Abridged Table of Atomic and Molar Masses of the Elements*

ELEMENT	SYMBOL	ATOMIC MASS(u) MOLAR MASS(g mol⁻¹)	ELEMENT	SYMBOL	ATOMIC MASS(u) MOLAR MASS(g mol⁻¹)
Aluminum	Al	26.98	Lithium	Li	6.941
Argon	Ar	39.95	Magnesium	Mg	24.31
Arsenic	As	74.92	Manganese	Mn	54.94
Barium	Ba	137.3	Mercury	Hg	200.6
Beryllium	Be	9.012	Neon	Ne	20.18
Boron	B	10.81	Nickel	Ni	58.69
Bromine	Br	79.90	Nitrogen	N	14.01
Calcium	Ca	40.08	Oxygen	O	16.00
Carbon	C	12.01	Phosphorus	P	30.97
Cesium	Cs	132.9	Potassium	K	39.10
Chlorine	Cl	35.45	Rubidium	Rb	85.47
Chromium	Cr	52.00	Silicon	Si	28.09
Cobalt	Co	58.93	Silver	Ag	107.9
Copper	Cu	63.55	Sodium	Na	22.99
Fluorine	F	19.00	Strontium	Sr	87.62
Gold	Au	197.0	Sulfur	S	32.06
Helium	He	4.003	Tin	Sn	118.7
Hydrogen	H	1.008	Titanium	Ti	47.88
Iodine	I	126.9	Tungsten	W	183.9
Iron	Fe	55.85	Vanadium	V	50.94
Krypton	Kr	83.80	Xenon	Xe	131.3
Lead	Pb	207.2	Zinc	Zn	65.38

*The values in this table are to four significant figures.

Cadet David Koenzti

Squadron 26

~~(scribbled out text)~~

~~(scribbled out text)~~

~~(scribbled out text)~~

SEC 3254

DR. J.M. BELLAMA

T.A. - L. YUAN

CHEMISTRY

Ronald J. Gillespie
McMASTER UNIVERSITY

David A. Humphreys
McMASTER UNIVERSITY

N. Colin Baird
UNIVERSITY OF WESTERN ONTARIO

Edward A. Robinson
UNIVERSITY OF TORONTO

ALLYN AND BACON, INC.
Boston · London · Sydney · Toronto

To Madge, Vivienne, and Jennifer
for their encouragement, support, and understanding

Library of Congress Cataloging-in-Publication Data
Main entry under title:

Chemistry.

Includes index.
1. Chemistry.   I. Gillespie, Ronald J. (Ronald James)
QD31.2.C39   1985   540   85-6009
ISBN 0-205-08416-8
ISBN (International) 0-205-08687-X

Printed in the United States of America

10 9 8 7 6 5 4 3 2   90 89 88 87 86

Production Administrator: Elaine Ober
Composition Buyer: Linda Cox
Manufacturing Buyer: Ellen Glisker
Cover Coordinator: Linda Dickinson
Cover Designer: Richard Hannus
Interior Designer: Deborah Schneck
Illustrators: William De Pippo, Avis Thomas, and
    Deborah Schneck
Experiment Photographs: Tom Bochsler
    Photography Limited

CREDITS
All photos by Tom Bochsler, except the following:
CHAPTER ONE: Figure 1.1, page 3: TOM STACK AND ASSO-CIATES.   Figure 1.2, page 8: Courtesy of Monsanto Electronic Materials Company.   Figure 1.4, page 9: Grant Heilman Photography (a); John Gerlach/TOM STACK AND ASSOCIATES (b). Figure 1.5, page 10: Grant Heilman Photography/Runk/Schoen-berger.   Figure 1.7, page 11: Courtesy of General Electric Research and Development Center.   CHAPTER TWO: Figure 2.1, page 42: *(Credits continue on page I-14, which represents a continuation of the copyright page.)*

BRIEF CONTENTS

CONTENTS

EXPERIMENTS

PREFACE

Among today's scientific and technological advances, from space exploration to cancer research, chemistry is at the cutting edge. The excitement of participating in this scientific adventure lies as much in the discovery of previously hidden properties and reactions of matter, as in the explanation of these phenomena by fundamental principles and theories. In writing this introductory textbook, it has been our ambition to communicate a sense of this excitement to the student. Features contributing to this end include: the approach to the material, aids to visualization, and emphasis on developing problem-solving skills.

Approach

Chemistry is based on observations and measurements of the properties and reactions of substances. Principles and theories are developed to systematize, rationalize, and explain these properties and reactions and to make predictions about compounds that have not yet been studied or prepared. In order to reflect these two fundamental aspects of chemistry, we have emphasized a balanced combination of the facts about the properties and reactions of substances—descriptive chemistry—and the principles and theories that explain these facts. Since descriptive chemistry and the principles and theories of chemistry are interdependent and neither can be fully appreciated without the other the descriptive chemistry is integrated with the theory and principles throughout the text. Hence the concrete facts inform the theoretical discussion at every stage.

We begin with some simple descriptive chemistry such as the composition of the atmosphere and the formation of oxides, which will be familiar to students from previous science courses or from everyday experience. We use this descriptive chemistry to introduce some basic principles, and then use these to discuss more descriptive chemistry and so on. In discussing principles we try, as far as possible, to use only those facts of descriptive chemistry that have already been introduced. We often return to these same facts in several different contexts, thus reinforcing the student's knowledge of the facts and underscoring their importance. In the descriptive material we have focused on substances that are common in everyday life, that are important to industry, and that illustrate particularly well the behavior of a given element.

Some chapters have descriptive titles because they emphasize the reactions and properties of a group of related elements, but the descriptive material is here used to introduce theories and principles in a natural manner. Other chapters have titles that indicate that they are primarily concerned with principles, such as those of thermochemistry, equilibrium, and reaction rates, but these chapters are based on descriptive material already introduced. For instance, the combustion of hydrocarbons provides important examples of energy-producing reactions. Hence, the chapter on thermochemistry logically follows the one in which the properties of hydrocarbons are described.

Important basic principles such as those concerning acid-base reactions, oxidation-reduction reactions, and precipitation reactions are introduced in Chapter 5, well before the quantitative discussion of these principles, because the basic ideas are essential for any meaningful discussion of chemical reactions. We have placed considerable emphasis on acid-base, oxidation-reduction, and precipitation reactions from Chapter 5 on, and by Chapter 14 the student is ready for a quantitative treatment.

This approach to important concepts and principles facilitates the student's assimilation of the material. The student is continually made aware that theories and principles are developed to explain observations and that it is important to have a knowledge of facts as well as of theories. Descriptive chemistry is made easier and more approachable when it is presented in small doses together with the appropriate theory. At the same time the discussion of theories and principles becomes less abstract when it is associated with the discussion and demonstration of properties and reactions. In this way the important concepts and principles are continually reinforced in the student's mind. Finally, the integration of descriptive chemistry with theory facilitates the efficient presentation of both, thus enabling us to deal with all the theories and principles normally covered in a first-year course, while including somewhat more reaction chemistry than is often possible.

Visualization

The experience of doing chemistry is visual as well as intellectual, and we have made every effort to convey both aspects to the student. A unique feature of this text is the use of color throughout. This feature serves several important purposes. A classical and still widely used method for detecting and following chemical reactions is the observation of color changes. Thus full color illustrations bring a reality to reactions that a chemical equation alone can never give. The many remarkable and often spectacular color changes that can be observed in chemical reactions produce an aesthetic enjoyment which will stimulate and reinforce the student's interest in the subject. The use of color has also enabled us to introduce another unique feature—over one hundred experiments are described, each illustrated with one or more color photos that were taken especially for this book. While many of these experiments are suitable for lecture demonstrations, they are not meant to be carried out by the student, nor are they intended to replace the laboratory manual. They are meant to illustrate the reactions discussed in the text, to make these reactions more meaningful to the student, and to aid the instructor in the presentation of this material. Instructions for performing these experiments safely are provided in the *Instructor's Manual*.

Problem Solving

Like any science, chemistry is an intellectual skill to be practiced, not simply a body of information to be memorized. Reading and observing need to be continually supplemented with active learning; hence, problem solving is emphasized throughout the book. Numerous worked examples in each chapter illustrate all the common types of problems that a first-year student is likely to meet. These include many quantitative problems as well as qualitative problems based on descriptive chemistry. The latter are particularly useful for testing the students' knowledge and understanding of reaction chemistry; in particular, their ability to recognize the main reaction types and to make reasonable, if not necessarily correct, predictions of the likely products of reactions involving the more common elements.

There are over 800 end-of-chapter problems organized according to the main sections of each chapter. Most chapters also have a section of miscellaneous problems that involve ideas or principles from more than one section of the chapter or from other chapters. A few problems that are of more than average difficulty are starred. Answers to selected odd-numbered problems are provided at the end of the text, and complete solutions to all the problems are available in the *Solutions Manual*.

Other Features

We use SI units almost exclusively. The change to SI is being made slowly, but inexorably. There is little doubt that today's students will be using SI to an ever increasing extent in scientific work and indeed in their everyday lives. However, we have made extensive use of atmospheres or torr (mm Hg) for pressure, liters for volume, and g cm^{-3} for density, because these units have a practical convenience and continue to be widely used by the scientific community. With these exceptions, SI units have a convenience and simplicity that lessen the difficulties students often encounter in solving numerical problems.

This book has been written with the aim of making the study of chemistry as simple and as enjoyable as possible for the student. It is written in a clear and straightforward style, and three preliminary editions have met with enthusiastic acceptance from the more than three thousand students who have used them. All key terms are highlighted in the text when first introduced and are summarized at the end of each chapter. Material that is of interest because of its relevance to everyday life and contemporary problems or is of an historical or biographical nature, has been placed in boxes or in short biographical sketches. In general this material is not meant for detailed study, but has been included to emphasize the relevance of chemistry to contemporary problems and to illustrate the human side of chemistry.

Organization

The twenty-six chapters follow a logical sequence that is compatible with most laboratory courses while allowing considerable flexibility in the order of presentation. The first eight chapters introduce many important ideas and principles that are best presented in the order of the text, but the remaining chapters can be reordered in several different ways. For example, instructors who wish to get to a quantitative discussion of equilibria earlier in the course might choose to cover Chapter 14 at any convenient time after Chapter 8. Note also that Chapter 25, Quantum Mechanics and the Chemical Bond, and Chapter 26, Thermodynamics, are not meant to be left until the end of the course. Because these two chapters contain material that is not essential to the understanding of the rest of the book, they may be introduced at any time the instructor considers appropriate. Some or all of the material in Chapter 25 could be covered together with Chapter 6 or at any later point in the course. Similarly, Chapter 26 could be combined with Chapter 12 or presented at any later time. In order to allow flexibility and to satisfy the needs of different instructors the book contains more material than will normally be covered in a first year course. Thus we have as far as possible delegated the more advanced and less essential material to the end of each chapter and to the later chapters so that it may be more easily omitted.

Supplementary Package

The ancillary items for *Chemistry* have been designed to complement the approach of the text, and to facilitate the learning of descriptive chemistry and the development of problem-solving skills.

- *Study Guide*: Provides a summary review, review questions, learning objectives, problem-solving strategies with worked examples, and a self test for each corresponding text chapter.
- *Test Bank*: Provides over 1,000 multiple choice questions. Available in printed or disk format.
- *Videotape*: Shows every text experiment in full color with voice commentary.
- *Slides*: Full color slides of all the text experiments.
- *Solutions Manual*: Provides detailed solutions to all end-of-chapter problems.
- *Laboratory Manual*: Offers an appropriate collection of laboratory experiments to complement *Chemistry*. The laboratory manual has its own instructor's manual.
- *Instructor's Manual*: Provides instructions and cautionary notes on performing the text's experiments, commentary notes on each chapter, an overview of topic organization, and numerous transparency masters of key figures in the text.

Acknowledgements

The following chemistry teachers reviewed all or parts of the manuscript at various stages in its development. We are grateful to all of them for their criticisms and suggestions.

Christopher Allen, *University of Vermont*
Richard Bader, *McMaster University*
Peter Barrett, *Trent University*

John Bauman, *University of Missouri—Columbia*
Jon Bellama, *University of Maryland*
Christopher Bender, *University of Lethbridge*

Edmund Benson, *Central Michigan University*
Clark Bricker, *University of Kansas*
Doyle Britton, *University of Minnesota*
Allan Burkett, *Dillard University*
Larry Byrd, *Glendale College*
Dewey Carpenter, *Louisiana State University*
James Carr, *University of Nebraska*
Glenn Crosby, *Washington State University*
Derek Davenport, *Purdue University*
Marcia Davies, *Creighton University*
Roger DeKock, *Calvin College*
Norman Duffy, *Kent State University*
Lawrence Epstein, *University of Pittsburgh*
Gordon Ewing, *New Mexico State University*
Fabian Fang, *California State College—Bakersfield*
Bert Fickerson, *Ventura College*
Suzanne Fortier, *Queens University*
Patrick Garvey, *Des Moines Area Community College*
Ronald Graham, *McMaster University*
Peter Gold, *The Pennsylvania State University*
Stephen Hall, *Southern Illinois University—Edwardsville*
Forrest Hentz, *North Carolina State University*
James Hill, *California State University—Sacramento*
Charles Howard, *University of Texas—San Antonio*
Harold Hunt, *Georgia Institute of Technology*
Paul Hunter, *Michigan State University*
Raul Imman, *Hudson Valley Community College*

Eugene Jekel, *Hope College*
David Katz, *Community College of Philadelphia*
Joseph Laposa, *McMaster University*
Philip Lamprey, *University of Lowell*
William Litchman, *University of New Mexico*
Edward Mercer, *University of South Carolina*
John Milne, *University of Ottawa*
Joseph Morse, *Utah State University*
Robert Nakon, *West Virginia University*
Judy Okamura, *San Bernardino Valley College*
Charles Owens, *University of New Hampshire*
Judith Poë, *University of Toronto*
Geoffrey Rayner-Canham, *Memorial University of Newfoundland*
Daniel Reger, *University of South Carolina*
Eugene Rochow, *Harvard University*
Anthony Saturno, *SUNY—Albany*
Donald Smith, *McMaster University*
Dennis Stynes, *York University*
James Thompson, *University of Toronto*
Larry Thompson, *University of Minnesota—Duluth*
Robert Thompson, *University of British Columbia*
Richard Tomlinson, *McMaster University*
James Ufford, *Concordia University*
Trina Valencick, *California State University—Los Angeles*
Gary Van Kempen, *Lansing Community College*
Glenn Vogel, *Ithaca College*
Donald Williams, *Hope College*

We also thank Bob Rogers, who gave us encouragement and support during the early stages of the writing.

It is a pleasure to acknowledge the enthusiastic support and assistance given to us by the staff at Allyn and Bacon. We especially wish to thank Jim Smith, whose support and enthusiasm for our project revived our spirits at a difficult time and enabled us to carry it through to completion; Elaine Ober, who led us so competently and efficiently through the intricacies and deadlines of production; Judy Fiske, Production Director, and Gary Folven, Editor-in-Chief.

We also owe a debt of gratitude to Tom Boschler, who took the outstanding color photos that illustrate the experiments—his patience and fortitude under sometimes trying circumstances were greatly appreciated; Edith Denham, who efficiently typed and retyped the many versions of the manuscript; and Carol Dada, who also typed portions of the manuscript and kept the authors organized.

CHEMISTRY

CHAPTER 1

STRUCTURE OF MATTER

Elements, Compounds, Mixtures, and Units of Measurement

The world in which we live is one of infinite variety and continual change. We inhabit the earth along with the trees, plants, fish, birds, and all other forms of life. All depend for their existence on the multitude of different materials that compose the solid surface of the earth, the water in the rivers, lakes, and oceans, and the gases that form the atmosphere. Ever since we learned to communicate by speech and by writing, we have attempted to describe the natural world of which we are a part: its form, its variety, its beauty, and its continual changes. Change is, perhaps, the most striking feature of the natural world. Mountains are worn down by the action of water and of the atmosphere on the rocks that compose them. New mountains are formed as a consequence of changes occurring beneath the surface of the earth. Leaves turn red in the fall, iron rusts, and wood burns. We are born, we breathe—inhaling oxygen and exhaling carbon dioxide—we grow, and we die.

Most of our ideas concerning the nature of the material world have developed over the past two hundred years as a result of careful systematic observations and experiments. Such studies of materials and their transformations constitute the modern science of chemistry. As our understanding of the transformations of materials has increased, we have learned to make many changes in naturally occurring substances. We extract iron ore from the ground and convert it to steel; we use steel to build drilling rigs to obtain petroleum from the ground; petroleum in turn is transformed into gasoline, plastics, and drugs. Indeed, during the past hundred years chemists have transformed the material world by creating a multitude of new substances, from synthetic fibers to antibiotics.

To the beginning student chemistry may sometimes seem to be an overwhelming jungle of formulas, equations, and theories. Remember, however, that behind all the formulas and theories lies the real world, the fascinating world of chemistry, in which we study the materials making up the earth and, in fact, the whole universe. All the formulas, equations, and theories that form part of modern chemistry have been invented by chemists only to help us understand the real world around us: its multitude of different substances, both living and nonliving, and the extraordinary variety of changes they undergo.

1.1 COMPOSITION OF MATTER

We can describe the universe, and all the changes occurring in it, in terms of two fundamental concepts: **matter** and **energy**. Matter is anything that occupies space and has mass. Water, air, rocks, and petroleum, for example, are matter, but heat and light are not; they are forms of energy. The many different kinds of matter are known as **substances**. Chemists are concerned with determining the composition and structure of substances, with finding out how the properties of substances depend on their composition and structure, and with understanding the changes substances undergo. The transformation of one or more substances into one or more different substances is called a **chemical reaction**. When iron rusts, it is transformed into iron oxide. When gasoline burns, it is transformed into carbon dioxide and water. These transformations are chemical reactions. Thus chemistry is the study of the compositon, structure, and properties of substances and the transformations (called chemical reactions) by which substances are changed into other substances.

The scope of chemistry is very broad; in fact, it overlaps with all the other natural sciences. The biologist studies the substances in living organisms; the geologist is concerned with the rocks and minerals that compose the earth; the astrophysicist is interested in the substances in the stars and interstellar space. This overlap between chemistry and other sciences has led to the development of many interdisciplinary fields such as biochemistry (the chemistry of living matter), geochemistry (the chemistry of rocks and minerals), and cosmochemistry (the chemistry of stars and interstellar space). Thus the study of chemistry is important not only for its own sake but also for a full understanding of many other sciences and their applications.

Elements and Compounds

Some of the earliest attempts to understand the natural world were made over two thousand years ago by the ancient Greeks, who proposed that everything in the world was composed of four *elements*, or basic substances, which combined to form rocks, plants, clouds, sunlight and all the other components of the universe. They identified these elements as earth, water, fire, and air. Sunlight, for example, appeared to them to be a mixture of fire and air. Ice, they thought, was water plus the hardness of earth.

Although these ideas might seem rather naive and even amusing to us today, they made several valuable contributions to our understanding of the world. One contribution was an awareness that the different states that a substance can take are important: Earth is a *solid*; water is a *liquid*; air is a *gas*. Another contribution was their recognition that fire, or *energy*, is important in the changes that substances undergo. For the moment let us focus on the important idea that even the most complex substances are made up of basic components called elements.

The concept of an *element* has changed over the centuries. It has come to mean a substance that cannot be broken down into other simpler substances by any chemical reaction. Earth obviously does not qualify as an element; even a superficial look at a sample of dirt shows that it is composed of many different substances (Figure 1.1). Nor does water qualify as an element. If we pass an electric current through water, we can separate it into the gases hydrogen and oxygen (Experiment 1.1). But nobody has succeeded in breaking down hydrogen or oxygen into other substances by chemical reactions. They therefore are considered to be elements, whereas the substances that they combine to form, such as water, are known as *compounds*. Today we recognize just over a hundred elements (Table 1.1).

It may surprise you to learn that the number of elements known today cannot be stated with complete certainty. In recent years new elements have been made by nuclear, not chemical, reactions, and in some cases only a few atoms have been made. In such cases establishing the identity of the atoms with complete certainty is sometimes difficult. Table 1.1 lists 103 elements that are known with complete certainty, and there are probably at least 4 more. Others are likely to be synthesized in the future. We will discuss only about 40 of the elements in this book.

Figure 1.1 Sample of Earth. Earth is composed of many different substances. It may contain sand, clay, rock, and decaying plant material, for example.

Atoms

What is the difference between elements and compounds? Once again, some of the ancient Greek philosophers provided the germ of the idea that is our present-day answer to this question. Although Plato and Aristotle believed that matter was continuous, Democritus argued that matter was composed of very small indivisible particles, a view later shared by the Roman poet Lucretius. However, not until the beginning of the nineteenth century did the Englishman John Dalton (Box 1.1) show that this idea could form the basis for understanding the nature of elements and compounds and the transformations that they undergo. Dalton called the fundamental particles of which matter is composed **atoms**, from the Greek word *atomos*, meaning "indivisible." He also made the following propositions:

- The atoms of any given element are identical. Fake now, Isotopes?
- The atoms of one element are different from those of another element.
- Atoms of two or more elements may combine in definite ratios to form compounds.
- Atoms remain unchanged in chemical reactions.

EXPERIMENT 1.1

Decomposition of Water by an Electric Current

Water can be decomposed into the elements hydrogen and oxygen by passing an electric current through it. Bubbles of hydrogen can be seen rising from the wire at the right and bubbles of oxygen from the wire at the left. The volume of hydrogen produced is twice the volume of oxygen. A few drops of sulfuric acid have been added to the water to increase the electrical conductivity.

Table 1.1 The Elements

ELEMENT	SYMBOL	ELEMENT	SYMBOL	ELEMENT	SYMBOL
Actinium	Ac	Hafnium	Hf	Promethium	Pm
Aluminum	**Al**	**Helium**	He	Protactinium	Pa
Americium	Am	Holmium	Ho	Radium	Ra
Antimony	Sb	**Hydrogen**	**H**	Radon	Rn
Argon	**Ar**	Indium	In	Rhenium	Re
Arsenic	As	**Iodine**	**I**	Rhodium	Rh
Astatine	At	Iridium	Ir	**Rubidium**	**Rb**
Barium	**Ba**	**Iron**	**Fe**	Ruthenium	Ru
Berkelium	Bk	**Krypton**	**Kr**	Samarium	Sm
Beryllium	Be	Lanthanum	La	Scandium	Sc
Bismuth	Bi	Lawrencium	Lr	Selenium	Se
Boron	**B**	**Lead**	**Pb**	**Silicon**	**Si**
Bromine	**Br**	**Lithium**	**Li**	**Silver**	**Ag**
Cadmium	Cd	Lutetium	Lu	**Sodium**	**Na**
Calcium	**Ca**	**Magnesium**	**Mg**	**Strontium**	**Sr**
Californium	Cf	**Manganese**	**Mn**	**Sulfur**	**S**
Carbon	**C**	Mendelevium	Md	Tantalum	Ta
Cerium	Ce	**Mercury**	**Hg**	Technetium	Tc
Cesium	**Cs**	Molybdenum	Mo	Tellurium	Te
Chlorine	**Cl**	Neodymium	Nd	Terbium	Tb
Chromium	**Cr**	**Neon**	**Ne**	Thallium	Tl
Cobalt	**Co**	Neptunium	Np	Thorium	Th
Copper	**Cu**	**Nickel**	**Ni**	Thulium	Tm
Curium	Cm	Niobium	Nb	**Tin**	**Sn**
Dysprosium	Dy	**Nitrogen**	**N**	**Titanium**	**Ti**
Einsteinium	Es	Nobelium	No	Tungsten	W
Erbium	Er	Osmium	Os	Uranium	U
Europium	Eu	**Oxygen**	**O**	**Vanadium**	**V**
Fermium	Fm	Palladium	Pd	**Xenon**	**Xe**
Fluorine	**F**	**Phosphorus**	**P**	Ytterbium	Yb
Francium	Fr	Platinum	Pt	Yttrium	Y
Gadolinium	Gd	Plutonium	Pu	**Zinc**	**Zn**
Gallium	Ga	Polonium	Po	Zirconium	Zr
Germanium	Ge	**Potassium**	**K**		
Gold	**Au**	Praseodymium	Pr		

Note: The elements discussed in this book are in bold type.

We can therefore define an **element** as a substance that contains only one kind of atom. A **compound** is a substance that contains two or more kinds of atoms combined in fixed proportions. For example, the element oxygen contains only oxygen atoms, and the element hydrogen contains only hydrogen atoms. But the compound water contains both hydrogen and oxygen atoms combined in the ratio of two hydrogen atoms for every one oxygen atom. We have seen that water can be broken down into hydrogen and oxygen; in other

John Dalton, the son of a poor weaver, was born in Cumberland, England. He first studied at a village school, and he made such rapid progress that by the age of twelve he became the teacher at the school. Seven years later he became a school principal. In 1793 he moved to Manchester, and he remained there for the rest of his life. He first taught mathematics, physics, and chemistry at a college. But finding that his teaching duties interfered with his scientific studies, he resigned from this post and supported himself by teaching mathematics and chemistry to private pupils. He never married, and he continued to live a very simple and modest life even after he became famous. Dalton's first scientific investigations were in meteorology, and he continued throughout his life to make daily observations on the temperature, barometric pressure, and rainfall. He described the nature of color blindness, of which he was a victim. A devout Quaker, Dalton always wore plain and somber clothes. Friends were therefore surprised when he wore a scarlet academic

robe on being presented to King William IV in 1832. To Dalton, however, it appeared a dull gray, and he wore it without concern.

Dalton put forward his atomic theory in 1803. Although he proposed that compounds were formed by the combination of atoms of different elements in small, whole-number ratios, he had no certain method for determining the ratios in which the different atoms combine. He assumed therefore that when only one compound of two elements A and B was known, it had the simplest possible formula, AB. On the basis of this assumption and from the masses of different elements that were found to combine, he was able to deduce relative atomic masses. He published the first table of such relative atomic masses. His assumptions about formulas of compounds were, however, not always valid. For example, he assumed that the formula of water was HO, which led to some of the atomic masses in his table being incorrect. In fact, not until 1858 did chemists solve the problem of the correct determination of molecular formulas and therefore of atomic masses. Nevertheless, the credit must go to Dalton for first putting the atomic theory on a quantitative basis and for laying the foundation for the rapid development of chemistry that followed.

words, the hydrogen atoms can be separated from the oxygen atoms. Not surprisingly, we can also combine hydrogen and oxygen to form water (Experiment 1.2).

Both the decomposition of water to hydrogen and oxygen and the combination of hydrogen and oxygen to give water are chemical reactions. (The arrow denotes a chemical reaction.)

$$\text{Hydrogen} + \text{Oxygen} \longrightarrow \text{Water}$$

and

$$\text{Water} \longrightarrow \text{Hydrogen} + \text{Oxygen}$$

A chemical reaction takes place when atoms are rearranged from their original combinations to form new combinations, the atoms themselves remaining unchanged.

The silvery white metal magnesium is an element. When it is heated in air, it burns with a brilliant white light, combining with the oxygen in the air to form a white powder, which is the compound magnesium oxide (Experiment 1.3). Magnesium also burns in steam and in carbon dioxide. When magnesium

1.1 COMPOSITION OF MATTER

Reaction of Hydrogen with Oxygen

Hydrogen burns in the air with a pale blue, almost colorless flame, combining with the oxygen of the air to form water. The water can be observed condensing on the outside of the cold flask. Here the flame is colored yellow by the sodium in the glass tube.

reacts with carbon dioxide, the magnesium atoms combine with the oxygen atoms of carbon dioxide to form magnesium oxide, leaving the element carbon, which consists only of carbon atoms. In this reaction we note a very important feature of chemical reactions: They are always accompanied by energy changes. The burning of magnesium is accompanied by the emission of heat and light.

The many experiments carried out since Dalton put forward his ideas have fully confirmed that matter consists of atoms. Dalton's proposals form the basis of what we now call the **atomic theory**. We are so used to the idea of atoms today that we cannot imagine that people ever had other ideas about the composition of matter. In the present century, however, scientists have shown that

Reactions of Magnesium

Magnesium burning in air. The white powder in the watch glass and the white smoke are magnesium oxide.

Magnesium burns in substances that normally extinguish flames. Here we see it burning in a flask of boiling water. White magnesium oxide is produced in a different reaction.

Magnesium also burns in carbon dioxide, giving white magnesium oxide and carbon. The carbon can be seen as black specks on the sides of the flask.

contrary to what Dalton believed, atoms are not indivisible in all circumstances. Rather, each atom consists of still smaller particles. As we shall see later, it is the structure of atoms that determines how two or more elements may combine to form a compound.

Symbols and Formulas

To simplify the representation of elements and compounds, chemists have agreed on an international set of symbols. The symbol for a given element nearly always consists of the first letter of its English name, frequently followed by one other letter (see Table 1.1). For some elements an abbreviation of the name in another language, usually Latin, is used. For example, sodium has the symbol Na, from the Latin *natrium.*

A compound is represented by a formula that indicates the elements that it contains and the relative number of atoms of each element. Water is represented by the formula H_2O because it contains two hydrogen atoms for every oxygen atom. *Methane*, a major component of natural gas, is a compound of the elements carbon and hydrogen and contains four hydrogen atoms for every carbon atom; it has the formula CH_4. Common table salt, *sodium chloride*, contains equal numbers of sodium and chlorine atoms and is represented by the formula NaCl. Carbon and oxygen atoms combine in a 1:2 ratio to form the compound carbon dioxide, which makes up a very small part of the earth's atmosphere. *Carbon dioxide* is represented by the formula CO_2. Carbon and oxygen atoms also combine in a 1:1 ratio to form *carbon monoxide*, a poisonous gas that is present in automobile exhaust; it has the formula CO. Thus two elements can form more than one compound. Many compounds contain more than two elements; for example, *sulfuric acid*, H_2SO_4, contains hydrogen, oxygen, and sulfur. Most compounds, however, contain only a few different elements.

Substances and Mixtures

All the different materials that we recognize around us either are mixtures of two or more substances or are single substances. Every substance is either an element or a compound. Magnesium is an example of a substance that is an element. Water and sodium chloride are examples of substances that are compounds. Seawater, however, is a mixture of water, sodium chloride, and many other substances.

Every substance has a unique set of properties that allows us to distinguish it from all other substances. Some of these are **physical properties**, the characteristics of a substance that we can observe when we study it in isolation. These properties include, for example, physical state at room temperature, melting or freezing point, boiling point, color, solubility in water, and electrical conductivity. Some physical properties of water are that it is a liquid at room temperature, that it freezes at 0°C and boils at 100°C, that it is colorless, and that it has a very small electrical conductivity.

A substance also has characteristic **chemical properties**: Under the same circumstances it will always react with, and change into, other substances in exactly the same way. Iron always rusts when exposed to air and water; magnesium burns in oxygen, carbon dioxide, or steam. If we were to carry out Experiment 1.1 with several different samples of water, every sample would undergo the same decomposition to give exactly the same relative amounts of hydrogen and oxygen.

Figure 1.2 Ingots of Ultrapure Silicon. These large ingots are cut into thin wafers and then into tiny chips for making the many kinds of semiconductors used in radios. pocket calculators, and computers.

A very important property of a substance is that it has a constant **composition**. If it is an element it contains only one kind of atom. If it is a compound every sample contains exactly the same relative numbers of atoms of each kind. Regardless of the source from which it is obtained, water always has two hydrogen atoms for each oxygen atom and can be represented by the formula H_2O. Carbon dioxide always has two oxygen atoms for each carbon atom; sodium chloride always has one sodium atom for each chlorine atom; and magnesium consists only of magnesium atoms.

If a given material is a single substance and not a mixture of substances, we often call it a **pure substance**. In nature very few substances occur in a pure form. One substance may be the major component of a given sample, but other substances, which we call *impurities*, are nearly always present. Thus *purity* is a relative term. No substance can be regarded as absolutely pure; no matter how carefully it has been purified, traces of other, contaminating substances always remain. A substance may be described as 99% pure when it contains 1% by mass of impurities and as 99.99% pure when it contains 0.01% of impurities. Such purities are sufficient for many purposes, and such substances are often described as pure.

Sometimes, though, a substance with a still greater purity is required for a particular purpose. For example, the silicon, Si, used in silicon chips has to be 99.999 99% pure (Figure 1.2). And often we need to know not only the purity of a substance but also the nature of the impurities present. The labels on the containers of many substances used in a chemical laboratory often list the impurities and their amounts (Figure 1.3).

HETEROGENEOUS MIXTURES Many common materials such as soil, rocks, concrete, and wood are mixtures. Different parts of these materials have different properties such as color and hardness (Figure 1.4). Such mixtures are said to be **heterogeneous**; both the properties and the composition of the material are nonuniform.

If we mix powdered sulfur with iron filings, the result is a heterogeneous mixture. We can still discern hard, dark grey iron particles and yellow sulfur powder when we examine the mixture. We can separate the iron and the sulfur by taking advantage of their different physical properties. One way to separate them is with a magnet: The iron filings are attracted, while the sulfur is not (Experiment 1.4).

In general, heterogeneous mixtures can be separated into their components by making use of their different physical properties—that is, by carrying out some **physical change**. Any change in which the amounts and nature of the substances present do not change is a physical change. Boiling and freezing water and dissolving sugar in water are physical changes.

The nonuniform composition of a heterogeneous mixture is not always easily visible. Milk might not appear to be heterogeneous, but under a microscope we can see that it consists of small droplets of fat suspended in a clear liquid (Figure 1.5).

HOMOGENEOUS MIXTURES A mixture that has uniform properties throughout is a **homogeneous mixture**, or a **solution**. If you stir a spoonful of sugar into a cup of water until all the sugar dissolves, you create a homogeneous mixture, or a solution. The sugar has become completely dispersed in the water so that the mixture is uniform. Any sample of a particular mixture has the same composition and the same physical properties as any other sample—color and

SODIUM CHLORIDE
MEETS A.C.S. SPECIFICATIONS

NaCl = 58.44

Minimum assay (after ignition)	99.9%

Maximum limits of impurities

Insoluble matter	0.003%
Free acid (HCl)	0.0018%
Free alkali	0.05 ml N/1%
Bromide and iodide (Br)	0.005%
Ferrocyanide [Fe(CN)$_6$]	0.0001%
Nitrate (NO$_3$)	0.0005%
Phosphate (PO$_4$)	0.0005%
Sulphate (SO$_4$)	0.002%
Ammonium (NH$_4$)	0.0005%
Arsenic (As)	0.00004%
Barium (Ba)	0.001%
Calcium group and	
magnesium (Ca)	0.004%
Iron (Fe)	0.0003%
Heavy metals (Pb)	0.0005%
Potassium (K)	0.01%

Figure 1.3 Label on Bottle of Sodium Chloride Indicating Impurities. This sodium chloride is described as 99.9% pure. It therefore contains a total of 0.1% of impurities, which are listed on the label.

EXPERIMENT 1.4

Mixtures and Compounds

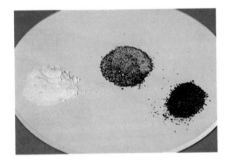

On the left is a pile of powdered sulfur. On the right is a pile of iron filings. In the middle is a mixture of the iron and sulfur.

The iron filings can be separated from the sulfur by using a magnet.

When the mixture is heated it glows brightly as the iron and the sulfur combine to form the compound iron sulfide.

Iron sulfide is grey-black and brittle, unlike either sulfur or iron. The iron in the compound iron sulfide cannot be separated from the sulfur by means of a magnet.

1.1 COMPOSITION OF MATTER

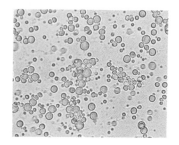

Figure 1.5 Milk Under a Microscope. The heterogeneity of milk is not obvious to the naked eye. But under a microscope we see that it consists of drops of oil suspended in a clear liquid.

sweetness, for example. Homogeneous mixtures, like heterogeneous mixtures, can be separated into their components by physical means. If you heat the mixture of sugar and water, the water evaporates and the sugar is left behind. A solution of salt in water can be separated into its two components in the same way.

Not all solutions are liquids. Pure, dry air is a solution—a homogeneous mixture of gases, principally nitrogen, oxygen, and argon. The gold used in jewelry is not pure but is a solid solution of copper or silver in gold.

How does a homogeneous mixture differ from a pure substance? The properties of a homogeneous mixture of a given composition do not vary from one part of the mixture to another, but mixtures do not have to have a constant composition or constant properties. Homogeneous mixtures, such as solutions of sugar in water, can have different compositions depending on how much sugar is dissolved in a given amount of water, and they will have correspondingly different sweetnesses. A mixture of CO_2 and CO can have any ratio of oxygen atoms to carbon atoms, from 2:1 to 1:1, depending on the relative amounts of the CO_2 and CO.

In contrast, a compound always has the same composition and properties. Water always has the composition expressed by the formula H_2O, and its color and freezing point are always the same. Moreover, water cannot be separated into its component elements by physical means but only by a chemical reaction that produces two new substances (hydrogen and oxygen). These new substances and the water from which they were obtained have completely different properties. Experiment 1.4 shows the differences between a mixture of iron and sulfur and the compound iron sulfide. Figure 1.6 summarizes the classification of matter and the ways in which one type of matter can be changed into another.

Molecules

Not only does a compound have a constant composition, being composed of fixed relative numbers of its component atoms, but these atoms have a definite spatial arrangement with respect to each other. The arrangement of the atoms

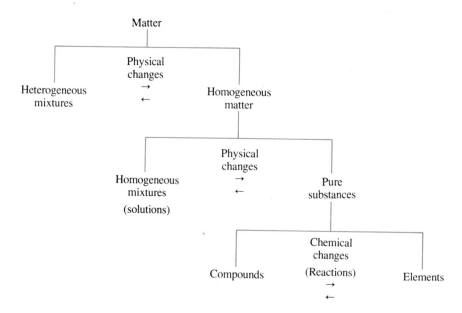

Figure 1.6 Classification of Matter. All matter is either homogeneous or heterogeneous. Homogeneous matter may be either a (pure) substance or a homogeneous mixture (solution). A substance may be either an element or a compound.

in a compound or element is known as its **structure.** Because the structure of a substance has a profound effect on its properties, one of the aims of modern chemistry is to relate the properties of a substance to its structure. For instance, the element carbon is known in several different forms, such as diamond and graphite, with very different properties (Figure 1.7). Diamond and graphite both consist of carbon atoms only; their different properties are due to the different ways these atoms are arranged.

The early chemists had no direct method of finding the structure of any given substance, although by studying its properties and reactions, they were sometimes able to deduce how its atoms were arranged. Only during the past sixty years have experimental methods been developed that enable chemists to determine the structures of substances directly. Most of these methods are based on the study of how different forms of electromagnetic radiation, such as visible light, infrared radiation, and X rays, interact with matter. Some of these methods are discussed later.

Experiments have shown that in many substances atoms are combined in small groups called **molecules.** Water, for example, consists of molecules in which two hydrogen atoms are joined to an oxygen atom. The water molecule has the formula H_2O. This is called a *molecular formula.* It shows the number of atoms of each kind in one molecule of the compound. Carbon monoxide consists of molecules composed of one carbon atom and one oxygen atom; its molecular formula is CO. Carbon dioxide consists of molecules composed of one carbon atom and two oxygen atoms; its molecular formula is CO_2.

Some elements are also composed of molecules. For example, oxygen, nitrogen, and hydrogen each have two atoms in their molecules, and their molecular formulas are O_2, N_2, and H_2, respectively. Some elements have bigger molecules; phosphorus consists of P_4 molecules and sulfur of S_8 molecules. Molecules such as H_2, N_2, O_2, and CO, which are composed of only two atoms, are known as **diatomic molecules**; molecules such as H_2O and CO_2 are called **triatomic molecules.** In general, any molecule containing more than two atoms is called a **polyatomic molecule.**

Not all substances are composed of molecules. For example, the compounds salt (sodium chloride) and chalk (calcium carbonate) do not have a molecular structure; neither do the elements magnesium and carbon. We will discuss the structures of these substances later. For the moment we will restrict our attention to those substances that consist of molecules.

Empirical and Molecular Formulas

The molecular formula of a substance is not always identical with the *simplest formula* that expresses the relative numbers of atoms of each kind in the substance. Since an element consists of only one kind of atom, the simplest formula for any element is simply the symbol for the element: O for oxygen, N for nitrogen, P for phosphorus, S for sulfur. The molecular formulas of these substances, however, are O_2, N_2, P_4, and S_8.

The composition of *ethane*, which is another component of natural gas in addition to methane, is expressed by the formula CH_3. For each carbon atom in ethane there are three hydrogen atoms. However, ethane consists of C_2H_6 molecules and therefore has the molecular formula C_2H_6. Hydrogen and oxygen combine to give a compound, hydrogen peroxide, which is a colorless liquid like water but has different chemical and physical properties. For example, it has a boiling point of 158°C. One of its characteristic chemical properties is

Figure 1.7 Diamond and Graphite. Both substances are forms of the same element, carbon, but they have very different properties. The "lead" in a lead pencil is made from graphite mixed with clay.

Properties of Water and Hydrogen Peroxide

Water (left) and hydrogen
peroxide (right) are both colorless
liquids. They are both compounds
composed of the elements
hydrogen and oxygen.

When a drop of blood is
added to water it slowly
mixes but no reaction is
observed. When a few drops
of blood are added to
hydrogen peroxide it
decomposes rapidly
producing bubbles of
oxygen which form a thick
foam that fills the beaker
and overflows the top.

When black solid
manganese dioxide is added
to water no reaction is
observed. When manganese
dioxide is added to
hydrogen peroxide it causes
hydrogen peroxide to
decompose, producing
bubbles of oxygen.

The oxygen that is evolved
ignites a glowing splint.

illustrated in Experiment 1.5. Hydrogen peroxide is composed of equal numbers
of hydrogen and oxygen atoms, and its composition can therefore be represented by the formula HO. However, its molecules each contain two hydrogen
atoms and two oxygen atoms, so its molecular formula is H_2O_2.

The simplest formula that correctly expresses the composition of a substance,
in terms of whole-number ratios of atoms, is called its **empirical formula**. Thus
the empirical formulas of water, hydrogen peroxide, ethane, sulfur, and phosphorus are H_2O, HO, CH_3, S, and P. Since we cannot have a fractional part
of an atom (half an atom is no longer an atom), empirical formulas are written
with integral numbers of atoms and not fractional numbers. The empirical
formulas of water and ethane are written as H_2O and CH_3, not as $HO_{1/2}$ and
$C_{1/3}H$.

The **molecular formula** of a substance tells us how many atoms of each kind
there are in one molecule of the substance. The molecular formulas of water,
hydrogen peroxide, ethane, sulfur, and phosphorus are H_2O, H_2O_2, C_2H_6, S_8,
and P_4. For some substances, such as H_2O, the empirical and molecular formulas are the same. For others, such as ethane, the molecular formula, C_2H_6,
is a whole-number multiple of the empirical formula, CH_3. Empirical and molecular formulas are compared in Table 1.2.

Table 1.2 Empirical and Molecular Formulas

SUBSTANCE	EMPIRICAL FORMULA	MOLECULAR FORMULA
Water	H_2O	H_2O
Hydrogen peroxide	HO	H_2O_2
Ethane	CH_3	C_2H_6
Sulfur	S	S_8
Phosphorus	P	P_4

Figure 1.8 summarizes the differences among mixtures, substances, compounds, and elements for hydrogen, oxygen, water, and hydrogen peroxide.

Many molecules, including most of those found in living organisms, are much larger that the simple molecules we have considered so far, and they have rather complicated structures. For example, sucrose (table sugar) has the molecular formula $C_{12}H_{22}O_{11}$; riboflavin (vitamin B_2) has the molecular formula $C_{17}H_{20}N_4O_6$; adenosine triphosphate (ATP) has the molecular formula $C_{10}H_{16}N_5O_{13}P_3$. Figure 1.9 shows how the atoms are connected in these molecules.

Structure and Shape of Molecules

Much detailed information on the arrangement of atoms in molecules has been obtained by the modern experimental methods developed in the past sixty years.

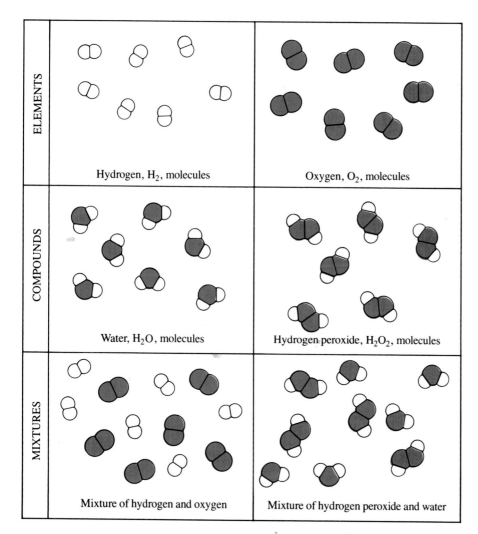

Figure 1.8 Mixtures, Substances, Compounds, and Elements.
Hydrogen, oxygen, water, and hydrogen peroxide are all substances. Hydrogen and oxygen are elements; water and hydrogen peroxide are compounds. A mixture of hydrogen peroxide and water is homogeneous; it is a solution. A mixture of hydrogen and oxygen is also homogeneous; it is a (gaseous) solution.

Figure 1.9 Structures of Sucrose, Riboflavin (Vitamin B₂), and Adenosine Triphosphate (ATP). These molecules are larger and have more complicated structures than most of the molecules with which we will be concerned in this book. But the principles that we discuss in later chapters that help us to understand how the atoms of simple molecules are held together and how their atoms are arranged in space also apply to these and still more complicated molecules.

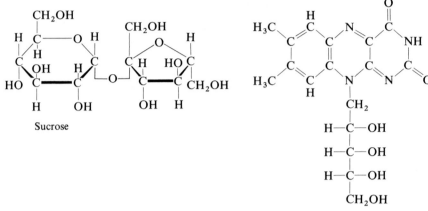

Sucrose

Riboflavin (Vitamin B₂)

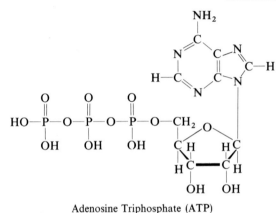

Adenosine Triphosphate (ATP)

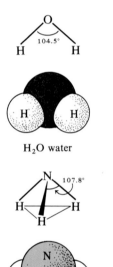

H₂O water

NH₃ ammonia

Figure 1.10 Shapes of Water and Ammonia Molecules. Although molecules are very small, the distances between the atoms and the arrangement of the atoms in space can be accurately determined by modern experimental methods.

For example, not only do we know, that there are two hydrogen atoms and an oxygen atom in a water molecule, but we also know the distance between the center of each hydrogen atom and the center of the oxygen atom and that the three atoms form an angle of 104.5° (Figure 1.10). The water molecule is described as *angular* because the three atoms are not in a straight line. The ammonia molecule has the shape of a *triangular pyramid*, with the nitrogen atom at the apex and the three hydrogen atoms at the corners of the pyramid's equilateral triangular base (Figure 1.10). The three angles at the apex of the pyramid are each 107.8°.

The shapes of some other molecules we have encountered so far in this chapter are shown in Figure 1.11. The shape of the methane molecule is a common shape for molecules but is less familiar in everyday experience. It is the *tetrahedron*. Each of the four faces of a tetrahedron is an equilateral triangle, so all the edges have the same length and all the angles between them are 60°,

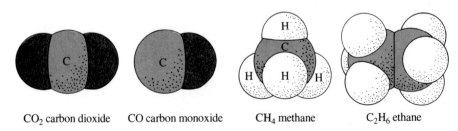

CO₂ carbon dioxide   CO carbon monoxide   CH₄ methane   C₂H₆ ethane

Figure 1.11 Shapes of Some Molecules.

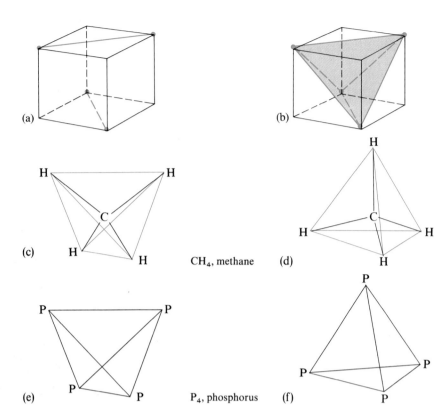

Figure 1.12 **The Tetrahedron.** The tetrahedron is a regular solid with four equivalent equilateral triangular faces and six equal edges. The shape of the tetrahedron is closely related to the shape of a cube. This relation can be seen by drawing a diagonal across one face of the cube and another diagonal at right angles across the opposite face of the cube as in (a). Then join the ends of the diagonals, as in (b), to form the tetrahedron. Two views of the methane molecule are shown in (c) and (d); two views of the phosphorus molecule are shown in (e) and (f).

(a)

(b)

(c) H   H   C   H   H CH_4, methane

(d) H   C   H   H   H

(e) P   P   P   P P_4, phosphorus

(f) P   P   P   P

as indicated in Figure 1.12. In the methane molecule the four hydrogen atoms are at the corners of the tetrahedron, and the carbon atom to which they are joined is at the center of the tetrahedron. Another molecule with a tetrahedral shape is the P_4 molecule. In this case there is a phosphorus atom at each corner of the tetrahedron, but there is no atom at the center.

1.2 UNITS OF MEASUREMENT

The distances between the atoms in a molecule are extremely small. In the water molecule the distance between the center of each hydrogen atom and the center of the oxygen atom is only 0.000 000 000 097 meter. We will also frequently encounter very large numbers in chemistry. There are, for example, 33 460 000 000 000 000 000 000 molecules in 1 gram of water.

Scientific Notation

To write out these very small or very large numbers in this way is inconvenient. **Scientific**, or *exponential*, **notation** is usually used for expressing very small and very large numbers. Scientific notation uses the form

$$N \times 10^n$$

where N is a number between $1.000\ldots$ and $9.999\ldots$, that is, a number with one nonzero digit before the decimal point. The number 33 460 000 000 000 000 000 000 is equal to 3.346 multiplied by 10 twenty-two times, that is,

$$3.346 \times 10 \times 10 \times 10 \times 10 \times 10 \times 10 \times 10 \times 10 \times 10 \times 10 \times 10$$
$$\times 10 \times 10 \times 10 \times 10 \times 10 \times 10 \times 10 \times 10 \times 10 \times 10 \times 10$$

In scientific notation this number is written as

$$3.346 \times 10^{22}$$

Thus there are 3.346×10^{22} molecules in 1 gram of water.

Similarly, the number 0.000 000 000 097 is equal to 9.7 divided by 10 eleven times, that is,

$$\frac{9.7}{10 \times 10 \times 10 \times 10 \times 10 \times 10 \times 10 \times 10 \times 10 \times 10 \times 10}$$

In scientific notation this number is written as

$$\frac{9.7}{10^{11}} \quad \text{or} \quad 9.7 \times \frac{1}{10^{11}}$$

and $1/10^{11}$ is written as 10^{-11}. Hence

$$0.000\,000\,000\,097 = 9.7 \times 10^{-11}$$

Thus the distance between a hydrogen atom and an oxygen atom in the water molecule is 9.7×10^{-11} meter.

In general, to convert a number to scientific notation, move the decimal point until there is only one nonzero digit in front of it. The number of places that the decimal point is moved gives the exponent of 10. For example,

$$3\,472\,809.0 \rightarrow 3.472\,809\,0 \times 10^{6}$$
$$\underbrace{\qquad}_{6 \text{ places}}$$

$$0.048\,729 \rightarrow 4.8729 \times 10^{-2}$$
$$\underbrace{\qquad}_{2 \text{ places}}$$

In writing numbers and doing calculations, be very careful to always use the correct number of **significant figures**. The number 0.048 729 has five significant figures, and it also has five significant figures when expressed in the form 4.8729×10^{-2}. Significant figures are discussed in Appendix A.

We have described the distance between the hydrogen and oxygen atoms in terms of meters. The meter is just one of many units of measurement. If we had been talking about the distance between two cities, we would have used miles or kilometers. If we were talking about the weight of this book, we probably would use kilograms or pounds and ounces. Unless a number is accompanied by a unit, it tells us nothing about the quantity measured.

Throughout history many units of measurement have been used, but in recent times two sets of units have become widely adopted. In most English-speaking countries length has traditionally been measured in inches, feet, yards, and miles; weight has been measured in ounces, pounds, and tons. In the rest of the world, however, lengths are commonly measured in meters or in decimal multiples or fractions of the meter, such as the kilometer (1000 meters) and the centimeter (0.01 meters). Masses are measured in grams or in decimal multiples or fractions of the gram, such as the kilogram (1000 grams) or the milligram (10^{-3} gram). Weight is very often (incorrectly) expressed in the same units (see Box 1.2).

The Metric System and SI Units

The system of units based on the meter and the gram is known as the **metric system**. This system was first adopted in France after the French Revolution (1789), and its common use quickly spread throughout Europe and beyond. In 1799 Thomas Jefferson tried to persuade the U.S. Congress to adopt the metric system, which he said would eventually become the standard for the world, but even today Americans resist changing from the English to the metric system.

Box 1.2

The *mass* of an object is the quantity that measures its resistance to a change in its state of rest or motion. For a change in the state of rest or of motion of an object, a force must be applied. An object of small mass such as a pebble needs only a small force to set it in motion if it is at rest or to change its motion if it is already moving. An object of large mass such as a space rocket needs a large force to set it in motion or to change its motion.

Mass can be measured by the force necessary to give an object a given acceleration. On earth we use the force of gravitational attraction of the earth for an object to measure its mass. We call this force the *weight* of the object; it depends on the mass of the object, the mass of the earth, and the distance of the object from the center of the earth.

A given object always has the same mass. But if it were taken to a planet with a mass different from that of the earth, it would be subject to a different gravitational force and would therefore have a different weight. On the surface of the earth, although the mass of the object and that of the earth are constant, the distance of an object from the center of the earth may vary slightly—from the top of a mountain to the bottom of a valley, for example—and therefore its weight will also vary slightly. Because the earth is not a perfect sphere but is slightly flattened at the poles, an object at one of the poles is slightly closer to the center of the earth than is an object at the equator. Thus a person weighing 140.0 lb at the equator would weigh 140.8 lb at the North Pole. So although the mass of an object is a constant, its weight depends on its location. The mass of an object is determined by comparison with a set of standard masses. Since the weights of two objects of equal mass are the same at any one place on the earth's surface, these objects balance each other when placed on the pans of a balance with arms of equal length.

In SI *mass* is measured in kilograms or convenient multiples of this unit, but *weight*, which is a force, is measured in newtons. In the English system weight is measured in pounds, while mass is measured in slugs, which is a unit you will not encounter in chemistry. Therefore we can quite correctly say that an object with a mass of 1 kg weighs 2.205 lb. But the terms *mass* and *weight* are frequently misused in everyday conversation, so we often speak of a weight of 1 kg rather than a mass of 1 kg. This misuse is unfortunate, but it is unlikely to cause any confusion if we understand the difference between mass and weight and recognize that *weight* is often used when *mass* is meant.

In science, however, the metric system was universally adopted, because the decimal relationship between different units of length and mass makes this system very convenient. Table 1.3 includes some metric system–English system equivalents.

Since 1899 international conferences have been held for the purposes of agreeing on systems of units, of providing accurate definitions of units, and of defining any new units that might be required. In 1960 the eleventh International Conference on Weights and Measures proposed some major changes to the metric system and suggested a new name for this modified metric system,

Table 1.3 Metric System–English System Equivalents

QUANTITY	METRIC AND ENGLISH UNITS
Length	1 meter = 1.094 yards
	2.540 centimeters = 1 inch
	1 kilometer =· 0.6214 miles
Mass	1 kilogram = 2.205 pounds[a]
	453.6 grams = 1 pound[a]
Volume	1 liter = 1.06 quarts (U.S.)
	1 cubic foot = 28.32 liters

[a] A pound is, strictly speaking, a unit of force (weight), not mass (see Box 1.2).

Table 1.4 The Seven SI Base Units

QUANTITY MEASURED	NAME OF UNIT	SYMBOL OF UNIT
Length	Meter	m
Mass	Kilogram	kg
Time	Second	s
Electric current	Ampere	A
Temperature	Kelvin	K
Amount of substance	Mole	mol
Luminous intensity	Candela	cd

the **International System of Units**. This name is abbreviated as **SI**, from the French *Système International d'Unités*. All the major countries in the world have agreed to adopt SI, not only for scientific purposes but also for everyday use.

The seven basic units of SI are given in Table 1.4. For the present we will consider in detail only length and mass. The other basic units will be considered as they are needed.

The Meter and Units of Length

The meter (m), the basic SI unit of length, is not convenient for expressing the dimensions of the water molecule or for many other purposes. Therefore a series of prefixes that represent various powers of ten have been defined in SI. These prefixes, which are listed in Table 1.5, allow us to reduce or enlarge the SI base units to an appropriate size. Figure 1.13 shows how these prefixes can be applied to the meter to describe a very wide range of lengths.

Table 1.5 Prefixes for Decimal Fractions and Multiples of SI Units

PREFIX	SYMBOL FOR PREFIX		SCIENTIFIC NOTATION
Exa	E	1 000 000 000 000 000 000	10^{18}
Peta	P	1 000 000 000 000 000	10^{15}
Tera	T	1 000 000 000 000	10^{12}
Giga	G	1 000 000 000	10^{9}
Mega	M	1 000 000	10^{6}
Kilo	k	1 000	10^{3}
Hecto	h	100	10^{2}
Deca	da	10	10^{1}
—	—	1	10^{0}
Deci	d	0.1	10^{-1}
Centi	c	0.01	10^{-2}
Milli	m	0.001	10^{-3}
Micro	μ	0.000 001	10^{-6}
Nano	n	0.000 000 001	10^{-9}
Pico	p	0.000 000 000 001	10^{-12}
Femto	f	0.000 000 000 000 001	10^{-15}
Atto	a	0.000 000 000 000 000 001	10^{-18}

Note: The more commonly used prefixes are printed in bold type.

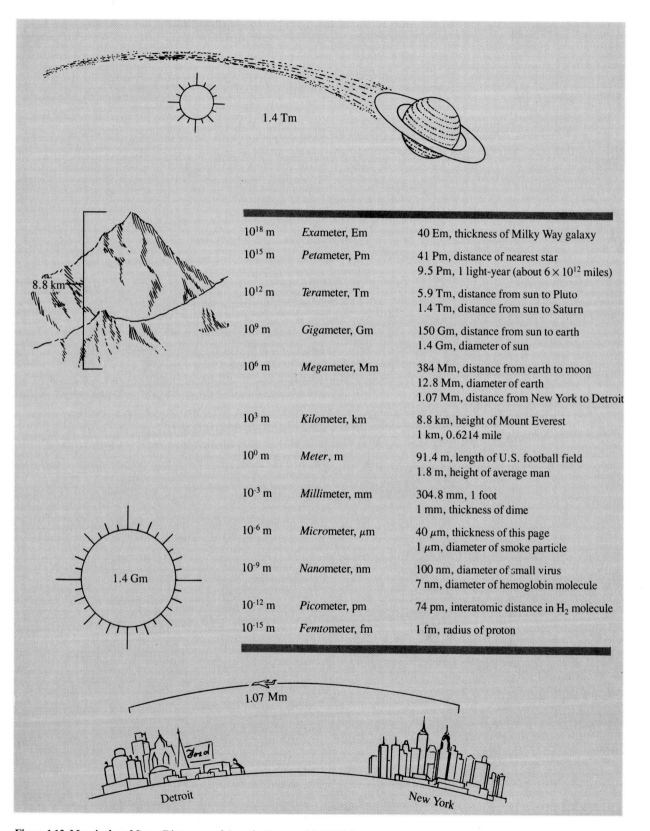

10^{18} m	*Exa*meter, Em	40 Em, thickness of Milky Way galaxy
10^{15} m	*Peta*meter, Pm	41 Pm, distance of nearest star 9.5 Pm, 1 light-year (about 6×10^{12} miles)
10^{12} m	*Tera*meter, Tm	5.9 Tm, distance from sun to Pluto 1.4 Tm, distance from sun to Saturn
10^{9} m	*Giga*meter, Gm	150 Gm, distance from sun to earth 1.4 Gm, diameter of sun
10^{6} m	*Mega*meter, Mm	384 Mm, distance from earth to moon 12.8 Mm, diameter of earth 1.07 Mm, distance from New York to Detroit
10^{3} m	*Kilo*meter, km	8.8 km, height of Mount Everest 1 km, 0.6214 mile
10^{0} m	*Meter*, m	91.4 m, length of U.S. football field 1.8 m, height of average man
10^{-3} m	*Milli*meter, mm	304.8 mm, 1 foot 1 mm, thickness of dime
10^{-6} m	*Micro*meter, μm	40 μm, thickness of this page 1 μm, diameter of smoke particle
10^{-9} m	*Nano*meter, nm	100 nm, diameter of small virus 7 nm, diameter of hemoglobin molecule
10^{-12} m	*Pico*meter, pm	74 pm, interatomic distance in H_2 molecule
10^{-15} m	*Femto*meter, fm	1 fm, radius of proton

Figure 1.13 Magnitudes of Some Distances and Lengths Expressed in SI Units.

The O–H distance in the water molecule, 9.7×10^{-11} m, can then be expressed as 97 picometers (pm) or as 0.097 nanometer (nm). We will normally use the picometer as the unit for molecular dimensions. The N–H distance in the ammonia molecule is 102 pm, and the P–P distance in the phosphorus, P_4, molecule is 221 pm.

Prior to the adoption of SI, chemists commonly used another unit, the *angstrom* (Å), equal to 10^{-16} m, for molecular dimensions. Although such non-SI units are being phased out, you should be familiar with the angstrom because it has been used very widely in the past and will undoubtedly continue to be used for some time in the future. The O–H distance of 97 pm in the water molecule is 0.97 Å. In general, the relationships between the angstrom (Å), the nanometer (nm), and the picometer (pm) are

$$1 \text{ nm} = 10 \text{ Å} \qquad \text{and} \qquad 1 \text{ pm} = 0.01 \text{ Å}$$

The Kilogram and Units of Mass

The standard mass of the metric system is a cylinder of corrosion-resistant, platinum-iridium alloy kept at the International Bureau of Weights and Measures in Sèvres, France. Its mass is defined as exactly 1 kilogram; all other masses are determined by comparison with this standard (Figure 1.14). The SI base unit of mass, the kilogram, is unusual because it already contains a prefix. Nevertheless, the standard prefixes are applied to the gram when larger or smaller units are needed. For example, the quantity 10^6 kg (1 million kg) is written as 1 Gg (gigagram), not as 1 Mkg (megakilogram). Figure 1.15 illustrates the wide range of masses that we encounter on earth. Mass and weight are often confused; the difference between them is explained in Box 1.2.

Conversion of Units

On many occasions we must convert from one set of units to another. For instance, we may need to convert one SI unit to another, English units to SI, or vice versa. A very convenient way of doing these conversions, as well as many other types of problems, is the **unit factor method**. To use this method, we write

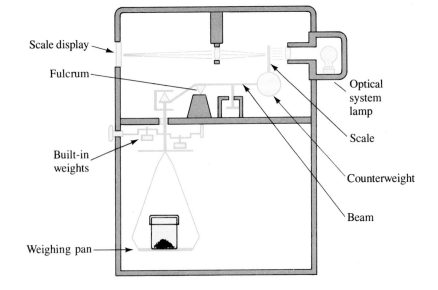

Figure 1.14 Chemical Balance.
The mass of an object is found by comparing its mass with standard masses (weights), using a chemical balance. A traditional balance consists of a beam supported at its midpoint with a pan at each end. A modern chemical balance has only one pan. The mass of the pan and a set of weights on the beam are balanced with a counterweight. When an object is added to the pan, weights are removed from the beam until it balances. Other modern balances operate electronically.

Scale display

Fulcrum

Built-in weights

Optical system lamp

Scale

Counterweight

Beam

Weighing pan

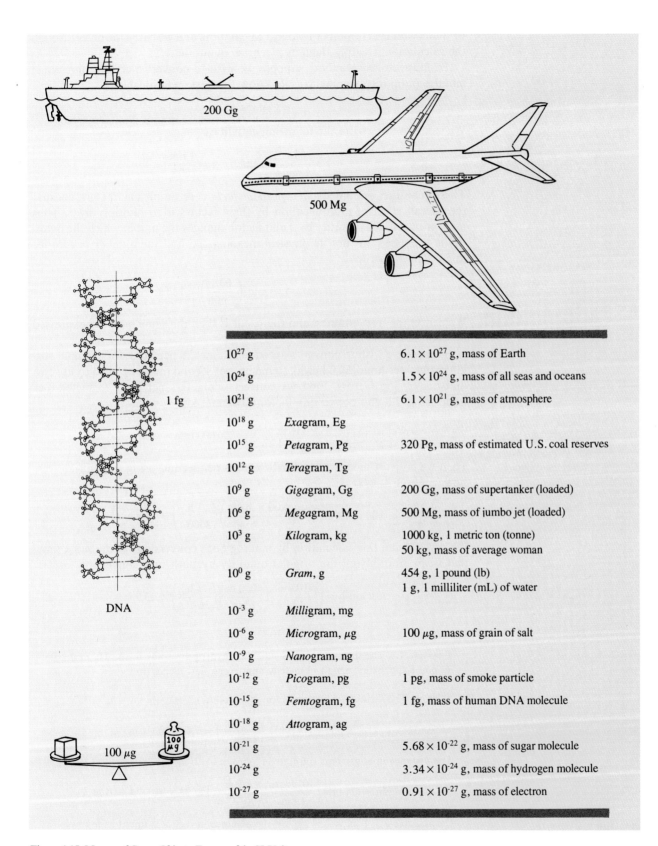

10^{27} g		6.1×10^{27} g, mass of Earth
10^{24} g		1.5×10^{24} g, mass of all seas and oceans
10^{21} g		6.1×10^{21} g, mass of atmosphere
10^{18} g	*Exa*gram, Eg	
10^{15} g	*Peta*gram, Pg	320 Pg, mass of estimated U.S. coal reserves
10^{12} g	*Tera*gram, Tg	
10^{9} g	*Giga*gram, Gg	200 Gg, mass of supertanker (loaded)
10^{6} g	*Mega*gram, Mg	500 Mg, mass of jumbo jet (loaded)
10^{3} g	*Kilo*gram, kg	1000 kg, 1 metric ton (tonne) 50 kg, mass of average woman
10^{0} g	*Gram*, g	454 g, 1 pound (lb) 1 g, 1 milliliter (mL) of water
10^{-3} g	*Milli*gram, mg	
10^{-6} g	*Micro*gram, μg	100 μg, mass of grain of salt
10^{-9} g	*Nano*gram, ng	
10^{-12} g	*Pico*gram, pg	1 pg, mass of smoke particle
10^{-15} g	*Femto*gram, fg	1 fg, mass of human DNA molecule
10^{-18} g	*Atto*gram, ag	
10^{-21} g		5.68×10^{-22} g, mass of sugar molecule
10^{-24} g		3.34×10^{-24} g, mass of hydrogen molecule
10^{-27} g		0.91×10^{-27} g, mass of electron

200 Gg

500 Mg

1 fg

DNA

100 μg

Figure 1.15 Masses of Some Objects Expressed in SI Units.

the units with every number used in a calculation, and we carry the units through the calculation, treating them as algebraic quantities.

To illustrate this method, suppose we wish to convert a certain number of minutes (min) into seconds (s). We know the basic relationship:

$$1 \text{ min} = 60 \text{ s}$$

We can therefore write the following equalities:

$$1 = \frac{60 \text{ s}}{1 \text{ min}} \quad \text{and} \quad 1 = \frac{1 \text{ min}}{60 \text{ s}}$$

These quantities are called *unit conversion factors*, or simply **unit factors**, because the overall effect of multiplication by these factors is to multiply by 1. Thus multiplication of a quantity by a unit factor changes the unit in which the quantity is expressed but not its physical meaning. To find the number of seconds in 3.0 min, we write

$$3.0 \text{ min} = (3.0 \text{ min})\left(\frac{60 \text{ s}}{1 \text{ min}}\right) = 180 \text{ s}$$

We cancel the unit minutes, and the result is then the same quantity expressed in seconds.

Suppose we wish to convert a speed of kilometers per hour, for example 50.0 kilometers per hour (50.0 km/h), to meters per second (m/s); we need to use two unit conversion factors. First we convert kilometers per hour to meters per hour using the unit conversion factor (1000 m/1 km)

$$50.0 \text{ km/h} = \left(\frac{50.0 \text{ km}}{1 \text{ h}}\right)\left(\frac{1000 \text{ m}}{1 \text{ km}}\right) = 50\,000 \text{ m/h} = 5.00 \times 10^4 \text{ m/h}$$

Then we convert meters per hour to meters per second using the unit conversion factor (1 h/3600 s)

$$5.00 \times 10^4 \text{ m/h} = \left(\frac{5.00 \times 10^4 \text{ m}}{1 \text{ h}}\right)\left(\frac{1 \text{ h}}{3600 \text{ s}}\right) = 13.9 \text{ m/s}$$

We can shorten this calculation by making both conversions at the same time; that is, by multiplying the original quantity by both unit conversion factors:

$$50.0 \text{ km/h} = \left(\frac{50.0 \text{ km}}{1 \text{ h}}\right)\left(\frac{1000 \text{ m}}{1 \text{ km}}\right)\left(\frac{1 \text{ h}}{3600 \text{ s}}\right) = 13.9 \text{ m/s}$$

Example 1.1 The distance between the two hydrogen atoms in the H_2 molecule is 74 pm. Express this length in meters, angstroms, and inches (in.).

Solution The distance in meters is found as follows:

$$74 \text{ pm} = (74 \text{ pm})\left(\frac{10^{-12} \text{ m}}{1 \text{ pm}}\right) = 7.4 \times 10^{-11} \text{ m}$$

The distance in angstroms is found as follows:

$$74 \text{ pm} = (74 \text{ pm})\left(\frac{10^{-12} \text{ m}}{1 \text{ pm}}\right)\left(\frac{1 \text{ Å}}{10^{-10} \text{ m}}\right) = 74 \times 10^{-2} \text{ Å} = 0.74 \text{ Å}$$

From Table 1.3, 1 in. = 2.54 cm. So the distance in inches is found as follows:

$$74 \text{ pm} = (74 \text{ pm})\left(\frac{10^{-10} \text{ cm}}{1 \text{ pm}}\right)\left(\frac{1 \text{ in.}}{2.54 \text{ cm}}\right) = 2.9 \times 10^{-9} \text{ in.}$$

Any number of unit factors may be used as required. In each case they are applied so that the units of the preceding factor cancel. Notice that for every equality such as

$$1 \text{ kg} = 2.205 \text{ lb}$$

there are two unit conversion factors. One is the reciprocal of the other.

$$\frac{1 \text{ kg}}{2.205 \text{ lb}} = \frac{2.205 \text{ lb}}{1 \text{ kg}} = 1$$

As an example of the use of the appropriate unit conversion factor, consider the following problem. If a woman weighs 140 lb, what is her mass in kilograms? We choose the unit factor that cancels the unit pounds:

$$140 \text{ lb} = (140 \text{ lb})\left(\frac{1 \text{ kg}}{2.205 \text{ lb}}\right) = 63.5 \text{ kg}$$

If we had used the other (incorrect) unit factor, we would have obtained

$$140 \text{ lb} = (140 \text{ lb})\left(\frac{2.205 \text{ lb}}{1 \text{ kg}}\right) = 309 \text{ lb}^2/\text{kg}$$

Because we see that the units do not cancel to give the desired units, we know that we have used the wrong unit factor.

Volume

Many other units are derived from the seven SI base units. One important derived unit is the unit of volume. If each side of a cube has a length of 1 m, its volume is 1 cubic meter (1 m^3). For many objects this unit of volume is rather large, so the cubic decimeter (dm^3) or, more commonly, the cubic centimeter (cm^3) is used instead.

Because 1 dm $= 10^{-1}$ m, we can write the unit conversion factor

$$\frac{10^{-1} \text{ m}}{1 \text{ dm}} = 1$$

Taking the cube of both sides of this equation gives the unit conversion factor $(10^{-1} \text{ m}/1 \text{ dm})^3 = 1$. Hence

$$1 \text{ dm}^3 = (1 \text{ dm}^3)\left(\frac{10^{-1} \text{ m}}{1 \text{ dm}}\right)^3 = (10^{-1} \text{ m})^3 = 10^{-3} \text{ m}^3$$

Similarly, since 1 cm $= 10^{-2}$ m, we can write the unit conversion factor

$$\frac{10^{-2} \text{ m}}{1 \text{ cm}} = 1$$

Hence

$$1 \text{ cm}^3 = (1 \text{ cm}^3)\left(\frac{10^{-2} \text{ m}}{1 \text{ cm}}\right)^3 = 10^{-6} \text{ m}^3$$

Prior to the development of SI, the liter (L) and the milliliter—10^{-3} L—(mL), were used to measure the volume of liquids and solutions. The liter was originally defined as the volume of 1 kg of water at the temperature of its maximum density (3.98°C), but it has since been redefined as exactly one-thousandth of a cubic meter, that is, 1 dm^3. Hence a milliliter is exactly 1 cm^3. The units liter and cubic decimeter, and milliliter and cubic centimeter, can therefore be used interchangeably. Most laboratory equipment for measuring volumes is graduated in liters and milliliters, and the use of these alternative names will continue for some time. We will use the units liter and milliliter extensively in this book.

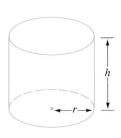

Example 1.2 A cylinder has a diameter of 5.00 cm and contains water to a height of 25.0 cm. Find the volume of water in cubic centimeters, cubic meters, and cubic inches (in.^{3}).

Solution The volume of a cylinder is given by the expression

$$\text{Volume} = \pi r^2 h$$

where r is the radius and h the height. For this example the diameter is 5.00 cm, so the radius r is 2.50 cm. Hence

$$\text{Volume of water} = \pi r^2 h = 3.142(2.50 \text{ cm})^2(25.0 \text{ cm}) = 491 \text{ cm}^3$$

Since 1 cm = 10^{-2} m, then $1 = 10^{-2}$ m/1 cm. Hence the volume in cubic meters is found as follows:

$$491 \text{ cm}^3 = (491 \text{ cm}^3)\left(\frac{10^{-2} \text{ m}}{1 \text{ cm}}\right)^3 = 491 \times 10^{-6} \text{ m}^3 = 4.91 \times 10^{-4} \text{ m}^3$$

Since 1 in. = 2.54 cm, then $1 = 1$ in/2.54 cm. Hence the volume in cubic inches is found as follows:

$$491 \text{ cm}^3 = (491 \text{ cm}^3)\left(\frac{1 \text{ in.}}{2.54 \text{ cm}}\right)^3 = 30.0 \text{ in.}^3$$

Density

The common statements that mercury is "heavier" than water or that iron is "heavier" than aluminum are inaccurate—or at least incomplete. What is actually being compared is not mass, which depends on the amount of a substance, but *mass per unit volume*, which is known as **density**.

$$\text{Density} = \frac{\text{Mass}}{\text{Volume}}$$

The mass of 1 cm^3 of water is 1.00 g, while the mass of 1 cm^3 of mercury is 13.6 g at 20°C. In other words, the density of water is 1.00 g cm^{-3}, and the density of mercury is 13.6 g cm^{-3}. The units for density are commonly written as g/cm^3 or g cm^{-3}. In the first expression the slash denotes "per." In the second expression 1/cm^3 has been written as cm^{-3}, so g/cm^3 becomes g cm^{-3}. Thus units are treated like algebraic quantities; for example, $1/x^2 = x^{-2}$, and similarly $1/\text{cm}^2 = \text{cm}^{-2}$.

Unlike mass and volume, the density of a substance is independent of the amount and the size of the sample. Density can therefore be used as an aid in distinguishing one pure substance from another. Table 1.6 lists the densities of some common substances.

Example 1.3 The density of liquid mercury is 13.6 g cm^{-3}. What is the mass of 1 L of mercury in grams, in kilograms, and in pounds?

Solution We have the relationship 1 L = 1000 cm^3, and since density = mass/volume, we can write mass = (volume)(density). Therefore the mass of 1 L of mercury is

Table 1.6 Densities of Various Substances at 20°C

SUBSTANCE	PHYSICAL STATE[a]	DENSITY $(g\ cm^{-3})$
Oxygen	g	0.001 43[b]
Hydrogen	g	0.000 090[b]
Alcohol (ethanol)	l	0.785
Benzene	l	0.880
Water	l	0.998
Magnesium	s	1.74
Salt (sodium chloride)	s	2.16
Aluminum	s	2.70
Iron	s	7.87
Copper	s	8.96
Silver	s	10.5
Lead	s	11.34
Mercury	l	13.6
Gold	s	19.32

[a] The abbreviations are g, gas; l, liquid; s, solid.
[b] At STP (standard temperature and pressure, that is 0°C and 1 atmosphere); see Chapter 3.

$(1\ L)(13.6\ g\ cm^{-3})$, or

$$(1\ L)\left(\frac{1000\ cm^3}{1\ L}\right)(13.6\ g\ cm^{-3}) = (1000\ cm^3)(13.6\ g\ cm^{-3})$$

$$= 1.36 \times 10^4\ g$$

The mass in kilograms is

$$1.36 \times 10^4\ g = (1.36 \times 10^4\ g)\left(\frac{1\ kg}{1000\ g}\right) = 13.6\ kg$$

From Table 1.3 we have the relationship 1 lb = 453.6 g, so the mass in pounds is

$$1.36 \times 10^4\ g = (1.36 \times 10^4\ g)\left(\frac{1\ lb}{453.6\ g}\right) = 30.0\ lb$$

Alternatively, we may use the relationship 1 kg = 2.205 lb:

$$13.6\ kg = (13.6\ kg)\left(\frac{2.205\ lb}{1\ kg}\right) = 30.0\ lb$$

1.3 SOLIDS, LIQUIDS, AND GASES

One of the most obvious properties of substances is that they can exist as *solids, liquids,* or *gases.* We call these the **three states of matter.** Many substances can exist in all three forms under appropriate conditions. When a gas is cooled, it eventually condenses to a liquid and finally freezes to a solid, but it remains the same substance. Water exists in all three forms on the earth's surface. Gaseous water (water vapor) is present in the atmosphere; liquid water is present in rivers, lakes, and oceans; and solid water (ice) is present in snow, in glaciers, and on the surfaces of frozen lakes and oceans.

Macroscopic Description

A **gas** is distinguished from other states of matter by two characteristic properties: (1) it is a fluid with no definite shape, and (2) it has no definite intrinsic volume but flows and expands to fill any container in which it is placed. If the volume of the container is reduced, the gas is easily compressed to the smaller volume. A **liquid** is also a fluid, but a given amount of liquid has its own definite volume. A liquid flows and takes the shape of a container, but it does not expand to completely fill a container of larger volume. In contrast, a **solid** is not a fluid; it does not flow. Any piece of a solid has a definite size and shape that does not depend on its container. Moreover, this shape can only be changed by exerting considerable forces on the solid. Unlike a gas, solids and liquids are only slightly compressible. A much greater force is needed to compress a liquid or a solid than a gas. A liquid usually has a much greater density than a gas, and a solid usually has a slightly greater density than the corresponding liquid.

These descriptions of solids, liquids, and gases are based on observations with our unaided senses. They are **macroscopic descriptions** of the different states of matter, that is, descriptions in terms of properties such as shape, fluidity, density, and hardness that we can recognize by using our unaided senses.

Microscopic Description

One of the main aims of chemistry—and, indeed, of all of the sciences—is to explain the properties of matter in terms of a theory or a model of the fundamental nature of matter. The model that is basic to modern chemistry is the atomic theory that was first clearly enunciated by Dalton, namely, that matter consists of very small particles called atoms. The atomic description of matter is a **microscopic description**, in contrast to the macroscopic description that we obtain with our unaided senses. In chemistry we are constantly striving to obtain a better understanding of the observed properties of matter in terms of the behavior of atoms and molecules.

We can use the element bromine as an example of the macroscopic and microscopic descriptions of the three states of matter. At room temperature bromine is a deep red brown liquid with a strong pungent odor. It causes severe burns on the skin. It freezes at $-7.2°C$ and boils at $58.8°C$. If a small amount of liquid bromine is sealed in a glass tube, we can freeze it to a brown solid by placing the tube in dry ice or an ice-salt mixture. If the tube is gently warmed, the bromine melts to a brown liquid. And if the warming is continued, the tube becomes filled with brown bromine gas (Figure 1.16).

What has happened at the atomic level that would explain these changes? Bromine consists of diatomic molecules, Br_2. The molecules are held together by forces of attraction that act between all molecules. These forces are called **intermolecular forces**; we will discuss them in detail in later chapters. In solid bromine the molecules are packed closely together in a regular manner, as Figure 1.16 illustrates. Although each molecule can move a little, its motion is restricted by the other molecules packed around it, and each individual molecule can do no more than oscillate and vibrate slightly around a fixed mean position. Because the molecules are packed closely together and cannot be easily displaced, the density of the solid is rather high (4.2 g cm^{-3}), and it is relatively hard and rigid.

As the temperature of solid bromine is increased, the molecules vibrate and oscillate more and more violently. Eventually, they break loose from their fixed

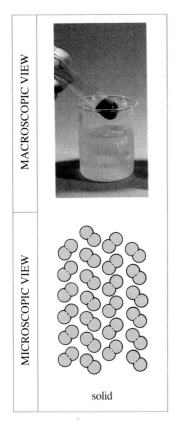

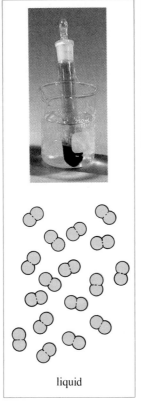

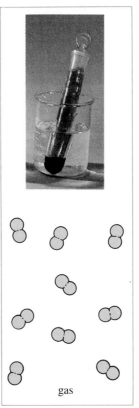

solid   liquid   gas

Figure 1.16 Three States of Bromine. At $-15°C$ bromine is a dark brown solid. The bromine molecules are packed closely together in a regular arrangement. At $0°C$ bromine is a dark brown liquid. The molecules are still packed closely together, but their arrangement is less regular, and they are constantly moving around each other. At $40°C$ brown bromine gas is seen clearly in the tube together with the liquid. In the gas bromine molecules are far apart and are moving rapidly so that they fill the whole of the tube. Because the molecules are far apart, the color of the gas is much less intense than the color of the liquid.

positions and move around each other. They remain packed rather closely together but in a random rather than a regular fashion. The solid melts and becomes a liquid, which is fluid because the molecules are now free to move around each other. Because of the increased motion of the molecules and their irregular packing, they take up a little more space than they did in the solid, and the density of the liquid (3.14 g cm^{-3}) is less than the density of the solid (4.2 g cm^{-3}).

As the temperature of the liquid is increased, the molecules move still more rapidly. And some of them move fast enough to fly off the surface of the liquid bromine, becoming bromine vapor, or gas. Eventually, at a temperature of $58.8°C$ all the bromine is converted to the gaseous state; in other words, bromine boils at $58.8°C$ at normal atmospheric pressure. When a gas is formed from a liquid, it is often called a **vapor**. Thus we often speak of bromine vapor, but there is no difference between a vapor and a gas.

In the gaseous state the bromine molecules are rather far apart and are moving rapidly and randomly through space. Because the molecules no longer stick closely together, the gas expands to fill any container in which it is placed. The density of the gas at the boiling point, $5.9 \times 10^{-3} \text{ g cm}^{-3}$, is much lower than the density of the liquid. One gram of gaseous bromine at normal atmospheric pressure and $58.8°C$ occupies a volume that is 500 times greater than the volume of 1 g of liquid bromine under the same conditions. Whereas the molecules in a liquid are packed closely together, the molecules in a gas are far apart. So the total volume occupied by the gas at ordinary pressure is very large compared with the space taken up by the molecules themselves.

Example 1.4 Ice has a density of 0.91 g cm^{-3} at $0°C$. Liquid water has a density of 1.00 g cm^{-3} at $25°C$. Steam has a density of $5.9 \times 10^{-4} \text{ g cm}^{-3}$ at $100°C$ and 1 atmosphere (atm) pressure. Calculate the volume of 100 g of water (a) as a solid, (b) as a liquid, and (c) as a gas. Also, calculate the ratios of these volumes.

Solution First, recall that density = mass/volume. Therefore

$$\text{Volume} = \frac{\text{Mass}}{\text{Density}}$$

(a)   Volume of 100 g ice $= \dfrac{100 \text{ g}}{0.91 \text{ g cm}^{-3}} = 110 \text{ cm}^3$

$$= (110 \text{ cm}^3)\left(\frac{1 \text{ L}}{1000 \text{ cm}^3}\right) = 0.11 \text{ L}$$

(b) Volume of 100 g water $= \dfrac{100 \text{ g}}{1.00 \text{ g cm}^{-3}} = 100 \text{ cm}^3 = 0.10 \text{ L}$

(c) Volume of 100 g steam $= \dfrac{100 \text{ g}}{5.9 \times 10^{-4} \text{ g cm}^{-3}} = 1.7 \times 10^5 \text{ cm}^3 = 170 \text{ L}$

The ratios of these volumes is as follows:

$$\text{ice} : \text{water} : \text{steam} = 0.11 : 0.10 : 170$$

Dividing by 0.10 gives

$$= 1.1 : 1.0 : 1700$$

Thus steam has a volume that is 1700 times as great as the volume of an equal mass of liquid water. This again emphasizes that the water molecules in steam are much further apart than they are in liquid water. Note also that water is a very unusual liquid in that the density of the liquid is greater than the density of the solid (ice). This unusual property of water is discussed in Chapter 13.

1.4 SOLUTIONS

A homogeneous mixture of two or more substances is usually known as a **solution**. Mixtures of gases are homogeneous and they are therefore solutions, but the term is usually employed to describe homogeneous mixtures of two or more liquids or of a liquid and one or more solids. When two substances are mixed to form a solution, one is said to dissolve in the other, or to be soluble in the other. We normally refer to the substance present in largest amount as the **solvent** and the other substances as **solutes**. If the solution consists of a solid in a liquid, however, the solid is always called the solute and the liquid the solvent. For simplicity we confine our discussion here to solutions obtained by mixing only two substances.

Types of Solutions

Liquids are probably the most familiar solvents. Many solid substances dissolve in particular liquids. For example, sodium chloride dissolves in water but not in gasoline, whereas paraffin wax dissolves in gasoline but not in water. Liquids may also dissolve in other liquids; for example, ethanol dissolves in water, and lubricating oil dissolves in gasoline. Liquids that are soluble in each other are said to be **miscible**. But not all liquids are soluble in each other.

Although ethanol dissolves in water in all proportions, both gasoline and carbon tetrachloride are insoluble in water (Experiment 1.6). Liquids that are insoluble in each other are said to be **immiscible**.

Gases also dissolve in many liquids. A solution of ammonia, NH_3, in water is commonly used as a household cleaner. Oxygen and nitrogen are soluble in water to a small extent. In fact, fish and other aquatic life depend on the small amount of dissolved oxygen in rivers, lakes, and oceans. Solutions in water are the most widely used solutions in chemistry; they are called **aqueous solutions**.

Gases may also dissolve in solids. For example, hydrogen is soluble in platinum. A solid may also form a solid solution with another solid. Sterling silver is a solution of copper in silver. Solid solutions of metals are generally called *alloys*, although an alloy may also be a heterogeneous mixture of metals.

Concentration

The term **concentration** is used to describe the amount of solute dissolved in a given quantity of solution. A solution having a relatively small concentration of solute is said to be **dilute**. A solution having a large concentration of solute is said to be **concentrated**.

One way of expressing concentration in a quantitative manner is as the **mass percentage** (mass percent) of the solute, which is the number of grams of solute in 100 g of solution. For example, a solution of 10 g of alcohol (ethanol) in 90 g of water (100 g of solution) is a 10% solution. For a solution containing two components (solute plus solvent),

$$(\text{Mass percent})_{\text{solute}} + (\text{Mass percent})_{\text{solvent}} = 100$$

Beer is typically 4% to 6% alcohol and wine 10% to 14%.

Immiscible Liquids and Solubility

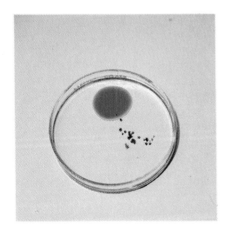

Water and carbon tetrachloride, CCl_4, are immiscible. Carbon tetrachloride forms a small pool when added to water. Solid iodine dissolves in the colorless carbon tetrachloride to give a violet solution but it is insoluble in water.

Copper sulfate dissolves in water to give a blue solution but it is insoluble in carbon tetrachloride, which remains colorless.

Example 1.5 An aqueous solution is prepared by adding 3.42 g of magnesium chloride, $MgCl_2$, and 2.63 g of sodium chloride, NaCl, to 88.20 g of water, H_2O. What are the concentrations, in mass percentage, of (a) NaCl, (b) $MgCl_2$, and (c) H_2O?

Solution First, we need to know the total mass of the solution:

$$\text{Total mass of solution} = \text{Mass NaCl} + \text{Mass } MgCl_2 + \text{Mass } H_2O$$
$$= 2.63 \text{ g} + 3.42 \text{ g} + 88.20 \text{ g} = 94.25 \text{ g}$$

(a) There is 2.63 g of NaCl in 94.25 g of solution; therefore

$$(2.63 \text{ g})\left(\frac{100 \text{ g}}{94.25 \text{ g}}\right) = 2.79 \text{ g in 100 g solution}$$

The concentration of NaCl is thus 2.79 mass percent.

(b) There is 3.42 g of $MgCl_2$ in 94.25 g of solution; thus

$$(3.42 \text{ g})\left(\frac{100 \text{ g}}{94.25 \text{ g}}\right) = 3.63 \text{ g in 100 g solution}$$

The concentration of $MgCl_2$ is 3.63 mass percent.

(c) There is 88.20 g of H_2O in 94.25 g of solution; therefore

$$(88.2 \text{ g})\left(\frac{100 \text{ g}}{94.25 \text{ g}}\right) = 93.58 \text{ g in 100 g solution}$$

So the concentration of H_2O is 93.6 mass percent.

Finally, as a check, we note that

$$\text{Mass percent NaCl} + \text{Mass percent } MgCl_2 + \text{Mass percent } H_2O$$
$$= 2.79 + 3.63 + 93.6 = 100.0\%$$

Example 1.6 Concentrated aqueous nitric acid has 69.0% by mass of HNO_3 and has a density of 1.41 g cm^{-3}. What volume of this solution contains 14.2 g of HNO_3?

Solution The 14.2 g of HNO_3 is contained in

$$\left(\frac{100 \text{ g aqueous solution}}{69.0 \text{ g } HNO_3}\right)(14.2 \text{ g } HNO_3) = 20.6 \text{ g concentrated aqueous } HNO_3$$

Using the relationship volume = mass/density, we have

$$\text{Volume} = \frac{20.6 \text{ g}}{1.41 \text{ g cm}^{-3}} = 14.6 \text{ cm}^3$$

Frequently, we need to give values for the concentrations of substances that are present in solutions in very small amounts. For example, just 2×10^{-5} g of copper per liter of water is believed to be lethal to fish. Such very small concentrations may be more conveniently expressed in *parts per million* (ppm). If there is 2×10^{-5} g of copper per liter of water, there is 2×10^{-5} g per 1000 g of water, since the density of water is 1.00 g cm^{-3}. There is therefore 0.02 g per 1 000 000 g of water, or 0.02 parts per million, that is, 0.02 ppm.

Example 1.7 Seawater contains 0.0064 g of dissolved oxygen, O_2, per liter. The density of seawater is 1.03 g cm^{-3}. What is the concentration of oxygen, in parts per million?

Solution   Mass 1 L seawater = Volume × Density

$$= (1 \text{ L})\left(\frac{10^3 \text{ cm}^3}{1 \text{ L}}\right)(1.03 \text{ g cm}^{-3}) = 1.03 \times 10^3 \text{ g}$$

$$\text{Mass } O_2 \text{ in } 10^6 \text{ g seawater} = (6.4 \times 10^{-3} \text{ g } O_2)\left(\frac{1 \times 10^6 \text{ g seawater}}{1.03 \times 10^3 \text{ g seawater}}\right)$$

$$= 6.2 \text{ g } O_2$$

There is 6.2 g of O_2 in 10^6 g of seawater. Therefore, the concentration of O_2 in seawater is 6.2 ppm.

The concentrations of gases present in the atmosphere in small traces are also often expressed in parts per million, but in this case volumes are used rather than masses. For example, 1 ppm of a trace gas is equivalent to dissolving 1 cm³ of the gas in 1 000 000 cm³, or 1000 L, of air. The concentration of helium in the atmosphere is 5 ppm. In an industrial city the concentration of sulfur dioxide, SO_2, in the atmosphere may be as high as 5 ppm or more, while in open country at some distance from a large city, the concentration of SO_2 may be less than 0.01 ppm.

Solubility

The extent to which a solid will dissolve in a liquid is limited. If sugar is stirred into water at a given temperature, increasing amounts may be dissolved up to a certain point, after which no more sugar will dissolve. A solution that contains the maximum amount of solute that can be dissolved at a particular temperature is said to be **saturated**. The concentration of solute in a saturated solution is called the **solubility** of the solute in that solvent. Solubilities are often expressed in grams of solute per 100 g of solvent, or, alternatively, in grams of solute per liter of solution.

EQUILIBRIUM   Once a saturated solution has been prepared at a given temperature, no more solute will dissolve. No matter how much additional solid is added and no matter how long it is left, the concentration of the solute remains unchanged. But this unchanged concentration does not mean that nothing is happening. In fact, solute continues to dissolve, but the amount that dissolves per second is exactly equal to the amount that simultaneously comes out of the solution as solid—that is, by the amount that crystallizes.

In other words, two processes are taking place: the solution process and its reverse, which is called **crystallization**. The solute dissolves and crystallizes at the same rate. In a saturated solution the number of solute molecules going into solution exactly balances, at all times, the number leaving the solution to form the solid. Thus the concentration of solute in a saturated solution and the amount of undissolved solute remain constant at a particular temperature. The saturated solution is said to be in **equilibrium** with the excess solid solute. This situation is an example of a **dynamic equilibrium**, in which two opposing processes occur at the same rate so that there is no overall or net change (Figure 1.17). Using the concept of equilibrium, we can redefine a *saturated solution* as one that is in equilibrium with excess undissolved solute.

SOLUBILITY AND TEMPERATURE   The solubility of a substance in a given solvent depends on the temperature. Sugar, for example, is more soluble in hot water than in cold. Although the solubility of many substances in water increases with increasing temperature, as Figure 1.18 illustrates, the solubility

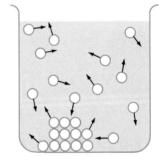

Figure 1.17 Dynamic Equilibrium in Saturated Solution. Solute molecules leave the crystal surface (dissolve) at the same rate as solute molecules attach themselves to the crystal surface (crystallize).

Figure 1.18 Temperature Dependence of Solubility. The effect of temperature on solubility varies from one substance to another. The solubility of potassium nitrate, KNO_3, increases very rapidly with increasing temperature. The solubility of sodium chloride, NaCl, hardly changes, while the solubility of sodium sulfate, Na_2SO_4, decreases with increasing temperature.

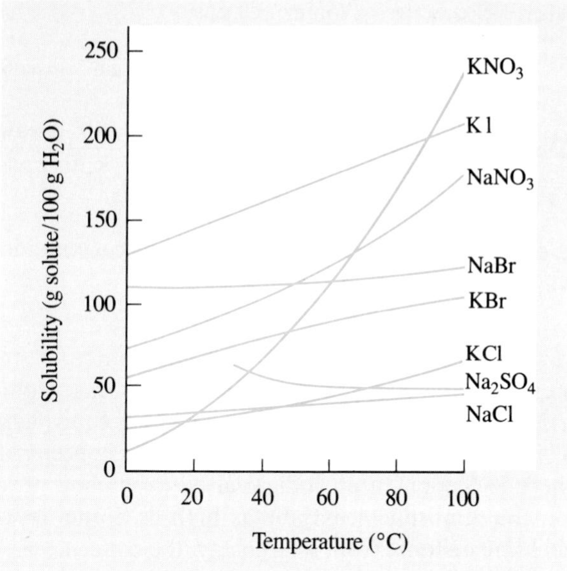

of some substances remains nearly constant or even decreases. The solubility in water of many gases such as carbon dioxide and oxygen also decreases with increasing temperature.

1.5 SEPARATION OF MIXTURES

The separation of mixtures into their component substances is an essential technique of experimental chemistry. If a reaction gives several products and we need one of them in a pure state, we must know how to separate the mixture. Chemists use many procedures for separating mixtures, but three of the most common are crystallization, distillation, and chromatography.

Crystallization

The **crystallization** method of purification depends on the change of solubility of a substance with a change in the temperature. Suppose, for example, that a sample of 80 g of potassium nitrate, KNO_3, contains 10 g of potassium chloride, KCl, and we want to obtain a pure sample of potassium nitrate. First, we dissolve the potassium nitrate–potassium chloride mixture in, for example, 100 g of water at a temperature of about 100°C. If we then allow the solution to cool slowly, it becomes saturated with potassium nitrate at about 50°C; then potassium nitrate starts to crystallize from the solution. If the solution is cooled to 0°C, 67 g of pure potassium nitrate crystallizes; only 13 g remains in solution because its solubility is 13 g in 100 g of water at 0°C. Since the solubility of potassium chloride at this temperature is 27 g in 100 g of water, all the 10 g of potassium chloride remains in solution.

Now we can separate solid potassium nitrate from the solution by the process of **filtration**, which is shown in Figure 1.19. The mixture of saturated solution and crystalline potassium nitrate is poured through a *filter funnel* containing either a porous paper known as *filter paper* or a *sintered glass disk*. The filter allows the solution to pass through but retains the solid potassium nitrate, which may then be washed free from adhering solution with small amounts of cold water and finally dried.

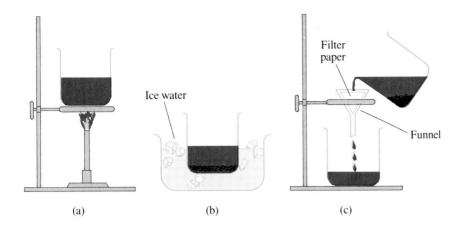

(a)   (b)   (c)

Figure 1.19 Crystallization and Filtration. (a) The solid is dissolved in a minimum amount of hot water (b) The solution is cooled until some solid crystallizes. (c) The mixture is poured through the filter funnel. The filter paper retains the solid, allowing the solution to pass through.

Crystallization does not separate the potassium nitrate and potassium chloride completely, since some of the potassium nitrate remains in solution with the potassium chloride. But we can, at least, obtain a large proportion of the originally impure potassium nitrate in a pure form. Crystallization is generally a rather efficient process for purification, particularly when the amounts of impurities are not too large and their solubilities are not very small. If one crystallization does not yield sufficiently pure material, it may be recrystallized.

Distillation

Distillation is a convenient method for separating the liquid from a solution of a solid in a liquid or for separating a solution of two or more liquids. The method is based on the differences in the **volatilities** of different substances, that is, differences in the ease with which they become gases. For example, pure water can be obtained from seawater in the apparatus shown in Figure 1.20. When seawater is heated, the water eventually boils. But at this temperature sodium chloride and the other salts are **nonvolatile**; they do not vaporize. Thus the water is completely converted to vapor, leaving a residue of the salts in the flask. The water vapor passes into the condenser, which is cooled by cold water running through its outer jacket. Thus the water vapor is cooled and changed

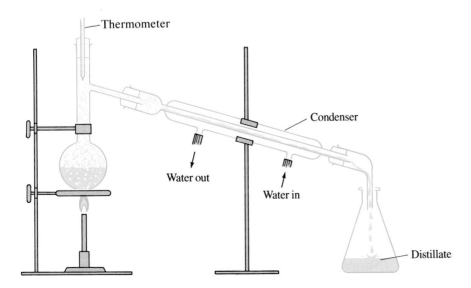

Figure 1.20 Distillation Apparatus. The nonvolatile component of the mixture remains in the distilling flask. The volatile component is converted to vapor, which passes into the water-cooled condenser. Here the vapor is converted to a liquid and flows into the receiving flask.

to liquid water. The process is known as **condensation**. The pure water runs out of the condenser into a receiver, where it is collected. The solid that remains in the flask after all the liquid has been removed contains whatever nonvolatile impurities were originally present. In the case of seawater the resulting solid is a mixture of sodium chloride and all the other substances present in seawater. Water that has been purified by distillation is commonly called *distilled water*. It is used in the laboratory whenever pure water is needed for preparing an aqueous solution.

Distillation can also be used to separate a solution of two or more liquids, although separation is not complete unless the two liquids have very different boiling points. For example, pure water may be obtained by distillation of a mixture of water and glycol, $C_2H_6O_2$ (a major component of antifreeze). The vapor entering the condenser is richer in water vapor than in glycol, which has a boiling point of 179°C. So the resulting condensed liquid, which is known as the **distillate**, contains a higher proportion of water and a lower proportion of glycol than the original mixture. The liquid remaining in the flask contains a higher proportion of glycol. Redistillation of the distillate further enriches it in the lower–boiling point component, and by repeated redistillations a sample of pure water can eventually be obtained.

Such a tedious and time-consuming procedure is avoided in practice by use of a *fractionating column* between the distillation flask and the condenser, as shown in Figure 1.21. The column is packed with some inert material, such as

Figure 1.21 Distillation with Fractionating Column. The column is packed with an inert material, such as glass beads. Vapor rises up the column and condenses on the beads. It is revaporized by the hot vapor rising up the column, and it condenses again further up in the column, and so on. The vapor becomes successively richer in the component of lower boiling point, and the pure, lower–boiling component leaves the top of the column and is reconverted to liquid in the condenser. The process of condensation and re-evaporation is equivalent to repeated distillations.

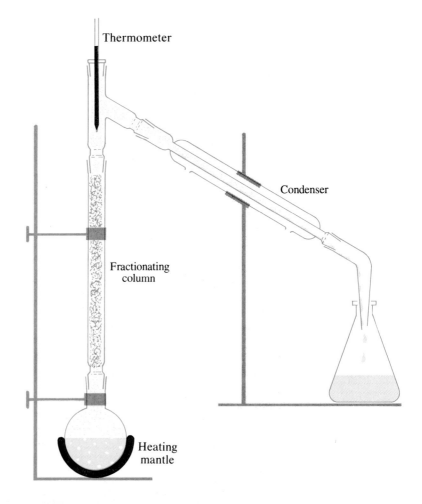

Thermometer

Condenser

Fractionating column

Heating mantle

glass beads, which does not react with the substances being distilled. The vapor ascending the column condenses on the glass beads. Then as this liquid trickles down the column, it is heated by the rising vapor. Some of it again vaporizes, to give a vapor still richer in the component with the lower boiling point. This vapor condenses further up in the column, is heated by the rising vapor, and again is partially vaporized, to give a vapor still richer in the lower–boiling point component. The vapor emerging from the top of a column of sufficient length condenses to a liquid, which is essentially the pure, lower–boiling point component. This process is known as **fractional distillation**.

Fractional distillation is important in many industrial processes. For example, petroleum (which is a complex mixture) is separated into gasoline, heating oil, and other materials by distillation in a fractionating column. Pure oxygen and pure nitrogen are obtained from air by first liquefying it at a low temperature and then carrying out a fractional distillation.

Chromatography

Another widely used technique for the separation of mixtures is **chromatography**. The name *chromatography* is based on the Greek word *chrōma*, for "color," since the method was first used for the separation of colored substances found in plants.

There are several different types of chromatography, but they are all based on differences in the solubilities of substances and on differences in their tendency to be **adsorbed**, that is, to stick to a solid surface. Experiment 1.7 illustrates the use of one type of chromatography, known as *paper chromatography*, to separate the different colored substances in ink.

Paper chromatography is useful for identifying substances, but it is not very useful for the separation of significant amounts of the substances in a mixture. Useful amounts of substances can be separated in a **chromatographic column**, such as the one shown in Figure 1.22.

Another very important type of chromatography that is useful for the separation and identification of volatile substances is **gas-liquid chromatography**. A

Paper Chromatography of Ink

A black ink line is drawn at the bottom of a long strip of absorbent paper. The paper is suspended in a mixture of ethanol and water.

As the solvent is drawn up the paper the ink separates into colored bands. Each color corresponds to a compound in the ink.

This shows the progress of the separation with time. The strips were removed from the solvent at different times during the separation.

1.5 SEPARATION OF MIXTURES

35

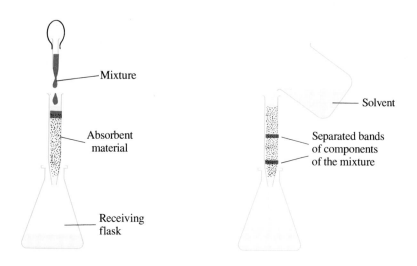

Figure 1.22 Chromatographic Column. The column is packed with an adsorbent material such as aluminum oxide, Al_2O_3. A solution of the mixture to be separated is poured into the top of the column. A suitable solvent is then poured slowly through the column. The solvent carries the different substances in the mixture down the column at different rates so that they separate into bands, which can be readily identified if they are colored. Each band can then be separately washed out of the column into the receiving flask.

Mixture

Absorbent material

Receiving flask

Solvent

Separated bands of components of the mixture

long tube is packed with a finely divided and inert solid. The surface of the solid is coated with a liquid that does not react with the substances to be separated (Figure 1.23). A stream of an inert gas such as helium, He, or nitrogen, N_2, passes continually through the tube, that is, through the tiny holes between the particles of the powdered solid. A mixture of substances to be separated for identification is injected into the stream of inert gas and is vaporized by heating. The vapor is swept down the tube by the inert gas. Those substances that are more soluble in the liquid lag behind, and the less soluble components of the mixture move more rapidly down the tube. Eventually, all the components separate from each other, and they emerge one by one from the end of the tube. They can then be separately identified.

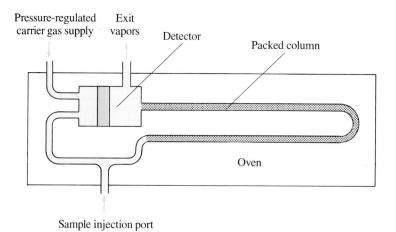

Pressure-regulated carrier gas supply

Exit vapors

Detector

Packed column

Oven

Sample injection port

Figure 1.23 Schematic Diagram of Gas-Liquid Chromatography Apparatus.
The detector is arranged to measure the difference in some property of the carrier gas alone and of the mixture of the carrier gas and the sample gas. Differences in thermal conductivity are fairly easy to measure and are often used for this purpose.

The development of the various forms of chromatography in the past twenty-five years has been one of the major advances in experimental chemistry. Separations previously considered impossible are now achieved easily. Perhaps the greatest impact of chromatography has been in biochemistry, where complex mixtures are frequently encountered.

IMPORTANT TERMS

An **aqueous solution** is a solution in water.

An **atom** is the smallest constituent of an element that retains the properties of the element.

A **chemical property** is any chemical change—that is, any chemical reaction—that a substance undergoes.

A **chemical reaction** is a process by which one or more substances are changed into one or more different substances; a process in which atoms are rearranged to form new combinations.

Chromatography is a physical method, based on the selective adsorption of gases or liquids on the surface of solids or on their relative solubilities in liquids, by which complex mixtures can be separated into their components.

A **compound** is a substance containing two or more kinds of atoms that are present in constant relative amounts. In other words, a compound always has the same constant chemical composition.

Condensation is a process in which a substance changes from the gaseous form (vapor) to the liquid form.

Crystallization is the formation of crystals from a saturated solution of a substance or from the molten state.

The **density** of a substance is its mass per unit volume:

$$\text{Density} = \frac{\text{Mass}}{\text{Volume}}$$

Diatomic molecules are molecules that contain only two atoms.

Distillation is a method of purifying liquids in which they are changed into a vapor and then condensed back to the liquid form.

Dynamic equilibrium is a situation in which a chemical or physical change and its reverse occur at the same rate so that there is no overall change in the system.

An **element** is a substance all of whose atoms are of the same kind.

A **formula** is a combination of element symbols and numerical subscripts that indicates either the relative numbers of each kind of atom in a substance (*empirical formula*) or the actual numbers of each kind of atom in a molecule (*molecular formula*).

A **heterogeneous mixture** of two or more materials has a nonuniform composition. Different parts of the material have different properties.

A **homogeneous mixture** of two or more substances has a uniform but variable composition and variable chemical and physical properties. A homogeneous mixture is also called a *solution*.

Macroscopic is a term used to describe objects that are large enough to be visible to the naked eye.

Microscopic is a term used to describe very small objects. Literally, it means "visible only under a microscope," but it is applied particularly to objects of atomic or molecular size.

Two liquids are said to be **miscible** when they form a homogeneous mixture or solution. Liquids that do not mix are said to be **immiscible**.

A **molecule** is a combination of atoms that has its own characteristic set of properties.

A **physical change** is any change in which the composition and amounts of each of the substances present do not change. In other words, a physical change is any change in which no chemical reaction occurs.

A **physical property** is any property of a substance that does not involve a chemical change. For example, melting point, boiling point, color, and density are physical properties.

A **polyatomic molecule** is a molecule that contains more than two atoms.

A **saturated solution** is a solution that contains the maximum amount of a substance (solute) that can be dissolved in a given amount of solvent at a specified temperature; a solution in which the dissolved solute is in equilibrium with the solid solute.

Solubility is the concentration of the solute in a saturated solution of that substance at a specified temperature.

A **solute** is the minor component of a solution or the component that is solid in the pure state.

Solution. See homogeneous mixture.

A **solvent** is the major component of a solution or the component that is liquid in the pure state.

A **substance** has a fixed, constant composition and a characteristic set of chemical and physical properties.

A **tetrahedron** is a regular solid with four identical, equilateral triangular faces, six sides of equal length, and four equivalent vertices.

A **vapor** is a gas. The term is usually used for a gas that has been, or is being, formed from a liquid.

A **volatile** liquid is one that readily forms a vapor, that is, a gas.

PROBLEMS

Elements, Compounds, Formulas, Mixtures, and Solutions

1. (a) What are the symbols for each of the following elements?

Hydrogen	Carbon	Oxygen	Nitrogen
Chlorine	Sodium	Potassium	Silicon
Iron	Nickel	Chromium	Krypton
Barium	Lead	Uranium	

(b) What is the name of each of the following elements?

He	Ne	F	Mg	Al	P	S
Ca	Fe	Mn	Co	Cu	Zn	As
K	Br	Ag	Pt	Au		

2. Classify each of the following as an element, a compound, or a mixture:

(a) Water   (b) Iron   (c) Beer   (d) Sugar

(e) Wine   (f) Salt   (g) Sulfur   (h) Milk

3. Give at least two physical properties (for example, color, physical state, electrical conductivity) of each of the following: water, sugar, mercury, copper, maple syrup, oxygen, glass, bromine.

4. Name a chemical property exhibited by each of the following: water, copper, iron, magnesium, carbon dioxide, hydrogen peroxide, hydrogen.

5. (a) Classify the following as pure substances or mixtures.

(b) Divide the pure substances into elements and compounds, and divide the mixtures into homogeneous and heterogeneous.

Nitrogen	Smog
Salad dressing	Gasoline
Iron	Cottage cheese
Mercury amalgam	14-carat gold
used in teeth fillings	Milk
Carbon dioxide	Black coffee
Wet sand	Oxygen
Carbon	Sodium chloride
Iodized table salt	Filtered seawater
Diamond	Distilled water
Nylon	Dime
Vegetable soup	

6. Classify the following as homogeneous or heterogeneous materials: seawater, soda water, black coffee, wood, snow, soil, clouds.

7. Suggest ways to separate the following mixtures:

Sugar and water	Sugar and powdered glass
Water and gasoline	Gasoline and kerosene
Iron filings and wood sawdust	Food coloring in water

* The asterisk denotes the more difficult problems.

8. Describe an experiment or give a simple physical property (not taste or smell) that would enable you to distinguish between each of the following:

(a) Two bottles of colorless liquid, one water and the other a solution of salt and water.

(b) Two samples of white powder, one chalk and the other baking soda.

9. Write empirical and molecular formulas for the following:

(a) An elemental form of arsenic containing four arsenic atoms.

(b) A compound of carbon and hydrogen consisting of molecules containing three carbon atoms and six hydrogen atoms.

(c) An oxide of phosphorus consisting of molecules containing four phosphorus atoms and ten oxygen atoms.

(d) A compound of xenon and fluorine consisting of molecules containing one xenon atom and four fluorine atoms.

10. What are the empirical formulas of substances having the following molecular formulas?

H_2O_2 H_2O Li_2CO_3 $C_2H_4O_2$ S_8

C_6H_{12} B_2H_6 O_3 S_3O_9 N_2O_3

Molecular Structure

11. The H–S–H bond angle in the molecule H_2S is very close to 90°. Calculate the H–H distance by using the Pythagorean theorem, given that the H–S distance is 134 pm.

12. Draw diagrams to scale to show the sizes and shapes of the following molecules:

(a) H_2, H–H = 74 pm; O_2, O–O = 121 pm; N_2, N–N = 109 pm; CO, C–O = 113 pm.

(b) CO_2, linear, C–O = 116 pm.

(c) O_3, angular, O–O = 128 pm, $\widehat{OOO}$ = 117°.

(d) As_4, tetrahedral, As–As = 244 pm.

13. What are the names of the shapes of the following molecules?

*14. By constructing a tetrahedron inside a cube, show by trigonometry that in a methane molecule, where the four hydrogen atoms are at the vertices of the tetrahedron and the carbon atom is at its center, each HCH angle is 109°28′.

*15. Determination of the molecular structure of water shows that it consists of angular H_2O molecules, with identical O–H distances and an HOH bond angle of 104.5°. If the H–H distance is observed to be 153.3 pm, what is the O–H distance?

Scientific Notation and Units

16. Express each of the following numbers in scientific (exponential) notation:

(a) 12 000 (b) 1 740 312.29

(c) 0.004 04 (d) −0.049 00

17. Express each of the following numbers without using scientific notation:

(a) 3.0×10^2 (b) 1.162×10^5

(c) 4.8×10^{-3} (d) -6.44×10^{-2}

18. Evaluate each of the following, after writing each number in scientific (exponential) notation:

(a) $(0.000\ 004)(0.001)(5000)$ (b) $\dfrac{0.000\ 12}{0.06}$

(c) $(200)^3(0.0009)^{1/2}$ (d) $\dfrac{0.02 \times 600 \times 50}{0.003 \times 500}$

(e) $(0.000\ 009)^{1/2}(20)^2$

19. What is the value of each of the following expressions? Give the answers in scientific (exponential) notation.

(a) $\dfrac{(6 \times 10^{-6})(3 \times 10^{14})}{(2 \times 10^3)(1 \times 10^6)}$

(b) $(6.022 \times 10^{23}) + (7.7 \times 10^{21})$

(c) $\dfrac{5000 \times 0.06}{0.0003}$

(d) $\dfrac{(3.6 \times 10^{-5})^{1/2}(0.000\ 12)}{3600}$

(e) $119.2 + (2.0412 \times 10^2) + (3.734 \times 10^{-4})$

20. Express each of the following in meters:

(a) 2.998×10^7 km (b) 143 pm

(c) 0.001 nm (d) 1.54 Å

21. Express each of the following in kilograms:

(a) 5.84×10^{-3} mg (b) 54.34 Mg

(c) 0.354 g (d) 1.673×10^{-21} g

22. Express each of the following in scientific notation and in SI base units:

(a) $0.000\ 199\ 8 \times 10^3$ Gm s^{-1}

(b) $(6.022 \times 10^{23})(1.673 \times 10^{-24})$ g

(c) $32\ 150$ Å min^{-1} (d) 22.41 mL

23. Using the conversion factors given in Table 1.3, express each of the following in SI units:

(a) 24 000 miles

(b) 150 pounds

(c) 14 pounds per square inch

(d) 60 miles per hour

24. Convert the following description of the moon to convenient metric units:

The moon revolves around the earth with a period of 27.32 days at a distance of 238 850 miles. The moon has a radius of 1081 miles, and its mass is estimated to be 8.1×10^{19} tons (1 ton = 2000 lb).

25. An automobile travels 25.4 miles on a gallon (U.S.) of gasoline. What is the gasoline consumption in liters per 100 km?

26. Convert the following description of the planet Jupiter to SI base units:

Jupiter revolves around the sun with a period of 11.86 years at a distance of approximately 4.84×10^8 miles. Jupiter has an average density of 1.330 g cm^{-3}.

27. Suggest reasons for and against converting to a decimal time system.

28. What is the unit factor that will convert each of the following:

(a) Inches to feet (b) Feet to inches

(c) Miles to kilometers (d) Kilometers to miles

(e) Cubic meters to cubic milliliters

(f) Square miles to square kilometers

(g) Square centimeters to square meters

29. Astronomical distances are measured in light-years (1 light-year is the distance traveled by light in 1 year; the speed of light is 3.00×10^8 m s^{-1}). The distance from Alpha Centauri to the earth is 4.0 light-years. What is the distance in kilometers and in miles?

30. The number of ways of playing the first ten moves in the game of chess is estimated at

169 518 829 544 000 000 000 000 000 000

Express this number in scientific notation.

31. By what powers of ten do each of the following prefixes multiply a basic SI unit?

(a) Kilo (b) Milli (c) Centi

(d) Pico (e) Micro

32. What are the prefixes for each of the following multiples of a unit?

(a) 10^3 (b) 10^{-1} (c) 10^{-2}

(d) 10^{-6} (e) 10^{-12}

33. (a) The populations of Canada, the United States, Australia, and the United Kingdom were, respectively, 23 845 000, 220 090 000, 14 510 000, and 55 819 000 in 1980. Express these populations in exponential notation.

(b) The areas of the four countries mentioned in part (a) are 9 976 139, 9 519 617, 7 686 849 and 244 013 km^2, respectively. Express these areas in exponential notation, and calculate the average population density per square kilometer.

Volume and Density

34. Using the data in Tables 1.3 and 1.6, calculate the volume of each of the following, in cubic centimeters:

(a) A crystal of sodium chloride having a mass of 1.34 mg.

(b) A pool of mercury having a mass of 21.34 g.

(c) 1.00 kg of benzene.

(d) 1.00 quart (U.S.) of water.

35. Using the data in Tables 1.3 and 1.6, calculate the masses of each of the following, in kilograms:

(a) 1 L of sulfuric acid (density = 1.84 $g\,cm^{-3}$).

(b) 25.00 mL of ethanol.

(c) 1.00 quart (U.S.) of water.

(d) A crystal of sodium chloride of volume 1.00 $in.^3$.

36. A bar of silver 40 cm long, 25 cm wide, and 15 cm deep has a mass of 157.5 kg. What is the density of silver?

37. What is the volume of 75.0 g of chloroform, $CHCl_3$, if the density of chloroform is 1.49 $g\,cm^{-3}$?

38. The density of air at ordinary atmospheric pressure and 25°C is 1.19 $g\,L^{-1}$. What is the mass, in kilograms, of the air in a room that measures $8.5 \times 13.5 \times 2.8$ m?

39. Rubbing alcohol (2-propanol) has a density of 6.56 lb $gallon^{-1}$. What is its density in grams per cubic centimeter?

40. Propanone, an important solvent also known as acetone, has a density of 0.785 $g\,cm^{-3}$. What is the mass of the contents of a 10 000-gallon tank car filled with acetone?

41. One piece of evidence for the fact that the core of the earth is composed largely of iron is its average density. Calculate the average density of the earth if its mass is 6.00×10^{24} kg and its average radius is 6.34×10^3 km.

Concentration

42. Express the concentrations of each of the following as mass percentages:

(a) Ethanol in a solution containing 5.34 g of ethanol and 121.51 g of water.

(b) Sodium chloride in a solution containing 18.12 g of sodium chloride in 250.0 g of aqueous solution.

(c) Both hydrochloric acid and sodium chloride in a solution containing 30.1 g of hydrochloric acid and 3.42 g of sodium chloride in 250.0 g of water.

(d) Phosphoric acid in 250 mL of an aqueous solution that contains 178 g of phosphoric acid and has a density of 1.35 $g\,cm^{-3}$.

(e) Water in an aqueous solution of nitric acid containing 21.35 g of nitric acid and 100 mL of water.

43. Vinegar is an aqueous solution of 5% by mass acetic acid. What mass of acetic acid is contained in 1 kg of vinegar?

44. How many grams of sulfuric acid are in 500 mL of concentrated sulfuric acid that is 95% sulfuric acid and 5% water by mass and has a density of 1.84 $g\,cm^{-3}$?

45. The density of a concentrated sodium hydroxide solution is 1.53 $g\,cm^{-3}$, and it contains 50% by mass of sodium hydroxide and 50% by mass of water. Calculate the volume of this solution that contains 20 g of sodium hydroxide.

46. The concentration of O_2 dissolved in a sample of seawater is 6.22 ppm, and the density of seawater is 1.03 $g\,cm^{-3}$. What is the mass and volume of seawater that contains 1.00 g of O_2?

CHAPTER 2
STOICHIOMETRY

Atomic Structure, Atomic Mass, The Mole, Formulas, and Equations

We have said that there are just over a hundred different elements and, there-fore, the same number of different kinds of atoms, but we have not yet discussed how atoms differ. To do this we need first to consider the structures of atoms. We will see that all atoms are made up of the same basic particles, but they contain different numbers of these particles. Second, we must consider the forces that hold these particles together.

An important property we must consider is an atom's mass. Dalton proposed that all the atoms of a given element have the same mass and that this mass is different from the mass of an atom of another element. We will see that this postulate has had to be slightly modified in the light of experimental evidence.

Finally, we will see how to use atomic masses to determine the formulas of substances and to calculate the amounts of substances taking part in chemical reactions. All the quantitive aspects of chemical composition and reactions are referred to collectively as **stoichiometry**, a term that is derived from Greek words meaning "element" and "measure."

2.1 ESSENTIALS OF ATOMIC STRUCTURE

The Hydrogen Atom: Electrons and Protons

The atom with the smallest mass and the simplest structure is that of hydrogen. The *hydrogen atom* consists of two electrically charged particles. One particle is a small, positively charged entity called the **proton**. The other particle is negatively charged and is called the **electron**. The charge on the electron is

1.6022×10^{-19} coulomb (C), and the charge on the proton is of equal magnitude but opposite sign. The hydrogen atom therefore has no net charge overall. In other words, it is electrically neutral.

The *mass* of the hydrogen atom is concentrated in the proton, which has a mass of 1.6726×10^{-27} kg; the electron has a mass of 9.1096×10^{-31} kg, which is only $\frac{1}{1836}$ of the mass of the proton. The negatively charged electron is attracted by the positively charged proton, and it moves around the proton. It is not, however, to be pictured as moving in a well-defined orbit—such as the earth's orbit around the sun—although this was an early model of the hydrogen atom. It is not possible to precisely describe the motion of the electron around the proton but as a result of this motion, the electron effectively fills a relatively large spherical space around the very small proton. Our current view of the structure of an atom dates back to 1910, when Ernest Rutherford (Figure 2.1) showed that all atoms consist of a very small, positively charged particle called a **nucleus** surrounded by one or more negatively charged electrons (Box 2.1). In the hydrogen atom the nucleus consists of a single proton.

For convenience we can think of the electron in the hydrogen atom not as a very small, negatively charged particle but as a cloud of negative charge surrounding the proton. This picture of the atom would be obtained if we could take a time exposure photograph of a hydrogen atom. The electron would appear not as a single dot but as a cloud of negative charge. The electron cloud in an isolated hydrogen atom is spherical, but it is not of uniform density. It is most dense near the center of the atom in the vicinity of the proton and rapidly diminishes in density as the distance away from the proton increases (See Figure 2.2 on page 44.)

Electrostatic Forces

The electron is held to the proton in the hydrogen atom because there is a force of attraction between them. We are familiar with the idea of a force from everyday experience. We exert a force when we kick a football or hit a tennis ball. A **force** acting on an object changes its velocity. The magnitude of the force is directly proportional both to the mass being moved and to the rate of change of its velocity, as shown in the following equation. Rate of change of

Figure 2.1 Ernest Rutherford (1871–1937).
Rutherford was born on a farm in New Zealand in 1871, the grandson of Scottish immigrants and the second of twelve children. On graduating from the University of New Zealand, he obtained a scholarship to Cambridge and began his research in the then new field of radioactivity. At the age of twenty-seven Rutherford was appointed professor of physics at McGill University, Montreal. In 1907 he moved to the University of Manchester, England, and in 1908 he was awarded the Nobel Prize in chemistry. Among his many discoveries were showing that alpha particles are the nuclei of helium atoms and finding the evidence that led to our present model of the atom. In 1917 Rutherford made another discovery of fundamental importance. By bombarding nuclei with subatomic particles, he changed the atoms of one element into atoms of another element, thus achieving the ancient dream of the alchemists. In 1919 he returned to Cambridge as head of the internationally renowned Cavendish Laboratory.

velocity is called **acceleration** and has the units meters per second per second, or meters per second squared (m s^{-2}).

$$\text{Force} = \text{Mass} \times \text{Acceleration} \qquad \text{or} \qquad F = ma$$

Box 2.1
RUTHERFORD AND THE NUCLEAR ATOM

Our present model of the atom as consisting of a very small nucleus, in which most of the mass of the atom is concentrated, surrounded by one or more electrons, which occupy a very much larger volume, dates from 1910. In that year, in Manchester, England, Ernest Rutherford and two of his collaborators, Hans Geiger and Ernest Marsden, a 20-year-old student, carried out an experiment on the scattering of alpha particles by thin metal foil. Rutherford had earlier found that some radioactive elements—that is, elements that have unstable nuclei—such as uranium, decompose by emitting positively charged particles, which he called alpha (α) particles, and electromagnetic radiation, or gamma (γ) rays. He later showed that alpha particles are the nuclei of helium atoms. They are emitted from unstable nuclei at very high speeds ($\approx 10^7 \text{ m s}^{-1}$). They can travel several centimeters through the air, or approximately 0.1 mm through solids, before they are stopped by collisions with atoms. They can pass right through very thin metal foil, but in so doing, they are deflected from their original path; they are said to be *scattered*. Rutherford was interested in obtaining information about the structure of atoms from this scattering of alpha-particles.

Rutherford and his collaborators observed that most of the alpha particles went straight through the metal foil with little or no deflection. A certain number, however, were scattered through large angles, and some even bounced right back toward the source. This result was very surprising, and Rutherford later remarked: "It was quite the most incredible event. . . . It was almost as surprising as if a gunner fired a shell at a piece of tissue and the shell bounced right back".

At the time, ideas about the structure of atoms were rather vague. The most popular model, proposed by J. J. Thompson in 1898, held that atoms consisted of electrons embedded in a uniform distribution of positively charged matter. Because the alpha particles are positively charged, they were expected to be attracted by the negatively charged electrons and repelled by the positively charged matter. But the electrons were not expected to deflect the high-speed alpha particles significantly; they have a much smaller mass than alpha particles and would simply be pushed out of the path of the alpha particles. A uniform distribution of positively charged matter was expected to deflect most of the alpha particles but only through rather small angles, because there is nowhere any large density of positive charge. This model of the atom was completely unable to explain the very large deflections that were observed for some alpha particles.

Rutherford was eventually led to the unexpected conclusion that almost all the mass and all the positive charge of an atom is concentrated in a very small particle situated at the center of the atom. He called this particle the nucleus. Since most of the volume occupied by an atom is filled only by a few electrons, most of the alpha particles pass through this space undeflected. It is only when an alpha particle comes very close to the heavy, positively charged nucleus of the atom that it is deflected through a large angle. The very few particles that make a more or less direct hit on the nucleus are bounced right back in the direction from which they came. Calculations of the number of alpha particles that would be expected to be scattered through any given angle on the basis of this model gave results in excellent agreement with the experimental observations and Rutherford's model.

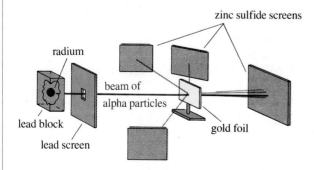

Rutherford's Apparatus for Studying Scattering of Alpha Particles. Alpha particles emitted by radium are blocked by a lead plate, but a small hole in the plate allows a narrow beam of particles to pass through. This beam then passes through a very thin foil of copper, silver, or gold and strikes a screen coated with zinc sulfide. A momentary flash is observed on the screen whenever it is struck by an alpha particle. (The screen in a television set works on the same principle). Thus the number of particles deflected at any angle can be found by counting the flashes on the screen when it is moved to different positions around the foil.

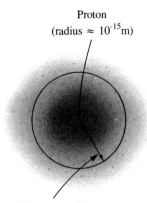

Proton
(radius ≈ 10^{-15}m)

Electron cloud
(radius ≈ 1.4×10^{-10}m)

Figure 2.2 Hydrogen Atom. If we could take a time exposure photograph of a hydrogen atom, its electron would, because of its rapid motion, appear as a cloud of negative charge surrounding the proton. This cloud has no precise boundary, but it is much larger than the very small proton. A radius of 1.4×10^{-10} m would include 90% of the electron density. Neutral atoms of other elements have more electrons and protons than hydrogen has, but in these atoms, too, the mass is concentrated in the nucleus, and the electrons account for most of the size of the atom.

A greater force is needed to change the velocity of a heavy object by a certain amount than is needed to change the velocity of a light object by the same amount. For a given mass a greater force is needed to accelerate it to a high velocity than to a low velocity in the same time interval.

The SI unit of force is called the **newton** (N). The newton is the force that, acting on a mass of 1 kg, causes an acceleration of 1 m s^{-2}:

$$1\ N = 1\ kg\,m\,s^{-2}$$

Only three types of forces are known in nature:

1. **Gravitational forces** act between all objects and are proportional to their masses.
2. **Electrical forces** act between charged objects and are proportional to their electric charges.
3. **Nuclear forces** act between subatomic particles and hold together the particles (protons and neutrons) of which a nucleus is composed.

Seemingly, there are many other types of forces, such as the mechanical force exerted by a tennis racket on a tennis ball or the force exerted by one atom on another when they collide. But all such forces are manifestations of electric forces between the charged particles of atoms. There are also magnetic forces that arise from charges in motion, but they are not fundamentally different from electric forces.

In chemistry only electrical forces are important. Gravitational forces are so much weaker than electrical forces that they can be completely ignored. For example, the electrical force of attraction between a proton and an electron is 10^{39} times stronger than the gravitational attraction between them. Nuclear forces are of importance only over extremely small distances, such as the distances between the particles inside the nucleus; they are completely negligible over larger distances.

COULOMB'S LAW   The electrical force between two charges of magnitude Q_1 and Q_2 is given by **Coulomb's law:**

$$F = k\left(\frac{Q_1 Q_2}{r^2}\right)$$

where r is the distance between the charges and k is a constant called the Coulomb constant. When the two charges are of opposite sign, the force is attractive; in other words, the charges are drawn to each other. When the two charges are of the same sign, the force is repulsive; that is, like charges repel each other.

THE COULOMB   The unit of charge is the **coulomb** (C). We will not give the formal definition of the coulomb, which is expressed in terms of magnetic forces. For our purposes it is more useful to think of the coulomb in terms of the number of individual elementary charges that add up to a coulomb of charge. All charges, both positive and negative, occur only in multiples of 1.602×10^{-19} C. The electron has a charge of -1.602×10^{-19} C and the proton has a charge of $+1.602 \times 10^{-19}$ C. A total of 6.241×10^{18} electrons have a charge of -1C and a total of 6.241×10^{18} protons have a charge of $+1$C. The amount of charge, 1.602×10^{-19} C, has been given a special name, the **electron charge**, and a special symbol, e:

$$e = 1.60 \times 10^{-19}\ C$$

The charge on an electron was first measured by the American physicist Robert Millikan (Figure 2.3 and Box 2.2).

Charges on particles such as the electron or proton are usually expressed as multiples of e and not in coulombs. Thus an electron is said to have a charge of -1, that is, a charge of $-e$, or -1.60×10^{-19} C. Similarly, a proton is said to have a charge of $+1$, that is $+e$, or $+1.60 \times 10^{-19}$ C.

The Helium Atom: Neutrons

After hydrogen, the next lightest atom is the atom of the element *helium*. It has two electrons surrounding a nucleus with a charge of $+2$; thus the atom is neutral. Like the nuclei of all atoms, the helium nucleus is very small and occupies only a tiny fraction of the total volume of the atom, although it contains most of the mass.

Because the charge on the helium nucleus is twice the charge of the proton, we might expect the helium nucleus to consist of two protons and to have a mass twice the mass of the proton. But it does not. Instead, the mass of the helium nucleus is very nearly *four times* the mass of the proton. Two protons account for the charge of $+2$ but only one-half of the mass, so there must be additional constituents in the helium nucleus. These additional particles are called **neutrons**. A neutron is a particle that has very nearly the same mass as the proton but zero charge.

Although the detailed structure of the nucleus is not fully understood, for our purposes we can consider the nuclei of all atoms to be made up of protons, each with a charge of $+1$, and a certain number of neutrons. *The charge on a*

Figure 2.3 Robert Millikan (1868–1953).

The son of a poor clergyman with a large family, Millikan grew up in the midwestern United States. As a second-year student at Oberlin College, he was asked to teach physics, a subject he had never studied. As a result of this experience, his future as a physicist was decided. He carried out his famous oil drop experiments (Box 2.2) at the University of Chicago. In 1921 he became professor of physics at the California Institute of Technology, where he established a world-renowned laboratory that did much important work in nuclear physics.

<hr>

Box 2.2

MILLIKAN'S DETERMINATION OF THE CHARGE OF AN ELECTRON

The charge of the electron was first accurately determined by the American physicist Robert Millikan at the University of Chicago during the period 1909–1913. Millikan allowed a fine spray of very small oil drops to fall between two metal plates. Air between the plates was irradiated with X rays, knocking electrons out of nitrogen and oxygen molecules. Some of these electrons collided with the oil drops and became attached to them, giving the oil drop a negative charge. The fall of the drops under gravity could then be stopped by charging the plates, one positively and the other negatively.

Millikan observed individual oil drops with a microscope and adjusted the charge on the plates until the force acting on the drop due to the charged plates just balanced the force due to gravity and a drop remained stationary. Knowing the charge on the plates and using Coulomb's law, Millikan was able to calculate the charge on the drop. On repeating the experiment for very many individual drops, he found that the charge on a drop was always -1.60×10^{-19} C or an integral multiple of this charge. In other words, he found that there is a smallest possible amount of electric charge, indicating that electric charge is not continuous but consists of particles, each having the same charge—the particles that we now call *electrons*. The oil drops were picking up one or more electrons and therefore always had the charge of one or more electrons. Millikan's experiments not only established an accurate value for the charge on the electron but also provided the first conclusive proof for the existence of electrons.

nucleus is determined by the number of protons that it contains. The mass of a nucleus is determined by the total number of protons and neutrons, which are sometimes collectively called **nucleons**. Thus the nucleus of a helium atom consists of two protons and two neutrons; it has a charge of $+2$ and a mass that is approximately four times the mass of a proton. The nucleus of a fluorine atom consists of nine protons and ten neutrons; it has a charge of $+9$ and a mass that is approximately 19 times the mass of the proton. Except for the nucleus of the hydrogen atom, all nuclei are composed of both protons and neutrons.

Atomic Numbers

We can summarize the constitution of any nucleus by just two numbers:

1. The number of protons in the nucleus is called the **atomic number**, Z.
2. The total number of protons and neutrons is called the **mass number** or the **nucleon number**. It determines the mass of the nucleus.

The difference between the mass number and the atomic number is equal to the number of neutrons in the nucleus.

By convention, the atomic number is written at the bottom left corner of the symbol of the atom, and the mass number is written at the top left corner. For example, we write ^1_1H, ^4_2He, ^7_3Li, and $^{12}_6C$. The symbol $^{12}_6C$ indicates that there is a total of 12 particles (nucleons) in the nucleus of a C atom, 6 of which are protons. Thus there must be $12 - 6 = 6$ neutrons.

Because each atom has a zero charge overall, its nuclear charge, which is $+Z$, must be balanced by the total charge of the surrounding electrons, each of which has a charge of -1. For example, a ^7_3Li atom, which has $Z = 3$, has a nuclear charge of $+3$ and 3 surrounding electrons, for a total charge of 0. A $^{12}_6C$ atom, $Z = 6$, has 6 electrons surrounding the nucleus, and $^{16}_8O$ has 8 electrons. Figure 2.4 shows the structure of a few atoms in a diagrammatic way. It is the nuclear charge—that is, the atomic number Z—that differentiates the atoms of one element from the atoms of another. Thus an **element** *may be defined as a substance whose atoms have the same atomic number.*

Isotopes

One further important aspect of nuclei must be considered. We have said that all the atoms of a given element have nuclei containing the same number of

---

Example 2.1 For each of the following, what is the mass number? How many protons are in the nucleus? How many neutrons? How many electrons are in the neutral atom?

(a) ^1_1H (b) ^4_2He (c) $^{19}_9F$ (d) $^{118}_{50}Sn$ (e) $^{238}_{92}U$

Solution

ATOM	MASS NUMBER (NUCLEON NUMBER)	NUMBER OF PROTONS IN NUCLEUS (ATOMIC NUMBER, Z)	NUMBER OF NEUTRONS IN NUCLEUS	NUMBER OF ELECTRONS
(a) ^1_1H	1	1	0	1
(b) ^4_2He	4	2	2	2
(c) $^{19}_9F$	19	9	10	9
(d) $^{118}_{50}Sn$	118	50	68	50
(e) $^{238}_{92}U$	238	92	146	92

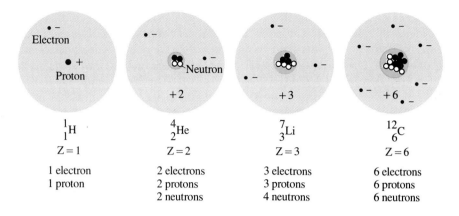

Figure 2.4 Structures of
Some Atoms.

These diagrams do not represent
the relative sizes of the electron
cloud and the nucleus. They are
meant to show only the numbers
of particles of each kind.

Electron	Neutron		
Proton	$+2$	$+3$	$+6$
$^{1}_{1}H$	$^{4}_{2}He$	$^{7}_{3}Li$	$^{12}_{6}C$
$Z = 1$	$Z = 2$	$Z = 3$	$Z = 6$
1 electron	2 electrons	3 electrons	6 electrons
1 proton	2 protons	3 protons	6 protons
	2 neutrons	4 neutrons	6 neutrons

protons and having the same charge. But the nuclei of all the atoms of a given element do not necessarily contain the same number of neutrons. For example, all oxygen atoms have an atomic number Z of 8, contain 8 protons, and have 8 electrons surrounding the nucleus; these features distinguish oxygen atoms from atoms of all the other elements. But three different kinds of oxygen atoms are found in nature: $^{16}_{8}O$ atoms with 8 protons and 8 neutrons in the nucleus, $^{17}_{8}O$ atoms with 8 protons and 9 neutrons in the nucleus, and $^{18}_{8}O$ atoms with 8 protons and 10 neutrons in the nucleus. All oxygen atoms have the same atomic number, $Z = 8$, and their nuclei have the same nuclear charge of $+8$, but they have different masses depending on the number of neutrons in their nuclei. Atoms of an element that have different masses are called **isotopes**.

Many of the elements have two or more naturally occurring isotopes. Even the simplest element, hydrogen, has two stable isotopes, $^{1}_{1}H$ and $^{2}_{1}H$. The nucleus of the lighter isotope is the proton; the nucleus of the heavier isotope contains a proton and a neutron. The mass of the $^{2}_{1}H$ isotope is nearly twice the mass of $^{1}_{1}H$. Isotopes are distinguished by the mass number in the symbol for the isotope, and they are not usually given different names. However, the $^{2}_{1}H$ isotope of hydrogen is an exception. It is called *deuterium* or *heavy hydrogen* and is sometimes given the symbol D.

Some elements have just a single naturally occurring isotope, for example, $^{19}_{9}F$ and $^{31}_{15}P$. But many have two or more isotopes, for example, helium, $^{3}_{2}He$ and $^{4}_{2}He$; lithium, $^{6}_{3}Li$ and $^{7}_{3}Li$; and oxygen, $^{16}_{8}O$, $^{17}_{8}O$, and $^{18}_{8}O$. In addition to naturally occurring isotopes, others can be made artificially by bombarding atoms with nuclear particles, such as protons, neutrons, and helium nuclei (Chapter 24). In these nuclear reactions nuclei are transformed into different nuclei—that is, atoms are transformed into different atoms. In chemical reactions, however, atoms remain unchanged; they simply form new combinations.

Because the atomic number is implied by the symbol of the element, it is often omitted. The isotopes of oxygen, for example, are often represented simply by the symbols ^{16}O, ^{17}O, and ^{18}O, and they are referred to as oxygen-16, oxygen-17, and oxygen-18.

The isotopes of an element are found not only in the element itself but also in all the compounds of the element. Naturally occurring water, for instance, consists mainly of molecules containing two $^{1}_{1}H$ atoms and an oxygen atom, some molecules containing a $^{1}_{1}H$ atom and a $^{2}_{1}H$ atom, and a very few molecules containing two $^{2}_{1}H$ atoms and an oxygen atom—in other words, H_2O, HDO, and D_2O molecules. Water that consists entirely of $^{2}_{1}H_2O$ (D_2O) molecules is called *deuterium oxide* or, more commonly, *heavy water*. It is used to slow down neutrons in some types of nuclear reactors.

2.1 ESSENTIALS OF
ATOMIC STRUCTURE

Molecules containing different isotopes of an element have very similar, but not identical, properties. The differences are more marked for ^{1}H and ^{2}H because the ratio of their masses is larger than the ratios for the isotopes of any other element. Ordinary water freezes at 0°C and boils at 100°C, whereas heavy water (^{2}H$_2$O) freezes at 3.82°C and boils at 101.42°C. Such differences in physical properties, even though they are very small, allow isotopically different molecules to be separated. For example, heavy water can be obtained by fractional distillation of ordinary water. It is more commonly obtained by electrolysis of water, that is, by passing an electric current through it. The $^1_1$H$_2$O molecules are split up into hydrogen and oxygen slightly faster than the $^2_1$H$_2$O molecules. Thus if a large quantity of water is electrolyzed until only a very small amount remains, this small amount consists almost entirely of $^2_1$H$_2$O molecules.

2.2 ATOMIC MASS

The mass of an atom is much too small to be conveniently expressed in kilograms or grams. Because the mass of the proton is 1.6726×10^{-27} kg, a unit of 10^{-27} kg would be a convenient size. However, this unit is not used. In the past, several different units have been used, but today the unit in which atomic masses are measured is defined by international agreement as $\frac{1}{12}$ the mass of one $^{12}_6$C atom. This unit is called the **unified atomic mass unit** and is given the symbol u. In other words, the mass of one $^{12}_6$C atom is defined as *exactly* 12 u. The masses of the electron, the neutron, the proton, and a number of lighter atoms on this scale are given in Table 2.1. An important advantage of this scale is that the atomic masses have values that are close to the mass (nucleon) number, the total number of protons and neutrons in the nucleus.

Average Atomic Mass

Because elements occur naturally as mixtures of their isotopes, the mass of a given number of atoms of an element depends on the relative abundance of isotopes in the sample. A hundred atoms of hydrogen, for example, could have a mass between 100.783 u and 201.410 u, depending on the relative numbers of ^{1}H and ^{2}H atoms in the sample. During the earth's long history, however, the isotopes of most elements have become thoroughly mixed with each other. As a result, the isotopic composition of most elements is constant throughout the surface of the earth; only a few elements have an isotopic composition that varies significantly from one sample of a substance containing the element to another sample. Wherever water and other hydrogen-containing compounds are found on the earth's surface they are composed of the same proportions of ^{1}H and ^{2}H atoms; 99.985% of the hydrogen atoms are ^{1}H atoms and 0.015% are ^{2}H atoms.

If we know the relative abundances—that is, the percentages—and masses of the isotopes of an element, we can calculate the average mass of an atom of the element. The masses and abundances of isotopes can be measured with a *mass spectrometer*, which is described in Box 2.3. Table 2.1 lists the masses and abundances of the isotopes of the elements with atomic numbers from 1 to 19. To calculate the average mass of an atom of an element, we simply multiply the fractional abundance of each isotope by its mass and then add the products. For example, Table 2.1 tells us that in a naturally occurring sample of hydrogen 99.985% of the atoms are ^{1}H atoms and 0.015% are ^{2}H atoms. Thus the fractional abundances are ^{1}H, 0.999 85 and ^{2}H, 0.000 15. Since ^{1}H has a mass of 1.007 83 u and ^{2}H has a mass of 2.014 10 u, the average mass of an atom of

Table 2.1 Mass and Natural Abundance of Some Fundamental Particles and Isotopes

PARTICLE OR ISOTOPE	MASS (u)	PERCENT NATURAL ABUNDANCE	ISOTOPE	MASS (u)	PERCENT NATURAL ABUNDANCE
Proton	1.007 28		$^{23}_{11}\text{Na}$	22.989 77	100
Neutron	1.008 66		$^{24}_{12}\text{Mg}$	23.985 04	78.7
Electron	0.000 548		$^{25}_{12}\text{Mg}$	24.985 84	10.2
$^{1}_{1}\text{H}$	1.007 83	99.985	$^{26}_{12}\text{Mg}$	25.986 36	11.1
$^{2}_{1}\text{H}$	2.014 10	0.015	$^{27}_{13}\text{Al}$	26.981 53	100
$^{3}_{2}\text{He}$	3.016 03	0.000 13	$^{28}_{14}\text{Si}$	27.976 93	92.18
$^{4}_{2}\text{He}$	4.002 60	99.999 87	$^{29}_{14}\text{Si}$	28.976 50	4.71
$^{6}_{3}\text{Li}$	6.015 12	7.42	$^{30}_{14}\text{Si}$	29.973 77	3.12
$^{7}_{3}\text{Li}$	7.016 00	92.58	$^{31}_{15}\text{P}$	30.973 77	100
$^{9}_{4}\text{Be}$	9.012 19	100	$^{32}_{16}\text{S}$	31.972 07	95.02
$^{10}_{5}\text{B}$	10.012 94	19.77	$^{33}_{16}\text{S}$	32.971 46	0.76
$^{11}_{5}\text{B}$	11.009 31	80.23	$^{34}_{16}\text{S}$	33.967 86	4.22
$^{12}_{6}\text{C}$	12.000 00	98.892	$^{36}_{16}\text{S}$	35.967 10	0.014
$^{13}_{6}\text{C}$	13.003 36	1.108	$^{35}_{17}\text{Cl}$	34.968 85	75.76
$^{14}_{7}\text{N}$	14.003 07	99.635	$^{37}_{17}\text{Cl}$	36.965 90	24.24
$^{15}_{7}\text{N}$	15.000 11	0.365	$^{36}_{18}\text{Ar}$	35.967 54	0.337
$^{16}_{8}\text{O}$	15.994 91	99.759	$^{38}_{18}\text{Ar}$	37.962 73	0.063
$^{17}_{8}\text{O}$	16.999 13	0.037	$^{40}_{18}\text{Ar}$	39.962 38	99.60
$^{18}_{8}\text{O}$	17.999 16	0.204	$^{39}_{19}\text{K}$	38.963 71	93.22
$^{19}_{9}\text{F}$	18.998 40	100	$^{40}_{19}\text{K}$	39.964 01	0.012
$^{20}_{10}\text{Ne}$	19.992 44	90.92	$^{41}_{19}\text{K}$	40.961 83	6.77
$^{21}_{10}\text{Ne}$	20.993 85	0.257			
$^{22}_{10}\text{Ne}$	21.991 38	8.82			

hydrogen can be found as follows:

Contribution of ^{1}H atoms	$0.999\,85 \times 1.007\,83\ \text{u} = 1.007\,68\ \text{u}$
Contribution of ^{2}H atoms	$0.000\,15 \times 2.014\,10\ \text{u} = 0.000\,30\ \text{u}$
Average of ^{1}H and ^{2}H atoms	$1.007\,98\ \text{u}$

Thus the average mass of one hydrogen atom is 1.007 98 u.

Example 2.2 Naturally occurring oxygen consists of a mixture of the isotopes ^{16}O, ^{17}O, and ^{18}O, in the relative abundances 99.759%, 0.037%, and 0.204%. Their masses are 15.994 91, 16.999 13, and 17.999 16 u. What is the average mass of an oxygen atom?

Solution To find the average mass, we calculate the masses contributed by each isotope and then add these amounts. It is convenient first to convert the percent abundances to fractional abundances. For example, percent abundance 99.759 = fractional abundance 0.997 59.

MASS NUMBER	MASS		FRACTIONAL ABUNDANCE		MASS × ABUNDANCE
^{16}O	15.994 91 u	×	0.997 59	=	15.9564 u
^{17}O	16.999 13 u	×	0.000 37	=	0.0063 u
^{18}O	17.999 16 u	×	0.002 04	=	0.0367 u
Average mass per O atom					15.9994 u

2.2 ATOMIC MASS

In 1913 the British physicist J. J. Thomson was working at the Cavendish laboratory in Cambridge, England, on methods to determine the masses of individual atoms. For this work he developed an instrument known as the *mass spectrometer*. It is based on the principle that if a force is applied at right angles to the path of a moving object, the force will change the object's direction of motion. A light object will be deflected from its original path more than a heavy object.

The accompanying figure gives a diagram of a mass spectrometer. One or more electrons are removed from atoms to produce positively charged particles known as positive *ions*. A beam of these ions is passed through a magnetic field, which exerts a force on the positively charged ions at right angles to their direction of motion. If the beam contains atoms of different masses, the heavier atoms are deflected less than the lighter atoms, and the beam splits up into separate beams, one for each atom of different mass. From the amount by which the beam is deflected, the mass of the atom can be calculated.

In his first experiments Thomson found that a sample of pure neon gas gave two deflected beams, which showed that neon contains atoms of two different masses. This experiment was the first demonstration of the existence of isotopes. Thomson recorded the two beams on a photographic plate. From the relative intensities of the two lines on the plate, he obtained the relative abundances of the two isotopes.

The simple instrument invented by Thomson has been developed into the very accurate, modern mass spectrometer, which is capable of measuring the abundance and masses of isotopes with great precision. Almost all the values given in the table of atomic masses have been determined in this way. Today the mass spectrometer is used primarily for the analysis of new compounds. A compound is converted to a mixture of positively charged atoms and molecules (ions), and from their masses one can usually deduce the composition of the compound.

In the accompanying diagram of a mass spectrometer the substance to be studied is introduced in the form of a gas at A. It passes into the region between the two metal plates B and C, which have a large electric potential between them. High-energy electrons, which are injected at D, collide with the atoms or molecules of the gas. If the gas is a simple monatomic gas such as neon or argon, one or more electrons are knocked out of the atoms to give positively charged atoms (called ions) such as Ne^+, Ar^+, and Ar^{2+}. If the gas is composed of molecules, such as H_2, positively charged molecules such as H_2^+ are produced. In addition, molecules may be split into smaller molecules and into atoms by the force of the collisions. In turn, these atoms and molecules also lose one or more electrons, to give positive ions.

The positively charged atoms and molecules are attracted toward the negatively charged plate C and are accelerated to high speeds. Some of the positive ions pass through the slit in the plate to form a narrow beam which then passes between the poles of a powerful magnet. The magnet deflects the beam into a circular path whose curvature depends on the mass of the ions and on their charge. The lighter ions are deflected more than the heavier ions, and the more highly charged ions are deflected more than the ions with a single positive charge.

With a given strength of magnetic field, only ions with a given value of Q/m, the ratio of charge to mass, pass through the slit E to reach the plate F, where the arrival of the positive ions is detected electronically. When the magnetic field is varied, ions with different values of Q/m can be focused onto the detector. From the current produced at the detector by the arrival of the positive ions, the relative amounts of each of the ions of different mass can be found. Thus in the case of the atoms of a given element the relative abundance of the isotopes can be found.

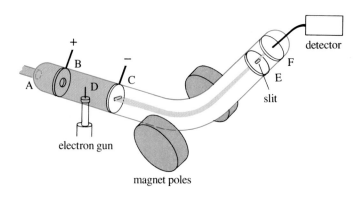

electron gun

magnet poles

detector

Example 2.3 Chlorine has two isotopes, $^{35}_{17}Cl$ and $^{37}_{17}Cl$, which have masses of 34.968 85 u and 36.965 90 u, respectively. Its average atomic mass is 35.453 u. What are the percent abundances of the two isotopes?

Solution Let the fractional abundance of $^{35}_{17}Cl$ be x. Then the fractional abundance of $^{37}_{17}Cl$ is $1 - x$. Thus we have

$$(34.968\ 85\ u)x + (36.965\ 90\ u)(1 - x) = 35.453\ u$$

$$(34.969\ u - 36.966\ u)x + 36.966\ u = 35.453\ u$$

$$x = \frac{36.966\ u - 35.453\ u}{36.966\ u - 34.969\ u} = 0.7576$$

Hence the abundance of $^{35}_{17}Cl$ is 75.76% and the abundance of $^{37}_{17}Cl$ is 24.24%.

J. J. Thomson

The average atomic masses of the elements are usually referred to simply as *atomic masses* and frequently (although strictly speaking, incorrectly) as *atomic weights*. Often they are expressed on a relative basis, which avoids the use of the atomic mass unit. The *relative average atomic mass* of an element is defined as the average mass of one of its atoms *relative to* $\frac{1}{12}$ the mass of a ^{12}C atom. Relative average atomic masses have no units; they are simply numbers. The numerical values of average atomic masses expressed in atomic mass units and as relative atomic masses are the same, and the difference between the two has no great practical significance.

A table of atomic masses appears in Appendix B. Notice that the accuracy with which atomic masses are quoted—that is, the number of significant figures—varies. The more constant the isotopic composition of an element, the greater is the number of significant figures cited in the atomic mass. For example, the atomic mass of sulfur, which has the isotopes ^{32}S, ^{33}S, ^{34}S, and ^{36}S, is given as 32.06 u. Only four significant figures are quoted because the isotopic composition of sulfur, and therefore its average atomic mass, varies slightly, depending on its source. In contrast, the atomic mass of fluorine, which has only one naturally occurring isotope, ^{19}F, is given as 18.998 403 u. For most purposes, including most of the calculations in this book, values of the average atomic masses rounded off to four significant figures are adequate. A table of atomic masses rounded off to four significant figures is given on the inside front cover.

Conservation of Mass and Energy

We might think that we could calculate the mass of an atom from the masses of its electrons, protons, and neutrons. But we cannot. Consider, for example, a ^4_2He atom. The sum of the masses of its protons, neutrons, and electrons is

$$
\begin{array}{ll}
2\ \text{protons} = 2 \times 1.007\ 28\ u = & 2.014\ 56\ u \\
2\ \text{neutrons} = 2 \times 1.008\ 66\ u = & 2.017\ 32\ u \\
2\ \text{electrons} = 2 \times 0.000\ 55\ u = & \underline{0.001\ 10\ u} \\
& 4.032\ 98\ u
\end{array}
$$

The total is greater than the experimentally determined mass of the helium ^4_2He atom, which is 4.002 60 u. This discrepancy illustrates a rule, not an exception: The mass of any atom is always slightly less than the sum of the masses of the constituent electrons, protons, and neutrons. This difference is called the **mass defect**.

The mass defect arises because a very large amount of energy is released when protons and neutrons combine to form a nucleus. The amount of energy is so large that it has a significant mass equivalent, which is given by the equation

$$E = mc^2$$

where E is energy, m is mass, and c is the velocity of light (3.00×10^8 m s^{-1}). This equation, proposed by Albert Einstein in 1905 and subsequently verified in many experiments, is one of the most important and fundamental equations in all science. It gives the relationship between the two fundamental concepts, mass and energy, that we use to describe the universe.

However, we are not directly concerned with this relationship in most of chemistry, although it is of great importance when we are dealing with *nuclear reactions* (Chapter 24). These reactions are changes in which nuclei are split or are fused. Very large energy changes are associated with nuclear reactions, which are accompanied by significant changes in mass. For instance, we have seen that when a helium nucleus is formed from two neutrons and two protons, there is a significant decrease in the total mass. In contrast, *chemical reactions* involve only the rearrangement of atoms. The energy changes associated with these rearrangements of atoms are too small for the accompanying changes in mass to be detectable. For example, when 1000 metric tons (10^6 kg) of gasoline are burned, 4.2×10^9 joules of energy are released. According to Einstein's equation, this energy is equivalent to only 5×10^{-10} kg, which is much too small to detect experimentally.

Not surprisingly, therefore, one of the earliest laws of chemistry was the **law of conservation of mass**, which is as follows:

Mass can neither be created nor destroyed.

This law was first formulated more than two hundred years ago by the French chemist Antoine Lavoisier (Figure 2.5). By carefully weighing the substances involved in a number of different reactions. Lavoisier showed that there was no detectable change in their *total* mass. He postulated that this result was true for all chemical reactions, a hypothesis that subsequently has been confirmed in countless experiments.

We can therefore state that in a chemical reaction the sum of the masses of the substances that react together (the *reactants*) is equal to the sum of the masses of the substances produced (the *products*). In other words, in any chemical reaction the total mass of the system is conserved. In terms of the atomic theory the **law of conservation of mass** indicates that

In a chemical reaction atoms are neither created nor destroyed; they are merely rearranged.

Similarly, although energy is changed from one form to another during a chemical reaction, the total energy of the reacting system and its surroundings is constant. In other words, the **law of conservation of energy** applies to chemical reactions. It states

Energy can neither be created nor destroyed but only changed from one form to another.

Strictly speaking, only the sum of the mass and energy remains constant, but in chemical reactions the mass changes are so small that they can be ignored. We cannot ignore them for processes involving changes in nuclei, however, in which the energy changes are very large and there are significant accompanying mass changes. In such nuclear reactions only the sum of the mass and energy remains constant.

Figure 2.5 Antoine Laurent Lavoisier (1743–1794). Lavoisier was the son of a wealthy Parisian lawyer, who hoped that his son would follow him in that profession. Although Lavoisier qualified as a lawyer in 1764 he became interested in science and he devoted much of his life to research in chemistry. He is shown here with his wife, Marie-Anne Pierrette, who assisted him. Before he was twenty-five he had lectured to the French Academy on such diverse subjects as the divining rod, hypnotism, and the construction of invalid chairs. He was fascinated by the phenomenon of burning and, rich enough to have the best in apparatus and chemicals, he worked tirelessly to try to understand it. He was the first to clearly recognize that air consists of two gases, one of which supports combustion and the other (nitrogen) which does not. Lavoisier called this gas oxygen and recognized that it was an element. He was one of the first chemists to recognize the importance of accurate quantitative measurements, and his accurate determination of the masses of substances formed in reactions led him to the law of conservation of mass. He was also one of the first to demonstrate clearly the nature of elements and compounds. In a textbook that he wrote he gave a table of elements that included many that we would recognize today. It was a great tragedy that he was arrested and guillotined during the French revolution.

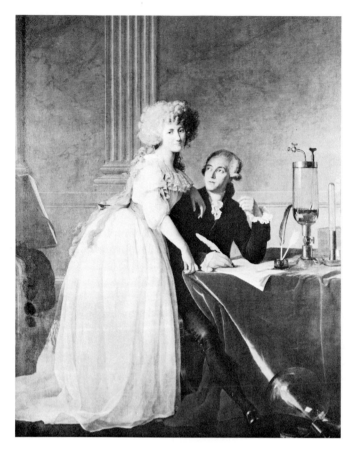

Work and Energy

We have talked a lot about energy, and we have a qualitative idea of what we mean by energy from everyday experience. But we have not yet defined energy in a precise way. We will do so now.

When a force displaces an object, we say that **work** is done on the object. The work done, w, is defined as the magnitude of the force, F, multiplied by the magnitude of the displacement, d:

$$w = F \times d$$

The SI unit of work is the **newton meter** (N m), which is called a **joule** (J):

$$1\ \text{J} = 1\ \text{N m} = 1\ \text{kg m}^2\ \text{s}^{-2}$$

The capacity to do work is called **energy**. Gasoline, for example, possesses energy because when it is burned, it can do the work of moving a car. We measure energy by the work done, and thus energy, like work, is measured in joules.

In practice, it is convenient to distinguish different forms of energy, such as heat energy, light energy, electric energy, and chemical energy. We will discuss several of these different types of energy in later chapters. They are all different forms of kinetic energy and/or potential energy:

- *Kinetic energy* is the energy possessed by an object by virtue of its motion. For a body of mass m with a speed v, the kinetic energy $(KE) = \frac{1}{2}mv^2$.
- *Potential energy* is the energy possessed by an object by virtue of its position in relation to another object that exerts a force upon it.

If we hold a stone above the ground, it has a certain potential energy, which depends on its height above the ground because of the gravitational force between the earth and the stone. If we let the stone fall, its potential energy is converted to kinetic energy as it accelerates toward the ground. When the stone hits the ground, it may do work, for example, by making a hole in the ground.

2.3 THE MOLE

One of the most important uses of the table of atomic masses is to find the relative masses of substances taking part in a chemical reaction and the masses of the products formed.

Formula Mass and Molecular Mass

The mass of a molecule is the sum of the masses of its component atoms, that is, the sum of the masses of the atoms in the molecular formula. This mass is called the **molecular mass**. For example, the molecular formula of hydrogen peroxide is H_2O_2. Therefore the molecular mass is $2(1.008\ u) + 2(16.00\ u) = 34.02\ u$.

Many substances do not consist of molecules, and for these substances we cannot calculate a molecular mass. We can instead calculate a **formula mass**, which is simply the sum of the masses of the atoms in the empirical formula. The empirical formula of sodium chloride is NaCl. Therefore the formula mass of sodium chloride is $22.99\ u + 35.45\ u = 58.44\ u$. We can also calculate a formula mass for substances which exist as molecules. If the empirical formula differs from the molecular formula, the empirical formula mass will differ from the molecular mass. In the case of hydrogen peroxide, for example, the empirical formula is HO and the empirical formula mass is therefore $1.008\ u + 16.00\ u = 17.01\ u$ which is one-half the molecular mass.

> **Example 2.4** What is the empirical formula mass of ethane? What is its molecular mass?
>
> **Solution** The empirical formula of ethane is CH_3 and the molecular formula is C_2H_6. Thus the empirical formula mass is $12.01\ u + 3(1.008\ u) = 15.03\ u$. The molecular mass is $2(12.01\ u) + 6(1.008\ u) = 30.07\ u = 2$ (empirical formula mass).

The Avogadro Constant

In studying reactions in the laboratory, we are normally concerned with large numbers of atoms and molecules rather than with single atoms and molecules. Moreover, we normally measure the masses of substances in grams rather than in atomic mass units. We therefore need a convenient way of translating the information provided by chemical formulas and the table of atomic masses into practical units that apply to convenient amounts of substances.

The formula of CO tells us that one atom of oxygen combines with one atom of carbon to give one molecule of carbon monoxide.

For 1 atom of C and 1 atom of O:

$$C + O \quad\quad \text{gives}\quad CO$$

1 atom C + 1 atom O   gives   1 molecule CO

1 atomic mass C + 1 atomic mass O $\longrightarrow$ 1 molecular mass CO

12.01 u C + 16.00 u O $\longrightarrow$ 28.01 u CO

For 100 *atoms of* C *and* 100 *atoms of* O:

$$100 \text{ atoms C} + 100 \text{ atoms O} \longrightarrow 100 \text{ molecules CO}$$

$$100(\text{atomic mass C}) + 100(\text{atomic mass O}) \longrightarrow 100(\text{molecular mass CO})$$

$$1201 \text{ u C} + 1600 \text{ u O} \longrightarrow 2801 \text{ u CO}$$

These numbers of atoms still correspond to extremely small amounts of carbon and oxygen. Normally, in any reaction carried out in the laboratory, we would be dealing with much larger numbers of atoms. We could, for example, consider the reactions of **6.022×10^{23}** atoms of carbon with the same number of oxygen atoms to give carbon monoxide molecules.

For 6.022×10^{23} *atoms of* C *and* 6.022×10^{23} *atoms of* O:

$$6.022 \times 10^{23} \text{ atoms C} + 6.022 \times 10^{23} \text{ atoms O} \longrightarrow 6.022 \times 10^{23} \text{ molecules CO}$$

$$(6.022 \times 10^{23} \times 12.01 \text{ u}) \text{ C} + (6.022 \times 10^{23} \times 16.00 \text{ u}) \text{ O} \longrightarrow (6.022 \times 10^{23} \times 28.01 \text{ u}) \text{ CO}$$

The number 6.022×10^{23} is a very large number. If we had 6.022×10^{23} marbles, they would cover the surface of the earth to a depth of 2 km. Why did we choose this particular very large number? We did so because 6.022×10^{23} u is 1.000 g. Hence

$$6.022 \times 10^{23} \times 12.01 \text{ u} = 12.01 \text{ g}$$

$$6.022 \times 10^{23} \times 16.00 \text{ u} = 16.00 \text{ g}$$

$$6.022 \times 10^{23} \times 28.01 \text{ u} = 28.01 \text{ g}$$

Thus for the reaction

$$C + O \longrightarrow CO$$

12.01 g of carbon reacts with 16.00 g of oxygen to give 28.01 g of carbon monoxide. In other words, we can write

$$\underset{12.01 \text{ g}}{\text{C}} + \underset{16.00 \text{ g}}{\text{O}} \longrightarrow \underset{28.01 \text{ g}}{\text{CO}}$$

Thus 6.022×10^{23} is the number of atoms in 12.01 g of carbon, or the number of atoms in 16.00 g of oxygen, or the number of molecules in 28.01 g of carbon monoxide. Using the relationship

$$1 \text{ g} = 6.022 \times 10^{23} \text{ u}$$

we can express the atomic mass unit in grams:

$$1 \text{ u} = \frac{1 \text{ g}}{6.022 \times 10^{23}} = 1.66 \times 10^{-24} \text{ g}$$

The number 6.022×10^{23} is called the **Avogadro constant**, after the Italian scientist Amedeo Avogadro (1776–1856); it is given the symbol N:

$$N = 6.022 \times 10^{23}$$

You might naturally wonder how the value of such a large number could possibly be determined; its determination will be discussed in Chapters 10 and 16. The presently accepted, most accurate value of the Avogadro constant is $6.022\,045 \times 10^{23}$, which gives a value of $1.660\,566 \times 10^{-24}$ g for 1 u. Values of $N = 6.022 \times 10^{23}$ and $1 \text{ u} = 1.661 \times 10^{-24}$ g are sufficiently accurate for our purposes in this book.

We may summarize our discussion as follows:

- 1 C atom has a mass of 12.01 u.
- 6.022×10^{23} C atoms have a mass of 12.01 g.
- 1 O atom has a mass of 16.00 u.
- 6.022×10^{23} O atoms have a mass of 16.00 g.
- 1 CO molecule has a mass of 28.01 u.
- 6.022×10^{23} CO molecules have a mass of 28.01 g.

Because of the very special significance of 6.022×10^{23} atoms or molecules, the amount of any substance containing this number of atoms or molecules is called a **mole** (abbreviated mol) of the substance. More formally, according to SI:

The mole is the amount of any substance that contains as many elementary entities as there are atoms in exactly 0.012 kg (12 g) of carbon-12. When the mole is used, the elementary entities must be stated. They may be atoms, molecules, ions, electrons, or other entities.

One mole of carbon-12 atoms has a mass of 12.00 g, and 1 mol of (naturally occurring) carbon atoms has a mass of 12.01 g. One mole of oxygen atoms has a mass of 16.00 g, and 1 mol of oxygen molecules (O_2) has a mass of 32.00 g.

One mole of atoms has a mass in grams numerically equal to the atomic mass in unified atomic mass units:

Atomic mass C = 12.01 u	1 mol C atoms has a mass of 12.01 g
Atomic mass O = 16.00 u	1 mol O atoms has a mass of 16.00 g
Atomic mass Al = 26.98 u	1 mol Al atoms has a mass of 26.98 g

One mole of molecules has a mass in grams numerically equal to the molecular mass in atomic mass units:

Molecular mass CO = 28.01 u	1 mol CO molecules has a mass of 28.01 g
Molecular mass CH_4 = 16.04 u	1 mol CH_4 molecules has a mass of 16.04 g

When we are dealing with a compound such as sodium chloride, which does not exist in the solid state in the form of molecules, we speak of a mole of empirical formula units:

1 mol NaCl (empirical formula units) has a mass of 22.99 g + 35.45 g = 58.44 g

Figure 2.6 shows 1 mol of the elements mercury (Hg), lead (Pb), copper (Cu), iron (Fe), and sulfur (S atoms and S_8 molecules), and 1 mol of the compounds water (H_2O), sugar ($C_{12}H_{22}O_{11}$), and sodium chloride (NaCl).

Figure 2.6 The Mole. Front row, left to right: The beakers contain one mole of each of the elements copper, lead, mercury, and iron. Each beaker contains 6.022×10^{23} atoms. Back row, left to right: The beakers contain 1 mol of sugar, $C_{12}H_{22}O_{11}$, (6.022×10^{23} molecules), 1 mol water, H_2O, (6.022×10^{23} molecules), 1 mol sulfur, S_8, (6.0222×10^{23} S_8 molecules) 1 mol sulfur, S, (6.022×10^{23} S atoms), 1 mol sodium chloride, NaCl, (6.022×10^{23} empirical formula units).

Example 2.5 How many moles of aluminum atoms are in 50.00 g of aluminum?

Solution We can summarize the problem as follows:

50.00 g Al = ? mol Al atoms

Since the atomic mass of aluminum is 26.98 u, 1 mol of aluminum atoms has a mass of 26.98 g. Then we can set up the unit conversion factor:

$$\frac{1 \text{ mol Al atoms}}{26.98 \text{ g Al}} = 1$$

which enables us to convert grams to moles. Therefore

$$(50.00 \text{ g Al}) \left(\frac{1 \text{ mol Al atoms}}{26.98 \text{ g Al}} \right) = 1.853 \text{ mol Al atoms}$$

Example 2.6 How many moles of carbon dioxide molecules are in 10.00 g of carbon dioxide?

Solution We can summarize the problem as follows:

$$10.00 \text{ g CO}_2 = ? \text{ mol CO}_2 \text{ molecules}$$

The molecular mass of carbon dioxide is 12.01 u + 2(16.00 u) = 44.01 u. So 1 mol of carbon dioxide molecules has a mass of 44.01 g. We can set up the unit conversion factor

$$\frac{1 \text{ mol CO}_2 \text{ molecules}}{44.01 \text{ g CO}_2} = 1$$

which enables us to convert grams to moles. Therefore

$$(10.00 \text{ g CO}_2)\left(\frac{1 \text{ mol CO}_2 \text{ molecules}}{44.01 \text{ g CO}_2}\right) = 0.2272 \text{ mol CO}_2 \text{ molecules}$$

Molar Mass

The mass of 1 mol of a substance is called its **molar mass**. The molar mass of hydrogen *atoms* is the mass of 1 mol of hydrogen *atoms*, which is 1.008 g mol^{-1}. The molar mass of hydrogen *molecules* is the mass of 1 mol of hydrogen *molecules*, H_2, which is 2(1.008) = 2.016 g mol^{-1}. The molar mass of water molecules, H_2O, is the mass of 1 mol of water molecules, or 18.02 g mol^{-1}. The molar mass of NaCl is the mass of 1 mole of NaCl empirical formulas or 58.44 g mol^{-1}.

When using moles and molar quantities such as molar mass, we must, strictly speaking, specify the nature of the entities being considered (atoms, molecules, empirical formulas, electrons). Nevertheless, we can speak of the molar mass of water without ambiguity, because the fact that water consists of H_2O molecules is well known. We might also speak of the molar mass of oxygen and assume that it is 32.00 g mol^{-1}, since oxygen normally consists of O_2 molecules. We must be careful to specify the entities, however, if the substances concerned are not in their ordinary or familiar states and there could be some doubt about the entity concerned. For example, if we were concerned with an experiment or with a calculation involving oxygen *atoms*, we would have to be careful to specify atoms, because the molar mass of oxygen atoms is 16.00 g mol^{-1}, not 32.00 g mol^{-1}.

2.4 CHEMICAL CALCULATIONS: COMPOSITIONS AND FORMULAS

How much carbon is in a kilogram of sugar given that the molecular formula of sugar (sucrose) is $C_{12}H_{22}O_{11}$? If we have prepared a substance that is made up of 56.8% carbon, 28.4% oxygen, 8.28% nitrogen, and 6.56% hydrogen, what is its empirical formula? Atomic masses, formulas, and the mole are the basic tools we need to solve problems like these, as we will now see.

Percentage Composition

From the formula of a compound we can find out how much of each of the elements is present in a given amount of the compound. Iron is obtained from iron oxide, Fe_2O_3, which is found as the ore hematite. From this formula we can calculate how much iron would be obtained, at least in principle, from a given amount of the iron oxide. This calculation is done most conveniently by making use of the idea of **percentage composition**, which is the percentage of the total mass of a compound contributed by each element. The percentage by mass of an element in a compound is as follows:

$$\text{Percent mass of element in compound} = \frac{\text{Mass of element in 1 molecular formula or 1 empirical formula of compound}}{\text{Molecular mass or empirical formula mass of compound}} \times 100\%$$

$$= \frac{\text{Mass of element in 1 mol compound}}{\text{Molar mass of compound}} \times 100\%$$

Let us calculate the percentage composition of iron oxide, Fe_2O_3:

$$\text{Percent iron} = \frac{\text{Mass of iron in 1 mol } Fe_2O_3}{\text{Molar mass } Fe_2O_3} \times 100\%$$

The molar mass of Fe_2O_3 is $(2 \times 55.85 \text{ g}) + (3 \times 16.00 \text{ g}) = 159.70$ g. Since 1 mole of Fe_2O_3 contains 2 mol of Fe atoms, the mass of Fe is 2×55.85 g Fe $= 111.7$ g Fe. Therefore

$$\text{Percent iron} = \frac{111.7 \text{ g}}{159.7 \text{ g}} \times 100\% = 69.94\%$$

We calculate the percentage of oxygen in the same way. One mole of Fe_2O_3 contains 3 mol of O atoms. So

$$\text{Percent oxygen} = \frac{3 \times 16.00 \text{ g}}{159.7 \text{ g}} = 30.06\%$$

Or since Fe_2O_3 contains only iron and oxygen, we could have found the percentage of oxygen more simply by subtracting 69.94% from 100%, that is,

$$\text{Percent oxygen} = 100.00\% - 69.94\% = 30.06\%$$

Now we are in a position to solve a problem such as the following: How much iron could be obtained from 10.00 kg of Fe_2O_3? Since the percentage of iron in Fe_2O_3 is 69.94%, there are 69.94 kg of Fe in 100.0 kg of Fe_2O_3. Hence the mass of iron in 10.00 kg of Fe_2O_3 is

$$(10.00 \text{ kg } Fe_2O_3)\left(\frac{69.94 \text{ kg Fe}}{100.0 \text{ kg } Fe_2O_3}\right) = 6.99 \text{ kg Fe}$$

Example 2.7 How much nitrogen is needed to make 10.00 kg of ammonia?

Solution One molecule of ammonia, NH_3, contains one N atom, or 1 mole of NH_3 contains 1 mol of N atoms. Thus

$$\text{Percent N} = \frac{\text{Mass of 1 mol N}}{\text{Mass of 1 mol } NH_3} = \frac{14.01 \text{ g}}{14.01 \text{ g} + 3(1.008 \text{ g})} \times 100\% = 82.25\%$$

The mass of nitrogen needed to make 10.00 kg of NH_3 is therefore

$$(10.00 \text{ kg } NH_3)\left(\frac{82.25 \text{ kg N}}{100.0 \text{ kg } NH_3}\right) = 8.225 \text{ kg}$$

Determination of Empirical Formulas

If we know the percentage composition of a compound from the results of analysis of the compound, we can obtain its empirical formula by reversing the calculations just described. Determining the empirical formula is important in the investigation of any new compound.

For example, water is 11.19% hydrogen and 88.81% oxygen by mass. From this data we can determine the empirical formula of water. The simplest way to proceed is to assume that we have a 100.00-g sample of water. The percentage composition then tells us that 100.00 g of water contains 11.19 g of hydrogen atoms and 88.81 g of oxygen atoms. From the table of atomic masses we find that 1 mol of hydrogen atoms has a mass of 1.008 g, and 1 mol of oxygen atoms has a mass of 16.00 g. Now we can write unit conversion factors so that the mass of hydrogen can be converted to moles of H atoms and the mass of oxygen can be converted to moles of O atoms. Because 1 mol H atoms has a mass of 1.008 g

$$\frac{1 \text{ mol H atoms}}{1.008 \text{ g H}} = 1$$

Therefore

$$11.19 \text{ g H} = (11.19 \text{ g H})\left(\frac{1 \text{ mol H atoms}}{1.008 \text{ g H}}\right) = 11.10 \text{ mol H atoms}$$

Similarly,

$$\frac{1 \text{ mol O atoms}}{16.00 \text{ g O}} = 1$$

Therefore

$$88.81 \text{ g O} = (88.81 \text{ g O})\left(\frac{1 \text{ mol O atoms}}{16.00 \text{ g O}}\right) = 5.55 \text{ mol O atoms}$$

Thus in water the ratio of moles of hydrogen atoms to moles of oxygen atoms is 11.10:5.55.

Since a mole of atoms always contains the same number of atoms ($N = 6.022 \times 10^{23}$), *the ratio of moles of atoms in a compound is also the ratio of the numbers of atoms.* Hence the ratio of hydrogen atoms to oxygen atoms is 11.10:5.55. We can convert both numbers to integers by dividing each by the smaller of the two numbers, 5.55:

$$\frac{11.10}{5.55} = 2.00 \quad \text{and} \quad \frac{5.55}{5.55} = 1.00$$

Thus the ratio of H atoms to O atoms is 2:1, and the empirical formula of water is therefore H_2O.

Example 2.8 A white compound is formed when phosphorus burns in air. Analysis shows that the compound is composed of 43.7% P and 56.3% O by mass. What is the empirical formula of the compound?

Solution The simplest procedure is to assume that we have a 100.0-g sample of the compound. According to the percentage composition given, 100.0 g of the compound contains 43.7 g of P atoms and 56.3 g of O atoms. Using the atomic masses of P and O, we can convert these masses to moles by using the appropriate unit conversion factors:

$$43.7 \text{ g P} = (43.7 \text{ g P})\left(\frac{1 \text{ mol P}}{30.97 \text{ g P}}\right) = 1.41 \text{ mol P}$$

$$56.3 \text{ g O} = (56.3 \text{ g O})\left(\frac{1 \text{ mol O}}{16.00 \text{ g O}}\right) = 3.52 \text{ mol O}$$

Thus the ratio of moles of P atoms to moles of O atoms is 1.41:3.52. Hence the ratio of P atoms to O atoms is also 1.41:3.52. We now divide by the smaller number, 1.41, so that one number in the ratio is 1:

$$\text{Ratio of P atoms to O atoms} = \frac{1.41}{1.41} : \frac{3.52}{1.41} = 1.00:2.50$$

This ratio gives us the empirical formula $PO_{2.5}$. But because we cannot have half an oxygen atom, we must multiply by the smallest factor that will make both numbers integral. In this case the appropriate factor is 2. So, finally, we have

$$\text{Ratio of P atoms to O atoms} = 2.00:5.00$$

The empirical formula is therefore P_2O_5.

Mole ratios do not always work out to exactly whole numbers, both because the numbers used in the calculations are rounded off and because there may be experimental errors in the percentage composition. When the experimental errors are small, there is little doubt about the appropriate whole numbers, as in the following example.

Example 2.9 The compound adrenaline is released in the human body in times of stress; it increases the body's metabolic rate. Like many compounds in living systems, it is composed of carbon, hydrogen, oxygen, and nitrogen. It was found by experiment to have the composition 56.8% C, 6.50% H, 28.4% O, and 8.28% N. What is the empirical formula of adrenaline?

Solution As in Example 2.9, we will assume that we are dealing with 100.0 g of the compound. For convenience we set up this type of calculation in a table.

	CARBON	HYDROGEN	OXYGEN	NITROGEN
Mass	56.8 g	6.50 g	28.4 g	8.28 g
Moles of atoms	$(56.8 \text{ g})\left(\frac{1 \text{ mol}}{12.01 \text{ g}}\right)$	$(6.56 \text{ g})\left(\frac{1 \text{ mol}}{1.008 \text{ g}}\right)$	$(28.4 \text{ g})\left(\frac{1 \text{ mol}}{16.00 \text{ g}}\right)$	$(8.28 \text{ g})\left(\frac{1 \text{ mol}}{14.01 \text{ g}}\right)$
	= 4.73	= 6.45	= 1.78	= 0.591
Mole ratio (atom ratio)	$\frac{4.73}{0.591}$:	$\frac{6.45}{0.591}$:	$\frac{1.78}{0.591}$:	$\frac{0.591}{0.591}$
	8.00 :	10.9 :	3.01 :	1.00

Thus the empirical formula of adrenaline is $C_8H_{11}O_3N$.

The percentage compositions from which we calculated the empirical formulas in the preceding examples are obtained from the results of experiments. There are two types of experiment by which the empirical formulas of a compound can be determined. Either the compound is synthesized from its elements, or it is broken down into its elements or into compounds of those elements that are of known composition. The following examples illustrate how the calculations may be carried out when the results are not given as a percentage composition.

Example 2.10 When the element antimony, Sb, is heated in air with excess sulfur, a reaction occurs to give a compound containing only antimony and sulfur. On further heating, excess sulfur is burnt off, forming gaseous sulfur dioxide, SO_2, and the substance left is a pure compound of antimony and sulfur. In one experiment 2.435 g of antimony was used, and the mass of the pure compound of antimony and sulfur was found to be 3.397 g. What is the empirical formula of the antimony-sulfur compound?

Solution From the law of conservation of mass we have

$$\text{Mass compound} = \text{Mass sulfur} + \text{Mass antimony}$$

$$\text{Mass sulfur} = \text{Mass compound} - \text{Mass antimony}$$

$$= 3.397 \text{ g} - 2.435 \text{ g} = 0.962 \text{ g}$$

Thus we see that 2.435 g of antimony combines with 0.962 g of sulfur to give the antimony-sulfur compound.

Next, we proceed as in previous examples to find the ratio of atoms of each element in the compound.

	ANTIMONY	SULFUR
Mass	2.435 g	0.962 g
Moles of atoms	$(2.435 \text{ g})\left(\dfrac{1 \text{ mol}}{121.8 \text{ g}}\right) = 0.0200$	$(0.962 \text{ g})\left(\dfrac{1 \text{ mol}}{32.06 \text{ g}}\right) = 0.0300$
Mole ratio (atom ratio)	$\dfrac{0.0200}{0.0200}$:	$\dfrac{0.0300}{0.0200}$
	1.00 :	1.50

This result corresponds to the formula $SbS_{1.5}$, but since we cannot have 1.5 S atoms, we must multiply by a suitable factor to obtain whole numbers. In this case the factor 2 is required. Thus the relative numbers of atoms are 2.00 Sb and 3.00 S. The empirical formula is therefore Sb_2S_3.

In Example 2.10 we found the empirical formula of antimony sulfide by synthesizing it from the elements. More commonly, the composition of a substance is obtained not by synthesizing it in this way but by carrying out a reaction that converts it into products of known formulas and, hence, known compositions. If the masses of these products obtained from a known mass of the original compound are determined, its composition can then be found.

For example, many of the carbon compounds known as organic compounds contain only carbon and hydrogen or only carbon, hydrogen, and oxygen. The composition of these organic compounds is often determined by burning them in an excess of dry oxygen gas, that is, sufficient oxygen to convert all the hydrogen in the compound to water vapor and all the carbon in the compound to carbon dioxide. The gases are then passed through a tube that contains a substance that absorbs the water and through another tube that contains a substance that absorbs the carbon dioxide, as shown in Figure 2.7. These tubes are weighed both before and after the experiment. The increase in the mass of the first tube is the mass of water absorbed, and the increase in mass of the second tube is the mass of carbon dioxide absorbed. From this information one can calculate the amounts of carbon and hydrogen in the original compound and thus determine its empirical formula, as the following example demonstrates.

2.4 CHEMICAL CALCULATIONS: COMPOSITIONS AND FORMULAS

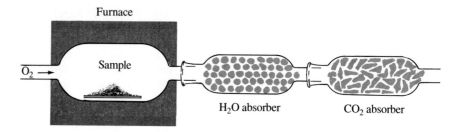

Furnace

Sample

O$_2$ →

H$_2$O absorber

CO$_2$ absorber

Figure 2.7 Combustion Apparatus for Analyzing Organic Compounds.
A weighed sample of the organic compound is burned in a stream of oxygen. The
reaction produces gaseous H$_2$O and CO$_2$. These gases then pass through a series
of tubes. One tube contains a substance that absorbs H$_2$O; the substance in another
tube absorbs CO$_2$. By comparing the masses of these tubes before and after the
reaction, one can determine the masses of hydrogen and of carbon present in the
compound that was burned.

Example 2.11 Ascorbic acid (vitamin C) is known to contain only C, H, and O. A
6.49-mg sample was burned in an apparatus like the one in Figure 2.7. The increased
weights of the absorption tubes showed that 9.74 mg of CO$_2$ and 2.64 mg of H$_2$O were
formed. What is the empirical formula of ascorbic acid?

Solution We first find the masses of C and H in the sample of ascorbic acid from the
masses of CO$_2$ and H$_2$O. For this calculation we need the molar masses of CO$_2$ and
H$_2$O:

$$\text{Molar mass CO}_2 = 12.01 + 2(16.00) = 44.01 \text{ g}$$

$$\text{Molar mass H}_2\text{O} = 2(1.008) + 16.00 = 18.02 \text{ g}$$

Then we have

$$\text{Mass C} = (9.74 \times 10^{-3} \text{ g CO}_2)\left(\frac{12.01 \text{ g C}}{44.01 \text{ g CO}_2}\right) = 2.66 \times 10^{-3} \text{ g C}$$

$$\text{Mass H} = (2.64 \times 10^{-3} \text{ g H}_2\text{O})\left(\frac{2.016 \text{ g H}}{18.02 \text{ g H}_2\text{O}}\right) = 0.295 \times 10^{-3} \text{ g H}$$

The mass of oxygen in the compound cannot be found from the masses of CO$_2$
and H$_2$O because some of this oxygen comes from the O$_2$ that was used to burn the
compound. But since we know that ascorbic acid contains only C, H, and O, the mass
of oxygen can be found by subtraction:

$$\text{Mass O} = 6.49 \text{ mg ascorbic acid} - (2.66 \text{ mg C} + 0.30 \text{ mg H})$$

$$= 3.53 \text{ mg} = 3.53 \times 10^{-3} \text{ g}$$

Now that we have the mass of C, H, and O in a given mass of ascorbic acid, we can
proceed to find the empirical formula in the usual way.

	C	H	O
Mass	2.66×10^{-3} g	0.295×10^{-3} g	3.53×10^{-3} g
Moles	$(2.66 \times 10^{-3} \text{ g})\left(\frac{1 \text{ mol}}{12.01 \text{ g}}\right)$	$(0.295 \times 10^{-3} \text{ g})\left(\frac{1 \text{ mol}}{1.01 \text{ g}}\right)$	$(3.53 \times 10^{-3} \text{ g})\left(\frac{1 \text{ mol}}{16.00 \text{ g}}\right)$
	$= 2.21 \times 10^{-4}$	$= 2.92 \times 10^{-4}$	$= 2.21 \times 10^{-4}$
Mole ratio (atom ratio)	$\dfrac{2.21 \times 10^{-4}}{2.21 \times 10^{-4}}$	$\dfrac{2.92 \times 10^{-4}}{2.21 \times 10^{-4}}$	$\dfrac{2.21 \times 10^{-4}}{2.21 \times 10^{-4}}$
	1.00	1.32	1.00

Finally, we convert these values to integers. Multiplying by 2 does not give integral values, so we try multiplying by 3, which gives

3.00	3.96	3.00

Thus the empirical formula of ascorbic acid is $C_3H_4O_3$.

If in addition to the percentage composition or the analytical data for a molecular compound, we also know its molar mass, we may find its molecular formula as well as its empirical formula, as Example 2.12 shows. Some methods for determining molar mass are described in later chapters.

Example 2.12 An organic compound contains 58.8% C, 9.80% H, and 31.4% O, and it has a molar mass of 204 g mol^{-1}. What are the empirical and molecular formulas of the compound?

Solution Assume that we have 100.0 g of the compound.

	C	H	O
Mass	58.8 g	9.80 g	31.4 g
Moles	$\dfrac{58.8 \text{ g}}{12.01 \text{ g mol}^{-1}} = 4.90 \text{ mol}$	$\dfrac{9.80 \text{ g}}{1.008 \text{ g mol}^{-1}} = 9.72 \text{ mol}$	$\dfrac{31.4 \text{ g}}{16.00 \text{ g mol}^{-1}} = 1.96 \text{ mol}$
Mole ratio (atom ratio)	$\dfrac{4.90}{1.96}$	$\dfrac{9.72}{1.96}$	$\dfrac{1.96}{1.96}$
	2.50	4.96	1.00

Since one of these values is not close to a whole number, we multiply by a suitable factor, which in this case is 2, giving

5.00	9.92	2.00

Thus the empirical formula of the compound is $C_5H_{10}O_2$. The empirical formula mass is therefore

$$5(12.01) + 10(1.008) + 2(16.00) = 102.1 \text{ g mol}^{-1}$$

The given molar mass of 204 g mol^{-1} is twice the empirical formula mass. Therefore the molecular formula is twice the empirical formula and is $C_{10}H_{20}O_4$.

2.5 CHEMICAL CALCULATIONS: REACTIONS

In this section we will see how we can answer a question such as: How much oxygen is needed to burn 1 kg of methane to give carbon dioxide and water? The first step in solving such a problem is to write down the chemical equation for the reaction.

Chemical Equations

A **chemical equation** is a shorthand description of a reaction that gives the formulas for all the reactants and all the products. For example,

$$\text{Reactants} \qquad \text{Products}$$
$$2H_2 + O_2 \longrightarrow 2H_2O$$
$$C + O_2 \longrightarrow CO_2$$
$$CH_4 + 2O_2 \longrightarrow CO_2 + 2H_2O$$

Most importantly, an equation must be consistent with the law of conservation of mass, which tells us that atoms are neither created nor destroyed in chemical reactions. The total numbers of atoms of each element must be the same in the products and in the reactants; that is, they must be the same on both sides of the equation. Thus two molecules (or four atoms) of hydrogen react with one molecule (or two atoms) of oxygen to give two water molecules, in which there are a total of four hydrogen atoms and two oxygen atoms. Every chemical equation must *balance* in this way.

Often in the formula of an element or a compound in an equation, we add a symbol to indicate whether the substance is in the gaseous, liquid, or solid state. The designations customarily used are (s) for solid, (l) for liquid, (g) for gas, and (aq) for substances that are in solution in water, that is, in aqueous solution. Thus the equation

$$CH_4 + 2O_2 \longrightarrow CO_2 + 2H_2O$$

could be written more completely as

$$CH_4(g) + 2O_2(g) \longrightarrow CO_2(g) + 2H_2O(g)$$

or as

$$CH_4(g) + 2O_2(g) \longrightarrow CO_2(g) + 2H_2O(l)$$

depending on whether the water was in the gaseous state or in the liquid state.

Balancing equations is not difficult, but practice is required. As an example, consider the equation representing the burning of methane in oxygen to give carbon dioxide and water. We first write down the reactants and products as follows:

$$CH_4 + O_2 \longrightarrow CO_2 + H_2O \qquad \text{(Unbalanced)}$$
$$\text{Reactants} \qquad\qquad \text{Products}$$

This equation is called the unbalanced equation because it indicates the reactants and products but not the relative numbers of molecules of each. Note that before any equation can be balanced, all the reactants and products and their correct formulas must be known. If all this information is not available, then a correct, balanced equation for a reaction cannot be written.

The simplest procedure for balancing an equation is to consider first those elements that appear the least frequently in the equation. In this example hydrogen and carbon appear in only two formulas each, while oxygen appears three times. So we begin by balancing the numbers of carbon and hydrogen atoms. All the carbon in methane, CH_4, must be converted to carbon dioxide, CO_2, because no other product contains carbon. Thus one molecule of CH_4 must give one molecule of CO_2, as already written. Each molecule of CH_4, however, contains four hydrogen atoms; and since all the hydrogen ends up in water molecules, two water molecules, H_2O, must be produced for each methane molecule. Hence we must place a 2 in front of the formula for water, to give

$$CH_4 + O_2 \longrightarrow CO_2 + 2H_2O \qquad \text{(Unbalanced)}$$

Now we can balance the remaining element, oxygen. We note that there are four oxygen atoms on the right side of the equation: two in the CO_2 molecule and two in the two H_2O molecules. We must therefore place a 2 in front of the formula for oxygen, O_2, so that we have four oxygen atoms on each side of the equation:

$$CH_4 + 2O_2 \longrightarrow CO_2 + 2H_2O \qquad \text{(Balanced)}$$

We can now make a final check of the numbers of each kind of atom on both sides of the equation. We find that we have one C atom, four H atoms, and four O atoms on both sides of the equation. The equation is therefore balanced.

Example 2.13 Butane, C_4H_{10}, is one of the gaseous components of petroleum. When liquified under pressure, it is sold as bottled gas for use as a fuel. When it is burned in sufficient oxygen, O_2, the only products are carbon dioxide, CO_2, and water, H_2O. Write a balanced equation to describe this reaction.

Solution First, we write an unbalanced equation showing the correct formulas of all the reactants and products:

$$C_4H_{10} + O_2 \longrightarrow CO_2 + H_2O \qquad \text{(Unbalanced)}$$

Since O atoms appear in three of these formulas, whereas C and H atoms appear in only two, we balance C and H first. There are 4 C atoms in C_4H_{10}, and CO_2 is the only product containing carbon. So 4 CO_2 molecules must be formed:

$$C_4H_{10} + O_2 \longrightarrow 4CO_2 + H_2O \qquad \text{(Unbalanced)}$$

The 10 H atoms in C_4H_{10} must appear in the products as 5 H_2O molecules:

$$C_4H_{10} + O_2 \longrightarrow 4CO_2 + 5H_2O \qquad \text{(Unbalanced)}$$

We now have a total of 13 O atoms on the right-hand side. The equation could be balanced by using a coefficient of $\frac{13}{2}$ in front of O_2 on the left-hand side:

$$C_4H_{10} + \tfrac{13}{2} O_2 \longrightarrow 4CO_2 + 5H_2O$$

We would then have 13 O atoms on each side of the equation. Customarily, however, we do not write fractional coefficients in equations, because they could be interpreted as meaning fractions of a molecule. One-half of an oxygen molecule is an oxygen atom, which has properties quite different from those of an oxygen molecule. Ambiguity is avoided by multiplying both sides of the equation by 2 to obtain the final balanced equation:

$$2C_4H_{10} + 13O_2 \longrightarrow 8CO_2 + 10H_2O$$

A final check should be made by counting the numbers of atoms of each kind on both sides.

$$2C_4H_{10} + 13O_2 \longrightarrow 8CO_2 + 10H_2O$$

	C	H	O	C	H	O
Number of atoms	8	20	26	8	20	26

Although this equation looks complicated, we balanced it rather easily because we used a logical, systematic approach.

Some remarks on incorrect ways of balancing equations might be helpful. The equation for the reaction between hydrogen and oxygen to give water,

$$H_2 + O_2 \longrightarrow H_2O$$

cannot be balanced by writing a 2 after the O in H_2O,

$$H_2 + O_2 \longrightarrow H_2O_2$$

This equation is balanced, but it is *not* the equation for the reaction between hydrogen and oxygen to give *water*. The product has been changed to hydrogen peroxide, and this equation is now for a *different reaction*.

The equation should also not be balanced as

$$H_2 + O_2 \longrightarrow H_2O + O$$

Although this equation is also balanced, another product, namely O atoms, has been introduced, which was not originally specified. In balancing an equation, we can make no change in the nature of either the reactants or the products. Coefficients must be inserted to balance the equation without introducing new products or reactants. The balanced equation for this reaction is

$$2H_2 + O_2 \longrightarrow 2H_2O$$

Finally, we note that equations can only be balanced after all the products are known. Incomplete equations such as

$$H_2 + O_2 \longrightarrow$$

cannot be balanced in an unambiguous way because no product is specified. This equation might be completed in three ways:

$$H_2 + O_2 \longrightarrow H_2O_2$$
$$H_2 + O_2 \longrightarrow H_2O + O$$
$$2H_2 + O_2 \longrightarrow 2H_2O$$

These equations represent different reactions, and only the last equation is the balanced equation for the reaction between hydrogen and oxygen to produce water.

Being able to predict the probable products of a reaction is important, and much of this book is devoted to providing the background information that will help you to do this. But predicting products has nothing to do with balancing equations. The products may be given; they may be determined by experiment; or they may be predicted. But only after the products have been specified in some way can we balance an equation.

Calculations Using Chemical Equations

From the balanced chemical equation and a knowledge of atomic masses, we can find the masses of the reactants and the products in any reaction. For example, from the atomic masses of hydrogen and oxygen and the equation

$$2H_2 + O_2 \longrightarrow 2H_2O$$

we know that

$$2 \text{ molecules } H_2 + 1 \text{ molecule } O_2 \longrightarrow 2 \text{ molecules } H_2O$$
$$\qquad 4.032 \text{ u} \qquad\qquad 32.00 \text{ u} \qquad\qquad\qquad 36.03 \text{ u}$$

Normally, we are not concerned with such small numbers of molecules. As we have seen, we may, much more conveniently, consider the reaction of 2 mol of hydrogen ($2 \times 6.022 \times 10^{23}$ molecules) with 1 mol of oxygen (6.022×10^{23} molecules) to give 2 mol of water ($2 \times 6.022 \times 10^{23}$ molecules):

$$2H_2 \qquad + \qquad O_2 \qquad\qquad 2H_2O$$

2 mol	1 mol	2 mol
$2(6.022 \times 10^{23})$ molecules	6.022×10^{23} molecules	$2(6.022 \times 10^{23})$ molecules
4.032 g	32.00 g	36.03 g

We conclude that 4.032 g of hydrogen reacts with 32.00 g of oxygen to give 36.03 g of water.

We could also change to any other convenient mass unit, such as tons (1 ton = 2000 lb) or metric tons (tonnes) (1 metric ton = 10^6 g = 10^3 kg). Thus we could write

$$2H_2 \quad + \quad O_2 \quad \longrightarrow \quad 2H_2O$$

$$\text{4.032 tons} \quad \text{32.00 tons} \qquad \text{36.03 tons}$$

The *relative* masses are the same in *any* units. We may also divide or multiply these amounts by any convenient factor. For example, dividing by 4.032, we have

$$2H_2 \quad + \quad O_2 \quad \longrightarrow \quad 2H_2O$$

$$\text{1.000 ton} \quad \text{7.936 tons} \qquad \text{8.936 tons}$$

Thus we see that a chemical equation can be interpreted in two important ways. First, it gives the number of atoms and molecules taking part in the reaction and the corresponding masses in atomic mass units. Second, it gives the number of moles taking part in the reaction, with the corresponding masses in grams or in other convenient mass units. The balancing of chemical equations and the calculation of the amounts of reactants and products taking part in reactions are important exercises with which you should become familiar.

Example 2.14 How many grams of chlorine, Cl_2, are needed to react completely with 0.245 g of hydrogen, H_2, to give hydrogen chloride, HCl? How much HCl is formed?

Solution First, we write the balanced equation:

$$H_2 + Cl_2 \longrightarrow 2HCl$$

Next, we calculate the number of moles of H_2:

$$(0.245 \text{ g H}_2)\left(\frac{1 \text{ mol H}_2}{2.016 \text{ g H}_2}\right) = 0.122 \text{ mol}$$

Then we find the number of moles of Cl_2 that react with 0.122 mol of H_2. Since 1 mol of Cl_2 reacts with 1 mol of H_2, we need

$$(0.122 \text{ mol H}_2)\left(\frac{1 \text{ mol Cl}_2}{1 \text{ mol H}_2}\right) = 0.122 \text{ mol Cl}_2$$

Then we convert 0.122 mol of Cl_2 to grams:

$$(0.122 \text{ mol Cl}_2)\left(\frac{70.90 \text{ g Cl}_2}{1 \text{ mol Cl}_2}\right) = 8.65 \text{ g Cl}_2$$

The amount of HCl produced must equal the total amount of H_2 and Cl_2 consumed in the reaction, which is 8.65 + 0.245 = 8.90 g HCl. Or we could calculate the amount of HCl by making use of the appropriate unit conversion factors:

$$(0.122 \text{ mol H}_2)\left(\frac{2 \text{ mol HCl}}{1 \text{ mol H}_2}\right)\left(\frac{36.46 \text{ g HCl}}{1 \text{ mol HCl}}\right) = 8.90 \text{ g HCl}$$

From the balanced equation   From the molar mass of HCl

Example 2.15 The first step in obtaining elemental zinc from its sulfide ore, ZnS, involves heating it in air to obtain the oxide of zinc, ZnO, and sulfur dioxide, SO_2. Calculate the mass of O_2 required to react with 7.00 g of ZnS. How much SO_2 is produced?

Solution First, we write the unbalanced equation:

$$ZnS + O_2 \longrightarrow ZnO + SO_2 \quad \text{(Unbalanced)}$$

Next, we balance the equation. Since Zn and S already are balanced, the only element that remains to be considered is O. In the unbalanced equation there are two O atoms on the left side and three on the right. Thus the equation can be balanced by a coefficient of $\frac{3}{2}$ in front of O_2:

$$ZnS + \tfrac{3}{2}O_2 \longrightarrow ZnO + SO_2$$

Multiplying all coefficients by 2 eliminates fractions and gives

$$2ZnS + 3O_2 \longrightarrow 2ZnO + 2SO_2$$

$$\text{2 mol} \qquad \text{3 mol} \qquad \text{2 mol} \qquad \text{2 mol}$$

Since the formula mass of ZnS is $65.38 + 32.06 = 97.44$ u and the molar mass is therefore 97.44 g, we can convert the mass of ZnS to moles of ZnS:

$$(7.00 \text{ g ZnS})\left(\frac{1 \text{ mol ZnS}}{97.44 \text{ g ZnS}}\right) = 0.0718 \text{ mol ZnS}$$

From the balanced equation we can write

$$\frac{3 \text{ mol } O_2}{2 \text{ mol ZnS}} = 1$$

Hence

$$(0.0718 \text{ mol ZnS})\left(\frac{3 \text{ mol } O_2}{2 \text{ mol ZnS}}\right) = 0.108 \text{ mol } O_2$$

Finally, we convert to the mass of O_2:

$$(0.108 \text{ mol } O_2)\left(\frac{32.00 \text{ g } O_2}{1 \text{ mol } O_2}\right) = 3.45 \text{ g } O_2$$

With a little experience we can carry out the calculation in one step, using the appropriate unit conversion factors as follows:

$$? \text{ g } O_2 = (7.00 \text{ g ZnS})\left(\frac{1 \text{ mol ZnS}}{97.44 \text{ g ZnS}}\right)\left(\frac{3 \text{ mol } O_2}{2 \text{ mol ZnS}}\right)\left(\frac{32.00 \text{ g } O_2}{1 \text{ mol } O_2}\right) = 3.45 \text{ g } O_2$$

Similarly, the amount of sulfur dioxide produced is

$$(7.00 \text{ ZnS})\left(\frac{1 \text{ mol ZnS}}{97.44 \text{ g ZnS}}\right)\left(\frac{2 \text{ mol } SO_2}{2 \text{ mol ZnS}}\right)\left(\frac{64.06 \text{ g } SO_2}{1 \text{ mol } SO_2}\right) = 4.60 \text{ g } SO_2$$

Limiting Reactant

Often substances that react with each other are not present in exactly the proportions stated in the balanced equation. For example, if 2 mol each of hydrogen and oxygen are mixed and a spark is passed through the mixture, water is formed. According to the equation

$$2H_2 + O_2 \longrightarrow 2H_2O$$

2 mol of hydrogen react with only 1 mol of oxygen, and 1 mol of oxygen there-

fore remains unreacted. In this example hydrogen is said to be the **limiting reactant** because its concentration becomes zero and the reaction therefore stops before all the other reactant, that is, the oxygen, is used up. The amount of hydrogen present initially limits the amount of the product that is formed.

Example 2.16 Three moles of sulfur dioxide, SO_2, react with 2 mol of oxygen, O_2, to give sulfur trioxide, SO_3.

(a) Which is the limiting reactant?

(b) What is the maximum amount of SO_3 that can be formed?

(c) How much of one of the reactants remains unreacted?

Solution

(a) We must first write the balanced equation for the reaction. The unbalanced equation is

$$SO_2 + O_2 \longrightarrow SO_3 \qquad \text{(Unbalanced)}$$

If we multiply SO_2 by 2 and SO_3 by 2, the equation is balanced:

$$2SO_2 + O_2 \longrightarrow 2SO_3$$

From this equation we can see that 2 mol of SO_2 give 2 mol of SO_3 and that 1 mol of O_2 gives 2 mol of SO_3. But we have 3 mol SO_2 and 2 mol O_2, so the corresponding amounts of SO_3 are

$$(3 \text{ mol } SO_2)\left(\frac{2 \text{ mol } SO_3}{2 \text{ mol } SO_2}\right) = 3 \text{ mol } SO_3$$

$$(2 \text{ mol } O_2)\left(\frac{2 \text{ mol } SO_3}{1 \text{ mol } O_2}\right) = 4 \text{ mol } SO_3$$

Whichever reactant gives the smallest amount of product is the limiting reactant. In this case we see that SO_2 is the limiting reactant.

(b) The maximum amount of product that can be obtained is the amount formed by the limiting reactant. Thus the maximum amount of SO_3 that can be obtained is 3 mol.

(c) From the equation we see that the reaction of 3 mol of SO_2 to give 3 mol of SO_3 uses 1.5 mol of O_2. Since there are initially 2 mol of O_2, the amount of O_2 that remains unreacted is $(2 - 1.5)$ mol $= 0.5$ mol.

Now that we have examined the principles underlying a limiting reactant problem, we may consider a more common type of problem in which the masses of the reactants are given rather than the numbers of moles of reactants.

Example 2.17 When zinc and sulfur are heated together, they react to form zinc sulfide, according to the equation

$$Zn + S \longrightarrow ZnS$$

Suppose 12.00 g of zinc is heated with 7.00 g of sulfur.

(a) Which is the limiting reactant?

(b) What is the maximum amount of ZnS that can be formed?

(c) How much of one of the elements remains unreacted?

Solution

(a) We convert the mass of each reactant to moles:

$$(12.00 \text{ g Zn})\left(\frac{1 \text{ mol Zn}}{65.38 \text{ g Zn}}\right) = 0.184 \text{ mol Zn}$$

$$(7.50 \text{ g S})\left(\frac{1 \text{ mol S}}{32.06 \text{ g S}}\right) = 0.234 \text{ mol S}$$

We see from the equation that 1 mol of Zn reacts with 1 mol of S. Therefore there is more S than required; thus Zn is the limiting reactant.

(b) The amount of product formed depends on the amount of the limiting reactant, Zn, and not on the amount of S. The equation shows that 1 mol of Zn gives 1 mol of ZnS. Therefore 0.184 mol of Zn gives 0.184 mol ZnS. The mass of ZnS formed is thus

$$(0.184 \text{ mol ZnS})\left(\frac{97.44 \text{ g ZnS}}{1 \text{ mol ZnS}}\right) = 17.9 \text{ g ZnS}$$

(c) The excess amount of sulfur is 0.234 mol − 0.184 mol = 0.050 mol S. We convert this result to grams:

$$(0.050 \text{ mol S})\left(\frac{32.06 \text{ g S}}{1 \text{ mol S}}\right) = 1.60 \text{ g excess S}$$

2.6 MOLAR CONCENTRATIONS

Because so many reactions are carried out in solution, it is convenient to be able to measure out a given amount of a substance by taking a known volume of a solution of known concentration. There are a number of ways of expressing the concentration of a solution. For instance, we previously defined the concentration of a solution in terms of grams of solute in 100 g of solution (mass percent). But often expressing concentrations in terms of moles of solute in a given volume of solution is more convenient. Thus the **molarity** of a solute in solution is defined as the number of moles of solute contained in 1 L (1 dm^3) of solution:

$$\text{Molarity} = \frac{\text{Moles of solute}}{\text{Liters of solution}}$$

The units of molarity are thus moles per liter (mol L^{-1}), or moles per cubic decimeter (mol dm^{-3}). **Molar concentration** is given the symbol M (M = mol L^{-1} or mol dm^{-3}).

Example 2.18 A solution is prepared by dissolving 20.36 g of sodium chloride, NaCl, in sufficient distilled water to give a 1.000 L of solution. What is the molarity of NaCl in the solution?

Solution   Molar mass NaCl = (22.99 + 35.45) = 58.44 g

To obtain the number of moles of NaCl, we multiply the mass of NaCl (20.36 g) by the appropriate unit conversion factor:

$$(20.36 \text{ g NaCl})\left(\frac{1 \text{ mol NaCl}}{58.44 \text{ g NaCl}}\right) = 0.3484 \text{ mol NaCl}$$

This is the amount of NaCl dissolved in 1.000 L of solution. Thus

$$\text{Molarity NaCl} = 0.3484 \text{ mol L}^{-1} = 0.3484 \text{ } M$$

Example 2.19 Suppose 20.36 g of NaCl is dissolved in sufficient distilled water to give a solution with a volume of 250 mL. What is the molar concentration of NaCl?

Solution As in Example 2.18, moles of NaCl = 0.3484, but the volume of the solution is now 0.250 L.

Thus

$$\text{Molarity NaCl} = \frac{0.348 \text{ mol}}{0.250 \text{ L}} = 1.39 \text{ mol L}^{-1} = 1.39 \text{ } M$$

Volumes of solutions are measured in the laboratory in several ways. Some of the more common apparatus for measuring the volume of solutions are shown in Figure 2.8. If we measure a given volume of a solution of known concentration, we must be able to find how much solute this volume of solution contains. This amount may be calculated as shown in the following example.

Example 2.20 How many moles and how many grams of sodium hydroxide, NaOH, are in 25.0 mL of a 0.500M NaOH solution?

Solution A 0.500M NaOH solution contains 0.500 mol of NaOH in 1.000 L, or 1000 mL, of solution:

$$0.500M \text{ NaOH} = \frac{0.500 \text{ mol NaOH}}{1000 \text{ mL solution}}$$

We can use this ratio as a conversion factor to convert volume of solution to moles:

$$(25.0 \text{ mL solution})\left(\frac{0.500 \text{ mol NaOH}}{1000 \text{ mL solution}}\right) = 0.0125 \text{ mol NaOH}$$

Now we can convert this result to grams:

$$(0.0125 \text{ mol NaOH})\left(\frac{40.00 \text{ g NaOH}}{1 \text{ mol NaOH}}\right) = 0.500 \text{ g NaOH}$$

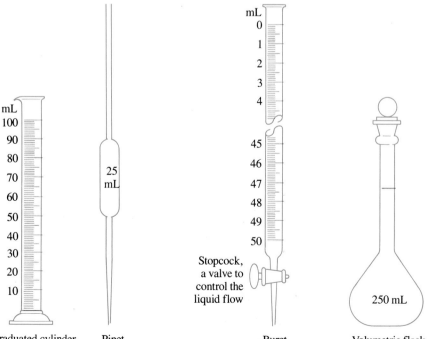

Figure 2.8 Apparatus for Measuring Volumes of Solutions.

Graduated cylinder   Pipet   Buret   Volumetric flask

2.6 MOLAR CONCENTRATIONS

71

(a)

(b)

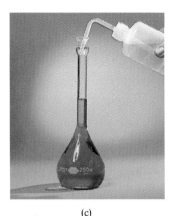

(c)

Figure 2.9 Using a Volumetric (Graduated) Flask. An accurately weighed amount of solute is added to the flask (a). Distilled water is added and the flask is shaken to dissolve the solute. More distilled water is added until the level of the solution reaches the mark on the neck of the flask.

Note that molarity is defined as moles of solute in a given volume of *solution*. Thus a $1M$ solution of sodium hydroxide *cannot* be prepared by adding 1 mol of sodium hydroxide (40.00 g) to 1 L of water. Because of the additional volume occupied by the sodium hydroxide, the total volume of the solution would not be exactly 1.000 L, and therefore the concentration of the solution would not be exactly $1.000M$. A solution of known molarity is usually prepared by using a *volumetric flask* (Figure 2.8). This flask has a graduation mark on the neck; when the flask is filled with a solution exactly to this point, the volume of the solution, for example, 250 or 500 mL, is precisely known. The procedure for making up a solution of known concentration by using such a flask is illustrated in Figure 2.9.

To make up a certain volume of a solution of known concentration, we must know what mass of the solute will be needed. The required amount may be calculated as shown in the following example.

Example 2.21 How many grams of sodium sulfate, Na_2SO_4, are required to prepare 250 mL of $0.500M$ Na_2SO_4 solution?

Solution This problem can be restated as follows: How many grams of Na_2SO_4 are in 250 mL of $0.500M$ Na_2SO_4 solution? A $0.500M$ Na_2SO_4 solution contains 0.500 mol in 1 L of solution. To find the number of moles of Na_2SO_4 in 250 mL, we use the conversion factor 0.500 mol/1000 mL. Therefore 250 mL solution contains

$$(250 \text{ mL})\left(\frac{0.500 \text{ mol}}{1000 \text{ mL}}\right) = 0.125 \text{ mol}$$

Now we can convert to grams, using the unit conversion factor

$$\frac{142.0 \text{ g Na}_2\text{SO}_4}{1 \text{ mol Na}_2\text{SO}_4} = 1$$

Therefore

$$(0.125 \text{ mol Na}_2\text{SO}_4)\left(\frac{142.0 \text{ g Na}_2\text{SO}_4}{1 \text{ mol Na}_2\text{SO}_4}\right) = 17.8 \text{ g Na}_2\text{SO}_4$$

Or we can do the calculation in one step, as follows:

$$? \text{ g Na}_2\text{SO}_4 = (250 \text{ mL})\left(\frac{0.500 \text{ mol Na}_2\text{SO}_4}{1000 \text{ mL}}\right)\left(\frac{142.0 \text{ g Na}_2\text{SO}_4}{1 \text{ mol Na}_2\text{SO}_4}\right) = 17.8 \text{ g Na}_2\text{SO}_4$$

Acids such as sulfuric acid are available commercially as concentrated aqueous solutions. These acids must be diluted by adding them to more water in order to prepare the more dilute solution that might be required in the laboratory.

Example 2.22 What volume of concentrated, aqueous sulfuric acid, which is 98% H_2SO_4 by mass and has a density of 1.84 g cm^{-3}, is required to make 10.0 L of $0.200M$ H_2SO_4 solution?

Solution We find the number of moles of H_2SO_4 in 10.0 L of $0.200M$ H_2SO_4 solution:

$$(10.0 \text{ L})\left(\frac{0.200 \text{ mol H}_2\text{SO}_4}{1 \text{ L solution}}\right) = 2.00 \text{ mol H}_2\text{SO}_4$$

Thus we require sufficient concentrated acid to give 2.00 mol of H_2SO_4, or

$$(2.00 \text{ mol } H_2SO_4)\left(\frac{98.08 \text{ g } H_2SO_4}{1 \text{ mol } H_2SO_4}\right) = 196 \text{ g } H_2SO_4$$

Since the concentrated (conc) acid is 98% by mass, we need

$$(196 \text{ g } H_2SO_4)\left(\frac{100 \text{ g conc acid}}{98 \text{ g } H_2SO_4}\right) = 200 \text{ g conc acid}$$

The density of the concentrated acid is $1.84 \text{ g cm}^{-3} = 1.84 \text{ g mL}^{-1}$. So we need

$$(200 \text{ g conc acid})\left(\frac{1 \text{ mL conc acid}}{1.84 \text{ g conc acid}}\right) = 109 \text{ mL conc acid}$$

With experience we can do this problem in one step, as follows:

$$? \text{ g conc acid} = (10.0 \text{ L})\left(\frac{0.200 \text{ mol } H_2SO_4}{1 \text{ L solution}}\right)\left(\frac{98.08 \text{ g } H_2SO_4}{1 \text{ mol } H_2SO_4}\right)$$

$$\times \left(\frac{100 \text{ g conc acid}}{98 \text{ g } H_2SO_4}\right)\left(\frac{1 \text{ mL conc acid}}{1.84 \text{ g conc acid}}\right)$$

$$= 109 \text{ mL conc acid}$$

Many reactions are carried out by using aqueous solutions of known concentration. The amount of a product of a reaction can be found from the volumes of the solutions of the reactants and their concentrations, as shown in the following example.

Example 2.23 An aqueous solution of sodium sulfate, Na_2SO_4, reacts with an aqueous solution of barium chloride, $BaCl_2$, to give a precipitate of insoluble barium sulfate, $BaSO_4$. Suppose 250 mL of $0.500M$ Na_2SO_4 solution is added to an aqueous solution of 10.00 g of $BaCl_2$. How many moles and how many grams of $BaSO_4$ are obtained?

Solution First, we must write the balanced equation for the reaction. The unbalanced equation is

$$BaCl_2 + Na_2SO_4 \longrightarrow BaSO_4 + NaCl \qquad \text{(Unbalanced)}$$

The Ba, S, and O are already balanced. Because there are 2 Na and 2 Cl on the left, we need 2 formula units of NaCl on the right. The balanced equation is

$$BaCl_2 + Na_2SO_4 \longrightarrow BaSO_4 + 2NaCl$$

Next, since we know the amounts of both reactants, we must find which is the limiting reactant.

$$\left(\frac{0.500 \text{ mol } Na_2SO_4}{1 \text{ L}}\right)\left(\frac{1 \text{ L}}{1000 \text{ mL}}\right)(250 \text{ mL}) = 0.125 \text{ mol } Na_2SO_4$$

$$(10.00 \text{ g } BaCl_2)\left(\frac{1 \text{ mol } BaCl_2}{208.2 \text{ g}}\right) = 0.0480 \text{ mol } BaCl_2$$

The balanced equation shows us that 1 mol of $BaCl_2$ reacts with 1 mol of Na_2SO_4, so $BaCl_2$ is the limiting reactant. The equation also shows that 1 mol of $BaCl_2$ gives 1 mol of $BaSO_4$, so the mass of $BaSO_4$ formed is:

$$(0.0480 \text{ mol } BaCl_2)\left(\frac{1 \text{ mol } BaSO_4}{1 \text{ mol } BaCl_2}\right)\left(\frac{233.4 \text{ g } BaSO_4}{1 \text{ mol } BaSO_4}\right) = 11.20 \text{ g } BaSO_4$$

IMPORTANT TERMS

The **atomic mass** of an atom is the mass of the atom expressed in atomic mass units.

The **atomic mass unit**, u, is $\frac{1}{12}$ the mass of a $^{12}_6C$ atom.

The **atomic number**, Z, of an element is the number of protons in the nucleus of an atom of the element; it is equal to the number of electrons surrounding the nucleus in the neutral atom.

The Avogadro constant is the number of elementary entities (atoms, molecules, electrons, and so on) in 1 mol of entities; $N = 6.022 \times 10^{23}$ mol^{-1}.

A **chemical reaction** is a process in which atoms are rearranged to form new substances. Atoms are neither created nor destroyed; the total number of each kind of atom is conserved.

Coulomb's law states that the force, F, between two charges, Q_1 and Q_2, is proportional to the product of the charges and inversely proportional to the square of the distance, r, between the charges: $F = k(Q_1Q_2/r^2)$.

An **element** is a substance composed of atoms of only one kind, all of which have the same atomic number Z.

The **empirical formula** of a substance is the simplest formula that expresses the relative numbers of atoms as whole numbers.

The **empirical formula mass** is the mass obtained by adding up the masses of all the atoms in the empirical formula of a substance.

Isotopes are atoms with the same atomic number Z but with different masses.

The **law of conservation of mass** states that the total mass of the products of a chemical reaction is equal to the total mass of the reactants; atoms are neither created nor destroyed.

A **limiting reactant** is the particular reactant among two or more reactants whose amount determines the amount of product that may be formed.

The **mass defect** is the difference between the mass of an atom and the sum of the masses of its constituent particles (protons, neutrons, and electrons).

The **mass number** (*nucleon number*) is the sum of the number of protons and the number of neutrons in an atom.

Molarity is the concentration of a solute in solution, expressed in moles per liter ($M = $ mol L^{-1}).

Molar mass is the mass of 1 mol of a substance.

A **mole** is the amount of a substance that contains as many elementary entities (for example, atoms, molecules, empirical formula units) as there are atoms in 0.012 kg (12 g) of carbon-12; 1 mol $= 6.022 \times 10^{23}$ entities.

A **molecular formula** shows the numbers of atoms of each kind in a molecule of a substance. It is always some integral multiple of the empirical formula.

The **molecular mass** is the mass obtained by adding the masses of all the atoms in the molecular formula of a substance.

Nucleon number. See *mass number*.

Stoichiometry is the term used to refer to all the quantitative aspects of chemical composition and chemical reactions.

PROBLEMS

Atoms, Isotopes, Atomic Mass

1. What is the number of protons, neutrons, and electrons in each of the following atoms?

^2_1H $^{19}_9F$ $^{40}_{20}Ca$ $^{112}_{48}Cd$ $^{117}_{50}Sn$ $^{131}_{54}Xe$

2. Complete the accompanying table.

ATOM SYMBOL	MASS NUMBER	ATOMIC NUMBER	NUMBER OF PROTONS	NUMBER OF ELECTRONS	NUMBER OF NEUTRONS
^9_4Be	9	4	4	4	5
$^{15}_7N$	—	—	—	—	—
$^{18}_8O$	—	—	—	—	—
—	—	—	6	6	6
—	23	11	—	—	—

---

* The asterisk denotes the more difficult problems.

3. Complete the accompanying table.

ATOM SYMBOL	MASS NUMBER	ATOMIC NUMBER	NUMBER OF PROTONS	NUMBER OF ELECTRONS	NUMBER OF NEUTRONS
$^{24}_{12}Mg$					
$^{106}_{47}Ag$					
$^{137}_{56}Ba$					

4. Which elements have atoms containing nuclei with the following numbers of protons?

5   9   32   54   92

5. Give the symbols for each of the following isotopes:

(a) Atomic number 19, mass number 40.

(b) Atomic number 14, mass number 30.

(c) Atomic number 18, mass number 40.

(d) Atomic number 7, mass number 15.

6. (a) From the mass and approximate radius of the proton quoted in this chapter, determine the approximate density, in grams per cubic centimeter, of matter in the proton. (The volume of a sphere of radius r is $4\pi r^3/3$.)

(b) What is the approximate density of matter in a hydrogen atom, assuming its radius to be 140 pm.

(c) Compare the values obtained in parts (a) and (b) with the density of water. Explain the differences.

7. Write the formulas for each of the isotopically different molecules of water. The isotopes of hydrogen and oxygen are ^1H, ^2H, ^{16}O, ^{17}O, and ^{18}O.

*8. Bromine atoms and chlorine atoms combine to give bromine chloride molecules with the molecular formula BrCl. Bromine chloride is found to consist of molecules with approximate masses 114, 116, and 118 u, and it is known that chlorine has just two isotopes, with mass numbers 35 and 37.

(a) Deduce the possible isotopes of bromine, and write symbols for each of them.

(b) Give a formula for each of the possible kinds of BrCl molecules.

9. Boron has two isotopes with masses of 10.012 94 and 11.009 31 u and abundances of 19.77% and 80.23%. What is the average atomic mass of boron?

10. (a) What information is given by the symbol $^{23}_{12}Mg$?

(b) From the data in the accompanying table, calculate the average atomic mass of magnesium.

MASS NUMBER	ABUNDANCE (%)	ATOMIC MASS (u)
24	78.60	23.993
25	10.11	24.994
26	11.29	25.991

11. Define *mass number* and *atomic number*, and explain what is meant by the term *isotope*. Naturally occurring copper has an atomic mass of 63.54 u and contains two isotopes with masses of 62.9298 and 64.9278 u. What is the abundance of each isotope?

12. Gallium, Ga, has two isotopes of atomic mass 68.926 and 70.926 u. The atomic number of gallium is 31. How many protons and neutrons are present in the nucleus of each isotope? Write symbols for each. What is the natural abundance of each isotope if the atomic mass of gallium is 69.72?

13. Uranium has an atomic mass of 238.03 and consists of ^{235}U, mass = 235.044, and ^{238}U, mass = 238.051. The ^{235}U isotope is used in nuclear power reactors. What is the percentage abundance of ^{235}U in natural uranium?

The Mole

14. The charge on an electron is 1.6022×10^{-19} C. What is the total charge on 1 mol of electrons? What is the total charge on 1 mol of protons?

15. How many atoms of mercury are in 1 g of the metal? (The atomic mass of mercury is 200.6 u.)

16. How many molecules of O_2 are in 2.00 g of O_2? If the O_2 molecules were completely split into O atoms, how many moles of O atoms would be obtained?

17. (a) Assume the human body contains 6×10^{13} body cells, and the earth's population is 4×10^9. Approximately how many moles of living human body cells are there on earth?

(b) Assume the human body is 80% water. Calculate the number of molecules of water that are in the body of a person who has a mass of 65 kg.

18. Carbon monoxide, CO, taken into the lungs reduces the ability of blood to transport oxygen. It is fatal if its concentration reaches 2.38×10^{-4} g L^{-1}. Calculate the number of CO molecules that must be emitted by an automobile exhaust in order to produce a fatal concentration in a garage of volume 150 m^3.

19. Assume table salt, NaCl, and sugar, $C_{12}H_{22}O_{11}$, cost about the same per kilogram. Which is cheaper, a mole of sugar or a mole of salt?

20. Using the atomic masses given in the table on the inside front cover, calculate the molar masses of each of the following compounds:

H_2O H_2O_2 NaCl $MgBr_2$ CO

CO_2 CH_4 C_2H_6 NH_3 HCl

21. (a) How many moles of SO_2 are in a 0.0280-g sample of SO_2?

(b) What mass of SO_2 contains exactly 3 mol of this compound?

22. The average atomic mass of carbon is 12.01 u. Find the number of moles of carbon in

(a) 1.000 g of carbon,

(b) 12.00 g of carbon,

(c) 1.500×10^{21} atoms of carbon.

23. What is the molecular mass of benzene, C_6H_6? How many molecules of benzene are in 1 cm^3 of benzene, given that its density is 0.880 g cm^{-3}? How many moles of atoms are produced if each benzene molecule is completely split into atoms?

Empirical Formulas and Composition

24. What is the elemental composition, as mass percentage, of each of the following?

H_2O NaCl C_2H_6 $MgBr_2$ CO_2

25. What is the elemental composition, in mass percentage, of each of the following?

NH_3 Cl_2 NaOH C_2H_6O $C_6H_5NO_2$

26. What mass of oxygen is required to produce 3.40 g of nitric acid HNO_3, from its elements?

27. Ammonium sulfate, $(NH_4)_2SO_4$, is often used as an agricultural fertilizer. What is the percentage of nitrogen by mass in this compound? What mass of ammonium sulfate is required to supply 100 g of nitrogen?

28. What is the percentage by mass of sulfur in the compound Sb_2S_3? What mass of sulfur is contained in 28.4 g of the compound? What mass of the compound contains 64.4 g of sulfur?

29. A compound contains 7.00% C and 93.00% Br. What is its empirical formula and its empirical formula mass?

30. An oxide of nitrogen is found to contain 3.04 g of nitrogen and 6.95 g of oxygen. Its molecular mass is determined to be 91 u. What is its empirical formula and its molecular formula?

31. A compound was analyzed and found to contain 21.7% C, 9.6% O, and 68.7% F. What is its empirical formula and its empirical formula mass?

32. When 0.210 g of a hydrocarbon (containing only carbon and hydrogen) was burned completely in oxygen, 0.660 g of carbon dioxide was collected. What is the empirical formula of the hydrocarbon?

33. Caffeine is a compound containing 49.5% carbon, 5.2% hydrogen, 28.8% nitrogen, and 16.6% oxygen by mass. What is its empirical formula?

34. A 0.100-mol sample of a compound of carbon, hydrogen, and nitrogen was burned in oxygen and produced 26.4 g of carbon dioxide, CO_2, 6.30 g of water, H_2O, and 4.60 g of nitrogen dioxide, NO_2. What is the empirical formula of the compound?

35. A 3.62-g sample of a compound containing C, H, and O only was burnt in air and produced 5.19 g of CO_2 and 2.83 g of H_2O. What is the empirical formula of the compound?

36. A 3.10-g sample of a compound containing only C, H, and O yielded 4.40 g of CO_2 and 2.70 g of H_2O. What is its empirical formula? If the molar mass is 62.1 u, what is its molecular formula?

37. Analysis of nicotine gave 74.0% C, 8.7% H, and 17.3% N by mass, and the molar mass of nicotine is found to be 162 g. What is the empirical formula of nicotine? What is its molecular formula?

38. Anthracene is a hydrocarbon with the composition 94.33% C and 5.67% H. What is its empirical formula? What is its empirical formula mass?

39. On being heated in air, a 2.862-g sample of a red copper oxide reacted to give 3.182 g of black copper oxide. On being heated in hydrogen, the black copper oxide reacted to give 2.542 g of pure copper. What are the empirical formulas of the two oxides?

40. When a 6.20-g sample of a compound containing only S, H, and C, reacted with an excess of Cl_2, 21.9 g of HCl and 30.8 g of CCl_4 were obtained. What is the empirical formula of the compound?

41. Aspirin has the empirical formula $C_9H_7O_4Na$. A 1.00-g sample of aspirin gave 1.96 g of carbon dioxide on combustion. A tablet containing only aspirin and magnesium hydroxide (as an antacid) had a mass of 2.00 g. On combustion it gave 1.80 g of carbon dioxide. What is the mass percentage of aspirin in the tablet?

42. A compound of sulfur and fluorine is analyzed and is found to contain 70.3% fluorine by mass. What is the empirical formula for this compound?

43. A sample of hydrated lithium sulfate contains an unknown amount of water. Its composition may be expressed by the formula $Li_2SO_4(H_2O)_x$, where x is unknown. When 3.25 g of the hydrated lithium sulfate was heated, all the water was removed and 2.80 g of anhydrous lithium sulfate, Li_2SO_4, was obtained. Find x in the formula.

44. By analysis, a compound with the formula $KClO_x$ is found to contain 28.9% by mass of chlorine. What is the value of the integer x.

45. Polychlorinated biphenyls (PCBs), now known to be environmental pollutants, are a group of compounds all having the general empirical formula $C_{12}H_mCl_{10-m}$, where m is an integer. What is the value of m for, and thus the empirical formula of, the PCB that contains 58.9% by mass chlorine?

Balancing Equations

46. Which of the following equations are not correctly balanced? Balance any that are incorrect.

(a) $2SO_2 + H_2O + O_2 \longrightarrow 2H_2SO_4$

(b) $CH_3OH + 2O_2 \longrightarrow CO_2 + 2H_2O$

(c) $H_2O_2 \longrightarrow H_2O + O_2$

(d) $H_2SO_4 + KOH \longrightarrow KHSO_4 + H_2O$

47. Balance each of the following equations:

(a) $S + O_2 \longrightarrow SO_3$

(b) $C_2H_2 + O_2 \longrightarrow CO + H_2O$

(c) $Na_2CO_3 + Ca(OH)_2 \longrightarrow NaOH + CaCO_3$

(d) $Na_2SO_4 + H_2 \longrightarrow Na_2S + H_2O$

(e) $Cu_2S + O_2 \longrightarrow Cu_2O + SO_2$

(f) $Cu_2O + Cu_2S \longrightarrow Cu + SO_2$

(g) $Cu + H_2SO_4 \longrightarrow CuSO_4 + H_2O + SO_2$

(h) $B + SiO_2 \longrightarrow Si + B_2O_3$

48. Write balanced chemical equations for each of the following reactions:

(a) Elemental phosphorus, P_4, burnt in excess oxygen to give phosphoric oxide, P_4O_{10}.

(b) The reaction of sodium metal with water to give sodium hydroxide, NaOH and hydrogen gas, H_2.

(c) The formation of dinitrogen monoxide, N_2O, and water, H_2O, by heating ammonium nitrate, NH_4NO_3.

(d) The formation of lead monoxide, PbO, nitrogen dioxide, NO_2, and oxygen, O_2 by heating lead nitrate, $Pb(NO_3)_2$.

49. Balance each of the following equations:

(a) $Na_2SO_4(s) + C(s) \longrightarrow Na_2S(s) + CO_2(g)$

(b) $Cl_2(aq) + H_2O(l) \longrightarrow HCl(aq) + HOCl(aq)$

(c) $PCl_3(l) + H_2O(l) \longrightarrow H_3PO_3(aq) + HCl(aq)$

(d) $NO_2(g) + H_2O(l) \longrightarrow HNO_3(aq) + NO(g)$

(e) $Mg_3N_2(s) + H_2O(l) \longrightarrow Mg(OH)_2(s) + NH_3(g)$

Reactions

50. Ammonia gas, NH_3, reacts with hydrogen chloride gas, HCl, to produce the white solid ammonium chloride,

NH_4Cl. Write a balanced equation for this reaction, and calculate the mass of HCl that reacts with 0.20 g of NH_3.

51. Hydrochloric acid, an aqueous solution of HCl, reacts with a solution of sodium hydroxide, NaOH, to give sodium chloride, NaCl, and water, H_2O. Write the balanced equation for this reaction. What mass of NaOH reacts with 3.00 g of HCl? What mass of NaCl is obtained?

52. Phosphoric acid, H_3PO_4, combines with calcium hydroxide, $Ca(OH)_2$, to produce water, H_2O, and calcium phosphate, $Ca_3(PO_4)_2$. Write a balanced equation for this reaction, and calculate the mass of $Ca(OH)_2$ that reacts with 30.0 g of H_3PO_4. What mass of $Ca_3(PO_4)_2$ is produced?

53. Phosphorus trichloride, PCl_3, is a colorless liquid made by passing a stream of chlorine gas, Cl_2, over phosphorus and condensing it in a cooled, dry receiver. What mass of phosphorus is required to yield 100 g of PCl_3?

54. Write the balanced equation for the reaction of chlorine, Cl_2, with silicon dioxide (silica), SiO_2, and carbon, C, to give silicon tetrachloride, $SiCl_4$, and carbon monoxide, CO. What mass of $SiCl_4$ can be obtained from 15.0 g of SiO_2?

55. A sample of an oxide of barium of unknown composition and with a mass of 5.53 g gave 5.00 g of pure BaO and 0.53 g of O_2 when heated. What is the empirical formula of the oxide of barium of unknown composition?

56. When sulfur dioxide reacts with oxygen and water, sulfuric acid is produced:

$$2SO_2 + O_2 + 2H_2O \longrightarrow 2H_2SO_4$$

What mass of O_2 and what mass of H_2O react with 0.32 g of SO_2? What mass of H_2SO_4 is produced?

***57.** Dry boric oxide, B_2O_3, reacts with magnesium powder on strong heating to give a mixture of magnesium oxide, MgO, and magnesium boride, Mg_3B_2. Write the balanced equation for the reaction. Magnesium boride reacts with dilute aqueous hydrochloric acid, HCl, to give a hydride of boron of formula B_4H_{10}. What is the maximum possible yield of B_4H_{10}, starting with 10.00 g of B_2O_3?

***58.** A sample of 1.000 kg of impure limestone containing 74.2% by mass calcium carbonate, $CaCO_3$, and 25.8% impurities is heated until the carbonate is completely decomposed to calcium oxide, CaO, and carbon dioxide, CO_2. What mass of CO_2 is produced?

Limiting Reactant

59. If aqueous solutions containing silver nitrate, $AgNO_3$, and calcium chloride, $CaCl_2$, are mixed, solid silver chloride, AgCl, is formed. The other product, calcium nitrate, $Ca(NO_3)_2$, remains in solution. Write a balanced equation for the reaction. The original solutions contain 0.010 mol of $AgNO_3$ and 0.010 mol of $CaCl_2$, respectively. Calculate the maximum mass of AgCl that can be produced.

60. A strip of zinc metal is placed in an aqueous solution containing copper chloride, $CuCl_2$, and a reaction occurs to produce copper, Cu, and zinc chloride, $ZnCl_2$:

$$Zn + CuCl_2 \longrightarrow ZnCl_2 + Cu$$

The solution originally contains 2.0 g of $CuCl_2$, and the zinc strip added weighs 2.0 g also. Calculate the mass of elemental copper produced.

61. A 5.00-g sample of antimony, Sb, and a 1.00-g sample of sulfur, S, are heated together to produce Sb_2S_3. What is the maximum possible yield of Sb_2S_3 in moles and in grams?

62. When sulfur dioxide reacts with oxygen in the presence of a catalyst and then with water, sulfuric acid is produced:

$$2SO_2 + O_2 + 2H_2O \longrightarrow 2H_2SO_4$$

If 5.6 mol of SO_2 reacts with 4.8 mol of O_2 and a large excess of water, what is the maximum number of moles of H_2SO_4 that can be formed?

63. Determine the mass of $BaSO_4$ produced by mixing an aqueous solution containing 6.00 g of K_2SO_4 with an aqueous solution containing 8.00 g $Ba(NO_3)_2$, given that the reaction is

$$K_2SO_4 + Ba(NO_3)_2 \longrightarrow BaSO_4 + 2KNO_3$$

64. Determine the mass of $AlCl_3$ produced and the mass of Al or HCl that remains unreacted when 2.70 g of Al is allowed to react with 4.00 g of HCl. The unbalanced equation for the reaction is

$$Al + HCl \qquad AlCl_3 + H_2$$

65. Magnesium hydroxide, $Mg(OH)_2$, and phosphoric acid, H_3PO_4, react to form magnesium phosphate, $Mg_3(PO_4)_2$, and water, H_2O. What mass of magnesium phosphate is produced when 7.00 g of magnesium hydroxide reacts with 9.00 g of phosphoric acid?

Molar Concentrations

66. What masses of each of the following would be required to give

(a) 1 L of $1M$ solution,

(b) 250 mL of $0.025M$ solution?

$NaCl \qquad H_2SO_4 \qquad HCl \qquad Na_2SO_4$

67. A 12.00-g sample of potassium permanganate, $KMnO_4$, was dissolved in sufficient water to give 2.00 L of solution. What is the molarity of $KMnO_4$?

68. (a) What mass of glucose, $C_6H_{12}O_6$, must be dissolved in water to prepare 0.25 L of a $0.10M$ solution?

(b) What volume of the resulting solution contains 0.010 mol of glucose?

69. How many milliliters of concentrated sulfuric acid (concentration 98% by mass of H_2SO_4; density of 1.842 g mL^{-1}) are required to prepare 500 mL of a $0.175M$ solution of sulfuric acid?

*70. How would the following be prepared?

(a) A volume of 6.3 L of $0.003M$ $Ba(OH)_2$ from $0.10M$ $Ba(OH)_2$ solution.

(b) A volume of 750 mL of $0.25M$ $Cr_2(SO_4)_3$ from a solution containing 35 mass percent $Cr_2(SO_4)_3$ (density 1.412 g cm^{-3}).

*71. (a) What mass of solid KOH is needed to prepare 1.5 L of $0.532M$ KOH solution?

(b) What mass of a 50 mass percent aqueous solution of perchloric acid, $HClO_4$ (density 1.410 g mL^{-1}), is needed to prepare 600 mL of $0.1M$ perchloric acid solution?

(c) What mass of an 85 mass percent aqueous solution of phosphoric acid, H_3PO_4 (density 1.689 g mL^{-1}), is needed to prepare 2.5 L of $1.5M$ phosphoric acid solution?

*72. Concentrated nitric acid, HNO_3, is 69 mass percent and has a density of 1.41 g mL^{-1}. What volume of concentrated nitric acid is needed to prepare 250 mL of $0.1M$ nitric acid?

Miscellaneous

73. A fusion reaction that may be usable for power generation in the future involves two deuterium nuclei:

$$2{}_1^2H \longrightarrow {}_2^3He + 1 \text{ neutron}$$

Suppose that 1.00 g of deuterium reacts in this way.

(a) What is the change in mass that accompanies this reaction?

(b) What is the energy equivalent to this mass loss?

(c) What is the mass of water that must be processed to extract 1.00 g of ${}_1^2H$? (Naturally occurring hydrogen contains 0.015% ${}_1^2H$.)

CHAPTER 3
THE ATMOSPHERE AND
THE GAS LAWS

Oxygen, Nitrogen, and Hydrogen

We live on the surface of the earth, immersed in a mixture of gases called the atmosphere. Not only is the atmosphere essential to life, but it also plays a vital role in determining the temperature of the earth and in producing the weather. It is also the source of several elements of great industrial importance. The atmosphere has played an important role in the history of chemistry. Air was the first gas to be studied, and these studies produced the first scientific laws.

In this chapter we first examine the nature of the atmosphere and, in particular, the gaseous elements oxygen and nitrogen, which are its two principal components. We also consider another important gaseous element, hydrogen. Then we turn to the general properties of all gases. The physical behavior of all gases is very nearly the same. We will see that, unlike solids and liquids, all gases behave in much the same way with changing conditions, such as pressure and temperature, and we will see why this is.

3.1 THE ATMOSPHERE

The **atmosphere** is a mixture of gases held to earth by gravity. This gaseous envelope is most dense at sea level and thins rapidly with increasing altitude. Almost all the atmosphere (99%) lies within 30 km of the earth's surface. Except for variable amounts of water vapor, this lowest layer of the atmosphere, which we call air, has a remarkably constant composition.

Pure, dry air consists largely of oxygen and nitrogen; they make up 99% of its volume (Table 3.1). Among the other components are the *noble gases*, helium, neon, argon, krypton, and xenon, of which argon is the most abundant.

Table 3.1 Composition of Dry Air

COMPONENT	FORMULA	PERCENT BY VOLUME
Nitrogen	N_2	78.084
Oxygen	O_2	20.948
Argon	Ar	0.934
Carbon dioxide	CO_2	0.031 4
Neon	Ne	0.001 82
Helium	He	0.000 52
Methane	CH_4	0.000 2
Krypton	Kr	0.000 11
Hydrogen	H_2	0.000 05
Nitrous oxide	N_2O	0.000 05
Xenon	Xe	0.000 008

Carbon dioxide, although it is only present in very small amounts (0.3%), is nevertheless important to life. Plants synthesize from carbon dioxide and water the complex substances they need in order to grow and reproduce in the process called *photosynthesis*. The atmosphere retains its constant composition up to a height of about 80 km, and this lowest layer of the atmosphere is therefore known as the **homosphere**. Although the homosphere has the same composition throughout, its temperature varies over a wide range, as illustrated in Figure 3.1.

Above about 80 km the atmosphere is no longer mixed up by air currents, and it separates into layers with different compositions (Figure 3.1). This part of the atmosphere is therefore called the **heterosphere**. Over the aeons the lighter atoms and molecules have gradually diffused further away from the earth's surface because they are subject to a smaller gravitational attraction. Therefore, the gases with larger molecular masses are closer to the surface of the earth, and those with smaller molecular masses are further away.

The lowest layer of the heterosphere is the *molecular nitrogen layer*, con-

Figure 3.1 The Atmosphere: Homosphere and Heterosphere.
The homosphere divides into three layers on the basis of temperature. In the *troposphere*—the layer which supports life—the temperature decreases continuously with height, reaching a minimum of $-50°C$. In the *stratosphere* the temperature rises to $+80°C$ because ultraviolet light energy is absorbed by oxygen molecules, which are split into oxygen atoms and then react with oxygen molecules to form ozone ($O_2 \longrightarrow O + O$ followed by $O + O_2 \rightleftarrows O_3$). In the *mesophere* the temperature decreases again to about $-100°C$ because there are too few oxygen molecules to absorb enough energy to raise the temperature. The *heterosphere* divides into four layers on the basis of chemical composition: molecular nitrogen, atomic oxygen, helium, and atomic hydrogen.

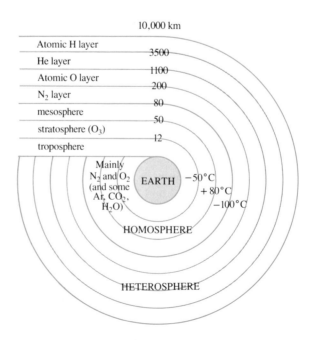

sisting almost entirely of nitrogen molecules, N_2. Very little molecular oxygen is present in the heterosphere because it is dissociated into oxygen atoms by intense ultraviolet radiation from the sun, and the next layer of the heterosphere, the *atomic oxygen layer*, consists largely of oxygen atoms. Above the oxygen layer is the *helium layer*, consisting almost entirely of helium atoms. The uppermost layer consists of the lightest atoms of all, hydrogen atoms. It is called the *atomic hydrogen layer*. There is no definite upper boundary to this layer, but at a height of approximately 10 000 km the density of hydrogen atoms is about the same as the density of interplanetary space.

Most of the hydrogen and the helium that might originally have been part of the atmosphere of the earth have been lost to outer space. Considerably larger amounts of hydrogen and helium are found in the atmospheres of the four larger, outer planets of the solar system: Jupiter, Saturn, Uranus, and Neptune. These planets have masses from 17 to 318 times that of the earth, and they exert a much greater gravitational attraction on their atmospheres. In contrast, the moon has lost all its atmosphere; because its mass is only $\frac{1}{6}$ the mass of the earth, it exerts a correspondingly smaller gravitational attraction.

The atmosphere has not always contained the large amount of oxygen that is present today. It is probable that at the time life first appeared on earth, more than 3000 million (3×10^9) years ago, there was little, if any, oxygen in the atmosphere. Early forms of bacteria obtained hydrogen from hydrogen sulfide in order to synthesize the compounds that they needed for growth and reproduction. Later the blue-green algae developed. They contained chlorophyll, the substance that enabled them to carry out photosynthesis and to utilize the most abundant source of hydrogen, namely water, liberating oxygen in the process. All the higher plants that subsequently developed made use of chlorophyll to carry out photosynthesis. The oxygen they produced has accumulated over the millenia to form the oxygen-rich atmosphere that we know today.

3.2 THE ABUNDANCE OF THE ELEMENTS

Although nitrogen is the most abundant element in the atmosphere, oxygen is by far the most abundant element in the whole of the earth's crust. The crust consists of the relatively thin layer of solid rocks called the *lithosphere*, the oceans, which are called the *hydrosphere*, and the *atmosphere* (Figure 3.2). Oxygen, in the form of its many compounds, constitutes almost half the earth's crust by mass. Indeed, there are more oxygen atoms than the total of all other kinds of atoms (Table 3.2). In contrast, nitrogen constitutes only 0.03% of the earth's crust. This striking difference is a reflection of the fact that oxygen is a very reactive element and forms many compounds. It is a major component of almost all rocks and, of course, water. Nitrogen is a much less reactive element and is found in far fewer compounds in the lithosphere.

After oxygen the next most abundant element in the earth's crust is silicon; together, silicon and oxygen compose 75% of the crust. Only 10 elements (O, Si, Al, Fe, Ca, Na, K, Mg, H, and Ti) together comprise 99.2% of the mass of the earth's crust. They are all light elements. Fifteen of the 20 most abundant elements have atomic numbers below 21. This is consistent with the belief that the elements are built up in stars from protons, neutrons, and electrons, forming first hydrogen and helium and then other, heavier elements. In fact, hydrogen and helium are by far the most abundant elements in the universe as a whole. Because they are so light, hydrogen and helium in their elemental forms have diffused away from the surface of the earth and are found only in the upper

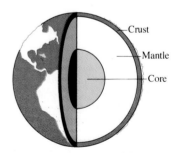

Figure 3.2 Structure of the Earth. We do not have reliable information about the composition of the deep interior of the earth, but the *core*, with a 3500 km (2200 miles) radius, is believed to be composed mainly of iron. Surrounding the core is fluid material called the *mantle*. It is 2900 km (1800 miles) thick and is thought to be composed primarily of silicon, oxygen, iron, calcium, and magnesium. The very thin outer, solid layer is the *crust*. It is 10–50 km (8–26 miles) thick and is composed of the hydrosphere, lithosphere, and atmosphere.

Table 3.2 Abundance of the 20 Most Common Elements in the Earth's Crust (Atmosphere, Hydrosphere and Lithosphere)

ELEMENT	Z	MASS PERCENT		ATOM PERCENT
Oxygen	8	49.4	} 75.2	55.1
Silicon	14	25.8		16.3
Aluminum	13	7.5		5.0
Iron	26	4.7		1.5
Calcium	20	3.4		1.5
Sodium	11	2.6	} 99.2	2.0
Potassium	19	2.4		1.1
Magnesium	12	1.9		1.4
Hydrogen	1	0.9		15.4
Titanium	22	0.6		0.2
Chlorine	17	0.2		
Phosphorus	15	0.12		
Manganese	25	0.10		
Carbon	6	0.08		
Sulfur	16	0.06		
Argon	18	0.04		
Nitrogen	7	0.03		
Rubidium	37	0.03		
Strontium	38	0.03		
Fluorine	9	0.03		

layers of the atmosphere. Helium forms no compounds and is a very rare element on the surface of the earth; it occurs only in very small amounts in the atmosphere and in association with oil and natural gas. Hydrogen, however, is relatively common in the form of its many compounds. Indeed, on an atom percentage basis it is the third most common element.

The data in Table 3.2 refer to the earth's crust only. They do not take into account the material of the core, which is believed to be mainly iron, nor the material of the fluid mantle that surrounds the core, which is believed to contain mainly iron, calcium, magnesium, silicon, and oxygen.

3.3 OXYGEN AND NITROGEN

Oxygen was not recognized as a distinct substance until the late eighteenth century. In fact, before then nobody knew that air was a mixture of gases. Experiments by the English chemist Joseph Priestley (Figure 3.3) and the French chemist Antoine Lavoisier in the latter part of the eighteenth century showed that when a substance burns in air, or when a metal is heated in air, it combines with part of the air (Box 3.1). Lavoisier called the reactive component of air, oxygen. The other, much less reactive component, was later called nitrogen.

Oxygen: Oxides and Oxidation

A colorless gas at ordinary temperatures, oxygen condenses to a blue liquid at $-183°C$ and freezes to a pale blue solid at $-218°C$. It consists of diatomic O_2 molecules. Oxygen is obtained on a large scale by liquefying air and then dis-

Figure 3.3 Joseph Priestley (1733–1804).
The son of a Nonconformist minister, Priestley was born in Yorkshire, England. As a young man he had many interests and he taught himself several languages, including Arabic and Hebrew, as well as philosophy and science. He had radical religious beliefs and he eventually became a Unitarian Minister. In 1766 he met Benjamin Franklin who was on a visit to London to attempt to settle the dispute over taxation between the American colonists and the British government. This meeting inspired him to begin his own research in electricity, in which field he did much important work. He then turned to chemistry. As a result of his experiment on the decomposition of mercury oxide by heat he discovered oxygen (Box 3.1). He was the first to collect gases over mercury, and he thus discovered several water-soluble gases, including ammonia, hydrogen chloride, and sulfur dioxide. An outspoken man with very liberal religious and political views, he was sympathetic to the American and French revolutions and was viewed with suspicion by the conservative British majority. In 1791 his house and laboratory were burned down by an angry mob. Priestley managed to escape, went into hiding for a time, and eventually emigrated to the United States. He spent the last ten years of his life in relative seclusion in Pennsylvania.

tilling it to separate the components. Very pure oxygen can be obtained by electrolysis, that is, by passing an electric current through water (Experiment 1.1). Small amounts of oxygen can be made in the laboratory by heating potassium chlorate, $KClO_3$, with manganese dioxide, which behaves as a catalyst:

$$2KClO_3 \xrightarrow{\text{heat}} 2KCl + 3O_2$$

Box 3.1

DISCOVERY OF OXYGEN

For much of the eighteenth century chemists believed that anything that burned contained a substance called *phlogiston*. Burning supposedly released the phlogiston. In August 1774 Priestley used a lens to focus sunlight on mercury oxide, HgO; mercury metal and a gas formed. When Priestley tested the gas with a burning candle, he found that the candle burned more vigorously in the gas than in ordinary air. Priestley had discovered oxygen but he called the gas *dephlogisticated air*. He believed that a candle burned less brightly in ordinary air because ordinary air contained some phlogiston. His experiment led to the downfall of the theory, although Priestley himself never gave up his belief in the theory.

In October that year he visited Lavoisier's laboratory in Paris and told him about the experiment. Lavoisier immediately began an experiment of his own in which he heated mercury in a sealed, air-filled flask in a furnace. Nothing remarkable happened on the first day. On the second day small red specks began to appear on the surface of the mercury, and as time passed, the number of red particles increased. At the end of 12 days Lavoisier extinguished the furnace and examined the gas left in the flask. If the phlogiston theory were correct, the mass of the mercury should have decreased as it released phlogiston. But Lavoisier found that in fact the mass of the mercury had increased. Contrary to the phlogiston theory, a burning material does not release a substance into the air; instead, it combines with part of the air.

Lavoisier next heated the red solid that had formed on the surface of the mercury. He obtained pure mercury and a gas that supported combustion much better than air, just as Priestley had done. Unable to find any reaction in which this gas was broken down into simpler substances, he concluded that it was an element, and he named it oxygen. This reaction of mercury with oxygen to give mercury oxide, HgO, and its decomposition back to the elements can be represented by the equation

$$2Hg + O_2 \rightleftharpoons 2HgO$$

A **catalyst** is a substance that increases the rate of a reaction without changing the nature of the products. Manganese dioxide and blood are catalysts for the decomposition of hydrogen peroxide (Experiment 1.5). We will discuss in Chapter 18 how a catalyst operates.

Oxygen reacts directly with most other elements. The reactions with many elements are quite slow at room temperature; frequently, no reaction at all is apparent. However, when the temperature is raised, the rate of reaction increases. Reactions of the elements with oxygen liberate heat, which is often sufficient to keep the temperature high enough that the reaction continues rapidly without the need to supply more heat. The element is said to be undergoing **combustion**, or burning. Sulfur, for example, shows no tendency to react with oxygen at room temperature. However, if sulfur is heated until it melts and is then placed in oxygen, it reacts rapidly, burning with a bright blue flame to form sulfur dioxide, a colorless gas with a pungent odor (Experiment 3.1). Magnesium does not react with oxygen at ordinary temperatures, but when heated, it burns rapidly, emitting a brilliant white light.

The products of the reactions of the elements with oxygen are known as **oxides**. Table 3.3 lists some typical reactions in which elements are converted to their oxides.

Only a very few elements do not react directly with oxygen. They are the noble gases, He, Ne, Ar, Kr, Xe, and Rn, and the metals gold, Au, and platinum, Pt. Nevertheless, oxides of some of these elements—for example, XeO_3, XeO_4, PtO_2, and Au_2O_3—can be made from other compounds of these elements.

Many compounds are converted to the oxides of their constituent elements on reaction with oxygen. For example, the compounds of carbon and hydrogen, called **hydrocarbons**, burn in oxygen to give carbon dioxide and water. Propane,

Reactions of Metals and Nonmetals with Oxygen

Burning magnesium burns even more violently in oxygen, forming a white smoke of solid particles of magnesium oxide, MgO.

If steel wool is first heated in a flame it ignites and burns vigorously in oxygen, giving a shower of sparks and forming brown solid iron oxide, Fe_2O_3.

White phosphorus ignites spontaneously in oxygen and burns with a very bright flame, forming a dense white smoke of solid P_4O_{10}.

If sulfur is warmed until it melts, it burns in oxygen with a bright blue flame, producing the pungent smelling gas sulfur dioxide, SO_2.

Table 3.3 Reactions of Some Elements with Oxygen

ELEMENT	REACTION WITH OXYGEN	OXIDE
Copper, a reddish metal	$2Cu + O_2 \longrightarrow 2CuO$	Copper oxide, a black solid, insoluble in water
Mercury, a silvery liquid metal	$2Hg + O_2 \longrightarrow 2HgO$	Mercury oxide, a red solid, insoluble in water
Magnesium, a silvery metal	$2Mg + O_2 \longrightarrow 2MgO$	Magnesium oxide, a white solid, insoluble in water
Sulfur, a yellow solid	$S + O_2 \longrightarrow SO_2$	Sulfur dioxide, a colorless and pungent-smelling gas, soluble in water
Phosphorus, a red solid	$4P + 5O_2 \longrightarrow P_4O_{10}$	Phosphoric oxide, a white solid, soluble in water
Hydrogen, a colorless gas	$2H_2 + O_2 \longrightarrow 2H_2O$	Water
Carbon (graphite), a black solid	$C + O_2 \longrightarrow CO_2$	Carbon dioxide, a colorless gas, slightly soluble in water

C_3H_8, reacts with oxygen according to the equation

$$C_3H_8 + 5O_2 \longrightarrow 3CO_2 + 4H_2O$$

A simple *test for oxygen* is to hold a glowing splint of wood in a tube of the gas suspected to be oxygen. In oxygen the combustion of wood is much faster than it is in air, and the wood splint bursts into flame.

All these reactions in which an element or a compound combines with oxygen to form oxides are examples of **oxidation reactions**. (A more general definition of oxidation will be discussed in Chapter 5.) Oxidation reactions were among the first reactions to be carefully and quantitatively studied by chemists (Box 3.1). Priestley and Lavoisier carried out many important studies of combustion, which led to the discovery of oxygen.

Exothermic and Endothermic Reactions

Burning or combustion—and, indeed, many other oxidation reactions—are accompanied by the liberation of energy in the form of heat. All reactions that occur with the liberation of heat are known as **exothermic reactions**. Experiment 3.2 provides another example of an exothermic reaction. If a reaction occurs with the absorption of heat, it is called an **endothermic reaction**.

Since the combustion of hydrocarbons is associated with the liberation of a considerable amount of heat, hydrocarbons are often used as fuels. Natural gas, which is mainly methane, CH_4, is often used in homes for heating and cooking. Propane, C_3H_8, and butane, C_4H_{10}, which are available compressed in cylinders, are used for the same purposes and also, for example, for camp stoves. Gasoline and diesel fuel, which are mixtures of a large number of hydrocarbons, are burned in internal combustion engines to provide the necessary power for automobiles, airplanes, trains, and ships. The combustion of acetylene, C_2H_2, is the basis of the oxyacetylene torch, which gives a very hot flame which can be used for welding metals.

Oxidation reactions form the basis for life. Animals inhale oxygen from the atmosphere, and this oxygen is absorbed in the lungs by the hemoglobin of

The Exothermic Reaction Between Iron and Sulfur

A mixture of iron and sulfur (see Experiment 1.4) reacts to form iron sulfide FeS when it is heated.

If the tube is removed from the flame it continues to glow very brightly as the heat evolved in the reaction maintains the high temperature of the mixture.

the blood and distributed to different parts of the body, where it is used for the oxidation of many different substances. These oxidation reactions are exothermic and they provide the energy that is needed to maintain life.

Reduction

Since most elements combine with oxygen, and since oxygen is very abundant on the surface of the earth, many elements, particularly metals, are found in the form of their oxides (for example, H_2O, Al_2O_3, Fe_2O_3, MnO_2, CuO). Frequently, these oxides are major sources of the elements, which can be obtained by removing the oxygen in a process known as **reduction**. Iron is made industrially by reducing the oxide Fe_2O_3 with carbon monoxide, CO, which combines with the oxygen to form carbon dioxide, CO_2:

$$Fe_2O_3(s) + 3CO(g) \longrightarrow 2Fe(s) + 3CO_2(g)$$

Copper oxide, CuO, can similarly be reduced with carbon and also with hydrogen, (see Experiment 3.3).

$$2CuO(s) + C(s) \longrightarrow 2Cu(s) + CO_2(g)$$
$$CuO(s) + H_2(g) \longrightarrow Cu(s) + H_2O(l)$$

Any reaction in which oxygen is removed either partially or completely from a compound is called a reduction reaction. Again, a more general definition of reduction is given in Chapter 5.

Nitrogen

Although nitrogen forms the major part of the atmosphere, it is not a very abundant element on the earth because very little occurs in the form of solid compounds in the lithosphere (Table 3.2). Nevertheless, nitrogen, together with carbon, oxygen, and hydrogen, is one of the principal elements found in all living matter. It is a colorless, odorless, tasteless gas consisting of N_2 molecules. It boils at $-196°C$ and freezes at $-210°C$. Like oxygen it can be obtained from the air by liquefaction and distillation.

Nitrogen is an unreactive element. It has very few reactions at ordinary temperatures, but it becomes somewhat more reactive at high temperatures. On heating, it reacts with hydrogen to give ammonia,

$$N_2 + 3H_2 \longrightarrow 2NH_3$$

with oxygen to give nitrogen monoxide,

$$N_2 + O_2 \longrightarrow 2NO$$

and with a few metals to give nitrides, such as magnesium nitride,

$$3Mg + N_2 \longrightarrow Mg_3N_2$$

EXPERIMENT 3.3

Reduction of Copper Oxide

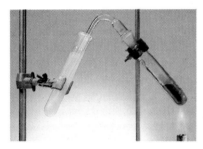

When a black mixture of powdered copper oxide and carbon is heated, the copper oxide is reduced to red copper and the carbon is oxidized to carbon dioxide. The formation of carbon dioxide is shown by passing the evolved gases through lime water—an aqueous solution of calcium hydroxide, $Ca(OH)_2$. A white precipitate of calcium carbonate is formed.

$$Ca(OH)_2(aq) + CO_2(g) \longrightarrow CaCO_3(s) + H_2O(l)$$

The close up clearly shows the shiny red-brown copper produced.

Black copper oxide can be reduced to copper by heating it in a stream of hydrogen. The water that is formed can be condensed in a U-tube and cooled in a freezing mixture of ice and salt.

This shows the surface of the black copper oxide covered with red copper after reduction with hydrogen.

The liquid condensed in the U-tube can be shown to be water by adding a few drops of it to white anhydrous copper sulfate, $CuSO_4$, which becomes bright blue due to the formation of $CuSO_4 \cdot 5H_2O$, copper sulfate pentahydrate.

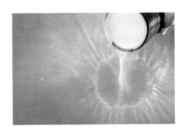

Liquid Nitrogen

Since the only widely available source of nitrogen is the nitrogen of the atmosphere, all nitrogen compounds have to be ultimately obtained from this source. The process of converting the nitrogen of the air into useful compounds is known as **nitrogen fixation**. This process is extremely important because of the great need for nitrogen fertilizers. The only practical method of large-scale nitrogen fixation is the preparation of ammonia by the reaction of nitrogen with hydrogen. In the industrial process known as the **Haber process**, nitrogen and hydrogen are heated together at about 400°C at a high pressure, and in the presence of a catalyst, to give ammonia.

Certain bacteria found in the soil and in the root nodules of leguminous plants, such as peas and beans, are able to carry out nitrogen fixation at ordinary temperatures. In other words, they convert the nitrogen of the atmosphere into nitrogen compounds that plants can assimilate.

3.4 HYDROGEN

Hydrogen is the most abundant element in the universe. Interstellar space is very sparsely filled with atoms that are predominantly hydrogen. The stars consist mostly of hydrogen, and the uppermost layer of the atmosphere is composed of hydrogen atoms. Elsewhere in the earth's crust hydrogen is not present as the free element, but hydrogen compounds are very common. Hydrogen is the third most abundant element on an atom basis, and the ninth most abundant on the basis of mass (Table 3.2). It is found in water, hydrocarbons, proteins, carbohydrates, and almost all the other substances in living organisms.

Hydrogen is a colorless, odorless, tasteless, nonpoisonous gas consisting of H_2 molecules. It combines readily with many other elements. It has a very low boiling point ($-252.8°C$) and a very low melting point ($-259.1°C$). A simple test for hydrogen is to bring a flame or glowing splint of wood to the open end of a test tube containing the gas suspected to be hydrogen. If the gas is hydrogen, a characteristic "pop" is heard as H_2 and O_2 combine rapidly to form H_2O.

Industrial Production of Hydrogen

Water is by far the cheapest and most abundant source of hydrogen; all the large-scale methods of making hydrogen are based on the removal of oxygen from water. Removal of oxygen from water can be done, for example, by combining the oxygen with carbon or with metals such as iron to form oxides.

In the production of hydrogen on an industrial scale, steam is passed over coke, an impure form of carbon, at about 1000°C:

$$C(s) + H_2O(g) \longrightarrow CO(g) + H_2(g)$$

Carbon is oxidized to carbon monoxide, and water is reduced to hydrogen. This mixture of CO(g) and $H_2(g)$ is known as *water gas*, an important industrial fuel.

If pure H_2 is needed, it is separated from the CO by mixing the water gas with steam and passing the mixture over a catalyst at 500°C, to convert CO to CO_2:

$$[CO(g) + H_2(g)] + H_2O(g) \longrightarrow 2H_2(g) + CO_2(g)$$
$$\text{Water gas}$$

A catalyst is needed here because otherwise the reaction would be much too slow to be useful. Because carbon dioxide is much more soluble in water than is hydrogen, it is easily removed from the H_2–CO_2 mixture by passing the mixture, under pressure, into water.

Hydrogen is also made industrially by passing steam over heated iron:

$$3Fe(s) + 4H_2O(g) \longrightarrow Fe_3O_4(s) + 4H_2(g)$$

In this reaction iron is *oxidized* to Fe_3O_4—it adds oxygen—and the steam is *reduced* to hydrogen—it loses oxygen. In the laboratory the reduction of water with a metal can be conveniently demonstrated by the reaction of steam with heated magnesium (Experiment 1.3):

$$Mg(s) + H_2O(g) \longrightarrow MgO(s) + H_2(g)$$

Very pure hydrogen is made by the **electrolysis** of water. When an electric current is passed through water, it is decomposed into hydrogen and oxygen, which can be collected separately (Experiment 1.1):

$$2H_2O(l) \xrightarrow{\text{Electric current}} 2H_2(g) + O_2(g)$$

Because electrolysis uses large amounts of electrical energy, this method is a rather expensive way of producing hydrogen. A related process is the manufacture of sodium hydroxide, $NaOH$, chlorine, and hydrogen by the electrolysis of an aqueous sodium chloride solution (see Chapter 16).

Another increasingly important source of hydrogen is methane, CH_4, the major component of natural gas. When methane mixed with steam is passed over a heated nickel catalyst, a mixture of carbon monoxide and hydrogen, called *synthesis gas*, is produced:

$$CH_4(g) + H_2O(g) \longrightarrow CO(g) + 3H_2(g)$$

Synthesis gas is the starting material for the industrial production of a number of important compounds (see Chapter 19).

Compounds of Hydrogen

Under appropriate conditions hydrogen, H_2, combines with most elements to form compounds. Compounds of hydrogen with metals are known as **hydrides**. Examples include sodium hydride, NaH, calcium hydride, CaH_2, and aluminum hydride, AlH_3. Many common compounds of hydrogen, particularly those with nonmetals, have special names, such as methane, CH_4, ammonia, NH_3, and water, H_2O.

WATER The reaction of hydrogen with oxygen to form water is a very exothermic reaction, releasing 286 kJ per mole of H_2O formed. A practical use of this reaction is in the oxyhydrogen torch, where the reaction generates temperatures up to 2800°C, which are useful for welding materials that have high melting points. The combustion of hydrogen and oxygen is also used to fuel rockets (Box 3.2). Mixtures of H_2 and O_2 are explosive, particularly when the $H_2:O_2$ ratio is approximately 2:1.

AMMONIA The reaction of nitrogen with hydrogen to form ammonia, NH_3, occurs much less readily than the reaction between hydrogen and oxygen. We have seen that in the Haber process for the manufacture of ammonia, nitrogen and hydrogen are heated together at a high pressure and temperature in the presence of a catalyst:

$$N_2(g) + 3H_2(g) \longrightarrow 2NH_3(g)$$

Ammonia is a colorless gas (boiling point = -33.4°C) with a characteristic, irritating odor. It is very soluble in water. Ammonia solutions are widely used as household cleaning agents.

Hydrogen is an important rocket fuel. A primary consideration for a rocket fuel is that its mass be as small as possible for a given amount of energy produced. Hydrogen was used in the *Saturn V* rocket that enabled the first astronauts to land on the moon, and it is the main fuel in the space shuttle rockets. Both the hydrogen and the oxygen needed to burn the hydrogen are carried on the rocket in liquid form.

Hydrogen also has attractive features as a fuel for more general use. But its use at present is hampered by the difficulty of safely storing, transporting, and distributing such a highly flammable, potentially explosive material. In addition, because there are no natural sources of hydrogen, energy from another source must be expended in order to obtain H_2 from hydrogen compounds such as water. Hydrogen is said to be an energy carrier rather than an energy source. Both sunlight and excess electric energy from nuclear reactors are being studied as possible sources of energy for producing hydrogen from water. Thus widespread use of hydrogen as a fuel will be economical only when hydrogen can be produced rather cheaply.

If the problems of safety and economics are resolved, hydrogen may eventually be delivered to homes and industry by pipelines, as natural gas is delivered today. Hydrogen might even be used to fuel vehicles. For this purpose it might be stored as a liquid at very low tem-peratures or as a solid metal hydride that, when heated, decomposes to hydrogen and metal. Hydrogen has already been used on a trial basis as a fuel for jet airplanes.

Columbia *Space Shuttle. The rocket that launched the shuttle used liquid hydrogen as a fuel. It was stored in a tank 40 m long and 8.4 m in diameter. Liquid oxygen for burning the hydrogen was stored in a similar tank.*

One of the major uses of ammonia is for the manufacture of *fertilizers*. Plants require nitrogen, but they cannot use nitrogen from the atmosphere directly. Instead, they take up nitrogen compounds from the soil. When the same land is used repeatedly for crops, the nitrogen compounds in the soil are depleted. Farmers therefore add nitrogen-containing compounds known as fertilizers. Liquid ammonia stored at a low temperature can be added directly to the soil, but more commonly, it is converted to other nitrogen-containing compounds, such as ammonium sulfate, $(NH_4)_2SO_4$, ammonium hydrogen phosphate, $(NH_4)_2HPO_4$, and urea $(NH_2)_2CO$.

METHANE   Although methane, CH_4, can be made by the direct reaction of carbon and hydrogen, this is not an important reaction as ample supplies of methane are available from natural gas. A very small amount of methane (0.0002%) is found in the atmosphere. It is formed by the bacterial decomposition of vegetable matter under water, and it is produced by some animals, such as cows, during the digestion of plant material.

Methane is a colorless gas with a very low melting point ($-182°C$) and a very low boiling point ($-162°C$). It burns readily in air in an exothermic reaction:

$$CH_4(g) + 2O_2(g) \longrightarrow CO_2(g) + 2H_2O(g)$$

Methane, particularly in its impure form, natural gas, is widely used as a fuel.

3.5 PHYSICAL PROPERTIES OF GASES AND THE GAS LAWS

We have seen that several important elements and compounds are gases under ordinary conditions. The study of gases has been important since the early days of science. Air was the first gas to be studied, and it was studied long before scientists understood that it is a mixture of a number of different elements and compounds. That air is, in fact, a mixture caused no problems, because, as we will see, all gases behave approximately in the same way; thus in many ways air behaves like a single substance.

Any given sample of a gas can be described in terms of four fundamental properties: *mass*, *volume*, *pressure*, and *temperature*. The investigation of these properties of air led to the discovery of the quantitative relationships between them. The statements of these quantitative relationships constitute the earliest scientific laws (see Box 3.3).

Box 3.3

THE SCIENTIFIC METHOD: LAWS, HYPOTHESES, AND THEORIES

Boyle's law, Charles's law, and the law of conservation of mass are statements that in each case summarize a large number of observations. The fundamental activity of science is making *observations* of the world around us. If observations are made under carefully controlled conditions, they can be repeated by any other person who has the appropriate equipment. In this way the facts of nature are established. When a large number of observations have been made and a number of facts established, regularities and consistencies in a set of facts may enable one to make a concise statement or give a mathematical equation that summarizes the observed facts. Such a statement is known as a scientific *law*, for example, Boyle's law. Another familiar example is the law of gravity, which summarizes the numerous observations that show that separate masses attract each other.

Once a law has been established, a scientist asks the question: "Why does nature behave in the way that is summarized by the law?" Why, for example, is the volume of gas inversely proportional to the pressure, as stated by Boyle's law? A tentative answer to such a question is known as a *hypothesis*. If it is to be useful, a hypothesis must suggest new experiments that will either verify or refute the hypothesis. A hypothesis that continually withstands such tests develops into a *theory*. The theory that provides an explanation for Boyle's law is the kinetic molecular theory. A theory is a model of nature that enables us to better understand our observations.

Theories are, however, only tentative. A theory continues to be useful only as long as we fail to find any experimental facts that cannot be accounted for by the theory. But only one fact that the theory cannot explain will cause the theory to be modified or replaced by a new theory. Dalton's atomic theory continues to be very useful today, but it has been modified from its original form in that we no longer believe that an atom is indivisible nor that all the atoms of an element have the same mass.

Observations that have been verified by repeated experiments will never be changed, but the theories invented to explain these observations may well be replaced or at least modified in the future. In this sense the facts are more important than the theories. Thus it is a mistake to believe that if one knows all the laws and theories that are derived from experimental observations one need not know the experimental facts. New theories can only be developed by those who have a wide knowledge of the facts relating to a particular field, particularly those facts that have not been satisfactorily accounted for by existing theories.

The deduction of laws and the development of hypotheses and theories that lead to predictions—which in turn must be tested by experiment, which then lead to new observations—is a continuous and never-ending process. This process is known as the *scientific method*. Although in principle it appears to be a logical process, science does not, in fact, advance in a completely organized and logical manner. Success in scientific research often depends on the ability to observe and interpret the unexpected. The experiment "that does not work" can be the clue to an important discovery. New theories are generally not developed in an entirely logical and planned manner but through a slow and tortuous process in which many incorrect hypotheses may be made before an adequate theory is formulated. Theories may depend as much on flashes of insight and inspiration as on logical argument.

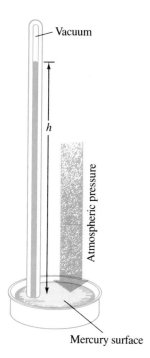

Vacuum

h

Atmospheric pressure

Mercury surface

Figure 3.4 Toricellian, or Mercury, Barometer.
Atmospheric pressure may be measured by the height, *h*, of the column of mercury. Average atmospheric pressure at sea level supports a column of mercury 760 mm high. At the same pressure a column of water, which is less dense than mercury, would be 10.3 m high.

Pressure of the Atmosphere

From the time of Aristotle (384–322 B.C.) to the time of Descartes (1596–1650), philosophers thought that a *vacuum*–a space that is empty of all matter—could not exist. They believed that "Nature abhors a vacuum." Then in 1643 Italian physicist Evangelista Torricelli (1608–1647) carried out a classic experiment in which he fulfilled his goal "not simply to produce a vacuum but to make an instrument which shows the mutations of the air, now heavier and dense, now lighter and thin." In other words, his aim was to demonstrate that the atmosphere exerts a pressure and that this pressure varies.

Torricelli used what is called a *Torricellian* or *mercury barometer*, a long glass tube about 1 m long, closed at one end and filled with mercury (Figure 3.4). The open end is temporarily closed, inverted, and immersed in a dish of mercury. When the end of the tube under the surface is opened, the mercury in the tube falls until it has a height of approximately 760 mm above the level of the mercury in the dish. The height of the mercury in the tube is always approximately the same, whatever the diameter of the tube. The space at the top of the tube is empty except for a minute amount of mercury vapor, which can be neglected for all practical purposes; this empty space constitutes a vacuum and exerts zero pressure. The pressure of the atmosphere acting on the surface of the mercury in the dish pushes the mercury in the dish down and therefore up into the tube. When the downward pressure exerted by the mercury in the tube is exactly balanced by the atmospheric pressure that holds the mercury up in the tube, the height of the mercury in the tube remains constant.

The height of the mercury column supported by the atmosphere decreases with height above the surface of the earth, and it varies somewhat with the atmospheric conditions. The average height of the mercury column at sea level is 760 mm, and this value is called the **standard atmospheric pressure** (1 atmosphere, atm).

Any instrument that can measure the pressure of the atmosphere is known as a **barometer**, and the type first devised by Torricelli is known as a Torricellian, or mercury, barometer. Any liquid could, in principle, be used in the tube. Because of its high density, mercury gives a column that has a convenient height. Water, which has a much lower density, would give a column 10.3 m high! Because the mercury barometer has been widely used, atmospheric pressure is often expressed simply in terms of the height of the mercury column. Thus we speak of an atmospheric pressure of 760 mm of mercury, or 760 mm Hg.

Since pressure is force per unit area, however, it should be measured in appropriate units such as pounds per square inch (psi), or in the SI unit of pascals (Pa). The *pascal* is defined as 1 newton per square meter ($N\ m^{-2}$):

$$1\ Pa = 1\ N\ m^{-2} \quad \text{or} \quad 1\ kPa = 10^3\ N\ m^{-2}$$

A pressure of 760 mm Hg is a pressure of 101.33 kPa. Another unit of pressure is the *torr*, named after Torricelli:

$$1\ torr = 1\ mm\ Hg$$

The pressure of gases is also very commonly measured in atmospheres (atm). The *standard atmosphere* was originally defined as the pressure that will support a column of mercury 760 mm high at 0°C and at sea level. The standard atmosphere is now defined as 101.33 kPa. Thus we have the following relationships between the various units for measuring pressure:

$$1\ atm = 101.33\ kPa = 760\ mm\ Hg = 760\ torr$$

Although the atmosphere is not an SI unit, it is a convenient unit for many purposes; we will use it often in this book.

Pressure and Volume: Boyle's Law

We are all familiar with the fact that gases are compressible. When the pressure on a certain amount of gas is increased, as in a bicycle tire pump, the volume of the gas decreases; the greater the pressure, the smaller the volume is. In 1660 English chemist Robert Boyle (1627–1691) studied the effects of pressure on the volume of air by using the apparatus shown in Figure 3.5. He found that if he doubled the pressure, the volume of the air was halved; if the pressure was increased four times, the volume was decreased to one-quarter of its original value. In other words, Boyle found that, in general, the volume, V, of a given mass of air is inversely proportional to its pressure, P, if the temperature, T, is held constant:

$$V \propto \frac{1}{P} \quad \text{or} \quad V = \text{Constant} \times \frac{1}{P} \quad (T \text{ constant})$$

This relationship has been found to hold for *any gas*. It can also be stated in the following form:

Pressure times volume is constant for a given amount of gas at a constant temperature:

$$PV = \text{Constant} \quad (T \text{ constant})$$

This relationship is known as **Boyle's law**.

To compare the same gas sample at constant temperature under different pressure and volume conditions, we can write Boyle's law conveniently as

$$P_1 V_1 = P_2 V_2 \quad (T \text{ constant})$$

If the initial pressure and volume of a given quantity of gas are initially P_1 and V_1, and the pressure is changed to P_2, then the new volume, V_2, is given by this relationship. Figure 3.6 shows the relationship between P and V graphically.

Boyle's law expresses quantitatively the important fact that a gas is compressible. The more a gas is compressed, the denser it becomes, because the

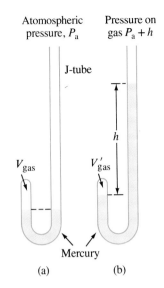

Figure 3.5 Boyle's Apparatus.
Boyle used this apparatus to study the volume and pressure of a gas sample. One end of the J tube is closed, trapping air at the end of the tube. (a) When the height of the mercury is the same in the open and the closed parts of the tube, the pressure exerted on the gas is equal to the atmospheric pressure. (b) The pressure of the gas is increased by adding mercury to the tube. Then the pressure exerted on the gas is equal to h (the difference in the heights of the two mercury surfaces) plus the atmospheric pressure, and the volume of the gas is smaller.

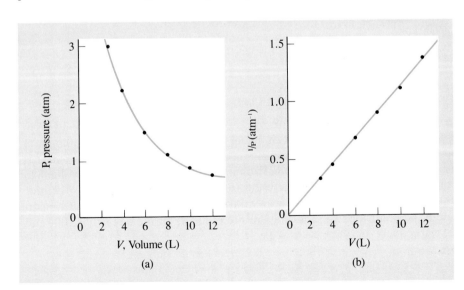

Figure 3.6 Boyle's Law.
Boyle's law gives the relationship between pressure and volume at constant temperature. If the temperature is held constant, the volume of a given amount of gas is inversely proportional to its pressure. (a) A plot of P against V gives a curve (a hyperbola). (b) A plot of $1/P$ against V or $1/V$ against P gives a straight line.

same number of molecules and the same mass occupy a smaller volume. For instance, the air at the surface of the earth is compressed by the mass of air resting above it; so the higher the altitude, the less the air is compressed. As a result, the density and the pressure of the air decrease with increasing altitude. Thus at 2500 m (8000 ft) in the Rocky Mountains, the pressure is only 0.75 atm, and at 8000 m (26 000 ft) in the Himalayas, the world's highest mountains, the atmospheric pressure is only 0.47 atm.

At high altitudes the amount of oxygen that the body takes in during each breath is considerably decreased. This decreased amount of oxygen makes any exertion very difficult and causes the weakness and headaches known as altitude sickness. Mountain climbers who tackle Everest and other very high peaks must undergo a long period of acclimatization at high altitude to allow the body to adapt to the low oxygen pressure, or they must carry cylinders of oxygen for breathing. For the same reasons jet aircraft, which fly at altitudes up to 10 000 m, must be pressurized and equipped with emergency oxygen in case the pressurization should fail.

Example 3.1 Calculate the volume occupied by a sample of hydrogen at a pressure of 3.00 atm if it has a volume of 6.20 L at a pressure of 1.05 atm.

Solution We will use subscript 1 to denote the original conditions of the hydrogen and subscript 2 to denote the final conditions. Thus we have

$$P_1 = 1.05 \text{ atm} \qquad V_1 = 6.20 \text{ L}$$
$$P_2 = 3.00 \text{ atm} \qquad V_2 = ? \qquad \text{(To be found)}$$

We can rearrange Boyle's law,

$$P_1 V_1 = P_2 V_2$$

to obtain an expression for the unknown volume, V_2, as follows:

$$V_2 = \frac{P_1 V_1}{P_2}$$

$$= \frac{1.05 \text{ atm} \times 6.20 \text{ L}}{3.00 \text{ atm}} = 2.17 \text{ L}$$

Thus the sample of hydrogen gas occupies 2.17 L when the pressure is increased to 3.00 atm.

Temperature and Volume: Charles's Law

In 1787, a hundred years after Boyle's work, Jacques Charles (1746–1823) in France investigated the effect of changing the temperature of a given amount of air while holding the pressure constant. He found that air expands when it is heated and that whatever the initial volume of the gas, the ratio of its volume in boiling water to its volume in ice water is constant. Experiment shows that this ratio is 1.366 for air and is very nearly the same for *all* gases. Since the temperature of boiling water and the temperature of ice water are defined as 100° and 0°, respectively, on the Celsius scale,

$$\frac{V_{\text{in boiling water}}}{V_{\text{in ice}}} = \frac{V_{100°C}}{V_{0°C}} = 1.366$$

Therefore

$$V_{100°C} = 1.366V_{0°C} = (1 + 0.366)V_{0°C}$$
$$= V_{0°C} + 0.366V_{0°C}$$

In other words, the volume of a gas at 100°C is greater than its volume at 0°C by 0.366 of its volume at 0°C. Moreover, experiment shows that the expansion of a gas is uniform; thus for every 1° increase in the temperature, the volume of a gas increases by $0.366/100 = 1/273$ of its volume at 0°. So the volume of a gas is a linear function of its Celsius temperature, as Figure 3.7 illustrates.

THE KELVIN TEMPERATURE SCALE   Because the volume of a gas decreases in a linear fashion from 100° to 0°C, we expect it to continue to decrease in the same way, that is, by 1/273 of its volume at 0°C for every degree that it is cooled below 0°. This expectation leads us to the surprising conclusion that the volume of the gas should be zero at −273°C and should become negative at a lower temperature (see Figure 3.7). Since we can attach no meaning to a negative volume, we are forced to assume that we cannot obtain a lower temperature than −273°C, which is therefore called the **absolute zero of temperature**.

In practice, all gases condense to liquids and solids at temperatures above −273°C, so no gas can, in fact, be cooled until it has zero volume. But the idea of a lowest possible temperature—that is, an absolute zero of temperature—is exceedingly important. Instead of arbitrarily choosing the melting point of ice as the zero of the temperature scale, as is done on the Celsius scale, we can logically, and also conveniently, choose the absolute zero as the zero of a temperature scale. This choice of zero is the basis of the **Kelvin temperature scale**, first suggested by the British scientist Lord Kelvin (1824–1907). According to accurate measurements, the absolute zero of temperature is −273.15°C. Thus 0 K = −273.15°C, and the Kelvin (K) scale is related to the Celsius scale by the expression

$$T = t + 273.15$$

where T is the temperature on the Kelvin scale and t is the temperature on the Celsius scale. In other words, temperatures on the Celsius scale are converted to temperatures on the Kelvin scale simply by adding 273.15. Notice that, by convention, the degree sign (°) is not used when we are expressing temperatures on the Kelvin scale. The unit on the Kelvin scale is the kelvin (K), and a temperature such as 100 K is read as "one hundred kelvins."

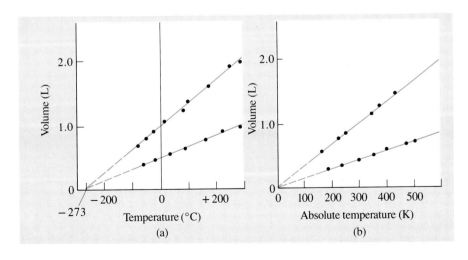

Figure 3.7 Temperature and Volume.
(a) For a fixed quantity of gas at constant pressure, a 1°C change in the temperature alters the volume of the gas by $\frac{1}{273}$ of its volume at 0°C and the volume extrapolates to zero at −273°C. The volume of a gas is a linear function of its Celsius temperature, but volume and temperature on the Celsius scale are not directly proportional. (b) The volume of a fixed quantity of gas at constant pressure is directly proportional to the absolute (Kelvin) temperature.

CHARLES'S LAW If the temperature is expressed on the Kelvin scale, the volume of a gas is directly proportional to the temperature, which is not true if the temperature is measured on the Celsius scale (Figure 3.7):

$$V \propto T \quad \text{or} \quad V = \text{Constant} \times T \quad (P \text{ constant})$$

$$\frac{V}{T} = \text{Constant}$$

or

$$\frac{V_1}{V_2} = \frac{T_1}{T_2} \quad (P \text{ constant})$$

where T is the temperature on the Kelvin scale. These expressions summarize **Charles's law**:

> **At constant pressure the volume of a given mass of gas is proportional to its temperature on the Kelvin scale.**

Hot-air balloons are an interesting application of Charles's law (see Box 3.4).

Example 3.2 To what temperature must a sample of nitrogen, which has a volume of 900 mL at 25°C, be cooled in order for it to occupy a volume of 350 mL?

Solution Using subscript 1 for initial conditions and subscript 2 for final conditions, we have

$$V_1 = 900 \text{ mL} \qquad t_1 = 25°C$$
$$V_2 = 350 \text{ mL} \qquad t_2 = ? \qquad \text{(To be determined)}$$

Since the volume-temperature relationship (Charles's law) involves Kelvin scale temperature, we convert t_1 to T_1, rounding off 273.15 to three significant figures:

$$T_1 = 273 + 25 = 298 \text{ K}$$

Box 3.4

BALLOONS

Balloons have long been used for meteorological observation and other purposes. The simplest and earliest balloons were filled with hot air. Because gases expand when heated (Charles's law), the hot air in a balloon is less dense than cool atmospheric air. Hence the cool air displaced by a hot-air balloon has a greater mass than the air in the balloon. The balloon therefore rises in the atmosphere, much as a cork released under water rises to the surface. Since the density of the atmosphere decreases with increasing height, the balloon rises until the mass of the air displaced by the balloon equals the mass of the balloon and its load, at which point the balloon floats in the atmosphere, just as a cork floats on the surface of water.

The first ascent in a hot-air balloon was made in France in 1783. In 1803 the French scientist Joseph Gay-Lussac, a balloon pioneer, set a record by ascending to 23 000 ft (7 km). He used balloon ascents to carry out experiments that included studies of the composition of the atmosphere and variations in the earth's magnetic field. In recent years hot-air ballooning has been revived as a sport.

Balloons filled with hydrogen represented a considerable improvement on hot-air balloons. Because hydrogen has very low density, much less than the density of air, balloons filled with hydrogen rise to a considerable height. Weather balloons can reach heights of approximately 40 km.

By the 1930s a logical development of the hydrogen balloon, the airship, was providing regular transportation across the Atlantic. Instead of just drifting in the wind, airships were driven by engine-powered propellors and included cabins for passengers. The disastrous fire that destroyed the German airship *Hindenburg* in 1937 marked the end of the airship era, but the possibility of building airships that use helium, which is nonflammable but expensive, has attracted new interest. Although they would be too slow to compete with jets, helium airships might have other uses, such as transporting timber in forestry operations.

Now we can use Charles's law:

$$\frac{V_1}{V_2} = \frac{T_1}{T_2}$$

Rearranging and substituting the known values gives

$$T_2 = \frac{V_2 T_1}{V_1} = \frac{350 \text{ mL} \times 298 \text{ K}}{900 \text{ mL}} = 116 \text{ K}$$

The required temperature is 116 K, which on the Celsius scale is

$$t_2 = T_2 - 273 = -157°C$$

Standard Temperature and Pressure (STP)

Since the volume of a gas changes with both pressure and temperature, stating that a certain gas sample has a particular volume is not sufficient; the pressure and the temperature must also be specified. To simplify comparisons the volume of a given sample of gas is normally reported at 0°C (273.15 K) and 1 atm (101.33 kPa); these conditions are known as **standard temperature and pressure**, abbreviated **STP**.

Combined Gas Law

By combining Boyle's law, which states that at constant temperature $V \propto 1/P$, and Charles's law, which states that at constant pressure $V \propto T$, we obtain a relationship between the volume, temperature, and pressure of a given amount of gas:

$$V \propto \frac{T}{P} \quad \text{or} \quad V = \text{Constant} \times \frac{T}{P}$$

Rearranging this relationship, we find that $PV/T = \text{constant}$ for a given mass of gas, or

$$\frac{P_1 V_1}{T_1} = \frac{P_2 V_2}{T_2}$$

This equation is known as the **combined gas law**.

If the volume of a given mass of gas is kept constant, that is, $V_1 = V_2$, then

$$\frac{P_1 V_1}{T_1} = \frac{P_2 V_1}{T_2} \quad \text{or} \quad \frac{P_1}{T_1} = \frac{P_2}{T_2}$$

Thus

$$\frac{P}{T} = \text{Constant} \quad \text{or} \quad P = \text{Constant} \times T$$

In other words, the pressure is proportional to the absolute temperature.

Example 3.3 If a sample of oxygen gas has a volume of 425 mL at 70°C and a pressure of 0.950 atm, what will its volume be at STP?

Solution We summarize the data given in the statement of the problem as follows:

$P_1 = 0.950$ atm	$P_2 = 1.00$ atm
$V_1 = 425$ mL	$V_2 = ?$ (To be found)
$T_1 = 343$ K (70°C)	$T_2 = 273$ K

Because both pressure and temperature are varying, we use the combined gas law,

$$\frac{P_1 V_1}{T_1} = \frac{P_2 V_2}{T_2}$$

which we can rearrange to obtain an expression for the unknown volume, V_2:

$$V_2 = \frac{P_1 V_1 T_2}{P_2 T_1}$$

Substituting the values given we have:

$$V_2 = \frac{0.950 \text{ atm} \times 425 \text{ mL} \times 273 \text{ K}}{1.00 \text{ atm} \times 343 \text{ K}} = 321 \text{ mL}$$

3.6 KINETIC MOLECULAR THEORY OF GASES

Because a given mass of a substance occupies much more space as a gas than as a liquid, and because a gas is much more compressible than a liquid, we can conclude that the molecules in a gas are far apart and that much of a gas consists of empty space. But how can something that is mostly empty space exert a pressure on its surroundings?

We can explain this property and many other properties of a gas by an extension of the atomic theory known as the **kinetic molecular theory**. It assumes that not only do all substances consist of atoms or molecules, but also these atoms and molecules are in a constant state of motion. The kinetic molecular theory for gases is based on the following four assumptions:

1. A gas is composed of molecules that are far apart from each other in comparison with their own dimensions. Most of the volume occupied by a gas is empty space.

2. The gas molecules are in constant random motion. Each molecule continues to move in a straight line, unless it collides with another molecule or with a wall of the container.

3. The molecules exert no force on each other or on the container, except when they collide with each other or with the walls of the container. These collisions are *elastic*; that is, one molecule may gain energy and the other may lose energy, but their total energy remains constant. As a result, the total energy of all the molecules remains constant.

4. The average kinetic energy of the molecules of a gas is proportional to the absolute temperature. The kinetic energy of a molecule is equal to $\frac{1}{2}mv^2$, where m is its mass and v is its speed.

Explaining the Gas Laws

We know from experiment that the pressure of a gas is doubled if we reduce the volume by one-half. According to the kinetic molecular theory, the pressure exerted by a gas on the walls of its container results from the continual bombardment of the walls by the fast-moving molecules. Every time a molecule collides with a wall, it exerts a force on it. Since pressure is force per unit area, the pressure is proportional to the number of collisions per unit area in a given time. Suppose we decrease the volume of a given mass of gas to one-half its original value. There will then be twice as many molecules in any given volume of the gas, and hence they will make twice as many collisions per unit area with the walls of the container. Consequently, the pressure will be doubled (Figure 3.8). Thus the kinetic theory provides a simple explanation of Boyle's law, that pressure and volume are inversely proportional.

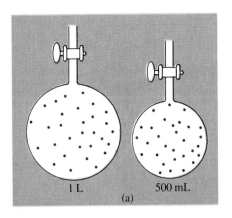 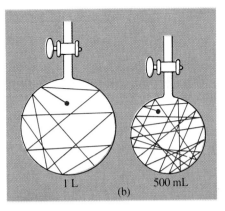

1 L   500 mL   1 L   500 mL

(a)   (b)

Figure 3.8 Volume and Pressure.
(a) If the volume of gas is halved while the temperature and number of molecules are held constant, the molecules are packed closer and the density is doubled. (b) On the average, a molecule now hits the wall twice as often. The total number of impacts with the wall is therefore doubled, which doubles the pressure.

The kinetic theory also explains why the volume of a gas increases with increasing temperature. Molecules in motion possess kinetic energy, $\frac{1}{2}mv^2$. A basic assumption of the kinetic theory is that the kinetic energy of the molecules of a gas is proportional to the absolute temperature. As the temperature increases, the molecules move faster; they make more collisions with the walls, and since they have a greater kinetic energy, they exert a greater force on the walls.

Suppose that a gas is confined in a container of variable volume, such as the vessel with a movable piston shown in Figure 3.9. As the temperature is raised, the increased number and force of the collisions of the gas molecules with the piston exerts a greater pressure on the piston than the pressure exerted by the atmosphere outside. The piston therefore moves outward, increasing the volume of the gas. This increase in volume reduces the number of collisions of the gas molecules with the piston and therefore reduces the pressure. Thus the piston moves outward until the pressure of the gas again equals the atmospheric pressure. The volume of a gas increases as the temperature increases if the pressure is held constant, as is observed experimentally. A more detailed mathematical treatment shows that the volume is directly proportional to the absolute temperature, thus providing an explanation for Charles's law.

The assumption that the average kinetic energy of the molecules of a gas is proportional to the temperature is not so much an assumption as a definition of absolute temperature. It applies not only to gases but also to liquids and solids, even though their molecules are packed more closely together. Suppose that a gas is confined in a container that is initially at a higher temperature than the gas. In the collisions between the slower molecules of the gas and the

Figure 3.9 Effect of Temperature on Volume.
When the temperature is raised, the molecules move faster and make more collisions and more energetic collisions with the walls of the vessel. The pressure inside the vessel is increased. Since the pressure inside the vessel is greater than the pressure outside, the piston moves outward, increasing the volume and decreasing the pressure in the vessel. The volume continues to increase, reducing the pressure exerted by the molecules in the vessel, until the pressure of the gas is once again equal to the atmospheric pressure. Thus the volume increases with increasing temperature.

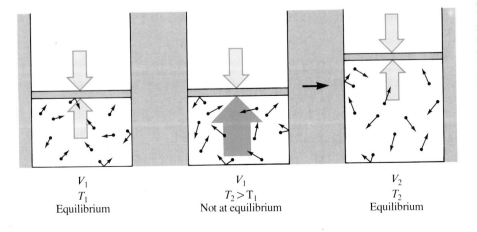

V_1 V_1 V_2
T_1 $T_2 > T_1$ T_2
Equilibrium   Not at equilibrium   Equilibrium

faster molecules of the container walls, the gas molecules are speeded up and those of the container walls are slowed down. The average kinetic energy of the molecules of the container walls therefore decreases, and the average kinetic energy of the gas molecules increases. Energy flows from the container to the gas; this flow of energy is called **heat**. We say that heat flows from the hot container to the cooler gas. Thus we see that temperature is simply a way of describing the average kinetic energy of the molecules of a substance; it is a property of a substance that is proportional to the kinetic energy of its molecules.

Since as a gas is cooled, the average speed and the average kinetic energy of its molecules decrease, a temperature will be reached at which the average speed and kinetic energy of the molecules are zero. Clearly, the speed and the kinetic energy of the molecules cannot be decreased further, and their temperature must be the lowest possible temperature. This temperature is the absolute zero, 0 K.

Explaining the States of Matter: Intermolecular Forces

In fact, no gas can be cooled to 0 K, because all substances become liquids or solids before this temperature is reached. This change of state results from attractive forces that act between all molecules. These forces are called **intermolecular forces**. In a substance that is a gas at ordinary temperature, these forces are relatively weak, and their influence on the motion of the molecules is negligible. But if a gas is cooled to low temperatures, the average speed of its molecules is greatly reduced. Even weak intermolecular forces then cause the molecules to stick together to some extent. As a result, they no longer move independently and randomly, as is assumed by the kinetic molecular theory. When the temperature is low enough for a sufficiently large number of molecules to stick together, the gas changes to a liquid.

In a liquid the molecules are packed rather closely, so a liquid has a relatively high density and is not very compressible. But the molecules still have enough kinetic energy to move around each other. When the temperature is lowered still further, the molecules no longer have sufficient energy to be able to jostle past each other. Each one becomes trapped in a hole formed by the surrounding molecules, and it can only rotate and oscillate about a fixed mean position. The molecules then usually take up the regular ordered arrangement that is characteristic of most solids.

The fact that all substances become liquids or solids at sufficiently low temperatures shows that there must be attractive forces between all molecules. Thus the assumption of the kinetic molecular theory of gases that there are no forces between molecules is not correct, but it is valid to a very good approximation for substances that are gases at ordinary temperatures. For substances that are gases even at low temperatures, such as O_2, H_2, N_2, and Ar, the intermolecular forces must be very weak. In contrast, a substance such as sulfur, which consists of S_8 molecules and is a solid at room temperature, melting at 112.8°C, must have relatively strong intermolecular forces.

Because the volume of the molecules of a gas is very small compared with the space that they occupy, and because the intermolecular forces are almost negligible, the properties of gases are essentially independent of the nature of the molecules. In other words, all gases behave in the same way. They all obey the combined gas law, to a reasonable approximation. In contrast, in a solid and in a liquid the molecules are held close together by intermolecular forces. The properties of solids and liquids therefore depend very much on the sizes and shapes of the molecules of which they are composed and on the strength of the intermolecular forces. Thus the properties of liquids and solids cannot be described, even approximately, by a single equation like the combined gas law.

3.7 THE IDEAL GAS LAW

We have seen that the pressure of a gas is proportional to the number of collisions of gas molecules per unit area of the vessel wall in a given time. If the volume and the temperature of a vessel are constant, the number of collisions per unit area in a given time is proportional to the number of molecules. Hence the pressure is proportional to the number of molecules. Since 1 mol of molecules always contains the same number of molecules (6.022×10^{23}), *the pressure is proportional to the number of moles of molecules, n, or*

$$P \propto n \quad \text{(at constant } V \text{ and } T) \quad \text{or} \quad \frac{P}{n} = \text{Constant} \quad \text{(At constant } V \text{ and } T)$$

In previous sections we have seen that

$$PV = \text{Constant} \quad \text{(At constant } T)$$

$$\frac{V}{T} = \text{Constant} \quad \text{(At constant } P)$$

Combining these relationships gives us the equation

$$\frac{PV}{nT} = \text{Constant} \quad \text{or} \quad PV = nRT$$

where R is a constant called the **gas constant**. If pressure is expressed in atmospheres, volume in liters, and temperature in kelvins, R has the value $0.0821 \text{ atm L mol}^{-1} \text{K}^{-1}$. If the pressure, volume, and temperature are expressed in SI units, the value of R is $8.31 \text{ kPa dm}^3 \text{ mol}^{-1} \text{K}^{-1}$.

This equation is known as the **ideal gas equation**, or the **ideal gas law**. It is obeyed closely, but not exactly, by nearly all gases. Different gases differ slightly in their behavior because the size of the molecules is not completely negligible compared with the distances between them and because there are intermolecular forces. A hypothetical gas in which the molecules have zero volume and in which there are no intermolecular forces would obey the ideal gas equation exactly and is known as an **ideal gas**. For the calculations in this book we may assume that real gases behave like an ideal gas. The ideal gas equation applies accurately to all gases at very low pressures and high temperatures, that is, when the molecules are far apart and are moving at high speeds. It is a very good approximation under most conditions, but it becomes less exact at very high pressures and at very low temperatures.

At very high pressures the volume of the molecules is no longer negligible compared with the space between the molecules, and therefore the volume of the gas is somewhat greater than expected from Boyle's law. At very low temperatures the molecules move very slowly and have a small average kinetic energy. Even weak intermolecular forces cause them to stick together to some extent, so the volume of the gas is somewhat less than predicted by Charles's law.

If we know any three of the variables P, V, n, and T, which describe the physical state of a gas, we may calculate the fourth by using the ideal gas equation.

Example 3.4 Calculate the number of moles of hydrogen in a 100-cm^3 sample at a temperature of 300 K and a pressure of 750 mm Hg.

Solution Either of the two values of the gas constant may be used provided that appropriate units are used for P, V, and T. If we use $R = 0.0821$ atm L mol^{-1} K^{-1}, then we convert the units of P to atmospheres and the units of V to liters:

$$V = (100 \text{ cm}^3)\left(\frac{1 \text{ L}}{1000 \text{ cm}^3}\right) = 0.100 \text{ L}$$

$$P = (750 \text{ mm Hg})\left(\frac{1 \text{ atm}}{760 \text{ mm Hg}}\right) = 0.987 \text{ atm}$$

$$T = 300 \text{ K}$$

Rearranging the ideal gas law, we obtain

$$n = \frac{PV}{RT} = \frac{0.987 \text{ atm} \times 0.100 \text{ L}}{0.0821 \text{ atm L mol}^{-1} \text{ K}^{-1} \times 300 \text{ K}} = 4.01 \times 10^{-3} \text{ mol}$$

The calculation is only slightly different if we use SI units. In this case $R = 8.31$ kPa dm^3 mol^{-1} K^{-1}:

$$V = (100 \text{ cm}^3)\left(\frac{1 \text{ dm}}{10 \text{ cm}}\right)^3 = 0.100 \text{ dm}^3$$

$$P = (750 \text{ mm Hg})\left(\frac{101.3 \text{ kPa}}{760 \text{ mm Hg}}\right) = 100 \text{ kPa}$$

$$T = 300 \text{ K}$$

$$n = \frac{PV}{RT} = \frac{100 \text{ kPa} \times 0.100 \text{ dm}^3}{8.31 \text{ kPa dm}^3 \text{ mol}^{-1} \text{ K}^{-1} \times 300 \text{ K}} = 4.01 \times 10^{-3} \text{ mol}$$

Example 3.5 Calculate the pressure exerted by 3.00 g of N_2 gas in a container of volume 2.00 L at a temperature of $-23°C$.

Solution The volume is 2.00 L. The temperature given must be converted to kelvins:

$$T = 273 + t = 273 + (-23) = 250 \text{ K}$$

The mass of N_2 must be converted to moles of N_2, using its molar mass of 28.02 g:

$$n = (3.00 \text{ g } N_2)\left(\frac{1 \text{ mol } N_2}{28.02 \text{ g } N_2}\right) = 0.107 \text{ mol } N_2$$

Now we can obtain the value for the pressure P from the ideal gas law:

$$P = \frac{nRT}{V} = \frac{0.107 \text{ mol} \times 0.0821 \text{ L atm mol}^{-1} \text{ K}^{-1} \times 250 \text{ K}}{2.00 \text{ L}} = 1.10 \text{ atm}$$

Determination of Molar Mass

As shown in Example 3.4, we can use the ideal gas law to find the number of moles of a gas if we can measure V, P, and T. If we also know the mass of the gas, we can calculate its molar mass. A simple procedure for finding the molar mass is as follows:

1. Weigh an evacuated flask of known volume.
2. Fill it, at a known temperature and pressure, with the gas whose molar mass is to be determined; weigh the flask again.
3. Calculate the increase in the mass of the flask; this increase equals the mass of the gas.

The **molar mass**, M, is given by

$$M = \frac{\text{Mass of gas}}{\text{Number of moles}} = \frac{m}{n}$$

Example 3.6 A flask of 0.300-L volume was weighed after it had been evacuated. It was then filled with a gas of unknown molar mass at 1.00 atm pressure and a temperature of 300 K. The increase in the mass of the flask was 0.977 g.

(a) What is the molar mass of the gas?

(b) Assuming that the gas molecules contain only sulfur and oxygen, what is the molecular formula of the gas?

Solution

(a) We first find the number of moles of the gas:

$$n = \frac{PV}{RT} = \frac{1.00 \text{ atm} \times 0.300 \text{ L}}{0.0821 \text{ atm L mol}^{-1}\text{K}^{-1} \times 300 \text{ K}} = 0.0122 \text{ mol}$$

Then we can find the molar mass M:

$$M = \frac{m}{n} = \frac{0.977 \text{ g}}{0.0122 \text{ mol}} = 80.1 \text{ g mol}^{-1}$$

Thus the molar mass is 80.1 g mol^{-1}.

(b) The molecular formula could be either SO_3 ($M = 80.06$ g mol^{-1}) or S_2O ($M = 80.12$ g mol^{-1}). Any other combinations of sulfur and oxygen have molar masses quite different from 80 g.

For a substance that is not a gas at room temperature, we can find the molar mass by a modification of the method just described if the substance has a relatively low boiling point, for example, below 100°C. A flask containing only air is weighed. A sample of the liquid to be studied is added, and the flask is heated, usually in a bath of boiling water (see Figure 3.10). Heating is continued until all the liquid has evaporated and the large amount of vapor thus formed has completely displaced all the air from the flask. The flask then contains

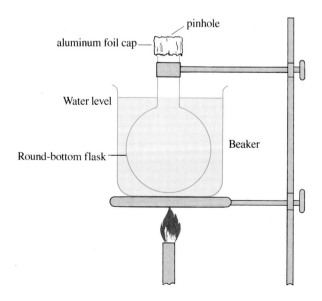

pinhole

aluminum foil cap

Water level

Round-bottom flask

Beaker

Figure 3.10 Determination of Molar Mass of a Volatile Liquid.

only the vapor of the substance under investigation at 100°C and atmospheric pressure.

The flask is then closed to the atmosphere and cooled to room temperature. The vapor condenses to a liquid, its mass remaining unchanged. The pressure in the flask is now low, and when we open the flask, air rushes in until the pressure in the flask is again atmospheric.

The flask is then reweighed. The difference between the mass of the flask now and its original mass is the mass of the condensed vapor, which is the mass of the vapor that filled the flask at 100°C and atmospheric pressure. We thus have all the information required to calculate the molar mass, as the following example shows.

Example 3.7 Benzene, a colorless liquid, is another example of the large class of compounds containing only carbon and hydrogen that are called hydrocarbons. Benzene has the empirical formula CH. A flask of volume 247.2 cm³ had a mass of 25.201 g when it contained only air. A sample of benzene was added to the flask and heated to 100°C. The benzene vaporized and drove all the air from the flask. The flask was then cooled to room temperature, opened to the atmosphere, and weighed. It had a mass of 25.817 g. The barometric pressure was 742 mm Hg. Calculate the molar mass and the molecular formula of benzene.

Solution First, we find the number of moles of benzene in the flask. We convert the information given to appropriate units:

$$V = (247.2 \text{ cm}^3)\left(\frac{1 \text{ L}}{1000 \text{ cm}^3}\right) = 0.2472 \text{ L}$$

$$T = (273 + 100) \text{ K} = 373 \text{ K}$$

$$P = (742 \text{ mm Hg})\left(\frac{1.00 \text{ atm}}{760 \text{ mm Hg}}\right) = 0.976 \text{ atm}$$

$$n = \frac{PV}{RT} = \frac{0.976 \text{ atm} \times 0.2472 \text{ L}}{0.0821 \text{ atm L mol}^{-1} \text{ K}^{-1} \times 373 \text{ K}} = 7.88 \times 10^{-3} \text{ mol}$$

Mass of benzene = (Mass of flask + air + condensed vapor) − (Mass of flask + air)

$$= 25.817 \text{ g} - 25.201 \text{ g} = 0.616\text{g}$$

Thus the molar mass is

$$M = \frac{m}{n} = \frac{0.616 \text{ g}}{7.88 \times 10^{-3} \text{ mol}} = 78.2 \text{ g mol}^{-1}$$

The empirical formula CH corresponds to a molar mass of $(12.01 + 1.008)$ g mol^{-1} = 13.02 g mol^{-1}. The experimentally determined molar mass is six times larger:

$$\frac{78.2 \text{ g mol}^{-1}}{13.02 \text{ g mol}^{-1}} = 6.01$$

Hence the molecular formula of benzene must be C_6H_6.

Gas Density

We have seen that the determination of the molar mass of a gaseous substance depends on finding the mass of gas that occupies a known volume at a known P and T. Since density = mass/volume, or $d = m/V$, we can find the molar mass of a gaseous substance if we know its density, pressure, and temperature.

Using the relationship

$$\text{Number of moles} = \frac{\text{Mass}}{\text{Molar mass}} \quad \text{or} \quad n = \frac{m}{M}$$

we may express the ideal gas law, $PV = nRT$, in the form

$$PV = \frac{m}{M}RT \quad \text{or} \quad P = \frac{m}{VM}RT$$

Now since

$$d = \frac{m}{V}$$

we have

$$P = \frac{d}{M}RT \quad \text{or} \quad M = \frac{dRT}{P}$$

Using this equation, we can find the molar mass M of a gaseous substance from its density at a given temperature and pressure. Or in general, we may find any one of the variables P, M, T, and d if we know the other three.

Example 3.8 The density of a gas was found to be 2.06 g L^{-1} at STP. What is its molar mass?

Solution At STP

$$P = 1.00 \text{ atm} \quad T = 273 \text{ K}$$

Substituting in the equation $M = dRT/P$, we have

$$M = \frac{(2.06 \text{ g L}^{-1})(0.0821 \text{ atm L mol}^{-1} \text{ K}^{-1})(273 \text{ K})}{1.00 \text{ atm}} = 46.2 \text{ g mol}^{-1}$$

Molar Volume of a Gas: Avogadro's Law

From the ideal gas equation we can show that at the same temperature and pressure n moles of any gas occupy the same volume. The volume of 1 mol of a substance is called its **molar volume.** The molar volume of an ideal gas at STP is given by

$$V = \frac{nRT}{P} = \frac{1 \text{ mol} \times 0.0821 \text{ atm L mol}^{-1} \text{ K}^{-1} \times 273 \text{ K}}{1 \text{ atm}} = 22.4 \text{ L}$$

In fact, though, the molar volumes of gases are not exactly equal. As Table 3.4 shows, the experimentally measured molar volumes at STP are very close to, but not exactly equal to, 22.4 L. In other words 6.022×10^{23} molecules of any gas occupy almost the same volume, namely, 22.4 L at STP.

In general, we can state the following, to a reasonable approximation:

Equal volumes of gases at the same temperature and pressure contain equal numbers of molecules.

This statement is known as **Avogadro's Law**. It was first suggested in 1811 by the Italian scientist Amedeo Avogadro (1776–1856) long before the kinetic theory of gases had been proposed.

Table 3.4 Molar Volumes of Some Gases

GAS	MOLAR VOLUME AT STP (L mol^{-1})
Hydrogen, H_2	22.43
Neon, Ne	22.44
Oxygen, O_2	22.39
Nitrogen, N_2	22.40
Carbon dioxide, CO_2	22.26

Avogadro put forward his hypothesis in order to explain the following observation made by the French chemist Jean-Louis Gay-Lussac (1778–1850) in 1808:

When gases combine at constant temperature and pressure, the volumes of the reactants and of the gaseous products are always in the ratio of small whole numbers.

This statement is known as the **law of combining volumes**. Gay-Lussac reached this conclusion by means of careful experiments in which he obtained results, in many cases accurate to better than 1%, such as the following:

$$2 \text{ volumes hydrogen} + 1 \text{ volume oxygen} \longrightarrow 2 \text{ volumes water vapor}$$

$$1 \text{ volume hydrogen} + 1 \text{ volume chlorine} \longrightarrow 2 \text{ volumes hydrogen chloride}$$

$$2 \text{ volumes carbon monoxide} + 1 \text{ volume oxygen} \longrightarrow 2 \text{ volumes carbon dioxide}$$

The law of combining volumes can be explained on the basis of Avogadro's law and our present knowledge concerning the formulas of the reactants and the products. For the reaction between hydrogen and chloride we can write the equation

$$H_2 + Cl_2 \longrightarrow 2HCl$$

which in terms of molecules means

$$1 \text{ molecule} + 1 \text{ molecule} \longrightarrow 2 \text{ molecules}$$

or

$$n \text{ molecules} + n \text{ molecules} \qquad 2n \text{ molecules}$$

According to Avogadro's law, equal numbers of molecules occupy the same volume under the same conditions. So

$$1 \text{ volume} + 1 \text{ volume} \longrightarrow 2 \text{ volumes}$$

in agreement with Gay-Lussac's observations (see Figure 3.11).

Similarly, for the reaction between hydrogen and oxygen to give water vapor, we have

$$2H_2 + O_2 \longrightarrow 2H_2O$$

Therefore

$$2 \text{ molecules} + 1 \text{ molecule} \longrightarrow 2 \text{ molecules}$$

$$2n \text{ molecules} + n \text{ molecules} \longrightarrow 2n \text{ molecules}$$

and using Avogadro's law, we have

$$2 \text{ volumes} + 1 \text{ volume} \longrightarrow 2 \text{ volumes}$$

However, we have the benefit of knowing the formulas of all the reactants and products, and we know that the kinetic theory provides an explanation

Figure 3.11 Law of Combining Volumes and Avogadro's Law.

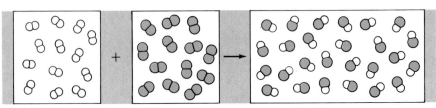

x H$_2$ molecules x Cl$_2$ molecules $2x$ HCl molecules

for Avogadro's law. Things were not so clear in 1811. Most chemists, following Dalton, thought that the elements were composed of single atoms. Moreover, they had no simple way of finding the correct formulas of compounds. They assumed the simplest formula in the absence of other evidence, for example, HO for water. Consequently Dalton wrote

$$H + Cl \longrightarrow HCl \quad \text{and} \quad H + O \longrightarrow HO$$

On the basis of these formulas, Avogadro's hypothesis leads to the following predictions:

$$H + Cl \longrightarrow HCl$$
1 volume + 1 volume $\longrightarrow$ 1 volume

and

$$H + O \longrightarrow HO$$
1 volume + 1 volume $\longrightarrow$ 1 volume

which are not in accordance with Gay-Lussac's observations. Avogadro therefore made the revolutionary proposal that some elements consist of diatomic molecules. Then it was easy to see how one volume of hydrogen could react with one volume of chlorine to give *two* volumes of hydrogen chloride, as we showed earlier.

We can also conclude that the formula of water must be H_2O. Gay-Lussac's experiments showed that:

$$\text{Hydrogen} + \text{oxygen} \longrightarrow \text{Water vapor}$$
2 volumes   1 volume   2 volumes

According to Avogadro's law,

$$2 \text{ molecules} + 1 \text{ molecule} \longrightarrow 2 \text{ molecules}$$

Then if the formulas of hydrogen and oxygen are H_2 and O_2, we can see that the formula of water must be H_2O:

$$2H_2 + O_2 \longrightarrow 2H_2O$$

However, Dalton was unable to accept that atoms of the same kind might combine to form molecules. So he rejected Avogadro's ideas, and he tried to show that Gay-Lussac's experiments were not accurate. Not until 1860, almost fifty years later, did Avogadro's countryman Stanislao Cannizzaro succeed in convincing chemists of the correctness of Avogadro's ideas. He did so by showing that Avogadro's ideas together with Gay-Lussac's law of combining volumes provided a method for determining correct formulas of compounds. The use of incorrect formulas, such as HO for water, had led to the deduction of incorrect and inconsistent atomic masses, which were a cause of much confusion in chemistry at the time. Cannizzaro was able to deduce correct formulas and a consistent and reasonably correct table of atomic masses, which did much to clear up the confusion. Cannizzaro's work paved the way for another important step forward in chemistry: the formulation of the periodic table by Mendeleev, which we describe in the next chapter.

Example 3.9 The gas propane, C_3H_8, burns in oxygen to give carbon dioxide and water. Write the balanced equation for the reaction. How many liters of carbon dioxide are formed at 25°C and 1 atm if 5.00 L of propane at 25°C and 1 atm is burned in oxygen?

Solution The balanced equation is

$$C_3H_8 + 5O_2 \longrightarrow 3CO_2 + 4H_2O$$

We see from the equation that 1 L of C_3H_8 gives 3 L of CO_2. Hence

$$V_{CO_2} = (5.00 \text{ L } C_3H_8)\left(\frac{3 \text{ L } CO_2}{1 \text{ L } C_3H_8}\right) = 15.0 \text{ L } CO_2$$

Example 3.10 How many liters of carbon dioxide are formed at 1.00 atm and 900°C if 5.00 L of propane at 10.0 atm and 25°C is burned in excess air?

Solution In this case we need to take account of the fact that the carbon dioxide is not formed at the same temperature and pressure as the propane. We begin by finding the volume that the propane would occupy at 900°C and 1.00 atm pressure, the conditions cited for the formation of carbon dioxide. We convert the temperatures given to kelvins:

$$25°C = 273 + 25 = 298 \text{ K} \qquad (T_1)$$
$$900°C = 273 + 900 = 1173 \text{ K} \qquad (T_2)$$

Now we use the combined gas law to find the volume of the propane:

$$V_2 = \frac{V_1 P_1 T_2}{P_2 T_1}$$

$$V(C_3H_8, 900°C, 1 \text{ atm}) = 5.00 \text{ L} \times \frac{10.00 \text{ atm}}{1.00 \text{ atm}} \times \frac{1173 \text{ K}}{298 \text{ K}} = 197 \text{ L}$$

The equation for the reaction (Example 3.9) indicates that 1 L of C_3H_8 gives 3 L of CO_2. Hence

$$V_{CO_2} = (197 \text{ L } C_3H_8)\left(\frac{3 \text{ L } CO_2}{1 \text{ L } C_3H_8}\right) = 591 \text{ L } CO_2$$

Example 3.11 Suppose 5.00 L of oxygen at 150°C and 1 atm pressure is mixed with 5.00 L of hydrogen at 150°C and 1 atm pressure and an electric spark is passed through the mixture. What volume of steam is formed at 150°C and 1 atm pressure? What is the total volume of the gaseous mixture after the reaction?

Solution The equation for the reaction is

$$2H_2 + O_2 \longrightarrow 2H_2O$$
$$\text{2 vol} \quad \text{1 vol} \qquad \text{2 vol}$$

The important thing to recognize is that this problem is a limiting reactant problem—we have equal volumes of oxygen and hydrogen, but two volumes H_2 react with only one volume O_2. Thus 5.00 L of hydrogen reacts with only 2.50 L of oxygen. Thus H_2 is the limiting reactant, and the equation shows that 5.00 L of hydrogen gives 5.00 L of steam.

We see that 2.50 L of oxygen remains unreacted. So the total volume of the gaseous mixture at the end of the reaction is 5.00 L of steam + 2.50 L of oxygen, or 7.50 L.

Dalton's Law of Partial Pressure

In a mixture of gases that do not react with each other, each molecule moves independently, just as it would in the absence of molecules of other kinds. Each gas distributes itself uniformly throughout the container. The molecules strike the walls with the same frequency and force—therefore exert the same pressure—as they do when no other gas is present. In other words, the pressure exerted by any one gas in a mixture is the same as it would be if the gas occupied the container by itself. This pressure is called the **partial pressure** of the gas. Furthermore:

In a mixture of gases the total pressure exerted is the sum of the pressures that each gas would exert if it were present alone under the same conditions.

Thus the total pressure of a gas mixture is given by

$$P_{total} = p_1 + p_2 + p_3 + \cdots$$

where p_1, p_2, and so on represent the partial pressures of each gas in the mixture. This law was formulated by John Dalton in 1803 and is known as **Dalton's law of partial pressures**.

We can apply the ideal gas equation to any particular gas in the mixture. Thus for gas 2, for example, $p_2 V = n_2 RT$, where V is the *total* volume of the gas mixture. And in general, for gas i, we have $p_i V = n_i RT$. We see that the partial pressure of a gas is proportional to the number of molecules of the gas in the mixture, or $p_i \propto n_i$. For example, in the atmosphere 78% of the molecules are nitrogen, 21% are oxygen, and 1% are argon. If the total pressure is 1.00 atm, the nitrogen has a partial pressure of 0.78 atm, the oxygen has a partial pressure of 0.21 atm, and the argon has a partial pressure of 0.01 atm.

Example 3.12 A 2.00 L flask contains 3.00 g of carbon dioxide, CO_2, and 0.10 g of helium, He, and it has a temperature of 17°C. What are the partial pressures of CO_2 and He? What is the total pressure exerted by the gas mixture?

Solution According to Dalton's law, we can treat each gas as if it alone occupied the container. First, we apply the ideal gas law to CO_2 to find its partial pressure. To do so, we need to find n, the number of moles of CO_2:

$$n_{CO_2} = (3.00 \text{ g } CO_2)\left(\frac{1 \text{ mol } CO_2}{44.01 \text{ g } CO_2}\right) = 0.0682 \text{ mol } CO_2$$

Now we use the ideal gas law:

$$p_{CO_2} = \frac{n_{CO_2} RT}{V} = \frac{(0.0682 \text{ mol})(0.0821 \text{ atm L mol}^{-1} \text{ K}^{-1})(290 \text{ K})}{2.00 \text{ L}}$$

$$= 0.812 \text{ atm}$$

Similarly,

$$n_{He} = (0.10 \text{ g He})\left(\frac{1 \text{ mol He}}{4.003 \text{ g He}}\right) = 0.025 \text{ mol He}$$

$$p_{He} = \frac{(0.025 \text{ mol})(0.0821 \text{ atm L mol}^{-1} \text{ K}^{-1})(290 \text{ K})}{2.00 \text{ L}} = 0.30 \text{ atm}$$

Thus the partial pressures of CO_2 and He are 0.81 atm and 0.30 atm. The total pressure is given by

$$P_{total} = p_{CO_2} + p_{He} = 0.81 \text{ atm} + 0.30 \text{ atm} = 1.11 \text{ atm}$$

Stoichiometry of Reactions Involving Gases

The ideal gas law enables us to calculate the volumes of gaseous reactants or products in reactions, as shown in Examples 3.13 and 3.14 that follow.

Example 3.13 What volume of oxygen, O_2, at STP can be obtained by heating 10.00 g of potassium chlorate, $KClO_3$, in the presence of manganese dioxide, MnO_2, as a catalyst?

Solution The equation for the reaction is

$$2KClO_3 \longrightarrow 2KCl + 3O_2$$

$$2 \text{ mol} \qquad\qquad\qquad 3 \text{ mol}$$

Because 1 mol of an ideal gas occupies 22.4 L at STP, we can also write

$$2KClO_3 \longrightarrow 2KCl + \quad 3O_2$$

$$2 \text{ mol} \qquad\qquad\qquad 3 \times 22.4 \text{ L}$$

Therefore

$$V = (10.00 \text{ g } KClO_3)\left(\frac{1 \text{ mol } KClO_3}{122.5 \text{ g } KClO_3}\right)\left(\frac{3 \times 22.4 \text{ L } O_2}{2 \text{ mol } KClO_3}\right) = 2.74 \text{ L } O_2$$

Example 3.14 In the production of water gas, which is used as a fuel, coke is heated in steam. What volume of water gas at a temperature of 20°C and 100 atm pressure can be obtained by heating 1.00 tonne (10^6 g) of coke in steam at 1000°C?

Solution The equation for the reaction is

$$C + H_2O \longrightarrow CO + H_2$$

Thus 1 mol of C gives 1 mol of CO and 1 mol of H_2, that is, a total of 2 mol of gas (water gas). Hence 1.00 tonne C gives

$$10^6 \text{ gC}\left(\frac{1 \text{ mol C}}{12.01 \text{ g C}}\right)\left(\frac{2 \text{ mol gas}}{1 \text{ mol C}}\right) = 16.7 \times 10^4 \text{ mol gas}$$

Now we can use the ideal gas equation to calculate the volume of the gas:

$$V = \frac{nRT}{P} = \frac{(16.7 \times 10^4 \text{ mol})(0.0821 \text{ atm L mol}^{-1} \text{ K}^{-1})(293 \text{ K})}{100 \text{ atm}}$$

$$= 4.02 \times 10^4 \text{ L water gas}$$

3.8 DIFFUSION AND EFFUSION

If we place a drop of bromine in a bottle containing only air, the bromine evaporates, forming a brown gas (vapor) that is seen initially at the bottom of the bottle. Gradually, the bromine molecules become evenly mixed with the air, and the color of the gas in the bottle becomes a uniform light brown. Similarly, we smell substances such as perfume or coffee because molecules of the perfume or coffee mix with other molecules in the air and eventually reach our noses. This mixing of one gas with another is called **diffusion**, and it illustrates that the molecules of a gas are in motion, as assumed by kinetic theory.

A closely related phenomenon is called **effusion**. Effusion is the process by which a gas escapes from a container through a very small opening. Like diffusion, effusion results from the random motions of the molecules of a gas. For example, a toy balloon gradually deflates because the apparently solid rubber skin of the balloon actually contains very small holes through which the gas in the balloon escapes. If a molecule is traveling toward a hole, and if it makes no further collisions, it passes through and escapes from the container. Because the pressure inside the balloon is greater than the pressure of the atmosphere outside, more molecules pass out through the hole than pass into the balloon from the air outside, and the pressure in the balloon slowly decreases.

Molecular Speeds

How fast do gases diffuse or effuse? If we open a bottle of ammonia, several seconds pass before the smell reaches our noses even if we are only about a meter away.

The rate of diffusion of a gas will clearly depend on how fast the molecules of the gas are moving. The kinetic theory tells us that if the temperature is constant, the average energy and the average speed of the molecules are also constant. But since the molecules of a gas are moving randomly and colliding with each other, they cannot all have the same speed or the same kinetic energy. When one molecule collides with another, generally one will be speeded up and the other slowed down. Thus the speeds and the energies of individual molecules change constantly, and they vary over a wide range.

The distribution of energies and speeds can be obtained from experimental measurements, as shown in Figure 3.12, or from a detailed treatment of kinetic theory. The distribution of molecular speeds in oxygen at 25°C and at 1000°C is shown in Figure 3.13. Notice that many molecules have speeds that are close to the average value, but some have very low speeds and a few have very high speeds.

The average kinetic energy ($\frac{1}{2}mv^2$) of the molecules of a gas depends only on the temperature; the average kinetic energy of all gases is the same at the same temperature. If two gases at the same temperature had *different* average kinetic energies, then when we mixed the gases, collisions between the molecules of the two gases would increase the average kinetic energy of one gas and decrease the average kinetic energy of the other gas. In other words, energy (heat) would flow from one gas to the other. However, we know that no heat flows if two gases are at the same temperature. Hence two gases at the same temperature must have the same average kinetic energy, or

$$\frac{1}{2}m_1v_1{}^2 = \frac{1}{2}m_2v_2{}^2$$

Therefore

$$\frac{v_1}{v_2} = \sqrt{\frac{m_2}{m_1}}$$

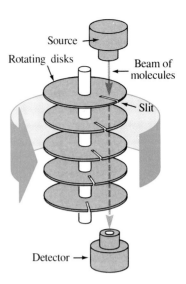

Figure 3.12 Determination of Molecular Speeds.
Molecular speeds can be measured by passing a beam of molecules through a velocity selector. This device consists of a series of disks mounted on a common axis. Each disk has a slit, and each slit is displaced by the same angle from the preceding slit. When the disks are rotated, the slits arrive successively in the same angular position. If a molecule moves the distance between the disks in the same time that it takes the next disk to arrive in the same angular position, the molecule will pass through each slit and arrive at the detector. Molecules moving at all other speeds will not arrive at a disk simultaneously with a slit and will therefore not pass through. From the speed of rotation of the disks, the speed of the molecules arriving at the detector can be calculated. Then if the speed of rotation of the disks is varied, the number of molecules arriving at the detector for each different speed can be found. Curves such as those in Figures 3.13 and 3.14 can then be plotted.

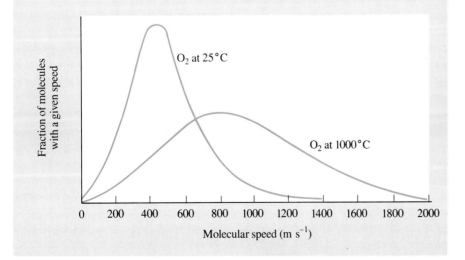

Figure 3.13 Molecular Speeds of Oxygen. Molecules have a wide range of speeds, but a large fraction have speeds close to the average. The distribution of speeds changes if the temperature changes. At 1000°C the average speed of oxygen molecules is greater than the average speed at 25°C, and the speeds have a wider distribution.

The average speed of the molecules of a gas is inversely proportional to the square root of the molecular mass. Molecules with a low molecular mass have higher average speeds than those with a high molecular mass, as Figure 3.14 illustrates.

Graham's Law

We can reasonably assume that the rate at which a gas effuses from a small hole is proportional to the average speed of its molecules. Hence if r_1 is the rate at which gas 1 effuses from a small hole and if r_2 is the rate at which gas 2 effuses from the same hole, we may write

$$\frac{r_1}{r_2} = \frac{v_1}{v_2} = \sqrt{\frac{m_2}{m_1}} = \sqrt{\frac{M_2}{M_1}}$$

where m_1 and m_2 are the molecular masses and M_1 and M_2 are the molar masses. The expression

$$\frac{r_1}{r_2} = \sqrt{\frac{M_2}{M_1}}$$

is **Graham's law**. This law states the following:

> **The rate of effusion of a gas is inversely proportional to the square root of its molar mass.**

Hydrogen therefore effuses more rapidly than other gases (Experiment 3.4).

We similarly expect that the rate of diffusion of a gas into another gas will depend on the speed of its molecules and that Graham's law therefore also applies to diffusion.

> **The rate of diffusion of a gas is inversely proportional to the square root of its molar mass.**

The dependence of the rate of diffusion on molar mass is demonstrated in Experiment 3.5. The laws of effusion and diffusion were first deduced by the Scottish chemist Thomas Graham (1805-1869) from the results of his experiments.

Example 3.15

(a) Compare the rates of diffusion of NH_3 and HCl.

(b) If Experiment 3.5 is carried out in a tube 1.00 m long, at what position in the tube do you expect to see the formation of a white cloud of NH_4Cl?

Figure 3.14 Distribution of Molecular Speeds for Some Gases at 25°C. Light gases such as hydrogen and helium have higher average speeds and a wider distribution of speeds than heavier gases such as oxygen and nitrogen.

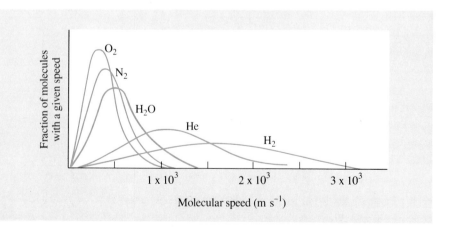

Rates of Effusion

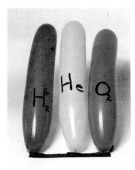
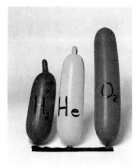

Three balloons of the same size are filled with hydrogen, helium, and oxygen.

After a few hours the balloons have decreased in size because the gases effuse through tiny holes in the porous rubber skin of the balloon. The hydrogen balloon is smaller than the helium balloon which is in turn smaller than the oxygen balloon. Thus, the rates of effusion are $H_2 > He > O_2$.

A large beaker containing hydrogen is inverted over a porous clay cylinder. The hydrogen molecules effuse through the holes in the porous cylinder faster than the oxygen and nitrogen molecules inside effuse out, creating an excess pressure inside the cylinder and the flask. This excess pressure forces the colored water out of the flask through a fine jet to give the fountain effect shown.

Solution

(a) The molar mass of NH_3 is 17.03 g mol^{-1}, and the molar mass of HCl is 36.46 g mol^{-1}. From Graham's law we have ("diff" in the subscript stands for "diffusion"):

$$\frac{r_{diff}\ NH_3}{r_{diff}\ HCl} = \sqrt{\frac{M_{HCl}}{M_{NH_3}}} = \sqrt{\frac{36.46}{17.03}} = 1.46$$

(b) The NH_3 diffuses 1.46 times more rapidly than HCl. Therefore the distance moved by NH_3 down the tube is 1.46 times the distance moved by HCl in the same time:

$$\frac{d_{NH_3}}{d_{HCl}} = \frac{r_{diff}\ NH_3}{r_{diff}\ HCl} = 1.46$$

$$d_{NH_3} = 1.46 d_{HCl}$$

When the NH_4Cl cloud forms

$$d_{NH_3} + d_{HCl} = 1.00\ m$$

So that

$$1.46 d_{HCl} + d_{HCl} = 1.00\ m$$

or

$$d_{HCl} = \frac{1.00\ m}{2.46} = 0.41\ m = 41\ cm$$

Thus the NH_4Cl cloud forms at 41 cm from the HCl end of the tube, or at 59 cm from the NH_3 end.

3.8 DIFFUSION AND EFFUSION

Graham's Law of Diffusion

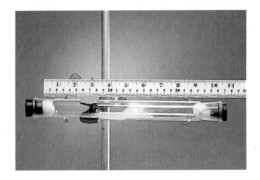

Swabs soaked in concentrated aqueous solutions of HCl and NH_3 are placed simultaneously in opposite ends of a long glass tube. Gaseous HCl and NH_3 diffuse down the tube toward each other. When they meet, a white cloud of solid ammonium chloride, NH_4Cl, is produced. Because NH_3 has a lower molecular mass than HCl it diffuses faster and the white cloud forms in the position shown.

Graham's law provides us with another method for finding the molar mass of a gas. By comparing the rate of diffusion or effusion of a gas of unknown molar mass with that for a gas of known molar mass, we can calculate the unknown molar mass, as shown in the next example.

Example 3.16 A gaseous compound known to contain only carbon, oxygen, and sulfur was found to effuse from a porous container at a rate of 100 mL in 5.00 min. Molecular oxygen effused from the same container under the same conditions at a rate of 137 mL in 5.00 min. What is the molar mass of the gas? What is its molecular formula?

Solution The molar mass is found as follows ("eff" in the subscript stands for "effusion"):

$$\frac{r_{\text{eff}} \, O_2}{r_{\text{eff}} \, \text{gas}} = \frac{137}{100} = \sqrt{\frac{M_{\text{gas}}}{M_{O_2}}} = \sqrt{\frac{M_{\text{gas}}}{32.00 \text{ g mol}^{-1}}}$$

$$M_{\text{gas}} = \left(\frac{137}{100}\right)^2 (32.00 \text{ g mol}^{-1}) = 60.1 \text{ g mol}^{-1}$$

If the gas contains C, O, and S, it must have the molecular formula COS, for which $M = 60.07 \text{ g mol}^{-1}$.

If we look at the horizontal scale of Figure 3.14, we see that the actual speeds of molecules are very high. The average speed of oxygen molecules at 25°C is 450 m s^{-1}, or 1620 km h^{-1}. Yet the rates of diffusion demonstrated by Experiment 3.5 are quite slow because the molecules of a gas are constantly colliding with each other. At 1 atm pressure a molecule makes about 10^{10} collisions a second with other molecules. Any one molecule therefore has a very tortuous path, since its direction is changed in every collision. The average distance traveled by a molecule between collisions is known as the *mean free path*. The mean free path of oxygen molecules at 0°C and 1 atm is only 60 nm, or 6×10^{-9} m. Thus although an oxygen molecule is moving at high speed, the large number of collisions it makes means that it is constantly changing direction, so its resultant motion in any particular direction is very slow. The rate of diffusion increases with decreasing pressure, because a molecule makes fewer collisions at lower pressures, and therefore diffuses more rapidly, than at higher pressures.

IMPORTANT TERMS

Avogadro's law states that equal volumes of gases at the same temperature and pressure contain equal numbers of molecules.

A **barometer** is an instrument that measures the pressure of the atmosphere.

Boyle's law states that the volume of a given amount of gas at constant temperature is inversely proportional to the pressure.

Charles's law states that the volume of a given amount of gas at constant pressure is directly proportional to the absolute (Kelvin) temperature.

Combustion is an exothermic reaction that is self-sustaining.

Dalton's law of partial pressures states that in a mixture of gases the total pressure is the sum of the partial pressures of the gases present.

Diffusion is the process by which one gas mixes with another.

Effusion is the process by which a gas escapes from a container through a very small hole.

An **endothermic reaction** is a reaction that absorbs heat.

An **exothermic reaction** is a reaction that liberates heat.

The **gas constant** is the constant R in the ideal gas equation, $PV = nRT$. It has the value 0.0821 atm L mol^{-1} K^{-1} or 8.31 kPa dm^3 mol^{-1} K^{-1}.

Graham's law states that the rate of effusion (or diffusion) of a gas is inversely proportional to the square root of its molecular mass (or molar mass).

A **hydride** is a compound of a metallic element and hydrogen.

An **ideal gas** is one whose behavior can be predicted by the ideal gas equation.

The **ideal gas equation** relates the pressure, volume, temperature, and number of moles of a gas through the expression $PV = nRT$, where R is the *gas constant*.

Intermolecular forces are the forces of attraction that exist between all molecules.

The **kinetic molecular theory** assumes that the atoms and molecules of which all substances are composed are in constant motion at any temperature above 0 K.

The **law of combining volumes** states that when gases combine at constant temperature and pressure, the volumes of the reactants and of the gaseous products are always in the ratio of small whole numbers.

Nitrogen fixation is a process by which atmospheric nitrogen is converted to useful compounds such as ammonia.

An **oxidation reaction** is a reaction in which an element or compound combines with oxygen. (A more general definition is given in Chapter 5).

An **oxide** is a compound of an element and oxygen.

The **partial pressure** of any one gas in a mixture of gases is the pressure that the gas would exert if it occupied the container by itself. The total pressure exerted by a gas mixture is the sum of the partial pressures of all the gases in the mixture.

A **reduction reaction** is a reaction in which oxygen is removed partially or completely from a compound. (See Chapter 5 for a more general definition).

Standard temperature and pressure (STP) are 273.15 K or 0°C and 101.33 kPa or 1 atm or 760 mm Hg.

PROBLEMS

The Atmosphere: Oxygen, Nitrogen, and Hydrogen

1. What species are present in the different layers of the heterosphere? Explain the origin of these species and why they have separated into different layers.

2. Describe the reactions by which ozone is formed in the stratosphere. What is the biological significance at the earth's surface of the ozone layer in the atmosphere? (See Figure 3.1.)

3. (a) What are the two most abundant elements in the earth's crust?

(b) What is the most abundant metal in the earth's crust?

(c) What is the most abundant element in the universe?

4. Write a balanced equation to describe the reaction that occurs when

(a) Sulfur burns in air.

(b) Magnesium burns in air.

(c) Methane burns in air.

5. Write a balanced equation to describe each of the following:

(a) The reduction of Fe_2O_3 to Fe with CO.

(b) The reduction of Fe_2O_3 to Fe with H_2.

(c) The reduction of CuO to Cu with CO.

(d) The reduction of H_2O with Mg.

6. Write a balanced equation for the reaction of steam at high temperature with each of the following:

(a) Iron   (b) Magnesium   (c) Methane

7. Write a balanced equation to describe the reaction of nitrogen at high temperature with each of the following:

(a) Hydrogen   (b) Oxygen   (c) Magnesium

8. Explain what is meant by a *catalyst*. Give two examples of reactions in which a catalyst is used.

9. Explain why the CO in a mixture of CO and H_2 must be converted to CO_2 before it can be easily separated from the H_2.

10. (a) Describe a simple test that you could use to confirm that a test tube full of a colorless gas contained hydrogen.

(b) Describe another test that you could use to confirm that another tube contained oxygen.

(c) Describe what is observed in a and b, and give equations for any reactions that occur.

11. Under the action of lightning some of the nitrogen and oxygen in the air react to form nitrogen monoxide, NO, which subsequently oxidizes further and reacts with water to form nitric acid, HNO_3, which forms a dilute solution in rainwater.

(a) Balance the equation

$$N_2 + O_2 + H_2O \longrightarrow HNO_3.$$

(b) Deduce the maximum mass of nitric acid that can be formed from 0.76 g of nitrogen, 0.24 g of oxygen (the constituents of 1 g of air), and excess water.

12. Write balanced equations for three reactions by which hydrogen can be produced from water.

Boyle's Law

13. A cylinder in a gasoline engine initially has a volume of 0.50 L and contains gases that exert a pressure of 1.0 atm. What is the pressure of the gases when the volume of the cylinder is reduced to 0.20 L and there is no change in the amount or temperature of the gaseous mixture?

14. A meteorological balloon had a volume of 150 L when filled with hydrogen at a pressure of 1.00 atm. To what volume will the balloon expand when it rises in the atmosphere to a height of 2500 m (8000 ft), where the atmospheric pressure is only 0.75 atm, assuming that the temperature remains constant?

15. A sample of hydrogen gas has a pressure of 0.98 atm when it is confined in a bulb with a volume of 2.00 L. A tap connects the bulb to another bulb that has a volume of 5.00 L and has been completely evacuated. What will the pressure be in the two bulbs after the tap is opened?

Charles's Law

16. A balloon filled with helium has a volume of 1.60 L at a pressure of 1.00 atm at 25.0°C. What will the volume of the balloon be if it is cooled in liquid nitrogen that has a temperature of −196°C while the pressure remains constant at 1.00 atm?

17. A sample of nitrogen has a volume of 400 mL at 100°C. At what temperature will it have a volume of 200 mL if the pressure does not change?

Combined Gas Law

18. The temperature of a closed vessel containing air at a pressure of 1.02 atm is raised from 20°C to 200°C. What will the pressure be inside the vessel if the volume of the vessel increases by 10% from 20° to 200°C?

19. A sample of helium occupies a volume of 10.0 L at 25°C and a pressure of 850 mm Hg. What volume will it occupy at STP?

20. What will be the volume of the meteorological balloon described in Problem 14 at a height of 2500 m if the temperature at that altitude is −10°C and if the balloon was filled at a temperature of 29°C?

21. A sample of oxygen occupies a volume of 0.840 L at a pressure of 0.450 atm at 37°C. What will the pressure be if it is cooled to −13°C and compressed to a volume of 0.150 L?

22. A metal cylinder, which has a safety valve that opens at a pressure of 100 atm, is to be filled with nitrogen gas and heated to 300°C. What is the maximum pressure to which it can be filled at 25°C?

The Ideal Gas Law and Molar Volume

23. A 1.00-L evacuated flask is to be filled with CO_2 gas at 300°C and a pressure of 500 mm Hg by placing a piece of dry ice [$CO_2(s)$] in the flask. What mass of dry ice should be used?

24. A NASA scientist proposed that barium peroxide, BaO_2, be used on a space capsule to supply emergency oxygen. The BaO_2 decomposes on heating according to the equation

$$2BaO_2(s) \longrightarrow 2BaO(s) + O_2(g)$$

(a) What mass of BaO_2 would be needed to supply enough oxygen to fill a 10,000-L space capsule to 1.00 atm of pressure at 25°C?

(b) How long would this oxygen last if the crew, operating at 20°C and 1 atm, used $1.00\,L\,s^{-1}$ in respiration?

25. The atmosphere of Venus is mostly carbon dioxide. At the surface the temperature is about 800°C and the pressure is about 75 atm. In the event that an inhabitant (if any!) of Venus took these values as STP for Venus, what value would the Venusian find for the molar volume of an ideal gas?

26. AT STP the mass of 1.00 L of gas is 1.89 g. What mass of the gas will occupy 1.00 L at 200°C and a pressure of 1.25 atm?

27. What mass of gaseous hydrogen chloride, HCl, is needed to exert a pressure of 0.240 atm in a container of volume 250 mL at 37°C?

28. As a publicity stunt a water bed retailer in New Brunswick, Canada, filled a water bed bag with helium and floated it above his store (unfortunately, he did not tie it down well enough, and it escaped its moorings and has not been seen since). Calculate the mass of helium required to fill the water bed bag at a pressure of 1.03 atm at 23°C if its dimensions were 2.00 m × 1.50 m × 0.20 m.

29. The density of a gas at STP is 1.62 g L^{-1}. What is its density at 302°C and a pressure of 0.950 atm?

30. An anesthetic gas used to relax patients contains oxygen mixed with dinitrogen monoxide, N_2O. The mixture has a density of 1.482 g L^{-1} at 25°C and 0.980 atm. What is the percentage by mass of N_2O in the gas?

31. The molar volume at STP of an ideal gas is 22.4 L. Calculate the molar volume for a pressure of 3.40 atm and a temperature of $-70°C$.

Determination of Molar Mass

32. A sample of a noble gas with a mass of 0.20 g exerted a pressure of 0.48 atm in a container of volume 0.26 L at 27°C. Is the gas helium, neon, argon, krypton, or xenon?

33. Oxygen can exist not only as a diatomic molecule, O_2, but also as ozone. What is the molecular formula for ozone, given that at the same pressure and temperature it has a density 1.50 times that of O_2?

34. A gas at a pressure of 740 mm Hg at 20°C and occupying a volume of 1.00 L has a mass of 1.134 g. What is its molar mass? If the empirical formula of the gas is CH_2, what is its molecular formula?

35. A sample of a gas has a volume of 2.00 L and a mass of 2.57 g at 37°C and a pressure of 785 torr. What is the molar mass of the gas?

36. A gas has a density of 1.275 g L^{-1} at 18°C and 750 mm Hg pressure.

(a) What is the molecular mass of the gas?

(b) How many molecules are in 0.010 mL of the gas under these conditions?

37. A gas has a density of 1.402 g L^{-1} at 20°C and 740 mm Hg pressure. What is the molar mass of the gas?

38. Several grams of a volatile liquid were heated at 100°C in a flask of volume 350 mL until all the air had been driven out. It was then cooled to room temperature, allowing air to reenter the flask. The flask then had a mass of 0.750 g greater than the mass of the empty (full-of-air) flask. If the pressure is 0.980 atm, what is the molar mass of the liquid?

39. A volatile liquid was found to have the composition 62.04% carbon, 10.41% hydrogen, and 27.55% oxygen, by mass. At 100°C and 1.00 atm pressure 440 mL of the compound had a mass of 1.673 g.

(a) What is the molar mass of the compound?

(b) What is its molecular formula?

Partial Pressures

40. One liter each of oxygen, nitrogen, and hydrogen gases, all originally at 1.00 atm pressure, are forced into a single evacuated 2.00-L container. What is the resulting pressure if the temperature is unchanged? What is the partial pressure of each gas in the mixture?

41. A mixture of 0.20 g of He and 0.20 g of H_2 is contained in a vessel of volume 225 mL at 27°C.

(a) What is the partial pressure of each gas?

(b) What is the total pressure exerted by the mixture?

42. On a certain day in Los Angeles the NO concentration in the atmosphere was 0.94 ppm. The atmospheric pressure was 750 mm Hg, and the temperature was 30°C.

(a) What was the partial pressure of the NO?

(b) What was the number of NO molecules per cubic meter?

Graham's Law

43. Which effuses faster, molecular nitrogen, N_2, or molecular oxygen, O_2? What is the ratio of their rates of effusion?

44. Calculate the ratio of the rates of diffusion of hydrogen gas, H_2, and heavy hydrogen, D_2.

45. The original separation of ^{235}U (required for atomic bombs) from ^{238}U was achieved by exploiting the slight difference in effusion rates of $^{235}UF_6$ and $^{238}UF_6$. Calculate the ratio of the rates of effusion of $^{235}UF_6$ and $^{238}UF_6$, assuming that the mass of each isotope is exactly equal to mass number.

46. The average speed of He atoms is 0.707 mile per second at 25°C. What will be the average speed of N_2 molecules at the same temperature?

47. Suppose Experiment 3.5 is carried out with NH_3 and HBr, which react to give a white cloud of NH_4Br. Predict at what point in a tube of length 1.00 m the white cloud will be formed.

Stoichiometry of Gas Reactions

48. An important reaction in the production of nitrogen fertilizers is the oxidation of ammonia, NH_3, to nitrogen monoxide, NO:

$$4NH_3(g) + 5O_2(g) \xrightarrow{500°C} 4NO(g) + 6H_2O(g)$$

How many liters of O_2, measured at 25°C and 0.896 atm, must be used to produce 500 L of NO at 500°C and 740 mm Hg?

49. How many liters of oxygen, O_2, are required to burn 1.00 L of (a) methane, and (b) hexane, C_6H_{14}, to carbon dioxide and water if all gases are initially at the same temperature and pressure?

50. Determine the amount, in liters, of CO_2 at STP that can be obtained by heating 1.00 kg of $CaCO_3$, which decomposes according to the equation

$$CaCO_3(s) \longrightarrow CaO(s) + CO_2(g)$$

51. How many liters of oxygen at 25°C and 740 mm Hg pressure are needed to burn completely 5.00 g of magnesium?

52. Calcium hydride, $CaH_2(s)$, reacts with water to give hydrogen, $H_2(g)$, and calcium hydroxide, $Ca(OH)_2(s)$.

    (a) Write a balanced equation for this reaction.

    (b) How many grams of CaH_2 are needed to prepare 10.0 L of $H_2(g)$ at STP?

53. What mass of $KClO_3(s)$ would have to be decomposed by heating to prepare 5.00 L of $O_2(g)$ at STP?

54. What volume of hydrogen at STP would be required to

    (a) reduce 1 kg of zinc oxide, ZnO, to zinc.

    (b) form 1 kg of water on burning in oxygen.

    (c) form 1 kg of lithium hydride, LiH.

Miscellaneous

55. The anesthetic cyclopropane is a gas containing only carbon and hydrogen. Suppose 0.55 L of cyclopropane at 120°C and 0.90 atm reacts with O_2 to give 1.65 L of CO_2 and 1.65 L of $H_2O(g)$ at the same temperature and pressure. What is the molecular formula of cyclopropane?

56. Calculate the average volume occupied by each molecule in an ideal gas at 27°C and 1 atm pressure. Find the actual volume of a molecule, and compare this volume with the volume effectively occupied by the molecule in the gas. Assume the molecules to be spherical and to have a radius of 100 pm.

57. Calculate the factor by which water expands when converted to vapor at 100°C at 1 atm pressure, given that the density of liquid water at 100°C is 0.96 g cm^{-3}.

58. A 50.0-mL bulb has a mass of 67.6259 g when evacuated and 67.8883 g when filled with xenon gas at 25°C and 1.00 atm pressure. What is the density of xenon under these conditions? What would the mass of the same bulb be after it is filled with a mixture of 35 volume percent oxygen (density = 1.30 g L^{-1}) and 65 volume percent xenon at the same temperature and pressure?

CHAPTER 4

THE PERIODIC TABLE AND CHEMICAL BONDS

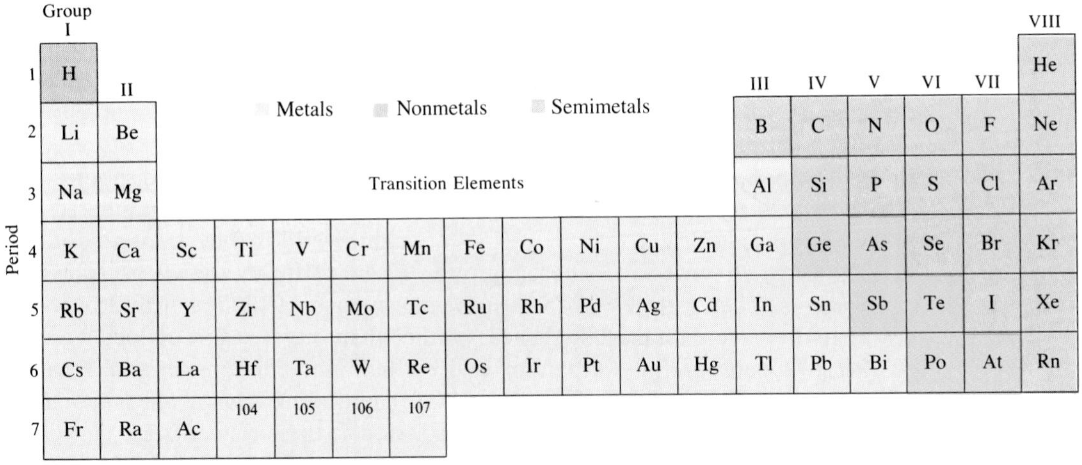

Group I	II											III	IV	V	VI	VII	VIII
																	He
H				Metals	Nonmetals	Semimetals											
Li	Be											B	C	N	O	F	Ne
Na	Mg				Transition Elements							Al	Si	P	S	Cl	Ar
K	Ca	Sc	Ti	V	Cr	Mn	Fe	Co	Ni	Cu	Zn	Ga	Ge	As	Se	Br	Kr
Rb	Sr	Y	Zr	Nb	Mo	Tc	Ru	Rh	Pd	Ag	Cd	In	Sn	Sb	Te	I	Xe
Cs	Ba	La	Hf	Ta	W	Re	Os	Ir	Pt	Au	Hg	Tl	Pb	Bi	Po	At	Rn
Fr	Ra	Ac	104	105	106	107											

Period (vertical label, rows 1–7)

The great variety of behavior of the elements and their compounds was as bewildering to early chemists as it must sometimes seem to students at the beginning of a course in chemistry. We have seen that oxygen is a very reactive element that combines readily with almost all the other elements. In contrast, nitrogen is a rather unreactive element that combines directly with far fewer elements than oxygen does. Another difference between elements that became apparent as soon as chemists began to determine the composition of compounds is that the elements have different combining powers. Nitrogen combines with three hydrogen atoms to form ammonia, NH_3; oxygen combines with two hydrogen atoms to form water, H_2O; fluorine combines with only one hydrogen atom to form hydrogen fluoride, HF; and neon does not form compounds with hydrogen—or, indeed, any other element.

These differences in reactivity and in combining power must be related in some way to the structures of the atoms, which in turn determine how the atoms of different elements are held together in a compound. We take up these important matters in this chapter.

As they increased their knowledge of both elements and compounds, nineteenth-century chemists began to recognize that certain elements, for example, sodium and potassium, have very similar properties, while other elements, such as sodium and chlorine, have very different properties. Attempts to classify the elements in terms of the similarities and differences in their properties culminated in the formulation of the **periodic table** by Dmitri Mendeleev (Box 4.1) in 1869.

Mendeleev listed the elements in order of their atomic masses and found that elements with similar properties recurred in a periodic manner in this list. Because of this regularity in properties, he was able to arrange the elements, listed in order of increasing atomic mass, in the form of a table in which similar elements fell in the same column. A modern form of the table is given in Figure 4.1. In this periodic table the elements are arranged in order of atomic number rather than atomic mass. The order of atomic mass is identical with the order of atomic number, except in a very few cases. The atomic number is the more fundamental property since it determines the number of protons and therefore the number of electrons in an atom, which in turn determines its chemical behavior. The table given in Figure 4.1 does not differ substantially from that proposed by Mendeleev, although his table contained only the 60 elements known at that time.

The formulation of the periodic table marked the beginning of a new era in chemistry. It led to a much greater understanding of the properties of the elements, and it remains today a most important working tool for the chemist. Learning the general form of the periodic table and the positions of at least the first 20 elements is essential for understanding chemistry.

The periodic table has eight principal vertical columns, called **groups**, and seven main horizontal rows, known as **periods**. The groups are numbered I to VIII from left to right, and the periods are numbered 1 to 7 from top to bottom. The elements that fall between groups II and III are known as **transition elements**. There are ten of these elements in each of periods 4 and 5, which therefore have a total of 18 elements rather than only 8, as in periods 2 and 3. Period 6 between La (lanthanum) and Hf (hafnium) contains an additional 14 elements, making a total of 32. These 14 elements are known as the *lanthanides*. There is a similar group of elements in period 7; these elements are called the *actinides*. To give the table a convenient size and shape, the lanthanides and actinides are usually set off separately at the bottom of the table.

Metals and Nonmetals

A diagonal line that runs across the periodic table from top left to bottom right indicates one of the broadest classifications of the elements: the division into metals and nonmetals. Metals have the following characteristic physical properties:

- They conduct heat very well.
- They have high electrical conductivities that increase with decreasing temperature.
- They have a high reflectivity and a shiny metallic luster.
- They are malleable and ductile; that is, they can be beaten out into sheets or foil and pulled out into thin wires without breaking.

Box 4.1

Mendeleev was born in Tobolsk, Siberia, and was the youngest of a family of seventeen. After his father died, his mother, determined that Dmitri should have the best possible education, moved the family to St. Petersburg (now Leningrad). In 1856 Mendeleev obtained a master's degree in chemistry and then taught at the University of St. Petersburg. In 1867 he was appointed professor of inorganic chemistry.

While giving his course of lectures, he felt the need for a new textbook, and he began writing what was to become a famous and widely used textbook, *Principles of Chemistry*. He realized that the order and system that was then becoming apparent in organic chemistry was lacking in inorganic chemistry, and in attempting to remedy this situation, he was led to formulate the periodic table.

To consider that the properties of the elements were in some way related to their atomic masses was an imaginative idea, since the structures of atoms were unknown at the time. Moreover, to bring certain elements into the correct group from the point of view of their chemical properties, he was bold enough to reverse the order of some pairs of elements and to predict that their atomic masses were incorrect. Some of these predictions were correct, but others were not, because we now know that the fundamental basis of the periodic table is atomic number rather than atomic mass. Also, to keep elements in the correct groups according to their chemical prop-

erties, he left gaps in his table, making the prediction that they would be filled by elements that were still undiscovered at that time. From the trends that he observed among the properties of related elements, he predicted the properties of these undiscovered elements.

At first the periodic table attracted little attention. Then, when his predictions concerning undiscovered elements were fulfilled in considerable detail by the successive discoveries of gallium (1874), scandium (1879), and germanium (1885), chemists began to realize that in the periodic table they had a tool of the greatest value. From that time on Mendeleev was recognized as one of the foremost scientists of the day.

Mendeleev was a versatile genius who was interested in many fields of science, both pure and applied. He worked on many problems associated with Russia's natural resources, such as coal, salt, and petroleum, and in 1876 he visited the United States to study the Pennsylvania oil fields. He invented an accurate barometer; he made an ascent in a balloon to study a total eclipse of the sun; and he was interested in the possibility of air travel.

As a professor at the University of St. Petersburg, Mendeleev was unavoidably involved in the political turmoil that affected all nineteenth-century Russia. Although in the middle of the century universities had been left comparatively free to carry out their research and teaching as they saw fit, the government came to suspect that academic freedom encouraged political unrest. Many repressive measures were taken against the universities, and the students reacted with demonstrations and riots. The universities were frequently closed for long periods, and many students were exiled to Siberia. In 1890 Mendeleev resigned from the university because of the government's oppressive treatment of students and the lack of academic freedom. Fortunately, he still had friends at the czar's court, and he was appointed director of the Bureau of Weights and Measures, where he continued to carry out important research.

Mendeleev was a colorful character as well as a genius. A characteristic always noticed in the portraits of Mendeleev is his enormous head of hair. Biographers say that he only had it cut once a year in the spring and that he would not deviate from this custom even when he had an audience with the czar.

· With the exception of mercury (melting point −39°C), they are solids at room temperature.

· They emit electrons when they are exposed to radiation of sufficiently high energy or when they are heated. These two effects are known as the *photoelectric effect* and the *thermionic effect*, respectively.

Figure 4.1 Periodic Table.

The elements are arranged in order of increasing atomic number, which usually (but not always) matches the order of increasing atomic mass. This arrangement places elements with similar properties in the same column. The atomic numbers are noted above the symbols of the elements; average atomic masses appear below each symbol. The solid line divides the metals on the left from the nonmetals on the right.

Group → I, II, III, IV, V, VI, VII, VIII

Legend: atomic number (top), symbol, atomic mass (bottom). Example:
1
H
1.008

Period 1

I	VIII
1 **H** 1.008	2 **He** 4.003

Period 2

I	II	III	IV	V	VI	VII	VIII
3 **Li** 6.941	4 **Be** 9.012	5 **B** 10.81	6 **C** 12.01	7 **N** 14.01	8 **O** 16.00	9 **F** 19.00	10 **Ne** 20.18

Period 3

I	II	III	IV	V	VI	VII	VIII
11 **Na** 22.99	12 **Mg** 24.31	13 **Al** 26.98	14 **Si** 28.09	15 **P** 30.97	16 **S** 32.06	17 **Cl** 35.45	18 **Ar** 39.95

Transition Elements / Periods 4–7

19 **K** 39.10	20 **Ca** 40.08	21 **Sc** 44.96	22 **Ti** 47.90	23 **V** 50.94	24 **Cr** 52.00	25 **Mn** 54.94	26 **Fe** 55.85	27 **Co** 58.93	28 **Ni** 58.70	29 **Cu** 63.55	30 **Zn** 65.38	31 **Ga** 69.72	32 **Ge** 72.59	33 **As** 74.92	34 **Se** 78.96	35 **Br** 79.90	36 **Kr** 83.80
37 **Rb** 85.47	38 **Sr** 87.62	39 **Y** 88.91	40 **Zr** 91.22	41 **Nb** 92.91	42 **Mo** 95.94	43 **Tc** (98)	44 **Ru** 101.1	45 **Rh** 102.9	46 **Pd** 106.4	47 **Ag** 107.9	48 **Cd** 112.4	49 **In** 114.8	50 **Sn** 118.7	51 **Sb** 121.8	52 **Te** 127.6	53 **I** 126.9	54 **Xe** 131.3
55 **Cs** 132.9	56 **Ba** 137.3	57 **La** 138.9	72 **Hf** 178.5	73 **Ta** 180.9	74 **W** 183.9	75 **Re** 186.2	76 **Os** 190.2	77 **Ir** 192.2	78 **Pt** 195.1	79 **Au** 197.0	80 **Hg** 200.6	81 **Tl** 204.4	82 **Pb** 207.2	83 **Bi** 209.0	84 **Po** (209)	85 **At** (210)	86 **Rn** (222)
87 **Fr** (223)	88 **Ra** (226.0)	89 **Ac** (227)	104	105	106	107											

Lanthanides (Period 6)

58 **Ce** 140.1	59 **Pr** 140.9	60 **Nd** 144.2	61 **Pm** (145)	62 **Sm** 150.4	63 **Eu** 152.0	64 **Gd** 157.3	65 **Tb** 158.9	66 **Dy** 162.5	67 **Ho** 164.9	68 **Er** 167.3	69 **Tm** 168.9	70 **Yb** 173.0	71 **Lu** 175.0

Actinides (Period 7)

90 **Th** 232.0	91 **Pa** (231)	92 **U** 238.0	93 **Np** (244)	94 **Pu** (242)	95 **Am** (243)	96 **Cm** (247)	97 **Bk** (247)	98 **Cf** (251)	99 **Es** (252)	100 **Fm** (257)	101 **Md** (258)	102 **No** (259)	103 **Lr** (260)

Nonmetallic elements generally have the following properties:

- They are poor conductors of heat.
- They are insulators; that is, they are very poor conductors of electricity.
- They do not have a high reflectivity or a shiny metallic appearance.
- They may be gases, liquids, or solids at room temperature.
- In the solid form they are generally brittle and fracture easily under stress, rather than being malleable and ductile.
- They do not exhibit the thermionic or photoelectric effects.

Not every metal or nonmetal exhibits all the listed properties, and there are other exceptions to these generalizations. For example, the metal chromium is quite brittle, and graphite, a form of the nonmetal carbon, is a fairly good conductor of electricity. Furthermore, the elements in the vicinity of the borderline have properties that are intermediate between those of a metal and those of a nonmetal. Germanium, antimony, and tellurium are examples. They are usually called **semimetals**, or **metalloids**.

If we divide all the elements into metals and nonmetals, 86 are metals and only 21 are nonmetals. But these numbers do not reflect the relative importance of the two classes of elements. Most nonmetals are rather common and have many important compounds; quite a few metals are rare and have few important compounds. Of the 20 most abundant elements in the earth's crust, 10 are nonmetals, including the two most abundant elements, oxygen and silicon (see Table 3.2).

Families of Elements

The elements can be further classified into a number of families whose members all have rather similar properties. These families constitute some of the groups of the periodic table. The resemblances between the properties of the elements in a given group are most marked at the sides of the table, that is, in groups I, II, VI, VII, and VIII. These groups have been given special names.

GROUP I: THE ALKALI METALS, Li, Na, K, Rb, AND Cs   The group I elements are all metals and have most of the typical metallic properties; for example, they are shiny solids and good conductors of heat and electricity. Unlike most metals, they are soft enough to be cut with a knife and have relatively low melting points. They react with hydrogen to form solid *hydrides* that have the general formula MH, where M represents an alkali metal:

$$2M(s) + H_2(g) \longrightarrow 2MH(s)$$

The alkali metals also react with chlorine to form *chlorides*, which are all colorless solids with the formula MCl:

$$2M(s) + Cl_2(g) \longrightarrow 2MCl(s)$$

The alkali metals react vigorously with water, producing hydrogen and a solution of a metal hydroxide (see Experiment 4.1). For example,

$$2Na(s) + 2H_2O(l) \longrightarrow 2NaOH(aq) + H_2(g)$$

GROUP II: THE ALKALINE EARTH METALS, Be, Mg, Ca, Sr, AND Ba   Like the alkali metals, the group II elements are all solids with typical metallic properties. Compared with the alkali metals, they are harder, melt at higher temperatures,

Reactivity of the Alkali Metals

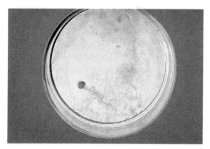

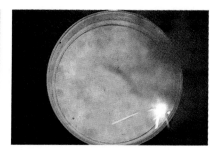

Sodium reacts so vigorously with water that it skates over the surface as hydrogen is evolved. A few drops of the indicator bromothymol blue have been added to the water. The blue trails are due to the sodium hydroxide formed in the reaction, which makes the solution basic as shown by the change in the color of the indicator.

The heat evolved in the reaction melts the sodium, which forms a small shiny metallic bead that floats on the water.

Potassium reacts even more violently, skating very rapidly over the surface leaving blue trails as the indicator changes color. The heat of the reaction ignites the hydrogen. The potassium imparts a characteristic lilac color to the flame. Steam can be seen rising from the water as it is heated by the burning potassium.

and are somewhat less reactive. For example, although magnesium reacts with steam (Experiment 1.1), it does not react with cold water, and calcium reacts much more slowly than either sodium, or potassium:

$$Ca(s) + 2H_2O(l) \longrightarrow Ca(OH)_2(s) + H_2(g)$$

The alkaline earth metals form solid hydrides with the general formula MH_2 and solid chlorides with the general formula MCl_2:

$$M(s) + H_2(g) \longrightarrow MH_2(s) \quad \text{and} \quad M(s) + Cl_2(g) \longrightarrow MCl_2(s)$$

GROUP VI: THE CHALCOGENS, O, S, Se, AND Te   Except for oxygen, which is a gas, the elements of group VI are solids. Oxygen and sulfur are typical nonmetals. Selenium and particularly tellurium have some metallic properties, the latter is, in fact, usually classified as a semimetal. The group VI elements react with hydrogen to form the compounds H_2O, H_2S, H_2Se, and H_2Te, which, with the exception of water, are all gases at room temperature.

GROUP VII: THE HALOGENS, F, Cl, Br, AND I   The halogens are all typical nonmetals. Although their physical forms differ—fluorine and chlorine are gases, bromine is a liquid, and iodine is a solid at room temperature—each consists of diatomic molecules: F_2, Cl_2, Br_2, and I_2. The halogens all react with hydrogen to form gaseous compounds, with the formulas HF, HCl, HBr, and HI, all of which are very soluble in water. The halogens all react with metals to give *halides*. Examples include sodium fluoride, NaF, potassium chloride, KCl, magnesium bromide, $MgBr_2$, and sodium iodide, NaI.

GROUP VIII: THE NOBLE GASES, He, Ne, Ar, Kr, Xe, AND Rn   These elements are all gases at room temperature and are typical nonmetals. They are all *monatomic*; that is, they consist of single atoms that are not combined to form molecules. They are very inert, and the first noble gas compound was not made until 1962. Krypton, xenon, and radon react with fluorine to form fluorines, but they do not react with oxygen or any other elements.

Example 4.1 Write balanced equations for the following reactions: potassium with bromine; cesium with water; strontium with iodine; calcium with water.

Solution Potassium is an alkali metal; therefore the equation is

$$2K + Br_2 \longrightarrow 2KBr$$

Cesium is an alkali metal; therefore the equation is

$$2Cs + 2H_2O \longrightarrow 2CsOH + H_2$$

Strontium is an alkaline earth metal; therefore the equation is

$$Sr + I_2 \longrightarrow SrI_2$$

Calcium is an alkaline earth metal; therefore the equation is

$$Ca + 2H_2O \longrightarrow Ca(OH)_2 + H_2$$

Valence

A very important property of an element that is related to its position in the periodic table is its combining power for other elements, a property called its **valence**. For example, with very few exceptions, one atom of hydrogen combines with no more than one atom of any other element; thus we find the molecules H_2, HF, HCl, and H_2O. Hydrogen is therefore said to have a valence of 1.

An oxygen atom, however, combines with two hydrogen atoms to give the water molecule, H_2O. A nitrogen atom combines with three hydrogen atoms to give the ammonia molecule, NH_3. A carbon atom combines with four hydrogen atoms to give the methane molecule, CH_4. Thus oxygen, nitrogen, and carbon are said to have valences of 2, 3, and 4, respectively.

All elements within a group have the same valence.

For example, all the alkali metals (Group I) have a valence of 1; all the elements in Group II have a valence of 2. The valences of the main group elements are shown in Table 4.1. From these valences we can immediately write the empirical formulas for the compounds of these elements with hydrogen:

LiH	BeH_2	BH_3	CH_4	NH_3	H_2O	HF
NaH	MgH_2	AlH_3	SiH_4	PH_3	H_2S	HCl

Like hydrogen, fluorine and chlorine have a valence of 1. Hence we may write the empirical formulas of the compounds that they form with the elements of periods 2 and 3, as follows:

Period 2	Fluorides	LiF	BeF_2	BF_3	CF_4	NF_3	F_2O	F_2
	Chlorides	LiCl	$BeCl_2$	BCl_3	CCl_4	NCl_3	Cl_2O	FCl
Period 3	Fluorides	NaF	MgF_2	AlF_3	SiF_4	PF_3	SF_2	ClF
	Chlorides	NaCl	$MgCl_2$	$AlCl_3$	$SiCl_4$	PCl_3	SCl_2	Cl_2

Table 4.1 Valences of Main Group Elements

Group	I	II	III	IV	V	VI	VII	VIII
Element	H							He
	Li	Be	B	C	N	O	F	Ne
	Na	Mg	Al	Si	P	S	Cl	Ar
Valence	1	2	3	4	3	2	1	0

Oxygen, with a valence of 2, combines with two atoms of an element with a valence of 1, such as hydrogen, and with one atom of an element with a valence of 2, such as beryllium, Be, to form the compound BeO. In general, in a compound A_yB_z the values of y and z are such that

$$y \times \text{valence of A} = z \times \text{valence of B}$$

On the basis of this rule, the empirical formulas of the oxides of the elements in periods 2 and 3 of the periodic table may be written as follows:

Period 2 Oxides Li_2O BeO B_2O_3 CO_2 N_2O_3 O_2 OF_2

Period 3 Oxides Na_2O MgO Al_2O_3 SiO_2 P_2O_3 SO Cl_2O

The valences we have discussed are the normal valences of these elements. Some elements, particularly those in periods 3–7 of Groups V–VIII, have additional, higher valences. For example, phosphorus has a valence of 5 as well as 3. It forms a compound with the formula PCl_5 as well as one with the formula PCl_3, and it forms an oxide with the empirical formula P_2O_5 and one with the empirical formula P_2O_3. Sulfur has higher valences of 4 and 6; it forms the oxides SO_2 and SO_3 as well as SO, which is an unstable and unimportant compound. These additional valences are discussed in Chapter 7.

> **Example 4.2** Predict the empirical formulas of the sulfides of C, Si, Na, Mg, and Cl.
>
> **Solution** We can write the empirical formula of the sulfide of carbon as C_yS_z. Since carbon is in Group IV its valence is 4; since sulfur is in Group VI its valence is 2. Hence
>
> $$y \times 4 = z \times 2$$
>
> and therefore
>
> $$\frac{y}{z} = \frac{2}{4} = \frac{1}{2}$$
>
> Thus the empirical formula is CS_2. Alternatively since sulfur has the same valence as oxygen the formulas can be predicted by analogy with the formulas of the oxides just given. They are CS_2, SiS_2, Na_2S, MgS, and Cl_2S.

4.2 ELECTRON ARRANGEMENTS

Why do similar properties, such as a particular valence, occur periodically with increasing atomic number? An obvious answer is that the properties depend on the structure of the atoms of an element and that the structure in turn varies in a regular way with atomic number. Thus to understand the periodic variation in properties, we need to amplify our description of atomic structure.

The Shell Model

A model of the atom that accounts well for the periodic variation in the properties of the elements is the **shell model**. According to this model, the electrons surrounding a nucleus in an atom are arranged in successive spherical layers called **shells**. The hydrogen atom consists of a nucleus surrounded by a rapidly moving electron that effectively fills a spherical space surrounding the nucleus. The helium atom similarly has two electrons in a similar sphere around the nucleus. The next element, lithium, also has two electrons in a sphere surrounding the nucleus, but the third electron occupies a spherical layer, or shell, that is mostly outside the sphere containing the first two electrons. In the succeeding elements, beryllium, boron, and so on, electrons continue to enter this outer shell until it contains eight electrons. Thus neon ($Z = 10$), which has a total of ten electrons, has two electrons in the inner shell and eight electrons in the second shell.

In sodium ($Z = 11$), the element following neon, the additional electron begins a third shell. Electrons continue to fill this shell up to the element argon ($Z = 18$), which has inner shells containing, respectively, two electrons and eight electrons, with a further eight electrons in the third shell. In the next element, potassium ($Z = 19$), a fourth shell is commenced. The electron arrangements for the first 20 elements are shown in Table 4.2. The successive electron shells are designated by $n = 1$ for the first shell, $n = 2$ for the second shell, and so on. The outer shell of an atom is called its **valence shell**, because it is the electrons in this shell that determine the valence of the atom.

Figure 4.2 shows a simple representation of the shell structures of the first 18 elements. As we will see in Chapter 6, we cannot locate electrons very precisely. Thus the boundaries of each shell are rather fuzzy, and they overlap each other to some extent, as shown in Figure 4.3. Nevertheless, on average, the electrons in the $n = 1$ shell are considerably closer to the nucleus than are those in the $n = 2$ shell, and those in the $n = 3$ shell are still further from the nucleus, and so on.

Table 4.2 Shell Structure of Atoms of First 20 Elements

PERIOD	Z	ELEMENT	NUMBER OF ELECTRONS IN EACH SHELL			
			$n = 1$	2	3	4
1	1	H	1			
	2	He	2			
2	3	Li	2	1		
	4	Be	2	2		
	5	B	2	3		
	6	C	2	4		
	7	N	2	5		
	8	O	2	6		
	9	F	2	7		
	10	Ne	2	8		
3	11	Na	2	8	1	
	12	Mg	2	8	2	
	13	Al	2	8	3	
	14	Si	2	8	4	
	15	P	2	8	5	
	16	S	2	8	6	
	17	Cl	2	8	7	
	18	Ar	2	8	8	
4	19	K	2	8	8	1
	20	Ca	2	8	8	2

Note: The boxes indicate the valence shell.

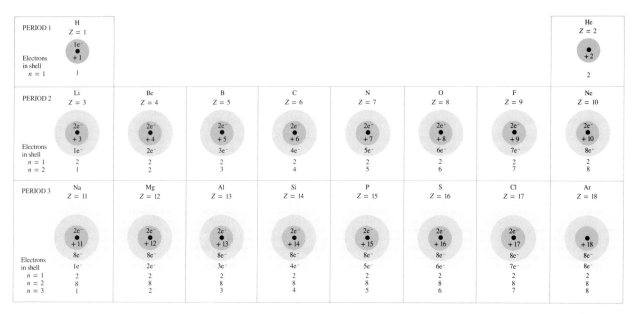

Figure 4.2 Shell Model of First 18 Elements. These diagrams show how electrons are arranged in shells; they do not indicate the relative size of shells or atoms. Each element within a group has a different number of shells, but they have the same number of electrons in their outermost shell.

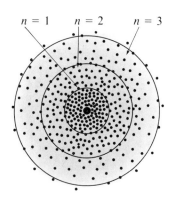

Figure 4.3 Approximate Representation of the Electron Density in the Argon Atom. The electron density decreases with increasing size of the shell; it is greatest in the $n = 1$ shell, less in the $n = 2$ shell, and least in the $n = 3$ shell. The shells overlap, however, so the electron density does not decrease abruptly from one shell to the next.

We see from Table 4.2 and Figure 4.2 that the shell model provides an explanation for the periodic variation in the properties of the elements and for the general form of the periodic table. Each time a new shell commences, we have an element with one electron in its valence shell. These elements therefore all have similar properties; they are the alkali metals Li, Na, and K. The elements following the alkali metals all have two electrons in their valence shell; they are the alkaline earth metals Be, Mg, and Ca. The halogens, F and Cl, each have seven electrons in their valence shell.

Elements in the same group of the periodic table have the same number of electrons in their valence shell.

For a main group element (Groups I–VIII) the number of valence shell electrons is equal to the group number. The only exception is helium, in Group VIII, which has only two electrons in its valence shell.

Ionization Energies

Direct experimental evidence for the shell model is provided by ionization energies. The **ionization energy** of an element is the energy needed to remove an electron from an atom of the element in the gas phase. Thus the ionization energy of an element M is the energy required for the process

$$M(g) \longrightarrow M^+(g) + e^-$$

When one or more electrons are removed from a neutral atom, a charged atom called a **positive ion** is formed. Some atoms may also add electrons forming negatively charged atoms known as **negative ions**.

One method of measuring ionization energies, the **electron impact method**, is described in Box 4.2. Atoms are bombarded with fast-moving electrons. If these electrons have sufficient energy, they will, on colliding with an atom, knock out one or more of the atom's electrons. The bombarding electrons can be given

The apparatus used in the electron impact method for determining ionization energies consists of a sealed tube containing a heated metal filament called the *cathode*, which emits electrons and is kept at a negative potential with respect to a wire grid at a positive potential. Another metal plate called the *anode* is kept at a slight negative potential with respect to the grid but positive with respect to the cathode.

The tube is filled with the atoms to be studied in the form of a gas at low pressure. Electrons emitted by the cathode are accelerated toward the grid. Many of the electrons pass through the grid. If they have sufficient energy to overcome the small decelerating potential between the grid and the anode, they reach the anode, causing a current to flow in the external circuit, where it can be measured with a suitable meter.

The energy of the bombarding electrons can be increased by increasing the voltage through which they are accelerated. They collide with atoms on their way down the tube, but until they have enough energy to knock electrons out of any atoms with which they collide, the bombarding electrons simply bounce off with their energy unchanged and continue on their way down the tube to the anode. However, when it has sufficient energy, a bombarding electron will knock an electron out of an atom with which it collides, thereby losing all its own energy. This electron then has too little energy to get from the grid to the anode so the number of electrons reaching the anode falls markedly. Thus the meter registers a correspondingly large decrease in the current.

From the voltage used to accelerate the electrons their energy can be calculated. Hence from the voltage at which a sudden reduction in the current is noted on the meter, the energy that must be supplied to the bombarding electrons in order to ionize the gaseous atoms can be found. This is the ionization energy.

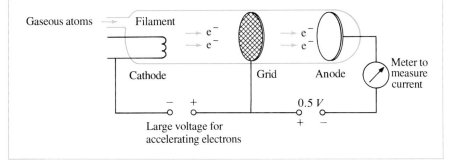

any desired energy by accelerating them through an electric potential that can be appropriately adjusted. The ionization of a hydrogen atom by electron bombardment can be summarized as follows:

$$H(g) + e^- \longrightarrow H^+(g) + 2e^-$$

A bombarding electron must have a minimum energy of 2.17×10^{-18} J in order to knock an electron out of a hydrogen atom:

$$H(g) + 2.18 \times 10^{-18} \text{ J} \longrightarrow H^+(g) + e^-$$

Thus 2.18×10^{-18} J is the ionization energy of one hydrogen atom. It is more convenient, however, to consider the total energy needed to remove an electron

4.2 ELECTRON
ARRANGEMENTS

129

from each of the hydrogen atoms in 1 mol of hydrogen atoms, which is

$$\left(\frac{2.18 \times 10^{-18} \text{ J}}{1 \text{ atom}}\right)\left(\frac{6.022 \times 10^{23} \text{ atoms}}{1 \text{ mol}}\right) = 1.31 \times 10^6 \text{ J mol}^{-1}$$

$$= 1.31 \text{ MJ mol}^{-1}$$

This value is the ionization energy of 1 mol of hydrogen atoms, and we will quote ionization energies in this form.

The ionization energies of the first 20 elements are listed in Table 4.3. They vary in a regular periodic manner with atomic number, as Figure 4.4 illustrates. We can understand these variations by using the shell model in conjunction with Coulomb's law.

Recall from Chapter 2 that according to Coulomb's law, the electrostatic force, F, between two charges, Q_1 and Q_2, is proportional to the magnitude of each charge and inversely proportional to the square of the distance, r, between them:

$$F = k\frac{Q_1 Q_2}{r^2}$$

The electrons of hydrogen and helium are both in the first, $n = 1$, shell, and they are therefore at approximately the same distance from the nucleus. But the nuclear charge of helium $(+2)$ is greater than that of hydrogen $(+1)$. Therefore the force of attraction between the nucleus and the electrons in the helium atom is greater than the force of attraction between the nucleus and the electron in the hydrogen atom. In other words, removing one of the electrons of the helium atom should take more energy than removing the electron in the hydrogen atom. And as Table 4.3 shows, helium has a higher ionization energy than hydrogen.

We see, however, that for lithium, $Z = 3$, a further increase in the ionization energy with increasing nuclear charge is not observed, and the ionization energy is very much less than that of hydrogen or helium. This observation is consistent with the shell model, which indicates that the third electron occupies a second shell outside the first shell. This electron is at a greater distance from the nucleus than are the two electrons in the inner shell. Since according to Coulomb's law, the electrostatic force between two charges depends inversely on the square of the distance between them, r^2, the force with which a nucleus attracts an electron

Table 4.3 Ionization Energies of First 20 Elements

Z	ELEMENT	IONIZATION ENERGY (MJ mol^{-1})	SHELL n	Z	ELEMENT	IONIZATION ENERGY (MJ mol^{-1})	SHELL n
1	H	1.31 ⎫	1	11	Na	0.50 ⎫	
2	He	2.37 ⎭		12	Mg	0.74	
3	Li	0.52 ⎫		13	Al	0.58	
4	Be	0.90		14	Si	0.79	
5	B	0.80		15	P	1.01	3
6	C	1.09		16	S	1.00	
7	N	1.40 ⎬ 2		17	Cl	1.25	
8	O	1.31		18	Ar	1.52 ⎭	
9	F	1.68		19	K	0.419 ⎫	4
10	Ne	2.08 ⎭		20	Ca	0.590 ⎭	

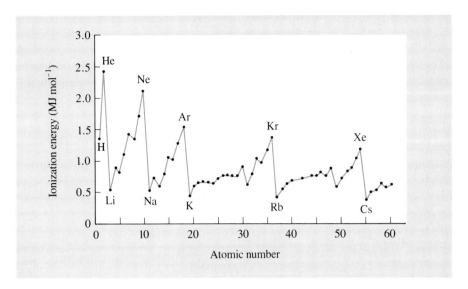

drops off rapidly, as the distance between the electron and the nucleus increases. Accordingly, the ionization energy of the outer electron in lithium is much smaller than the ionization energy of either hydrogen or helium.

Moreover, the electron in the valence shell of lithium is repelled by the two inner electrons. Thus the resultant charge acting on the outer electron is not the nuclear charge of $+3$, but rather the charge of the inner core of the atom, consisting of the nucleus and the two inner-shell electrons, which is $+3 - 2 = +1$. The nucleus plus the completed inner shells of electrons constitute the **core** of an atom. The overall charge on the core is called the **core charge**. *The core charge is equal to the atomic number Z minus the total number of electrons in the inner shells* (see Figure 4.5).

Thus the outer-shell electrons experience the charge of the core rather than the full charge of the nucleus. The inner electrons that surround the nucleus are said to *shield* the nucleus. In fact, because the shells overlap each other and because there is some repulsion between the electrons in the valence shell, the

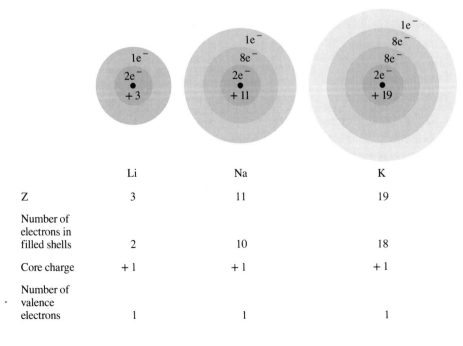

Figure 4.5 Core Charge and Shell Structure. The resultant force acting on a valence shell electron is the sum of an attraction due to the nucleus and a repulsion due to all the inner-shell electrons. The core charge takes both of these forces into account.

	Li	Na	K
Z	3	11	19
Number of electrons in filled shells	2	10	18
Core charge	$+1$	$+1$	$+1$
Number of valence electrons	1	1	1

effective charge acting on a valence shell electron differs somewhat from the core charge. This effective charge acting on a valence shell electron is called the **effective nuclear charge**. Since there is no simple way to obtain values for the effective nuclear charge we will use the core charge as a basis for qualitative explanations of the properties of atoms. It is only an approximation, but it is adequate for our purposes.

> **Example 4.3** Which has a higher core charge, a magnesium atom or an oxygen atom?
>
> **Solution** The electron arrangement of magnesium is 2, 8, 2, and it has atomic number $Z = 12$. The core charge is $+12 - (2 + 8) = +2$. The electron arrangement of oxygen is 2, 6, and its atomic number is $Z = 8$. The core charge is $+8 - 2 = +6$. So oxygen has a higher core charge than magnesium.

Like lithium, the other elements in period 2 have their outermost electrons in the second shell, but their core charges are greater than that of lithium. For example, beryllium has two electrons in its outer shell, an inner shell of two electrons, and a nuclear charge of $+4$; thus beryllium has a core charge of $+4 - 2 = +2$. Similarly, boron has a core charge of $+3$, carbon has a core charge of $+4$, and so on. The core charge is simply equal to the group number. As the core charge increases across the second period, the ionization energies of the elements also increase, except for two small irregularities that are discussed in Chapter 6. The ionization energy reaches a maximum at neon, at which point the second shell is complete.

The element after neon, sodium, has a nuclear charge of $+11$ and ten electrons in its inner shells, so the core charge is only $+1$. Thus, not surprisingly, the ionization energy of sodium is much lower than that of neon. But it is lower than that of lithium, too, which also has a core charge of $+1$. Because the outermost electron in sodium is in the third shell from the nucleus, it is further from the nucleus than the outer electron in lithium and thus has a lower ionization energy.

As the core charge increases throughout the rest of period 3, from sodium to argon, the ionization energies increase correspondingly . Then for potassium, the first element in period 4, the ionization energy again drops, as we would predict, because potassium has a core charge of $+1$ and an outer electron in the fourth shell from the nucleus. The ionization energy of potassium is therefore even lower than that of sodium.

We may summarize the dependence of the ionization energy of an element on its position in the periodic table by the following two statements:

1. With one or two minor exceptions, the ionization energy increases across each period of the periodic table as the core charge increases. In other words, as we move across a period, an increasing amount of energy is needed to remove an electron from the outermost (valence) shell.

2. The ionization energy decreases down any group of the periodic table as the distance of the valence shell electrons from the nucleus increases, with the increasing number of shells, while the core charge remains constant.

> **Example 4.4** Which element in each of the following pairs would be expected to have the higher ionization energy:
>
> (a) Li, Na;   (b) Na, Mg;   (c) N, F;   (d) O, S.
>
> Answer this question by reference to the periodic table but not to Table 4.3.

Solution

(a) Since Li comes above Na in Group I, Li has the higher ionization energy.

(b) Since Mg follows Na in period 3, Mg has the higher ionization energy.

(c) Since F comes after N in period 2, F has the higher ionization energy.

(d) Since O comes above S in Group VI, O has the higher ionization energy.

Electron Affinities

We mentioned earlier that an electron may be added to many atoms to form a negative ion. For most elements energy is released in this process; this energy is called the **electron affinity** of the atom:

$$X + e^- \longrightarrow X^- + \text{Energy (electron affinity)}$$

For example, the electron affinity of the H atom is $72.77 \text{ kJ mol}^{-1}$. Thus 72.77 kJ is released when 1 mol of H atoms combines with 1 mol of electrons to give 1 mol of H^- ions:

$$H + e^- \longrightarrow H^- + 72.77 \text{ kJ mol}^{-1}$$

For a few elements (see Table 4.4) energy is required to add an electron—the process is endothermic rather than exothermic. In such cases the electron affinity has a negative value.

The force with which an electron is pulled into the valence shell of an atom depends on the core charge and is expected to increase with increasing core charge, from left to right in any period. Thus we expect the nonmetals in Groups VI and VII to have higher electron affinities than the metals in Groups I and II. And so they do, although there is no steady increase in the electron affinity from left to right in a period. We noticed similar small irregularities in the ionization energies of the elements, which will be explained in Chapter 6.

Sizes of Atoms

Another very important property that varies in a regular, periodic manner with atomic number is atomic size. The valence shell of an atom has no precise outer boundary; the electron density of an atom decreases rapidly to a very

Table 4.4 Electron Affinities of Elements (kJ mol^{-1})

H							
72.8							
Li	Be	B	C	N	O	F	Ne
60	<0	27	122	<0	141	328	<0
Na	Mg	Al	Si	P	S	Cl	Ar
53	−14	43	134	72	200	349	<0
K	Ca				Se	Br	Kr
48	<0				195	324	<0
Rb	Sr				Te	I	Xe
47	<0				190	295	<0
Cs	Ba						
46	<0						

Note: <0 indicates a negative value.

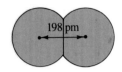

Cl₂
Bond length = 198 pm
rCl = radius of Cl atom
= 99 pm

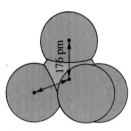

CCl₄
Bond length = 176 pm
$r_C + r_{Cl}$ = 176 pm
r_C = 77 pm

Figure 4.6 Atomic Radii and Bond Lengths. The length of a bond is the sum of the atomic (covalent) radii of the two atoms that are bonded together.

small value (see Figures 2.1 and 4.3), but it does not become exactly zero even at a large distance from the nucleus. Therefore it is impossible to give an exact value for the size of a free atom. However, the distance between the nuclei of any pair of atoms *in a molecule* can be measured rather accurately (see Chapter 10).

The distance between the two nuclei in the H_2 molecule, which is known as the H–H **bond length**, is 74 picometers (pm). If we take one-half this distance, that is, 37 pm, we have a measure of the size of the H atom, which is known as the **atomic**, or **covalent**, **radius** of hydrogen. The distance between the two Cl nuclei in the Cl_2 molecule, the Cl–Cl bond length, is 198 pm. The atomic radius of Cl is one-half of this distance, that is, 99 pm. Other atomic radii can be found in a similar manner.

We cannot obtain values for atomic radii directly from the length of a bond between two dissimilar atoms because we do not know how to divide the distance into two unequal lengths. The length of the C–Cl bond in carbon tetrachloride has been found by experiment to be 176 pm. This distance is the sum of the atomic radii of Cl and C; in other words,

$$r_C + r_{Cl} = \text{C–Cl bond length}$$

If we use the value of 99 pm for the atomic radius of Cl that we calculated from the Cl_2 bond length, we can calculate the atomic radius of C:

$$r_C = \text{C–Cl bond length} - r_{Cl} = 176\ \text{pm} - 99\ \text{pm} = 77\ \text{pm}$$

This calculation assumes that the radius of a Cl atom, or indeed of any atom, is the same in all molecules, and this assumption has been shown in many experiments to be approximately true. The relationships between atomic radii and bond lengths are illustrated in Figure 4.6.

The relative sizes of the atoms of the main group elements, as indicated by the values of their covalent radii, are shown in Figure 4.7. We note two important trends:

1. *As we move down any group, the atoms become progressively larger.* This trend is consistent with the fact that an electron shell is added when we pass from one element to the next in a group.

Figure 4.7 Covalent Radii. The sizes of atoms generally increase from top to bottom within a group and decrease from left to right within a period. The radii here are given in picometers. No values are given for He, Ne, and Ar because no compounds of these elements are known.

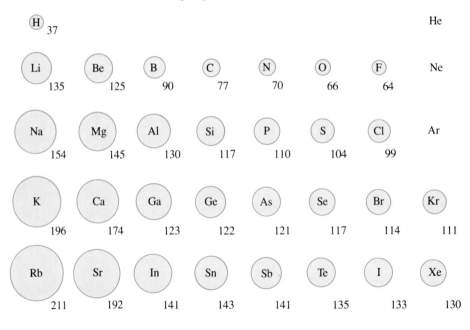

2. *As we move across any period,* the atoms become progressively smaller. Although the outer electrons in each of the atoms in the same period are in the same valence shell, they are acted upon by an increasing core charge. This increasing core charge pulls the valence electrons in toward the core, thus decreasing the size of the atom, which becomes progressively smaller across a period.

Figure 4.8 shows the periodic variation of the atomic radii of the elements with atomic number. The alkali metal atoms are at the maxima of the plot, and the halogen and noble gas (Kr and Xe) atoms are at the minima. No values can be given for the radii of He, Ne, and Ar because no molecules containing these atoms are known. Thus it has not been possible to measure the length of any bonds involving these atoms.

4.3 CHEMICAL BONDS AND LEWIS STRUCTURES

We are now in a position to consider the important question of how atoms are held together, as they are, for example, in compounds such as NaCl and H_2O and in the molecules of some elements such as H_2 and Cl_2. Whenever atoms are held strongly together, we say that there is a **chemical bond** between them. Chemical bonds, then, are strong forces of attraction between atoms.

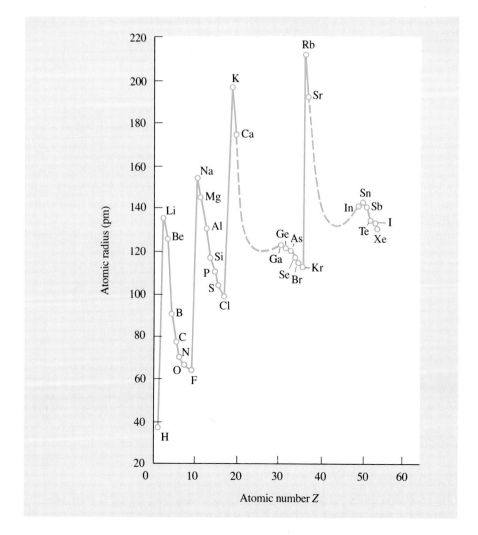

Figure 4.8 Periodic Variation of Covalent Radii. The largest atoms in any period are the alkali metals; the smallest are the noble gases.

Figure 4.9 Gilbert Newton Lewis (1875–1946).

Lewis was born in Massachusetts, but in 1884 his family moved to Lincoln, Nebraska, where he received little formal schooling. He began his college career at the University of Nebraska, but he transferred to Harvard where he obtained his Ph.D. in 1899. After a period of advanced studies in Germany he joined the faculty of the Massachusetts Institute of Technology in 1905. In 1912 he became professor of chemistry at the University of California, Berkeley. Under his guidance the chemistry department at Berkeley gained international recognition.

Lewis's inquisitive, imaginative mind led him to make important contributions in several areas of chemistry. He was the first to propose that atoms could be held together by sharing an electron pair, and he proposed the electron dot (Lewis) structures that we use in this chapter. He made outstanding contributions to thermodynamics (the study of energy changes) and was the coauthor of a textbook that had a profound influence on the teaching of thermodynamics. In addition, he proposed a new definition of acids and bases (which we discuss in Chapter 9) and was the first to prepare and study pure heavy water, ^2H_2O (D_2O).

We have previously stressed that the only forces that are important in chemistry are electrical forces, specifically the forces between negatively charged electrons and positively charged nuclei. Because the electrons in the completed inner shells of an atom are held much more strongly than the electrons in the outer shell, the inner-shell electrons are not generally involved when atoms combine. In discussing chemical bonds, we will therefore be concerned only with the valence shell electrons. The American chemist Gilbert Lewis (Figure 4.9) introduced very convenient symbols, called **electron dot symbols**, or **Lewis symbols**, in which the electrons in the valence shell of an atom are indicated by dots surrounding the symbol of the element.

There is a simple relationship between the number of electrons in the valence shell and the common valence of the element. The valence of an atom in Groups I to IV is equal to the number of electrons in the valence shell, that is, to the group number. The valence of an atom in Groups IV to VIII is equal to 8 minus the number of electrons in the valence shell. This relationship is illustrated in Table 4.5 for the elements of the second period.

Octet Rule

The valences of neon and argon, both of which have eight electrons in their valence shell, are 0; these elements are not known to form any compounds, and the elements themselves are monatomic. Thus a valence shell containing eight electrons appears to be a specially stable arrangement. This observation led Lewis to suggest the following rule:

In compound formation an atom tends to gain or lose electrons or to share electrons until there are eight electrons in its valence shell.

Table 4.5 Electron Arrangements and Valences of Main Group Elements

GROUP	I	II	III	IV	V	VI	VII	VIII
Number of valence shell electrons	1	2	3	4	5	6	7	8
Valence	1	2	3	4	3	2	1	0
Period 2 element	Li·	·Be·	·Ḃ·	·Ċ·	·N̈·	:Ö·	:F̈·	:N̈e:
Empirical formulas of some typical compounds	LiCl LiH	BeCl$_2$ BeH$_2$	BCl$_3$ BH$_3$	CCl$_4$ CH$_4$	NCl$_3$ NH$_3$	OCl$_2$ H$_2$O	FCl HF	— —

A valence shell containing eight electrons is called an **octet**, and the generalization is known as the **octet rule**. We will see that it enables us to give a simple explanation for the normal valences of the elements.

Ionic Bonds

Two obvious ways in which an atom can obtain a valence shell of eight electrons are to lose all its outer electrons or to gain additional electrons. For example, sodium reacts with chlorine to form sodium chloride. We can see in Table 4.2 that sodium has the electron arrangement 2, 8, 1, with only one electron in its valence shell. If a sodium atom loses this electron, a positive sodium ion, Na$^+$, is formed. This ion has the electron arrangement 2, 8, with eight electrons in its valence shell. If chlorine, which has the electron arrangement 2, 8, 7, gains an electron, a negative chloride ion, Cl$^-$, is formed. This ion has the electron arrangement 2, 8, 8. Thus when sodium reacts with chlorine, each sodium atom loses an electron to form a sodium ion, Na$^+$, and each chlorine atom gains an electron to form chloride ion, Cl$^-$. We can summarize this reaction very conveniently in terms of Lewis symbols:

$$\text{Na·} + \text{·C̈l:} \longrightarrow (\text{Na}^+)(\text{:C̈l:}^-)$$

The representation of sodium chloride as $(\text{Na}^+)(\text{:C̈l:}^-)$ is called the **Lewis structure** (or electron dot structure) of sodium chloride.

Since opposite charges attract each other, there is an attractive force between the sodium and chloride ions. Thus solid sodium chloride, NaCl, consists of equal numbers of Na$^+$ ions and Cl$^-$ ions held together by electrostatic attraction (Figure 4.10). The electrostatic attraction between oppositely charged ions is called an **ionic bond**.

We have seen, in Experiment 1.3, that magnesium oxide can be made by burning magnesium in air or oxygen. If magnesium, which has a valence shell with two electrons, loses those two electrons, it becomes the positive ion Mg^{2+}, which has an outer shell with eight electrons. If oxygen, which has six electrons in its valence shell, gains two electrons, it becomes the oxide ion, O^{2-}, which has a valence shell with eight electrons. Magnesium oxide consists of Mg^{2+} and O^{2-} held together by electrostatic attraction, that is, by ionic bonds. The Lewis structure of magnesium oxide is written as

$$(\text{Mg}^{2+})(\text{:Ö:}^{2-})$$

Magnesium also forms the compound magnesium chloride, MgCl$_2$, which occurs, for example, together with sodium chloride, in the sea. This compound consists of magnesium ions, Mg^{2+}, and twice as many chloride ions, Cl$^-$. It is

Figure 4.10 The Structure of Sodium Chloride. The arrangement of sodium ions and chloride ions in solid sodium chloride is shown.

4.3 CHEMICAL BONDS AND LEWIS STRUCTURES

137

represented by the Lewis structure

$$(Mg^{2+})(\text{:}\overset{\cdot\cdot}{\underset{\cdot\cdot}{Cl}}\text{:}^-)_2$$

Compounds such as Na^+Cl^-, $Mg^{2+}O^{2-}$, and $Mg^{2+}(Cl^-)_2$, which are composed of ions held together by electrostatic attraction (ionic bonds), are called **ionic compounds**.

Notice that both sodium and magnesium, which exist as positive ions in NaCl, MgO, and $MgCl_2$, are metals, whereas chlorine and oxygen, which form the negative ions in these compounds, are nonmetals. In general, the ionization energy of a valence electron of a metal is relatively low, so the electron is rather easily removed, leaving a positive ion. Moreover, although the metals have space in their valence shells for more electrons, they have little tendency to attract extra electrons because of their small core charges. We see from Table 4.4 that they have only small electron affinites. Except for the noble gases, the nonmetallic elements also have space for additional electrons in their valence shells, and since they have high core charges, they tend to strongly attract and hold on to additional electrons. As a result, electrons are normally transferred from metallic to nonmetallic elements, forming positive metal ions and negative nonmetal ions held together by ionic bonds.

The valence of an element that forms an ionic compound is equal to the number of electrons that it loses to form a positive ion or to the number of electrons that it gains to form a negative ion. In other words the valence is equal to the charge on the ion. An ionic compound consists of an arrangement of a large number of positive and negative ions, but it has no overall charge, so the total charge on the positive ions is equal to the total charge on the negative ions. Thus in Na^+Cl^- and $Mg^{2+}O^{2-}$ there are equal numbers of positive and negative ions, but in $Mg^{2+}(Cl^-)_2$ there are twice as many negative Cl^- ions as positive Mg^{2+} ions.

The formula of an ionic compound such as sodium chloride may be written as NaCl or as Na^+Cl^-. Because ionic compounds are normally solids consisting of a regular arrangement of ions of opposite charge, the formulas of ionic compounds are always empirical formulas. No individual molecules can be recognized in a solid ionic compound (see Figure 4.10), and therefore no molecular formula can be assigned to these compounds. The empirical formulas of ionic compounds may be derived from the charges of the common ions of the main group elements, which are listed in Table 4.6.

Table 4.6 Common Monatomic Ions of Main Group Elements

			GROUP			
I	II	III	IV	V	VI	VII
Li^+	Be^{2+}			$\text{:}\overset{\cdot\cdot}{N}\text{:}^{3-}$	$\text{:}\overset{\cdot\cdot}{\underset{\cdot\cdot}{O}}\text{:}^{2-}$	$\text{:}\overset{\cdot\cdot}{\underset{\cdot\cdot}{F}}\text{:}^-$
Na^+	Mg^{2+}	Al^{3+}		$\text{:}\overset{\cdot\cdot}{P}\text{:}^{3-}$	$\text{:}\overset{\cdot\cdot}{\underset{\cdot\cdot}{S}}\text{:}^{2-}$	$\text{:}\overset{\cdot\cdot}{\underset{\cdot\cdot}{Cl}}\text{:}^-$
K^+	Ca^{2+}					$\text{:}\overset{\cdot\cdot}{\underset{\cdot\cdot}{Br}}\text{:}^-$
Rb^+	Sr^{2+}					$\text{:}\overset{\cdot\cdot}{\underset{\cdot\cdot}{I}}\text{:}^-$
Cs^+	Ba^{2+}					

Note: All these ions, except Li^+ and Be^{2+}, have eight electrons in their outer shell and have the same electron arrangements as the noble gases Ne to Xe. Both Li^+ and Be^{2+} have only two electrons in the $n = 1$ shell, as in the noble gas He.

Notice that the ions N^{3-}, O^{2-}, F^-, Na^+, Mg^{2+}, and Al^{3+} all have the same electron arrangement, namely, that of neon, 2, 8. Atoms and ions that have the same electron arrangement are said to be **isoelectronic**. They differ only in their nuclear charge, which ranges from $+7$ for N^{3-} to $+13$ for Al^{3+}.

Example 4.5 Give the empirical formulas and draw Lewis structures for the ionic compounds formed by each of the following pairs of elements: Na, O; Mg, Br; K, S; Al, F; Na, P.

Solution We can find the charge on each of the ions formed by these elements from Table 4.6 or directly from the position of the element in the periodic table. From the charges on the ions we can deduce the ratio of the numbers of positive and negative ions in the compound and hence write the empirical formula and the Lewis structure. The accompanying table lists the results.

ELEMENTS	IONS	EMPIRICAL FORMULA	LEWIS STRUCTURE
Na, O	Na^+, $:\overset{..}{\underset{..}{O}}:^{2-}$	Na_2O	$(Na^+)_2(:\overset{..}{\underset{..}{O}}:^{2-})$
Mg, Br	Mg^{2+}, $:\overset{..}{\underset{..}{Br}}:^-$	$MgBr_2$	$(Mg^{2+})(:\overset{..}{\underset{..}{Br}}:^-)_2$
K, S	K^+, $:\overset{..}{\underset{..}{S}}:^{2-}$	K_2S	$(K^+)_2(:\overset{..}{\underset{..}{S}}:^{2-})$
Al, F	Al^{3+}, $:\overset{..}{\underset{..}{F}}:^-$	AlF_3	$(Al^{3+})(:\overset{..}{\underset{..}{F}}:^-)_3$
Na, P	Na^+, $:\overset{..}{\underset{..}{P}}:^{3-}$	Na_3P	$(Na^+)_3(:\overset{..}{\underset{..}{P}}:^{3-})$

Covalent Bonds

We have seen that when a metal such as magnesium, from which the valence shell electrons are easily removed, combines with a nonmetal such as chlorine, which has a strong tendency to gain an additional electron, an ionic compound is formed in which oppositely charged ions are held by electrostatic attraction. But how are the two identical atoms held in the chlorine molecule, Cl_2? If one chlorine atom were to gain an electron to become Cl^-, the other would have to lose an electron to become Cl^+. Because of its high core charge, the chlorine atom holds on to its valence electrons rather strongly, and it has a high ionization energy. In any case Cl^+ would have only six, rather than eight, electrons in its valence shell. Lewis suggested in 1916 that since each chlorine atom in Cl_2 requires an additional electron to complete its octet they may be thought of as *sharing* a pair of electrons. Thus he viewed the formation of the Cl_2 molecule in the following way

$$:\overset{..}{\underset{..}{Cl}}\cdot \; + \; \cdot\overset{..}{\underset{..}{Cl}}: \; \longrightarrow \; :\overset{..}{\underset{..}{Cl}}:\overset{..}{\underset{..}{Cl}}:$$

Shared pair of electrons

By sharing a pair of electrons, both atoms effectively acquire eight electrons in their valence shells (Figure 4.11).

The atoms in the molecules of many compounds formed between the nonmetals can be thought of as being held together in the same way. For example, phosphorus in Group V has a valence of 3 and forms the chloride PCl_3. Phosphorus needs three more electrons to complete its octet, and it can obtain these electrons by sharing a pair of electrons with each of three chlorine atoms:

$$\cdot\overset{..}{P}\cdot \; + \; 3\cdot\overset{..}{\underset{..}{Cl}}: \; \longrightarrow \; :\overset{..}{\underset{..}{Cl}}:\overset{..}{P}:\overset{..}{\underset{..}{Cl}}: \\ :\overset{..}{\underset{..}{Cl}}:$$

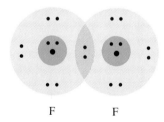

F   F

Figure 4.11 Formation of a Covalent Bond by Sharing of an Electron Pair. By sharing a pair of electrons, each F atom effectively acquires a valence shell of eight electrons in the F_2 molecule.

The diagrams

$$:\overset{\cdot\cdot}{\underset{\cdot\cdot}{Cl}}\cdot\overset{\cdot\cdot}{\underset{\cdot\cdot}{Cl}}: \quad\quad \text{and} \quad\quad :\overset{\cdot\cdot}{\underset{\cdot\cdot}{Cl}}:\overset{\cdot\cdot}{P}:\overset{\cdot\cdot}{\underset{\cdot\cdot}{Cl}}:$$
$$:\overset{\cdot\cdot}{\underset{\cdot\cdot}{Cl}}:$$

which show the arrangement of the electrons in the Cl_2 and PCl_3 molecules, are **Lewis structures.**

Whenever two atoms are held by a pair of shared electrons, we say that they have a **covalent bond** between them. In short, a covalent bond may be described as a pair of shared electrons. The Cl_2 molecule has one covalent bond. The PCl_3 molecule has three covalent bonds.

We can describe the bond in the hydrogen molecule, H_2, in the same way:

$$H\cdot + \cdot H \longrightarrow H\cdot H$$

We note, however, that hydrogen has only one electron in the first ($n = 1$) shell; it only needs one more electron to fill this shell and thus to have the same electron arrangement as the noble gas helium. Hydrogen and helium are therefore exceptions to the octet rule, because only two electrons are needed to fill the first shell. The ions Li^+ and Be^{2+} also have a filled ($n = 1$) shell of two electrons.

The bonding in many molecules formed by hydrogen with other nonmetallic elements can be described in the same way. We can derive the Lewis structure for the water molecule as follows:

$$2H\cdot + :\overset{\cdot\cdot}{\underset{\cdot}{O}}\cdot \longrightarrow :\overset{\cdot\cdot}{\underset{\cdot\cdot}{O}}\cdot H$$
$$H$$

Each hydrogen atom has effectively filled its valence shell with two electrons, and the oxygen atom has filled its valence shell with eight electrons.

It is sometimes convenient when you are learning to write Lewis structures to distinguish between the electrons that originally belonged to different atoms, as we have done above. However, because all electrons are identical, we do not usually make this distinction. We normally draw the Lewis structures of Cl_2, PCl_3, H_2, and H_2O as follows, where all the electrons are represented by identical dots:

$$:\overset{\cdot\cdot}{\underset{\cdot\cdot}{Cl}}:\overset{\cdot\cdot}{\underset{\cdot\cdot}{Cl}}: \quad :\overset{\cdot\cdot}{\underset{\cdot\cdot}{Cl}}:\overset{\cdot\cdot}{P}:\overset{\cdot\cdot}{\underset{\cdot\cdot}{Cl}}: \quad H:H \quad :\overset{\cdot\cdot}{\underset{\cdot\cdot}{O}}:H$$
$$:\overset{\cdot\cdot}{\underset{\cdot\cdot}{Cl}}: \quad\quad\quad\quad H$$

For the compounds of hydrogen with carbon, nitrogen, oxygen, and fluorine, we can write the following Lewis structures for the molecules:

$$\begin{array}{cccc} H & & & \\ H:\overset{\cdot\cdot}{C}:H & H:\overset{\cdot\cdot}{N}:H & :\overset{\cdot\cdot}{\underset{\cdot\cdot}{O}}:H & :\overset{\cdot\cdot}{\underset{\cdot\cdot}{F}}:H \\ H & H & H & \end{array}$$

$$\text{Methane} \quad \text{Ammonia} \quad \text{Water} \quad \text{Hydrogen fluoride}$$

The valence of hydrogen is 1, because it needs only one electron to complete its valence shell of two electrons. The valence of the other nonmetallic elements is determined by the number of electrons that they need to fill their valence shells with eight electrons. In other words, it is equal to 8 minus the number of electrons in the valence shell.

Hydrogen can also form ionic compounds. In particular, its compounds with metals, such as sodium, are ionic and contain the hydride ion, H^-, which has a filled shell of two electrons:

$$Na\cdot + \cdot H \longrightarrow (Na^+)(H:^-)$$
$$\text{Sodium hydride}$$

Molecules are frequently represented by simplified forms of the Lewis structures. In particular, a bonding electron pair shared between two atoms is usually represented by a line drawn between the two atoms. In this convention the Lewis structures of water, ammonia, methane, and carbon tetrachloride molecules are written in the following manner:

$$
:\!\ddot{O}\!-\!H \qquad H\!-\!\ddot{N}\!-\!H \qquad H\!-\!\underset{|}{\overset{|}{C}}\!-\!H \qquad :\!\ddot{Cl}\!-\!\underset{\ddot{Cl}:}{\overset{:\ddot{Cl}:}{\underset{|}{\overset{|}{C}}}}\!-\!\ddot{Cl}:
$$

Depending on the number of bonds that it forms, an atom with an octet of electrons may have one or more pairs of electrons that are not forming bonds; these pairs are called **unshared pairs**, **nonbonding pairs**, or **lone pairs**. In the ammonia molecule the nitrogen atom has one unshared pair of electrons. In the water molecule the oxygen atom has two unshared pairs of electrons. The chlorine atoms in carbon tetrachloride each have three unshared pairs of electrons.

Example 4.6 Draw the Lewis structure for silicon tetrafluoride, $SiCl_4$, a colorless, volatile liquid.

Solution Silicon is in group IV of the periodic table and has the Lewis symbol $\cdot\dot{Si}\cdot$. Silicon therefore can accommodate four more electrons in its valence shell to complete its octet. Chlorine is in group VII and has the Lewis symbol $:\ddot{Cl}\cdot$; it requires one more electron to complete its octet. Silicon can therefore use each of its four electrons to form four shared pairs with four chlorine atoms, as follows:

$$
\cdot\dot{Si}\cdot + 4:\ddot{Cl}\cdot \longrightarrow :\ddot{Cl}\!:\!\underset{:\ddot{Cl}:}{\overset{:\ddot{Cl}:}{Si}}\!:\!\ddot{Cl}: \qquad \text{or} \qquad :\ddot{Cl}\!-\!\underset{:\ddot{Cl}:}{\overset{:\ddot{Cl}:}{\underset{|}{\overset{|}{Si}}}}\!-\!\ddot{Cl}:
$$

Exceptions to the Octet Rule

Although as we have seen, the octet rule is very useful in enabling us to understand how both covalent and ionic compounds are formed, it has many exceptions. One is obvious and has already been mentioned: The rule cannot apply to hydrogen, which has a valence shell that is filled by only two electrons. The elements on the left-hand side of the periodic table also depart from the octet rule in their covalent compounds. Strict application of the octet rule would predict that the elements in groups I, II, and III should not form covalent compounds because they have fewer than four electrons in their valence shells and cannot therefore complete their octets by electron sharing. However, these elements do form some covalent molecules. Boron trifluoride, BF_3, and boron trichloride, BCl_3, are two examples. In forming these molecules, boron uses all three electrons to form covalent bonds:

$$
\cdot\dot{B}\cdot + 3:\ddot{F}\cdot \longrightarrow :\ddot{F}\!-\!\underset{:\ddot{F}:}{\overset{}{\underset{|}{B}}}\!-\!\ddot{F}:
$$

but it still has only six electrons in its valence shell. It is therefore an exception to the octet rule.

We have mentioned that some elements such as phosphorus and sulfur have "higher" valences in addition to the normal valences predicted from the position of the element in the periodic table. For example, phosphorus forms the compound PCl_5 as well as the compound PCl_3. As we will discuss in Chapter 7, the phosphorus atom in PCl_5 is another exception to the octet rule.

Polyatomic Ions

As we just mentioned, the valence shell of the boron atom in the boron trifluoride molecule is incomplete and can accommodate two more electrons. Thus it has a strong tendency to undergo reactions in which it acquires these two additional electrons. For example, BF_3 combines readily with a fluoride ion, F^-, to form the negatively charged molecule (negative ion) tetrafluoroborate, BF_4^-:

$$
\begin{array}{ccc}
\ddot{\:F}\!: & & \left[\begin{array}{c}\ddot{\:F}\!:\end{array}\right]^- \\
| & & | \\
:\!\ddot{F}\!-\!B \;\; + \;\; :\!\ddot{F}\!:^- \;\longrightarrow\; :\!\ddot{F}\!-\!B\!-\!\ddot{F}\!: \\
| & & | \\
:\!\ddot{F}\!: & & :\!\ddot{F}\!:
\end{array}
$$

In this reaction boron forms a fourth covalent bond by sharing one of the unshared electron pairs of the fluoride ion.

The tetrafluoroborate ion, BF_4^-, is an example, of which we will meet many in later chapters, of an ion that is not a single charged atom, like Cl^- or Na^+. It is, rather, a molecule consisting of two or more atoms joined by covalent bonds but with an overall charge. It is called a **polyatomic** (or molecular) **ion**. Another example is the ammonium ion,

$$
\left[\begin{array}{c}
H \\
| \\
H\!-\!N\!-\!H \\
| \\
H
\end{array}\right]^+
$$

These polyatomic ions form ionic compounds in the same way as monatomic ions. Ammonium chloride, $NH_4^+Cl^-$, and sodium tetrafluoroborate, $Na^+BF_4^-$, are examples of such compounds:

$$
\left[\begin{array}{c}
H \\
| \\
H\!-\!N\!-\!H \\
| \\
H
\end{array}\right]^+ :\!\ddot{C}l\!:^- \qquad Na^+ \left[\begin{array}{c}
:\!\ddot{F}\!: \\
| \\
:\!\ddot{F}\!-\!B\!-\!\ddot{F}\!: \\
| \\
:\!\ddot{F}\!:
\end{array}\right]^-
$$

Formal Charge

Where is the charge located on a polyatomic ion? We can obtain an approximate answer to this question by assuming that the electrons of each bond are shared equally between the two atoms that are bonded. Thus for the ammonium ion,

$$
\left[\begin{array}{c}
H \\
\cdot\cdot \\
H\!:\!N\!:\!H \\
\cdot\cdot \\
H
\end{array}\right]^+
$$

we assign one electron of each bond (blue) to each hydrogen and one electron of each bond (black) to the nitrogen.

Of the total of eight electrons in the ion we have assigned one to each of the hydrogen atoms and four to the nitrogen atom. Since each hydrogen atom has a core charge of $+1$ and one electron, it has a zero charge. The nitrogen atom has a core charge of $+5$; it has four electrons assigned to it, so it has a charge of $+5 - 4 = +1$. The positive charge of the ammonium ion is formally located on the nitrogen atom; this charge can be indicated in the Lewis structure as follows:

$$\begin{array}{c} \overset{..}{\underset{..}{H}} \\ H\!:\!\overset{\oplus}{\underset{\cdot\cdot}{N}}\!:\!H \\ \overset{..}{H} \end{array}$$

The $+1$ charge on the nitrogen atom is called its **formal charge**.

Notice that the ammonium ion and the methane molecule have the same Lewis structures. They have the same number of nuclei and the same number of electrons arranged in the same way. However, carbon has a core charge of $+4$, and since it is assigned four electrons, it has a zero charge:

$$\begin{array}{cc} \overset{..}{H} & \overset{..}{H} \\ H\!:\!\overset{\oplus}{N}\!:\!H \qquad & H\!:\!\overset{..}{C}\!:\!H \\ \overset{..}{H} & \overset{..}{H} \end{array}$$

As another example, consider the hydroxide ion, OH^-, which we find in compounds such as sodium hydroxide, $NaOH$. First, we draw the Lewis structure. Oxygen has six electrons, $:\overset{..}{O}\cdot$, and hydrogen, $\cdot H$, has one. We can form one covalent bond,

$$:\overset{..}{\underset{\cdot}{O}}\cdot + \cdot H \longrightarrow :\overset{..}{\underset{\cdot\cdot}{O}}\!:\!H$$

We can then add another electron to give the negative charge:

$$:\overset{..}{\underset{\cdot\cdot}{O}}\!:\!H + e^- \longrightarrow [:\overset{..}{\underset{\cdot\cdot}{O}}\!:\!H]^-$$

To find the formal charges, we assign one of the electrons of the covalent bond to oxygen. The oxygen atom also has six more electrons—three unshared pairs—which are not shared with any atom and which therefore are all assigned to the oxygen atom, giving a total of seven electrons to oxygen (black). Since oxygen has a core charge of $+6$, it has a formal charge of $+6 - 7 = -1$. We assign one of the electrons of the covalent bond to hydrogen (blue), which therefore has a zero formal charge. Thus we write the Lewis structure as follows:

$$\overset{\ominus}{}:\overset{..}{\underset{\cdot\cdot}{O}}\!:\!H$$

It is important to be able to correctly assign formal charges in Lewis structures, and you should always assign the formal charges when you write a Lewis structure.

The procedure that we have used for assigning formal charges can be summarized by the following rule:

$$\begin{pmatrix} \text{Formal} \\ \text{charge} \end{pmatrix} = \begin{pmatrix} \text{Core} \\ \text{charge} \end{pmatrix} - \begin{pmatrix} \text{Number of} \\ \text{unshared electrons} \end{pmatrix} - \frac{1}{2}\begin{pmatrix} \text{Number of} \\ \text{shared electrons} \end{pmatrix}$$

Alternatively, since the core charge is equal to the group number, and since $\frac{1}{2}$(number of shared electrons) = number of bonds, we have

$$\begin{pmatrix} \text{Formal} \\ \text{charge} \end{pmatrix} = \begin{pmatrix} \text{Group} \\ \text{number} \end{pmatrix} - \begin{pmatrix} \text{Number of} \\ \text{unshared electrons} \end{pmatrix} - \begin{pmatrix} \text{Number} \\ \text{of bonds} \end{pmatrix}$$

Example 4.7 Assign formal charges to each of the atoms in the BF_4^- ion.

Solution The Lewis structure is

$$\left[\begin{array}{c} :\overset{..}{\underset{..}{F}}: \\ :\overset{..}{F}\!:\!B\!:\!\overset{..}{F}: \\ :\underset{..}{\overset{..}{F}}: \end{array} \right]^-$$

Boron has four shared pairs of electrons. We assign one electron of each of these shared pairs to boron, giving boron four electrons (blue). Since boron is in group III and has a core charge of $+3$, the formal charge on boron is $+3 - 4 = -1$. Each fluorine atom has three unshared pairs and is also assigned one electron from the pair that it shares with boron, making a total of seven electrons (black) assigned to each fluorine atom. Since fluorine is in group VII and has a core charge of $+7$, each fluorine has a formal charge of $+7 - 7 = 0$. Thus we can write the Lewis structure with formal charges as follows:

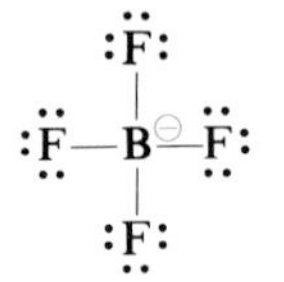

$$
\begin{array}{c}
:\ddot{F}: \\
| \\
:\ddot{F}-\overset{\ominus}{B}-\ddot{F}: \\
| \\
:\ddot{F}:
\end{array}
$$

Alternatively, using the rule just given, we have the following:

ATOM	GROUP NUMBER	NUMBER OF UNSHARED ELECTRONS	NUMBER OF BONDS	FORMAL CHARGE
Boron	3	0	4	$3 - 0 - 4 = -1$
Fluorine	7	6	1	$7 - 6 - 1 = 0$

We note that the number of covalent bonds formed by an atom depends on its formal charge. A neutral nitrogen atom forms three bonds, as in NH_3, but a positively charged nitrogen atom forms four bonds, as in NH_4^+

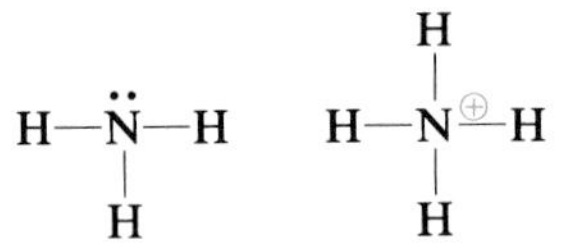

$$
\begin{array}{cc}
 & H \\
 & | \\
H-\overset{\cdot\cdot}{N}-H \qquad & H-\overset{\oplus}{N}-H \\
| & | \\
H & H
\end{array}
$$

A positively charged nitrogen has four electrons and is isoelectronic with a carbon atom:

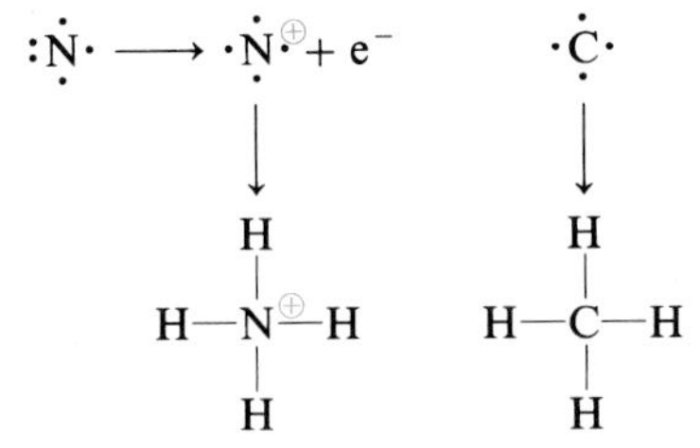

$$
:\dot{N}\cdot \longrightarrow \cdot\overset{\oplus}{N}\cdot + e^- \qquad \cdot\dot{C}\cdot
$$

$$
\begin{array}{cc}
H & H \\
| & | \\
H-\overset{\oplus}{N}-H & H-C-H \\
| & | \\
H & H
\end{array}
$$

Thus both a neutral carbon atom and a positively charged nitrogen atom have four electrons and can form four bonds. They both have a valence of 4. A neutral nitrogen atom has a valence of 3.

Multiple Bonds

When two atoms share one electron pair, the bond is called a **single bond**. Some atoms can also form **double bonds**, which consist of two pairs of shared electrons. For example, carbon normally forms four covalent bonds and oxygen two covalent bonds; therefore, the CO_2 molecule must have two covalent bonds between each oxygen atom and the carbon atom. Each oxygen atom shares *two* pairs of electrons with the carbon atom, so there are two double bonds in the CO_2 molecule:

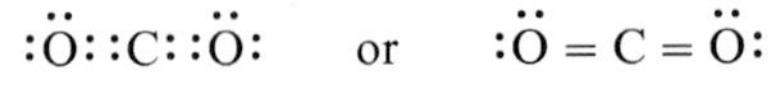

$$
:\ddot{O}::C::\ddot{O}: \qquad \text{or} \qquad :\ddot{O} = C = \ddot{O}:
$$

A carbon atom can also form double bonds with other carbon atoms. Carbon-carbon double bonds are found in many molecules. For example, the hydrocarbon ethene (ethylene) has the following Lewis structure:

$$\underset{H}{\overset{H}{>}}C::C\underset{H}{\overset{H}{<}} \quad \text{or} \quad \underset{H}{\overset{H}{>}}C=C\underset{H}{\overset{H}{<}}$$

Here each carbon atom shares an electron pair with each of two hydrogen atoms, and it shares two electron pairs with the other carbon atom.

Sometimes, three pairs of electrons may be shared between two atoms, thereby forming a **triple bond**. Examples are found in the hydrocarbon ethyne (acetylene), C_2H_2, and in the nitrogen molecule, N_2:

$$H—C\equiv C—H \qquad :N\equiv N:$$

Ethyne   Nitrogen

Example 4.8 Draw the Lewis structure for the highly poisonous gas hydrogen cyanide, HCN.

Solution The Lewis symbols for the atoms are $H\cdot$, $\cdot\dot{C}\cdot$, and $\cdot\ddot{N}\cdot$. Hydrogen can form one bond; carbon, four; and nitrogen, three. We first form the HC bond:

$$H\cdot + \cdot\dot{C}\cdot \longrightarrow H:\dot{C}\cdot$$

We see that we can then form three bonds between carbon and nitrogen to give the final structure:

$$H:\dot{C}\cdot + \cdot\ddot{N}: \longrightarrow H:C:::N: \qquad \text{or} \qquad H—C\equiv N:$$

We have seen that Lewis's idea that a covalent bond is formed by a pair of shared electrons enables us to understand the number of bonds formed by an atom in a covalent compound and therefore to rationalize the formulas of covalent compounds. But we have not explained *why* a pair of shared electrons is able to hold two atoms together. We will take up this important topic in Chapter 6, where we will see that our current understanding of chemical bonding goes far beyond the simple ideas on which Lewis structures are based. Nevertheless, Lewis structures are a very useful way of approximately describing the arrangement of the electrons in molecules and polyatomic ions. We will make extensive use of them in this book.

IMPORTANT TERMS

The **alkali metals** are the elements of group I.

The **alkaline earth metals** are the elements of group II.

The **bond length** is the distance between the nuclei of two atoms that have a chemical bond between them.

The **chalcogens** are the elements of group VI.

A **chemical bond** is said to exist between any two atoms that are held strongly together.

The **core** of an atom is the nucleus plus the completed inner electron shells.

The **core charge** of an atom is equal to the atomic number Z minus the total number of electrons in the completed inner shells.

A **covalent bond** consists of a pair of electrons shared between two atoms.

The **covalent** (or atomic) **radius** is one-half the distance between two atoms of the same kind held together by a covalent bond.

A **double bond** consists of two pairs of shared electrons.

The **formal charge** of an atom in a molecule or ion is given by either of the following equivalent rules:

$$\begin{pmatrix}\text{Formal}\\\text{charge}\end{pmatrix} = \begin{pmatrix}\text{Core}\\\text{charge}\end{pmatrix} - \begin{pmatrix}\text{Number of}\\\text{unshared electrons}\end{pmatrix}$$

$$-\frac{1}{2}\begin{pmatrix}\text{Number of}\\\text{shared electrons}\end{pmatrix}$$

or

$$\begin{pmatrix}\text{Formal}\\\text{charge}\end{pmatrix} = \begin{pmatrix}\text{Group}\\\text{number}\end{pmatrix} - \begin{pmatrix}\text{Number of}\\\text{unshared electrons}\end{pmatrix}$$

$$-\begin{pmatrix}\text{Number}\\\text{of bonds}\end{pmatrix}$$

A **group** is a vertical column of the periodic table.

The **halogens** are the elements of group VII.

There is said to be an **ionic bond** between two oppositely charged ions that are held together by their electrostatic attraction.

Ionic compounds are composed of oppositely charged ions held together by electrostatic attraction, that is by ionic bonds.

The **ionization energy** is the energy needed to remove an electron from an atom in the gas phase.

A **Lewis (electron dot) symbol** is the symbol of an element surrounded by a number of dots equal to the number of electrons in its valence shell.

A **negative ion** is a negatively charged atom or molecule. Negative ions may be formed by the addition of one or more electrons to a neutral atom or molecule.

The **noble gases** are the elements of group VIII.

An **octet** is a valence shell containing eight electrons.

The **octet rule** states that when an atom forms a compound, it gains, loses, or shares electrons to obtain eight electrons in its valence shell.

A **period** is a horizontal row of the periodic table.

A **positive ion** is a positively charged atom or molecule. Positive ions may be formed by the removal of one or more electrons from a neutral atom or molecule.

A **semimetal (metalloid)** is an element whose properties are intermediate between those of a metal and those of a nonmetal.

A **shell** is one of the successive layers of electrons around an atom.

A **single bond** is formed by one pair of shared electrons.

A **transition element** is one of the ten elements in each period between groups II and III in the periodic table, excluding the 14 lanthanide and 14 actinide elements.

A **triple bond** is formed by three shared pairs of electrons.

Unshared pairs are electron pairs that are not involved in bonding (nonbonding pairs or lone pairs).

Valence is the combining power of an element. In ionic compounds it is equal to the charge on an ion. In covalent compounds it is the number of bonds formed by a neutral atom of the element.

The **valence shell** is the outer electron shell of an atom.

PROBLEMS

Periodic Table

1. Find each of the following elements in the periodic table. State whether it is a metal, a semimetal (metalloid), or a nonmetal, and give the name of the element.

    (a) He   (b) P   (c) K   (d) Ca

    (e) Te   (f) Br   (g) Al   (h) Sn

2. Identify three pairs of elements whose atomic masses do not follow the order of their atomic numbers. Explain why.

3. The element with atomic number 22 forms crystals that melt at 1668°C, and the liquid element boils at 3313°C. The crystals are hard, conduct heat and electricity, can be drawn into thin wires, and emit electrons when exposed to ultraviolet radiation. On the basis of these properties, classify the element as a metal or a nonmetal. Which element is it?

4. Refer to the periodic table and identify the following:

    (a) The element that is in Group III and in the fifth period.

    (b) An element that has properties similar to those of sulfur.

    (c) A very reactive metal in the sixth period.

    (d) The halogen in the fifth period.

    (e) The alkaline earth metal in the fourth period.

5. Write balanced equations for the following reactions (if any):

    (a) Magnesium with steam

    (b) Sulfur with hydrogen

    (c) Sodium with iodine

    (d) Potassium with water

    (e) Hydrogen with chlorine

    (f) Neon with water

6. Refer to the periodic table, and predict the valences of the following elements:

    Ba   Cs   Ga   Te   Ar   Ge   Sb.

* The asterisk denotes the more difficult problems.

Give the empirical formulas of the compounds that they are expected to form on reaction with (a) hydrogen, (b) oxygen, (c) fluorine.

7. You are given the information that francium, Fr, is an alkali metal; tin, Sn, is in the same group of the periodic table as carbon; astatine, At, is a halogen; and radon, Rn, is a noble gas. Write formulas for their hydrides, if any. What are the expected products of the reaction, if any, of these elements with chlorine?

8. Predict the formulas of the chlorides formed by elements in Group III of the periodic table.

9. Write the empirical formulas of the compounds that the elements of the second period may be expected to form (a) with sulfur and (b) with nitrogen.

10. Elements A, X, Y, and Z form the fluorides AF_2, XF_3, YF_4, and ZF. In which groups of the periodic table do you expect to find the elements A, X, Y, and Z?

11. Classify each of the following elements as a metal or as a nonmetal, assign each to its appropriate group in the periodic table, and write the formula for its hydride:

C	Ca	He	B	Cl
Li	O	F	P	Mg

12. Indium oxide contains 82.7% indium by mass. Indium occurs naturally in ores containing zinc oxide, ZnO, and it was therefore originally assumed to have the formula InO. Calculate the atomic mass of indium on this basis. Predict its location in the periodic table. Explain why this location is not a suitable position for indium. Mendeleev suggested that the formula of indium oxide must be In_2O_3. Calculate the atomic mass of indium on this basis and find its actual position in the periodic table.

*13. An element A has a melting point of 845°C, is silvery white, and dissolves in water with the evolution of hydrogen. It reacts with hydrogen chloride to give gaseous hydrogen and a solid chloride with empirical formula ACl_2. When 0.230 g of element A was reacted completely with water, 144.1 mL of hydrogen gas, measured at 25°C and a pressure of 740 mm Hg, was evolved. The chloride was analyzed by dissolving 0.1456 g of ACl_2 in water and adding aqueous silver nitrate solution in order to precipitate all the chlorine as silver chloride, AgCl. The weight of dry AgCl obtained was 0.3760 g. What is the element A, and what is its atomic mass?

Shell Model

14. The atomic numbers of phosphorus, carbon, and potassium are, respectively, 15, 6, and 19. Without reference to any other material, predict (a) the number of electrons in each shell for each element and (b) the core charge of each element.

15. What is meant by the *valence shell* of an atom? How many electrons are there in the valence shell of (a) boron, (b) a halogen, (c) helium, (d) neon, (e) magnesium?

16. Determine the number of electrons in each of the following. In each case describe the arrangement of the electrons in shells.

(a) A neutral chlorine atom.

(b) A negatively charged chlorine atom, that is, a chloride ion.

(c) A silicon atom.

(d) A positively charged neon atom, that is, the positive ion, Ne^+

17. What are the core charges of the following atoms and ions?

(a) C   (b) Mg   (c) Mg^{2+} (d) Si

(e) O   (f) O^{2-} (g) S^{2-}

Ionization Energies

18. Explain what is meant by the *ionization energy* of an element. Why is there a tendency for ionization energies to increase in going from left to right along any period of the periodic table, and to decrease in going from top to bottom of any group?

19. What is the order of ionization energies for the atoms F, Ne, and Na?

20. Without consulting a table of ionization energies, arrange the following atoms in order of increasing value of their ionization energy:

Ba   Cs   F   S   As

21. How much energy is needed to convert 1.00 g of sodium atoms to Na^+ ions in the gaseous state?

22. Explain why the ionization energy of helium in higher than that for any other neutral atom.

Covalent Radii

23. Why do the covalent radii of the atoms in any row of the periodic table generally decrease in going from left to right, and increase in any group in going from top to bottom of the periodic table? Why are no values quoted for the covalent radii of He, Ne, or Ar?

24. The bond length in F_2 is 138 pm, and the covalent radius of C is 77 pm. Predict the carbon-fluorine bond lengths in CF_4.

25. From the data in the accompanying table, calculate the covalent radius of each atom. What do you expect the bond lengths in Br_2, BrCl, and I_2 to be?

MOLECULE	BOND LENGTH (pm)
Cl_2	198
CCl_4	176
CBr_4	194
CI_4	215

26. Without consulting Figure 4.6 or the periodic table, predict which atom in each of the following pairs is expected to have the largest covalent (atomic) radius:

(a) F, Cl   (b) B, C   (c) C, Si

(d) P, Al   (e) Si, O

Lewis Structures

27. Write Lewis (electron dot) symbols for each of the following elements:

K	Ca	B	Sn	Sb
Te	Br	Xe	As	Ge

28. On the basis of the octet rule and the position of the element in the periodic table, predict the charge on the ion formed by each of the following elements:

(a) Mg   (b) Rb   (c) Br

(d) S   (e) Al   (f) Li

29. Draw Lewis structures for each of the following ionic compounds:

(a) LiCl   (b) Na_2O (c) AlF_3

(d) CaS   (e) $MgBr_2$

30. Four elements have the following Lewis symbols:

$$A\cdot \quad \cdot\dot{D}\cdot \quad \cdot\dot{E}\cdot \quad :\dot{G}\cdot$$

(a) Place the elements in the appropriate group of the periodic table.

(b) Which elements do you expect to form ions? What do you expect the charge on each of the ions to be?

31. Predict the empirical formula of the ionic compound formed by each of the following pairs of elements:

(a) Li, S   (b) Ca, O   (c) Mg, Br

(d) Na, H   (e) Al, I

32. Predict the empirical formula of the ionic compound formed by each of the following pairs of elements:

(a) Ca, I   (b) Be, O   (c) Al, S

(d) Ca, Br   (e) Rb, Se   (f) Ba, O

33. What is the empirical formula of the compound composed of each of the following pairs of ions?

(a) NH_4^+, PO_4^{3-} (b) Fe^{3+}, O^{2-}

(c) Cu^+, O^{2-} (d) Al^{3+}, SO_4^{2-}

34. What is the empirical formula of each of the following compounds?

(a) Barium iodide

(b) Aluminum chloride

(c) Calcium oxide

(d) Sodium sulfide

(e) Aluminum oxide

35. What is the Lewis (electron-dot) symbol for each of the following atoms and ions? Find two pairs of isoelectronic species in this list.

(a) Ca   (b) Ca^{2+} (c) Ne   (d) O^{2-}

(e) S^{2-} (f) Cl   (g) Ar

36. Draw Lewis structures for the following molecules and ions:

H_2	HCl	HI	PH_3	H_2S
Br^-	Na^+	SiF_4	C_3H_8	F_2O

37. In one of its several different forms the element sulfur consists of S_6 molecules in which the sulfur atoms are joined in a ring. Draw a Lewis structure for S_6.

38. Draw the Lewis structures for the following ions, and assign formal charges to the appropriate atoms:

NH_4^+	NH_2^-	H_3O^+	H_2F^+
PH_4^+	BH_4^-	PCl_4^+	

39. Draw Lewis structures for each of the following molecules:

(a) H_2CO (b) P_2

(c) HNNH   (d) H_2NNH_2

*40. Draw a Lewis structure for PCl_5, and explain why it is an exception to the octet rule.

CHAPTER 5
THE HALOGENS

Introduction to Oxidation-Reduction and Acid-Base Reactions

Group
I

1	H																	He
		II		Metals	Nonmetals	Semimetals						III	IV	V	VI	VII		VIII
2	Li	Be											B	C	N	O	F 19.00	Ne
3	Na	Mg			Transition Elements								Al	Si	P	S	Cl 35.45	Ar
4	K	Ca	Sc	Ti	V	Cr	Mn	Fe	Co	Ni	Cu	Zn	Ga	Ge	As	Se	Br 79.90	Kr
5	Rb	Sr	Y	Zr	Nb	Mo	Tc	Ru	Rh	Pd	Ag	Cd	In	Sn	Sb	Te	I 126.9	Xe
6	Cs	Ba	La	Hf	Ta	W	Re	Os	Ir	Pt	Au	Hg	Tl	Pb	Bi	Po	At (210)	Rn
7	Fr	Ra	Ac	104	105	106	107											

Period

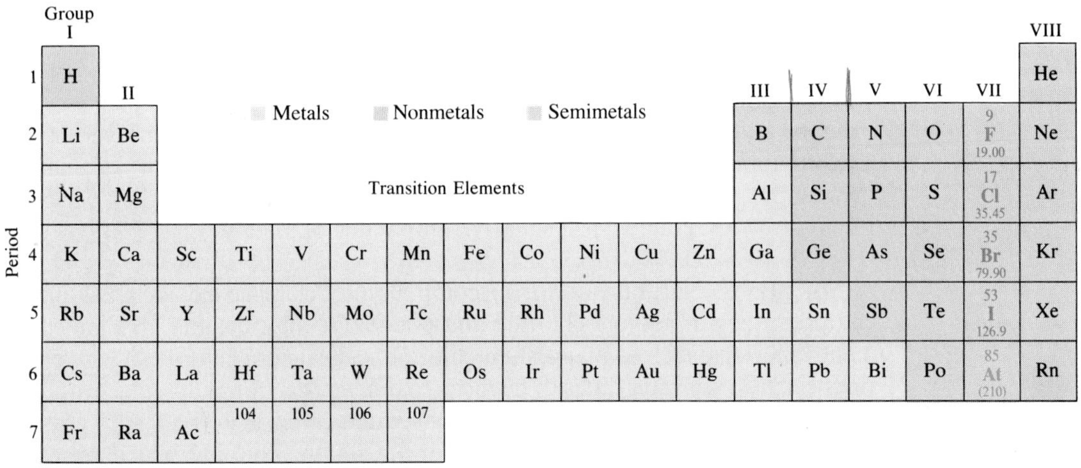

The elements of Group VII, fluorine, chlorine, bromine, iodine, and astatine, are collectively called the **halogens**. Astatine is an extremely rare, radioactive element, and we do not discuss it here. The other halogens are well-known nonmetals that form many important compounds. For example, chlorine is used to purify water supplies; the hypochlorite ion, OCl^-, is the active component of household bleach; fluoride ion is added to the water in many cities and to many toothpastes to help prevent tooth decay; iodine is used as an antiseptic.

The halogens and their compounds provide many examples of both ionic and covalent substances. Discussion of the halogens will enable us to expand our knowledge and understanding of chemical bonding. The reactions of the halogens illustrate two very important types of chemical reactions: oxidation-reduction reactions and acid-base reactions. We begin our study of these two reactions in this chapter.

These elements are all composed of diatomic molecules in which the two atoms are held together by a single covalent bond; for example, the Lewis structure for chlorine is

$$:\ddot{C}l - \ddot{C}l:$$

If we go from top to bottom in the group, the atoms follow the usual trend and become progressively larger. Therefore the molecules also increase in size, in the order $F_2 < Cl_2 < Br_2 < I_2$. The halogens are among the most reactive of the elements; this reactivity decreases from fluorine to iodine. The compounds of the halogens with another element are called fluorides, chlorides, bromides, and iodides, or, in general, **halides**; examples are sodium chloride, NaCl, and phosphorus trifluoride, PF_3.

Table 5.1 summarizes several other physical properties of the halogens. As the size of their molecules increases with increasing atomic number, so do their melting points and boiling points. At room temperature fluorine and chlorine are gases, whereas bromine is a red-brown liquid, and iodine is a black solid (see Experiment 5.1). These facts indicate that the strength of intermolecular forces increases with increasing atomic number and molecular size, in the order $F_2 < Cl_2 < Br_2 < I_2$. This relationship between molecular size, and the strength of intermolecular forces is typical of many substances, for reasons that we discuss in Chapter 13.

Sources and Uses

Chlorine is the most abundant of the halogens, making up about 0.20% of the earth's crust (Table 3.2). Fluorine composes only 0.03% of the earth's crust; bromine and iodine are far less abundant. Because the halogens are very reactive, combining directly with many other elements and reacting with many compounds, they are not found in the free state on the earth. They occur in the form of compounds, particularly compounds of metals.

CHLORINE Naturally occurring compounds of chlorine include alkali metal chlorides, such as sodium chloride and potassium chloride, and alkaline earth metal chlorides, such as magnesium chloride and calcium chloride. These chlorides occur in seawater and salt deposits that have been formed by the evaporation of ancient seas. The total amount of sodium chloride in the earth's crust is truly enormous. The oceans alone contain 3% sodium chloride, which amounts

Table 5.1 Some Properties of the Halogens

	MELTING POINT (°C)	BOILING POINT (°C)	COLOR	BOND LENGTH (pm)	COVALENT (ATOMIC) RADIUS (pm)
F_2	−223	−187	Pale yellow	144	64[a]
Cl_2	−102	−35	Yellow-green	198	99
Br_2	−7	58	Red-brown	228	114
I_2	113	183	Black	266	133

[a] The covalent radius of F, which is an average obtained from a large number of fluorine compounds, is smaller than the value of 72 pm that would be obtained from the F_2 bond length. The F—F bond is longer and weaker than expected. Repulsions between the nonbonding electron pairs, which are closer together than in the other halogen molecules, appear to be responsible for the length and weakness of this bond.

Physical Properties of the Halogens

Chlorine is a pale yellow-green gas at room temperature. The tube inside the flask contains dry ice (solid carbon dioxide, $-78°C$). Yellow, liquid chlorine (b.p. $-35°C$) is condensing on the cold tube.

Bromine is red-brown liquid that is sufficiently volatile to fill the flask with vapor.

When solid iodine is gently heated it forms a violet vapor, which can be seen recrystallizing on the cold flask on top of the beaker.

to a total of approximately 4.6×10^{19} kg of sodium chloride. This amount would form a solid cube of sides 277 km, or 172 miles.

In the amount produced, chlorine ranks as the eighth most important substance produced by the chemical industry. It is used for bleaching wood pulp in the manufacture of paper, for bleaching textiles, for making plastics, such as polyvinyl chloride (PVC), insecticides, and dry-cleaning agents, and for the manufacture of bromine. Many communities treat their water supply with chlorine to kill bacteria. The hypochlorite ion, OCl^-, is used for the same purpose in swimming pools.

BROMINE Metal bromides occur in small amounts in seawater and in salt deposits as well as in the water from mineral springs. In the past seawater was used as a source of bromide ion, but the concentration of bromide in the oceans is only 65–70 ppm. Israel today produces a significant amount of bromine from the waters of the Dead Sea, which contains 4500–5000 ppm of bromine. In the United States most commercially produced bromine is obtained from subterranean *brines*, which are concentrated aqueous solutions of chloride and bromide that can be pumped to the surface. At the present time the most economical brine deposits are in Arkansas; they contain 3800–5000 ppm of bromide ion.

One of the most important uses for bromine is for the manufacture of bromine compounds used as gasoline additives. Bromine compounds are also widely used as pesticides and for treating plastic materials and textiles to make them fireproof. A considerable amount of silver bromide is used in the manufacture of photographic film.

IODINE The concentration of the iodide ion in seawater and underground brines is much lower than that of bromide. At one time seaweed (kelp), which

5.1 THE HALOGENS

151

concentrates the iodine in seawater, was an important source. When the seaweed is burned, its ashes contain as much as 1% of iodide ion. Until recently the most important commercial source of iodine was sodium iodate, $NaIO_3$, which occurs in small amounts in large deposits of sodium nitrate found in Chile. Underground brines containing up to 100 ppm iodide have recently been discovered in Japan and in Midland, Michigan. These are now the world's major source of iodine.

Iodine does not have as many important uses as the other halogens. Silver iodide is used in the manufacture of photographic film. Certain antiseptics are organic compounds containing iodine; a solution of the element in alcohol (tincture of iodine) is also a good, although rather old-fashioned, antiseptic. Iodine is an essential element in our diet because it is part of the structure of the growth-regulating hormone thyroxine, which is produced by the thyroid gland. Lack of iodine in our diet leads to enlargement of the thyroid gland, a condition called goiter. The necessary iodine can be obtained from fish, from sea salt, and from iodized table salt, which contains 0.01% of sodium iodide or potassium iodide.

FLUORINE Although it is less abundant than chlorine, fluorine is widely distributed in minerals such as fluorspar, CaF_2; cryolite, Na_3AlF_6, and fluorapatite, $Ca_5(PO_4)_3F$. Plants absorb small amounts of calcium from the soil and this calcium eventually passes into our bones and teeth, which are mainly hydroxyapatite, $Ca_5(PO_4)_3(OH)$. When OH^- is replaced by F^-, a harder, more acid-resistant layer of $Ca_5(PO_4)_3F$ is formed on the surface of teeth. Because this hard layer protects teeth from decay, most cities now add sodium fluoride to their water supply. A somewhat less effective protection is given by toothpaste containing fluoride ion.

Reactions with Nonmetals

The halogens combine readily with almost all the other nonmetals to give compounds that consist of covalent molecules.

HYDROGEN HALIDES Hydrogen reacts with each of the halogens producing the **hydrogen halides** HF, HCl, HBr, and HI. For example,

$$H_2(g) + Cl_2(g) \longrightarrow 2HCl(g)$$

The reaction between hydrogen and fluorine is violent under all conditions. The reaction of hydrogen with chlorine is slow in the dark but becomes explosive in sunlight or other bright light or on heating to 250°C. Bromine reacts more slowly with hydrogen, and iodine reacts still more slowly; neither of these reactions goes to completion. Thus the reactivity of the halogens toward hydrogen decreases from fluorine to iodine.

Example 5.1 Suppose we wish to convert 100 mL of Cl_2 completely to HCl at 350°C and 1 atm pressure.

(a) What volume of H_2 is required?

(b) How many moles of HCl will be formed?

Solution The reaction is

$$H_2(g) + Cl_2(g) \longrightarrow 2HCl(g)$$

(a) According to Gay-Lussac's law of combining volumes (Chapter 3),

$$1 \text{ volume } H_2 + 1 \text{ volume } Cl_2 \longrightarrow 2 \text{ volumes HCl}$$

Therefore 100 mL of H_2 are required for complete reaction with 100 mL of Cl_2 at 350°C and 1 atm pressure.

(b) The reaction produces 200 mL (0.200 L) of HCl, measured under the same conditions of temperature and pressure (350°C and 1 atm). We can calculate the number of moles of HCl by using the ideal gas law:

$$n = \frac{PV}{RT} = \frac{1.00 \text{ atm} \times 0.200 \text{ L}}{0.0821 \text{ atm L mol}^{-1} \text{ K}^{-1} \times 623 \text{ K}} = 3.91 \times 10^{-3} \text{ mol}$$

The hydrogen halides are all colorless gases that have pungent odors, and are all very soluble in water (see Experiment 5.2). Some of their properties are summarized in Table 5.2. The bond lengths of the molecules increase steadily from HF to HI as the size of the halogen atom increases. The melting points and boiling points increase from HCl to HI as the intermolecular forces become stronger with increasing atomic number and increasing size of the halogen atom. Hydrogen fluoride, however, has unexpectedly high melting and boiling points compared with the other halogen halides; these unusual properties will be discussed in Chapter 13.

NONMETAL HALIDES   When chlorine is heated with carbon, carbon tetrachloride is produced:

$$C(s) + 2Cl_2(g) \longrightarrow CCl_4(l)$$

When red phosphorus is gently heated in a slow stream of chlorine, liquid phosphorus trichloride, PCl_3, is formed (see Figure 5.1):

$$2P(s) + 3Cl_2(g) \longrightarrow 2PCl_3(l)$$

If an excess of chlorine is used, pale yellow solid phosphorus pentachloride, PCl_5, is also produced.

EXPERIMENT 5.2

Solubility of Hydrogen Chloride in Water

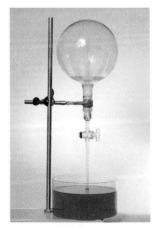

The flask contains colorless hydrogen chloride gas. When the tap is opened hydrogen chloride dissolves in the water, reducing the pressure in the flask so that water is drawn into the flask from the dish below, and creating a fountain.

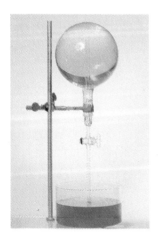

A few drops of bromothymol blue have been added to the water in the dish. The color changes to yellow as the water enters the flask because hydrogen chloride forms an acid solution in water. Hydrogen chloride is so soluble in water that the flask fills almost completely.

Table 5.2 Some Properties of the Hydrogen Halides

	MELTING POINT (°C)	BOILING POINT (°C)	BOND LENGTH (pm)	SOLUBILITY IN WATER AT 10°C (g L^{-1})
HF	−83	20	92	Miscible
HCl	−112	−85	127	780
HBr	−87	−67	141	2100
HI	−51	−35	161	2340

When a stream of dry chlorine is passed over molten sulfur, an orange-red liquid is formed. It is a mixture of two chlorides of sulfur: S_2Cl_2, disulfur dichloride, which is a foul-smelling, orange liquid, and SCl_2, sulfur dichloride, which is a deep red liquid:

$$2S(s) + Cl_2(g) \longrightarrow S_2Cl_2(l)$$
$$S(s) + Cl_2(g) \longrightarrow SCl_2(l)$$

Many other nonmetal halides such as PF_3, CF_4, CBr_4, OF_2, and ICl can be made similarly. They are all covalent molecular compounds. The shapes of some nonmetal halides and their Lewis structures are given in Figure 5.2. Table 5.3 summarizes the properties of some nonmetal chlorides. Most of them are gases or liquids with low boiling points. The boiling points and melting points increase, with increasing atomic number, as we go down each group, so that a few of the halides of the heavier elements are solids.

Except for carbon tetrachloride, these nonmetal chlorides all fume in moist air and react with water, in many cases rather vigorously. Carbon tetrachloride is the best known and most widely used of these nonmetal chlorides. It is exceptional in that it is insoluble in water and does not react with it (see Experiment 5.3). The reason for this difference between CCl_4 and the other non-metal halides is discussed in Chapter 7. Carbon tetrachloride dissolves many of the other nonmetal halides and many other substances such as Br_2 and I_2 (see Experiment 1.6); it is therefore used as a solvent. Because it dissolves grease, it has been used as a cleaning fluid in dry cleaning. But recently CCl_4 has been replaced by other solvents since it was found to be toxic.

Figure 5.1 Apparatus for Preparing Nonmetal Chlorides. The nonmetal chlorides PCl_3, S_2Cl_2, and SCl_2 may be prepared in this apparatus by passing chlorine over the heated element. The chlorine is first dried by passing it over anhydrous calcium chloride.

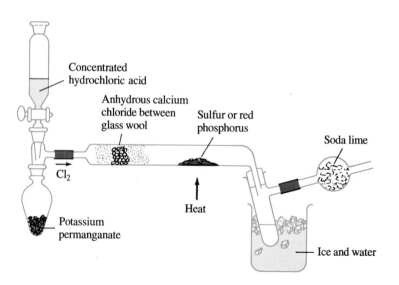

Concentrated hydrochloric acid

Anhydrous calcium chloride between glass wool

Sulfur or red phosphorus

Soda lime

Cl_2

Heat

Potassium permanganate

Ice and water

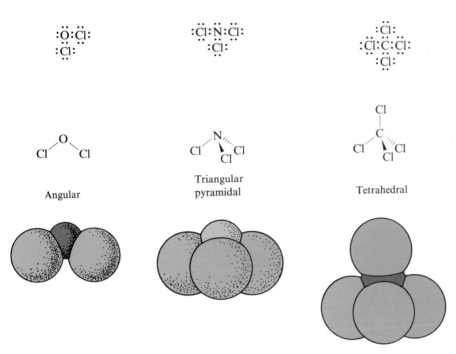

Figure 5.2 Lewis Structures and Molecular Shapes of Nonmetal Halides. Halides of group VI elements, such as Cl_2O, are all angular molecules like H_2O. The halides of group V, such as PCl_3 and NCl_3, are all triangular pyramidal molecules; they have the same shape as NH_3. The halides of group IV are tetrahedral, with the group IV atom at the center and a halogen atom at each of the four corners of the tetrahedron; CCl_4 is an example.

Angular

Triangular pyramidal

Tetrahedral

Reactions with Metals: Metal Halides

The halogens react with many metals to form halides (see Experiment 5.4). Some typical reactions are the following:

$$2Na(s) + Cl_2(g) \longrightarrow 2NaCl(s)$$

$$Ca(s) + F_2(g) \longrightarrow CaF_2(s)$$

$$Mg(s) + Cl_2(g) \longrightarrow MgCl_2(s)$$

Table 5.3 Properties of Some Chlorides of Nonmetals

PERIOD	CHLORIDE	MELTING POINT (°C)	BOILING POINT (°C)	BOND LENGTH (pm)
2	CCl_4	−23	77	176
	NCl_3	−27	71	173
	OCl_2	−20	4	169
	FCl	−154	−101	163
3	$SiCl_4$	−68	57	201
	PCl_3	−91	74	204
	SCl_2	−122	59	200
	Cl_2	−102	−35	198
4	$GeCl_4$	−50	86	209
	$AsCl_3$	−16	130	216
	$SeCl_2$[a]	—	—	—
	$BrCl$	−66	10	214
5	$SnCl_4$	−33	114	232
	$SbCl_3$	73	223	238
	$TeCl_2$	209	327	234
	ICl	27	97	230

[a] $SeCl_2$ is an unstable compound that has never been isolated in the pure state.

Behavior of CCl₄ and SiCl₄ with Water

When carbon tetrachloride is added to water it sinks to the bottom forming a colorless layer. A few drops of bromothymol blue indicator have been added to the water to give a green color. Silicon tetrachloride in the vial on the right is also a colorless liquid.

When silicon tetrachloride is added to water it reacts, forming a white precipitate of silicic acid and HCl which changes the color of the indicator to yellow. The HCl forms white fumes of ammonium chloride when the glass stopper of a bottle of a concentrated aqueous ammonia solution is held over the beaker.

The compounds formed in such reactions have quite different physical properties and chemical reactivity from the halides of the nonmetals. Some of the physical properties of the fluorides and chlorides of the alkali and alkaline earth metals are summarized in Table 5.4. They are all colorless crystalline solids (see Figure 5.3) with high melting points. In contrast, the nonmetal halides are generally gases, liquids, or solids of low melting point, as indicated in Table 5.5.

Reactions of Metals with Chlorine

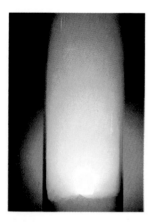

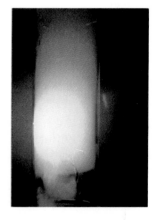

Sodium metal heated until it melts burns with a intense yellow flame when inserted into a jar of chlorine. A white smoke of sodium chloride particles is formed.

When steel wool is ignited in air and then inserted into a jar of chlorine it continues to burn, forming dense red-brown fumes of FeCl₃ which deposit on the sides of the jar. The sparks are white hot iron.

Small pieces of antimony react immediately, becoming white hot when dropped into a jar of chlorine. They can be seen here bouncing from the bottom of the jar. A white smoke of antimony trichloride SbCl₃ is formed.

Burning magnesium continues to burn with a bright white flame when inserted into a jar of chlorine gas. White solid magnesium chloride is formed.

Table 5.4 Some Properties of Alkali and Alkaline Earth Metal Fluorides and Chlorides

COMPOUND	MELTING POINT (°C)	BOILING POINT (°C)	SOLUBILITY IN WATER AT 25°C (g L⁻¹)
LiF	845	1681	1.3
NaF	995	1704	40.1
KF	856	1501	1020[a]
RbF	775	1408	1310
CsF	682	1250	3700[a]
LiCl	610	1382	850[a]
NaCl	808	1465	360
KCl	772	1407	350
RbCl	777	1381	940
CsCl	645	1300	1900
BeF_2	—	800[b]	5500
MgF_2	1263	2227	0.13
CaF_2	1418	2500	0.016
SrF_2	1400	2460	0.12
BaF_2	1320	2260	1.6
$BeCl_2$	405	488	720[a]
$MgCl_2$	714	1418	550
$CaCl_2$	772	>1600	830[a]
$SrCl_2$	875	1250	560[a]
$BaCl_2$	1350	—	370[a]

[a] Solubilities of hydrates $KF \cdot 2H_2O$, $CsF \cdot H_2O$, $LiCl \cdot H_2O$, $BeCl_2 \cdot 4H_2O$, $CaCl_2 \cdot 6H_2O$, $SrCl_2 \cdot 6H_2O$, $BaCl_2 \cdot 2H_2O$.
[b] Sublimes (changes directly from solid to gas without forming a liquid).

Figure 5.3 Sodium Chloride Crystals. Sodium chloride forms colorless cubic crystals. These crystals can be seen by looking at table salt under a microscope. Large crystals are sometimes found in salt deposits; they are called halite or rock salt crystals. The cubic shape of these crystals is determined by the arrangement of the ions in the structure of the crystal (see Figure 5.7).

As we have seen in Chapter 4, the metal halides are ionic compounds consisting of oppositely charged ions held together by electrostatic attraction. The formation of these ionic compounds can be represented by Lewis structures, as follows:

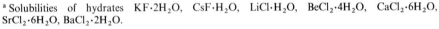

The bonds in the metal halides are ionic, while those in the nonmetal halides are covalent. These are two extreme types of bonds; the bonds in many substances have an intermediate character. The nature of a bond between two atoms depends primarily on the *electronegativity* of the atoms. We take up this important concept in the next section.

Table 5.5 Melting Points (°C) of Some Chlorides

METAL CHLORIDES: IONIC SOLIDS			NONMETAL CHLORIDES: COVALENT MOLECULAR LIQUIDS AND GASES			
LiCl	$BeCl_2$	BCl_3	CCl_4	NCl_3	OCl_2	FCl
610	405	−107	−23	−27	4	−101
NaCl	$MgCl_2$		$SiCl_4$	PCl_3	SCl_2	Cl_2
808	714		−68	−91	−122	−102

[a] Under pressure; sublimes at 180°C.

5.2 ELECTRONEGATIVITY

In a molecule such as H_2, F_2, or Cl_2 the electron pair of the covalent bond is shared equally between the two atoms because they are identical and therefore attract the electrons of the bond equally strongly. However, in molecules such as HF, HCl, or ClF, in which the two atoms are not the same, one atom attracts electrons more strongly than the other. The ability of an atom in a molecule to attract the electrons of a covalent bond to itself is called its **electronegativity**. The greater the electronegativity of an atom, the more strongly it attracts the electrons of a bond. Electronegativity is a qualitative concept and is not a quantity that can be directly measured experimentally.

According to Coulomb's law, the force of attraction exerted by the positively charged core of an atom on an electron in its valence shell is given by

$$F = k\frac{Q_1 Q_2}{r^2} = k\frac{(Z_{core}e)(-e)}{r^2}$$

where $Q_1 = +Z_{core}e$, the core charge, $Q_2 = -e$, the charge on an electron, and r is the distance of the electron from the core. We see that this force is proportional to the core charge, $Z_{core}e$, and that it decreases with increasing distance of the electron from the core. We might expect that we could calculate a value for the electronegativity of an atom from this expression. However, the resultant force on any one electron in the valence shell depends not only on the attraction exerted on it by the positive core but also on the repulsions due to all the other electrons, which cannot be calculated in an exact manner.

Although the electronegativity of an atom cannot be measured experimentally, nor calculated accurately, electronegativity is a very useful concept, and approximate values have been estimated by several methods. Figure 5.4 gives values for the electronegativities of the elements of the main groups.

Electronegativity and the Periodic Table

As we would expect from Coulomb's law, the electronegativity of an atom increases with increasing core charge and decreases with increasing distance of

Figure 5.4 Electronegativities of Main Group Elements. Across any period the electronegativity increases from left to right. Generally, the electronegativity decreases down any group from top to bottom, although there are some exceptions to this trend among the heavier elements.

Period	Group I	II	III	IV	V	VI	VII	VIII
1	H 2.2							He —
2	Li 1.0	Be 1.5	B 2.0	C 2.5	N 3.1	O 3.5	F 4.1	Ne —
3	Na 1.0	Mg 1.2	Al 1.3	Si 1.7	P 2.1	S 2.4	Cl 2.8	Ar —
4	K 0.9	Ca 1.0	Ga 1.8	Ge 2.0	As 2.2	Se 2.5	Br 2.7	Kr 3.1
5	Rb 0.9	Sr 1.0	In 1.5	Sn 1.7	Sb 1.8	Te 2.0	I 2.2	Xe 2.4

Electronegativity increases

Electronegativity decreases

the valence shell from the core, that is, with increasing atomic size. Thus there are two important trends in electronegativity within the periodic table:

1. Electronegativity increases across a period as the core charge increases.
2. Electronegativity generally decreases from top to bottom in a group as atomic size increases and the bond electrons are further from the nucleus.

Figure 5.4 summarizes these trends.

The nonmetals on the right-hand side of the periodic table have high electronegativities, and the metals on the left-hand side have low electronegativities. To a good approximation, elements with an electronegativity, χ (Greek letter chi), of 2.0 or greater are nonmetals. Elements with an electronegativity χ of less than 2.0 are metals. Values are not given for helium, neon, and argon because there are no known compounds of these elements. The halogens are among the most electronegative elements—indeed, fluorine is the most electronegative of all the elements ($\chi = 4.1$). The second most electronegative element is oxygen ($\chi = 3.5$), followed by krypton and nitrogen ($\chi = 3.1$) and chlorine ($\chi = 2.8$). The decrease in electronegativity on descending a group is largest in group VII and smallest in group I. This decrease is interrupted in the fourth period for groups III, IV, V, and VI. This interruption is a consequence of the fact that there are ten transition elements between calcium and gallium (Chapter 21).

Because electronegativities cannot be determined in a quantitative manner, and because different methods of obtaining electronegativities give somewhat different values, the electronegativity values in Figure 5.4 have only a limited quantitative significance. Moreover, the electronegativity of a given element does not have a truly constant value: It varies somewhat from one molecule to another, depending on the number and the nature of the other atoms that are bonded to it. Nevertheless, electronegativity is an important and useful concept, as we will see in the next section and on many other occasions.

Polar Bonds

If two atoms forming a diatomic molecule have exactly the same electronegativity, then the bonding electron pair will be shared exactly equally between the two atoms. In other words, the electron density of the bonding electron pair is distributed equally and symmetrically between the two atoms, as shown in Figure 5.5(a). When the two atoms forming the bond are identical, in molecules such as H_2, F_2 and Cl_2, for example, the sharing of the bonding electron density is always equal and symmetrical.

In contrast, if the atoms in a molecule have different electronegativities, the bonding electron pair is shared unequally, as is the case in HCl. The bonding electron pair is not shared equally between the H atom and the Cl atom because chlorine has a greater electronegativity then hydrogen. Therefore the chlorine atom attracts the bonding electron pair more strongly than the hydrogen atom does. The electron density of the bonding electron pair is not distributed equally between the two nuclei but is located more on the chlorine side of the molecule than the hydrogen side. In other words, the chlorine atom has a greater share of the bonding pair than the hydrogen atom, as shown in Figure 5.5(b). As a result, the chlorine atom, which has a little more electron density in HCl than in the isolated atom, becomes slightly negatively charged. And the hydrogen atom, which has a little less electron density in HCl than in the isolated atom, becomes slightly positively charged. That is,

$$\overset{\delta+}{H}\!-\!\overset{\delta-}{Cl}$$

Figure 5.5 Electron Density Distributions in Polar and Nonpolar Molecules. (a) In a nonpolar molecule, such as H_2 or Cl_2, the electron density of the bonding pair is shared equally between the two atoms. (b) In a polar molecule, such as HCl, the electron density of the bonding pair is shared unequally; the more electronegative chlorine atom acquires a greater share. It therefore has more electron density than it needs to balance its nuclear charge, giving it a small negative charge. The less electronegative hydrogen atom has less density than it needs to balance its nuclear charge; it therefore has a small positive charge.

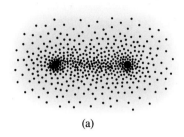

(a)

$\delta +$ $\delta -$

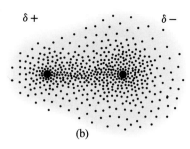

(b)

These charges are of equal magnitude but opposite sign, and the overall charge on the molecule remains zero. Such small charges are denoted by the symbols $\delta +$ (delta plus) and $\delta -$ (delta minus), which indicate some fraction of the charge of one electron.

A diatomic molecule such as HCl, in which the atoms carry small charges of equal magnitude but opposite sign, is called a **polar molecule**. The bond is called a **polar covalent bond**. When the difference in the electronegativities of the two atoms forming a bond is zero, the charges on the two atoms are zero. The bonding electron pair is shared equally between the two atoms, and the bond is said to be a **nonpolar bond**, as in H_2, F_2, and Cl_2. The greater the difference in the electronegativities of two atoms forming a bond, the more polar is the bond; that is, the greater are the charges on its atoms.

Example 5.2 Which of the following molecules have polar bonds and which have nonpolar bonds? For the polar molecules indicate the charges on the atoms.

H_2 FCl N_2 HCl O_2 ICl Na_2

Solution

H_2	Nonpolar	
FCl	Polar	$\overset{\delta -}{F}$—$\overset{\delta +}{Cl}$ since $\chi(F) > \chi(Cl)$
N_2	Nonpolar	
HCl	Polar	$\overset{\delta +}{H}$—$\overset{\delta -}{Cl}$ since $\chi(Cl) > \chi(H)$
O_2	Nonpolar	
ICl	Polar	$\overset{\delta +}{I}$—$\overset{\delta -}{Cl}$ since $\chi(Cl) > \chi(I)$
Na_2	Nonpolar	

Ionic Bonds

If we consider the series of bonds F—F, O—F, N—F, C—F, B—F, Be—F, and Li—F, we expect the bond polarity to increase from zero in F_2 to a rather large value in LiF, as the electronegativity difference increases (see Figure 5.6). Indeed, the sharing of the electron pair between Li and F is so unequal that the electron pair is acquired almost completely by the fluorine atom. As a result, the fluorine atom has a negative charge that is very nearly equal to -1. The fluorine atom has gained an electron to become the fluoride ion, F^-, and the lithium atom has lost an electron to become the lithium ion, Li^+. An electron has effectively been transferred from Li to F.

Thus lithium fluoride consists of positive lithium ions and negative fluoride ions held together by the electrostatic attraction between oppositely charged ions. Thus we no longer think of a lithium atom and a fluorine atom as being bonded by a shared electron pair. Rather, we think of the bond as arising

Period 2 fluorides	LiF	BeF$_2$	BF$_3$	CF$_4$	NF$_3$	OF$_2$	F$_2$
Electronegativity difference	3.1	2.6	2.1	1.6	1.0	0.6	0
Bond polarity	Highly polar $\text{Li}^+\ \ :\!\ddot{\text{F}}\!:^-$		Polar $\delta+\ \ \delta-$ $\text{C}\ :\ \ddot{\text{F}}\!:$				Non-polar $:\!\ddot{\text{F}}\ :\ \ddot{\text{F}}\!:$
Bond pair	Very unequally shared		Unequally shared				Equally shared
Type of bond	Ionic		Polar covalent				Covalent

Figure 5.6 Variation of Bond Polarity Across Periodic Table. The bonds in the fluorides of the second-period elements become increasingly polar from right to left. Fluorine, F$_2$, is nonpolar; OF$_2$, NF$_3$, CF$_4$, and BF$_3$ are covalent molecules with increasingly polar bonds; LiF and BeF$_2$ have highly polar bonds and are usually described as ionic. In the solid state LiF and BeF$_2$ consist of an infinite three-dimensional array of positive and negative ions.

from the electrostatic attraction between oppositely charged ions, and the bond is described as an **ionic bond**. An ionic bond can therefore be thought of as the extreme case of a polar bond, as indicated in Figure 5.6.

PREDICTING THE NATURE OF BONDS   It would be convenient if electronegativity values could be used to predict whether a particular bond is nonpolar covalent, polar covalent, or ionic. Unfortunately, the values are too approximate to enable us to make such predictions with much confidence. If the electronegativity difference is zero, then the bond is certainly nonpolar covalent. If the electronegativity difference is 1.0 or smaller, the bond is almost certainly polar covalent. Examples include the bonds in PCl$_3$ (0.7), SCl$_2$ (0.4), and NF$_3$ (1.0). If the electronegativity difference is greater than 2.0, the bond is almost certainly ionic. Examples include the bonds in NaF (3.1) and K$_2$O (2.6). For electronegativity differences between 1.0 or 2.0, however, predictions are difficult to make. For example, AsF$_3$ (1.9) is a molecular compound with polar covalent bonds, but NaCl (1.8) is an ionic compound composed of Na$^+$ and Cl$^-$ ions.

A simpler and more reliable guide to the nature of the bonds in a compound of the main group elements is the statement that: *compounds of Group I and Group II metals with a nonmetal usually have ionic bonds whereas compounds of two nonmetals usually have covalent bonds.*

Example 5.3 Classify the bonds between the following pairs of atoms as ionic, polar covalent, or nonpolar covalent:

LiH   LiF   CH   NH   OH   NN   RbBr   SiH   Ca

Which of the polar covalent bonds would be expected to have the greatest polarity?

Solution The classification can be made on the basis of the following generalizations:

- Bonds between a group I or a group II metal atom and a nonmetal atom are ionic.

- Bonds between two different nonmetal atoms are polar covalent.

- Bonds between two identical nonmetal atoms are nonpolar covalent.

LiH	Ionic	NN	Nonpolar covalent
LiF	Ionic	RbBr	Ionic
CH	Polar covalent	SiH	Polar covalent
NH	Polar covalent	Ca	Ionic
OH	Polar covalent		

The OH bond has the greatest electronegativity difference (1.4) and is therefore the most polar of these polar covalent bonds.

How do we know that a substance such as lithium fluoride or sodium chloride has ionic bonds? One important piece of evidence is that such substances conduct an electric current when melted.

ELECTRICAL CONDUCTIVITY   If copper wires connected to a current source and a light bulb are dipped into molten lithium chloride, the bulb lights up (see Experiment 5.5). Molten (liquid) lithium chloride is therefore capable of conducting an electric current. Any substance through which an electric current will pass when an electric potential (voltage) is applied to it is said to be an *electrical conductor* and to have an **electrical conductivity**.

An electric current is composed of moving electric charges. In the copper wires the current is conducted by electrons (as we will see in Chapter 9). In molten lithium chloride the current is conducted by positive lithium ions and negative chloride ions, which move in opposite directions through the liquid under the influence of the applied potential. In contrast, many liquid nonmetal halides, such as CCl_4, PCl_3, and SCl_2, do not conduct an electric current; they are nonconductors, or *insulators*.

When a metal halide is dissolved in water, it gives a solution that also conducts an electric current. In the solution the ions separate from each other and move independently through the solution. In contrast, substances that are soluble in water but consist of covalent molecules, such as alcohol (ethanol) and sugar (sucrose), do not give conducting solutions. Substances that dissolve in water to give conducting solutions are called **electrolytes**. Substances that dissolve in water to give nonconducting solutions are called **nonelectrolytes**.

Other evidence that metal halides are ionic comes from their structures, which we consider in the next section.

5.3 IONIC CRYSTALS

We have seen that the ionic metal halides are usually hard crystalline solids with rather high melting points. In contrast, the covalent nonmetal halides are usually gases, liquids, or, in some cases, solids with low melting points. This

EXPERIMENT 5.5

Electrical Conductivity of Molten Lithium Chloride

The crucible contains lithium chloride, which is heated to its melting point (610°C) to give a colorless liquid. The light bulb connected to a circuit with graphite electrodes shines brightly, showing that molten lithium chloride is an electrical conductor.

If the heating is stopped and the electrodes are removed from the molten lithium chloride, a plug of solid lithium chloride forms between the electrodes. The bulb does not light, showing that solid lithium chloride is not an electrical conductor.

striking difference in properties is a consequence of their different structures. We consider now the structure of ionic substances.

Sodium chloride consists of positive sodium ions and negative chloride ions. When one sodium atom reacts with one chlorine atom to form a sodium ion and a chloride ion,

$$\cdot Na + \cdot \ddot{\underset{\cdot\cdot}{Cl}}: \longrightarrow Na^+ : \ddot{\underset{\cdot\cdot}{Cl}}:^-$$

these two ions are held together by electrostatic attraction, giving a diatomic molecule in which the bond is ionic. Such a molecule is sometimes called an **ion pair**. This ionic molecule, or ion pair, occurs only in gaseous sodium chloride at very high temperature.

When a large number of positive ions and negative ions are formed, each positive ion tends to surround itself with as many negative ions as possible, and each negative ion tends to surround itself with as many positive ions as it can. The result is a structure consisting of a regular pattern of alternate positive and negative ions in which no individual molecules can be distinguished. In the structure of sodium chloride (see Figure 5.7), each sodium ion is surrounded by six chloride ions, and each chloride ion is surrounded by six sodium ions. The six chloride ions are arranged in the form of an **octahedron** around each Na^+ ion. Similarly, the six Na^+ ions surrounding each Cl^- ion have an octahedral arrangement. Like the tetrahedron, the octahedron is the basis of the structures of many molecules and crystals (see Figure 5.8). This regular arrangement continues indefinitely in three dimensions throughout the whole crystal. The structure is therefore described as an *infinite three-dimensional structure*. Thus an **ionic crystal** is composed of an infinite array of positive and negative ions in which no individual molecules can be distinguished.

Melting or vaporizing an ionic substance requires that the strong ionic bonds, which extend throughout the crystal, be broken, which takes a considerable amount of energy. Ionic substances, such as the alkali and alkaline earth metal halides, therefore generally have high melting points and boiling points, as we

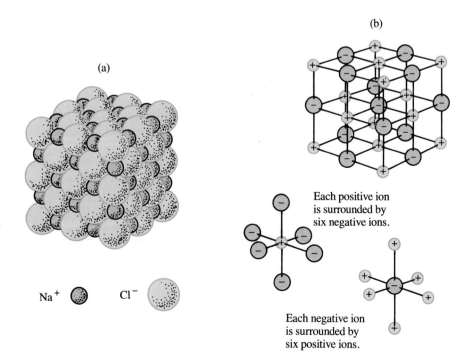

(a)

(b)

Na$^+$ ⬤   Cl$^-$ ⬤

Each positive ion is surrounded by six negative ions.

Each negative ion is surrounded by six positive ions.

Figure 5.7 Structure of Sodium Chloride. The regular array of sodium ions and chloride ions shown in (a) extends indefinitely throughout the crystal. The crystal can therefore be regarded as one huge molecule. No individual NaCl molecules can be recognized in the structure. Each sodium ion is surrounded by six chloride ions, which have an octahedral arrangement. And each chloride ion is surrounded by six sodium ions, which also have an octahedral arrangement. This arrangement is most easily seen in the expanded view of the structure shown in (b). Here the ions have been separated from each other but their arrangement has been maintained. The lines joining the ions do not represent bonds; they are included only to assist in visualizing the structure. The attraction of one ion for others of opposite sign is exerted in all directions. In contrast, covalent bonds form only in specific directions.

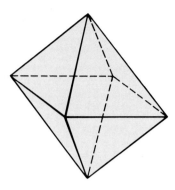

Figure 5.8 Octahedron. The octahedron is a very important shape found commonly in molecules and crystal structures. It is a regular polyhedron with eight equilateral triangular faces, twelve equivalent edges, and six equivalent corners.

have seen in Table 5.4. In contrast, the nonmetal halides, such as CCl_4 and PCl_3, consist of molecules in which the atoms are held together by strong covalent bonds, but the molecules themselves are only held together by weak intermolecular forces. Only a small amount of energy is needed to overcome the weak forces *between* the molecules, so the nonmetal halides are generally gases and liquids with low melting points (Table 5.3).

When a covalent molecular compound, such as CCl_4 or H_2O, is melted and then vaporized, the molecules are free to move relative to each other. But they remain intact because their atoms are held together by strong covalent bonds. Ice, liquid water, and steam all consist of H_2O molecules; solid, liquid, and gaseous carbon tetrachloride all consist of CCl_4 molecules. The regular, infinite, three-dimensional structure of sodium chloride or any other ionic substance cannot, however, be retained in the liquid or gaseous state. Gaseous sodium chloride, which is only formed at a very high temperature, consists of small molecules, such as Na^+Cl^- and $(Na^+)_2(Cl^-)_2$, in which the bonding is predominantly ionic.

The Sodium Chloride and Cesium Chloride Structures

Most of the alkali metal halides and many other ionic compounds have the same arrangement of ions as sodium chloride has. They are said to have the *sodium chloride structure* (see Figure 5.7). However, cesium chloride and a few other alkali metal halides have the *cesium chloride structure* (see Figure 5.9), in which each Cs^+ is surrounded by eight Cl^- ions in a *cubic* arrangement, and each Cl^- is surrounded by eight Cs^+ ions in a cubic arrangement. As in the NaCl structure, no individual CsCl molecules can be distinguished. The fact that each positive ion is surrounded by eight Cl^- ions, rather than six Cl^- ions, as in the NaCl structure, is a consequence of the larger size of the Cs^+ ion. A positive ion attracts as many negative ions as can pack around it. There is room for eight Cl^- ions around a Cs^+ ion, but for only six Cl^- ions around the smaller Na^+ ion.

Sizes of Ions: Ionic Radii

The sizes of ions can be obtained from the structures of crystals as determined by X ray crystallography (described in Chapter 10). Some values for the radii of the *cations* (positively charged ions) formed by the alkali and alkaline earth metals and of the *anions* (negatively charged ions) formed by the halogens and the Group VI elements are given in Table 5.6. By comparison with the covalent (atomic) radii in Figure 4.7, we see that positive ions are smaller and negative ions are larger than the corresponding covalently bound atoms. Removal of the valence electrons from an alkali or alkaline earth metal atom corresponds to the removal of its outer shell, so the positive ion that is formed is considerably smaller than the neutral atom. Addition of one or two electrons to a neutral atom to form a negative ion increases the number of repulsions between the valence shell electrons and causes the valence shell to expand.

Because anions are usually larger than cations, the size of a cation determines the number of anions of a given size that can be packed around it. Thus the structure of an ionic crystal depends primarily on the ratio of the sizes of the anion and the cation. As we have seen, CsCl has a different structure from NaCl because of the larger size of the Cs^+ ion.

Note that within a period the alkaline earth M^{2+} ion is **isoelectronic** with the alkali metal M^+ ion; that is, it has the same electron arrangement. But

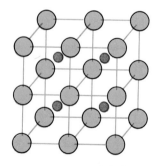

Figure 5.9 Structure of Cesium Chloride. Each cesium ion is surrounded by eight chloride ions in a cubic arrangement, and each chloride ion is surrounded by eight cesium ions in a cubic arrangement. This arrangement of ions extends throughout the crystal. No individual cesium chloride molecules can be distinguished.

Table 5.6 Ionic Radii (pm)

PERIOD	I	II	VI	VII
			GROUP	
1	Li^+	Be^{2+}	O^{2-}	F^-
	74	35	140	135
2	Na^+	Mg^{2+}	S^{2-}	Cl^-
	102	72	185	180
3	K^+	Ca^{2+}	Se^{2-}	Br^-
	138	100	195	195
4	Rb^+	Sr^{2+}	Te^{2-}	I^-
	149	116	220	215
5	Cs^+	Ba^{2+}		
	170	136		

because M^{2+} has a greater nuclear charge than M^+, its electrons are pulled closer to the nucleus, and the M^{2+} ion is therefore smaller than the M^+ ion. For example, Na^+ and Mg^{2+} both have the electron arrangement, 2, 8; but Mg^{2+} ($Z = 12$) has a smaller radius (72 pm) than Na^+ ($Z = 11$), which has a radius of 102 pm. Similarly, within a period the Group VI X^{2-} ions and the halogen X^- ions are isoelectronic, but the X^- ion has a greater nuclear charge and therefore a smaller radius than the X^{2-} ion. For example, O^{2-} and F^- both have the electron arrangement, 2, 8, but O^{2-} ($Z = 8$) has a radius of 140 pm, whereas F^- ($Z = 9$) has a radius of 135 pm.

Example 5.4 In each of the following pairs, choose the larger ion:

(a) Na^+, F^- (b) Na^+, K^+ (c) F^-, Cl^- (d) Na^+, Mg^{2+} (e) S^{2-}, Cl^-

Solution

(a) The ions Na^+ and F^- are isoelectronic; they have the same electron arrangement, 2, 8. But these electrons are attracted by different nuclear charges. The atomic number Z of Na is 11, therefore the charge on the nucleus of both an Na atom and an Na^+ ion is $+11$. The atomic number of F is 9; therefore the charge on the nucleus of both F and F^- is $+9$. The electrons of Na^+ are attracted by a greater nuclear charge than are those of F^- and are therefore pulled in closer to the nucleus. Thus F^- is larger than Na^+.

(b) Although K^+ (2, 8, 8) has a nuclear charge of $+19$, and Na^+ (2, 8) has a nuclear charge of $+11$, the K nucleus is shielded by an additional shell of eight electrons. The effective charge acting on the outer shell of eight electrons is $+19 - 8 - 2 = +9$ for K^+ and is $+11 - 2 = +9$ for Na^+. Thus the electrons are attracted by the same charge, but since K^+ has three shells while Na^+ has only two, K^+ is the larger ion.

(c) The same type of argument we used in part (b) shows that Cl^- is the larger ion.

(d) The ions Na^+ and Mg^{2+} are isoelectronic; they have the electron arrangement 2, 8. But Mg^{2+} has a nuclear charge of $+12$ whereas Na^+ has a nuclear charge of $+11$. So the electrons of Na^+ are less strongly attracted than are those of Mg^{2+}, and therefore Na^+ is the larger ion.

(e) The same type of argument we used in part (d) shows that S^{2-} is the larger ion.

In the reaction between sodium and chlorine to give sodium chloride, sodium atoms lose electrons to become sodium ions, and chlorine molecules gain electrons to become chloride ions. We can therefore conveniently consider the reaction to take place in two steps:

$$Na \longrightarrow Na^+ + e^-$$
$$Cl_2 + 2e^- \longrightarrow 2Cl^-$$

The electrons produced in the first reaction must be used up in the second reaction, because no electrons appear in the overall reaction.

To obtain the balanced equation for the overall reaction, we must multiply the first reaction by 2 and add it to the second reaction so that the electrons produced in the first reaction are used up in the second reaction:

$$2Na + Cl_2 \longrightarrow 2(Na^+Cl^-)$$

The overall reaction is an **electron transfer reaction**: Electrons are transferred from sodium to chlorine. Such a reaction is commonly called an **oxidation-reduction reaction**.

We previously defined oxidation as the addition of oxygen to an element or compound and reduction as the removal of oxygen from a compound. The formation of sodium oxide, Na_2O, from sodium and oxygen can be represented by the equation

$$4Na + O_2 \longrightarrow 2(Na^+)_2O^{2-}$$

This equation can similarly be split into two halves to show that it is an electron transfer reaction. Each sodium atom loses an electron to form a sodium ion:

$$Na \longrightarrow Na^+ + e^-$$

And each oxygen molecule gains four electrons to form two oxide ions:

$$O_2 + 4e^- \longrightarrow 2O^{2-}$$

Combining the two equations

$$4(Na \longrightarrow Na^+ + e^-)$$
$$O_2 + 4e^- \longrightarrow 2O^{2-}$$

gives the overall equation

$$4Na + O_2 \longrightarrow 2(Na^+)_2O^{2-}$$

In its reactions with both chlorine and oxygen, sodium undergoes the same basic reaction, namely, loss of an electron. It is said to have been *oxidized* to Na^+. *An atom is said to have been oxidized if it loses electrons.* **Oxidation** is the loss of electrons. The chlorine and the oxygen are said to have been *reduced*. **Reduction** is the gain of electrons. *In any oxidation-reduced reaction one substance is oxidized and another is reduced.* Electrons are transferred from the substance that is oxidized to the substance that is reduced.

According to this more general definition of oxidation, many substances other than oxygen can oxidize other substances. They are said to be **oxidizing agents**. Any substance that tends to gain electrons—for example, nonmetals such as oxygen, sulfur, and the halogens—may behave as an oxidizing agent. Any substance from which electrons can be readily removed, such as metals, may behave as a **reducing agent**. An *oxidizing agent is a substance that can gain electrons, and a reducing agent is a substance that can give up electrons.*

Oxidation-reduction reactions are a very important general class of chemical reactions, which we will examine in more detail in Chapter 7. In general, the reaction between a metal and a nonmetal is an oxidation-reduction reaction in which the metal behaves as a reducing agent and is oxidized and the nonmetal behaves as an oxidizing agent and is reduced.

The Halogens as Oxidizing Agents

The halogens are all strong oxidizing agents, but their strength decreases in the series $F_2 > Cl_2 > Br_2 > I_2$. When an aqueous solution of chlorine, Cl_2, is added to a colorless aqueous solution of sodium bromide, NaBr, the solution rapidly becomes orange-red because of the formation of molecular bromine in the reaction (see Experiment 5.6):

$$Cl_2 + 2Br^- \longrightarrow 2Cl^- + Br_2$$

This reaction is an oxidation-reduction reaction in which bromide ion is oxidized to bromine,

$$2Br^- \longrightarrow Br_2 + 2e^-$$

while chlorine is reduced to chloride ion,

$$Cl_2 + 2e^- \longrightarrow 2Cl^-$$

Similarly, if an aqueous solution of chlorine is added to a colorless aqueous solution of an iodide, or if gaseous chlorine is passed into an iodide solution, the solution becomes brown because of the formation of molecular iodine:

$$Cl_2 + 2I^- \longrightarrow 2Cl^- + I_2$$

Chlorine oxidizes iodide ion to iodine. Iodine is insoluble in water, and the solution, in fact, contains the brown triiodide ion, I_3^-, formed by combination

Oxidation-Reduction Reactions of the Halogens

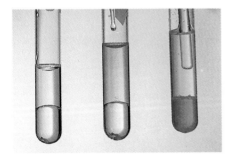

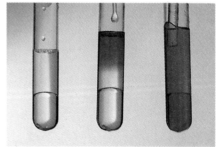

Left: The top layer is an aqueous solution of NaBr. The bottom layer is carbon tetrachloride, CCl_4. Center: When a few drops of a dilute aqueous solution of Cl_2 are added, the NaBr solution becomes yellow because Br_2 is formed. Right: Bromine is much more soluble in carbon tetrachloride than in water. On stirring, it is extracted into the carbon tetrachloride, forming a brown solution.

Left: The top layer is an aqueous solution of NaI. The bottom layer is CCl_4. Center: When a few drops of an aqueous solution of Cl_2 are added, iodine, I_2, is formed. Iodine reacts with iodide ion, I^-, to form a brown solution of the triiodide ion, I_3^-. Right: On stirring, I_2 is extracted into the CCl_4, forming a violet solution.

of molecular iodine and an iodide ion:

$$I_2 + I^- \longrightarrow I_3^-$$

If this solution is shaken with carbon tetrachloride, which is insoluble in water, molecular iodine is removed from I_3^- and forms a violet solution in the carbon tetrachloride.

If bromine is added to an aqueous solution of an iodide, iodine is formed according to the equation

$$Br_2 + 2I^- \longrightarrow 2Br^- + I_2$$

However, if bromine is added to an aqueous solution of a chloride, no reaction is observed. In other words, the reaction

$$Cl_2 + 2Br^- \longrightarrow 2Cl^- + Br_2$$

proceeds from left to right but not from right to left. Whereas chlorine can oxidize bromide and iodide, bromine can oxidize only iodide. Iodine, the weakest oxidizing agent among the halogens, oxidizes neither Br^- nor Cl^-.

Similar reactions occur between the halogens and the hydrogen halides in the gas phase (see Experiment 5.7). For example,

$$Cl_2(g) + 2HI(g) \longrightarrow I_2(s) + 2HCl(g)$$

These simple experiments show that the order of oxidizing strengths is $Cl_2 > Br_2 > I_2$. Fluorine is a much stronger oxidizing agent than chlorine. Fluorine, for example, oxidizes water to oxygen:

$$2F_2 + 2H_2O \longrightarrow 4H^+(aq) + 4F^-(aq) + O_2(g)$$

In contrast, chlorine and bromine dissolve largely unchanged in water, and iodine is insoluble. Fluorine reacts so vigorously with many substances that it can only be handled in the laboratory by using special apparatus and taking suitable precautions.

The order of oxidizing strengths $F_2 > Cl_2 > Br_2 > I_2$ is not unexpected when we recall that the electronegativities of F, Cl, Br, and I also decrease in the same sequence. The electronegativity of a halogen is a measure of the tendency of a halogen atom to attract the electrons of a covalent bond and therefore to

Oxidation of Hydrogen Iodide by Chlorine

Chlorine is separated from twice its volume of hydrogen iodide by a glass plate.

When the plate is removed, chlorine oxidizes hydrogen iodide to iodine. Both the violet vapor and the black solid can be seen. Some brown iodine monochloride, ICl, is also observed on the sides of the jar.

acquire a partial negative charge. It is therefore not the same as the oxidizing strength of a halogen, which is a measure of the tendency of the diatomic molecule X_2 to acquire two electrons to become two X^- ions.

$$X_2 + 2e^- \longrightarrow 2X^-$$

Nevertheless, both the oxidizing strength and the electronegativity of the halogens are related to the tendency of halogen atoms to acquire electrons. Therefore, not surprisingly, as the electronegativity of X decreases, the oxidizing strength of X_2 also decreases in the order $F_2 > Cl_2 > Br_2 > I_2$. The ease with which halide ions can be oxidized increases in the order $F^- < Cl^- < Br^- < I^-$, which is the order of their strengths as reducing agents.

Example 5.5 In each of the following reactions, identify the oxidizing agent, the reducing agent, the species (molecule or ion) that is oxidized, and the species that is reduced:

(a) $Ca(s) + Br_2(l) \longrightarrow Ca^{2+}(Br^-)_2(s)$

(b) $4Li(s) + O_2(g) \longrightarrow 2(Li^+)_2O^{2-}(s)$

(c) $Fe(s) + S(s) \longrightarrow Fe^{2+}S^{2-}(s)$

(d) $2Fe^{2+}(aq) + Cl_2(g) \longrightarrow 2Fe^{3+}(aq) + 2Cl^-(aq)$

Solution

(a) The Ca loses electrons to become Ca^{2+}. It is therefore a reducing agent and is itself oxidized:

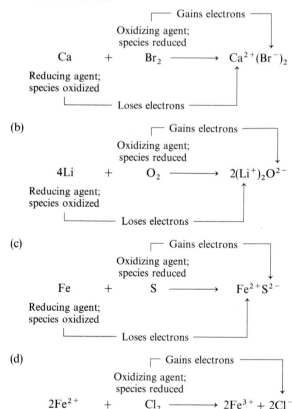

Preparation of the Halogens

Since the halogens occur naturally as halide ions, they are prepared by oxidation of halide ion to the free halogen. Although in principle Cl^- could be oxidized by fluorine, F_2, this method is impractical because fluorine is expensive to produce and reacts vigorously with water and many other substances. In the industrial manufacture of chlorine, chloride ion is oxidized by directly removing electrons in a process called **electrolysis**, in which an electric current is passed through molten sodium chloride or a concentrated aqueous solution of sodium chloride:

$$2Cl^-(aq) \xrightarrow{\text{Electrolysis}} Cl_2 + 2e^-$$

The manufacture of chlorine by this process is discussed in detail in Chapter 16.

In the laboratory chloride ion can be oxidized to gaseous chlorine, Cl_2, by using a strong oxidizing agent. A suitable oxidizing agent is manganese dioxide, MnO_2 (see Experiment 5.8). The reaction is carried out in an acidic solution:

$$MnO_2(s) + 2Cl^-(aq) + 4H^+(aq) \longrightarrow Mn^{2+}(aq) + Cl_2(g) + 2H_2O(l)$$

As we have seen, bromide ion is readily oxidized to bromine by chlorine. Because chlorine is manufactured in very large amounts and is relatively inexpensive, it is used to prepare bromine from the bromide ion present in subterranean brines:

$$Cl_2(g) + 2Br^-(aq) \longrightarrow Br_2(l) + 2Cl^-(aq)$$

Chlorine and steam are passed through the brine. The gaseous bromine that is formed is carried off from the solution with steam and excess chlorine. It is condensed from this mixture and purified by distillation. Iodine can be obtained from an aqueous iodide solution by the same method.

5.5 AQUEOUS SOLUTION OF ELECTROLYTES

Solutions of Ionic Compounds

When ionic compounds dissolve in water, the ions separate from each other and mix with the water molecules. A solution of sodium chloride in water

EXPERIMENT 5.8

Preparation of Chlorine

Chlorine is being prepared by adding concentrated hydrochloric acid from the funnel to solid manganese dioxide, MnO_2, in the flask and gently warming the mixture. The $Cl_2(g)$, which is heavier than air, is collected by upward displacement of air.

consists simply of sodium ions, Na^+, and chloride ions, Cl^-, mixed with H_2O molecules. If potassium fluoride is dissolved in water, the solution contains potassium ions, K^+, and fluoride ions, F^-. If we mix equal amounts of solutions of NaCl and KF of equal concentrations, we obtain a solution containing equal concentrations of Na^+, K^+, Cl^-, and F^-. We cannot say that we have a solution of sodium chloride and potassium fluoride because we could have prepared the same solution by mixing equal amounts of solutions of sodium fluoride and potassium chloride. We would again obtain a solution containing Na^+, K^+, Cl^-, and F^-. Thus in solutions of mixtures of ionic compounds the individual compounds lose their identity.

Precipitation Reactions

If we were to evaporate a solution containing equal amounts of Na^+, K^+, Cl^-, and F^-, would we obtain a mixture of all four possible compounds, NaF, NaCl, KF, and KCl, or would we obtain only some of them? The answer depends on the solubilities of the four possible compounds. If we allow the solution to evaporate, sodium fluoride, in fact, will crystallize first, as shown in Figure 5.10(a), since it has the smallest solubility (see Table 5.4).

If we start with solutions of KF and NaCl, we can represent this reaction by the equation

$$KF(aq) + NaCl(aq) \longrightarrow NaF(s) + KCl(aq)$$

where we have represented each substance by its empirical formula. The remaining solution still contains some Na^+ and F^- ions together with all the K^+ and Cl^- ions. No reaction occurs on simply mixing the solutions. When the solution is concentrated by evaporation until the solubility of NaF is exceeded, Na^+ ions combine with F^- ions to form a **precipitate** of insoluble NaF:

$$Na^+(aq) + F^-(aq) \longrightarrow NaF(s)$$

This reaction is an example of a **precipitation reaction**.

Precipitation of sodium fluoride only occurs from a rather concentrated solution, because sodium fluoride is relatively soluble. More common examples of precipitation reactions involve substances that have very small solubilities; they precipitate, therefore, even from dilute solutions.

Although the chlorides of the alkali and alkaline earth metals are all soluble in water, the chlorides of some metals such as silver and lead have very small solubilities. Thus if we mix dilute solutions of a soluble silver salt such as silver nitrate, $AgNO_3$, and sodium chloride, silver chloride precipitates immediately (Figure 5.10b):

$$AgNO_3(aq) + NaCl(aq) \longrightarrow AgCl(s) + NaNO_3(aq)$$

Since these are all ionic compounds, we can also write the equation in terms of the ions:

$$Ag^+(aq) + NO_3^-(aq) + Na^+(aq) + Cl^-(aq) \longrightarrow (Ag^+Cl^-)(s) + Na^+(aq) + NO_3^-(aq)$$

We see that the ions Na^+ and NO_3^- appear on both sides of the equation; they take no part in the reaction and hence are said to be **spectator ions**. We can cancel them from the equation, leaving

$$Ag^+(aq) + Cl^-(aq) \longrightarrow (Ag^+Cl^-)(s)$$

This equation is called a **net ionic equation**. It is also frequently written as

$$Ag^+(aq) + Cl^-(aq) \longrightarrow AgCl(s)$$

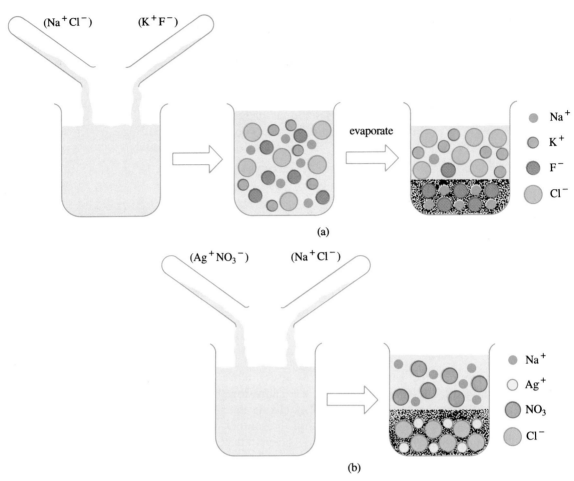

(Na^+Cl^-) (K^+F^-)

evaporate

Na$^+$
K$^+$
F$^-$
Cl$^-$

(a)

$(Ag^+NO_3^-)$ (Na^+Cl^-)

Na$^+$
Ag$^+$
NO$_3$
Cl$^-$

(b)

Figure 5.10 Precipitation reactions.
(a) When sodium chloride and potassium fluoride are dissolved in water, a
solution containing Na^+, K^+, Cl^-, and F^- is obtained. On evaporation the
solution eventually becomes saturated in sodium fluoride, which then separates
from the solution. Sodium fluoride is the least soluble of the four compounds
NaF, KF, KCl, and NaCl that might be formed. (b) When a solution of sodium
chloride is added to a solution of silver nitrate, a precipitate of silver chloride
forms immediately, because silver chloride has a very low solubility.

where the ions in the solid are not shown. This equation represents the reaction
that occurs when a solution of *any* soluble silver salt is mixed with a solution
of *any* soluble chloride, for example,

$$AgClO_4(aq) + KCl(aq) \longrightarrow AgCl(s) + KClO_4(aq)$$

as we would see if we were to write this equation in the ionic form and cancel
the spectator ions.

Example 5.6 Write the net ionic equation for the formation of a precipitate of insoluble barium sulfate, $BaSO_4$, when an aqueous solution of barium chloride, $BaCl_2$,
is mixed with an aqueous solution of sodium sulfate, Na_2SO_4.

Solution In terms of empirical formulas the equation is

$$BaCl_2(aq) + Na_2SO_4(aq) \qquad BaSO_4(s) + 2NaCl(aq)$$

The corresponding ionic equation is

$$Ba^{2+}(aq) + 2Cl^-(aq) + 2Na^+(aq) + SO_4^{2-}(aq) \longrightarrow$$
$$(Ba^{2+}SO_4^{2-})(s) + 2Na^+(aq) + 2Cl^-(aq)$$

Canceling the spectator ions Na^+ and Cl^- gives the net ionic equation:

$$Ba^{2+}(aq) + SO_4^{2-}(aq) \longrightarrow (Ba^{2+}SO_4^{2-})(s)$$

or

$$Ba^{2+}(aq) + SO_4^{2-}(aq) \longrightarrow BaSO_4(s)$$

Silver bromide and silver iodide are also insoluble, and precipitates of silver bromide and silver iodide are obtained when silver ion is added to aqueous solutions containing bromide or iodide ions. Since most other silver salts are soluble, the formation of these precipitates is a good test for chloride, bromide, or iodide ions (see Experiment 5.9).

Solutions of Hydrogen Halides: The Hydronium Ion

Hydrogen chloride and the other hydrogen halides are typical molecular covalent substances that have very low melting and boiling points and are nonconductors of electricity in both the liquid and the solid states. But the hydrogen halides are very soluble in water and form solutions that are very good conductors of electricity. Ions must be formed when the hydrogen halides are dissolved in water. What are these ions?

If a silver nitrate solution is added to an aqueous solution of hydrogen chloride, a white precipitate of silver chloride is obtained:

$$Ag^+(aq) + Cl^-(aq) \longrightarrow AgCl(s)$$

EXPERIMENT 5.9

Precipitation Reactions

 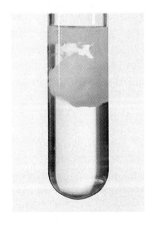 

When an aqueous NaCl solution is added to an aqueous $AgNO_3$ solution a white precipitate of insoluble AgCl is formed.

When an aqueous solution of NaI is added to an aqueous solution of $AgNO_3$ a pale yellow precipitate of insoluble AgI is formed.

When an aqueous solution of NaI is added to an aqueous solution of $Pb(NO_3)_2$ a bright yellow precipitate of insoluble PbI_2 is formed.

This reaction shows that one of the ions in the hydrogen chloride solution is the chloride ion. If a chloride ion is formed from hydrogen chloride, a hydrogen ion (a proton) must also be formed:

$$H:\overset{..}{\underset{..}{Cl}}:(aq) \longrightarrow H^+(aq) + :\overset{..}{\underset{..}{Cl}}:^-(aq)$$

A proton has an empty valence shell and attracts electrons very strongly. In aqueous solution it combines with a water molecule to form the **hydronium ion**, H_3O^+:

$$H^+ + :\overset{..}{\underset{H}{O}}:H \longrightarrow H:\overset{..\oplus}{\underset{H}{O}}:H$$

Thus the net reaction that occurs when hydrogen chloride gas dissolves in water is the transfer of a hydrogen ion, H^+, from a hydrogen chloride molecule to a water molecule, with the formation of the hydronium ion, H_3O^+, and a chloride ion, Cl^-:

$$:\overset{..}{\underset{..}{Cl}}-H + :\overset{..}{\underset{H}{O}}-H \longrightarrow :\overset{..}{\underset{..}{Cl}}:^- + H-\overset{..\oplus}{\underset{H}{O}}-H$$

Because of the large electronegativity differences between hydrogen and chlorine and between hydrogen and oxygen, both hydrogen chloride and water are polar molecules:

$$:\overset{\overset{\delta-}{..}}{\underset{..}{Cl}}-\overset{\delta+}{H}\quad\overset{2\delta-}{:\overset{..}{O}}-\overset{\delta+}{H} \longrightarrow :\overset{..}{\underset{..}{Cl}}:^- + H-\overset{..\oplus}{\underset{H}{O}}-H$$
$$\underset{\delta+H}{}$$

The positive hydrogen atom of the hydrogen chloride molecule is attracted to the negative oxygen atom of the water molecule. A hydrogen ion (proton) then moves from the chlorine atom to the oxygen atom of the water molecule, leaving behind the electrons that it originally shared with the chlorine atom to give a $:\overset{..}{\underset{..}{Cl}}:^-$ ion. The hydrogen ion attaches itself to one of the unshared pairs of electrons of the oxygen atom, forming an additional O—H bond and converting the water molecule to a hydronium ion, H_3O^+.

If a concentrated solution of HCl in water is cooled to $-15°C$, colorless crystals of the compound $H_3O^+Cl^-$ separate. This is hydronium chloride, an ionic compound that is similar to other ionic chlorides such as sodium chloride and calcium chloride. It differs from these simple ionic compounds only in that it has a polyatomic cation instead of a simple monatomic ion such as Na^+ or Ca^{2+}. The H_3O^+ ion and the NH_3 molecule have the same Lewis structure,

$$H-\overset{..}{\underset{H}{N}}-H \qquad H-\overset{..\oplus}{\underset{H}{O}}-H$$

in which both the nitrogen atom and the oxygen atom form three covalent bonds. In their valence shells they both have three shared (bonding) pairs of electrons and one unshared pair of electrons. The H_3O^+ also has the same triangular pyramidal shape as the NH_3 molecule.

In the formation of the hydronium ion from a hydrogen ion and a water molecule, the positive charge on the hydrogen ion (proton) is transferred to the oxygen atom of the water molecule:

$$H^+ + :\overset{..}{\underset{H}{O}}-H \longrightarrow H-\overset{..\oplus}{\underset{H}{O}}-H$$

Triangular Pyramidal Molecules

When the hydrogen ion becomes attached to one of the unshared pairs of electrons on the oxygen atom, it acquires a share in this electron pair and thus effectively acquires one electron. It therefore loses its charge. The oxygen atom at the same time effectively loses an electron when one of its unshared pairs is converted to a shared pair. Thus the oxygen atom becomes positively charged; it has a formal charge of $+1$. This formal charge can also be found by following the procedure for finding formal charges given in Chapter 4:

Formal charge = Group number − Number of bonds − Number of unshared electrons
$$= +6 - 3 - 2 = +1$$

Thus the charge on the hydronium ion may be considered to be located on the oxygen atom. However, oxygen is in fact more electronegative than hydrogen. The oxygen atom will therefore have more than a half share in each bonding pair and so will have a smaller charge than $+1$, while each hydrogen will acquire a small positive charge. The positive charge is thus somewhat spread out over the whole ion. However, for many purposes we can conveniently consider the positive charge to be located on the oxygen atom, as indicated by the formal charge.

5.6 ACIDS AND BASES

Solutions of hydrogen chloride and the other hydrogen halides in water share certain properties with the aqueous solutions of a number of other substances. These properties have long been used to identify these substances as **acids**. Some of these characteristic properties are as follows:

- Aqueous solutions of acids have a sour taste. We are familiar with the taste of vinegar or lemon juice. Vinegar is an aqueous solution of acetic acid, $C_2H_4O_2$; lemon juice is an aqueous solution containing citric acid, $C_6H_8O_7$.
- Acids have the ability to change the colors of substances called *indicators*. For example, litmus, a substance extracted from certain lichens, has a characteristic red color in aqueous acid solutions. Grape juice and tea also act as indicators (see Experiment 5.10).
- When acids are added to a solution of a soluble carbonate such as sodium carbonate, Na_2CO_3, or to an insoluble solid carbonate such as calcium carbonate, $CaCO_3$, carbon dioxide is evolved. The solution bubbles vigorously—it is said to effervesce.
- Aqueous solutions of acids react with many metals, such as zinc and magnesium, producing hydrogen, which bubbles off from the solution. The solution effervesces.

These common characteristic properties of aqueous solutions of acids result from the fact that, like a solution of hydrogen chloride in water, they all contain the hydronium ion, H_3O^+. These properties of aqueous acid solutions are the properties of the hydronium ion.

An acid has at least one hydrogen atom attached to an electronegative atom or group of atoms. We may write the formula of an acid as HA, where A is an electronegative atom or group of atoms (such as Cl or NO_3). The HA bond is therefore polar

$$\overset{\delta+}{H}-\overset{\delta-}{A}$$

and the hydrogen atom has a small positive charge. This positively charged H atom is attracted to the negatively charged oxygen atom of a water molecule and then H^+ is transferred to one of the unshared electron pairs of the water molecule to form H_3O^+.

Reactions of Acids

Left: Pieces of the metal zinc
Right: When dilute hydrochloric
acid is added bubbles of $H_2(g)$
are evolved.

Left: Marble chips ($CaCO_3$).
Right: When dilute
hydrochloric acid is added
bubbles of $CO_2(g)$ are
evolved.

Citric acid in lemon juice
changes the color of tea,
which behaves as an indicator.

We can represent the reaction of an acid with water by the general equation

$$HA + H_2O \longrightarrow H_3O^+ + A^-$$

For example,

$$HCl + H_2O \longrightarrow H_3O^+ + Cl^-$$

$$HNO_3 + H_2O \longrightarrow H_3O^+ + NO_3^-$$

In each of these reactions the acid gives up an H^+, that is, a proton. Water in turn accepts a proton. The reaction is thus a **proton transfer reaction**; one substance donates a proton and the other accepts a proton. In 1923 Danish chemist Johannes Brønsted (1879–1947) and English chemist Thomas Lowry (1874–1936) proposed that an **acid** may be defined as a **proton donor**. A **proton acceptor** is called a **base**. A proton transfer reaction is therefore called an **acid-base reaction**. Because an acid and a base may be defined in other ways a proton donor is often called a Brønsted acid (or a Brønsted-Lowry acid) and a proton acceptor is often called a Brønsted base (or a Brønsted-Lowry base).

When an acid is dissolved in water, there is an acid-base reaction in which water behaves as a base, accepting a proton from the acid:

$$HA(aq) + H_2O \longrightarrow H_3O^+(aq) + A^-(aq)$$

This equation is sometimes written in the form

$$HA(aq) \longrightarrow H^+(aq) + A^-(aq)$$

and the acid is commonly said to *ionize* in water. Remember, however, that in this equation $H^+(aq)$ is merely an abbreviation for H_3O^+. There is no evidence for the existence of free hydrogen ions, that is free protons, in any solutions. Because the proton is a bare nucleus and is therefore extremely small, it can approach very closely to an unshared electron pair and is then held very strongly by the electron pair. Consequently, a proton formed in water becomes attached to an unshared pair of electrons on the oxygen atom of a water molecule, forming the hydronium ion, H_3O^+.

The reaction of acids with carbonate ions, CO_3^{2-}, is another example of an acid-base reaction. In this case the base is the carbonate ion. It can accept protons from two H_3O^+ ions to form carbonic acid, H_2CO_3:

$$CO_3^{2-}(aq) + 2H_3O^+(aq) \longrightarrow H_2CO_3(aq) + 2H_2O$$

Carbonic acid is, however, unstable and decomposes to water and carbon dioxide;

$$H_2CO_3(aq) \longrightarrow H_2O(l) + CO_2(g)$$

which bubbles off from the solution.

However, the reactions of metals with the hydronium ion are not acid-base reactions. In the reaction of magnesium with an aqueous acid,

$$Mg(s) + 2H_3O^+(aq) \longrightarrow Mg^{2+}(aq) + H_2(g) + 2H_2O$$

each magnesium atom loses two electrons to become Mg^{2+}:

$$Mg \longrightarrow Mg^{2+} + 2e^-$$

It is therefore oxidized. At the same time hydrogen in the hydronium ion is reduced to molecular hydrogen. That this is a reduction can be seen more clearly if we write the equation in the form

$$Mg(s) + 2H^+(aq) \longrightarrow Mg^{2+}(aq) + H_2(g)$$

Each hydrogen ion acquires an electron to become a hydrogen atom, and two hydrogen atoms form a hydrogen molecule:

$$2H^+(aq) + 2e^- \longrightarrow H_2(g)$$

The reactions of metals with the hydronium ion in aqueous acid solutions are therefore oxidation-reduction reactions, and $H^+(aq)$ is an oxidizing agent.

Strong and Weak Acids

Hydrogen chloride reacts completely with water; it is completely converted to the corresponding ions:

$$HCl(g) + H_2O \longrightarrow H_3O^+(aq) + Cl^-(aq)$$

Acids that react completely with water in this way are called **strong acids**. They are completely ionized and their solutions have a high conductivity. Both HBr and HI are also strong acids.

In contrast, the reaction of HF with water,

$$HF(aq) + H_2O \longrightarrow H_3O^+(aq) + F^-(aq)$$

is incomplete. This is shown by the fact that the electrical conductivity of a solution of HF is much smaller than that of a solution of HCl of the same concentration. There are therefore fewer ions in the HF solution than in the HCl solution. Much of the HF in the solution remains as un-ionized HF molecules. An acid that is incompletely ionized in an aqueous solution is called a **weak acid**.

The incomplete ionization of HF in water does not mean, however, that the reaction of HF with water simply stops after a certain amount of the HF has ionized to form F^-. In fact, this reaction continues indefinitely. However, at the same time a reaction occurs between the products H_3O^+ and F^- to give back the original reactants HF and H_2O. The reaction between HF and H_2O is called the **forward reaction**. The reaction between H_3O^+ and F^-, to give HF and H_2O, is called the **reverse reaction**. When the rate of the reverse reaction is equal to the rate of the forward reaction, a state of dynamic equilibrium is reached. There is then no resultant change in the concentrations of HF, H_3O^+, H_2O, or F^-; all four of these species are present in equilibrium. The two reactions proceed simultaneously but at the same rate, so the reaction appears to have stopped. The dynamic nature of the equilibrium is emphasized by writing the equation for the reaction with two arrows, one pointing in each direction:

$$HF(aq) + H_2O \rightleftharpoons H_3O^+(aq) + F^-(aq)$$

The concept of dynamic equilibrium is extremely important in chemistry. All reactions eventually reach a state of equilibrium unless products or reactants are removed during the reaction. In the ionization of strong acids such as HCl, HBr, and HI, there is, in principle, a reverse reaction. But the concentrations of un-ionized HCl, HBr, or HI that are present at equilibrium are infinitesimally small, so we may say that the reaction has proceeded to completion. Equilibrium in chemical reactions is discussed in more detail in Chapter 14.

Table 5.7 lists some common strong and weak acids. Among the weak acids is carbonic acid, H_2CO_3, which is formed in small amounts when carbon dioxide is dissolved in water. Most of the carbon dioxide remains unchanged, but a small equilibrium amount of carbonic acid is formed:

$$CO_2(aq) + H_2O \rightleftharpoons H_2CO_3(aq)$$

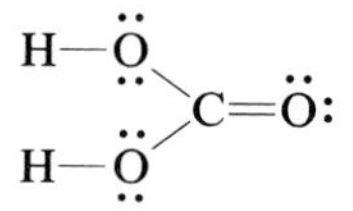

Carbonic acid is only ionized to a small extent according to the equation

$$H_2CO_3 + H_2O \rightleftharpoons H_3O^+ + HCO_3^-$$

In this case two separate equilibria are set up: the formation of carbonic acid from carbon dioxide and the ionization of carbonic acid.

Another weak acid is hypochlorous acid, HOCl. This acid is formed in a solution of chlorine in water. In such a solution most of the chlorine remains as Cl_2 molecules, but small equilibrium amounts of HCl and the weak acid HOCl are formed:

$$Cl_2(aq) + H_2O(l) \rightleftharpoons HCl(aq) + HOCl(aq)$$

Table 5.7 Some Common Acids

STRONG ACIDS		WEAK ACIDS	
Hydrochloric acid	HCl	Hydrofluoric acid	HF
Hydrobromic acid	HBr	Carbonic acid	H_2CO_3
Hydroiodic acid	HI	Phosphoric acid	H_3PO_4
Sulfuric acid	H_2SO_4	Acetic acid	CH_3CO_2H
Nitric acid	HNO_3	Hypochlorous acid	HOCl
Perchloric acid	$HClO_4$	Boric acid	H_3BO_3
		Hydrocyanic acid	HCN

Acetic acid, $C_2H_4O_2$, is another example of a weak acid (see Experiment 5.11). Vinegar is a dilute solution of acetic acid in water. The ionization of acetic acid may be written as

$$H-\overset{\overset{\displaystyle H}{|}}{\underset{\underset{\displaystyle H}{|}}{C}}-\overset{\overset{\displaystyle \ddot{O}:}{\|}}{C}-\ddot{\underset{\cdot\cdot}{O}}-H + :\overset{\overset{\displaystyle H}{|}}{\underset{\underset{\displaystyle H}{|}}{\ddot{O}}}-H \rightleftharpoons H-\overset{\overset{\displaystyle H}{|}}{\underset{\underset{\displaystyle H}{|}}{C}}-\overset{\overset{\displaystyle \ddot{O}:}{\|}}{C}-\ddot{\underset{\cdot\cdot}{O}}:^{\ominus} + H-\overset{\overset{\displaystyle H}{|}}{\ddot{O}}^{\oplus}-H$$

The negative ion formed in this reaction is called the *acetate ion*. Acetic acid is a member of a series of acids, called **carboxylic acids**. These acids are further discussed in Chapter 19.

Only one of the four hydrogen atoms in the acetic acid molecule is acidic. Carbon has only a very slightly greater electronegativity than hydrogen. Thus C—H bonds normally have only a very small polarity, and the hydrogen atoms have only a very small positive charge. Hence they have no tendency to be transferred to a water molecule. These three hydrogens are therefore said to be nonacidic. The hydrogen atoms in methane and almost all other hydrocarbons are also not acidic in water.

There are only a very few molecules in which a hydrogen atom attached to carbon is acidic in aqueous solution. One example is hydrogen cyanide, $H—C\equiv N:$, which is a weak acid in aqueous solution.

Electrical Conductivity of Solutions of Hydrogen Chloride and Acetic Acid

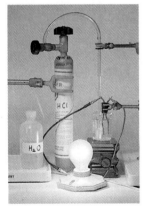 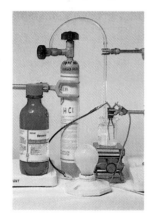 

Two metal electrodes connected to a power supply are dipped into a beaker of water. When HCl(g) is passed into the water the light bulb in the circuit glows brightly. Water behaves as a base, accepting protons from HCl(g) to give H_3O^+(aq) and Cl^-(aq).

When HCl(g) is passed into benzene, C_6H_6, it dissolves but the light bulb does not light—the solution is not conducting. Hydrogen chloride does not ionize in benzene because benzene is too weak a base to accept a proton from HCl. Hydrogen chloride dissolves in benzene as covalent HCl molecules.

The beaker contains a 0.10 M solution of acetic acid in water. The light bulb glows dimly, showing that the solution has only a small conductivity. Acetic acid is a weak acid and is only slightly ionized in water.

5.6 ACIDS AND BASES

Naming Binary Acids

The solutions formed when the hydrogen halides are dissolved in water are given special names:

Hydrogen halide	Formula	Name of aqueous solution
Hydrogen fluoride	HF	Hydrofluoric acid
Hydrogen chloride	HCl	Hydrochloric acid
Hydrogen bromide	HBr	Hydrobromic acid
Hydrogen iodide	HI	Hydroiodic acid

The name is derived by adding the prefix *hydro-* to the name of the halide and replacing *-ide* by *-ic acid*. The same system is used for an aqueous solution of hydrogen cyanide:

Hydrogen cyanide	HCN	Hydrocyanic acid

Bases

A **base** is a proton acceptor. Water behaves as a base, accepting a proton from an acid to form H_3O^+. Ammonia, NH_3, is another base. It is a polar molecule like water, and the positively charged hydrogen of the hydrogen chloride molecule is attracted to the negatively charged nitrogen of the ammonia molecule and transferred to its unshared pair of electrons:

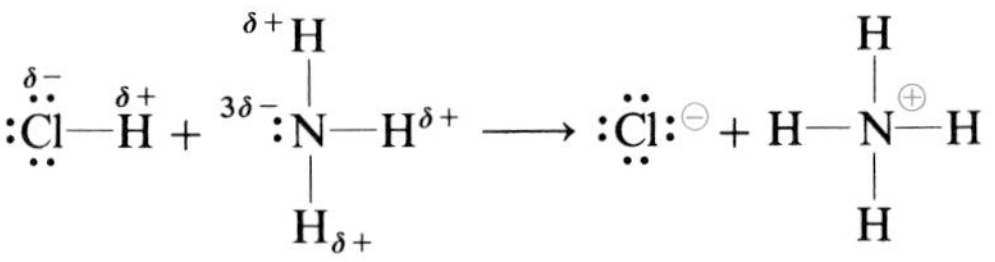

$$:\overset{\delta-}{\underset{..}{Cl}}-\overset{\delta+}{H} + {}^{3\delta-}:N-H^{\delta+} \longrightarrow :\overset{..}{\underset{..}{Cl}}:{}^{\ominus} + H-\overset{H}{\underset{H}{N}}-H$$

The products of this reaction are the ammonium ion and the chloride ion. Both hydrogen chloride and ammonia are gases. When they are allowed to mix, a dense white cloud of solid ammonium chloride is formed (see Experiment 3.5).

If the two gases are first dissolved in water, the reaction may then be written

$$H_3O^+(aq) + Cl^-(aq) + NH_3(aq) \longrightarrow NH_4^+(aq) + Cl^-(aq) + H_2O$$

The chloride ion is a spectator ion and takes no part in the reaction. So the equation may be simplified to

$$H_3O^+(aq) + NH_3(aq) \longrightarrow NH_4^+(aq) + H_2O$$

In this case the acid is H_3O^+ rather than HCl.

Ammonia also reacts with water in a proton transfer reaction in which water acts as an acid and ammonia as a base. The reaction is incomplete, however, and the following equilibrium is set up:

$$H_2O + NH_3(aq) \rightleftharpoons NH_4^+(aq) + OH^-(aq)$$
$$\text{acid} \qquad \text{base}$$

The products of this reaction are the ammonium ion, NH_4^+, and the **hydroxide ion**, OH^-. This reaction can be represented in terms of Lewis structures in the following way:

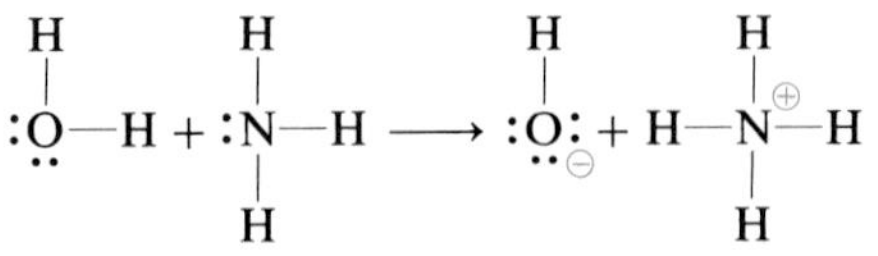

$$:\overset{H}{\underset{..}{O}}-H + :\overset{H}{\underset{H}{N}}-H \longrightarrow :\overset{H}{\underset{..}{O}}:{}^{\ominus} + H-\overset{H}{\underset{H}{N}}-H$$

A proton is transferred from the water molecule to the unshared pair of electrons on the nitrogen atom of the ammonia molecule. Since the reaction of ammonia with water is incomplete, ammonia is described as a **weak base**.

A base, B, is a proton acceptor. A base in water, therefore, gives rise to the hydroxide ion:

$$B + H_2O \longrightarrow BH^+ + OH^-$$

If this reaction is complete, the base is strong. A **strong base** is fully ionized in water to give OH^-. A **weak base** is incompletely ionized in water to give OH^-.

An example of a strong base in water is the oxide ion, O^{2-}. The oxide ion accepts a proton from a water molecule and is completely converted to the hydroxide ion:

$$O^{2-}(aq) + H_2O \longrightarrow OH^-(aq) + OH^-(aq)$$

The alkali metal oxides, for example, $(Na^+)_2O^{2-}$, which are ionic compounds, are strong bases in water:

$$(Na^+)_2O^{2-}(s) + H_2O \longrightarrow 2Na^+(aq) + 2OH^-(aq)$$

The product of this reaction is a solution of sodium hydroxide, NaOH. All of the alkali metals form similar hydroxides. They can also be prepared by the reaction of the alkali metal with water, as we saw in Experiment 4.1. For example,

$$2K(s) + H_2O \longrightarrow 2K^+(aq) + 2OH^-(aq) + H_2(g)$$

The alkali metal hydroxides are all ionic compounds composed of alkali metal cations and hydroxide ions, for example, potassium hydroxide, K^+OH^-. The alkali metal hydroxides are soluble in water, and they behave as strong bases because they give a quantitative yield of hydroxide ion:

$$K^+OH^-(s) \longrightarrow K^+(aq) + OH^-(aq)$$

A solution of ammonia in water is often called ammonium hydroxide, even though only a small fraction of the ammonia is converted to NH_4^+ and OH^-:

$$NH_3(aq) + H_2O \rightleftharpoons NH_4^+(aq) + OH^-(aq)$$

Another strong base in water is the hydride ion, H^-:

$$H^- + H_2O \longrightarrow H_2 + OH^-$$

Thus sodium hydride reacts with water to give hydrogen and a solution of sodium hydroxide:

$$NaH(s) + H_2O \longrightarrow NaOH(aq) + H_2(g)$$

All bases in water give rise to the hydroxide ion, OH^-, just as all acids give rise to the hydronium ion, H_3O^+. Aqueous solutions of bases therefore have several common properties by which they have long been recognized; these are the properties of the hydroxide ion, OH^-. These *basic solutions* are also often called *alkaline solutions*; their properties include the following:

- Basic (alkaline) solutions have an unpleasant, bitter taste.
- Basic (alkaline) solutions have the ability to change the colors of indicators and, in particular, to reverse the effect of acids on these indicators.
- Basic (alkaline) solutions have the ability to destroy or neutralize the properties of acids.

Acid-Base Reactions

A reaction between an acid and a base is called a **neutralization reaction**. The reaction of an aqueous solution of HCl with an aqueous solution of NaOH is an example:

$$NaOH(aq) + HCl(aq) \longrightarrow NaCl(aq) + H_2O(l)$$

Another example is

$$KOH(aq) + HBr(aq) \longrightarrow KBr(aq) + H_2O(l)$$

The products of these reactions between acids and bases in aqueous solution are a **salt** and water. A salt consists of a positive ion (cation) derived from a base and a negative ion (anion) derived from an acid. The salts formed in the above reactions are the alkali metal halides, sodium chloride, Na^+Cl^-, and potassium bromide, K^+Br^-.

Since NaOH consists of Na^+ and OH^- ions and KOH consists of K^+ and OH^- ions, and since HCl and HBr ionize in water to give the H_3O^+, Cl^-, and Br^- ions, we may rewrite the above equations as

$$Na^+(aq) + OH^-(aq) + H_3O^+(aq) + Cl^-(aq) \longrightarrow Na^+(aq) + Cl^-(aq) + 2H_2O$$

and

$$K^+(aq) + OH^-(aq) + H_3O^+(aq) + Br^-(aq) \longrightarrow K^+(aq) + Br^-(aq) + 2H_2O$$

Canceling the spectator ions in both cases, we obtain the equation

$$H_3O^+(aq) + OH^-(aq) \longrightarrow 2H_2O$$

This equation describes the neutralization reaction that occurs when *any* acid reacts with *any* base in aqueous solution.

A common laboratory procedure is the determination of the concentration of a solution of a base by allowing it to react with a solution of an acid of known concentration, or vice versa. The procedure is called a **titration**. A solution of the acid is placed in a buret, and it is run into a known volume of the solution of the base until just sufficient acid solution has been added to react with all the base. This point is determined by observation of the change in color of a suitable indicator (see Figure 5.11).

Example 5.7 In a titration 25.00 mL of sodium hydroxide, NaOH, solution was neutralized by 32.72 mL of hydrochloric acid, HCl. The HCl solution had a concentration of 0.129M. Find the concentration of the NaOH solution.

Solution First, we must write the balanced equation for the reaction:

$$NaOH(aq) + HCl(aq) \longrightarrow NaCl(aq) + H_2O$$

Now, since we know both the volume and the concentration of the HCl solution, we can determine the amount of HCl that took part in the reaction:

$$\text{Amount HCl} = (32.72 \text{ mL})\left(\frac{1 \text{ L}}{1000 \text{ mL}}\right)\left(\frac{0.129 \text{ mol}}{1 \text{ L}}\right) = 4.22 \times 10^{-3} \text{ mol HCl}$$

The balanced equation shows that 1 mol of HCl reacts with 1 mol of NaOH. Therefore

$$\text{Amount NaOH} = 4.22 \times 10^{-3} \text{ mol NaOH}$$

This amount of NaOH was in 25.00 mL of the NaOH solution. Therefore the concentration of the NaOH solution is

Figure 5.11. Acid-Base Titration.
(a) A measured volume of a solution of a base is added to a flask using a pipet. (b) A solution of an acid is run into the flask from a buret until just enough has been added to react completely with the base, as shown by the color change of a suitable indicator.

(a) A known volume of a sample of base solution is placed in a flask using a pipet.

(b) Acid is added from a buret until there is just enough to completely react with the base, as shown by the color change of a suitable indicator.

$$\left(\frac{4.22 \times 10^{-3} \text{ mol NaOH}}{25.00 \text{ mL}}\right)\left(\frac{1000 \text{ mL}}{1 \text{ L}}\right) = 0.169 \text{ mol L}^{-1} = 0.169 M$$

Once we understand the principles, we can solve this problem more quickly in one step, as follows (concn is the abbreviation for concentration, and soln is the abbreviation for solution):

$$\text{Concn NaOH soln} = (32.72 \text{ mL HCl})\left(\frac{1 \text{ L}}{1000 \text{ mL}}\right)\left(\frac{0.129 \text{ mol HCl}}{1 \text{ L HCl}}\right)$$

$$\times \left(\frac{1 \text{ mol NaOH}}{1 \text{ mol HCl}}\right)\left(\frac{1}{25.00 \text{ mL}}\right)\left(\frac{1000 \text{ mL}}{\text{L}}\right)$$

$$= 0.169 \text{ mol L}^{-1} = 0.169 \ M$$

Still more time can be saved by working in millimoles (10^{-3} mol) rather than in moles. A $1M$ solution contains 1 mmol in 1 mL. Hence a $0.129M$ solution contains 0.129 mmol in 1 mL. Therefore

$$\text{Concn NaOH soln} = (32.72 \text{ mL HCl})\left(\frac{0.129 \text{ mmol HCl}}{1 \text{ mL}}\right)$$

$$\times \left(\frac{1 \text{ mol NaOH}}{1 \text{ mol HCl}}\right)\left(\frac{1}{25.00 \text{ mL}}\right)$$

$$= 0.169 \text{ mmol mL}^{-1} = 0.169 \text{ mol L}^{-1} = 0.169 M$$

Conjugate Acid-Base Pairs

When an acid, HA, gives up a proton, the anion, A^-, that is formed is a base, because it can add a proton to re-form the acid HA:

$$HA \rightleftharpoons A^- + H^+$$

The anion A^- is called the **conjugate base** of the acid HA, and HA and A^- are described as a **conjugate acid-base pair**. Examples include the following:

$$HCl \rightleftharpoons Cl^- + H^+$$
$$H_2O \rightleftharpoons OH^- + H^+$$
$$NH_4^+ \rightleftharpoons NH_3 + H^+$$
$$H_3O^+ \rightleftharpoons H_2O + H^+$$

$$\text{Acid} \qquad \text{Conjugate base}$$

When an acid is a cation, such as NH_4^+, its conjugate base is a neutral molecule, such as NH_3.

Similarly, when a base B accepts a proton, the cation, BH^+, that is formed is an acid, because it can lose a proton to give back the base B:

$$B + H^+ \rightleftharpoons BH^+$$

The cation BH^+ is called the **conjugate acid** of the base B, and BH^+ and B are described as a conjugate acid-base pair. For example,

$$H_2O + H^+ \rightleftharpoons H_3O^+$$
$$NH_3 + H^+ \rightleftharpoons NH_4^+$$
$$OH^- + H^+ \rightleftharpoons H_2O$$
$$F^- + H^+ \rightleftharpoons HF$$

$$\text{Base} \qquad\qquad \text{Conjugate acid}$$

If the base is an anion, such as F^-, the conjugate acid is a neutral molecule, such as HF.

Example 5.8

(a) What is the conjugate acid of each of the following: CN^-, Br^-, and PH_3?

(b) What is the conjugate base of each of the following: HF, $HClO_4$, NH_3, and NH_4^+?

Solution

(a) The conjugate acid is the species obtained by adding a proton to a base. Thus we have

$$CN^- + H^+ \longrightarrow HCN$$
$$Br^- + H^+ \longrightarrow HBr$$
$$PH_3 + H^+ \longrightarrow PH_4^+$$

$$\text{Base} \qquad\qquad \text{Conjugate acid}$$

(b) The conjugate base is the species formed by removing a proton from an acid. Thus we have

$$HF \longrightarrow F^- + H^+$$
$$HClO_4 \longrightarrow ClO_4^- + H^+$$
$$NH_3 \longrightarrow NH_2^- + H^+$$
$$NH_4^+ \longrightarrow NH_3 + H^+$$

Acid   Conjugate base

Since a base can obtain a proton only from an acid, and an acid can give up a proton only if there is a base to accept it, every acid-base reaction involves two conjugate acid-base pairs:

$$AH + B \rightleftharpoons BH^+ + A^-$$
$$Acid_1 + Base_2 \rightleftharpoons Acid_2 + Base_1$$

$Acid_1$ and $base_1$ (AH and A^-) form one conjugate acid-base pair; $acid_2$ and $base_2$ (BH^+ and B) constitute another. For example,

$$Acid_1 + Base_2 \rightleftharpoons Acid_2 + Base_1$$
$$HF + H_2O \rightleftharpoons H_3O^+ + F^-$$
$$H_2O + CN^- \rightleftharpoons HCN + OH^-$$
$$H_2O + NH_3 \rightleftharpoons NH_4^+ + OH^-$$

We see from these examples that water can act both as an acid and as a base. It accepts a proton from an acid, such as HCl, forming the conjugate acid H_3O^+:

$$HF + H_2O \rightleftharpoons H_3O^+ + F^-$$

But it also donates a proton to an ammonia molecule, forming its conjugate base OH^-:

$$H_2O + NH_3 \rightleftharpoons NH_4^+ + OH^-$$

Many molecules and ions can exhibit both acid and base properties. They are said to be *amphiprotic*, or *amphoteric*. **Amphiprotic** means having the ability to either accept or donate a proton. **Amphoteric** is a more general term that means having the ability to act either as an acid or as a base.

The existence of amphiprotic substances such as water serves to emphasize the relative nature of the terms *acid* and *base*. It is not strictly correct to state that water *is* an acid or that it *is* a base. Water may be said to *behave as an acid* or to *behave as a base*, depending on the circumstances. In the presence of HCl water behaves as a base, but in the presence of NH_3 it behaves as an acid.

Since water is amphiprotic, a proton exchange can occur between two water molecules. In this reaction one water molecule acts as an acid and the other as a base:

$$H_2O + H_2O \rightleftharpoons H_3O^+ + OH^-$$

$Acid_1$ $Base_2$ $Acid_2$ $Base_1$

This reaction is known as the **self-ionization**, or **autoprotolysis, of water**. The concentrations of the ions H_3O^+ and OH^- in equilibrium with pure water are very small, but nevertheless, the self-ionization of water is important in a quantitative treatment of acid-base reactions in water, as we will see in Chapter 14.

Strengths of Acids and Bases

STRONG ACIDS The strong acids listed in Table 5.7 are all completely ionized when dissolved in a large amount of water. Their reaction with water goes to completion. For example,

$$HCl + H_2O \longrightarrow H_3O^+ + Cl^-$$
$$HClO_4 + H_2O \longrightarrow H_3O^+ + ClO_4^-$$

Since all these acids are completely converted to H_3O^+, any intrinsic differences in their strengths cannot be detected. Their strengths are said to be leveled by the base, water, and this effect is referred to as the **leveling effect** of water. The strongest acid that can exist in a dilute aqueous solution is H_3O^+. Any acid that is intrinsically stronger than H_3O^+ is quantitatively converted to H_3O^+ in water.

The conjugate bases of strong acids are such weak bases that they have no tendency to remove a proton from an H_3O^+ ion. They therefore have no tendency to remove a proton from a water molecule, which is a much weaker acid than H_3O^+. In other words, they have no base properties in water. The equilibrium

$$Cl^- + H_2O \rightleftharpoons OH^- + HCl$$

lies very far over on the left-hand side. Although formally they may be described as bases because they have the potential to add a proton, anions such as Cl^-, Br^-, NO_3^- and ClO_4^- are much too weak to act as bases in water (see Figure 5.12).

STRONG BASES Similar considerations apply to strong bases. They are completely converted to the hydroxide ion in water, and any intrinsic differences

Figure 5.12. Strengths of Acids and Bases in Water.

ACIDS						
An acid donates a proton to form a conjugate base				A base accepts a proton to form a conjugate acid		
	Acid		Conjugate Base		Base	Conjugate Acid
Strong acids 100% dissociated in water / Completely converted to H_2O^+	$HClO_4$ + $H_2O \rightarrow H_3O^+$ + ClO_4^-			Too weak to behave as bases in water	ClO_4^-	
	HCl + $H_2O \rightarrow H_3O^+$ + Cl^-				Cl^-	
	HNO_3 + $H_2O \rightarrow H_3O^+$ + NO_3^-				NO_3^-	
Strongest acid	H_3O^+ + $H_2O \rightleftharpoons H_3O^+$ + H_2O			Weakest base	H_2O + $H_2O \rightleftharpoons OH^-$ + H_3O^+	
Weak acids	HF + $H_2O \rightleftharpoons H_3O^+$ + F^-			Weak bases	F^- + $H_2O \rightleftharpoons OH^-$ + HF	
	CH_3CO_2H + $H_2O \rightleftharpoons H_3O^+$ + $CH_3CO_2^-$				$CH_3CO_2^-$ + $H_2O \rightleftharpoons OH^-$ + CH_3CO_2H	
	NH_4^+ + $H_2O \rightleftharpoons H_3O^+$ + NH_3				NH_3 + $H_2O \rightleftharpoons OH^-$ + NH_4^+	
Weakest acid	H_2O + $H_2O \rightleftharpoons H_3O^+$ + OH^-			Strongest base	OH^- + $H_2O \rightleftharpoons OH^-$ + H_2O	
Too weak to behave as acids in water	NH_3			Strong bases 100% dissociated in water / Completely converted to OH^-	NH_2^- + $H_2O \rightarrow OH^-$ + NH_3	
	H_2				H^- + $H_2O \rightarrow OH^-$ + H_2	
	OH^-				O^{2-} + $H_2O \rightarrow OH^-$ + OH^-	

Range of acid and base strengths in water

in their strengths cannot be detected. For example,

$$O^{2-} + H_2O \longrightarrow OH^- + OH^-$$

$$H^- + H_2O \longrightarrow H_2 + OH^-$$

The strongest base that can exist in a dilute aqueous solution is the hydroxide ion, OH^-. Water therefore also levels the strengths of strong bases, just as it levels the strengths of strong acids. The range of acid-base strengths that is accessible in aqueous solution therefore lies between the strong acid H_3O^+ and the strong base OH^-.

Since the reactions of strong bases with water proceed completely to the right, the reverse reactions show no tendency to proceed. Thus the conjugate acids of O^{2-} and H^-, in other words, OH^- and H_2, have no tendency to donate a proton to an OH^- ion. They also therefore have no tendency to donate a proton to the much weaker base, H_2O. In other words, neither OH^- nor H_2 behave as acids in water (see Figure 5.12).

WEAK ACIDS AND BASES   A weak acid such as HF is incompletely ionized. In other words, it is only partially converted to H_3O^+, and the following equilibrium is set up:

$$HF + H_2O \rightleftharpoons H_3O^+ + F^-$$

In this case the reverse reaction does occur; F^- accepts a proton from H_3O^+ to a limited extent, and so it behaves as a base. The F^- ion also accepts a proton from an H_2O molecule but to an even more limited extent, to form very small equilibrium amounts of HF and OH^-:

$$F^- + H_2O \rightleftharpoons HF + OH^-$$

Thus the conjugate base of a weak acid is a weak base in water.

Similarly, NH_4^+, the conjugate acid of the weak base, NH_3,

$$NH_3 + H_2O \rightleftharpoons NH_4^+ + OH^-$$

is a weak acid in water:

$$NH_4^+ + H_2O \rightleftharpoons H_3O^+ + NH_3$$

In a series of acids of increasing strength in water, the strength of the conjugate base decreases until finally, for a strong acid in water, the conjugate base shows no base properties at all in water. Similarly, in a series of bases of increasing strength in water, the conjugate acids decrease in strength until finally, for a strong base, the conjugate acid has no acid properties at all in water (see Figure 5.12).

WHY IS HF A WEAK ACID?   Since fluorine is the most electronegative of the halogens, and since therefore HF is the most polar of the hydrogen halides, it seems surprising that HF is a weak acid, whereas HCl, HBr, and HI are strong acids. Clearly, the electronegativity of the halogen is not the only factor involved in determining the acid strength. Many other factors such as the strengths of the bonds in the hydrogen halides and the strengths of the interactions between the HF molecules and H_2O molecules and between the halide ions and H_2O molecules also play a role. We cannot give a complete discussion here of what is, in fact, a relatively complicated question, but we will have more to say about it in Chapters 7 and 13.

5.7 REACTION TYPES

In this chapter we have met three very important types of reactions:

1. In *oxidation-reduction reactions* electrons are transferred from one species (molecule or ion) to another species. For example, in the reaction

$$2Br^-(aq) + Cl_2(aq) \longrightarrow Br_2(aq) + 2Cl^-(aq)$$

electrons are transferred from Br^- ions (the reducing agent) to Cl_2 molecules (the oxidizing agent).

2. In *acid-base reactions* hydrogen ions (protons), H^+, are transferred from one species (molecule or ion) to another species. For example, in the reaction

$$HBr(aq) + H_2O(l) \longrightarrow H_3O^+(aq) + Br^-(aq)$$

protons are transferred from HBr molecules (the acid) to H_2O molecules (the base).

3. In *precipitation reactions* an insoluble solid is formed from a solution of soluble substances. For example, in the reaction

$$KBr(aq) + AgNO_3(aq) \longrightarrow AgBr(s) + KNO_3(aq)$$

insoluble silver chloride, AgCl, is formed on mixing aqueous solutions of KBr and $AgNO_3$.

It is important that you are able to recognize these three types of reactions. You will meet them many times in the following chapters, where you will also encounter several other types of reactions.

Example 5.9 Classify each of the following reactions as an oxidation-reduction, acid-base, or precipitation reaction.

(a) $Mg(s) + F_2(g) \longrightarrow Mg^{2+}(F^-)_2(s)$

(b) $2Ca(s) + O_2(g) \longrightarrow 2CaO(s)$

(c) $Pb(NO_3)_2(aq) + 2NaCl(aq) \longrightarrow PbCl_2(s) + 2NaNO_3(aq)$

(d) $HNO_3(aq) + NH_3(aq) \longrightarrow NH_4^+(aq) + NO_3^-(aq)$

(e) $2Na(s) + S(s) \longrightarrow Na_2S(s)$

(f) $H_3O^+(aq) + CO_3^{2-}(aq) \longrightarrow HCO_3^-(aq) + H_2O$

(g) $Zn(s) + 2HCl(aq) \longrightarrow Zn^{2+}(aq) + 2Cl^-(aq) + H_2(g)$

Solution

(a) It is an oxidation-reduction reaction. The Mg loses electrons to become Mg^{2+}; it is oxidized:

$$Mg \longrightarrow Mg^{2+} + 2e^-$$

The F_2 gains electrons to become $2F^-$; it is reduced:

$$F_2 + 2e^- \longrightarrow 2F^-$$

(b) It is an oxidation-reduction reaction. Calcium oxide, CaO, is a compound of a group II metal with a group VI nonmetal. It is therefore ionic, $Ca^{2+}O^{2-}$. Thus Ca loses electrons to become Ca^{2+} and is oxidized, while O_2 gains electrons to become $2O^{2-}$ and is reduced.

(c) It is a precipitation reaction. An insoluble solid, $PbCl_2$, is formed from an aqueous solution of two soluble substances, $Pb(NO_3)_2$ and NaCl.

(d) It is an acid-base reaction. Nitric acid, HNO_3, loses a proton to form the nitrate ion, NO_3^-, and NH_3 gains a proton to give the ammonium ion, NH_4^+. Thus it is a proton transfer, or acid-base, reaction.

(e) It is an oxidation-reduction reaction. Sodium sulfide, Na_2S, is a compound of a group I metal and a group VI nonmetal and is therefore ionic. Thus Na loses an electron to become Na^+, and S gains two electrons to become S^{2-}. It is an electron transfer, or oxidation-reduction, reaction.

(f) It is an acid-base reaction. The H_3O^+ ion loses a proton to become H_2O; it is an acid. The carbonate ion, CO_3^{2-}, gains a proton to become the hydrogen carbonate ion, HCO_3^-; CO_3^{2-} is a base.

(g) It is an oxidation-reduction reaction. Although this reaction involves the acid HCl(aq), it is not an acid-base reaction. The proton in HCl is not transferred to a base. In fact, Zn loses two electrons to become Zn^{2+}; it is oxidized. And two H^+ (H_3O^+) ions gain two electrons to become $H_2(g)$:

$$2H^+ + 2e^- \longrightarrow H_2$$

IMPORTANT TERMS

An **acid** (Brønsted acid) is a proton donor. In aqueous solution all acids transfer a proton to a water molecule, to give the hydronium ion, H_3O^+.

An **acid-base reaction** is a reaction in which protons are transferred; it is a proton transfer reaction.

An **amphiprotic substance** is one that can both add and donate a proton.

An **amphoteric substance** is one that can behave as both an acid and a base.

Autoprotolysis (self-ionization) is an acid-base reaction in which one solvent molecule acts as an acid, donating a proton to another solvent molecule, which acts as a base.

A **base** (Brønsted base) is a proton acceptor. In aqueous solution a base accepts a proton from a water molecule, forming the hydroxide ion, OH^-.

A **conjugate acid** is the acid formed by the addition of a proton to a base.

A **conjugate base** is the base formed by the loss of a proton from an acid.

Electronegativity is the ability of an atom in a molecule to attract the electrons of a covalent bond to itself.

Equilibrium is reached in a reaction when the rate of the forward reaction equals the rate of the back (reverse) reaction so that there is no change in the overall concentrations of the reactants and products.

The **halogens** are the elements fluorine, chlorine, bromine, iodine and astatine which form group VII of the periodic table.

The **hydronium ion**, H_3O^+, is the ion that is formed by all *acids* in aqueous solution. It is the strongest acid that can exist in water.

The **hydroxide ion**, OH^-, is the ion that is formed by all *bases* in aqueous solution. It is the strongest base that can exist in water.

An **ionic bond** is a bond that results from the electrostatic attraction between oppositely charged ions.

An **ionic crystal** is a crystal that is composed of an infinite array of positive and negative ions and in which no individual molecules can be distinguished.

A **neutralization reaction** is a reaction between an acid and a base.

A **nonpolar bond** is a bond formed between atoms of equal electronegativity.

An **octahedron** is a regular solid with 8 equivalent equilateral triangular faces, 12 equivalent edges, and 6 equivalent vertices.

Oxidation is a process in which electrons are lost.

An **oxidation-reduction reaction** is a reaction in which electrons are transferred; it is an electron transfer reaction.

An **oxidizing agent** is a substance that has a tendency to gain electrons.

A **polar bond** is a covalent bond between two atoms of different electronegativity in which the atom of higher electronegativity has a partial negative charge and the other atom has a partial positive charge; a bond in which the bond electrons are unequally shared.

A **precipitation reaction** is one in which an insoluble substance separates from solution.

A **reducing agent** is a substance that has a tendency to lose electrons.

Reduction is a process in which electrons are gained.

A **salt** is an ionic compound consisting of a positive ion (cation) derived from a base and a negative ion (anion) derived from an acid. It is the product of the neutralization of a base such as NaOH with an acid such as an aqueous solution of HCl.

Spectator ions are ions that are present in a solution but which do not take part in any reactions that are occurring.

A **strong acid** in water is an acid that is quantitatively converted to H_3O^+.

A **strong base** in water is a base that is quantitatively converted to OH^-.

A **titration** is a procedure in which a measured volume of a solution of a reactant is added to a known volume of a solution of a second reactant until the reaction is complete, as indicated by a color change produced by a reactant or by a product, or by a suitable indicator.

A **weak acid** in water is an acid that is only partially converted to H_3O^+.

A **weak base** in water is a base that is only partially converted to OH^-.

PROBLEMS

Reactions and Equations

1. Write a balanced equation for each of the following reactions:

(a) The reaction of chlorine with phosphorus to give phosphorus trichloride.

(b) The reaction of chlorine with sulfur to give (i) SCl_2 and (ii) S_2Cl_2.

(c) The reaction of fluorine with carbon to give CF_4.

(d) The reaction of bromine with arsenic to give arsenic tribromide.

2. Write a balanced equation for each of the following:

(a) The reaction of hydrogen bromide with water.

(b) The reaction of carbon dioxide with water.

(c) The reaction of ammonia with water.

(d) The reaction of chlorine with water.

(e) The reaction of fluorine with water.

3. Write a balanced equation for each of the following:

(a) The reaction of calcium with chlorine.

(b) The reaction of aluminum with bromine.

(c) The reaction of water with chlorine.

(d) The reaction of phosphorus with chlorine.

(e) The reaction of phosphorus with iodine.

Electronegativity

4. What are the two principal properties of an atom that determine its electronegativity?

5. Describe how the electronegativities of the elements change in going along any period from the left-hand side to the right-hand side of the periodic table and going down any group. Explain these variations.

6. In each of the following pairs of elements, choose the element of higher electronegativity without referring to Figure 5.4:

* The asterisk denotes the more difficult problems.

(a) F, Cl   (b) F, O   (c) P, S   (d) C, Si

(e) O, P   (f) Br, Se   (g) P, Al

Ionic, Covalent, and Polar Bonds

7. Classify the bonds in each of the following substances as ionic, nonpolar covalent, or polar covalent:

Cl_2 PCl_3 $LiCl$ ClF $MgCl_2$ S_2Cl_2

8. Classify the bonds in each of the following substances as ionic, nonpolar covalent, or polar covalent:

Li_2O MgO O_2 SO_2 Cl_2O NO

9. Classify each of the following molecules as polar or nonpolar:

HBr I_2 ClF H_2

10. Explain why NaCl forms a solid ionic compound at room temperature rather than a gas consisting of Na^+Cl^- molecules (ion pairs).

11. Suggest a reason why ions with charges greater than 3 are rarely found in ionic compounds.

12. Explain, on the basis of structure, why F_2 is a gas, whereas LiF is a crystalline solid at room temperature.

13. Without referring to Table 5.6, select the largest ion in each of the following pairs:

(a) K^+, Ca^{2+} (b) S^{2-}, Cl^- (c) Cl^-, K^+

(d) Na^+, Li^+ (e) I^-, Br^-

14. Without referring to Table 5.6 and Figure 4.6, select the largest species in each of the following pairs:

(a) Cl, Cl^- (b) Na, Na^+

(c) Mg^{2+}, Na (d) Cl^-, K^+

15. Give the empirical formula of the ionic compound formed by each of the following pairs of elements:

(a) Ca, I   (b) Be, O   (c) Al, S

(d) Mg, Br   (e) Rb, Se   (f) Ba, O

16. Write formulas for compounds composed of the following ions:

(a) NH_4^+, PO_4^{3-} (b) Fe^{3+}, O^{2-}

(c) Cu^+, O^{2-} (d) Al^{3+}, SO_4^{2-}

17. Write formulas for each of the following compounds:

Barium iodide   Aluminum chloride

Ammonium perchlorate   Calcium nitrate

*18. The density of sodium chloride is 2.17 g cm^{-3}. How many Na^+ ions and how many Cl^- ions are in an NaCl crystal that is a cube of size 1 mm?

19. Draw Lewis structures for each of the following molecules:

$$PCl_3 \quad ClF \quad LiF(g) \quad SiF_4$$

Oxidation-Reduction Reactions

20. Would you expect Cl_2 to react with I^-, or I_2 to react with Cl^-? In the reaction that does proceed, identify the oxidizing agent, the reducing agent, the substance oxidized, and the substance reduced.

21. Complete and balance each of the following equations. If no reaction occurs, write NR.

(a) $Cl_2 + KI(aq)$ (b) $I_2 + NaCl(aq)$

(c) $Br_2 + NaI(aq)$ (d) $F_2 + H_2O(l)$

22. Suggest a reason why samples of crystalline sodium iodide are sometimes pale yellow, although both Na^+ and I^- are colorless.

23. In each of the following reactions, identify the oxidizing agent, the reducing agent, the species oxidized, and the species reduced:

(a) $Rb(s) + I_2(s) \longrightarrow RbI(s)$

(b) $4Al(s) + 3O_2(g) \longrightarrow 2(Al^{3+})_2(O^{2-})_3(s)$

(c) $Cu^{2+}(aq) + 3I^-(aq) \quad Cu^+I^-(s) + I_2(s)$

(d) $Zn(s) + S(s) \longrightarrow Zn^{2+}S^{2-}(s)$

(e) $Mg(s) + 2HCl(aq) \longrightarrow MgCl_2(aq) + H_2(g)$

24. Suppose 2.45 g of magnesium is burned completely in chlorine. What mass of magnesium chloride is formed?

25. If you are given colorless aqueous solutions of a chloride, a bromide, and an iodide, explain how you could identify each solution. Write balanced equations for the reactions.

Precipitation Reactions

26. An aqueous solution of silver nitrate contains 2.50 g $AgNO_3$. To this solution sufficient barium chloride, $BaCl_2$, solution is added to precipitate all the silver as insoluble silver chloride, AgCl. What mass of dry silver chloride is obtained?

27. In attempting to prepare a $0.1M$ solution of silver nitrate in tap water, a student claimed that he could not dissolve all the silver nitrate because the solution remained cloudy. Explain the student's problem. The solubility of $AgNO_3$ is 245 g in 100 g of water at 25°C.

28. Using compounds of bromine, give at least one example of each of the following types of reactions:

(a) Acid-base   (b) Oxidation-reduction

(c) Precipitation

Acid-Base Reactions

29. Describe three properties that you could use to decide whether a given aqueous solution was acidic. What property would you *not* use in investigating an unknown solution?

30. For each of the following acids, write an equation for its reaction with water, and indicate whether the acid is strong or weak:

$$HNO_3 \quad H_3PO_4 \quad HOCl \quad H_2SO_4 \quad HF$$

31. Write an equation to show how each of the following bases is ionized in aqueous solution. Which are strong bases and which are weak bases?

$$Na_2O \quad KOH \quad NH_3 \quad KNH_2$$

32. Write the simplest equation that describes the neutralization of any acid by any base in aqueous solution. Suppose 250 mL of $1M$ HCl solution is neutralized with 500 mL of $0.5M$ NaOH solution. What is the concentration of NaCl in the resulting solution? If it is evaporated to dryness, what mass of NaCl is obtained?

33. An acid, HZ, is described as strong, while another acid, HY, is described as weak in aqueous solution. Write equations for the ionization of the two acids in water. What can be said about the base strengths of Z^- and Y^-? What species are present in aqueous solutions of the sodium salts of these anions?

34. What volume of $0.100M$ HCl is required to react completely with 5.00 g of calcium hydroxide, $Ca(OH)_2$?

35. The primary active ingredient in an antacid is $NaAl(OH)_2CO_3$, which reacts with excess acid in the stomach according to the equation

$$NaAl(OH)_2CO_3(s) + 4HCl(aq) \longrightarrow$$
$$NaCl(aq) + AlCl_3(aq) + 3H_2O(l) + CO_2(g)$$

What mass of $NaAl(OH)_2CO_3$ is required to react with 2.00 L of $0.120M$ HCl?

36. Give the formulas of the conjugate bases of the following acids:

(a) HF   (b) HNO_3 (c) $HClO_4$

(d) H_2O (e) H_3O^+

37. Give the formulas of the conjugate acids of the following bases:

(a) NH_3 (b) CH_3NH_2 (c) OH^- (d) H_2O

38. Write the formulas and draw the Lewis structures of each of the following:

(a) Hydronium ion   (b) Ammonium ion

(c) Hydroxide ion

39. Write equations for the ionization of the following acids in aqueous solution. Indicate which of the reactants and which of the products in these reactions are acids and which are bases.

HCl CH_3CO_2H $HClO_4$ H_2SO_4

$HOCl$ NH_4^+ H_2O

40. Write balanced equations for the reactions that occur when sodium hypochlorite is dissolved in water and when ammonium chloride is dissolved in water. Are the resulting solutions acidic, basic, or neutral?

41. What is the strongest acid species and the strongest base species that can exist in aqueous solution? The hydride ion, H^-, and the oxide ion, O^{2-}, are both strong bases in water; write balanced equations for the reactions of LiH, CaH_2, Li_2O, and CaO with water.

42. What volume of $0.124M$ HBr solution is required to neutralize 25.00 mL of $0.107M$ NaOH solution?

43. How much $0.115M$ KOH solution is required to react completely with 100.0 mL of $0.211M$ HF solution?

*44. Element A is a metal and element B is a liquid. They react to give a compound of formula AB, which is a colorless, crystalline solid that melts at 760°C and vaporizes at 1380°C. The solid is a nonconductor, but when molten and when it is dissolved in water, it conducts electricity. The addition of an aqueous solution of silver nitrate to a solution of 0.543 g of AB dissolved in water gave 0.857 g of a pale yellow, insoluble compound. When chlorine was bubbled through an aqueous solution of AB, the solution turned brown. Identify the elements A and B, and explain your reasoning.

Miscellaneous

45. Astatine (At) is the last member of the halogen family. Predict the following for At:

(a) Physical state.

(b) Ionization energy and size (compared with iodine).

(c) The formula of a compound between astatine and bromine.

(d) The acid strength of HAt in water.

(e) The shape of the PAt_3 molecule.

(f) The type of bond in KAt.

(g) Equations for the reactions of At with Na, Ca. P, and H_2.

46. Bleaching powder, $Ca(OCl)_2$, is the calcium salt of hypochlorous acid, HOCl. When carbon dioxide, CO_2, is passed into an aqueous solution of this salt, HOCl and insoluble calcium carbonate, $CaCO_3$, are formed. Write an equation for the reaction. In an experiment that started with 8.00 g of $Ca(OCl)_2$, the final solution of HOCl in water had a volume of 200 mL. Assuming that the reaction goes to completion, what is the concentration of the HOCl solution obtained?

47. Javex is a solution of sodium hypochlorite, NaOCl, and sodium chloride, NaCl, obtained by passing chlorine, Cl_2, into sodium hydroxide, NaOH, solution. Write a balanced equation for the reaction. How many grams of NaOCl could be obtained by passing 1.00 L of Cl_2 into excess of an NaOH solution?

48. Radium, a radioactive element of group II of the periodic table, was discovered by Marie Curie in France in 1898. When it was allowed to react with dilute hydrochloric acid, a pure anhydrous yellow-white salt, melting at 1000°C, was obtained. After the salt had been purified by fractional crystallization, analysis showed that it contained 76.1% radium. What is the atomic mass of radium?

49. Which of the following reactions are acid-base reactions, and which are oxidation-reduction reactions? For the acid-base reactions, identify the acid and the base. For the oxidation-reduction reactions, identify the oxidizing agent and the reducing agent.

(a) $Cl_2(aq) + 2I^-(aq) \longrightarrow 2Cl^-(aq) + I_2(s)$

(b) $HCl(aq) + H_2O \longrightarrow Cl^-(aq) + H_3O^+(aq)$

(c) $Zn(s) + 2HCl(aq) \longrightarrow ZnCl_2(aq) + H_2(g)$

(d) $HCO_3^-(aq) + H_3O^+(aq) \longrightarrow CO_2(aq) + 2H_2O$

CHAPTER 6

ENERGY LEVELS, ELECTRON CONFIGURATIONS, AND THE COVALENT BOND

We have seen in Chapters 4 and 5 that Lewis structures and the concepts of ionic and covalent bonds are very useful in enabling us to understand the compositions and properties of substances. We have described two important types of substances—ionic compounds and covalent molecular substances. Ionic compounds consist of oppositely charged ions held together by electrostatic attraction. Examples are sodium chloride, Na^+Cl^-, magnesium chloride, $Mg^{2+}(Cl^-)_2$, and magnesium oxide, $Mg^{2+}O^{2-}$. Covalent molecular substances consist of molecules in which the atoms are held together by covalent bonds. Examples are water, H_2O, chlorine, Cl_2, and hydrogen chloride, HCl. Although we have described how oppositely charged ions are held by electrostatic attraction, we have not yet answered the question "How does a pair of shared electrons hold two atoms together in a covalent bond?" To answer this question, and to generally extend our understanding of chemical bonds, we need to consider the properties of electrons and their arrangement in atoms in more detail.

Experiments carried out in the early years of this century showed that the laws of motion that apply to ordinary-sized objects do not apply to electrons. A new theory was developed to account for the behavior of electrons and other very small particles; it is known as **quantum mechanics**. We will examine only some of the very basic ideas of quantum mechanics here, but even this limited discussion will give us a better understanding of the behavior of electrons in atoms and molecules and therefore of chemical bonds. Quantum mechanics had its origins in attempts to explain the nature of light and its interactions with matter, so we begin with a discussion of light.

6.1 LIGHT AND ELECTROMAGNETIC WAVES

For thousands of years philosophers and scientists have argued about the nature of light. Isaac Newton (1642–1727) believed that light consisted of a stream of particles. Dutch physicist Christian Huygens (1629–1695) believed that light was a type of wave motion. The dispute over the nature of light was apparently resolved by the work of Scottish physicist James Clerk Maxwell (1831–1879). Maxwell showed that all the then-known properties of light could be accounted for by means of equations based on the hypothesis that light is an electromagnetic wave. An **electromagnetic wave** consists of electric and magnetic fields that oscillate in directions perpendicular to each other and perpendicular to the direction in which the wave is traveling, as Figure 6.1 illustrates. From his equations Maxwell was able to calculate a value for the speed of light that agreed exactly with the experimentally measured value of 3.00×10^8 m s^{-1}.

All waves can be described in terms of their velocity, frequency, wavelength, and amplitude (see Figure 6.2). The **wavelength**, λ (Greek letter lambda), is the distance between successive wave *crests*, or points of equal displacement on successive waves. The **frequency**, ν (Greek letter nu), is the number of wave crests that pass a given point in 1 second. It has the units second^{-1}, which is given the special name **hertz (Hz)**:

$$1 \text{ Hz} = 1 \text{ s}^{-1}$$

A frequency of 10 Hz means that ten wave crests pass a given point in 1 s. If there are ν waves per second moving past a given point, and if the length of each wave is λ, the distance traveled by the wave in 1 s is $\lambda\nu$, which is its **speed**, v:

$$v = \lambda\nu$$

Light and all other types of electromagnetic radiation have a constant speed of 3.00×10^8 m s^{-1} in a vacuum; this speed is given the symbol c:

$$c = \lambda\nu = 3.00 \times 10^8 \text{ m s}^{-1}$$

The **amplitude**, A, of a wave is the height of a crest or the depth of a trough. The energy per unit volume stored in a wave is proportional to A^2. In the case of light the **intensity**, or brightness, of light is proportional to A^2.

Electromagnetic Spectrum

The complete range of electromagnetic waves is called the **electromagnetic spectrum** (see Figure 6.3). When we speak of visible light, or the *visible spectrum*, we are referring to radiation with wavelengths in the range 4×10^{-7} to 7.5×10^{-7} m. Our eyes are sensitive only to this very small part of the complete

Figure 6.1 Electromagnetic waves. Electromagnetic waves consist of electric and magnetic fields that oscillate perpendicular to each other and to the direction in which the wave travels.

CHAPTER 6
ENERGY LEVELS,
ELECTRON
CONFIGURATIONS, AND
THE COVALENT BOND

194

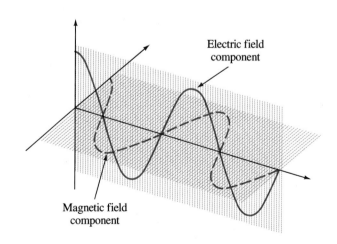

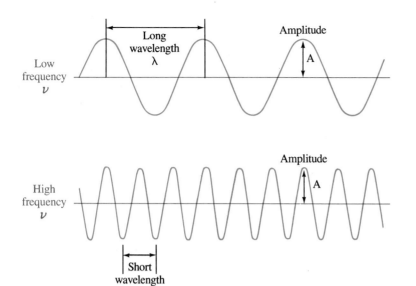

Figure 6.2 Properties of Waves. Any wave can be described by its frequency, wavelength, amplitude, and speed. The speed of a wave is the product of its wavelength and frequency. The energy of a wave is related to its amplitude. For example, the greater the height of an ocean wave, the greater is its destructive power.

electromagnetic spectrum. X rays have wavelengths as short as 10^{-13} m, and ultraviolet, visible, and infrared radiation have increasingly longer wavelengths, in the range of 10^{-8} to 10^{-4} m. In contrast, radio waves have wavelengths as long as 1 km or more.

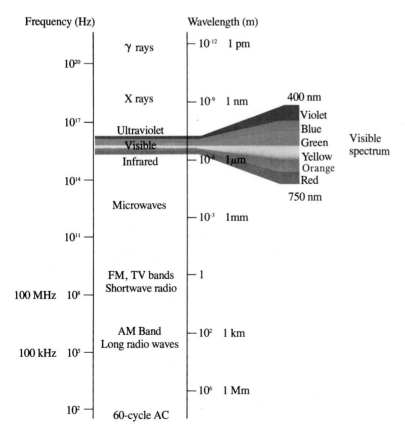

Figure 6.3 Electromagnetic Spectrum. All electromagnetic waves have the same speed in a vacuum, but their wavelengths and frequencies vary. Thus electromagnetic waves with a low frequency have a long wavelength; those with a high frequency have a short wavelength. Electromagnetic radiation with a wavelength of about 400 to 750 nm is detectable by the human eye and thus constitutes visible light.

6.1 LIGHT AND
ELECTROMAGNETIC WAVES

White light from the sun or an incandescent light bulb consists of all the wavelengths in the visible spectrum. When white light is passed through a glass prism, as shown in Figure 6.4, it is spread out into a band of colors ranging from long-wavelength red light to shorter-wavelength violet light.

Example 6.1 A radio station broadcasts on a frequency of 900 kHz. What is the wavelength of the electromagnetic radiation emitted by the transmitter?

Solution We can rearrange the equation $c = v\lambda$ to give $\lambda = c/v$. Then inserting the values of c and v and converting kilohertz (10^3 s^{-1}) to hertz (s^{-1}) we have

$$\lambda = \frac{c}{v} = \left(\frac{3.00 \times 10^8 \text{ m s}^{-1}}{900 \text{ kHz}}\right)\left(\frac{1 \text{ kHz}}{10^3 \text{ s}^{-1}}\right)$$

$$= 3.33 \times 10^2 \text{ m} = 0.333 \text{ km}$$

Example 6.2 The colors that make up visible light range in wavelength from 400 to 750 nm (violet to red). What is the corresponding range of frequencies, in hertz?

Solution Violet light has a frequency of

$$v = \frac{c}{\lambda} = \frac{3.00 \times 10^8 \text{ m s}^{-1}}{400 \times 10^{-9} \text{ m}} = 7.50 \times 10^{14} \text{ Hz}$$

Red light has a frequency of

$$v = \frac{c}{\lambda} = \frac{3.00 \times 10^8 \text{ m s}^{-1}}{750 \times 10^{-9} \text{ m}} = 4.00 \times 10^{14} \text{ Hz}$$

Interference and Diffraction

The view that light is an electromagnetic wave is supported by experimental observations that show that light behaves like other waves, such as waves on a water surface. In particular, light exhibits the property known as **interference**. If waves are generated at two points close together, the waves spreading out from these points interfere with each other and produce an *interference pattern*.

Figure 6.4 Visible Spectrum
Because light travels more slowly in glass than in air, the direction in which the light travels is changed when it enters a prism and again when it leaves the prism. The amount of bending of the light beam depends on the wavelength of the light. As a result, when ordinary white light, such as sunlight, is passed through a glass prism, it is spread out into a band of its constituent wavelengths, which we perceive as a band of colors.

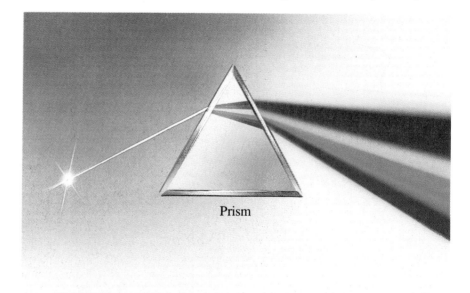

Prism

CHAPTER 6
ENERGY LEVELS,
ELECTRON
CONFIGURATIONS, AND
THE COVALENT BOND

196

In general, the two waves travel a different distance to reach the same point. At some points, therefore, the crests of the two waves will arrive together, producing a wave of larger magnitude. At other points the crest of one wave will arrive at the same time as a trough of another wave arrives. The two waves cancel, so the resultant wave has zero amplitude at that point. When light is passed through two very narrow slits that are very close together, an interference pattern consisting of a series of alternately light and dark lines is produced on a screen placed in front of the two slits (Figure 6.5).

Waves bend around obstacles in their path, a property known as **diffraction**. Thus when waves pass through an opening, they tend to bend around its edges and spread out as they emerge through the opening. The extent of this diffraction depends on the size of the opening and the wavelength of the waves. If light is passed through a slit or hole that is large compared with the wavelength of the light, a sharp image is produced on a screen. But if light is passed through a very narrow slit or a very small hole, a blurred image consisting of light and dark regions is produced. This image is an interference or *diffraction pattern* (Figure 6.6).

Because interference and diffraction patterns are only produced by slits, holes, and other objects comparable in size to the wavelength of light, they are not readily observed. Light produces sharp images of ordinary-sized objects; it therefore appears to travel in straight lines, like a beam of particles.

Particle Nature of Light

For several decades after Maxwell described electromagnetic waves, the wave theory seemed to explain all the observed phenomena related to the transmission of light. But when attempts were made to understand the interaction of light and matter, difficulties arose. The most striking effect of light that is not in accord with the wave theory is the photoelectric effect. This effect was discovered by German physicist Heinrich Hertz (1857–1894) in 1888 during the course of his experiments on generating radio waves. His name is commemorated in the unit for the frequency of a wave.

PHOTOELECTRIC EFFECT When light, particularly ultraviolet light, shines on certain metallic surfaces, electrons are emitted from the surface (see Figure

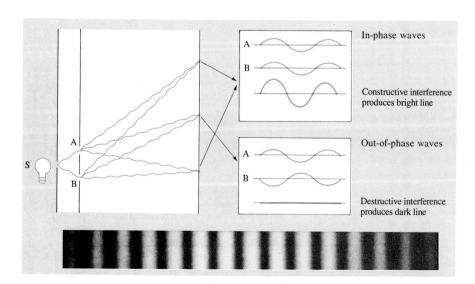

Figure 6.5 Interference Pattern of Light Waves Passing Through Two Slits. Light passing through two narrow, closely spaced slits, produces an interference pattern. When the waves arrive in phase, reinforcing each other, they produce a bright band. When the waves arrive out of phase, they cancel each other, creating a dark band. To obtain an interference pattern with visible light, the slits should be less than 0.1 mm wide and less than 0.1 mm apart.

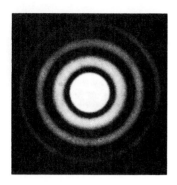

Figure 6.6 Diffraction of Light.
When light passes through a very small hole, the light spreads out beyond the hole, giving a fuzzy image with light and dark regions. This image is called a diffraction pattern.

6.7). This phenomenon is known as the **photoelectric effect**, and the emitted electrons are called **photoelectrons**. Experiments show the following:

1. For each metal there is a frequency of light, v_0, below which no electrons are emitted (see Figure 6.8)
2. At frequencies v, greater than v_0, the kinetic energy of the emitted electrons is proportional to $v - v_0$ (see Figure 6.8).
3. The number of electrons emitted is proportional to the light intensity, but the kinetic energy of the electrons is independent of the light intensity.
4. Electrons are emitted from the surface almost instantaneously (less than 10^{-9} s after the surface is illuminated), even with very low-intensity light.

These observations could not be explained by the wave theory.

The explanation of the photoelectric effect was given by Albert Einstein in 1905 (see Box 6.1). He proposed that a beam of light consists of small bundles of energy called light **quanta**, or **photons**, and that the energy, E, of a photon is proportional to its frequency, v, or

$$E = hv$$

where h is a constant called the **Planck constant**.

A photon striking the metal surface imparts all its energy to an electron and disappears from existence. The electron may then have sufficient energy to escape from the metal. Part of its energy, ϕ (Greek letter phi), is used to overcome the attractive forces holding it in the metal, and the remainder appears as the kinetic energy, KE, of the emitted electron:

$$\begin{pmatrix} \text{Energy} \\ \text{of photon} \end{pmatrix} = \begin{pmatrix} \text{Energy needed} \\ \text{for electron to} \\ \text{escape from metal} \end{pmatrix} + \begin{pmatrix} \text{Kinetic energy} \\ \text{of emitted electron} \end{pmatrix}$$

or

$$hv = \phi + KE$$

Figure 6.7 Photoelectric Effect.

(a) Detecting the photoelectric effect: The photoelectrons ejected from the irradiated metal plate are attracted to the positive collection electrode at the other end of the tube. The current that results is measured with an ammeter. (b) Measuring the kinetic energy of the photoelectrons: The polarity of the electrodes is reversed. The negative stopping electrode repels the photoelectrons, and if it is made negative enough, no photoelectrons reach the electrode. From the voltage needed to stop the electrons, their kinetic energy may be calculated.

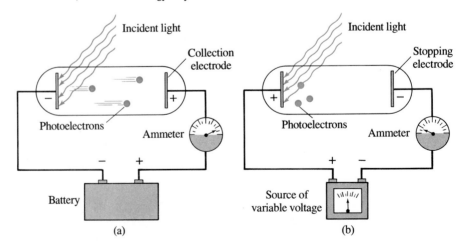

CHAPTER 6
ENERGY LEVELS,
ELECTRON
CONFIGURATIONS, AND
THE COVALENT BOND

198

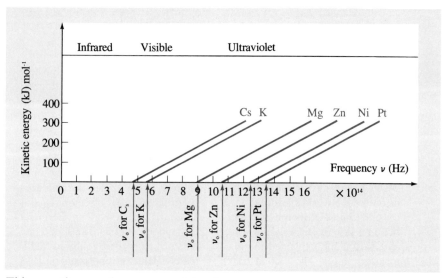

Figure 6.8 Kinetic Energy of Photoelectrons. The variation of kinetic energy with the frequency of the incident light is shown for several metals. No electrons are emitted below a frequency v_0 that is characteristic for each metal. The slope of each line is the same; it is equal to h, the Planck constant.

This equation can be rearranged to give

$$KE = hv - \phi$$

A plot of KE against v for any metal is a straight line of slope h (see Figure 6.8). The value of the Planck constant can be determined from the slope of these plots. The experimentally determined value of h is 6.626×10^{-34} J s.

When the frequency of the light is less than the critical frequency, v_0, the energy, hv, absorbed by an electron is less than the energy, ϕ, needed for an electron to escape from the metal, so no electrons are emitted. If the frequency of the light is increased so that it is equal to the critical frequency v_0, an electron that absorbs a photon has just enough energy, $\phi = hv_0$, to escape from the metal. But

$$KE = hv_0 - \phi = hv_0 - hv_0 = 0$$

so that the electron has zero kinetic energy. With increasing frequency of the light, electrons are emitted with increasing kinetic energy.

$$KE = hv - \phi = hv - hv_0 = h(v - v_0)$$

So the kinetic energy is proportional to $v - v_0$. Thus observations (1) and (2) are accounted for. The more intense the light, the more photons reach the metal, and therefore the more electrons are emitted. But for a constant frequency they all have the same energy. Thus observation (3) is accounted for. When a photon gives all its energy to an electron, the electron is emitted immediately. If the light is of low intensity, only a few electrons will be emitted, but there will be no delay in their emission. Thus observation (4) is accounted for. Einstein's revolutionary proposal that light consists of photons therefore accounts for all the features of the photoelectric effect, whereas they could not be accounted for in terms of the wave theory. For example, since the energy of an electromagnetic wave depends on its amplitude but not on its frequency, the kinetic energy of the emitted electrons should be independent of the frequency, and any frequency of light should cause electrons to be emitted. But as we have seen, the kinetic energy of the electrons *does* depend on the frequency, and below a certain frequency no electrons are emitted.

We have seen previously that the metals that most easily lose electrons, as indicated by their reactions and as measured by their ionization energies, are the alkali metals. Thus not surprisingly, the alkali metals exhibit the photo-

Einstein was born in Germany in the old city of Ulm on the Danube. As a child, he was so slow at learning that his parents feared he might be retarded. At high school he disliked the harsh discipline, and when his family emigrated to Milan, Einstein left his school and joined them. He applied for admission to the Swiss Federal Polytechnical School in Zurich but was refused because he did not have a high school diploma and he failed the entrance examination, although he did very well in mathematics and physics. He then spent two years at a small college and finally was able to enter the Polytechnical School in Zurich. He did not particularly impress his teachers, and after graduation he had difficulty finding employment.

After taking several part-time positions, Einstein went to work as a junior official in the Swiss Patent Office in Berne. The work appears to have left Einstein lots of time to think about theoretical physics. In 1905, at the age of twenty-six, he published three articles, any one of which would have established him as one of the world's leading physicists. The first proposed that light has a particlelike, as well as a wavelike, nature and explained the photoelectric effect. The second paper explained Brownian motion, the random erratic motion of very small particles suspended in a liquid, as being due to collisions with the rapidly moving molecules of the liquid. The third showed that ideas of absolute space and time had to be replaced by the concept that space and time are relative to each other (the theory of relativity). It was in this paper that Einstein derived the famous equation $E = mc^2$.

Einstein reached his revolutionary conclusions by means of rather simple but uncompromising logic based on experimental observations. Remarkably, he did all this work without any contact with other important physicists of the time.

After the publication of these papers, the University of Zurich offered him a position, and Einstein quickly became an important figure in the world of theoretical physics. In 1914 he was persuaded to move to Berlin as the head of the physics department of the world-famous Kaiser-Wilhelm Institute. Despite his prestigious position he was not entirely happy under the militaristic Prussian rulers of Germany, but he continued to work in Germany throughout World War I and the difficult years that followed. Ultimately, Hitler's repression of Jews forced Einstein to leave. He arrived in New York in October 1933, and he stayed in the United States until his death in 1955.

Albert Einstein is universally recognized as the greatest physicist of our age. Some say that if someone else had discovered the theory of relativity, his other work would have made him the second greatest physicist of his time. His ideas radically changed our concepts of space and time. From the 1905 publication of the theory of relativity until the end of his long life, he concentrated on one main task: the attempt to find a single unifying theory that would explain all physical events.

electric effect at lower frequencies than any other metals. Long-wavelength, low-frequency, red light causes the photoelectric effect in cesium, whereas for magnesium, for example, higher-frequency violet light is needed (see Figure 6.8).

Example 6.3 When light with a wavelength of 300 nm falls on the surface of sodium, electrons with a kinetic energy of 1.68×10^5 J mol^{-1} are emitted. What is the minimum energy needed to remove an electron from sodium? What is the maximum wavelength of light that will cause a photoelectron to be emitted?

Solution The energy of a 300-nm photon is given by

$$E = h\nu = \frac{hc}{\lambda} = \left(\frac{6.63 \times 10^{-34}\ \text{J s}}{300 \times 10^{-9}\ \text{m}}\right)\left(\frac{3.00 \times 10^8\ \text{m}}{1\ \text{s}}\right)$$

$$= 6.63 \times 10^{-19}\ \text{J}$$

CHAPTER 6
ENERGY LEVELS,
ELECTRON
CONFIGURATIONS, AND
THE COVALENT BOND

200

The energy of 1 mol of photons is

$$\left(\frac{6.63 \times 10^{-19} \text{ J}}{1 \text{ photon}}\right)\left(\frac{6.022 \times 10^{23} \text{ photons}}{1 \text{ mol photons}}\right) = 3.99 \times 10^5 \text{ J mol}^{-1}$$

The minimum energy needed to remove a mole of electrons from sodium is

Energy of photons − Kinetic energy of photoelectrons

$$= 3.99 \times 10^5 \text{ J mol}^{-1} - 1.68 \times 10^5 \text{ J mol}^{-1}$$

$$= 2.31 \times 10^5 \text{ J mol}^{-1}$$

The minimum energy needed to remove one electron from sodium is therefore

$$\left(\frac{2.31 \times 10^5 \text{ J}}{1 \text{ mol}}\right)\left(\frac{1 \text{ mol electrons}}{6.022 \times 10^{23} \text{ electrons}}\right) = 3.84 \times 10^{-19} \text{ J per electron}$$

Thus the maximum wavelength of a photon that will cause the emission of an electron from sodium is

$$\lambda = \frac{hc}{E} = (6.63 \times 10^{-34} \text{ J s})\left(\frac{3.00 \times 10^8 \text{ m}}{1 \text{ s}}\right)\left(\frac{1}{3.84 \times 10^{-19} \text{ J}}\right)$$

$$= 5.18 \times 10^{-7} \text{ m} = 518 \text{ nm}$$

This wavelength is in the green region of the spectrum.

PHOTOCHEMICAL REACTIONS   The description of light in terms of photons is also needed to explain chemical reactions that are caused by the absorption of light. These reactions are known as **photochemical reactions**. We saw in Chapter 5 that a bright light will initiate the reaction between hydrogen and chlorine, which then occurs with explosive violence.

A chlorine molecule is dissociated into two chlorine atoms when a photon of sufficient energy, $E = h\nu$, is absorbed:

$$Cl_2 + h\nu \longrightarrow 2Cl$$

A chlorine atom then reacts with a hydrogen molecule,

$$Cl + H_2 \longrightarrow HCl + H$$

to form a hydrogen chloride molecule and a hydrogen atom. The hydrogen atom then reacts with a chlorine molecule to form another hydrogen chloride molecule and another chlorine atom:

$$H + Cl_2 \longrightarrow HCl + Cl$$

This chlorine atom can react with another hydrogen molecule, and so these two successive reactions can continue indefinitely.

The overall reaction is the sum of these two reactions:

$$\begin{array}{r} Cl + H_2 \longrightarrow HCl + H \\ \underline{H + Cl_2 \longrightarrow HCl + Cl} \\ H_2 + Cl_2 \longrightarrow 2HCl \end{array}$$

The very reactive chlorine and hydrogen atoms are used up as fast as they are formed and do not appear as products of the overall reaction.

As Example 6.4 shows, blue-green light or light of higher frequency provides enough energy to dissociate a chlorine molecule, but light of lower frequency does not. Thus even an intense red light will not initiate the reaction between hydrogen and chlorine.

Many other reactions are caused by the absorption of light by molecules. In the upper atmosphere many photochemical reactions involving nitrogen and oxygen molecules occur as a result of the absorption of ultraviolet radiation from the sun; the formation of ozone (Figure 3.1) is an example.

Example 6.4 The energy needed to dissociate a chlorine molecule into chlorine atoms is 243 kJ mol^{-1}. What is the maximum wavelength of light that will initiate the hydrogen-chlorine reaction?

Solution The energy needed to dissociate one chlorine molecule is

$$E = \left(\frac{243 \text{ kJ}}{1 \text{ mol}}\right)\left(\frac{1 \text{ mol}}{6.022 \times 10^{23} \text{ molecules}}\right)\left(\frac{1000 \text{ J}}{1 \text{ kJ}}\right)$$

$$= 4.035 \times 10^{-19} \text{ J (molecule)}^{-1}$$

Now since one photon is needed to dissociate one chlorine molecule, the photon must have a minimum energy of 4.035×10^{-19} J. It must therefore have a minimum frequency of

$$\nu = \frac{E}{h} = \frac{4.035 \times 10^{-19} \text{ J}}{6.63 \times 10^{-34} \text{ J s}} = 6.09 \times 10^{14} \text{ s}^{-1}$$

We can now convert frequency to wavelength:

$$\lambda = \frac{c}{\nu} = \left(\frac{3.00 \times 10^8 \text{ m}}{1 \text{ s}}\right)\left(\frac{1 \text{ s}}{6.09 \times 10^{14}}\right)$$

$$= 4.93 \times 10^{-7} \text{ m} = (4.93 \times 10^{-7} \text{ m})\left(\frac{10^9 \text{ nm}}{1 \text{ m}}\right)$$

$$= 493 \text{ nm}$$

This value is the maximum wavelength of light that will dissociate a chlorine molecule; it is in the green region of the spectrum.

Since light behaves in some experiments as if it consisted of photons, but nevertheless also exhibits wavelike behavior, as in the phenomena of diffraction and interference, we have to accept the idea that light behaves both like particles and like waves. Which behavior we observe depends on the kind of experiment being carried out. When light is passed through two adjacent slits, we observe an interference pattern that is characteristic of waves; when light interacts with matter and there is a transfer of energy from the light to electrons, we observe the particlelike behavior of light. This dual nature of light is something that is not familiar from everyday experience, but experiment shows us that this is how light behaves.

6.2 ATOMIC SPECTRA

Studies of the absorption and emission of electromagnetic radiation by substances provide important information about the structures of atoms and molecules. We consider now the spectra of atoms, that is, **atomic spectra**. The range of frequencies (in wavelengths) of light emitted by an atom is called its **spectrum** (plural, spectra).

When the alkali metals or compounds of the alkali metals are heated to a high temperature in a very hot flame, they impart distinctive colors to the flame that are characteristic of a particular metal—for example, red for lithium, yellow for sodium, and lilac for potassium (Experiment 6.1). If the light emitted by the

CHAPTER 6
ENERGY LEVELS,
ELECTRON
CONFIGURATIONS, AND
THE COVALENT BOND

202

Colored Flames and Atomic Spectra

Compounds of the alkali metals give characteristic colors to a flame. The color arises
from alkali metal atoms that are raised to an excited state by the high temperature of
the flame and then return to the ground state, emitting light of a characteristic frequency.
The flames shown here were produced by holding a platinum wire on which a small amount
of the alkali metal chloride had been placed in the flame of a bunsen burner.

Lithium—red   Sodium—yellow   Potassium—lilac

flame is passed first through a slit (to obtain a narrow beam) and then through
a glass prism, a spectrum is obtained. Unlike the continuous spectrum obtained
from sunlight, this spectrum consists of only a few sharp lines, as Figure 6.9
shows. Thus light emitted by an alkali metal flame does not contain all the frequencies in the visible region but only a few frequencies that are characteristic
of the particular metal. Such a spectrum is called an emission **line spectrum**.

If an electric discharge is passed through helium, the gas glows with a characteristic blue-violet color. If the light is passed through a prism, the spectrum

**Figure 6.9 Atomic (Line) Spectra
of (top to bottom) Hydrogen,
Sodium, Helium, and Neon.**

obtained again consists of only a limited number of sharp lines. It is a line spectrum like those given by the alkali metals (see Figure 6.9).

What is the origin of these spectra? When an alkali metal is heated to a high temperature in a flame, some of its atoms are raised to high energy states by collisions with fast-moving atoms and molecules in the flame. When an alkali metal compound is heated in a flame, it is decomposed to give alkali metal atoms, which are similarly in high energy states. When an electric discharge is passed through helium gas, the very fast-moving electrons of the discharge collide with helium atoms, transfering some of their energy to the helium atoms and thus raising them to a higher energy state. Atoms that have been raised to a high energy state are said to be in an **excited state**. The normal, unexcited state of an atom is called its **ground state**. In an excited state of an atom one or more of its electrons are at a greater average distance from the nucleus than in the ground state. An excited atom is unstable, and may lose some or all of its excess energy by emitting light. This light constitutes the spectrum of the atoms (Figure 6.9).

Quantization of Energy

The fact that excited atoms emit only certain frequencies of light leads to the very important conclusion that an electron in an atom cannot have just any arbitrary energy but only certain definite energies. If an electron in an excited state has energy E_2, whereas in the ground state it has energy E_1, then the energy emitted when the atom returns from the excited state to the ground state is $E_2 - E_1$. This energy is emitted as a single photon with a frequency v:

$$E_2 - E_1 = hv$$

Each frequency in the emitted light, and thus each line in the spectrum, corresponds to the difference in energy between two different states (see Figure 6.10). For example, if the atom is in a different excited state, with energy E_3, and returns to the ground state, the energy $E_3 - E_1$ will be emitted in the form of a photon of light of frequency v', where $hv' = E_3 - E_1$.

From the observed spectra of atoms we conclude that each atom has a set of definite energy levels, E_1, E_2, E_3, and so on (Figure 6.10). An atom may have a rather large number of energy levels, but it cannot have an energy that does not correspond to one of these levels. A set of energy levels is rather like a set of shelves, each one at a certain height. An object can be placed on any one of the shelves but not at any position between the shelves. Thus the object can have the potential energy corresponding to the height of any particular shelf, but it cannot have any energy between these values.

Any system that can only have certain definite energies is said to be **quantized**. All microscopic systems, such as atoms and molecules, are quantized. The energies of everyday macroscopic objects are also quantized, but the separation between the energy levels is too small to be observed. As a result, we may treat the energy of a macroscopic system as being continuous instead of quantized.

Any microscopic system has associated with it a certain number of energy

Figure 6.10 Energy Levels. Each atom has a characteristic set of energy levels. An atom emits light when an electron returns from an excited state to a lower energy state. Each frequency in the line spectrum therefore corresponds to the difference in energy between two energy levels of the atom.

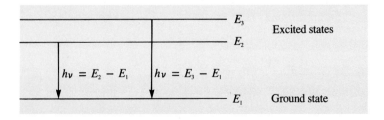

levels, E_1, E_2, E_3, ... The electron shells that we described in Chapter 4 correspond to the different energy levels. Starting with the shell closest to the nucleus, we numbered the shells, $n = 1, 2, 3, ...$. These numbers are called **quantum numbers**.

Spectrum of the Hydrogen Atom

From the spectrum of an atom we can deduce the energy levels for the electrons in the atom. As we might expect, the simplest atom, hydrogen, has a simpler spectrum than any other atom. The spectrum has lines in the visible, ultraviolet, and infrared regions; they can all be observed by recording the spectrum on suitable photographic film. In the region of the spectrum shown in Figure 6.11, three series of lines are observed; they are named after their discoverers. All the lines in the spectrum can be accounted for by the simple set of energy levels in Figure 6.11. Each energy level is designated by a quantum number, n, which can have any positive integral value, that is, $n = 1, 2, 3, ...$, where $n = 1$ refers to the lowest energy, $n = 2$ to the next energy level, and so on.

The *Lyman series* of lines in the ultraviolet region have the highest frequencies of any in the hydrogen spectrum and correspond to the largest energy changes. These lines result from electrons moving from different excited states to the ground state ($n = 1$). There is a line corresponding to the transition from $n = 2$ to $n = 1$, from $n = 3$ to $n = 1$, from $n = 4$ to $n = 1$, and so on.

An electron in an upper level can also undergo a transition to the $n = 2$ level, for example, from the $n = 3$ to the $n = 2$ level, from $n = 4$ to $n = 2$, and so on. These transitions produce the *Balmer series* of lines in the visible part of the spectrum. Transitions to the $n = 3$ level produce the *Paschen series* of lines in the infrared part of the spectrum. Other series of lines correspond to transitions to higher energy levels; these lines are weak and difficult to observe.

In short, one set of energy levels can account for *all* the lines observed in the atomic spectrum of hydrogen. The energy of each level depends in a very simple way on the value of the quantum number n and is given by the expression

$$E = -\frac{1312}{n^2}$$

where E is in kilojoules per mole.

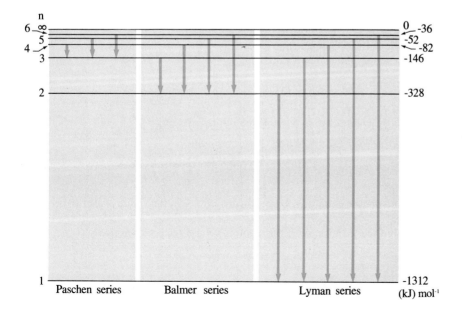

Figure 6.11 Energy Levels and Transitions for Hydrogen Atom. Each line in the spectrum of the hydrogen atom is produced by a photon emitted when the electron returns from an excited state to a lower energy level. The energy of the photon, and hence the frequency of the light emitted, depends on the difference in energy between the two energy levels. For example, the $n = 1$ state has an energy of $-1312 \text{ kJ mol}^{-1}$; the $n = 2$ state has an energy of -328 kJ mol^{-1}. Thus the energy change associated with a transition from the $n = 2$ to the $n = 1$ level is $-328 - (-1312) = 984 \text{ kJ mol}^{-1}$, which is the energy of the photons producing one of the lines in the Lyman series.

The energy of the electron in the hydrogen atom is measured from the energy of the highest excited state ($n = \infty$) in which the electron is completely separated from the atom—in other words, in which the atom is ionized. This level is taken to have zero energy. Then since all the other energy levels have less energy, they have negative energy values. The energy of the lowest level, that is, the ground state, of the hydrogen atom, for which $n = 1$, is

$$E = \frac{-1312}{1^2} \text{ kJ mol}^{-1} = -1312 \text{ kJ mol}^{-1}$$

The energy needed to move this electron to the highest ($n = \infty$) level is 1312 kJ mol^{-1}; this value is the ionization energy of the hydrogen atom (Chapter 4). The energy of the first excited state of the hydrogen atom, for which $n = 2$, is

$$E = \frac{-1312}{2^2} \text{ kJ mol}^{-1} = -328.0 \text{ kJ mol}^{-1}$$

The energy of the $n = 3$ state is $-1312/3^2 = -145.8$ kJ mol^{-1}, and so on.

To find the frequency or wavelength of the light emitted when an electron moves from a higher to a lower energy level, we first find the energy emitted:

$$E = E_{\text{initial}} - E_{\text{final}} = \left(\frac{-1312}{n_{\text{initial}}^2}\right) - \left(\frac{-1312}{n_{\text{final}}^2}\right) \text{ kJ mol}^{-1}$$

$$= 1312\left(\frac{1}{n_{\text{final}}^2} - \frac{1}{n_{\text{initial}}^2}\right) \text{ kJ mol}^{-1}$$

Then we convert this energy to a frequency or wavelength, as shown in the next example.

Example 6.5 Find the energy emitted when an electron drops from the $n = 4$ level to the $n = 2$ level in the hydrogen atom. If this energy is emitted as a photon, what is the wavelength of the photon?

Solution We have

$$\Delta E = 1312\left(\frac{1}{n_f^2} - \frac{1}{n_i^2}\right) \text{ kJ mol}^{-1}$$

where ΔE (Δ is the Greek capital letter delta) represents the energy difference between the initial and final states. Substituting $n_i = 4$ and $n_f = 2$ gives

$$\Delta E = 1312\left(\frac{1}{2^2} - \frac{1}{4^2}\right) = 246.0 \text{ kJ mol}^{-1}$$

The energy emitted by *one* hydrogen atom is

$$E = \left(\frac{246.0 \text{ kJ}}{1 \text{ mol}}\right)\left(\frac{1 \text{ mol}}{6.022 \times 10^{23} \text{ atoms}}\right)\left(\frac{1000 \text{ J}}{1 \text{ kJ}}\right)$$

$$= 4.09 \times 10^{-19} \text{ J (atom)}^{-1}$$

Since one photon is emitted by one hydrogen atom, this value is also the energy of the photon.

To find the wavelength of this photon, we use the relationships

$$E_{\text{photon}} = h\nu \qquad \text{and} \qquad \lambda\nu = c$$

Thus

CHAPTER 6
ENERGY LEVELS,
ELECTRON
CONFIGURATIONS, AND
THE COVALENT BOND

206

$$E_{photon} = \frac{hc}{\lambda} \quad \text{or} \quad \lambda = \frac{hc}{E_{photon}}$$

$$\lambda = \frac{(6.63 \times 10^{-34} \text{ J s})(3.00 \times 10^8 \text{ m s}^{-1})}{4.09 \times 10^{-19} \text{ J}}$$

$$= 4.86 \times 10^{-7} \text{ m} = 486 \text{ nm}$$

Because they have more than one electron, other elements have more complicated atomic spectra than hydrogen; however, they can all be accounted for in terms of a set of energy levels characteristic of each element. The atomic spectrum of an element is unique; it therefore provides a very convenient method for identifying that element. For example, the sodium spectrum has two characteristic intense lines in the yellow region of the spectrum, which enable sodium and its compounds to be recognized very easily.

The emission of light by excited atoms also has some practical uses in everyday life. The yellow light of certain street and highway lamps is emitted by excited sodium atoms produced by bombardment of the sodium atoms in sodium vapor by the fast-moving electrons of an electric discharge. This type of lamp produces less heat than an incandescent light bulb and therefore wastes less energy. Another type of street lamp, the mercury lamp, gives a blue-green light, which arises from excited mercury atoms produced in mercury vapor by an electric discharge. The red light emitted by excited neon atoms is also familiar from neon signs, which are made from tubes that contain neon gas through which an electric discharge is passed.

When an electric discharge is passed through neon, the gas glows with a characteristic red color.

Bohr's Model of the Hydrogen Atom

We have seen that the electron in a hydrogen atom can only have certain energies; its energy is quantized. But where is the electron situated when it has a certain energy? Or more precisely, what path does the electron of a given energy take as it moves around the nucleus? The electron must be moving; otherwise, the electrostatic attraction of the nucleus would pull the electron into the nucleus.

In 1913 Danish physicist Niels Bohr (1885–1962) postulated that the electron moves around the nucleus in any one of a limited number of circular orbits similar to the orbits of the planets around the sun. Each orbit corresponds to an allowed energy level. The orbit closest to the nucleus has the lowest energy and $n = 1$. The other orbits at successively greater distances from the nucleus have higher energies and correspond to $n = 2$, $n = 3$, and so on. Using this model, Bohr found that the energy of an electron in a given orbit is

$$E = -\frac{2.178 \times 10^{-18}}{n^2} \text{ J}$$

This expression gives the energy of the electron in a single hydrogen atom.
Expressing this energy on a molar basis, we have

$$E = \left(-\frac{1}{n^2}\right)\left(\frac{2.178 \times 10^{-18} \text{ J}}{1 \text{ atom}}\right)\left(\frac{6.022 \times 10^{23} \text{ atoms}}{1 \text{ mol}}\right)\left(\frac{1 \text{ kJ}}{1000 \text{ J}}\right)$$

$$= \frac{-1312}{n^2} \text{ kJ mol}^{-1}$$

Figure 6.12 Louis Victor, Prince de Broglie (1892–1977).

This expression is identical with the relationship deduced from the spectrum of the H atom. In other words, Bohr's model correctly predicts the spectrum of the hydrogen atom.

Although Bohr's theory works well for the hydrogen atom and represented a considerable advance in the understanding of atomic spectra, it did not correctly predict the spectra of other atoms. Obviously, Bohr's model must have a flaw, and we will see what that is in the next section.

6.3 WAVE PROPERTIES OF PARTICLES

In 1924 French physicist Louis de Broglie (Figure 6.12) made the bold suggestion that since light exhibits both wave and particle characteristics, particles like electrons, protons, and atoms should show wavelike properties. De Broglie went further and proposed that the wavelength associated with a particle of mass m and speed v is

$$\lambda = \frac{h}{mv}$$

The product mv is called **momentum**. It is usually given the symbol p ($p = mv$). So the above equation can be written as

$$\lambda = \frac{h}{p}$$

He was led to propose this equation because from Einstein's theory of relativity the momentum of a photon is given by the expression

$$p = \frac{h}{\lambda}$$

And he proposed that the same equation should apply to the waves associated with a particle.

Diffraction of Electrons

When de Broglie put forward his hypothesis, there were no experimental observations to support it, and it was not taken seriously. Then in 1927 Clinton Davisson in the United States and George Thomson in Britain independently demonstrated that beams of electrons are diffracted by crystals. Figure 6.13 shows a diffraction pattern produced by passing a beam of electrons through very thin aluminum foil. Subsequently, diffraction patterns were observed for other particles such as neutrons, protons, and even neutral atoms.

Figure 6.13 Diffraction pattern produced by passing a beam of fast moving electrons through thin aluminum foil. The foil consists of many tiny aluminum crystals. In each crystal the atoms are packed together so that there are many regular, parallel rows of atoms. These act like a set of parallel slits, and the electron waves passing through them give a diffraction pattern.

Example 6.6 The mass of an electron is 9.11×10^{-31} kg. What is the wavelength of an electron with a speed of 6.12×10^6 m s^{-1}?

Solution We use the equation

$$\lambda = \frac{h}{mv}$$

The value of the Planck constant h is 6.63×10^{-34} J s. We must convert this constant to basic SI units by using the unit conversion factor $1\ \text{kg m}^2\,\text{s}^{-2}/1\ \text{J}$ in order to obtain the wavelength in meters. We then have

$$\lambda = \frac{h}{mv} = \frac{6.63 \times 10^{-34}\ \text{J s}}{(9.11 \times 10^{-31}\ \text{kg}) \times (6.12 \times 10^6\ \text{m s}^{-1})}\left(\frac{1\ \text{kg m}^2\,\text{s}^{-2}}{1\ \text{J}}\right)$$

$$= 1.19 \times 10^{-10}\ \text{m} = 119\ \text{pm}$$

Thus electrons moving with a speed of 6.12×10^6 m s^{-1} have a wavelength of 119 pm, which, as shown in Figure 6.3, is comparable with the wavelength of X rays.

Example 6.6 shows that the wavelength of a rapidly moving electron is in the X ray region and is considerably shorter than that of visible light. This fact is put to good use in the electron microscope, as Figure 6.14 shows.

The wave behavior of matter can only be observed for very small particles such as electrons and atoms. For an ordinary-sized object the associated wavelength is much too small to be of any significance. For example, the wavelength of a 1500-kg car moving at a speed of 30 m s^{-1} can be calculated from de Broglie's equation: It is only 1.5×10^{-38} m.

Thus we have seen that in certain aspects of their behavior small moving objects, such as electrons, resemble particles, while in others they resemble waves. Similarly, in some aspects of its behavior light resembles particles, and in other aspects it resembles waves. Neither the wave model nor the particle model can be used exclusively to describe either matter or light. Some phenomena are best described by one model, while other phenomena are best described by the other model.

Uncertainty Principle

Whereas a particle is localized in space and its position can be accurately defined, a wave is spread out in space. If an electron has both wave and particle properties, can its position in space be defined? If it is a particle, we should be able to find precisely where it is located. But if it is a wave, we cannot do so. By considering the various ways in which one might try, in principle, to find the position of an electron, German physicist Werner Heisenberg (Figure 6.15) came to the conclusion in 1927 that there are definite limitations on the accuracy with which one can define the position of an electron. He showed that:

There is an uncertainty Δx in the position and an uncertainty Δp in the momentum of a particle: If one is known more accurately, the other becomes more uncertain.

This statement is known as the **uncertainty principle**.

We can get an idea of how Heisenberg arrived at the uncertainty principle by considering the problem of determining the position of an electron in an atom. We know that we can determine the position of the moon in its orbit around the earth by observing the light from the sun that is reflected to the earth by the moon. If we imagine attempting to determine the position of an electron in the same way, we must remember that because of diffraction effects, we can determine the position of an object only with an accuracy of the order of magnitude of the wavelength of the light used.

To determine the position of the electron with as great an accuracy as possible, we might consider using light of very short wavelength. However, the shorter the wavelength, the greater is the momentum of each photon, because $p = h/\lambda$. Even if we were to suppose that, in principle at least, we could observe the electron by bouncing just one photon off it, momentum would be transferred from the photon to the electron during collision, and thus the momentum of the electron would become uncertain. If we increase the wavelength of the photon in order to decrease its momentum, and therefore reduce the momentum transferred to the electron, then we can no longer define the position of the electron with as much precision. Neither the position nor the momentum of an electron can be determined simultaneously with absolute precision. Thus although Bohr's theory was clearly along the right lines, it was in error in proposing that an electron could be located in a precisely defined orbit.

The idea that an object has a wavelength or that an electron cannot be precisely located may seem contrary to common sense. But common sense is based

Figure 6.14 The Electron Microscope. The size of an object that can be distinguished in an ordinary microscope is limited by the wavelength of visible light. Because diffraction blurs the image, a clear and accurate image of an object that is smaller than the wavelength of the light used cannot be obtained. Thus the lower limit to the size of objects that can be distinguished in an ordinary optical microscope is about 500 nm. However, objects as small as 0.1 nm can be observed in a microscope that uses beams of electrons instead of visible light, because the wavelength of electrons traveling at high speed is much shorter than that of visible light. Here we see an electron microscope photograph of a piece of graphite. The bright bands are layers of carbon atoms that are only 341 pm apart. This corresponds to a magnification of about 15 million times.

Figure 6.15 Werner Heisenberg (1901–1976).

6.3 WAVE PROPERTIES OF PARTICLES

on everyday experience with everyday objects; electrons are many orders of magnitude smaller than these objects and are far removed from everyday experience. Still, if you find these ideas of theoretical physics disconcerting, you are far from alone. The great experimental physicist Ernest Rutherford (Box 2.1) wrote:

> I was brought up to look at the atom as a nice hard fellow, red or grey in colour, according to taste. In order to explain the facts however, the atom cannot be regarded as a sphere of matter but rather as a sort of wave motion of a peculiar kind. The theory of wave (quantum) mechanics, however bizarre it may appear— and it is so in some respects—has the astonishing virtue that it works and works in detail, so that it is now possible to understand and explain things which looked almost impossible in earlier days.

The ideas of Einstein, de Broglie, Bohr, Heisenberg, and others led to the development, during the period 1924–1928, of a new theory that could describe the behavior of electrons in atoms and molecules. This theory is called **quantum mechanics**. No attempt is made in this theory to precisely describe the position or the path of an electron. Instead, the theory gives the *probability* of finding an electron at some specified point in an atom or a molecule. An electron is frequently described by a probability cloud, which is densest at those points where the electron is most likely to be found and less dense where it is less likely to be found. A probability cloud is also often called an electron density cloud. We imagine that the electron is spread out over the atom or molecule and that the density of the electron cloud is greatest where the electron is most likely to be found. We have already described atoms in this way in Chapters 2 and 4.

The very basic ideas of quantum mechanics that we have described are the essential background that we need for all the chemistry discussed in this book. The discussion of quantum mechanics and some of its applications to chemical bonding is, however, continued in Chapter 25 for those who wish to delve a little more deeply into this subject.

We are now in a position to look in more detail at the arrangement of electrons in atoms and to amplify the shell model of the atom that we described in Chapter 4.

6.4 ELECTRON CONFIGURATIONS

We have seen that atomic spectra provide clear evidence that the energies of the electrons in an atom are quantized, and we have looked at the energy levels of the single electron in a hydrogen atom in some detail. What are the energy levels associated with the electrons in other atoms? Their spectra are more complicated than the spectrum of the hydrogen atom because they have more electrons occupying more energy levels.

The form of the periodic table and the values of ionization energies led us to conclude in Chapter 4 that the electrons in an atom are arranged in shells. Because of the increasing distance of successive shells from the nucleus, less energy is needed to remove an electron from the $n = 2$ shell of a given atom than from the $n = 1$ shell, and still less energy is needed to remove an electron from the $n = 3$ shell. The electrons in an $n = 2$ shell are held less tightly than those in the $n = 1$ shell, and those in the $n = 3$ shell are held still less tightly, and so on. But do all the electrons in a given shell have precisely the same energy? That is, are they all in the same energy level? In order to answer this question, we must consider ionization energies in a little more detail.

CHAPTER 6
ENERGY LEVELS,
ELECTRON
CONFIGURATIONS, AND
THE COVALENT BOND

210

Ionization Energies and Energy Levels

We described in Chapter 4 how ionization energies may be measured by the *electron impact method*, in which atoms in the gas phase are bombarded with fast-moving electrons. An alternative, and generally more accurate, method is known as **photoelectron spectroscopy**; this method uses photons to knock electrons out of atoms. Electrons obtained in this way are called *photoelectrons*. In general, electron impact experiments give a value for the ionization energy of the electron that is most easily removed from the atom—in other words, the ionization energy for an electron in the highest occupied energy level of the ground state. Photoelectron spectroscopy provides much more detailed information on ionization energies.

Figure 6.16 shows a photoelectron spectrometer. Atoms are irradiated with photons of known fixed frequency and therefore of known energy. If the energy of the photons is less than the ionization energy of the atoms, no electrons are knocked out of the atoms; that is, no photoelectrons are obtained. If the energy of the photons is equal to or greater than the ionization energy, an electron is ejected from each atom that interacts with a photon. The difference between the energy of the photon and the ionization energy of the atom appears as kinetic energy of the photoelectron. The ionization energy, IE, is equal to the energy of the photon, $h\nu$, minus the kinetic energy, KE, of the photoelectron:

$$IE = h\nu - KE$$

The kinetic energy of the photons is measured in the photoelectron spectrometer as shown in Figure 6.16. The ionization energy of the atom is then calculated from the preceding expression.

If photons of sufficient energy are used, an electron may be ejected from *any* of the energy levels of an atom. Thus if we bombard a large number of atoms with photons of sufficient energy, electrons of several different kinetic energies will usually be obtained.

HELIUM The photoelectrons obtained from helium, using radiation of a single frequency, all have the same kinetic energy. Thus both the electrons occupying the $n = 1$ shell of the helium atom have the same energy. In other words, there is only one energy level for the $n = 1$ shell, and both electrons are in this level. From the known energy of the photons and the measured kinetic energy of the photoelectrons, the ionization energy of an electron in this level is found to be 2.37 MJ mol^{-1}. The energy of this level is therefore 2.37 MJ mol^{-1} less than the energy of an ionized helium atom; it has an energy of

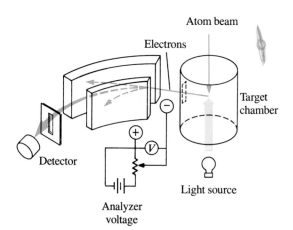

Figure 6.16 Photoelectron Spectrometer.
A beam of atoms is directed into the target chamber. In the target chamber, which is kept at very low pressure, the atoms are irradiated with ultraviolet light or X rays of a known, fixed frequency. The kinetic energies of the ejected electrons are measured by passing them between two curved plates. A voltage is applied between the plates, creating an electric field which deflects the electrons into curved paths. Only electrons of one particular kinetic energy will have their paths curved just the right amount to enable them to pass through a narrow slit and reach the detector. The detector counts the number of electrons arriving in a given time. Varying the voltage between the plates allows electrons of different energies to reach the detector. Thus the number of electrons of each energy that arrive per second can be found.

-2.37 MJ mol^{-1}, with the energy of the ionized helium atom being taken to be zero. This value is the same as that obtained from electron impact experiments (Table 4.2).

NEON When neon atoms are irradiated with light of a sufficiently high frequency, electrons with three different kinetic energies are obtained. This result shows that the electrons in a neon atom occupy three energy levels. A given atom may lose an electron from any one of these levels if the energy of the photon with which it interacts is large enough.

The electrons with the lowest kinetic energy must require the largest amount of energy for their removal. Therefore, these electrons must come from the shell in which the electrons are most tightly held, which is the $n = 1$ shell. These electrons have an ionization energy of 84.0 MJ mol^{-1}; therefore the $n = 1$ shell has just one level, with an energy of -84.0 MJ mol^{-1}.

The electrons with higher kinetic energies need much less energy for their removal, and they must come from the $n = 2$ shell of the neon atoms. Since these electrons have two different kinetic energies, the electrons in the $n = 2$ shell must have two slightly different energies. The ionization energies for these two electrons are found from the measured kinetic energies and the known energy of the photons to be 4.68 and 2.07 MJ mol^{-1}. Therefore, these electrons came from energy levels having energies of -4.68 and -2.07 MJ mol^{-1}. These levels are labeled 2s and 2p, respectively. The number 2 indicates that the electrons are in the $n = 2$ shell. The label s always indicates the lowest energy level in a given shell, and p denotes the next highest energy level. Thus the $n = 2$ shell consists of the 2s and 2p energy levels. The $n = 1$ shell consists of a single 1s energy level.

Example 6.7 When a beam of neon atoms was irradiated with a beam of X rays of wavelength 0.2291 nm, electrons with kinetic energies of 438.2 MJ mol^{-1}, 517.5 MJ mol^{-1}, and 520.1 MJ mol^{-1} were obtained. Calculate the corresponding ionization energies of neon.

Solution First, we need the energy of the X ray photons. For one photon $E = h\nu$ and $\nu = c/\lambda$, so

$$E = \frac{hc}{\lambda} = \left(\frac{6.626 \times 10^{-34} \text{ J s}}{0.2291 \times 10^{-9} \text{ m}}\right)\left(\frac{2.998 \times 10^{8} \text{ m}}{1 \text{ s}}\right)$$

$$= 8.671 \times 10^{-16} \text{ J per photon}$$

For 1 mol of photons

$$E = \left(\frac{8.671 \times 10^{-16} \text{ J}}{1 \text{ photon}}\right)\left(\frac{6.022 \times 10^{23} \text{ photons}}{1 \text{ mol}}\right)$$

$$= 5.222 \times 10^{8} \text{ J mol}^{-1} = 522.2 \text{ MJ mol}^{-1}$$

Now, we can use the relationship between the energy of the photons, E_{photons}, the kinetic energy of the electrons, KE, and their ionization energy, IE,

$$IE = E_{\text{photons}} - KE$$

to find the ionization energies:

$$IE = 522.2 \text{ MJ mol}^{-1} - 438.2 \text{ MJ mol}^{-1} = 84.0 \text{ MJ mol}^{-1}$$

$$IE = 522.2 \text{ MJ mol}^{-1} - 517.5 \text{ MJ mol}^{-1} = 4.7 \text{ MJ mol}^{-1}$$

$$IE = 522.2 \text{ MJ mol}^{-1} - 520.1 \text{ MJ mol}^{-1} = 2.1 \text{ MJ mol}^{-1}$$

CHAPTER 6
ENERGY LEVELS,
ELECTRON
CONFIGURATIONS, AND
THE COVALENT BOND

212

Energy level diagrams for the helium and neon atoms are given in Figure 6.17. The value of the ionization energy of neon obtained from electron impact experiments is 2.07 MJ mol^{-1} and corresponds to the removal of a 2p electron, namely, the most easily removed electron. In general, the ionization energies obtained from electron impact experiments (Table 4.3) are for the most easily removed electron in each case—in other words, for an electron in the highest occupied energy level.

Periods 1, 2, and 3

The ionization energies of the electrons in the atoms from hydrogen to argon are given in Table 6.1 and in Figure 6.18. We see that lithium has two ionization

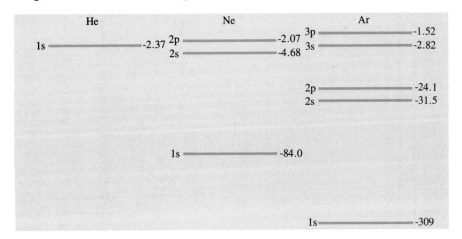

Figure 6.17 Energy levels for helium, neon, and argon atoms obtained by photoelectron spectroscopy. (The energies are in MJ mol^{-1}).

Table 6.1 Ionization Energies (MJ mol^{-1}) for First 21 Elements

ELEMENT	1s	2s	2p	3s	3p	3d	4s
H	1.31						
He	2.37						
Li	6.26	0.52					
Be	11.5	0.90					
B	19.3	1.36	0.80				
C	28.6	1.72	1.09				
N	39.6	2.45	1.40				
O	52.6	3.04	1.31				
F	67.2	3.88	1.68				
Ne	84.0	4.68	2.08				
Na	104	6.84	3.67	0.50			
Mg	126	9.07	5.31	0.74			
Al	151	12.1	7.19	1.09	0.58		
Si	178	15.1	10.3	1.46	0.79		
P	208	18.7	13.5	1.95	1.06		
S	239	22.7	16.5	2.05	1.00		
Cl	273	26.8	20.2	2.44	1.25		
Ar	309	31.5	24.1	2.82	1.52		
K	347	37.1	29.1	3.93	2.38		0.42
Ca	390	42.7	34.0	4.65	2.90		0.59
Sc	433	48.5	39.2	5.44	3.24	0.77	0.63

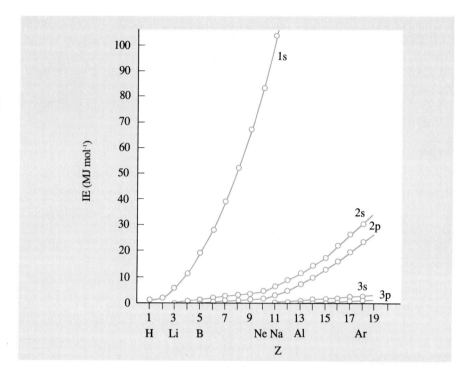

energies: one for the two electrons in the 1s level and the second one for one electron in the 2s level. Similarly, beryllium has two ionization energies corresponding to two electrons in the 1s level and two in the 2s level. However, boron and all the following elements up to neon have three ionization energies, showing that there must be a second energy level in the $n = 2$ shell, namely, the 2p level. In boron one electron is in this level; in carbon two electrons are in this level. The number of electrons in the 2p level increases from element to element up to six in neon, when the 2p level is filled and the $n = 2$ shell is complete.

The arrangement of the electrons for the elements of the first two periods can therefore be written as follows:

H	$1s^1$
He	$1s^2$
Li	$1s^2 2s^1$
Be	$1s^2 2s^2$
B	$1s^2 2s^2 2p^1$
C	$1s^2 2s^2 2p^2$
N	$1s^2 2s^2 2p^3$
O	$1s^2 2s^2 2p^4$
F	$1s^2 2s^2 2p^5$
Ne	$1s^2 2s^2 2p^6$

These representations are called **electron configurations**. The energy levels of the neon atom are shown in Figure 6.17.

The element following neon is sodium. It is in the third period and has an electron in the $n = 3$ shell. Sodium is an alkali metal like lithium, and therefore it has a similar electron configuration, with just one electron in its valence shell. It has four ionization energies and the configuration $1s^2 2s^2 2p^6 3s^1$. Similarly, magnesium has four ionization energies corresponding to the 1s, 2s, 2p, and 3s levels. But the next element, aluminum, has five ionization energies; its 3s level

CHAPTER 6
ENERGY LEVELS,
ELECTRON
CONFIGURATIONS, AND
THE COVALENT BOND

214

is filled with two electrons and the remaining electron in the $n = 3$ shell must be in the 3p level. Like the 2p level, the 3p level can contain a maximum of six electrons; electrons are added to this level until it is filled at argon. Thus the electron configurations of the elements from sodium to argon are as follows:

$$
\begin{array}{ll}
\text{Na} & 1s^2 2s^2 2p^6 3s^1 \\
\text{Mg} & 1s^2 2s^2 2p^6 3s^2 \\
\text{Al} & 1s^2 2s^2 2p^6 3s^2 3p^1 \\
\text{Si} & 1s^2 2s^2 2p^6 3s^2 3p^2 \\
\text{P} & 1s^2 2s^2 2p^6 3s^2 3p^3 \\
\text{S} & 1s^2 2s^2 2p^6 3s^2 3p^4 \\
\text{Cl} & 1s^2 2s^2 2p^6 3s^2 3p^5 \\
\text{Ar} & 1s^2 2s^2 2p^6 3s^2 3p^6
\end{array}
$$

The existence of the s and p energy levels in the $n = 2$ and $n = 3$ shells accounts for some of the small irregularities in the ionization energy of the most easily removed electron that we saw in Table 4.3 and Figure 4.4 but did not explain. We expect a steady increase in the ionization energy of an $n = 2$ shell electron from lithium to neon as the core charge increases, but this increase is interrupted at boron and at oxygen. The ionization energy of the electron that is most easily removed from boron (0.80 MJ mol^{-1}) is slightly lower, not higher, than that for beryllium (0.90 MJ mol^{-1}), even though boron has a higher core charge than beryllium. We now see that this energy is lower because the most easily removed electron comes from the 2s level ın beryllium, but the most easily removed electron in boron comes from the 2p level. Electrons in the 2p level of boron have a slightly lower ionization energy than the 2s electrons of beryllium, despite the increase in the core charge.

A similar situation occurs in the third period. The ionization energy of the most easily removed electron in aluminum (0.58 MJ mol^{-1}) is less than the ionization energy of the preceding element magnesium (0.74 MJ mol^{-1}). The electron removed from aluminum is a 3p electron, whereas the electron removed from magnesium is a more tightly held 3s electron. The similar anomaly observed between group V and group VI atoms of the same period is explained shortly.

Periods 4 to 7

The element following argon is another alkali metal, potassium, and the next element is the alkaline earth metal calcium. Each of these elements has six different ionization energies, one of which has a very small value and therefore corresponds to an electron from the $n = 4$ shell. We may assign these elements the following electron configurations:

$$
\begin{array}{ll}
\text{K} & 1s^2 2s^2 2p^6 3s^2 3p^6 4s^1 \\
\text{Ca} & 1s^2 2s^2 2p^6 3s^2 3p^6 4s^2
\end{array}
$$

The next element, scandium, has seven ionization energies. However, the ionization energy for the most easily removed electron is higher than that for a calcium 4s electron. If this electron were being removed from the 4p level, then by analogy with boron and aluminum it should have a lower ionization energy than that for the 4s electron in calcium. Instead, the electron is in a third level of the $n = 3$ shell, the 3d level, which is higher in energy than the 3p level and may accommodate up to ten electrons. The elements from scandium to zinc all have electrons in a 3d level; these elements are the transition elements of the fourth period. Thus the electron configuration of scandium is $1s^2 2s^2 2p^6 3s^2 3p^6 3d^1 4s^2$; that of zinc is $1s^2 2s^2 2p^6 3s^2 3p^6 3d^{10} 4s^2$.

The fact that electrons in the ground states of potassium and calcium occupy the 4s level, rather than the 3d level, shows that for these two elements the 3d level is of higher energy than the 4s level. Although the *average* energy of the electrons in all the energy levels of the $n = 3$ shell is lower than that of electrons in the $n = 4$ shell, the energies of the different levels in the two shells overlap for some elements. In general, for the $n = 3, 4, 5, \ldots$ shells the energy levels of neighboring shells overlap (see Figure 6.19), which complicates the electron configurations of the elements in periods 4 to 7.

Following zinc, electrons enter the 4p level, starting with gallium and finishing with krypton. Thus krypton has the valence shell electron configuration $4s^2 4p^6$, which is like those of argon, $3s^2 3p^6$, and neon, $2s^2 2p^6$. The electron configurations are as follows:

Ga	$1s^2 2s^2 2p^6 3s^2 3p^6 3d^{10} 4s^2 4p^1$
Ge	$1s^2 2s^2 2p^6 3s^2 3p^6 3d^{10} 4s^2 4p^2$
As	$1s^2 2s^2 2p^6 3s^2 3p^6 3d^{10} 4s^2 4p^3$
Se	$1s^2 2s^2 2p^6 3s^2 3p^6 3d^{10} 4s^2 4p^4$
Br	$1s^2 2s^2 2p^6 3s^2 3p^6 3d^{10} 4s^2 4p^5$
Kr	$1s^2 2s^2 2p^6 3s^2 3p^6 3d^{10} 4s^2 4p^6$

The six elements from Ga to Kr resemble the six elements from Al to Ar in which the 3p level is progressively filled, and the six elements from B to Ne in which the 2p level is filled.

Electron configurations of all the elements are given in Table 6.2. For convenience only the energy levels higher than those in the preceding noble gas are shown in full in the table. Thus the electron configuration of sodium, $1s^2 2s^2 2p^6 3s^1$ is written as $[Ne] 3s^1$, and the electron configuration of bromine, $1s^2 2s^2 2p^6 3s^2 3p^6 4s^2 3d^{10} 4p^5$ is written as $[Ar] 3d^{10} 4s^2 4p^5$. We will be concerned primarily with the electron configurations of the first 36 elements.

GENERAL RULES   The order in which the various energy levels are occupied can be remembered by making use of the diagram in Figure 6.20. For the first 36 elements the order is 1s, 2s, 2p, 3s, 3p, 4s, 3d, 4p. Each of the energy levels in a given shell is often called a **subshell**. Thus the $n = 2$ shell has a 2s and a 2p subshell, and the $n = 3$ shell has a 3s, a 3p, and a 3d subshell.

Figure 6.19 Energy Levels. The energy of the shells increases with increasing n, and the energy of the subshells increases in the order s, p, d, f. When $n \geq 3$, there is some overlap between the energy levels of one shell and the next highest shell.

CHAPTER 6
ENERGY LEVELS,
ELECTRON
CONFIGURATIONS, AND
THE COVALENT BOND

216

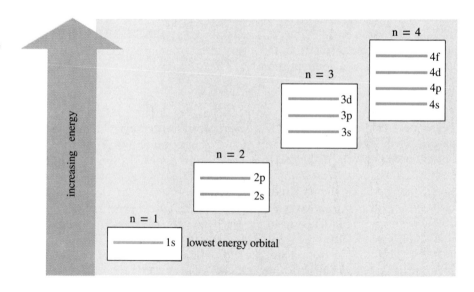

Table 6.2 Electron Configurations of Elements

ATOMIC NUMBER	ELEMENT	ELECTRON CONFIGURATION	ATOMIC NUMBER	ELEMENT	ELECTRON CONFIGURATION
1	H	$1s^1$	45	Rh	$[Kr]\,4d^85s^1$
2	He	$1s^2$	46	Pd	$[Kr]\,4d^{10}$
3	Li	$[He]\,2s^1$	47	Ag	$[Kr]\,4d^{10}5s^1$
4	Be	$[He]\,2s^2$	48	Cd	$[Kr]\,4d^{10}5s^2$
5	B	$[He]\,2s^22p^1$	49	In	$[Kr]\,4d^{10}5s^25p^1$
6	C	$[He]\,2s^22p^2$	50	Sn	$[Kr]\,4d^{10}5s^25p^2$
7	N	$[He]\,2s^22p^3$	51	Sb	$[Kr]\,4d^{10}5s^25p^3$
8	O	$[He]\,2s^22p^4$	52	Te	$[Kr]\,4d^{10}5d^25p^4$
9	F	$[He]\,2s^22p^5$	53	I	$[Kr]\,4d^{10}5s^25p^5$
10	Ne	$[He]\,2s^22p^6$	54	Xe	$[Kr]\,4d^{10}5s^25p^6$
11	Na	$[Ne]\,3s^1$	55	Cs	$[Xe]\,6s^1$
12	Mg	$[Ne]\,3s^2$	56	Ba	$[Xe]\,6s^2$
13	Al	$[Ne]\,3s^23p^1$	57	La	$[Xe]\,5d^16s^2$
14	Si	$[Ne]\,3s^23p^2$	58	Ce	$[Xe]\,4f^15d^16s^2$
15	P	$[Ne]\,3s^23p^3$	59	Pr	$[Xe]\,4f^3\quad6s^2$
16	S	$[Ne]\,3s^23p^4$	60	Nd	$[Xe]\,4f^4\quad6s^2$
17	Cl	$[Ne]\,3s^23p^5$	61	Pm	$[Xe]\,4f^5\quad6s^2$
18	Ar	$[Ne]\,3s^23p^6$	62	Sm	$[Xe]\,4f^6\quad6s^2$
19	K	$[Ar]\,4s^1$	63	Eu	$[Xe]\,4f^7\quad6s^2$
20	Ca	$[Ar]\,4s^2$	64	Gd	$[Xe]\,4f^75d^16s^2$
21	Sc	$[Ar]\,3d^14s^2$	65	Tb	$[Xe]\,4f^9\quad6s^2$
22	Ti	$[Ar]\,3d^24s^2$	66	Dy	$[Xe]\,4f^{10}\quad6s^2$
23	V	$[Ar]\,3d^34s^2$	67	Ho	$[Xe]\,4f^{11}\quad6s^2$
24	Cr	$[Ar]\,3d^54s^1$	68	Er	$[Xe]\,4f^{12}\quad6s^2$
25	Mn	$[Ar]\,3d^54s^2$	69	Tm	$[Xe]\,4f^{13}\quad6s^2$
26	Fe	$[Ar]\,3d^64s^2$	70	Yb	$[Xe]\,4f^{14}\quad6s^2$
27	Co	$[Ar]\,3d^74s^2$	71	Lu	$[Xe]\,4f^{14}5d^16s^2$
28	Ni	$[Ar]\,3d^84s^2$	72	Hf	$[Xe]\,4f^{14}5d^26s^2$
29	Cu	$[Ar]\,3d^{10}4s^1$	73	Ta	$[Xe]\,4f^{14}5d^36s^2$
30	Zn	$[Ar]\,3d^{10}4s^2$	74	W	$[Xe]\,4f^{14}5d^46s^2$
31	Ga	$[Ar]\,3d^{10}4s^24p^1$	75	Re	$[Xe]\,4f^{14}5d^56s^2$
32	Ge	$[Ar]\,3d^{10}4s^24p^2$	76	Os	$[Xe]\,4f^{14}5d^66s^2$
33	As	$[Ar]\,3d^{10}4s^24p^3$	77	Ir	$[Xe]\,4f^{14}5d^76s^2$
34	Se	$[Ar]\,3d^{10}4s^24p^4$	78	Pt	$[Xe]\,4f^{14}5d^96s^1$
35	Br	$[Ar]\,3d^{10}4s^24p^5$	79	Au	$[Xe]\,4f^{14}5d^{10}6s^1$
36	Kr	$[Ar]\,3d^{10}4s^24p^6$	80	Hg	$[Xe]\,4f^{14}5d^{10}6s^2$
37	Rb	$[Kr]\,5s^1$	81	Tl	$[Xe]\,4f^{14}5d^{10}6s^26p^1$
38	Sr	$[Kr]\,5s^2$	82	Pb	$[Xe]\,4f^{14}5d^{10}6s^26p^2$
39	Y	$[Kr]\,4d^15s^2$	83	Bi	$[Xe]\,4f^{14}5d^{10}6s^26p^3$
40	Zr	$[Kr]\,4d^25s^2$	84	Po	$[Xe]\,4f^{14}5d^{10}6s^26p^4$
41	Nb	$[Kr]\,4d^45s^1$	85	At	$[Xe]\,4f^{14}5d^{10}6s^26p^5$
42	Mo	$[Kr]\,4d^55s^1$	86	Rn	$[Xe]\,4f^{14}5d^{10}6s^26p^6$
43	Tc	$[Kr]\,4d^55s^2$	87	Fr	$[Rn]\,7s^1$
44	Ru	$[Kr]\,4d^75s^1$	88	Ra	$[Rn]\,7s^2$

Table 6.2 Electron Configurations of Elements (*continued*)

ATOMIC NUMBER	ELEMENT	ELECTRON CONFIGURATION	ATOMIC NUMBER	ELEMENT	ELECTRON CONFIGURATION
89	Ac	$[Rn]\, 6d^1 7s^2$	97	Bk	$[Rn]\, 5f^9\quad 7s^2$
90	Th	$[Rn]\, 6d^2 7s^2$	98	Cf	$[Rn]\, 5f^{10}\quad 7s^2$
91	Pa	$[Rn]\, 5f^2 6d^1 7s^2$	99	Es	$[Rn]\, 5f^{11}\quad 7s^2$
92	U	$[Rn]\, 5f^3 6d^1 7s^2$	100	Fm	$[Rn]\, 5f^{12}\quad 7s^2$
93	Np	$[Rn]\, 5f^4 6d^1 7s^2$	101	Md	$[Rn]\, 5f^{13}\quad 7s^2$
94	Pu	$[Rn]\, 5f^6\quad 7s^2$	102	No	$[Rn]\, 5f^{14}\quad 7s^2$
95	Am	$[Rn]\, 5f^7\quad 7s^2$	103	Lr	$[Rn]\, 5f^{14} 6d^1 7s^2$
96	Cm	$[Rn]\, 5f^7 6d^1 7s^2$			

The arrangements of electrons in the shells and the subshells (energy levels) of an atom are summarized in Table 6.3. These arrangements are governed by several rules, as follows:

1. In the $n = 1$ shell there is only one subshell (1s); in the $n = 2$ shell there are two subshells (2s and 2p); and in the $n = 3$ shell there are three subshells (3s, 3p, and 3d). In general, *the number of subshells in any shell is equal to the quantum number n.*

2. The maximum number of electrons that can occupy the $n = 1$ shell is 2 ($1s^2$); in the $n = 2$ shell it is 8 ($2s^2 2p^6$); and in the $n = 3$ shell it is 18 ($3s^2 3p^6 3d^{10}$). In general, *the maximum number of electrons in a shell is $2n^2$.* For example, for $n = 3$, the maximum number of electrons is $2 \times 3^2 = 18$.

3. *The number of electrons in an s subshell is 2, in a p subshell is 6, in a d subshell is 10, in an f subshell is 14, and so on.* We will be concerned only with the s, p, and d subshells.

ENERGY LEVELS AND THE PERIODIC TABLE   Figure 6.20 shows the relationship between the periodic table and the electron configurations. The alkali and alkaline earth metals are called the *s block elements*, because the highest-energy, most easily removed electrons for these elements are s electrons. The elements on the right, which are mainly nonmetals, are called the *p block elements* because the highest-energy electrons for these elements are p electrons. The elements in the middle, the transition elements, are called the *d block elements*.

Figure 6.20 Periodic Table and Electron Configurations. The labels indicate the energy level being filled with electrons as we move through the table.

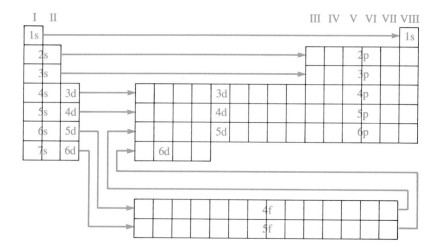

CHAPTER 6
ENERGY LEVELS,
ELECTRON
CONFIGURATIONS, AND
THE COVALENT BOND

218

Table 6.3 Arrangement of Electrons in Shells and Subshells

SHELL n	ENERGY LEVELS	NUMBER OF ELECTRONS	TOTAL NUMBER OF ELECTRONS IN SHELL ($2n^2$)
1	1s	2	2
2	2s	2 ⎫	8
	2p	6 ⎭	
3	3s	2 ⎫	18
	3p	6 ⎬	
	3d	10 ⎭	
4	4s	2 ⎫	32
	4p	6 ⎪	
	4d	10 ⎬	
	4f	14 ⎭	

Example 6.8 Write the electron configuration for an arsenic atom ($Z = 33$).

Solution We first find arsenic in the periodic table. It is in period 4 of Group V. Using the periodic table in Figure 6.20 and working from top to bottom, we move across each period writing down the occupancies of the energy levels until we come to the element arsenic:

$$
\begin{array}{ll}
\text{Period 1} & 1s^2 \\
\text{Period 2} & 2s^2 2p^6 \\
\text{Period 3} & 3s^2 3p^6 \\
\text{Period 4} & 4s^2 3d^{10} 4p^3
\end{array}
$$

By writing these down in order we obtain the complete electron configuration, $1s^2 2s^2 2p^6 3s^2 3p^6 4s^2 3d^{10} 4p^3$.

Another method is to remember (1) the order in which the energy levels are filled (namely, 1s, 2s, 2p, 3s, 3p, 4s, 3d, 4p) and (2) the maximum number of electrons that can be accommodated in each level (namely, 2 in s, 6 in p, 10 in d). Then fill the energy levels until the number of electrons is equal to the atomic number of arsenic, $Z = 33$, which gives the same result: $1s^2 2s^2 2p^6 3s^2 3p^6 4s^2 3d^{10} 4p^3$.

Electron configurations are also often written with the energy levels listed strictly in order of the quantum number n; in this case we have $1s^2 2s^2 2p^6 3s^2 3p^6 3d^{10} 4s^2 4p^3$. Both ways of writing electron configurations are acceptable.

We might next ask, "How are the electrons arranged within each energy level?" To understand these arrangements, we must first consider two more properties of electrons, electron spin and the Pauli exclusion principle.

Electron Spin

Several experiments carried out in the 1920s (Box 6.2) showed that an electron has magnetic properties; in other words, when it is placed in a magnetic field, it behaves rather like a small bar magnet. The magnetic properties of the electron can be understood if we think of it as a charged sphere rotating about an axis through its center (Figure 6.21). A rotating charge generates a magnetic field, and therefore the electron behaves like a magnet.

An ordinary bar magnet may be placed in any orientation in a magnetic field. But its energy is lowest when it is pointing in the direction of the field,

Box 6.2

EXPERIMENTAL EVIDENCE FOR ELECTRON SPIN

Experiments carried out by Otto Stern and Walter Gerlach in Germany beginning in 1921 provided the first direct evidence for electron spin. They passed a beam of alkali metal or silver atoms between the poles of a magnet designed to give a very nonuniform field. In each case the beam of atoms was split into two. Since the atoms were not charged, they could be deflected only if they behaved like magnets.

A magnet situated in a nonuniform magnetic field experiences a resultant force that displaces it, because one pole is situated in a stronger magnetic field than the other pole. In a uniform magnetic field both poles experience the same force, and there is no resultant force to displace the magnet. The amount by which a moving magnet is deflected from its original path in a nonuniform field depends on the orientation of the magnet. When it is lined up along the direction in which the field is changing the most (the greatest field gradient), it experiences the greatest force. When it is at right angles to this direction (zero field gradient), it experiences no deflecting force.

A beam of tiny magnets having random orientations, when entering the magnetic field, would be spread out into a continuous band. The fact that a beam of alkali metal or silver atoms is split up only into two beams shows not only that the atoms are magnetic but also that the atomic magnets can only have two orientations with respect to the magnetic field.

The explanation of the magnetic behavior of these atoms was given by two Dutch physicists, George Uhlen-beck and Sam Goudsmit. To explain certain fine details of atomic spectra, they proposed in 1925 that an electron has spin. The alkali metal atoms and the silver atom each have one electron in their valence shells. It is the spin of this single unpaired electron that is responsible for the magnetic properties of these atoms.

The Experimental Arrangement for the Stern-Gerlach Experiment. (a) A beam of atoms is passed through an inhomogeneous magnetic field produced by the specially shaped pole pieces. (b) The expected result for magnets that can take up any orientation with respect to the field: the beam of atoms is spread out uniformly. (c) The experimental result for silver atoms which have only one unpaired electron in the valence shell. The beam of atoms is split up into two distant beams. This result shows that the magnetic moment due to the single valence shell electron can only take up two orientations with respect to the field.

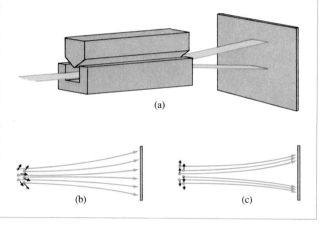

(a)

(b)

(c)

as when a compass needle, which is free to rotate, points toward the magnetic north pole. The electron, however, differs from a compass needle in that it can have only two orientations in a magnetic field, either along the direction of the field or in the opposite direction. When the axis of rotation is pointing in the direction of the field, the electron has a lower energy than when the axis is pointing in the opposite direction. The energy of the electron in a magnetic field is quantized, and there are only two corresponding energy levels. In contrast, the energy of a compass needle in the earth's magnetic field is not quantized. The compass needle can have any orientation, and there is essentially an infinite number of corresponding energy levels.

When the electron moves from the lower to the upper of its two energy levels, the orientation of the axis of the spinning electron changes from being along the direction of the magnetic field to being in the opposite direction. In other words, the direction of rotation of the electron is reversed. The difference in energy between the two levels for an electron in a magnetic field is very small. In the field of a strong electromagnet the difference in energy between the two orientations of the electron is about 10^6 times smaller than the ionization energy of an electron in an atom. Thus it takes very little energy to reverse the spin of an electron. In a system containing more than one electron, any two

CHAPTER 6
ENERGY LEVELS,
ELECTRON
CONFIGURATIONS, AND
THE COVALENT BOND

220

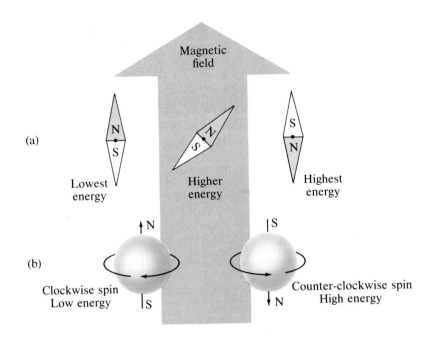

(a)

Magnetic field

Lowest energy
Higher energy
Highest energy

(b)

Clockwise spin
Low energy

Counter-clockwise spin
High energy

Figure 6.21 Electron Spin. A spinning electron generates a magnetic field and behaves like a bar magnet. However, unlike a compass needle, it can only have two orientations when placed in a magnetic field. Either its spin axis points in the direction of the magnetic field (clockwise spin), or its spin axis points in the opposite direction (counter-clockwise spin). (a) A compass needle has an infinite number of possible orientations in a magnetic field, three of which are shown here. (b) An electron has only two possible orientations in a magnetic field. Its orientation is quantized.

electrons may be spinning in the same direction—we say that they have the same spin—or in opposite directions—we say that they have opposite spins.

The Pauli Exclusion Principle

The arrangement of the electrons in a given energy level is determined by their spins and by a fundamental property of electrons that is called the Pauli exclusion principle. This principle was proposed by Swiss physicist Wolfgang Pauli (1900–1958) in 1925 to explain certain features of the spectra of atoms and to account for the fact that the electrons in an atom do not all crowd into the ls level. The electron arrangements that we deduced from ionization energies and from the periodic table show that the capacity of any energy level to accommodate electrons is limited to two electrons in an s level and six electrons in a p level.

We have seen that the electrons in an atom cannot be thought of as circulating around the nucleus in precisely defined orbits. Rather, each electron has a certain distribution around the nucleus—it has a high probability of being found in certain regions of the space around the nucleus and a low probability of being found in other regions. The probability distribution of an electron is described by a mathematical function called an *orbital*. The exact form of this function for a particular electron can be obtained from quantum mechanics but it is not important to us.

The **Pauli exclusion principle** states that:

Only two electrons can be described by the same orbital and these two electrons must have opposite spin.

This means that two electrons with the same spin cannot have the same distribution in space. Thus an important consequence of the exclusion principle is that:

Electrons with the same spin keep apart in space whereas electrons of opposite spin may occupy the same region of space.

Thus we may consider that the electrons in the same energy level and having the same spin keep as far apart as possible by each separately occupying a different region of the space around the nucleus. Each of these regions of space corresponds approximately to the orbital of a particular electron. Each orbital can then contain two electrons provided that the electrons have opposite spin. Thus the Pauli exclusion principle can also be stated in the form:

No orbital can accommodate more than two electrons and these two electrons must have opposite spin.

An orbital can be conveniently represented by a box in which we place arrows pointing either up or down to represent the spin of the electron:

$$\boxed{\downarrow} \qquad \boxed{\uparrow}$$

A 1s or 2s energy level can contain only two electrons; it therefore has only one orbital, represented by a single box. This orbital may contain either one electron, which may have its spin either up or down, or two electrons, which must have opposite spins:

1s or 2s orbital containing one electron $\boxed{\uparrow}$ or $\boxed{\downarrow}$

1s or 2s orbital containing two electrons $\boxed{\uparrow\downarrow}$

Since the 2p energy level may contain up to six electrons, there must be three 2p orbitals, which may be represented by three boxes, each of which may contain up to two electrons:

Six electrons in the 2p orbitals $\boxed{\uparrow\downarrow\,|\,\uparrow\downarrow\,|\,\uparrow\downarrow}$

HUND'S RULE The electrostatic repulsion energy between two electrons is a minimum when they are as far apart as possible. The minimum-energy arrangement for two or more electrons in the same energy level is the arrangement in which they have the same spin and are therefore forced to occupy separate orbitals, that is, separate regions of space. Their average distance apart is thus greater than if they had opposite spins and could occupy the same orbital. Thus:

In the lowest-energy (ground state) electron configuration electrons in the same energy level as far as possible occupy separate orbitals and have the same spin.

This statement is known as **Hund's rule**.

Following this rule, we can represent the electron configurations of the elements from lithium to neon by the orbital diagrams in Figure 6.22. When the number of electrons in an energy level does not exceed the number of available orbitals, then, following Hund's rule, all the electrons occupy separate orbitals and have the same spin, as in the cases of boron, carbon, and nitrogen. When there are more electrons than available orbitals, then at least some of the orbitals must contain two electrons of opposite spin, as in the case of oxygen, fluorine, and neon.

When two electrons occupy the same orbital, the repulsion between them decreases their ionization energy. This decrease shows up in the ionization energies of the 2p electrons (Tables 4.3 and 6.1). We would expect oxygen to have a higher ionization energy than nitrogen because of the greater core charge of oxygen. But the ionization energy for the most easily removed electron of oxygen (1.31 MJ mol^{-1}) is slightly smaller than that of nitrogen (1.40 MJ mol^{-1}). In oxygen two electrons must occupy the same orbital, whereas all the electrons are in separate orbitals in nitrogen. Similarly, sulfur has a lower ionization

CHAPTER 6
ENERGY LEVELS,
ELECTRON
CONFIGURATIONS, AND
THE COVALENT BOND

222

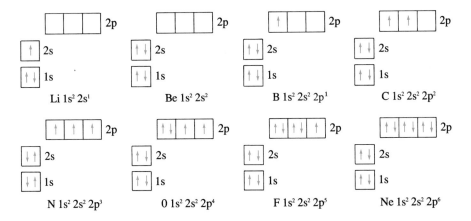

Figure 6.22 Orbital Diagrams for the Elements Lithium to Neon. These diagrams are the ground state configurations. In accordance with Hund's rule, electrons in the same energy level as far as possible occupy separate orbitals and have the same spin. When electrons must occupy the same orbital, they have opposite spins, as the arrows indicate.

Li $1s^2 2s^1$ Be $1s^2 2s^2$ B $1s^2 2s^2 2p^1$ C $1s^2 2s^2 2p^2$

N $1s^2 2s^2 2p^3$ O $1s^2 2s^2 2p^4$ F $1s^2 2s^2 2p^5$ Ne $1s^2 2s^2 2p^6$

energy than phosphorus because in sulfur (but not in phosphorus) two electrons occupy the same 3p orbital.

Notice that although the orbital diagrams in Figure 6.22 are drawn to show the orbitals in order of increasing energy, they do not show the relative energies of the orbitals in a quantitative manner. In fact, the difference in energy between the 1s and 2s orbitals is much greater than that between the 2s and 2p orbitals. Furthermore, all the energies vary from one atom to another, as shown by the ionization energies in Table 6.1. So box diagrams are frequently written out on one line, with the boxes in order of increasing energy. For example, for carbon we have

1s   2s   2p

The orbital diagrams in Figure 6.22 represent ground state electron configurations. There are many other possible electron configurations for these atoms, but they all represent higher-energy excited states. The following diagrams show possible excited state electron configurations of boron, carbon, and nitrogen:

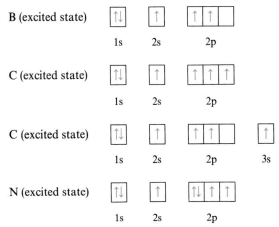

B (excited state)   1s   2s   2p

C (excited state)   1s   2s   2p

C (excited state)   1s   2s   2p   3s

N (excited state)   1s   2s   2p

Example 6.9 Write the electron configuration and draw an orbital diagram for the ground state of each of the following atoms and ions:

O   F⁻   Na   Na⁺   P   S

Solution First, we write the electron configuration:

$$O \qquad 1s^2 2s^2 2p^4$$

Then we draw one box for each s level and three boxes for each p level:

$$\text{1s} \qquad \text{2s} \qquad \text{2p}$$

Now we place two electrons of opposite spin in each of the 1s and 2s levels:

$$\boxed{\uparrow\downarrow} \quad \boxed{\uparrow\downarrow} \quad \boxed{}\boxed{}\boxed{}$$
$$\text{1s} \qquad \text{2s} \qquad \text{2p}$$

Then we place one electron in each of the 2p orbitals, following Hund's rule:

$$\boxed{\uparrow\downarrow} \quad \boxed{\uparrow\downarrow} \quad \boxed{\uparrow}\boxed{\uparrow}\boxed{\uparrow}$$
$$\text{1s} \qquad \text{2s} \qquad \text{2p}$$

Finally, we add the fourth electron to one of the 2p orbitals.

$$O \quad \boxed{\uparrow\downarrow} \quad \boxed{\uparrow\downarrow} \quad \boxed{\uparrow\downarrow}\boxed{\uparrow}\boxed{\uparrow}$$
$$\text{1s} \qquad \text{2s} \qquad \text{2p}$$

The same procedure gives the other orbital diagrams:

F^- $\quad 1s^2 2s^2 2p^6 \qquad \boxed{\uparrow\downarrow} \quad \boxed{\uparrow\downarrow} \quad \boxed{\uparrow\downarrow}\boxed{\uparrow\downarrow}\boxed{\uparrow\downarrow}$

$\qquad\qquad\qquad\qquad\quad\; \text{1s} \qquad \text{2s} \qquad \text{2p}$

Na $\quad 1s^2 2s^2 2p^6 3s^1 \qquad \boxed{\uparrow\downarrow} \quad \boxed{\uparrow\downarrow} \quad \boxed{\uparrow\downarrow}\boxed{\uparrow\downarrow}\boxed{\uparrow\downarrow} \quad \boxed{\uparrow}$

$\qquad\qquad\qquad\qquad\quad\; \text{1s} \qquad \text{2s} \qquad \text{2p} \qquad \text{3s}$

Na^+ $\quad 1s^2 2s^2 2p^6 \qquad \boxed{\uparrow\downarrow} \quad \boxed{\uparrow\downarrow} \quad \boxed{\uparrow\downarrow}\boxed{\uparrow\downarrow}\boxed{\uparrow\downarrow}$

$\qquad\qquad\qquad\qquad\quad\; \text{1s} \qquad \text{2s} \qquad \text{2p}$

P $\quad 1s^2 2s^2 2p^6 3s^2 3p^3 \qquad \boxed{\uparrow\downarrow} \quad \boxed{\uparrow\downarrow} \quad \boxed{\uparrow\downarrow}\boxed{\uparrow\downarrow}\boxed{\uparrow\downarrow} \quad \boxed{\uparrow\downarrow} \quad \boxed{\uparrow}\boxed{\uparrow}\boxed{\uparrow}$

$\qquad\qquad\qquad\qquad\quad\; \text{1s} \qquad \text{2s} \qquad \text{2p} \qquad \text{3s} \qquad \text{3p}$

S $\quad 1s^2 2s^2 2p^6 3s^2 3p^4 \qquad \boxed{\uparrow\downarrow} \quad \boxed{\uparrow\downarrow} \quad \boxed{\uparrow\downarrow}\boxed{\uparrow\downarrow}\boxed{\uparrow\downarrow} \quad \boxed{\uparrow\downarrow} \quad \boxed{\uparrow\downarrow}\boxed{\uparrow}\boxed{\uparrow}$

$\qquad\qquad\qquad\qquad\quad\; \text{1s} \qquad\qquad\; \text{2p} \qquad \text{3s} \qquad \text{3p}$

SHAPES OF ORBITALS The electron density distribution that depicts an electron occupies a certain region of space and has a certain shape. The orbital occupied by the electron has this same shape. The single electron in the hydrogen atom occupies all the space close to the hydrogen nucleus, and it has a spherical distribution. Thus the 1s orbital has a spherical shape. The electron density, however, is not uniform in this sphere—it is greatest close to the nucleus and decreases continuously with increasing distance from the nucleus. In helium two electrons occupy the 1s orbital—they both have the same distribution in space—so the electron distribution of the helium atom is also spherical. All other s orbitals also have a spherical shape. The shapes of p and d orbitals are more complex and are discussed in Chapter 25.

CHAPTER 6
ENERGY LEVELS,
ELECTRON
CONFIGURATIONS, AND
THE COVALENT BOND

224

Having discussed the electron configurations of the elements in some detail, we are now in a position to look a little more closely at the covalent bond. And we can ask *why* a shared pair of electrons can bind two atoms together. We first consider the hydrogen molecule.

Hydrogen Molecule

As two hydrogen atoms approach each other, the nucleus of one atom exerts an attractive force on the electron of the other atom. Each hydrogen atom has space in its 1s orbital for a second electron, of opposite spin to the first electron. Therefore if the electrons of the two hydrogen atoms have opposite spins, both electrons can occupy both orbitals simultaneously; the electron densities of the two atoms can overlap each other. As a result, the electron density increases in the region where they overlap, that is, in the region between the two nuclei (see Figure 6.23). It is basically this increased electron density that holds the nuclei together by electrostatic attraction.

As the two hydrogen atoms come together, the electron density between the two nuclei increases, and the attractive force that this electron density exerts on the nuclei increases correspondingly. But as the nuclei and electrons come closer together, there is also an increase in the nucleus-nucleus and electron-electron repulsions. At a certain distance between the nuclei these repulsive forces equal the attractive forces between the nuclei and electrons, and the total

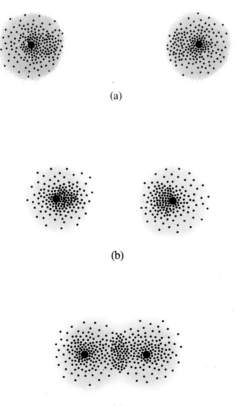

(a)

(b)

(c)

Figure 6.23 Electron Density Changes in the Formation of an H_2 Molecule. Because the hydrogen atom has space for another electron in its 1s orbital, the electron densities of two hydrogen atoms can overlap. The resulting increased electron density between the two hydrogen nuclei holds the two nuclei together by electrostatic attraction. (a) There is only a very weak attraction between two hydrogen atoms when they are a large distance apart. (b) As the distance between atoms decreases, the atoms interact. (c) The increased electron density between hydrogen nuclei holds the atoms together by electrostatic attraction.

energy of the system is then a minimum (see Figure 6.24). If the nuclei were to come closer, the repulsive forces would dominate, and the total energy of the system would therefore increase. The internuclear separation at which the energy is a minimum is the bond length of the hydrogen molecule in its ground state.

Bonding Orbitals

We can think of the two electrons in H_2 as now belonging to the molecule as a whole rather than to each separate atom. The two electrons may be said to occupy a **molecular orbital** that surrounds both nuclei. Since the two electrons constitute the covalent bond between the two H atoms, this orbital may also be called a **bonding orbital**. Thus we can represent the formation of the H_2 molecule diagrammatically as follows:

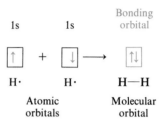

Any pair of atomic orbitals on different atoms can be combined in the same way to give a bonding orbital, if each atomic orbital is occupied by only one electron. For example, the F atom has the electron configuration $1s^2 2s^2 2p^5$; it therefore has one 2p orbital that contains only one electron. This 2p orbital may be combined with the 1s orbital of an H atom to form a bonding orbital that describes approximately the space occupied by the bonding electron pair in the HF molecule. Using orbital diagrams, we can represent the formation of the HF molecule as follows:

The singly occupied 2p orbital of one F atom may also be combined with the singly occupied 2p orbital of another F atom to give an approximate description of the bonding in the F_2 molecule, as shown by the following orbital diagram:

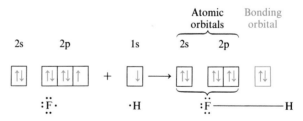

CHAPTER 6
ENERGY LEVELS,
ELECTRON
CONFIGURATIONS, AND
THE COVALENT BOND

226

Valence and Electron Configurations

We can describe any single bond in terms of a bonding orbital formed by the combination of two atomic orbitals. Thus the number of bonds formed by an

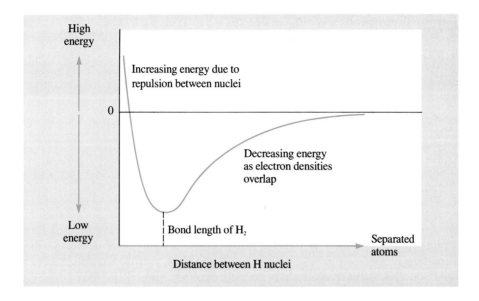

Figure 6.24 Energy Changes in the Formation of an H_2 Molecule. The curve shows the change in energy as the distance between two hydrogen nuclei changes. The distance at which the energy is a minimum is the distance between the nuclei in the H_2 molecule, 74 pm.

atom depends on the number of electrons in singly occupied orbitals, that is, on the number of unpaired electrons. Table 6.4 shows the valences we would predict on this basis for the elements in the second period.

These valences are in accordance with experiment and with the positions of the elements in the periodic table for Li, N, O, F, and Ne but not for Be, B, and C, which have valences of 2, 3, and 4 not 0, 1, and 2, as predicted from the number of unpaired electrons. However, if we were to excite an electron from the 2s orbital to a vacant 2p orbital, the number of unpaired electrons in Be, B, and C would correspond to the observed valence in each case, as shown in Table 6.5. In this case an electron is said to have been *promoted* from the 2s orbital to a 2p orbital. This excited state of the atom is often called the **valence state**.

Table 6.4 Ground State Electron Configurations and Predicted Valences for Second-Period Elements

ELEMENT	1s	2s	2p	NUMBER OF UNPAIRED ELECTRONS	PREDICTED VALENCE	OBSERVED VALENCE
Li	↑↓	↑		1	1	1
Be	↑↓	↑↓		0	0	2
B	↑↓	↑↓	↑	1	1	3
C	↑↓	↑↓	↑ ↑	2	2	4
N	↑↓	↑↓	↑ ↑ ↑	3	3	3
O	↑↓	↑↓	↑↓ ↑ ↑	2	2	2
F	↑↓	↑↓	↑↓ ↑↓ ↑	1	1	1
Ne	↑↓	↑↓	↑↓ ↑↓ ↑↓	0	0	0

Table 6.5 Valence State Electron Configurations and Predicted Valences for Be, B, and C

ELEMENT	1s	2s	2p	NUMBER OF UNPAIRED ELECTRONS	PREDICTED VALENCE	OBSERVED VALENCE
Be	↑↓	↑	↑ □ □	2	2	2
B	↑↓	↑	↑ ↑ □	3	3	3
C	↑↓	↑	↑ ↑ ↑	4	4	4

We can then imagine that in the formation of the methane molecule each of the singly occupied orbitals of the carbon atom is used to form a bonding orbital with a hydrogen 1s orbital. But it is not correct to think that in the formation of methane from a carbon atom and four hydrogen atoms, this excited state of the carbon atom is actually formed. Rather, the four bonding orbitals are formed directly from the carbon atom in its ground state and the four hydrogen atoms.

Nevertheless, imagining the formation of methane in two stages is often convenient. First, the promotion of a 2s electron into a vacant 2p orbital gives the electron configuration $2s^1 2p^3$ (the valence state), which is followed by the formation of four bonding orbitals from these carbon orbitals and the four hydrogen 1s orbitals. A certain amount of energy is needed to promote the 2s electron to a 2p orbital. But because the formation of a C—H bond liberates a considerable amount of energy this is more than compensated for by the energy evolved in the formation of two additional C—H bonds.

The advantage of thinking about the process in this way is that we can easily count the number of unpaired electrons in the valence state, and we can therefore easily predict how many bonds the atom will form. The important thing for us to note is that Be, B, and C all form as many bonds as possible; that is, they use *all* their valence shell electrons in bond formation.

For Li, N, O, and F promotion of a 2s electron to a 2p orbital does not change the number of unpaired electrons, so promotion does not allow the formation of additional bonds. In the case of Ne, since all the orbitals are filled, an electron cannot be promoted from a 2s orbital to a 2p orbital. Thus neon has no unpaired electrons and forms no compounds. Promotion of an electron to an $n = 3$ orbital would need more energy than could be obtained by bond formation.

Similar considerations apply to the elements of period 3. For example, silicon in group IV has a valence of 4 in compounds such as $SiCl_4$ and SiH_4. We can imagine that these compounds are formed via the valence state of silicon, which has the electron configuration $3s^1 3p^3$ with four unpaired electrons, rather than via the ground state $3s^2 3p^2$ configuration, which has only two unpaired electrons.

Why Solids and Liquids Are Difficult to Compress

Comparing the interaction of two helium atoms with the interaction of two hydrogen atoms is instructive. At relatively large distances intermolecular forces are responsible for a very weak attraction between two helium atoms. But when the two atoms are brought closer together, the electrons of one helium atom

CHAPTER 6
ENERGY LEVELS,
ELECTRON
CONFIGURATIONS, AND
THE COVALENT BOND

228

cannot be drawn into the region occupied by the electrons of the other because the orbitals of both atoms already contain two electrons of opposite spin.

According to the Pauli exclusion principle, neither orbital can accommodate additional electrons. If the atoms are pushed close together, the electron clouds of each atom distort in order to prevent the overlap of the two filled orbitals, as shown in Figure 6.25. This distortion leads to an increase in the electron density *outside* the nuclei rather than between the nuclei. The increased electron density outside the region between the nuclei exerts a force that pulls the nuclei apart. Because of this force and the nucleus-nucleus repulsion, two helium atoms repel each other if they are pushed close together. Thus because of the operation of the Pauli exclusion principle, two helium atoms cannot form a stable molecule.

Two hydrogen *molecules* behave in the same way. At relatively large distances they attract each other very weakly, but if pushed close together, they repel each other because their charge clouds cannot overlap. In fact, all atoms and molecules with filled orbitals behave in this way; they attract each other weakly at large distances but repel each other strongly at short distances. This repulsion is the reason that liquids and solids are very difficult to compress.

If a helium atom is described as consisting of two particlelike electrons moving around a nucleus, it is mostly empty space. But helium atoms behave like solid objects because the two electrons in the helium atom, or the two electrons in any orbital, exclude all other electrons. This situation is more easily understood in terms of the wave picture of an electron than the particle picture. We can imagine how an electron wave might occupy all of some given region of space. The Pauli exclusion principle is a statement in quantum mechanical terms of the observed solidity of matter. It describes a fundamental property of electrons, namely, that two electrons of opposite spin may occupy a region of space, called an orbital, and no other electrons can penetrate into this space

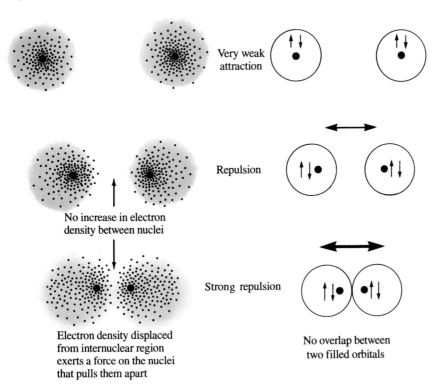

No increase in electron density between nuclei

Electron density displaced from internuclear region exerts a force on the nuclei that pulls them apart

Very weak attraction

Repulsion

Strong repulsion

No overlap between two filled orbitals

Figure 6.25 Interaction of Two Helium Atoms. If two helium atoms are brought close together, there is no increase in the electron density between the nuclei. Electron density is, in fact, pushed away from the nuclei, because the two charge clouds cannot overlap. The electron density that has been pushed away from the internuclear region pulls the nuclei apart. Two helium atoms that are brought together therefore repel each other rather than attract each other to form a molecule.

to a significant extent. Hence all atoms and molecules that have filled orbitals behave as essentially solid objects that cannot overlap or interpenetrate each other to any significant extent.

In contrast, atoms and molecules that have one or more electrons in singly occupied orbitals can interact readily with other atoms and molecules that have electrons in singly occupied orbitals to form bonds. Thus almost all atoms and molecules with one or more unpaired electrons are very reactive. For example, under normal conditions H atoms, F atoms, and O atoms exist only in the combined state. Free H atoms *are* found in relatively large numbers in interstellar space, but this is because they are so far apart that they rarely encounter another atom with which they could combine and because any H_2 molecules that form are decomposed back into H atoms by radiation from the sun. The only unreactive free atoms are those of the noble gases, because they have all their electrons in filled orbitals.

Almost all stable molecules have an even number of electrons. In other words, all their electrons are in filled orbitals, and they are therefore much less reactive than free atoms. Molecules containing an odd number of electrons are known, but like most free atoms, they are almost all extremely reactive, because they have an incompletely filled orbital—that is, an unpaired electron. A few molecules with an odd number of electrons are relatively unreactive; these molecules are discussed in Chapter 17.

IMPORTANT TERMS

A **bonding orbital** extends over two or more nuclei in a molecule. The electron density in such an orbital is concentrated between the nuclei and serves to hold them together.

Diffraction is the term used to describe the spreading out of waves emerging from a slit or other aperture or the bending of waves around an obstacle placed in their path.

An **electron configuration** is a particular arrangement of electrons in the orbitals of an atom.

Electron spin is a concept used to explain the magnetic properties of electrons. It is imagined that a spherical electron can spin around its own axis, thus generating a magnetic field.

In an **excited state** an atom has one or more of its electrons in energy levels other than the lowest available level.

The **frequency** of a wave is the number of wave crests that pass a given point in one second.

In the **ground state** of an atom all the electrons are in their lowest-possible energy levels.

Hund's rule states that in the lowest energy (ground state) electron configuration electrons in the same energy level as far as possible occupy separate orbitals and have the same spins.

Interference occurs between waves when waves from two different sources arrive at the same point.

A **molecular orbital** is an orbital that extends over an entire molecule.

Momentum is the product of the mass and the velocity of an object, or $p = mv$.

An **orbital** is a function that describes the distribution of an electron in an atom or molecule. It may be considered to be the region of space occupied by a single electron or by two electrons of opposite spin.

The **Pauli exclusion principle** may be stated in a number of different but equivalent ways. For example: No orbital can contain more than two electrons, which must have opposite spins, or, electrons with the same spin keep apart in space, whereas electrons of opposite spin may occupy the same region of space.

A **photochemical reaction** is a reaction that is caused by the absorption of light by a molecule.

The **photoelectric effect** is the emission of electrons by metals when light with a frequency greater than a certain minimum frequency shines upon them.

In **photoelectron spectroscopy** photons are used to knock electrons out of atoms. From the energy of the photons and the kinetic energies of the electrons thus obtained, the ionization energies of the electrons in an atom can be determined, and its electron configuration can be deduced.

A **photon** is a quantum of light; its energy is given by $E = hv$, where v is the frequency of the light and h is the Planck constant.

A **quantized** system is one that can have only certain energies. The energies of the electrons in atoms are quantized.

A **quantum** (plural, quanta) is a small packet of energy.

A **spectrum** is the range of frequencies of radiation emitted by an atom, a heated filament, the sun, or other energy source. A *line spectrum* consists of only a few sharp lines; it consists of radiation of only a few definite wavelengths.

Subshell is an alternative name for each of the energy levels associated with a particular shell.

The **uncertainty principle** states that there is an uncertainty

$$c = 3.00 \times 10^8 \text{ m/s} \qquad h = 6.626 \times 10^{-34} \text{ Js}$$

in the position, Δx, and an uncertainty in the momentum, Δp, of any particle such that if one is decreased, the other is increased.

The **wavelength** is the distance between successive wave crests or points of equal displacement on successive waves.

$$\lambda = \frac{c}{\nu} \qquad E_{photon} = \frac{hc}{\lambda} \qquad E = 1312\left(\frac{1}{m^2} - \frac{1}{n^2}\right) \text{ kJ/mol}$$

PROBLEMS

Electromagnetic Radiation

1. Radio station CBC in Toronto, Ontario, broadcasts its FM signal at 94.1 MHz and its AM signal at 740 kHz. What are the wavelengths of these signals in meters?

2. Mercury vapor lamps used for street and highway lighting emit the atomic spectrum of mercury. One of the lines in this spectrum is in the blue region and has a wavelength of 435.8 nm. Express the wavelength in (a) meters, (b) micrometers, (c) angstroms. What is the frequency of this line?

3. Citizens band (CB) radio operates at a frequency of 27.3 MHz. What is the wavelength of the radiowave?

4. Calculate the range of frequencies associated with the various parts of the electromagnetic spectrum from the following wavelengths.

(a) Radio (1 km to 30 cm)

(b) Microwave (30 cm to 2 mm)

(c) Far infrared (2 mm to 30 μm)

(d) Near infrared (30 μm to 710 nm)

(e) Visible (710 nm to 400 nm)

(f) Ultraviolet (400 nm to 4 nm)

(g) X rays (4 nm to 30 pm)

(h) γ rays (30 pm to 0.1 pm)

5. A helium-neon laser produces light of wavelength 633 nm. What is the frequency of this light?

6. A sodium vapor street lamp emits radiation of wavelength 589.2 nm. What is the frequency of this radiation?

7. The atomic spectrum of lithium has a strong red line at 670.8 nm. What is the energy of each photon of this wavelength? What is the energy of a mole of these photons?

8. The ionization energy of potassium is 0.42 MJ mol^{-1}. What is the maximum wavelength of light that will ionize a potassium atom in the gas phase?

9. Using the wavelengths listed in Problem 4, calculate the energy range of the photons of each type of radiation. What type of radiation is needed to dissociate a gaseous H_2 molecule into H atoms (dissociation energy $= 434.0$ kJ mol^{-1}), a gaseous O_2 molecule (dissociated energy $= 498$ kJ mol^{-1}), a gaseous Cl_2 molecule (dissociated energy $= 243$ kJ mol^{-1}), and a gaseous F_2 molecule (dissociation energy $= 159$ kJ mol^{-1})?

10. Nitrogen dioxide, NO_2, is one of the components of photochemical smog. The energy required to dissociate NO_2 molecules into NO molecules and O atoms is 305 kJ mol^{-1}. What is the maximum wavelength of light that can cause this dissociation? What type of radiation is it? If the minimum wavelength of light that reaches the earth's surface at sea level is 320 nm, does the dissociation of NO_2 into NO and O occur on the surface of the earth near sea level?

Photoelectric Effect

11. The longest wavelength of light that will cause an electron to be emitted from a lithium atom is 520 nm. Gaseous lithium atoms are irradiated with light of wavelength 360 nm. What is the kinetic energy of the emitted electrons, in kJ mol^{-1}?

12. Photons of minimum energy 496 kJ mol^{-1} are required to ionize sodium atoms. Calculate the lowest frequency of light that will ionize a sodium atom. What is the color of this light? If light of energy 600 kJ mol^{-1} is used, what is the velocity of the emitted electrons?

13. When light with a wavelength of 470.0 nm falls on the surface of potassium metal, electrons are emitted with a velocity of 6.4×10^4 m s^{-1}.

(a) What is the kinetic energy of the emitted electrons?

(b) What is the energy of a 470.0 nm photon?

(c) What is the minimum energy needed to remove an electron from potassium metal?

The Hydrogen Atom

14. What wavelength of light is emitted when an electron moves from the $n = 6$ to the $n = 2$ energy level in the hydrogen atom? In what region of the spectrum is the corresponding line found?

$$4.11 \times 10^{-7} \text{ m}$$

15. Calculate the energy required to excite an electron from the $n = 2$ to the $n = 4$ level of atomic hydrogen. What is the maximum wavelength of light that causes this excitation?

16. A line in the Lyman series of the spectrum of atomic hydrogen has a wavelength of 103 nm. In what energy level is the electron in the excited atoms that give rise to this line?

3

17. How much energy is needed to ionize a hydrogen atom starting (a) from the ground state and (b) from the $n = 2$ state? In each case, find the maximum wavelength of light that ionizes the hydrogen atom.

Matter Waves

18. What is the wavelength of a neutron moving at a speed of 4.21×10^3 m s^{-1}? The mass of a neutron is 1.67×10^{-27} kg.

19. A beam of neutrons has a wavelength of 3.00×10^{-10} m. What is the speed of the neutrons?

20. What wavelength is associated with a neutron moving with a speed of 1.00×10^2 km s^{-1}? What wavelength is associated with an electron moving with a speed of 5.00×10^7 m s^{-1}?

21. Major league pitchers can throw a baseball at a maximum speed of about 95 miles per hour. What is the de Broglie wavelength associated with a 5.0-oz baseball thrown at that speed?

Ionization Energies

22. Arrange the following species in order according to the energy needed to remove the most easily removed electron:

$$\text{He} \quad \text{Li}^+ \quad \text{Ar} \quad \text{Ne} \quad \text{Be}^{2+}$$

23. The ionization energy of helium is 2.37 MJ mol^{-1}. What is the kinetic energy of the photoelectrons produced when helium is irradiated with radiation of wavelength 40.0 nm?

Electron Configurations

24. What is the total number of orbitals associated with the $n = 3$ level?

25. Which of the following atomic orbital designations are not possible?

(a) 6s   (b) 1p   (c) 4d   (d) 2d

26. Explain, with reference to the Pauli exclusion principle, why beryllium cannot have the electron configuration $1s^4$.

27. Without referring to Table 6.2, decide which of the following electron configurations are not allowed by the Pauli exclusion principle. Explain why.

(a) $1s^2 2s^2 2p^4$ (b) $1s^2 2s^2 2p^6 3s^3$
(c) $1s^2 3p^1$ (d) $1s^2 2s^2 2p^6 3s^2 3p^{10}$

28. Without referring to Table 6.2, select the electron configurations that are not ground state configurations:

(a) $1s^2 2p^1$ (b) $1s^2 2s^2 3p^2$
(c) $1s^2 2s^2 2p^5$ (d) $1s^2 2s^2 3p^6 3d^3$

29. Without referring to Table 6.2, write the ground state electron configurations for each of the following atoms:

(a) K   (b) Al
(c) Cl   (d) Ti $(Z = 22)$
(e) Zn $(Z = 30)$ (f) As $(Z = 33)$

30. Write the ground state electron configuration for each of the following atoms and ions:

(a) Be   (b) N   (c) F   (d) Mg
(e) Cl   (f) Ne$^+$ (g) Al^{3+}

31. Without referring to Table 6.2, draw orbital box diagrams for each of the following atoms:

(a) P   (b) Ca   (c) V $(Z = 23)$ (d) Br $(Z = 35)$

32. Referring to a periodic table if necessary, identify the elements that possess each of the following electron configurations:

(a) $1s^2 2s^1$ (b) $1s^2 2s^2 2p^3$
(c) [Ar] $4s^2$ (d) [Ar] $4s^2 3d^2$
(e) [Ar] $4s^2 3d^{10} 4p^3$ (f) [Kr] $5s^2 4d^{10} 5p^5$
(g) [Xe] $6s^2$

33. Use Hund's rule to decide which of the following electron configurations are not ground states:

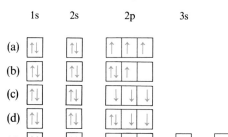

34. Identify the atoms that have the ground state or excited state electron configurations in Problem 33.

Miscellaneous

35. Explain why two hydrogen atoms attract each other when brought close together and combine to form a molecule, whereas two helium atoms repel each other when brought close together and do not form a molecule.

36. Write the ground state electron configuration of boron. How many unpaired electrons are there? How many covalent bonds would you expect boron to form? How do you explain the fact that boron forms the molecule BF_3 and the ion BF_4^-?

37. Write the ground state electron configuration of silicon. How many unpaired electrons are there? What valence would you predict for silicon? How do you explain the fact that silicon forms the compounds $SiCl_4$ and SiH_4?

38. Which of the following molecules and ions have an odd number of electrons?

(a) NO_2 (b) O_3 (c) SO_2
(d) ClO_2 (e) O_2^- (f) NO_2^-

39. Which of the following atoms have unpaired electrons in their ground states?

(a) Li   (b) Be   (c) O   (d) Ne
(e) Na   (f) Ca   (g) P   (h) Zn

CHAPTER 7

PHOSPHORUS AND SULFUR: TWO NONMETALS

Oxidation Numbers and Oxoacids

Group I																	VIII
1 H	II											III	IV	V	VI	VII	He
2 Li	Be		Metals	Nonmetals	Semimetals							B	C	N	O	F	Ne
3 Na	Mg			Transition Elements								Al	Si	15 P 30.97	16 S 32.06	Cl	Ar
4 K	Ca	Sc	Ti	V	Cr	Mn	Fe	Co	Ni	Cu	Zn	Ga	Ge	As	Se	Br	Kr
5 Rb	Sr	Y	Zr	Nb	Mo	Tc	Ru	Rh	Pd	Ag	Cd	In	Sn	Sb	Te	I	Xe
6 Cs	Ba	La	Hf	Ta	W	Re	Os	Ir	Pt	Au	Hg	Tl	Pb	Bi	Po	At	Rn
7 Fr	Ra	Ac	104	105	106	107											

Period

Among the 20 most abundant elements in the earth's crust (Table 3.2), 10 are nonmetals: O, Si, H, Cl, P, C, S, Ar, N, and F. We have already discussed oxygen and hydrogen (Chapter 3) and fluorine and chlorine (Chapter 5). In this chapter we consider two more common nonmetals, phosphorus and sulfur. They are neighboring elements in the third period. Phosphorus is in Group V and sulfur is in Group VI. They are relatively abundant elements, twelfth and fifteenth, respectively. The elements and their compounds have many important practical uses. Sulfuric acid, H_2SO_4, is produced by the chemical industry in a larger quantity than any other substance. Enormous quantities of phosphorus compounds are used as fertilizers, and phosphorus plays a vital role in life processes. Indeed, after carbon, hydrogen, oxygen, and nitrogen, phosphorus and sulfur are the two most abundant nonmetals in living matter.

In addition to the valences of 2 for sulfur and 3 for phosphorus, which can be predicted from their positions in the periodic table, these elements have other important valences. For example, sulfur forms the oxides SO_2 and SO_3. We will see that these additional valences can be explained in terms of the electron configurations we described in Chapter 6. To classify the relatively large number of different compounds formed by these elements, we introduce the concepts of oxidation state and oxidation number, thereby extending and making more quantitative our discussion of oxidation-reduction reactions. Among the compounds of these elements are two important acids—phosphoric acid, H_3PO_4, and sulfuric acid, H_2SO_4—that are examples of an important class of acids, the oxoacids. We discuss the structures and strengths of these acids in this chapter, thereby extending our knowledge of acids and bases and their reactions.

7.1 OCCURRENCE AND PRODUCTION OF ELEMENTS

Sulfur

Sulfur has been known since the beginning of recorded history. In the Bible it is called *brimstone*, which means "the stone that burns." It has been known for so long because in volcanic regions it is found as the free element on the earth's surface, although in small amounts. Important deposits of sulfur are found at depths of 300 m or more in association with sodium chloride and the mineral anhydrite—calcium sulfate, $CaSO_4$. Reduction of calcium sulfate by bacteria probably formed the elemental sulfur in these deposits.

Today the most productive deposits are in Louisiana and Texas. Until recently, they produced about 80% of the world's sulfur, but this percentage is declining. When these sulfur deposits were first discovered, they were considered inaccessible. They are covered by layers of sand, gravel, and mud, which would have made normal mining difficult and dangerous because of the possibility of excavations collapsing. Moreover, the deposits contain considerable quantities of the poisonous gases SO_2 and H_2S.

These problems were solved by the development of the **Frasch process**. A bore is made down to the sulfur, and three concentric pipes are pushed down the bore (see Figure 7.1). Superheated water (150°C) under pressure is pumped down one pipe, melting the sulfur around the bottom of the pipe. Compressed air is then sent down the inner pipe, and a bubbly froth of air, water, and molten sulfur is forced to the surface through the third pipe. At the surface the molten sulfur solidifies. The sulfur obtained by this process is 99.5% pure and is suitable for most commercial purposes.

Compounds of sulfur are also widespread. There are many sulfide minerals such as pyrite, FeS_2, galena, PbS, and chalcocite, Cu_2S, and several common sulfates such as gypsum, $CaSO_4 \cdot 2H_2O$. The sulfides are important sources of metals such as lead and copper. During the smelting operations that extract the metal from the sulfide ore, sulfur dioxide is produced:

$$Cu_2S + O_2 \longrightarrow 2Cu + SO_2$$

When this sulfur dioxide is allowed to escape into the atmosphere, severe pollution and acid rain can result. For example, in the past sulfur dioxide produced by the nickel smelters at Sudbury, Ontario, destroyed the vegetation for many miles around. The landscape was so denuded of vegetation that it was used as a training ground for the astronauts who made the first landing on the moon.

Controls on SO_2 emission are now being introduced, and at least some of the SO_2 is removed from the gases evolved in smelting operations. One method

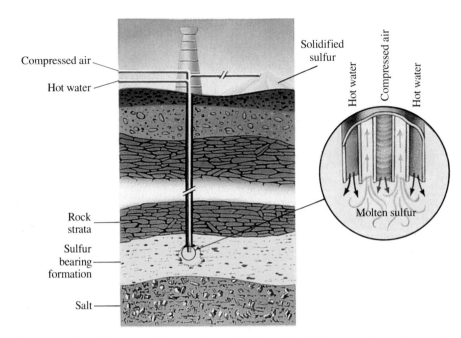

Compressed air

Hot water

Solidified sulfur

Hot water

Compressed air

Hot water

Rock strata

Sulfur bearing formation

Salt

Molten sulfur

Figure 7.1 Extraction of Sulfur by Frasch Process. A bore is made by a drilling rig similar to the rigs used for drilling oil wells; then three concentric pipes are pushed down the bore. Superheated water at about 150°C is forced down the outer pipe. When it has melted the sulfur around the end of the pipe, compressed air is used to blow the mixture of hot water and molten sulfur to the surface. The sulfur, which is immiscible with and denser than water, flows into enormous

for removing SO_2 uses the reaction of SO_2 with H_2S to produce elemental sulfur:

$$2H_2S + SO_2 \longrightarrow 3S + 2H_2O$$

Another source of sulfur of increasing importance is the H_2S present in some natural gas wells; this compound is removed by using its reaction with SO_2. Considerable quantities of sulfur are now produced as a by-product of the purification of natural gas and of smelting operations, so that the Frasch process is becoming less important.

Phosphorus

Phosphorus is too reactive to occur in nature in the free state. Most of the phosphorus in the earth's crust occurs as the minerals fluorapatite, $Ca_5(PO_4)_3F$, hydroxyapatite, $Ca_5(PO_4)_3(OH)$, and chlorapatite, $Ca_5(PO_4)_3Cl$, which are collectively known as phosphate rock. These minerals are all ionic compounds containing calcium ions, Ca^{2+}, phosphate ions, PO_4^{3-} and fluoride, F^-, chloride, Cl^-, or hydroxide, OH^-, ions. Phosphate rock and sulfur are two of the basic substances on which a very large part of modern inorganic chemical industry is based. Other substances of basic importance are salt, $NaCl$, limestone, $CaCO_3$, and sand, SiO_2. These compounds are the starting substances for the manufacture of a wide variety of important products.

Hydroxyapatite is the main constituent of the bones and teeth of animals. Fluoridation of water leads to the replacement of hydroxyapatite, $Ca_5(PO_4)_3(OH)$ in teeth enamel by fluorapatite, $Ca_5(PO_4)_3F$, which is less basic and more resistant to attack by acids. Complex organic compounds of phosphorus are essential constituents of many proteins and of nerve and brain tissue. Phosphate groups are an essential part of the structure of DNA, the storehouse of genetic information, and they play a vital role in the transfer of energy in our bodies. Thus phosphorus is an essential component of our diet. An adult excretes daily phosphorus compounds containing about 2 g of phosphorus. Indeed, phosphorus was discovered by German alchemist Hennig Brand, who obtained it by strongly heating the residue obtained on the evaporation of urine (see Figure 7.2).

Figure 7.2 Discovery of Phosphorus. German alchemist Hennig Brand discovered phosphorous in 1669 when he strongly heated the residue left by the evaporation of urine and observed the striking blue-green light that is emitted by freshly distilled phosphorus. The strong heating of the urine had decomposed its organic compound, producing carbon, which reduced the phosphate present in urine to phosphorus.

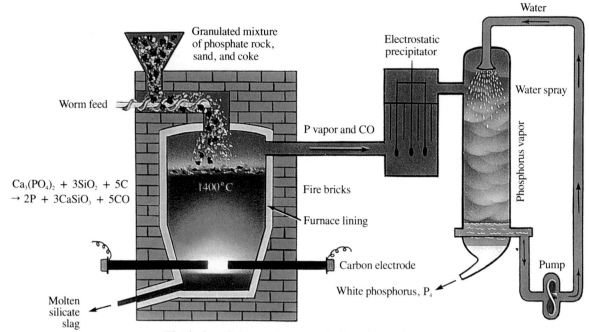

Ca$_3$(PO$_4$)$_2$ + 3SiO$_2$ + 5C
→ 2P + 3CaSiO$_3$ + 5CO

Figure 7.3 Preparation of Phosphorus. A mixture of phosphate rock, coke, and silica sand is fed into the top of the furnace and heated by a large electric current that is passed between two large graphite rods in the lower part of the furnace. The phosphorus vapor passes out of the furnace through an electrostatic precipitator to remove dust and is then condensed by a spray of water at about 70°C. The liquid phosphorus is run off at the bottom of the spray tower into large storage tanks.

The industrial manufacture of phosphorus is based on the reaction that led to the discovery of phosphorus, namely, the reduction of phosphate by carbon. A mixture of phosphate rock, coke, and silica sand is fed into an electric furnace and heated to 1400°–1500°C by passing a large electric current through it (see Figure 7.3). The reactions that occur in the molten mixture are complex but may be summarized by two equations:

$$2Ca_3(PO_4)_2(l) + 6SiO_2(l) \longrightarrow P_4O_{10}(g) + 6CaSiO_3(l)$$
$$P_4O_{10}(g) + 10C(s) \longrightarrow P_4(g) + 10CO(g)$$

The oxide, P$_4$O$_{10}$, formed in the first reaction is reduced to elemental phosphorus by carbon in the second. The gaseous phosphorus emerging from the furnace is condensed by a spray of warm water, and the liquid is pumped to storage tanks through steam-heated pipes, which keep it above its melting point of 44°C.

7.2 PROPERTIES OF THE ELEMENTS

Sulfur can be obtained in several different solid forms. When sulfur is allowed to crystallize at room temperature from a suitable solvent such as carbon disulfide, it is obtained in the form of brilliant yellow crystals of *orthorhombic sulfur* (Figure 7.4a). If orthorhombic sulfur is heated to 95.5°C, it changes to another crystalline form, *monoclinic sulfur*. However, the rate of this transformation is quite slow, and it is simpler to obtain monoclinic sulfur by cooling molten sulfur. Monoclinic sulfur crystallizes from molten sulfur at 119.3°C in the form of long needles (Figure 7.4b). But monoclinic sulfur is not stable at room temperature; it slowly changes back to orthorhombic sulfur.

Allotropes

The different forms of an element are called **allotropes**. Thus orthorhombic sulfur and monoclinic sulfur are two allotropes of sulfur. Their names refer to the different structures of their crystals. Both crystals contain S$_8$ molecules, which consist of a zigzag ring of eight sulfur atoms. The arrangement of these S$_8$

(a)   (b)

Figure 7.4 Crystals of Orthorhombic and Monoclinic Sulfur.
(a) Orthorhombic sulfur is the
stable allotrope of sulfur at room
temperature. It occurs naturally,
sometimes in the form of large
crystals such as the crystal shown
here. (b) The needle-like crystals of
monoclinic sulfur are stable only
above 119°C. They are sometimes
found in the vicinity of volcanoes
where they are formed from the
hot gases emitted by the volcano.

molecules in orthorhombic sulfur is shown in Figure 7.5. The arrangement of
the molecules in monoclinic sulfur is slightly different.

Another form of sulfur is obtained by quickly cooling molten sulfur by pouring it into cold water. A brown rubbery material known as *plastic sulfur* is
obtained, as described in Experiment 7.1. It is not stable and slowly becomes
brittle as it is transformed back into crystalline orthorhombic sulfur.

Plastic sulfur consists of very long chains of sulfur atoms rather than S_8
rings. When liquid sulfur is heated to about 160°C, one of the S—S bonds in
some of the rings break, and the rings open to give chains of eight sulfur atoms:

EXPERIMENT 7.1

Plastic Sulfur

Sulfur is heated until it melts
to form an orange liquid.

On further heating the color
becomes dark red and the
liquid becomes very viscous.

When the liquid is rapidly
cooled by pouring it into
cold water, a rubbery brown
solid—plastic sulfur—is
formed.

Plastic sulfur is elastic, like
rubber.

Figure 7.5 The Structure of Orthorhombic Sulfur. Both orthorhombic and monoclinic sulfur consist of crown shaped rings of eight sulfur atoms. In orthorhombic sulfur these molecules are packed together as shown here. In monoclinic sulfur the arrangement of the S_8 molecules is slightly different.

The sulfur atoms at each end of the chain have only seven valence electrons. Therefore they have a strong tendency to attract an additional electron and are very reactive. An S_8 chain reacts with an S_8 ring causing it to open up and thereby forming a 16-atom chain. But the sulfur atoms at the ends of the chain still have only seven electrons, so the process continues, leading to the formation of very long chains of thousands of atoms. As the temperature is raised from 160° to 190°, the chains become longer and more tangled, and the liquid becomes increasingly thick and sticky, like maple syrup or molasses. It is described as being very **viscous**. A liquid like water that flows easily has a low **viscosity**, while a liquid like molasses or heavy engine oil that flows slowly is said to have a high viscosity.

The behavior of liquid sulfur on heating is unusual. Most liquids become less viscous on heating, because the increased thermal motion of the molecules enables them to move past each other more easily. Liquid sulfur, however, becomes more viscous. When this viscous liquid sulfur is cooled rapidly—for example, by pouring it into water, with which it does not react—the long tangled chains do not have time to rearrange to the more stable cyclic S_8 molecules. So the rubbery solid that is obtained is not crystalline. Its molecules do not have the ordered arrangement that is characteristic of a crystalline solid. They still retain the random arrangement that is characteristic of liquids. Solids that are not crystalline are called **amorphous solids**. Plastic sulfur exhibits properties similar to those of rubber, which also consists of long chain molecules. Long chain molecules tend to coil up into compact shapes. If plastic sulfur or rubber is stretched, the coiled-up molecules straighten out a little, and the material stretches. If the stretching force is removed, the molecules tend to resume their coiled-up structure and the material contracts again (see Figure 7.6).

There are still other, less important allotropes of sulfur that contain rings of six, seven, twelve, and more sulfur atoms.

Like sulfur, phosphorus also has several allotropes. The best-known allotrope is *white phosphorus*, which is the form obtained when phosphorus vapor is condensed. It is a colorless crystalline solid, with a melting point of 44°C, but it rapidly becomes white and opaque unless stored under nitrogen in the dark. White phosphorus is very toxic. The vapor causes decay of cartilage and bones, particularly of the nose and jaw.

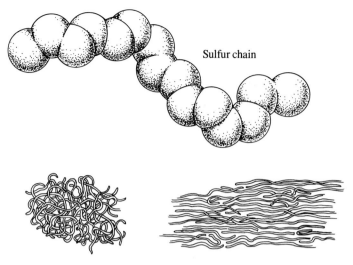

Sulfur chain

Coiled-up molecules

Stretched chains

Figure 7.6 Plastic Sulfur. Plastic sulfur is an amorphous (noncrystalline) allotrope of sulfur consisting of long chains of sulfur atoms. These chains have an irregular, disordered arrangement rather than the regular arrangement characteristic of crystalline substances. The chains tend to coil up and form a tangled mass. When a force is applied, the molecules straighten out a little but coil up again when the force is removed. So the solid can stretch and contract like rubber.

White phosphorus reacts with oxygen at room temperature, emitting a bluish green light (see Experiment 7.2). Some of the energy produced in the reaction is emitted as radiation. This phenomenon is known as **chemiluminescence**, which is the emission of the energy released by a reaction as light rather than as heat. At temperatures above 40°C the oxidation becomes quite rapid, and phosphorus ignites. So that the danger of fire is avoided, white phosphorus is stored away from contact with the air; usually, it is stored under water, in which it is insoluble. It is quite soluble, however, in certain other solvents such as carbon disulfide, CS_2. White phosphorus consists of tetraatomic P_4 molecules that have a tetrahedral structure (Figure 7.7a). Each of the phosphorus atoms forms three bonds to neighboring atoms and has an unshared pair of electrons.

There are several other allotropes of phosphorus; the best-known are *red phosphorus* and *black phosphorus*. Red phosphorus can be obtained by heating

EXPERIMENT 7.2

Oxidation of White Phosphorus: Chemiluminescence

When water containing a piece of white phosphorus is boiled, the jet of steam coming from the tube at the top of the flask glows in a subdued light with a blue-green color, as the phosphorus vapor in the steam is oxidized by the oxygen in the air. Some of the energy produced by the oxidation of phosphorus is in the form of light rather than heat.

7.2 PROPERTIES OF THE ELEMENTS

239

Figure 7.7 Structures of White and Black Phosphorus. (a) White phosphorus consists of tetrahedral P_4 molecules. (b) Black phosphorus also has a pyramidal arrangement of three bonds around each phosphorus atom, but the atoms are connected to form a corrugated sheet of atoms that extends indefinitely in two dimensions—in other words, a giant two-dimensional molecule. In the crystal of black phosphorus these sheets of atoms are stacked one upon another. When black phosphorus is vaporized, this structure is broken up and P_4 molecules are formed.

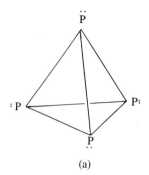

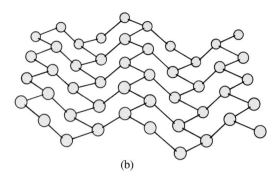

(a) (b)

Red and white phosphorus

white phosphorus in the absence of air at atmospheric pressure. Black phosphorus is obtained by heating white or red phosphorus under a very high pressure. Both allotropes have much higher melting points and boiling points than white phosphorus. For example, red phosphorus sublimes (goes directly to vapor) at 417°C and only melts under pressure at 540°C. Neither red nor black phosphorus ignites spontaneously in air, and they are much less reactive and less poisonous than white phosphorus. These differences result from differences in their structures. Red and black phosphorus do not consist of P_4 molecules but have polymeric, giant-molecule structures (Figure 7.7b). To melt and vaporize a giant-molecule structure requires that covalent bonds be broken, so the melting point and boiling point are high. The vapor obtained from both red and black phosphorus consists of P_4 molecules; when the vapor is condensed, white phosphorus is obtained.

Multiple Bonding

Although both phosphorus and nitrogen are in Group V, nitrogen occurs as N_2 molecules containing a triple bond, whereas the allotropes of phosphorus all have single bonds. Similarly, oxygen exists as diatomic molecules with a double bond between the oxygen atoms, whereas elemental sulfur normally occurs as an eight-atom ring with only single bonds. The S_2 molecule with a double bond can be formed, but only in gaseous sulfur at very high temperatures.

These differences between nitrogen and phosphorus and between oxygen and sulfur reflect an important contrast between the second-period elements and those in the following periods. The second-period elements—especially carbon, nitrogen, and oxygen—have a much greater tendency to form double and triple bonds. In fact, in all but a very few cases at least one of the elements involved in a double or triple bond is C, N, or O. This tendency is not completely understood. It seems that the only elements that commonly form stable molecules with multiple bonds are the small, highly electronegative elements.

7.3 COMPOUNDS OF SULFUR

Oxides

From its position in the periodic table, we might expect sulfur to have a valence of 2 and to form an oxide with the formula SO. This oxide is formed in very small amounts as a very unstable, reactive molecule when an electric discharge is passed through sulfur dioxide.

When sulfur is heated in air, it ignites at 350°C and burns with a blue flame to produce *sulfur dioxide*, SO_2 rather than sulfur monoxide, SO (Experiment 3.1). Sulfur dioxide is a colorless gas with a pungent, choking odor. It destroys bacteria and is used as a preservative in the storage of fruits, such as apples, and in the preparation of dried fruits, such as prunes and apricots. It condenses to a liquid at $-10°C$; at $20°C$ it can be liquefied by a pressure of about 3 atm. It is usually sold as a liquid under pressure in metal cylinders.

When sulfur dioxide is heated with oxygen, in the presence of finely divided platinum metal or vanadium pentoxide, V_2O_5, another oxide, sulfur trioxide, SO_3, is formed:

$$2SO_2 + O_2 \longrightarrow 2SO_3$$

The platinum or vanadium pentoxide acts as a *catalyst* for the reaction. Recall that a *catalyst* is a substance that increases the rate at which a reaction occurs but is not used up in the reaction. Sulfur trioxide condenses to a colorless liquid at $44.5°C$ and freezes to transparent crystals at $16.8°C$. It forms dense white fumes in the air.

Higher Valences of Sulfur and Phosphorus

The formulas SO_2 and SO_3 show that sulfur can exhibit valences of 4 and 6 in addition to the expected valence of 2, which it has in compounds such as H_2S and SCl_2. Indeed, sulfur forms many compounds in which it exhibits these *higher valences*. These compounds include the sulfite ion, SO_3^{2-}, the sulfate ion, SO_4^{2-}, sulfuric acid, H_2SO_4, and the fluorides SF_4 and SF_6.

Similarly, from its position in the periodic table we expect phosphorus to have a valence of 3, which it does in the oxide P_4O_6 (empirical formula P_2O_3) and the chloride PCl_3. But phosphorus also forms an oxide P_4O_{10} (empirical formula P_2O_5), the chloride PCl_5, and phosphoric acid, H_3PO_4. In each of these compounds phosphorus has a valence of 5.

To understand the higher valences of phosphorus and sulfur, we must look at their electron configurations (Table 7.1). They are just like those of the corresponding elements of the second period, nitrogen and oxygen, except that the electrons occupy 3s and 3p rather 2s and 2p orbitals.

Thus sulfur has the following ground state electron configuration:

3s	3p

$\boxed{\uparrow\downarrow}$ $\boxed{\uparrow\downarrow\,|\,\uparrow\,|\,\uparrow}$

Like oxygen, sulfur is therefore expected to form two covalent bonds as it does in H_2S. However, the $n = 3$ shell also has five 3d orbitals which are not occupied in the ground state of sulfur:

3s	3p	3d

$\boxed{\uparrow\downarrow}$ $\boxed{\uparrow\downarrow\,|\,\uparrow\,|\,\uparrow}$ $\boxed{\;|\;|\;|\;|\;}$

If an electron is promoted from a 3p orbital to one of the 3d orbitals we obtain the excited or valence state configuration:

3s	3p	3d

$\boxed{\uparrow\downarrow}$ $\boxed{\uparrow\,|\,\uparrow\,|\,\uparrow}$ $\boxed{\uparrow\,|\;|\;|\;|\;}$

In this valence state there are four electrons in singly occupied orbitals, that is, four unpaired electrons. These electrons can be used to form four covalent bonds, as in SO_2 and SF_4.

Table 7.1 Ground State Electron Configurations of Third-Period Elements

Na	[Ne] $3s^1$
Mg	[Ne] $3s^2$
Al	[Ne] $3s^2 3p^1$
Si	[Ne] $3s^2 3p^2$
P	[Ne] $3s^2 3p^3$
S	[Ne] $3s^2 3p^4$
Cl	[Ne] $3s^2 3p^5$
Ar	[Ne] $3s^2 3p^6$

Promotion of both a 3s and a 3p electron to a 3d orbital gives another valence state in which there are six electrons in singly occupied orbitals:

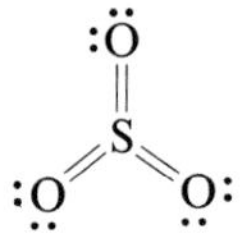

These six unpaired electrons can be used to form six covalent bonds, as in the compounds SO_3 and SF_6.

Thus, because the $n = 3$ shell can contain a maximum of 18 rather than 8 electrons as in the $n = 2$ shell, sulfur can form compounds such as SO_2 and SF_4 in which it has 10 electrons in its valence shell and SO_3 and SF_6 in which it has 12 electrons in its valence shell. As we have already discussed in Chapter 6 in the case of carbon, these excited or valence states of the sulfur atom are not formed before compound formation begins but rather four or six bonding orbitals are formed as the oxygen or fluorine atoms combine with the sulfur atom. Nevertheless, it is convenient to think of these higher valence compounds as being formed from the corresponding valence state because we can then readily see how many bonds will be formed.

Phosphorus has the following ground state configuration:

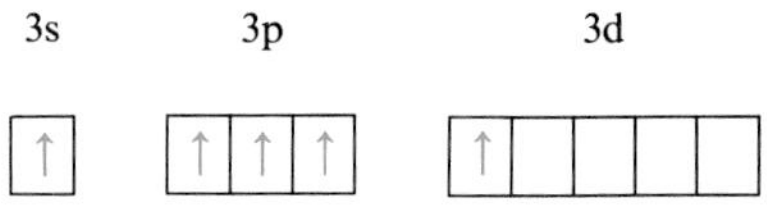

This valence state has five unpaired electrons. Phosphorus in this state can therefore form five covalent bonds, as in PCl_5 and P_4O_{10}.

If an electron is promoted from the 3s orbital to a 3d orbital, the electron configuration becomes

In writing Lewis structures for the higher-valence compounds of sulfur and phosphorus we must remember that they do not obey the octet rule; they may have 10 or 12 electrons in their valence shells. For example, in the case of SO_2 we can use 2 of the 6 sulfur electrons to form two bonds to an oxygen atom and 2 more electrons to form two bonds to the second oxygen atom, which leaves the 2 remaining electrons as an unshared pair. Thus the sulfur atom in SO_2 has a pair of nonbonding electrons and four shared pairs of electrons, making a total of five pairs, or 10 electrons, in its valence shell:

$$:\ddot{S}: + 2:\ddot{O}\cdot \longrightarrow :\ddot{O}::\ddot{S}::\ddot{O}: \quad \text{or} \quad :\ddot{O}=\ddot{S}=\ddot{O}:$$

In the molecule SO_3 sulfur uses all 6 of its valence electrons in the formation of three double bonds to three oxygen atoms:

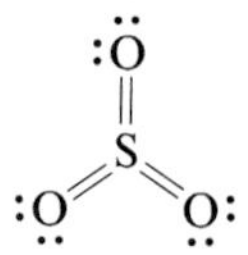

It therefore has six pairs, or 12 electrons, in its valence shell.

Gaseous sulfur trioxide consists partly of this molecule and partly of a trimeric molecule, S_3O_9, which has the ring structure in which each sulfur atom forms six bonds. The liquid consists very largely of S_3O_9 molecules. Sulfur trioxide reacts with water to give sulfuric acid, H_2SO_4, in which sulfur again forms six bonds:

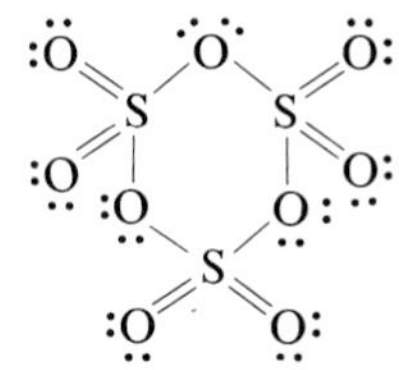

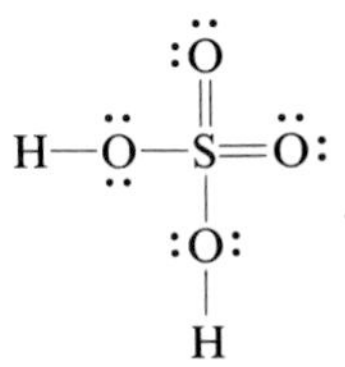

In both SO_2 and SO_3, and indeed in all its other covalent compounds, oxygen forms only two bonds and always has a valence shell of eight electrons. The valence shell of oxygen, the $n = 2$ shell, is completely filled by eight electrons, and oxygen always obeys the octet rule. In fact, the only elements that invariably obey the octet rule in compound formation are carbon, nitrogen, oxygen, and fluorine. All heavier elements obey the octet rule only in the compounds in which they exhibit their normal valence. In compounds in which they have higher valences, the valence shell may contain 10, 12, or even more electrons.

Phosphorus, sulfur, and other elements in periods 3–7 normally exhibit their higher valences only in combination with very electronegative elements such as oxygen, fluorine, and chlorine. To understand why, recall that a d orbital has a higher energy than an s or a p orbital in the same shell, and an electron in a d orbital is at a correspondingly greater distance from the nucleus. Energy must be supplied to move the electrons to a higher energy level, that is, to pull the electron away from the nucleus. Only the atoms of elements with high electronegativities, such as fluorine, oxygen, and chlorine, appear to be able to attract the valence shell electrons of phosphorus and sulfur sufficiently strongly to promote one or more of these electrons from an s or p orbital into a d orbital.

Industrial Preparation of Sulfuric Acid: Contact Process

Sulfuric acid may be described as the world's most important industrial chemical; it is produced in larger quantities than any other substance and has many uses. The uses of sulfuric acid are so important and so varied that it has been said that a country's sulfuric acid production is a good measure of its industrial development. The U.S. production in 1980 was 40 million tons. About half of this amount is used for the production of phosphate fertilizers. Other uses include the manufacture of paints, dyes, explosives, detergents, synthetic fibers, and other chemicals.

The industrial preparation of sulfuric acid involves three reactions that we have already encountered:

1. The burning of sulfur or a metal sulfide in air to give SO_2:

$$S + O_2 \longrightarrow SO_2$$
$$CuS + O_2 \longrightarrow Cu + SO_2$$

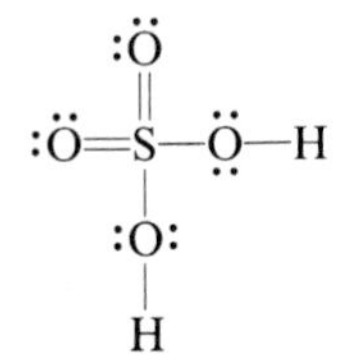

$$\begin{array}{c} \overset{\displaystyle :\ddot{O}}{\underset{\displaystyle :\ddot{O}:}{\overset{\|}{:\ddot{O}=S}-\ddot{O}}-H} \\ \phantom{:\ddot{O}=S}| \\ \phantom{:\ddot{O}=S}H \end{array}$$

2. The oxidation of SO_2 to SO_3:

$$2SO_2 + O_2 \xrightarrow{\text{Catalyst}} 2SO_3$$

3. The combination of SO_3 with water to give H_2SO_4:

$$SO_3 + H_2O \longrightarrow H_2SO_4$$

The oxidation of SO_2 to SO_3 is very slow at ordinary temperatures. To increase the reaction rate this oxidation is carried out at approximately 400°C in the presence of a catalyst, usually vanadium pentoxide, V_2O_5 (see Experiment 7.3). The process is known as the **contact process** because the reaction takes place when SO_2 and O_2 molecules come into contact on the surface of the solid V_2O_5 catalyst. The SO_3 is not allowed to react directly with water, because this is a violent reaction that produces a dense mist of H_2SO_4 droplets. Instead, the gaseous SO_3 is absorbed in 98% H_2SO_4, and water is added at a controlled rate to keep the concentration of H_2SO_4 at approximately 98%. This solution is the acid that is normally sold as *concentrated sulfuric acid*.

If no water is added, the concentration of the sulfuric acid rises to 100%, and then a series of *polysulfuric acids* are formed:

$$H_2SO_4 + SO_3 \longrightarrow H_2S_2O_7$$

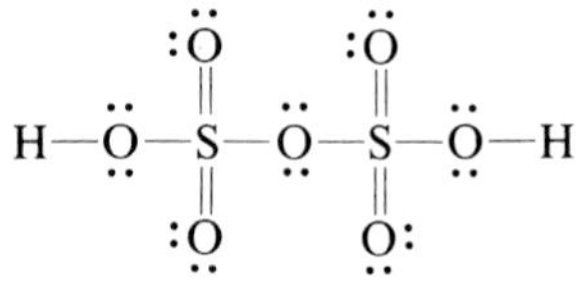

Disulfuric acid

$$H_2SO_4 + 2SO_3 \longrightarrow H_2S_3O_{10}$$

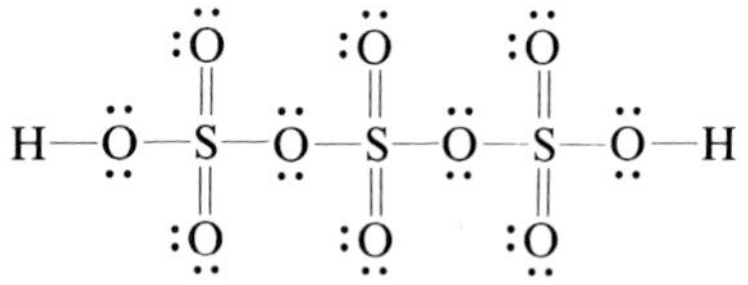

Trisulfuric acid

A mixture of sulfuric acid and polysulfuric acids is often called *oleum* or *fuming sulfuric acid*.

EXPERIMENT 7.3

Preparation of SO₃: Contact Process

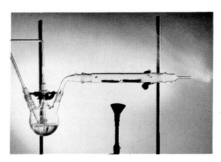

Colorless SO_2 and O_2 gases, dried by passing them through concentrated H_2SO_4, are passed over hot V_2O_5 catalyst. The SO_3 forms dense white fumes when it comes into contact with the moisture in the atmosphere. The SO_3 vapor condenses to form colorless, needle-shaped crystals on the inside of a cool, dry flask.

In the presence of a trace of moisture SO_3 forms white, needlelike crystals of β-SO_3, which is actually a mixture of long chain polysulfuric acids,

$$H-\overset{..}{\underset{..}{O}}\left(\overset{\overset{..}{\underset{}{O}}}{\underset{\underset{..}{O}}{\overset{\|}{\underset{\|}{S}}}}-\overset{..}{\underset{..}{O}}\right)_{\!\! n}H$$

where n is very large ($\approx 10^5$). The dense white fumes formed by SO_3 in (damp) air consist of this solid (see Experiment 7.3).

Properties and Reactions of Sulfuric Acid: Le Châtelier's Principle

SOLUTIONS IN WATER   Sulfuric acid dissolves in water with the evolution of a large amount of heat. Consequently, the dilution of concentrated sulfuric acid must be carried out with care. If water is added to the concentrated acid, the heat of the reaction may be sufficient to raise the temperature of the water to its boiling point and cause drops of the acid to be thrown violently out of the container. The only safe procedure is to add the concentrated acid slowly, with constant stirring, to a large amount of cold water. The heat is produced by the very exothermic reaction of sulfuric acid with water.

Sulfuric acid is a strong acid and is fully ionized:

$$H_2O + H_2SO_4 \longrightarrow H_3O^+ + HSO_4^-$$

Since sulfuric acid has two OH groups, it has two ionizable hydrogen atoms and it can donate a second proton to another water molecule:

$$HSO_4^- + H_2O \longrightarrow H_3O^+ + SO_4^{2-}$$

An acid that has two ionizable hydrogen atoms is known as a **diprotic acid**.

The anions formed in these two reactions are the hydrogen sulfate ion, HSO_4^-, and the sulfate ion, SO_4^{2-}. The HSO_4^- ion is a weak acid and its reaction with water is not complete. The Lewis structures for the hydrogen sulfate ion and the sulfate ion are

$$H\overset{..}{\underset{..}{O}}-\overset{\overset{\overset{..}{O}:}{\|}}{\underset{\underset{:\underset{..}{O}:^{\ominus}}{|}}{S}}=\overset{..}{\underset{..}{O}}: \quad \text{and} \quad {}^{\ominus}:\overset{..}{\underset{..}{O}}-\overset{\overset{\overset{..}{O}:}{\|}}{\underset{\underset{:\underset{..}{O}:^{\ominus}}{|}}{S}}=\overset{..}{\underset{..}{O}}:$$

A solution of sulfuric acid in water contains H_3O^+, HSO_4^-, and SO_4^{2-}. The acidic properties of such a solution are those of H_3O^+, and they are the same therefore as those of an aqueous solution of any strong acid such as hydrochloric acid. When we speak of hydrochloric acid we always mean a solution of hydrogen chloride in water. Hydrogen chloride is a gas, which condenses to a liquid only at $-85°C$. The pure liquid is therefore not commonly encountered and is not of any great importance.

The situation is very different, however, for sulfuric acid. The name *sulfuric acid* is itself ambiguous since it is used both for the pure substance H_2SO_4 and for solutions of H_2SO_4 in water. Unlike pure liquid HCl, pure liquid H_2SO_4 is easily obtained. It is a colorless liquid with a melting point of 10.4°C. The 98% H_2SO_4 that is produced industrially and is commonly used in the laboratory is a mixture of water and sulfuric acid, but since the latter is in a large excess, it is best regarded as a solution of water in sulfuric acid rather than as a solution of sulfuric acid in water. It contains H_2SO_4 molecules and some H_3O^+ and HSO_4^-, but no un-ionized H_2O molecules. Its properties are quite similar to those of the 100% pure H_2SO_4 that consists only of H_2SO_4 molecules.

Concentrated (98%) sulfuric acid has four important properties, which are utilized in a number of important ways:

1. It has a high boiling point of 338°C.
2. It is a strong acid.
3. It is a dehydrating agent.
4. It is a strong oxidizing agent.

First we consider some uses of sulfuric acid that depend on its boiling point.

PREPARATION OF HYDROGEN CHLORIDE   Concentrated sulfuric acid reacts with sodium chloride to give hydrogen chloride, which is evolved as a gas from the reaction mixture:

$$NaCl(s) + H_2SO_4(l) \rightleftharpoons HCl(g) + NaHSO_4(s)$$

In this reaction sulfuric acid *protonates*—that is, adds a proton to—the chloride ion:

$$Cl^- + H_2SO_4 \rightleftharpoons HCl + HSO_4^-$$

In principle, this reaction should come to equilibrium, but since hydrogen chloride is a gas, if the reaction is not carried out in a closed container, hydrogen chloride is evolved as it is formed and is lost from the reaction mixture. Thus more and more hydrogen chloride is formed until the reaction goes nearly to completion (see Experiment 7.4).

PREPARATION OF NITRIC ACID   Nitric acid, HNO_3, can be prepared in a similar manner by heating sodium nitrate with concentrated sulfuric acid:

$$NaNO_3(s) + H_2SO_4(l) \rightleftharpoons HNO_3(l) + NaHSO_4(s)$$

In this reaction sulfuric acid protonates the nitrate ion:

$$NO_3^- + H_2SO_4 \rightleftharpoons HNO_3 + HSO_4^-$$

Unlike hydrogen chloride, which is a gas, nitric acid is a liquid with a boiling point of 86°C. It can be distilled from the reaction mixture by heating and is collected as a pale yellow liquid. Because sulfuric acid has a very high boiling point, it does not distill at the same time (see Experiment 7.4).

LE CHÂTELIER'S PRINCIPLE   The preparations of hydrogen chloride and nitric acid are examples of the effect of removing one of the substances taking part in an equilibrium. In any chemical equilibrium the reaction between the reactants to give the products, the forward reaction, is proceeding at the same rate as the reaction between the products to give back the reactants. If one of the products is removed, or if its concentration is decreased in some way, then the rate of the back reaction is decreased. The rate of the forward reaction is unchanged and is therefore greater than the rate of the back reaction. Hence the concentration of the products increases until equilibrium is once again established.

$Ba(NO_3)_2 + Na_2SO_4 \rightarrow Ba SO_4 \downarrow + 2Na NO_3$

Preparation of Volatile Acids

When sodium nitrate is heated with concentrated sulfuric acid, nitric acid is formed. The nitric acid distills into the test tube and collects as a pale yellow liquid. Some brown NO_2 vapor is also formed by decomposition of the nitric acid in the hot flask.

This is a close-up view of the pale yellow liquid nitric acid. Pure nitric acid is colorless but here it is colored yellow by the small amount of nitrogen dioxide, NO_2, which was formed.

When concentrated sulfuric acid is added to sodium chloride, hydrogen chloride gas is evolved, causing effervescence. When the $HCl(g)$ comes in contact with a swab soaked in aqueous ammonia solution that is held at the mouth of the tube, white fumes of solid ammonia chloride are formed.

In the preparation of nitric acid it is removed from the equilibrium mixture by vaporizing it on heating the solution; more nitric acid is then formed until equilibrium is again established. Continued heating causes a continual removal of nitric acid, which therefore leads to the continual formation of more nitric acid. In this way the reaction can be driven very nearly to completion.

In general, for any system in chemical equilibrium, the effect of a change in the concentration of any of the reacting substances is to cause the equilibrium to shift so as to minimize the concentration change.

Thus if the concentration of one of the products is decreased, then more product is formed. Conversely, if the concentration of one of the reactants is decreased, then more reactant will form and there will be less product at equilibrium. This statement is a special case of a general principle first stated by French chemist Henri le Châtelier (1850–1936), which describes how systems at equilibrium react to changes in any of the factors that affect the equilibrium. **Le Châtelier's principle** is as follows:

When any of the conditions that affect the position of a dynamic equilibrium are changed, the position of the equilibrium shifts so as to minimize the effect of the change.

These conditions include the concentrations of the reactants and the products, the temperature, and the pressure. In this chapter we will consider several examples of the effect of changing the concentrations of the reactants or products.

7.3 COMPOUNDS OF SULFUR

247

SULFATES Sulfuric acid is a strong diprotic acid which reacts with bases to form two series of salts, the hydrogen sulfates and the sulfates:

$$H_2SO_4(aq) + KOH(aq) \longrightarrow KHSO_4(aq) + H_2O$$
$$\text{Potassium hydrogen sulfate}$$

$$KHSO_4(aq) + KOH(aq) \longrightarrow K_2SO_4(aq) + H_2O$$
$$\text{Potassium sulfate}$$

The overall reaction is

$$H_2SO_4(aq) + 2KOH(aq) \longrightarrow K_2SO_4(aq) + 2H_2O$$

Metal hydrogen sulfates and sulfates are ionic compounds containing metal ions and HSO_4^- or SO_4^{2-}. Common soluble sulfates include $Na_2SO_4 \cdot 10H_2O$, $(NH_4)_2SO_4$, $MgSO_4 \cdot 7H_2O$ (Epsom salts), $CuSO_4 \cdot 5H_2O$, $ZnSO_4 \cdot 7H_2O$, and some sulfates containing two different positive ions, such as $(NH_4)_2Fe(SO_4)_2 \cdot 6H_2O$ and $KAl(SO_4)_2 \cdot 12H_2O$ (alum), which are known as double salts.

Some insoluble sulfates—such as $CaSO_4 \cdot 2H_2O$ (gypsum), $BaSO_4$ (barite), and $PbSO_4$—occur as minerals. Barium sulfate is very insoluble, and its formation as a white precipitate on addition of an acidic aqueous solution of $BaCl_2$ to a solution indicates the presence of sulfate in that solution:

$$Ba^{2+}(aq) + SO_4^{2-}(aq) \longrightarrow BaSO_4(s)$$

SULFURIC ACID AS A DEHYDRATING AGENT Water is completely ionized in solution in sulfuric acid. It behaves as a strong base:

$$H_2O + H_2SO_4 \longrightarrow H_3O^+ + HSO_4^-$$
$$\text{Concentrated 98\%}$$

This property makes sulfuric acid a very good **dehydrating agent**. Gases that do not react with sulfuric acid, such as O_2, N_2, CO_2, and SO_2, may be dried by bubbling them through concentrated sulfuric acid.

Hydrated salts, such as $CuSO_4 \cdot 5H_2O$, lose their water of crystallization when stored in a closed container (*desiccator*) with concentrated sulfuric acid. A small amount of water vapor is in equilibrium with the hydrated salt, and as this vapor is absorbed by the sulfuric acid, more of the salt gives up its water of crystallization. Thus slowly all the water is removed:

$$CuSO_4 \cdot 5H_2O \rightleftharpoons CuSO_4 + \qquad 5H_2O$$
$$\text{Blue} \qquad\qquad \text{White} \qquad \text{Absorbed by sulfuric acid}$$
$$(H_2O + H_2SO_4 \longrightarrow H_3O^+ + HSO_4^-)$$

This reaction is another example of the operation of Le Châtelier's principle. The copper sulfate that results is a white powder and is called *anhydrous* copper sulfate. It can be reconverted to the blue hydrated salt simply by addition of water. This reaction can be used as a test for water.

The tendency of sulfuric acid to combine with water is so strong that it will remove hydrogen and oxygen in a 2:1 atomic ratio from many compounds that do not contain water in a molecular form. For example, it removes hydrogen and oxygen as water from carbohydrates and many other organic compounds, leaving behind a charred residue of carbon. Wood, paper, starch, cotton, and sugar (sucrose) are all dehydrated in this way (Experiment 7.5).

$$C_{12}H_{22}O_{11} + 11H_2SO_4 \longrightarrow 12C(s) + 11H_3O^+ + 11HSO_4^-$$

Sulfuric Acid as a Dehydrating Agent

Concentrated sulfuric acid is added to sugar, $C_{12}H_{22}O_{11}$.

The concentrated sulfuric acid removes water from the sugar to form black carbon and steam.

The steam mixes with the carbon to form a black pillar which rises high above the beaker as the mixture heats and expands.

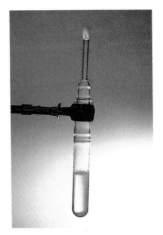

Concentrated sulfuric acid dehydrates methanoic (formic) acid, HCO_2H, to form carbon monoxide, CO, which can be ignited and burns with a blue flame.

Methanoic acid (formic acid) is dehydrated to give carbon monoxide:

$$HCO_2H + H_2SO_4 \longrightarrow CO + H_3O^+ + HSO_4^-$$

Ethene, C_2H_4, may be prepared by heating ethanol, C_2H_5OH, with sulfuric acid:

$$C_2H_5OH + H_2SO_4 \longrightarrow C_2H_4 + H_3O^+ + HSO_4^-$$

SULFURIC ACID AS AN OXIDIZING AGENT Concentrated aqueous H_2SO_4 is a strong oxidizing agent. For example, it will oxidize Br^- and I^- to Br_2 and I_2, respectively (see Experiment 7.6):

$$2Br^- + 5H_2SO_4(l) \longrightarrow Br_2(g) + SO_2(g) + 2H_3O^+(soln) + 4HSO_4^-(soln)$$

We use solution (soln) here, and in some of the following equations, to indicate that the SO_4^{2-} is in solution. But it is a solution in concentrated H_2SO_4 and not a solution in water (aqueous solution, aq). Because Br^- is oxidized by concentrated H_2SO_4, the reaction of a bromide with H_2SO_4 cannot be used to prepare HBr in the same way as HCl can be prepared from a metal chloride (see Experiment 7.4). Although some HBr is formed, a large amount of Br_2 is formed at the same time.

When carbon is heated with concentrated sulfuric acid, it is oxidized to carbon dioxide:

$$C(s) + 2H_2SO_4(l) \longrightarrow CO_2(g) + 2H_2O(l) + 2SO_2(g)$$

When copper is heated with concentrated sulfuric acid, it is oxidized to Cu^{2+}:

$$Cu(s) + 5H_2SO_4(l) \longrightarrow Cu^{2+}(soln) + 4HSO_4^-(soln) + 2H_3O^+(soln) + SO_2(g)$$

In all these cases we note that the other product of the reaction is sulfur dioxide. Sulfuric acid is reduced to sulfur dioxide.

Methanoic acid
(formic acid)

Ethanol

Ethene

7.3 COMPOUNDS OF SULFUR

Sulfuric Acid as an Oxidizing Agent

Concentrated sulfuric acid is added to white solid sodium bromide and white solid sodium iodide.

The sulfuric acid oxidizes the sodium bromide to reddish-brown bromine and the sodium iodide to violet iodine.

Hydrochloric acid and aqueous solutions of sulfuric acid dissolve some metals, but they do not dissolve copper. In these cases the other product of the reaction is hydrogen and not sulfur dioxide. The reactions of aqueous acids with metals are reactions of the hydronium ion, in which it is reduced to hydrogen, as, for example, in the reaction of magnesium with dilute aqueous sulfuric acid:

$$Mg(s) + 2H_3O^+(aq) + SO_4^{2-}(aq) \longrightarrow Mg^{2+}(aq) + SO_4^{2-}(aq) + H_2(g) + 2H_2O(l)$$

Or, more simply, canceling spectator ions, we have

$$Mg(s) + 2H_3O^+(aq) \longrightarrow Mg^{2+}(aq) + H_2(g) + 2H_2O(l)$$

When concentrated sulfuric acid behaves as an oxidizing agent, the reaction is different. Sulfuric acid molecules, not hydronium ions, are the oxidizing agent, and the reduction product is not hydrogen but sulfur dioxide or, in some reactions, other sulfur compounds such as hydrogen sulfide. Copper is not oxidized by the hydronium ion, but it is oxidized by sulfuric acid, which is a stronger oxidizing agent than the hydronium ion.

We recognize the conversion of copper metal to copper ion and the conversion of bromide ion to bromine as oxidations because they involve the loss of electrons. But it is not so clear that the conversion of sulfuric acid to sulfur dioxide is a reduction, because it is not obvious that this reaction involves a gain of electrons. Similarly, although the conversion of carbon to carbon dioxide is an oxidation in the original sense of addition of oxygen, it is not obvious that this reaction involves a loss of electrons. To help us to understand how these reactions may be regarded as oxidations and reductions in terms of electron loss and gain, and to help us identify the substances that are oxidized and reduced, we use the concept of oxidation numbers.

Oxidation Numbers

The **oxidation number** is a measure of the extent of oxidation of an element in its compounds and is assigned by using a set of simple rules. The free element is the standard state from which the extent of oxidation and reduction is measured. *An element in any of its allotropic forms is assigned an oxidation number of 0.*

A singly charged positive ion formed by the loss of one electron from a neutral atom is assigned an oxidation number of $+1$. Thus sodium in Na^+ has an oxidation number of $+1$. A doubly charged positive ion formed by the loss of two electrons from an atom is assigned an oxidation number of $+2$, and so on. Singly charged, monatomic, negative ions, such as Cl^-, are assigned an oxidation number of -1, and doubly charged negative ions, such as O^{2-}, have an oxidation number of -2. *The oxidation number of a monatomic ion is equal to its charge.*

But how do we assign oxidation numbers in covalent compounds such as SO_2 and polyatomic ions such as SO_4^{2-}? Consider the reaction

$$H_2 + Cl_2 \longrightarrow 2HCl$$

Because chlorine is more electronegative than hydrogen, hydrogen chloride is a polar molecule. The chlorine atom has a small negative charge, and the hydrogen atom has a small positive charge:

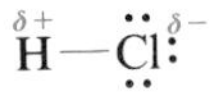

$$\overset{\delta+}{H}-\overset{\delta-}{\underset{..}{\overset{..}{Cl}}}:$$

There has been some transfer of electron density from hydrogen to chlorine; the hydrogen has been oxidized and the chlorine reduced, but less than one electron has been transferred. Assigning fractional oxidation numbers to the hydrogen and chlorine atoms would be too complicated and would not be very useful. So we define the oxidation number as the charge that *would be* associated with a particular atom *if* both electrons of each bond were transferred completely to the atom of higher electronegativity. Thus in hydrogen chloride the bonding electron pair is assigned to chlorine because it is more electronegative than hydrogen. Consequently, the Cl atom is considered to be Cl^- and is assigned an oxidation number of -1. The H atom then is thought of as having lost an electron, becoming H^+, and is assigned an oxidation number of $+1$:

$$\overset{+1}{H}:\overset{-1}{\underset{..}{\overset{..}{Cl}}}:$$

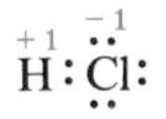

Similarly, in PCl_3 the shared electrons are assigned to Cl, and the oxidation number of each Cl atom is -1. The oxidation number of P, which is thought of as having lost three electrons, is $+3$:

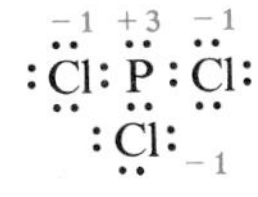

$$:\overset{-1}{\underset{..}{\overset{..}{Cl}}}:\overset{+3}{P}:\overset{-1}{\underset{..}{\overset{..}{Cl}}}:$$
$$:\underset{-1}{\underset{..}{\overset{..}{Cl}}}:$$

In assigning oxidation numbers in covalent compounds, we imagine electrons to be transferred from one atom to another, but no electrons are added to or removed from the molecule. Thus the oxidation numbers of all the atoms must add up to zero, as is the case in HCl and PCl_3. Similarly, in any polyatomic ion the sum of the oxidation numbers must equal the charge on the ion. We therefore have the following rule:

In any neutral molecule the sum of the oxidation numbers is zero; in an ion the sum of the oxidation numbers is equal to the charge on the ion.

Because fluorine is the most electronegative element in any covalent or ionic compound, it always has an oxidation number of -1. Because the other halogens also have high electronegativities, they too have an oxidation number of -1 in most, but not all, of their compounds. Similarly, in its compounds oxygen is normally assigned an oxidation number of -2. In its covalent compounds hydrogen is usually less electronegative than the atom to which it is attached, so hydrogen usually has an oxidation number $+1$.

We can assign oxidation numbers for atoms in a great many compounds if we combine these generalizations about the halogens, oxygen, and hydrogen with the rule that the sum of the oxidation numbers of the atoms must equal zero for a neutral compound or the charge on an ion. For example, in sulfur dioxide each oxygen has an oxidation number of -2. The sum of their oxidation numbers is -4, and since the sum of all the oxidation numbers must be zero, the oxidation number of the sulfur atom is $+4$:

$$\overset{-2\ +4\ -2}{O\ S\ O}$$

In sulfuric acid, H_2SO_4, the sum of the oxidation numbers of hydrogen and oxygen is

$$2H = 2 \times (+1) = +2$$
$$4O = 4 \times (-2) = \underline{-8}$$
$$-6$$

Therefore since the sum of the oxidation numbers must be zero, sulfur in this case has an oxidation number of $+6$:

$$\overset{+1}{H_2}\quad \overset{+6}{S}\quad \overset{-2}{O_4}$$
$$2(+1) + (+6) + 4(-2) = 0$$

In the singly charged hydrogen sulfate ion, HSO_4^-, the sum of the oxidation numbers must be -1:

$$\overset{+1}{H}\quad \overset{+6}{S}\quad \overset{-2}{O_4}$$
$$(+1) + (+6) + 4(-2) = -1$$

One situation that has not been covered by the previous rules is that in which an atom is bonded to another of identical electronegativity and, in particular, *when it is bonded to one of the same kind. In this case the bond electrons are divided equally between the two atoms.* Thus in the case of the H_2 molecule each atom is assigned one electron, and therefore each H atom has an oxidation number of zero. Similar arguments apply to the eight sulfur atoms in S_8 and the four phosphorus atoms in P_4. As previously stated, *the oxidation number of an atom in any of the allotropic forms of an element is 0.*

Another possibility arises when there is a bond between two like atoms in a molecule that also contains other atoms. For example, in hydrogen peroxide, H_2O_2, each hydrogen is assigned an oxidation number of $+1$. And from the rule that the sum of the oxidation numbers must be zero, each oxygen atom must have an oxidation number of -1. Alternatively, we see that if the electrons of the O—O bond are divided equally between the two oxygen atoms, each oxygen is assigned seven electrons. It would then have a charge of -1 and therefore an oxidation number of -1. This compound is one of the rare cases in which oxygen does not have an oxidation number of -2:

$$\overset{+1\ -1\ -1\ +1}{H:\ddot{O}:\ddot{O}:H}$$

In summary, the rules for assigning oxidation numbers are:

1. In any of the allotropic forms of an element each atom is assigned an oxidation number of 0.

2. Fluorine, the most electronegative element, has an oxidation number of -1 in all its compounds.

3. Oxygen usually has oxidation number -2 in its compounds. There are two important exceptions:
   - When oxygen is combined with fluorine, as in F_2O, its oxidation number is $+2$.
   - In compounds containing oxygen-oxygen bonds, such as hydrogen peroxide, the oxidation number of oxygen is -1.

4. The halogens other than fluorine have an oxidation number of -1 except when they are combined with a more electronegative element, that is, with oxygen or a more electronegative halogen.

5. Hydrogen in its compounds usually has an oxidation number of $+1$. The only exceptions occur in metal hydrides, in which hydrogen has an oxidation number of -1, as, for example, in Na^+H^-.

6. The sum of the oxidation numbers of all the atoms in a neutral compound is 0 and in an ion is equal to the charge on the ion. As a result, the oxidation number of a monatomic ion is equal to its charge, for example, $+2$ for Cu^{2+} and -1 for Cl^-.

Example 7.1 Assign oxidation numbers (abbreviated as O.N.) to each of the elements in the following compounds:

(a) NaCl   (b) BaF_2 (c) CO_2 (d) SCl_2 (e) H_2S

Solution

(a) NaCl is ionic: Na^+Cl^-. The O.N.s are equal to the charges. Therefore O.N.(Na) = $+1$; O.N.(Cl) = -1.

(b) BaF_2 is ionic: $Ba^{2+}(F^-)_2$. Therefore O.N.(Ba) = $+2$; O.N.(F) = -1.

(c) Each O has O.N. = -2, and therefore C has O.N. = $+4$.

(d) Each Cl has O.N. = -1, and therefore S has O.N. = $+2$.

(e) Each H has O.N. = $+1$, and therefore S has O.N. = -2.

Example 7.2 Assign oxidation numbers to each of the atoms in the following molecules and ions:

(a) HS^- (b) S_2Cl_2 (c) SO_4^{2-} (d) HSO_3^-

Solution

(a) The sum of the oxidation numbers must be -1. Hydrogen has O.N. = $+1$. So $+1 + $ O.N.(S) = -1, hence O.N.(S) = -2.

(b) Chlorine has O.N. = -1. So

$$2(-1) + 2[\text{O.N.(S)}] = 0$$

Hence O.N.(S) = $+1$. This oxidation state is unusual for sulfur. It arises because the molecule has an S—S bond:

$$:\overset{..}{\underset{..}{Cl}}-\overset{..}{\underset{..}{S}}-\overset{..}{\underset{..}{S}}-\overset{..}{\underset{..}{Cl}}:$$

The S—S bond electrons are shared equally between the S atoms, and the S—Cl bond electrons are assigned to Cl. Thus each S atom is assigned five electrons so each has a charge of $+1$. Hence O.N.(S) = $+1$.

7.3 COMPOUNDS OF SULFUR

253

(c) Oxygen has O.N. $= -2$. Then we have $4(-2) + \text{O.N.(S)} = -2$. Hence O.N.(S) $= +6$.

(d) Taking the oxidation numbers of oxygen and hydrogen to be -2 and $+1$, respectively, we have

$$3(-2) + (+1) + \text{O.N.(S)} = -1$$
$$\text{O.N.(S)} = +4$$

Obviously, oxidation numbers do not correspond to the actual distribution of charge in a molecule or polyatomic ions. They are arbitrary numbers that enable us to follow the state of oxidation of an element in its reactions and its compounds and to keep count of electrons in oxidation-reduction reactions. Thus oxidation numbers resemble formal charges in being arbitrary, but note that the two numbers are not the same. Oxidation numbers are assigned by assuming that the electrons of bonds are shared *unequally* between atoms, while formal charges are found by assuming that bonding electrons are shared *equally* between atoms. In other words, oxidation numbers are based on the assumption that all bonds can be regarded as ionic, while formal charges are based on the assumption that all bonds are purely covalent and have no polarity.

For the chloride ion, ammonium ion, and sulfate ion, the formal charges and oxidation numbers are as follows:

Formal charges

Oxidation numbers

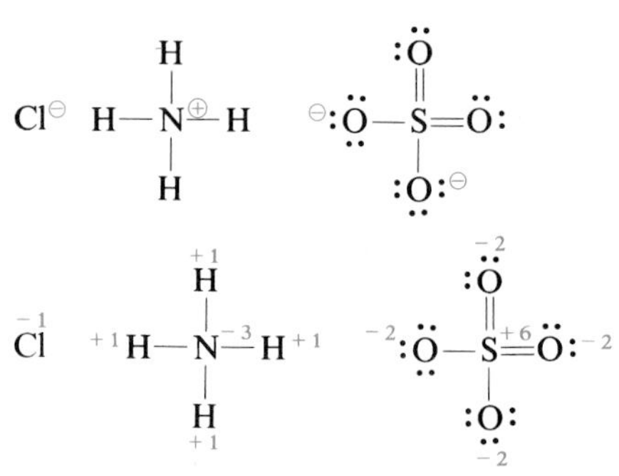

For the chloride ion the formal charge is equal to the oxidation number. For the polyatomic ions NH_4^+ and SO_4^{2-}, the formal charges give a reasonable, although approximate, description of the distribution of charge in the ion, but the oxidation numbers do not—and they are not intended for this purpose. In the ammonium ion the positive charge of the ion is on the nitrogen atom, and in the sulfate ion the double negative charge is distributed between two oxygen atoms. The formal charges are part of a correctly drawn Lewis structure. The oxidation numbers tell us the state of oxidation of each atom in a molecule of polyatomic ion.

Let us now see how oxidation numbers are useful to us in describing the compounds of sulfur and their reactions.

Oxidation States of Sulfur

The common oxidation numbers found for sulfur in its compounds are $+6$, $+4$, $+2$, 0, and -2, as shown in Figure 7.8. These numbers describe the states of oxidation of sulfur. When it has an oxidation number of $+6$, as in H_2SO_4, sulfur is said to be in the $+6$ oxidation state, which is the most highly oxidized state of sulfur. When it has an oxidation number of -2, as in H_2S, sulfur is said to be in the -2 oxidation state, which is the least highly oxidized, or the most highly reduced, state of sulfur. In its elemental forms, S_n (usually S_8), sulfur is in

Figure 7.8 Classification of Compounds of Sulfur by Oxidation States.

Oxidation state	Compounds
+6	SO_3, H_2SO_4, SO_4^{2-}, $H_2S_2O_7$
+4	SO_2, HSO_3^-, SO_3^{2-}
+2	SCl_2
0	S_8 and all other forms of elemental sulfur
−2	H_2S, S^{2-}

Oxidation Oxidation number increases

Reduction Oxidation number decreases

the 0 oxidation state. When sulfur is converted to H_2S, it is reduced to its lowest (-2) oxidation state. When it is oxidized to H_2SO_4, it is converted to its highest ($+6$) oxidation state. Note that the ground state of sulfur in which there are two unpaired electrons gives rise to both the -2 and the $+2$ oxidation states in compound formation depending on whether sulfur is combined with a less or a more electronegative element. But the valence states in which there are four or six unpaired electrons only give rise to the $+4$ and $+6$ oxidation states, because only electronegative elements such as fluorine and oxygen can cause the promotion of electrons to the 3d level.

The concept of oxidation numbers leads to another useful definition of oxidation and reduction: **Oxidation** *can be defined as any change in which the oxidation number of an atom increases.* **Reduction** *may be defined as any change in which the oxidation number of an atom decreases.*

By considering the reaction between copper and sulfuric acid, we can see that these definitions are equivalent to our previous definitions of oxidation as electron loss and reduction as electron gain. We can split the overall reaction,

$$Cu(s) + 5H_2SO_4(l) \longrightarrow Cu^{2+}(soln) + 4HSO_4^-(soln) + 2H_3O^+(soln) + SO_2(g)$$

into two **half-reactions**: one for oxidation and the other for reduction:

$$Cu \longrightarrow Cu^{2+} + 2e^- \qquad \text{Oxidation}$$

The oxidation number of copper changes from 0 to $+2$; that is, it increases by two, which corresponds to the loss of two electrons. Also,

$$5H_2SO_4(l) + 2e^- \longrightarrow 4HSO_4^-(soln) + 2H_3O^+(soln) + SO_2(g) \qquad \text{Reduction}$$

The oxidation number of sulfur decreases from $+6$ in H_2SO_4 to $+4$ in SO_2, which corresponds to the gain of two electrons. Adding these two half-reactions and canceling two electrons from both sides gives the overall equation for the reaction.

This same equation for the reduction of sulfuric acid to sulfur dioxide applies to all reactions in which sulfuric acid behaves as an oxidizing agent and is reduced to sulfur dioxide. We can therefore use it to write the overall equation for the oxidation of bromide ion to bromine by concentrated sulfuric acid. The half-reaction for the oxidation of bromide ion to bromine is

$$2Br^- \longrightarrow Br_2 + 2e^-$$

If we add this equation to the equation for the reduction of sulfuric acid to sulfur dioxide, namely

$$5H_2SO_4(l) + 2e^- \longrightarrow 4HSO_4^-(soln) + 2H_3O^+(soln) + SO_2(g)$$

7.3 COMPOUNDS OF SULFUR

we obtain the complete equation for the oxidation of bromide to bromine by concentrated sulfuric acid:

$$2Br^-(soln) + 5H_2SO_4(l) \longrightarrow Br_2(l) + 2H_3O^+(soln) + 4HSO_4^-(soln) + SO_2(g)$$

Example 7.3 Write balanced equations for the following reactions:

(a) The reaction of zinc with dilute aqueous sulfuric acid to give hydrogen and zinc sulfate, $ZnSO_4$.

(b) The reaction of concentrated sulfuric acid with sodium iodide.

(c) The reaction of concentrated sulfuric acid with silver to give silver sulfate, Ag_2SO_4, and sulfur dioxide.

(d) The reaction of an aqueous solution of sulfuric acid with solid magnesium hydroxide.

Solution

(a) In this reaction H_3O^+, (H^+) is reduced to H_2 and Zn is oxidized to Zn^{2+}, just as in the reaction with magnesium. The equation is similar:

$$H_2SO_4(aq) + Zn(s) \longrightarrow ZnSO_4(aq) + H_2(g)$$

or

$$H_2SO_4(aq) + Zn(s) \longrightarrow Zn^{2+}(aq) + SO_4^{2-}(aq) + H_2(g)$$

The equation could also be written as

$$2H_3O^+(aq) + SO_4^{2-}(aq) + Zn(s) \longrightarrow Zn^{2+}(aq) + SO_4^{2-}(aq) + H_2 + 2H_2O$$

to show that sulfuric acid, which is a strong acid, is ionized in aqueous solution. However, in this form it looks more complicated, and so it is not usually written this way unless there is a special reason to do so.

(b) We know that I^- is oxidized to I_2 by concentrated H_2SO_4. The half-reaction is

$$2I^- \longrightarrow I_2 + 2e^-$$

We cannot predict the reduction product of H_2SO_4 with certainty, but we can reasonably suppose that H_2SO_4 is reduced to SO_2 as it is by Br^-, in which case the half-reaction is

$$5H_2SO_4(l) + 2e^- \longrightarrow 2H_3O^+(soln) + 4HSO_4^-(soln) + SO_2(g)$$

So the overall reaction is

$$2I^- + 5H_2SO_4(l) \longrightarrow I_2(s) + 2H_3O^+(soln) + 4HSO_4^-(soln) + SO_2(g)$$

(c) Silver is oxidized to Ag^+ in $(Ag^+)_2(SO_4^{2-})$, so the oxidation half-reaction is

$$Ag \longrightarrow Ag^+ + e^-$$

This reaction is combined with the half-reaction for the reduction of H_2SO_4 to SO_2 to give

$$2Ag(s) + 5H_2SO_4(l) \longrightarrow$$
$$2Ag^+(soln) + 2H_3O^+(soln) + 4HSO_4^-(soln) + SO_2(g)$$

(d) The alkaline earth hydroxides are bases. Thus H_2SO_4 reacts with $Mg(OH)_2$ to give $MgSO_4$ and water. The equation is

$$Mg(OH)_2(s) + H_2SO_4(aq) \longrightarrow MgSO_4(aq) + 2H_2O$$

or

$$Mg(OH)_2(s) + H_2SO_4(aq) \longrightarrow Mg^{2+}(aq) + SO_4^{2-}(aq) + 2H_2O$$

The +4 Oxidation State: Sulfur Dioxide and Sulfites

In addition to sulfur dioxide, the most important compounds in which sulfur is in the +4 oxidation state are the metal **sulfites**. Sulfur dioxide is very soluble in water (9.4 g L^{-1} at STP), and it has generally been assumed that *sulfurous acid* is found in such solutions. The reaction would be

$$H_2O + SO_2 \longrightarrow H_2SO_3$$

In fact, there is no firm evidence that the molecule H_2SO_3 exists. Nevertheless, the anions HSO_3^- and SO_3^{2-} are formed in small amounts:

$$SO_2(aq) + 2H_2O \rightleftharpoons H_3O^+ + HSO_3^-$$
$$HSO_3^- + H_2O \rightleftharpoons H_3O^+ + SO_3^{2-}$$

Thus sulfur dioxide behaves as a weak diprotic acid in water.

The sulfite and hydrogen sulfite ions have the following Lewis structures:

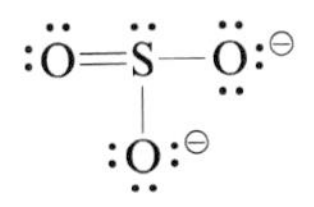

Metal sulfites are all soluble in water. Addition of an aqueous acid (H_3O^+) to a solution of a sulfite shifts the above equilibria to the left, and gaseous SO_2 is evolved:

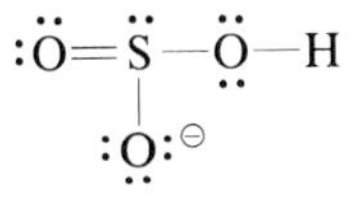

$$SO_3^{2-}(aq) + 2H_3O^+(aq) \longrightarrow SO_2(g) + 3H_2O$$

The reaction is driven to completion by the escape of the gaseous SO_2 (Le Châtelier's principle). Small amounts of SO_2 can be prepared in the laboratory in this way.

Both SO_2 and SO_3^{2-} are good reducing agents. They are oxidized to SO_3 and sulfate ion, SO_4^{2-}, by many oxidizing agents, although, as we have seen, the reaction with O_2 is very slow in the absence of a catalyst. For example, an orange-red solution of Br_2 in water is decolorized by SO_2 or by a sulfite solution, because Br_2 is reduced to Br^- (see Experiment 7.7):

$$Br_2(aq) + SO_2(aq) + 2H_2O(l) \longrightarrow 2Br^-(aq) + SO_4^{2-}(aq) + 4H^+(aq)$$

Both Cl_2 and I_2 are reduced in a similar way.

The reducing properties of an aqueous solution of sulfur dioxide make it a useful bleaching agent. For example, colored compounds in wool, silk, paper, straw, and sponges become colorless when reduced by sulfur dioxide (see Experiment 7.7). This reaction may be reversed when the material is exposed to air and light; the colorless, reduced compound is then oxidized back to the original colored compound. For example, oxidation causes newspaper, which is made from bleached wood pulp, to yellow when left out in the sun.

Because the sulfur in SO_2 is in an intermediate oxidation state, it may either be oxidized to the +6 state or reduced to a lower oxidation state. Thus SO_2 can act not only as a reducing agent but also as an oxidizing agent. For example, it is an oxidizing agent in the important reaction with H_2S used in the purification of natural gas:

$$2\overset{-2}{H_2S} + \overset{+4}{SO_2} \longrightarrow 3\overset{0}{S} + 2H_2O$$

In this reaction sulfur in the −2 oxidation state in H_2S is oxidized to sulfur in the 0 oxidation state by SO_2 in which sulfur is in the +4 oxidation state.

Sulfur Dioxide as a Reducing Agent

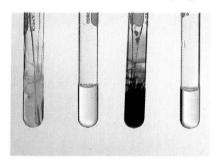  

When sulfur dioxide is passed into an orange-red aqueous solution of bromine it reduces it to colorless bromide ion (left). When sulfur dioxide is passed into a purple potassium permanganate solution it reduces it to a colorless solution of Mn^{2+} (right).

Sulfur dioxide reduces the red pigment in a dampened rose to a colorless substance.

The rose can be oxidized back to a red color by exposing it to chlorine.

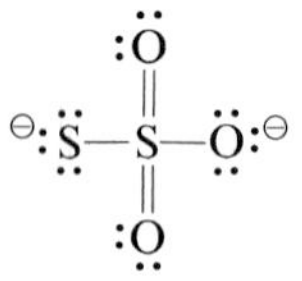

:Ö:
‖
⁻:Ṡ—Ṡ—Ö:⁻
‖
:Ö:

Thiosulfate ion

The sulfur in SO_2 is, at the same time, reduced to the 0 oxidation state.

When an aqueous solution of a sulfite is boiled with sulfur, the thiosulfate ion, $S_2O_3^{2-}$, is formed:

$$S(s) + SO_3^{2-}(aq) \longrightarrow S_2O_3^{2-}(aq)$$

The Lewis structure for this ion is the same as that of the sulfate ion, except that a sulfur atom replaces one of the oxygen atoms. The replacement of an oxygen atom by a sulfur atom in a compound is often denoted by the prefix *thio-*, as in thiosulfate.

When dilute acid is added to an aqueous solution of a thiosulfate, it is decomposed to sulfur and sulfur dioxide:

$$S_2O_3^{2-}(aq) + 2H_3O^+(aq) \longrightarrow S(s) + SO_2(g) + 3H_2O(l)$$

This reaction is the reverse of the reaction by which the thiosulfate ion is formed. Addition of acid converts the small equilibrium amount of SO_3^{2-} to H_2O and SO_2 and thus drives the equilibrium back to the left:

$$S(s) + SO_3^{2-}(aq) \rightleftharpoons S_2O_3^{2-}(aq)$$

$$\downarrow +2H^+(aq)$$

$$H_2O(l) + SO_2(g)$$

This reaction is another example of the operation of Le Châtelier's principle.

The −2 Oxidation State: Hydrogen Sulfide and Metal Sulfides

HYDROGEN SULFIDE Unlike water, which is a liquid at room temperature, H_2S is a gas (melting point −85.6°C; boiling point −60.7°C). It has a powerful, unpleasant odor, and it is very poisonous. Hydrogen sulfide is soluble in water; it is a weak diprotic acid:

$$H_2S + H_2O \rightleftharpoons H_3O^+ + HS^- \qquad \text{Hydrogen sulfide ion}$$

$$HS^- + H_2O \rightleftharpoons H_3O^+ + S^{2-} \qquad \text{Sulfide ion}$$

Hydrogen sulfide is a good reducing agent; it is usually oxidized to elemental sulfur. For example, if an aqueous solution is exposed to air, a precipitate of finely divided sulfur is slowly formed:

$$2H_2S + O_2 \longrightarrow 2S + 2H_2O$$

The same reaction occurs when H_2S burns in air; sulfur is deposited on a cold surface held near the flame (see Experiment 7.8). Some SO_2 is formed at the same time:

$$2H_2S + 3O_2 \longrightarrow 2H_2O + 2SO_2$$

If H_2S is passed into an aqueous solution of bromine, Br_2, the red-brown color of the Br_2 disappears as it is reduced to Br^-. At the same time a milky precipitate of finely divided sulfur is formed:

$$H_2S(g) + Br_2(aq) \longrightarrow 2H^+(aq) + 2Br^-(aq) + S(s)$$

Chlorine and iodine are reduced in the same way.

In the laboratory H_2S is easily prepared by the action of an acid on a metal sulfide, such as FeS:

$$H_2SO_4(aq) + FeS(s) \longrightarrow FeSO_4(aq) + H_2S(g)$$

EXPERIMENT 7.8

Properties of Hydrogen Sulfide

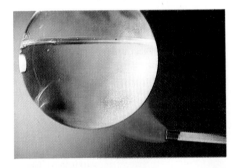

Hydrogen sulfide burns in air with a blue flame, depositing yellow sulfur on a cool surface.

When H_2S is passed into a colorless aqueous solution of zinc sulfate, $ZnSO_4$, a white precipitate of insoluble zinc sulfide, ZnS, is formed.

When H_2S is passed into a colorless aqueous solution of lead nitrate, $Pb(NO_3)_2$, a dark brown precipitate of lead sulfide, PbS, is formed.

7.3 COMPOUNDS OF SULFUR

259

Because H_2S is a weak acid, the sulfide ion, S^{2-}, is a weak base, which reacts with H_3O^+ to give its conjugate acid HS^-, which, in turn, gives its conjugate acid, H_2S. The overall equation is

$$S^{2-}(aq) + 2H_3O^+(aq) \longrightarrow H_2S(g) + 2H_2O(l)$$

SULFIDES AND POLYSULFIDES　The **sulfides** of the alkali and alkaline earth metals are colorless solids that are soluble in water. The sulfides of most other metals are often colored and insoluble or only very slightly soluble in water. They are precipitated when H_2S is passed through a solution of a salt of the metal (see Experiment 7.8):

$$Zn^{2+}(aq) + H_2S(g) \longrightarrow ZnS(s) + 2H^+(aq)$$
$$Pb^{2+}(aq) + H_2S(g) \longrightarrow PbS(s) + 2H^+(aq)$$

Many insoluble metal sulfides occur in nature and are important ores. Examples include FeS, CuS, Ag_2S, HgS, and PbS.

A colorless solution of an alkali or alkaline earth metal sulfide such as Na_2S can dissolve considerable amounts of sulfur to give a bright yellow solution containing a mixture of polysulfides:

$$S^{2-} + S \longrightarrow S_2^{2-} \qquad \text{Disulfide} \qquad {}^{\ominus}\!:\!\ddot{S}\!-\!\ddot{S}\!:^{\ominus}$$

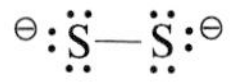

$$S^{2-} + 2S \longrightarrow S_3^{2-} \qquad \text{Trisulfide} \qquad {}^{\ominus}\!:\!\ddot{S}\!-\!\ddot{S}\!-\!\ddot{S}\!:^{\ominus}$$

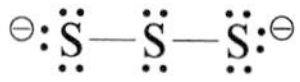

$$S^{2-} + 3S \longrightarrow S_4^{2-} \qquad \text{Tetrasulfide} \qquad {}^{\ominus}\!:\!\ddot{S}\!-\!\ddot{S}\!-\!\ddot{S}\!-\!\ddot{S}\!:^{\ominus}$$

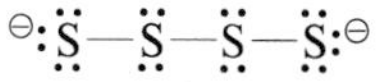

These polysulfide ions consist of chains of sulfur atoms. The common mineral pyrite, FeS_2, is a salt of the disulfide ion, S_2^{2-}.

7.4 COMPOUNDS OF PHOSPHORUS

The most important oxidation states of phosphorus are $+5$, $+3$, 0, and -3. Figure 7.9 gives examples of compounds in which phosphorus is in these oxidation states.

Oxides

When phosphorus is burnt in air or oxygen, two different oxides may be formed; they have the empirical formulas P_2O_3 and P_2O_5 (see Experiment 7.9). Their molecular masses indicate that their molecular formulas are P_4O_6 and P_4O_{10},

Figure 7.9 Classification of Compounds of Phosphorus by Oxidation State.

Oxidation state	Compounds
+5	P_4O_{10}, H_3PO_4, PO_4^{3-}
+3	P_4O_6, H_3PO_3, HPO_3^{2-}
0	P_4
−3	PH_3, PH_4^+

Oxidation　Oxidation number increases

Reduction　Oxidation number decreases

Reaction of White Phosphorus with Oxygen

When a piece of white phosphorus is inserted into a flask of oxygen it ignites, burns, and produces a white smoke of P_4O_{10}.

As the phosphorus burns it produces a brilliant white light.

When the reaction is complete, the inside of the flask is coated with a layer of white P_4O_{10}.

respectively. If the supply of oxygen is limited, P_4O_6, tetraphosphorus hexaoxide, is the major product. In excess oxygen, P_4O_{10}, tetraphosphorus decaoxide, is formed:

$$P_4(s) + 3O_2(g) \longrightarrow P_4O_6(s)$$
$$P_4(s) + 5O_2(g) \longrightarrow P_4O_{10}(s)$$

Figure 7.10 shows how the structures of these oxides are related to that of the P_4 tetrahedron. In P_4O_6 phosphorus exhibits its normal valence of 3 and is in the $+3$ oxidation state. Each phosphorus forms three bonds and has one unshared pair to complete an octet of electrons. The structure of P_4O_{10} is similar, except that an additional oxygen atom is bonded to each phosphorus atom by a double bond. Each phosphorus atom uses all its five valence electrons in the valence state configuration $3s^1 3p^3 3d^1$ to form five covalent bonds; phosphorus is in the $+5$ oxidation state.

The oxide P_4O_{10} reacts vigorously with excess water, liberating a considerable amount of heat and forming H_3PO_4, *orthophosphoric acid* (usually called simply *phosphoric acid*):

$$P_4O_{10}(s) + 6H_2O(l) \longrightarrow 4H_3PO_4(aq)$$

If a limited quantity of water is used, *metaphosphoric acid*, HPO_3, is formed:

$$P_4O_{10}(s) + 2H_2O(l) \longrightarrow 4HPO_3(s)$$

The oxide P_4O_{10} has such a strong tendency to react with water that it is a very powerful dehydrating agent. It can even remove water from H_2SO_4. When a mixture of concentrated H_2SO_4 and P_4O_{10} is heated, sulfur trioxide, SO_3, is formed:

$$2H_2SO_4(l) + P_4O_{10}(s) \longrightarrow 2SO_3(g) + 4HPO_3(s)$$

7.4 COMPOUNDS OF PHOSPHORUS

261

Figure 7.10 Structures of P₄, P₄O₆, and P₄O₁₀. In P_4O_6 each edge of the P_4 tetrahedron is bridged by an oxygen atom. The structure of P_4O_{10} is the same, except that an additional oxygen atom is bonded to each phosphorus atom by a double bond.

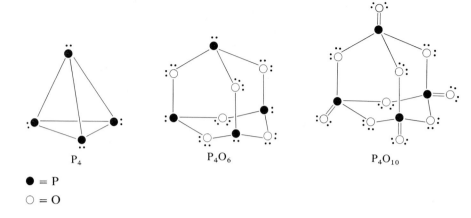

$$\bullet = P$$
$$\circ = O$$

$$P_4 \qquad P_4O_6 \qquad P_4O_{10}$$

Oxacids and Their Salts

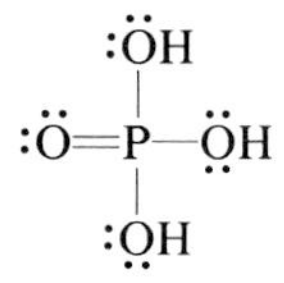

PHOSPHORIC ACID   Phosphoric acid is a colorless solid with a melting point of 42°C. A concentrated (82% by mass) solution in water is usually used in the laboratory.

Phosphoric acid is **triprotic** because it has three OH groups and thus three ionizable hydrogen atoms. It ionizes in water in three stages:

$$H_3PO_4 + H_2O \rightleftharpoons H_3O^+ + H_2PO_4^-$$
$$H_2PO_4^- + H_2O \rightleftharpoons H_3O^+ + HPO_4^{2-}$$
$$HPO_4^{2-} + H_2O \rightleftharpoons H_3O^+ + PO_4^{3-}$$

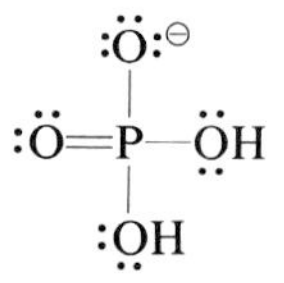

Even the first of these ionizations is incomplete. Thus unlike sulfuric acid, which is a strong acid, phosphoric acid is a weak acid.

Notice that there is no relationship between the number of ionizable hydrogen atoms and the strength of an acid. Hydrochloric acid, HCl, which has one ionizable hydrogen, and H_2SO_4, which has two ionizable hydrogens, are both strong acids. But carbonic acid, H_2CO_3, which has two ionizable hydrogens, and H_3PO_4, which has three ionizable hydrogens, are both weak acids.

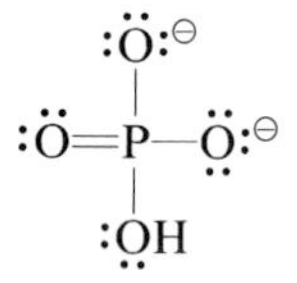

In its reactions with bases H_3PO_4 forms three series of salts. These salts contain the dihydrogen phosphate ion, $H_2PO_4^-$, the hydrogen phosphate ion, HPO_4^{2-}, and the phosphate ion, PO_4^{3-}. For example, 1 mol of H_3PO_4 reacts with 1, 2, or 3 mol of potassium hydroxide, KOH, to give KH_2PO_4, K_2HPO_4, or K_3PO_4, respectively:

$$KOH + H_3PO_4 \longrightarrow KH_2PO_4 + H_2O$$
$$2KOH + H_3PO_4 \longrightarrow K_2HPO_4 + 2H_2O$$
$$3KOH + H_3PO_4 \longrightarrow K_3PO_4 + 3H_2O$$

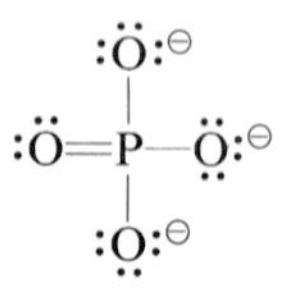

Example 7.4 What volume of 0.113M KOH solution must be added to 50.0 mL of 0.105M H_3PO_4 solution to prepare a solution of K_3PO_4?

Solution 50.0 mL of 0.105M H_3PO_4 contain

$$50.0 \text{ mL} \left(\frac{1 \text{ L}}{1000 \text{ mL}} \right) (0.105 \text{ mol L}^{-1}) = 5.25 \times 10^{-3} \text{ mol } H_3PO_4$$

1 mol H_3PO_4 reacts with 3 mol KOH to form 1 mol K_3PO_4

(side figures, left column:)

:ÖH
:Ö=P—ÖH
:ÖH
Phosphoric acid

:Ö:⊖
:Ö=P—ÖH
:ÖH
Dihydrogen phosphate ion

:Ö:⊖
:Ö=P—Ö:⊖
:ÖH
Hydrogen phosphate ion

:Ö:⊖
:Ö=P—Ö:⊖
:Ö:⊖
Phosphate ion

Therefore, we require

$$3 \times 5.25 \times 10^{-3} \text{ mol KOH} = 15.75 \times 10^{-3} \text{ mol KOH}$$

That is,

$$(15.75 \times 10^{-3} \text{ mol KOH})\left(\frac{1 \text{ L solution}}{0.113 \text{ mol KOH}}\right)\left(\frac{1000 \text{ mL}}{1 \text{ L}}\right)$$

$$= 139 \text{ mL KOH solution}$$

PHOSPHATES AS FERTILIZERS Phosphorus is essential for plant life. Although most soils contain an appreciable amount of phosphate, it is often present in the form of insoluble salts and hence is unavailable to plants. Phosphate rock can be ground up and added to phosphorus-deficient soil, but it is not soluble enough to be a very good fertilizer. A better fertilizer can be obtained by treating phosphate rock with aqueous sulfuric acid to convert it to the more soluble calcium dihydrogen phosphate. The amount and concentration of the acid are adjusted so that all the water is used up in the formation of hydrated salts:

$$Ca_3(PO_4)_2(s) + 2H_2SO_4(aq) + 5H_2O(l) \longrightarrow Ca(H_2PO_4)_2 \cdot H_2O(s) + 2CaSO_4 \cdot 2H_2O(s)$$

If this equation is written in the ionic form and all the spectator ions are eliminated, we see that the reaction is simply

$$PO_4^{3-}(aq) + 2H^+(aq) \longrightarrow H_2PO_4^-(aq)$$

in which phosphate ion, which is a base, combines with hydrogen ions from the sulfuric acid. The resulting solid mixture of hydrated calcium sulfate (gypsum), $CaSO_4 \cdot 2H_2O$, and hydrated calcium dihydrogen phosphate, $Ca(H_2PO_4)_2 \cdot H_2O$, is called *superphosphate of lime*.

A more concentrated phosphorus fertilizer called *triple phosphate* is made by treating phosphate rock with phosphoric acid to give a product that is essentially pure calcium dihydrogen phosphate and contains no calcium sulfate. The phosphoric acid is made by treating phosphate rock with a larger quantity of sulfuric acid than is needed for the production of superphosphate of lime:

$$Ca_3(PO_4)_2(s) + 3H_2SO_4(aq) + 6H_2O(l) \longrightarrow 2H_3PO_4(aq) + 3CaSO_4 \cdot 2H_2O(s)$$

The insoluble calcium sulfate is removed by filtration, leaving phosphoric acid, which is then used to treat more phosphate rock to form calcium dihydrogen phosphate (triple phosphate), $Ca(H_2PO_4)_2$:

$$Ca_3(PO_4)_2 + 4H_3PO_4 \longrightarrow 3Ca(H_2PO_4)_2$$

Another important fertilizer is made by treating phosphoric acid with ammonia:

$$H_3PO_4 + NH_3 \longrightarrow NH_4H_2PO_4$$

The product, ammonium dihydrogen phosphate, $NH_4^+H_2PO_4^-$, supplies the soil with nitrogen as well as phosphorus.

POLYMERIC PHOSPHORIC ACIDS AND CONDENSATION Phosphoric acid readily undergoes **condensation**, the reaction of two or more molecules to form a larger molecule by the elimination of small molecules, such as water. Two hydroxyl groups on different molecules frequently eliminate a molecule of water in this way; the product is a new molecule containing a bridging oxygen atom:

$$X—OH + HO—X \longrightarrow X—O—X + H_2O$$

This process can produce polymeric chains or rings. For example, when phosphoric acid is heated, it loses water and condenses to give diphosphoric acid (pyrophosphoric acid), $H_4P_2O_7$:

$$HO-\overset{\overset{\textstyle :O}{\|}}{\underset{\underset{\textstyle :OH}{|}}{P}}-OH + HO-\overset{\overset{\textstyle :O}{\|}}{\underset{\underset{\textstyle :OH}{|}}{P}}-OH \longrightarrow HO-\overset{\overset{\textstyle :O}{\|}}{\underset{\underset{\textstyle :OH}{|}}{P}}-O-\overset{\overset{\textstyle :O}{\|}}{\underset{\underset{\textstyle :OH}{|}}{P}}-OH + H_2O$$

Continued heating leads to the formation of triphosphoric acid (often called tripolyphosphoric acid), $H_5P_3O_{10}$, and acids with still longer chains.

Sulfuric acid undergoes a similar reaction on heating to give disulfuric acid (pyrosulfuric acid) and longer-chain polysulfuric acids:

$$:O=\overset{\overset{\textstyle :O}{\|}}{\underset{\underset{\textstyle :OH}{|}}{S}}-OH + HO-\overset{\overset{\textstyle :O}{\|}}{\underset{\underset{\textstyle :OH}{|}}{S}}=O: \longrightarrow :O=\overset{\overset{\textstyle :O}{\|}}{\underset{\underset{\textstyle :OH}{|}}{S}}-O-\overset{\overset{\textstyle :O}{\|}}{\underset{\underset{\textstyle :OH}{|}}{S}}=O: + H_2O$$

The interconversion of triphosphates, diphosphates, and phosphates, such as adenosine triphosphate (ATP) and adenosine diphosphate (ADP), is very important in many biochemical processes. These reactions occur at body temperature under the influence of catalysts known as enzymes.

Further dehydration of phosphoric acid leads eventually to metaphosphoric acid:

$$H_3PO_4 \longrightarrow HPO_3 + H_2O$$

As we have seen, HPO_3 can also be made by addition of a limited amount of water to P_4O_{10}.

The formula HPO_3 is the empirical formula of metaphosphoric acid, which is in fact polymeric. It has many different forms depending on the degree of polymerization. Thus the general formula is $(HPO_3)_n$, where $n = 3, 4, 5, 6, \ldots$. These molecules are rings, such as $(HPO_3)_3$ and $(HPO_3)_4$, or very long chains, $(HPO_3)_n$. Salts of the metaphosphoric acids, *metaphosphates*, are used extensively as water softeners.

PHOSPHOROUS ACID AND PHOSPHITES   When P_4O_6 reacts with water, it forms phosphorous acid, H_3PO_3:

$$P_4O_6 + 6H_2O \longrightarrow 4H_3PO_3$$

Phosphorous acid is a white solid with a melting point of 74°C. In aqueous solution it behaves as a weak acid. It has only two replaceable hydrogen atoms; it is a diprotic acid. At first sight we might expect phosphorous acid to be triprotic, like phosphoric acid. But H_3PO_3 has only two OH groups; the third hydrogen is bonded directly to phosphorus. The P—H bond is nonpolar, because hydrogen and phosphorus have essentially the same electronegativity ($\chi_P = 2.1, \chi_H = 2.2$), so this third hydrogen does not ionize in water. Phosphorous acid is, therefore, diprotic and forms two series of salts, for example, sodium dihydrogen phosphite, NaH_2PO_3, and sodium hydrogen phosphite, Na_2HPO_3.

Just as SO_2 and sulfites are rather easily oxidized to the +6 oxidation state, so phosphorous acid and the phosphites, in which phosphorus is in the +3 oxidation state, are readily oxidized to phosphates, in which phosphorus is in the +5 oxidation state. They are therefore good reducing agents. For example, phosphorous acid precipitates metallic copper from a solution of a copper salt:

$$Cu^{2+}(aq) + H_3PO_3(aq) + 3H_2O(l) \longrightarrow Cu(s) + H_3PO_4(aq) + 2H_3O^+(aq)$$

Metaphosphates

Phosphorous acid

Phosphides and Phosphine

The lowest oxidation state of phosphorus is the -3 state. When phosphorus is heated with some metals, **phosphides**, such as calcium phosphide, Ca_3P_2, and aluminum phosphide, AlP, are formed. If these phosphides are treated with dilute aqueous acid, *phosphine*, PH_3, is formed:

$$Ca_3P_2(s) + 6HCl(aq) \longrightarrow 2PH_3(g) + 3CaCl_2(aq)$$

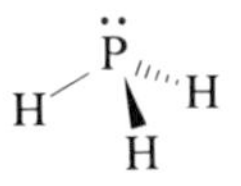

Phosphine is an unpleasant-smelling toxic gas. It has a pyramidal structure like ammonia. It is much less soluble in water than ammonia, and it is an exceedingly weak base. In other words, the equilibrium

$$PH_3 + H_2O \rightleftharpoons PH_4^+ + OH^-$$

lies far to the left.

Very few salts of the tetrahedral phosphonium ion, PH_4^+, are known. Phosphonium iodide, PH_4I, is an example. It reacts with water, liberating phosphine:

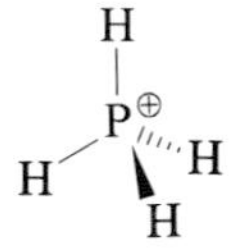

$$PH_4^+ + H_2O \rightleftharpoons H_3O^+ + PH_3$$

Because PH_3 is an exceedingly weak base, its conjugate acid, PH_4^+, is a strong acid. Thus the equilibrium lies far to the right.

Phosphine

Phosphonium ion

7.5 STRUCTURE AND ACID-BASE CHARACTER

In Chapter 5 and in this chapter we have described a number of different acids and bases. We are now in a position to give some answers to these questions: What makes a substance an acid or a base? What determines the strength of an acid or a base? We first consider the nonmetal hydrides.

Nonmetal Hydrides

As Figure 7.11 shows, acid-base strengths vary in a regular way in the periodic table. In Groups IV to VII the acid strength of the hydrides increases from left to right and the base strengths decrease.

Figure 7.11 Acid and Base Strengths of Nonmetal Hydrides.

	Group			
	IV	V	VI	VII
Period 2	CH_4	NH_3	H_2O	HF
	No acid or base properties	Weak base	Amphoteric	Weak acid
3	SiH_4	PH_3	H_2S	HCl
	No acid or base properties	Weak base	Weak acid	Strong acid

Acid strength increases →

Base strength increases →

Base strength increases (vertical)

Acid strength increases (vertical)

We can explain this trend by comparing electronegativities. As the electronegativity of X in the molecule HX increases, the X—H bond becomes more polar, and the tendency of the molecule to donate a proton increases. In other words, *the acid strength of HX increases with increasing electronegativity of X*. The ability of a molecule to behave as a base depends on its having one or more unshared electron pairs that can accept a proton from an acid. With increasing electronegativity an atom holds onto its unshared electron pairs more tightly and has a decreasing tendency to share them with a proton. Therefore *the base strength of HX decreases as the electronegativity of X increases*.

Consider the hydrides shown in Figure 7.11. Methane has nonpolar C—H bonds and therefore has no acidic properties. Since it has no unshared electron pairs, methane also cannot behave as a base. Ammonia, however, has an unshared pair of electrons, and it does behave as a base. Since nitrogen has a higher electronegativity than hydrogen, the N—H bonds in ammonia are more polar than the bonds in methane, but they are still not polar enough for ammonia to behave as an acid in water. The O—H bonds in the water molecule are more polar, and, indeed, water is a very weak acid. In addition, water has two unshared electrons pairs and also behaves as a weak base, accepting protons from acids such as HCl. Water therefore has both acid and base properties. Because fluorine is still more electronegative than oxygen, the H—F bond is more polar than the H—O bonds in water; therefore HF is a stronger acid than H_2O. Although the HF molecule has three unshared electron pairs, these electron pairs are held very strongly by the highly electronegative fluorine atom, and it has very little tendency to share them with a proton. The HF molecule therefore does not behave as a base in water.

Note that the base strength of a molecule does not depend on how many unshared electron pairs it has but only on the availability of such pairs for sharing with a hydrogen ion. Note also that although the water molecule has two unshared pairs of electrons, it only adds one proton to give the hydronium ion, H_3O^+. The positive charge thus produced on the oxygen atom strongly attracts the remaining unshared pair and makes it much less available for sharing with another proton. Thus the ion H_4O^{2+} has never been observed.

In period 3 the basicity of the hydrides again decreases, and their acidity increases from group IV to group VII (see Figure 7.11). But the hydrides of the third-period elements are less basic and more acidic than those in the second period, even though the period 3 elements are less electronegative than the corresponding period 2 elements. Clearly, electronegativity is not the only factor determining acid and base strength. What else, then, is important?

An important difference between the elements in these periods is that the atoms of the third-period elements are considerably larger (Figure 4.7). Because the hydrogen is farther from the nucleus of the larger atom, the attraction between the departing H^+ and the negative ion that remains is correspondingly smaller. Hence it is easier to remove a hydrogen atom attached to a larger atom than to a smaller atom. So the *acid strength of a molecule HX increases with increasing size of X*. Thus HCl is a stronger acid than HF, and H_2S is a stronger acid than H_2O.

The size of an atom also influences base strengths. The unshared electron pair on a large atom such as phosphorus in PH_3 is more spread out and less localized than the unshared pair on the much smaller nitrogen atom in NH_3. It therefore exerts a smaller attraction on a proton. Hence PH_3 is a weaker base than NH_3. In general, *the basicity of an atom with one or more unshared electrons pairs decreases with increasing size of the atom.*

Oxoacids

Phosphoric acid, phosphorous acid, and sulfuric acid are members of an important class of compounds known as **oxoacids**. Oxoacids have one or more OH groups attached to an electronegative atom. They have the general formula $XO_m(OH)_n$, where $m = 0, 1, 2, \ldots$ and $n = 1, 2, 3, \ldots$. For example, H_3PO_4 can be written as $PO(OH)_3$, where $m = 1$ and $n = 3$; H_2SO_4 can be written as $SO_2(OH)_2$, where $m = 2$ and $n = 2$.

Other important oxoacids include perchloric acid, $HClO_4$ or $ClO_3(OH)$, and nitric acid, HNO_3 or $NO_2(OH)$.

The oxoacids $XO_m(OH)_n$ have one or more hydroxyl groups. But we know that some compounds containing OH groups, such as NaOH and $Ca(OH)_2$, are ionic hydroxides, Na^+OH^- and $Ca^{2+}(OH^-)_2$, and are therefore strong bases. What, then, determines if a compound containing OH groups is an acid or a base?

The structures and acid-base properties of the hydroxides of the elements of the second and third periods are shown in Figure 7.12. We see that the hydroxides of metals are bases, while the hydroxides of nonmetals are acids. More specifically, we see that the acid strength of the hydroxides increases from left to right across each period, and the base strength decreases. The alkali metal hydroxides on the left are strong bases, $Be(OH)_2$ and $Al(OH)_3$ are amphoteric, and the nonmetal hydroxides are acids, their strength increasing from left to right. In period 2 boric acid, $B(OH)_3$, and carbonic acid, $CO(OH)_2$ are weak, and nitric acid, $NO_2(OH)$, is strong. In period 3 silicic acid, $Si(OH)_4$, and phosphoric acid, $PO(OH)_3$, are weak, while sulfuric acid, $SO_2(OH)_2$, and perchloric acid, $ClO_3(OH)$, are strong.

If the atom X in XOH has a low electronegativity, the X—O bond is polar,

Figure 7.12 Acid and Base Strengths of Hydroxyl Compounds.

	Group I	Group II	Group III	Group IV	Group V	Group VI	Group VII
Period 2	LiOH	$Be(OH)_2$	$B(OH)_3$	$[C(OH)_4]$	$[N(OH)_5]$		
				$-H_2O$	$-2H_2O$		
	Li^+OH^-	$Be(OH)_2$	$B(OH)_3$	$O{=}C(OH)_2$	$O{=}N^{\oplus}(O^{\ominus})-OH$		
	Strong base	Amphoteric	Weak acid	Weak acid	Strong acid		
Period 3	NaOH	$Mg(OH)_2$	$Al(OH)_3$	$Si(OH)_4$	$[P(OH)_5]$	$[S(OH)_6]$	$[Cl(OH)_7]$
					$-H_2O$	$-2H_2O$	$-3H_2O$
	Na^+OH^-	$Mg^{2+}(OH^-)_2$	$Al(OH)_3$	$Si(OH)_4$	$O{=}P(OH)_3$	$(O{=})_2S(OH)_2$	$(O{=})_3Cl{-}OH$
	Strong base	Weak base	Amphoteric	Weak acid	Weak acid	Strong acid	Strong acid

Acid strength increases →

← Base strength increases

with the oxygen having a negative charge and X a positive charge:

$$\overset{\delta+}{X}-\overset{\delta-}{O}-H$$

When this is the case, XOH ionizes to give X^+ and OH^- as in NaOH and $Mg(OH)_2$. When X has a high electronegativity, the X—O bond is less polar but the O—H bond becomes more polar

$$X-\overset{\delta-}{O}-\overset{\delta+}{H}$$

Hence there is an increased tendency for the hydrogen to be donated as a proton to a base to give XO^- and BH^+. Thus acid strength increases with increasing electronegativity of X, and base strength decreases, as shown in Figure 7.12.

The oxoacids (the hydroxides of the nonmetals) often do not have the maximum number of hydroxyl groups that would be expected from their valences. Instead, pairs of —OH groups are often replaced by a doubly bonded oxygen, =O. They can be imagined as being derived from the hypotnetical hydroxides with the maximum number of OH groups by loss of one or more water molecules. For example,

$$HO-\underset{\underset{\textstyle OH}{|}}{\overset{\overset{\textstyle OH}{|}}{C}}-OH \longrightarrow O{=}C\overset{\diagup OH}{\diagdown OH} + H_2O$$

and

$$\underset{HO}{\overset{HO}{>}}\underset{\underset{OH}{|}}{\overset{\overset{OH}{|}}{S}}\underset{OH}{\diagdown} \longrightarrow O{=}\underset{\underset{\textstyle OH}{|}}{\overset{\overset{\textstyle O}{||}}{S}}-OH + 2H_2O$$

In the oxoacids $XO_m(OH)_n$ acid strength increases with the number of doubly bonded oxygen atoms on the central atom X, that is, with the value of m. The electronegative oxygen atom attracts electrons from X, which in turn causes X to attract electrons from other atoms more strongly. In other words, the more oxygen atoms that are attached to X, the more its electronegativity is increased, which in turn increases the strength of the acid. In general, for an acid $XO_m(OH)_n$, if $m = 0$ or 1 the acid is weak, but if $m = 2$ or 3 the acid is strong. We see from Figure 7.12 that boric acid and silicic acid, with $m = 0$, and phosphoric acid and carbonic acid, with $m = 1$, are weak. But nitric acid and sulfuric acid, with $m = 2$, and perchloric acid, with $m = 3$, are strong.

This correlation of acid strength with the number of doubly bonded oxygen atoms also implies that acid strength increases with the increasing oxidation state of the central atom. Sulfurous acid, H_2SO_3 (oxidation number S $= +4$), is a weak acid, while sulfuric acid, H_2SO_4 (oxidation number S $= +6$), is a strong acid. Hypochlorous acid, ClOH (oxidation number Cl $= +1$), is a weak acid, while perchloric acid, $HClO_4$ (oxidation number Cl $= +7$), is a strong acid.

Example 7.5 Classify the following acids as either strong or weak:

$$HNO_2 \qquad HBrO_4 \qquad Si(OH)_4 \qquad H_2SeO_4$$

Solution

HNO_2	$NO(OH)$	$m = 1$	Weak
$HBrO_4$	$BrO_3(OH)$	$m = 3$	Strong
$Si(OH)_4$		$m = 0$	Weak
H_2SeO_4	$SeO_2(OH)_2$	$m = 2$	Strong

Some of the compounds we have encountered so far have names that give no clue to their composition; water, ammonia, methane, and acetic acid are examples. If all compounds were named in such an unsystematic way, chemists would have to memorize hundreds of thousands of names. Fortunately, the names of most compounds are related to their composition in a systematic way. Now that we have examined oxidation states as well as electronegativity, we can introduce the methods of naming inorganic compounds; the names of organic compounds are discussed in later chapters.

Binary Compounds

The simplest compounds are **binary compounds**; they are composed of only two elements. We name them by giving first the name of the less electronegative (more metallic) element and then the stem of the name of the other element (almost always a nonmetal), to which the suffix *ide* has been added. The following examples give the names of some binary compounds:

NaCl	Sodium chloride	$CaCl_2$	Calcium chloride
KBr	Potassium bromide	Na_2S	Sodium sulfide
MgO	Magnesium oxide	Mg_3N_2	Magnesium nitride
HBr	Hydrogen bromide		

When two elements form more than one binary compound, we need some way of distinguishing the two compounds. The simplest method uses a Greek prefix to indicate the numbers of each atom in the formula. The prefixes are *mono* for one, *di* for two, *tri* for three, *tetra* for four, *penta* for five, and *hexa* for six. By this method we have the following names:

Carbon monoxide	CO	Carbon dioxide	CO_2
Sulfur dioxide	SO_2	Sulfur trioxide	SO_3
Phosphorus trichloride	PCl_3	Phosphorus pentachloride	PCl_5
Sulfur tetrafluoride	SF_4	Tetraphosphorus hexaoxide	P_4O_6

The prefix *mono* is generally omitted before the first element in the name.

Alternatively, we can distinguish two compounds of the same elements by specifying the oxidation state of the less electronegative element with a Roman numeral. For example, phosphorus in PCl_3 has an oxidation number of $+3$, while in PCl_5 it has an oxidation number of $+5$. Thus according to this system, the names of PCl_3 and PCl_5 are phosphorus(III) chloride and phosphorus(V) chloride. Other examples are as follows:

SO_2	Sulfur(IV) oxide	SO_3	Sulfur(VI) oxide
$FeCl_2$	Iron(II) chloride	$FeCl_3$	Iron(III) chloride

This method is more commonly used for the compounds of metals than for compounds between nonmetals. Thus the name phosphorus trichloride is more commonly used than phosphorus(III) chloride.

Note that when the stoichiometry of a compound is not indicated in its name, as, for example, in aluminum chloride, we must know that aluminum is in group III and has a valence of 3 in order to write the correct formula, $AlCl_3$. The names aluminum trichloride or aluminum(III) chloride are more specific, but they are not always used.

Binary Acids

The aqueous solutions of nonmetal hydrides that behave as acids are named by adding the prefix *hydro* and the suffix *ic* to the stem of the name of the nonmetal:

Binary hydride	Formula	Name of aqueous solution
Hydrogen fluoride	HF	Hydrofluoric acid
Hydrogen chloride	HCl	Hydrochloric acid
Hydrogen bromide	HBr	Hydrobromic acid
Hydrogen iodide	HI	Hydroiodic acid
Hydrogen cyanide	HCN	Hydrocyanic acid

This method of naming is commonly used only for the aqueous solutions of the hydrogen halides.

Oxoacids

If an element forms only one oxoacid, it is named by adding *ic acid* to the stem of the name of the element. For example:

$$H_2CO_3 \quad \text{Carbonic acid}$$
$$H_3BO_3 \quad \text{Boric acid}$$

When an element has more than one oxidation state, it may form more than one oxoacid. If the element forms only two oxoacids, the acid containing the element in the higher oxidation state is given the suffix *ic*; the acid having the element in the lower oxidation state is given the ending *ous*. Thus we have

$$H_3PO_4 \quad \text{Phosphoric acid}$$
$$H_3PO_3 \quad \text{Phosphorous acid}$$

The name of the anion derived from the *ic* acid ends in *ate*, while the name of the anion derived from the *ous* acid ends in *ite*:

H_2SO_4	Sulfuric acid	SO_4^{2-}	Sulfate
H_2SO_3	Sulfurous acid	SO_3^{2-}	Sulfite
HNO_3	Nitric acid	NO_3^-	Nitrate
HNO_2	Nitrous acid	NO_2^-	Nitrite
H_3PO_4	Phosphoric acid	PO_4^{3-}	Phosphate
H_2CO_3	Carbonic acid	CO_3^{2-}	Carbonate

A few elements form oxoacids in more than two oxidation states. In this case the prefixes *hypo* and *per* are used to designate a lower and a higher oxidation state, respectively. The oxoacids of chlorine, of which we have already briefly encountered HClO and $HClO_4$, provide a good example:

HClO	Hypochlorous acid	ClO^-	Hypochlorite
$HClO_2$	Chlorous acid	ClO_2^-	Chlorite
$HClO_3$	Chloric acid	ClO_3^-	Chlorate
$HClO_4$	Perchloric acid	ClO_4^-	Perchlorate

Many acids are *polyprotic*; they have more than one ionizable hydrogen and therefore give two or more anions. The three anions derived from phosphoric acid are dihydrogen phosphate, $H_2PO_4^-$, hydrogen phosphate, HPO_4^-, and phosphate, PO_4^{3-}. When it reacts with sodium hydroxide, phosphoric acid gives the following salts:

NaH_2PO_4	Sodium dihydrogen phosphate
Na_2HPO_4	Disodium hydrogen phosphate (or sodium hydrogen phosphate)
Na_3PO_4	Trisodium phosphate (or sodium phosphate)

Example 7.6 Name each of the following compounds:

(a) $ZnCO_3$ (b) SiF_4 (c) Na_2SO_3 (d) $HBrO_4$ (e) Li_3N

Solution

(a) Zinc carbonate   (b) Silicon tetrafluoride   (c) Sodium sulfite

(d) Perbromic acid   (e) Lithium nitride

Example 7.7 Write the formulas of each of the following compounds:

(a) Phosphorus triiodide   (b) Barium iodide

(c) Magnesium phosphate   (d) Manganese (IV) oxide

(e) Calcium hydrogen carbonate   (f) Hypobromous acid

Solution

(a) PI_3.

(b) The formula is BaI_2. It is not BaI because barium is in Group II and has a valence of 2 and forms the Ba^{2+} ion. Since BaI_2 is ionic we can also write the formula as $Ba^{2+}(I^-)_2$.

(c) The formula is $Mg_3(PO_4)_2$. Magnesium is in Group III and therefore forms the Mg^{2+} ion. Thus we can write the formula as $(Mg^{2+})_3(PO_4^{3-})_2$. The sum of the positive charges must equal the sum of the negative charges, so the formula is *not* $MgPO_4$.

(d) The formula is MnO_2. The Roman numeral IV indicates that manganese is in the oxidation state $+4$ and has an oxidation number of $+4$. Therefore the formula must be MnO_2, assuming, as is usual, that oxygen has oxidation number -2.

(e) The formula is $Ca(HCO_3)_2$. The hydrogen carbonate ion is HCO_3^-, and calcium forms the Ca^{2+} ion.

(f) The formula is $HOBr$. This compound is an oxoacid of bromine in which bromine is in the $+1$ oxidation state. It is named in the same way as $HOCl$, hypochlorous acid.

IMPORTANT TERMS

Allotropes are different forms of an element, such as white, red, and black phosphorus.

A **binary compound** is composed of only two elements.

Condensation is the reaction of two or more molecules to form a larger molecule with the elimination of small molecules, such as water.

A **diprotic acid** has two ionizable hydrogen atoms. A **triprotic acid** has three ionizable hydrogen atoms.

A **half-reaction** describes either the oxidation or the reduction part of an oxidation-reduction reaction. Electrons appear on the right side of the equation describing the oxidation half-reaction and on the left side of the equation describing the reduction half-reaction.

Le Châtelier's principle states that when one of the conditions that affect the position of a dynamic equilibrium is changed, the position of the equilibrium shifts in such a way as to minimize the effect of the change.

Oxidation is any change in which the oxidation number of an atom increases.

The **oxidation number** is a measure of the extent of oxidation of an element in its compounds. It is assigned by using a set of rules. For a monatomic ion it is equal to the charge on the ion. For an atom in a covalent molecule it is equal to the charge that the atom would have if both electrons of each covalent bond were assigned to the more electronegative element.

Oxoacids have one or more OH groups attached to an electronegative atom. They have the general formula $XO_m(OH)_n$.

Reduction is any change in which the oxidation number of an atom decreases.

PROBLEMS

Sulfur

1. Name and briefly describe three allotropes of sulfur. Describe and explain what is observed when sulfur is heated to its boiling point.

2. A solution is known to contain either SO_3^{2-} or SO_4^{2-} (but not both). Suggest two tests that could be used to identify the anion.

3. Write balanced equations for the reactions of sulfuric acid with each of the following. In each case, state whether the reaction occurs at room temperature or only at high temperature and whether dilute or concentrated sulfuric acid should be used.

(a) $CaSO_4 \cdot 2H_2O(s)$

(b) An aqueous solution of NaOH

(c) An aqueous solution of $BaCl_2$

(d) Ethanol, C_2H_5OH (e) NaF(s)

(f) $NaNO_3(s)$ (g) Carbon(s)

4. Write a balanced equation for each of the following:

(a) A reaction in which sulfuric acid acts as an oxidizing agent.

(b) A reaction in which sulfuric acid acts as an acid.

(c) A reaction in which sulfuric acid acts as a dehydrating agent.

5. Draw box diagrams showing the valence state electron configurations of sulfur when it has valences of 2, 4, and 6. Explain why sulfur cannot have a valence of 8. Explain why oxygen cannot have valences of 4 and 6.

6. Give two examples of the use of concentrated sulfuric acid for the preparation of more volatile acids. Why is this method not suitable for the preparation of hydrogen iodide?

7. (a) Benjamin Disraeli, the British statesman, said: "There is no better barometer to show the state of an industrial nation than the figure representing the annual consumption of sulfuric acid per head of the population." Explain, by means of specific examples of the use of sulfuric acid, why this statement is still largely true.

(b) Suggest a reason why the industrial process for making sulfuric acid is called the contact process.

(c) Explain why sulfuric acid should always be diluted by adding the acid to water and not the other way round.

8. A compound containing only iron and sulfur was analyzed by heating a sample in air to convert all the sulfur to sulfur dioxide. A 0.4203-g sample of the compound produced 173.6 mL of SO_2, measured at a pressure of 750 mm Hg and 25°C. What is the empirical formula of the compound?

* The asterisk denotes the more difficult problems.

9. (a) What property of sulfuric acid is illustrated by its reaction with copper?

(b) Suppose 50.0 g of pure H_2SO_4 is heated with copper. What volume of SO_2 is produced at STP?

(c) How much pure H_2SO_4 is needed to make 50.0 mL of a solution having a concentration of 0.0500 mol L^{-1}? Will this solution react with copper? Explain.

10. Some well water and springwater is contaminated with small amounts of hydrogen sulfide, which causes it to have a bad smell. Hydrogen sulfide can be removed by treating the water with chlorine, which oxidizes the H_2S to sulfur. Write the balanced equation for this reaction. The H_2S content of water from a particular source is 22 ppm. Calculate the amount of chlorine needed to remove the H_2S from 5000 L of water.

11. One method that can be used for the removal of SO_2 from the flue gases of power plants is to react it with an aqueous solution of H_2S. One product of the reaction is sulfur. Write a balanced equation for the reaction. What volume of H_2S at 25°C and 750 mm Hg is required to remove the SO_2 formed by burning 1 metric ton (10^6 g) of coal containing 4.00% sulfur by mass? What mass of elemental sulfur is produced?

Phosphorus

12. Explain the following properties of the allotropes of phosphorus in terms of their structures:

(a) White phosphorus has a lower melting point than black phosphorus.

(b) White phosphorus dissolves in CS_2, but red phosphorus does not.

13. Write an equation for the combustion of PH_3 in air to give P_4O_{10}.

14. Show by means of equations how phosphate rock can be used as a source of phosphoric acid, sodium phosphate, white phosphorus, and superphosphate-of-lime fertilizer.

15. One of the components of a match head is phosphorus sulfide, P_4S_3. Propose a structure for P_4S_3 based on the structures of P_4 and P_4O_6.

16. Suggest possible explanations for the fact that PCl_5 has been made, but PH_5 and NCl_5 have not been made.

17. Striking a wooden match involves the combustion of phosphorus sulfide, P_4S_3, to produce a white smoke of P_4O_{10} and gaseous sulfur dioxide, SO_2. Write a balanced equation for the reaction. Calculate the volume of SO_2 at 20°C and 772 mm Hg pressure that results from the combustion of 0.157 g of P_4S_3.

*18. Dry hydrogen fluoride, HF, reacts with P_4O_{10} to give a gas containing 29.8% P, 54.8% F, and 15.4% O. The gas has a density of 4.46 g L^{-1} at STP. What is the molecular formula of the gas? Draw a possible Lewis structure for this compound, and give the oxidation numbers of each of the elements.

*19. When aluminum phosphide, AlP, was dissolved in dilute sulfuric acid, a gas X with a density of 1.531 g L^{-1} at STP was evolved. When the solution was evaporated to dryness and the solid product was recrystallized from water and heated to dryness at a high temperature, a salt containing 15.8% Al, 28.1% S, and 56.1% O was obtained. When 0.158 g of the gas X reacted completely with hydrogen iodide, 0.752 g of a white crystalline compound that contained 19.1% P, 2.5% H, and 78.4% I was obtained. What was the gas? Write balanced equations for all the reactions.

Oxidation-Reduction

20. Define *oxidation* and *reduction* in terms of oxidation numbers.

21. What are the principal oxidation states of phosphorus? Give two examples for each oxidation state.

22. What are the oxidation numbers of the atoms in the following substances?

(a) PH_3 (b) AsH_3 (c) PF_3
(d) $K^+MnO_4^-$ (e) SiO_2 (f) TeF_6
(g) PH_4^+

23. What are the oxidation numbers for each of the following:

(a) Carbon in CO
(b) Silicon in SiO_2
(c) Phosphorus in $Ca_5(PO_4)_3(OH)$
(d) Sulfur in SF_6
(e) Sulfur in S_2^{2-}
(f) Phosphorus in $H_2PO_3^-$

24. What are the oxidation numbers of each of the elements in the following substances?

(a) H_2O_2 (b) Li_2O (c) PH_4I
(d) $NaClO_4$ (e) OF_2 (f) CaS_2
(g) AlP   (h) S_8 (i) $Ba(OCl)_2$

25. When a metal bromide is heated with concentrated sulfuric acid, the bromide ion is oxidized to bromine and the sulfuric acid is reduced to sulfur dioxide. Write a balanced equation for this reaction. When sulfur dioxide is bubbled through an aqueous solution of bromine, the bromine is reduced to bromide ion and the sulfur dioxide is oxidized to sulfate. Write a balanced equation for this reaction. Explain why this reaction proceeds in one direction under one set of conditions and in the reverse direction under another set of conditions.

26. Write equations for the following half-reactions:

(a) The oxidation of H_2S to S in aqueous solution.
(b) The oxidation of SO_2 to SO_4^{2-} in aqueous solution.

(c) The oxidation of H_3PO_3 to H_3PO_4 in aqueous solution.

Lewis Structures

27. Draw Lewis structures for each of the following species:

(a) H_2SO_4 (b) $H_2S_2O_7$
(c) SO_4^{2-} (d) HSO_3^-

28. Draw Lewis structures for each of the following species:

(a) SO_2 (b) SO_3 (c) SO_3^{2-}
(d) $S_2O_3^{2-}$ (e) $S_2O_7^{2-}$

29. Draw Lewis structures for each of the following:

(a) H_3PO_4 (b) H_3PO_3 (c) PH_4^+
(d) HPO_4^{2-} (e) PCl_3

Acids and Bases

30. Classify the following acids as either strong or weak:

(a) $B(OH)_3$ (b) Phosphorous acid
(c) H_2SeO_3 (d) Nitric acid

31. Explain the acid and base properties of the following hydrides:

CH_4	NH_3	H_2O	HF
No acid or base properties	Weak base	Very weak base	No base properties in water
	No acid properties in water	Very weak acid	Weak acid

32. Explain the acid and base properties of the following hydrides:

SiH_4	PH_3	H_2S	HCl
No acid or base properties	Very weak base	No base properties in water	No base properties in water
	No acid properties in water	Weak acid	Strong acid

33. Explain why HCl is a stronger acid than HF and why H_2S is a stronger acid than H_2O.

34. Explain why the anion $H_{m-1}XO_n^-$ is always a weaker acid than the parent acid H_mXO_n (for example, HSO_4^- versus H_2SO_4).

35. On the basis of the composition of the oxoacids of carbon and nitrogen, explain why there is no oxoacid of oxygen.

36. What is the relationship between the following, mostly unknown hydroxides, $B(OH)_3$, $C(OH)_4$, $N(OH)_5$, $Si(OH)_4$, $P(OH)_5$, $S(OH)_6$, and the known oxoacids of these elements?

37. Explain why KOH and $Ca(OH)_2$ are bases, whereas $Si(OH)_4$ is an acid.

Nomenclature

38. Write the formulas for each of the following compounds:

(a) Hydrofluoric acid   (b) Phosphoric acid

(c) Phosphorous acid   (d) Hypochlorous acid

39. Write the formulas for each of the following compounds:

(a) Calcium sulphate

(b) Phosphorus pentabromide

(c) Phosphorus(III) iodide

(d) Ammonium hydrogen phosphate

(e) Calcium nitride

(f) Sulfur tetrafluoride

(g) Chromium(III) chloride

40. Name each of the following compounds:

(a) K_3PO_4 (b) CF_4 (c) $ZnCO_3$

(d) $Ca(OCl)_2$ (e) $CaSO_3$

41. Name each of the following compounds:

(a) K_2SeO_4 (b) H_2Te (c) Na_2S_4

(d) FeS_2 (e) $RbHSO_4$ (f) P_4O_6

42. What is the name of each of the following acids?

(a) H_2SO_4 (b) H_3PO_4 (c) H_3PO_3

(d) H_2CO_3 (e) HNO_3 (f) H_3BO_3

(g) H_4SiO_4 (h) $HClO$ (i) $HClO_4$

43. Write a balanced equation for the reactions of each of the following acids with sodium hydroxide in aqueous solution, and name each of the salts that could be obtained from the solutions.

(a) Nitric acid, HNO_3

(b) Sulfuric acid, H_2SO_4

(c) Phosphoric acid, H_3PO_4

(d) Phosphorous acid, H_3PO_3

(e) Diphosphoric acid, $H_4P_2O_7$

Miscellaneous

44. What species are present in aqueous solutions of the following? Give both names and formulas.

(a) Sulfuric acid   (b) Sulfur dioxide

(c) Phosphoric acid   (d) Carbonic acid

45. Classify each of the following reactions in aqueous solution as an acid-base reaction or an oxidation-reduction reaction, and balance the equations. In the case of the acid-base reactions, identify the acid-base conjugate pairs. In the case of the oxidation-reduction reactions, identify the elements that are oxidized and those that are reduced.

(a) $CO_3^{2-} + H_2SO_4 \longrightarrow HSO_4^- + HCO_3^-$

(b) $HSO_4^- + H_2O \rightleftharpoons H_3O^+ + SO_4^{2-}$

(c) $Cu + 2H_2SO_4 \longrightarrow$
$$Cu^{2+} + SO_4^{2-} + H_2O + SO_2$$

(d) $2Br^- + 2H_2SO_4 \longrightarrow$
$$Br_2 + SO_4^{2-} + SO_2 + H_2O$$

(e) $CaCO_3 + H_3O^+ \longrightarrow Ca^{2+} + HCO_3^- + H_2O$

(f) $Mg + H_3O^+ \longrightarrow Mg^{2+} + H_2 + H_2O$

(g) $Ca_3P_2 + 6H_2O \longrightarrow 3Ca(OH)_2 + 2PH_3$

46. Each of the following sets of compounds has a common structural feature. In each case, explain what it is.

(a) P_4, P_4O_6, and P_4O_{10}

(b) $(HPO_3)_3$ and $(SO_3)_3$

(c) PO_4^{3-}, HPO_3^{2-}, PH_4^+, SO_4^{2-}, and $S_2O_3^{2-}$

47. Draw orbital box diagrams for the electron configurations of arsenic and selenium. Predict the formulas of the hydrides of these elements. Arsenic forms an unstable chloride, $AsCl_5$, and selenium forms the fluorides SeF_4 and SeF_6. Give the orbital box diagrams for the valence states of the elements on which these higher-valence compounds are based.

48. Explain in terms of Le Châtelier's principle why sulfuric acid and phosphoric acid can be used to prepare HCl from sodium chloride but nitric acid cannot be used for this purpose.

*49. Phosphine may be prepared by heating white phosphorus with sodium hydroxide.

$$P_4(s) + 3NaOH(aq) + 3H_2O \longrightarrow PH_3(g) + 3NaH_2PO_2$$

(a) Give the formula of the acid from which the salt NaH_2PO_2 can be obtained.

(b) Name this acid.

(c) The acid is monoprotic. Draw a Lewis structure for the acid.

(d) Assign oxidation numbers to phosphorus in each of its compounds in the equation.

(e) What species is oxidized, and what species is reduced?

(f) Suggest why NH_3 cannot be prepared by a similar reaction from nitrogen.

(g) How many grams of phosphorus are required to give 100 mL of phosphine at 0.95 atm and 25°C?

*50. A yellow solid, A, was heated in air to give a colorless gas, B, which was dissolved in water to give a solution, C. Another sample of the yellow solid A was mixed with iron powder and heated to produce a black solid, D. Dilute hydrochloric acid was added to the solid D, and a colorless gas, E, was formed. This gas was bubbled through solution C, and a finely divided pale yellow precipitate, F, was formed. Identify A, B, C, D, E, and F, and write balanced equations for all the reactions.

CHAPTER 8
MOLECULAR GEOMETRY

VSEPR Theory, Lewis Structures,
Resonance Structures, and
Dipole Moments

Molecules have a fascinating variety of different shapes. Some are long and thin, some are round, some are flat, some are rings, and others are spirals. We have already mentioned the shapes of several simple molecules. Water is an angular molecule; ammonia is a pyramidal molecule; methane has a tetrahedral shape, as does the P_4 molecule; and the S_8 molecule is a ring with a crown shape. Larger molecules may have much more complicated shapes. One of the most complex molecules that has been studied is the DNA molecule, which has an overall, beautifully simple, double spiral shape, although the detailed structure of each of the long spiral chains is very complex (see Figure 8.1). However, we will see that the arrangements of atoms even in very complex molecules are based on some very simple principles.

The properties of a substance are determined in the first instance by the kinds of atoms it contains, that is, by its composition. But hardly less important is the geometric arrangement of the atoms in a molecule and the nature and the strength of the bonds between them. The shape of the DNA molecule is intimately related to its function of transmitting information from one generation to the next. Again, we only have to recall the difference in the properties of allotropes, such as white and black phosphorus or orthorhombic and plastic sulfur, to appreciate the importance of the arrangement of the atoms and the bonds in a substance in determining its properties.

Some molecules can have more than one shape, and they can change their shape—sometimes easily, sometimes with difficulty. A fascinating example is provided by the molecule retinal, which is found in the retina of the eye. It changes

**Figure 8.1 Two Models of the
DNA Molecule.** The DNA
molecule consists of two long
chains that spiral around each
other and are linked by weaker
interactions at various places, as
indicated by the dotted lines.

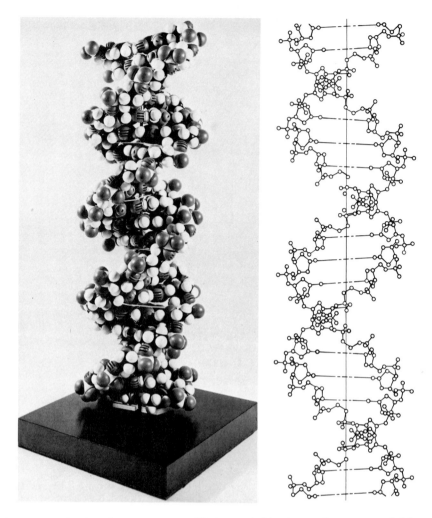

shape when it absorbs a quantum of light, and this change in structure initiates
the electric signal transmitted to the brain (see Figure 8.2).

We have seen that *ionic bonds are nondirectional*; an ion attracts to itself as
many ions of opposite charge as can be packed around it. The structure of an
ionic crystal is therefore determined primarily by the relative sizes of the ions
of which it is composed. In contrast, *covalent bonds are directional*; an atom
forms covalent bonds only in certain specific directions. **Molecular geometry**—
that is, the geometric arrangement of the atoms in molecules and in covalently
bonded crystals—depends on the directions of the covalent bonds formed by
each atom.

Although chemists realized as long ago as the middle of the past century
that the atoms in molecules have definite arrangements in space, they had no
direct methods of determining the structures of molecules. They were able to
formulate some qualitative ideas concerning molecular geometry by studying
substances that have the same formula but have different properties. They concluded that the reason for these different properties is that the same atoms may
be arranged in different ways to give different molecules (see Box 8.1).

The general arrangement of the atoms in a large number of molecules was
deduced in this way, but not until the present century were methods developed
by which the distances between the atoms in a molecule could actually be mea-

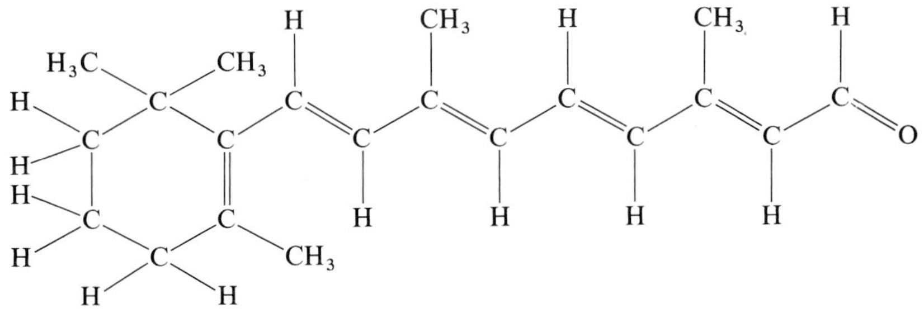

trans-retinal

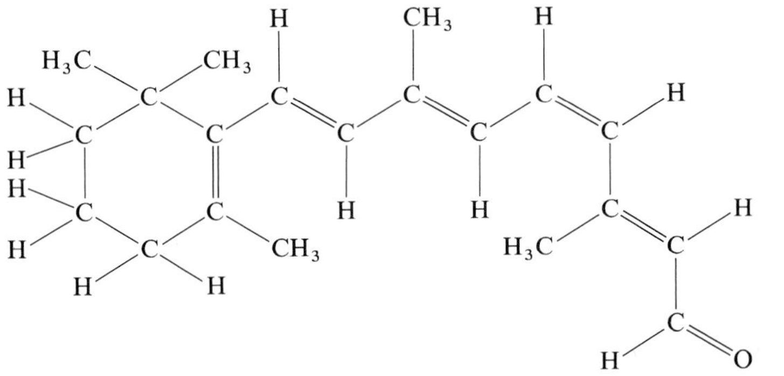

cis-retinal

Figure 8.2 Molecular Shape and Vision. Retinal, $C_{20}H_{27}O$, is an oxidized form of Vitamin A (retinol), $C_{20}H_{29}O$. It has two different shapes, called *cis* and *trans*. In the cells of the eye that are responsible for vision, *cis*-retinal is combined with a protein called opsin to give a substance called rhodopsin. When rhodopsin absorbs light, it decomposes to opsin and retinal, but the retinal is in the trans form. In some way not yet understood, the change in shape of the retinal molecules initiates the signal transmitted to the brain.

Box 8.1

HOW CHEMISTS DEDUCED SHAPES OF MOLECULES BEFORE
THE INVENTION OF X RAY CRYSTALLOGRAPHY

If we were to successively replace the hydrogen atoms in methane by halogen atoms, we could make the molecules CH_3F, CH_2FCl, and $CHFClBr$. Suppose that these molecules have a *square planar shape*; we predict that the following different molecules would exist:

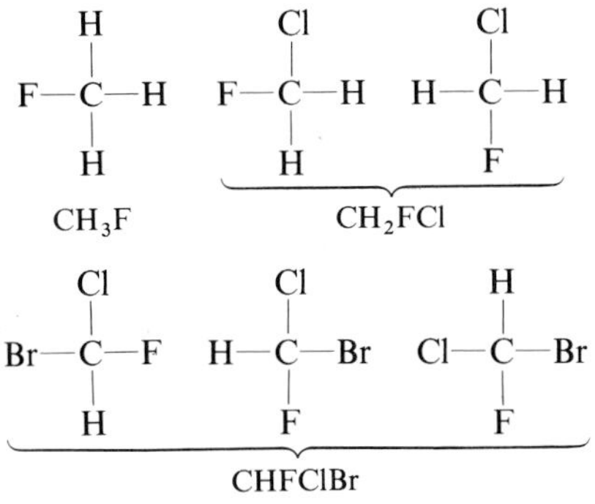

On the basis of a square planar shape, we expect only one type of molecule with the formula CH_3F but two molecules with the formula CH_2FCl and three molecules

with the formula $CHFClBr$. In fact, there is only one substance with the formula CH_3F and only one with the formula CH_2FCl, but there are two substances with the formula $CHFClBr$. Evidence like this led chemists to conclude that the four bonds formed by a carbon atom do not have a planar arrangement but, rather, a tetrahedral arrangement, because this arrangement leads to the prediction of the correct number of compounds with the same formula (*isomers*):

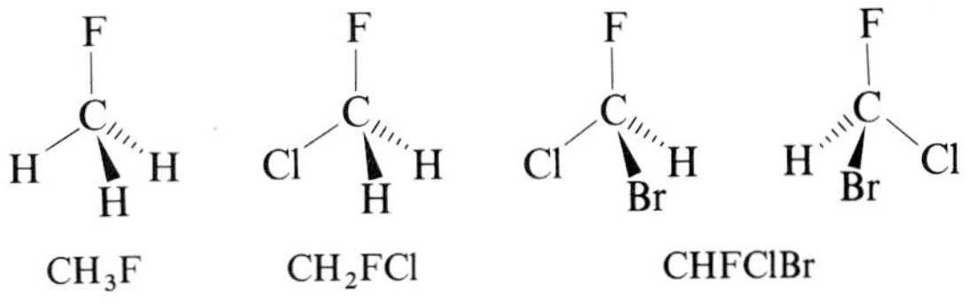

The two isomers of $CHFClBr$ differ in a rather subtle way; they are like right- and left-handed gloves. Right- and left-handed molecules, called *chiral molecules*, can be recognized by the fact that they rotate the plane of polarization of light in opposite directions.

Cl—Be—Cl

O
H   H

CH₄

H
C
H   H
H

sured and the geometry of a molecule completely defined. Most of these modern methods are based on the interaction of different kinds of electromagnetic radiation with molecules. The most important and generally useful of these methods is the diffraction of X rays by crystalline solids, which will be described in Chapter 10. Other methods, which are particularly applicable to gases, are the diffraction of a beam of electrons (electron diffraction) and the absorption of infrared and microwave radiation by molecules (infrared and microwave spectroscopy).

By means of these different experimental methods it is often possible to determine distances between the atoms in a molecule or in a crystalline solid to an accuracy of 0.1 pm, or better, and the angles between bonds (bond angles) to 0.1°. The measurement of distances to such an accuracy would appear to imply that atoms are very rigidly fixed in position in molecules. However, the atoms in a molecule are actually vibrating around a mean position. So it is the distance between their mean positions that is measured. If the vibrations are large, then one cannot measure the distances with great accuracy. In some molecules atoms are actually continually moving from one position to another. The study of these intriguing molecules, however, must be left to more advanced courses.

We will be concerned in this chapter with the results of the experimental determination of the shapes of simple molecules and with the principles that govern their shapes. The water molecule has an angular shape, but the beryllium chloride, BeCl₂, molecule has a linear shape. The methane molecule is tetrahedral, not planar. But boron trifluoride, BF₃, does have a planar shape, while nitrogen trifluoride, NF₃, has a pyramidal shape like ammonia. We use several different types of models to help us visualize the shapes of molecules. These are described in Box 8.2.

Box 8.2

To visualize the shapes of molecules, you will find it very helpful to have a set of molecular models. These models do not have to be elaborate; they can be constructed even from gum drops and toothpicks. Relatively inexpensive sets are available commercially.

Stick models use plastic or metal forms consisting of a number of spokes radiating from a central point. Plastic tubes can be pushed onto the spokes to represent the bonds to other atoms and nonbonding electron pairs. The atoms are not directly represented in these models. Alternatively, plastic or wooden balls, drilled with holes at appropriate angles, can be used to represent atoms; wooden or plastic rods on stiff springs can be used to represent bonds. This kind of model is often called a *ball-and-stick model*. It represents only the directions of the bonds in space, not the relative sizes of the atoms. Nevertheless, a ball-and-stick model gives a reasonably good representation of the general shape of a molecule and the relative orientations of the bonds.

Space-filling models are more elaborate. Since bonded atoms share electron density, they come together more closely than do non-bonded atoms; thus the atoms in a molecule are not spherical. In space-filling models they are represented by spheres with one or more slices cut off at suitable angles; each of the flat faces thus formed has a fastener, which enables it to be connected to similar spheres. These models give no direct representation of bonds, but compared with stick models, they give a much more accurate representation of the shape of a molecule and of the relative sizes of its atoms.

What determines the different shapes of these apparently similar molecules? A simple theory enables us to understand and predict the directions in which covalent bonds are formed and therefore to understand the shapes of molecules. It is called the VSEPR theory (valence shell electron pair repulsion theory).

8.1 VALENCE SHELL ELECTRON PAIR REPULSION (VSEPR) THEORY

The basic idea of the **VSEPR theory** is as follows:

> **The geometrical arrangement of the bonds around an atom in a molecule depends on the total number of electron pairs in the valence shell of the atom, including both bonding pairs and nonbonding (unshared) pairs.**

In a molecule AX_n, where A is a central atom to which n other atoms X are attached by single electron pair bonds, the arrangement of the atoms X around the central atom A depends on the total number of electron pairs in the valence shell of the central atom A. To find the number of electron pairs in the valence shell of the central atom in a molecule AX_n, we fiirst draw its Lewis structure:

$$:\ddot{C}l:Be:\ddot{C}l: \qquad :\ddot{F}:\ddot{B}:\ddot{F}: \qquad H:\overset{H}{\underset{H}{C}}:H \qquad H:\overset{}{\underset{H}{N}}:H \qquad :\overset{}{\underset{H}{O}}:H$$

We then count the number of electron pairs in the valence shell of the central atom. In $BeCl_2$ there are only two electron pairs in the valence shell of the beryllium atom. In BF_3 there are three electron pairs in the valence shell of boron. All three of the molecules H_2O, NH_3, and CH_4 have four electron pairs in the valence shell of the central atom.

Two pairs of electrons have a linear arrangement, three pairs an equilateral triangular arrangement, and four pairs a tetrahedral arrangement, as shown in Table 8.1. These arrangements are adopted because electron pairs have a strong

Table 8.1 Arrangements of Two, Three, and Four Electron Pairs in a Valence Shell. The electron pairs in a valence shell adopt the arrangement that maximizes the distance between any two pairs. Column (a) shows the electron pairs arranged as far apart as possible on the surface of a sphere. Column (b) shows the angles between any two pairs of electrons. Column (c) shows an approximate representation of the electron density distribution of each pair of electrons, that is, the orbital occupied by each pair of electrons.

NUMBER OF ELECTRON PAIRS	SHAPE	ARRANGEMENT OF ELECTRON PAIRS		
		(a)	(b)	(c)
2	Linear		180°	
3	Triangular		120°	
4	Tetrahedral		109.5°	

NH_3

tendency to keep as far apart from each other as possible. This tendency for electron pairs to keep apart in space—that is, to behave as if they repelled each other—has its origin in the Pauli exclusion principle.

As we have seen in Chapter 6, pairs of electrons of opposite spin occupy different regions of space. Thus the electron pairs in a valence shell share the space available to them, keeping as far apart as possible. In other words, *the electron pairs in a valence shell adopt the arrangement that maximizes the angle between any two pairs.* Thus two pairs are 180° apart, three pairs are in a plane 120° apart, and four pairs have a tetrahedral arrangement 109.5° apart.

It is not immediately obvious that the tetrahedral arrangement is the one that maximizes the distances between the four electron pairs. But one can show, by making models or by calculation, that any other arrangement, such as a square planar arrangement, brings at least some of the electron pairs closer together. A model that is very convenient for showing the arrangements of electron pairs in a valence shell is the electron pair sphere model described in Experiment 8.1. It assumes that the space occupied by a pair of electrons, either bonding or nonbonding, can be approximately represented by a sphere.

Valence Shells with Four Electron Pairs

When there are four electron pairs in the valence shell of a central atom, they always have a tetrahedral arrangement. But whether a molecule is itself tetrahedral depends on whether all four electron pairs are forming bonds.

EXPERIMENT 8.1

Electron Pair Arrangements: Styrofoam Sphere Models

Electron pair arrangements can be demonstrated by joining pairs or threes of styrofoam spheres by elastic bands held in place by small nails or toothpicks. Each sphere represents a pair of electrons, or more exactly the space occupied by a pair of electrons. The elastic band provides a force of attraction between the spheres that corresponds to the electrostatic attraction between the electrons and a positive core situated at the mid-point of the elastic band. The spheres naturally adopt the arrangements shown. If they are forced into some other arrangement—such as the square planar arrangement for four spheres—they immediately adopt the tetrahedral arrangement when the restraining force is removed.

Two spheres have a linear arrangement

Three spheres adopt a triangular arrangement

Four spheres adopt a tetrahedral arrangement

Five spheres adopt a trigonal bipyramidal arrangement
Six spheres adopt an octahedral arrangement.

TETRAHEDRAL MOLECULES   All molecules of the type AX_4, such as CH_4, CCl_4, SiH_4, and SiF_4, have four electron pairs in the valence shell of the central atom. All four pairs are bonding pairs. Therefore all these molecules have a tetrahedral structure (see Figure 8.3). The angle between any pair of bonds in a tetrahedral AX_4 molecule is 109.5°. Other examples of tetrahedral AX_4 species are polyatomic ions such as NH_4^+ and BF_4^- (see Figure 8.3).

MOLECULES WITH ONE OR MORE NONBONDING PAIRS ON THE CENTRAL ATOM   Molecules such as NH_3, H_2O, and HF also have four electron pairs in the valence shell of the central atom, but there are nonbonding electron pairs as well as bonding pairs. The NH_3 molecule has three bonding pairs and one nonbonding pair; the H_2O molecule has two bonding and two nonbonding electron pairs; and the HF molecule has one bonding and three nonbonding pairs. The four electron pairs always have a tetrahedral arrangement, and the shapes of these molecules can therefore be derived as shown in Figure 8.4.

AX_3E MOLECULES   In the NH_3 molecule nitrogen has three bonding pairs and one nonbonding electron pair (E) in its valence shell. These four pairs have a tetrahedral arrangement. Three of the four corners of a tetrahedron are therefore occupied by hydrogen atoms; the fourth is occupied by a nonbonding pair. Three hydrogen atoms form an equilateral triangle, and the nitrogen atom completes a pyramid on this triangular base. The molecule therefore has a *triangular pyramidal shape*. The angle between any pair of bonds is predicted to be the same as in a tetrahedral molecule, namely 109.5°.

Figure 8.3 Some Tetrahedral AX_4 Molecules and Ions.
In depicting tetrahedral and related molecules on a two-dimensional page, we use the convention shown in the lower part of the figure. The tetrahedron is oriented so that two of the bonds are in the plane of the paper. Of the remaining two bonds, one projects forward out of the plane and is denoted by a solid wedge that becomes wider toward the forward atom; the other projects behind the plane of the paper and is denoted by a dashed wedge that becomes narrower toward the atom behind the plane of the page.

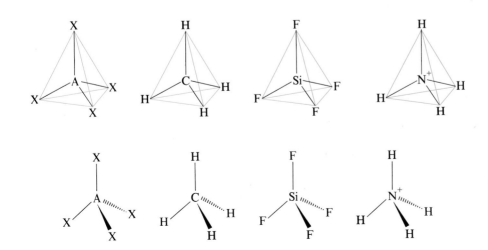

Figure 8.4 The Shapes of the CH₄, NH₃, H₂O, and HF molecules. The shapes of these molecules are based on the tetrahedral arrangement of four pairs of electrons. We can derive the shape of any of these molecules by placing the atoms bonded to the central atom at the corners of the tetrahedron; any remaining corners are occupied by unshared pairs of electrons. In each case there is a tetrahedral arrangement of four electron pairs in the valence shell of the central atom. But the shapes of these molecules are described by the arrangement of the atoms. Thus only CH₄ is described as tetrahedral. In the ammonia molecule, NH₃, the three hydrogen atoms are at the base and the nitrogen atom is at the apex of a triangular pyramid. Water is an angular molecule and HF, which is diatomic, is necessarily linear. Notice that NH₃ is not a tetrahedron because only one face—the face containing the three H atoms—is an equilateral triangle; the other faces have one angle of 109.5°.

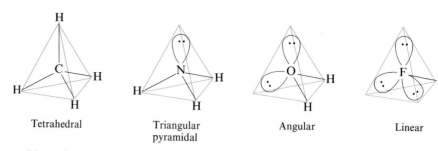

Tetrahedral   Triangular pyramidal   Angular   Linear

Note that a triangular pyramid is not a tetrahedron, because the faces of a triangular pyramid are not, in general, equilateral triangles. In the NH₃ molecule only the face containing the three hydrogen atoms is an equilateral triangle. We describe a molecule such as NH₃ as an AX₃E molecule, where E represents a nonbonding electron pair; there are three bonding electron pairs and one nonbonding electron pair in the valence shell of A. Other molecules of this type include NF₃, PCl₃, and ions such as H₃O⁺. All these molecules have similar Lewis structures and a triangular pyramidal shape similar to that of the NH₃ molecule, as shown in Figure 8.5.

AX₂E₂ MOLECULES   In the water molecule the oxygen atom has two bonding and two nonbonding electrons pairs (E) in its valence shell. It is therefore described as an AX₂E₂ molecule. The two hydrogen atoms occupy two of the corners of a tetrahedron, and the two nonbonding pairs occupy the other two corners. The water molecule is thus *angular*, with a predicted bond angle of 109.5°. All molecules of the AX₂E₂ type have similar Lewis structures and the same angular shape. Other examples include F₂O, SCl₂ and NH₂⁻ (Figure 8.6).

AXE₃ MOLECULES   The fluorine atom in the HF molecule has one bonding pair and three nonbonding pairs in its valence shell. The hydrogen atom is therefore at one corner of a tetrahedron, and the other three corners are occupied

Figure 8.5 Shapes of Some AX₃E Molecules. All AX₃E molecules have a triangular pyramidal shape.

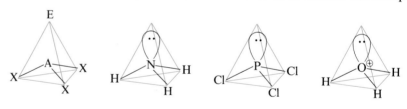

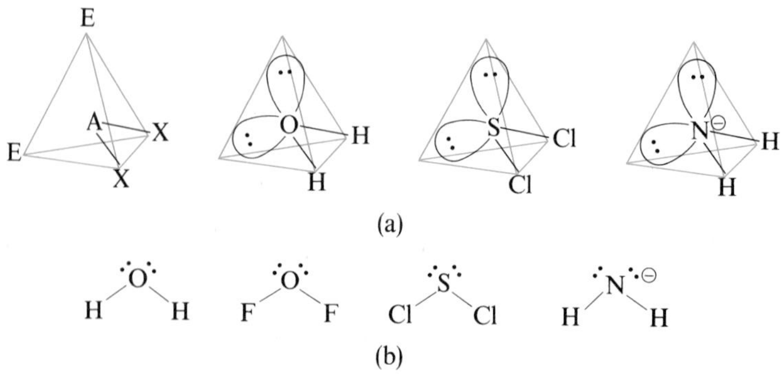

Figure 8.6 Angular AX₂E₂ Molecules. (a) The molecules are oriented with the nonbonding electron pairs in the plane of the paper. (b) It is usually more convenient to depict these molecules with the bonds in the plane of the paper. The nonbonding pairs then project forward and behind the paper, but this is not usually explicitly shown.

(a)

(b)

by bonding pairs, as shown in Figure 8.4. The molecule is necessarily *linear* since there are only two atoms.

Models of AX_3E, AX_2E_2, and AXE_3 molecules can be made with balloons, as shown in Experiment 8.2.

Example 8.1 Classify the following molecules and ions as AX_4, AX_3E, AX_2E_2, or AXE_3, and describe their shapes.

NCl_3 $\quad$ PF_3 $\quad$ H_2F^+ $\quad$ OH^- $\quad$ $AsCl_3$ $\quad$ PH_4^+ $\quad$ SiO_4^{4-} $\quad$ OCl_2

Solution

	Lewis structure	Class of molecule	Shape
NCl_3	 :Cl:N:Cl: :Cl:	AX_3E	Triangular pyramidal
PF_3	 :F:P:F: :F:	AX_3E	Triangular pyramidal
H_2F^+	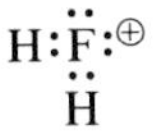 H:F:⊕ H	AX_2E_2	Angular
OH^-	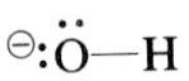 ⊖:Ö—H	AXE_3	Linear
$AsCl_3^*$	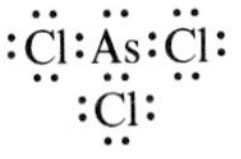 :Cl:As:Cl: :Cl:	AX_3E	Triangular pyramidal
PH_4^+	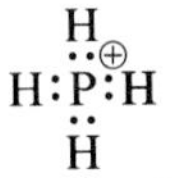 H H:P:H H	AX_4	Tetrahedral
SiO_4^{4-}	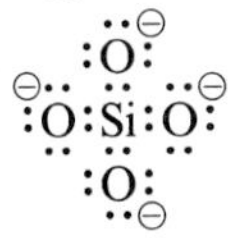 :O:⊖ ⊖:O:Si:O:⊖ :O:⊖	AX_4	Tetrahedral
OCl_2	 :O:Cl: :Cl:	AX_2E_2	Angular

* As is in Group V, below P.

ISOELECTRONIC MOLECULES Molecules and ions that have the same number of nuclei and the same number of electrons are said to be **isoelectronic**. Isoelectronic species have the same shapes. For example, BH_4^-, CH_4, and NH_4^+ all have five nuclei and ten electrons, two core electrons ($1s^2$) and eight valence electrons. The eight valence electrons are arranged as four pairs around the core of the central atom. All three are tetrahedral AX_4 molecules:

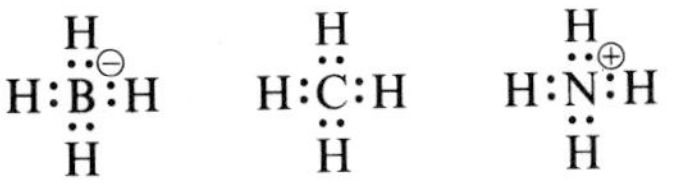

H⊖ H H⊕
H:B:H $\quad$ H:C:H $\quad$ H:N:H
H $\qquad$ H $\qquad$ H

Similarly, BF_4^-, CF_4, and NF_4^+ all have five nuclei and 34 electrons. They have the same Lewis structures and the same tetrahedral shape:

:F: $\quad$:F: $\quad$:F:
:F:B⊖:F: $\quad$:F:C:F: $\quad$:F:N⊕:F:
:F: $\quad$:F: $\quad$:F:

8.1 VALENCE SHELL
ELECTRON PAIR
REPULSION (VSEPR) THEORY

283

Electron Pair Arrangements: Balloon Models

The long orange balloons represent bonding electron pairs and the round blue balloons represent the larger space occupied by the nonbonding or unshared pairs of electrons.

A tetrahedral arrangement of four bonding electron pairs—a tetrahedral AX_4 molecule, such as CH_4.

A tetrahedral arrangement of three bonding pairs and one nonbonding electron pair—a pyramidal AX_3E molecule, such as NH_3.

A tetrahedral arrangement of two bonding pairs and two nonbonding electron pairs—an angular AX_2E_2 molecule, such as H_2O.

The molecules H_3O^+ and NH_3 have four nuclei and ten electrons. They have the same Lewis diagrams and the same triangular pyramidal shape. Both H_2O and NH_2^- have three nuclei and ten electrons and the same angular shape.

Molecules with the same number of *valence* electrons are often also described as isoelectronic, and we will find it convenient to use this extended definition of *isoelectronic*. Thus we would regard SiF_4, $SiCl_4$, PCl_4^+, and $AlCl_4^-$ as being isoelectronic with CF_4 and BF_4^-. They are all tetrahedral molecules with the same Lewis structures.

Valence Shells with Two or Three Electron Pairs

Molecules in which the central atom has four electron pairs in its valence shell are by far the most common, but in some molecules there are only three valence shell pairs on the central atom. For example, when boron forms three covalent bonds, as in BF_3 or BCl_3, it has only three electron pairs in its valence shell. Similarly, aluminum, which is also in group III, has three electron pairs in its valence shell in $AlCl_3$, and beryllium, which is in group II, forms a few molecules in which it has only two electron pairs in its valence shell.

AX_3 MOLECULES The BF_3, BCl_3, and $AlCl_3$ molecules are examples of AX_3 molecules, since the central atom is bonded to three other atoms and has no unshared electrons pairs. As shown in Table 8.1, the arrangement that keeps three electron pairs as far apart as possible is the planar triangular arrangement. Hence *AX_3 molecules all have a planar triangular structure* with equal 120° bond angles (see Figure 8.7).

We should note that X does not necessarily represent a single atom; it may represent a group of atoms, one of which is covalently bonded to the central atom. For example, $B(OH)_3$, boric acid, is an AX_3 molecule. A single atom or a group of atoms, such as OH, that is attached to a central atom is called a **ligand**. Thus AX_3 is a general formula for any molecule in which three ligands are at-

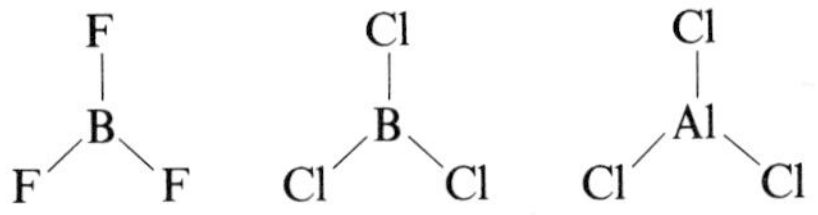

F—B(F)(F) and Cl—B(Cl)(Cl) and Cl—Al(Cl)(Cl) structures shown at top.

tached by covalent bonds to a central atom that does not have any unshared electron pairs. The three covalent bonds always have a planar triangular arrangement. Thus in $B(OH)_3$ the three B—O bonds have a triangular arrangement with a bond angle of 120° (see Figure 8.8).

However, the *overall* shape of $B(OH)_3$ depends also on the arrangements of the bonds within the ligands. Each oxygen atom has an AX_2E_2 geometry: The two bonds formed by each oxygen atom are therefore predicted to have an angular arrangement. Two possible geometries for $B(OH)_3$ are shown in Figure 8.8. Other overall shapes are also possible for this molecule; they differ in the relative orientations of the bonds on different oxygen atoms. The arrangements of electron pairs given in Table 8.1 enable us to predict the geometry of the bonds around each of the atoms in a molecule but not the arrangements of the bonds around one atom *relative* to those around another atom. This topic is discussed in Chapter 11.

Figure 8.8 Two Possible Shapes for the B(OH)₃ Molecule. The bonds around the boron atom have a planar triangular AX_3 arrangement, and the bonds around the oxygen atoms have angular AX_2E_2 arrangements. The overall shape of $B(OH)_3$ depends on the relative orientation of the bonds to the oxygen atoms; this cannot be predicted by the VSEPR theory. Two possibilities for the overall shape of $B(OH)_3$ are shown here.

AX₂E MOLECULES   Tin, which is in group IV, forms the chloride $SnCl_2$. Since tin has four electrons in its valence shell, we can write the Lewis structure as follows:

$$\cdot\ddot{Sn}\cdot + 2\cdot\ddot{Cl}: \longrightarrow :\ddot{Cl}:\ddot{Sn}:\ddot{Cl}:$$

There are three electron pairs in the valence shell of the tin, one of which is a nonbonding pair. Since these three electron pairs have a triangular arrangement, the two bonding pairs have an angular arrangement. Thus the $SnCl_2$ molecule has an angular shape with a bond angle of 120°, as shown in Figure 8.9. It is described as an AX_2E molecule.

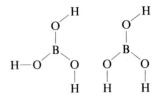

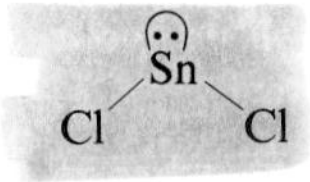

Figure 8.9 SnCl₂, an Angular AX₂E Molecule.

AX₂ MOLECULES   A few molecules of beryllium, such as $BeCl_2$, have only two electron pairs in the valence shell of the beryllium atom. They belong to the AX_2 class of molecules. As shown in Table 8.1, two electron pairs keep as far apart from each other as possible by adopting a linear arrangement, that is, by forming a bond angle of 180° (see Figure 8.10). The molecular shapes we have discussed to this point are summarized in Table 8.2.

Figure 8.10 BeCl₂, a Linear AX₂ Molecule.

Valence Shells with More Than Four Electron Pairs

In Chapter 7 we saw that when period 3 elements form compounds in their high oxidation states, they have more than four pairs of electrons in their valence shells. For example, in PCl_5 phosphorus has a valence shell of 10 electrons; in SF_6 the valence shell of sulfur has 12 electrons. To predict the shapes of molecules with five or six ligands, such as PCl_5 and SF_6, we must consider the arrangements of five and six electron pairs (see Experiment 8.3 and Table 8.3).

The arrangement that keeps five electron pairs as far apart as possible is the *trigonal bipyramid*, which consists of two triangular pyramids that share a face (see Table 8.3). It has six equivalent triangular faces, nine edges, and 5 vertices. The PCl_5 molecule has this shape. Although the faces are all equivalent triangles, they are not necessarily equilateral triangles as in the tetrahedron and octahedron.

8.1 VALENCE SHELL ELECTRON PAIR REPULSION (VSEPR) THEORY

Table 8.2 Electron Pair Arrangements and Shapes of Molecules.

NUMBER OF ELECTRON PAIRS	ARRANGEMENT OF ELECTRON PAIRS	NUMBER OF NONBONDING ELECTRON PAIRS	CLASS OF MOLECULE	ARRANGEMENT OF BONDS OR SHAPE OF MOLECULE	EXAMPLES	PREDICTED BOND ANGLE
2	Linear	0	AX_2	Linear X—A—X	BeH_2, $BeCl_2$	180°
3	Equilateral triangular	0	AX_3	Equilateral triangular 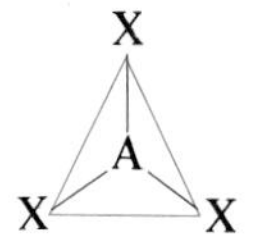	BF_3, $B(OH)_3$, $AlCl_3$	120°
		1	AX_2E	Angular 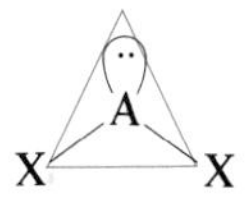	$SnCl_2$	120°
4	Tetrahedral	0	AX_4	Tetrahedral 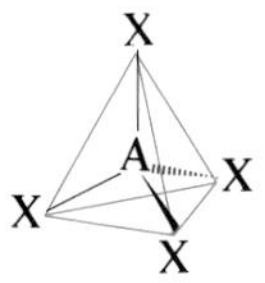	CH_4, CCl_4, SiH_4, $NH_4{}^+$	109.5°
		1	AX_3E	Triangular pyramidal 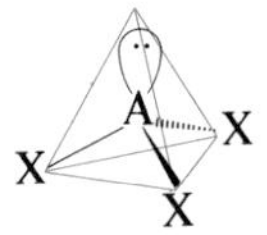	NH_3, NF_3, PCl_3, H_3O^+	109.5°
		2	AX_2E_2	Angular 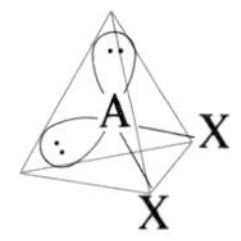	H_2O, F_2O, SCl_2	109.5°
		3	AXE_3	Linear 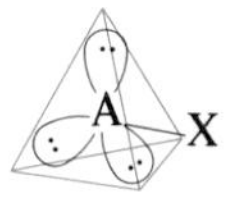	HF, HCl	—

The arrangement that keeps six electron pairs as far apart as possible is the octahedron, which is a regular solid with 6 equivalent vertices, 12 equivalent edges, and 8 equilateral triangular faces (see Table 8.3). The SF_6 molecule has this shape.

Hybrid Orbitals

We saw in Chapter 6 that the four bonding orbitals corresponding to the four C—H bonds in methane may be considered to be formed from the carbon atom in the valence state in which it has an unpaired electron in the 2s orbital and

AX$_6$, AX$_5$, AX$_4$, and AX$_3$ molecules: Balloon Models

If six balloons are tied together they automatically adopt an AX$_6$ octahedral arrangement.

If one of the balloons is popped, the remaining five adopt an AX$_5$ trigonal bipyramidal arrangement.

If another balloon is popped, the remaining four adopt an AX$_4$ tetrahedral arrangement.

If a third balloon is popped, the remaining three adopt an AX$_3$ triangular arrangement.

each of the three 2p orbitals and four hydrogen 1s orbitals.

2s   2p

C valence state   ↑   ↑ ↑ ↑

We have seen in this chapter that the four bonding electron pairs in the methane molecule are equivalent and have a tetrahedral arrangement around the core of

Table 8.3 Arrangements of Five and Six Electron Pairs. Five electron pairs have a triangular pyramidal arrangement; therefore AX$_5$ molecules have this shape. Six electron pairs have an octahedral arrangement; therefore AX$_6$ molecules have this shape.

NUMBER OF ELECTRON PAIRS	ARRANGEMENT		EXAMPLE
5	Trigonal bipyramid		Cl Cl—P—Cl Cl Cl
6	Octahedron		F F—S—F F F

8.1 VALENCE SHELL ELECTRON PAIR REPULSION (VSEPR) THEORY

the carbon atom. Thus the four bonding orbitals occupied by these four electron pairs have a tetrahedral arrangement. Because they are derived from one 2s and three 2p orbitals these four equivalent bonding orbitals are called sp^3 bonding or *hybrid orbitals*. Hybrid orbitals are discussed in more detail in Chapter 25. For the moment we need only know that four tetrahedrally arranged valence shell electron pairs are described as occupying four sp^3 hybrid orbitals. In general the four electron pairs in AX_4, AX_3E, and AX_2E_2 molecules are said to occupy four tetrahedrally arranged sp^3 hybrid orbitals.

The three bonding orbitals in an AX_3 molecule such as BF_3 may be considered to be derived from the valence state configuration of boron in which it has three unpaired electrons and a singly occupied 2p orbital on each fluorine atom.

	2s	2p
B valence State	↑	↑ ↑

Thus they are called sp^2 bonding or hybrid orbitals. Similarily the bonding orbitals in $BeCl_2$ are described as sp orbitals.

We saw in Chapter 7 that in their higher valence states sulfur and phosphorus utilize d orbitals. Thus the bonding orbitals in a trigonal bipyramidal molecule such as PF_5 are described as sp^3d hybrids, and in an AX_6 molecule such as SF_6 as sp^3d^2 hybrids.

The names of the hybrid orbitals occupied by the bonding electron pairs in molecules with two to six electron pairs in their valence shells are summarized in Table 8.4.

Table 8.4 Bonding or Hybrid Orbitals

CLASS OF MOLECULE	BONDING OR HYBRID ORBITAL	ARRANGEMENT OF ORBITALS	
AX_2	sp	Linear	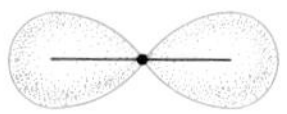
AX_3	sp^2	Triangular	
AX_4	sp^3	Tetrahedral	
AX_5	sp^3d	Trigonal bipyramidal	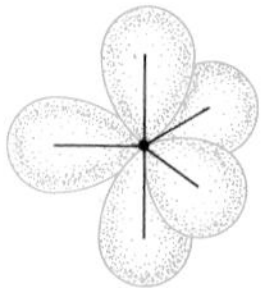
AX_6	sp^3d^2	Octahedral	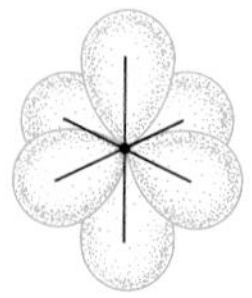

Bond Angles

Because the shapes of AX_4, AX_3E, and AX_2E_2 molecules are all based on a tetrahedral arrangement of four electron pairs, the bond angles in molecules of this type are all predicted to be the same as in a tetrahedral molecule such as CH_4, namely 109.5°. This prediction is correct for all molecules of the AX_4 type such as CCl_4, SiH_4, and NH_4^+. But it is not exactly correct for AX_3E and AX_2E_2 molecules in which the bond angles are generally smaller than 109.5°. For example the NH_3 and H_2O molecules have bond angles of 107.5° and 104.5°, respectively.

EFFECT OF NONBONDING PAIRS These small differences from the predicted angles are due to the fact that the four electron pairs in the valence shell are not all equivalent; some are bonding pairs and some are nonbonding pairs. A nonbonding pair takes up more space in the valence shell than a bonding pair, as shown in Experiment 8.2 and Figure 8.11 for the ammonia molecule. A bonding pair is situated between two atomic cores—in this case the nitrogen and the hydrogen—and is attracted by the positive charge of both cores. In contrast, the nonbonding pair is attracted by only one positive core, in this case that of nitrogen. The greater the positive charge attracting an electron pair, the smaller is the volume occupied by the electron pair. Hence a bonding pair that is associated with two atomic cores occupies a smaller volume than a nonbonding pair that is under the influence of only one atomic core. *Nonbonding pairs occupy more space than bonding pairs.*

The difference in the space occupied by nonbonding and bonding electron pairs affects bond angles at any atom with one or more nonbonding electron pairs in its valence shell. Four equivalent bonding pairs adopt a perfect tetrahedral arrangement, and all molecules of the type AX_4 with four identical ligands X have bond angles of exactly 109.5°. However, because nonbonding electron pairs take up more space in the valence shell than bonding pairs, the angles between the nonbonding pairs in AX_3E and AX_2E_2 molecules are larger than 109.5° and the angles between the bonding pairs are smaller than 109.5°. Thus in all AX_3E and AX_2E_2 molecules we consistently find bond angles less than the tetrahedral angle. Some examples are given in Table 8.5.

An alternative statement of the difference between bonding electron pairs (B) and nonbonding electron pairs (NB) is that nonbonding pairs appear to repel bonding pairs more strongly than they repel each other. In short, NB—B repulsions are greater than B—B repulsions. This again leads to the conclusion that in AX_3E and AX_2E_2 molecules the bond angles will be less than 109.5°.

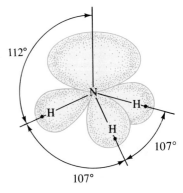

Figure 8.11 Bonding and Nonbonding Electron Pairs. Nonbonding pairs are larger than bonding pairs. Therefore angles involving nonbonding pairs are larger than those involving bonding pairs. So bond angles in AX_3E and AX_2E_2 molecules are always less than the tetrahedral angle, as shown here for ammonia.

Table 8.5 Bond Angles in Some AX_3E and AX_2E_2 Molecules

AX_3E MOLECULES		AX_2E_2 MOLECULES	
NH_3	107°	H_2O	104°
NF_3	102°	F_2O	103°
PH_3	93°	H_2S	92°
PI_3	102°	SCl_2	100°
PBr_3	101°	SF_2	98°
PCl_3	100°		
PF_3	97°		

EFFECT OF ELECTRONEGATIVITY Even in AX_4 molecules, if all the ligands X are not identical, the bond angles deviate from the tetrahedral angle because the space occupied by a bonding electron pair depends to some extent on the ligand. For example, in difluoromethane, CH_2F_2, the HCH bond angle is 112° and the FCF bond angle is 108°. Some other examples are given in Figure 8.12.

Notice that there is a trend in the data given in Figure 8.12 and in Table 8.5: *Bond angles usually decrease with an increase in the electronegativity of the ligands.* For example, in the phosphorus trihalides PX_3 the bond angle decreases from PI_3 to PF_3 as the electronegativity of the ligand increases. A more electronegative ligand exerts a greater force of attraction on the bonding electron pair than does a less electronegative ligand, thereby decreasing the volume of space occupied by the bonding pair. Thus the bonding pair takes up less space in the valence shell, and any angle made by this bond is correspondingly smaller.

Double and Triple Bonds

We can derive the shapes of molecules containing multiple bonds from the electron pair arrangements in Table 8.1. In the ethene molecule, C_2H_4, the two carbon atoms share two pairs of electrons; that is, they are joined by a **double bond**. Figure 8.13 shows a ball-and-stick model of ethene. The four tetrahedrally arranged electron pairs in the valence shell of each carbon atom are represented by four tetrahedrally arranged sticks. To represent two carbon atoms joined by a double bond, two of these sticks must be bent, as shown in Figure 8.13. The double bond can be described as consisting of two bent bonds. **A bent bond** is a bond in which the maximum electron density of the bond pair does not lie on the straight line between the atoms that it bonds. We see from the model that all six atoms lie in the same plane. So ethene is a planar molecule. Figure 8.13 also shows an electron pair sphere model, which emphasizes the tetrahedral arrangement of the electron pairs around each carbon and shows the planarity of the molecule.

Ethyne (acetylene) has the formula C_2H_2; its Lewis structure shows that three electron pairs are shared between two carbon atoms, forming a **triple bond**. The ball-and-stick model in Figure 8.14 shows that three of the bonds on each carbon atom must be bent to join the two carbon atoms. The triple bond can therefore be described as consisting of three bent bonds. We see from the model

Figure 8.13 Ethene Molecule. The arrangement of electron pairs around each carbon atom is tetrahedral. Two pairs are shared between the carbon atoms, giving the molecule an overall planar shape.

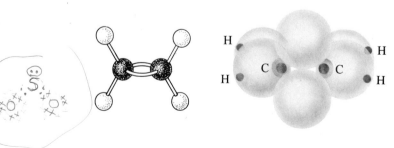

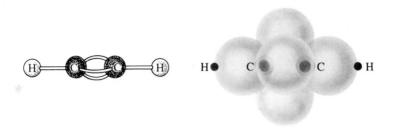

Figure 8.14 Ethyne Molecule.
The arrangement of electron pairs
around each carbon atom is
tetrahedral. Three pairs are shared
between the two carbon atoms,
giving the molecule an overall linear
shape.

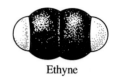

Ethyne

that the molecule is linear; that is, all four atoms lie along the same straight line. Figure 8.14 also shows an electron pair sphere model, which emphasizes the tetrahedral arrangement of electron pairs around each carbon atom and shows that the molecule is linear.

We can derive the shapes of other molecules with multiple bonds in a similar fashion, as Figure 8.15 shows. Thus methanal, CH_2O (also called formaldehyde), has a planar triangular structure with an arrangement of bonds around the carbon atom, just like that of the carbon atom in ethene. Carbon dioxide, in which carbon forms two double bonds, and hydrogen cyanide, in which carbon forms one single and one triple bond, are both linear molecules.

We must be careful not to interpret our bent-bond model of the double bond as meaning that there are two quite distinct regions of electron density with a "hole" in the middle. Each stick in fact represents a large electron density cloud, and these overlap sufficiently that the *total* electron density has a maximum along the C—C direction. There is one large cloud of electron density in which the individual electron pairs cannot be distinguished. Similar considerations apply to the three electron pairs of a triple bond (see Figure 8.16).

We can now consider an alternative and simpler method for deriving the shapes of molecules containing multiple bonds. The two electron pairs of a double bond or the three electron pairs of a triple bond form one large electron cloud that maximizes its distance from the charge clouds of other bonds or nonbonding electron pairs. Thus to derive the arrangements of ligands around a given atom, we can ignore the exact nature of the bonds—that is, whether they are single, double, or triple. We simply count the number of ligands X and the number of nonbonding pairs E and use Table 8.1 to find the shape of the molecule. For example, ethene has three ligands attached to each carbon atom and no nonbonding electron pairs; thus the geometry around each carbon atom may be described as AX_3. In other words there is a planar triangular arrangement of bonds around each carbon atom.

H ⨯ C ⦂⦂ C ⨯ H

Figure 8.15 Some Molecules with Multiple Bonds.
Each molecule has four electron pairs in a tetrahedral
arrangement around each of the carbon, nitrogen, and, oxygen
atoms. Sharing two or three pairs in double and triple bonds
leads to the shapes shown: CH_2O, methanal, is planar triangular;
CO_2, N_2, and HCN are linear.

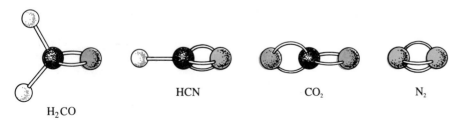

H_2CO HCN CO_2 N_2

8.1 VALENCE SHELL
ELECTRON PAIR
REPULSION (VSEPR) THEORY

291

Figure 8.16 Representation of Double and Triple Bonds. The overlap of the electron densities of the charge clouds corresponding to the two individual bent bonds forms a large charge cloud. (a) This charge cloud corresponds to one of the bonds making up the double bond. (b) This charge cloud corresponds to the other half of the double bond. (c) The total electron density of the combined charge clouds. The maximum is along the C—C direction, and the individual electron pair densities cannot be distinguished. We can represent this bonding more simply by using the electron pair sphere model and allowing the electron pair spheres to overlap to form a larger and very approximately spherical double or triple electron pair cloud. (d) Two electron pairs are shown in a double bond. (e) Three electron pairs are shown in a triple bond looking down the CC axis.

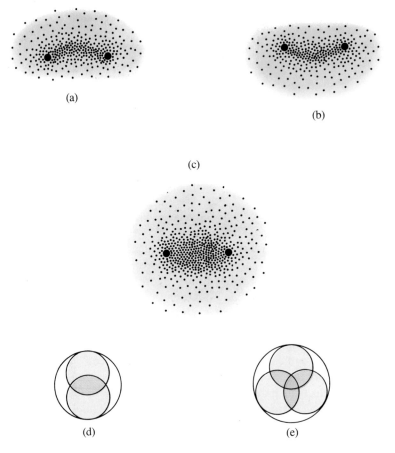

(a)

(b)

(c)

(d)

(e)

The Lewis structures for CO_2 and SO_2 are

$$:\ddot{O}=C=\ddot{O}: \qquad :O \overset{\overset{\displaystyle \ddot{S}}{\diagup \diagdown}}{} O:$$

Thus CO_2 is an AX_2 molecule, and SO_2 is an AX_2E molecule. We have seen that an AX_2 molecule is linear; thus CO_2 is linear (Figure 8.17a). In an AX_2E molecule the electron pairs on the central atom have a triangular arrangement but

Figure 8.17 Multiple Bonds in the CO_2, SO_2, and HCN Molecules.
(a) CO_2—an AX_2 molecule. Since the carbon atom has two double bonds and no unshared electron pairs, CO_2 can be classified as an AX_2 molecule. (b) SO_2—an AX_2E molecule. Since the sulfur atom has two double bonds and an unshared electron pair, SO_2 can be classified as an AX_2E molecule. (c) HCN—an AX_2 molecule. Since HCN has a single bond and a triple bond on the central carbon atom, HCN can be classified as an AX_2 molecule.

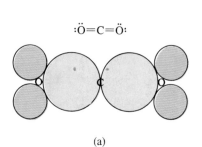

○ Double bond
● Triple bond
○ Nonbonding pair

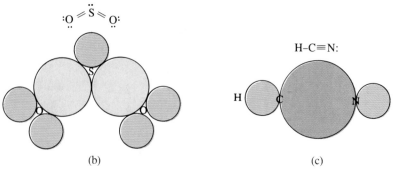

(a)

(b)

(c)

one of the three positions is occupied by a nonbonding pair, so an AX_2E molecule is angular. Thus SO_2 is an angular molecule with a bond angle close to $120°$ (see Figure 8.17b).

An important class of ions containing multiple bonds are the anions such as carbonate, CO_3^{2-}, and sulfate, SO_4^{2-}, derived from oxoacids. We can derive their shapes in the manner just outlined. For example, from the Lewis structure for the carbonate ion we see that it is an AX_3 molecule. Accordingly, it has a planar triangular structure. The Lewis structure of the nitrate ion, NO_3^-, similarly shows that it is an AX_3 molecule. It therefore also has a planar triangular AX_3 structure. The sulfate and phosphate ions are both AX_4 molecules with four ligands (two double and two single bonds in SO_4^{2-} and one double and three single bonds in PO_4^{3-}) and no unshared pairs of electrons. They are therefore tetrahedral in shape. The sulfite ion, SO_3^{2-}, has three ligands and one unshared pair of electrons. Therefore it is an AX_3E molecule and has a triangular pyramidal shape like ammonia.

Table 8.6 summarizes the shapes of molecules and ions containing double and triple bonds.

Since the total volume occupied by the two pairs of electrons of a double bond is greater than that occupied by a single bonding pair, we expect deviations from ideal bond angles in all molecules having both single and double bonds. For molecules having an AX_3 geometry around a central atom, the ideal bond angles for these equivalent electron pairs are $120°$. We therefore expect angles involving a double bond to be somewhat greater than $120°$ and angles involving only single bonds to be somewhat less than $120°$. For AX_4 molecules we expect

Table 8.6 Shapes of Molecules Containing Double and Triple Bonds.

TOTAL NUMBER OF LIGANDS AND NONBONDING PAIRS	ARRANGEMENT	NUMBER OF LIGANDS	NUMBER OF NONBONDING PAIRS	CLASS OF MOLECULE	SHAPE OF MOLECULE	EXAMPLES
2	Linear	2	0	AX_2	Linear	$O{=}C{=}O$ $H{-}C{\equiv}N$
3	Triangular	3	0	AX_3	Triangular	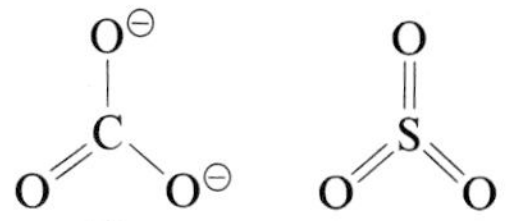
		2	1	AX_2E	Angular	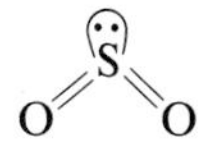
4	Tetrahedral	4	0	AX_4	Tetrahedral	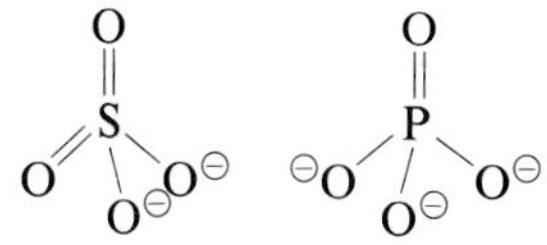
		3	1	AX_3E	Triangular pyramidal	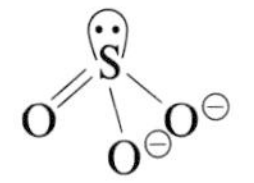
		2	2	AX_2E_2	Angular	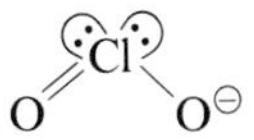

angles involving double bonds to be somewhat greater than the tetrahedral angle of 109.5° and angles involving only single bonds to be somewhat smaller than 109.5°. Some examples are given in Figure 8.18.

To predict the departure of the bond angle in a molecule such as SO_2 from the ideal angle of 120°, we need to know which occupies the larger volume, the nonbonding pair or the two pairs of a double bond. Unfortunately there is no simple way to decide which will have the largest volume. But, by considering the observed structures of suitable molecules, we can see that they often occupy a similar volume. For example, the bond angle in SO_2, which has two double bonds and a nonbonding pair, is 119°, which is very close to the ideal angle of 120° for an AX_2E molecule.

Example 8.2 Classify the following molecules and ions as AX_2, AX_2E, AX_3, and so on, and describe their shapes:

$$SeO_3^{2-} \quad ClO_4^- \quad CS_2 \quad H_3PO_3$$

Selenium is in Group VI.

Solution

	Lewis structure	Class of molecule	Shape
SeO_3^{2-}	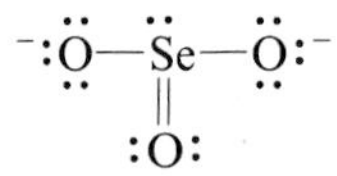	AX_3E	Triangular pyramidal
ClO_4^-	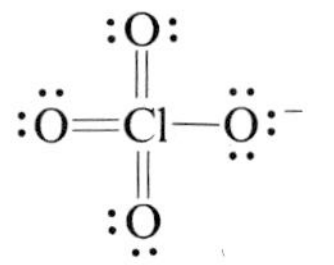	AX_4	Tetrahedral
CS_2	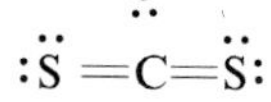 $\ddot{S}=C=\ddot{S}$	AX_2	Linear
H_3PO_3	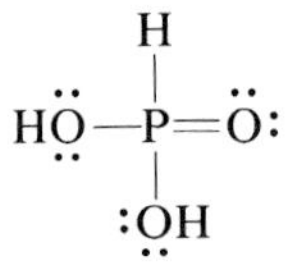	AX_4	Tetrahedral

Figure 8.18 Bond Angles in Some Molecules Containing Double Bonds.
The ideal bond angles in AX_3 and AX_4 molecules are 120° and 109.5°, respectively. Because the electron cloud of a double bond is larger than that of a single bond, angles involving double bonds are larger than those between single bonds. Therefore in AX_3 and AX_4 molecules involving double bonds, the angles involving double bonds are larger than the ideal angles, and those involving single bonds only are smaller than the ideal angles. According to the bent bond model of an $X_2A=Y$ molecule, the angle between the single bonds is 109.5° and that between the single bonds and a double bond is 125°. The observed angles are intermediate between the limiting values predicted by the two models.

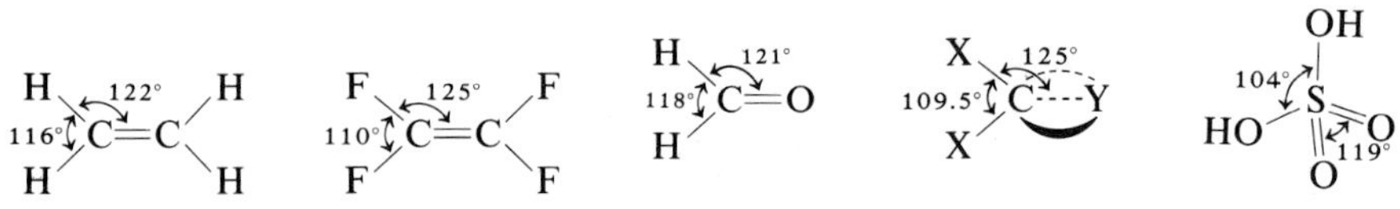

Ethene   Tetrafluoroethene   Methanal   Bent bond model   Sulfuric acid

Lewis structures are the simplest and commonest method used to show how the electrons are arranged in molecules and ions, and as we have seen, they are the starting point for the prediction of the geometry of molecules. You have already learned how to draw Lewis structures for simple molecules, and you have encountered a few more complicated cases such as the sulfate and sulfite ions. It is important to be able to draw correct Lewis structures for any of the molecules and ions that are discussed in this book.

Knowing the valences of the elements and taking account of the octet rule, when it is valid, we can often arrive at a correct Lewis structure quite easily. However, some cases are more difficult, and it is for dealing with such cases that we now describe a general set of rules for drawing Lewis structures. With practice you may not always need to work systematically through these rules, but they are useful when you are uncertain how to proceed. The rules are summarized below:

1. Draw a diagram of the molecule or polyatomic ion, showing the atoms connected by single bonds.

2. Add the number of valence electrons on each atom in the molecule to find the total number of valence electrons. If the molecule is charged—that is, if it is a polyatomic ion—add one electron for each negative charge and subtract one electron for each positive charge.

3. Subtract the number of electrons needed to form the single bonds between the atoms, and use the remainder to complete octets around each atom except hydrogen. If there are insufficient electrons to complete all the octets, complete those of the more electronegative atoms first. Then assign formal charges.

4. If any atom still has an incomplete octet, convert nonbonding electron pairs to bonding pairs. In other words, use nonbonding pairs to form double or triple bonds until each atom has an octet. Then reassign formal charges.

5. If rule 4 creates additional formal charges, use the structure given by rule 3.

6. If the central atom is from period 3 and later periods, the octet rule does not apply. Form additional multiple bonds in order to remove as many formal charges as possible.

To illustrate the application of the rules, we now use them to deduce the Lewis structures of PCl_3 and the carbonate ion CO_3^{2-}. First, we deduce the Lewis structure for PCl_3:

1. Before any Lewis structure can be drawn, we must know how the atoms are connected in the molecule. This information can only be obtained with complete certainty by experiment. If we do not know the arrangement of the atoms, often we can deduce a probable arrangement of the atoms as follows: (a) The structure assumed must be consistent with the common valences of the elements; (b) in a molecule with the formula AX_n the unique atom A is usually the central atom to which the ligands X are attached; (c) atom A is also normally the least electronegative atom or the atom with the highest valence. The arrangement of the atoms in PCl_3 is known to be

$$Cl-P-Cl$$
$$|$$
$$Cl$$

We could in any case have deduced this arrangement because it is the only one consistent with a valence of 3 for P and 1 for Cl and because P is the least electronegative of the atoms in the molecule and it is the unique atom.

2. The number of electrons in the valence shell of an atom is given by its electron configuration or simply by its position in the periodic table:

Phosphorus (group V)	$5e^-$	1P	$5e^-$
Chlorine (group VII)	$7e^-$	3Cl	$21e^-$
			$\overline{26e^-}$

3. The three single bonds in the structure use up 6 of these electrons. Thus $26 - 6 = 20$ electrons remain to be allocated. So 6 electrons are added to each Cl and 2 to the P atom to complete their octets:

$$:\ddot{\underset{..}{Cl}}-\ddot{P}-\ddot{\underset{..}{Cl}}:$$
$$\underset{..}{:\ddot{Cl}:}$$

This structure uses up all 20 electrons. Each atom has a zero formal charge.

4. All atoms have an octet, so no further adjustment is necessary. Rules 5 and 6 are not applicable in this case.

Now we deduce the Lewis structure for CO_3^{2-}.

1. The arrangement of the atoms in CO_3^{2-} is

$$\begin{array}{c} O \\ | \\ O-C-O \end{array}$$

2. The number of electrons is found as follows:

Carbon (group IV)	$4e^-$	1C	$4e^-$
Oxygen (group VI)	$6e^-$	3O	$18e^-$
Negative charge			$2e^-$
			$\overline{24e^-}$

3. Six of these electrons are used for the three single bonds, so $24 - 6 = 18$ electrons remain to be distributed. Three pairs of electrons can be added to each O (the most electronegative atom) to give

$$\begin{array}{c} :\ddot{O}:^{\ominus} \\ | \\ {}^{\ominus}:\ddot{O}-\underset{\oplus}{C}-\ddot{O}:^{\ominus} \end{array}$$

thus using up all 18 electrons. Formal charges are then assigned. Each O atom (core charge $+6$) is assigned $6 + 1 = 7$ electrons and therefore has a -1 formal charge. The C atom (core charge $+4$) is assigned 3 electrons and therefore has a $+1$ formal charge.

4. In this structure C does not have an octet. One of the nonbonding electron pairs is therefore moved from an O to form a double bond with C:

$$\begin{array}{ccc} :\ddot{O}:^{\ominus} & & :\ddot{O}:^{\ominus} \\ | & & | \\ {}^{\ominus}:\ddot{O}-\underset{\oplus}{C}-\ddot{O}:^{\ominus} & \longrightarrow & :O=C-\ddot{O}:^{\ominus} \end{array}$$

Rules 5 and 6 are not applicable in this case.

Example 8.3 Draw the Lewis structure for the nitrate ion, NO_3^-.

Solution

1. The arrangement of the atoms is known to be

$$\begin{array}{c} O \\ | \\ O-N-O \end{array}$$

It would in any case have been a plausible assumption to put the less electronegative and unique atom N in the center of the three O atoms.

2. The number of electrons is found as follows:

Nitrogen (group V)	$5e^-$	1N	$5e^-$
Oxygen (group VI)	$6e^-$	3O	$18e^-$
One negative charge			$1e^-$
			$24e^-$

3. Six of these electrons are used for the three single bonds, so $24 - 6 = 18$ remain to be distributed. Three pairs can be added to each O to give

$$\overset{\displaystyle :\!\ddot{O}\!:^{\ominus}}{\underset{}{\overset{|}{\underset{}{{}^{\ominus}:\!\ddot{O}\!-\!\underset{\text{\textcircled{+2}}}{N}\!-\!\ddot{O}\!:^{\ominus}}}}}$$

Each O has a -1 formal charge, and the N (core charge $+5$) has a $+2$ formal charge.

4. In this structure the N atom does not have an octet; therefore one of the non-bonding electron pairs on O is moved to form a double bond to N:

$$\overset{:\!\ddot{O}\!:^{\ominus}}{\underset{{}^{\ominus}:\!\ddot{O}\!-\!\underset{\text{\textcircled{+2}}}{N}\!-\!\ddot{O}\!:^{\ominus}}{|}} \longrightarrow \overset{:\!\ddot{O}\!:^{\ominus}}{\underset{:\!\ddot{O}\!=\!\underset{\oplus}{N}\!-\!\ddot{O}\!:^{\ominus}}{|}}$$

Rules 5 and 6 are not applicable in this case.

If we had not known the structure of NO_3^-, we might have arranged the atoms in the following way: O—N—O—O. If we follow the rules, we arrive at the Lewis structure:

$$:\!\ddot{O}\!=\!\ddot{N}\!-\!\ddot{O}\!-\!\ddot{O}\!:^{\ominus}$$

This is a correct Lewis structure and is, in fact, a known ion called peroxonitrite. It is not, however, the nitrate ion. There are two ions with the formula NO_3^-: the nitrate ion and the peroxonitrite ion, each with a corresponding Lewis structure. This example emphasizes the need to know how the atoms are arranged before we can draw a correct Lewis structure for a molecule or ion.

For atoms that can have more than eight electrons in their valence shell, we need to use rule 6. As an example, let us consider the sulfate ion, SO_4^{2-}.

1. The S atom is in the middle of four O atoms:

$$\overset{\displaystyle O}{\underset{\displaystyle O}{\overset{|}{O\!-\!S\!-\!O}}}$$

Note that S is the less electronegative atom and the unique atom.

2. The number of electrons is found as follows:

Sulfur (group VI)	$6e^-$	1S	$6e^-$
Oxygen (group VI)	$6e^-$	4O	$24e^-$
Negative charges			$2e^-$
			$32e^-$

3. The four bonds use up 8 of these electrons, leaving 24 to distribute. These 24 can be allocated 6 to each O to complete the octet on each atom:

$$\overset{\displaystyle :\!\ddot{O}\!:^{\ominus}}{\underset{\displaystyle :\!\ddot{O}\!:_{\ominus}}{\overset{|}{{}^{\ominus}:\!\ddot{O}\!-\!\underset{\text{\textcircled{2+}}}{S}\!-\!\ddot{O}\!:^{\ominus}}}}$$

The S atom has a formal charge of $+2$, and each O atom has a formal charge of -1.

4. Each atom has an octet, so no more bonds need to be formed.

5. Rule 5 is not applicable.

6. The structure in step 3 is a correct Lewis structure, but we can draw a better structure by applying rule 6, which recognizes the fact that S is not restricted to eight electrons in its valence shell. The formal charges on the S atom and two of the O atoms can be removed by forming two double bonds. This step does not change the number of electrons in the valence shell of each of the O atoms, but it increases the number in the valence shell of S to 12. We then obtain

$$\overset{\ddot{\text{O}}}{\underset{\underset{\ddot{\text{O}}:^{\ominus}}{\mid}}{\overset{\parallel}{:\text{O}=\text{S}-\ddot{\text{O}}:^{\ominus}}}}$$

This is the structure we gave in Chapter 7.

In general, when two or more Lewis structures can be written, the structure with the fewest formal charges is the preferred structure. Removing formal charges moves electrons from negatively charged atoms toward positively charged atoms, thereby decreasing the energy of the structure (see Box 8.3).

Example 8.4 Draw the Lewis structure of SO_2.

Solution

1. The arrangement of atoms is O—S—O.

2. There are 18 electrons and two bonds. Therefore 14 electrons are available to complete the octets.

3. Completing the octets, we have $^{\ominus}:\ddot{\text{O}}-\overset{\overset{2+}{\cdots}}{\text{S}}-\ddot{\text{O}}:^{\ominus}$

4. Forming a double bond, we obtain $:\ddot{\text{O}}=\overset{\oplus}{\text{S}}\ddot{\text{O}}:^{\ominus}.$

5. Rule 5 is not applicable.

6. The structure in step 4 is a correct Lewis structure. But since S is a third-period element, we can eliminate the formal charges by forming another double bond, thereby giving S a valence shell of ten electrons. The preferred Lewis structure is therefore $:\text{O}=\overset{\cdot\cdot}{\text{S}}=\text{O}:$.

Note that the purpose of a Lewis structure is to show how the electrons are arranged in bonding pairs and nonbonding pairs in an ion or a molecule. Although the shape of a molecule can be predicted from the Lewis structures, the Lewis structure need not show the shape. Thus the Lewis structure of SO_2 may be correctly drawn as

$$:\ddot{\text{O}}=\ddot{\text{S}}=\ddot{\text{O}}: \quad \text{or} \quad \overset{\ddot{\text{S}}}{\underset{:\text{O}}{\diagup}}\overset{}{\underset{\text{O}:}{\diagdown}}$$

The Lewis structure of a more complicated molecule is often most easily drawn without attempting to show the shape. Thus we can draw the Lewis structure of SO_4^{2-} as follows:

$$\overset{\ddot{\text{O}}}{\underset{\underset{:\ddot{\text{O}}:^{\ominus}}{}}{:\ddot{\text{O}}::\text{S}:\ddot{\text{O}}:^{\ominus}}} \quad \text{or} \quad \overset{\ddot{\text{O}}}{\underset{\underset{:\ddot{\text{O}}:^{\ominus}}{\mid}}{\overset{\parallel}{:\ddot{\text{O}}=\text{S}-\ddot{\text{O}}:^{\ominus}}}}$$

We can show that it is tetrahedral by means of a separate diagram, which clearly shows the shape and to which we often add other information such as bond lengths and bond angles:

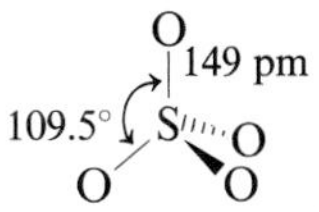

Example 8.5 Draw the Lewis structure of ozone, O_3.

Solution

1. O—O—O.

2. 18 electrons, 2 bonds, 14 electrons to distribute.

3. $\overset{\ominus}{:}\ddot{\text{O}}—\overset{\overset{\scriptsize\textcircled{2+}}{..}}{\text{O}}—\ddot{\text{O}}\overset{..}{:}^{\ominus}$

4. $:\ddot{\text{O}}=\overset{..}{\text{O}}\overset{\oplus}{—}\overset{..}{\ddot{\text{O}}}:^{\ominus}$. In this case we cannot form a second double bond because oxygen cannot exceed the octet.

Another possible arrangement of the atoms for the O_3 molecule is

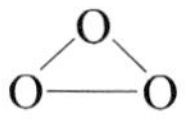

Completing the Lewis structures gives

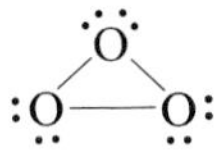

This is a correct Lewis structure but for a molecule that is not known. It is not the Lewis structure of ozone, which does not have a cyclic structure with 60° bond angles but rather an angular structure with a bond angle of 117°. Ozone is an AX_2E molecule for which a bond angle of approximately 120° would be predicted. This example emphasizes again that one must know how the atoms are connected before a Lewis structure can be written.

Another atom that does not always obey the octet rule is boron. Consider BCl_3.

1. In BCl_3 the boron atom is in the center of the three chlorine atoms:

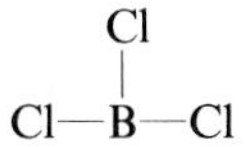

2. There are a total of $3 + 21 = 24$ electrons.

3. Six electrons are used for the three bonds, therefore 18 remain. Six can be added to each chlorine atom to complete their octets:

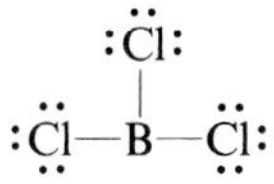

This structure leaves the boron atom without an octet.

4. Following rule 4 we would write the structure

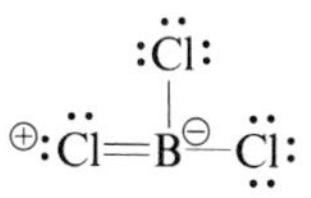

However, we find that we have created formal charges. This structure is expected to be less stable than that in which all the atoms have a zero formal charge, particularly since there is a positive charge on the very electronegative chlorine atom and a negative charge on the much less electronegative boron. We therefore apply rule 5 and return to the structure obtained in step 3.

When we draw Lewis structures, it is useful to be able to recognize the possible combinations of bonds, nonbonding pairs, and formal charges for common atoms. Because carbon, nitrogen, oxygen, and fluorine always obey the octet rule, they always have four electron pairs in their valence shell. So the only possibilities are

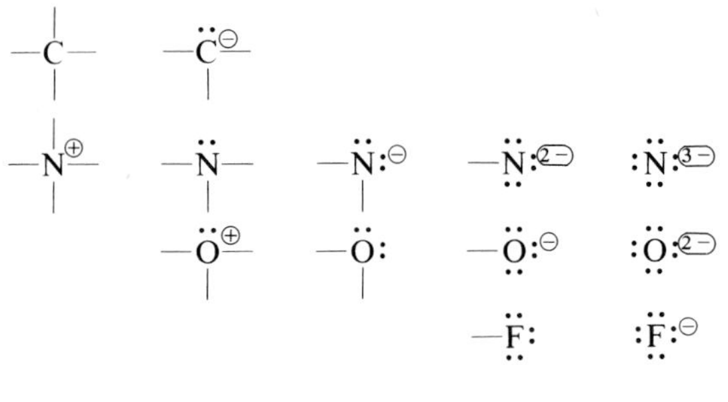

8.3 RESONANCE STRUCTURES

The Lewis structure of the planar triangular carbonate ion,

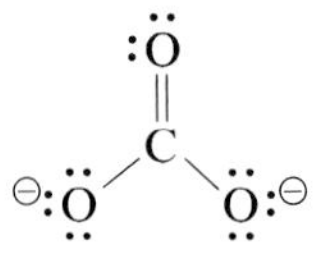

indicates that the central carbon atom forms one double bond and two single bonds to oxygen atoms. We therefore expect one of the bonds in this ion to be shorter than the other two and two of the angles to be larger than the third. However, determination of the structures of this ion by X ray crystallography has shown that all three bonds have the same length and that the bond angles are all exactly 120°:

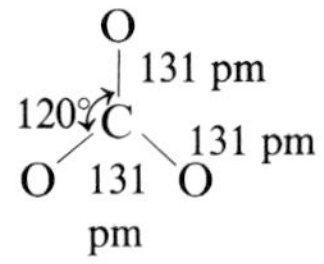

Similarly, the Lewis structures of the phosphate and sulfate ions have both single and double bonds. Yet again, researchers have found experimentally that all the bonds are the same length and all the bond angles are equal; these ions have a regular tetrahedral shape:

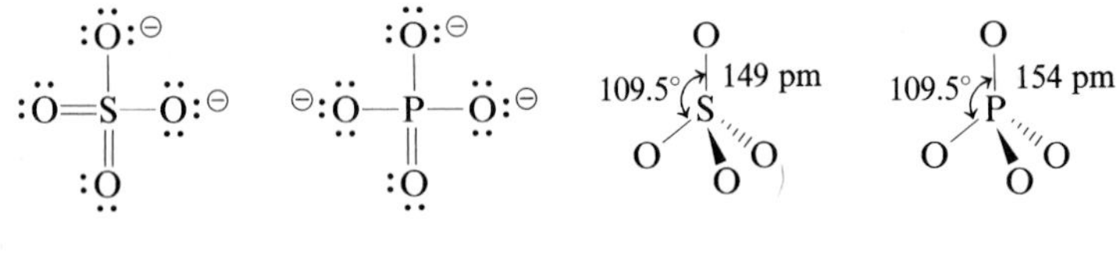

Delocalized Electrons

Clearly, the Lewis structures do not give a completely accurate description of the arrangement of the electrons in these ions. Let us see why they do not by returning to the derivation of the Lewis structure of the carbonate ion. Rule 3 leads to the structure

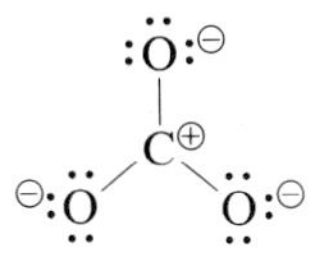

There are then three ways to complete the octet on carbon, following rule 4:

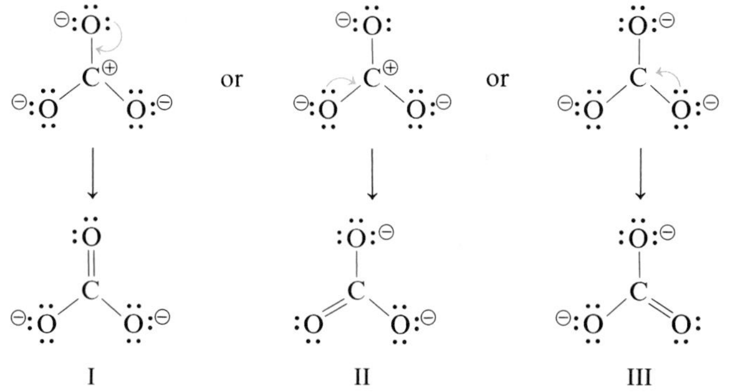

So there are three possible Lewis structures, I, II, and III. The carbon atom cannot discriminate among the three oxygen atoms. It does not accept a complete electron pair from any one of them but, rather, one-third of a pair of electrons from each of the three oxygen atoms:

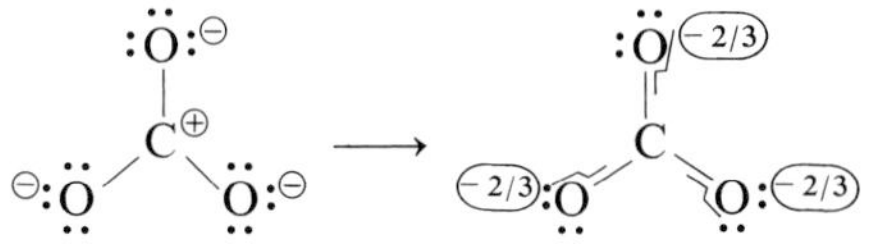

Each of these three electron pairs is partially a nonbonding pair and partially a bonding pair. They are denoted in the structure by ⌐. On the average two-thirds of each of these electron pairs is left on oxygen and one-third of each pair contributes to each C—O bond. Three times one-third of an electron pair, in other words, a total of one electron pair, is donated by the oxygen atoms to be shared with the carbon atom. This is equivalent to the carbon atom gaining one electron and each oxygen losing one-third of an electron. Thus the charge on carbon becomes $+1 + (-1) = 0$ and the charge on each oxygen becomes $-1 - (-\frac{1}{3}) = -\frac{2}{3}$.

Clearly, we cannot write a single, conventional Lewis structure to show this arrangement of electrons in the carbonate ion because Lewis structures depict electrons as being completely localized either as bond pairs or as nonbonding pairs. In fact, electron clouds tend to spread out so as to be associated with as many nuclei as possible. In the carbonate ion three of the electron pairs may be described as partially nonbonding and partially bonding.

We get around this problem by representing the arrangement of the electrons by a combination of several Lewis structures—in the case of the carbonate ion by the structures I, II, and III, which are called **resonance structures**. We indicate that they are resonance structures by means of double-headed arrows between them:

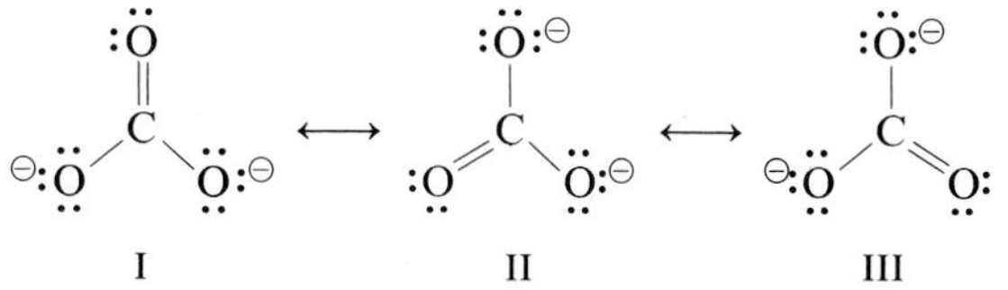

Box 8.3
MORE ON LEWIS STRUCTURES

Some of the Lewis structures that we described in Chapter 7 and in this chapter may differ from those that you learned in another chemistry course in high school or elsewhere. For example, we have given the structures

whereas you may have learned the structures

for these molecules. So you may be wondering if the structures that you learned previously are wrong.

They are not wrong, but they are not the best structures that can be given for these molecules. It is important to remember that Lewis structures are only very approximate representations of the electron distributions in molecules. This electron distribution is a continuous charge cloud in which individual electrons cannot be recognized. Determining this electron distribution experimentally is very difficult, although it has been done in a few cases by X ray crystallography. In relatively simple molecules the electron distribution can be calculated with some accuracy on the basis of better models for the electron distribution than are provided by Lewis structures. However, we cannot discuss these calculations at the level of this book. We have to make do with Lewis structures, which are, in any case, widely used by chemists because of their simplicity. We must be aware of their limitations, however. A continuous electron distribution is only very crudely represented by dots and lines representing bonding and nonbonding electron pairs.

The structures originally proposed by Lewis have subsequently been modified in two important ways to make them more consistent with experimental observations, particularly bond lengths.

First, to account for the fact, for example, that the three CO bonds in the carbonate ion, CO_3^{2-}, all have the same length, the concept of *resonance* has been introduced. As we have seen, this is a way of allowing for the fact that in many molecules not all electrons can be regarded as strictly localized nonbonding pairs associated with only one atom or as bonding pairs associated with just two atoms. Some electrons may behave as both bonding and nonbonding pairs, and some bonding pairs may be delocalized over more than two atoms.

Second, to account for the existence of molecules such as PCl_5 and SF_6 and for the short SO bonds in SO_2, SO_3, and SO_4^{2-}, we recognized that the octet rule is not necessarily valid for the elements of period 3 and beyond. These elements may have more than 8 electrons in their valence shell in some of their compounds, because the $n = 3$ shell can accommodate a maximum of 18 electrons. The structures that you may have learned previously, such as IV, V, and VI are the structures that Lewis wrote for these molecules and ions and they are therefore correct Lewis structures. Indeed, we deduce their structures following our rules 1 to 4, but these structures have subsequently been modified so that they are in better agreement with experimental data, such as bond lengths, and because it has been recognized that elements such as sulfur do not need to obey the octet rule in all their compounds. We allow for this modification in rule 6.

Thus the structure III and the equivalent resonance structures are preferred to the structure VI because the second structure has many formal charges and is not consistent with the length of the SO bonds. Both structures are only a very crude representation of the electron distribution in these molecules, but most chemists now consider that the double-bond structures for SO_2, SO_3, and SO_4^{2-}, for example, are better representations of the electron distributions than Lewis's original structures.

If we wish, we can go one step further and say that SO_4^{2-}, for example, is best represented by a combination—that is, a resonance hybrid—of all the possible Lewis structures. This combination gives more weight to those structures that we consider to be of greatest importance—that is, the structures of lower energy. For SO_4^{2-} we would then have the following resonance structures, in order of decreasing importance:

| 6 equivalent structures | 4 equivalent structures | 1 structure |

However, this begins to get too complicated to be worthwhile; we are trying to push our simple model beyond its limits. We will normally just use the single structure or the equivalent resonance structures that most nearly represent the electron distribution as far as we can tell from the available experimental information.

The carbonate ion is not correctly described by any one of these structures, but it may be described by a combination, called a **resonance hybrid**, of the three structures. One of the three electron pairs on each oxygen is nonbonding in two of the structures and bonding in the other structure. In other words, each of these three electron pairs is partially bonding and partially nonbonding. All three bonds are equivalent and each may be described as a $1\frac{1}{3}$ bond; each is said to have a **bond order** of $1\frac{1}{3}$.

Bond Order

The **order of a bond** *is defined as the total number of electron pairs that constitute the bond.* A triple bond has a bond order of 3, a double bond has a bond order of 2, and a single bond has a bond order of 1. In the carbonate ion the bonds are intermediate between single and double bonds and have a bond order of $1\frac{1}{3}$. The observed bond length of 131 pm for each bond is consistent with this bond order; it is intermediate between that for the CO double bond in methanal (121 pm) and that for the CO single bond in methanol (143 pm):

121 pm	143 pm
Methanal	Methanol
(formaldehyde)	

To find the bond order from a set of resonance structures, we take the same bond in each of the resonance structures, sum the bond order for this bond in all the structures, and then divide by the number of structures. To determine the bond order of the carbonate ion, we might use the top CO bond in structures I, II, and III. The sum of the bond orders is $2 + 1 + 1 = 4$. The number of structures is 3, so the average bond order is $\frac{4}{3}$, or $1\frac{1}{3}$. Note that it does not matter which bond we select; for any bond the result is the same.

The formal charge on each oxygen atom in the carbonate ion may be obtained by a similar procedure. The sum of the charges on any given oxygen atom for all three structures is $0 - 1 - 1 = -2$. The average charge on oxygen is then obtained by dividing -2 by the number of structures; it is $-\frac{2}{3}$. We see that the -2 charge of the ion is spread over the three oxygen atoms.

It is often said that there is **resonance** between the three structures I, II, and III. However, we must be careful not to interpret such a statement as meaning that resonance is a phenomenon. There is no rearrangement of electrons corresponding to the change of one structure to another. The molecule never has any of the three individual Lewis structures, all of which localize the electrons. It has only one structure in which some of the electrons are less localized—in other words, more spread out—than is represented by any single resonance structure. This delocalization of some of the electrons can be represented by a combination of resonance structures.

We use resonance structures because no individual Lewis structure is capable of accurately representing the arrangement of electrons in some molecules. Lewis structures depict each electron pair as either a bonding pair or a nonbonding pair, and they do not allow for the possibility that a pair of electrons may be partially bonding and partially nonbonding. The concepts of resonance and of resonance structures are necessary because of the inadequacies of Lewis diagrams for representing the structures of some molecules (see Box 8.3). We emphasize again that *resonance is a concept and not a phenomenon*. A phenomenon is some change in nature that we can observe. A concept is a model or theory proposed by us to explain phenomena that we observe.

Whenever we can write more than one equivalent Lewis structure for a polyatomic ion or molecule, as in the case of the carbonate ion, we may suspect that some of the electrons are more delocalized than any one of the individual structures indicates. When we can only write one Lewis structure, as in the case of methane, we may reasonably suppose that the electrons are localized approximately in the manner indicated by the Lewis diagram.

The Lewis structure for the phosphate ion has one double bond and three single PO bonds. The following four resonance structures may be written:

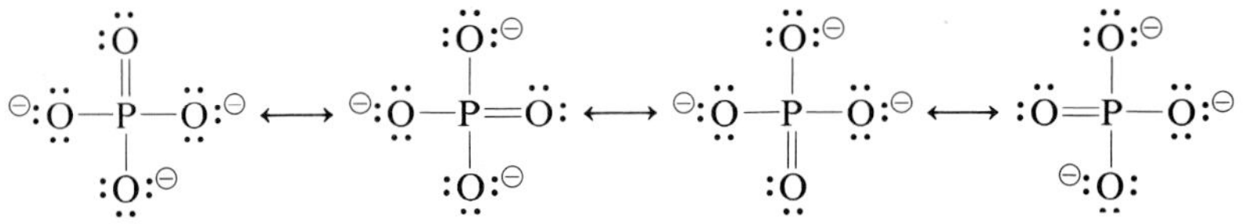

They show that each bond has an order of $1\frac{1}{4}$. Each oxygen atom has an electron pair that is 25% a bonding pair and 75% a nonbonding pair. Each oxygen has a charge of $-\frac{3}{4}$, because a total of three negative charges are shared between four oxygen atoms. Or we may sum the charge on any one oxygen for all the structures $(-1 - 1 - 1 + 0 = -3)$ and divide by the total number of structures, to give $-\frac{3}{4}$.

Example 8.6 The sulfate ion has a tetrahedral structure in which all four SO bonds have the same length (149 pm). Draw resonance structures for the sulfate ion, and determine the bond order and the charge on each oxygen atom.

Solution The Lewis structure of the sulfate ion is

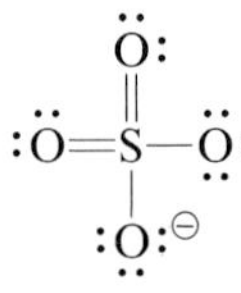

There are six possible combinations of double and single bonds, as follows:

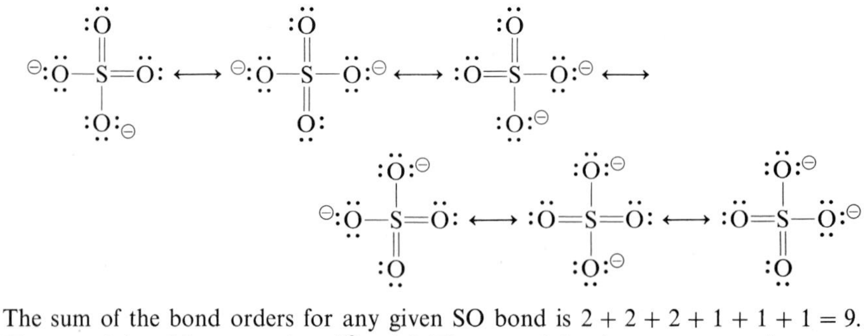

The sum of the bond orders for any given SO bond is $2 + 2 + 2 + 1 + 1 + 1 = 9$. Hence each bond has an order of $\frac{9}{6} = 1.5$. The sum of the charges on any given oxygen atom is $0 + 0 + 0 - 1 - 1 - 1 = -3$. Hence each oxygen atom has a charge of $-\frac{3}{6} = -\frac{1}{2}$.

The sulfite ion, SO_3^{2-}, is an AX_3E molecule with a triangular pyramidal shape. It has three equal bonds of length 151 pm and equal bond angles of 106°.

It can be described by the following three resonance structures:

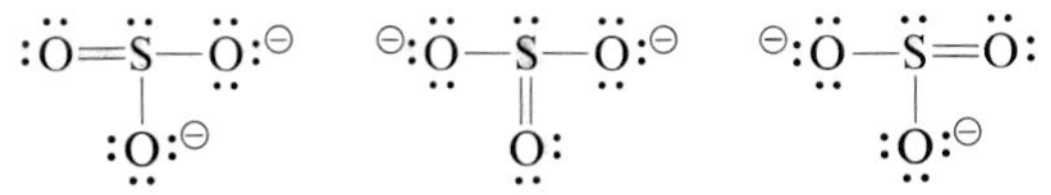

$$:\ddot{O}=\underset{|}{\overset{\cdots}{S}}-\ddot{O}:^{\ominus} \qquad ^{\ominus}:\ddot{O}-\underset{\|}{\overset{\cdots}{S}}-\ddot{O}:^{\ominus} \qquad ^{\ominus}:\ddot{O}-\underset{|}{\overset{\cdots}{S}}=\ddot{O}:$$

The bonds have an order of $1\frac{1}{3}$, and each oxygen has a charge of $-\frac{2}{3}$.

The sulfur–oxygen bonds in the following molecules and ions illustrate the relationship between bond order and bond length:

Structure	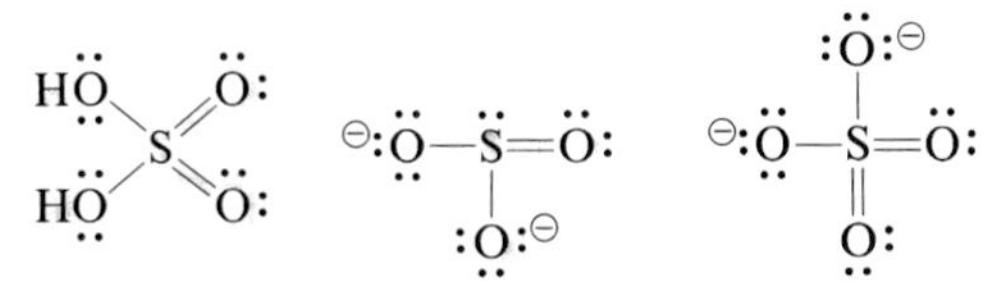 HO∖S∕O: with HO∕ ∖O: (S)	$^{\ominus}:\ddot{O}-\ddot{S}=\ddot{O}:$ with $:\ddot{O}:^{\ominus}$	$^{\ominus}:\ddot{O}-\ddot{S}=\ddot{O}:$ with :Ö:⁻ above and O: below
Bond length (pm)	157 (S—O)	151	149
Bond order	1.00	1.33	1.50

Structure	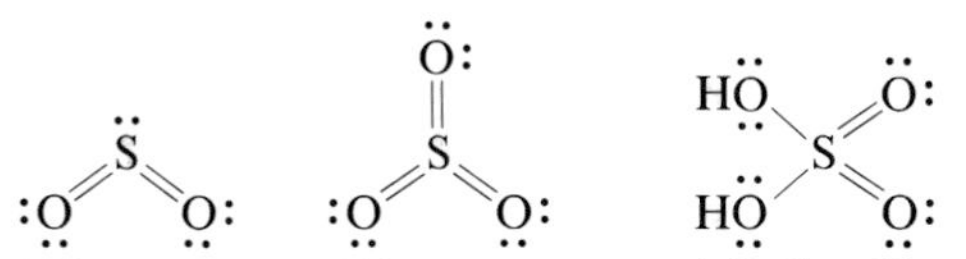 :O═S═O:	O: ═S═ with :O ∕ ∖ O:	HO∖S∕O: with HO∕ ∖O:
Bond length (pm)	143	143	142 (S═O)
Bond order	2	2	2

In general, bond length decreases as bond order increases. The more electron pairs that form the bond, the more strongly they hold the two nuclei together.

Example 8.7 Draw resonance structures for the methanoate (formate) ion, HCO_2^-. What is the formal charge on each O atom? What is the bond order of the CO bonds? Predict an approximate value for the CO bond length.

Solution

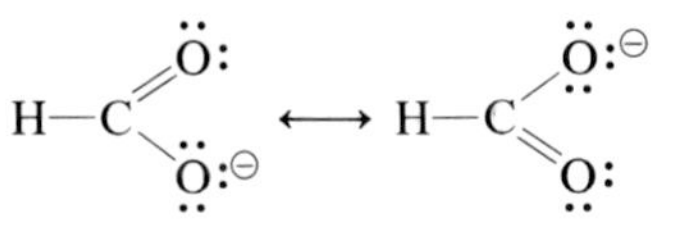

$$H-C\overset{\displaystyle\ddot{O}:}{\underset{\displaystyle\ddot{O}:^{\ominus}}{}} \longleftrightarrow H-C\overset{\displaystyle\ddot{O}:^{\ominus}}{\underset{\displaystyle\ddot{O}:}{}}$$

The formal charge on each O atom is $-\frac{1}{2}$. The CO bond order is $\frac{3}{2} = 1.5$. The bond length is predicted to be intermediate between that of CO_3^{2-} (bond order 1.33, bond length 131 pm) and methanal CH_2O (bond order 2.0, bond length 121 pm). Taking the mean of these two bond lengths as a rough approximation, we predict a value of 126 pm. The observed value in sodium methanoate is 127 pm.

8.4 MOLECULAR POLARITY AND DIPOLE MOMENTS

The geometric arrangement of the bonds in a molecule and the electronegativities of the atoms determine the distribution of charge in the molecule. The distribution of charge in a molecule has an important effect on its properties and on the forces between molecules (intermolecular forces), as we will discuss in Chapter 13.

We have seen that when two atoms in a diatomic molecule have different electronegativities, the bond is polar covalent; HCl is an example. Because it has a greater electronegativity than H and because the bond electrons are shared unequally between Cl and H, the Cl atom in HCl has a small negative charge ($\delta-$) and the H atom has a small positive charge ($\delta+$):

$$\overset{\delta+}{H}-\overset{\delta-}{Cl}$$

Two separated, equal, and opposite charges constitute a **dipole**.

A dipole is described quantitatively by its **dipole moment**, μ, which is defined as the product of the magnitude of the charge, Q, at each end of the dipole and the distance, r, between the charges:

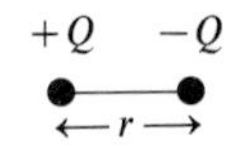

$$+Q \quad\quad -Q$$
$$\bullet\!\!-\!\!-\!\!-\!\!-\!\!-\!\!\bullet$$
$$\longleftarrow r \longrightarrow$$
$$\mu = Qr$$

In SI units the charge Q is measured in coulombs; the distance r is measured in meters. So the dipole moment is measured in coulomb meters, C m.

As Figure 8.19 describes, the dipole moment of a molecule is found experimentally by measuring how the substance composed of these molecules affects the ability of a capacitor to store electric charge. Table 8.7 lists the dipole moments of several molecules.

Note that the dipole moment of the hydrogen halides increases with an increase in the difference in electronegativity between the halogen and hydrogen because the charges on the atoms increase with increasing electronegativity difference.

A molecule that has a dipole moment is described as **polar**. All *diatomic* molecules with polar covalent bonds have dipole moments and thus are polar. But not all polyatomic molecules with polar covalent bonds are polar. The geometry as well as the polarity of its bonds determines whether a polyatomic molecule is polar.

Consider SO_2 and CO_2. Because oxygen has a greater electronegativity than either carbon or sulfur, we expect both a CO bond and an SO bond to be polar. But we see from Table 8.7 that SO_2 has a dipole moment of 5.42×10^{-30} C m, whereas the dipole moment of CO_2 is 0. The important difference between the two molecules lies in their geometry. The CO_2 molecule is a linear AX_2 molecule, whereas SO_2 is an angular AX_2E molecule:

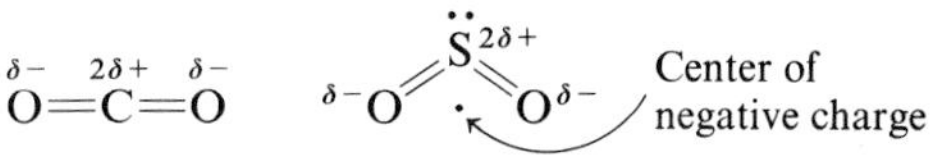

$$\overset{\delta-}{O}\!\!=\!\!\overset{2\delta+}{C}\!\!=\!\!\overset{\delta-}{O}$$

Linear AX_2

Angular AX_2E — Center of negative charge

If a molecule has a dipole moment, the center of the negative charges does not coincide with the center of the positive charges. In SO₂ the center of the two

Figure 8.19 Measurement of Dipole Moment of a Molecule.
A substance is placed between the plates of a capacitor, and the plates are charged as shown. (a) Nonpolar molecules orient themselves randomly between the plates. (b) In contrast, polar molecules, with a dipole moment, tend to line up with their + ends pointing toward the negative plate and their − ends pointing toward the positive plate. This lining up of the polar molecules enables the plates to accept a greater electric charge. The greater the dipole moment, the greater is the charge that the plates can hold, that is, the greater the capacitance of the capacitor in (b).

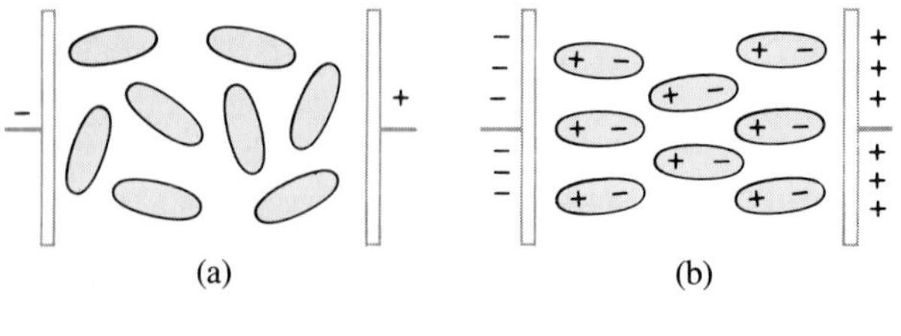

(a)   (b)

negative charges on oxygen is at a point midway between the two oxygens and does not coincide with the positive charge on sulfur. The molecule therefore has a dipole moment and is described as polar. In contrast, in the linear CO_2 molecule the center of the two negative charges on oxygen coincides with the positively charged carbon atom, so the molecule has no dipole moment; it is described as a nonpolar.

We could arrive at the same conclusion by considering each bond as a dipole that has it own dipole moment and then taking the vector sum of the two dipole moments to obtain the dipole moment for the molecule. Thus for CO_2 and SO_2 we have

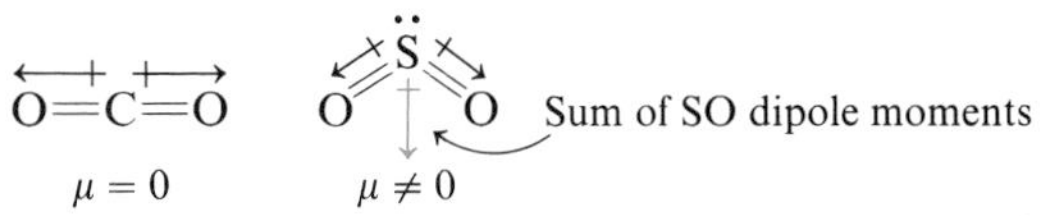

The two SO dipoles make an angle of 120° to each other and they do not cancel each other; their vector sum gives a resultant dipole pointing along the bisector of the OSO angle. In contrast, the two CO dipoles are of equal magnitude and point in opposite directions, so they cancel each other.

The water molecule, which is an angular AX_2E_2 molecule, is also a polar molecule with a dipole moment:

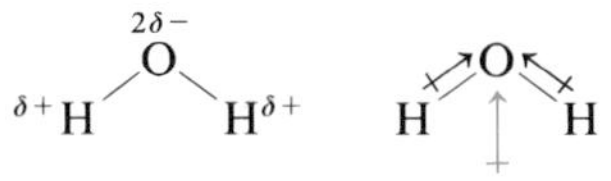

For a triatomic molecule consisting of a central atom A with two identical ligands X, the measurement of the dipole moment can tell us immediately whether the molecule is linear or angular. If it has a dipole moment it is angular not linear. AX_2 molecules do not have a dipole moment, AX_2E and AX_2E_2 do.

Example 8.8 Which of the following molecules will have a dipole moment and which will not?

$$CS_2 \quad BeCl_2 \quad SnCl_2 \quad H_2S \quad SCl_2$$

Solution First, we draw the Lewis structures for the molecules, and then we classify them as AX_2 linear, AX_2E angular, or AX_2E_2 angular.

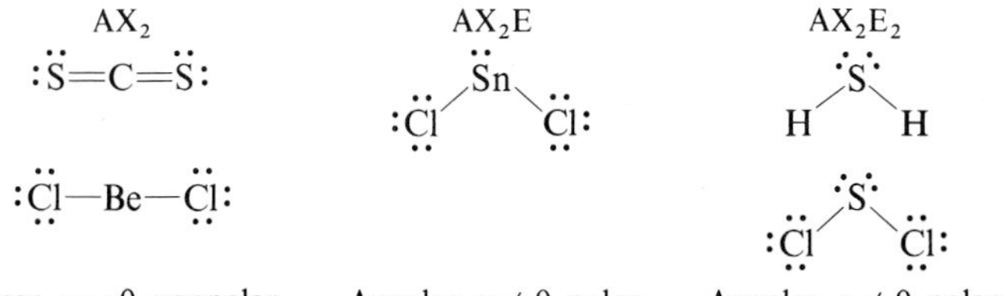

The linear CS_2 and $BeCl_2$ molecules have no dipole moment and are nonpolar, while $SnCl_2$, H_2S, and SCl_2, which are all angular molecules, have dipole moments and are polar.

We can similarly compare molecules in which three ligands are attached to a central atom. For example, Figure 8.20 compares PCl_3 and BCl_3. As the

Table 8.7 Dipole Moments of Some Molecules

MOLECULE	DIPOLE MOMENT (UNITS OF 10^{-30} Cm)
HF	6.36
HCl	3.43
HBr	2.63
HI	1.27
H_2O	6.17
H_2S	3.12
NH_3	4.88
PCl_3	2.30
BCl_3	0
SO_2	5.42
CO_2	0

8.4. MOLECULAR POLARITY AND DIPOLE MOMENTS

307

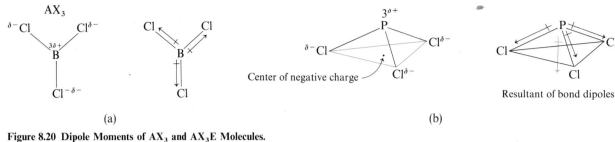

AX_3

Center of negative charge

Resultant of bond dipoles

(a)   (b)

Figure 8.20 Dipole Moments of AX_3 and AX_3E Molecules.
(a) In BCl_3 the center of the negative charges on the Cl atoms coincides with the
positive charge on the B atom. Therefore the molecule has no dipole moment.
In other words, the three BCl bond dipoles, which have a symmetrical planar
arrangement, have a vector sum of zero. (b) In PCl_3 the center of the negative
charges does not coincide with the positive charge on the phosphorus atom. So the
molecule has a dipole moment and is polar. The vector sum of the three bond dipoles
in PCl_3 is a dipole that has its positive end on the P atom and points toward the
midpoint of the triangular base of the pyramid.

sum of CCl_a and CCl_b dipoles

Cl_b

Cl_a

C

Cl_c

Cl_d

sum of CCl_c and CCl_d dipoles

**Figure 8.21 Dipole Moment of
CCl_4—an AX_4 Molecule.** The sum
of the upper C—Cl dipoles is a
vector that points from the C atom
to the midpoint of the upper face of
the cube; the sum of the lower
C—Cl dipoles points toward the
midpoint of the lower face of the
cube. These two vectors have equal
magnitudes and point in opposite
directions; therefore their sum is
zero. The center of the four negative
charges coincides with the positive
charge on the C atom. The dipole
moment of CCl_4 is zero.

figure shows, PCl_3, which is an AX_3E molecule with a triangular pyramidal
shape, has a dipole moment ($\mu = 2.30 \times 10^{-30}$ C m); BCl_3, which is a planar
AX_3 molecule, does not ($\mu = 0$). In general, AX_3 molecules, like AX_2 molecules,
are nonpolar; AX_3E molecules, like AX_2E and AX_2E_2 molecules, have dipole
moments and are polar. Molecules that have one or more nonbonding electron
pairs in the valence shell of the central atom have a dipole moment, while those
that do not have a nonbonding electron pair have no dipole moment ($\mu = 0$).

Finally, if we look at AX_4 molecules, we again see the importance of symmetry in determining whether a molecule is polar. Figure 8.21 shows CCl_4, a
tetrahedral AX_4 molecule. The C—Cl bonds are polar, but the center of the
four negative charges on the Cl atoms lies in the center of the tetrahedron and
thus coincides with the positive charge on the C atom. Therefore the molecule
has no dipole moment and is nonpolar. We reach the same conclusion if we
consider the bond dipoles; they have a vector sum of zero. This can be seen
if we represent the tetrahedron by four corners of a cube and consider the sum
of the vectors two at a time, as shown in Figure 8.21.

Because of their symmetrical structures, AX_2, AX_3, and AX_4 molecules have
no dipole moment—if the ligands are identical. But if the ligands are not the
same, even these molecules will be polar. For example, CF_2Cl_2 has a dipole
moment because the sum of the C—F dipole moments is not equal to the sum
of the C—Cl dipole moments (see Figure 8.22).

**Figure 8.22 Dipole Moment of a Tetrahedral Molecule with
Nonidentical Ligands.**
When the ligands are not identical, as in CF_2Cl_2, an AX_4 molecule
will have a dipole moment. The sum of the two C—F dipoles in
CF_2Cl_2 is greater than the sum of the two C—Cl dipoles. Therefore
they do not cancel, and there is a resultant molecular dipole moment.

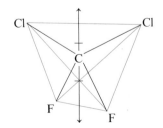

IMPORTANT TERMS

In a **bent bond** the maximum of the electron density does not lie on the straight line between two bonded atoms.

A **bond angle** is the angle formed by two bonds to the same atom.

A **bond dipole** is formed by the partial charges on two atoms that are bonded together.

The **bond length** is the distance between the nuclei of two atoms that are bonded together.

The **bond order** is the number of electron pairs involved in a bond between two atoms. For molecules and ions that are described by resonance structures, the bond order may be fractional.

A **bonding pair** is a pair of electrons that is shared between two atoms and that bonds them together.

A **dipole** is formed by two separated charges of equal magnitude and opposite sign.

The **dipole moment**, μ, is the product of the magnitude of the charges forming a dipole and the distance between the charges: $\mu = Qr$.

A **double bond** is a covalent bond of order 2; a bond formed by the sharing of two electron pairs.

A **ligand** is a single atom, or a group of atoms, such as OH, that is attached to a central atom.

A **nonbonding pair** is a pair of electrons in the valence shell of an atom that is not forming a bond; also called an *unshared pair* or a *lone pair*.

A **nonpolar bond** is formed between two atoms that have the same electronegativity. A **nonpolar molecule** is a molecule that has no dipole moment.

A **polar bond** is a bond formed between two atoms with different electronegativities. A **polar molecule** is a molecule that has a dipole moment.

Resonance is a method for describing the structures of molecules that cannot be adequately represented by a single Lewis structure because the electrons are not as localized as shown in a Lewis structure. Such a molecule may be represented by a combination of two or more Lewis structures called **resonance structures**.

A **triple** bond is a covalent bond of order 3; a bond formed by the sharing of three electron pairs.

PROBLEMS

Lewis Structures

1. Draw Lewis structures for these ions:

S_2^{2-} SO_3^{2-} ClO^- ClO_4^-

2. In which of the following compounds does the central atom not obey the octet rule?

OF_2 SF_2 SF_4 SO_2 PCl_3 BCl_3

3. Draw Lewis structures for the following species. In each case the central atom is a sulfur atom.

$S_2O_3^{2-}$ SO_3F^- SOF_2 SO_2F_2 SNF_3

4. Draw Lewis structures for the following molecules and ions, and categorize them according to the AX_nE_m nomenclature. In each case the first atom in the formula is the central atom.

CO_2 CS_2 SO_2 PO_4^{3-}* SO_3^{2-} ClO_4^- BO_3^{3-}

5. Draw Lewis structures for the following species:

CN^- NO_2^- O_3

6. Draw Lewis structures for the following oxoacids:

HNO_2 HNO_3 H_2SO_4 $HClO_3$ H_2CO_3

7. Dinitrogen monoxide, N_2O, is one of several oxides of nitrogen. Draw Lewis structures for both the atomic arrangements NON and NNO. By comparing the formal charges in the two Lewis structures, decide which structure is the more probable.

8. In addition to ammonia, there are two other molecules containing nitrogen and hydrogen. They are hydrazine, N_2H_4, and diazine, N_2H_2. Draw Lewis structures for these two molecules.

VSEPR Theory

9. Draw diagrams of the following molecules and give a value for the XAX bond angle in each case:
   (a) A linear AX_2 molecule
   (b) An equilateral triangular AX_3 molecule
   (c) A tetrahedral AX_4 molecule

10. Draw diagrams to illustrate the geometries of the following types of molecules, and name the shape in each case:

AX_3E AX_2E_2 AXE_3 AX_2E AXE_2 AX_2

11. Describe each of the following molecules and ions using the AX_nE_m nomenclature, and hence name their shapes:

H_2O H_3O^+ PCl_3 BCl_3 SiH_4 BH_4^-
H_2S SCl_2 NH_4^+ BeH_2 BeH_4^{2-}

* The asterisk denotes the more difficult problems.

12. Draw Lewis structures for the molecules formed by the elements Be to F with chlorine. Describe each molecule using the AX_nE_m nomenclature, and draw a diagram to show the shapes in each case.

*13. Draw a cube and connect appropriate corners to form a tetrahedron. Join the midpoint of the cube (and the tetrahedron) to each corner of the tetrahedron. These lines represent the bonds in a tetrahedral molecule such as CH_4. Calculate the angle between the bonds by trigonometry.

Resonance Structure

14. Write all of the possible equivalent Lewis structures, that is, the resonance structures, for each of the following molecules and ions. In each case, find the charge on each oxygen atom and the CO bond order:

CH_3OH HCO_2^- CO_3^{2-} H_2CO CO

15. Draw resonance structures for each of the following, and find the bond orders:

NO_3^- O_3 PO_4^{3-} SO_4^{2-} ClO_4^-

16. Would you expect the Cl—O bond in ClO^- to be longer or shorter than that in ClO_4^-? Explain why.

17. The HNO_3 molecule has two different NO bond lengths of 121 pm and 140 pm. The NO_3^- ion has a bond length of 122 pm. Account for these different bond lengths in terms of the Lewis structures for HNO_3 and NO_3^-.

18. The observed PO bond lengths in P_4O_{10} are 160 and 140 pm. On the basis of these two values, account for the following observed PO bond lengths:

PO_4^{3-} 154 pm HPO_3^{2-} 151 pm $P_2O_7^{4-}$ 152 and 161 pm

Dipole Moments

19. Predict which of the following molecules will have a dipole moment:

BF_3 NF_3 $BeCl_2$ SCl_2 I_2
ICl CCl_4 $CHCl_3$ CH_2Cl_2

20. The neutral molecule ACl_3 is found to have a dipole moment. Predict the group of the periodic table to which the central atom A belongs. What would your answer be if ACl_3 had no dipole moment ($\mu = 0$)?

21. Indicate the expected polarity of the bonds in the following molecules. State which molecules are expected to have a dipole moment.

SO_2 CO_2 H_2O NH_3 SO_3 BeH_2

22. The ozone molecule has a dipole moment of 1.8×10^{-30} C m. Explain how the dipole moment could be used to distinguish between the two possible Lewis structures given in Example 8.5.

23. Draw diagrams to show the charge distribution in each of the following molecules:

NF_3 OF_2 H_2S PH_3

24. Arrange the following groups of molecules in order of increasing dipole moment:
   (a) HF, HI, HBr, HCl   (b) AsH_3, NH_3, PH_3
   (c) Cl_2O, F_2O, H_2O (d) H_2O, H_2Te, H_2S, H_2Se

25. Explain the following statements:
   (a) The dipole moment of F_2O is much less than that of H_2O, even though both molecules have similar bond angles.
   (b) The dipole moment of CS_2 is zero, while OCS has a measurable dipole moment.

CHAPTER 9

SOME COMMON METALS: ALUMINUM, IRON, COPPER, AND LEAD

Metallic Bonding, Oxidation-Reduction Equations, Lewis Acids and Bases

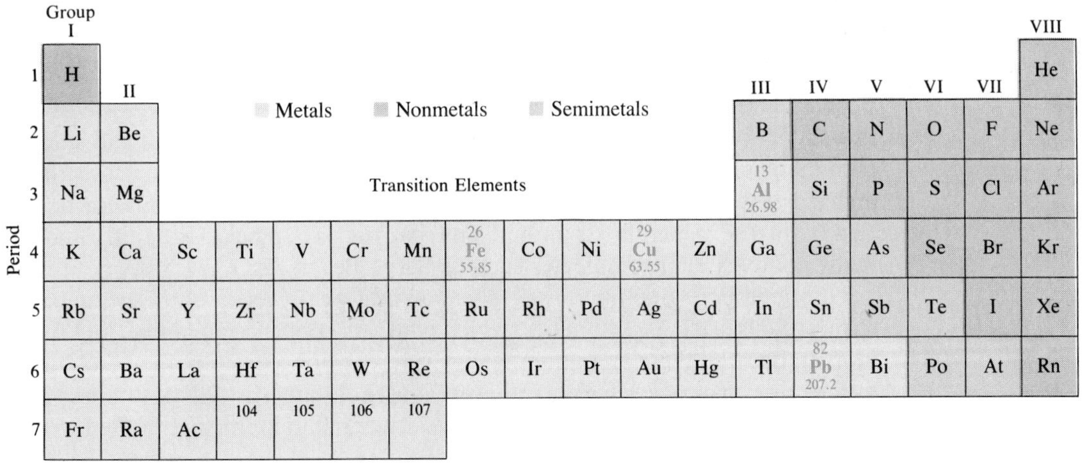

Group I																	VIII
1 H	II											III	IV	V	VI	VII	He
2 Li	Be		Metals		Nonmetals		Semimetals					B	C	N	O	F	Ne
3 Na	Mg			Transition Elements								13 Al 26.98	Si	P	S	Cl	Ar
4 K	Ca	Sc	Ti	V	Cr	Mn	26 Fe 55.85	Co	Ni	29 Cu 63.55	Zn	Ga	Ge	As	Se	Br	Kr
5 Rb	Sr	Y	Zr	Nb	Mo	Tc	Ru	Rh	Pd	Ag	Cd	In	Sn	Sb	Te	I	Xe
6 Cs	Ba	La	Hf	Ta	W	Re	Os	Ir	Pt	Au	Hg	Tl	82 Pb 207.2	Bi	Po	At	Rn
7 Fr	Ra	Ac	104	105	106	107											

Period

In Chapters 5 and 7 we described the chemistry of the halogens, phosphorus, and sulfur, which are all nonmetals. Of the more than one hundred elements about eighty are metals. This chapter is devoted to four typical metals: aluminum, iron, copper, and lead. These metals are all produced in very large quantities and have many important and often familiar uses. We first describe the reactions by which the metals are obtained from their naturally occurring compounds. Then we discuss the structures of metals, that is, the arrangements of the atoms in the solids and the nature of the forces holding them together—metallic bonds. We will see that the metals have particularly simple structures, which we will use later as a basis for the discussion of the structures of many other solids. Finally, we describe some of the reactions of the metals and their compounds. Consideration of these reactions will enable us to further extend our ideas on acid-base and oxidation-reduction reactions.

9.1 PROPERTIES AND USES OF METALS

Aluminum is in Group III below boron in the periodic table. Lead is the last element in Group IV. In this group the properties of the elements change gradually from those of carbon, a typical nonmetal, to those of lead, a typical metal. Copper and iron are members of the first transition series in the fourth period.

For many practical purposes **alloys** are often more useful than pure metals. An alloy is a solid that has metallic properties and that is usually composed of two or more metals; it may also be composed of a metal and a nonmetal, such as carbon, silicon, or nitrogen. In a few cases alloys are compounds, but more frequently they are solid solutions or simply mixtures. Alloys are often far stronger and harder than their constituent metallic elements. The alloys of copper include *brass*, an alloy of copper and zinc, and *bronze*, an alloy of copper and tin.

Some of the properties of these four elements are summarized in Table 9.1. All four show the typical properties of metals; they are good *conductors of heat and electricity*, they are *malleable* and *ductile*, and they have a shiny surface that reflects light, a so-called *metallic luster*. This is not always apparent, because many metals become coated with a layer of oxide or carbonate when they are exposed to the air. If the metal is cut or the surface layers are scraped away, the bright shiny surface is exposed.

Aluminum

Because aluminum is a rather reactive metal, it is not found in nature in the free state; however, its compounds are very common. After oxygen and silicon, aluminum ranks as the third most abundant element on the earth's surface, and it is the most abundant of all the metals. It occurs as aluminum oxide, Al_2O_3, in the minerals corundum and bauxite, as spinel, $MgAl_2O_4$, and as many complex aluminosilicates such as beryl, $Be_3Al_2Si_6O_{18}$.

Although pure aluminum is a rather soft and weak metal, it has several hard, strong alloys, such as duralumin, which contains copper, manganese, and magnesium. Because these aluminum alloys have very low densities, they find many uses in aircraft, space vehicles, and even household utensils. Billions of beer and soft drink cans are manufactured each year from aluminum, and in the form of foil it is used as a wrapping material.

Other applications of aluminum are based on the fact that it is an excellent

Table 9.1 Some Properties of Aluminum, Iron, Copper, and Lead

ELEMENT	Z	ELECTRON CONFIGURATION	ABUNDANCE (%)	MELTING POINT (°C)	BOILING POINT (°C)	DENSITY (g cm^{-3})
Aluminum	13	[Ne] $3s^2 3p^1$	7.5	660	2467	2.71
Iron	26	[Ar] $3d^6 4s^2$	4.7	1535	2750	7.86
Copper	29	[Ar] $3d^{10} 4s^1$	0.0058	1083	2567	8.97
Lead	82	[Xe] $4f^{14} 5d^{10} 6s^2 6p^2$	0.002	327	1740	11.40

conductor of electricity. Although a wire of aluminum has only one-third the conductivity of a copper wire of the same diameter, the aluminum wire is much lighter because of the low density of aluminum. Since it is also considerably cheaper than copper, aluminum is being used increasingly for electric wiring and is widely used in high-voltage transmission lines.

Iron

Iron ranks as the fourth most abundant element in the earth's crust and the second most abundant metal (see Table 3.2). It is found only in very small amounts on the earth's surface in the free state, usually in association with nickel. But it is believed to form a large fraction of the earth's core, and it is commonly found in meteorites. Most of the iron in the earth's crust is combined with oxygen, silicon, and sulfur. Important iron minerals include the oxides *hematite*, Fe_2O_3, and *magnetite*, Fe_3O_4, and the carbonate $FeCO_3$, *siderite*. The sulfide *pyrite*, FeS_2, also called "fool's gold" because of a superficial resemblance to gold, is another abundant iron mineral. But it is not used as a source of iron because of the difficulty of removing the large amount of sulfur.

Pure iron is a rather soft silvery metal that very few people have ever seen. Obtaining it free of carbon, oxygen, nitrogen, phosphorus, sulfur, and other metals is difficult; and pure iron so obtained has no use. However, the alloys of iron have a greater industrial importance than any other metal. Almost all iron is produced in the form of steel, which is an alloy of iron containing carbon and certain transition metals, such as, nickel, chromium, manganese, and vanadium. Steel is much harder and stronger than pure iron. It is used in buildings, bridges, ships, cars, machinery of all kinds, and many other familiar items.

Copper

Copper is a shiny red metal that is resistant to corrosion. Very few metals are found on the surface of the earth in the uncombined state, and then only in small amounts. Among these copper, silver, and gold are by far the most important. They were the earliest known metals because they are found in the free state, whereas other metals must be obtained from their compounds, and processes for so obtaining them were only slowly discovered (see Box 9.1).

Because of their availability as the free elements and their resistance to corrosion, these metals have been used since antiquity for coins, and they are called the *coinage metals*. However, as the price of these metals has increased dramatically in recent times, their use in coins has decreased. Today gold and silver are not so used at all, and even copper is being replaced by other metals.

Because copper has a very high electrical conductivity, second only to silver, and because it is rather soft and very ductile and thus easily drawn out into thin wires, one of its most important uses is for electric wiring. Elemental copper occurs in amounts too small to be a useful source of the metal, which is therefore obtained from its minerals, which include the sulfides, *chalcocite*, Cu_2S, and *chalcopyrite*, $CuFeS_2$, the oxide *cuprite*, Cu_2O, and the hydroxide carbonate, *malachite*, $Cu_2(CO_3)(OH)_2$.

Lead

Pure lead is a soft metal with a high density and a relatively low melting point. When freshly cut, it has a silvery luster. But when it is exposed to the air, a surface layer of oxide and carbonate forms, and the lead turns a dull blue-gray.

An iron-nickel meteorite

Pyrite

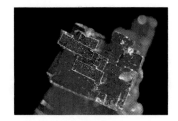

Cuprite

9.1 PROPERTIES AND USES OF METALS

313

Articles of copper, silver, and gold have been found in the remains of the earliest civilizations. These objects were made from the naturally occurring metals simply by shaping them with stone hammers and chisels. Other metals were not available until people had learned how to extract a metal from one of its ores; we do not know how this momentous discovery was first made. The ores of copper, lead, tin, and mercury are easily reduced to the metal by heat or by heating with charcoal. Thus it seems likely that the formation of a metal was first observed when a fire was built on an ore-bearing rock or when an ore was dropped accidentally into a fire. One plausible suggestion is that copper was obtained accidentally from its ore malachite, $Cu_2(CO_3)(OH)_2$. In early Egypt malachite, which is green, was used for facial adornment. When it is heated, it decomposes to give copper oxide, which, when heated with carbon, is reduced to the metal. If malachite had been accidentally dropped into a charcoal fire, metallic copper would have been formed.

Before 3000 B.C., perhaps because copper and tin ores frequently occur together, people discovered that copper could be hardened by alloying it with tin. The resulting bronze could be forged into utensils, tools, and weapons so superior to others then available that this discovery produced a major revolution in technology, initiating the period known as the Bronze Age.

The next great step forward was made about 1000 B.C. with the discovery of iron smelting, the production of iron from iron ore. This discovery marked the beginning of the Iron Age. The rise and fall of early civilizations was closely tied to progress and discoveries in metallurgy.

By the time of Christ, people had learned how to make and to use the metals iron, copper, silver, tin, gold, mercury, and lead. But the nature of the processes involved remained an enigma until the end of the eighteenth century. Until then, the extraction of metals from their ores was an art and certainly not a science. Lack of understanding of the processes involved in making metals encouraged the belief that one metal might be changed into another, leading to the alchemists' futile attempts to transmute lead into gold.

All the metals known to early humans, except iron, are rare elements, but they either occur in the uncombined state or could be obtained easily from their ores. The most abundant metal, aluminum, is difficult to obtain from its ores and was not prepared as the free element until 1825; a practical method for the large-scale industrial production of aluminum was not devised until 1886.

Galena

Although it is not abundant, lead is easily extracted from its most important ore, *galena*, PbS. As a result, it has been known for a long time. In ancient Rome lead pipes were used for the water supply of villas; buildings such as the Pantheon had roofs sheathed in lead; and even wine casks were lined with lead. Some historians have claimed that the use of lead by Romans was so extensive that many Romans, particularly the rich governing class, suffered from lead poisoning, which led to both madness and sterility, and that lead poisoning was one cause of the decline of the Roman Empire. In the Middle Ages lead was used extensively in the building of the great cathedrals, for roofs, gutters, and stained glass windows.

Today lead is used for making storage batteries, for covering electric cables, and as an important component of several alloys such as type metal and solder. Large quantities are also used for the production of the gasoline additive tetraethyl lead, $Pb(C_2H_5)_4$. This use is now declining, since the health hazards associated with "leaded" gasoline have been recognized.

9.2 METALLURGY: EXTRACTION OF METALS

Metallurgy is the science of metals. It had its origins in the development of methods of extracting metals from their ores. An **ore** is a mineral deposit that may be profitably treated for the extraction of one or more metals.

Most metals occur in the earth's crust in an oxidized form, that is, as positive ions. The most important ores of many metals are either oxides or sulfides. Thus if we wish to obtain a metal from one of its ores, a reducing agent must be used. In order of increasing cost, the more commonly used reducing agents are carbon (usually in the form of coke), carbon monoxide, hydrogen, free electrons (in electrochemical processes; see Chapter 16), and other metals, for example, sodium and aluminum.

Extraction of Metals from Oxide Ores

Often metals can be extracted from their oxide ores simply by heating the oxide with carbon, usually in the form of coke, or with carbon monoxide. This process is used, for example, with cuprite, Cu_2O:

$$Cu_2O + C \longrightarrow 2Cu + CO$$

A carbonate ore such as malachite can be treated in the same way because it decomposes when heated to give CuO:

$$Cu_2(CO_3)(OH)_2 \longrightarrow 2CuO + CO_2 + H_2O$$

Since the majority of the rocks on the earth's surface are composed of silicon dioxide, SiO_2 (silica), and silicates (substances containing oxoanions of silicon such as SiO_3^{2-}), these substances are almost always present in an ore, and they must be removed before or during the reduction of the ore. Various mechanical methods are used, but an important chemical method is to add a material called a *flux* that combines with the silica and the silicates to produce a material called a *slag*. The slag is liquid at the temperature of the molten metal and does not dissolve in the molten metal; it forms a separate liquid layer, which can easily be removed. For example, when oxide ores of copper are reduced, *limestone*, or calcium carbonate, $CaCO_3$, is added to the ore along with the coke. When it is heated, the calcium carbonate decomposes to calcium oxide, CaO, which combines with silicon dioxide (silica) to give molten calcium silicate slag, $CaSiO_3$. This slag is insoluble in the molten copper and therefore can be separated from it:

$$CaCO_3(s) \longrightarrow CaO(s) + CO_2(g)$$
$$CaO(s) + SiO_2(s) \longrightarrow CaSiO_3(l)$$

Aluminum is sufficiently reactive to be used for reducing oxides of some less reactive metals to the metal. Some manganese and chromium are produced industrially by using aluminum powder as the reducing agent:

$$3MnO_2(s) + 2Al(s) \longrightarrow 3Mn(s) + 2Al_2O_3(s)$$

In the **thermite process** aluminum powder is used for producing small quantities of molten iron for special purposes such as the welding and repair of railway lines:

$$Fe_2O_3(s) + 2Al(s) \longrightarrow Al_2O_3(s) + 2Fe(l)$$

This reaction is strongly exothermic, producing sufficient heat to melt the iron (see Experiment 9.1).

Manufacture of Iron and Steel

In prehistoric times iron was prepared simply by mixing iron ore with charcoal and heating the mixture in a fired clay pot. The modern process uses the same reaction. Iron ore, coke (carbon), and limestone, $CaCO_3$, are added at the top of

The Reduction of Iron(III) Oxide to Iron Using Aluminum as the Reducing Agent: The Thermite Process

The crucible contains a mixture of Fe_2O_3 and aluminum powder. A small amount of a mixture of potassium chlorate and sugar is placed on the top of the Fe_2O_3–Al mixture and a few drops of concentrated sulfuric acid are added to start the reaction.

The heat of the strongly exothermic reaction of H_2SO_4 with the $KClO_3$–sugar mixture ignites the Fe_2O_3–Al mixture, which reacts violently in a strongly exothermic reaction. A shower of white hot sparks is emitted and the crucible becomes red hot.

A ball of white hot iron can be seen glowing in the bottom of the crucible.

When the crucible has cooled, a magnet can be used to pick up the ball of iron

a blast furnace while preheated air (or pure oxygen) is blown in at the bottom (see Figure 9.1). The reducing agent is the carbon monoxide that is produced from the coke at the high temperature in the furnace:

$$2C(s) + O_2(g) \longrightarrow 2CO(g)$$

The iron ore, which is an impure mixture of Fe_2O_3, Fe_3O_4, and silicates, is reduced as it moves down the furnace—first to FeO and then, in the lower and hottest part of the furnace, to liquid iron, which runs down to the bottom of the furnace. The overall process is

$$Fe_2O_3 + 3CO \longrightarrow 2Fe + 3CO_2$$

Meanwhile, the limestone decomposes to CaO, which combines with silica impurities in the ore to form a layer of molten slag, $CaSiO_3$, which collects on top of the iron. This slag protects the molten iron from reoxidation. Solid slag is used in building materials such as cement, concrete, and cinder blocks.

The product of the blast furnace is called *pig iron*. It contains about 4% carbon, 2% silicon, and small amounts of other elements such as phosphorus and sulfur. Most of the carbon and nearly all the other impurities are removed by melting

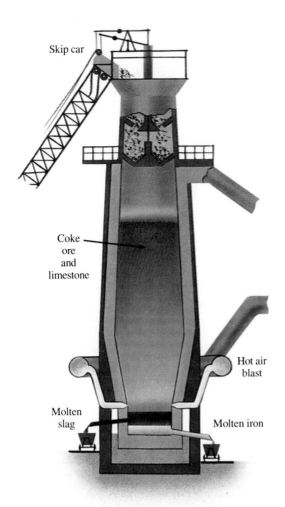

Skip car

Coke
ore
and
limestone

Hot air
blast

Molten
slag

Molten iron

Figure 9.1 Blast Furnace.
Iron ore, coke, and limestone are added at the top of the furnace, and air is blown
in at the bottom. The oxygen of the air reacts with the hot coke to form carbon
monoxide, which passes up the furnace and reduces the iron oxide as it moves
down, first to iron oxide, FeO, and finally to iron. Liquid iron collects at the
bottom of the furnace covered by a layer of molten slag, $CaSiO_3$. The iron is tapped
off at the bottom of the furnace and allowed to solidify, forming pig iron.

the pig iron and blowing oxygen through it for a short time (Figure 9.2) in a
process called the basic oxygen process. The impurities are converted to their
oxides by this treatment and either are evolved as gases, such as SO_2 and CO_2,
or form a slag with added calcium oxide:

$$Si \xrightarrow{O_2} SiO_2 \xrightarrow{CaO} CaSiO_3$$

$$P_4 \xrightarrow{O_2} P_4O_{10} \xrightarrow{CaO} Ca_3(PO_4)_2$$

The product of this process is *steel*, of which there are many different kinds.
They differ primarily in the amount of carbon that they contain, from less than
0.2% up to 1.5%. Addition of other metals such as nickel, chromium, manganese,
tungsten, and vanadium gives a very large number of alloy steels, of which stainless steel is the most familiar; it typically contains 18% chromium and 8%
nickel.

Figure 9.2 Basic Oxygen Furnace for the Production of Steel. A typical basic oxygen furnace is charged with about 200 tons of molten pig iron, 100 tons of scrap iron, and 20 tons of limestone. A stream of hot oxygen gas is blown through the mixture to oxidize the impurities, which either escape as their gaseous oxides or form a slag with the calcium oxide formed from the limestone.

Extraction of Metals from Sulfide Ores

Sulfide ores are first heated strongly in air to remove the sulfur as sulfur dioxide, SO_2, and to form the oxide. The process is known as **roasting**. It produces enormous quantities of sulfur dioxide, much of which is allowed to escape into the atmosphere and is a major source of acid rain. For example, galena, PbS, is roasted in the air to give the oxide, which is then reduced to lead with coke in a small blast furnace:

$$2PbS(s) + 3O_2(g) \longrightarrow 2PbO(s) + 2SO_2(g)$$
$$2PbO(s) + C(s) \longrightarrow 2Pb(l) + CO_2(g)$$

The copper sulfide ores, Cu_2S and $CuFeS_2$, are first roasted in a limited supply of air. This roasting converts the iron to FeO and removes some of the sulfur as SO_2 but leaves the copper as Cu_2S:

$$2CuFeS_2(s) + 4O_2(g) \longrightarrow Cu_2S(s) + 2FeO(s) + 3SO_2(g)$$

The FeO forms a slag of $FeSiO_3$ with any silica that is present or is added, and this slag is removed, leaving a relatively pure molten copper sulfide, Cu_2S. Air is then blown through the molten Cu_2S, converting it to the metal:

$$Cu_2S(l) + O_2(g) \longrightarrow 2Cu(l) + SO_2(g)$$

This copper is then purified by electrolysis to produce the very pure copper that is needed to make electrical conductors.

9.3 STRUCTURE OF METALS

Copper, aluminum, iron, and lead—and indeed all metals except mercury—are crystalline solids at room temperature. Usually, a metal consists of many tiny crystals not visible to the eye, but they can often be seen under the microscope. Occasionally, small gold and silver crystals are found in nature (see Figure 9.3).

Close-Packed Arrangements of Spheres: Hexagonal and Cubic Close-Packed Structures

The crystalline form of metals implies that the atoms in the structure are packed together in a regular manner. Because metals consist of identical spherical atoms, their structures are relatively simple: They are based on the two different ways that spheres may be packed together as closely as possible. If we place identical spherical balls in a slightly inclined tray and gently push them together as closely as possible, we obtain a **close-packed arrangement** (see Box 9.2 and Experiment 9.2). The spheres are arranged in straight rows that make angles of 60° with each other. Each sphere is surrounded by a regular hexagonal arrangement of six other spheres. Box 9.2 also shows how the rows of atoms define the edges of "two-dimensional" crystals of different shapes. Although these two-dimensional crystals may have various shapes, the angles between the edges are always 60° or 120°.

When layers of close-packed atoms are stacked so that the atoms of one layer nestle in the hollows between the atoms in the adjacent layer, a three-dimensional, close-packed structure is obtained. Each sphere is in contact with six spheres in its own layer, three spheres in the layer above, and three spheres in the layer below. Thus each sphere has a total of 12 nearest neighbors (see Box 9.2).

There are two ways in which close-packed layers can be stacked, and thus there are two closely related but distinct *close-packed structures*. In **hexagonal close packing** the spheres in the third layer lie directly over the spheres of the first layer. The sequence of layers in the hexagonal close-packed structure is therefore described as ABABAB.... In **cubic close packing** the spheres in the third layer do *not* lie above the spheres of the first layer (see Box 9.2). The sequence of layers in the cubic close-packed structure is therefore described as ABCABCABC.... It is not until the fourth layer is reached that the atoms lie directly above those in the first layer.

Body-Centered Cubic Structure

A majority of the metals have one of these two close-packed structures. Most of the remaining metals have a third type of structure, called the **monatomic body-centered cubic structure**. To understand this structure, consider first a single layer of spheres packed in a square arrangement (see Box 9.2). Although

Figure 9.3 Crystals of Metals.
(a) Gold and silver are sometimes found naturally in crystalline forms. (b) This section through a piece of zinc that has been allowed to solidify very slowly shows a mass of interlocking crystals.

(a)
(b)

Close Packing of Spheres

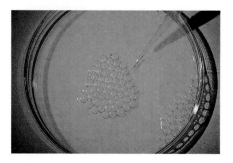

When a fine jet of air is bubbled through a soap solution a layer of soap bubbles forms on the surface of the solution.

Each bubble is surrounded by a hexagonal arrangement of six more bubbles giving a two-dimensional close-packed arrangement.

the spheres are packed in a regular way, they are not close-packed; each sphere has only four rather than six close neighbors, and a given number of spheres occupy a greater total area than do the same number in the close-packed arrangement. If layers of this type are stacked directly one on another, a simple cubic-packed arrangement is obtained. If such an arrangement is expanded slightly so that the spheres no longer quite touch each other, an additional sphere can be inserted in the middle of each cube. This arrangement is the monatomic body-centered cubic structure, in which each atom is in contact with only 8 neighbors rather than 12, as in a close-packed structure (Box 9.2).

Box 9.2
PACKING OF SPHERES

Close-packed layer of spheres

Each sphere has six close neighbors surrounding it in a hexagonal arrangement. The rows of atoms define the shapes of different "two-dimensional" crystals, which are outlined by thick lines.

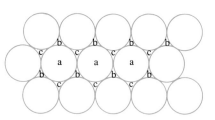

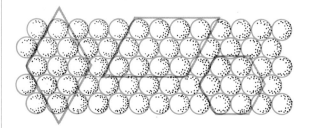

These spheres may be placed over only three of the depressions, either the depressions b or the depressions c. Here we have placed them over the depressions b.

Hexagonal and cubic close-packing

Each sphere in a close-packed layer is surrounded by six depressions into which the spheres of a second layer may be placed.

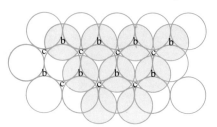

A third layer of spheres may be added in two ways. If we place the spheres of the third layer over the depressions a in the second layer, they lie directly above the spheres in the first layer. If we call the bottom layer an A layer and the second layer a B layer, then since the third layer is an exact replica of the first layer, we may also label it an A layer. Thus we have an arrangement in which the layers alternate in position, and so it is an ABABAB . . . arrangement, which is called hexagonal close packing.

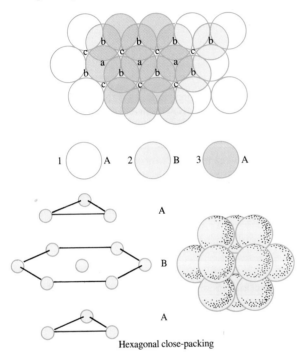

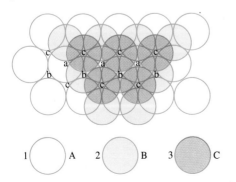

Hexagonal close-packing

If the third layer of spheres is placed not over the depressions a in the second layer but over the depressions c, then the spheres in the third layer do not lie above spheres in the first layer. Only when a fourth layer is placed over the depressions a do they lie directly above the spheres in the first layer. This arrangement is therefore different from the ABABA . . . arrangement. It is an ABCABCA . . . arrangement, which is called cubic close-packing.

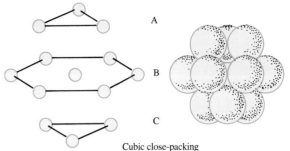

Cubic close-packing

Body-centered cubic packing

Most metals that do not have a close-packed structure have a monatomic body-centered cubic structure. Imagine a single layer of spheres with square packing as shown below.

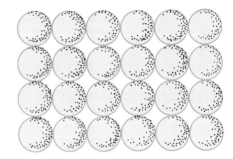

If square packed layers are stacked directly upon each other, a simple cubic packing is obtained.

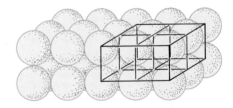

There are rather large holes in the middle of each cube, but they are not large enough to accommodate another sphere. However, if each cube is expanded by 15% so that the spheres no longer touch, another sphere can be inserted in the middle of each cube. In this arrangement each sphere is in contact with 8 other spheres at the corners of a cube rather than with 12 other spheres, as in the hexagonal and cubic close-packed arrangements.

In the two close-packed arrangements, the spheres occupy 74% of the total volume; the rest of the space is the interstices ("holes") between the spheres. In the body-centered cubic structure the spheres occupy only 68% of the total volume. Although this structure is not close-packed, it may be described as nearly close-packed. In addition to the 8 close neighbors of an atom in this structure, 6 more neighbors are not much further away, making a total of 14 close and nearly close neighbors.

Copper, aluminum, and lead have the cubic close-packed structure, whereas iron has the body-centered cubic structure. Common metals having the hexagonal close-packed structure include zinc and magnesium. A few metals have different, more complicated structures.

9.4 METALLIC BONDING

Metals in general have low ionization energies; one, two, and sometimes three electrons can be rather easily removed to give positive ions. Because of their small core charges, metals have little tendency to accept electrons to form negative ions. Hence we do not expect a metal to have an ionic structure consisting of positive and negative ions. Covalent bonding also is not possible in solid metals, because a metal atom does not have enough valence electrons to form covalent bonds to all of its 12 or 8 neighboring atoms. A sodium atom, for example, has only one valence electron. A sodium atom could therefore form a localized covalent bond to only one of its neighbors (see Figure 9.4). Even in aluminum each atom could form bonds to only 3 of the neighboring atoms.

Metals, in fact, have a type of bonding that differs from both ionic and covalent bonding; it is called **metallic bonding**. We will describe metallic bonding initially by using sodium as an example. Sodium is simple to discuss because it has only one valence electron, but the model that we will develop applies to all metals.

The 3s orbital of a sodium atom can overlap with the 3s orbital of another sodium atom to form a covalent bond and give a diatomic Na_2 molecule, as shown in Figure 9.5(a) just as the 1s orbital of a hydrogen atom overlaps with that of another hydrogen atom in the formation of the H_2 molecule. But the Na_2 molecule is found only in the gas phase. In the solid state the 3s orbital of a sodium atom overlaps with those of all its close neighbors, as shown in Figure 9.5(b), so the electron density is spread out among all the sodium atoms (Figure 9.5c).

Figure 9.4 Covalent Bond Representation of Sodium Crystal. A sodium atom can form a bond to only one of its four nearest neighbors in the two-dimensional representation of sodium shown here.

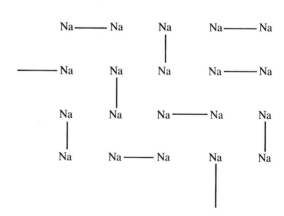

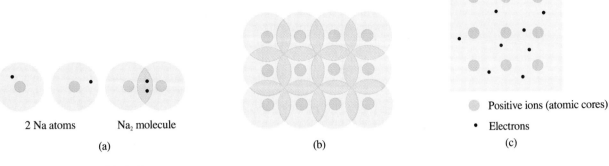

2 Na atoms   Na₂ molecule

(a)   (b)   (c)

○ Positive ions (atomic cores)

· Electrons

Figure 9.5 Metallic Bond in Sodium.
(a) The 3s orbitals of two isolated sodium atoms overlap in the formation of a diatomic Na_2 molecule. (b) In solid sodium the 3s orbital of any one sodium atom overlaps with those of all its neighbors. (c) Thus the electron density is spread out among all the sodium atoms to form a cloud of freely moving electrons—a negative charge cloud or electron gas. The electrostatic attraction between the electron cloud and the positive ions constitutes the metallic bond.

Each sodium atom in metallic sodium may be considered to have given up its single 3s electron to give a sodium ion, Na^+, and a free electron. These free electrons form a cloud of negative charge, a kind of electron gas, that occupies the space between the positive ions and can move freely between them. The electrostatic attraction between the cloud of negative electrons and the positive ions holds them together. The **metallic bond** may be defined as the force of attraction between a mobile charge cloud (electron gas) and positively charged atomic cores. In any metal each atom may be considered to have contributed some or all of its valence electrons to this mobile electron cloud. For example, aluminum consists of Al^{3+} ions held together by an electron cloud made up of three electrons from each atom (see Figure 9.6).

As we have seen before, resonance structures are used to describe the distribution of electrons in ions and molecules when they are not strictly localized,

Figure 9.6 Metallic Bonds in Sodium, Magnesium, and Aluminum.
The number of electrons contributed to the metallic bond increases from one per sodium atom, to two for each magnesium atom, to three for each aluminum atom. Thus the strength of the metallic bond increases from sodium to magnesium to aluminum.

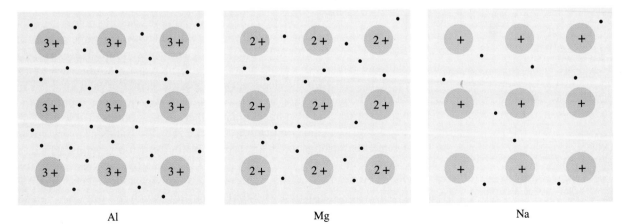

Al   Mg   Na

as implied by a normal Lewis structure. A metal represents an extreme case of electron delocalization; the electrons may be regarded as being delocalized over the whole metal crystal. We could, in principle, represent a metal by a very large number of resonance structures, such as the structures given in Figure 9.4, each one having a different arrangements of bonds, but this model is clearly not very practical. The delocalized electron cloud model is much more convenient, and it provides a satisfactory explanation for most of the properties of metals.

Just as many compounds have bonds that are intermediate in character between covalent and ionic, so other compounds have bonds that can be described as intermediate between metallic and ionic or between metallic and covalent. Each of the three types of chemical bond—covalent, ionic, and metallic—represents an extreme limiting case. In most compounds the bonds have an intermediate character, although it is often convenient to classify compounds according to their predominant bonding type.

Properties of Metals

The charge cloud (electron gas) model of metallic bonding is only a very simplified description of a metal. Nevertheless, we can use this model to provide an elementary explanation of most of the typical properties of metals.

MELTING POINT AND STRENGTH OF THE METALLIC BOND   Consider the three metals sodium, magnesium, and aluminum in period 3. Since sodium has one electron in its valence shell, magnesium two, and aluminum three, there are twice as many electrons in the charge cloud of magnesium and three times as many in the charge cloud of aluminum as in sodium (see Figure 9.6). Since the core charge also increases from sodium, $+1$, to aluminum, $+3$, the strength of the attraction between the atomic cores and the charge cloud increases from sodium to magnesium to aluminum. Therefore we expect the strength of the metallic bonding to increase in the same order.

To melt a metal, the atoms must be separated slightly so that they have sufficient space to move around each other. The stronger the bonds between the atoms, the more energy will be needed to separate them and therefore the higher will be the melting point. Other factors such as the structure of the metal also affect the melting point, but melting point does give us a rough indication of the strength of the metallic bonding. The melting points of the alkali and alkaline earth metals and aluminum are given in Table 9.2. In accordance with our model, the melting points increase in the series sodium, magnesium, aluminum, and in general, Group II metals have higher melting points than the group I metals. The melting point decreases with increasing atomic size down both Groups I and II, although magnesium is an exception. We expect the strength of the metallic bond to decrease with increasing atomic size as the average distance between the electron cloud and the center of each atomic core increases and the electrostatic force between them decreases correspondingly.

Note that copper and iron have much higher melting points (Table 9.1) than the Group I and II metals, indicating that the metallic bonds are very strong in these elements. Strong bonds are typical of the transition metals, which, as we will see in Chapter 21, can contribute as many as five electrons per atom to the charge cloud. The melting point of lead, which has four valence electrons, is unexpectedly low compared with that of barium, which has only two valence electrons. The unexpectedly low value reminds us that we can take the melting point only as a rough guide to the strength of the metallic bond.

Table 9.2 Melting Points (°C) of Some Metals

Li	Be	
180	1280	
Na	Mg	Al
98	650	660
K	Ca	
64	838	
Rb	Sr	
39	770	
Cs	Ba	
29	725	

ELECTRICAL CONDUCTIVITY   Since the valence electrons in a metal are free to move between the positively charged ions, they migrate readily under the influence of an electric potential. In other words, an electric current, which consists of a stream of negative electrons, will flow through a metal under the influence of an applied electric potential, provided, for example, by a battery. This type of conduction, which is due to the movement of electrons, is called **metallic conduction**.

A metal conducts electricity in both the solid and liquid states. In contrast, most ionic substances conduct electricity only in the liquid state, and covalent substances do not conduct either in the solid or in the liquid state. They are said to be *insulators*. In a solid ionic crystal the ions are not free to move; in the liquid state the ions can move, and the current consists of moving positive and negative ions. A covalent solid or liquid has no ions, and the electrons are not free to move because they are held firmly to the atoms as bonding or non-bonding pairs of electrons.

Whereas the conductivity of an ionic conductor decreases with decreasing temperature, the conductivity of a metallic conductor increases as the temperature is decreased. As a molten ionic substance or a solution of an ionic substance is cooled, the speeds of the ions decrease. Thus under the influence of an applied electric potential, the distance moved by the ions in a given time decreases with decreasing temperature; hence the conductivity decreases. In a metal, on the other hand, the small and extremely light electrons move at very high speeds, even at low temperatures. They are hindered in their motion by collisions with the vibrating positive ions (see Figure 9.7). Because the amplitude of the vibrations of the ions decreases with decreasing temperature, they interfere to a lesser extent with the motion of the electrons. Thus the conductivity of the metal increases as the temperature is decreased.

Silver is the most highly conducting metal, followed by the other coinage metals, copper and gold, and then aluminum.

HEAT CONDUCTION   The transfer of heat energy results from collisions of the fast-moving particles in the hotter part of a material with the slower-moving particles in the cooler part. The slower particles are speeded up by these colli-

Figure 9.7 Effect of Temperature on Electrical Conductivity.
The movement of electrons through a metal is impeded by their collisions with the atomic cores. The atomic cores are vibrating, and the amplitude of the vibrations increases with increasing temperature. The greater the amplitude of these vibrations, the greater is the chance that an electron will collide with an atomic core and that its motion through the solid will be impeded. With decreasing temperature the amplitude of the vibrations of the atomic cores decreases, and the chance of an electron colliding with a core decreases correspondingly. Hence the electrons are less impeded by the atomic cores, and the conductivity increases correspondingly.

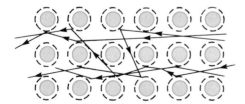

High temperature   Low temperature

sions, and the temperature of this part of the material therefore rises. At a given temperature all the particles—electrons and nuclei—in a solid have the same average kinetic energy. But because electrons have a much smaller mass than nuclei, they move at much higher speeds than the nuclei. Thus the free electrons move rapidly from one part of a metal to another, thus making metals very good conductors of heat. Covalent and ionic substances are generally poor conductors of heat because they have no free electrons and because their atoms and ions move only relatively slowly and to a limited extent about fixed positions.

Copper tubing

MECHANICAL PROPERTIES Because the electrons that form the bonds in a metal are constantly moving and are not fixed in position between any particular atoms, the atoms in a solid metal can be moved with respect to each other without breaking the bonds between them. The mobile bonding electrons readily move to adjust to the changing positions of the atoms. For this reason a metal can be distorted in shape relatively easily; that is, it is malleable and ductile. For example, steel can be rolled out into thin sheets for making automobiles and household appliances, and copper can be drawn out into thin wires for use in electric circuits. Most covalent and ionic solids cannot be distorted in this way; they break in two or more pieces if they are placed under stress in an attempt to alter their shape.

LUSTER Light falling on a metal surface causes the loosely held electrons of the metal to vibrate at the frequency of the light. A vibrating charge, such as an electron, emits electromagnetic waves, and therefore it emits light at the same frequency as the incident light. Thus light is reflected from a metal surface, which therefore has a shiny appearance.

THERMIONIC EFFECT The free electrons in a metal have a distribution of speeds just like the molecules of a gas (Chapter 3). When a metal is heated, the speeds of the electrons increase; at a certain temperature a few acquire enough energy to escape from the metal. If they are attracted away from the metal by a positive potential, a flow of electrons—in other words, an electric current—is obtained. This phenomenon is known as the **thermionic effect**. In a television picture tube the electrons emitted by a heated metal wire are accelerated and focused into a beam, which moves across the coated end of the tube to produce the visible pattern that is the television picture.

9.5 REACTIONS AND COMPOUNDS OF ALUMINUM, IRON, COPPER, AND LEAD

In this section we describe some of the reactions of the metals aluminum, iron, copper, and lead and some of their more important compounds. Very many of these reactions are either acid-base or oxidation-reduction reactions. In considering these reactions, we outline a systematic method for balancing oxidation-reduction equations, and we introduce the concept of Lewis acids and bases. First, we consider the reactions of the metals with acids.

Reactions with Acids

Many metals react with *dilute* aqueous solutions of acids; hydrogen is evolved and a solution of an ionic salt of the metal is formed. For example, aluminum and iron react with hydrochloric acid in this way (see Experiment 9.3):

$$Fe(s) + 2HCl(aq) \longrightarrow FeCl_2(aq) + H_2(g)$$

$$2Al(s) + 6HCl(aq) \longrightarrow 2AlCl_3(aq) + 3H_2(g)$$

Lead is much less reactive than aluminum or iron, but it does dissolve in hot, concentrated hydrochloric acid in a similar reaction:

$$Pb(s) + 2HCl(aq) \longrightarrow PbCl_2(aq) + H_2(g)$$

Because lead chloride is rather insoluble in cold water, it is precipitated as a white solid if the concentrated HCl solution is diluted with water.

Aqueous solutions of the more soluble $FeCl_2$ and $AlCl_3$ must be evaporated to obtain the solid salts. The crystals obtained from these solutions contain water molecules in addition to the metal ion and the chloride ion. They have the formulas $FeCl_2 \cdot 4H_2O$ and $AlCl_3 \cdot 6H_2O$. The salts are said to be **hydrated**, and the water that they contain is called **water of crystallization**. In most cases the water molecules are bonded to the metal ion to form hydrated ions such as $Fe(H_2O)_4^{2+}$ and $Al(H_2O)_6^{3+}$. The positive charge of the metal ion attracts the negatively charged oxygen atom of the polar water molecule. Thus several water molecules, frequently four to six, become attached to the metal ion. This can be shown by writing the formulas of the salts as $[Fe(H_2O)_4]Cl_2$ and $[Al(H_2O)_6]Cl_3$, although they are often written $FeCl_2 \cdot 4H_2O$ and $AlCl_3 \cdot 6H_2O$.

The six water molecules in $Al(H_2O)_6^{3+}$ have an octahedral arrangement around the aluminum ion. The bonds between the water molecules and the

Reactions of Metals with Acids

Reactions with dilute hydrochloric acid. Left to right: Copper does not react. Aluminum dissolves with the formation of hydrogen and a solution of $AlCl_3$. Lead does not react. Iron dissolves with the formation of hydrogen and a solution of $FeCl_2$.

Reactions with nitric acid. Left: Copper dissolves in concentrated nitric acid, forming a dark green solution of $Cu(NO_3)_2$ and brown NO_2 gas. Right: Copper dissolves in dilute nitric acid, forming a blue solution of $Cu(NO_3)_2$ and colorless NO gas.

Lead dissolves in concentrated nitric acid, forming a colorless solution of $Pb(NO_3)_2$ and colorless NO(g) which reacts with the oxygen in the air to give brown $NO_2(g)$.

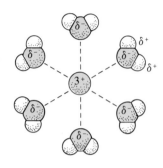

$Al(H_2O)_6^{3+}$—ionic bonding

$Al(H_2O)_6^{3+}$—covalent bonding

aluminum ion can be described in two ways: (1) The bond can be considered to arise from the electrostatic attraction between the negatively charged oxygen atom of the polar water molecule and the positive aluminum ion; or (2) the bonds can be considered to be covalent bonds formed by a water molecule donating an electron pair to the empty valence shell of the aluminum ion.

The actual bonding between the aluminum ion and a water molecule is best regarded as being intermediate between these two extremes. In other words, an electron pair on an oxygen atom is donated to the aluminum ion, but this donation is only partial, so the resulting bond is intermediate between a covalent bond and an ionic bond. The formal charge on aluminum is $+3$ in the ionic description and -3 in the covalent description. The real situation is intermediate between these two extremes.

Aluminum and iron react in a similar way with other acids, such as HBr and H_2SO_4, to give hydrogen and the corresponding salt of the acid:

$$2Al(s) + 6HBr(aq) \longrightarrow 2AlBr_3(aq) + 3H_2(g)$$

$$Fe(s) + H_2SO_4(aq) \longrightarrow FeSO_4(aq) + H_2(g)$$

Aluminum bromide crystallizes from solution as the hydrate $AlBr_3 \cdot 6H_2O$, that is, $[Al(H_2O)_6]Br_3$. Iron(II) sulfate, $FeSO_4$, crystallizes as $FeSO_4 \cdot 7H_2O$, which is best written as $[Fe(H_2O)_6]SO_4 \cdot H_2O$, since only six of the water molecules are strongly attached to the Fe^{2+} ion.

Since acids such as HCl and H_2SO_4 are ionized to give H_3O^+ in dilute aqueous solution, the reaction between a divalent metal and an aqueous acid may be described by the general equation:

$$M(s) + 2H_3O^+(aq) \longrightarrow M^{2+}(aq) + H_2(g) + 2H_2O$$

For example,

$$Fe(s) + 2H_3O^+(aq) \longrightarrow Fe^{2+}(aq) + H_2(g) + 2H_2O$$

This equation is frequently written in the simpler form

$$M(s) + 2H^+(aq) \longrightarrow M^{2+}(aq) + H_2(g)$$

However, we must remember that H^+ is always associated with a water molecule to form H_3O^+. The essential reaction in each case is the oxidation of the metal to M^{2+} and the reduction of H^+ (H_3O^+) to hydrogen, H_2. There is a transfer of electrons from the metal to H^+ (H_3O^+). Similar equations can be written for the formation of M^+ and M^{3+}.

Copper, lead, and some other metals do not react with dilute aqueous acid solutions because the hydronium ion, H_3O^+, is not a sufficiently strong oxidizing agent to oxidize these metals. Copper reacts only with aqueous acids in which the undissociated molecular acid, HA, or its anion, A^-, is a stronger oxidizing agent than H_3O^+. Thus as we saw in Chapter 7, copper does not react with dilute sulfuric acid, but it is dissolved by hot concentrated sulfuric acid, which is reduced to sulfur dioxide. Lead reacts similarly to give lead sulfate:

$$Pb(s) + 2H_2SO_4(l) \longrightarrow PbSO_4(aq) + SO_2(g) + 2H_2O(l)$$

Nitric acid, HNO_3, and the nitrate ion, NO_3^-, are both strong oxidizing agents, like concentrated H_2SO_4. Copper and lead react with both dilute and concentrated nitric acid. The reactions are complex, and the products depend on the concentration of the acid and the temperature (see Experiment 9.3).

We may represent the reactions of copper with nitric acid by the following simplified equations (the abbreviation "dil" represents "dilute"):

$$Cu(s) + 4HNO_3(aq, conc) \longrightarrow Cu(NO_3)_2(aq) + 2NO_2(g) + 2H_2O$$

$$3Cu(s) + 8HNO_3(aq, dil) \longrightarrow 3Cu(NO_3)_2(aq) + 2NO(g) + 4H_2O$$

In these reactions molecular nitric acid, HNO_3, and the nitrate ion, NO_3^-, not H_3O^+, are the oxidizing agents. Copper is oxidized to copper(II) nitrate. Nitric acid, in which nitrogen is in the $+5$ oxidation state, is reduced to NO_2, in which it is in the $+4$ oxidation state, or to NO, in which it is in the $+2$ oxidation state.

Balancing Oxidation-Reduction Equations

The somewhat complicated equations describing the reactions between copper and nitric acid are not simple to balance unless we use a systematic approach. If we know all the reactants and the products for an oxidation-reduction reaction, then we can balance the equation by following five steps, as we show now for the oxidation of copper by concentrated nitric acid.

1. *Write down the reactants and products of the reaction and inspect them to determine which elements have undergone changes in their oxidation numbers.* We have

$$\underset{\text{Reactants}}{Cu + HNO_3} \longrightarrow \underset{\text{Products}}{Cu(NO_3)_2 + NO_2}$$

For reactions in aqueous solution, it is convenient to rewrite the equation in the ionic form:

$$Cu + H_3O^+ + NO_3^- \longrightarrow Cu^{2+} + 2NO_3^- + NO_2$$

Examining the oxidation numbers, we find

$$\overset{+5\,-2}{N\,O_3^-} \xrightarrow{\text{Reduction}} \overset{+4\,-2}{N\,O_2}$$

$$\overset{0}{Cu} \xrightarrow{\text{Oxidation}} \overset{+2}{Cu^{2+}}$$

Only N and Cu have undergone changes in oxidation numbers.

2. *Write a half equation for the reduction process and another half equation for the oxidation process. Balance all atoms except O and H, and then balance the oxidation numbers by adding the requisite number of electrons to the appropriate side of the equation.* In our example we have

$$\text{Reduction} \quad \overset{+5}{NO_3^-} + e^- \longrightarrow \overset{+4}{NO_2}$$

$$\text{Oxidation} \quad \overset{0}{Cu} \longrightarrow \overset{+2}{Cu^{2+}} + 2e^-$$

Nitrogen is balanced, and in the change from NO_3^- to NO_2 the oxidation number of N changes from $+5$ to $+4$, so one electron must be involved in the reaction. Similarly, Cu is balanced, and the oxidation number of Cu changes from 0 for the metal to $+2$ for Cu^{2+}. Two electrons must be added to the right side of the equation for this half-reaction. At this point the half equations are not necessarily balanced with respect to charges or H and O atoms. The next step is to *balance charges.* How we do this depends on whether the reaction is taking place in an acidic solution or a basic solution. In an acidic aqueous solution H^+ (H_3O^+) ions are always present, and in a basic aqueous solution OH^- ions are always present. Thus for acidic solutions we add H^+ (H_3O^+) to balance charges, and for basic solutions we add OH^- to balance charges.

3. *Balance each half-reaction for charge.* In the half equation for the reduction of NO_3^- to NO_2, there are two negative charges on the left side of the equation and no charges on the right side, since NO_2 is a neutral molecule. The reaction is in *acidic* solution, so H_3O^+ ions can participate in the reaction. For simplicity we will write H_3O^+ as H^+. We balance charges by adding the requisite number of H^+ to the appropriate side of the equation:

$$NO_3^- + e^- + 2H^+ \longrightarrow NO_2$$

The equation is now balanced with respect to the changes in both the oxidation numbers and the charges. The equation $Cu \rightarrow Cu^{2+} + 2e^-$ is already balanced.

4. *Balance each half equation with respect to H and O atoms.* If the H and O atoms do not balance, because we are concerned with aqueous solutions, we can add an appropriate number of H_2O molecules to either side of the equation. The left side of our equation has two more H atoms and one more O atom than the right side; it can be balanced by adding one H_2O to the right side:

$$NO_3^- + e^- + 2H^+ \longrightarrow NO_2 + H_2O$$

Having balanced charges by adding H^+ to the appropriate side of the equation, atoms can be balanced by adding the requisite number of H_2O molecules to the other side of the equation. The half-reaction for reduction of NO_3^- to NO_2 is now balanced in all respects. In general, for reactions carried out in acidic aqueous solution, H_2O molecules and H_3O^+ (H^+) ions may participate in the reaction as either reactants or products.

5. *Multiply the two half equations by appropriate coefficients so that when the two half equations are added, the electrons on one side of the equation cancel those on the other side.* In our example this step can be accomplished by adding the equation for the oxidation of Cu to twice the equation for the reduction of NO_3^-:

$$Cu \longrightarrow Cu^{2+} + 2e^-$$
$$\underline{2NO_3^- + 2e^- + 4H^+ \longrightarrow 2NO_2 + 2H_2O}$$
$$Cu + 2NO_3^- + 4H^+ \longrightarrow Cu^{2+} + 2NO_2 + 2H_2O$$

This equation is the required balanced equation for the oxidation-reduction reaction.

The final equation should be checked to see whether it is balanced for atoms and for charges:

	Left side	Right side
Cu atoms	1	1
N atoms	2	2
O atoms	6	6
H atoms	4	4
Charge	2+	2+

We see that atoms and charges balance.

We can also check changes in the oxidation numbers:

Oxidation number change in element(s) reduced	Oxidation number change in element(s) oxidized
$2N(+5) \longrightarrow 2N(+4) = -2$	$Cu(0) \longrightarrow Cu^{2+}(+2) = +2$

We see that the total *decrease* in the oxidation number of N is equal to the total *increase* in the oxidation number of Cu, so the overall sum of the oxidation number changes is zero.

The equation can also be written in terms of the empirical and molecular formulas of the reactants and products. In this case addition of two NO_3^- ions to each side of the equation gives

$$Cu + 4NO_3^- + 4H^+ \longrightarrow Cu^{2+} + 2NO_3^- + 2NO_2 + 2H_2O$$

or

$$Cu + 4HNO_3 \longrightarrow Cu(NO_3)_2 + 2NO_2 + 2H_2O$$

Example 9.1 Copper reacts with dilute nitric acid to give nitrogen monoxide, NO. Write a balanced equation for the reaction.

Solution

$$Cu + H_3O^+ + NO_3^- \longrightarrow Cu^{2+} + NO$$
$$\underbrace{\qquad\qquad\qquad}_{\text{Reactants}} \qquad \underbrace{\qquad\qquad}_{\text{Products}}$$

We have

Oxidation $\quad \overset{0}{Cu} \longrightarrow \overset{+2}{Cu}{}^{2+}$

Reduction $\quad \overset{+5}{N}O_3^- \longrightarrow \overset{+2}{N}O$

Adding electrons, we have

Oxidation $\quad \overset{0}{Cu} \longrightarrow \overset{+2}{Cu}{}^{2+} + 2e^-$

Reduction $\quad \overset{+5}{N}O_3^- + 3e^- \longrightarrow \overset{+2}{N}O$

and the oxidation equation is balanced. But the reduction equation is not balanced. We first balance charges by adding H^+:

$$NO_3^- + 3e^- + 4H^+ \longrightarrow NO$$

We then balance atoms by adding $2H_2O$ to the right-hand side of the equation:

$$NO_3^- + 3e^- + 4H^+ \longrightarrow NO + 2H_2O$$

The equations for both half-reactions are now balanced.
We add three times the first equation to twice the second equation to eliminate the electrons:

$$
\begin{array}{rcl}
3Cu & \longrightarrow & 3Cu^{2+} + 6e^- \\
2NO_3^- + 6e^- \quad + 8H^+ & \longrightarrow & 2NO \quad + 4H_2O \\
\hline
3Cu \quad + 2NO_3^- + 8H^+ & \longrightarrow & 3Cu^{2+} + 2NO \ + 4H_2O
\end{array}
$$

This equation is the balanced equation for the oxidation-reduction reaction between Cu and dilute HNO_3 to give Cu^{2+} and NO.
Finally, we check for atoms and charges:

Left side	*Right side*
3Cu	3Cu
2N	2N
6O	6O
8H	8H
6+	6+

And we check for oxidation number changes:

$$2N(+5) \longrightarrow 2N(+2) = -6 \quad \text{and} \quad 3Cu(0) \longrightarrow 3Cu^{2+}(+2) = +6$$

The balanced equation for any oxidation-reduction reaction can be obtained in the same way, provided we know all the reactants and all the products. In *acidic solutions* we balance charges by adding H^+ as appropriate. In *basic solutions* only insignificant concentrations of H^+ ions are present; so we balance charges by adding an appropriate number of OH^- to the equation. In general, in a basic solution H_2O molecules and OH^- ions may take part in a reaction as either reactants or products.

Example 9.2 Sodium nitrate reacts with zinc in sodium hydroxide solution to give ammonia and the tetrahydroxozinc(II) ion, $Zn(OH)_4^{2-}$. Write a balanced equation for this reaction.

Solution

$$Zn + NO_3^- \longrightarrow Zn(OH)_4^{2-} + NH_3$$

$$\underset{\text{Reactants}}{\qquad} \qquad \underset{\text{Products}}{\qquad}$$

$$\text{Oxidation} \quad \overset{0}{Zn} \longrightarrow \overset{+2}{Zn}(OH)_4^{2-} + 2e^-$$

We balance charges by adding $4OH^-$ to the left side:

$$Zn + 4OH^- \longrightarrow Zn(OH)_4^{2-} + 2e^-$$

Atoms are then also balanced. For the reduction half-reaction we have

$$\text{Reduction} \quad \overset{+5}{NO_3^-} + 8e^- \longrightarrow \overset{-3}{NH_3}$$

We balance charges by adding $9OH^-$ to the right side of the equation:

$$NO_3^- + 8e^- \longrightarrow NH_3 + 9OH^-$$

We balance atoms by adding $6H_2O$ to the left side of the equation:

$$NO_3^- + 8e^- + 6H_2O \longrightarrow NH_3 + 9OH^-$$

The electrons are eliminated by adding four times the first equation to the second equation:

$$4Zn + 16OH^- \longrightarrow 4Zn(OH)_4^{2-} + 8e^-$$
$$\underline{NO_3^- + 8e^- + 6H_2O \longrightarrow NH_3 + 9OH^-}$$
$$4Zn + NO_3^- + 6H_2O + 7OH^- \longrightarrow 4Zn(OH)_4^{2-} + NH_3$$

Checking the overall equation for atoms and charges, we have

Left side	Right side
4Zn	4Zn
1N	1N
16O	16O
19H	19H
8−	8−

Checking for oxidation number changes we have

$$4Zn(0) \longrightarrow 4Zn(+2) = +8 \quad \text{and} \quad N(+5) \longrightarrow N(-3) = -8$$

The steps to follow for balancing all oxidation-reduction equations are:

1. Write down the reactants and products of the reaction, and inspect them to determine which elements have undergone changes in their oxidation numbers.

2. Write half equations representing the changes in the oxidized and reduced species, and balance the oxidation number changes by adding the requisite number of electrons to the appropriate side of each equation.

3. Balance each half equation for charge by adding the requisite number of hydrogen ions or hydroxide ions to the appropriate side of the equation.

4. Balance each half equation with respect to atoms by adding the requisite number of water molecules to the appropriate side of the equation.

5. Multiply the two half equations by appropriate coefficients so that when the two half equations are added, the electrons on one side of the equation cancel those on the other side.

6. Check the final equation to see that it is balanced for atoms and for charges.

There are several ways of balancing oxidation-reduction equations. Sometimes, one is more convenient than the others, but beginning students are advised to adopt one method and stick to it. We will illustrate a variation in the method we have described by again using the reaction between copper and concentrated nitric acid as an example.

1. Write the reactants and products of the reaction and inspect them to determine which elements have undergone changes in their oxidation numbers:

$$\text{Oxidation} \quad \overset{0}{\text{Cu}} \longrightarrow \overset{+2}{\text{Cu}}{}^{2+}$$

$$\text{Reduction} \quad \overset{+5}{\text{NO}_3^-} \longrightarrow \overset{+4}{\text{NO}_2}$$

2. Write two half equations and balance them for atoms. In an acid solution we may use hydrogen ions, H^+, and H_2O molecules as necessary:

$$\text{Oxidation} \qquad\qquad \text{Cu} \longrightarrow \text{Cu}^{2+}$$

$$\text{Reduction} \quad \text{NO}_3^- + 2H^+ \longrightarrow \text{NO}_2 + H_2O$$

3. Balance for charges by adding electrons as necessary:

$$\text{NO}_3^- + 2H^+ + e^- \longrightarrow \text{NO}_2 + H_2O$$

$$\text{Cu} \longrightarrow \text{Cu}^{2+} + 2e^-$$

At this point we can check that the number of electrons in each equation is equal to the change in oxidation number.

4. Multiply the equations by appropriate coefficients so that, on adding, we have the same number of electrons on each side, and they cancel:

$$\begin{aligned}
\text{Cu} &\longrightarrow \text{Cu}^{2+} + 2e^- \\
2\text{NO}_3^- + 2e^- + 4H^+ &\longrightarrow 2\text{NO}_2 + 2H_2O \\
\hline
\text{Cu} + 2\text{NO}_3^- + 4H^+ &\longrightarrow \text{Cu}^{2+} + 2\text{NO}_2 + 2H_2O
\end{aligned}$$

The final equation should be checked as before for atom balance, for charge balance, and for oxidation number changes.

Example 9.3 Hot, concentrated H_2SO_4 reacts with Cu to give $CuSO_4$ and SO_2. Write the equation for this reaction using the preceding method.

Solution

$$\text{Cu} + H_2SO_4 \longrightarrow \text{CuSO}_4 + SO_2$$

or writing it in ionic form, we have

$$\text{Cu} + 2H^+ + SO_4^{2-} \longrightarrow \text{Cu}^{2+} + SO_4^{2-} + SO_2$$

1. We see that Cu is oxidized to Cu^{2+}:

$$\overset{0}{\text{Cu}} \longrightarrow \overset{+2}{\text{Cu}}{}^{2+}$$

Sulfuric acid is reduced to SO_2:

$$2H^+ + \overset{+6}{S}O_4^{2-} \longrightarrow \overset{+4}{S}O_2$$

2. Balance for atoms.

$$Cu \longrightarrow Cu^{2+}$$
$$4H^+ + SO_4^{2-} \longrightarrow SO_2 + 2H_2O$$

3. Balance for charges by adding electrons:

$$Cu \longrightarrow Cu^{2+} + 2e^-$$
$$4H^+ + SO_4^{2-} + 2e^- \longrightarrow SO_2 + 2H_2O$$

4. Adding the two equations gives the overall balanced equation:

$$Cu + 4H^+ + SO_4^{2-} \longrightarrow Cu^{2+} + SO_2 + 2H_2O$$

Adding one SO_4^{2-} to each side, we can write the equation in the following alternative form:

$$Cu + 2H_2SO_4 \longrightarrow CuSO_4 + SO_2 + 2H_2O$$

The final equation should be checked for atom balance, charge balance, and oxidation number changes.

Reactivity of metals

On the basis of our discussion of the reactions of metals with dilute aqueous acids and our previous observation of the reactions of the alkali and alkaline earth metals with water, we can classify the common metals according to their reactivity toward the hydronium ion, H_3O^+, that is, the ease with which they are oxidized by H_3O^+. The alkali metals and some of the alkaline earth metals are extremely reactive and are so readily oxidized that they react even with water in which the concentration of H_3O^+ is very small. Other metals such as Zn, Fe, and Al are less reactive, but they are oxidized by H_3O^+ in aqueous acid solutions. Metals such as Cu, Ag, and Au are still less reactive and are not oxidized by H_3O^+:

Na, K, Ca	Mg, Zn, Al, Fe, Pb	Cu, Ag, Au
Very reactive	Less reactive	Unreactive
React with water	React with H_3O^+ in aqueous acids	Do not react with H_3O^+ in aqueous acids

This is called the *activity series* of the metals. The metals on the left are strong reducing agents. Those on the right are weak reducing agents. In Chapter 16 we will see how to put this series on a quantitative basis.

Oxidation States of Aluminum, Iron, Copper, and Lead

As we saw in Chapter 7, classifying the compounds of an element according to the oxidation state of the element in the compound is very convenient. Table 9.3 shows the principal oxidation states of aluminum, iron, copper, and lead. Because they are metals, they have only positive oxidation states. Aluminum has only one oxidation state, $+3$, while the other metals each have two common oxidation states, $+2$ and $+3$ for iron, $+1$ and $+2$ for copper, and $+2$ and $+4$ for lead. In the following sections we consider some of the important compounds of these elements, principally the oxides, the hydroxides, and the halides.

Table 9.3 Oxides and Chlorides of Aluminum, Iron, Copper, and Lead

OXIDATION STATE	Al	Fe	Cu	Pb
+1			Cu_2O; $CuCl$	
+2		FeO; $FeCl_2$; Fe_3O_4[a]	CuO; $CuCl_2$	PbO; $PbCl_2$; Pb_3O_4[a]
+3	Al_2O_3; $AlCl_3$	Fe_2O_3; $FeCl_3$		
+4				PbO_2; $PbCl_4$

Note: Except for $PbCl_4$, w   '⁻ⁿt liquid, all these compounds are solids and have three-dimensional network str   ⁻ntly ionic bonding. In the gas phase, however, $FeCl_3$ and $AlCl_3$ are co⁻

[a] Fe_3O_4 and Pb_3O_4 a   metal atoms in two different oxidation states.

Compounds of A

Aluminum is ir   figuration [Ne] $3s^2 3p^1$. The majority of al   contain Al^{3+}, formed by the loss of the thr   there are also some important covalent co

The gro   aluminum is

3p

It has   would therefore expect aluminum to be un   and carbon, aluminum forms com-poun   ell electrons. So we can conveniently imagi   not from the ground state of aluminum but fr   ctron has been promoted to a 3p orbital and in whicn   unpaired electrons:

3s   3p

[Ne]   ↑   ↑ ↑

By sharing these electrons, it forms three covalent bonds. The fact that aluminum forms covalent as well as ionic bonds reflects the fact that it lies near the border-line between the metals and the nonmetals of the periodic table.

ALUMINUM OXIDE Although aluminum is rather reactive, it does not cor-rode in the atmosphere because the surface of the metal quickly becomes covered by a thin, hard layer of aluminum oxide, Al_2O_3. In nature aluminum oxide, also called *alumina*, occurs as *bauxite*, $Al_2O_3 \cdot xH_2O$, and in the anhydrous form, Al_2O_3, as *corundum*. To a first approximation, Al_2O_3 can be regarded as con-sisting of the ions Al^{3+} and O^{2-}. The force between two ions at a given distance is, according to Coulomb's law, proportional to the product of their charges. Thus if we neglect differences in the sizes of the ions, the force between an Al^{3+} and an O^{2-} would be six times that between singly charged ions such as Na^+ and Cl^-. For this reason Al_2O_3 is very hard and has a very high melting point (approximately 2000°C). However, the small highly charged Al^{3+} attracts elec-trons very strongly. It therefore attracts the electron pairs of adjacent O^{2-} ions so that they are partially shared with the Al^{3+} ion. In other words, the bonds

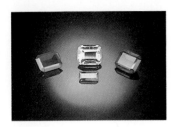

have some covalent character; they may be described as very polar covalent bonds.

Transparent crystals of corundum containing traces of other metal ions, which impart colors to the crystals, occur naturally; they are used as gemstones. For example, sapphires are corundum crystals that have a blue color because a small number of Al^{3+} ions are replaced by Fe^{2+}, Fe^{3+}, and Ti^{3+}. Rubies are red because a small number of Al^{3+} ions are replaced by Cr^{3+} ions. Artificial rubies and sapphires are produced by melting Al_2O_3 with small amounts of other suitable oxides in the flame of an oxyhydrogen torch. Because of the great hardness of Al_2O_3, such synthetic jewels are also used as bearing points in watches and other precision instruments (Figure 9.8).

Figure 9.8 Synthetic Gemstones. Ruby, diamond, sapphire, and emerald. Ruby and sapphire are impure forms of corundum. Emerald is an impure form of beryl, $Be_3Al_2Si_6O_{18}$.

Like many other metal oxides, aluminum oxide is insoluble in water. Its insolubility can be attributed to the very strong ionic bonding in the solid that results from the high charges on the ions.

ALUMINUM CHLORIDE, $AlCl_3$, AND ALUMINUM BROMIDE, $AlBr_3$ When aluminum is heated in chlorine or hydrogen chloride, aluminum chloride is formed:

$$2Al(s) + 3Cl_2(g) \longrightarrow 2AlCl_3(s)$$
$$2Al(s) + 6HCl(g) \longrightarrow 2AlCl_3(s) + 3H_2(g)$$

Aluminum reacts vigorously with liquid bromine at room temperature to form aluminum bromide, $AlBr_3$ (see Experiment 9.4):

$$2Al(s) + 3Br_2(l) \longrightarrow 2AlBr_3(s)$$

Aluminum chloride and aluminum bromide are white crystalline substances. They are very **hygroscopic**, strongly absorbing water from the atmosphere and eventually dissolving in the absorbed water. When heated at atmospheric pressure, $AlCl_3$ sublimes at 183° without melting. Aluminum chloride may be purified by heating it until it sublimes and then condensing the vapor, thus leaving behind all the less volatile impurities.

Solid aluminum chloride is an ionic solid consisting of Al^{3+} and Cl^-. But the gas, which is obtained when aluminum chloride sublimes, consists of covalent Al_2Cl_6 molecules:

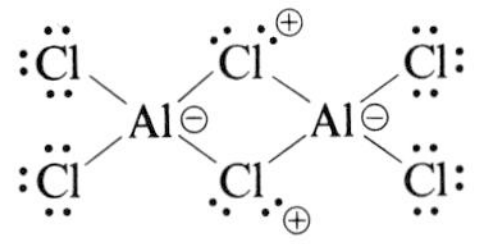

Both aluminum atoms have a tetrahedral AX_4 geometry, so the overall geometry of the molecule is

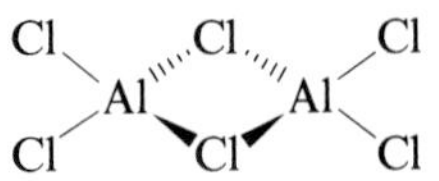

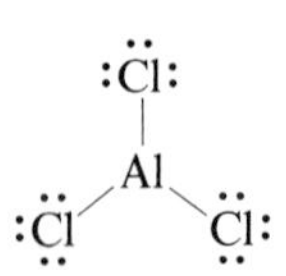

Above 750°, Al_2Cl_6 molecules dissociate into $AlCl_3$ molecules, which have a planar triangular AX_3 geometry.

Unlike $AlCl_3$, aluminum bromide is not ionic in the solid state but consists of Al_2Br_6 molecules, just like aluminum chloride in the gaseous state. At high temperatures in the gas phase Al_2Br_6 molecules dissociate to $AlBr_3$ molecules.

In the $AlCl_3$ molecule the Al atom has only six electrons in its valence shell and thus has space for more electrons. In the Al_2Cl_6 molecule each Al atom completes its octet by accepting an electron pair from a Cl atom. When an Al—Cl bond is formed in this way, formal charges are generated, + on Cl

The Preparation of Aluminum bromide and Aluminum Chloride

Aluminum pellets can be seen on the watch glass on top of the beaker, which contains bromine.

The aluminum pellets are tipped into the bromine and after a few minutes the aluminum begins to react vigorously with the bromine. The aluminum pellets ignite, burning with a bright flame, and skate around the surface of the bromine.

The product of the reaction is a white solid, aluminum bromide, which coats the beaker and the watch glass.

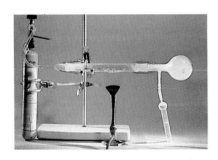

When alumimum is heated in a stream of chlorine, it forms aluminum chloride, which condenses as a white powder in the flask on the right. The reaction is exothermic, and round drops of red-hot molten aluminum (mp 660°C) can be seen in the tube. The yellow-green color is due to the chlorine.

and − on Al, because a pair of electrons initially belonging only to a Cl atom is shared with the Al atom. In general, whenever an Al atom forms four bonds, it must carry a formal negative charge; whenever a Cl atom forms two covalent bonds, it must carry a formal positive charge.

In forming a covalent chloride in the vapor state, aluminum is behaving like a nonmetal. This is consistent with its position in the periodic table on the borderline between metals and nonmetals.

LEWIS ACIDS AND BASES The formation of Al_2Cl_6 by the donation of an unshared electron pair on a Cl atom into the valence shell of the Al atom is an example of a type of behavior that is common for molecules in which the central atom has only six or fewer electrons in its valence shell. Such molecules often combine with other molecules or ions by accepting an electron pair to complete an octet of electrons in their valence shells. Thus $AlCl_3$ also combines with a Cl^-

9.5 REACTIONS AND COMPOUNDS OF ALUMINUM, IRON, COPPER, AND LEAD

337

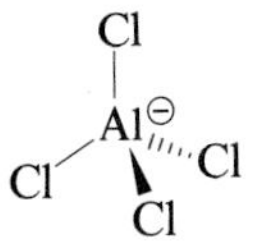

ion to give the tetrahedral $AlCl_4^-$ ion:

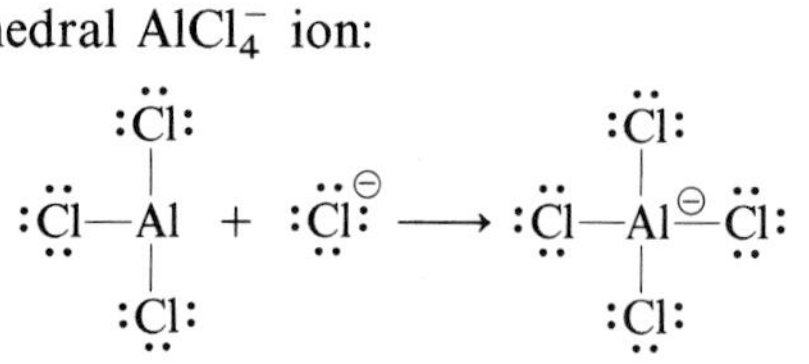

$$:\ddot{C}l:$$

(Left margin structure showing $AlCl_4^-$ with Cl atoms around Al with negative charge)

$$:\ddot{C}l-Al \;+\; :\ddot{C}l:^{\ominus} \longrightarrow :\ddot{C}l-Al^{\ominus}\ddot{C}l:$$

Boron, the first element in group III, forms covalent trihalides, such as boron trichloride and boron trifluoride, in which the boron atom has only six electrons in its valence shell. Unlike aluminum chloride, boron trichloride and boron trifluoride do not form dimeric molecules, but they do combine with a chloride ion or a fluoride ion to give the tetrahedral BCl_4^- and BF_4^- ions and with other neutral molecules, such as ammonia, that have unshared electron pairs:

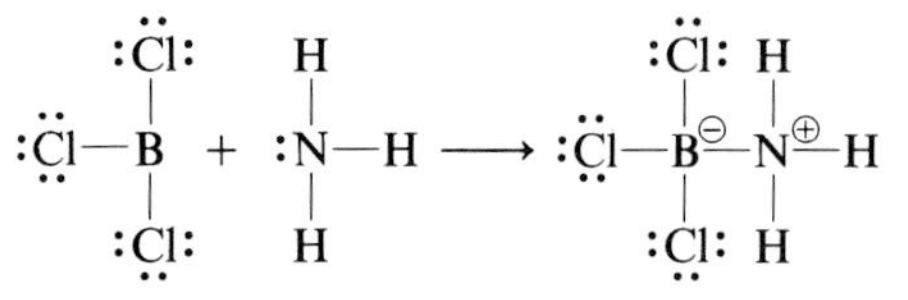

$$:\ddot{C}l-B \;+\; :N-H \longrightarrow :\ddot{C}l-B^{\ominus}-N^{\oplus}-H$$

Such a compound between two neutral molecules is called an **adduct**, in this case the boron trichloride–ammonia adduct.

A proton adds to a molecule or ion with an unshared pair of electrons (a base), forming a covalent bond in an acid-base reaction:

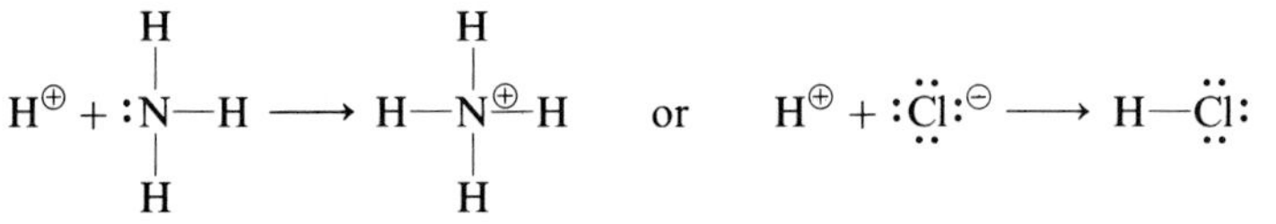

$$H^{\oplus} + :N-H \longrightarrow H-N^{\oplus}-H \quad \text{or} \quad H^{\oplus} + :\ddot{C}l:^{\ominus} \longrightarrow H-\ddot{C}l:$$

The molecules BF_3 and $AlCl_3$ behave in a similar manner by accepting an electron pair from another molecule or ion to form a covalent bond:

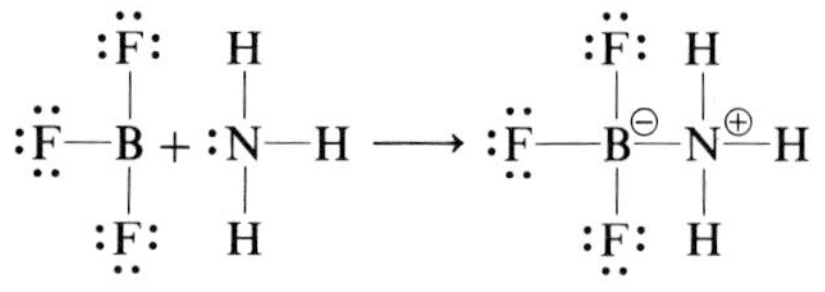

$$:\ddot{F}-B \;+\; :N-H \longrightarrow :\ddot{F}-B^{\ominus}-N^{\oplus}-H$$

Lewis (Figure 4.9) therefore suggested that all molecules and ions that combine with other molecules and ions by accepting an electron pair should be regarded as acids. Acids thus defined are called **Lewis acids**, to differentiate them from acids as defined by Brønsted and Lowry. Molecules that provide an electron pair are called **Lewis bases**. *A Lewis acid is an electron pair acceptor, and a Lewis base is an electron pair donor.* Bases are the same according to both definitions; in both cases the prerequisite for base behavior is the availability of an unshared electron pair.

In forming Al_2Cl_6, each $AlCl_3$ molecule behaves as both a Lewis acid and a Lewis base. The Al atom accepts an electron pair, and a Cl atom donates an electron pair.

ACIDIC AND BASIC OXIDES Another example of a Lewis acid-base reaction is provided by the reaction in which slag is formed when ores are reduced to obtain metals. The silica impurity in the ore reacts with lime, CaO, to form calcium silicate, $CaSiO_3$:

$$CaO + SiO_2 \longrightarrow CaSiO_3$$

Or since CaO and $CaSiO_3$ are ionic,

$$O^{2-} + SiO_2 \longrightarrow SiO_3^{2-}$$

Similar examples include

$$Na_2O + CO_2 \longrightarrow Na_2CO_3 \quad \text{or} \quad O^{2-} + CO_2 \longrightarrow CO_3^{2-}$$

and

$$MgO + SO_3 \longrightarrow MgSO_4 \quad \text{or} \quad O^{2-} + SO_3 \longrightarrow SO_4^{2-}$$

In each case an unshared pair of electrons of the oxide ion of an ionic metal oxide forms a bond with the central atom of a nonmetal oxide:

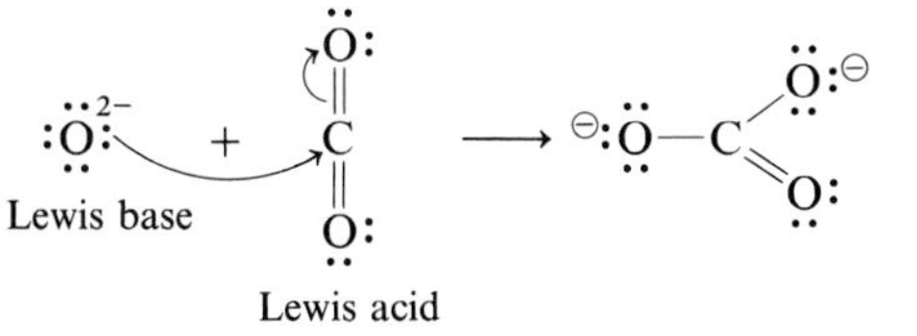

Because the C atom in CO_2 has four pairs of electrons in its valence shell, we might think that it would not be able to accept an electron pair. But one of the electron pairs of a CO double bond shifts to the O atom to make room for the incoming electron pair from the oxide ion. The reaction between an oxide ion and SO_3 is similar:

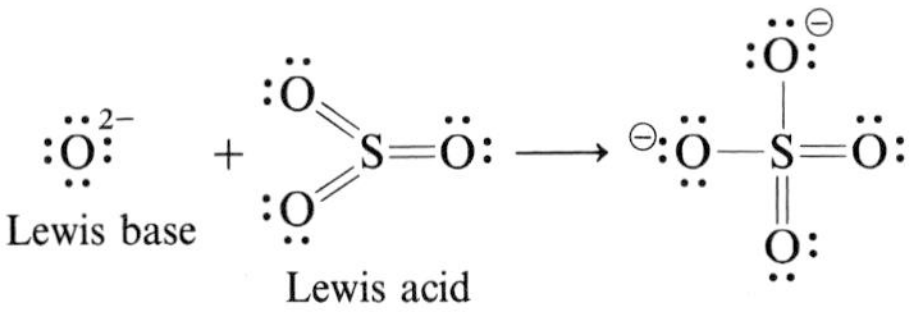

In summary, the oxide ion is a Lewis base, and ionic metal oxides are therefore called **basic oxides**. Covalent nonmetal oxides are Lewis acids and are called **acidic oxides**. An acidic oxide reacts with a basic oxide to form a salt:

$$CaO \;+\; SiO_2 \;\longrightarrow\; CaSiO_3$$
$$CaO \;+\; CO_2 \;\longrightarrow\; CaCO_3$$

| Basic oxide | Acidic oxide | Salt |
| Lewis base | Lewis acid | |

The acid and base character of oxides can also be seen in their reactions in aqueous solution. Basic oxides react with water to give basic solutions; for example,

$$Na_2O(s) + H_2O \longrightarrow 2Na^+(aq) + 2OH^-(aq)$$

Acidic oxides react with water to give an acidic solution; for example,

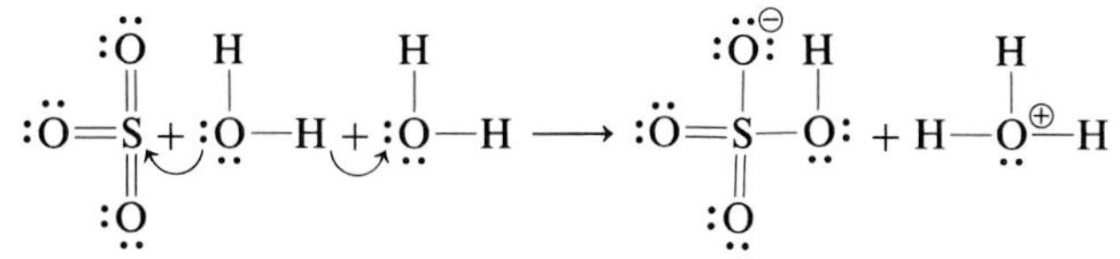

This reaction is another example of a Lewis acid-base reaction. Here the Lewis acid SO_3 accepts an electron pair from the Lewis base H_2O, forming an S—O bond. At the same time the H_2O molecule loses a proton to another H_2O molecule to give HSO_4^- and H_3O^+.

Because they react with water to give acids, acidic oxides are often called **acid anhydrides**. Thus an acidic oxide can be thought of as being derived from an acid by the removal of water—although, in practice, in some cases it is very difficult to dehydrate the acid to give the oxide. For example,

$$H_2SO_4 - H_2O \longrightarrow SO_3 \quad \text{(Occurs only with difficulty)}$$

$$H_2CO_3 - H_2O \longrightarrow CO_2 \quad \text{(Occurs readily)}$$

We saw in Chapter 7 that P_4O_{10} reacts with excess water to give orthophosphoric acid, H_3PO_4:

$$P_4O_{10} + H_2O \longrightarrow H_3PO_4 \quad \text{(Unbalanced)}$$

Balancing P, we write

$$P_4O_{10} + H_2O \longrightarrow 4H_3PO_4 \quad \text{(Unbalanced)}$$

Balancing H and O, we have

$$P_4O_{10} + 6H_2O \longrightarrow 4H_3PO_4 \quad \text{(Balanced)}$$

With a limited amount of water, metaphosphoric acid is the product:

$$P_4O_{10} + 2H_2O \longrightarrow 4HPO_3$$

Thus P_4O_{10} is the anhydride of phosphoric acid.

Similarly, the anhydride of nitric acid is the oxide N_2O_5:

$$N_2O_5 + H_2O \longrightarrow 2HNO_3 \qquad 2HNO_3 - H_2O \longrightarrow N_2O_5$$

Visualizing the dehydration of nitric acid as follows is helpful:

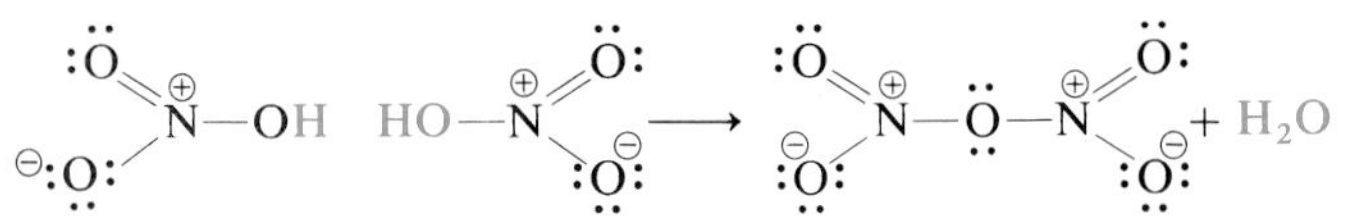

Thus acidic oxides react with water to give an oxoacid, and they react with an oxide ion to give the anion of the corresponding acid. For example,

$$SO_3 + H_2O \longrightarrow H_2SO_4$$

and

$$SO_3 + O^{2-} \longrightarrow SO_4^{2-}$$

Some oxides such as MgO and SiO_2 react only very slowly and incompletely with water. Their acid-base properties are, however, shown by their reactions with acids and bases. For example, MgO reacts with hydrochloric acid to give $MgCl_2$; it is therefore a basic oxide:

$$MgO(s) + 2HCl(aq) \longrightarrow MgCl_2(aq) + H_2O$$

And silicon dioxide reacts with an aqueous solution of NaOH to give Na_2SiO_3; it is therefore an acidic oxide:

$$SiO_2(s) + 2NaOH(aq) \longrightarrow Na_2SiO_3(aq) + H_2O$$

Some oxides, such as aluminum oxide, Al_2O_3, have both acid and base properties; they are called **amphoteric oxides**. Although corundum is very inert and insoluble in both acids and bases, other forms of Al_2O_3 that have slightly different structures are soluble in both acidic and basic aqueous solutions:

$$Al_2O_3(s) + 6HCl(aq) \longrightarrow 2Al^{3+}(aq) + 6Cl^-(aq) + 3H_2O$$

$$Al_2O_3(s) + 2NaOH(aq) + 3H_2O \longrightarrow 2Na^+(aq) + 2Al(OH)_4^-$$

In general, the oxides of the main group metals are basic. They are ionic, and the oxide ion is a Lewis base (and a Brønsted-Lowry base). However, oxides of elements such as aluminum that are near the borderline between metals and nonmetals in the periodic table may be amphoteric. The oxides of some of the transition metals may be basic, amphoteric, or acidic, depending on the oxidation state of the metal, as we will discuss in Chapter 21. The relationship between the acid and base character of an oxide and the position of the element in the periodic table is shown in Figure 9.9.

Basic oxides	Amphoteric oxides	Acidic oxides			Oxides with no acidic or basic properties	
Li_2O	BeO	B_2O_3	CO_2	N_2O_3	O_2	F_2O
				N_2O_5		
Na_2O	MgO	Al_2O_3	SiO_2	P_2O_3	SO_2	Cl_2O
				P_2O_5	SO_3	Cl_2O_7

Figure 9.9 Acid and Base Properties of Oxides of Periods 2 and 3. On the left of the periodic table the oxides are basic. On the right they are acidic (except O_2 and F_2O).

Example 9.4 Write equations for the reactions of the oxide ion with the following acidic oxides:

(a) SiO_2 (b) N_2O_5 (c) P_4O_{10}

Solution

(a) The equation is $SiO_2 + O^{2-} \longrightarrow SiO_3^{2-}$.

(b) The anhydride of HNO_3 is N_2O_5, so N_2O_5 reacts with an oxide ion to give two nitrate ions, NO_3^-:

$$N_2O_5 + O^{2-} \longrightarrow 2NO_3^-,$$

(c) The anhydride of phosphoric acid is P_4O_{10}. Since there are several different phosphoric acids, there are several possibilities; for example,

$$P_4O_{10} + 6O^{2-} \longrightarrow 4PO_4^{3-} \text{ (Orthophosphate)}$$
$$P_4O_{10} + 2O^{2-} \longrightarrow 4PO_3^- \text{ (Metaphosphate)}$$

Example 9.5 Write equations for the reactions of the following oxides with water. In each case state whether the oxide is acidic or basic.

(a) CO_2 (b) BaO   (c) N_2O_5 (d) Rb_2O

Solution

(a) Carbon dioxide reacts with water to give the weak acid, carbonic acid:

$$H_2O + CO_2(aq) \rightleftharpoons H_2CO_3(aq)$$
$$H_2CO_3(aq) + H_2O \rightleftharpoons H_3O^+(aq) + HCO_3^-(aq)$$

Since CO_2 is a nonmetal oxide, it is acidic.

(b) Group I and II metal oxides are basic:

$$BaO(s) + H_2O \longrightarrow Ba(OH)_2(aq)$$

9.5 REACTIONS AND COMPOUNDS OF ALUMINUM, IRON, COPPER, AND LEAD

(c) The oxide N_2O_5 is a nonmetal oxide; it is therefore acidic. It reacts with water to give HNO_3:

$$N_2O_5(s) + H_2O \longrightarrow 2HNO_3(aq)$$

$$HNO_3(aq) + H_2O \longrightarrow H_3O^+(aq) + NO_3^-(aq)$$

(d) Rubidium is in group I, so Rb_2O is a basic oxide. It reacts with water to give a basic solution:

$$Rb_2O(s) + H_2O \longrightarrow 2RbOH(aq)$$

ALUMINUM SALTS   Hydrated aluminum chloride can also be prepared by dissolving aluminum or the oxide Al_2O_3 in hydrochloric acid:

$$2Al(s) + 6HCl(aq) \longrightarrow 2AlCl_3(aq) + 3H_2(g)$$

$$Al_2O_3(s) + 6HCl(aq) \longrightarrow 2AlCl_3(aq) + 3H_2O(l)$$

Other salts can be prepared similarly by using the appropriate acid; for example,

$$2Al(s) + 3H_2SO_4(aq) \longrightarrow Al_2(SO_4)_3(aq) + 3H_2(g)$$

Aluminum sulfate crystallizes from aqueous solution as the hydrate $Al_2(SO_4)_3 \cdot 18H_2O$. Six water molecules are associated with each aluminum ion, and six more form part of the crystal structure. If an equimolar mixture of potassium sulfate and aluminum sulfate is allowed to crystallize, a salt with the composition $KAl(SO_4)_2 \cdot 12H_2O$ is obtained as large octahedral-shaped crystals (see Experiment 9.5). This salt is called **alum**, or, more specifically potassium alum. Other related salts such as ammonium alum, $(NH_4)Al(SO_4)_2 \cdot 12H_2O$, can be prepared. They are all ionic solids consisting of ammonium, potassium, or another singly charged cation, aluminum ions, sulfate ions, and water molecules.

The formation of the hydrated aluminum ion, $Al(H_2O)_6^{3+}$, provides another example of a Lewis acid-base reaction. The aluminum ion behaves as a Lewis acid, accepting not one but six electron pairs from water molecules. In general, because they have empty valence shells, metal ions, particularly those with high charges, behave as electron pair acceptors, or Lewis acids.

EXPERIMENT 9.5

Growing Alum Crystals

A crystal of potassium alum growing in a saturated solution.

Octahedral crystals of potassium alum.

ALUMINUM HYDROXIDE   If hydroxide ion is added to a solution of an aluminum salt, a white gelatinous precipitate of aluminum hydroxide, $Al(OH)_3$, is obtained:

$$Al^{3+}(aq) + 3OH^-(aq) \longrightarrow Al(OH)_3(s)$$

Except for the hydroxides of the alkali metals and barium, most metal hydroxides are insoluble in water. If excess OH^- ion is added to the precipitate of $Al(OH)_3$, it dissolves again because of the formation of the soluble tetrahydroxoaluminum ion, $Al(OH)_4^-$:

$$Al(OH)_3(s) + OH^-(aq) \longrightarrow Al(OH)_4^-(aq)$$

This reaction is analogous to the reaction between $AlCl_3$ and Cl^- to give $AlCl_4^-$. It is another example of Lewis acid-base behavior, in which $Al(OH)_3$ is the Lewis acid and OH^- is the Lewis base. The tetrahydroxoaluminate ion can also be obtained by dissolving aluminum in aqueous hydroxide solution:

$$2Al(s) + 2OH^-(aq) + 6H_2O \longrightarrow 2Al(OH)_4^-(aq) + 3H_2(g)$$

Aluminum hydroxide also dissolves in acids to give an aluminum salt:

$$Al(OH)_3(s) + 3HCl(aq) \longrightarrow AlCl_3(aq) + 3H_2O(l)$$

Thus aluminum hydroxide behaves as both a base and an acid. It is amphoteric, like the oxide.

Like most metal hydroxides, except those of the alkali metals, aluminum hydroxide decomposes when strongly heated to give the oxide and water:

$$2Al(OH)_3(s) \longrightarrow Al_2O_3(s) + 3H_2O(g)$$

Compounds of Lead

Lead is in Group IV and has the electron configuration $[Xe]\ 4f^{14}5d^{10}6s^26p^2$, with a valence shell consisting of the four $6s^26p^2$ electrons. It has an electron configuration similar to that of carbon, $2s^22p^2$. Like carbon, it forms a number of compounds from a valence state electron configuration with four unpaired electrons:

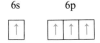

In these compounds, which are predominately covalent and resemble the corresponding compounds of carbon and silicon, lead is in the $+4$ oxidation state.

However, we have pointed out previously that elements in the central groups of the periodic table resemble each other less than those in the groups at either end. The elements in Group IV all have a common valence of 4, but there is a distinct trend in the group from the typically nonmetallic element carbon at the top to the typically metallic element lead at the bottom. Because of the large size of the lead atom, its valence shell electrons are much less strongly held than are those of carbon. Thus lead loses two electrons from its ground state configuration quite readily to form the Pb^{2+} ion in which lead is in the $+2$ oxidation state:

Tetrahydroxoaluminate ion

These Pb(II) compounds are typical ionic salts like those formed by other metals. In contrast, the less numerous Pb(IV) compounds are much less ionic and are best described as covalent.

OXIDES   When lead is heated in air, it forms the yellow oxide Pb(II)O, *litharge*, which contains lead in the $+2$ oxidation state. If lead is heated in oxygen, or if PbO is strongly heated in air, it is further oxidized to Pb_3O_4, *red lead*:

$$6PbO(s) + O_2(g) \xrightarrow{\text{heat}} 2Pb_3O_4(s)$$

The oxide Pb_3O_4 is the basis of a red paint that is widely used for protecting iron and steel structures against rust; apparently, it forms an oxidized, unreactive layer on the surface of the iron. Red lead, Pb_3O_4, contains both Pb(II) and Pb(IV) and may be written as $Pb(II)_2Pb(IV)O_4$.

Red lead reacts with concentrated nitric acid to give soluble lead(II) nitrate and lead dioxide, $Pb(IV)O_2$, a brown insoluble solid in which lead is in the $+4$ oxidation state:

$$Pb_3O_4(s) + 4HNO_3(aq, conc) \longrightarrow 2Pb(NO_3)_2(aq) + PbO_2(s) + 2H_2O(l)$$

The reaction of Pb_3O_4 with HNO_3 is consistent with the formulation of this oxide as $Pb(II)_2Pb(IV)O_4$, because the Pb(II) atoms are converted to soluble $Pb(II)(NO_3)_2$, lead nitrate, leaving the Pb(IV) as the insoluble oxide PbO_2. Lead dioxide is an important component of lead storage batteries.

LEAD(II) CHLORIDE, PbCl₂   Boiling, concentrated hydrochloric acid slowly dissolves lead to form $PbCl_2$:

$$Pb(s) + 2HCl(aq) \longrightarrow PbCl_2(aq) + H_2(g)$$

Lead(II) chloride is more easily prepared by addition of a soluble chloride to a solution of a lead(II) salt (see Experiment 9.6):

$$Pb^{2+}(aq) + 2Cl^-(aq) \longrightarrow PbCl_2(s)$$

It is only sparingly soluble in cold water but is more soluble in hot water. When the hot solution is cooled, white needlelike crystals of $PbCl_2$ separate out.

LEAD(IV) CHLORIDE, PbCl₄   Lead also forms the rather unstable lead(IV) chloride (lead tetrachloride), $PbCl_4$. This is a yellow liquid that decomposes when warmed to give chlorine and $PbCl_2$:

$$PbCl_4 \longrightarrow PbCl_2 + Cl_2$$

The tetrachloride is a covalent compound consisting of tetrahedral $PbCl_4$ molecules, analogous to carbon tetrachloride, CCl_4, and silicon tetrachloride, $SiCl_4$.

LEAD(II) IODIDE, PbI₂   Lead(II) iodide can be precipitated from a solution of a Pb(II) salt. The precipitate is somewhat soluble in hot water. When the solution is cooled, sparkling golden yellow crystals are obtained (see Experiment 9.6):

$$Pb^{2+}(aq) + 2I^-(aq) \longrightarrow PbI_2(s)$$

LEAD(II) HYDROXIDE, Pb(OH)₂   When an alkali metal hydroxide is added to a solution of a soluble Pb(II) salt, a white precipitate, $Pb(OH)_2$, is obtained:

$$Pb^{2+} + 2OH^- \longrightarrow Pb(OH)_2(s)$$

Reactions of $Pb^{2+}(aq)$

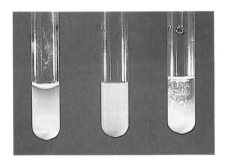

Left to right: Pb^{2+} forms an insoluble precipitate of white $PbSO_4$ with SO_4^{2-}, an insoluble precipitate of yellow PbI_2 with I^-, and a white insoluble precipitate of $PbCl_2$ with Cl^-.

The precipitate of PbI_2 (like $PbCl_2$) dissolves when the water is heated to boiling.

On cooling, the solution PbI_2 slowly precipitates as shiny golden-yellow spangles.

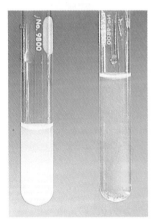

Left: When an aqueous NaOH solution is added to a solution of a soluble Pb^{2+} salt, a white insoluble precipitate of $Pb(OH)_2$ is formed. Right: When an excess of sodium hydroxide is added, the white precipitate dissolves to give a solution of $Pb(OH)_4^{2-}$.

This precipitate is soluble in excess OH^- to give $Pb(OH)_4^{2-}$:

$$Pb(OH)_2(s) + 2OH^-(aq) \longrightarrow Pb(OH)_4^{2-}(aq)$$

Lead hydroxide is also soluble in aqueous acids; it is amphoteric:

$$Pb(OH)_2(s) + 2HCl(aq) \longrightarrow PbCl_2(aq) + 2H_2O$$

Compounds of Iron

Iron is a member of the first transition metal series of elements. A characteristic feature of these elements is that they have several different oxidation states.

As Table 9.4 shows, most of the elements in the first transition series have two electrons in the 4s orbital, which they can lose to form M^{2+} ions, such as Fe^{2+}. These elements also have oxidation states in which they utilize some or all of their 3d electrons in addition to the 4s electrons. In particular, iron has a common $+3$ oxidation state.

OXIDES We have already mentioned two oxides of iron: the minerals hematite, Fe_2O_3, and magnetite, Fe_3O_4. The latter, as its name implies, is magnetic. When iron is heated strongly in oxygen, it burns to give Fe_3O_4. Iron(II) oxide, FeO, can be made by reducing Fe_2O_3 with hydrogen at 300°:

$$Fe_2O_3(s) + H_2(g) \longrightarrow 2FeO(s) + H_2O(g)$$

Table 9.4 Electron Configurations of Elements of First Transition Series

Sc	$[Ar]\, 3d^1 4s^2$
Ti	$[Ar]\, 3d^2 4s^2$
V	$[Ar]\, 3d^3 4s^2$
Cr	$[Ar]\, 3d^5 4s^1$
Mn	$[Ar]\, 3d^5 4s^2$
Fe	$[Ar]\, 3d^6 4s^2$
Co	$[Ar]\, 3d^7 4s^2$
Ni	$[Ar]\, 3d^8 4s^2$
Cu	$[Ar]\, 3d^{10} 4s^1$
Zn	$[Ar]\, 3d^{10} 4s^2$

9.5 REACTIONS AND COMPOUNDS OF ALUMINUM, IRON, COPPER, AND LEAD

The oxides FeO and Fe_2O_3 exemplify the two important oxidation states of iron, namely $+2$ and $+3$. The oxide Fe_3O_4 contains iron in both oxidation states and can be written as $Fe(II)Fe(III)_2O_4$.

IRON(II) CHLORIDE, $FeCl_2$ As previously mentioned, iron reacts with aqueous solutions of acids to form iron(II) salts. With hydrochloric acid iron(II) chloride, $FeCl_2$, is formed. It crystallizes as the tetrahydrate, $FeCl_2 \cdot 4H_2O$:

$$Fe(s) + 2HCl(aq) \longrightarrow FeCl_2(aq) + H_2(g)$$

IRON(II) SULFATE Iron dissolves in sulfuric acid to form iron(II) sulfate;

$$Fe(s) + H_2SO_4(aq) \longrightarrow FeSO_4(aq) + H_2(g)$$

which crystallizes from solution as $FeSO_4 \cdot 7H_2O$. Six water molecules are arranged octahedrally around Fe^{2+}, and there is an additional water molecule forming part of the crystal structure. Hydrated Fe(II) salts are generally pale green.

IRON(II) HYDROXIDE, $Fe(OH)_2$ When an alkali metal hydroxide is added to a solution of an Fe(II) salt, a white precipitate, $Fe(OH)_2$, is obtained:

$$Fe^{2+}(aq) + 2OH^-(aq) \longrightarrow Fe(OH)_2(s)$$

The precipitate rapidly becomes a dirty green and finally brown, because the oxygen in the air oxidizes iron(II) hydroxide to iron(III) hydroxide, $Fe(OH)_3$ (see Experiment 9.7).

IRON(II) SULFIDE, FeS When iron and sulfur are heated together, black FeS is obtained (see Experiment 3.2). It is often used for the preparation of hydrogen sulfide by treating it with an aqueous acid; for example,

$$FeS(s) + H_2SO_4(aq) \longrightarrow FeSO_4(aq) + H_2S(g)$$

EXPERIMENT 9.7

Reactions of Fe^{2+}(aq)

Left: When NaOH(aq) is added to a solution of a soluble Fe^{2+} salt, a dirty white precipitate of $Fe(OH)_2$ is formed. Center: In a few minutes this becomes a dark brown-green color as the $Fe(OH)_2$ is oxidized by the air to $Fe(OH)_3$. Right: If aqueous hydrogen peroxide is added to the $Fe(OH)_2$ precipitate it is immediately oxidized to red-brown $Fe(OH)_3$.

When an Fe^{2+} solution is added to a purple solution of potassium permanganate, it is immediately decolorized as purple MnO_4^- is reduced to colorless Mn^{2+}. An aqueous solution of Mn^{2+} is actually pale pink, but here the solution is too dilute for the color to be seen.

When hydrogen sulfide is passed through a solution of an iron(II) salt, the reverse reaction occurs and iron(II) sulfide is precipitated

$$Fe^{2+}(aq) + H_2S(g) \longrightarrow FeS(s) + 2H^+(aq)$$

The reversibility of this reaction provides another example of Le Châtelier's principle. Excess hydrogen sulfide and a low H^+ concentration cause iron(II) sulfide to precipitate. But a high H^+ concentration and a low hydrogen sulfide concentration cause iron(II) sulfide to dissolve with the formation of hydrogen sulfide.

IRON(II) DISULFIDE   Another important sulfide of iron, which occurs widely as a mineral, is pyrite, FeS_2. It is an ionic compound consisting of Fe^{2+} and disulfide ions, $[:\ddot{S}\!-\!\ddot{S}:]^{2-}$ (see page 313).

IRON(III) CHLORIDE, $FeCl_3$ When iron is heated in chlorine, $FeCl_3$ is obtained. Like $AlCl_3$, the deep red-black $FeCl_3$ sublimes on heating and forms covalent dimeric molecules, Fe_2Cl_6, in the vapor state. Also like $AlCl_3$, it behaves as a Lewis acid and combines with Cl^- to give the tetrahedral $FeCl_4^-$ ion:

$$FeCl_3 + Cl^- \longrightarrow FeCl_4^-$$

Iron(III) chloride is soluble in water, and it crystallizes as yellow $FeCl_3 \cdot 6H_2O$.

IRON(III) HYDROXIDE, $Fe(OH)_3$ When an alkali metal hydroxide is added to a solution of $FeCl_3$, a brown gelatinous precipitate of $Fe(OH)_3$ is obtained. When this is heated, it loses water to give iron(III) oxide, Fe_2O_3:

$$2Fe(OH)_3 \longrightarrow Fe_2O_3 + 3H_2O$$

RUSTING OF IRON   Although iron, particularly in the form of steel, is a very useful metal, it has the disadvantage that it corrodes or rusts rather easily when exposed to the atmosphere. Rust is brown hydrated iron(III) oxide, $Fe_2O_3 \cdot H_2O$. Both water and oxygen are required to form rust; iron does not rust in dry air or in water that is free of O_2.

The initial step in aqueous solutions is

$$2Fe(s) + 4H^+(aq) + O_2(g) \longrightarrow 2Fe^{2+}(aq) + 2H_2O$$

In the second step Fe^{2+} is oxidized by oxygen to insoluble Fe_2O_3 in which iron is in the $+3$ oxidation state.

$$4Fe^{2+}(aq) + O_2(g) + 4H_2O \longrightarrow 2Fe_2O_3(s) + 8H^+(aq)$$

The overall reaction, the sum of these two reactions, is

$$4Fe(s) + 3O_2(g) \longrightarrow 2Fe_2O_3(s)$$

Aluminum also forms an oxide in the air, but it is in the form of a tough continuous skin over the surface of the metal and thus protects it from further reaction with oxygen. Thus although aluminum is a rather reactive metal, it does not corrode. In contrast, iron(III) oxide does not form a continuous skin over the surface of the iron. Rather, it flakes off, exposing more metal to attack by oxygen, and therefore iron continues to rust.

Iron may be protected against corrosion in various ways. It may be coated with a thin layer of a metal, such as tin, that is not attacked by the atmosphere

Rust

9.5 REACTIONS AND COMPOUNDS OF ALUMINUM, IRON, COPPER, AND LEAD

347

or with a thin layer of chromium, which is protected itself by a thin layer of the oxide, or with a layer of zinc on which an adhering, insoluble layer of zinc carbonate hydroxide, $Zn_2CO_3(OH)_2$, is formed. Iron that is protected by a layer of zinc is called *galvanized iron*. Painting iron also protects it; red lead paint is often used. Many alloys of iron such as stainless steels, which contain nickel and chromium, are also resistant to corrosion.

OXIDATION AND REDUCTION REACTIONS OF IRON   Because they can be oxidized to iron(III) compounds, iron(II) compounds can behave as reducing agents. An important reaction in volumetric analysis is the reduction of permanganate ion, MnO_4^-, to manganous ion by Fe^{2+} in an acidic solution:

$$5Fe^{2+}(aq) + MnO_4^-(aq) + 8H^+(aq) \longrightarrow 5Fe^{3+}(aq) + Mn^{2+}(aq) + 4H_2O$$

Iron(III) compounds can behave as oxidizing agents because they can be reduced to iron(II) compounds. For example, I^- is oxidized to I_3^- by Fe^{3+}:

$$2Fe^{3+}(aq) + 3I^-(aq) \longrightarrow 2Fe^{2+}(aq) + I_3^-(aq)$$

We saw in Chapter 5 that the strength of the halide ions as reducing agents increases in the order $F^- < Cl^- < Br^- < I^-$. Of the halide ions only I^- is strong enough to reduce Fe^{3+} to Fe^{2+}. Thus FeI_3 cannot be prepared, although FeF_3, $FeCl_3$, and $FeBr_3$ can.

Example 9.6 Derive the equation for the reduction of MnO_4^- to Mn^{2+} by Fe^{2+}.

Solution

$$\text{Oxidation} \quad Fe^{2+} \longrightarrow Fe^{3+} + e^-$$
$$\text{Reduction} \quad MnO_4^- \longrightarrow Mn^{2+}$$

Assigning oxidation numbers, we have

$$\overset{+7}{Mn}O_4^- \longrightarrow \overset{+2}{Mn}^{2+}$$

Therefore we add five electrons to the left side:

$$MnO_4^- + 5e^- \longrightarrow Mn^{2+}$$

Next, we add $8H^+$ to the left side to balance charges:

$$MnO_4^- + 8H^+ + 5e^- \longrightarrow Mn^{2+}$$

Finally, we add $4H_2O$ to the right side to balance atoms:

$$MnO_4^- + 8H^+ + 5e^- \longrightarrow Mn^{2+} + 4H_2O$$

Adding this equation to $5(Fe^{2+} \rightarrow Fe^{3+} + e^-)$ gives the required equation:

$$5Fe^{2+} + MnO_4^- + 8H^+ \longrightarrow 5Fe^{3+} + Mn^{2+} + 4H_2O$$

Compounds of Copper

Copper, like iron, is a transition element. It has two common oxidation states, $+1$ and $+2$. The compounds of copper are predominately ionic and contain Cu^+ or Cu^{2+}. The electron configuration of copper is $[Ar]\, 3d^{10}4s^1$. It loses the 4s electron to give Cu^+, and it may also lose one of the 3d electrons to give Cu^{2+}.

COPPER(II) COMPOUNDS   When copper is heated strongly in air or oxygen, black *copper(II) oxide*, CuO, is formed. It can also be prepared by heating copper(II) nitrate:

$$2Cu(NO_3)_2(s) \longrightarrow 2CuO(s) + 4NO_2(g) + O_2(g)$$

Most metal nitrates, except those of the alkali metals, decompose in a similar way on heating. For example,

$$2Pb(NO_3)_2(s) \longrightarrow 2PbO(s) + 4NO_2(g) + O_2(g)$$

Copper(II) nitrate can be obtained as blue hydrated crystals, $Cu(NO_3)_2 \cdot 6H_2O$, by evaporating a solution made by dissolving copper in nitric acid.

Example 9.7 Assign oxidation numbers to all the elements in the reaction

$$2Cu(NO_3)_2 \longrightarrow 2CuO + 4NO_2 + O_2$$

and find which elements are oxidized and which are reduced.

Solution

$$\overset{+2\,+5\,-2}{2Cu(NO_3)_2} \longrightarrow \overset{+2\,-2}{2CuO} + \overset{+4\,-2}{4NO_2} + \overset{0}{O_2}$$

The oxidation state of copper remains $+2$ throughout the reaction. Nitrogen oxidizes some of the oxygen from the -2 state to the element (oxidation state 0) and is itself reduced to the $+4$ state.

The most common salt of copper(II) is *copper(II) sulfate*, $CuSO_4$. It can be prepared by dissolving CuO in aqueous H_2SO_4. It crystallizes from water in large blue crystals of the hydrate $CuSO_4 \cdot 5H_2O$. White $CuSO_4$, formed when the hydrate is heated strongly, is called *anhydrous* copper(II) sulfate. It can be used to detect small amounts of water in solvents such as alcohol and ether; when it is added to these liquids, it becomes blue if water is present because the hydrate $CuSO_4 \cdot 5H_2O$ is formed (see Experiment 3.3).

Brown *copper(II) chloride* is obtained when copper is heated in an excess of chlorine gas:

$$Cu(s) + Cl_2(g) \longrightarrow CuCl_2(s)$$

Copper(II) chloride is also formed when CuO reacts with hydrochloric acid:

$$CuO(s) + 2HCl(aq) \longrightarrow CuCl_2(aq) + H_2O(l)$$

When the solution is evaporated, green crystals of the hydrate $CuCl_2 \cdot 2H_2O$ are obtained.

Copper(II) sulfide is a black solid formed by heating copper with excess sulfur or by passing H_2S into a solution of a Cu^{2+} salt (see Experiment 9.8):

$$Cu(s) + S(s) \longrightarrow CuS(s)$$
$$Cu^{2+}(aq) + H_2S(g) \longrightarrow CuS(s) + 2H^+(aq)$$

Copper(II) hydroxide is formed as a pale blue precipitate when OH^- is added to a solution of Cu^{2+} salt (see Experiment 9.8):

$$Cu^{2+}(aq) + 2OH^-(aq) \longrightarrow Cu(OH)_2(s)$$

This precipitate dissolves in an aqueous NH_3 solution to give a deep blue solution of the tetraamminecopper(II) ion, $Cu(NH_3)_4^{2+}$. If $Cu(OH)_2$ is heated,

Reactions of Cu^{2+}(aq)

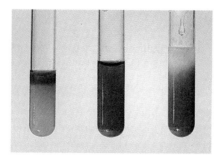

When an aqueous NH_3 solution is added to an aqueous Cu^{2+} solution, a pale blue precipitate of $Cu(OH)_2$ is obtained. This precipitate dissolves in an excess of aqueous ammonia solution with the formation of the soluble dark blue $Cu(NH_3)_4^{2+}$ complex ion.

Left: When an aqueous solution of KI is added to a blue Cu^{2+} solution, I^- is oxidized to brown I_3^-, Cu^{2+} is reduced to Cu^+, and a white precipitate of CuI is formed. Center: When sufficient KI has been added to react with all the Cu^{2+}, the solution is brown due to I_3^- and contains a white precipitate of CuI. Right: When a solution of sodium thiosulfate, $Na_2S_2O_3$, is added, it reduces the I_3^- to colorless I^-, leaving the white precipitate of CuI.

When H_2S(g) is bubbled into a blue solution of a soluble Cu^{2+} salt, a black precipite of copper (II) sulfide, CuS, is formed.

black CuO is formed:

$$Cu(OH)_2(s) \longrightarrow CuO(s) + H_2O$$

COPPER(I) COMPOUNDS If CuO is heated strongly with copper filings or copper powder, red *copper(I) oxide*, Cu_2O, is formed:

$$CuO(s) + Cu(s) \longrightarrow Cu_2O(s)$$

White *copper(I) chloride*, CuCl, is formed when HCl gas is passed over heated copper:

$$2Cu(s) + 2HCl(g) \longrightarrow 2CuCl(s) + H_2(g)$$

Black *copper(I) sulfide*, Cu_2S, can be obtained simply by heating copper(II) sulfide, CuS:

$$2CuS(s) \longrightarrow Cu_2S(s) + S(s)$$

It can also be made by heating the requisite amounts of copper and sulfur:

$$2Cu(s) + S(s) \longrightarrow Cu_2S(s)$$

Copper(I) iodide can be obtained by the reduction of Cu^{2+} with I^- in a reaction similar to that between Fe^{3+} and I^- (see Experiment 9.8):

$$2Cu^{2+}(aq) + 5I^-(aq) \longrightarrow 2CuI(s) + I_3^-(aq)$$

Copper (I) iodide is insoluble in water. The only copper(I) compounds that can be made in aqueous solution are insoluble compounds like CuI, because Cu^+ is not stable in aqueous solution; it reacts to give Cu and Cu^{2+}:

$$2Cu^+(aq) \longrightarrow Cu^{2+}(aq) + Cu(s)$$

In this reaction half of the copper(I) is oxidized to copper(II) and the other half is reduced to copper(0), that is, elemental copper. This type of reaction in which an element oxidizes itself to give a more oxidized and a more reduced form is called a **disproportionation reaction**.

Example 9.8 In an experiment 1.50 g of copper was dissolved in concentrated nitric acid, and the solution was evaporated to dryness to give blue crystals. When these crystals were heated strongly, a black powder was obtained. When the black powder was dissolved in dilute aqueous sulfuric acid, a blue solution was obtained, which on evaporation yielded blue crystals. Write balanced equations for all the reactions, and calculate the maximum amount of the final product that can be obtained.

Solution

$$Cu(s) + 4HNO_3(aq) \longrightarrow Cu(NO_3)_2(aq) + 4NO_2(g) + 2H_2O$$

$$Cu(NO_3)_2(aq) \xrightarrow{\text{Evaporate}} Cu(NO_3)_2 \cdot 6H_2O(s)$$
$$\text{Blue crystals}$$

$$Cu(NO_3)_2 \cdot 6H_2O \xrightarrow{\text{Heat}} CuO(s) + 2NO_2(g) + O_2(g) + 6H_2O(g)$$
$$\text{Black}$$

$$CuO(s) + H_2SO_4(aq) \longrightarrow CuSO_4(aq) + H_2O$$

$$CuSO_4(aq) + 5H_2O \xrightarrow{\text{Evaporate}} CuSO_4 \cdot 5H_2O(s)$$

$$(1.50 \text{ g Cu})\left(\frac{1 \text{ mol Cu}}{63.5 \text{ g Cu}}\right)\left(\frac{1 \text{ mol Cu(NO}_3)_2}{1 \text{ mol Cu}}\right)\left(\frac{1 \text{ mol CuO}}{1 \text{ mol Cu(NO}_3)_2}\right)\left(\frac{1 \text{ mol CuSO}_4}{1 \text{ mol CuO}}\right)$$

$$\left(\frac{1 \text{ mol CuSO}_4 \cdot 5H_2O}{1 \text{ mol CuSO}_4}\right)\left(\frac{249.6 \text{ g}}{1 \text{ mol CuSO}_4 \cdot 5H_2O}\right) = 5.90 \text{ g CuSO}_4 \cdot 5H_2O$$

Summary

The large number of reactions and compounds mentioned in this section may appear confusing at first sight. But there are many similarities between the reactions and compounds of the four metals, and it will make it easier to remember and understand these reactions if we look at these similarities. The important compounds of the metals that we have considered are the oxides, the hydroxides, the halides, and the sulfides, together with the sulfates and nitrates in a few cases. These compounds are summarized in Table 9.5. The oxides, hydroxides, and sulfides are insoluble in water, whereas the halides, sulfates, and nitrates are generally soluble. (Exceptions are $PbCl_2$, PbI_2, and $PbSO_4$.) Insoluble compounds can usually be prepared by precipitation. For example,

$$Al^{3+}(aq) + 3OH^-(aq) \longrightarrow Al(OH)_3(s)$$
$$Fe^{2+}(aq) + S^{2-}(aq) \longrightarrow FeS(s)$$

The oxide ion, O^{2-}, is too strong a base to exist in water, so oxides cannot be prepared in this way, but they can be prepared by direct combination of the elements. They can also be prepared by dehydrating a hydroxide; for example,

$$Cu(OH)_2(s) \xrightarrow{\text{heat}} CuO(s) + H_2O(s)$$

Sulfides can similarly be prepared by direct combination. Soluble compounds such as halides and sulfates cannot be prepared by precipitation, but they can be prepared by the reaction of the oxide and in some cases the metal with an aqueous solution of the appropriate acid:

$$Fe(s) + H_2SO_4(aq) \longrightarrow Fe^{2+}(aq) + SO_4^{2-}(aq) + H_2(g)$$

$$CuO(s) + H_2SO_4(aq) \longrightarrow Cu^{2+}(aq) + SO_4^{2-}(aq) + H_2O(l)$$

Halides can also be prepared by direct combination of the elements; for example,

$$Fe(s) + 3Cl_2(g) \longrightarrow 2FeCl_3(s)$$

Because the halogens are strong oxidizing agents, direct combination usually gives a halide of the metal in one of its higher oxidation states, for example Fe(III) and Cu(II). The methods of preparation are summarized in Table 9.6.

Table 9.5 Compounds of Aluminum, Iron, Lead, and Copper

	OXIDES	HYDROXIDES	SULFIDES	CHLORIDES	SULFATES
Al(III)	$Al_2O_3(s)$	$Al(OH)_3(s)$	—	$AlCl_3(s)$	$Al_2(SO_4)_3(s)$
Fe(II)	$FeO(s)$ $Fe_3O_4(s)$	$Fe(OH)_2(s)$	$FeS(s)$	$FeCl_2(s)$	$FeSO_4(s)$
Fe(III)	$Fe_2O_3(s)$	$Fe(OH)_3(s)$	—	$FeCl_3(s)$	$Fe_2(SO_4)_3(s)$
Pb(II)	$PbO(s)$ $Pb_3O_4(s)$	$Pb(OH)_2(s)$	$PbS(s)$	$PbCl_2(s)$	$PbSO_4(s)$
Pb(IV)	$PbO_2(s)$	—	—	$PbCl_4(l)$	—
Cu(I)	$Cu_2O(s)$	—	$Cu_2S(s)$	$CuCl(s)$	—
Cu(II)	$CuO(s)$	$Cu(OH)_2(s)$	$CuS(s)$	$CuCl_2(s)$	$CuSO_4(s)$

Table 9.6 Preparative Methods for the Compounds of Aluminum, Lead, Iron, and Copper

DIRECT COMBINATION OF ELEMENTS	PRECIPITATION	METAL + ACID	OXIDE + ACID
Oxides	Hydroxides	Sulfates	Sulfates
Sulfides	Sulfides	Nitrates	Nitrates
Halides		Halides	Halides

IMPORTANT TERMS

An **acidic oxide** is an oxide of a nonmetal that dissolves in water to give an acidic solution or that reacts with a base or a basic oxide to give a salt.

An **alloy** is a solid that has metallic properties and is usually composed of two or more metals; it may also be composed of a metal and a nonmetal such as carbon, silicon, or nitrogen.

An **alum** is a double salt, $MAl(SO_4)_2 \cdot 12H_2O$, where M is an alkali metal ion or the ammonium ion.

An **amphoteric oxide** is an oxide that has both acid and base properties.

A **basic oxide** is an oxide of a metal that dissolves in water to give a basic solution or that reacts with an acid or an acidic oxide to give a salt.

In **body-centered cubic packing** each spherical atom is surrounded by eight neighbors that are arranged at the corners of a cube.

Cubic close packing is an arrangement of spherical atoms to give close-packed layers in which each atom is surrounded by a hexagon of nearest neighbors. The layers are stacked on each other in an ABCABC... arrangement so that the atoms in any layer lie directly above those three layers below.

A **disproportionation reaction** is one in which an element in a given oxidation state reacts to give compounds containing the element in two different oxidation states.

Hexagonal close packing is an arrangement of spherical atoms in close-packed layers in which each atom is surrounded by a hexagon of nearest neighbors. The layers are stacked one on another in an ABAB ... arrangement so that the atoms in alternate layers are directly above each other.

A **hydrated salt** is one that crystallizes from water with molecules of *water of crystallization.*

A **hygroscopic compound** is one that will absorb water directly from the atmosphere.

A **Lewis acid** is an electron pair acceptor. A **Lewis base** is an electron pair donor.

A **metallic bond** is the type of bond that holds the atoms together in a metal. It is the force of attraction between a mobile charge cloud (electron gas) and positively charged atomic cores.

Water of crystallization. See *hydrated salt.*

PROBLEMS

Metallurgy

1. List four different reducing agents that may be used to prepare metals from their compounds. Give balanced equations for the preparation of copper, iron, aluminum, and lead, using a different reducing agent in each case.

2. A copper ore contains 1.60% Cu as Cu_2S. What volume of SO_2 measured at 20°C and 760 mm Hg is produced when 1.00 metric ton (10^3 kg) of the ore is roasted?

3. What is the minimum mass of limestone, containing 96% $CaCO_3$, required in a blast furnace charged with 2.00 metric tons of iron ore containing 12.2% SiO_2?

4. (a) What is the difference between a mineral and an ore? How do you account for the fact that the common ores are oxides, sulfides, and carbonates?

(b) Give equations to show how metals may be extracted from each of the types of ore mentioned in part (a).

(c) Write equations to show the extraction of a metal from an oxide, using reduction by another metal and reduction by a nonmetal.

5. (a) What quantity of an ore containing 30% Fe_2O_3 is required to produce 10^6 metric tons of steel?

(b) How many kilograms of iron are present in each metric ton of the ore?

(c) Calculate the mass of coke (assuming that it is pure carbon and that only CO is formed) that is required to reduce 1 metric ton of the ore.

6. Describe the function of each of the following in the manufacture of pig iron:

Fe_3O_4 Coke   CO   CaO

Properties and Structures of Metals

7. How does the metallic character of the elements vary through the periodic table? Describe the metallic character of the elements in the third period from sodium to argon and in group IV from carbon to lead.

8. Draw diagrams to show (a) closest packing and (b) square packing of spheres in a single, two-dimensional layer. What is the number of nearest neighbors of each sphere in the close-packed layer and in the square-packed layer?

9. How would you describe the type of bonding in an Na_2 molecule and in solid sodium? What is the essential difference between the bonding in the two cases?

10. Why is a metal a good conductor of heat and electricity? Why does the electric conductivity increase with decreasing temperature?

11. Draw a diagram of a body-centered cubic structure to show the eight nearest neighbors and the six next-nearest neighbors to a particular atom in the structure.

12. The metals gold, silver, copper, tin, lead, and iron are mentioned in the Bible, but lithium, sodium, potassium, and aluminum were not isolated as elements until the nineteenth century. What differences in properties between the "biblical" elements and the alkali metals and aluminum account for the late discovery of the latter metals?

13. Considering the properties of the metals discussed in this chapter, which would you consider the best choice for (a) making the main structural parts of a space vehicle and (b) making a statue for the town square? Justify your choice by citing some advantages and disadvantages for each metal. Indicate any advantage of using alloys of these metals.

14. Explain the following statements:

(a) Copper was one of the earliest known metals.

(b) Copper-covered roofs acquire a green color.

(c) Copper is used in "silver" coins.

(d) Most slightly soluble copper salts dissolve readily in aqueous ammonia.

(e) Although aluminum is a rather reactive metal, it can be used for cooking utensils.

(f) Aluminum did not become common until the beginning of this century.

Reactions of Metals with Acids

15. Name two metals that react with dilute hydrochloric acid. What are the products of the reactions? Name two metals that do not react with a dilute hydrochloric acid solution.

16. Write equations for the following reactions:

(a) Aluminum with dilute sulfuric acid.

(b) Silver with hot concentrated sulfuric acid.

Oxidation-Reduction

17. Write balanced equations for the following oxidation-reduction reactions carried out in acidic aqueous solution:

(a) $SO_3^{2-}(aq) + MnO_4^-(aq) \longrightarrow SO_4^{2-}(aq) + Mn^{2+}(aq)$

(b) $H_2O_2(aq) + MnO_4^-(aq) \longrightarrow Mn^{2+}(aq) + O_2(g)$

(c) $Zn(s) + NO_3^-(aq) \longrightarrow Zn^{2+}(aq) + N_2O(g)$

(d) $P(s) + NO_3^-(aq) \longrightarrow H_2PO_4^-(aq) + NO(g)$

(e) $NO_3^-(aq) + I_2(s) \longrightarrow IO_3^-(aq) + NO_2(g)$

18. Write balanced equations for the following oxidation-reduction reactions carried out in basic aqueous solution:

(a) $MnO_4^-(aq) + I^-(aq) \longrightarrow MnO_2(s) + IO^-(aq)$

(b) $Br_2(aq) \longrightarrow Br^-(aq) + BrO_3^-(aq)$

(c) $I^-(aq) + ClO^-(aq) \longrightarrow IO_3^-(aq) + Cl^-(aq)$

(d) $SO_3^{2-}(aq) + MnO_4^-(aq) \longrightarrow SO_4^{2-}(aq) + MnO_2(s)$

19. Assign oxidation numbers to each of the atoms in the following:

$Al_2O_3 \cdot xH_2O$ $AlCl_4^-$ SiO_2 SiO_3^{2-}
$Pb(OH)_4^{2-}$ $PbCl_4$ Fe_2O_3 $Cu(NH_3)_4^{2+}$

Acidic and Basic Oxides

20. What are the anhydrides of the following acids?

H_2SO_4 HNO_3 H_2CO_3 $HClO_4$

21. Write equations for the reactions of the following oxides with water. Classify the oxides as acidic or basic.

K_2O SrO SO_2 SO_3
CO_2 P_4O_6 Cl_2O_7

22. Which oxide would you describe as being more basic, Al_2O_3 or SiO_2? Which oxide would you describe as being more acidic, SiO_2 or SO_2?

23. Write balanced equations for the following reactions:

(a) The reaction of Li_2O with SiO_2.

(b) The reaction of Na_2O with N_2O_5.

(c) The reaction of CaO with P_4O_{10}.

Compounds of Aluminum, Iron, Copper, and Lead

24. Describe and write balanced equations for the reactions that occur when a solution of NaOH is added to aqueous solutions of the following:

$AlCl_3$ $CuSO_4$ $FeSO_4$ $Pb(NO_3)_2$

25. Write Lewis structures for the following molecules and ions:

$AlCl_3$ Al_2Cl_6 $Al(H_2O)_6^{3+}$

26. Give three examples, taken from this chapter, of metal hydroxides that are decomposed on heating to give the oxide. In each case, write the balanced equation for the reaction.

27. Solutions of $CuSO_4$, $Al_2(SO_4)_3$, $Pb(NO_3)_2$, and $FeSO_4$ were prepared but not labeled. Describe some simple tests that could be used to identify the solutions.

28. Write balanced equations for the following reactions:

(a) The reaction of powdered aluminum with aqueous sodium hydroxide to give sodium aluminate, $NaAl(OH)_4$, and hydrogen.

(b) The reaction of aluminum oxide with carbon and chlorine to give aluminum trichloride and carbon monoxide.

(c) The decomposition of ammonium alum when it is heated to give ammonia, sulfuric acid, aluminum oxide, and water.

29. Give the names and formulas of all the iron and copper salts that may be produced from the following acids:

HNO_3 H_2SO_4 H_3PO_4

30. Give the formula and a systematic name for each of the following:

Limestone   Alumina   Magnetite   Pyrite
Coke   Red lead   Rust

31. Describe how iron is produced in a blast furnace and how it is converted to steel by the basic oxygen process.

32. Describe the reactions that occur in the corrosion of iron.

33. What is slag? Why is it formed in a blast furnace?

34. What is the mass percentage of aluminum in common clay, assuming a composition of $H_2Al_2(SiO_4)_2 \cdot H_2O$?

35. A sample of 0.250 g of aluminum metal was heated in a stream of dry chlorine gas and gave a white hygroscopic solid containing only aluminum and chlorine. The white solid weighed 1.236 g and changed from solid to vapor at a temperature above 183°C. The vapor gave a pressure of 720 torr when contained in a 210-mL flask at a temperature of 250°C. Find the empirical formula and the molecular formula of aluminum chloride.

36. A sample of 1.00 g of soft solder (a lead/tin alloy) was treated with concentrated nitric acid and warmed. After the reaction was completed, water was added and insoluble $SnO_2 \cdot 2H_2O$ was filtered off and dried. It had a mass of 0.778 g. What is the mass percentage of lead in the solder?

*__37.__ A sample of 2.00 g of lead was strongly heated in oxygen and converted to a red powder. This powder was then treated with concentrated nitric acid to give a colorless solution and a brown powder, which was filtered off and dried. A solution of potassium iodide was added to the colorless solution, and a bright yellow precipitate was obtained. Write equations for all the reactions, and identify all the substances formed. What is the maximum amount of the brown powder and of the yellow precipitate that can be formed?

*__38.__ A sample of 1.50 g of iron was heated in chlorine to give a deep red-black solid. This solid was dissolved in water, and when a solution of sodium hydroxide was added, a gelatinous brown precipitate was obtained. When this precipitate was strongly heated, it was converted to a red-brown powder. Write equations for all the reactions, and identify all the products. What is the maximum amount of the red-brown powder that can be obtained?

CHAPTER 10
THE SOLID STATE

The vast majority of substances are solids under ordinary conditions. The importance in our daily lives of metals, glass, concrete, plastics, silicon chips, and gemstones hardly needs emphasizing.

In earlier chapters we have given considerable emphasis to gases and to solutions, for two important reasons. First, chemical reactions occur much more readily between gases and between substances in solution than between solids, because in the gas state and in solution molecules are free to move and to collide with each other. Second, unlike solids, all gases have similar properties; they obey the ideal gas law, at least approximately. This is because they all have the same simple structure, with widely separated, rapidly moving molecules between which there are only relative weak forces.

In contrast, the forces holding the atoms and molecules together in a solid are relatively strong, and the structures of solids are varied and often complex. As a result, different types of solids often have very different properties. We have already described some of the differences between molecular, ionic, and metallic solids. In this chapter we give a more comprehensive account of the different types of solids and their properties.

10.1 RELATIONSHIP BETWEEN STRUCTURE AND PROPERTIES: DIAMOND AND GRAPHITE

The properties of a solid depend on its composition and its structure. The allotropes of an element, which, of course, all have the same composition, provide striking illustrations of the effect of structure on properties. We have only to recall the differences between white and red phosphorus or between orthorhombic and plastic sulfur to appreciate the profound effect of structure on properties. A still more striking example is provided by diamond and graphite, the allotropes of carbon.

Diamond and graphite are so different that in the absence of chemical analysis, we would find it difficult to believe that they are forms of the same element.

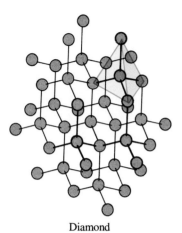

Diamond

Figure 10.1 Structure of Diamond.
Each carbon atom forms four
bonds to neighboring carbon
atoms. The four bonds formed by
each carbon atom have a
tetrahedral AX_4 arrangement. A
three-dimensional network of
covalent bonds extends
throughout the crystal.

Pure diamond forms beautiful, transparent, colorless, very hard, and highly
refractive crystals. In contrast, graphite is a soft, black substance that is used
as a lubricant, and mixed with varying amounts of clay, it is used as the "lead"
in lead pencils. Diamond has a much higher density (3.53 g cm^{-3}) than graphite
(2.25 g cm^{-3}) and is an electrical insulator, whereas graphite is a fairly good
conductor. At ordinary temperatures diamond is inert. It must be heated to
800°C to react even with oxygen. Graphite, on the other hand, reacts at room
temperature with several acids and other oxidizing agents.

In diamond each carbon atom forms bonds to four neighboring carbon atoms,
which have a tetrahedral arrangement; in other words, each portion of the diamond structure has an AX_4 geometry. Each carbon atom is bonded to four other
carbon atoms, each of which in turn is bonded to four more carbon atoms, and
so on (see Figure 10.1). The distance between each pair of adjacent carbon atoms
is 154 pm. The structure extends throughout the diamond crystal, which may
be regarded, therefore, as one giant covalent molecule.

In contrast, graphite consists of planar sheets of carbon atoms in which each
carbon atom is surrounded by only three other carbon atoms. Each carbon atom
has a planar triangular AX_3 arrangement of bonds (see Figure 10.2a). The distance between adjacent carbon atoms in each sheet is 142 pm. The layers are
stacked one on another with a separation of 334 pm (see Figure 10.2b). This distance between the sheets is much greater than the length of a single carbon-
carbon bond. There are therefore no chemical bonds between the sheets of
carbon atoms but only relatively weak intermolecular forces.

Although a carbon atom has four valence electrons and normally forms four
bonds, in graphite each carbon atom is bonded to only three other atoms. We
therefore expect that one of the three bonds will be a double bond. This double
bond can be formed to any of the three neighboring carbon atoms. Thus for
each carbon atom there are three possible bond arrangements, and for each
layer of graphite we can write three resonance structures (see Figure 10.3). This
description of the bonding implies that one of the electron pairs associated with
each carbon atom is not localized in a bond between two particular carbon
atoms but must be considered to be delocalized. An alternative model is to

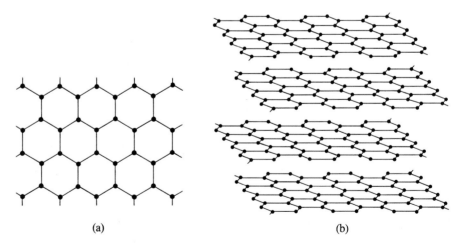

(a)   (b)

Figure 10.2 Structure of Graphite.
(a) Graphite is composed of flat sheets of carbon atoms in which
each carbon atom is bonded to three neighboring carbon atoms
in a planar AX_3 arrangement. (b) These sheets of carbon atoms
are stacked one on another and held together by relatively weak
intermolecular forces.

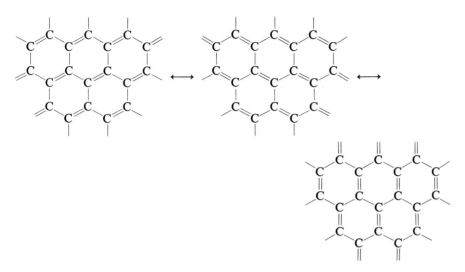

Figure 10.3 Bonding in Graphite.
There are three possible arrangements for one double bond and two single bonds
to each carbon atom:

Therefore there are three possible arrangements of the double bonds and single
bonds in a sheet of carbon atoms, that is, three possible Lewis structures. The
structure consists of carbon atoms joined by three single bonds, so each carbon
atom has an AX_3 geometry. Also, the electrons of the double bonds are not localized
in one position but are spread over the whole sheet of carbon atoms. They are
delocalized, rather like the electrons in a metal, except that they are spread over a
sheet of carbon atoms in two dimensions only, rather than over three-dimensions,
as in a metal. According to this model, each carbon atom forms three localized
covalent bonds and contributes a single electron to a delocalized electron cloud, that
is, to a metallic bond.

consider that each carbon atom forms three single bonds and contributes one
electron to a delocalized charge cloud of electrons extending over the whole
layer. Thus in graphite there are two types of bonding: localized covalent bonds
that link the carbon atoms in planar sheets and metallic bonding due to the
delocalized electrons.

Not surprisingly, therefore, graphite is a shiny black substance with a metallic
luster, and it is a fairly good conductor of electricity. It is used for electrodes in
high-temperature electric furnaces such as the furnace used in the preparation
of phosphorus (Chapter 7) and in some flashlight batteries. But unlike the true
metals described in Chapter 9, graphite has a network of localized bonds in
addition to the metallic bonding, and the metallic bonding extends only in two
dimensions rather than three. Therefore graphite is a good electrical conductor
in the planes of carbon atoms but it is much less conducting in the direction
perpendicular to these planes. The sheets of carbon atoms are not held together
by covalent bonds or metallic bonds but by relatively weak intermolecular
forces. Because the forces between the sheets of carbon atoms are not very
strong, they slide over each other rather easily. Thus graphite is a soft substance that feels quite slippery (see Box 10.1) and is a useful lubricant.

Graphite is slightly more stable than diamond, but the rate at which diamond
is converted to graphite at room temperature is almost infinitely slow. At ordinary temperatures the strongly bonded atoms have far too little energy to
enable them to undergo the very drastic rearrangement needed to convert diamond into graphite. But at a very high pressure of 10^5 atm diamond becomes

Box 10.1
DIAMOND, GRAPHITE, AND HARDNESS

Although we understand what is meant by *hardness*, it is not a property that can be defined simply or precisely. Several scales of hardness and instruments for measuring hardness have been proposed. For example, mineralogists use the *scratch test*: Any substance that scratches another and is not scratched by it is said to be harder than the second substance. On this basis ten minerals have been ranked on a hardness scale called the *Mohs scale*, after Friedrich Mohs (1773–1839), a German mineralogist:

1. Talc, $Mg_3Si_4O_{10}(OH)_2$ (softest)
2. Gypsum, $CaSO_4 \cdot 2H_2O$
3. Calcite, $CaCO_3$
4. Fluorite, CaF_2
5. Fluorapatite, $Ca_5(PO_4)_3F$
6. Orthoclase, $KAlSi_3O_8$
7. Quartz, SiO_2
8. Topaz, $Al_2SiO_4F_2$
9. Corundum, Al_2O_3
10. Diamond, C (hardest)

Diamond is the hardest substance on this scale, whereas graphite has a hardness between 1 and 2. Graphite is almost as soft as talc, which is the main constituent of talcum powder.

more stable than graphite. Graphite can therefore be converted to diamond by subjecting it to a very high pressure and by using a very high temperature of 1200°C plus a catalyst. The synthetic diamonds produced in this way are not necessarily of gem quality but they have many important uses, for example, in oil well drilling bits, in abrasive wheels, and in glass and metal cutting tools.

The structures of solids can be determined with considerable accuracy from the patterns produced when a beam of X rays is diffracted by a crystal (see Box 10.2). This technique, known as **X ray crystallography**, was invented and developed by William and Lawrence Bragg (Figure 10.4). Because the properties of solids depend so strongly on their structure, it is very useful to classify solids according to their structures. This we do in the next section.

Figure 10.4 W. Lawrence Bragg (1890–1971) and William H. Bragg (1862–1942).

William Bragg left England in 1885, where he had graduated from the University of Cambridge, to become professor of physics at the University of Adelaide, Australia, at the age of twenty-three. While at Adelaide he made some important discoveries in the field of radioactivity. His son Lawrence was born in 1890. The family returned to England when William Bragg was offered a position as professor of physics at the University of Leeds, England, in 1909. In 1912 he became intrigued with von Laue's discovery that a beam of X rays could be diffracted by a crystal, because, like many other physicists, he believed that X rays consisted of fast-moving particles. He discussed this discovery with his son Lawrence, who had graduated from the University of Adelaide at the age of eighteen and was studying at the University of Cambridge. During the summer of 1912 Lawrence worked out the theory of the diffraction of X rays by crystals, deducing what is now known as the Bragg equation. Father and son became convinced that X rays were indeed a form of electromagnetic radiation, and working together, they went on to deduce the structures of several ionic crystals. Thus they founded the science of X ray crystallography, to which they were both to devote a major part of their scientific lives. In 1915 they jointly received the Nobel Prize for physics for their work on X rays and the determination of the structures of crystals. They are the only father-and-son team to have received this honor, and Lawrence is also the youngest person to have received it.

Box 10.2
X RAY DIFFRACTION AND STRUCTURES OF SOLIDS

Before 1912 chemists had no direct way of determining how the atoms are arranged in a molecule and in crystal solids. The discovery by German physicist von Laue, in 1912, that a beam of X rays is diffracted by a crystalline solid led to the rapid development of a new branch of science called *X ray crystallography*, which had an enormous effect on the development of chemistry. By studying the diffraction of X rays by crystalline solids, we can obtain precise information about the arrangement of the atoms in solids.

We cannot see atoms and molecules even with the most powerful optical microscope because the wavelength of visible light is much longer than their dimensions, so light waves pass over them undisturbed, much as ocean waves are undisturbed by a small floating cork. But if the wavelength of the waves is made comparable to the dimensions of the object, a diffraction pattern, due to interference between waves scattered by different parts of the object, is produced, as we described in Chapter 6. X rays have wavelengths comparable to the distances between the atoms and ions in solids. For example, the wavelength of a typical X ray beam is 154 pm, while the distance between the sodium ion and the chloride ion in a sodium chloride crystal is 281 pm. Thus when a beam of X rays impinges on a crystal of sodium chloride or any other crystalline solid, a diffraction pattern is produced. Although the diffraction pattern is not a simple image of the arrangement of the atoms in the crystal, it can be interpreted, at least in principle, to give complete information of the atomic arrangement in the crystal.

The theory of X ray diffraction by a crystal was first worked out by Lawrence Bragg. He showed that when X rays are scattered by a plane of atoms, interference occurs so that an intense beam comes from the plane of

Diffraction pattern produced by a sodium chloride crystal

atoms at the same angle as the incident beam, just as if the X ray beam had been reflected by a mirror. In a crystal there is also interference between the X rays "reflected" by one plane of atoms and those reflected by parallel planes. So a diffracted beam is obtained from such a set of parallel planes only for certain angles of incidence, given by the Bragg equation

$$n\lambda = 2d \sin \theta$$

where λ is the wavelength of the X rays, d is the distance between successive planes of atoms in the crystal, θ is the angle of reflection, which is equal to the angle of incidence, and $n = 1, 2, 3, \ldots$.

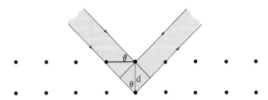

The two reflected X ray beams are in phase when the difference in path length equals $n\lambda$; that is, when $n\lambda = 2d \sin \theta$

If we measure the angle at which a reflected X ray beam is obtained from a crystal, we can use the Bragg equation to calculate the distance between the planes of atoms giving rise to the reflection. Even in a crystal with a very simple structure there are a number of different sets of parallel planes of atoms, each of which gives rise to a characteristic reflection. By using the Bragg equation, we can determine the distance between the planes in each set, and thus we can build up a picture of the complete crystal structure. When the crystal contains two or more different kinds of atoms, different sets of planes have a different composition. In sodium chloride, for example, the planes may consist of sodium ions only, of chloride ions only, or of both sodium and chloride ions. The intensity of the beam scattered by a given set of planes depends on the kinds of atoms in those planes. So it is possible, by measuring the intensities of the reflections, to identify the different atoms in the crystal.

Except for the very simplest structures, it is generally necessary to measure several thousand reflections. Collecting all this information and determining the structure from it was, until fairly recently, a long and tedious process. However, instrumentation has now been developed that automatically records all the reflections. Moreover, the enormous amount of data and the rather complex calculations involved can readily be handled by a modern computer, so a structure can often be determined in just a few days. The structures of increasingly large and complex molecules, such as proteins, are now being routinely determined by X ray crystallography.

10.2 CLASSIFICATION OF SOLIDS

Most solids are **crystalline**; that is, they have a regular periodic arrangement of their atoms or ions, which is reflected in the fact that they form crystals with flat faces and definite angles between the faces. We will be mainly concerned with crystalline solids in this chapter. But as we will see, some solids do not have a regular periodic arrangement of their atoms; they are called **amorphous solids**. A simple classification of the most important types of solids is given in Table 10.1. We first divide solids into two main types, molecular solids and network solids.

Molecular solids consist of individual molecules (or in the case of the noble gases, individual atoms) held together by relatively weak intermolecular forces. Some molecular substances, such as I_2, S_8, P_4, and P_4O_6, are solids at room temperature. But because of the weakness of intermolecular forces, many are liquids, including H_2O, Br_2, CCl_4, and H_2SO_4, and a large number are gases, for example, CO_2, CH_4, HCl, and NH_3. Molecular solids are often soft and have no great mechanical strength. Also, they are usually soluble in several different solvents, because the forces holding the molecules together are relatively weak.

Network solids do not contain finite individual molecules. They have a continuous network of atoms or ions in which each atom or ion is bound strongly to its neighbors, and the regular network extends indefinitely throughout the crystal. Each crystal may be thought of as a molecule of almost infinite size.

We can distinguish three main types of network solids depending on the nature of the bonds between the atoms or ions: covalent solids, ionic solids, and metallic solids. In **covalent network solids** the atoms are held together by covalent bonds, and the covalently bound network of atoms extends throughout the crystal. Diamond is an example (see Figure 10.1). In **ionic solids** oppositely charged ions are held together by their mutual electrostatic attraction, that is, by ionic bonds. The regular arrangement of ions extends continuously throughout the crystal, as in sodium chloride (Figure 5.7). In **metallic solids** positive ions are held together in a regular arrangement by free electrons, that is, by a mobile electron cloud. The regular arrangement of positive ions extends throughout the crystal.

No sharp dividing line can be drawn between ionic and covalent network solids. When the electronegativity difference between the atoms is small, the bonds can be regarded as covalent. If the electronegativity difference is large, the bonds are polar. If the electronegativity difference is very large, the bonds can be regarded as ionic, as in sodium chloride. Again, in many solids with metallic properties the bonding may not be purely metallic but may be intermediate between covalent and metallic or ionic and metallic.

Table 10.1 Classification of Solids

| | MOLECULAR | NETWORK | | | | | |
		1D CHAINS	2D LAYERS (SHEETS)		3D NETWORKS		
Structural units	Molecules	Atoms	Atoms	Ions of opposite charge	Atoms	Ions of opposite charge	Positive ions and electrons
Types of interaction between units	Intermolecular forces	Covalent bonds	Covalent bonds	Ionic bonds	Covalent bonds	Ionic bonds	Metallic bonds
Examples	CH_4, CO_2, HCl, NH_3, I_2, S_8, P_4	S_nSe_n $(SO_3)_n$ $(PO_3^-)_n$	C(graphite) P(black)	$MgCl_2$	C(diamond) SiO_2 SiC	NaCl MgO CaF_2	Ca Fe Al

Properties of Molecular and Network Solids

MELTING POINT AND BOILING POINT   Molecular substances usually consist of the same molecules in the solid, liquid, and gaseous states. Molecular solids are converted to the corresponding liquid and gas at relatively low temperatures because only a small amount of energy is needed to overcome the weak intermolecular forces. All molecular substances either have a melting point below 400°C or decompose below this temperature, in which case they cannot be obtained in the liquid or gaseous state.

In contrast, many covalent, metallic, and ionic network solids have very high melting points and still higher boiling points. For example, the melting point of the covalent solid diamond is 3600°C, the melting point of iron is 1540°C, and the melting point of the ionic solid magnesium oxide is 2800°C. These substances have high melting points because a large amount of energy is needed to break the strong covalent, metallic, or ionic bonds in order to destroy the regular structure of the network solid, allowing the atoms to move relative to each other to form a liquid. Not all ionic solids have very high melting points, however. For example, potassium nitrate melts at 334°C, and potassium hydrogen sulfate, $KHSO_4$, melts at 210°C. The melting point depends on the strength of the ionic bonds. In accordance with Coulomb's law, the strength of ionic bonds increases with the charge on the ions and with decreasing distance between the ions. Thus the strongest bonds are between small highly charged ions such as Mg^{2+} and O^{2-}, whereas the bonds between large, singly charged ions such as K^+ and NO_3^- or HSO_4^- are relatively weak.

If a network solid such as silicon dioxide, SiO_2, is converted to a gas, it undergoes a profound change in structure. In the solid state each silicon atom forms four single bonds to four different oxygen atoms, which surround it in a tetrahedral arrangement (see Figure 10.5). In the gas, silicon dioxide consists of linear SiO_2 molecules, similar to CO_2. In solid sodium chloride, each Cl^- is surrounded by six Na^+ ions, and each Na^+ is surrounded by six Cl^- ions. In contrast, in the gas there are small clusters of two or more ions—in other words, small molecules such as NaCl and Na_2Cl_2.

The melting points of metals cover a wide range, from 3400°C for tungsten to as low as −39°C for mercury. Mercury is the only metal that melts below room temperature, but cesium (mp = 29°C) and gallium (mp = 30°C) melt just above room temperature. There are two important reasons that metals may have relatively low melting points. First, metallic bonds are not broken when a metal melts. The metal ions are held together by a mobile electron cloud, which can adjust to the movement of the positive ions and still keep them strongly bound together. Second, some metallic bonds are relatively weak because only a small number of electrons are available for metallic bonding. For example, in the alkali metal cesium the metallic bonds are formed by only one electron from each metal atom.

In summary, the melting points of molecular solids are usually low, often well below room temperature and always below 400°C. In contrast, network solids usually have high melting points, often as high as several thousand degrees; with very few exceptions network solids melt above room temperature.

SOLUBILITY   Because the forces holding the molecules together are relatively weak, molecular solids are often soluble in molecular liquids. We have seen that phosphorus and sulfur are soluble in carbon disulfide and that bromine and iodine are soluble in carbon tetrachloride.

In contrast, all covalent, most metallic, and many ionic network solids are insoluble in all solvents, including water. For example, silica, SiO_2, and diamond

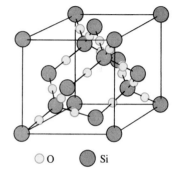

○ O   ● Si

Figure 10.5 Structure of Silicon Dioxide, SiO_2. Silicon dioxide (silica) crystallizes in a number of different forms. Each Si atom is surrounded by a tetrahedron of O atoms, each of which in turn is bonded to another Si atom. In the different forms of silica the SiO_4 tetrahedra are arranged in different ways. The structure shown here is the form of silica called β-cristobalite. Note that the O atom of each SiO_4 tetrahedron is shared with another Si atom, so each Si atom has a half share in four O atoms and the overall composition in SiO_2.

10.2 CLASSIFICATION OF SOLIDS

361

are insoluble in water—and, indeed, in most other solvents. Metals such as copper, iron, and aluminum are also insoluble in water and other solvents. Among the many insoluble ionic compounds are the majority of metal oxides, such as MgO, Al_2O_3, CuO, and Fe_2O_3. Because of the strength of the forces holding the atoms or ions together in these substances, they cannot separate into individual atoms or ions, which is necessary if they are to dissolve to form a solution. But many ionic solids are soluble in water because the attractive forces between the ions and the polar water molecules are strong enough to break the ions away from the crystal and pull them into solution. Although metals are generally insoluble in almost all solvents, many metals are soluble in mercury and other metals.

One-, Two-, and Three-Dimensional Networks

We can also classify network solids according to the type of network rather than the type of bonding. We can distinguish three types of network, one-dimensional (1D), two-dimensional (2D), and three-dimensional (3D).

A **three-dimensional network solid** is one in which strong bonds extend indefinitely in three dimensions so that the whole crystal may be regarded as one giant, almost infinite molecule. Examples include diamond (covalent bonding), zinc sulfide (polar covalent bonding), sodium chloride (ionic bonding), and copper (metallic bonding) (see Figure 10.6).

A **two-dimensional network solid** has a network that extends indefinitely in only two dimensions; therefore the atoms or ions form an infinite layer, or sheet, of atoms. Examples include graphite (covalent and metallic bonding), black phosphorus (covalent bonding), and magnesium chloride (ionic bonding) (see Figure 10.7). The sheets of atoms are not necessarily just one atom thick, as in

Figure 10.6 Three-Dimensional Network Solids. The bonds in these solids extend throughout the crystal, which may be regarded therefore as one giant molecule. The bonding may be covalent, as in diamond, polar covalent, as in zinc sulfide (sphalerite), ionic, as in sodium chloride, or metallic, as in copper.

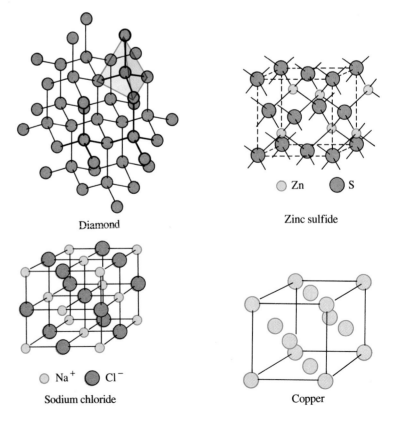

Diamond

Zinc sulfide

○ Zn ● S

○ Na^+ ● Cl^-

Sodium chloride

Copper

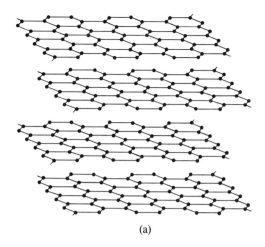

(a)

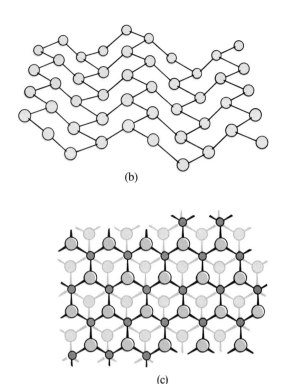

(b)

(c)

Figure 10.7 Two-Dimensional Network Solids, or Sheet Molecules.
These solids have a network of bonded atoms that extends in two
dimensions only, forming sheets of atoms or giant two-dimensional
molecules. The sheets are stacked on each other in the solids. (a) The
bonding in the sheets may be covalent, as in graphite, which consists
of planar, two-dimensional carbon atoms. (b) The bonding in black
phosphorus is also covalent; it consists of buckled two-dimensional
sheets of phosphorus atoms. (c) The bonding may also be ionic, as in
magnesium chloride, which consists of Mg^{2+} ions in the octrahedral
holes between planar layers of Cl^- ions.

graphite; they may have two or more layers of atoms or ions, but they do not
extend indefinitely in the third dimension. The sheets are stacked one on another and are held together by relatively weak intermolecular forces.

Recall that three-dimensional covalent and ionic network solids, such as
diamond and aluminum oxide, are very hard, because strong bonds must be
broken if atoms are displaced. In contrast, two-dimensional solids are often quite
soft, because the sheets of atoms can slide over each other relatively easily. They
also tend to form flaky, platelike crystals, as found, for example, for graphite
and mica. Two-dimensional network solids are intermediate between three-
dimensional solids, in which all the atoms are held together by strong bonds,
and molecular solids, in which the atoms are held together in small groups
(molecules) by strong bonds and the molecules are held together by weak intermolecular forces. A two-dimensional network solid consists of extremely large
(virtually infinite), sheetlike molecules held together in the crystal by intermolecular forces.

One-dimensional network solids consist of infinitely long chain molecules held
together by weak intermolecular forces. These chains may be arranged in a
regular fashion to give a crystalline solid, as in grey selenium, or the polymeric
form of sulfur trioxide, $(SO_3)_n$, and certain asbestos minerals (Figure 10.8). The
crystals of such solids are often thin and fibrous. Frequently, however, long chain
molecules get tangled up with each other and form an amorphous solid, as in
the case of plastic sulfur. Polyethylene and many other familiar plastics consist
of long chain molecules forming an amorphous solid.

Amorphous Solids

Amorphous solids lack the regular, repeating arrangement of atoms or ions that
is characteristic of crystalline solids, but they do possess the most obvious characteristic property of solids, namely, that they are hard and rigid and do not

10.2 CLASSIFICATION OF
SOLIDS

363

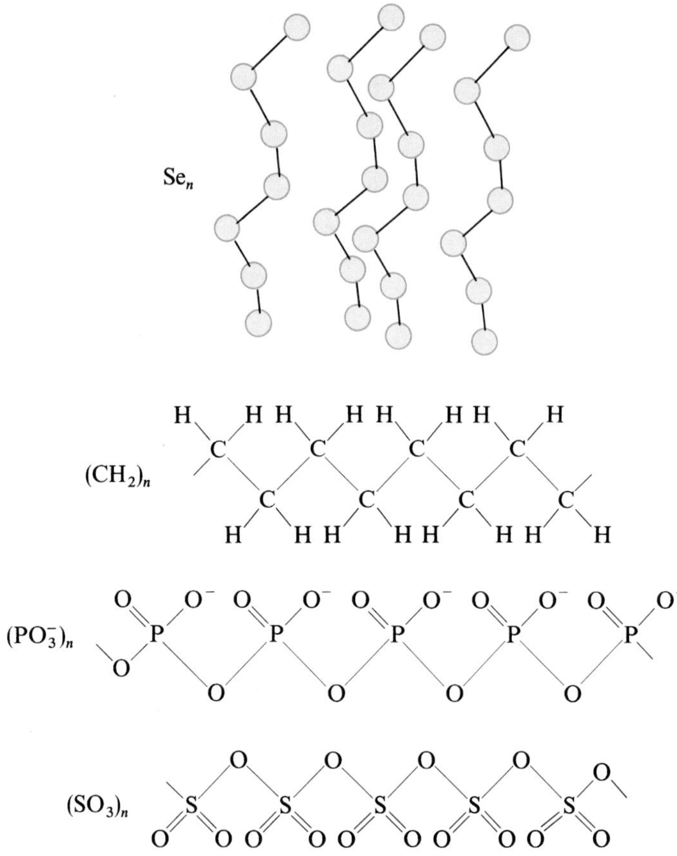

Figure 10.8 Some Chain Molecules that Form One-Dimensional Network Solids. Spiral chains of selenium atoms are arranged in a regular manner to give a crystalline solid. Other long chain molecules such as $(SO_3)_n$ and $(PO_3^-)_n$ may also be arranged in a regular manner to give crystalline solids, but many long chain molecules such as $(CH_2)_n$ form only amorphous or partially crystalline solids.

flow. However, they do not form crystals because they have a random, disordered arrangement of atoms rather than the regular, periodic arrangement characteristic of a crystalline solid. Many synthetic plastics or polymers form amorphous, noncrystalline solids. Silicon dioxide (silica), SiO_2, crystallizes as the mineral quartz, but if molten silica is cooled rapidly, it sets to a transparent, noncrystalline solid that we call glass. Each silicon atom is surrounded by a tetrahedral arrangement of four oxygen atoms, but in the amorphous form these tetrahedra are joined in a random rather than a regular manner. Figure 10.9 illustrates, by a two-dimensional diagram, the difference between an ordered crystalline structure and a disordered amorphous structure.

Amorphous solids do not have a sharp melting point as crystalline solids do. When heated, they soften gradually and become more fluid without exhibiting any sharp melting point, because the solid already has the random arrangement characteristic of the liquid state. When amorphous solids are heated, the motion of the atoms increases, and there is a smooth transition from the solid to the liquid state through an intermediate stage that may be described as a very soft solid or a very viscous liquid. In contrast, in a crystalline solid the regular arrangement of the atoms must change to the random arrangement of the liquid. This change can only occur when the atoms of the solid acquire enough energy to be able to move past each other. The regular arrangement then changes abruptly to the random arrangement characteristic of the liquid, and a sharp melting point is observed. Ordinary glass, which is a complex metal silicate, is an amorphous solid. The fact that it softens over a temperature range and does not melt sharply enables it to be molded and blown into a variety of shapes.

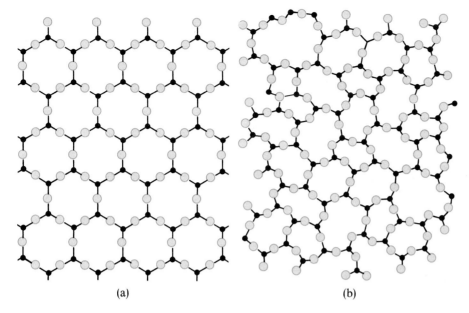

Figure 10.9 Two-Dimensional Models for a Crystalline Solid and an Amorphous Solid. (a) In the crystalline solid the arrangement of the atoms is regular throughout the crystal. (b) In the amorphous solid the arrangement of the atoms is irregular. But in both cases there is an equilateral triangular arrangement of the ○ atoms around the ● atoms.

(a)   (b)

Generally, the amorphous state of a solid is less stable than the crystalline state, and the amorphous state may change spontaneously to the crystalline state. We saw in Chapter 7 that plastic sulfur changes within a few hours to crystalline orthorhombic sulfur. But the transformation of an amorphous solid to a crystalline solid may be extremely slow, because at ordinary temperatures the atoms do not have sufficient kinetic energy to undergo the complicated rearrangement that may be necessary for the transformation. Glass objects from ancient civilizations sometimes have a milky appearance because very tiny crystals have grown over the centuries.

10.3 LATTICES

The atoms, ions, or molecules of a crystalline substance are arranged in a regular manner and form a repeating, three-dimensional pattern. Because there are an enormous number of different structures, we need a systematic method of describing and classifying them. We base our classification of structure on the concept of the **space lattice**, which is a regular, repeating arrangement of points in space. But before considering space lattices, we will examine two-dimensional arrangements of points called *two-dimensional lattices* or nets.

Two-Dimensional Lattices, or Nets

Two-dimensional patterns are familiar from the designs on many wallpapers, carpets, and tiled floors. Any two-dimensional pattern can be described by a **net**, or **two-dimensional lattice**, which is a regular repeating arrangement of points in a plane. Figure 10.10(a) provides an example.

To construct the net corresponding to this two-dimensional pattern, we lay tracing paper on it and put a dot at some chosen point of the pattern, for example, at the eye of each white horse (Figure 10.10b). The regular arrangement of points that we obtain in this way is the net on which the pattern is based. The whole pattern can be constructed by placing the white horse, which is called the *motif* of the pattern, in the same position with respect to each point of the net.

There are five types of two-dimensional lattices, as Figure 10.11 shows. They are the hexagonal, square, rectangular, rhombic, and parallelogram lattices.

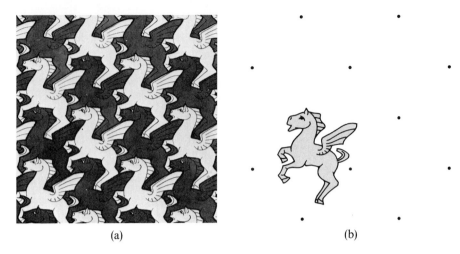

(a)

(b)

Figure 10.10 Two-Dimensional Pattern.
(a) This two-dimensional pattern is by the Dutch artist M. C. Escher. (b) This two-dimensional lattice or net is the basis of the pattern. It may be constructed by placing a dot at the position of the eye of each white horse.

They differ in the symmetry of the arrangement of the points; the hexagonal lattice has the most symmetrical arrangement of points and the parallelogram has the least symmetrical.

Unit Cells

Because of the regular arrangement of the points in a net or a two-dimensional lattice, we need only describe a small part of the lattice in order to specify it completely. We choose four points and connect them to give a parallelogram (see Figure 10.11). This figure is known as a **unit cell**. We can generate the complete lattice by repeatedly moving the unit cell in the directions of its edges by a distance equal to the cell edge.

Figure 10.11 The Five Two-Dimensional Lattices or Nets.
(a) The hexagonal lattice: the unit cell is a rhombus with equal sides and an angle of 60°. (b) The square lattice: the unit cell is a square. (c) The rectangular lattice: the unit cell is a rectangle. (d) The rhombic lattice, or centered-rectangular lattice: the unit cell that is normally chosen is the centered rectangle. The alternative primitive cell is a rhombus with equal sides and an angle that is not equal to 60° or 90°. (e) The parallelogram lattice: the unit cell is a parallelogram with unequal sides and an angle that is not 90°.

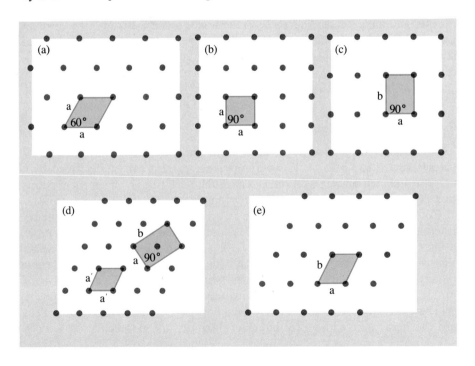

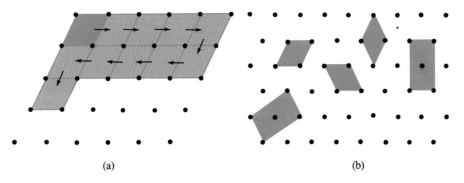

Figure 10.12 Unit Cells. (a) A unit cell is constructed by choosing four points of the lattice and connecting them to give a four-sided figure with pairs of parallel sides. The complete lattice can be generated by repeatedly moving the unit cell in the directions of its edges by a distance equal to the cell edge. (b) A unit cell can be chosen in many different ways. The most convenient cell is the smallest cell that has the full symmetry of the lattice. Some cells have additional points in the center.

(a)   (b)

As Figure 10.12 shows, unit cells for any given lattice can be chosen in many different ways, and may include some cells that have interior points. The most convenient cell is the smallest cell that shows the full symmetry of the lattice. For the square, rectangular, and parallelogram lattices the unit cells normally chosen are the square, the rectangle, and the parallelogram. Any other cells are either larger or are not as symmetrical as the lattice. For the hexagonal lattice the unit cell is a rhombus with an angle of 60°. For the rhombic lattice a rectangular cell with an interior point is normally chosen. A cell with an interior point is called a **centered cell**—in this case a *centered rectangular cell*. The other four cells that do not contain any interior points are known as **primitive cells**.

To describe any two-dimensional pattern, we must specify the unit cell by the lengths of its edges and the angle between them, and we must specify the motif and its position in the cell. Figure 10.13 shows some imaginary two-dimensional structures of molecular, ionic, and metallic crystals. The description of a metallic crystal (Figure 10.13a) is particularly simple since the motif is a single atom that is placed at a lattice point. Thus a close-packed layer of metal atoms can be described by a hexagonal lattice with a motif consisting of a single atom at a lattice point. The simple two-dimensional ionic crystal shown in Figure 10.13(b) can be described by a square lattice. In this case the motif consists of two ions,

Figure 10.13 Structures of Some Imaginary Two-Dimensional Crystals.
(a) This metallic crystal can be described by a hexagonal lattice. The motif is a single metal atom that is placed at a lattice point. (b) This ionic crystal can be described by a square lattice. The motif is a single anion and cation. If the anion is placed at a lattice point, the cation is at the center of the unit cell. (c) This molecular crystal can be described by a parallelogram lattice. The motif is the CO_2 molecule, with the C atom placed at a lattice point.

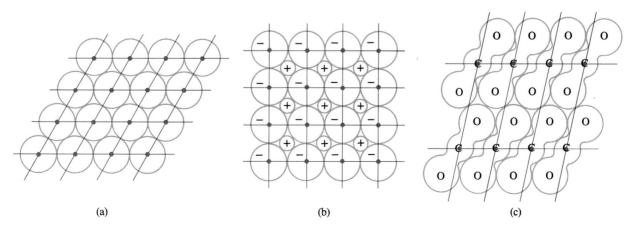

(a)   (b)   (c)

a negative ion and a positive ion. If the negative ion is placed at a lattice point, the positive ion is located in the middle of the square unit cell. In a molecular crystal the motif is one or more molecules (Figure 10.13c).

Space Lattices

A regular three-dimensional arrangement of points in space is called a *space lattice*. Just as there are only 5 possible two-dimensional lattices, there are only 14 possible three-dimensional space lattices. They are known as *Bravais lattices* after the French mathematician who first described them. The most symmetrical of the 14 three-dimensional lattices are the 3 cubic lattices shown in Figure 10.14. Many structures can be described in terms of the primitive cubic lattice and two centered versions of this lattice, the body-centered cubic lattice and the face-centered cubic lattice.

The unit cell of a space lattice is obtained by joining eight points to give a solid with pairs of parallel faces—a parallelepiped. The unit cell of the **primitive cubic lattice** is a cube. The unit cell of the **body-centered cubic lattice** is a cube with an additional point in the middle of the cube. The unit cell of a **face-centered cubic lattice** is a cube with an additional point in the center of each face.

As we will see in the next section, it is important to recognize how many lattice points are associated with each unit cell. The primitive cubic cell has a point at each corner, but each point is shared with seven other unit cells that meet at that point. Thus only one-eighth of the point belongs to any particular cell. Since the unit cell has eight corners, then $8 \times \frac{1}{8} = 1$ point is associated with each primitive unit cell. The body-centered cubic cell has an additional point at its center, which is not shared with other cells. Thus there are two lattice points associated with this unit cell. The face-centered cubic unit cell has an additional point at the center of each face, each of which is shared between two adjacent unit cells. Since the cube has six faces, $6 \times \frac{1}{2} = 3$ additional points are added to the unit cell, to make a total of four points. These conclusions are summarized in Table 10.2.

We can now describe the structures of some different types of network solids in terms of their space lattices.

Figure 10.14 The Three-Cubic Lattices. (a) The primitive cubic lattice. One unit cell is outlined with heavy lines. (b) The unit ceils of the primitive, body-centered, and face-centered cubic lattices. The body-centered cell has an additional point at the center of the cube. The face-centered cell has additional points at the center of each face.

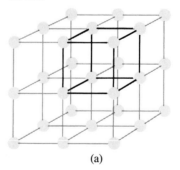

(a)

Primitive cubic   Body-centered cubic   Face-centered cubic

(b)

Table 10.2 Cubic Unit Cells

TYPE OF CELL	NUMBER OF POINTS AT CORNERS	NUMBER OF POINTS IN FACES	NUMBER OF POINTS IN CENTER OF CUBE	TOTAL
Primitive	$8 \times \frac{1}{8}$	0	0	1
Body-centered	$8 \times \frac{1}{8}$	0	1	2
Face-centered	$8 \times \frac{1}{8}$	$6 \times \frac{1}{2}$	0	4

10.4 METALLIC CRYSTALS

Because they consist of a network of single identical atoms, metals generally have very simple structures. We discussed the structures of metals in Chapter 9 in terms of close packing of spherical atoms. As Figure 10.15 shows, almost all the metals have a cubic close-packed, hexagonal closed-packed, or monatomic body-centered cubic structure. We can now describe these structures in terms of their space lattices.

The Iron (Monatomic Body-Centered Cubic) Structure

Recall that we can describe any structure in terms of the appropriate space lattice and a motif. The structure of iron is based on the body-centered cubic lattice. For a monatomic element such as iron the motif is just a single atom, and it can be most conveniently situated at a lattice point. Thus the unit cell of the iron structure has an iron atom at each point of the body-centered cubic lattice (see Figure 10.16). Since the unit cell contains two lattice points, it contains two iron atoms.

We can reach the same conclusion by counting the contents of a cell. As shown in Figure 10.16(a), there are eight corner atoms that are shared between eight unit cells plus the atom in the center of the cell. Therefore there are

$$8 \text{ corner atoms} \times \tfrac{1}{8} \text{ atom per cell} = \tfrac{8}{8} = 1 \text{ atom}$$

$$1 \text{ center atom per cell} = 1 \text{ atom}$$

$$\text{Total} = 2 \text{ atoms}$$

Although it might appear that the lattice point in the middle of the cell is different from the lattice points at the corners, such is not the case. In any space

Figure 10.15 Structures of the Metals. All the metals except Mn, Hg, Ga, In, Ge, and Sn have one of the three types of structures as shown in the key.

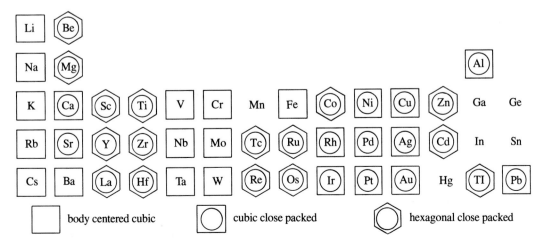

body centered cubic   cubic close packed   hexagonal close packed

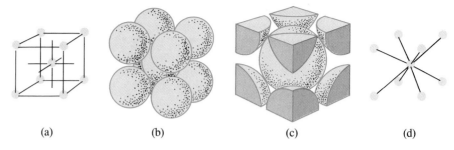

Figure 10.16 Iron (Monatomic Body-Centered Cubic) Structure. (a) The lattice is body-centered cubic. (b) One atom (the motif) is situated at each lattice point. (c) There are two atoms per unit cell. (d) Each atom has a cubic eight coordination.

(a)   (b)   (c)   (d)

lattice every point is equivalent to every other point. In this case every point is surrounded by eight other points in the form of a cube. Thus every iron atom is surrounded by eight other iron atoms in a cubic arrangement. The number of neighboring atoms that are in contact with a given atom is called the **coordination number** of that atom. Each iron atom has a coordination number of eight.

Iron is sometimes described as having a body-centered cubic structure. But we should remember that the description of any structure requires a lattice plus a motif. In this case the motif is a single iron atom; thus the arrangement of the iron atoms is the same as that of the points of the body-centered cubic lattice. The structure is correctly described as a *monatomic* body-centered cubic structure, because there is a single atom at each point of the body-centered cubic lattice. Other metals that have this structure include the alkali metals, barium, chromium, and vanadium.

Example 10.1 The length of the unit cell edge in the iron structure is 286 pm. What is the radius of the iron atom?

Solution In calculations on structures involving cubic unit cells, it is useful to know the relationships between a cube edge, a face diagonal, and a body diagonal. These relationships can be obtained from trigonometry or from the Pythagorean theorem.

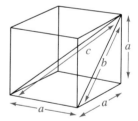

If the length of the unit cell edge is a, if b is a face diagonal, and if c is a body diagonal, then

$$b^2 = a^2 + a^2 = 2a^2 \quad \text{or} \quad b = \sqrt{2}\,a$$
$$c^2 = a^2 + b^2 = 3a^2 \quad \text{or} \quad c = \sqrt{3}\,a$$

From Figure 10.16 we see that for the iron structure the length of the body diagonal c is equal to $4r$, where r is the radius of the iron atom. Hence

$$r = \frac{c}{4} = \frac{\sqrt{3}}{4}\,a = \frac{\sqrt{3} \times 286}{4}\,\text{pm} = 124\,\text{pm}$$

The radius of the iron atom is 124 pm.

Example 10.2 Vanadium has the iron (monatomic body-centered cubic) structure. The length of the edge of the unit cell is found by X ray diffraction to be 305 pm. What is the density of vanadium?

Solution We can find the density by finding the mass and the volume of a unit cell:

$$\text{Volume} = (\text{Cell edge})^3 = (3.05 \times 10^{-8} \text{ cm})^3$$

The iron structure has two atoms per unit cell. Therefore the mass of the unit cell is the mass of two vanadium atoms:

$$\text{Mass of 2 mol of vanadium atoms} = 2 \times 50.94 \text{ g}$$

$$\text{Hence, mass of two vanadium atoms} = \frac{2 \times 50.94 \text{ g}}{6.022 \times 10^{23}}$$

$$\text{Density} = \frac{\text{Mass}}{\text{Volume}} = \frac{2 \times 50.94 \text{ g}}{6.022 \times 10^{23} \times 3.05^3 \times 10^{-24} \text{ cm}^3} = 5.96 \text{ g cm}^{-3}$$

Alternatively, using the unit factor method, we have

$$\text{Density} = (2 \text{ atoms V}) \left(\frac{1 \text{ mol V}}{6.022 \times 10^{23} \text{ atoms}}\right) \left(\frac{50.94 \text{ g}}{1 \text{ mol V}}\right) \left(\frac{1}{3.05 \times 10^{-8} \text{ cm}}\right)^3$$

$$= 5.96 \text{ g cm}^{-3}$$

The Copper (Cubic Close-Packed) Structure

In Chapter 9 we described copper as having a cubic close-packed structure. We can now describe it in more detail in terms of its space lattice, which is face-centered cubic (see Figure 10.17a). The motif is a single copper atom that is located at each lattice point (Figure 10.17b). Because there are four lattice points associated with the face-centered cubic unit cell, there are four copper atoms per unit cell (Figure 10.17c). We can also reach this conclusion by counting the copper atoms inside a unit cell. Those at the corners are shared between eight unit cells, while those at the faces are shared between two cells. Therefore

$$8 \text{ corner atoms} \times \tfrac{1}{8} \text{ atom per cell} = \tfrac{8}{8} = 1 \text{ atom}$$

$$6 \text{ face atoms} \times \tfrac{1}{2} \text{ atom per cell} = \tfrac{6}{2} = 3 \text{ atoms}$$

giving a total of four atoms. This structure is the cubic close-packed structure, which, as we saw in Chapter 9, consists of close-packed layers stacked one on another in the sequence ABCABCABC . . . .

The close-packed layers of atoms in this structure are not immediately apparent because they are not parallel to the faces of the unit cell. Rather, they are perpendicular to the body diagonals of the cell, as shown in Figure 10.18. The name *cubic* close packing refers to the fact that the structure is based on a face-centered cubic lattice. The motif is a single atom situated at each lattice point, so it may also be called the *monatomic face-centered cubic structure*.

Many other metals, including Al, Ca, Sr, Co, Ni, Pb, Ag, and Au, have the cubic close-packed structure. The noble gases Ne, Kr, Ar, and Xe, which consist of single atoms, also have this structure in the solid state.

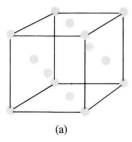

(a)

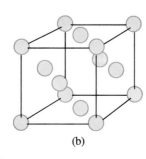

(b)

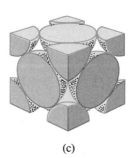

(c)

Figure 10.17 Copper (Cubic Close-Packed) Structure. (a) The lattice is face-centered cubic. (b) The motif is a single copper atom situated at a lattice point. (c) There are four atoms per unit cell.

Figure 10.18 Cubic Close-Packed Structure. (a) Two unit cells are shown, so that we can see that the 12 nearest neighbors of the atom 0 are atoms 1 to 12. (b) The close-packed layers ABC are perpendicular to the body diagonal of the face-centered cubic cell. (c) In the close-packed layers ABC, layer B contains the atom 0 surrounded by atoms 1 to 6; layer A contains atoms 7, 8, and 9; and layer C contains atoms 10, 11, and 12. (d) The cubic close-packed structure showing the removal of successive layers of atoms, exposing the close-packed layers perpendicular to the body diagonal.

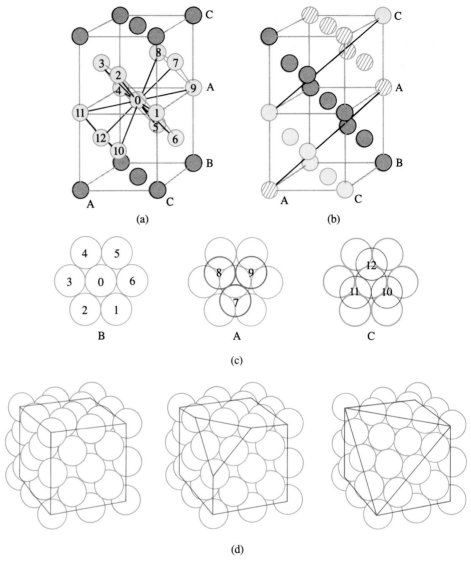

Example 10.3 Silver has a structure based on a cubic lattice. The edge of the unit cell is found to have a length of 408 pm by X ray diffraction. The density of silver is 10.6 g cm^{-3}. How many atoms of silver are there in the unit cell? What type of cubic lattice is the structure of silver based on?

Solution To find how many atoms of silver there are in the unit cell, we will need to find the mass of the unit cell. We can find the mass from the density and the volume:

$$\text{Volume of unit cell} = (\text{Cell edge})^3 = (408 \text{ pm})^3 = (4.08 \times 10^{-8} \text{ cm})^3$$

$$\text{Mass of unit cell} = \text{Density} \times \text{Volume}$$
$$= 10.6 \text{ g cm}^{-3} \times 4.08^3 \times 10^{-24} \text{ cm}^3$$

If the cell contains x atoms, its mass is

$$x(\text{Atomic mass}) = \frac{x(107.9 \text{ g})}{6.022 \times 10^{23}}$$

Hence we have

$$\frac{x(107.9 \text{ g})}{6.022 \times 10^{23}} = 10.6 \times 4.08^3 \times 10^{-24} \text{ g}$$

$$x = 4.02$$

There are four atoms per unit cell. Therefore it is the unit cell of a face-centered cubic lattice.

Alternatively, using the unit factor method, we have

$$\frac{\text{Number of atoms}}{\text{Unit cell}}$$

$$= (408 \text{ pm})^3 \left(\frac{10^{-10} \text{ cm}}{1 \text{ pm}}\right)^3 \left(\frac{10.6 \text{ g Ag}}{1 \text{ cm}^3}\right) \left(\frac{1 \text{ mol Ag}}{107.9 \text{ g Ag}}\right) \left(\frac{6.022 \times 10^{23} \text{ atoms}}{1 \text{ mol Ag}}\right)$$

$$= 4.02 \text{ atoms}$$

The Hexagonal Close-Packed Structure

Some other metals, such as magnesium, titanium, and zinc, have the second type of close-packed structure—hexagonal close-packed. As we saw in Chapter 9, this structure consists of close-packed layers stacked one on another in the sequence ABABAB . . . It is based not on a cubic lattice but on a hexagonal lattice (see Figure 10.19). The unit cell is not cubic but has a less symmetrical shape.

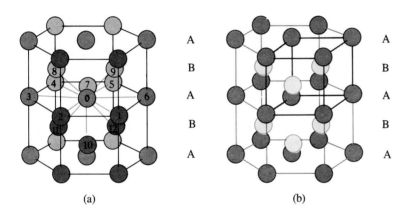

(a)   (b)

Figure 10.19 Hexagonal Close-Packed Structure. (a) The 12 nearest neighbors of atom 0 are atoms 1 to 12. (b) The close-packed layers A and B. A unit cell is shown in heavy lines (it is not a cubic cell).

Example 10.4 By considering the volume occupied by the atoms in appropriate unit cells, show that 74% of the total volume is occupied by atoms in a close-packed structure, whereas only 68% of the volume is occupied by atoms in a monatomic body-centered cubic structure, which is therefore not quite close-packed.

Solution We can solve this problem by comparing the total volume of the atoms in a unit cell with the volume of the cell. First, we consider close-packed structures. The packing in the hexagonal close-packed (ABABAB) structure is related to that of a cubic close-packed structure (ABCABC) simply by shifting planes of atoms, so it does not matter which we select. The cubic close-packed structure has a face-centered cubic unit cell (Figure 10.19), in which there are four atoms per unit cell. For an atom of radius r the volume is given by $\frac{4}{3}\pi r^3$, so four atoms have a volume of $\frac{16}{3}\pi r^3$.

Looking at any face of the cube, we have

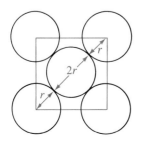
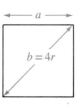

so that $4r = b = \sqrt{2}\,a$ (see Example 10.1), or $a = 2\sqrt{2}\,r$. Thus the volume of the unit cell is $a^3 = (2\sqrt{2}\,r)^3 = 16\sqrt{2}\,r^3$, and therefore

$$\text{Percent occupied space} = \frac{\text{Total volume of atoms}}{\text{Volume of unit cell}} \times 100\% = \frac{\frac{16}{3}\pi r^3}{16\sqrt{2}\,r^3} \times 100\% = 74\%$$

Now consider the monatomic body-centered cubic structure. The unit cell shown in Figure 10.18 contains two atoms. For an atom of radius r the volume of two atoms is $2 \times \frac{4}{3}\pi r^3 = \frac{8}{3}\pi r^3$.

Considering a body diagonal plane of the cube, we have

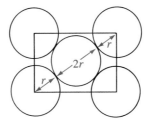 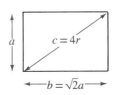

As we saw in Example 10.1,

$$c = 4r = \sqrt{3}\,a \qquad \text{or} \qquad a = \frac{4r}{\sqrt{3}}$$

Hence the volume of the unit cell is $a^3 = [64/(3\sqrt{3})]r^3$. Finally,

$$\text{Percent occupied space} = \frac{\text{Total volume of atoms}}{\text{Volume of unit cell}} \times 100\%$$

$$= \frac{\frac{8}{3}\pi r^3}{[64/(3\sqrt{3})]r^3} \times 100\% = 68\%$$

Accurate determination of the size of the unit cell for a metal structure provides a method for the experimental determination of the Avogadro constant, as shown in the next example.

Example 10.5 Chromium has the iron (monatomic body-centered cubic) structure. Its density is 7.19 g cm^{-3}, and the length of the edge of the unit cell is 288.4 pm. Use these data to calculate a value for the Avogadro constant.

Solution To solve this problem, we first find the mass of the unit cell from its volume and density:

$$\text{Mass of unit cell} = \text{Volume} \times \text{Density}$$

$$= (2.884 \times 10^{-8} \text{ cm})^3 \times 7.19 \text{ g cm}^{-3}$$

$$= 2.884^3 \times 7.19 \times 10^{-24} \text{ g}$$

The unit cell of the iron structure contains two atoms. The mass of two Cr atoms is therefore $2.884^3 \times 7.19 \times 10^{-24}$ g. The mass of 2 mol of chromium atoms is

$$2.884^3 \times 7.19 \times 10^{-24} N \text{ g} = 2(\text{molar mass Cr}) = 2 \times 52.00 \text{ g}$$

Hence

$$\text{Avogadro constant } N = 6.03 \times 10^{23}$$

This value is not exactly equal to the more accurate value (6.022×10^{23}) that we have been using, because the density was given only to three significant figures.

10.5 COVALENT NETWORK CRYSTALS

The structures of covalent crystals are determined primarily by the geometry of the bonds formed by each atom, as in the diamond structure.

The Diamond Structure

The diamond structure was briefly described at the beginning of this chapter. We can now describe it in terms of its space lattice, which is face-centered cubic. The unit cell is shown in Figure 10.20. A carbon atom may be located at each lattice point, and there are also four more carbon atoms located at a position one-fourth of the way along each body diagonal of the unit cell. Since there are four lattice points associated with the unit cell of a face-centered cubic lattice, there are a total of $4 + 4 = 8$ carbon atoms in each unit cell of the diamond structure. Each atom in this structure has an AX_4 geometry and forms four covalent bonds to four nearest neighbors in a tetrahedral arrangement around it (Figure 10.20). This network of tetrahedrally coordinated carbon atoms extends throughout the crystal.

Among the other elements of group IV, silicon, germanium, and one form of tin have the diamond structure.

Layer Structures

Arsenic and phosphorus each form three bonds with a triangular pyramidal AX_3E geometry. One allotrope of each element contains the tetraatomic molecules P_4 or As_4; another form of each element has a layer structure. The layer structure of As can be derived by breaking all the vertical bonds between the buckled sheets in the diamond structure (Figure 10.20). This procedure gives buckled layers of atoms in which the atoms are all joined in hexagonal rings, as Figure 10.21(a) shows. The structure of black phosphorus is very similar, but the hexagonal rings are joined to each other in a slightly different manner (Figure 10.21b).

10.6 IONIC CRYSTALS

We have seen that the structures of metallic crystals are determined by the tendency of identical metal atoms to pack together as closely as possible and that the structures of covalent crystals are determined by the geometry of the covalent

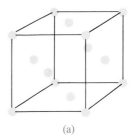

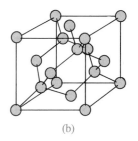

(a)

(b)

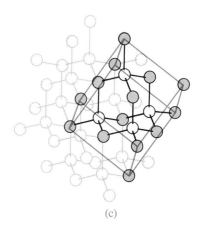

(c)

Figure 10.20 Diamond Structure.
(a) The lattice is face-centered cubic. (b) There is an atom at each lattice point and at one-fourth the distance along each body diagonal. (c) This shows another view of the structure in which a body diagonal of the outlined unit cell is in a vertical position. The atoms at the face-centered lattice points are shaded; the other four atoms are shown as open circles.

Figure 10.21 Layer Structures.
(a) The structure of arsenic is based on a pyramidal AX_3E geometry at each arsenic atom. The three bonds at each arsenic atom link the atoms into six-membered rings. (b) The structure of black phosphorus is similar, except that the six-membered rings are joined to each other in a slightly different way.

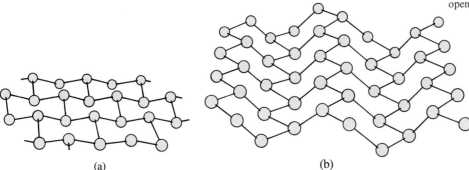

(a)

(b)

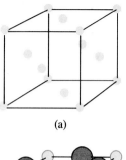

(a)

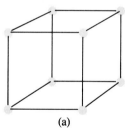

○ Na⁺ ● Cl⁻

(b)

Figure 10.22 Sodium Chloride Structure. (a) The lattice is face-centered cubic. (b) There is a positive ion at each lattice point and a negative ion to the right of each positive ion at a distance equal to one-half of the cell edge. The unit cell contains four cations and four anions.

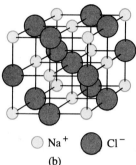

(a)

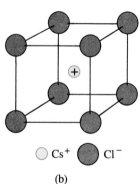

○ Cs⁺ ● Cl⁻

(b)

Figure 10.23 Cesium Chloride Structure. (a) The lattice is a primitive cubic lattice. (b) The anions are at the lattice points, and the cation is at the center of the cell. The unit cell contains one cation and one anion.

bonds formed by each atom. In contrast, the structures of ionic crystals are determined by the tendency of each ion to attract to itself as many oppositely charged ions as possible. As a consequence, all ionic compounds have network structures.

Ionic crystals that are composed of equal numbers of ions of opposite charge have the simplest structures. They have the general formula MX, where M is a positive ion and X is a negative ion. These structures are called **1:1 structures**. We have briefly described some simple structures of this type in Chapter 5, and we can now consider them in more detail in terms of their space lattices.

The Sodium Chloride Structure

Figure 10.22 shows the unit cell of the sodium chloride structure, which is based on a face-centered cubic lattice. A sodium ion may be placed at each lattice point, and there is a chloride ion associated with each sodium ion at a distance of one-half the unit cell edge away. Since the unit cell of the face-centered cubic lattice has four lattice points associated with it, there are four Na^+ ions and four Cl^- ions in the unit cell, in other words, four NaCl formula units per unit cell. Each positive ion is surrounded by an octahedral arrangement of six negative ions, and each negative ion is surrounded by an octahedral arrangement of six positive ions. Each ion has a coordination number of 6.

The sodium chloride structure is very common. It is found for all the alkali metal halides (except CsCl, CsBr, and CsI), the oxides and sulfides of Mg, Ca, Sr, and Ba, and the oxides of a number of transition metals such as iron(II) oxide, FeO.

The Cesium Chloride Structure

Cesium chloride, cesium bromide, and cesium iodide have a structure that is different from sodium chloride, called the cesium chloride structure. In this structure, each Cs^+ ion is surrounded by a cubic arrangement of eight Cl^- ions, and each Cl^- ion is surrounded by a cubic arrangement of eight Cs^+ ions (see Figure 10.23). The lattice is the primitive cubic lattice. If the Cl^- ions are placed at each lattice point, then there is a Cs^+ at the center of each cell. Since this is a primitive cell, it has only one lattice point associated with it. There is therefore only one Cs^+ and one Cl^- associated with each unit cell, or in other words, there is one CsCl formula unit per unit cell.

The Sphalerite Structure

A third important 1:1 structure is the structure of sphalerite, one form of zinc sulfide, ZnS. In this structure the ions are only four-coordinated. The structure is based on the face-centered cubic lattice, and the unit cell is shown in Figure 10.24. If the sulfide ions are placed at the lattice points, there are zinc ions a fourth of the distance along each body diagonal. This structure is the same as the diamond structure, except that alternate atoms are zinc and sulfur. Since there are four lattice points associated with the face-centered cubic lattice, each unit cell contains four zinc ions and four sulfide ions, or four ZnS formula units. By connecting each ion to its nearest neighbors, we see that each ion has a tetrahedral coordination. Because of the large size of the sulfide ion compared with the size of the zinc ion, only four, rather than six or eight, sulfide ions can be packed around a zinc ion.

It is only a rather crude approximation to describe ZnS as an ionic crystal.

Many crystals that are commonly described as consisting of doubly charged ions have some covalent character, particularly when the difference in the electronegativities of the two elements is not very large. In the present case $\chi_{Zn} = 1.6$ and $\chi_S = 2.5$, so the difference is only 0.9. In other words, there are covalent bonds between Zn and S, although the bonds have a considerable polarity. Whenever a small highly charged positive ion is situated next to a large negative ion, we expect some electron density to be transferred from the electron pairs of the negative ion to the empty valence shell of the positive ion, giving some covalent character to the bonds. Assuming that there are polar covalent bonds between Zn and S, they would each have four electron pairs in their valence shells and would therefore be expected to have an AX_4 geometry. Thus the structure of ZnS is consistent with a polar covalent description of the bonding.

Many substances, including CuCl, CuBr, CuI, BeS, HgS, and SiC, possess the sphalerite structure. The amount of covalent character in the bonds varies from one substance to another; in SiC the bonds are almost nonpolar covalent bonds.

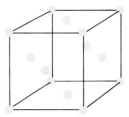

(a)

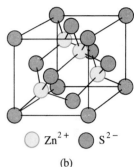

  Zn^{2+} ⬤ S^{2-}

(b)

Figure 10.24 Sphalerite (ZnS) Structure. (a) The lattice is face-centered cubic. (b) There are sulfide ions, S^{2-}, at the lattice points and zinc ions, Zn^{2+}, a fourth of the distance along each body diagonal. The unit cell contains four zinc ions and four sulfide ions.

Example 10.6 Potassium iodide, KI, has a cubic unit cell with a cell edge of 705 pm. The density of KI is 3.12 g cm^{-3}. How many K^+ ions and I^- ions are contained in the unit cell? What is the structure of KI?

Solution We can find the number of formula units per unit cell by finding the mass of the unit cell from its volume and density:

$$\text{Volume of unit cell} = (7.05 \times 10^{-8} \text{ cm})^3$$

$$\text{Mass of unit cell} = \text{Volume} \times \text{Density}$$

$$= (7.05 \times 10^{-8})^3 \text{ cm}^3 \times 3.12 \text{ g cm}^{-3}$$

$$= 1.09 \times 10^{-21} \text{ g}$$

This mass is the mass of a certain number of K^+ and I^-. We can find the total mass of one K^+ and one I^- from the molar mass of KI:

$$\text{Molar mass KI} = 166.0 \text{ g}$$

$$\text{Mass of one empirical formula} = \frac{166.0 \text{ g}}{6.022 \times 10^{23}} = 2.76 \times 10^{-22} \text{ g}$$

$$\text{Number of empirical formulas per unit cell} = \frac{\text{Mass of unit cell}}{1 \text{ empirical formula mass}}$$

$$= \frac{1.09 \times 10^{-21} \text{ g}}{2.76 \times 10^{-22} \text{ g}} = 3.95$$

There are four empirical formulas per unit cell, that is, four K^+ and four I^- ions per unit cell. Thus there are probably four lattice points in the unit cell, which is therefore a face-centered cubic lattice. We conclude that KI has the NaCl structure.

Example 10.7 The density of CsCl is 3.99 g cm^{-3}.

(a) From this value and the structure of CsCl (Figure 10.23), calculate the length of the edge of the unit cell.

(b) What is the distance between a Cs^+ ion and a neighboring Cl^- ion?

(c) The radius of Cl^- is 180 pm. What is the radius of Cs^+?

Solution

(a) We first calculate the mass of the unit cell. The unit cell contains one Cs^+ and one Cl^-. Hence

$$\text{Mass of unit cell} = \text{Mass}(Cs^+ + Cl^-)$$

$$= \frac{132.9 + 35.5}{6.022 \times 10^{23}} \text{ g} = 2.796 \times 10^{-22} \text{ g}$$

Next, we calculate the volume of the cell from the relationship

$$\text{Volume} = \frac{\text{Mass}}{\text{Density}} = \frac{2.796 \times 10^{-22}}{3.99 \text{ g cm}^{-3}} \text{ g} = 70.1 \times 10^{-24} \text{ cm}^3$$

If we let the edge of the unit cell be a, then the volume of the unit cell is a^3. Hence

$$a^3 = 70.1 \times 10^{-24} \text{ cm}^3$$

and therefore

$$a = 4.12 \times 10^{-8} \text{ cm} = 412 \text{ pm}$$

(b) We see from Figure 10.23 that the Cs^+ ion at the center of the unit cell is in contact with eight Cl^- ions at the corners. Therefore the distance between the center of the Cs^+ ion and the center of a Cl^- ion is the distance from the center of the unit cell to one corner, that is, half the length of a body diagonal. We saw in Example 10.1 that the length of the body diagonal is $\sqrt{3}\,a$. Therefore the distance between a Cs^+ and a Cl^- is

$$\frac{\sqrt{3}a}{2} = \frac{\sqrt{3} \times 412}{2} = 357 \text{ pm.}$$

(c) Let r_{Cl^-} be the radius of Cl^- and r_{Cs^+} be the radius of Cs^+. Then

$$r_{Cl^-} + r_{Cs^+} = 357 \text{ pm}$$

If $r_{Cl^-} = 180$ pm, then $r_{Cs^+} = 177$ pm.

Ionic Structures Based on Close Packing of Anions

Many ionic crystal structures can also be described as close-packed arrangements of the larger anions with the smaller cations occupying the holes in the close-packed arrangement. Consider sodium chloride and sphalerite, which both have structures based on the face-centered cubic lattice. If the anions are large enough to touch each other, they have a cubic close-packed arrangement. The holes in such a cubic close-packed structure are of two types, **tetrahedral holes**, which have four neighboring anions in a tetrahedral arrangement, and **octahedral holes**, which are surrounded by six anions in an octahedral arrangement.

As shown in Figure 10.25, there are four octahedral holes per unit cell—one in the center of the unit cell and $12 \times \frac{1}{4} = 3$ holes situated at the middle of each edge of the unit cell. Thus we can describe the sodium chloride structure as a cubic close-packed arrangement of chloride ions, with all the octahedral holes occupied by sodium ions. There are eight tetrahedral holes per unit cell situated in a cubic arrangement in the interior of the unit cell. We can describe the sphalerite structure as a cubic close-packed arrangement of sulfide ions with half the tetrahedral holes occupied by cations.

The next simplest types of ionic crystal structures are the **1:2 and 2:1 structures**. Two of these structures that can be described in terms of the face-centered cubic lattice or cubic close packing of one of the ions are the fluorite and the antifluorite structures.

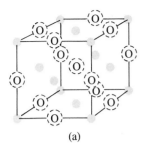 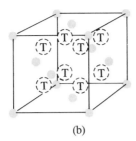 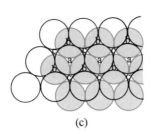

(a)   (b)   (c)

Figure 10.25 Tetrahedral and Octahedral Holes in a Cubic Close-Packed Structure. (a) The octahedral holes in the face-centered cubic lattice of a cubic close-packed structure. (b) The tetrahedral holes in the face-centered cubic lattice for a cubic close-packed structure. (c) Tetrahedral holes a and b and octahedral holes c between two close-packed layers.

The Fluorite Structure

Fluorite is the mineral name for calcium fluoride, CaF_2. The fluorite structure is shown in Figure 10.26. The calcium ions are located at the face-centered cubic lattice points and therefore have a cubic close-packed arrangement. The fluoride ions then occupy *all* of the eight tetrahedral holes. There are therefore eight fluoride ions and four calcium ions in the unit cell, which is consistent with the empirical formula CaF_2. Each fluoride ion is surrounded by four calcium ions in a tetrahedral arrangement, while each calcium ion is surrounded by eight fluoride ions, which have a cubic arrangement. There are four calcium ions and eight fluoride ions per unit cell.

Other ionic compounds that have the fluorite structure include SrF_2, BaF_2, PbF_2, and $BaCl_2$.

The Antifluorite Structure

Several M_2X compounds have a structure closely related to the fluorite structure; it is called the antifluorite structure. In this structure the smaller cations occupy the position of the fluoride ions, and the larger anions occupy the positions of the calcium ions in the fluorite structure. The following oxides and sulfides of the alkali metals, Li_2O, Na_2O, K_2O, Rb_2O, Li_2S, K_2S, and Rb_2S, have this structure. In these structures the oxide or sulfide ions have a cubic close-packed arrangement, with the cations occupying all the tetrahedral holes.

Ionic Radii

The number of negative ions (anions) that can crowd around a given positive ion (cation) depends on the relative sizes of the two ions, that is, on the **ratio** of the **radius** of the positive ion to that of the negative ion, or r_+/r_-. Table 10.3 gives the possible range of radius ratios for tetrahedral four coordination, octahedral six coordination, and cubic eight coordination. It shows that if we start with a very small cation and gradually increase its size, then when the radius of the cation r is 0.225 times the radius of the anion, r_-, that is, when $r_+ = 0.225r_-$ or $r_+/r_- = 0.225$, we can place just four anions in a tetrahedral

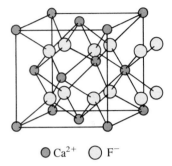

$\bigcirc$ Ca^{2+} $\bigcirc$ F^-

Figure 10.26 Fluorite (CaF_2) Structure

The lattice is face-centered cubic. The calcium ions are at the lattice points and have a cubic close-packed arrangement. The fluoride ions occupy all the tetrahedral holes.

Fluorite

Table 10.3 Radius Ratios

COORDINATION NUMBER	RADIUS RATIO r_+/r_-	COORDINATION POLYHEDRON
4	0.225–0.414	Tetrahedron
6	0.414–0.732	Octahedron
8	0.732–1.00	Cube

10.6 IONIC CRYSTALS

379

arrangement around the cation, with each anion touching its three neighbors. If the size of the cation is increased further, the anions no longer touch each other but there is insufficient room for additional cations until $r_+ = 0.414r_-$. Six anions can then be grouped around the cation in an octahedral arrangement, with each anion just touching its four neighbors. If the size of the cation is further increased, the anions no longer touch each other; but when $r_+ = 0.732r_-$, eight anions can be grouped around the cation in a cubic arrangement. Note that $r_+ = 0.225r_-$ is the radius of a tetrahedral hole, and $r_+ = 0.414r_-$ is the radius of an octahedral hole in a close-packed arrangement of anions.

Example 10.8 Calculate the minimum radius ratios for the coordination of a cation by (a) eight, (b) six, and (c) four anions.

Solution

(a) For eight coordination we have the structure shown in Figure 10.23 (cesium chloride structure). The limiting case is when the anions at the corners of the cube just touch each other and the central cation. Thus the length of the side of the cube is $a = 2r_-$, and the length of the body diagonal is $c = 2r_- + 2r_+$. However, as we saw in Example 10.1, the relationship between the length a of the edge of a cube and the body diagonal c is

$$c = \sqrt{3}\, a$$

Thus for limiting eight coordination we have

$$2r_- + 2r_+ = 2\sqrt{3}\, r_-$$

Therefore

$$\frac{r_+}{r_-} = 0.732$$

(b) For six coordination we have the structure shown in Figure 10.22 (sodium chloride structure). The limiting case occurs when the cation at the center of a face touches the four surrounding anions and these anions touch each other. Thus the cations form a square (length of edge is $2r_-$) in which the length of the diagonal is $2r_+ + 2r_-$, which from the Pythagorean theorem is also $2\sqrt{2}r_-$ (see Example 10.1):

$$2r_+ + 2r_- = 2\sqrt{2}r_-$$

Therefore

$$\frac{r_+}{r_-} = 0.414$$

(c) We saw in Chapter 1 that a tetrahedron can be constructed by placing two pairs of points at opposite ends of the diagonals of the opposite faces of a cube. Thus two anions at opposite ends of a face diagonal will just touch each other and a cation situated at the center of the cube. So we have

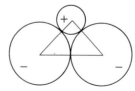

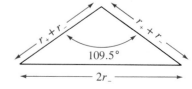

$$\sin \frac{109.5°}{2} = \frac{r_-}{r_+ + r_-} = 0.817 \quad \text{or} \quad \frac{r_+}{r_-} = 0.224$$

These radius ratios are useful in enabling us to understand the structures of ionic crystals, but to use them, we need to know the radii of ions. The distances between the centers of adjacent ions in an ionic crystal can be measured by X ray diffraction with considerable accuracy. Values of these interionic distances for the alkali halides are given in Table 10.4. These distances vary in the expected manner; that is, we expect the halide ions to increase in size from F^- to I^- and the alkali metal ions to increase in size from Li^+ to Cs^+. But unfortunately, there is no completely reliable way of dividing up these distances to give values characteristic of the individual ions.

Table 10.4 Interionic Distances (pm) in Alkali Halide Crystals

	Li^+	Na^+	K^+	Rb^+	Cs^+
F^-	201	231	266	282	300
Cl^-	257	281	314	327	356
Br^-	275	298	329	343	371
I^-	300	323	353	366	395

Chemists, physicists, and geologists have all suggested different ways of dividing up interionic distances to obtain the radii of ions. A recent set of values based on the analysis of a very large number of crystal structures is given in Table 10.5. The values in this table—or, indeed, any other table of ionic radii—must be regarded as approximate since they vary somewhat with the method by which they are determined. Such a set of ionic radii assumes that an ion always has exactly the same size, in other words, that each ion behaves as a hard sphere. But ions, like neutral atoms, are somewhat compressible. They may be compressed to different extents in different crystals, which means that the radius of an ion may vary somewhat from one crystal to another.

Nevertheless, the values in Table 10.5 are reasonably accurate for most crystals containing these ions. As we saw in Chapter 5, both cations and anions increase in size down any group of the periodic table. Comparison of the ionic

Table 10.5 Ionic Radii (pm)

Li^+	Be^{2+}	Al^{3+}	Ti^{2+}	V^{2+}	Cr^{2+}	Mn^{2+}	Fe^{2+}	Co^{2+}	Ni^{2+}	Cu^{2+}	Zn^{2+}	O^{2-}	F^-
74	35											140	135
Na^+	Mg^{2+}	Al^{3+}										S^{2-}	Cl^-
102	72	53										185	180
K^+	Ca^{2+}	Sc^{3+}	Ti^{2+}	V^{2+}	Cr^{2+}	Mn^{2+}	Fe^{2+}	Co^{2+}	Ni^{2+}	Cu^{2+}	Zn^{2+}	Se^{2-}	Br^-
138	100	73	86	79	82	82	77	74	70	73	75	195	195
Rb^+	Sr^{2+}											Te^{2-}	I^-
149	116											220	215
Cs^+	Ba^{2+}												
170	136												

radii in Table 10.5 with the atomic radii in Figure 4.7 shows that anions are, in general, larger than the corresponding neutral atoms, while cations are, in general, smaller than the corresponding neutral atoms.

From the ionic radii in Table 10.5 and the radius ratios for different coordination numbers given in Table 10.3, we might expect to be able to predict the coordination number for any given pair of ions and therefore to make a prediction about the crystal structure. Some examples of such predictions for 1:1 structures are given in Table 10.6.

Although the ionic radius ratios are consistent with the coordination numbers of the ions in many crystal structures, there are also a large number of exceptions, only a few of which are shown in Table 10.6. Cesium chloride, CsBr, and CsI have radius ratios consistent with the eight coordination observed in their structures, and many of the other alkali halides that have the NaCl structure have radius ratios consistent with six coordination. However, others have the NaCl structure, even though their radius ratios are too large, as in the case of RbCl, or too small, as in the case of LiI. Although ZnS has a radius ratio consistent with the four coordination found in its structure, BeS also has this structure, despite the very small value of its radius ratio.

These exceptions to predictions made from the radius ratios are not very surprising in view of the somewhat approximate nature of the ionic radii. These exceptions also lead us to suspect that there are other factors that influence the structures of ionic crystals. For example, we know that many "ionic crystals" have considerable covalent character in their bonding. The structures of such substances are influenced by the geometry of the covalent bonds, as is certainly the case for ZnS and BeS, where each atom forms a set of four tetrahedral AX_4 bonds.

Table 10.6 Radius Ratios and 1:1 Structures

r_+/r_-	COORDINATION NUMBER	STRUCTURE	EXAMPLES			
			COMPOUND	r_+/r_-	COMPOUND	r_+/r_-
0.225–0.414	4	Sphalerite	ZnS	0.41	BeS	0.19
0.414–0.732	6	Sodium chloride	LiCl	0.41	LiI	0.34
			MgO	0.51	LiBr	0.38
			NaBr	0.52	KCl	0.77
			CaS	0.54	RbCl	0.83
			FeO	0.55	BaO	0.97
			NaCl	0.57		
			MnO	0.59		
			KBr	0.71		
			CaO	0.71		
0.732–1.00	8	Cesium chloride	CsI	0.79		
			CsBr	0.87		
			CsCl	0.94		

IMPORTANT TERMS

Amorphous solids have a random, disordered arrangement of atoms rather than the regular periodic arrangement characteristic of crystalline solids.

A **body-centered cubic cell** has a lattice point at each corner of a cube and one in the center of the cube, giving a total of two points.

The **coordination number** of an ion is the number of ions of opposite sign that are in contact with it in the crystal structure.

Crystalline solids have a regular periodic arrangement of their atoms or ions. They form crystals with flat faces and definite characteristic angles between the faces.

A **face-centered cubic cell** has one lattice point at each corner of a cube and one at the center of each face, giving a total of four points.

Molecular solids consist of individual molecules held together by relatively weak intermolecular forces.

A **net**, or **two-dimensional lattice**, is a regular arrangement of points in a plane.

Network solids have a continuous network of atoms or ions in which each atom or ion is bound strongly to its neighbors, and the regular network extends indefinitely throughout the crystal.

A **primitive cubic unit cell** has lattice points at each corner of a cube. It has one point associated with it.

The **radius ratio** is the ratio of the radius of the cation to the radius of the anion in an ionic crystal.

A **space lattice** is a regular, three-dimensional arrangement of points in space.

The **unit cell** of a two-dimensional lattice is obtained by choosing four points of the lattice and joining them to give a parallelogram. The unit cell of a three-dimensional lattice is obtained by joining eight points of the lattice to give a parallelepiped. Many different unit cells are possible. The cell usually chosen is the smallest that shows the full symmetry of the lattice. It is possible, by repeatedly moving the cell in the directions of its edges by a distance equal to the cell edge, to generate the complete lattice.

PROBLEMS

Diamond and Graphite

1. Describe the structure of diamond and the nature of the bonding.

2. Describe the structure of graphite and the nature of the bonding.

3. Account for the physical properties of diamond and graphite in terms of their structures.

Types of Solids

4. Which of the following solids are molecular solids and which are network solids?

C S_8 CO_2 SiO_2 P_4O_6 NaCl   MgO   Al

5. Explain why molecular solids usually melt below 400°C, whereas network solids often have much higher melting points.

6. (a) What type of force holds the molecules together in orthorhombic sulfur?

(b) What type of bond holds the atoms together in black phosphorus?

(c) What type of bond holds the ions together in magnesium oxide?

(d) What type of bond holds the atoms together in silicon dioxide?

7. Explain why polymeric sulfur trioxide, $(SO_3)_n$, forms needlelike crystals. Explain why plastic sulfur is a soft rubbery solid.

8. What is an amorphous solid? Give an example. Why does an amorphous solid not have a sharp melting point?

9. Explain in terms of their structures why each of the following pairs of substances have very different chemical and physical properties:

(a) Silica and carbon dioxide

(b) Oxygen and sulfur

10. Explain what is meant by a two-dimensional network solid. Give an example.

Lattices

11. What is a space lattice?

12. What is a unit cell?

13. Describe the three cubic unit cells, and draw a diagram of each.

14. How many lattice points are associated with each of the following?

(a) A primitive cubic unit cell

(b) A body-centered cubic unit cell

(c) A face-centered cubic unit cell

Structures of Solids

15. What are the three common structures of metals? In each case give the corresponding space lattice and the motif.

16. Describe the structures of each of the following: copper, sodium, and diamond.

17. Describe the structures of each of the following ionic compounds: KCl, BaO, and CuCl.

18. Explain the difference between a cubic close-packed structure and a hexagonal close-packed structure.

19. Draw a diagram to show the relationship of the close-packed layers in a cubic close-packed structure to the unit cell.

20. Aluminum has a structure based on a cubic lattice. The length of the edge of the unit cell has been found to be 405 pm at 25°C. The density of aluminum is 2.70 g cm^{-3} at 25°C. How many atoms of aluminum are there in the unit cell? On what type of cubic lattice is the structure of aluminum based?

21. The density of platinum is 21.5 g cm^{-3}, and platinum has a cubic close-packed structure with a unit cell edge of 392 pm. Calculate the atomic mass of platinum by using this data.

22. Aluminum has a cubic close-packed structure. The length of the unit cell edge is 405 pm. Calculate the length of the diagonal of a face of the unit cell and hence find the radius of an aluminum atom.

23. The compound CuCl has the sphalerite (zinc sulfide) structure. Its density is 3.41 g cm^{-3}.

(a) What is the length of the edge of the unit cell?

(b) What is the shortest distance between a Cu^{+} and a Cl^{-} ion?

(c) The radius of the Cl^{-} ion is 180 pm. What is the radius of the Cu^{+} ion?

24. At 24 K neon has a cubic lattice with a unit cell edge of 450 pm. The density of the solid at this temperature is 1.45 g cm^{-3}. How many atoms of neon are in the unit cell? What type of cubic lattice does crystalline neon have? What is the radius of the neon atom?

25. Solid krypton has a cubic close-packed structure with a cell edge of 559 pm.

(a) Sketch the unit cell, showing the positions of the atoms.

(b) How many atoms of krypton are in the unit cell?

(c) Calculate the density of solid krypton.

(d) Find the radius of a krypton atom.

(e) Compare the radius of krypton found here with the covalent radius given in Table 4.8, and explain the difference between the two values.

26. Gold has a cubic close-packed structure and a density of 9.329 g cm^{-3}. The length of the unit cell edge is found by X ray crystallography to be 407 pm. Calculate a value for the Avogadro constant.

27. From the structure of sodium chloride, the NaCl distance of 281 pm, and the density of 2.165 g cm^{-3}, calculate a value for the Avogadro constant.

28. Copper has a density of 8.930 cm^{-3}. The edge of the unit cell has a length of 361.5 pm. Calculate a value for the Avogadro constant.

29. Calcium fluoride, CaF$_2$, has a cubic lattice. The length of an edge of the unit cell is 546.3 pm. The density of CaF$_2$ is 3.180 g cm^{-3}. How many formula units of CaF$_2$ are there per unit cell?

30. Potassium fluoride has the NaCl structure and a density of 2.481 g cm^{-3}.

(a) What is the length of the edge of the unit cell?

(b) What is the distance between a potassium ion and a neighboring fluoride ion?

31. On the basis of the diagram of the unit cell of β-cristobalite, a form of SiO$_2$, given in Figure 10.5, identify the space lattice, and describe the positions of the atoms in the unit cell.

CHAPTER 11

CARBON: INORGANIC COMPOUNDS AND HYDROCARBONS

Although living matter is composed almost entirely of water and an enormous variety of carbon compounds, carbon is only the fourteenth most abundant element. Only 0.08% of the earth's crust is carbon, and about half of this is in the form of the carbonate ion, CO_3^{2-}. The most common metal carbonates are calcium carbonate, $CaCO_3$, in its various forms such as chalk, limestone, and marble; magnesium carbonate, $MgCO_3$; and dolomite, $CaMg(CO_3)_2$. The remaining carbon is present in vegetable and animal matter, in coal and petroleum, and in the atmosphere and the oceans as carbon dioxide.

There are an enormous number of carbon compounds, and the vast majority of them are classified as **organic compounds**. At one time it was believed that many compounds of carbon could only be obtained from organic, or living, matter; hence these compounds were called organic compounds. Compounds occurring in the nonliving or mineral world were called **inorganic compounds**. Today we know that carbon compounds found in living matter can be synthesized from substances that are not found in living matter. Nevertheless, we retain the name *organic* to denote the compounds of carbon with the exception of CO_2, carbonates, and a few other compounds traditionally regarded as inorganic compounds. With a very few exceptions we may define **organic chemistry** as the chemistry of carbon compounds and **inorganic chemistry** as the chemistry of all the other elements and their compounds.

Very nearly all organic compounds contain hydrogen, and they can be regarded as being derived from the very large number of compounds that contain carbon and hydrogen only, the **hydrocarbons**, by replacing hydrogen atoms by other atoms or groups of atoms. Modern civilization is almost totally dependent on hydrocarbons, which occur in the earth's crust as *natural gas* and *petroleum*. Natural gas, gasoline, diesel fuel, domestic heating oil, and industrial

Offshore drilling allows us to reach the large deposits of oil that lie under water.

fuel oil, all of which are mixtures of hydrocarbons obtained by the distillation of petroleum, provide our major source of energy. Hydrocarbons are also the starting materials for the synthesis of a wide variety of organic compounds, ranging from drugs to plastics. We could well be said to be living in the Petroleum Age. In this chapter we consider the structures and properties of the most important types of hydrocarbons. Other organic compounds derived from the hydrocarbons are discussed in Chapter 19.

The hydrocarbons provide numerous examples of *isomers*—substances that have the same molecular formula but different structures and hence different properties—and we describe some of the more important types of isomers. We will see that the very large number of hydrocarbons and therefore the enormous number of organic compounds arise from the ability of carbon to form stable chains and rings of carbon atoms, which can be of almost any size.

Before we describe the hydrocarbons, we discuss the element carbon and the more important of its relatively small number of inorganic compounds. The division between the inorganic and organic compounds of carbon is, in any case, an artificial one. In the process of photosynthesis carbon dioxide and water are converted by plants to organic compounds. We convert organic compounds in our bodies back to carbon dioxide and water. Indeed, life itself appears to have originated from simple inorganic compounds such as H_2, H_2O, NH_3, and HCN as well as CH_4, which were present in the original atmosphere of our planet. Probably these compounds were combined under the action of lightning discharges or ultraviolet radiation to form the amino acids and other organic compounds that form the basis of life.

11.1 ALLOTROPES OF CARBON

We have already described in Chapter 10 the two most important allotropes of carbon, *diamond* and *graphite*. They illustrate in a striking way how the properties of a solid depend on its structure. There are several other forms of elemental carbon such as carbon black, charcoal, and coke that appear to be amorphous but, in fact, consist mainly of extremely small crystals of graphite.

Carbon black is a pure form of soot that is deposited when hydrocarbons are burned in a very limited supply of air. For example,

$$2C_2H_2(g) + O_2(g) \longrightarrow 4C(s) + 2H_2O(g)$$

Ethyne

Carbon black has a very intense color and is used in large quantities as a pigment for paint, paper, and printer's ink and to reinforce and color the rubber used in automobile tires.

Charcoal is made by heating wood and other organic materials to a high temperature in the absence of air. We are familiar with the fact that charcoal is much lighter than the wood from which it is made; in other words, it appears to have a very low density. The density is low because charcoal is extremely porous; it has a structure resembling a sponge but with holes that are too small to be visible to the eye. This porous structure means that it has a very large surface area relative to its volume. This large surface area can adsorb considerable quantities of other substances. Charcoal that has been thoroughly cleaned by heating with steam is known as *activated charcoal*. It has many applications, such as removing unburnt hydrocarbons from automobile exhaust, unpleasant and dangerous gases from the air, and impurities from water (see Experiment 11.1). Many municipal water treatment plants pass water through beds of activated charcoal.

Adsorption by Activated Charcoal

The flask contains bromine vapor. The dish contains charcoal.

When the charcoal is added to the flask, the color of the bromine vapor begins to diminish rapidly.

Within a few minutes no more bromine vapor can be seen because it has been completely adsorbed by the charcoal.

When charcoal is added to an aqueous solution of potassium permanganate (left), the purple $KMnO_4$ is absorbed by the charcoal and only colorless water remains (right).

Coke is made by heating coal in the absence of air. Coal is a very complex material consisting mainly of hydrocarbons and other organic compounds. It contains 60%–90% C together with H, O, N, S, Al, Si, and some other elements. When coal is heated to a high temperature, it decomposes, producing a variety of gaseous and liquid products. The mixture of gases, which is largely methane and hydrogen, is known as *coal gas*. The mixture of liquid products, which includes many hydrocarbons and other organic compounds, is called *coal tar*. The solid residue contains 90%–98% C and is known as coke. Coke is used in enormous quantities in industry as a reducing agent for the production of metals, phosphorus, and other substances.

11.2 INORGANIC COMPOUNDS OF CARBON

Carbon, the first element in group IV, is a typical nonmetal, although it is much less reactive than some nonmetals such as phosphorus and the halogens. In this respect it resembles nitrogen and sulfur, which are also rather unreactive at ordinary temperatures. Carbon does, however, react with oxygen, many metal oxides, and many other elements and compounds at high temperatures. As we saw in Chapter 9, it is a very useful reducing agent.

Carbon Monoxide

When carbon is burned in a limited supply of air, carbon monoxide, CO, is obtained:

$$2C(s) + O_2(g) \longrightarrow 2CO(g)$$

Industrially, carbon monoxide mixed with hydrogen is made on a large scale by passing steam over red-hot coke:

$$C(g) + H_2O(g) \longrightarrow CO(g) + H_2(g)$$

This mixture of hydrogen and carbon monoxide is called *water gas*.

Another increasingly important method of making carbon monoxide is the high-temperature reaction of methane (natural gas) with steam in the presence of a catalyst:

$$CH_4(g) + H_2O(g) \xrightarrow{\text{Catalyst}} CO(g) + 3H_2(g)$$

When prepared in this way, the mixture of carbon monoxide and hydrogen is known as *synthesis gas*, because of its importance as the basic material from which many organic compounds are synthesized. It is also used as a fuel.

Carbon monoxide is a colorless, odorless, and tasteless gas that has only a slight solubility in water. It is very toxic because it combines with hemoglobin in the blood, thus preventing hemoglobin from carrying out its function as an oxygen carrier. It is particularly dangerous because it is odorless and therefore not easily detected.

In carbon monoxide carbon is apparently divalent, and we might therefore think that it is formed from carbon in its ground state, $1s^2 2s^2 2p^2$, in which it has two unpaired electrons rather than from the valence state, $1s^2 2s^1 2p^3$, in which it has four unpaired electrons. However, the CO bond in carbon monoxide has a length of 113 pm and is much shorter than the CO double bond in methanal, H_2CO (121 pm), which is in turn shorter than the CO single bond in methanol, CH_3OH (143 pm). Hence carbon monoxide is best represented by the Lewis structure

$$:\overset{\ominus}{C}\equiv\overset{\oplus}{O}:$$

in which there is a triple bond and in which both atoms have an octet of electrons, rather than as $:C=\overset{..}{O}:$, in which carbon has only six electrons in its valence shell. Because oxygen is much more electronegative than carbon, the electrons of the triple bond are not shared equally between the oxygen and the carbon atoms; the oxygen atom has a considerably greater share than the carbon atom. Thus the actual charges on the atoms in the molecule are much smaller than the formal charges in the Lewis structure.

Carbon monoxide is a reducing agent because carbon is in the $+2$ oxidation state. Carbon monoxide is oxidized to carbon dioxide, in which carbon is in the $+4$ oxidation state. It burns in air to form carbon dioxide:

$$2CO(g) + O_2(g) \longrightarrow 2CO_2(g)$$

It reduces steam at high temperature, giving an equilibrium mixture of CO_2 and H_2:

$$CO(g) + H_2O(g) \rightleftharpoons CO_2(g) + H_2(g)$$

As we saw in Chapter 9, CO reduces metal oxides to the metal. For example,

$$Fe_2O_3(s) + 3CO(g) \longrightarrow 2Fe(s) + 3CO_2(g)$$

At about 400°C carbon monoxide reacts with chlorine in the presence of charcoal, which acts as a catalyst:

$$:\overset{\ominus}{C}\equiv\overset{\oplus}{O}: + Cl_2 \longrightarrow \underset{:\overset{..}{Cl}}{\overset{:\overset{..}{Cl}}{}}\!\!\diagdown\!\!\!\diagup C=\overset{..}{O}:$$

In this reaction the triple bond is converted to a double bond. The product oxocarbon dichloride, which has a planar AX_3 structure, is commonly called carbonyl chloride or phosgene. It is a very poisonous gas that has been used in warfare.

Carbon monoxide adds hydrogen at high temperature in the presence of suitable catalysts to give methanal (formaldehyde) and methanol, in which the triple bond is converted first to a double bond and then to a single bond:

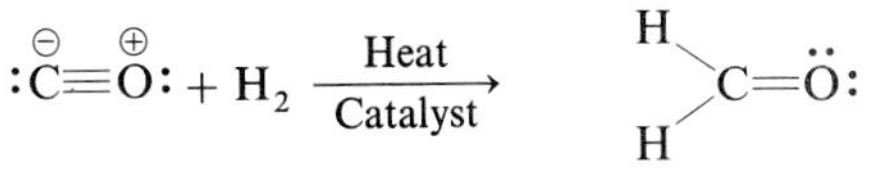

$$:C{\equiv}\overset{\oplus}{O}: + H_2 \xrightarrow[\text{Catalyst}]{\text{Heat}} \underset{H}{\overset{H}{>}}C{=}\overset{..}{O}:$$

Methanal (formaldehyde)

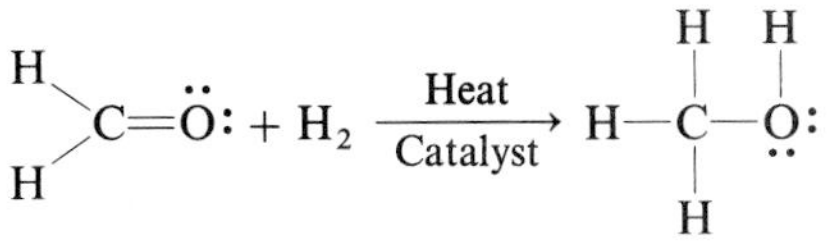

$$\underset{H}{\overset{H}{>}}C{=}\overset{..}{O}: + H_2 \xrightarrow[\text{Catalyst}]{\text{Heat}} H{-}\overset{\overset{H}{|}}{\underset{\underset{H}{|}}{C}}{-}\overset{..}{\underset{..}{O}}:$$

Methanol

These are important industrial reactions for the preparation of these substances.

Carbon Dioxide

The combustion of carbon and carbon-containing compounds in excess oxygen leads to the formation of carbon dioxide, CO_2. It is also produced by heating many metal carbonates such as calcium carbonate:

$$CaCO_3(s) \longrightarrow CaO(s) + CO_2(g)$$

Beer and wine making also produce carbon dioxide as a by-product because the fermentation of a sugar such as glucose gives ethanol and carbon dioxide:

$$C_6H_{12}O_6(aq) \longrightarrow 2C_2H_5OH(aq) + 2CO_2(g)$$

Preparation and Properties of Carbon Dioxide

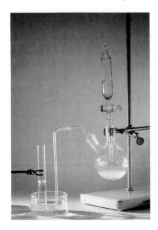

Carbon dioxide is generated by the reaction of hydrochloric acid with solid sodium carbonate.

Carbon dioxide is prepared by the reaction of hydrochloric acid with solid calcium carbonate. The gas is collected by bubbling it into an inverted tube filled with water that is standing in a dish of water. The gas rises to the top of the tube, displacing the water into the dish.

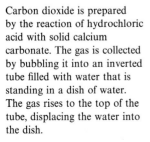

Because carbon dioxide is heavier than air, it pours from the tube. Because carbon dioxide does not support combustion, it extinguishes the candle flame.

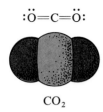

$$:\ddot{O}=C=\ddot{O}:$$

CO_2

Solid CO_2

In the laboratory we can prepare small amounts of carbon dioxide by adding a dilute aqueous acid to a metal carbonate (Experiment 11.2):

$$CaCO_3(s) + 2H^+(aq) \longrightarrow Ca^{2+}(aq) + CO_2(g) + H_2O(l)$$

Carbon dioxide is a linear AX_2 molecule with two double bonds. It is a colorless gas with a very slight odor. When cooled, it forms a white solid (dry ice) that sublimes at $-78°C$ at 1 atm pressure. Liquid carbon dioxide can be obtained only under pressure.

Carbon dioxide dissolves readily in water to give a solution that contains a small equilibrium amount of carbonic acid, H_2CO_3:

$$CO_2(aq) + H_2O \rightleftharpoons H_2CO_3(aq)$$

This solution is commonly called soda water; it has a slightly acidic taste.

Carbonic acid (Experiment 11.3) is a weak diprotic acid that is partially ionized in two stages:

$$H_2CO_3 + H_2O \rightleftharpoons H_3O^+ + HCO_3^-$$
$$HCO_3^- + H_2O \rightleftharpoons H_3O^+ + CO_3^{2-}$$

Carbonic acid and its two ions, HCO_3^- and CO_3^{2-}, are represented by the following Lewis structures:

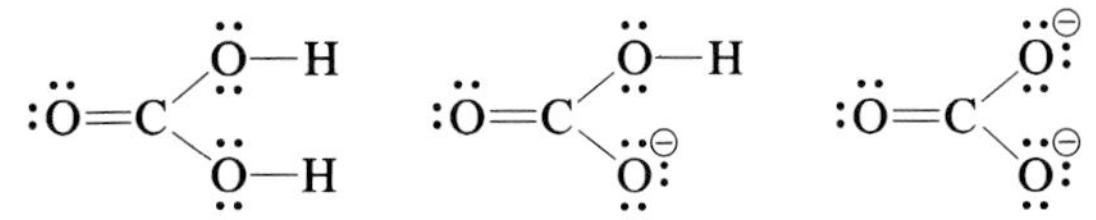

EXPERIMENT 11.3

Carbon Dioxide Behaves as an Acid in Water

A piece of solid carbon dioxide (dry ice) is dropped into water to which a small amount of bromothymol blue indicator has been added.

As the CO_2 dissolves in the water it forms carbonic acid, which changes the color of the indicator from blue to yellow. The white smoke is a mist of water droplets condensed from the air by the cold CO_2 gas.

The original blue color of the indicator can be restored by adding sodium hydroxide solution, which converts the carbonic acid to sodium carbonate.

We can write two equivalent resonance structures for the hydrogen carbonate ion:

$$HO-C{\overset{\ddot{O}:}{\underset{\ddot{\ddot{O}}:^{\ominus}}{\Bigg<}}} \quad \text{and} \quad HO-C{\overset{\ddot{O}:^{\ominus}}{\underset{\ddot{O}:}{\Bigg<}}}$$

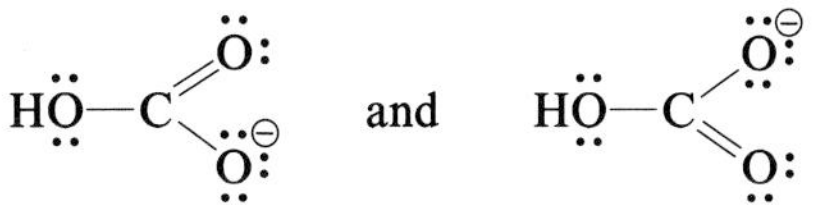

The three equivalent resonance structures for the carbonate ion were given in Chapter 8. Carbonic acid, the hydrogen carbonate ion, and the carbonate ion all have a planar AX_3 geometry at the carbon atom.

Because carbon dioxide does not burn or easily support combustion of other substances and because it has a higher density than air, it is used as a fire extinguisher. It sinks down on the fire, forming a blanket of carbon dioxide that excludes air, extinguishing the fire (see Experiment 11.2). Only a few very reactive metals such as sodium, potassium, and magnesium will burn in carbon dioxide. For example, a previously ignited piece of magnesium will burn in carbon dioxide with a spluttering flame to produce white magnesium oxide and black carbon particles (Experiment 1.3):

$$CO_2(g) + 2Mg(s) \longrightarrow 2MgO(s) + C(s)$$

In photosynthesis carbon dioxide is combined with water to form many organic compounds (see Box 11.1).

Carbon Disulfide

Carbon combines with sulfur on heating to give carbon disulfide, CS_2:

$$C + 2S \longrightarrow CS_2$$

Carbon disulfide is a linear AX_2 molecule with double bonds, just like carbon dioxide:

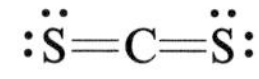

$$:\ddot{S}{=}C{=}\ddot{S}:$$

Carbon disulfide is a toxic, flammable liquid, but it is a useful solvent for sulfur and rubber.

Tetrachloromethane (Carbon Tetrachloride)

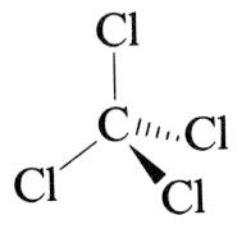

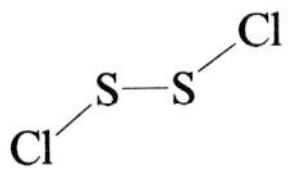

When carbon disulfide is heated with chlorine, tetrachloromethane, CCl_4, and disulfur dichloride, S_2Cl_2, are formed:

$$CS_2(g) + 3Cl_2(g) \longrightarrow CCl_4(g) + S_2Cl_2(g)$$

The two products can be separated by distillation. Other ways of preparing tetrachloromethane are described in Chapter 19. Tetrachloromethane is a useful solvent. Because it is a good solvent for oils and grease, it may be used as a cleaning agent. But it should be used with suitable precautions because it is toxic and the liquid can pass through the skin.

Hydrogen Cyanide

When heated to approximately 1000°–1200°C in the presence of a catalyst, a mixture of methane and ammonia burns in air to give hydrogen cyanide in an exothermic reaction:

$$2CH_4 + 2NH_3 + 3O_2 \longrightarrow 2HCN + 6H_2O$$

The carbon atoms on earth are continually being recycled through various compounds, producing what is called the *carbon cycle*. Although carbon dioxide constitutes only 0.05% of the earth's atmosphere by volume, it plays a vital role in this cycle and generally on earth.

There is a fine balance in nature between the processes that produce carbon dioxide and those that use it up. Plants take up carbon dioxide and use it in *photosynthesis*. In this process they use the energy provided by sunlight to convert carbon dioxide and water into carbohydrates and other organic compounds:

$$xCO_2 + yH_2O \longrightarrow C_x(H_2O)_y + xO_2$$

Animals consume these carbon-containing compounds and convert them to CO_2 in exothermic reactions with O_2; these reactions provide the energy needed to keep the body at its normal temperature and to enable the bodily functions to be carried out. The CO_2 is exhaled into the atmosphere by animals and used by plants. In addition, CO_2 in the atmosphere is in equilibrium with an enormous amount of CO_2 dissolved in the oceans and lakes. Some of this CO_2 is converted by marine animals into calcium carbonate, $CaCO_3$, the main component of their shells. The shells are eventually converted into limestone, which represents an enormous store of carbon on the earth. When the limestone weathers under the action of rain and surface water, some of the CO_2 is released into the atmosphere. Carbon dioxide is also produced by the decay of plants as well as by the respiration of animals.

Human activities are altering the balance established by nature's carbon cycle. The combustion of fossil fuels and industrial activities such as smelting add CO_2 to the atmosphere. Moreoever, the amount of CO_2 removed from the atmosphere by photosynthesis is being reduced by the continued destruction of vast forested areas, particularly the tropical jungles. As a result of these activities, the CO_2 concentration in the atmosphere has been increasing slowly in the past hundred years, and the rate of increase is accelerating.

This accumulation of CO_2 in the atmosphere may be slowly increasing earth's temperature, because CO_2 absorbs some of the energy radiated from the earth's surface and converts it to heat. This phenomenon is known as the *greenhouse effect*. Although an increase in the earth's average temperature might appear to be a benefit to the many people who live in temperate climates with cold winters, it could have very serious consequences. For example, it would lead to extensive melting of the polar ice caps, which would cause extensive flooding of coastal areas, including major cities such as New York and Washington. Although this disastrous consequence seems at first sight to be inevitable, the greenhouse effect may, in fact, be counterbalanced by other effects, such as an increase in the amount of dust in the atmosphere from industrial operations. Dust produces a cooling effect by reducing the amount of sunlight that reaches the surface of the earth.

Ammonia and methane also react in the absence of air. The reaction is endothermic and requires a temperature of 1200°–1300°C and a platinum catalyst:

$$CH_4 + NH_3 \xrightarrow[\text{Catalyst}]{1200°C} HCN + 3H_2$$

Hydrogen cyanide is a colorless liquid that boils just above room temperature (25.6°C). It has the odor and taste of bitter almonds and is highly toxic. The HCN molecule has a triple CN bond. It has a linear geometry as expected for an AX_2 molecule.

$$H—C≡N:$$

Hydrogen cyanide behaves as a weak acid in water. An aqueous solution is called *hydrocyanic acid*. Its salts are the cyanides, such as sodium cyanide, NaCN. They contain the cyanide ion, $:C≡N:^{\ominus}$, which is isoelectronic with $:\overset{\ominus}{C}≡\overset{\oplus}{O}:$ and $:N≡N:$.

Hydrogen cyanide has many important applications in the plastics industry. Sodium cyanide is used for the extraction of gold and silver from their ores.

When an alkali metal cyanide is fused with sulfur, a *thiocyanate* is formed; for example,

$$S + NaCN \longrightarrow NaSCN$$

The thiocyanate ion, SCN^-, has the Lewis structure $^{\ominus}:\overset{..}{\underset{..}{S}}—C≡N:$. The CN bond length of 115 pm is almost exactly the same as in the cyanide ion, CN^-

The Carbon Cycle in Nature. *Carbon dioxide in the atmosphere and dissolved in the ocean is converted by photosynthesis, P, into carbohydrates in plants. Oxygen is released in the process. Animals on land and in the sea use the carbohydrates for food, F. Respiration, R, and decomposition D, of animals and plants uses up oxygen and produces carbon dioxide. In the oceans carbon dioxide reacts with water to give carbonate ion, CO_3^{2-}, which combines with calcium ion present in the sea to form deposits of calcium carbonate as sediments, S. The shells of certain aquatic animals also consist mainly of calcium carbonate* *and are formed from the calcium and carbonate ions in the ocean. These shells accumulate on the ocean floor, contributing to the sediment of calcium carbonate. Other sediments are formed when dead organic matter accumulates on the bed of lakes and seas in the absence of oxygen. These accumulations ultimately lead to coal and oil deposits. Combustion, C, of coal and oil regenerates carbon dioxide. The combustion of huge amounts of coal and oil in recent times has upset the balance in the carbon cycle, and the amount of carbon dioxide in the atmosphere is slowly increasing.*

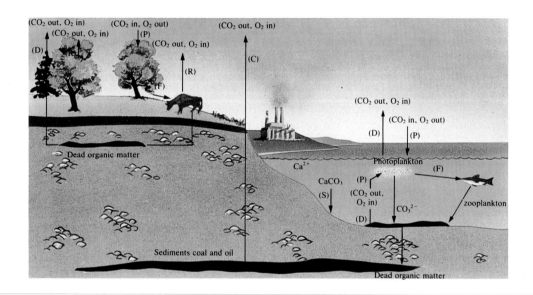

(116 pm). The thiocyanate ion may be used as a sensitive test for iron(III), since it combines with Fe^{3+} to form $Fe(SCN)^{2+}$, which has a characteristic intense red color. The analogous ion containing oxygen rather than sulfur is also known. It is called the *cyanate* ion, $^{\ominus}:\!\ddot{O}\!-\!C\!\equiv\!N:$.

Carbides

In addition to its compounds with nonmetals such as oxygen, sulfur, and nitrogen, carbon reacts with many metals. The compounds of carbon with the alkali and alkaline earth metals contain the carbide ion, C_2^{2-}; examples are Na_2C_2 and CaC_2. The carbide ion is isoelectronic with the nitrogen molecule and the cyanide ion and has a triple bond, $:\!\overset{\ominus}{C}\!\equiv\!\overset{\ominus}{C}\!:$.

Commercially, calcium carbide is made by heating lime with coke in a furnace:

$$CaO + 3C \xrightarrow{\text{Heat}} CaC_2 + CO$$

Calcium carbide has a structure similar to that of sodium chloride (see Figure 11.1). It reacts with water to give ethyne (acetylene), C_2H_2:

$$CaC_2 + 2H_2O \longrightarrow Ca(OH)_2 + C_2H_2$$

11.2 INORGANIC
COMPOUNDS OF CARBONS

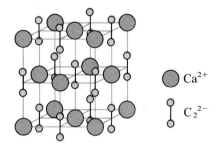

Figure 11.1 Structure of Calcium Carbide. This structure is very similar to that of sodium chloride. The lattice is face-centered, but the unit cell is not cubic; it is longer in the vertical direction because of the space needed to accommodate the linear carbide ion.

These carbides are salts of ethyne, which is too weak an acid to ionize to a significant extent in water. Therefore, the carbide ion, C_2^{2-}, is a strong base in water. It removes protons from water to give the undissociated acid C_2H_2, which bubbles off as a gas:

$$C_2^{2-}(s) + 2H_2O(l) \longrightarrow C_2H_2(g) + 2OH^-(aq)$$

There are many other metal carbides that do not contain the C_2^{2-} ion. Often they have unexpected formulas, such as iron carbide, Fe_3C, which is called *cementite* and is an important component of some steels. In many of these metal carbides the small carbon atoms occupy holes in the close-packed metal structure. We cannot easily explain the bonding and structures of these compounds by the simple bonding models used in this book; they are treated in advanced texts of inorganic chemistry.

$$H\!-\!C\!\equiv\!C\!-\!H$$

SILICON CARBIDE   With more electronegative elements such as silicon, carbon forms a covalent rather than an ionic carbide. *Silicon carbide*, SiC, is a covalent compound with the diamond structure, except that each alternate atom is a silicon atom rather than a carbon atom (see Figure 11.2). Silicon carbide (9 on the Mohs scale) is almost as hard as diamond and is used as an abrasive; for example, carbide sandpaper and grinding wheels are coated with silicon carbide.

Multiple Bonding in the Compounds of Carbon, Nitrogen, and Oxygen

As we stated in Chapter 7, carbon, nitrogen, and oxygen have a much greater tendency to form multiple bonds than any other elements. For example, the following molecules and ions all have multiple bonds:

$$:\!\overset{\ominus}{C}\!\equiv\!\overset{\oplus}{O}\!: \qquad :\!\overset{\ominus}{C}\!\equiv\!N\!: \qquad :\!N\!\equiv\!N\!: \qquad :\!\overset{\ominus}{C}\!=\!\overset{\ominus}{C}\!:$$

$$:\!\overset{..}{O}\!=\!C\!=\!\overset{..}{O}\!: \qquad :\!\overset{..}{S}\!=\!C\!=\!\overset{..}{S}\!: \qquad {}^{\ominus}\!:\!\overset{..}{\underset{..}{S}}\!-\!C\!\equiv\!N\!:$$

The heavier elements in groups IV, V, and VI, namely silicon, phosphorus, and sulfur, form very few analogous compounds. If compounds with the same formulas are known, they usually have completely different structures that do not involve multiple bonds. White phosphorus is composed of tetrahedral P_4 molecules in which the bonds are single (Chapter 7), whereas nitrogen consists of N_2 molecules. Diatomic $:\!P\!\equiv\!P\!:$ molecules are found only in phosphorus vapor at high temperatures. As we saw in Chapter 10, silicon dioxide has a covalent network structure with single bonds, whereas carbon dioxide consists of CO_2 molecules. When an element such as sulfur or phosphorus does form a multiple bond, it is almost invariably with oxygen or nitrogen, as in SO_2 and SO_3.

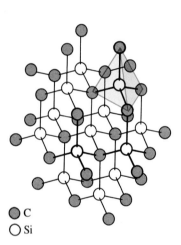

○ C
○ Si

Figure 11.2 Structure of Silicon Carbide. The structure is very similar to that of diamond; each alternate atom in the diamond structure is replaced by a silicon atom. Each carbon atom and each silicon atom has tetrahedral AX_4 geometry in this three-dimensional, covalently bonded network structure.

CHAPTER 11
CARBON: INORGANIC
COMPOUNDS AND
HYDROCARBONS

11.3 ALKANES

One large and important group of hydrocarbons is the **alkanes**. They have single bonds between the carbon atoms and the general formula C_nH_{2n+2}, where n has all integral values from 1 up to a very large number. The simplest alkanes are *methane*, CH_4, which is the major constituent of natural gas; *ethane*, C_2H_6, a minor constituent of natural gas; *propane*, C_3H_8; and *butane*, C_4H_{10}, both of which are used as fuels. Methane has $n = 1$, ethane has $n = 2$; and butane has $n = 4$. Some of the higher members of the series are important constituents of gasoline, for example, octane ($n = 8$), C_8H_{18}, and nonane ($n = 9$), C_9H_{20}.

The alkanes up to butane are gases at ordinary temperatures and pressures; the higher members are liquids and solids. The melting points and the boiling points of the alkanes increase with increasing molecular size (see Table 11.1) Alkanes with more than 16 carbon atoms are waxy solids. Paraffin wax is a mixture of solid alkanes.

Structures of Alkanes

The Lewis structures for methane, ethane, and propane are as follows:

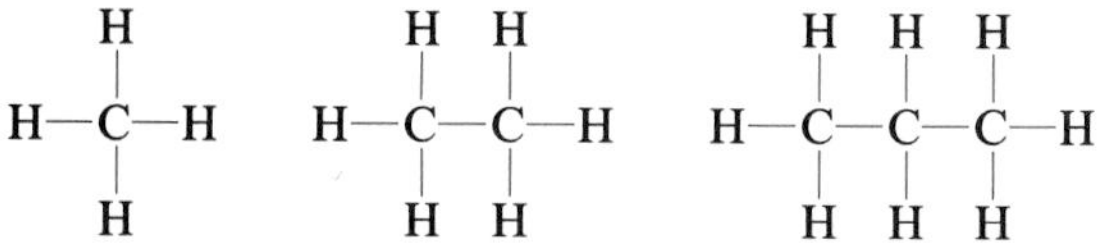

Methane	Ethane	Propane
CH_4	C_2H_6	C_3H_8

These structures are also called **structural formulas**. Frequently, structural formulas are further simplified by omitting the bond lines to the hydrogen atoms and collecting together the hydrogen atoms bonded to a given carbon atom. For example,

$$CH_4 \qquad CH_3-CH_3 \qquad CH_3-CH_2-CH_3$$

In these molecules each carbon atom forms four tetrahederally arranged bonds, some to hydrogen atoms and some to other carbon atoms (see Figure 11.3).

Table 11.1 Boiling Points and Melting Points of *n*-Alkanes, C_nH_{2n+2}

n	NAME	BOILING POINT (°C at 1 atm)	MELTING POINT (°C)	FORMULA
1	Methane	-162	-183	CH_4
2	Ethane	-89	-183	CH_3CH_3
3	Propane	-42	-188	$CH_3CH_2CH_3$
4	Butane	0	-138	$CH_3(CH_2)_2CH_3$
5	Pentane	36	-130	$CH_3(CH_2)_3CH_3$
6	Hexane	69	-95	$CH_3(CH_2)_4CH_3$
7	Heptane	98	-91	$CH_3(CH_2)_5CH_3$
8	Octane	126	-57	$CH_3(CH_2)_6CH_3$
9	Nonane	151	-54	$CH_3(CH_2)_7CH_3$
10	Decane	174	-30	$CH_3(CH_2)_8CH_3$
20	Eicosane	343	37	$CH_3(CH_2)_{18}CH_3$
30	Triacontane	446	66	$CH_3(CH_2)_{28}CH_3$

Figure 11.3 Structures of Methane, Ethane, and Propane. (a) Ball-and-stick models. (b) Space-filling models.

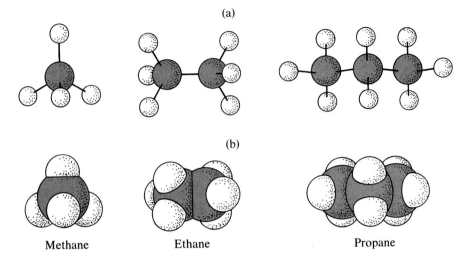

(a)

(b)

Methane   Ethane   Propane

Continuing the series, we expect the structural formulas for butane, C_4H_{10}, and pentane, C_5H_{12}, to be

$$\begin{array}{ccccccc} & H & H & H & H & & \\ | & | & | & | & | & & \\ H- & C- & C- & C- & C- & H & \\ | & | & | & | & | & & \\ & H & H & H & H & & \end{array} \quad \text{or} \quad CH_3-CH_2-CH_2-CH_3$$

$$\begin{array}{ccccccc} & H & H & H & H & H & \\ | & | & | & | & | & | & \\ H- & C- & C- & C- & C- & C- & H \\ | & | & | & | & | & | & \\ & H & H & H & H & H & \end{array} \quad \text{or} \quad CH_3-CH_2-CH_2-CH_2-CH_3$$

These alkanes are often called *straight-chain alkanes*, because all the carbon atoms are connected in one continuous chain. In fact, the carbon chains are not straight but have a zigzag shape with an angle between the C—C bonds at each carbon atom of approximately 109°, as Figure 11.4 shows.

Structural Isomers

Two different substances with slightly different properties have the molecular formula C_4H_{10}; one has a boiling point of 0°C and the other has a boiling point

Figure 11.4 Structure of Pentane.
Pentane is called a straight-chain alkane because all the carbon atoms are connected in a continuous chain. But the chain actually has a zigzag shape; the angles between the bonds at each carbon atom are approximately 109°. (a) Ball-and-stick model. (b) Space-filling model. (c) Structural formula.

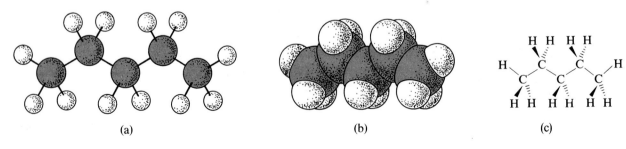

(a)   (b)   (c)

of $-10°C$. Similarly, there are three substances with the formula C_5H_{12}; they have boiling points of 10°C, 28°C, and 36°C.

There are two different butanes and three different pentanes, because carbon atoms in an alkane with four or more carbon atoms can be joined in different ways. The four carbon atoms of C_4H_{10} may be joined consecutively in one chain as we have seen. But we can draw another structure in which there is a continuous chain of only three carbon atoms; the fourth carbon is attached to the middle carbon of the chain, forming a **side chain** (Figure 11.5):

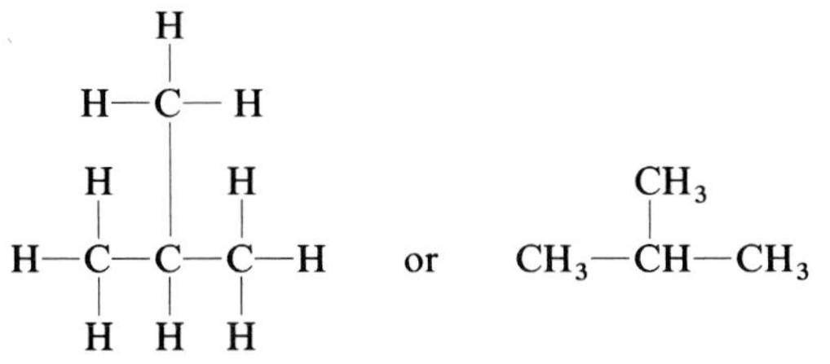

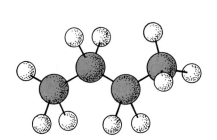

Isobutane

This is an example of a **branched-chain alkane**. It is called *isobutane*, whereas that with a single four-carbon chain is called normal butane, or *n-butane*.

Notice that the middle carbon in isobutane is attached to three other carbons, whereas all the carbons in *n*-butane are bonded at most to two others. Both butanes have the same molecular formula, C_4H_{10}, and the same molar mass; but since their structures differ, they have slightly different properties. Molecules that have the same molecular formula but different structures are called **structural isomers**. In other words, structural isomers have the same number of each kind of atom, but the atoms are arranged in different ways.

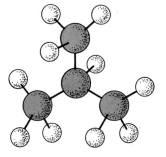

Figure 11.5 *n*-Butane and Isobutane. (a) Ball-and-stick models. (b) Space-filling models.

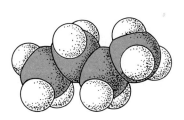

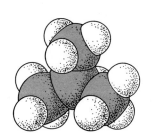

n-Butane

Isobutane

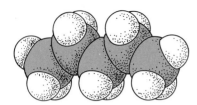

n-Pentane

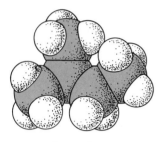

Isopentane

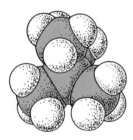

Neopentane

For pentane three unique arrangements of carbon atoms are possible. Therefore, there are three isomers with the following structures:

$$CH_3-CH_2-CH_2-CH_2-CH_3 \qquad CH_3-\overset{\overset{\displaystyle CH_3}{|}}{CH}-CH_2-CH_3 \qquad CH_3-\overset{\overset{\displaystyle CH_3}{|}}{\underset{\underset{\displaystyle CH_3}{|}}{C}}-CH_3$$

n-Pentane (bp = 36°C)   Isopentane (bp = 28°C)   Neopentane (bp = 10°C)

We might think that there is a fourth possibility:

$$CH_3-CH_2-\overset{\overset{\displaystyle CH_3}{|}}{CH}-CH_3$$

But this structure is identical with isopentane, as we can see by rotating the structure about an axis passing through the middle C—C bond,

$$CH_3-CH_2-\overset{\overset{\displaystyle CH_3}{|}}{CH}-CH_3 \longrightarrow CH_3-\overset{\overset{\displaystyle CH_3}{|}}{CH}-CH_2-CH_3$$

or by comparing models of the two structures.

Structural isomers of alkanes differ in the way in which the atoms are connected together. For example, the carbon atoms in *n*-pentane are joined in a continuous chain, whereas in isopentane the chain is not continuous but is branched. Remember that structural formulas do not give any information about the three-dimensional geometry of molecules; they merely show how the atoms are connected. Thus we can write isopentane as

$$\underset{\underset{\displaystyle CH_3}{|}}{CH_2}-\overset{\overset{\displaystyle CH_3}{|}}{CH}-CH_3 \quad \text{or} \quad CH_3-CH_2-\overset{\overset{\displaystyle CH_3}{|}}{CH}-CH_3 \quad \text{or} \quad \overset{\overset{\displaystyle CH_3}{|}}{\underset{\underset{\displaystyle CH_2-CH_3}{|}}{CH}}-CH_3$$

All three structures represent the same molecule. They do not represent different structural isomers because the atoms are all connected in the same way. It is irrelevant whether the angle between two C—C bonds is shown as 90° or as 180°, because structural formulas are not meant to give any information on bond angles.

Following the pentanes, C_5H_{12}, we have the hexanes, C_6H_{14}, the heptanes, C_7H_{16}, and so on. As the number of carbon atoms increases, the number of isomers increases rapidly. There are 3 isomeric pentanes, 5 hexanes, 9 heptanes, and 75 decanes, $C_{10}H_{22}$. The names and formulas of the first ten *n*-alkanes are given in Table 11.1, together with their boiling points and melting points.

It takes practice to be able to recognize whether two structural formulas that look different actually represent different isomers and to find the correct number of isomers of a particular alkane. To identify the isomers systematically, first find the longest continuous chain of carbon atoms, and then look for the location of the side chains along the principal chain. *n*-Pentane has a continuous chain of five carbon atoms. Isopentane has a continuous chain of four carbon atoms with one CH_3 side chain. Neopentane has a continuous chain of only three carbon atoms with two CH_3 side chains. No other arrangements of five carbon atoms are possible.

Example 11.1 What are the structures of the five isomers of hexane, C_6H_{14}?

Solution First, we write a structure with a continuous chain of six carbon atoms:

$$CH_3-CH_2-CH_2-CH_2-CH_2-CH_3$$
$$(1)$$

This structure is the only possibility for a six-carbon chain; simply bending the chain cannot give a different isomer.

Second, we consider structures containing a continuous chain of five carbon atoms. There are two such isomers:

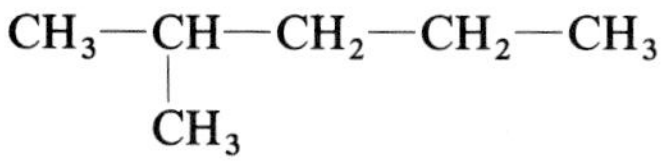

$$CH_3-\underset{\underset{CH_3}{|}}{CH}-CH_2-CH_2-CH_3 \quad \text{and} \quad CH_3-CH_2-\underset{\underset{CH_3}{|}}{CH}-CH_2-CH_3$$

$$(2) \qquad\qquad\qquad\qquad (3)$$

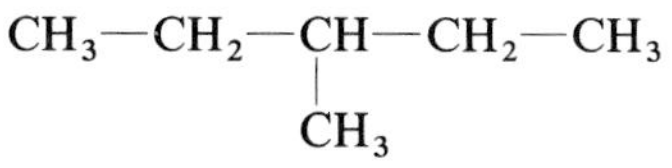

And there are no others. For example,

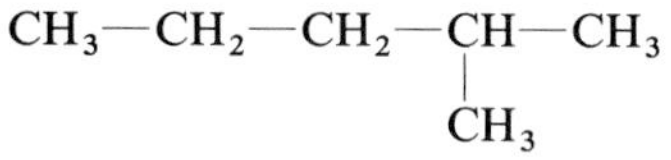

$$CH_3-CH_2-CH_2-\underset{\underset{CH_3}{|}}{CH}-CH_3$$

is the same as (2), because there is a $-CH_3$ side chain attached to the second carbon atom of a continuous chain of five C atoms.

We must also consider structures in which there is a continuous chain of four C atoms. There are two isomers of this kind:

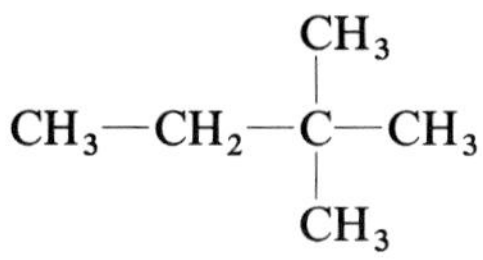

$$CH_3-CH_2-\underset{\underset{CH_3}{\overset{\overset{CH_3}{|}}{|}}}{C}-CH_3 \quad \text{and} \quad CH_3-\underset{\underset{CH_3}{|}}{CH}-\underset{\underset{CH_3}{|}}{CH}-CH_3$$

$$(4) \qquad\qquad\qquad\qquad (5)$$

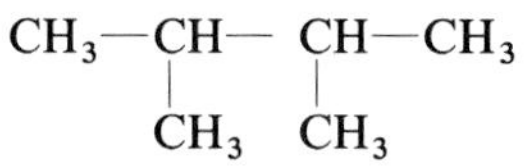

We could try writing structures with a continuous chain of only three carbon atoms, but this does not lead to any different structures. For example, in the structure

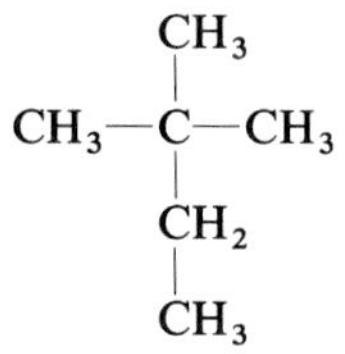

$$CH_3-\underset{\underset{CH_2}{\overset{\overset{CH_3}{|}}{|}}}{\underset{\underset{CH_3}{|}}{C}}-CH_3$$

the longest continuous chain contains four C atoms. The structure is the same as (4). We have simply drawn a structure that at first sight looks different but, in fact, is not.

Conformations

Although we know that the four bonds around each carbon atom in ethane have a tetrahedral arrangement, this does not tell us how the C—H bonds at one end of the molecule are oriented with respect to those at the other end. Indeed, it might seem that there could be two ethane molecules with rather different overall shapes, as shown in Figure 11.6. In one form of the molecule, the C—H bonds at one end of the molecule are directly opposite the C—H bonds at the other end. The C—H bonds are said to have an **eclipsed arrangement**. In the other form the C—H bonds at one end of the molecule, rather than being exactly opposite the bonds at the other end, lie exactly between them. The C—H bonds are said to have a **staggered arrangement**.

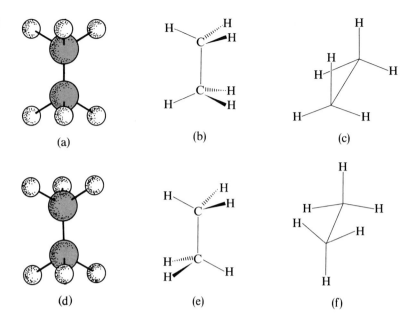

(a) (b) (c)

(d) (e) (f)

Different arrangements of atoms that can be converted into one another by rotation about single bonds are called **conformations**. Ethane can have an eclipsed conformation, a staggered conformation, and an infinite number of conformations between these two that are called **skew conformations**.

As we might expect, there are small differences in energy between the different conformations. The eclipsed conformation has a slightly higher energy and is therefore slightly less stable than the staggered conformation. But because the energy difference between the different conformations is small, even at low temperatures the conformation of an ethane molecule is continually changing. In effect, one end of the molecule is continually rotating with respect to the other end. We say that there is **free rotation** around the C—C bond. Any orientation of the C—H bonds at one end with respect to those at the other end is possible. This free rotation is easily demonstrated by using a ball-and-stick model of ethane.

Reactions of Alkanes

Alkanes are rather unreactive compounds. At ordinary temperatures they do not react with oxidizing agents such as oxygen, chlorine, bromine, or sulfuric acid. They do not react with reducing agents such as hydrogen, and they have no acid or base properties. An important reason for this lack of reactivity is that both the C—H bond and the C—C bond are strong, and a rather large amount of energy is needed to break them. Only at relatively high temperatures do the molecules have enough energy for the C—H or C—C bonds to be broken when molecules collide. Also, unlike nitrogen in ammonia, for example, the carbon atom has no unshared pairs of electrons to which a hydrogen ion or other molecules can be added—alkanes are not bases. Unlike boron and silicon, carbon has no space in its valence shell for additional electrons—alkanes are not Lewis acids. Moreover, the C—H bond has very little polarity—alkanes are not Brönsted-Lowry acids.

By far the most important reaction of alkanes is their reaction with oxygen at elevated temperatures. If oxygen is in excess, complete combustion occurs to give carbon dioxide and water; for example,

$$C_3H_8(g) + 5O_2(g) \longrightarrow 3CO_2(g) + 4H_2O(g)$$

The combustion of any alkane is a strongly exothermic reaction, and that is why alkanes are used as fuels.

Alkanes also react with the halogens at high temperatures or in a photochemical reaction (see Chapter 6) to give products in which one or more of the hydrogen atoms are replaced by halogen atoms:

$$CH_4 + Cl_2 \longrightarrow CH_3Cl + HCl$$
$$\text{Chloromethane}$$

$$C_2H_6 + Cl_2 \longrightarrow C_2H_5Cl + HCl$$
$$\text{Chloroethane}$$

When alkanes are heated by themselves to a sufficiently high temperature, they decompose to a mixture of molecules including smaller alkanes, other hydrocarbons such as ethene, and hydrogen. For example,

$$C_4H_{10} \xrightarrow{\text{Heat}} CH_2{=}CH_2 + C_2H_6$$

$$C_2H_6 \xrightarrow{\text{Heat}} CH_2{=}CH_2 + H_2$$

In the petroleum industry such reactions are called **cracking reactions**, because they make small molecules out of larger ones.

Petroleum and Natural Gas

Animal and vegetable matter that has been covered with other deposits and then subjected to enormous pressure and high temperature for millions of years has been transformed into a complex mixture of alkanes and other hydrocarbons. The more volatile alkanes in this mixture constitute *natural gas*, and the liquid mixture of all the others is called *petroleum*.

Natural gas consists chiefly of methane and small amounts of ethane, propane, and the two butanes. It also contains small amounts of helium. The propane and butane can be separated by compressing and cooling the gas until the propane and the butanes are liquefied. The liquefied propane and butane is sold as bottled gas.

Petroleum is normally separated by *fractional distillation* (see Figure 11.7), not into pure hydrocarbons but into mixtures of hydrocarbons, called fractions, each of which boils over a certain limited temperature range. The particular fractions that are collected depend partly on the source of the petroleum and partly on the proposed uses of the fractions. Typical fractions are shown in Table 11.2. So that the demand for gasoline can be met, much of the kerosene and higher–boiling point fractions are decomposed in cracking reactions to form the shorter-chain alkanes of gasoline.

The industrialized world has come to rely heavily on petroleum as an energy source. The invention of the internal combustion engine (see Box 11.2) led to the replacement of coal-burning steam engines in trains and ships with the diesel engine and to the development of the private automobile. These developments have in turn led to an increasing demand for petroleum. Hydrocarbons are also valuable as the raw materials for the manufacture of plastics and many other materials that have come to be a part of modern life. Despite the discovery of new deposits of petroleum, the world's total petroleum resources are limited and will last for only a relatively short time. Estimates of this time vary from 30 years to several hundred years. One of the challenges facing humanity is to learn how to utilize, efficiently and safely, alternative energy sources, such as coal, nuclear power, and solar energy, and to find alternative raw materials for all the substances now manufactured from petroleum.

Figure 11.7 Oil Refinery Distillation Column. Petroleum is heated with superheated steam at the bottom of a tall distillation column. Most of the petroleum is vaporized; the higher–boiling point components condense at a low point in the column, and the lower–boiling point components move toward the top of the column. Fractions of different compositions are taken from the column at different heights.

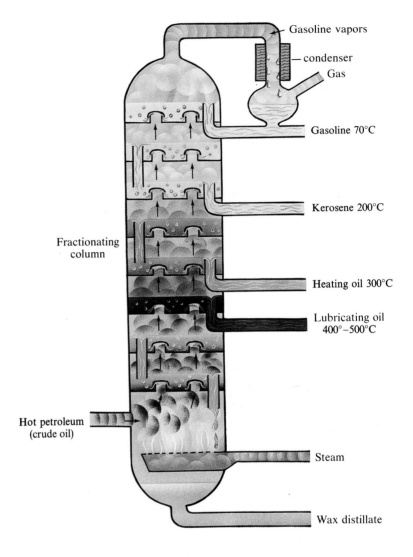

Gasoline vapors

condenser

Gas

Gasoline 70°C

Kerosene 200°C

Fractionating column

Heating oil 300°C

Lubricating oil 400°–500°C

Hot petroleum (crude oil)

Steam

Wax distillate

Table 11.2 Typical Fractions Obtained in Distillation of Petroleum

FRACTION	BOILING POINT (°C)	COMPOSITION	USES
Gas	Up to 20°	Alkanes from CH_4 to C_4H_{10}	Synthesis of other carbon compounds; fuel
Petroleum ether	20°–70°	C_5H_{12}, C_6H_{14}	Solvent; gasoline additive for cold weather
Gasoline	70°–180°	Alkanes from C_6H_{14} to $C_{10}H_{22}$	Fuel for gasoline engines
Kerosene	180°–230°	$C_{11}H_{24}$, $C_{12}H_{26}$	Jet engine fuel
Light gas oil	230°–305°	$C_{13}H_{28}$ to $C_{17}H_{36}$	Fuel for furnaces and for diesel engines
Heavy gas oil and light lubricating distillate	305°–405°,	$C_{18}H_{38}$ to $C_{25}H_{52}$	Fuel for generating stations; lubricating oil
Lubricants	405°–515°	Higher alkanes	Thick oils, greases, and waxy solids; lubricating grease; petroleum jelly
Solid residue			Pitch or asphalt for roofing and road material

Box 11.2
INTERNAL COMBUSTION ENGINE

The old-fashioned steam engine is an *external combustion engine*, because the fuel is burned outside the engine and the heat produced is used to convert water into steam. A wide variety of fuels could be used, of which coal was the most common. But steam engines are inconveniently heavy and require a driving fluid (water) that must be carried around. For all forms of transportation, from private automobiles to trains and airplanes, the steam engine has been replaced by the *internal combustion engine* in which the fuel is burned inside the engine. There are two important types of internal combustion engine: the diesel engine and the gasoline engine.

In the internal combustion engine, the hydrocarbon fuel is burned in air, yielding mainly carbon dioxide and water as products. For example, for the representative alkane nonane the principal reaction is

$$C_9H_{20} + 14O_2(g) \longrightarrow 9CO_2(g) + 10H_2O(g)$$

Some oxides of nitrogen are also formed by reactions that involve the nitrogen in the air.

In the operation of the gasoline engine a mixture of gasoline and air is sucked into a cylinder as a piston descends, thereby creating a partial vacuum by increasing the cylinder volume. The valve on the cylinder then closes and the piston returns, compressing the air-fuel mixture. An electric spark is then passed through the compressed air-fuel mixture to ignite it. Since the number of moles of gaseous products exceeds the number of moles of reactants, and since the reaction products are much hotter than the reactants because the reaction is exothermic, the gas pressure on the piston increases, forcing its descent and thus delivering power to the engine.

The burning of the fuel-air mixture in the gasoline engine must occur at a rate that delivers a smooth thrust to the descending piston. Too rapid a reaction causes a distinct explosive noise known as engine knocking or pinging, and some of the power is wasted. Most of the hydrocarbons obtained by distillation of petroleum are unbranched alkanes, which tend to explode too rapidly and cause knock. The highly branched alkane isooctane causes little knocking and is arbitrarily given an octane rating of 100:

Isooctane, C_8H_{18}

At the other end of the scale is *n*-heptane which causes considerable knocking and is given an octane rating of 0. The octane rating of any fuel is then established by determining the ratio of isooctane and *n*-heptane needed to produce the same amount of knocking. For example, a 50–50 mixture has an octane rating of 50.

The octane number of gasoline is increased by heating it and passing it over a catalyst (catalytic reforming) to convert some of the less-branched alkanes to more highly branched isomers. For example,

The octane rating of gasolines can also be increased by using various additives. In recent years the most common additive was tetraethyl lead, $Pb(C_2H_5)_4$. But because lead compounds are toxic, leaded gasoline is gradually being phased out.

A diesel engine is similar to a gasoline engine, except that the fuel-air mixture is more highly compressed so that its temperature rises sufficiently to ignite the mixture without the use of a spark. Thus unlike the more familiar gasoline engine, the diesel engine does not have any spark plugs.

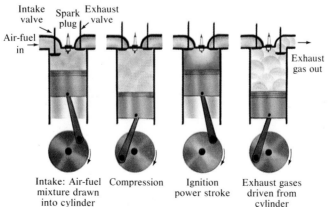

Intake: Air-fuel mixture drawn into cylinder | Compression | Ignition power stroke | Exhaust gases driven from cylinder

Four-Cycle Internal Combustion Engine.

When alkanes are strongly heated, they decompose in cracking reactions to give hydrogen and simpler hydrocarbons. One of the most important products of these reactions is ethene (ethylene), C_2H_4:

$$C_2H_6 \longrightarrow C_2H_4 + H_2$$
$$C_4H_{10} \longrightarrow 2C_2H_4 + H_2$$

Ethene is the first member of a series of hydrocarbons with the general formula C_nH_{2n}; these hydrocarbons contain one carbon-carbon double bond and are known as the **alkenes**. In the same way that the ending -*ane* in alkane tells us that a compound contains all single bonds, the ending -*ene* is used to signify the presence of a carbon-carbon double bond, C=C.

Ethene

In the *ethene* molecule the two carbon atoms share two pairs of electrons, that is, they are joined by a double bond:

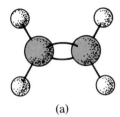

Figure 11.8 shows two models of the ethene molecule. Since the four bonds around carbon have a tetrahedral orientation, the two carbon-carbon bonds must be bent to join the two carbon atoms. The molecule has a rigid structure, and all six atoms lie in the same plane. Each carbon atom may be described as having an AX_3 geometry; therefore there is a planar triangular arrangement of the three attached atoms around each C atom (see Chapter 8, Figure 8.13). Because the two electron pairs of the double bond together take up more space than the C—H bond pairs, we expect the CCH bond angle to be slightly larger than 120° and the HCH angle to be correspondingly smaller. In fact, the HCH bond angle is 116° and the CCH angle is 122°.

In contrast to alkanes, which are rather unreactive, alkenes such as ethene react with a variety of substances. For example, in the presence of suitable catalysts, such as finely divided nickel or platinum, hydrogen adds to the carbon-carbon double bond, converting the double bond into a single bond and yielding ethane:

$$H_2C{=}CH_2 + H_2 \xrightarrow{\text{Catalyst}} H_3C{-}CH_3$$

Ethene   Ethane

Other reagents such as the halogens and the hydrogen halides also add to a carbon-carbon double bond. For example,

$$H_2C{=}CH_2 + Br_2 \longrightarrow \begin{array}{c} CH_2{-}CH_2 \\ | \quad\quad | \\ Br \quad\, Br \end{array}$$

Dibromoethane

$$H_2C{=}CH_2 + HCl \longrightarrow \begin{array}{c} CH_2{-}CH_2 \\ | \quad\quad | \\ H \quad\, Cl \end{array}$$

Chloroethane

Alkenes, which contain less hydrogen than the corresponding alkanes, are sometimes called **unsaturated hydrocarbons**, while the alkanes are called **saturated hydrocarbons**. Various molecules such as H_2, Br_2, and HCl can be added

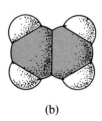

Figure 11.8 Structure of Ethene. (a) A ball-and-stick model. (b) A space-filling model. Both models show that the molecule is planar.

to unsaturated hydrocarbons to give compounds derived from saturated hydrocarbons, that is, compounds that do not have a double bond.

Like all hydrocarbons, ethene burns in air or oxygen:

$$C_2H_4 + 3O_2 \longrightarrow 2CO_2 + 2H_2O$$

In the laboratory we can prepare ethene by dehydrating ethanol by heating it with concentrated sulfuric acid:

$$C_2H_5OH + H_2SO_4 \longrightarrow C_2H_4 + H_3O^+ \cdot HSO_4^-$$

In 1984 ethene was the fifth most important industrial chemical produced in the United States in terms of the amount manufactured per year. It is the starting material for the synthesis of many other important substances. Almost half the ethene produced is polymerized to give *polyethylene* (polythene):

$$\frac{n}{2} \, H_2C{=}CH_2 \xrightarrow{\text{Catalyst}} (CH_2)_n$$

Ethene   Polyethylene

In a **polymerization reaction** a large number of small molecules (*monomers*) are combined to form a single very large molecule (*polymer*). In the formation of polyethylene one of the bonds of the carbon-carbon double bond breaks, and new bonds form between the individual molecules:

$$\cdots + H_2C{=}CH_2 + H_2C{=}CH_2 + H_2C{=}CH_2 + H_2C{=}CH_2 + \cdots$$

$$\downarrow$$

$$\cdots CH_2{-}CH_2{-}CH_2{-}CH_2{-}CH_2{-}CH_2{-}CH_2{-}CH_2\cdots$$

Thousands of ethene molecules are joined together to form one molecule of polyethylene. Polymers are discussed in more detail in Chapter 23.

Other Alkenes

The second member of the alkene series is *propene*, C_3H_6. It has three carbon atoms like propane, but one of the carbon-carbon bonds is a double bond. It is also known as *propylene*:

$$\underset{H}{\overset{H}{\diagdown}}C{=}C\underset{CH_3}{\overset{H}{\diagup}} \qquad \text{or} \qquad CH_2{=}CH{-}CH_3$$

Figure 11.9 shows two models of the molecule. Propene can be regarded as a derivative of ethene, in which one of the H atoms has been replaced by a —CH_3 group. It does not matter which of the four H atoms in the planar $CH_2{=}CH_2$ molecule is replaced by a —CH_3 group; exactly the same molecule results. The chief industrial use of propene is in the production of *polypropylene*, a polymer

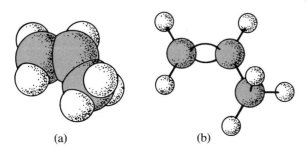

(a)   (b)

Figure 11.9 Structure of Propene.
(a) A space-filling model. (b) A ball-and-stick model.

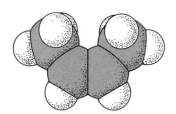

cis-2-butene

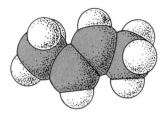

trans-2-butene

used in the manufacture of packing materials and synthetic fibers for ropes and carpets.

Butene, C_4H_8, contains four carbon atoms and one C=C bond. There are four isomers of butene:

(1)   (2)

(3)   (4)

Structures (1), (2), and (3) are structural isomers because their atoms are connected in different ways. But (3) and (4) differ in a more subtle way: They both have one —CH_3 group on each carbon atom and they differ only in the position of these groups relative to each other. These isomers exist because *there is no free rotation around a* C=C *bond*. In order for rotation to occur around a double bond, one of the two bonds in the double bond would have to be broken, which requires a considerable amount of energy.

Isomers in which all the atoms are connected in the same way and that differ only with respect to the positions of identical groups in space are known as **geometric isomers**. Structure (3) is named the **trans isomer**, and (4) is called the **cis isomer**. *Trans* means across (as in transatlantic). In *trans*-butene the CH_3 groups are across from each other, while in *cis*-butene the CH_3 groups are on the *same* side of the double bond.

There are also many alkenes containing two or more double bonds. They are called alka*dienes*, alka*trienes*, and so on. One very important compound of this type is *butadiene*, CH_2=CH—CH=CH_2 (see Figure 11.10). When buta-

Figure 11.10 Structures of Butadiene, Isoprene, and Their Polymers. Butadiene and isoprene (2-methylbutadiene) are dienes; they have two double bonds. When they are polymerized, one double bond per molecule is lost. Polybutadiene is a synthetic rubber, and polyisoprene is natural rubber.

Butadiene

Polybutadiene (a synthetic rubber)

Isoprene

Polyisoprene (natural rubber)

diene is polymerized, a synthetic rubber is produced. The product is similar in many respects to natural rubber, which is obtained from the sap of the rubber tree and is a polymer of *isoprene*. Isoprene is very similar to butadiene but has the hydrogen atom on one of the inner carbon atoms replaced by a —CH$_3$ group (see Figure 11.10).

Most of the world's rubber is now synthetic rubber. The large-scale production of synthetic rubber from butadiene began during World War II when the Japanese occupied the rubber plantations of Southeast Asia, and Britain and the United States were deprived of their principal source of natural rubber.

Pure polyisoprene and polybutadiene are sticky substances of little strength that are not very useful except as components of adhesives and cements. They are transformed into materials of greater strength and elasticity by *vulcanization*, a process in which the polymers are heated with small amounts of sulfur. Vulcanization causes the separate chains to be linked by chains of sulfur atoms. Other materials such as finely divided carbon and various pigments are often added, as in the manufacture of automobile tires.

11.5 ALKYNES

Hydrocarbons that contain a carbon-carbon triple bond are called **alkynes**. They have the general formula C$_n$H$_{2n-2}$. The simplest alkyne is *ethyne*, C$_2$H$_2$, which is commonly called *acetylene*.

As Figure 11.11 shows, ethyne is a linear molecule in which both carbon atoms have an AX$_2$ geometry. It is an important industrial chemical that is being made in increasing amounts by the high-temperature thermal decomposition (cracking) of ethane:

$$C_2H_6 \xrightarrow{\text{Heat}} C_2H_2 + 2H_2$$

It was formerly made by the reaction of calcium carbide with water (see Experiment 11.4):

$$CaC_2 + 2H_2O \longrightarrow Ca(OH)_2 + C_2H_2$$

Ethyne burns in excess air or oxygen with a very hot bright flame:

$$2C_2H_2 + 5O_2 \longrightarrow 4CO_2 + 2H_2O$$

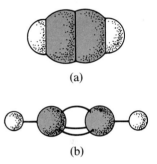

Figure 11.11 Structure of Ethyne. Ethyne is a linear AX$_2$ molecule. (a) Space-filling model. (b) Ball-and-stick model.

EXPERIMENT 11.4

Ethyne

Ethyne is being prepared by the reaction of calcium carbide, CaC$_2$, with water. It burns with a smoky yellow flame, depositing soot (carbon) on the surface of the evaporating dish.

In a limited supply of air or oxygen it gives a smoky flame because of the formation of carbon (Experiment 11.4).

One early form of portable lighting was the acetylene lamp in which a constant supply of ethyne was produced by water slowly dripping onto calcium carbide. A major use for ethyne is in the oxyacetylene torch for welding and cutting metals.

Ethyne is a rather reactive hydrocarbon, and it is a very important raw material for making other substances. Like ethene, ethyne readily adds the halogens, hydrogen halides, and hydrogen to the triple bond. These reactions occur in two stages. In the first stage one molecule of reactant is added to the triple bond to give a molecule with a double bond. Addition of a second molecule of reactant gives a compound containing a carbon-carbon single bond. For example,

$$H-C\equiv C-H \xrightarrow{Cl_2} CHCl=CHCl \xrightarrow{Cl_2} CHCl_2-CHCl_2$$
$$\text{Dichloroethene} \qquad\qquad \text{Tetrachloroethane}$$

$$H-C\equiv C-H \xrightarrow{HCl} CH_2=CHCl \xrightarrow{HCl} CH_3-CHCl_2 \text{ and } CH_2Cl-CH_2Cl$$
$$\text{Chloroethene} \qquad\qquad \text{Dichloroethane}$$

Ethyne and other alkynes, like the alkenes, are described as unsaturated hydrocarbons.

11.6 ADDITION AND ELIMINATION REACTIONS

Hydrocarbons provide us with simple examples of two important types of reactions: addition and elimination reactions. An **addition reaction** is a reaction in which two molecules combine to give a third molecule. Two important types are additions to a multiple bond and additions to an atom that can accept more electrons in its valence shell.

The addition reactions of the alkenes and the alkynes are typical examples of addition to a multiple bond:

$$H_2C=CH_2 + Cl_2 \longrightarrow CH_2Cl-CH_2Cl$$
$$HC\equiv CH + Br_2 \longrightarrow CHBr=CHBr$$

Similarly, carbon dioxide adds water to give carbonic acid, and sulfur trioxide adds water to give sulfuric acid:

$$O=S\!\!\begin{array}{c}\diagup O \\ \diagdown O\end{array} + H_2O \longrightarrow \begin{array}{c}O\diagdown \\ \diagup S \diagdown \\ O\end{array}\!\!\begin{array}{c}OH \\ \\ OH\end{array}$$

The reaction of chlorine with phosphorus trichloride to give phosphorus pentachloride illustrates an addition to an atom that can accept more electrons in its valence shell:

$$PCl_3 + Cl_2 \longrightarrow PCl_5$$

Some reactions can be classified in more than one way. For example, we recognize that this last reaction is also an oxidation-reduction reaction in which the oxidation number of the phosphorus increases from $+3$ to $+5$.

An **elimination reaction** is the opposite of an addition reaction. It involves the decomposition of a molecule into two molecules, one of which, normally the smaller one, is said to be eliminated. For example, the cracking reaction of ethane involves the elimination of a molecule of hydrogen and the formation of ethene:

$$C_2H_6 \longrightarrow C_2H_4 + H_2$$

When PCl_5 is heated, it decomposes to give PCl_3 and Cl_2. We say that chlorine is eliminated from the PCl_5:

$$PCl_5 \longrightarrow PCl_3 + Cl_2$$

The preparation of ethene from ethanol by heating with concentrated sulfuric acid involves the elimination of a molecule of water, which combines with the sulfuric acid:

$$C_2H_5OH \xrightarrow{\text{H}_2\text{SO}_4} C_2H_4 + H_2O$$

Reactions in which water is removed from a molecule are also frequently called **dehydration reactions** (see Chapter 7).

11.7 CYCLOALKANES

Another important series of hydrocarbons is the **cycloalkanes**. They have $(CH_2)_n$ chains in which the two ends are joined, forming a closed loop or ring. The simplest members of this series are the following:

	Melting point (°C)	Boiling point (°C)
Cyclopropane, C_3H_6	−127	−33
Cyclobutane, C_4H_8	−91	13
Cyclopentane, C_5H_{10}	−95	49
Cyclohexane, C_6H_{12}	7	81

An unusual feature of the structures of cyclopropane and cyclobutane is that the bond angles are unexpectedly small. If the C—C bonds are drawn as straight lines, the angles between them are only 60° and 90°, respectively. Since the angles between four single bonds on a carbon atom are always close to 109.5°, the C—C bonds must be bent, as shown in Figure 11.12.

Figure 11.12 Structures of Cycloalkanes.
So that an angle of approximately 109° is preserved between the bonds at the carbon atoms, the bonds must be bent. The amount of bending is greatest in cyclopropane, less in cyclobutane, and negligible in cyclopentane. (The internal angle of a regular pentagon is 108°, which is very close to the tetrahedral angle.)

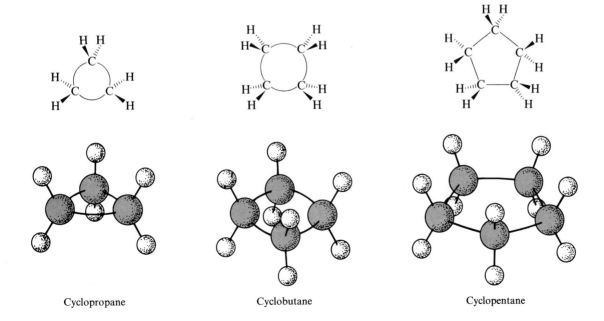

Cyclopropane   Cyclobutane   Cyclopentane

We have seen that the saturated hydrocarbons (alkanes) are in general unreactive compounds, whereas unsaturated hydrocarbons (alkenes and alkynes) react readily with reagents such as hydrogen, the halogens, and the hydrogen halides. Most of the cycloalkanes behave like typical alkanes and are relatively unreactive, but cyclopropane and cyclobutane are more reactive than the other cycloalkanes. For example, cyclopropane adds hydrogen in the presence of a catalyst to give *n*-propane, just as ethene reacts with hydrogen to give ethane:

$$\underset{H_2C-CH_2}{\overset{CH_2}{\diagup \diagdown}} + H_2 \xrightarrow{\text{Catalyst}} \underset{\textit{n}\text{-Propane}}{CH_3-CH_2-CH_3}$$

Cyclopropane also reacts with HBr or Br_2:

$$\underset{H_2C-CH_2}{\overset{CH_2}{\diagup \diagdown}} + HBr \longrightarrow \underset{\text{1-Bromopropane}}{CH_3-CH_2-CH_2Br}$$

This reaction is very similar to the reaction of propene with HBr:

$$CH_3-CH{=}CH_2 + HBr \longrightarrow \underset{\text{2-Bromopropane}}{CH_3-CHBr-CH_3}$$

Cyclobutane is not as reactive as cyclopropane. For example, unlike cyclopropane, it does not react with Br_2 or HBr at room temperature. Addition of hydrogen, which produces *n*-butane, must also be carried out at a higher temperature than for cyclopropane. Cyclopentane, cyclohexane, and the higher cycloalkanes do not add HBr, Br_2 or H_2; they have properties very similar to those of the noncyclic alkanes.

In 1855 the German chemist Johann Baeyer (1835–1917) proposed that the difference between the properties of cyclopropane and cyclobutane and the other alkanes was due to *ring strain*. He postulated that there would be strain in any cyclic molecule containing carbon atoms if the structure required that the bond angles deviate significantly from the tetrahedral angle of 109.5°. He applied this idea to both cyclopropane and to ethene to account for their reactivities toward reagents such as bromine.

Both ethene and cyclopropane have bent bonds, and the strain can be related to this bending of the bonds. When bromine is added to either molecule, the strain is removed and an unstrained molecule is formed in which there are no bent bonds. A bent bond is weaker than a normal straight bond because the maximum electron density is displaced from the straight line between the atomic cores, decreasing its effectiveness in pulling the cores together. When the bond electron density is displaced to one side, it is also more exposed to attack by another molecule, which is another reason that ethene and cyclopropane are more reactive than the alkanes.

Ring strain is greatest in ethene, which could be regarded as the first member of the cycloalkanes, and decreases progressively in cyclopropane and cyclobutane as the bonds become less bent. In cyclopentane the CCC bond angles of 108° are quite close to the tetrahedral angle (109.5°), and the molecule is virtually unstrained (see Figure 11.12).

If cyclohexane were a planar molecule, the bonds would again be strained but in the opposite sense, because the bond angles would have to be 120°, which is appreciably larger than the normal bond angle of 109°. However, normal tetrahedral bond angles can be maintained if cyclohexane adopts a nonplanar conformation (see Figure 11.13).

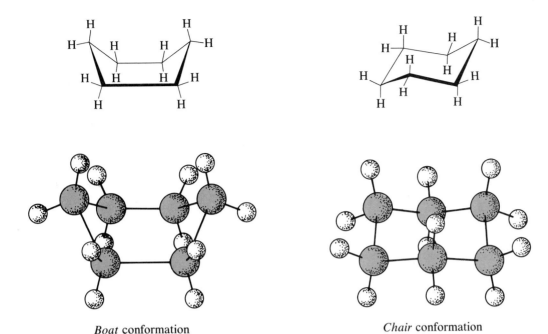

Boat conformation

Chair conformation

Cyclohexane

Figure 11.13 Structure of Cyclohexane.
Cyclohexane is not planar, because a planar arrangement would require 120° bond angles at each carbon atom. There are two nonplanar conformations of cyclohexane, called the *boat conformation* and the *chair conformation*. The bond angles at each carbon atom are approximately 109°.

11.8 NAMING OF HYDROCARBONS

The naming of individual compounds and of classes of related compounds has been an important but somewhat controversial subject during the development of chemistry. The names of compounds, particularly those discovered or first prepared many years ago, were chosen for a variety of reasons. Many names, particularly of substances that have been known for a long time, contain no structural information. For example, acetic acid, CH_3CO_2H, was named from the Latin word *acetum*, for "sour wine or vinegar." With the enormous and ever-increasing number of known compounds, particularly organic compounds, devising a systematic way of naming them has become essential so that their formulas and their structures can be deduced from their names. The most complete and useful rules were evolved during several international conferences; they are known as the *IUPAC* (International Union of Pure and Applied Chemistry) *rules*. We will show how to use these rules for naming hydrocarbons.

The naming of hydrocarbons is of particular importance because their names form the basis for naming all other organic compounds. The names and formulas of some of the alkanes are given in Table 11.1. For the first four hydrocarbons the common (nonsystematic) names are used:

CH_4 CH_3-CH_3 $CH_3-CH_2-CH_3$ $CH_3-CH_2-CH_2-CH_3$

Methane   Ethane   Propane   Butane

The higher members, beginning with pentane, are named systematically by denoting the number of carbon atoms with a numerical prefix (*pent-*, *hex-*, *hept-*, and so on) and adding the ending *-ane* to classify the compound as an alkane.

Thus we have

$$CH_3—CH_2—CH_2—CH_2—CH_3 \quad \text{and} \quad CH_3—CH_2—CH_2—CH_2—CH_2—CH_3$$

Pentane   Hexane

Alkanes are classified as *continuous-chain* (that is, unbranched) if all the carbon atoms in the chains are linked to no more than two other carbon atoms. They are called *branched-chain* if one or more carbon atoms are linked to more than two other carbon atoms. Thus

$$CH_3—CH_2—CH_2—CH_2—CH_2—CH_3$$

Continuous-chain hydrocarbon

$$\underset{\displaystyle CH_3—CH—CH—CH_3}{\overset{\displaystyle CH_3 \quad CH_3}{|\quad\quad|}} \qquad \underset{\underset{\displaystyle CH_3}{\displaystyle |}}{\overset{\overset{\displaystyle CH_3}{\displaystyle |}}{CH_3—C—CH_2—CH_3}}$$

Branched-chain hydrocarbons

The possibility of having branched-chain alkanes that are structural isomers of the continuous-chain alkanes begins with butane ($n = 4$).

The rules for the systematic naming of hydrocarbons are as follows:

1. The *longest* continuous chain of carbon atoms is taken as the parent hydrocarbon. Single carbon atoms or shorter chains of carbon atoms may then be attached to this longest chain at various places. They substitute for hydrogen atoms of the parent hydrocarbon and are therefore called **substituent groups**. Thus the following hydrocarbon is regarded as a substituted pentane, rather than as a substituted butane, because the longest continuous chain has five carbon atoms:

$$\overset{\displaystyle CH_3}{\underset{\displaystyle CH_3—CH—CH—CH_3}{\overset{\displaystyle |}{\underset{\displaystyle |\quad\quad|}{CH_3 \quad CH_2}}}}$$

2. The substituent groups attached to the main chain are named by replacing the ending *-ane* of the alkane by *-yl*. Hence the substituent groups are known as **alkyl groups**; the simplest examples are methyl, $—CH_3$, and ethyl, $—C_2H_5$.

3. The parent hydrocarbon is numbered starting from the end of the chain, and the substituent groups are assigned numbers corresponding to their position on the chain. The direction of numbering is chosen so as to give the lowest numbers to the side-chain substituents. When there are two or more substituents of the same kind, the number is denoted by the prefixes *di* (2), *tri* (3), *tetra* (4), *penta* (5), and so on. Thus the hydrocarbon illustrated above is 2,3-dimethylpentane:

$$\underset{\text{2,3-Dimethylpentane}}{\overset{\displaystyle _5CH_3}{\underset{\displaystyle _1CH_3—_2CH—_3CH—CH_3}{\overset{\displaystyle |}{\underset{\displaystyle CH_3 \quad _4CH_2 \atop |\quad\quad|}{}}}}} \quad not \quad \underset{\text{3,4-Dimethylpentane}}{\overset{\displaystyle _1CH_3}{\underset{\displaystyle _5CH_3—_4CH—_3CH—CH_3}{\overset{\displaystyle |}{\underset{\displaystyle CH_3 \quad _2CH_2 \atop |\quad\quad|}{}}}}}$$

Normally, the structure of this hydrocarbon would be written as follows:

$$\underset{\displaystyle _1CH_3—_2CH—_3CH—_4CH_2—_5CH_3}{\overset{\displaystyle CH_3 \quad CH_3}{|\quad\quad|}}$$

Example 11.2 What are the systematic names of the following alkanes?

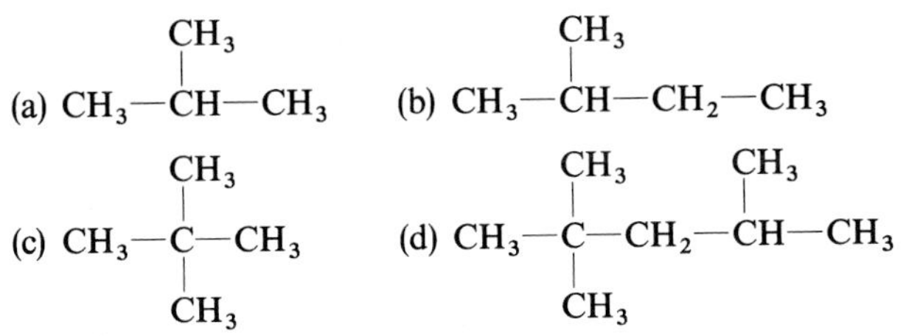

(a) CH$_3$—CH—CH$_3$ with CH$_3$ above the CH

(b) CH$_3$—CH—CH$_2$—CH$_3$ with CH$_3$ above the CH

(c) CH$_3$—C—CH$_3$ with CH$_3$ above and CH$_3$ below the C

(d) CH$_3$—C—CH$_2$—CH—CH$_3$ with CH$_3$ above and below the C, and CH$_3$ above the CH

Solution

(a)

CH$_3$
$_1$CH$_3$—$_2$CH—$_3$CH$_3$

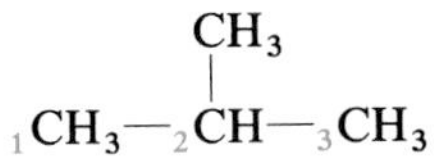

We have previously called this compound, isobutane. But according to the systematic rules, it is named as a derivative of propane, since the longest continuous chain has three carbon atoms. And since the methyl group is on the second carbon atom in the chain, it is 2-methylpropane.

(b)

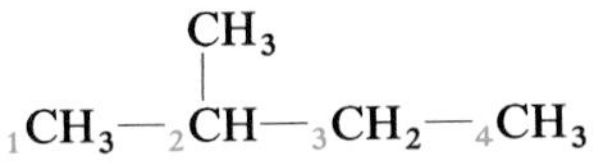

CH$_3$
$_1$CH$_3$—$_2$CH—$_3$CH$_2$—$_4$CH$_3$

This alkane is named as a derivative of butane and numbered from the left (so that the substituent —CH$_3$ group comes at carbon atom 2). It is 2-methylbutane.

CH$_3$
$_1$CH$_3$—$_2$C—$_3$CH$_3$
CH$_3$

(c)

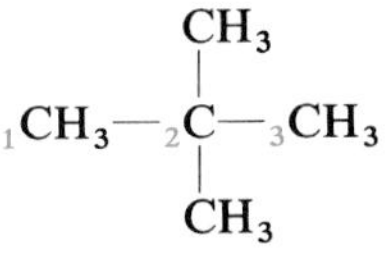

This alkane is a derivative of propane and has two methyl substituents at carbon atom 2. It is named 2,2-dimethylpropane. Note that when there is more than one of the same kind of substituent attached to the same carbon atom, we repeat the number of this carbon atom as many times as there are attached groups.

CH$_3$ CH$_3$
$_1$CH$_3$—$_2$C—$_3$CH$_2$—$_4$CH—$_5$CH$_3$
CH$_3$

(d)

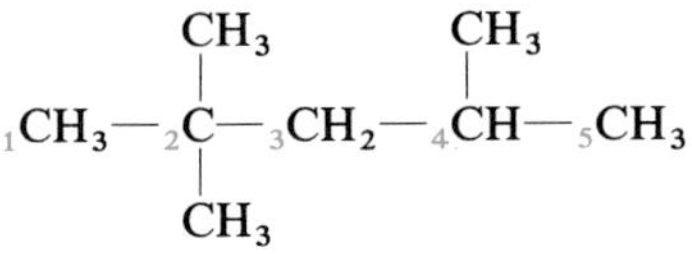

Whichever end of the chain we number from, there are substituents at carbon atoms 2 and 4, but we choose the numbering that gives the most *highly substituted* carbon—that is, the carbon with the greatest number of substituent groups—the lowest number; this is carbon atom 2. The compound is 2,2,4-trimethylpentane.

Now that we have seen how to write the names of alkanes given their structural formulas, we can readily draw the structural formulas from the systematic name.

Example 11.3 Draw the structures of the following compounds:

(a) Hexane   (b) 2-Methylpentane   (c) 3-Methylpentane
(d) 2,2-Dimethylbutane   (e) 2,3-Dimethylbutane

Solution

(a) According to systematic nomenclature, hexane has a six-carbon continuous chain:

$$CH_3—CH_2—CH_2—CH_2—CH_2—CH_3$$

(b) The longest continuous chain of 2-methylpentane contains 5 C atoms, and there is a methyl ($—CH_3$) substituent at C :

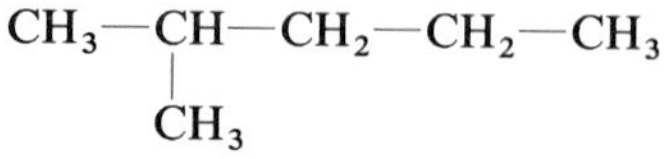

$$CH_3—\overset{\displaystyle |}{\underset{\displaystyle CH_3}{CH}}—CH_2—CH_2—CH_3$$

(c) The longest continuous chain of 3-methylpentane contains 5 C atoms, and there is a $—CH_3$ substituent at C .

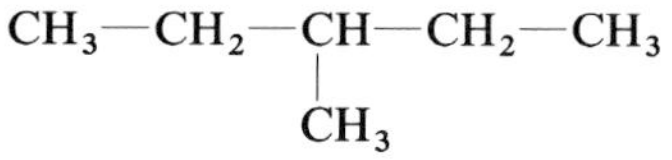

$$CH_3—CH_2—\overset{\displaystyle |}{\underset{\displaystyle CH_3}{CH}}—CH_2—CH_3$$

(d) The longest continuous chain of 2,2-dimethylbutane contains 4 C atoms, and there are two methyl substituents at C .

$$CH_3—\overset{\displaystyle CH_3}{\overset{\displaystyle |}{\underset{\displaystyle |}{\underset{\displaystyle CH_3}{C}}}}—CH_2—CH_3$$

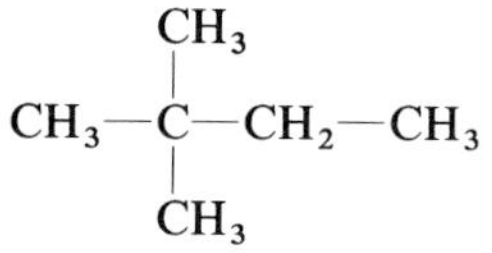

(e) The longest continuous chain of 2,3-dimethylbutane contains 4 C atoms, and there are methyl substituents at C and C .

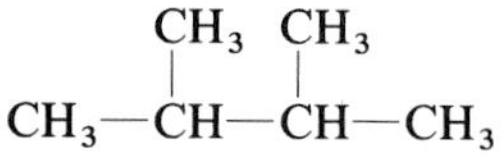

$$CH_3—\overset{\displaystyle CH_3}{\overset{\displaystyle |}{CH}}—\overset{\displaystyle CH_3}{\overset{\displaystyle |}{CH}}—CH_3$$

We have encountered all of these five compounds previously. They are the structural isomers of hexane, C_6H_{14}.

We can also use the systematic method for naming compounds to decide whether two structures are in fact different structural isomers. For example, we saw previously that the two structures

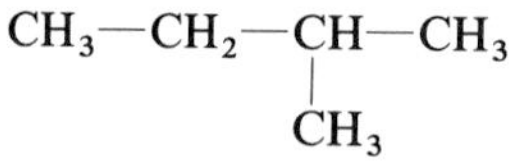

$$CH_3—CH_2—\overset{\displaystyle |}{\underset{\displaystyle CH_3}{CH}}—CH_3 \quad \text{and} \quad CH_3—\overset{\displaystyle |}{\underset{\displaystyle CH_3}{CH}}—CH_2—CH_3$$

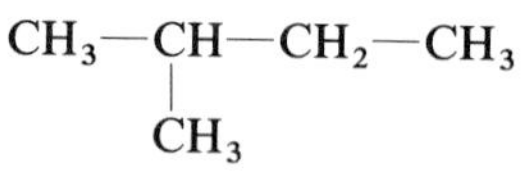

were identical by rotating one structure about an axis passing through the center of the middle C—C bond. However, if we name each of them systematically, we achieve the same result. They are the same compound because they have the same name, 2-methylbutane.

Two more rules are needed for naming hydrocarbons that contain multiple bonds.

4. Hydrocarbons containing one or more double bonds are *alkenes*. The longest continuous chain *containing a double bond* is given the name of the corresponding alkane but with the termination *-ane* changed to *-ene*. The chain is numbered so that the *first carbon atom* of the double bond is indicated by the *lowest possible number*. This number is sometimes omitted if the chain is so short that no ambiguity thereby arises; for example, $CH_2{=}CH—CH_3$, 1-propene, may be called simply propene.

	Systematic name	*Common name*

$CH_2{=}CH_2$ Ethene   Ethylene

$CH_2{=}CH{-}CH_3$ Propene   Propylene

$CH_2{=}CH{-}CH_2{-}CH_3$ 1-Butene

$CH_3{-}CH{=}CH{-}CH_3$ 2-Butene

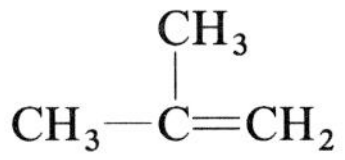

$$CH_3{-}\overset{\displaystyle CH_3}{\underset{|}{C}}{=}CH_2 \qquad \text{2-Methylpropene} \qquad \text{Isobutylene}$$

When compounds contain more than one C=C bond, a numerical prefix (*di-*, *tri-*, and so on) is used before the *-ene* suffix. For example,

$$CH_2{=}CH{-}CH{=}CH_2$$
1,3-Butadiene

Example 11.4 Name the following compounds:

(a) 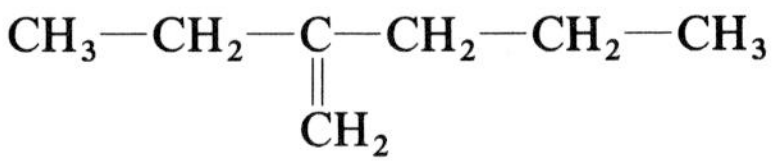 $CH_3{-}CH_2{-}\underset{\underset{\displaystyle CH_2}{\|}}{C}{-}CH_2{-}CH_2{-}CH_3$ (b) 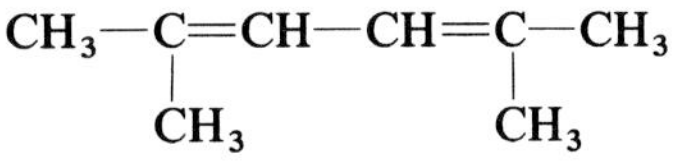 $CH_3{-}\underset{\underset{\displaystyle CH_3}{|}}{C}{=}CH{-}CH{=}\underset{\underset{\displaystyle CH_3}{|}}{C}{-}CH_3$

Solution (a) The name is 2-ethyl-1-pentene:

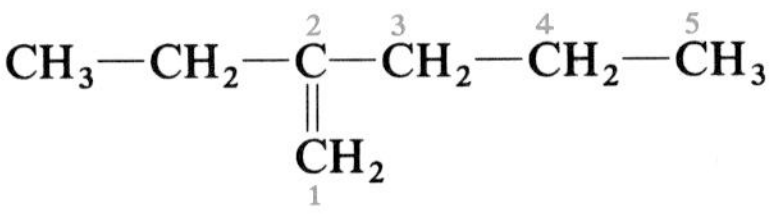

$$CH_3{-}CH_2{-}\overset{2}{\underset{\underset{\displaystyle \underset{1}{CH_2}}{\|}}{C}}{-}\overset{3}{CH_2}{-}\overset{4}{CH_2}{-}\overset{5}{CH_3}$$

(b) The name is 2,5-dimethyl-2,4-hexadiene:

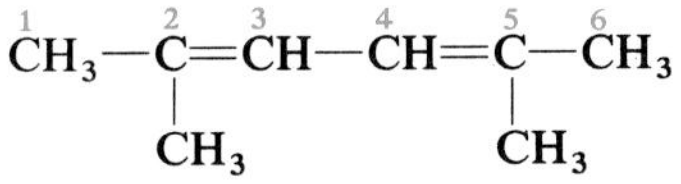

$$\overset{1}{CH_3}{-}\overset{2}{\underset{\underset{\displaystyle CH_3}{|}}{C}}{=}\overset{3}{CH}{-}\overset{4}{CH}{=}\overset{5}{\underset{\underset{\displaystyle CH_3}{|}}{C}}{-}\overset{6}{CH_3}$$

5. Hydrocarbons that contain triple bonds are *alkynes*. Their names are obtained by replacing the *-ane* ending of the corresponding alkane by *-yne*. The numbering system used for locating the triple bonds and any substituents is the same as that for the alkenes:

$$H{-}C{\equiv}C{-}H \qquad CH_3{-}C{\equiv}C{-}H \qquad CH_3{-}C{\equiv}C{-}CH_3$$

  Ethyne (acetylene)   Propyne (methylacetylene)   2-Butyne (dimethylacetylene)

A final rule is needed for cyclic alkanes.

6. Cycloalkanes are named by adding the prefix *cyclo-* to the name of the corresponding noncyclic alkane having the same number of carbon atoms as the ring:

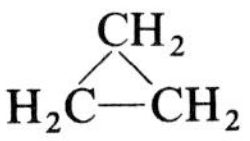

$$\underset{\text{Cyclopropane}}{\overset{\displaystyle CH_2}{H_2C{-}CH_2}} \qquad \underset{\text{Cyclobutane}}{\begin{matrix} H_2C{-}CH_2 \\ | \qquad | \\ H_2C{-}CH_2 \end{matrix}}$$

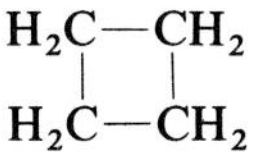

The carbon atoms are numbered around the ring so as to give the lowest possible numbers for the substituent positions. For example,

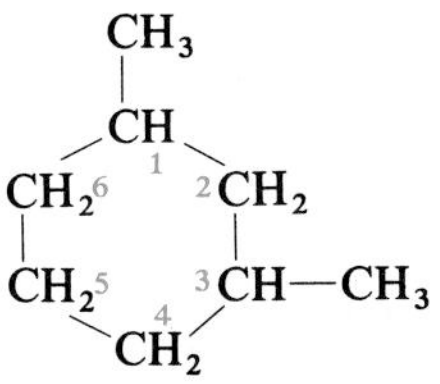

$$\begin{matrix} CH_3 \\ | \\ CH \\ \diagup \quad \diagdown \\ CH_2{}^6 \qquad {}^2CH_2 \\ | \qquad\qquad | \\ CH_2{}^5 \qquad {}^3CH{-}CH_3 \\ \diagdown \quad {}_4 \quad \diagup \\ CH_2 \end{matrix}$$
1,3-Dimethylcyclohexane

The systematic names for $CH_2{=}CH_2$ and $H{-}C{\equiv}C{-}H$ are ethene and ethyne, respectively. But the common and long established names, ethylene and acetylene, continue to be widely used. Although chemists appreciate the utility of the IUPAC rules for naming more complex hydrocarbons, like most people, they are reluctant to change familiar things. They continue to use the well-established names for many simple and well-known compounds. Although it would be logical and in some ways simpler to use only IUPAC nomenclature through this book, you would be at a serious disadvantage in reading newspapers, magazines, and other books if you did not know the nonsystematic names that are widely used. Throughout the history of chemistry old names have slowly given way to new, more systematic names. For example, hydrochloric acid was at one time known as muriatic acid, but this name is now very rarely used. We may suppose that ethene will gradually replace ethylene as the name for C_2H_4.

11.9 BENZENE AND THE ARENES

Another important class of hydrocarbons are the **arenes**, of which the simplest member is *benzene*, C_6H_6. Benzene and other arenes were first discovered among the products obtained when coal is heated to a high temperature in the absence of air. The products are *coke* and a complex mixture of liquids, called *coal* tar, that are mainly hydrocarbons, including benzene and other arenes, and a mixture of gases—mainly CH_4, H_2, and CO—called *coal gas* (see Experiment 11.5).

Coal is no longer the major source of benzene. Substantial amounts of benzene are present in petroleum from some sources, but large amounts are also made by *catalytic reforming*. In this process a mixture of hydrogen and alkanes from C_6H_{14} to $C_{10}H_{22}$ are passed over suitable catalysts heated to a high temperature. Many complex reactions occur in which alkanes are converted to cycloalkanes and cycloalkanes are converted to benzene and other arenes. For example,

$$C_6H_{14} \longrightarrow H_2 + \underset{\substack{CH_2 \\ CH_2\quad CH_2 \\ CH_2\quad CH_2 \\ CH_2}}{} \longrightarrow 4H_2 +$$

Many of the arenes and their derivatives have pleasant odors; therefore they are often called *aromatic hydrocarbons*. Many are toxic and some are carcinogenic. The volatile arenes are very flammable and burn with a yellow sooty flame. In contrast, alkanes, alkenes, and alkynes in an adequate amount of air burn with a bluish flame, leaving little carbon residue. Benzene is a colorless liquid that boils at 80°C and freezes at 5.5°C. It is less dense than water and is insoluble in water.

Structure of Benzene

Although benzene was first prepared by Michael Faraday in 1814, its structure remained a mystery until 1866. It was not easy to see how the formula C_6H_6 could be reconciled with a valence of 4 for carbon and a valence of 1 for hydrogen, and several structures were proposed. The structure that we now know to be nearly correct was suggested by Friedrich August Kekulé (1829–1896) in

Production of Coal Tar and Coal Gas

When coal is strongly heated it decomposes to give a liquid distillate, which consists of a complex mixture of organic compounds (coal tar) and a mixture of gases (coal gas)—mainly CH_4, H_2, and CO—which are burning at the exit tube.

1866. He proposed that the benzene molecule is a six-membered ring containing alternate double and single bonds:

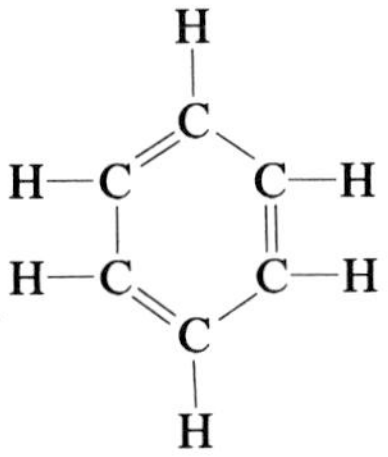

Although this structure seemed to account for the molecular formula, it was not generally accepted for a long time, because there were difficulties associated with this structure. For example, we have seen that a typical alkene adds halogens or hydrogen halides to the double bond. But although benzene apparently has three double bonds, it shows no tendency to add the halogens or the hydrogen halides (see Experiment 11.6). Another problem is that the Kekulé structure would predict that replacement of two *adjacent* hydrogen atoms of benzene by other groups would produce two structural isomers. For example,

Cyclohexane, Cyclohexene, and Benzene

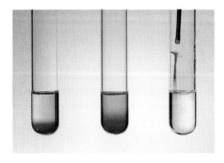

When bromine is added to cyclohexane (left) and benzene (center) it dissolves to give a red-brown solution. When bromine is added to cyclohexene (right) it reacts immediately, adding to the double bond to form dibromo-cyclohexane, which is colorless.

11.9 BENZENE AND THE ARENES

417

replacing two hydrogen atoms by methyl groups would be expected to give

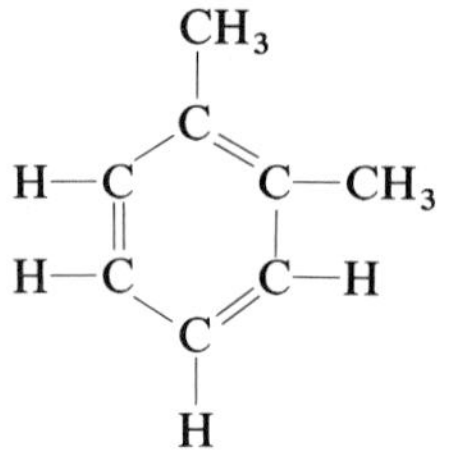

and

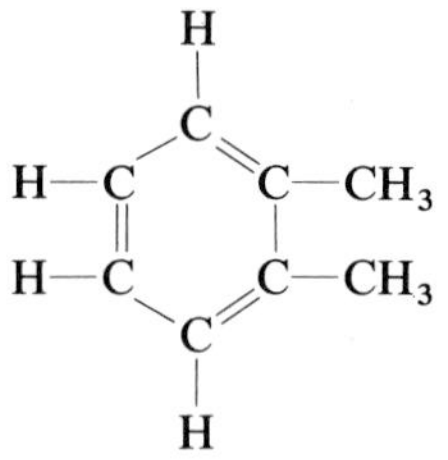

In fact, only one substance is obtained.

To overcome these problems, Kekulé suggested that the double bonds are not fixed in position but can be arranged in two ways and that one structure rapidly changes to the other and vice versa. He further suggested that this freedom of the double bonds to change position was responsible for their decreased reactivity.

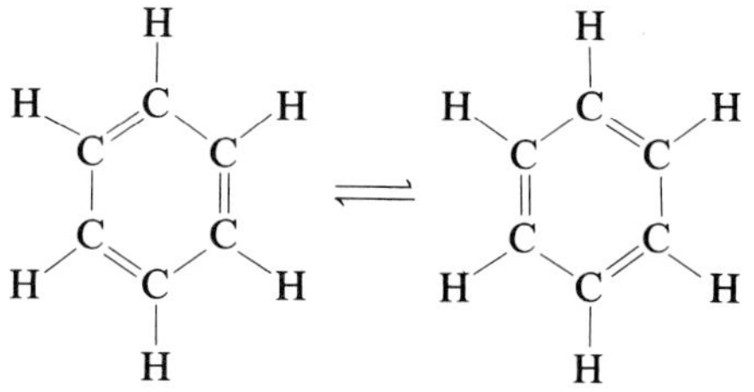

These structures are called the **Kekulé structures** for benzene. For simplicity just the C—C bonds are often shown, with the C and H atoms and the C—H bonds omitted:

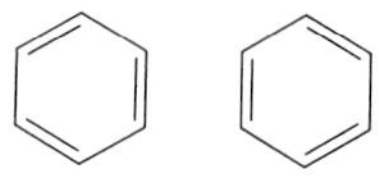

In fact, spectroscopic studies show that benzene does not have two structures that are rapidly interconverting but only one structure that is a regular planar hexagon with all six C—C bonds the same length and all the bond angles 120°. By comparison with the C—C bond lengths in ethane and ethene, we would predict that the three double bonds of a Kekulé structure should have a length of approximately 134 pm and the single bonds have a length of approximately 154 pm. In fact, all the bonds have the same length, 140 pm.

We now recognize the two Kekulé structures as two resonance structures that, taken together, describe the fact that three of the electron pairs are not as localized as the Kekulé structure would suggest. If we calculate the bond order of any of the C—C bonds as described in Chapter 8, taking the sum of the bond orders for any particular bond in the two structures ($1 + 2 = 3$),

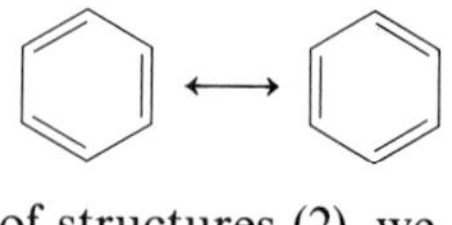

and dividing by the number of structures (2), we obtain a bond order for each bond of 1.5. The observed bond length of 140 pm is consistent with this bond order.

The three delocalized electron pairs are often depicted by a circle drawn inside the hexagon representing the six C—C bonds, which allows us to use one structural formula to represent benzene:

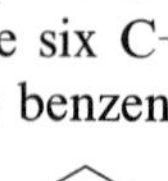

Each carbon atom forms three localized bonds with an AX_3 geometry, two to adjacent carbon atoms and one to a hydrogen atom. In addition, there are six delocalized electrons that are spread around the ring of six carbon atoms. Benzene is often represented simply by one Kekulé structure; we must then remember that the double-bond electrons are in fact delocalized around the ring.

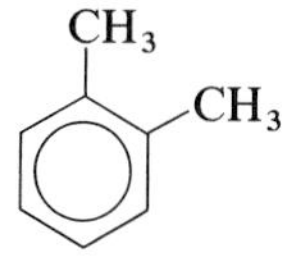

Derivatives of Benzene

Many other hydrocarbons can be derived from benzene by replacing one or more hydrogen atoms with alkyl groups. These compounds are all *arenes*. The simplest derivative is *methylbenzene*, or *toluene*:

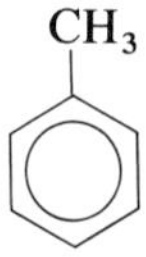

CH$_3$

Methylbenzene (toluene)
(bp = 111°C)

Substituting two methyl groups gives three *dimethylbenzenes* (*xylenes*) that are structural isomers. These three compounds differ only in the relative positions of the methyl groups (see Figure 11.14). According to the IUPAC rules of nomenclature, these isomeric molecules are named by numbering the carbon atoms of the ring so that the substituent groups have the lowest possible numbers:

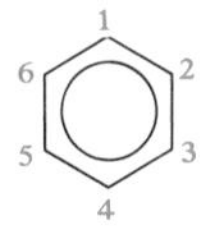

Another compound that has the same molecular formula as the dimethylbenzenes, C_8H_{10}, is ethylbenzene.

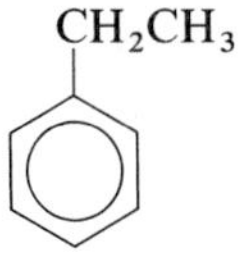

Example 11.5 How many isomers are there of trimethylbenzene, $C_6H_3(CH_3)_3$?

Solution We have seen that there are three ways of placing two substituent groups on a benzene ring. In this case we have three identical substituent groups. Only one compound results from placing them adjacent to each other, 1,2,3-trimethylbenzene. Another compound results from placing two groups adjacent to each other with the third group separated from this pair by one carbon atom, 1,2,4-trimethylbenzene. A final possibility is to have each of the —CH$_3$ groups separated from each other by one carbon atom, 1,3,5-trimethylbenzene.

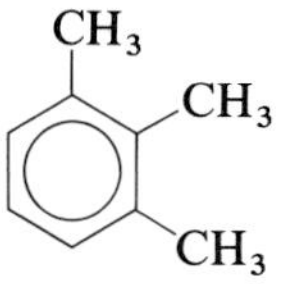

1,2,3-Trimethylbenzene 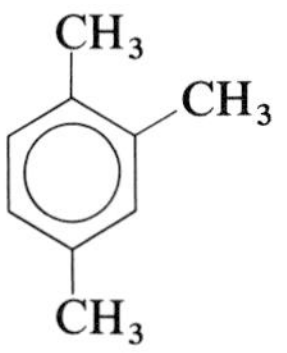   1,2,4-Trimethylbenzene 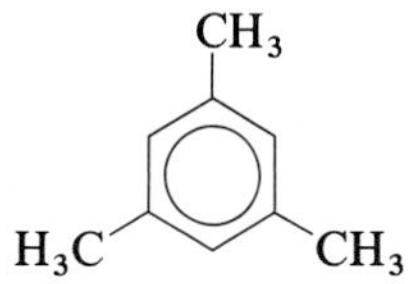   1,3,5-Trimethylbenzene

Arenes Containing Fused Rings

Naphthalene consists of two benzene rings fused together, that is, two rings sharing two carbon atoms. It has the formula $C_{10}H_8$ because the two carbon

CH$_3$
CH$_3$

1,2-Dimethylbenzene
o-Xylene

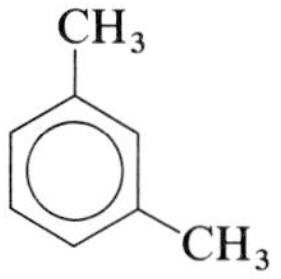

CH$_3$
CH$_3$

1,3-Dimethylbenzene
m-Xylene

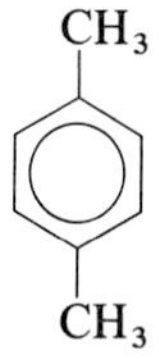

CH$_3$
CH$_3$

1,4-Dimethylbenzene
p-Xylene

Figure 11.14 The Three Isomeric Xylenes.

CH$_2$CH$_3$

Ethylbenzene

atoms at the point of fusion do not have hydrogen atoms bonded to them:

Naphthalene
(mp = 90°C)

A large number of arenes are known that contain different numbers of benzene rings fused in different ways. Typical examples include anthracene, pentacene, and coronene, which are shown in Figure 11.15. Although the vast majority of hydrocarbons are colorless, arenes that contain several fused rings are often colored. When the number of rings becomes essentially infinite, a planar infinite sheet of carbon atoms is obtained (Figure 11.15). Graphite, which is black, consists of these sheets stacked one on another (Figure 10.2).

The hydrogen atoms in hydrocarbons may be replaced by many other atoms, or groups of atoms called **functional groups**, to give a variety of organic compounds. We will discuss functional groups and some important types of organic compounds in Chapter 19.

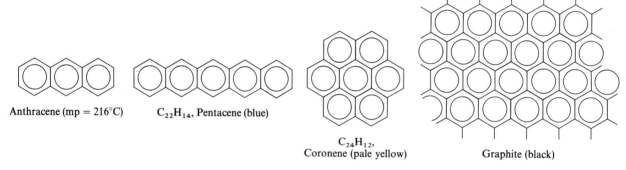

Anthracene (mp = 216°C) $C_{22}H_{14}$, Pentacene (blue)

$C_{24}H_{12}$,
Coronene (pale yellow)   Graphite (black)

Figure 11.15 Structures of Some Arenes with Fused Rings.
As the number of rings increases, the arenes become more intensely colored. Graphite may be regarded as an arene of essentially infinite size; it is black.

11.10 IDENTIFICATION AND ANALYSIS OF HYDROCARBONS

The hydrocarbons illustrate a common problem in organic chemistry, namely, that of identifying substances that have very similar compositions and very similar properties. For example, all the alkanes are rather unreactive, and except for the simplest members of the series, they have rather similar compositions. Therefore the identification of an alkane depends mainly on its physical properties, such as its boiling point, and on the determination of its molar mass. The molar mass of a volatile compound can be determined by measuring the volume occupied by a known mass of the gaseous compound at a known temperature and pressure, as we described in Chapter 3.

A modern method for determining molar mass is *mass spectrometry*. A substance in the gas state is bombarded with high-speed electrons. These electrons may knock an electron from a molecule of the substance to give a positive ion called the *parent ion* and they may also break up molecules to give smaller positive ions called *fragment ions*. The positive ions are accelerated by an electric field and passed into a mass spectrometer, where their masses can be determined by the amount by which they are deflected by a magnetic field, as described in Box 2.3. The mass of the parent ion gives the mass of the molecule from which it is derived, and the masses of the fragment ions enable the chemist to determine the nature and composition of the substance (see Figure 11.16).

Complex mixtures of very similar substances, such as hydrocarbons, present formidable problems for analysis. One useful method for volatile substances is *gas-liquid chromatography*, which we described for volatile liquids in Chapter 1 (Figure 1.23). Figure 11.17 shows the results achieved in the separation of a mixture of nine heptanes. As indicated in the figure, each isomer emerges from

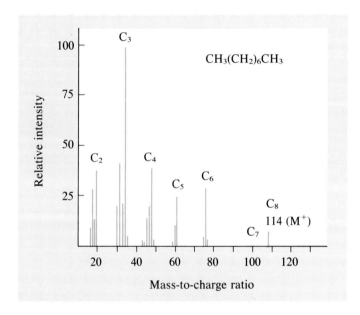

Figure 11.16 Mass Spectrum of Octane. The $C_8(M^+)$ peak is due to the parent ion. The other peaks are due to fragment ions.

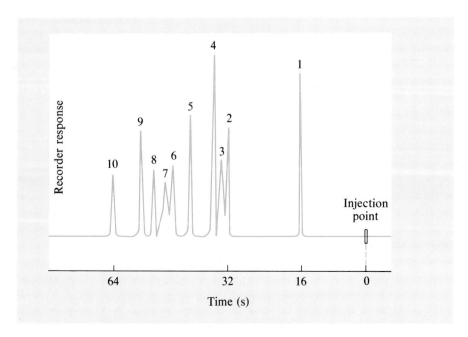

Figure 11.17 Gas Chromatogram Obtained from Mixture of Nine Heptane Isomers. Each of the isomers takes a different time to pass through the chromatographic column and is detected separately on emerging from the column. Peak 1 is a reference peak.

11.10 IDENTIFICATION AND ANALYSIS OF HYDROCARBONS

421

the long separating tube at a different time. The area under each peak corresponds to the relative amount of each substance in the mixture.

The chart in Figure 11.17 is called a **chromatogram**. The peaks in the chromatogram can be identified by comparison with a chromatogram obtained from a known mixture of substances. Alternatively, if the apparatus is large enough, samples of each substance can be collected as they emerge. They can then be separately identified by means of, for example, a mass spectrometer.

IMPORTANT TERMS

An **addition reaction** is a reaction in which two molecules combine to give a third molecule.

An **alkane** is a hydrocarbon in which all the C—C bonds are single bonds. Noncyclic alkanes have the general formula C_nH_{2n+2}. Cyclic alkanes have the general formula C_nH_{2n}.

An **alkene** is a hydrocarbon that has one or more carbon-carbon double bonds. Noncyclic alkenes with only one double bond have the general formula C_nH_{2n}.

An **alkyne** is a hydrocarbon that has one or more carbon-carbon triple bonds. Noncyclic alkynes with only one triple bond have the general formula C_nH_{2n-2}.

An **arene** is a hydrocarbon derived from benzene either by substituting its hydrogen atoms with alkyl groups or by fusing two or more benzene rings.

A **cis isomer** is a structural isomer of an alkene in which substituents at each end of the double bond are on the same side of the double bond.

Conformations are the different arrangements of the atoms and the groups in a molecule that are possible as a consequence of rotation about single bonds.

Cracking is a process in which alkanes are converted to shorter-chain alkanes and alkenes by heating.

Cycloalkanes are alkanes of general formula C_nH_{2n} in which the carbon chain forms a ring.

The **eclipsed conformation** of a molecule is one in which the atoms or groups attached to the ends of a single bond are in positions directly opposite each other.

An **elimination reaction** is the opposite of an addition reaction. It involves the decomposition of a molecule into two molecules, one of which, normally the smaller one, is said to be eliminated.

Free rotation is a property of a single bond that allows the atoms and the groups attached at one end of the bond

to rotate freely with respect to those at the other end.

Geometric isomers are isomers in which all the atoms are connected in the same way and that differ only in how identical groups are arranged in space.

Hydrocarbons are compounds containing only carbon and hydrogen.

Inorganic chemistry is the chemistry of the elements except carbon.

Inorganic compounds are compounds of all the elements except carbon.

A **Kekulé structure** is one of the resonance (Lewis) structures of benzene containing alternate carbon-carbon single and double bonds.

Organic chemistry is the chemistry of carbon and its compounds.

Organic compounds are the compounds of carbon. But there are a few compounds such as carbon dioxide that are normally regarded as inorganic compounds.

A **saturated hydrocarbon** is one in which every carbon atom is bonded to four other atoms by single bonds; an alkane or cycloalkane.

A **staggered conformation** is one in which the atoms or the groups attached to the ends of a single bond are in positions between those at the other end.

Structural isomers have the same molecular formula but they have different structures. In other words, they have the same number of different kinds of atoms but the atoms are arranged in different ways.

A **trans isomer** is an isomer of an alkene in which two substituent groups are on opposite sides of the double bond.

An **unsaturated hydrocarbon** is a hydrocarbon containing one or more double or triple bonds; an alkene or alkyne.

PROBLEMS

Inorganic Compounds of Carbon

1. Write a balanced equation for the reaction of carbon monoxide with each of the following:

(a) H_2 (b) O_2 (c) Cl_2 (d) H_2O (e) Fe_2O_3.

State the conditions under which each reaction occurs.

2. Write equations for the following:
(a) Two reactions by which CO may be prepared.
(b) Two reactions by which CO_2 may be prepared.
(c) One reaction by which CS_2 may be prepared.

* The asterisk denotes the more difficult problems.

3. Draw Lewis structures for each of the following molecules and ions:

   (a) CO_2 (b) CO   (c) CN^-

   (d) C_2^{2-} (e) HCN   (f) $COCl_2$

4. Two Lewis structures can be written for SC ⁻; only one of them is consistent with the experimental bond lengths. Draw both Lewis structures, and suggest a reason why one of these structures does not appear to be very important. (*Hint:* Consider the ease with which different atoms form multiple bonds.)

5. Two Lewis structures can be drawn for CS_2. By considering formal charges, predict which of these two structures is preferred.

6. When malonic acid, $CH_2(CO_2H)_2$, is heated with a large excess of the dehydrating agent P_4O_{10}, an oxide of carbon containing 53.0% C is obtained. This oxide boils at 6°C and freezes to a white solid at −111°C. A sample of 0.200 g of the oxide was found to occupy a volume of 74.3 mL at 27°C and 740 mm Hg pressure.

   (a) What is the molecular formula of the oxide?

   (b) Write an equation for its formation from malonic acid.

   (c) Suggest a Lewis structure for the oxide.

   (d) Would the molecule have a dipole moment?

7. When concentrated sulfuric acid is dripped onto carbon tetrabromide at 160°C, a compound containing 6.4% C, 85.0% Br, and 8.6% O is obtained. A sample of 0.940 g of the compound occupies a volume of 122 mL at 25°C and a pressure of 1.00 atm. What is the molecular formula of the compound? Draw a Lewis structure for the compound, and deduce its molecular geometry.

8. When aluminum oxide is heated with coke in an electric furnace, a yellow carbide of aluminum is obtained, which is stable up to 1400°C and reacts with water to give methane. A sample of 0.500 g of the carbide was reacted with excess water, and the methane obtained had a pressure of 1.02 atm when collected in a 250-mL bulb at 25°C. What is the empirical formula of the carbide? What mass of aluminum carbide is required to produce 20.0 L of methane at 25°C and a pressure of 1.00 atm?

9. Hydrocyanic acid is a weak acid. Write an equation for the ionization of hydrocyanic acid in water. Show the conjugate acid-base pairs. Would you expect an aqueous solution of sodium cyanide to be acidic, neutral, or basic? Write an equation for the reaction of CN^- with water. Identify the conjugate acid-base pairs.

10. Write balanced equations for the following reactions:

   (a) The reaction of sodium cyanide with sulfuric acid to give sodium hydrogen sulfate and hydrogen cyanide.

   (b) The reaction of sodium cyanide with concentrated sulfuric acid to give carbon monoxide and ammonium sulfate.

   (c) The combustion of hydrogen cyanide in air to give nitrogen, carbon dioxide, and water.

   (d) The reaction of carbon disulfide with a concentrated sodium hydroxide solution to give sodium thiocarbonate, Na_2CS_3, sodium carbonate, and water.

   (e) Which of the reactions in parts (a)–(d) are oxidation-reduction reactions?

11. When mercury(II) cyanide, $Hg(CN)_2$, is heated, a gaseous compound, X, containing only carbon and nitrogen is obtained. Analysis shows that it contains 46.2% C. A sample of 0.208 g of the gas occupies a volume of 126 mL at 100°C and 0.950 atm pressure.

   (a) What is the empirical and the molecular formula of the gas?

   (b) When the compound is reduced with hydrogen, ethylenediamine, $H_2N-CH_2-CH_2-NH_2$, is obtained. What is the arrangement of the atoms in X?

   (c) Draw a Lewis structure for the molecule and deduce its shape.

   (d) Would it have a dipole moment?

***12.** When carbon monoxide and sulfur vapor are passed through a heated tube, a colorless, odorless gas is obtained. The gas burns in excess oxygen to give CO_2 and SO_2. A sample of 0.246 g of this gas occupies a volume of 100 mL at 25°C and 1.00 atm pressure. When this sample was mixed with 200 mL of oxygen and burned completely, the final volume of gas was 250 mL at the initial temperature and pressure. When this mixture of gases was bubbled through a large volume of water, only 50 mL of the mixture remained undissolved at 25°C and 1.00 atm. A heated platinum wire decomposed the gas into sulfur and carbon monoxide without change of volume. A 25.00 mL sample of a 0.100M aqueous solution of the gas reacted completely with 50.00 mL of 0.200M potassium hydroxide to give a solution containing K^+ (aq), CO_3^{2-} (aq), and S^{2-} (aq). What is the formula of the gas? Would it have a dipole moment? Write balanced equations for each of the reactions.

Reactions of Hydrocarbons

13. What is a cracking reaction? Give an example.

14. Write balanced equations for the preparation of ethyne starting with calcium oxide, carbon, and water.

15. What is an addition reaction and an elimination reaction? Give two examples of each.

16. Starting with C_2H_2 and HCl, HBr, Cl_2, and Br_2, how could each of the following be prepared?

   (a) $CHCl_2-CHCl_2$ (b) $CH_2Cl-CHCl_2$

   (c) $CHBrCl-CHBrCl$ (d) $CH_2Cl-CHBrCl$

17. Why does cyclopropane have reactions that are more characteristic of an alkene than an alkane?

18. Give an example, in the form of a balanced equation, for each of the following reactions:

   (a) Combustion of an alkane.

   (b) An elimination reaction of an alkane.

(c) An elimination reaction of ethanol.

(d) An addition reaction of an alkene.

19. Write a balanced equation for each of the following reactions:

(a) The combustion of propane in an unlimited supply of air.

(b) The chlorination of ethene.

(c) The reaction of ethane with chlorine at high temperature to give chloroethane.

(d) The preparation of ethyne from calcium carbide.

(e) The polymerization of ethene.

(f) The cracking of ethane.

20. In each of the following pairs of hydrocarbons, which is the more reactive? Explain why.

(a) Butane and cyclobutane

(b) Cyclohexane and cyclohexene

(c) Benzene and cyclohexene

21. Give a short definition or description of each of the following terms. Give an appropriate example in each case.

(a) Cracking   (b) Polymerization

(c) Addition reaction   (d) Cyclic hydrocarbon

(e) Alkyl group   (f) Aromatic hydrocarbon

(g) Conformation

22. Explain why cyclopropane readily reacts with hydrogen in the presence of a catalyst (such as nickel or platinum) while cyclopentane does not. What is the product of the reaction of cyclopropane with hydrogen?

Molecular Geometry

23. Why is the C_2F_4 molecule planar?

24. Suggest an explanation for the bond angles in C_2F_4 and for why they differ from the bond angles in C_2H_4:

$$\overset{H}{\underset{H}{}}\,\overset{122°}{\underset{122°}{C}}{=}C\,\overset{H}{\underset{H}{}} \qquad \overset{F}{\underset{F}{}}\,\overset{125°}{\underset{125°}{C}}{=}C\,\overset{F}{\underset{F}{}}$$

with $116°$ and $110°$ labeled at the left carbons.

25. A, B, C, and D are four isomeric hydrocarbons with the molecular formula C_4H_6. Compound A has all its atoms in the same plane. Compounds A, B, and D have their carbon atoms in a continuous chain. Compound C is cyclic, and D contains an ethyl group. What are the structures of A, B, C, and D?

26. What is the chemical and structural evidence that leads us to formulate benzene as a regular hexagonal molecule containing equivalent carbon-carbon bonds of bond order $1\frac{1}{2}$?

27. Explain what is meant by a conformation. Draw diagrams of two conformations of ethane and two conformations of cyclohexane.

Names and Formulas of Hydrocarbons

28. What type of bond is present in an alkene but not in an alkane?

29. What is the difference between a saturated and an unsaturated hydrocarbon?

30. Classify each of the following as an alkane, an alkene, or an alkyne, and draw their Lewis structures:

(a) Methane   (b) Ethylene

(c) Methylacetylene   (d) Cyclobutane

(e) Cyclopropene

31. Draw the structural formula and give the IUPAC name for each of the nine isomers of heptane.

32. Give the molecular formula for each of the following:

(a) An alkane having 22 carbon atoms.

(b) An alkene having 19 carbon atoms and one double bond.

(c) An alkyne having 12 carbon atoms and one triple bond

(d) A cycloalkane having 10 carbon atoms.

33. Explain why 2-butene exists as *cis* and *trans* isomers while 2-butyne does not.

34. What is the IUPAC name of each of the following hydrocarbons?

(a) $CH_3{-}\overset{\overset{\displaystyle H}{|}}{\underset{\underset{\displaystyle CH_3}{|}}{C}}{-}CH_3$

(b) $CH_3{-}\overset{\overset{\displaystyle CH_3}{|}}{\underset{\underset{\displaystyle CH_2{-}CH_2{-}CH_2{-}CH_3}{|}}{CH}}{-}CH{-}CH_2{-}CH_2{-}CH_3$

(c) $CH_3{-}CH{=}CH_2$

(d) $CH_3{-}CH{=}C(CH_3)_2$

(e) $CH_3{-}CH_2{-}CH_2{-}CH{=}CH{-}CH_3$

(f) $H_3C{-}\overset{\overset{\displaystyle H}{|}}{\underset{\underset{\displaystyle H_2C{-}C(CH_3)_2}{|}}{C}}{-}\overset{\overset{\displaystyle H}{|}}{C}{-}CH_3$

(g) $CH_2{=}CH{-}CH_2{-}CH{=}CH{-}CH_3$

35. What is the structure of each of the following compounds?

(a) 2,2,4-Trimethylpentane

(b) 1,3-Cyclobutadiene

(c) 4,4-Dimethyl-1-pentyne

(d) *trans*-2-Pentene

(e) 1,3-Butadiene

(f) 1-Methyl-2,3-diethylbenzene

(g) 1,3,5-Trichlorobenzene

(h) *cis*-3,4-Dichloro-2-heptene

36. Draw the structures of as many isomers of C_5H_{10} as you can, and give the IUPAC name for each one.

37. Draw the structures and give the IUPAC name for each of the isomers of formula $C_2H_2Cl_2$.

38. Draw the structural formula of each of the following compounds:

(a) 1,3-Cyclopentadiene

(b) 1,3-Cyclohexadiene

(c) 3-Chloro-2,4-hexadiene

39. Which of the following names do not conform to the IUPAC rules? Rename those that do not conform.

(a) 2-Ethylbutane   (b) 3,3-Dimethylbutane

(c) 1-Ethylpropane   (d) 2,2-Dimethylpropane

40. Draw the structure and give the IUPAC name for all the isomers of C_4H_8.

41. In the following structural formula for benzene,

the circle represents six delocalized electrons. How many delocalized electrons are represented by the two circles in the following structural formula for naphthalene?

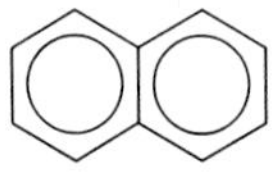

42. Which of the following pairs of molecules are isomers and which are not? Explain.

(a) $CH_3CH_2CH_2CH_3$ and $CH_3CH{=}CHCH_3$

(b) $CH_3(CH_2)_5CH(CH_3)_2$ and
    $CH_3(CH_2)_4CH(CH_3)CH_2CH_3$

(c) 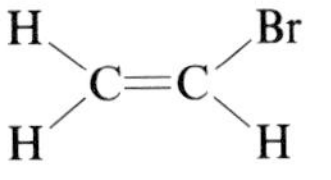 H₂C=CHBr (H, H / Br, H) and H₂C=CH₂ (H, H / Br, H)

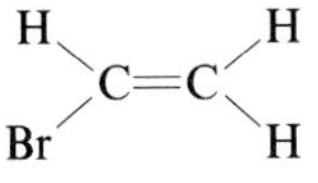

(d) 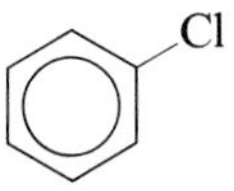 [monochlorobenzene] and [monochlorobenzene]

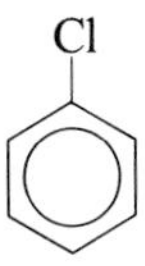

(e) 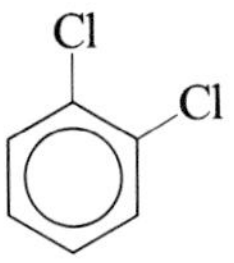 [1,2-dichlorobenzene] and [1,3-dichlorobenzene]

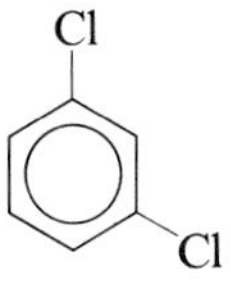

Miscellaneous

43. Classify each of the following reactions as an acid-base reaction, an oxidation-reduction reaction, an addition reaction, or an elimination reaction:

(a) $CO + Cl_2 \longrightarrow COCl_2$

(b) $CH_4 + H_2O \longrightarrow CO + 3H_2$

(c) $CO + H_2 \longrightarrow H_2CO$

(d) $CS_2 + 3Cl_2 \longrightarrow CCl_4 + S_2Cl_2$

(e) $CaC_2 + 2H_2O \longrightarrow Ca(OH)_2 + C_2H_2$

(f) $C_2H_6 \longrightarrow C_2H_4 + H_2$

44. Complete and balance the following equations

(a) C_6H_{10} $+ Br_2 \longrightarrow$
    Cyclohexene

(b) $C_3H_8 + O_2 \xrightarrow{\text{Heat}}$

(c) $C_2H_5OH + H_2SO_4 \xrightarrow{\text{Heat}}$

(d) $CH_4 + NH_3 \xrightarrow{\text{Heat}}$

45. What are the following?

(a) Natural gas   (b) Petroleum

(c) Coal tar   (d) Synthesis gas

46. What is a simple chemical test that would allow you to distinguish between the two gases in each of the following pairs? Describe what you would observe.

(a) C_2H_6 and C_2H_2 (b) CO_2 and C_3H_8

(c) CO and CH_4 (d) C_2H_4 and C_3H_8

47. A compound contains only carbon and hydrogen, and analysis shows it to contain 82.6% C. A flask of volume 199 mL was filled with the gaseous compound at a pressure of 750 mm Hg and a temperature of 25°C, and the amount of gas in the flask was found to be 0.470 g. What is the empirical formula of the compound, its molecular formula, and its molecular mass? From this information, can you write a unique structural formula?

48. A small amount of a liquid hydrocarbon was placed in a 250-mL flask and heated in a boiling water bath to 100°C. All the hydrocarbon was vaporized, and excess vapor was allowed to escape from the flask. The flask was then allowed to cool to room temperature. The mass of the flask and its contents was found to have increased by 0.687 g. The atmospheric pressure was 760 mm Hg. What is the molecular mass of the hydrocarbon? Analysis showed that it contained 85.6% C. What is the molecular formula? Suggest a possible structure for the hydrocarbon.

49. A sample of 0.200 g of a hydrocarbon containing 85.71% C occupies a volume of 95.3 mL at 0.921 atm and 27°C. Draw structures for the hydrocarbon consistent with the molecular formula, and give a correct name for each.

CHAPTER 12

THERMOCHEMISTRY

Over 90% of the world's energy at the present time is supplied by the combustion of fossil fuels: coal, natural gas, and petroleum. Only relatively small amounts are produced by hydroelectric power (4%) and nuclear reactors (5%), and an almost insignificant amount is obtained directly from the sun, although of course all our energy ultimately comes from the sun. The combustion of the hydrocarbons in fossil fuels is an exothermic process that releases energy in the form of heat.

We have already met many other examples of both exothermic and endothermic reactions. Almost all chemical reactions occur with either an absorption or a release of energy, depending on whether the energy stored in the products is greater or less than the energy stored in the reactants. The energy stored in elements and compounds is called **chemical energy**. We saw in Chapter 6 that when two atoms combine, the resulting molecule is more stable than the free atoms. In other words, the formation of a bond is an exothermic process. Thus if more energy is released by the formation of new bonds in a chemical reaction than is needed to break bonds, the reaction is exothermic. But if more energy is needed to break bonds than is released by the formation of new bonds, the reaction is endothermic. Chemical energy is therefore the energy associated with chemical bonds.

The energy released or absorbed in chemical reactions is most frequently observed in the form of heat, but it may also take other forms such as light and electrical energy. For example, we have seen that when magnesium burns in air, a considerable amount of energy is evolved as both heat and light (see Experiment 1.3). Another exothermic reaction is illustrated in Experiment 12.1.

Clearly, it is important to have quantitative information about the energy changes associated with chemical reactions. For instance, an engineer designing a space rocket must know how much energy is released in the reaction between hydrogen and oxygen in order to calculate how much liquid oxygen and liquid hydrogen must be carried in order to lift the rocket clear of the earth. An engineer must know how much fuel oil is needed to operate an electricity generating

An Exothermic Reaction

The colorless liquid glycerol (1,2,3-propanetriol), $C_3H_5(OH)_3$, is added to the purple-black solid, potassium permanganate, $KMnO_4$. The potassium permanganate oxidizes the glycerol to CO_2, H_2O, MnO_2, and K_2CO_3.

The reaction is sufficiently exothermic that the glycerol catches fire, burning with a flame that is colored lilac by the potassium of the $KMnO_4$.

station for a given period. A chemical engineer designing a chemical plant to operate as economically as possible needs to know how much heat each reaction in a process absorbs or evolves, so as to ensure that as far as possible the heat evolved in one reaction provides the heat necessary for another reaction. It is important for a biochemist to know the energy changes associated with the reactions in living cells to be able to understand the processes that occur in the cells and how a living organism transforms energy for its needs.

Thermochemistry, the quantitative study of heat changes, is the subject of this chapter. We will see how we can calculate the heat change associated with any reaction if we have certain information about each of the substances involved. We will see how we can find the energy associated with chemical bonds and how this information gives us a measure of the strength of a bond. Thermochemistry is part of a subject of much wider scope called *thermodynamics*, which is the science of the transformations of energy. We will consider some basic ideas of the science of thermodynamics, in particular, the concepts of heat and work and their relationship as summarized in the first law of thermodynamics.

12.1 HEAT CHANGES IN CHEMICAL REACTIONS

Heat is energy that is transferred as the result of a temperature difference. Heat always flows from a warmer object to a colder object. At the molecular level, when the molecules of the warmer object collide with those of the colder object, the molecules of the warmer object, which have a higher average kinetic energy, lose kinetic energy to those of the colder object. The average kinetic energy of the molecules of the warmer object decreases, and its temperature falls. The average kinetic energy of the molecules of the cooler object increases, and its temperature increases. Heat flows between the two objects until they have the same temperature.

The heat required to raise the temperature of an object or of a given amount of a substance by 1 kelvin is called the **heat capacity** of the object or of the particular amount of substance. The heat required to raise the temperature of 1 mole of a substance by 1 kelvin is the **molar heat capacity** of the substance. The molar heat capacity of liquid water is 75.4 J K^{-1} mol^{-1}.

Enthalpy

Most chemical reactions, including those in living matter, occur at constant pressure, usually a pressure of 1 atm. The heat changes accompanying chemical reactions are therefore usually measured at constant pressure. In other words, the pressure of the reactants and the pressure of the products are the same, usually 1 atm. The *heat absorbed in a reaction at constant pressure is called the* **enthalpy change** for the reaction; it is given the symbol ΔH (Δ is the Greek letter delta).

Endothermic reactions, in which heat is absorbed, have positive ΔH values, and exothermic reactions have negative ΔH values. For example, the equation and the associated enthalpy change for the combustion of CH_4 are

$$CH_4(g) + 2O_2(g) \longrightarrow CO_2(g) + 2H_2O(l) \qquad \Delta H = -890.4 \text{ kJ}$$

This reaction is *exothermic* (ΔH is negative); 890.4 kJ of heat are *evolved* for every mole of $CH_4(g)$ consumed in the reaction, at constant pressure. For the combination of 1 mol of H_2 with 1 mol of I_2 to give HI, we have

$$H_2(g) + I_2(s) \longrightarrow 2HI(g) \qquad \Delta H = 52.2 \text{ kJ}$$

This reaction is *endothermic* (ΔH is positive), and 52.2 kJ of heat are *absorbed* for every mole of H_2 and I_2 used up and for every 2 mol of HI produced in the reaction, at constant pressure.

We can consider that every substance has a property called enthalpy, H. Although we cannot measure the absolute enthalpy of any substance, this is of no concern to us because we are interested in the enthalpy changes in reactions, and these changes we can measure. The enthalpy change for a reaction is

$$\Delta H = \sum H_{\text{products}} - \sum H_{\text{reactants}}$$

where $\sum H_{\text{products}}$ is the sum of the enthalpies of the products and $\sum H_{\text{reactants}}$ is the sum of the enthalpies of the reactants.

When the total enthalpy of the products, $\sum H_{\text{products}}$, is greater than the total enthalpy of the reactants, $\sum H_{\text{reactants}}$, the enthalpy change, ΔH, is positive. An amount of heat ΔH is *absorbed* in the reaction. When the total enthalpy of the products is less than that of the reactants, the enthalpy change, ΔH, is negative. In other words, an amount of heat $-\Delta H$ is *evolved*. Figure 12.1 illustrates the enthalpy changes for an endothermic and an exothermic reaction.

Calorimetry

Figure 12.1 Enthalpy Changes in Exothermic and Endothermic Reactions. (a) Heat is absorbed in an endothermic reaction. (b) Heat is evolved in an exothermic reaction.

The measurement of the heat absorbed or evolved in chemical reactions is called **calorimetry**, after *caloric*, the old name for heat. The apparatus used for the measurement is known as a **calorimeter**. A very simple calorimeter can be made from two styrofoam coffee cups (Figure 12.2a). Because styrofoam is a very

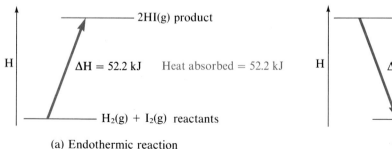

(a) Endothermic reaction

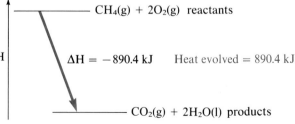

(b) Exothermic reaction

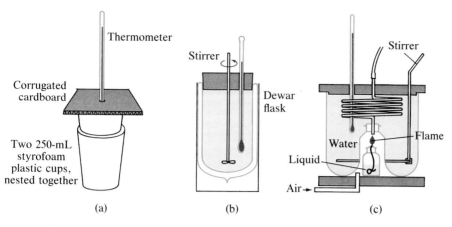

Figure 12.2 Calorimeters.
(a) A coffee cup calorimeter is suitable for a student laboratory experiment but not for accurate measurements. (b) A calorimeter of this type is suitable for making accurate measurements on reactions occurring in solution. (c) A flame calorimeter is used for measuring the enthalpies of combustion of gases and volatile liquids. The heat evolved in the combustion raises the temperature of the water in the calorimeter. When the heat capacity of water is known, the quantity of heat released in the reaction can be calculated.

good heat insulator, a coffee cup calorimeter absorbs only a very small amount of the heat produced in a reaction carried out inside it. Its temperature is, therefore, not increased significantly, and so very little heat is transmitted to the surroundings.

This simple calorimeter is most convenient for measuring the heat evolved in reactions carried out in dilute aqueous solution. Almost all the heat evolved in the reaction goes to raise the temperature of the solution. If the solution is dilute, its heat capacity does not differ significantly from that of water, and the amount of heat evolved or absorbed in the reaction can be calculated from the change in the temperature of the solution. A reaction that is easily carried out in a coffee cup calorimeter is the neutralization of an acid with a base. Some typical results are given in Example 12.1.

Example 12.1 A sample of 50 mL of a 0.20M solution of HCl was mixed with 50 mL of a 0.20M solution of NaOH in a coffee cup calorimeter. The initial temperature of both solutions was 22.2°C. After the two solutions were mixed, the temperature rose to 23.5°C. What is the enthalpy change for the reaction?

$$H_3O^+(aq) + OH^-(aq) \rightleftharpoons 2H_2O$$

Solution The total volume of the solution is 100 mL, and it has a mass of 100 g. The temperature rise was $23.5 - 22.2 = 1.3°C = 1.3$ K. The molar heat capacity of water is $75.4 \, J \, K^{-1} \, mol^{-1}$. Therefore

$$\left(\frac{75.4 \, J}{1.0 \, K \times 1.0 \, mol}\right)\left(\frac{1 \, mol}{18.02 \, g}\right)(100 \, g)(1.3 \, K) = 540 \, J$$

is needed to raise the temperature of 100 g of water (dilute solution) by 1.3 K. Hence the amount of heat produced by the reaction is 540 J.

The HCl solution contained

$$(0.20 \, mol \, L^{-1})(50 \, mL)\left(\frac{1 \, L}{1000 \, mL}\right) = 0.010 \, mol \, HCl$$

The NaOH solution similarly contained 0.010 mol NaOH. For the reaction of 1.0 mol NaOH with 1.0 mol HCl,

$$\text{Heat evolved} = (540 \text{ J})\left(\frac{1.0 \text{ mol}}{0.010 \text{ mol}}\right) = 54\,000 \text{ J} = 54 \text{ kJ}$$

Hence we may write

$$\text{HCl(aq)} + \text{NaOH(aq)} \longrightarrow \text{NaCl(aq)} + \text{H}_2\text{O(l)} \qquad \Delta H = -54 \text{ kJ}$$

or

$$\text{H}_3\text{O}^+(\text{aq}) + \text{OH}^-(\text{aq}) \longrightarrow 2\text{H}_2\text{O(l)} \qquad \Delta H = -54 \text{ kJ}$$

A coffee cup calorimeter does not give very accurate results because the Styrofoam absorbs a small amount of heat and some heat is lost to the surroundings. A calorimeter constructed from a vacuum flask, as shown in Figure 12.2(b), gives more accurate results. Careful experiments using this type of calorimeter have given the accurate value of $\Delta H = -56.02$ kJ for the enthalpy change for the neutralization reaction in aqueous solution. The equation

$$\text{H}_3\text{O}^+(\text{aq}) + \text{OH}^-(\text{aq}) \longrightarrow 2\text{H}_2\text{O} \qquad \Delta H = -56.02 \text{ kJ}$$

applies to the reaction of any strong acid with any strong base. For example,

$$\text{HClO}_4 + \text{KOH} \longrightarrow \text{KClO}_4 + \text{H}_2\text{O} \qquad \Delta H = -56.02 \text{ kJ}$$
$$\text{H}_2\text{SO}_4 + 2\text{KOH} \longrightarrow \text{K}_2\text{SO}_4 + 2\text{H}_2\text{O} \qquad \Delta H = 2(-56.02 \text{ kJ}) = -112.04 \text{ kJ}$$

For studying the enthalpies of combustion of gases and liquids, a flame calorimeter such as the one shown in Figure 12.2(c) can be used.

Example 12.2 When 0.510 g of ethanol was burned in oxygen in a flame calorimeter containing 1200 g of water, the temperature of the water rose from 22.46° to 25.52°C. What is the enthalpy change, ΔH, for the combustion of 1 mol of ethanol?

$$\text{C}_2\text{H}_5\text{OH} + 3\text{O}_2 \longrightarrow 2\text{CO}_2 + 3\text{H}_2\text{O}$$

Solution The increase in the temperature of the water is

$$25.52°\text{C} - 22.46°\text{C} = 3.06°\text{C} = 3.06 \text{ K}$$

The molar heat capacity of water is 75.4 J K^{-1} mol^{-1}. Therefore the amount of heat added to the water by the combustion of 0.510 g ethanol is

$$1200 \text{ g}\left(\frac{1 \text{ mol}}{18.02}\right)\left(\frac{75.4 \text{ J}}{1.0 \text{ K} \times 1.0 \text{ mol}}\right)(3.06 \text{ K}) = 15\,400 \text{ J} = 15.4 \text{ kJ}$$

The heat *evolved* in the combustion of 1 mol of ethanol (molar mass = 46.05 g) is

$$\left(\frac{15.4 \text{ kJ}}{0.510 \text{ g}}\right)\left(\frac{46.05 \text{ g}}{1 \text{ mol}}\right) = 1390 \text{ kJ mol}^{-1}$$

Therefore for the combustion of 1 mol of ethanol, $\Delta H = -1390$ kJ.

Standard Enthalpy Changes

Since reaction enthalpies depend somewhat on the conditions under which the reaction is carried out, we must specify some standard conditions. The **standard enthalpy** of reaction is the enthalpy change for a reaction when all the participating substances are in their standard states. A substance is in its *standard state* when it is at a pressure of 1 atm and at a specified temperature, which is usually chosen to be 25°C. All standard enthalpy changes in this book are at 25°C. Standard conditions are denoted by adding the superscript ° to the symbol ΔH, to give $\Delta H°$. For the combustion of 1 mol of methane under standard conditions, we write the equation

$$CH_4(g) + 2O_2(g) \longrightarrow CO_2(g) + 2H_2O(l) \qquad \Delta H° = -890.4 \text{ kJ}$$

This equation indicates that in the combustion of 1 mol of CH_4 in sufficient oxygen or air, 1 mol of CO_2 and 2 mol of H_2O are formed and 890.4 kJ of heat is released at a pressure of 1 atm and at 25°C.

It might seem surprising that the combustion of methane can apparently be carried out at 25°C. In fact, when methane is ignited, its temperature rises rapidly as it burns and produces heat, so the combustion is not in fact occurring at 25°. But if the reactants are at 25° at the start of the reaction, and the products cool to 25° at the conclusion of the reaction, all the heat produced in the reaction is eventually transferred to the calorimeter. It is immaterial that the reaction occurs at some temperature other than 25°, provided that the reactants are initially at 25° and the products are also at 25°.

Although we usually avoid fractional coefficients in writing equations for reactions, for convenience when we discuss reaction enthalpies, we often write the equation for the reaction of 1 mol of the reactant under consideration. Therefore the equation for the combustion of ethane may be written as

$$C_2H_6(g) + \tfrac{7}{2}O_2(g) \longrightarrow 2CO_2(g) + 3H_2O(l) \qquad \Delta H° = -1560.4 \text{ kJ}$$

This equation tells us that the enthalpy change for the combustion of 1 mol of ethane under standard conditions is -1560.4 kJ. The standard enthalpy of combustion of ethane is therefore $-1560.4 \text{ kJ mol}^{-1}$.

Note that *standard enthalpy changes, $\Delta H°$, are given for the reaction as written*. If we were to write the above equation in the more usual form, which involves 2 mol of ethane, we would have

$$2C_2H_6(g) + 7O_2(g) \longrightarrow 4CO_2(g) + 6H_2O(l)$$

and $\Delta H° = 2(-1560.4)\text{kJ} = -3120.8$ kJ.

Hess's Law

The combustion of ethane can be carried out directly, or alternatively, in the following steps (see Figure 12.3):

1. If ethane is strongly heated, it is decomposed to ethene, C_2H_4, and hydrogen; 136.2 kJ is absorbed for each mole of ethane decomposed:

$$C_2H_6(g) \longrightarrow C_2H_4(g) + H_2(g) \qquad \Delta H° = +136.2 \text{ kJ}$$

2. If ethene is burned in air, the reaction is very exothermic:

$$C_2H_4 + 3O_2(g) \longrightarrow 2CO_2(g) + 2H_2O(l) \qquad \Delta H° = -1410.8 \text{ kJ}$$

3. The combustion of hydrogen also releases heat:

$$H_2(g) + \tfrac{1}{2}O_2(g) \longrightarrow H_2O(l) \qquad \Delta H° = -285.8 \text{ kJ}$$

Together, these three reactions give the same result as the direct combustion of ethane. This result can be seen by adding the left and right sides of the equations and then canceling terms that appear on each side:

$C_2H_6(g) \longrightarrow C_2H_4(g) + H_2(g)$	$\Delta H° =$	$+136.2$ kJ
$C_2H_4(g) + 3O_2(g) \longrightarrow 2CO_2(g) + 2H_2O(l)$	$\Delta H° =$	-1410.8 kJ
$H_2(g) + \tfrac{1}{2}O_2(g) \longrightarrow H_2O(l)$	$\Delta H° =$	-285.8 kJ
$C_2H_6(g) + \tfrac{7}{2}O_2(g) \longrightarrow 2CO_2(g) + 3H_2O(l)$	$\Delta H° =$	-1560.4 kJ

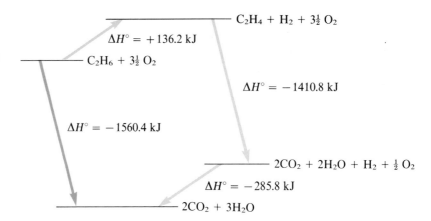

Figure 12.3 Enthalpy Changes in the Combustion of Ethane. The combustion of ethane to carbon dioxide and water in the gas phase can be carried out either directly or by the intermediate formation of ethene and hydrogen. The overall enthalpy change for the process is the same for the two routes. Directly, we have $\Delta H^\circ = -1560.4$ kJ; indirectly, we have $\Delta H^\circ = +136.2 - 1410.8 - 285.8 = -1560.4$ kJ

We find also that if we add the enthalpy changes for each reaction, we obtain a value equal to the experimentally determined value of the enthalpy change for the overall reaction. Thus the net ΔH° for the combustion of C_2H_6 in O_2 carried out indirectly by the formation of C_2H_4 and H_2 and the combustion of each of these separately is *exactly* the same as ΔH° for the direct, one-stage combustion.

The Russian chemist Germain Hess (1802–1850) found experimentally that a similar statement could be made for a large number of different reactions. In fact, this conclusion is valid for *any* chemical reaction and is expressed as **Hess's law**:

> **The enthalpy change in a reaction is the same regardless of the path by which the reaction occurs.**

In other words, the enthalpy change in a reaction is simply the difference in the enthalpies of the reactants and the products; it depends *only* on the enthalpies of the reactants and the products and is independent of the way in which the reactants are transformed into the products (see Figure 12.3).

Hess's law is a direct consequence of the law of conservation of energy. If Hess's law were not true, we could carry out an exothermic reaction and produce a certain amount of energy. Then we could convert the products back to the reactants by a different route that absorbed a smaller quantity of energy. The overall result would be that a certain amount of energy would have been created but with no overall change in the reactants, contrary to the law of conservation of energy.

We can use Hess's law to calculate enthalpy changes that are not easy to measure directly. *Enthalpies of combustion* are one of the simplest to measure experimentally, and they are often used to obtain values for other reaction enthalpies. For example, the enthalpy change for the formation of methane from carbon and hydrogen,

$$C(s) + 2H_2(g) \longrightarrow CH_4(g)$$

can be obtained from the ΔH° values for the combustion of carbon, hydrogen, and methane. These values are as follows:

(1) $\quad C(s) + O_2(g) \longrightarrow CO_2(g) \qquad\qquad \Delta H^\circ = -393.5$ kJ

(2) $\quad H_2(g) + \frac{1}{2}O_2(g) \longrightarrow H_2O(l) \qquad\qquad \Delta H^\circ = -285.8$ kJ

(3) $CH_4(g) + 2O_2(g) \longrightarrow CO_2(g) + 2H_2O(l) \qquad \Delta H^\circ = -890.4$ kJ

To obtain the equation for the formation of methane from carbon and hydrogen, we must multiply equation (2) by 2, add equation (1), and subtract equation (3). Or instead of subtracting equation (3), we may add its reverse. *The values of $\Delta H°$ must be combined in exactly the same way.* Reaction (3) is exothermic, therefore the reverse reaction is endothermic and we must change the sign of ΔH accordingly.

$$\begin{array}{ll} & \Delta H° \\ C(s) + O_2(g) \longrightarrow CO_2(g) & -393.5 \text{ kJ} \\ 2 \times [H_2(g) + \tfrac{1}{2}O_2(g) \longrightarrow H_2O(l)] & 2(-285.8 \text{ kJ}) \\ CO_2(g) + 2H_2O(l) \longrightarrow CH_4(g) + 2O_2(g) & +890.4 \text{ kJ} \\ \hline C(s) + 2H_2(g) + 2O_2(g) + CO_2(g) + 2H_2O(l) \longrightarrow & \\ \quad CO_2(g) + 2H_2O(l) + CH_4(g) + 2O_2(g) & -74.7 \text{ kJ} \end{array}$$

Canceling terms that appear on both sides gives

$$C(s) + 2H_2(g) \longrightarrow CH_4(g) \qquad -74.7 \text{ kJ}$$

Example 12.3 In the thermite reaction aluminum reacts with metal oxides at high temperature to generate the free metal and a large amount of heat (Chapter 9). Calculate the enthalpy change for the reaction

$$2Al(s) + Fe_2O_3(s) \longrightarrow Al_2O_3(s) + 2Fe(s)$$

from the enthalpy changes for the oxidation of iron and aluminum:

(1) $2Al(s) + \tfrac{3}{2}O_2(g) \longrightarrow Al_2O_3(s) \qquad \Delta H_1° = -1670 \text{ kJ}$

(2) $2Fe(s) + \tfrac{3}{2}O_2(g) \longrightarrow Fe_2O_3(s) \qquad \Delta H_2° = -824 \text{ kJ}$

Solution We add the reverse of equation (2) to equation (1) and cancel terms that appear on both sides:

$$\begin{array}{ll} 2Al(s) + \tfrac{3}{2}O_2(g) \longrightarrow Al_2O_3(s) & \Delta H° = -1670 \text{ kJ} \\ Fe_2O_3(s) \longrightarrow 2Fe(s) + \tfrac{3}{2}O_2(g) & \Delta H° = +824 \text{ kJ} \\ \hline 2Al(s) + Fe_2O_3(s) \longrightarrow 2Fe(s) + Al_2O_3(s) & \Delta H° = -846 \text{ kJ} \end{array}$$

We see that the reaction is highly exothermic, as we observed in Experiment 9.1.

Note that although the reactants in the thermite reaction have to be heated to initiate it, it is not an endothermic reaction. The initial heating is necessary only to raise the reactants to a sufficiently high temperature for the reaction to proceed at a reasonable speed. Once the reaction is proceeding, a large amount of heat is liberated, which raises the reactants to a higher temperature. The reaction is then further speeded up, and heat is liberated still more rapidly. The reacting mixture rapidly becomes hot enough to melt the iron that is formed.

Enthalpies of Formation

Because we can calculate enthalpy changes for reactions from the data for other reactions, we need to list the enthalpy changes for only a selected number of reactions. But what reactions should be listed? We must certainly have at least one reaction for each compound. The most useful way to tabulate the values of enthalpy changes is in terms of the standard enthalpies of formation of compounds. The **standard enthalpy of formation**, $\Delta H_f°$, is the $\Delta H°$ for the reaction in which 1 mol of a pure compound is formed from its elements in their most stable form and in their standard state.

Because an element can exist in the gaseous, liquid, and solid forms—and often in several different solid forms called allotropes—we must specify the form of the element on which standard enthalpies of formation are based. This is the most stable allotrope in its standard state, that is, at 1 atm and 25°C.

Thus for bromine, it is $Br_2(l)$, not $Br_2(g)$ or $Br(g)$; and for carbon, it is graphite, not diamond or $C(g)$.

Since to form the most stable form of an element in its standard state from the element in its most stable form in its standard state is to make no change at all, the corresponding standard enthalpy of formation is clearly zero. Thus by definition, *the enthalpy of formation of the most stable form of an element in its standard state is zero.* The standard states of the most stable forms of the elements are the arbitrary zero from which all enthalpy changes are measured. This choice is analogous to the choice of sea level as the arbitrary zero from which all land altitudes and ocean depths are measured. Thus the standard enthalpies of formation of carbon dioxide, water, and methane are the enthalpy changes for the following reactions at 25°C (298 K) and 1 atm:

$$C(graphite) + O_2(g) \longrightarrow CO_2(g) \qquad \Delta H_f^\circ(CO_2, g) = -393.5 \text{ kJ mol}^{-1}$$
$$H_2(g) + \tfrac{1}{2}O_2(g) \longrightarrow H_2O(l) \qquad \Delta H_f^\circ(H_2O, l) = -285.8 \text{ kJ mol}^{-1}$$
$$C(graphite) + 2H_2(g) \longrightarrow CH_4(g) \qquad \Delta H_f^\circ(CH_4, g) = -74.8 \text{ kJ mol}^{-1}$$

REACTION ENTHALPIES FROM ENTHALPIES OF FORMATION From the ΔH_f° values for the reactants and products of a reaction, we can calculate ΔH° for the reaction by using Hess's law. Consider the combustion of methane. This reaction may be carried out directly according to the equation

$$CH_4(g) + 2O_2(g) \longrightarrow CO_2(g) + 2H_2O(l) \qquad \Delta H_1^\circ \qquad \textbf{(1)}$$

Alternatively, at least in principle, the reaction may be carried out by first decomposing the reactants into the elements in their standard states (equation 2) and then recombining these elements to form the products (equation 3); see Figure 12.4:

$$CH_4(g) + 2O_2(g) \longrightarrow C(s) + 2H_2(g) + 2O_2(g) \qquad \Delta H_2^\circ \qquad \textbf{(2)}$$
$$C(s) + 2H_2(g) + 2O_2(g) \longrightarrow CO_2(g) + 2H_2O(l) \qquad \Delta H_3^\circ \qquad \textbf{(3)}$$

From Hess's law we have

$$\Delta H_1^\circ = \Delta H_2^\circ + \Delta H_3^\circ$$

Reaction (3) corresponds to the sum of the formation reactions for all the products from their elements. Thus

$$\Delta H_3^\circ = \text{Sum of the enthalpies of formation, } \Delta H_f^\circ, \text{ of all the products}$$
$$= (1 \text{ mol } CO_2)[\Delta H_f^\circ(CO_2, g)] + (2 \text{ mol } H_2O)[\Delta H_f^\circ(H_2O, l)]$$

Figure 12.4 Calculation of the Enthalpy Change for a Reaction. Any reaction can, in principle, be carried out by first decomposing all the reactants into the elements in their standard states and then combining the elements to give the products. From Hess's law, $\Delta H_1 = \Delta H_2 + \Delta H_3$. Hence, $\Delta H = \sum$ (enthalpies of formation of products) $- \sum$ (enthalpies of formation of reactions).

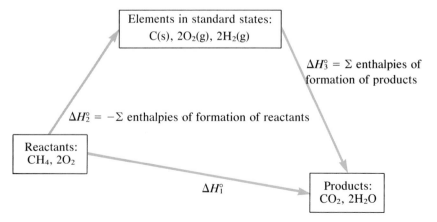

Elements in standard states:
$C(s)$, $2O_2(g)$, $2H_2(g)$

$\Delta H_3^\circ = \Sigma$ enthalpies of formation of products

$\Delta H_2^\circ = -\Sigma$ enthalpies of formation of reactants

Reactants: CH_4, $2O_2$

ΔH_1°

Products: CO_2, $2H_2O$

Reaction (2) is the reverse of the formation reactions for all the reactants. Thus

$$\Delta H_2^\circ = -[\text{Sum of the enthalpies of formation, } \Delta H_f^\circ, \text{ of all the reactants}]$$
$$= -\{(1 \text{ mol } CH_4)[\Delta H_f^\circ(CH_4, g)] + (2 \text{ mol } O_2)[\Delta H_f^\circ(O_2, g)]\}$$

Hence

$$\Delta H_1^\circ = \Delta H_2^\circ + \Delta H_3^\circ$$
$$= (\text{Sum of the enthalpies of formation, } \Delta H_f^\circ, \text{ of the products})$$
$$\quad -(\text{Sum of the enthalpies of formation, } \Delta H_f^\circ, \text{ of the reactants})$$
$$= (1 \text{ mol } CO_2)[\Delta H_f^\circ(CO_2, g)] + (2 \text{ mol } H_2O)[\Delta H_f^\circ(H_2O, g)]$$
$$\quad -(1 \text{ mol } CH_4)[\Delta H_f^\circ(CH_4, g)] - (2 \text{ mol } O_2)[\Delta H_f^\circ(O_2, g)]$$

Using the values given in Table 12.1, we have

$$\Delta H^\circ = (1 \text{ mol})(-393.5 \text{ kJ mol}^{-1}) + (2 \text{ mol})(-285.8 \text{ kJ mol}^{-1})$$
$$\quad - (1 \text{ mol})(-74.8 \text{ kJ mol}^{-1}) - (2 \text{ mol})(0)$$
$$= -890.3 \text{ kJ}$$

This approach is valid for any reaction, since any reaction can, in principle, be carried out by first decomposing all the reactants into their elements and then combining the elements to give the final products. Thus for any reaction

$$n_a A + n_b B + n_c C + \cdots \longrightarrow n_x X + n_y Y + \cdots$$
$$\Delta H^\circ = [n_x(\Delta H_f^\circ)_x + n_y(\Delta H_f^\circ)_y + \cdots] - [n_a(\Delta H_f^\circ)_a + n_b(\Delta H_f^\circ)_b + \cdots]$$
$$= \sum n_p(\Delta H_f^\circ)_p - \sum n_r(\Delta H_f^\circ)_r$$

where p stands for products and r for reactants and $\sum$ (Greek sigma) means *the sum of*. In other words,

$$\Delta H^\circ = (\text{Sum of standard enthalpies of formation of products})$$
$$\quad - (\text{Sum of standard enthalpies of formation of reactants})$$

Table 12.1 Standard Enthalpies of Formation

SUBSTANCE	ΔH_f° (kJ mol^{-1})	SUBSTANCE	ΔH_f° (kJ mol^{-1})
$CH_4(g)$	-74.5	$NO(g)$	$+90.3$
$C_2H_2(g)$	$+226.8$	$NO_2(g)$	$+33.2$
$C_2H_4(g)$	$+52.3$	$NH_3(g)$	-46.2
$C_2H_6(g)$	-84.7	$SO_2(g)$	-296.8
$C_3H_6(g)$ (cyclopropane)	$+53.3$	$SO_3(g)$	-395.7
$C_3H_8(g)$ (*n*-propane)	-103.8	$O_3(g)$	$+142.7$
$C_4H_8(g)$ (cyclobutane)	$+28.4$	$H_2SO_4(l)$	-814.0
$C_4H_{10}(g)$ (*n*-butane)	-126.1	$HNO_3(l)$	-174.1
$C_5H_{10}(g)$ (cyclopentane)	-78.4	$B_5H_9(s)$	$+73.2$
$C_5H_{12}(g)$ (*n*-pentane)	-146.4	$B_2O_3(s)$	-1273.5
$C_6H_6(l)$ (benzene)	$+49.0$	$NaOH(s)$	-425.6
$C_6H_{12}(g)$ (cyclohexane)	-123.3	$Na_2O_2(s)$	-511.7
$CO(g)$	-110.5	All elements	
$CO_2(g)$	-393.5	(most stable form)	0
$H_2O(l)$	-285.8	C(diamond)	$+1.90$
$H_2O(g)$	-241.8	H(g, atomic)	$+218.0$
		C(g, atomic)	$+716.7$

Notice that the ΔH_f° values listed in Table 12.1 are for the formation of one mol of the substance. If two or more moles of a substance are involved in a reaction, the corresponding ΔH_f° value must be multiplied by the appropriate factor n.

Notice also that the standard enthalpies of formation of the elements are zero only for their most stable form. For example, the standard enthalpy of formation of gaseous molecular hydrogen, H_2, is zero, but the standard enthalpy of formation of gaseous *atomic* hydrogen is the enthalpy change for the reaction

$$\tfrac{1}{2}H_2(g) \longrightarrow H(g, \text{atomic}) \qquad \Delta H^\circ = +218.0 \text{ kJ}$$

$$\Delta H_f^\circ(H, g, \text{atomic}) = +218.0 \text{ kJ mol}^{-1}$$

Similarly, the standard enthalpy of formation of solid carbon (graphite) is zero, but the standard enthalpy of formation of atomic carbon is the enthalpy change for the reaction

$$C(s, \text{graphite}) \longrightarrow C(g, \text{atomic}) \qquad \Delta H^\circ = +716.7 \text{ kJ}$$

$$\Delta H_f^\circ(C, g, \text{atomic}) = +716.7 \text{ kJ mol}^{-1}$$

The standard enthalpy of formation of diamond is also not zero; it is the enthalpy change for the reaction

$$C(\text{graphite}) \longrightarrow C(\text{diamond}) \qquad \Delta H^\circ = +1.90 \text{ kJ}$$

so that $\Delta H_f^\circ(\text{diamond}) = +1.90 \text{ kJ mol}^{-1}$.

Example 12.4 Find the standard enthalpy change for the oxidation of NH_3 according to the equation

$$4NH_3(g) + 5O_2(g) \longrightarrow 4NO(g) + 6H_2O(g)$$

from the standard enthalpies of formation given in Table 12.1.

Solution

$$\begin{aligned}
\Delta H^\circ &= \sum n_p(\Delta H_f^\circ)_p - \sum n_r(\Delta H_f^\circ)_r \\
&= (4 \text{ mol NO})(+90.3 \text{ kJ mol}^{-1} \text{ NO}) + (6 \text{ mol } H_2O)(-241.8 \text{ kJ mol}^{-1} \text{ } H_2O) \\
&\quad - [(4 \text{ mol } NH_3)(-46.2 \text{ kJ mol}^{-1} \text{ } NH_3) + (5 \text{ mol } O_2)(0)] \\
&= +361.2 \text{ kJ} - 1450.8 \text{ kJ} + 184.8 \text{ kJ} - 0 \\
&= -904.8 \text{ kJ}
\end{aligned}$$

Example 12.5 The hydrides of boron, which have unexpected formulas such as B_5H_9 and B_4H_{10}, have very high enthalpies of combustion. Some years ago, an extensive study was made of the hydrides of boron with a view to their utilization as rocket fuels. The hydride B_5H_9 ignites spontaneously in air with a green flash to produce solid B_2O_3 and water. What is the enthalpy of combustion of B_5H_9 under standard conditions?

Solution The unbalanced equation for the reaction is

$$B_5H_9(g) + O_2(g) \longrightarrow B_2O_3(s) + H_2O(l)$$

Two moles of B_5H_9 are needed for every 5 mol of B_2O_3 produced in order to balance the boron atoms. Hence the 18 hydrogen atoms must appear as 9 water molecules. There are then a total of 24 oxygen atoms on the right side of the equation, and so $12O_2$ molecules are needed on the left side. The balanced equation is

$$2B_5H_9(g) + 12O_2(g) \longrightarrow 5B_2O_3(s) + 9H_2O(l)$$

or for 1 mol of B_5H_9

$$B_5H_9(g) + 6O_2(g) \longrightarrow \tfrac{5}{2}B_2O_3(s) + \tfrac{9}{2}H_2O(l)$$

Using the standard enthalpies of formation given in Table 12.1, we have

$$\Delta H^\circ = \sum n_p (\Delta H_f^\circ)_p - \sum n_r (\Delta H_f^\circ)_r$$
$$= (\tfrac{5}{2}\,\text{mol})(-1273.5\,\text{kJ mol}^{-1})$$
$$+ (\tfrac{9}{2}\,\text{mol})(-285.8\,\text{kJ mol}^{-1}) - (1\,\text{mol})(73.2\,\text{kJ mol}^{-1})$$
$$= -4543\,\text{kJ}$$

The standard enthalpy of combustion of $B_5H_9(g)$ is $-4543\,\text{kJ mol}^{-1}$.

Example 12.6 When a little water is added to sodium peroxide, Na_2O_2, oxygen is liberated in a very exothermic reaction.

$$2Na_2O_2(s) + 2H_2O(l) \longrightarrow 4NaOH(s) + O_2(g)$$

How much heat is liberated when 10.0 g of Na_2O_2 reacts according to this equation?

Solution First we calculate ΔH° for the reaction:

$$\Delta H^\circ = \sum n_p (\Delta H_f^\circ)_p - \sum n_r (\Delta H_f^\circ)_r$$
$$= [4\,\Delta H_f^\circ(\text{NaOH, s}) + \Delta H_f^\circ(O_2, g)] - [2\,\Delta H_f^\circ(Na_2O_2, s) + 2\,\Delta H_f^\circ(H_2O, l)]$$

Substituting values of the standard enthalpies of formation from Table 12.1, we have

$$\Delta H^\circ = (4\,\text{mol})(-425.6\,\text{kJ mol}^{-1}) + 0.0 - (2\,\text{mol})(-511.7\,\text{kJ mol}^{-1})$$
$$- (2\,\text{mol})(-285.8\,\text{kJ mol}^{-1})$$
$$= -107.4\,\text{kJ}$$

The reaction is exothermic, and we see that the reaction of 2 mol of Na_2O_2 with water liberates 107 kJ of heat. Thus 10.0 g Na_2O_2 will liberate

$$(10.0\,\text{g }Na_2O_2)\left(\frac{1\,\text{mol }Na_2O_2}{77.98\,\text{g }Na_2O_2}\right)\left(\frac{107.4\,\text{kJ}}{2\,\text{mol }Na_2O_2}\right) = 6.89\,\text{kJ}$$

We can obtain some idea of the amount of heat 6.88 kJ represents by calculating its effect on 25.0 g of water initially at a temperature of 25.0°C. Since the molar heat capacity of water is 75.4 J K^{-1} mol^{-1}, 6.89 kJ would raise the temperature of 25.0 g of water by

$$6890\,\text{J}\left(\frac{(1\,\text{K})(1\,\text{mol})}{75.45\,\text{J}}\right)\left(\frac{18.02\,\text{g}}{1\,\text{mol}}\right)\left(\frac{1}{25.0\,\text{g}}\right) = 65.9\,\text{K}$$

that is, from 25.0° to 90.8°C

ENTHALPIES OF FORMATION FROM ENTHALPIES OF COMBUSTION   Direct determination of the enthalpies of formation of many compounds is difficult. For example, pure hydrocarbons are not easily obtained by directly combining carbon and hydrogen. In general, the enthalpy of formation of a hydrocarbon may be obtained from its experimentally measured enthalpy of combustion and the tabulated enthalpies of formation of the combustion products, CO_2 and H_2O, as we have already seen in the case of CH_4.

Example 12.7 The combustion of 1 mol of liquid benzene, $C_6H_6(l)$, at 25°C and 1 atm to produce $CO_2(g)$ and $H_2O(l)$ liberates 3267 kJ of heat when the products are at 25°C and 1.00 atm. What is the standard enthalpy of formation of $C_6H_6(l)$?

Solution The equation for the combustion of 1 mol of C_6H_6 is

$$C_6H_6(l) + 7\tfrac{1}{2}O_2(g) \longrightarrow 6CO_2(g) + 3H_2O(l)$$

We are given that the combustion of 1 mol of C_6H_6 at standard conditions liberates 3267 kJ of heat. Therefore $\Delta H^\circ = -3267$ kJ mol^{-1}. Then

$$\Delta H^\circ = \sum n_p(\Delta H_f^\circ)_p - \sum n_r(\Delta H_f^\circ)_r$$
$$-3267 \text{ kJ} = 6\,\Delta H_f^\circ(CO_2, g) + 3\,\Delta H_f^\circ(H_2O, l) - \Delta H_f^\circ(C_6H_6, l) - 7\tfrac{1}{2}\,\Delta H_f^\circ(O_2, g)$$

Using the values given in Table 12.1, we have

$$-3267 \text{ kJ} = (6 \text{ mol})(-393.5 \text{ kJ mol}^{-1}) + (3 \text{ mol})(-285.8 \text{ kJ mol}^{-1})$$
$$- 1 \text{ mol}[\Delta H_f^\circ(C_6H_6, l)] - 0$$

Therefore

$$(1 \text{ mol})[\Delta H_f^\circ(C_6H_6, l)] = [6(-393.5) + 3(-285.8) + 3267] \text{ kJ}$$
$$= (-2361.0 - 857.4 + 3267) \text{ kJ}$$
$$\Delta H_f^\circ(C_6H_6, l) = 49 \text{ kJ mol}^{-1}$$

The standard enthalpies of formation provide a measure of the relative amounts of chemical energy stored in a compound. If the standard enthalpy of formation of a compound is negative, the compound contains less stored energy than the standard states of the elements from which it is formed. The ΔH_f° for CO_2 is -393 kJ mol^{-1}, so CO_2 contains less stored energy than C(graphite) and O_2(g). If ΔH_f° is positive, the compound contains more stored energy than the elements from which it is formed. The ΔH_f° for ethyne, C_2H_2, is $+228.0$ kJ mol^{-1}; C_2H_2 therefore contains a large amount of stored energy. If it is decomposed to the elements, this energy is released as heat. When ethyne is burned in oxygen, the reaction is highly exothermic, because the energy stored in C_2H_2 is released together with the energy that is released in the formation of CO_2 and H_2O, which both contain less stored energy than the elements from which they are made. We express the same idea in a different way when we say that C_2H_2 is less stable than the elements from which it is formed, whereas CO_2 and H_2O are more stable than the elements from which they are formed.

12.2 BOND ENERGIES

Thermochemical data can give us useful information on the strengths of chemical bonds. For example, the *strength of the bond in a diatomic molecule is measured by the energy needed to dissociate the gaseous molecule into two atoms.* The energy needed could be supplied in various ways, for example, in the form of light or in the form of heat. Thus we could find the lowest frequency of light needed to dissociate the hydrogen molecule into two hydrogen atoms and then calculate the corresponding energy for the process from $\Delta E = h\nu$ (Chapter 6).

Alternatively, we can use the enthalpy change for the reaction:

$$H_2(g) \longrightarrow 2H(g, \text{atomic}) \qquad \Delta H^\circ = 436.0 \text{ kJ}$$

This value of 436.0 kJ is known as the **dissociation energy**, or **bond energy**, of the hydrogen molecule. The bond energy of a diatomic molecule is the energy needed to dissociate 1 mol of molecules into gaseous atoms. Notice that the enthalpy of formation of *1 mol of hydrogen atoms*, which is given in Table 12.1, is the ΔH° for the reaction:

$$\tfrac{1}{2}H_2(g) \longrightarrow H(g, \text{atomic}) \qquad \Delta H^\circ = \frac{436.0}{2} \text{ kJ} = 218.0 \text{ kJ}$$

that is,

$$\Delta H_f^\circ(H, g, atomic) = 218.0 \text{ kJ mol}^{-1}$$

The bond energies of diatomic molecules range from values of 149 kJ mol^{-1} and 155 kJ mol^{-1} for the rather weak bonds in I_2 and F_2 to higher values of 941 kJ mol^{-1} and 1070 kJ mol^{-1} for the very strong bonds in N_2 and CO.

The bond energies in polyatomic molecules are not quite so simple to determine. We consider two possible cases: first, when all the bonds are the same and, second, when there are two or more different kinds of bonds in the molecule.

Molecules with Only One Type of Bond

The energy needed to dissociate a CH_4 molecule completely into a C atom and four H atoms,

$$CH_4(g) \longrightarrow C(g) + 4H(g)$$

represents the energy needed to break all four C—H bonds. One-quarter of this value is the *average bond energy* for the C—H bond in methane. It is not easy to determine the enthalpy change for this reaction experimentally, but we can calculate it from the enthalpies of formation given in Table 12.1:

$$\Delta H^\circ = \Delta H_f^\circ(C, atomic) + 4\,\Delta H_f^\circ(H, atomic) - \Delta H_f^\circ(CH_4, g)$$

$$= (1 \text{ mol})(716.7 \text{ kJ mol}^{-1}) + (4 \text{ mol})(218.0 \text{ kJ mol}^{-1}) - (1 \text{ mol})(-74.5 \text{ kJ mol}^{-1})$$

$$= 1663.2 \text{ kJ}$$

Since this enthalpy change is the energy required to break four C—H bonds, the average C—H bond energy is one-quarter of this value, that is, 415.8 kJ mol^{-1}.

We have called this energy the *average* bond energy because if the four C—H bonds in CH_4 were to be broken one at a time, as in

$$
\begin{array}{lll}
CH_4 & \longrightarrow CH_3 + H & \Delta H_1^\circ \\
CH_3 & \longrightarrow CH_2 + H & \Delta H_2^\circ \\
CH_2 & \longrightarrow CH + H & \Delta H_3^\circ \\
CH & \longrightarrow C + H & \Delta H_4^\circ
\end{array}
$$

the values of ΔH_1°, ΔH_2°, ΔH_3°, and ΔH_4° would all be different and none of them would be equal to 415.8 kJ mol^{-1}. The four bonds in CH_4 are all equivalent, and the same amount of energy is required to break any one of them. But when one bond is broken, there is a rearrangement of the electron density. So the remaining bonds, although equivalent to each other, are slightly different from the bonds in CH_4. Similarly, the bonds in CH_2 and CH are not exactly the same as the bonds in CH_3 and CH_4. However, from Hess's law the sum of the energies needed to break the bonds one at a time, $\Delta H_1^\circ + \Delta H_2^\circ + \Delta H_3^\circ + \Delta H_4^\circ$, is equal to the energy needed to break them all at the same time, that is, 1663.2 kJ mol^{-1}.

Molecules with More Than One Type of Bond

When there are two or more kinds of bonds in a molecule, the determination of the bond energies is slightly more complicated and involves certain approximations. For example, from the appropriate enthalpies of formation we may obtain a value for the enthalpy change for the reaction

$$C_2H_6(g) \longrightarrow 2C(g, atomic) + 6H(g, atomic)$$

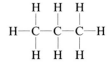

in which six C—H and one C—C bond are broken. Since we have two unknowns, namely, the C—H and C—C bond energies, we cannot determine values for both of them, and we have to make an assumption about one of them. If we assume, for example, that the C—H bond energy in C_2H_6 is the same as that in CH_4, we could then calculate a value for the C—C bond energy.

Example 12.8 From the data in Table 12.1, calculate a value for the enthalpy change for the reaction

$$C_2H_6(g) \longrightarrow 2C(g, \text{atomic}) + 6H(g, \text{atomic})$$

Then assuming that the C—H bond energy is the same as that in CH_4, calculate a value for the C—C bond energy in C_2H_6.

Solution

$$\Delta H^\circ = (2 \text{ mol})(716.7 \text{ kJ mol}^{-1}) + (6 \text{ mol})(218.0 \text{ kJ mol}^{-1}) - (1 \text{ mol})(-84.7 \text{ kJ mol})$$

$$= 2826.1 \text{ kJ}$$

Since C_2H_6 has six C—H bonds and one C—C bond, if $BE(\text{C—H})$ is the bond energy of a C—H bond and $BE(\text{C—C})$ is the bond energy of a C—C bond, then the energy needed to break all the bonds is

$$BE(\text{C—C}) + 6BE(\text{C—H}) = 2826.1 \text{ kJ}$$

If we now assume that $BE(\text{C—H})$ is the same as it is in CH_4, that is, $BE(\text{C—H}) = 415.8 \text{ kJ mol}^{-1}$, then

$$(1 \text{ mol})[BE(\text{C—C})] = (1 \text{ mol})(2826.1 \text{ kJ mol}^{-1}) - (6 \text{ mol})(415.8 \text{ kJ mol}^{-1})$$

$$BE(\text{C—C}) = 331.3 \text{ kJ mol}^{-1}$$

We could also calculate a value for the C—C bond energy from the enthalpy change for the reaction

$$C_3H_8(g) \longrightarrow 3C(g \text{ atomic}) + 8H(g \text{ atomic})$$

This enthalpy change can be calculated from the enthalpies of formation in Table 12.1 (see Example 12.8); it has the value 3998.6 kJ mol^{-1}. Propane has eight C—H and two C—C bonds; therefore

$$8BE(\text{C—H}) + 2BE(\text{C—C}) = 3998.6 \text{ kJ}$$

Then assuming that $BE(\text{C—H})$ is the same as it is in CH_4, that is, 415.8 kJ mol^{-1}, we obtain a value of 336.1 kJ mol^{-1} for $BE(\text{C—C})$. This value does not agree exactly with that obtained from C_2H_6 because we would not expect the C—H bond energy to be exactly the same in CH_4, C_2H_6, and C_3H_8, as we have assumed, nor would we expect a C—C bond to have the same bond energy in all compounds.

A third value for $BE(\text{C—C})$ can be obtained from data for diamond. The enthalpy change for the vaporization of 1 mol of carbon atoms is

$$C(\text{diamond}) \longrightarrow C(g, \text{atomic}) \qquad \Delta H^\circ = 714.8 \text{ kJ}$$

In this process all the C—C bonds in the solid are broken. Since each carbon atom forms four bonds, but each bond is shared between two atoms, the ΔH° of this process is a measure of the energy required to break $\frac{4}{2} = 2$ bonds per carbon. Thus we obtain $BE(\text{C—C}) = 714.8 \text{ kJ}/2 = 357.4 \text{ kJ mol}^{-1}$, which we can compare with the values of 336.1 kJ mol^{-1} from C_3H_8 and 331.3 kJ mol^{-1} from C_2H_6.

Table 12.2 Bond Energies

BOND	BOND ENERGY[a] (kJ mol^{-1})	BOND	AVERAGE BOND ENERGY[b] (kJ mol^{-1})	BOND	AVERAGE BOND ENERGY[b] (kJ mol^{-1})
H—H	436	O—O	138	C—F	485
H—F	565	N—N	159	O—Cl	205
H—Cl	431	N=N	418	N—Cl	201
H—Br	364	C—C	348	C—Cl	326
H—I	297	C=C	619	P—Cl	326
F—F	155	C≡C	812	S—Cl	276
Cl—Cl	239	O—H	463	C—O	335
Br—Br	190	N—H	389	C=O	707
I—I	149	C—H	413	C≡O	1070
O=O	494	P—H	318	C—N	293
N≡N	941	S—H	364	C=N	616
		O—F	184	C≡N	879

[a] These bond energies are the dissociation energies of diatomic molecules that have only one bond; they are therefore exact values.
[b] These bond energies are obtained, as described in the text, from molecules that contain more than one bond; they are average, not exact, values therefore.

Although these values differ somewhat, the average obtained from a large number of compounds gives a reasonably good approximation for the energy of the C—C bond in any substance. This value is called the *average bond energy* of the C—C bond. Values for the average bond energies of some common bonds are listed in Table 12.2. These values give a good idea of the relative strengths of different bonds. We see that the C—C and C—H bonds are among the strongest of the single bonds, whereas the N—N, O—O, and F—F bonds are relatively weak. That the properties of a given bond, such as the bond energy, do not vary greatly from one substance to another is a very important concept in chemistry. It enables us to make many predictions about the properties of substances from the knowledge gained from the study of other substances that contain the same bonds.

Double and Triple Bonds

From the heats of formation of ethene and ethyne the energy needed to dissociate the molecule completely into atoms can be calculated. If we assume that the C—H bonds have the same strength as those in alkanes, values for the bond energies of the C=C double bond and the C≡C triple bond can be obtained in the same manner as for C—C single bonds.

Example 12.9 From the value for the enthalpy of formation of ethene, C_2H_4, in Table 12.1 and the bond energy of the C—H bond in CH_4 (415.8 kJ mol^{-1}), calculate a value for the C=C bond energy.

Solution To find the C=C bond energy, we must first find the energy needed to decompose the C_2H_4 molecule completely into atoms. This energy is the enthalpy change for the reaction

$$C_2H_4(g) \longrightarrow 2C(g) + 4H(g)$$

Thus

$$\Delta H° = (2 \text{ mol})[\Delta H_f°(C, g)] + (4 \text{ mol})[\Delta H_f°(H, g)] - (1 \text{ mol})[\Delta H_f°(C_2H_4, g)]$$
$$= (2 \text{ mol})(716.7 \text{ kJ mol}^{-1}) + (4 \text{ mol})(218.0 \text{ kJ mol}^{-1}) - (1 \text{ mol})(52.3 \text{ kJ mol}^{-1})$$
$$= 2253.1 \text{ kJ}$$

This enthalpy change is the energy required to break four C—H bonds and one C=C bond. So

$$BE(C{=}C) + 4BE(C{-}H) = 2253.1 \text{ kJ}$$

If $BE(C{-}H)$ is the same as it is in CH_4, that is, $415.8 \text{ kJ mol}^{-1}$, then

$$BE(C{=}C) = 2253.1 \text{ kJ} - (4 \text{ mol})(415.8 \text{ kJ mol}^{-1}) = 589.9 \text{ kJ}$$

The average bond energies of the C=C and C≡C bonds given in Table 12.2 are mean values calculated from a number of different compounds. The values for some other double and triple bonds are also given in the table.

We see from Table 12.2 that a double bond is very approximately twice as strong as the corresponding single bond, and a triple bond is three times as strong. But notice that whereas a C≡C bond is considerably less than three times as strong as a C—C bond, the N≡N triple bond is considerably more than three times as strong as the corresponding single bonds. However, a discussion of these interesting differences would take us beyond the scope of this book. Because of the great strength of the N≡N triple bond, many reactions in which N_2 is formed are highly exothermic. Thus many nitrogen compounds decompose or burn in explosive reactions. Explosions are simply reactions in which a large amount of energy is released and a large volume of gaseous products is formed very rapidly (see Box 12.1 and Experiment 12.2).

EXPERIMENT 12.2

The Decomposition of Nitrogen Triiodide

The dark brown substance on the filter paper is a compound of nitrogen triiodide and ammonia, $NI_3 \cdot NH_3$.

When dry it is extremely shock sensitive and, when touched lightly with a feather, explodes violently, producing a cloud of purple-brown iodine vapor.

The explosion is violent enough to punch a hole through the filter papers and the asbestos mat on which they were resting.

Because of the great strength of the triple bond in the N_2 molecule, the decomposition of NI_3 to N_2 and I_2 is a strongly exothermic reaction. It is also very rapid so that a large amount of heat and a large volume of gaseous products are formed rapidly which produces a violent explosion.

An important application of chemical energy is the use of explosives. Any substance that undergoes a very rapid chemical reaction that is strongly exothermic—or that produces a large volume of gaseous products from a solid or a liquid—is potentially an explosive. The destructive power of an explosion is due to the shock wave caused by the rapid increase in volume from the gases formed or to the rapid expansion of the atmosphere as a consequence of the large amount of heat released in a short time, or to both of these circumstances.

The oldest known explosive is *gun powder*, which was used in ancient times in China, Arabia, and India. Gun powder is a mixture of approximately 75% KNO_3, 12% S, and 13% C. The products include a large volume of gases including CO_2, CO, and N_2, as well as a dense smoke that consists of fine particles of K_2CO_3, K_2SO_4, and K_2S.

For many purposes gun powder has been replaced by stronger explosives such as ammonium nitrate. When ammonium nitrate is detonated, it decomposes in a very exothermic reaction to give a large volume of gaseous products:

$$2NH_4NO_3(s) \longrightarrow 2N_2(g) + O_2(g) + 4H_2O(g)$$

The oxygen produced can also be used to oxidize other substances, thus increasing the energy released. A commonly used explosive for blasting in mines is composed of 95% NH_4NO_3 and 5% fuel oil. Ammonium nitrate is also used as a fertilizer. Normally, its use as a fertilizer is safe, because ammonium nitrate must be detonated before it will explode. However, the careless handling of large quantities of ammonium nitrate fertilizer can cause a massive explosion. For example, in 1947 a ship carrying ammonium nitrate fertilizer exploded and leveled a huge area of Texas City, Texas, claiming 576 lives.

If an explosive reaction is very rapid, the shock wave may travel at very high speeds, up to 6 km s^{-1}, and the explosive is classified as a high explosive. Trinitrotoluene (TNT), $C_7H_5O_6N_3$, and nitroglycerin, $C_3H_5O_9N_3$, are examples. The slower combustion that occurs in low explosives such as gun powder produces shock waves that travel at about 100 m s^{-1}.

The oxygen required for the very rapid combustion

*Alfred Nobel
(1833–1896)*

of high explosives cannot come from the air, because the oxidation is too rapid. For these high explosives the oxygen comes from the explosive itself. Often such explosives are mixed with other substances such as NH_4NO_3 in order to increase the amount of oxygen available. Many common explosives contain nitrogen compounds. Their combustion produces the oxides of nitrogen and molecular nitrogen, which is formed in a very exothermic process because of the high bond energy of the nitrogen molecule.

Nitroglycerin is a liquid that explodes 25 times as fast as gun powder and with three times the energy per gram. Although it was used as an explosive soon after it was first prepared in 1866, nitroglycerin is much too unstable and sensitive to shock to be handled safely. For transportation the containers were packed into a type of clay called kieselguhr in order to cushion them from shocks as much as possible. Alfred Nobel, a Swedish inventor, noticed that when nitroglycerin leaked from the containers, it was soaked up by the clay. The nitroglycerin-soaked clay is much more stable and less sensitive to shock than pure nitroglycerin. Nobel called this safer explosive *dynamite*, and he made enough money from manufacturing it to found the Nobel prizes in peace, literature, physics, chemistry, and physiology and medicine.

Approximate Reaction Enthalpies from Bond Energies

If the enthalpy change for a reaction is not known or cannot be measured experimentally, bond energies can be used to calculate an approximate value for the enthalpy change. The basic idea is simple. The heat absorbed in any reaction, ΔH, is the energy needed to break all the bonds in the reactants minus the energy evolved in the formation of the bonds in the products. If more energy is needed to break the bonds in the reactants than is obtained by forming the bonds in the products, then the reaction is endothermic. If less energy is needed to break all the bonds in the reactants than is evolved in the formation of the bonds in the products, then the reaction is exothermic (see Figure 12.5). In summary

$$\Delta H^\circ = \sum BE(\text{bonds broken}) - \sum BE(\text{bonds made})$$

or

$$\Delta H^\circ = \sum [\text{bond energies(reactants)}] - \sum [\text{bond energies(products)}]$$

Note that when we use bond energies, we find ΔH° by subtracting the values for the products from those for the reactants. When we use enthalpies of formation, however, we do the reverse—we subtract the values for the reactants from those for the products.

Example 12.10 Use the average bond energies in Table 12.2 to calculate an approximate ΔH° for the oxidation of HI(g) by Cl_2(g) (Experiment 5.7):

$$Cl_2(g) + 2HI(g) \longrightarrow I_2(g) + 2HCl(g)$$

Solution First, we dissociate the Cl_2 and HI molecules into free atoms, and then we combine the atoms to form the I_2 and HCl molecules.

Figure 12.5 Approximate Reaction Enthalpies from Bond Energies. Any reaction can, in principle, be carried out by first decomposing all the gaseous reactants into free atoms. This process is endothermic, and the heat needed is the sum of the bond energies of the reactants. Then the free atoms may be combined to give the products. This process is exothermic, and the heat evolved is the sum of the bond energies of the products. Hence $\Delta H = \sum [\text{bond energies(reactants)}] - \sum [\text{bond energies(products)}]$. Since the bond energies are approximate, this method gives only an approximate value for the enthalpy change of a reaction in the gas phase.

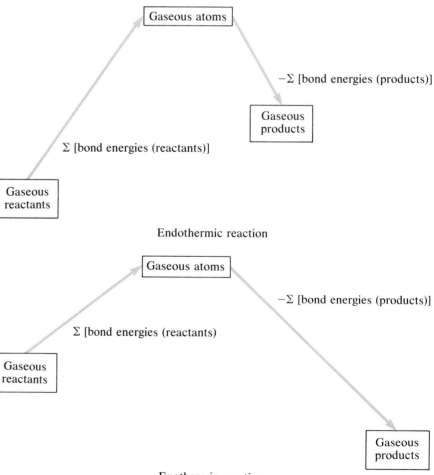

To break 1 mol of Cl_2 bonds requires 239 kJ. To break 2 mol of HI bonds requires 2 mol $\times$ 297 kJ mol^{-1} = 594 kJ. The total energy needed to break all the bonds in the reactants is therefore

$$239 \text{ kJ} + 594 \text{ kJ} = 833 \text{ kJ}$$

The energy evolved in the formation of 1 mol of I—I bonds is 149 kJ. The energy evolved in the formation of 2 mol of H—Cl bonds is 2 mol $\times$ 431 kJ mol^{-1} = 862 kJ. The total energy evolved in the formation of all the bonds in the products is

$$149 \text{ kJ} + 862 \text{ kJ} = 1011 \text{ kJ}$$

Thus for this reaction, 837 kJ is needed to break all the bonds in the reactants, but 1013 kJ is evolved in the formation of the new bonds in the products. Hence the overall reaction is exothermic:

$$\Delta H^\circ = 833 \text{ kJ} - 1011 \text{ kJ} = -178 \text{ kJ}$$

Example 12.11 A typical cracking reaction is

$$C_8H_{18}(g) \xrightarrow[\text{Heat}]{\text{Catalyst}} C_4H_8(g) + C_4H_{10}(g)$$
$$\text{Octane} \qquad\qquad \text{Butene} \quad \text{Butane}$$

Using bond energies, calculate an approximate value for the enthalpy change for this reaction.

Solution The C_8H_{18} molecule contains 7 C—C and 18 C—H bonds, whereas the products contain 5 C—C, 1 C=C, and 18 C—H bonds. Thus

$$\Delta H^\circ = 7BE(\text{C—C}) + 18BE(\text{C—H}) - [5BE(\text{C—C}) + BE(\text{C=C}) + 18BE(\text{C—H})]$$
$$= 2BE(\text{C—C}) - BE(\text{C=C})$$
$$= (2 \text{ mol})(348 \text{ kJ mol}^{-1}) - (1 \text{ mol})(619 \text{ kJ mol}^{-1}) = 77 \text{ kJ}$$

We see that the enthalpy change for the reaction is determined by the difference in energy of two C—C single bonds and a C=C double bond. Since the C=C double bond is not twice as strong as a C—C single bond, the reaction is endothermic.

Example 12.12 Estimate the standard enthalpy of formation of cyclohexane, $C_6H_{12}(g)$, from the bond energies in Table 12.2 and the enthalpies of formation of atomic hydrogen and atomic carbon given in Table 12.1.

Solution The standard enthalpy of formation of C_6H_{12} is the enthalpy change at 25°C and 1 atm for the reaction

$$6C(s, \text{graphite}) + 6H_2(g) \longrightarrow C_6H_{12}(g)$$

To determine the bond energy of the reactants, we convert graphite to carbon atoms in the gas phase and the hydrogen molecules to hydrogen atoms in the gas phase. Using the enthalpies of formation of atomic carbon and atomic hydrogen from Table 12.1, we have

$$6C(s, \text{graphite}) \longrightarrow 6C(g) \qquad \Delta H^\circ = 6 \times 716.7 \text{ kJ} = 4300 \text{ kJ}$$
$$6H_2(g) \longrightarrow 12H(g) \qquad \Delta H^\circ = 12 \times 218.0 \text{ kJ} = \underline{2616 \text{ kJ}}$$
$$6916 \text{ kJ}$$

To determine the bond energies of the products, we combine the 6 C atoms and the 12 H atoms to form a molecule of C_6H_{12}. Since C_6H_{12} has 6 C—C and 12 C—H bonds, the total energy evolved is

$$6BE(\text{C—C}) = 6 \text{ mol} \times 348 \text{ kJ mol}^{-1} = 2088 \text{ kJ}$$
$$12BE(\text{C—H}) = 12 \text{ mol} \times 413 \text{ kJ mol}^{-1} = 4956 \text{ kJ}$$
$$\overline{7044 \text{ kJ}}$$

Thus 6916 kJ is needed to break all the bonds in the reactants, and 7044 kJ is evolved in the formation of all the bonds in the product, C_6H_{12}. Therefore the enthalpy of formation of C_6H_{12} is estimated to be

$$\Delta H_f^\circ = 6916 \text{ kJ} - 7044 \text{ kJ} = -128 \text{ kJ}$$

The value for the enthalpy of formation of C_6H_{12} calculated in Example 12.12 by using bond energies agrees fairly well with the value determined experimentally, $-123.3 \text{ kJ mol}^{-1}$. Other experimental enthalpies of formation and values calculated from bond energies are compared in Table 12.3. In most cases the agreement is good. When there is a marked disagreement between the experimental and calculated values, we must look for some special explanation. As we will see, such discrepancies can provide us with valuable information about molecular structure.

Strain Energy

If we consider the values in Table 12.3 for the cyclic hydrocarbons from cyclopropane to cyclohexane, we see that the discrepancy between the calculated and experimental values of ΔH_f° decreases through this series. The discrepancy is negligible for cyclohexane but is rather large for cyclobutane and cyclopropane. The enthalpies of formation are less negative (more positive) than calculated. This means that the molecules are less stable than calculated from the bond energies, which suggests that one or more of the bonds in these molecules have smaller bond energies than given in Table 12.2. The difference in the calculated and experimental total bonding energies for these molecules is called the **strain energy**; it results mainly from the bending of the C—C bonds, which makes them weaker than normal single bonds.

The strain energies for cyclopropane, cyclobutane, and cyclopentane are 117, 113, and 28 kJ mol^{-1}, respectively. The strain energy per C—C bond is therefore $117/3 = 39 \text{ kJ mol}^{-1}$ for cyclopropane, $113/4 = 28 \text{ kJ mol}^{-1}$ for cyclobutane, and only 6 kJ mol^{-1} for cyclopentane. The strain energy decreases as the amount of bending decreases. The strain energy in cyclopentane is very small because in a planar cyclopentane ring the C—C bond angle is 108°, which implies an almost insignificant amount of bending of the C—C bonds, since the normal CCC tetrahedral angle is 109.5°.

Example 12.13 Use the average bond energies in Table 12.2 to calculate a value for the standard enthalpy of formation of cyclopropane. Compare this value with the value given in Table 12.1, and explain the difference.

Solution We need the standard enthalpy change for the reaction in which cyclopropane is formed from its elements in their standard states, that is, for the reaction

$$3\text{C(graphite)} + 3\text{H}_2(g) \longrightarrow \text{C}_3\text{H}_6(g)$$

We first convert graphite and molecular hydrogen to atoms, and then we combine the atoms to form cyclopropane:

$$3\text{C(graphite)} + 3\text{H}_2(g) \longrightarrow 3\text{C}(g) + 6\text{H}(g)$$

Table 12.3 Comparison of Enthalpies of Formation from Average Bond Energies with Experimental Values

COMPOUND	ΔH_f° CALCULATED FROM BOND ENERGIES (kJ mol^{-1})	ΔH_f° EXPERIMENTAL (kJ mol^{-1})	ΔH_f° EXPERIMENTAL $-\Delta H_f^\circ$ CALCULATED (kJ mol^{-1})
Methane, CH_4	-63	-75	-12
Ethane, C_2H_6	-85	-85	0
Propane, C_3H_8	-106	-104	$+2$
Butane, C_4H_{10}	-127	-126	$+1$
Pentane, C_5H_{12}	-148	-146	$+2$
Hexane, C_6H_{14}	-170	-167	$+3$
2-Methylpropane, C_4H_{10}	-127	-135	-8
2-Methylbutane, C_5H_{12}	-148	-154	-6
2,2-Dimethylpropane, C_5H_{12}	-148	-167	-19
2-Methylpentane, C_6H_{14}	-170	-175	-5
2,2-Dimethylbutane, C_6H_{14}	-170	-172	-2
Cyclopropane, C_3H_6	-64	$+53$	$+117$
Cyclobutane, C_4H_8	-89	$+28$	$+113$
Cyclopentane, C_5H_{10}	-106	-78	$+28$
Cyclohexane, C_6H_{12}	-128	-123	$+5$
Ethene, C_2H_4	$+52$	$+52$	0
Propene, C_3H_6	$+13$	$+20$	$+7$
1-Butene, C_4H_8	-8	0	$+8$
cis-2-Butene, C_4H_8	-8	-8	0
trans-2-Butene, C_4H_8	-8	-12	-4
Benzene, C_6H_6	$+229$	$+83$	-146

Using the values in Table 12.1, we have

$$\Delta H^\circ = (3 \text{ mol C})(716.7 \text{ kJ mol}^{-1}\text{ C}) + (3 \text{ mol H}_2)(436.0 \text{ kJ mol}^{-1}\text{ H}_2)$$
$$= 3458 \text{ kJ}$$

Now combining the atoms to form cyclopropane, we have

$$\Delta H^\circ = -3BE(\text{C}-\text{C}) - 6BE(\text{C}-\text{H}) = -1044 \text{ kJ} - 2478 \text{ kJ} = -3522 \text{ kJ}$$

	ΔH°
$3\text{C(graphite)} + 3\text{H}_2(\text{g}) \longrightarrow 3\text{C(g)} + 6\text{H(g)}$	$+3458$ kJ
$3\text{C(g)} + 6\text{H(g)} \longrightarrow \text{C}_3\text{H}_6(\text{g})$	-3522 kJ
$3\text{C(graphite)} + 3\text{H}_2(\text{g}) \longrightarrow \text{C}_3\text{H}_6(\text{g})$	-64 kJ

The value of ΔH_f° for cyclopropane from Table 12.1 is 53.3 kJ mol^{-1}. Thus cyclopropane has a higher energy than calculated from bond energies, by $53.3 - (-64) \text{ kJ} = 117$ kJ, because the bonds in cyclopropane are bent. Therefore 117 kJ is the strain energy of cyclopropane.

We see in Table 12.3 that there is also a discrepancy between ΔH_f (calculated) and ΔH_f (experimental) for benzene. The ΔH_f calculated from bond energies is $+229$ kJ mol^{-1}, but the observed value is only $+83$ kJ mol^{-1}. The formation of benzene from its elements is much less endothermic than calculated from bond energies; in other words, benzene is more stable than expected. This discrepancy arises because the calculated value is based on the Kekulé structure, which has alternate single and double bonds.

We determine the calculated value by first finding the energy needed to break all the bonds:

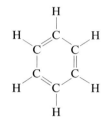

6 C—H bond	6×413 kJ =	2478 kJ
3 C—C bonds	3×348 kJ =	1044 kJ
3 C=C bonds	3×619 kJ =	1857 kJ
Total bond energy		5379 kJ

Then we calculate the enthalpy of formation as follows:

$$\Delta H°$$

6C(g) + 6H(g) $\longrightarrow$ C$_6$H$_6$(g)	Total bond energy =	-5379 kJ
6C(s, graphite) $\longrightarrow$ 6C(g)	6×716.7 kJ =	$+4300$ kJ
3H$_2$(g) $\longrightarrow$ 6H(g)	6×218 kJ =	$+1308$ kJ
6C(s, graphite) + 3H$_2$(g) $\longrightarrow$ C$_6$H$_6$(g)		$+229$ kJ

But we saw in Chapter 11 that all the bond lengths are equal and that benzene cannot be adequately described by a single structure with alternate single and double bonds. It can best be described by two resonance structures,

which show that six of the electrons are delocalized around the ring. When electrons in a molecule are delocalized, the molecule is more stable than it would be if the electrons were localized. Thus benzene is more stable than it would be if it had just one of the resonance structures, that is, a Kekulé structure.

The difference in energy between a benzene molecule with a Kekulé structure and an actual benzene molecule of $229 - 83$ kJ mol^{-1} = 146 kJ mol^{-1} is called the **resonance energy**, or **delocalization energy**, of benzene (see Figure 12.6).

It is sometimes said that benzene is *resonance-stabilized* or that benzene is more stable than expected because of resonance. Remember, however, that the resonance stabilization of a molecule is not a phenomenon. The concepts of resonance stabilization and resonance energy are necessary only because Lewis structures are an inadequate, approximate representation of the distribution of the electrons in some molecules such as benzene. A Lewis structure describes all the electrons in a molecule as being strictly localized in bonds and in unshared pairs, whereas in some molecules such as benzene not all the electrons are, in fact, localized in this way.

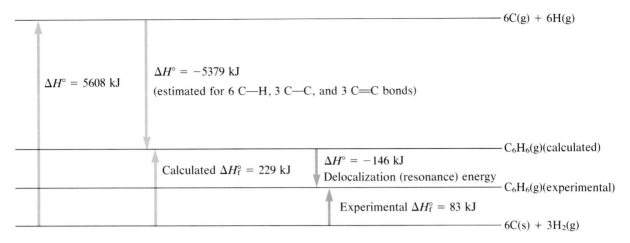

Figure 12.6 Delocalization (Resonance) Energy of Benzene.
The ΔH_f° of a Kekulé structure of benzene can be calculated from the ΔH_f° for
6 C(g) and 6 H(g), which is 5608 kJ, and the bond energies for 6 C—H, 3 C—C,
and 3 C=C bonds, which give −5379 kJ. So ΔH_f° (Kekulé structure) =
(5608 − 5379) kJ = 229 kJ. But the experimental ΔH_f° for C_6H_6(g) is only 83 kJ.
The difference 83 − 229 = −146 kJ is the delocalization (resonance) energy of
benzene.

12.4 FUELS AND FOODS

Before the Industrial Revolution in Europe wood and peat were the only significant fuels, and they were used primarily for heating and cooking. During the
latter half of the eighteenth century coal grew rapidly in importance because
of the increasing use of energy for power. Power was generated by burning coal
in a steam engine, which could be used for driving machinery, for generating
electricity, and for transportation. Today petroleum and natural gas provide
much larger amounts of energy. But as petroleum and natural gas stocks are
depleted, we are turning again to the use of coal and other fuels.

In assessing the suitability of a particular fuel for a particular purpose, not
only do we need to know the enthalpy of combustion, but we also must consider
several other factors. The cost of the fuel—in particular, the cost per kilojoule—
is of primary importance. Frequently, also, the enthalpy of combustion per
gram rather than per mole is important. For example, although liquid hydrogen
is very expensive, it gives more energy per unit mass than any other fuel. It is
therefore the preferred fuel for the Saturn rockets, since a rocket must lift its
fuel with it as it blasts off. With the increasing cost of petroleum fuels and rising
concern about atmospheric pollution, the use of liquid hydrogen for commercial
planes is becoming increasingly realistic. In other applications the volume of
the fuel that has to be carried by an automobile or by a plane, for example, is
of importance. Thus enthalpies of combustion of fuels are commonly quoted
either as enthalpies per gram or per liter (see Table 12.4).

Oxidation reactions are the major source not only of the energy for heating
and power production but also of the energy that sustains all living creatures.
We meet our energy needs by using the chemical energy that is released by the
oxidation of carbohydrates, proteins, and fats in food. These oxidation reactions
are typified by the oxidation of glucose:

$$C_6H_{12}O_6(s) + 6O_2(g) \longrightarrow 6CO_2(g) + 6H_2O(l) \qquad \Delta H^\circ = -2803.0 \text{ kJ}$$

In the body this reaction occurs in a relatively slow and controlled manner,

Table 12.4 Enthalpies of Combustion of Some Fuels and Pure Substances

SUBSTANCE	$(-\Delta H^\circ_{combustion})$ (kJ g^{-1})	$(-\Delta H^\circ_{combustion})$ (kJ L^{-1})	$-\Delta H^\circ_{combustion})$ (kJ mol^{-1})
Solids			
Wood[a]	≈25	—	—
Peat[a]	≈20	—	—
Lignite (brown coal)[a]	≈30	—	—
Bituminous coal (black coal)[a]	≈33	—	—
Anthracite[a]	≈35	—	—
Graphite	32.76	—	394.9
Glucose	15.57	—	2803.0
Liquids			
Gasoline		≈33 800	—
Kerosene		37 000	—
Diesel fuel		38 000	—
Heating oil		38 500	—
Hexane	48.30	31 897	4163.1
Octane	47.71	33 519	5450.5
Methanol, CH_3OH	22.67	17 944	726.5
Hydrogen (liquid, 20 K)	142.9	10 120	285.9
Gases			
Methane	55.64	39.72	890.3
Ethane	51.99	69.59	1559.8
Propane	50.45	99.04	2219.9
Coal gas		≈78	
Acetylene	49.98	57.98	1299.6
Ethylene	50.39	62.95	1411.0
Hydrogen	142.9	12.75	285.8

[a] These values are for the dried fuels; normally, these fuels contain some water.

Table 12.5 Enthalpies of Combustion of Various Foods

FOOD	ΔH° (kJ g^{-1})
White bread	−10.8
Potatoes	−2.9
Onion	−1.0
Apple	−1.9
Milk	−2.8
Cheese	−17.7
Butter	−33.3
Beef	−6.3
Sugar	−16.5
Beer	−1.5
Cake	−19.7

by a series of rather complicated steps. In contrast, the combustion of glucose outside the body occurs only at a high temperature, with the rapid evolution of a considerable amount of heat. The total energy made available in the complex series of reactions that occur in the body is exactly the same as would be obtained if the glucose were simply burned in air.

Therefore the energy that can be obtained from any food can be measured by burning it in oxygen in a calorimeter. It must first be dried, because most foods contain considerable quantities of water. Since foods are usually not pure substances, we cannot express the results as enthalpies of combustion per mole. They are usually expressed as the enthalpy change per gram; some typical average values for foods are given in Table 12.5. Often the energy content of foods is expressed in kilocalories (kcal) or Calories (Cal):

$$1 \text{ kcal} = 1 \text{ Cal} = 4.184 \text{ kJ}$$

12.5 THERMODYNAMICS

Thermochemistry is part of a much more general subject called *thermodynamics,* the science of transformations of energy. Most of this subject is beyond the scope of this book. However, in this section we take up an important principle of thermodynamics, the first law, which is fundamental to thermochemistry. We continue the discussion of thermodynamics in Chapter 26.

First Law of Thermodynamics

The **first law of thermodynamics** is just another name for the *law of conservation of energy*. As we noted in Chapter 2, this law is valid for mechanical and chemical systems, but it is not valid for any process in which atomic nuclei are broken down into fundamental particles or change into other nuclei, because these changes involve a significant conversion of mass to energy.

The first law of thermodynamics—that is the law of conservation of energy—may be stated in several ways. For example:

Energy can neither be created nor destroyed.

Or:

The energy of a system that is isolated from its surroundings is constant.

By a *system* we mean any part of the universe in which we are particularly interested. By the *surroundings* we mean all the rest of the universe, but it can usually be taken to mean just the immediate surroundings of a reaction vessel.

Energy can be transferred to a system in two fundamentally different ways: either by allowing heat to flow into the system or by doing work on the system. If the amount of energy transferred to a system as heat is q and if the amount of energy transferred to the system as work is w, then

$$\Delta E = q + w$$

This equation is a convenient mathematical expression of the first law of thermodynamics. Here E is called the *internal energy*.

If 10 kJ of energy is transferred to a system as heat, we write $q = 10$ kJ. If the same amount of energy is transferred as heat from the system to the surroundings, we write $q = -10$ kJ. The negative sign indicates loss of energy from the system.

If 15 kJ of energy is transferred to the system by doing work on the system, we write $w = +15$ kJ. If the system does 15 kJ of work on the surroundings, we write $w = -15$ kJ, because the system has lost 15 kJ of energy.

If we do 15 kJ of work on a system and it simultaneously loses 10 kJ of energy as heat to the surroundings, the internal energy of the system increases by

$$\Delta E = 15 \text{ kJ} - 10 \text{ kJ} = 5 \text{ kJ}$$

Heat and Work

At this point we can usefully examine the ideas of work and heat in a little more detail. Suppose a book with a mass of 1.0 kg is raised to a height of 10 cm. The work required is equal to the force acting on the book times the distance that it is raised, h. The force is $F = mg$, where m is the mass of the book and g is the acceleration due to gravity ($g = 9.8 \text{ m s}^{-2}$). The work done on the book is

$$w = Fh = mgh$$
$$= 1.0 \text{ kg} \times 9.8 \text{ m s}^{-2} \times 0.10 \text{ m}$$
$$= 0.98 \text{ kg m}^2 \text{ s}^{-2} = 0.98 \text{ J}$$

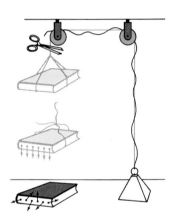

Thus 0.98 J of work is done on the book.

The potential energy of the book is therefore increased by the same amount, 0.98 J. If the book is allowed to fall back to the table, it could do this much work. If we were to attach it to a string running over a pulley and then to a weight, the book could raise the weight and thus do work on the weight. However, if the book is allowed to fall freely, it does no work. Its potential energy is converted to kinetic energy as it falls. And when it hits the table, this kinetic energy is converted to heat, which will slightly raise the temperature of the book and the table.

This example helps us to understand the essential difference between heat and work. When work is done on the book to lift it from the table, all its molecules are moved away from the table in an organized, coordinated way; we may say that work produces organized motion. When the book falls to the table, the force exerted by gravity causes all the molecules in the book to move in a coordinated, organized fashion toward the table. However, when the book hits the table, this organized molecular motion is transformed to the random or disorganized motion that is heat. The random, disorganized motion of the atoms and molecules of the book is increased, and its temperature is raised accordingly. We may say that heat is associated with random, disorganized motion, whereas work is associated with organized motion.

Internal Energy and State Functions

The internal energy of a system is the sum of the kinetic and potential energies of all the particles in the system. We cannot measure the absolute internal energy of a system, but we can measure the changes in the internal energy in terms of the work done on a system and the energy added to the system as heat.

The internal energy of a system is a characteristic property of the system; its value depends only on the conditions that are needed to specify the state of the system, such as composition, pressure, and temperature. Under exactly the same conditions a given system always has the same internal energy. The internal energy, E, is therefore called a **state function**. A state function is any property of a system that depends only on the state of the system and not on how that state was reached.

For a system going from some initial state (i) to some final state (f), the change in internal energy, ΔE, is given by

$$\Delta E = E_f - E_i$$

And since both E_f and E_i are state functions characteristic of the final state and the initial state, ΔE is always exactly the same no matter how the change from the initial state to the final state is carried out. In other words, it is completely independent of the route by which the change from initial state to final state took place.

A given system may change from an initial state to some final state by many routes. Different amounts of heat q may be added to the system, and different amounts of work w may be done on the system, provided only that the sum of q and w is constant, that is, that they satisfy the relation

$$\Delta E = q + w$$

The amount of heat added to the system, q, and the work done on the system, w, are not state functions. The initial and final states of a system may be likened to the balance in a bank account in which deposits and withdrawals are made in two currencies, namely, *heat* and *work*. The balance at the end of the month depends only on the sum of all the deposits and withdrawals. A particular balance could have been achieved in innumerable different ways; it is independent of the number and order of the deposits and withdrawals and the currency in which they were made, provided only that they have the same sum and therefore give the same final balance.

One mole of water at 25°C has a certain internal energy. One mole of ice at 0°C has a lower internal energy. The difference in their internal energies is always the same irrespective of the path by which the water at 25°C is converted to ice at 0°C. For example, the water could be cooled directly; that is, energy could be transferred as heat to the surroundings. Or it could first be boiled, in which case it does work in pushing away the atmosphere as it expands. Then the steam could be converted directly to ice by letting it come into contact with a surface held at 0°C, when energy would be transferred as heat from the steam to the surface (the surroundings).

The preceding discussion leads to another alternative statement of the *first law of thermodynamics*:

The change in energy, ΔE, accompanying the change of a system from an initial state to a final state is determined exactly by the initial and final states and is independent of the path by which change is effected.

Energy and Enthalpy

Suppose that we neutralize a solution of 1 mol of NaOH with a solution of 1 mol of HCl in a beaker open to the atmosphere. A certain amount of energy is produced as heat, q, which raises the temperature of the final solution. The solution therefore expands; that is, its volume increases. As it expands, the solution does work by pushing away the atmosphere.

The change in the energy of the system is not just $\Delta E = q$, though, because the system simultaneously loses energy by pushing back the atmosphere. If we could keep the volume of the system constant by carrying out the reaction in a vessel of fixed volume, it would then do no work against the surroundings, and

$$\Delta E = q \quad \text{(at constant volume)}$$

However, when reactions are carried out in vessels open to the atmosphere—that is, at constant pressure, the atmospheric pressure—then

$$\Delta E \neq q \quad \text{(at constant pressure)}$$

Because reactions are so often carried out at constant pressure, it is convenient to define a new property of the system H so that ΔH *is* equal to the energy transferred as heat. This property H is called the **enthalpy** of the system, and

$$\Delta H = q \quad \text{(at constant pressure)}$$

Thus as we stated earlier, the enthalpy change ΔH in a reaction equals the heat absorbed in the reaction at constant pressure. Using enthalpies of formation, we can, as we have seen, readily find the amount of energy that is required or is liberated as heat in any reaction that takes place at constant pressure. If we were to use internal energies rather than enthalpies, we would be faced with the complication of having to allow each time for the work done on or by the system.

We can find the relationship between ΔE and ΔH by considering a reaction carried out at constant pressure in a vessel fitted with a piston that allows the volume to change during the reaction, as shown in Figure 12.7. If the volume of the system increases by ΔV during a reaction, then the system does work $P\,\Delta V$ in pushing against the atmosphere. Therefore the work done *by* the system is $P\,\Delta V$. So the work done *on* the system is $-P\,\Delta V = w$. According to the first law of thermodynamics,

$$\Delta E = q + w$$

Hence

$$\Delta E = q - P\,\Delta V$$

$$q = \Delta E + P\,\Delta V \qquad \text{(at constant pressure)}$$

Now since $\Delta H = q$ (at constant pressure),

$$\Delta H = \Delta E + P\,\Delta V \qquad \text{(at constant pressure)}$$

or, in general

$$H = E + PV$$

Since P and V depend only on the state of the system and not on the path by which that state is reached, they are also state functions. Therefore the enthalpy H is a state function, like the internal energy E.

Because reactions are carried out at constant pressure more frequently than at constant volume, usually working with H rather than E is more convenient. Thus in tables of thermodynamic data ΔH rather than ΔE values are normally listed, as in Table 12.1. The usefulness of enthalpy was first recognized by the American scientist J. Willard Gibbs (Figure 12.8).

Reactions Carried Out at Constant Volume

Some reactions, particularly those involving gases, are often conveniently carried out in a sealed vessel, and therefore they occur at constant volume rather

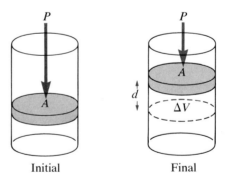

Initial   Final

Figure 12.7 Work Done in a Reaction at Constant Pressure.
Consider a reaction carried out in a reaction vessel with a tightly fitted piston, acted upon by an external pressure P, and in which there is an increase in volume, ΔV. Since pressure is force per unit area, the force acting on the piston is PA, where A is the area of the piston. If the piston moves out a distance, d, the work done by the system is force × distance = PAd. Now since Ad is the increase in the volume of the system, ΔV, the work done by the system is $PAd = P\,\Delta V$ at constant pressure P.

Figure 12.8 Josiah Willard Gibbs (1839–1903).
Gibbs, the son of a Yale professor, was the first person to be awarded a Ph.D. in science from an American university. After a period of study in France and Germany, he returned to New Haven in 1869. In 1871 he became professor of theoretical physics at Yale, retaining that position until his death. He is regarded by many as the most brilliant of native-born American scientists. He was a modest, reserved person who was said to be a poor teacher. He traveled very little, and he had almost no contact with the great scientists of the time in Europe. Working by himself, he applied the principles of thermodynamics to chemical reactions in a thorough and rigorous mathematical fashion. He published this work from 1876 to 1878 in the *Transactions of the Connecticut Academy of Sciences*. Unfortunately, this journal was not well known in Europe, and he wrote in such a concise and abstract style that most other scientists found his writing difficult to understand. Even Einstein said of a book that Gibbs wrote later, "It is a masterpiece but it is hard to read." It was not until the 1890s that his work was discovered by physical chemists in Europe. He was then universally recognized for his outstanding contribution to thermodynamics, and today his work remains the foundation of modern chemical thermodynamics.

than at constant pressure. A strong metal vessel, called a **bomb calorimeter**, is often used to measure the heat changes associated with reactions involving gases, particularly for measuring heats of combustion in which a large quantity of gaseous products may be formed at a high temperature (see Figure 12.9). For reactions in a bomb calorimeter the heat evolved gives the change in the internal energy, ΔE. If we wish to find the enthalpy change, ΔH, we must use the relationship

$$\Delta H = \Delta E + P\,\Delta V$$

as shown in the following example.

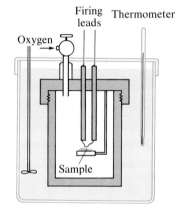

Figure 12.9 A Bomb Calorimeter.
A bomb calorimeter is particularly useful for measuring the enthalpy of combustion of solids. The reaction is initiated when a wire in contact with the solid is heated by passing an electric current through it. The reaction occurs at constant volume so that q, the heat absorbed, is equal to ΔE, the change in internal energy.

Example 12.14 When 1.000 g of propane was burned in excess oxygen in a bomb calorimeter, 52.50 kJ of heat was obtained at 25°C. Find ΔH for the reaction.

Solution The balanced equation for the reaction is

$$C_3H_8(g) + 5O_2(g) \longrightarrow 3CO_2(g) + 4H_2O(l)$$

The molar mass of propane is 44.09 g mol^{-1}. Hence

$$q = \left(\frac{-52.50\ \text{kJ}}{1.000\ \text{g C}_3\text{H}_8}\right)\left(\frac{44.09\ \text{g C}_3\text{H}_8}{1\ \text{mol C}_3\text{H}_8}\right) = -2315\ \text{kJ mol}^{-1}$$

Since the reaction was carried out at constant volume,

$$q = \Delta E = -2315\ \text{kJ mol}^{-1}$$

To find ΔH, we use the relation

$$\Delta H = \Delta E + P\,\Delta V$$

To find $P\,\Delta V$, we make use of the ideal gas law,

$$P\,\Delta V = \Delta nRT \qquad \text{(At constant pressure and temperature)}$$

We assume that the volume of any liquids and solids, in this case water, can be neglected. We note that 6 mol of gaseous reactants (1 mol C_3H_8 + 5 mol O_2) gives 3 mol of gaseous products (3 mol CO_2). So for this reaction $\Delta n = 3 - (5 + 1) = -3$ mol. Therefore

$$P\,\Delta V = (-3\ \text{mol})(8.31\ \text{J K}^{-1}\ \text{mol}^{-1})(298\ \text{K}) = -7.43\ \text{kJ}$$

$$\Delta H = \Delta E + P\,\Delta V = -2315 - 7.43\ \text{kJ} = -2322\ \text{kJ}$$

12.5 THERMODYNAMICS

455

We see from Example 12.14 that even for a gas reaction involving a relatively large volume change, the difference between ΔE and ΔH is quite small. For many purposes, particularly for reactions that do not involve large volume changes, we can justifiably ignore the difference between ΔH and ΔE.

Driving Force of Chemical Reactions

What makes a chemical reaction go? This question is one chemists and students have asked since chemical reactions first began to be studied. We might, perhaps, expect that a reaction would proceed spontaneously if the reacting system decreased in energy by transferring heat to its surroundings. However, this premise would lead us to conclude that all reactions are exothermic. Although many reactions that go to completion, such as the combustion of a hydrocarbon and the thermite reaction, are exothermic, there are also many endothermic reactions.

Clearly, some other factor must also be important. This factor is the tendency of any system composed of a large number of atoms and molecules to become more mixed up and more disordered with time as a consequence of their random motions. For example, two gases diffuse spontaneously into each other; that is, they become thoroughly mixed up with each other even though no energy changes are involved in the process. Also, a liquid evaporates spontaneously, despite the fact that this process is endothermic, because molecules in the gas state have a more disordered arrangement than in the liquid state. The disorder or "mixed-upness" of a system is called **entropy**. The *second law of thermodynamics* states that the entropy of a system and its surroundings increases in any spontaneous process. We will see several examples of the importance of entropy in later chapters and in Chapter 26 we discuss entropy and the second law of thermodynamics in detail.

IMPORTANT TERMS

The **average bond energy** of a bond in a polyatomic molecule is the average energy needed to break a bond of a particular type (for example a C—H bond).

The **bond energy** of a diatomic molecule is the energy needed to dissociate one mole of molecules in the gas state into atoms in the gas state.

A **calorimeter** is a device used to measure the quantity of heat absorbed or evolved during a chemical reaction or a physical change.

Calorimetry is the measurement of the heat absorbed or evolved in chemical reactions or other changes.

Endothermic reactions absorb heat from the surroundings.

Enthalpy is a state function defined by the expression $H = E + PV$.

The **enthalpy change ΔH** is the difference in the enthalpy of the final and initial states of a system. It is equal to the heat absorbed by the system at constant pressure, q_p.

The **standard enthalpy change $\Delta H°$** for a reaction is the enthalpy change for the reaction when all the reactants and the products are in their standard states.

The **standard enthalpy of formation** of a compound is the enthalpy change for the reaction in which one mole of the compound is formed from its elements in their standard states.

Exothermic reactions give off heat to the surroundings.

The **first law of thermodynamics** states that energy cannot be created or destroyed, or the energy of an isolated system is constant. In mathematical form, $\Delta E = q + w$, where ΔE is the change in internal energy, q is the amount of heat added to the system, and w is the amount of work done on the system. It is an alternative name for the **law of conservation of energy**.

Heat is energy that is transferred as a result of a temperature difference.

The **heat capacity** of an object or of a given amount of a substance is the amount of heat required to raise the temperature of the object or of the substance by 1 kelvin. **Molar heat capacity** is the amount of heat needed to raise the temperature of one mole of a substance by 1 kelvin.

Hess's law states that the enthalpy change for a reaction is the sum of the enthalpy changes for the individual steps of the reaction, or the enthalpy change for a reaction is independent of the path by which the reaction occurs.

The **internal energy E** is the sum of the kinetic and potential energies of all the particles in a system.

The **resonance** (or **delocalization energy**) of a molecule is the difference between the standard enthalpy of formation calculated on the basis of a structure containing localized bonds—for example, a Kekulé structure of benzene—and the experimental standard enthalpy of formation.

A **state function** is any property of a system that depends only on the state of the system and not on how that state was reached.

The **strain energy** of a substance whose molecules contain bent bonds is the difference between the experimental value of the standard enthalpy of formation of the substance and the standard enthalpy of formation calculated on the basis of bond energies for normal straight bonds.

The **surroundings** represent that portion of the universe with which a system interacts.

A **system** is a portion of the universe that is chosen for study.

Thermochemistry is the quantitative study of heat changes in chemical reactions.

PROBLEMS

Hess's Law

1. Calculate the value of ΔH° for the reaction

$$CuCl_2(s) + Cu(s) \longrightarrow 2CuCl(s)$$

given the following information:

$$Cu(s) + Cl_2(g) \longrightarrow CuCl_2(s) \qquad \Delta H^\circ = -206 \text{ kJ}$$
$$2Cu(s) + Cl_2(g) \longrightarrow 2CuCl(s) \qquad \Delta H^\circ = -36 \text{ kJ}$$

2. Given that

$$H_2(g) + F_2(g) \longrightarrow 2HF(g) \qquad \Delta H^\circ = -542 \text{ kJ}$$
$$2H_2(g) + O_2(g) \longrightarrow 2H_2O(l) \qquad \Delta H^\circ = -572 \text{ kJ}$$

calculate ΔH° for the reaction

$$2F_2(g) + 2H_2O(l) \longrightarrow 4HF(g) + O_2(g)$$

3. Calculate ΔH° for the reaction

$$2CO(g) + O_2(g) \longrightarrow 2CO_2(g)$$

given the following information:

$$C(graphite) + O_2(g) \longrightarrow CO_2(g) \quad \Delta H^\circ = -393.5 \text{ kJ}$$
$$2C(graphite) + O_2(g) \longrightarrow 2CO(g) \quad \Delta H^\circ = -110.5 \text{ kJ}$$

4. Calculate ΔH° for the reduction of FeO by CO,

$$FeO(s) + CO(g) \longrightarrow Fe(s) + CO_2(g)$$

given the following data:

$$Fe_3O_4(s) + CO(g) \rightarrow 3FeO(s) + CO_2(g) \, \Delta H^\circ = +38 \text{ kJ}$$
$$2Fe_3O_4(s) + CO_2(g) \rightarrow 3Fe_2O_3(s) + CO(g) \, \Delta H^\circ = +59 \text{ kJ}$$
$$Fe_2O_3(s) + 3CO(g) \rightarrow 2Fe(s) + 3CO_2(g) \, \Delta H^\circ = -28 \text{ kJ}$$

5. Calculate the enthalpy change for the reaction of nitrogen dioxide with water,

$$3NO_2(g) + H_2O(l) \longrightarrow 2HNO_3(aq) + NO(g)$$

given the following information:

$$2NO(g) + O_2(g) \longrightarrow 2NO_2(g) \qquad \Delta H^\circ = -173 \text{ kJ}$$
$$2N_2(g) + 5O_2(g) + 2H_2O(l) \longrightarrow$$
$$4HNO_3(aq) \qquad \Delta H^\circ = -255 \text{ kJ}$$
$$N_2(g) + O_2(g) \longrightarrow 2NO(g) \qquad \Delta H^\circ = +181 \text{ kJ}$$

Enthalpies of Formation and Combustion

6. What is ΔH° for the conversion of 2 mol of gaseous cyclohexane, C_6H_{12}, to 3 mol of gaseous cyclobutane, C_4H_8?

7. The overall reaction in photosynthesis is

$$6CO_2(g) + 6H_2O(l) \longrightarrow C_6H_{12}O_6(s) + 6O_2(g)$$

Given that ΔH_f° for glucose, $C_6H_{12}O_6(s)$, is -1273 kJ mol^{-1}, calculate the ΔH° of this reaction per mole and per gram of glucose. (In nature the large energy requirement for this reaction is supplied by light from the sun rather than as heat.)

8. What is the enthalpy change when 1 mol of $SO_3(g)$ combines with 1 mol of $H_2O(g)$ to yield 1 mol of $H_2SO_4(l)$?

9. Calculate ΔH_f° for ethyne, C_2H_2, from the standard enthalpy change of -312 kJ for the reaction

$$C_2H_2(g) + 2H_2(g) \longrightarrow C_2H_6(g)$$

and the standard enthalpy of formation of ethane, C_2H_6, given in Table 12.1.

10. The standard enthalpy of formation is 0 kJ mol^{-1} for $H_2(g)$, $N_2(g)$, $O_2(g)$, $F_2(g)$, and $Cl_2(g)$ but not for $Br_2(g)$ and $I_2(g)$. Explain this difference.

11. Calculate the enthalpy of formation of solid magnesium hydroxide from the following information:

$$2Mg(s) + O_2(g) \longrightarrow 2MgO(s) \qquad \Delta H^\circ = -1203.7 \text{ kJ}$$
$$MgO(s) + H_2O(l) \longrightarrow Mg(OH)_2(s) \qquad \Delta H^\circ = -36.7 \text{ kJ}$$
$$2H_2O(l) \longrightarrow 2H_2(g) + O_2(g) \qquad \Delta H^\circ = +572.4 \text{ kJ}$$

12. Many cigarette lighters contain liquid butane, C_4H_{10}, whose ΔH_f° is -127 kJ mol^{-1}. Calculate the heat given off by the combustion of 1 g of C_4H_{10} in the lighter, assuming that the final products are $CO_2(g)$ and $H_2O(g)$ at 298 K and 1 atm.

* The asterisk denotes the more difficult problems.

13. Calculate the enthalpy changes when $NO(g)$, $SO_2(g)$, and $O_2(g)$ are oxidized to $NO_2(g)$, $SO_3(g)$, and $O_3(g)$. Which reactions are exothermic, and which are endothermic?

14. The hydrocarbon propane, C_3H_8, is used as domestic fuel and as a fuel for automobiles in some areas. What is the standard enthalpy of formation of $C_3H_8(g)$, given that combustion of 1 g of gaseous C_3H_8 releases 46.3 kJ of heat when it is burned to $CO_2(g)$ and $H_2O(g)$ at 298 K and 1 atm?

15. Methanol, CH_3OH, is a potential future fuel supplement. Although it provides only half as much energy per liter as gasoline, it is clean burning and has a high octane number. It is made industrially from synthesis gas at high pressure, using a catalyst at about 300°C:

$$2H_2(g) + CO(g) \longrightarrow CH_3OH(l)$$

Use the following $\Delta H°$ (combustion) values to determine $\Delta H°$ for the synthesis of methanol:

$$CH_3OH(l) + \tfrac{3}{2}O_2 \longrightarrow CO_2(g) + 2H_2O(l)$$
$$\Delta H° = -726.6 \text{ kJ}$$

$$C(graphite) + \tfrac{1}{2}O_2(g) \longrightarrow CO(g) \qquad \Delta H° = -110.5 \text{ kJ}$$

$$C(graphite) + O_2(g) \longrightarrow CO_2(g) \qquad \Delta H° = -393.5 \text{ kJ}$$

$$H_2(g) + \tfrac{1}{2}O_2(g) \longrightarrow H_2O(l) \qquad \Delta H° = -285.8 \text{ kJ}$$

16. The standard enthalpies of combustion ($\Delta H°_{combustion}$) of C(graphite), $H_2(g)$, $C_2H_6(g)$, and $C_3H_8(g)$, in kilojoules per mole are -393.5, -285.8, -1559.8, and -2219.9, respectively. Calculate the enthalpy of formation of $C_2H_6(g)$ and $C_3H_8(g)$, and predict the values of $\Delta H°_f$ and $\Delta H°_{combustion}$ for butane, $C_4H_{10}(g)$.

17. The accompanying table gives the heat evolved for the combustion of a given mass of alcohol fuel at constant pressure. How much heat per mole, $\Delta H°_{combustion}$ is evolved for each alcohol? Explain the trend in the $\Delta H°_{combustion}$ values.

ALCOHOL	MASS (g)	HEAT EVOLVED (kJ)
Methanol, CH_3OH	3.20	71
Ethanol, C_2H_5OH	9.20	268
Propanol, C_3H_7OH	7.50	251
1-Butanol, C_4H_9OH	7.40	263
2-Butanol, C_4H_9OH	8.20	288

18. Calculate the standard enthalpy of formation, $\Delta H°_f$, for C_2H_5OH from its enthalpy of combustion to $CO_2(g)$ and $H_2O(l)$, which is $-1370 \text{ kJ mol}^{-1}$ at 25°C and 1 atm, and from the data given in Table 12.1.

19. Calculate the standard enthalpy of formation of propane, C_3H_8, from its standard enthalpy of combustion to $CO_2(g)$ and $H_2O(g)$ at 25°C, which is $-2044 \text{ kJ mol}^{-1}$, and the data given in Table 12.1.

Bond Energies

20. The standard enthalpies of formation of atomic O and atomic N are, respectively, 249 and 473 kJ mol^{-1}. What is the dissociation energy of the double bond in $O_2(g)$ and of the triple bond in $N_2(g)$?

21. Using the average C—H and C—C bond energies given in Table 12.2, estimate the $\Delta H°_f$ of ethane, $C_2H_6(g)$.

22. Calculate the average P—Cl bond energy in PCl_3, given the thermochemical information that follows. What is the average energy of the two additional PCl bonds in PCl_5? The $\Delta H°_f$ values, in kilojoules per mole, are as follows:

$$P(g) = +316.2 \qquad Cl(g) = +121.7$$
$$PCl_5(g) = -374.7 \qquad PCl_3(g) = -287.0$$

23. Calculate a value for the O—H bond energy in water from the accompanying data. Using this value, deduce a value for the O—O bond energy in hydrogen peroxide. Is the O—O bond strength more, or is it less, than half that in molecular oxygen? The $\Delta H°_f$ values, in kilojoules per mole, are as follows:

$$H(g) = +218.0 \qquad O(g) = +249.1$$
$$H_2O_2(g) = -136.4 \qquad H_2O(g) = -241.8$$

24. Calculate the carbon-oxygen bond energies in CO and CO_2, using the following $\Delta H°_f$ values:

$$CO(g) = -110.5 \text{ kJ mol}^{-1} \qquad CO_2(g) = -393.5 \text{ kJ mol}^{-1}$$
$$O(g) = 249.1 \text{ kJ mol}^{-1} \qquad C(g) = 716.7 \text{ kJ mol}^{-1}$$

Compare these energies with the CO bond energies given in Table 12.2, and discuss the differences.

25. Calculate the C=O bond energy in H_2CO from its $\Delta H°_f$ of $-115.9 \text{ kJ mol}^{-1}$ and the average C—H bond energy given in Table 12.2. Is the average C=O bond energy in CO_2 (see Problem 24) greater or less than this value? The $\Delta H°_f$ values, in kilojoules per mole, are as follows:

$$C(g) = +716.7 \qquad H(g) = +218.0 \qquad O(g) = +249.1$$

26. Use the bond energies given in Table 12.2 to estimate $\Delta H°$ for each of the following gas-phase reactions:

(a) $H_2S + Cl_2 \longrightarrow SCl_2 + H_2$

(b) $CH_4 + 2F_2 \longrightarrow CH_2F_2 + 2HF$

(c) $CH_2Cl_2 + CH_4 \longrightarrow 2CH_3Cl$

27. Use the bond energies given in Table 12.2 to estimate $\Delta H°$ for each of the following gas-phase reactions:

(a) $C_2H_2 + C_2H_6 \longrightarrow 2C_2H_4$

(b) $2H_2O_2 \longrightarrow 2H_2O + O_2$

(c) $CO + H_2 \longrightarrow H_2CO$

Resonance Energy and Strain Energy

28. The enthalpy of formation of the hydrocarbon naphthalene, $C_{10}H_8(g)$, is $+151$ kJ mol^{-1}. What is its resonance energy?

29. From the bond energy values given in Table 12.2, estimate values for the standard enthalpies of formation of cyclobutane and cyclohexane. Compare these values with the values given in Table 12.1, and comment on any differences.

First Law of Thermodynamics

30. What is $\Delta H°$ and what is $\Delta E°$ for the reaction in which 1 mol of ethane is burned in excess oxygen to give $CO_2(g)$ and $H_2O(l)$? What are the values of $\Delta H°$ and $\Delta E°$ if $H_2O(g)$ is formed?

31. How much heat will be evolved at 25°C if 2.42 g of *n*-butane is burned in excess oxygen in a bomb calorimeter?

32. The combustion of gaseous ethyne, C_2H_2, to give liquid water and gaseous carbon dioxide yields 1303 kJ at constant volume and 25°C. What is the enthalpy of combustion of ethyne at 25°C and 1 atm pressure?

Calorimetry

33. A 0.150-g sample of octane(l) was burned in a flame calorimeter (see Figure 12.2c). If the calorimeter contained 1500 g of water and the temperature of the water rose from 25.246°C to 26.386°C, what is the standard enthalpy of combustion of octane at 25°C?

34. A volume of 50.0 cm^3 of 0.400M NaOH solution was added to 20.0 cm^3 of 0.500M H_2SO_4 solution in a calorimeter of heat capacity 39.0 J K^{-1}. The temperature of the resulting solution rose by 3.60°C. What is the standard enthalpy of neutralization of $H_2SO_4(aq)$ with NaOH(aq)?

35. Calcium oxide (lime) reacts with water to give calcium hydroxide in an exothermic reaction:

$$CaO(s) + H_2O(l) \longrightarrow Ca(OH)_2(s)$$

A 5.40-g sample of calcium oxide was added to 500 mL of water in a calorimeter of heat capacity 350 J K^{-1}. The observed temperature increase was 2.60 K. What is the standard enthalpy change for the reaction of one mole of CaO(s) to give one mole of $Ca(OH)_2(s)$?

Miscellaneous

*36. By comparing the cost per joule of gasoline and electricity in your area, decide which form of energy is currently the best buy. Using the fact that 1 W equals 1 J s^{-1}, work out the number of joules in 1 kilowatt hour (kW h), the usual unit for which you can find a price quotes on utility bills. Assume that gasoline has the same enthalpy of formation as hexane, $C_6H_{14}(l)$, for which $\Delta H_f° = -199$ kJ mol^{-1} and whose density is 1.38 g mL^{-1}. Assume that gasoline burns completely to $CO_2(g)$ and $H_2O(l)$. Use the price of regular gasoline from local service stations.

*37. Estimate the enthalpy of vaporization of 1 mole of C(graphite), given that in the Lewis structure for a layer each carbon participates in one C=C and two C—C bonds. Use the bond energies given in Table 12.2.

*38. (a) Define standard enthalpy of formation, illustrating your answer by reference to $MgCO_3(s)$.

(b) When 0.203 g of magnesium was dissolved in an excess of dilute hydrochloric acid in a vacuum flask calorimeter, the temperature rose by 8.61 K. In a separate experiment, it was found that the calorimeter and its contents required 506 J to raise its temperature by 1.02 K. Calculate the heat released in the experiment. Hence find the enthalpy change for the reaction per mole of magnesium.

(c) In a similar experiment, solid magnesium carbonate reacted with an excess of dilute hydrochloric acid and the enthalpy change for the reaction was found to be -90.4 kJ per mole of magnesium carbonate. Use your result in (b) and the standard enthalpies of formation of $H_2O(l)$ and $CO_2(g)$ to find the standard enthalpy of formation of $MgCO_3(s)$.

CHAPTER 13

LIQUIDS, WATER, SOLUTIONS, AND INTERMOLECULAR FORCES

We discussed gases in Chapter 3 and solids in Chapter 10. We have left liquids to last because they are the most difficult to treat theoretically and are the least well understood. The simplifying assumptions that apply to the gaseous state—namely, that the volume of the atoms or molecules is negligible compared with the volume occupied by the gas, and that the forces between the molecules are also negligible—are not applicable to liquids. The molecules in a liquid are packed closely together, and the forces between them are relatively strong. In contrast to the properties of a gas, the properties of a liquid depend very much on the nature of the molecules—their size and shape—and on the forces between them. Moreover, the molecules in a liquid are not arranged in the regular patterns that we find in most solids, so it is much more difficult to describe the structure of a liquid than the structure of a solid.

Yet liquids are extremely important. Water is the most common liquid and the most common substance on earth. It covers 72% of the earth's surface. It is found naturally as a liquid, as a solid (snow and ice), and as a gas (water vapor) in the atmosphere. Water is essential to life; the human body is 65% water by mass, and a tomato is approximately 90% water.

In this chapter we discuss the properties of liquids, with particular emphasis on water. We also discuss the changes among the gas, liquid, and solid states of matter, called phase changes. Many chemical reactions are most conveniently carried out in solution, and water is the most common solvent for this purpose. The solubility of substances in water and other liquids is strongly dependent on the forces acting between the solvent molecules and between the solute molecules. We have frequently mentioned these intermolecular forces, and we now discuss their nature and strength in some detail. One type of intermolecular force, the hydrogen bond, is of particular importance in water and plays a vital role in living matter.

Not only is water a good solvent for many reactions, but it also undergoes many important reactions itself. Some of these reactions we have encountered in previous chapters. In this chapter we review these reactions, noting that water participates in both acid-base and oxidation-reduction reactions. Finally, we discuss some important physical properties of solutions—in particular, their freezing point, boiling point, and osmotic pressure.

13.1 NATURAL WATERS

Of the vast amount of water on the earth's surface, 97% is in the oceans. The remainder is fresh water, but 2.1% is in the form of ice caps and glaciers, and only 0.7% is readily available in rivers and lakes and as underground water. Water is continually being redistributed. It evaporates from the lakes and oceans into the atmosphere and returns in the form of rain and snow. This water runs off the surface or percolates through the ground and finally returns to the sea. In doing so, it dissolves many substances from the earth's crust, which accumulate in the oceans. This redistribution of water makes life on the land possible, and the dissolved substances that accumulate in the ocean are important for marine life.

Seawater

The 97% of all the earth's water that is found in the seas and oceans contains too much sodium chloride and other dissolved substances to be useful as drinking water or for many other purposes. Fresh water is obtained by distillation from seawater in Israel, Kuwait, and some other countries with a hot climate, but this process is expensive in most places because a large amount of energy is needed.

Although the oceans are not generally useful as a source of water, they are a vast storehouse of many other substances. Table 13.1 gives the concentrations of the most common of these. Each cubic meter of seawater contains 1.5 kg of dissolved substances, but the low concentrations of most of them makes any process for separating them expensive and uneconomical at present. However, two important substances are recovered commercially from seawater—sodium chloride and magnesium. In the past, bromine was also obtained from seawater, but it is now mostly obtained from underground brines (Chapter 5).

Sodium chloride is the most abundant dissolved substance in seawater, and much of it is obtained from this source. In hot climates seawater trapped in shallow ponds is allowed to evaporate in the sun until the solubility of sodium chloride is exceeded.

Magnesium is removed from seawater by adding lime, which precipitates magnesium hydroxide:

$$CaO(s) + H_2O \longrightarrow Ca^{2+}(aq) + 2OH^-(aq)$$
$$Mg^{2+}(aq) + 2OH^-(aq) \longrightarrow Mg(OH)_2(s)$$

The magnesium hydroxide is filtered off and converted to magnesium chloride with hydrochloric acid:

$$Mg(OH)_2(s) + 2HCl(aq) \longrightarrow MgCl_2(aq) + 2H_2O$$

The solution is evaporated to give solid magnesium chloride, which after being dried, is melted and then electrolyzed to give magnesium and chlorine (see Chapter 16).

Table 13.1 Major Constituents of Seawater (ppm)

Sodium, Na^+	10 561
Magnesium, Mg^{2+}	1 272
Calcium, Ca^{2+}	400
Potassium, K^+	380
Chloride, Cl^-	18 980
Sulfate, SO_4^{2-}	2 649
Hydrogen carbonate, HCO_3^-	142
Bromide, Br^-	65
Other substances	34

Purification of Water

The water needed for homes, agriculture, and industry is taken from lakes, rivers, and underground sources. Much of this water must be treated to remove bacteria and other dangerous impurities. Figure 13.1 illustrates a typical purification process. After a preliminary filtration, the water is allowed to stand in large tanks so that fine sand and other very small particles can settle. This process is aided by first making the water slightly alkaline by the addition of lime,

$$CaO(s) + H_2O(l) \longrightarrow Ca^{2+}(aq) + 2OH^-(aq)$$

then adding aluminum sulfate or alum to give a gelatinous precipitate of $Al(OH)_3$ (see Experiment 13.1):

$$Al^{3+}(aq) + 3OH^-(aq) \longrightarrow Al(OH)_3(s)$$

This precipitate settles out slowly and carries with it much of the suspended matter, including most of the bacteria. The water is then again passed through a filter and often through activated charcoal, which adsorbs most of the impurities that still remain. Then the water is sprayed into the air to speed up the oxidation of dissolved organic substances.

In the final stage of purification the water is treated with an oxidizing agent to complete the destruction of bacteria. Ozone, O_3, is very effective, but it must be generated on the site. In North America chlorine is most frequently used because it can be shipped in tanks in liquid form and dispensed from the tanks directly into the water supply. The hypochlorous acid formed by the chlorine in water is the effective sterilizing agent:

$$Cl_2(g) + 2H_2O(l) \longrightarrow HOCl(aq) + H_3O^+(aq) + Cl^-(aq)$$

A very small concentration of approximately 1 ppm of hypochlorous acid is sufficient to kill bacteria. This concentration is harmless to humans, although it adds a detectable taste to the water.

Even after this treatment water is not completely pure. It contains small amounts of dissolved salts, particularly sodium, potassium, magnesium, and

Figure 13.1 Water Purification System.

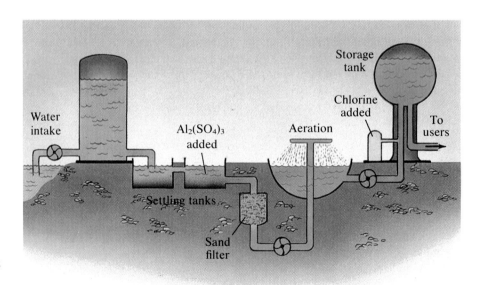

Purification of Water

The beaker on the left contains muddy water in which there is a large amount of suspended material. Ammonium hydroxide and aluminum chloride are added (center), producing a thick gelatinous precipitate of aluminum hydroxide which slowly settles to the bottom, carrying the suspended matter with it and leaving the water clear (right).

calcium, in the form of chlorides, sulfates, fluorides, and hydrogen carbonates. These salts have no harmful effects in the low concentrations usually present; indeed, they provide essential "minerals" for the body. Water also contains dissolved gases, in particular, oxygen, nitrogen, and carbon dioxide. The dissolved oxygen is essential to aquatic life.

13.2 PHASE CHANGES

With the exception of helium, all liquids, when cooled sufficiently at atmospheric pressure, freeze to form a solid. Also, all liquids are volatile to some extent; that is, they tend to evaporate to form gases. At atmospheric pressure, a liquid is completely transformed to gas at a particular temperature called its *boiling point*.

The term *vapor* is frequently used to refer to a gas that is formed by the evaporation of a liquid or a gas that is easily condensed to give a liquid. Thus one usually refers to water vapor or to gasoline vapor. However, there is no difference between a vapor and a gas; every vapor is a gas, and every gas can be condensed to a liquid. It is merely a matter of habit that we speak of water vapor rather than water gas or gaseous water.

The transformation of a substance from one state to another is called a **phase change**. Figure 13.2 summarizes the possible phase changes and the terms used to describe them.

Melting and Freezing

When a solid is heated, the kinetic energy of the atoms or molecules increases so that the molecules vibrate about their fixed positions more vigorously. With

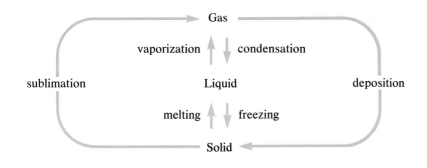

Figure 13.2 Phase Changes.

increasing temperature this motion eventually becomes sufficiently violent that the atoms or molecules break away from their fixed positions and the ordered arrangement of the solid is replaced by the random arrangement of the liquid. In other words, the solid melts. At the melting point of the solid, which is also the freezing point of the liquid, melting and freezing are occurring at the same rate, and the system is in dynamic equilibrium. Solid helium cannot be obtained at ordinary pressures because the motion of the very light helium atoms is so great that the solid simply shakes itself to pieces unless it is held together by increasing the pressure.

Because the energy of the molecules must be increased in order to break up their arrangement in a solid, melting (or fusion) is an endothermic process. The **molar enthalpy of fusion**, ΔH_{fus}, is the heat required to melt 1 mol of a substance at constant pressure. Values for the molar enthalpy of fusion of some solids are given in Table 13.2. A low molar enthalpy of fusion implies that the intermolecular forces are weak; such substances usually have low melting points. Conversely, a high enthalpy of fusion implies that the intermolecular forces are strong; such substances usually have high melting points.

A few substances exhibit an unusual state of matter that is intermediate between a liquid and a crystalline solid; it is called a *liquid crystal* (see Box 13.1).

Evaporation and Condensation

The molecules of a liquid have a distribution of energies similar to those in a gas (Figure 13.3). Molecules with enough kinetic energy to enable them to overcome the intermolecular forces holding them together may leave the liquid to form the vapor. This process is called **evaporation**. As the more energetic molecules leave the liquid, the average kinetic energy of the remaining molecules must decrease, so the temperature of the liquid must also decrease, unless heat is added to keep the temperature constant.

Vaporization is an endothermic process. The **molar enthalpy of vaporization**, ΔH_v, is the heat required to transform 1 mol of liquid into vapor at 1 atm pressure and at a specified temperature. Values of the enthalpy of vaporization at the boiling point of some liquids are given in Table 13.3. Like the enthalpy of fusion, the enthalpy of vaporization is a measure of the strength of intermolecular forces. Values range from as low as 0.9 kJ mol^{-1} for hydrogen, H_2, to 612 kJ mol^{-1} for carbon, C.

One of the important roles played by water in the body is that of maintaining

Table 13.2 Molar Enthalpies of Fusion

SUBSTANCE	FORMULA	ΔH_{fus} (kJ mol^{-1})	MELTING POINT (°C)
Hydrogen	H_2	0.12	−259
Methane	CH_4	0.94	−164
Mercury	Hg	2.3	−38.9
Tetrachloromethane	CCl_4	2.51	−22.9
Ethanol	C_2H_5OH	5.01	−114.6
Water	H_2O	6.0	0.0
Benzene	C_6H_6	10.6	5.5
Silver	Ag	11.3	961
Sodium chloride	NaCl	27.2	801

Box 13.1

LIQUID CRYSTALS

The displays in many calculators, wristwatches, readout meters, and temperature-measuring strips use substances called *liquid crystals*. These substances are unusual in that they have a structure intermediate between that of a liquid and a crystalline solid. In a liquid the molecules have a random arrangement, and they are able to move past each other. In a crystalline solid the molecules have an ordered arrangement and are in fixed positions. In a small temperature range just above the melting point, the molecules of a liquid crystal have an ordered arrangement, but they are able to move past each other, so the substance is a liquid. Molecules that form liquid crystals have rather special shapes; they are either long and cylindrical—rodlike—or large and flat—platelike. Rodlike molecules can form

arrangements like those shown in the accompanying figure. They can rotate around their own axis, and they can slide past each other, but they remain parallel to each other, like soda straws or matches in a long narrow box.

The practical applications of liquid crystals depend on their rather remarkable optical properties. Because of the ordered arrangement of the molecules, they can diffract light, just as the planes of atoms in a crystal diffract X rays (Box 10.2). Only light of a given wavelength will satisfy the Bragg relationship, and so only one of the wavelengths of white light is reflected by a liquid crystal, which therefore appears colored. As the temperature is changed, the distance between the layers of molecules changes, and therefore the color of the reflected light changes corres-

pondingly. Liquid crystals can therefore also be used as sensitive temperature-measuring devices. A liquid crystal film can be used to map the temperature on the surface of an object. One medical application is for locating veins by the slightly higher temperature of the skin above a vein.

The optical properties of liquid crystals are also affected by an electric field. When a thin film of a liquid crystal is placed between two electrodes and an electric potential is applied, a rearrangement of the structure occurs, and the transparent liquid crystal becomes opaque. This property is used in the number displays of digital watches and other instruments. Only very small potentials are needed, so these devices consume very little power.

 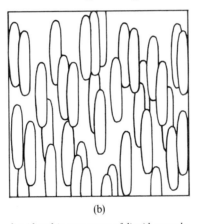 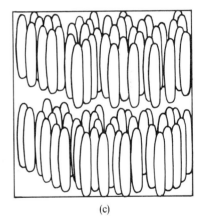

(a)　　　　　　　(b)　　　　　　　(c)

The arrangement of molecules in an ordinary liquid and in two types of liquid crystals. (a) An ordinary liquid. (b) A liquid crystal: the molecules can freely rotate and slip past each other, but retain their parallel orientation. (c) Another type of liquid crystal: the molecules are arranged in layers. They can rotate and move past each other in the layers, and the layers can also slide over each other, but the layers are retained.

the body temperature close to 37°C. The metabolic processes in the body generate heat, and any heat in excess of that required to maintain the body temperature must be dissipated. If the air temperature is lower than the body temperature, we lose heat by conduction. But if we are exercising vigorously or if the air temperature is very high, we cannot get rid of the heat rapidly enough by conduction. We then rely on the evaporation of water, that is, on sweating, to remove heat and maintain the body temperature. Because of its relatively high enthalpy of vaporization, water is very effective for this purpose.

13.2 PHASE CHANGES

465

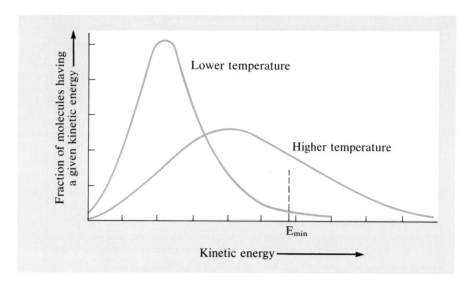

Figure 13.3 Distribution of Kinetic Energies of Molecules in a Liquid. The average kinetic energy of the molecules increases with increasing temperature. With increasing temperature a rapidly increasing number of molecules have more than the minimum energy, E_{min}, needed for a molecule to escape from the liquid into the vapor phase.

Table 13.3 Molar Enthalpies of Vaporization

SUBSTANCE	FORMULA	ΔH_v (kJ mol^{-1})	BOILING POINT (°C)
Hydrogen	H_2	0.9	−253
Methane	CH_4	10.4	−164
Pentane	C_5H_{12}	27.0	36.1
Tetrachloromethane	CCl_4	30.0	76.7
Benzene	C_6H_6	30.8	80.2
Ethanol	C_2H_5OH	38.6	78.5
Water	H_2O	40.7	100
Mercury	Hg	59.3	357
Sodium chloride	NaCl	207	1465
Carbon (graphite)	C	612	4830

Vapor Pressure

We are familiar with the fact that if water or gasoline is left in an open container or spilled on the ground, it gradually evaporates until all the liquid has been converted to vapor. But if a sufficient amount of liquid is present in a closed container, it does not all evaporate. Instead, an equilibrium is established between the liquid and the vapor, and a constant pressure is established in the container, as Figure 13.4 illustrates.

At first the amount of vapor increases, and the pressure that it exerts increases correspondingly. As the amount of vapor increases, the chance that a molecule in the vapor phase will collide with the surface of the water and return to the liquid phase also increases. Eventually, the number of molecules hitting the surface and returning to the liquid is equal to the number leaving. A state of dynamic equilibrium is reached, which can be summarized by the equation

$$H_2O(l) \rightleftharpoons H_2O(g)$$

No further changes in the amounts of water in the liquid phase and in the vapor phase are then observed, and the constant amount of vapor in the container exerts a constant pressure. The **vapor pressure** is the constant pressure

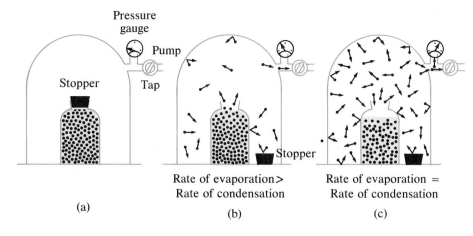

Figure 13.4 Vapor Pressure. (a) A bottle completely filled with water is placed in a container that can be evacuated by pumping out all the air. When all the air has been removed and the pressure gauge registers zero, the tap is turned to close the container, and the stopper of the flask is removed by remote control. (b) The water then begins to evaporate into the container, and the pressure gauge begins to register the pressure of the water vapor. As the amount of water vapor increases, the chance that a gaseous molecule in the vapor phase will collide with the surface of the water and return to the liquid phase also increases. (c) Eventually, the number of molecules returning to the liquid phase is equal to the number leaving. No further changes in the amounts of liquid and vapor are then observed. The vapor exerts a constant pressure (the vapor pressure of water at the temperature of the experiment), which is registered on the pressure gauge as a constant value.

exerted by the vapor above a liquid when equilibrium is established. The vapor pressure of a liquid is independent of the volume of the container in which it is confined, and it has a constant value at a given temperature (see Figure 13.5).

The vapor pressure of a liquid depends on the tendency of the molecules to escape into the vapor phase, which in turn depends on the strength of the intermolecular forces in the liquid. When the intermolecular forces are weak, the molecules escape easily, and the vapor pressure is high. Thus the vapor pressure of a liquid provides a measure of the magnitude of the intermolecular forces.

If a liquid is not confined in a container, it still exerts its vapor pressure, but the vapor diffuses away. The rate of condensation is thus always less than the rate of evaporation, and the liquid continues to evaporate until it has all been converted to vapor.

The vapor pressure of a liquid depends on the temperature, as Table 13.4 demonstrates. As the temperature increases, a larger fraction of the molecules

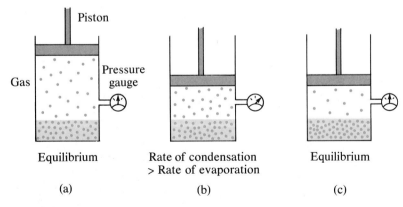

Figure 13.5 The Vapor Pressure of a Liquid Is Independent of the Volume of the Container.

(a) A liquid and its vapor are confined in a container with a tight-fitting piston. (b) If we push in the piston, thus decreasing the volume of the vapor, the concentration of the molecules in the gas phase will momentarily increase. In turn, the rate of collisions of gas molecules with the surface will increase, and thus the rate at which they return to the liquid will increase. Since this rate will be greater than the rate at which the molecules evaporate, the system will no longer be in equilibrium. Vapor will condense until the concentration of gas molecules has decreased to its original value. (c) The system is then again in equilibrium, and the vapor pressure again has its original value.

Table 13.4 Vapor Pressure of Water

T (°C)	P (mm Hg)	T (°C)	P (mm Hg)
0	4.6	25	23.8
5	6.5	30	31.8
10	9.2	40	55.3
15	12.8	60	147.4
20	17.5	80	355.1
		100	760.0

have sufficient energy to overcome the intermolecular forces and escape from the liquid phase. Thus the vapor pressure increases with increasing temperature.

Figure 13.6 shows how the vapor pressures of pentane, tetrachloromethane, and water vary with temperature. At any given temperature pentane has a higher vapor pressure than tetrachloromethane, which in turn has a higher vapor pressure than water. Thus the intermolecular forces are stronger in water than in tetrachloromethane, which in turn has stronger intermolecular forces than pentane.

BOILING POINT   When the temperature of a liquid is raised to the point at which the vapor pressure is equal to the atmospheric pressure, bubbles of vapor form in the liquid, and it is said to boil. A bubble of vapor can only form and expand in a liquid when the pressure in the bubble is sufficient to push the liquid away against the atmospheric pressure acting on the liquid—that is, when the vapor pressure of the liquid is equal to the atmospheric pressure. The **normal boiling point** of a liquid is defined as the temperature at which the vapor pressure equals 1 atm (760 mm Hg or 101.3 kPa). The temperature of the liquid cannot be raised above the boiling point, because if more heat is supplied, it simply causes bubbles to be formed more rapidly and to grow more rapidly.

Water boils at 100°C at 1 atm pressure; therefore 100°C is its normal boiling point. Tetrachloromethane, which has a higher vapor pressure, boils at 76.5°C at 1 atm pressure, and pentane boils at 36.1°C. If the atmospheric pressure is lower, as it is at high altitude, the boiling point of a liquid is correspondingly reduced. At an atmospheric pressure of 600 mm Hg, for example, water boils at the temperature at which its vapor pressure is 600 mm Hg, that is, 92°C. If we wish to purify, by distillation, a liquid that has an inconveniently high boiling point, we can lower the boiling point by carrying out the distillation under reduced pressure.

Example 13.1 Using Figure 13.6, estimate the boiling points of C_5H_{12}, CCl_4, and H_2O at a pressure of 400 mm Hg.

Solution If we draw a horizontal line on Figure 13.6 at a pressure of 400 mm Hg, we can read off the temperature at which this line cuts each vapor pressure curve. Thus we find these boiling points:

$$C_5H_{12}: 25°C \qquad CCl_4: 60°C \qquad H_2O: 80°C$$

Figure 13.6 Variation of Vapor Pressure with Temperature. The vapor pressure of a liquid increases with increasing temperature. The temperature at which the vapor pressure equals the pressure exerted by the atmosphere is the boiling point of the liquid.

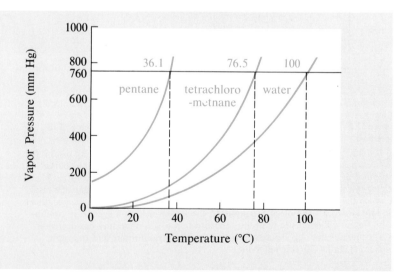

VAPOR PRESSURE AND PARTIAL PRESSURE   When a liquid evaporates in the presence of other gases, such as air, the total pressure of the gas phase is the sum of several partial pressures: the pressure caused by the vapor molecules and the pressure caused, for example, by the N_2 and O_2 of the air. The pressure produced by the water vapor is of importance when a gas is collected over water, as shown in Figure 13.7. The total pressure is the pressure of the gas plus the vapor pressure of water at the particular temperature. Thus to find the pressure of the gas, we must subtract the vapor pressure of water from the total pressure, as shown in Example 13.2.

Example 13.2   When 2.050 g of a mixture of magnesium metal and magnesium oxide was treated with an excess of dilute hydrochloric acid, 510 mL of hydrogen was collected over water at 20.0°C and a pressure of 742 mm Hg. How many moles of H_2 were produced? What was the percentage by mass of the Mg metal in the mixture?

Solution   The total pressure of 742 mm Hg is equal to the sum of the vapor pressure of water and the pressure of hydrogen:

$$P_{total} = p_{H_2} + p_{H_2O}$$

From Table 13.4 the vapor pressure of water at 20.0°C is 17.5 mm Hg. Therefore

$$p_{H_2} = P_{total} - p_{H_2O} = 742 - 17.5 = 724 \text{ mm Hg}$$

$$n_{H_2} = \frac{p_{H_2}V}{RT} = \frac{(724/760) \text{ atm} \times 0.510 \text{ L}}{0.0821 \text{ L atm mol}^{-1}\text{K}^{-1} \times 293 \text{ K}} = 0.0202 \text{ mol } H_2$$

Magnesium reacts with dilute HCl according to the equation

$$Mg + 2HCl \longrightarrow MgCl_2 + H_2$$

Magnesium oxide reacts with dilute HCl according to the equation

$$MgO + 2HCl \longrightarrow MgCl_2 + H_2O$$

so only the reaction with the Mg forms H_2. Hence 1 mol of Mg gives 1 mol of H_2, and 0.0202 mol of Mg gives 0.0202 mol of H_2. Therefore

$$(0.0202 \text{ mol Mg}) \left(\frac{24.31 \text{ g Mg}}{1 \text{ mol Mg}} \right) = 0.491 \text{ g Mg}$$

$$\text{Percent mass Mg in mixture} = \frac{0.491}{2.050} \times 100 = 24.0\%$$

VAPOR PRESSURE OF SOLIDS   A solid also exerts a vapor pressure, although this pressure is often very small because very few molecules in a solid normally

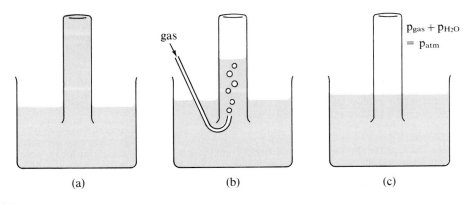

(a)   (b)   (c)

Figure 13.7 Collecting an Insoluble Gas over Water. (a) The inverted cylinder is filled completely with water. (b) The gas displaces the water in the cylinder. (c) The height of the cylinder is adjusted so that the water levels inside and outside the cylinder are equal. The total pressure in the cylinder, which is the sum of the partial pressure of the gas that has been collected, p_{gas}, and the vapor pressure of water, p_{H_2O}, at the temperature of the water, is then equal to the atmospheric pressure. Thus $p_{gas} = p_{atm} - p_{H_2O}$.

have enough energy to break loose from the solid to form a vapor. The process by which a solid is converted directly to vapor is called **sublimation**. If a solid has a relatively high vapor pressure, sublimation is a convenient procedure for purification. For example, we saw in Experiment 5.1 that iodine is easily sublimed; it can therefore be conveniently purified in this way. The vapor pressure of solid carbon dioxide (dry ice) is also rather high; it is equal to 1 atm at $-78°C$. In other words, at this temperature the solid is transformed directly to the vapor without melting. Liquid carbon dioxide can be obtained only under an external pressure of at least 5.2 atm; at this pressure carbon dioxide melts at $-57°C$.

13.3 INTERMOLECULAR FORCES

The fact that all gases condense to liquids at a sufficiently low temperature shows that there must be attractive forces between molecules that are strong enough to hold the molecules together in the liquid and solid states. These forces are called **intermolecular forces**.

Although it is convenient to distinguish several different types of intermolecular forces, they are all electrostatic. In fact, all the different types of interactions among atoms, molecules, and ions, whether we describe them as covalent bonds, ionic bonds, metallic bonds, or intermolecular forces, result from electrostatic attractions and repulsions between positive nuclei and negative electrons. Table 13.5 describes the various types of interactions between ions and molecules. We first briefly review interactions involving ions, and then we consider interactions between molecules.

Interactions Involving Ions

IONIC BONDS In ionic crystals the electrostatic attraction between oppositely charged ions is called **ionic bonding**. This ionic bonding persists when a crystal is melted, although the ions no longer have a completely ordered arrangement. The number and the arrangement of ions around any given ion

Table 13.5 Interionic and Intermolecular Forces

	TYPE OF INTERACTION	USUAL DESCRIPTION OF INTERACTION	DEPENDENCE OF FORCE ON DISTANCE
Ions and ions	Ion-ion	Ionic bonding	$1/r^2$
Ions and polar molecules	Ion-dipole		$1/r^3$
Ions and nonpolar molecules	Ion–induced dipole	Ion solvation	$1/r^5$
Polar molecules and polar molecules	Dipole–dipole		
Polar molecules and nonpolar molecules	Dipole–induced dipole	Intermolecular forces (van der Waals forces)	$1/r^7$
Nonpolar molecules and nonpolar molecules	Induced dipole–induced dipole	(London forces)*	

* The term *intermolecular (van der Waals) forces* applies to all types of attractions between molecules. The term *London forces* is used to describe only induced dipole-induced dipole forces—that is, the forces that act between nonpolar molecules.

are not constant, as they are in the solid, but they fluctuate with time. At very high temperatures ionic substances evaporate to form a vapor phase. For example, sodium chloride boils at 1465°C. Unlike the solid and the liquid, the vapor must contain small molecules consisting of two (or more) ions held together by their electrostatic attraction, although in many cases the bonds in these molecules have considerable covalent character.

Example 13.3 Account for the differences in the following boiling points:

$$\text{LiF: } 1717°C \qquad \text{Li}_2\text{O: } 2563°C \qquad \text{MgO: } 3260°C$$

Solution These compounds are all ionic. The force of attraction between the ions increases with increasing ionic charge, according to Coulomb's law. Therefore if the distance between the ions were the same in each case, we would expect the boiling points to increase in the order $(+1: -1)$, $(+1: -2)$, $(+2: -2)$. This is, in fact, the order observed. The anions O^{2-} and F^- have almost the same size, and the cations Mg^{2+} and Li^+ also have almost the same size (see Table 10.5), so the interionic distances are approximately equal.

ION-DIPOLE INTERACTIONS   In Chapter 8 we saw that many molecules have a **dipole moment**, because the center of positive charge does not correspond to the center of negative charge. Such molecules are said to be **polar**. A positive ion attracts the negative end of a polar molecule and repels the positive end. For example, a sodium ion attracts the negative oxygen atom of a water molecule and repels the positive hydrogen atoms. Consequently, a water molecule will orient itself next to a sodium ion in the following manner:

$$\text{Na}^+ \; {}^{2\delta-}\text{O} \underset{\diagdown \text{H}^{\delta+}}{\overset{\diagup \text{H}^{\delta+}}{}}$$

In this orientation the attraction between the sodium ion and the oxygen atom is greater than the repulsion between the sodium ion and the hydrogen atoms. As a result, there is an overall attraction between a sodium ion and a water molecule.

Any ion exerts an attraction on any polar molecule in the same manner. The force of attraction between an ion and a dipole varies with $1/r^3$, where r is the distance between the center of the ion and the midpoint of the dipole. It therefore decreases more rapidly with increasing distance than the force between two ions, which varies with $1/r^2$ (see Figure 13.8). The force is also usually weaker than that between two ions because, as we saw in Chapter 8, the charges on polar molecules are smaller than the charges on ions. The interaction between ions and polar molecules is important in solutions and is called **ion solvation**. In aqueous solution it is called **ion hydration**.

We saw in Chapter 9 that ion hydration can also be regarded as a Lewis acid-base interaction. Which description is the best depends on how much electron charge is donated from each water molecule to an ion. If very little electron density is donated, the interaction is best described as an ion-dipole interaction. But if there is a significant donation of charge, then it can be regarded as a Lewis acid-base interaction.

Because a negative charge repels electrons while a positive charge attracts electrons, the electron distribution of a nonpolar molecule is distorted when it is close to an ion, as Figure 13.8 illustrates. The center of negative charge no longer coincides with the center of positive charge, so a dipole is created. It is

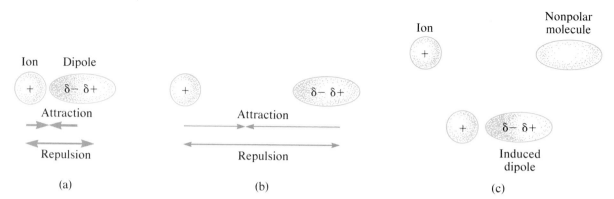

Figure 13.8 Ion-Molecule Interactions.
(a, b) Interaction between an ion and a polar molecule: When the ion and the polar molecule are close together, as in (a), the attractive force between the positive ion and the negative end of the dipole of the polar molecule is much greater than the repulsive force between the ion and the positive end of the dipole, because the positive end is farther from the ion than the negative end. The result is a strong attraction between the ion and the dipole. When the ion and the dipole are farther apart, as in (b), the positive and negative ends of the dipole are about equally far from the ion. Hence, the repulsion almost equals the attraction and the overall attraction is small. Thus the force of attraction between an ion and a polar molecule decreases rapidly with increasing distance between them. In particular, it decreases more rapidly than the force of attraction between two ions. In fact $F \propto 1/r^3$. (c) Interaction between an ion and a nonpolar molecule: The ion induces a dipole in the nonpolar molecule, which leads to an overall attraction between the ion and the molecule.

called an **induced dipole**. The attractive force between an ion and an induced dipole is proportional to the inverse fifth power of the distance between the ion and the dipole, that is, to $1/r^5$. *Ion-induced dipole interactions* are therefore important only when an ion and a nonpolar molecule are very close together.

Interactions Between Molecules

DIPOLE-DIPOLE FORCES   When two dipoles are arranged head to tail, they attract each other, because the sum of the attractive forces between opposite charges is greater than the sum of the repulsive forces between like charges. If two dipoles are arranged head to head or tail to tail, they repel each other. Since the head-to-tail arrangements have a lower energy than the head-to-head arrangements, the head-to-tail arrangements predominate in any collection of polar molecules. Therefore there is an overall attraction between such molecules (see Figure 13.9a).

All polar molecules therefore attract each other by dipole-dipole interactions. Dipole-dipole attractions are relatively weak compared with ion-ion attractions, because the charges on polar molecules are generally quite small. Moreover, this attractive force varies with $1/r^7$ and therefore falls off extremely rapidly with increasing distance. It is significant only when the two molecules are very close together (see Figure 13.9a).

DIPOLE–INDUCED DIPOLE FORCES   A polar molecule can induce a dipole in a nonpolar molecule. For example, as we see in Figure 13.9(b), the positive end of a polar molecule attracts electrons in an adjacent nonpolar molecule, thereby creating an induced dipole. Because of the motions of the molecules in a liquid, the induced dipole constantly fluctuates in magnitude and direction, but there is always a resultant attraction between a polar molecule and a nonpolar molecule. A dipole-induced dipole interaction is proportional to the inverse seventh power of their distance apart, that is, to $1/r^7$.

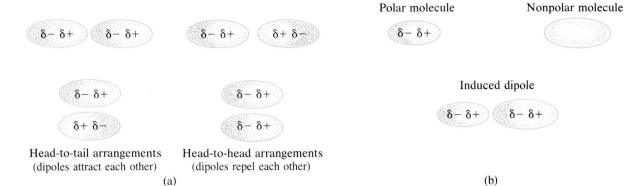

Polar molecule   Nonpolar molecule

δ− δ+

Induced dipole

δ− δ+   δ− δ+

Head-to-tail arrangements
(dipoles attract each other)
Head-to-head arrangements
(dipoles repel each other)

(a)   (b)

Figure 13.9 Intermolecular Forces: Dipole-Dipole and Dipole-Induced Dipole Forces.
(a) Two polar molecules can have many different relative arrangements. When they have a head-to-tail arrangement, their dipoles attract each other. When they have a head-to-head arrangement, the two dipoles repel each other. The head-to-tail arrangements have a lower energy, and are therefore more common, than the head-to-head arrangements. So there is an overall attraction between the polar molecules. (b) The dipole of a polar molecule can induce a dipole in a nonpolar molecule, which leads to a weak interaction between the polar molecule and the nonpolar molecule.

INDUCED-DIPOLE–INDUCED-DIPOLE FORCES: LONDON FORCES Finally, we know that there must be forces of attraction even between nonpolar molecules, since gases such as H_2, O_2, and Ar condense to liquids at very low temperatures. In 1926 German physicist Fritz London (1900–1954) proposed an explanation for these forces. They are therefore often called **London forces**. He postulated that they result from the fact that even a nonpolar molecule or an atom may have an instantaneous dipole moment. The electrons in a molecule are in constant motion, and their *average* distribution is such that in a nonpolar molecule the center of negative charge of the electrons coincides with the center of positive charge of the nuclei. But at any one instant the centers of positive and negative charges do not, in general, coincide, and the molecule will have an **instantaneous dipole moment**, which is constantly changing in magnitude and direction with the movement of electrons in the molecule (see Figure 13.10). The fluctuating dipole in one molecule influences the fluctuating dipole in an adjacent molecule, so two molecules have a head-to-tail arrangement of their fluctuating dipoles more often than a head-to-head arrangement. Because of this interaction between fluctuating dipoles on adjacent molecules, nonpolar molecules attract each other.

London showed that the force of attraction between two nonpolar molecules is inversely proportional to the seventh power of the distance and proportional

Figure 13.10 Intermolecular Forces (London Forces).
(a) The movement of the electrons in a nonpolar molecule gives rise to a small instantaneous dipole, which continually changes in magnitude and direction.
(b) This fluctuating dipole can induce a dipole in a neighboring molecule. This induced dipole changes in magnitude and direction following the changes in the fluctuating dipole in the first molecule. Thus there is always an attraction between the two molecules.

Attraction   Attraction

(a)   (b)

to a property of each molecule called the *polarizability*, α. Thus for two molecules A and B

$$F \propto \frac{\alpha_A \alpha_B}{r^7}$$

We see that London forces fall off very rapidly with distance, like the dipole-dipole and dipole–induced-dipole interactions.

POLARIZABILITY   The **polarizability** of an atom or a molecule is a measure of the ease with which the electrons and nuclei can be displaced from their average positions. The more easily the electrons and nuclei can be displaced, the greater the polarizability. The greater the polarizability, the greater the magnitude of the instantaneous dipole and therefore the stronger the London forces. Since the mass of an electron is much less than the mass of any nucleus, electrons are displaced much more easily; they therefore make the largest contribution to the polarizability.

The electrons that are most easily displaced are the valence electrons since they are furthest from the nucleus and are held less strongly than the other electrons. Therefore they make the greatest contribution to the polarizability, and to a first approximation we may consider that the polarizability is due to just these electrons. The force acting on the valence electrons depends on their distance from the nucleus and on the core charge. In any group of the periodic table the core charge is constant, so we expect polarizability to increase as atomic size increases from the top to the bottom of the group. For example, we see from the polarizabilities given in Table 13.6 that the polarizability of the hydrogen halides increases from HF to HI. The units of polarizability are the units of volume, cubic meters, m^3.

In a molecule a dipole is induced in each atom, and the total induced-dipole moment is the resultant of all these small dipoles. Hence large molecules with many atoms have larger polarizabilities than small molecules. For example, we saw in Table 11.1 that the boiling points of the alkanes, which are nonpolar molecules, increase regularly with increasing molecular size.

Table 13.6 Dipole Moments and Polarizabilities

SUBSTANCE	DIPOLE MOMENT μ (10^{-30} C m)	POLARIZABILITY α (10^{-30} m³)
He	0	2.5
Ar	0	20.1
H_2	0	9.9
N_2	0	22.1
CO	0.33	24.5
CO_2	0	33.3
HF	6.37	6.4
HCl	3.60	33.1
HBr	2.67	45.4
HI	1.40	68.5
H_2O	6.17	18.6
NH_3	4.90	27.8
CCl_4	0	127
CH_4	0	32.7
SO_2	5.42	54.6

Because of its smaller size, fluorine has a much smaller polarizability than chlorine. Molecular fluorides, therefore, usually have much lower boiling points than the corresponding chlorides. For example, the boiling point of CCl_4 is $+77°C$, whereas that of CF_4 is $-128°C$. In fact, the boiling points of fluorides are generally not much higher than those of the corresponding hydrides; the boiling point of methane is $-161°C$. Although fluorine has nine electrons, whereas hydrogen has only one, the fluorine electrons are held very tightly by the high core charge, and so the polarizability of fluorine is not much greater than that of hydrogen.

Molecular shape also plays a role in determining the strength of intermolecular forces. For example, the boiling point of pentane, $CH_3CH_2CH_2CH_2CH_3$, is 36.2°C, which is appreciably higher than that of its isomer, 2,2-dimethylpropane, $C(CH_3)_4$, which is 9.5°C. This difference is a consequence of the rapid diminution in the intermolecular (London) forces with distance. The long, thin pentane molecules can get close together by lying alongside each other, but two much more nearly spherical 2,2-dimethylpropane molecules can touch each other only at one point. Thus the centers of the 2,2-dimethylpropane molecules are necessarily further apart than those of the pentane molecules (see Figure 13.11). Because the strength of the London forces decreases rapidly with increasing distance, the forces between the pentane molecules are considerably stronger than those between 2,2-dimethylpropane molecules.

The boiling point increases in the series HCl $(-85°C)$, HBr $(-67°C)$, HI $(-35°C)$, despite the fact that the dipole moment decreases from HCl to HI. This increase is due to the increasing polarizability of the molecules in the series from HCl to HI (Table 13.6). We see that, particularly for larger molecules, the London forces make an important contribution to the overall intermolecular forces, even for molecules that have fairly large dipole moments.

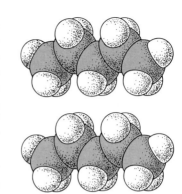

Two pentane molecules

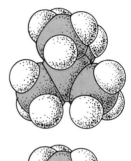

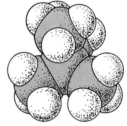

Two 2, 2-dimethylpropane molecules

Figure 13.11 Intermolecular Forces and Molecular Shape. The long, thin pentane molecules can get more atoms much closer together than the more spherical molecules of its isomer can.

Example 13.4 Which of the following substances is most likely to exist as a gas at room temperature and normal atmospheric pressure: PCl_3, Cl_2, $MgCl_2$, or Br_2? Why?

Solution The weakest intermolecular forces will be found in the substance with the smallest nonpolar molecules. Of the substances listed, $MgCl_2$ is ionic and PCl_3 consists of polar molecules. Both Cl_2 and Br_2 consist of nonpolar molecules, but since Cl_2 molecules are smaller than Br_2 molecules, they will be less polarizable and therefore have the smaller intermolecular forces. Thus Cl_2 is the substance most likely to be a gas at room temperature and normal atmospheric pressure. Even if the dipole-dipole forces in PCl_3 were small, we note that in any case it is a larger molecule than Cl_2 (three Cl atoms and one P atom rather than two Cl atoms), so we expect the intermolecular forces in PCl_3 to be larger.

Nonideal Behavior of Gases and van der Waals Forces

The existence of attractive forces between molecules is one important reason why real gases do not obey the ideal gas equation exactly (Chapter 3). The greater the intermolecular forces between the molecules of a gas, the more the gas deviates from the ideal gas law. Gases with very small intermolecular forces obey the ideal gas law most closely; small molecules with no dipole moment, such as He, Ne, H_2, and N_2, are examples. Larger molecules, and particularly those with dipole moments, such as SO_2 and HI, show greater deviations from the ideal gas law.

Dutch physicist Johannes van der Waals (1837–1923) was the first to study deviations from the ideal gas law in some detail. Hence intermolecular forces are frequently called **van der Waals forces**.

Van der Waals Radii

Although all atoms and molecules attract each other by intermolecular (van der Waals) forces, at very short intermolecular distances this attraction is opposed by a strong repulsion. According to the Pauli exclusion principle, completed shells of electrons cannot overlap. Thus when two molecules are pushed very close, the electron distributions distort so as to avoid overlapping. This distortion gives rise to a repulsion between the two molecules (Chapters 4 and 6).

Thus two molecules attract each other when they are fairly close together; but when they are very close together, their mutual repulsion is greater than the attraction, and they repel each other. At some particular distance apart the repulsive force equals the attractive force, and the total energy of the two molecules is a minimum. The two molecules may then be regarded as "touching" each other (see Figure 13.12). By measuring the distances between molecules that are touching each other in the solid state, we can obtain a set of radii, called **van der Waals radii**, that are a measure of the size of an atom in a molecule.

For example, from the structure of solid chlorine we can measure the following distances:

1. The distance between the chlorine nuclei in the *same* molecule. One-half this distance gives the covalent radius of the chlorine atom, as discussed in Chapter 4 (Figure 4.7). The covalent radius is a measure of the size of the chlorine atom when it is forming a bond.

2. The distance between the chlorine nuclei in two *adjacent* molecules. One-half this distance is the van der Waals radius of chlorine. This radius is the effective size of the chlorine atom in any direction in which it is *not* forming a bond (see Figure 13.12).

Van der Waals radii are considerably larger than covalent radii; for most atoms van der Waals radii are approximately twice the covalent radii (see Figure 13.12) and are approximately equal to the radii of the corresponding anions (Table 10.5).

Figure 13.12 Van der Waals Radii, Covalent Radii, and Ionic Radii.
(a) The van der Waals radius, r_{vdw}, is half the distance between two similar atoms in separate molecules in a solid. The molecules are close-packed in the solid but are not bonded together. (b) The covalent radius, r_c, is half the distance between two similar atoms joined by a covalent bond in the same molecule. (c) According to the electron sphere model of Cl_2, r_c is equal to the *radius* of the sphere occupied by the bonding electron pair, but r_{vdw} is equal to the *diameter* of the sphere occupied by a nonbonding pair, which shows why $r_{vdw} \approx 2r_c$ for the same atom. (d) In the electron sphere model of Cl^- the radius of the ion, r_{ion}, is also equal to the *diameter* of an electron pair. So for a monatomic anion $r_{ion} \approx r_{vdw}$ for the same atom.

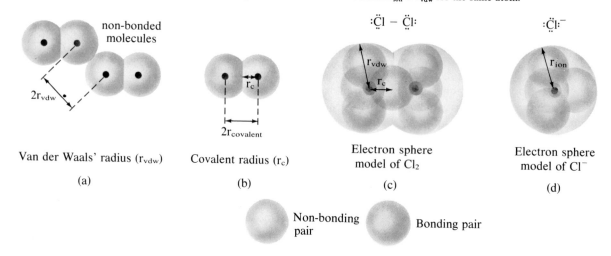

Table 13.7 Van der Waals Radii of Atoms (pm)

H	110	N	150	O	140	F	135
		P	190	S	185	Cl	180
		As	200	Se	200	Br	195
		Sb	220	Te	220	I	215

Some van der Waals radii for atoms are listed in Table 13.7. Like covalent radii, van der Waals radii are approximate and are not strictly constant from one molecule to another. They depend somewhat on how the molecules are packed together in the solid state.

13.4 SOME UNUSUAL PROPERTIES OF WATER

Although water is the most common liquid on the earth, it is also one of the most unusual. It has several properties that at first sight are unexpected and unlike those of most other liquids.

One unusual property is that the density of solid water (ice) is less than that of liquid water at the melting point. For almost all other substances the solid is more dense than the liquid. As a liquid is cooled, the kinetic energy of the molecules decreases, their movements become more restricted, and they pack more closely together. This contraction and increase in density continues until the liquid freezes to a solid, at which point the density increases abruptly. Because of their regular arrangement in the solid, the molecules are packed still more closely together. However, water behaves in a different manner when it is cooled. Its density increases in the normal way until a temperature of 3.98°C is reached, when its density is $1.000\ 00\ \text{g cm}^{-3}$. Then the density *decreases* very slightly with decreasing temperature to 0.00°C, when it has a density of $0.999\ 87\ \text{g cm}^{-3}$. Then when water freezes, its density again *decreases* abruptly, to $0.917\ \text{g cm}^{-3}$, rather than increasing.

This unusual variation of the density of water with temperature makes aquatic life possible in cold climates. When the temperature of the atmosphere falls to near or below 0°C, the water on the surface of lakes, ponds, and rivers cools first. Since it is then more dense, it sinks to the bottom, replacing warmer water, which in its turn is cooled and sinks to the bottom. But when all the water has been cooled to the temperature of maximum density (3.98°C), any further cooling produces a surface layer, which since it is less dense, remains on top, eventually freezing. Because ice is considerably less dense than water, the ice remains on the surface. Because ice is a poor conductor of heat, the water below the ice cools very slowly, and the ice layer thickens slowly. As a result, even a small pond rarely freezes solid, and large lakes never freeze completely. If ice were more dense than water, it would sink to the bottom of a lake, and eventually the whole lake would freeze. Clearly, fish could not survive under such conditions. In many parts of the world the ice at the bottom of a lake would not melt in the summer, and life on the lake bed would not be possible.

We are used to the fact that water freezes at 0°C and boils at 100°C, and so at first these properties do not seem unusual. But if we compare the boiling point and melting point of water with those of similar substances, we find that both are far higher than expected. For example, the boiling points of the hydrides of the elements of group VI decrease from H_2Te to H_2S as the size decreases, and therefore the polarizability of the molecules decreases, as Figure 13.13 illustrates. Extrapolation of these boiling points would lead us to expect

Figure 13.13 Boiling Points of Hydrides of Groups IV, V, VI, and VII. The boiling points of the nonpolar group IV hydrides decrease from SnH_4 with decreasing molecular size and, therefore, decreasing intermolecular (London) forces. Group VI hydrides have higher boiling points than hydrides of group IV because they are polar molecules, and therefore there are additional dipole-dipole attractions. Nevertheless, London forces make the largest contribution to the overall intermolecular forces. Therefore the boiling points decrease from H_2Te to H_2S with decreasing molecular size. But the boiling point of H_2O is very much higher than expected. This high boiling point is attributed to a strong additional intermolecular force called the *hydrogen bond*, which is not present in the group IV hydrides and is much weaker in H_2S, H_2Se, and H_2Te, if it is present at all. The hydrides NH_3 in group V and HF in group VII also have unexpectedly high boiling points, for the same reason.

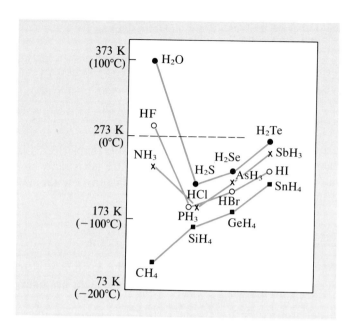

that water would have a boiling point of approximately $-100°C$, that is, about 200°C lower than the observed boiling point.

In contrast, the boiling points of the hydrides of the elements of Group IV, namely SnH_4, GeH_4, SiH_4, and CH_4, decrease in the expected manner with decreasing size and the consequent decrease in the strength of the induced-dipole–induced-dipole (London) forces (see Figure 13.13). The boiling points of the hydrides of group VI are somewhat higher than those of group IV because the group VI hydrides are polar molecules. Therefore in addition to the London forces there are also dipole-dipole forces.

But the dipole moment of water (6.17×10^{-30} C m) is quite inadequate to account for its exceptionally high boiling point. The dipole moment of water is only slightly larger than that of SO_2 ($\mu = 5.42 \times 10^{-30}$ C m), which also has a considerably greater polarizability than that of water (see Table 13.6), yet SO_2 has a boiling point of only $-10°C$. The exceptionally high boiling point of water must be due to an additional stronger force between H_2O molecules that is not present between SO_2 molecules nor between H_2S, H_2Se, or H_2Te molecules or the hydrides of group IV. This strong force must be of a different type from the intermolecular forces discussed so far. It is called the *hydrogen bond*.

13.5 HYDROGEN BOND

The hydrogen bond is found not only in water but rather generally in substances that have N—H, O—H, and F—H bonds. For example, although the boiling points of hydrogen fluoride and ammonia are not as high as that of water they are much higher than expected from the extrapolation of the boiling points of the other hydrides in the same groups (see Figure 13.13).

What, then, is a hydrogen bond? Like all other attractive forces between atoms and molecules, it is electrostatic in origin. Because water is a polar molecule, water molecules attract each other by dipole-dipole forces. But the comparison of the dipole moments of SO_2 and H_2O seems to show that dipole-dipole forces cannot account for the very strong attractions between water molecules. However, we must take into account the unique character of the hydrogen atom, which consists only of a proton and an electron and which has no inner shell of nonbonding electrons. When an H atom is bonded to a very electronegative atom, a significant amount of electron density is removed from the H atom. The proton of the H atom can then get very close to the negative end of a dipole, so the electrostatic attraction between them is unusually strong (see Figure 13.14). Indeed, the H atom of a water molecule gets so close to the O atom of another water molecule that it interacts almost exclusively with one particular unshared pair of the O atom. The interaction of a positively charged H atom with an unshared electron pair is strongest when the unshared electron pair is rather small and localized, as it is in the N, O, and F atoms. The larger and more diffuse unshared pairs on heavier atoms such as Cl, S, and P do not interact as strongly with positively charged H atoms.

We see therefore that a hydrogen bond is a dipole-dipole force that is exceptionally strong because the proton of a positively charged H atom that is attached to an electronegative atom can get very close to an unshared pair of electrons in another molecule. The proton is particularly strongly attracted by this unshared pair of electrons if it is a small localized pair such as is found on the small electronegative N, O, and F atoms. Briefly a **hydrogen bond** *may be defined as an intermolecular attraction in which a hydrogen atom that is bonded to an electronegative atom is attracted to an unshared electron pair on another small electronegative atom.*

By far the most common and strongest hydrogen bonds are therefore the following combinations, in which the hydrogen bond is depicted by a dashed line:

$$F—H---:F \qquad F—H---:O \qquad F—H---:N$$
$$O—H---:F \qquad O—H---:O \qquad O—H---:N$$
$$N—H---:F \qquad N—H---:O \qquad N—H---:N$$

The hydrogen bond is intermediate in strength between other intermolecular forces and the much stronger covalent and ionic bonds. The energies of most hydrogen bonds range from about 5 to 25 kJ mol^{-1}, whereas the energies of most covalent bonds range from 100 to over 500 kJ mol^{-1} (Chapter 12).

The relative weakness of the hydrogen bond compared with other bonds is also seen in its length. For example, Figure 13.15 shows that in ice the H---O hydrogen bond has a length of 177 pm, whereas the O—H covalent bond has

Figure 13.14 The Hydrogen Bond. (a) The H$_2$O and SO$_2$ molecules have similar dipole moments, but the force of attraction is much greater between H$_2$O molecules than between SO$_2$ molecules. The SO$_2$ molecule is larger than the H$_2$O molecule. Therefore the average distance between the dipoles is much larger for SO$_2$ than for H$_2$O. So the force between them is correspondingly weaker—if the distance is twice as great, the force will be eight times smaller, since $F \propto 1/r^3$. (b) Moreover, the positive core of the H atom (the proton) is surrounded by very little electron density—a total of less than one electron. Therefore the proton can get very close to an unshared electron pair of the O atom of another H$_2$O molecule, to which it is strongly attracted and with which it may share an electron density, thus forming a hydrogen bond.

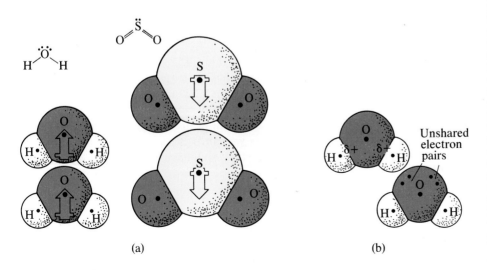

(a)   (b)

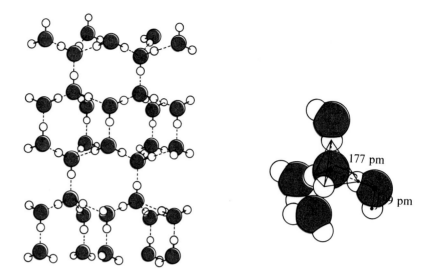

Figure 13.15 Structure of Ice. Each water molecule is surrounded by a tetrahedral arrangement of four other water molecules to which it is bound by hydrogen bonds. It is a three-dimensional network structure rather similar to the structures of diamond and silicon dioxide.

a length of 99 pm. The hydrogen bond can be regarded as being intermediate between a weak intermolecular force and a strong chemical bond. It resembles a chemical bond in that it is localized between a hydrogen atom and a specific unshared pair of electrons. In contrast, other intermolecular forces act between molecules as a whole rather than between specific atoms.

If we think of the hydrogen bond as a chemical bond, we might ask whether it is covalent or ionic. As is often the case, the answer to this question is not clear-cut. Since the valence shell of the hydrogen atom can contain only two electrons, we normally consider that the hydrogen atom can form only one covalent bond. However, when hydrogen is bonded to a very electronegative element, some of its electron density is pulled away by the electronegative atom, leaving the hydrogen deficient in electron density. It can therefore accept some electron density from the unshared electron pair to which it is attracted. Thus in some cases the hydrogen atom and the atom with the unshared pair share some electron density. In these cases the hydrogen bond can be considered to have some covalent character.

For most hydrogen bonds this covalent character is quite small, but a few hydrogen bonds have considerable covalent character. The strongest known hydrogen bond is in the very stable HF_2^- ion formed from an F^- ion and an HF molecule:

$$F^- + HF \longrightarrow F^- \text{---} H \text{---} F$$

This reaction has $\Delta H^\circ = -155 \text{ kJ mol}^{-1}$. Thus 155 kJ mol^{-1} is required to break this hydrogen bond, which therefore has a strength close to that of many weak covalent bonds. This hydrogen bond is also unusual in that it is symmetrical. The H atom is located exactly halfway between the two F atoms, as illustrated in Figure 13.16. In this ion we cannot distinguish a weak hydrogen bond and a stronger covalent bond, as we can between two water molecules. Both bonds have the same strength, and they each have 50% covalent character (see Figure 13.16).

Structure of Ice

The strong hydrogen bonding between water molecules accounts not only for the exceptionally high boiling and melting points of water but also for the

Figure 13.16 Hydrogen Bond in HF$_2^-$.
(a) The hydrogen bond in HF$_2^-$ is unusual. Unlike most hydrogen bonds, it is symmetrical; the H atom is exactly halfway between the two F atoms. The two halves of the bond must be identical. (b) The bonding in HF$_2^-$ can be represented by two resonance structures. (c) The negative charge is shared equally between the two F atoms.

structure and low density of ice and the expansion of water below 4°C. Ice has a network structure similar to that of diamond; each oxygen atom is surrounded by four other oxygen atoms in a tetrahedral arrangement (see Figure 13.15). Each oxygen has two hydrogen atoms covalently bound to it, and it has two unshared pairs of electrons, each of which forms a hydrogen bond with the hydrogen atoms of neighboring water molecules.

Each water molecule could have up to 12 nearest neighbors if they were packed closely together. But the number of neighbors of any water molecule is limited to 4 because each oxygen atom can form only two covalent O—H bonds and two O---H hydrogen bonds. As a result, the structure is an open one in which there is room for more molecules.

When ice melts, the regular structure collapses to some extent because some of the hydrogen bonds are broken, and consequently, some of the water molecules pack more closely together. Hence the density increases upon melting. However, only about 15% of the hydrogen bonds are broken when ice melts. Thus in liquid water at 0°C many water molecules are still hydrogen-bonded in groups that retain more or less the same structure they have in ice. With increasing temperature more hydrogen bonds are broken and more molecules pack more closely together, so the density continues to increase. Normally as a liquid is heated, the increased motion of the molecules causes them to take up more space. At 3.98°C this increased motion becomes more important than the continued collapse of the structure. So above 3.98°C the density of water begins to decrease in the normal way.

Hydrogen bonds are of widespread importance. For example, they hold long-chain protein molecules in certain definite shapes such as a spiral; the shape of a protein molecule plays an important role in its function. As we will see in the next section, hydrogen bonds also play an important role in determining the solvent properties of water.

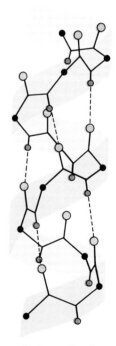

— — — Hydrogen bond

The spiral shape of a protein molecule

13.6 SOLUTIONS IN WATER AND OTHER SOLVENTS

Although water is a very good solvent, it is far from being a universal solvent. Metals, many metal oxides and salts, hydrocarbons, and some nonmetals such as carbon, iodine, and phosphorus are not soluble in water to any appreciable extent.

In the process of forming a solution, one substance mixes completely with another to form a homogeneous mixture. Two gases always form a gaseous mixture or solution. When two gases are placed in contact, the rapid, random motions of the molecules of each gas causes them to form a random, disordered mixture. However, when two liquids are brought into contact, they may or may not mix. Hydrocarbons are insoluble in water, whereas ethanol, C_2H_5OH, is soluble in water in all proportions.

Since the random motions of the molecules would be expected to cause the liquids to mix with each other, something must be preventing the mixing of the hydrocarbon molecules with the water molecules. It is the hydrogen bonds between water molecules that prevent the mixing. If hydrocarbon molecules were to mix with water molecules, many of the hydrogen bonds between water molecules would have to be broken. They are not broken because there are only weak, dipole-induced dipole forces between the nonpolar hydrocarbon molecules and the water molecules, and these forces are not strong enough to pull the water molecules apart. The strong attractions between water molecules due to hydrogen bonding thus prevent water molecules from mixing with hydrocarbon molecules (see Figure 13.17).

Ethanol, C_2H_5OH, is soluble in water because the OH group can form hydrogen bonds. Thus hydrogen bonds between ethanol molecules and water molecules replace some of the hydrogen bonds between water molecules. The energy needed to break the hydrogen bonds between water molecules is compensated by the energy evolved in the formation of hydrogen bonds between ethanol and water molecules, and so ethanol dissolves in water. Hydrocarbons are also soluble in each other. They mix easily because there are only weak London forces of attraction between hydrocarbon molecules.

These examples illustrate an often quoted generalization: "Like dissolves like." In other words, chemically similar substances are soluble in each other, whereas dissimilar substances are insoluble. Ethanol is like water in that both contain OH groups and are hydrogen-bonded. Different hydrocarbons are also

Figure 13.17 Solutions: Like Dissolves Like.
(a) Benzene and water are immiscible; the strong hydrogen bonds between the water molecules hold them together and prevent them from mixing with nonpolar C_6H_6 molecules. (b) Methanol, CH_3OH, is soluble in water in all proportions (miscible), because it forms strong hydrogen bonds with the water molecules. (c) Pentane, C_5H_{12}, is soluble in hexane, C_6H_{14}, because the forces between the molecules of both liquids and in a mixture are London forces of comparable strength.

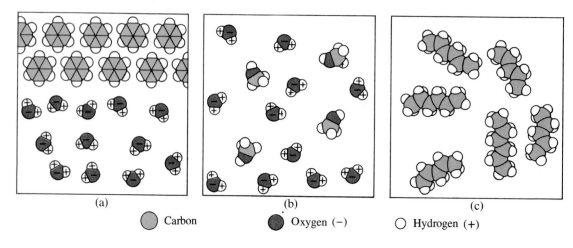

(a)   (b)   (c)

◯ Carbon   ● Oxygen (−)   ○ Hydrogen (+)

very similar to each other, and they are soluble in each other. But water and a hydrocarbon are not at all similar; one is polar and the other is nonpolar, and they are not soluble in each other.

We can apply the same considerations to understand the solubility of solids in liquids. Nonpolar substances, such as I_2 and S_8, are soluble in a nonpolar liquid such as tetrachloromethane, CCl_4, or toluene, $C_6H_5CH_3$, but they are insoluble in water, which is polar (Experiments 1.6 and 13.2). Sugar contains many OH groups that can form hydrogen bonds with water. It is therefore very soluble in water, but it is insoluble in CCl_4.

Substances having infinite network structures—such as metals, diamond, graphite, red phosphorus, silicon dioxide, and silicon carbide—are insoluble in water. They are insoluble in water because the metallic or covalent bonds holding the atoms together are much stronger than the hydrogen bonds between the H_2O molecules and thus keep the atoms from separating to mix with the H_2O molecules. Similarly, many ionic crystals, such as MgO, AgCl, CuO, CaF_2, and $BaSO_4$, are insoluble in water because the ionic bonds are too strong to allow the ions to separate and mix with the H_2O molecules.

However, many other ionic compounds, such as NaCl, KNO_3, $MgSO_4$, and $CuCl_2$, are soluble in water despite the strong forces holding the ions together. Why? There must be a relatively strong attractive force between the ions and the H_2O molecules so that the H_2O molecules can pull the ions away from the crystal. This force is the electrostatic attraction between an ion and a dipole (Table 13.5). Positive ions attract the negative end of the H_2O dipole, and consequently, each positive ion becomes surrounded by H_2O molecules, as we described in Chapter 9. Similarly, negative ions attract the positive end of the H_2O dipole, and consequently, each negative ion also becomes surrounded by H_2O molecules (see Figure 13.18). The ions are said to be *hydrated*. In the case of the F^- ion and oxoanions such as SO_4^{2-}, NO_3^-, CO_3^{2-}, and PO_4^{3-}, the ion-dipole interaction is enhanced by hydrogen bonding between H_2O molecules and F^- ions or the O atoms of the oxoanions. The layer of H_2O molecules surrounding cations and anions also keeps them relatively far apart, reducing the strength of the electrostatic attraction between the ions.

EXPERIMENT 13.2

Solubilities of Solids in Liquids

Left to right: Water, toluene, water, and toluene. Sulfur, which is nonpolar, is insoluble in the polar solvent, water, but is soluble in the nonpolar solvent, toluene. Sodium chloride, which is ionic, is soluble in the polar solvent, water, but is insoluble in the nonpolar solvent, toluene (methyl benzene).

13.6 SOLUTIONS IN WATER
AND OTHER SOLVENTS

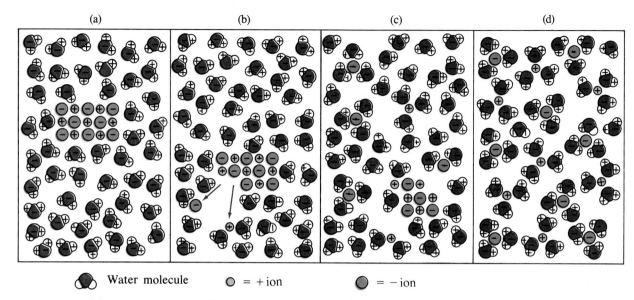

| (a) | (b) | (c) | (d) |

Water molecule   ⊕ = + ion   ● = − ion

Figure 13.18 Solution of Ionic Crystal. (a) A small crystal before it dissolves in the surrounding water. (b) Ions are pulled away from the crystal by the attraction of the polar water molecules. (c) Hydrated ions are formed. (d) The crystal has dissolved, giving a solution of hydrated positive and negative ions.

There are two important factors determining solubility. One is the tendency of the random, thermal motions of atoms and molecules to cause them to become mixed up with each other, as always happens in the case of gases. The second is the relative strengths of the attractive forces between the solute particles, the attractive forces between the solvent particles, and the attractive forces between the solute and the solvent particles. The first factor is an example of the tendency of all systems to become more mixed up or disordered as a result of the thermal motions of atoms and molecules—in other words, for the *entropy* to increase, as we discussed briefly in Chapter 12 (and as we discuss in more detail in Chapter 26). This tendency toward disorder is opposed by attractive forces between the atoms or molecules of the solvent or solute that may prevent them from mixing. The forces between the molecules in gases are always weak, so they mix easily. But the forces between the atoms and molecules in liquids and solids are often much stronger, so they may or may not mix easily, depending on the magnitude of these forces.

Enthalpy of Solution

The process of forming a solution may be either endothermic or exothermic. If more energy is needed to overcome the solvent-solvent and solute-solute attractions than is gained from the interactions between the solvent particles and the solute particles, then the process is endothermic (see Figure 13.19). *The heat absorbed when 1 mol of a solute dissolves in a solvent at constant pressure is called the* **molar enthalpy of solution**, $\Delta H^\circ_{solution}$. The enthalpy of solution depends on the concentration of the final solution, but it has a constant value for a sufficiently dilute solution, that is, if a large amount of solvent is used. If heat is absorbed when a solution is formed, the enthalpy of solution has a positive value. If heat is liberated when a solution is formed, the enthalpy of solution has a negative value.

Some enthalpies of solution in water are given in Table 13.8. The enthalpy of solution of sulfuric acid in water has a large negative value; sulfuric acid dissolves in water with the evolution of a large amount of heat. In the formation of an aqueous solution of a salt, there is a rather close balance between the energy needed to pull the ions apart and to pull the water molecules apart and

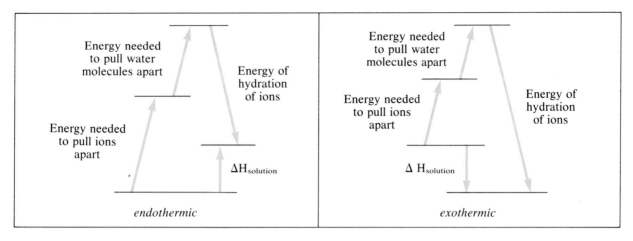

Figure 13.19 Enthalpy of Solution of Ionic Solid in Water.
The enthalpy of solution may be positive or negative, depending on the relative magnitudes of three energy terms: the energy needed to pull the ions apart, the energy needed to pull the solvent molecules apart, and the energy of hydration of the ions.

the energy gained from the hydration of the ions (see Figure 13.19). For most salts there is an overall small positive enthalpy of solution; the solution process is endothermic. But for a few salts $\Delta H^{\circ}_{\text{solution}}$ is negative (see Experiment 13.3).

Example 13.5 Ammonium nitrate has a reasonably large positive enthalpy of solution ($\Delta H = 26.4$ kJ mol^{-1}). This property is exploited in cold packs, which are used to treat athletic injuries. Cold packs contain bags of ammonium nitrate and water; these bags are broken when the pack is kneaded, allowing the ammonium nitrate to dissolve in the water. What is the final temperature reached when 10.0 g of ammonium nitrate dissolves in 50.0 g of water, initially at room temperature (20°C)?

Solution

Heat absorbed by dissolving 10.0 g $= 26.4$ kJ mol$^{-1} \left(\dfrac{1 \text{ mol}}{80.1 \text{ g}}\right)(10.0 \text{ g}) = 3300$ J

Molar heat capacity of water $= 75.4$ J K^{-1} mol^{-1}

Temperature decrease $= \left(\dfrac{3300 \text{ J}}{75.4 \text{ J K}^{-1} \text{ mol}^{-1}}\right)\left(\dfrac{18.02 \text{ g mol}^{-1}}{50.0 \text{ g}}\right) = 16$ K

The temperature is lowered from 20° to 4°C.

This calculation is only approximate since we have used the enthalpy of solution for a dilute solution, that is, for the addition of 1 mol of ammonium nitrate to a very large amount of water. When a concentrated solution is formed, the enthalpy of solution is somewhat smaller. The approximate calculated value nevertheless demonstrates that a considerable cooling effect may be achieved by dissolving some salts in water.

DEPENDENCE OF SOLUBILITY ON TEMPERATURE We saw in Chapter 1 that solubility depends on temperature: The solubility of most salts in water increases with increasing temperature. The effect of temperature depends on the enthalpy of solution and can be predicted by using Le Châtelier's principle. If the enthalpy of solution is positive, we can write

Salt + Water + Heat $\rightleftharpoons$ Solution

Table 13.8 Molar Enthalpies of Solution in Water

SOLUTE	$\Delta H^{\circ}_{\text{solution}}$ (kJ mol^{-1})
Na$_2$SO$_4 \cdot 10$H$_2$O	79
NaCl	3.9
Na$_2$SO$_4$	-23
CH$_3$CO$_2$Na	-18
H$_2$SO$_4$	-96
NaOH	-43
Na$_2$CO$_3 \cdot 10$H$_2$O	67
NH$_4$NO$_3$	26.4
LiCl	-37

13.6 SOLUTIONS IN WATER AND OTHER SOLVENTS

485

Enthalpy of Solution

Ammonium nitrate dissolves in water endothermically. The beaker contains water at room temperature, and the dish contains ammonium nitrate; there is a small pool of water on the wooden block. The ammonium nitrate is added to the water and the beaker is placed on the wooden block.

So much heat is absorbed that the temperature of the solution decreases to below 0°C as the ammonium nitrate dissolves and the water on the block freezes solid. The block can then be lifted with the beaker.

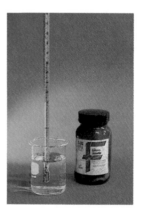

Lithium chloride dissolves exothermically in water.

In the experiment shown here, the temperature rose from 21°C to 46°C when lithium chloride was dissolved in the water in the beaker.

We see then that adding heat will shift the equilibrium to the right; in other words, the solubility increases with increasing temperature. If the enthalpy of solution is negative, we can write

$$\text{Salt} + \text{Water} \longrightarrow \text{Solution} + \text{Heat}$$

Then according to Le Châtelier's principle, the equilibrium will shift to the left on raising the temperature; in other words, the solubility decreases with increasing temperature. Because most salts have positive (endothermic) enthalpies of solution, their solubilities in water increase with increasing temperature, as we saw in Figure 1.18.

Hydration of the Hydronium Ion

Hydrogen bonds play an extremely important role in determining the properties of water. Not surprisingly, therefore, the hydronium ion, H_3O^+, in which the

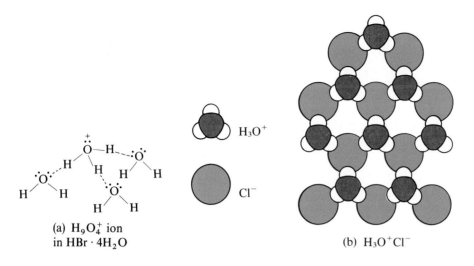

(a) $H_9O_4^+$ ion
in HBr · $4H_2O$

(b) $H_3O^+Cl^-$

electronegativity of the oxygen is increased by the presence of the formal positive charge, also forms strong hydrogen bonds. In dilute solution in water each of the three hydrogen atoms of the hydronium ion forms a hydrogen bond to a water molecule. Hence the hydrated hydronium ion, $H_3O^+(aq)$, can be written more specifically as $H_3O^+(H_2O)_3$, or $H_9O_4^+$ (see Figure 13.20a). Evidence for this and other hydrated forms of the hydronium ion is provided by the structures of **acid hydrates**.

Acids form many hydrates, such as HCl·H_2O, HCl·$2H_2O$, HCl·$4H_2O$, and HBr·$4H_2O$. These compounds are all ionic and contain the H_3O^+ ion or hydrated forms of the H_3O^+ ion. For example, HCl·H_2O has the ionic structure $H_3O^+Cl^-$ in which each hydrogen of H_3O^+ forms a hydrogen bond to a Cl^-, resulting in the layer structure shown in Figure 13.20(b). Similarly, $HClO_4$·H_2O has the ionic structure $H_3O^+ClO_4^-$, with hydrogen bonds between the H_3O^+ ion and the O atoms of the ClO_4^- ion. Although most acid hydrates have melting points below room temperature, $H_3O^+ClO_4^-$ has a melting point of 50°C, which emphasizes its close similarity to other ionic perchlorates such as $NH_4^+ClO_4^-$.

The dihydrate of perchloric acid, $HClO_4$·$2H_2O$, has the ionic structure $H_5O_2^+ClO_4^-$. The $H_5O_2^+$ ion has a symmetrical structure like the HF_2^- ion:

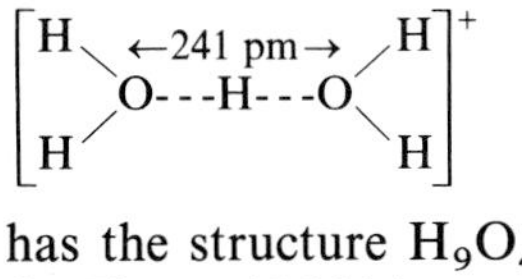

$$\left[\begin{array}{c} H \\ \diagdown \\ H \end{array} O \text{---H---} O \begin{array}{c} H \\ \diagup \\ H \end{array} \right]^+ \quad \leftarrow 241\ pm \rightarrow$$

The hydrate HBr·$4H_2O$ has the structure $H_9O_4^+Br^-$, in which the $H_9O_4^+$ ion has the structure shown in Figure 13.20(a).

13.7 REACTIONS OF WATER

So far we have discussed solutions in which there is no chemical reaction between the solvent and solute. In these cases the solute can be recovered unchanged from the solvent. For example, we can recover salt or sugar from water by heating the solution to evaporate the water. But often water is not an inert solvent, and in such cases the solution contains molecules or ions that differ from those in the original solute.

Because water is by far the most widely used solvent, reactions with water are among the most common reactions. In general, reactions with water are called **hydrolysis reactions**, and a substance that reacts with water is said to be **hydrolyzed**. Many of these reactions have been discussed in earlier chapters, but we can usefully summarize here the most important reactions of water.

Acid-Base Reactions

Many of the reactions of water depend on the fact that it is both a weak acid and a weak base. It can donate a proton to a base such as ammonia and thus behave as an acid:

$$H_2O + NH_3 \rightleftharpoons NH_4^+ + OH^-$$

Also, one of its unshared electron pairs can accept a proton from an acid such as hydrochloric acid, and thus it behaves as a base:

$$H_2O + HCl \longrightarrow H_3O^+ + Cl^-$$

Recall from Chapter 5 that water exerts a *leveling effect* on the strengths of acids and bases. All acids that are quantitatively converted to H_3O^+ in aqueous solution, such as HNO_3, HCl, and $HClO_4$, are said to be strong, and their strengths cannot be differentiated. Similarly, all bases that are quantitatively converted to OH^-, such as O^{2-} and NH_2^-, are said to be strong. The range of acid-base strengths that is available in water lies between the acid H_3O^+, the strongest acid that can exist in water, and OH^-, the strongest base that can exist in water. We take up this important property of water in a quantitative way in the next chapter.

The formation of hydrated metal ions in aqueous solution is an example of Lewis acid-base behavior. The water molecule, which is a Lewis base, can donate an unshared pair of electrons into the empty valence shell of a metal ion, which is a Lewis acid:

$$Al^{3+} + 6:\overset{\displaystyle \diagup H}{\underset{\displaystyle \diagdown H}{O}} \longrightarrow Al(OH_2)_6^{3+}$$

Oxidation-Reduction Reactions

Water is also an oxidizing agent and a reducing agent, although its oxidizing and particularly its reducing properties are rather weak. When it behaves as an oxidizing agent, it is reduced to hydrogen:

$$2H_2O + 2e^- \longrightarrow 2OH^- + H_2$$

We are familiar with its reaction with sodium and other reactive metals, which are strong reducing agents:

$$2Na + 2H_2O \longrightarrow 2NaOH + H_2$$

Similarly, water reacts with the hydride ion, H^-, which is oxidized to elemental hydrogen:

$$H^- + H_2O \longrightarrow H_2 + OH^-$$

This reaction is also an example of an acid-base reaction, since the hydride ion is a strong base that adds a proton to give hydrogen. At very high temperatures water, as steam, oxidizes carbon to carbon monoxide:

$$H_2O + C \longrightarrow CO + H_2$$

Water gas

Water is a rather weak reducing agent:

$$2H_2O \longrightarrow 4H^+ + O_2 + 4e^-$$

It will reduce only very strong oxidizing agents such as fluorine:

$$2F_2 + 2H_2O \longrightarrow 4HF + O_2$$

Reactions with Oxides

Water reacts with nonmetal oxides to give oxoacids. For example,

$$SO_3 + H_2O \longrightarrow H_2SO_4$$
$$CO_2 + H_2O \rightleftharpoons H_2CO_3$$
$$P_4O_{10} + 6H_2O \longrightarrow 4H_3PO_4$$

These oxides have polar bonds; the nonmetal atoms have a positive charge, and the oxygens have a negative charge. The reaction is initiated by attack of the negative oxygen atom of a water molecule on the positive nonmetal atom of the oxide. A proton is then transferred from the water molecule to a negative oxygen atom. For example

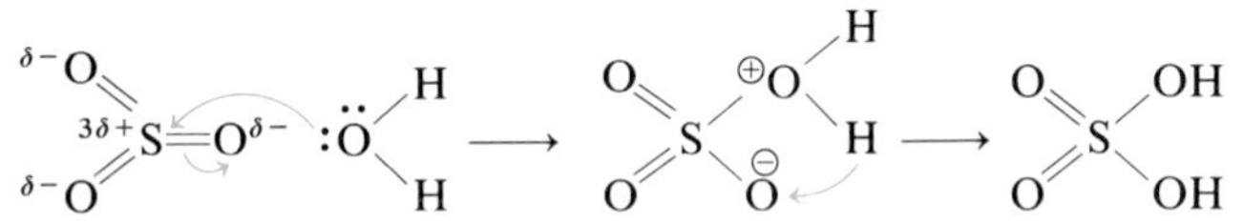

In these reactions water is a Lewis base, and the oxide is a Lewis acid. They are also examples of addition reactions. Water is added to a double bond converting it to two single bonds.

Water reacts with some metal oxides to give metal hydroxides. For example,

$$Na_2O + H_2O \longrightarrow 2NaOH$$

This is an acid-base reaction between the oxide ion, which is a strong base, and water, which is acting as an acid:

$$O^{2-} + H_2O \longrightarrow 2OH^-$$

Many metal oxides such as MgO and CuO do not react with water and are insoluble. In general, these oxides also have insoluble hydroxides. Therefore even if hydroxide forms on the surface of the oxide, there is no tendency for the reaction to go further.

Reactions with Halides

Most nonmetal halides react with water to give an oxoacid of the nonmetal and the hydrogen halide (see Experiments 5.3 and 13.4). For example,

$$PCl_3 + 3H_2O \longrightarrow HPO(OH)_2 + 3HCl$$
$$SiCl_4 + 4H_2O \longrightarrow Si(OH)_4 + 4HCl$$

These reactions resemble the reactions of water with the oxides of the nonmetals. The negatively charged O atom of the water molecule attacks the positively charged nonmetal atom. Then a H atom is transferred to the halogen atom, with the elimination of the hydrogen halide:

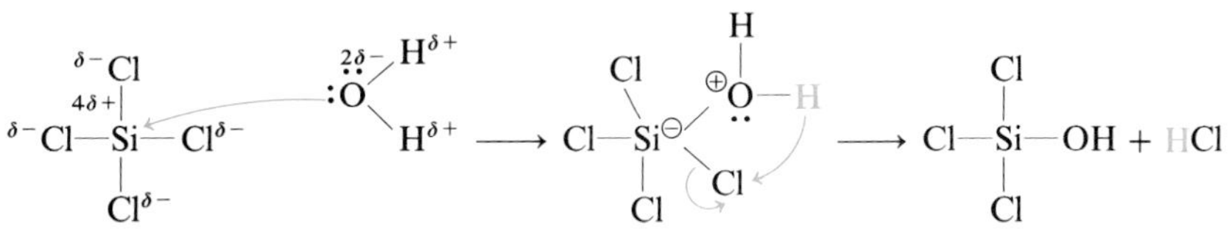

Reactions of Metal and Nonmetal Chlorides with Water

The three beakers contain water to which a small amount of methyl red indicator has been added. The dish on the left contains sodium chloride. The small bottle in the center contains the colorless liquid phosphorous trichloride, PCl_3, and the dish on the right contains the pale yellow solid phosphorous pentachloride, PCl_5.

When sodium chloride is added to the beaker on the left no change is observed. When PCl_3 is added to the beaker in the middle and PCl_5 to the beaker on the right the color of the indicator changes to red.

Sodium chloride dissolves in water without reacting. PCl_3 reacts rapidly with water to form phosphorous acid, H_3PO_3, and HCl, giving an acidic solution which changes the color of the indicator. PCl_5 reacts with water to form phosphoric acid, H_3PO_4, and HCl, giving an acidic solution which changes the color of the indicator.

The result is that OH^- replaces Cl^-, which adds H^+ to form HCl.

Repetition of this reaction leads eventually to the formation of silicic acid, $Si(OH)_4$. Again, the water molecule is behaving as a Lewis base and the metal halide as a Lewis acid. Silicon tetrachloride can behave as a Lewis acid because the valence shell of silicon can accommodate more than eight electrons. In contrast, the valence shell of carbon is restricted to eight electrons—carbon obeys the octet rule—and so carbon tetrachloride cannot accept an unshared pair of electrons from a water molecule. In other words, carbon tetrachloride cannot behave as a Lewis acid. Moreover, the C atom is completely surrounded by four large Cl atoms, which block the access of a water molecule to the carbon atom.

The reaction of water with chlorine to give hypochlorous acid can be understood on the basis of the same model:

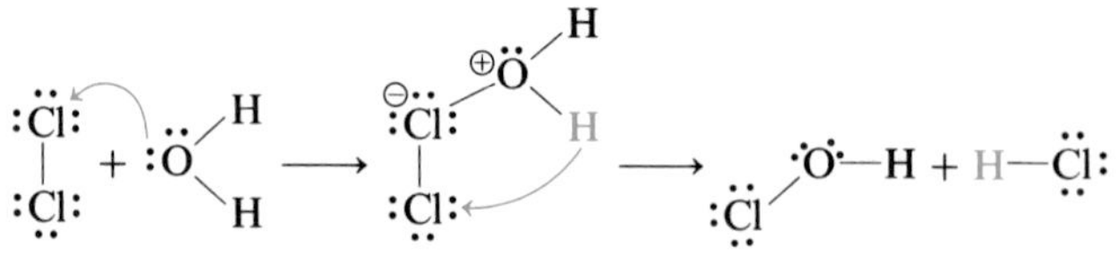

The overall equation is

$$Cl_2 + 2H_2O \rightleftharpoons HOCl + H_3O^+ + Cl^-$$

The reaction of bromine with water is similar, but iodine is insoluble in water. In contrast, fluorine oxidizes water to oxygen, as we have seen previously. In addition to the above reaction, chlorine also oxidizes water to oxygen, but very slowly:

$$2Cl_2 + 2H_2O \longrightarrow 4HCl + O_2$$

13.8 COLLIGATIVE PROPERTIES

Antifreeze is added to an automobile radiator to prevent water from freezing in the winter. Salt is placed on roads in the winter to prevent the formation of ice. Water continuously rises from the roots of trees to the leaves, even though the leaves may be as much as 100 m above the roots for very tall trees. Red blood cells placed in water burst.

At first, there does not appear to be much connection between these varied facts, but they all depend on some related properties of solutions called colligative properties. **Colligative properties** *are properties of solutions that depend on the number of solute and solvent molecules (or ions), but not on the nature of the solute.* They are the vapor pressure, the boiling point elevation, the freezing point depression, and the osmotic pressure.

Vapor Pressure of Solutions

Measurement of the vapor pressure of solutions shows that the vapor pressure of the solvent in a solution is less than that of the pure solvent. French chemist François Raoult (1830–1901) showed that the vapor pressure of the solvent in a solution is independent of the nature of the solute and depends only on its concentration. The vapor pressure of the solvent is given by the expression

$$p_{solvent} = x_{solvent} p_{solvent}^{\circ}$$

where $p_{solvent}$ is the vapor pressure of the solvent in the solution, $p_{solvent}^{\circ}$ is the vapor pressure of the pure solvent, and $x_{solvent}$, which is called the **mole fraction** of the solvent, is given by the expression

$$x_{solvent} = \frac{\text{moles solvent}}{\text{moles solvent} + \text{moles solute}}$$

The mole fraction has no units; it represents the fraction of the total number of molecules in a solution that are solvent molecules.

If the solute is nonvolatile, then the vapor pressure of the solution is equal to that of the solvent in the solution. Thus for a nonvolatile solute

$$p_{solution} = p_{solvent} = x_{solvent} p_{solvent}^{\circ}$$

A direct consequence of the lowering of the vapor pressure of the solvent in a solution is that the boiling point of the solution is higher than that of the pure solvent.

Elevation of the Boiling Point

Figure 13.21 shows that if the vapor pressure of a solution is lower than the vapor pressure of the pure solvent, then in order to increase the vapor pressure to 1 atm, we must increase its temperature to above the boiling point of the pure solvent. In other words, the boiling point of the solution is higher than that of the pure solvent.

The elevation of the boiling point is proportional to the number of moles of solute in a given amount of solvent. We can write

$$\Delta T = K_b m$$

where m is the number of moles of solute in 1 kg of solvent, called the **molality** of the solute, and K_b is the **boiling point elevation constant**. Values of K_b for some solvents are given in Table 13.9. Note that *molality* is not the same as *molarity. The molarity, M, of a solute is the number of moles of the solute in 1 L of solution*; the **molality**, *m, is the number of moles of solute in 1 kg of solvent.*

The elevation of the boiling point of a solvent by a solute can be used to determine the molar mass of the solute.

Figure 13.21 Elevation of Boiling Point and Depression of Freezing Point. A solution has a lower vapor pressure than the pure solvent; it must be heated to a *higher* temperature before its vapor pressure reaches the atmospheric pressure. The boiling point of a solution is greater than that of the solvent by ΔT_b. At the freezing point the vapor pressure of the solvent in a solution is lower than that of the pure solvent. Therefore the vapor pressure of the solid in equilibrium with the solution must also be lower. The vapor pressure of the solid decreases with decreasing temperature. So the pure solid and a solution are in equilibrium at a *lower* temperature than the solid and pure solvent. The freezing point of a solution is lower than that of the pure solvent by ΔT_f.

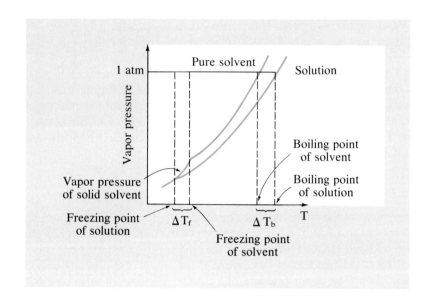

Example 13.6 A sample of 1.20 g of a nonvolatile organic compound was dissolved in 60.0 g of benzene. The boiling point of the solution was found to be 80.96°C. Pure benzene boils at 80.08°C. What is the molar mass of the solute?

Solution

$$\Delta T = 80.96°C - 80.08°C = 0.88°C$$

From Table 13.9 we have

$$K_b \text{ (benzene)} = 2.53°C \text{ kg mol}^{-1}$$

Therefore

$$m = \frac{\Delta T}{K_b} = \frac{0.88°C}{2.53°C \text{ kg mol}^{-1}} = 0.35 \text{ mol kg}^{-1}$$

The number of moles of solute in 60.0 g solvent is then found as follows:

$$\left[\frac{0.35 \text{ mol (solute)}}{1 \text{ kg (solvent)}}\right]\left(\frac{1 \text{ kg}}{1000 \text{ g}}\right)[60.0 \text{ g (solvent)}] = 0.021 \text{ mol (solute)}$$

Then the mass of 1 mol, the molar mass, can be calculated:

$$\text{Molar mass} = \frac{1.20 \text{ g}}{0.021 \text{ mol}} = 57 \text{ g mol}^{-1}$$

Table 13.9 Boiling Point Elevation Constants

SOLVENT	BOILING POINT (°C)	K_b (°C kg mol^{-1})
Water	100	0.52
Ethanol	79	1.19
Acetic acid	118	2.93
Benzene	80	2.53
Cyclohexane	81	2.79
Carbon disulfide	46	2.37
Carbon tetrachloride	76.5	5.03

Depression of the Freezing Point

At the freezing point of a solvent the solid and the liquid are in equilibrium. They must therefore have the same vapor pressure. If the solid had a lower vapor pressure, all the liquid would be transformed into solid—in other words, they could not be in equilibrium. If the vapor pressure of the solid were higher than that of the liquid, all the solid would be transformed into liquid. So at equilibrium the vapor pressure of the solid and liquid must be equal. The solid that separates from a solution is in most cases the pure solid solvent. Because the vapor pressure of a solution is less than that of the pure solvent, a solution can only be in equilibrium with the pure solid solvent if the vapor pressure of the solid solvent is decreased by decreasing its temperature. Thus a solution and the pure solid solvent are in equilibrium at a lower temperature than the pure liquid and the solid solvent. In other words, the freezing point of a solution is lower than that of the pure solvent as Figure 13.21 shows.

The relationship between the depression of the freezing point and the number of molecules of solute in 1 kg of solvent is similar to that for the boiling point elevation:

$$\Delta T = K_f m$$

Values for the **freezing point depression constant**, K_f, for some solvents are given in Table 13.10.

The measurement of the freezing point of a solution can be used, like the measurement of the boiling point, to determine the molar mass of a solute. If we know the empirical formula of a solute, we can then find its molecular formula, as the following example shows.

Table 13.10 Freezing Point Depression Constants

SOLVENT	FREEZING POINT (°C)	$K_f(°\text{C kg mol}^{-1})$
Water	0	1.86
Acetic acid	17	3.90
Sulfuric acid	10.4	5.98
Benzene	5.5	5.10
Cyclohexane	6.5	20.2

Example 13.7 A solution of 2.95 g of sulfur in 100 g of cyclohexane had a freezing point of 4.18°C. The freezing point of pure cyclohexane is 6.50°C. What is the molecular formula of sulfur in this solvent?

Solution
$$\Delta T = 6.50 - 4.18 = 2.32°\text{C}$$

From Table 13.10 we have

$$K_f = 20.2°\text{C kg mol}^{-1}$$

Therefore

$$m = \frac{\Delta T}{K_f} = \frac{2.32°\text{C}}{20.2°\text{C kg mol}^{-1}} = 0.115 \text{ mol kg}^{-1}$$

The number of moles of solute in 100 g of solvent is then found as follows:

$$\left[\frac{0.115 \text{ mol (solute)}}{1 \text{ kg (solvent)}}\right]\left(\frac{1 \text{ kg}}{1000 \text{ g}}\right)[100 \text{ g (solvent)}] = 0.0115 \text{ mol (solute)}$$

Since we know the mass of the solute, we can find the mass of 1 mol, that is, the molar mass:

$$\text{Molar mass} = \frac{2.95 \text{ g}}{0.0115 \text{ mol}} = 257 \text{ g mol}^{-1}$$

Hence the mass of one molecule of sulfur is 257 u.
Since the atomic mass of sulfur is 32.06 u, there must be

$$\frac{257 \text{ u molecule}^{-1}}{32.06 \text{ u atom}^{-1}} = 8.02 \text{ atoms molecule}^{-1}$$

In other words, there are eight sulfur atoms in a sulfur molecule. Thus the molecular formula of sulfur in solution in cyclohexane is S_8.

Example 13.8 How much ethylene glycol (1,2-ethanediol), $C_2H_6O_2$, must be added to 1.00 L of water so that the solution does not freeze above $-20.0°C$?

Solution From Table 13.10 the freezing point depression constant for water is $1.86°C\,kg\,mol^{-1}$. We can find the molality of a solution with a freezing point of $-20°C$ as follows:

$$m = \frac{\Delta T}{K_b} = \frac{20.0°C}{1.86°C\,kg\,mol^{-1}} = 10.8\ mol\,kg^{-1}$$

Since 1.00 L of water has a mass of 1.00 kg, we need 10.8 mol of ethylene glycol. Since the molar mass of ethylene glycol is $62.1\ g\,mol^{-1}$, we need

$$10.8\ mol \times 62.1\ g\,mol^{-1} = 670\ g\ ethylene\ glycol$$

Osmosis and Osmotic Pressure

Osmosis is another colligative property that is not only very useful for the measurement of molar mass but also is of great biological importance. A *membrane* is a thin liquid or solid film; paper, cellophane, and parchment are examples. The walls of the living cell are membranes. Most membranes are *semipermeable*; that is, they allow the passage of some molecules or ions but not others. Some important membranes allow the passage of water but not of ions or other solutes.

If a solution is separated from pure solvent by a semipermeable membrane that allows the passage of solvent molecules but not solute molecules, more solvent molecules pass through the semipermeable membrane from the solvent to the solution than pass in the other direction. Thus there is a net flow of solvent from the solute to the solution, which is called **osmosis** (Experiment 13.5).

The passage of the solvent through the membrane into the solution can be stopped if sufficient pressure is exerted on the solution (Figure 13.22). This pressure is called the **osmotic pressure**. It is osmotic pressure that drives the water up a tall tree, that keeps the stems of nonwoody plants rigid, and that causes a red blood cell to burst when placed in water.

EXPERIMENT 13.5

Osmotic Pressure

The carrot has been hollowed out and filled with a colored salt solution. A stopper with a long narrow glass tube is sealed into the top of the carrot. When the carrot is placed in the beaker of water, the solution rises to a considerable height up the tube. The cells of the carrot act as a semipermeable membrane, drawing water into the salt solution by osmosis, diluting the solution and increasing its volume, thus causing it to rise up the tube.

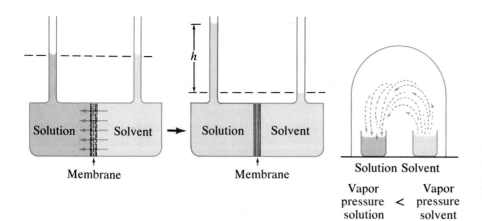

Figure 13.22 Osmotic Pressure.
When a solution and its solvent
are separated by a semipermeable
membrane, there is a transfer of
solvent from the pure solvent to
the solution. This transfer can be
stopped if a sufficient pressure is
exerted on the solution, as, for
example, from the the hydrostatic
pressure due to the increased
height of the solution on the left.
The flow of solvent through
a semipermeable membrane is
analogous to the flow of solvent
in the vapor phase that occurs
when separate samples of solvent
and solution are left in a closed
container. Because the vapor
pressure of the solvent is higher
than that of the solution, there
is an overall transfer of solvent
to the solution. The flow of solvent
continues until all the solvent has
been transferred to the solution.

Osmosis is a colligative property because the osmotic pressure of a solution depends on the number of solute molecules or ions but not on their nature. The osmotic pressure is related to the number of moles of solute by an equation that closely resembles the ideal gas equation:

$$\Pi V = nRT$$

where Π is the osmotic pressure, V is the volume of the solution, n is the number of moles of solute, T is the Kelvin temperature, and R is the gas constant.

The relationship of osmotic pressure to the other colligative properties can be understood by considering a pure solvent and a solution in separate vessels in an enclosed space. Because of its high vapor pressure, more solvent molecules leave the solvent than leave the solution in a given time interval, so there is a net transfer of solvent from pure solvent to solution. Similarly, when a solvent and a solution are separated by a semipermeable membrane, more solvent molecules pass from the solvent to the solution than from the solution to the solvent, so there is a net transfer of solvent from the pure solvent to the solution (see Figure 13.22).

Osmosis is a very sensitive method for measuring molar masses, and it is widely used for proteins, polymers, and other large molecules. Another advantage of using osmosis is that measurements can be made at room temperature, which is important for biological molecules that may not be stable at higher temperatures.

Example 13.9 An aqueous solution containing of 1.10 g of a protein in 100 mL of the solution had an osmotic pressure at 20°C of 3.93×10^{-3} atm. What is the molar mass of the protein?

Solution From the equation $\Pi V = nRT$ we calculate the number of moles of protein in the solution:

$$n = \frac{\Pi V}{RT} = \frac{(3.93 \times 10^{-3}\ \text{atm})(0.100\ \text{L})}{(0.0821\ \text{L atm mol}^{-1}\ \text{K}^{-1})(293\ \text{K})}$$

$$= 1.63 \times 10^{-5}\ \text{mol}$$

Hence

$$\text{Molar mass} = \frac{1.10\ \text{g}}{1.63 \times 10^{-5}\ \text{mol}} = 67\,500\ \text{g mol}^{-1}$$

Because of the very high molar mass, the solution is very dilute, 1.64×10^{-5} mol L^{-1}, and therefore the osmotic pressure is very small, only 3.93×10^{-3} atm. However, this pressure corresponds to a difference in height between the solvent and the solution of 2.99 cm, which is easily measured.

In contrast, the freezing point depression of this solution is only

$$\Delta T = 1.86°C \, kg \, mol^{-1} \times 1.63 \times 10^{-5} \, mol \, kg^{-1} = 0.000 \, 03°C$$

This freezing point depression is much too small to be measured with any accuracy. Thus freezing point depression, and similarly boiling point elevation, cannot be used for measuring the molar masses of very large molecules. But osmotic pressure is much more sensitive and is widely used for this purpose.

Example 13.10

(a) Analysis of hemoglobin from red blood cells showed that it contained 0.328% iron. What is the minimum molar mass of the hemoglobin?

(b) An aqueous solution containing 80.0 g of hemoglobin in 1.00 L had an osmotic pressure of 0.0260 atm at 4°C. What is the molar mass of hemoglobin?

(c) How many iron atoms are in a hemoglobin molecule?

Solution

(a) The analytical data show that 100 g of hemoglobin contains

$$0.328 \, g \, Fe = \frac{0.328 \, g}{55.85 \, g \, mol^{-1}} = 5.87 \times 10^{-3} \, mol \, Fe$$

One molecule of hemoglobin must contain at least one iron atom. Therefore in 100 g of hemoglobin there can be at most 5.87×10^{-3} mol of hemoglobin. Thus the minimum molar mass of hemoglobin is

$$\frac{100 \, g}{5.87 \times 10^{-3} \, mol} = 17\,000 \, g \, mol^{-1}$$

(b) We can find the number of moles of hemoglobin in solution as follows:

$$n = \frac{\Pi V}{RT} = \frac{(0.0260 \, atm)(1.00 \, L)}{(0.0821 \, atm \, L \, mol^{-1} \, K^{-1})(277 \, K)}$$

$$= 1.14 \times 10^{-3} \, mol$$

Hence the molar mass is

$$\frac{80.0 \, g}{1.14 \times 10^{-3} \, mol} = 70\,200 \, g \, mol^{-1}$$

(c) The actual molar mass is approximately four times the minimum molar mass assuming that the molecule contains only one iron atom. Therefore we conclude that the hemoglobin molecule contains four iron atoms.

Why Does a Solute Lower the Vapor Pressure of a Solvent?

We have seen that a solute dissolves in a solvent because the random, thermal motions of the solvent and solute molecules cause the molecules to become mixed up with each other. All systems that are not at equilibrium tend with time to become more mixed up, more chaotic, more random. Their entropy is said to increase. Solutions have a higher entropy than the pure solvent. The molecules of a vapor have more random motion that those of a liquid, so a vapor has a higher entropy than the corresponding liquid. Thus the tendency of a liquid to evaporate is another example of a system tending to increase its entropy. Because a solution has a higher entropy than a pure solvent, the difference in entropy between a solution and the vapor is less than that between the pure solvent and the vapor. Thus solvent molecules have a smaller tendency to leave a solution than to leave the pure solvent to become vapor. So the vapor pressure of a solution is less than that of the solvent.

Similarly, solvent molecules leave a pure solvent and pass through a semipermeable membrane into a solution because they have a greater entropy in the solution than in the pure solvent. They do not move in the reverse direction because this would lead to a decrease in entropy.

The equations that we have given for the lowering of the vapor pressure, the elevation of the boiling point, the depression of the freezing point, and the osmotic pressure can all be derived from thermodynamics, but these derivations are beyond the scope of this book.

Electrolyte Solutions

Colligative properties depend on the numbers of solvent and solute molecules, not on their nature. One mole of NaCl and 1 mol of HCl both give 2 mol of ions, $(Na^+ + Cl^-)$ and $(H_3O^+ + Cl^-)$, respectively, in solution. These solutes therefore give twice the freezing point depression, boiling point elevation, or osmotic pressure of a nonelectrolyte solution of the same concentration. If we call i the number of moles of solute particles produced in a solution by 1 mol of solute, then $i = 2$ for HCl and NaCl and $i = 3$ for Na_2SO_4. The equations for the colligative properties are then as follows:

Elevation of boiling point	$\Delta T = iK_b m$
Depression of boiling point	$\Delta T = iK_f m$
Osmotic pressure	$\Pi V = inRT$

The fact that salts and acids give larger than expected freezing point depressions in aqueous solution provided Swedish chemist Svante Arrhenius (1859–1927) with strong evidence for his theory that these substances are electrolytes that form ions in aqueous solution (see Figure 13.23).

Example 13.11 How many grams of NaCl must be added to 1.00 L (1.00 kg) of water to decrease the freezing point to $-6.00°C$?

Solution

$$NaCl(s) \xrightarrow{H_2O} Na^+(aq) + Cl^-(aq)$$

Therefore $i = 2$. Using the equation

$$\Delta T = iK_f m \quad \text{or} \quad m = \frac{\Delta T}{iK_f}$$

and substituting $i = 2$, $\Delta T = 6.00°C$, and $K_f = 1.86°C \, kg \, mol^{-1}$ (from Table 13.10), we have

$$m = \frac{6.00°C}{2 \times 1.86°C \, kg \, mol^{-1} \, (solvent)} = 1.61 \, mol \, kg^{-1} \, (solvent)$$

Therefore $1.61 \, mol \times 58.44 \, g \, mol^{-1} = 94.1 \, g$ NaCl must be added to $1.00 \, kg = 1.00 \, L$ of water.

Figure 13.23 Svante Arrhenius. Arrhenius was born near Uppsala, Sweden, in 1859, and became one of the most brilliant physical chemists of his time. He was the first to propose, in his Ph.D. thesis of 1884, that substances such as NaCl exist as ions in aqueous solution. This was a revolutionary suggestion, since the electron had not yet been discovered and chemists had great difficulty in understanding how sodium and chlorine atoms could become charged. Arrhenius barely passed his Ph.D. oral, and only after evidence for the electron had accumulated in the 1890s did Arrhenius's ideas concerning the nature of solutions of salts in water come to be generally accepted. Finally, in 1903, for the same thesis that had barely been accepted for the Ph.D., Arrhenius was awarded the Nobel Prize in chemistry. His other great contribution to chemistry was the idea of activation energy in reactions, which we consider in Chapter 18.

Note that calculations of this kind for electrolyte solutions only give approximate answers, because the electrostatic attractions between the ions prevent them from acting completely independently, except in very dilute solutions. Therefore the observed freezing point depression, elevation of boiling point, and osmotic pressure are, in fact, somewhat smaller than predicted.

IMPORTANT TERMS

A **colligative property** is a property of a solution that depends only on the number of solute particles (molecules or ions) and not on their nature.

A **hydrogen bond** is an intermolecular attraction in which a hydrogen atom that is bonded to an electronegative atom is attracted to an unshared electron pair on another electronegative atom.

An **induced dipole** results from the separation of charge produced in an atom or a molecule by a polar molecule in its vicinity.

An **instantaneous dipole moment** is the fluctuating, temporary dipole moment that results from the motions of the electrons in an atom or molecule.

Intermolecular forces are the forces of attraction between molecules.

London forces are intermolecular forces that result from fluctuating instantaneous dipoles. They are the only intermolecular forces between nonpolar molecules.

The **molality, m** of a solute in a solution is the number of moles of the solute in 1 kg of solvent.

The **molar enthalpy of fusion**, ΔH_{fus}, is the heat required to melt 1 mol of a solid substance at 1 atm pressure.

The **molar enthalpy of solution** is the heat absorbed at constant pressure when 1 mol of a solute dissolves in a solvent at a given temperature. The enthalpy of solution depends on the concentration of the final solution, but

it has a constant value for sufficiently dilute solutions, that is, for a sufficiently large amount of solvent.

The **molar enthalpy of vaporization**, ΔH_v, is the heat required to vaporize 1 mol of a liquid substance at 1 atm pressure.

The **mole fraction** of the solvent in a solution is the fraction of the total number of molecules in the solution that are solvent molecules.

Osmosis is the flow of solvent through a semipermeable membrane from a more dilute to a more concentrated solution.

Osmotic pressure is the pressure that must be exerted on a solution to prevent osmosis.

The **polarizability** of an atom or a molecule is a measure of the ease with which the electrons and nuclei can be displaced from their equilibrium positions by an external force so that a dipole is created.

Sublimation is the process by which a solid is converted directly to vapor.

Van der Waals force is an alternative name for intermolecular force.

The **van der Waals radius** is a measure of the size of an atom in the directions in which it is not forming a bond. It is equal to one-half the distance between adjacent atoms of the same kind in molecules that are not bonded together.

Vapor pressure is the pressure exerted by a vapor in equilibrium with a liquid or a solid.

PROBLEMS

Vapor Pressure and Enthalpy of Vaporization

1. Why is the molar enthalpy of sublimation of a solid always larger than the molar enthalpy of vaporization of the corresponding liquid?

2. (a) Why does the vapor pressure of a liquid increase with increasing temperature?

(b) At 20°C the vapor pressure of benzene is 75.0 mm and that of toluene is 50.0 mm. Which hydrocarbon is expected to have the higher boiling point at atmospheric pressure?

3. Before the refrigerator was invented, butter and milk were stored in porous clay pots standing in water. Why would such a system keep the contents cool even on the hottest day?

4. Calculate the standard enthalpy change for the reaction in which CH_4 is burned completely in air to give CO_2 and H_2O. Use the value obtained to find the amount of methane (natural gas) that would have to be burned to convert 1000 kg of liquid water at 100°C to steam.

Partial Pressure of Water Vapor

5. A student prepared a sample of O_2 in the laboratory by heating potassium chlorate, $KClO_3$, in the presence of manganese dioxide, MnO_2, as a catalyst. He collected the O_2 in a jar over water, as illustrated in Figure 13.7. After the levels of the water inside and outside the jar were adjusted so that they were the same, the volume of the gas in the jar was 365 mL. The atmospheric pressure was 745 torr, and the temperature was 21°C.

(a) What was the partial pressure of O_2 in the gas collected?

(b) What would the volume of dry O_2 have been at STP?

6. A sample of 5.00 L of air was saturated with water vapor at 25°C and bubbled through a concentrated H_2SO_4 solution. The H_2SO_4 solution was found to increase in mass by 0.115 g. What is the vapor pressure of water at 25°C, assuming that all the water vapor is removed from the air by the H_2SO_4?

7. What volume of gas can be collected over water at 25°C and a pressure of 754 torr when 0.540 g of LiH reacts completely with water? If the volume of the resulting LiOH(aq) solution is 50.5 mL, what is the LiOH concentration? What is the volume of dry H_2 at STP?

Intermolecular Forces

8. Elemental boron is almost as hard as diamond, and it has a melting point of 2300°C. It is a poor conductor of electricity at room temperature. Would you expect boron to be a molecular solid, an ionic solid, or a covalent network solid? Would you expect boron to be soluble in water?

9. The compound $SnCl_4$ is a liquid with a boiling point of 114°C and a melting point of $-33°C$. The compound $SnCl_2$ is a solid with a melting point of 246°C. Suggest an explanation for these differences in properties in terms of their structures and their intermolecular forces.

10. For each of the following, predict whether you would expect it to form a molecular solid or a network solid. What is the most important type of intermolecular force in each case? Give reasons for your answers.

(a) N_2 (b) H_2S (c) Cr   (d) CaO
(e) SiH_4 (f) SiO_2 (g) KOH   (h) H_2SO_4

11. Which substance of each of the following pairs would be expected to have the higher melting point? Give reasons for your answers.

(a) ClF or BrF   (b) BrCl or Cl_2
(c) KBr or BrCl   (d) K or Br_2

12. Why is the boiling point of chlorobenzene, C_6H_5Cl (132°C), higher than that of benzene, C_6H_6 (80°C)? Why is the boiling point of hexachlorobenzene, C_6Cl_6 (310°C), much higher than that of C_6H_6?

13. What are the intermolecular forces of attraction in each of the following?

I-I (a) I_2(s) *induced dipole*   (b) CaO(s) *ionic*   (c) CO_2(g)
(d) CH_3CN(l)   (e) HF(l)
dipole-dipole   *ionic*

14. The boiling points of the fluorides of the second-row elements are LiF, 1717°C; BeF_2, 1175°C; BF_3, $-101°C$; CF_4, $-128°C$; NF_3, $-120°C$; OF_2, $-145°C$; F_2, $-188°C$. Account for the enormous difference in the boiling points of LiF and BeF_2 and those of the other fluorides in terms of the nature and the strength of the intermolecular forces in these substances.

15. Why is the vapor pressure of water at 25°C less than the vapor pressure of gasoline at the same temperature?

16. Which substance in each of the following pairs will have the highest enthalpy of vaporization? Justify your choices.

(a) Cl_2O or H_2O
(b) CCl_4 or CBr_4
(c) Benzene, C_6H_6, or toluene, C_7H_8
(d) Methanol, CH_3OH, or dimethyl ether, CH_3OCH_3
(e) acetic acid, CH_3CO_2H, or acetyl chloride, CH_3COCl

17. Which of the following properties would you expect to depend on the strength of intermolecular forces: boiling point, density, molecular mass, solubility, viscosity, covalent radius, electronegativity, covalent bond strength?

18. Make qualitative predictions about the solubility of the following:

(a) HCl(g) in water and in pentane, C_5H_{12}
(b) Water in liquid HF and in gasoline
(c) Chloroform, $CHCl_3$, in water and in tetrachloromethane, CCl_4
(d) Naphthalene, $C_{10}H_8$, in water and in benzene, (C_6H_6)
(e) N_2(g) and HCN(g) in water
(f) Benzene, C_6H_6, in toluene (methylbenzene), C_7H_8, and in water

19. What type of intermolecular force is most important in each of the following substances?

(a) Ar   (b) Cl_2 (c) LiF(l)
(d) $AlCl_3 \cdot 6H_2O$ (e) CH_3OH (f) H_2SO_4
(g) HCl   (h) C_2F_6

20. Explain the following observations: Sulfuric acid, $(HO)_2SO_2$, boils at 320°C, fluorosulfuric acid, $(HO)SO_2F$, boils at 170°C, and sulfuryl fluoride, SO_2F_2, has a boiling point of $-55°C$.

21. Explain the following phenomena:

(a) Water freezes from the top down.

(b) Two of the components of gasoline are n-octane (straight chain) and isooctane (branched chain). Both have the formula C_8H_{18}, but n-octane has the higher boiling point.

(c) Ammonia, NH_3, shows a bigger deviation from the ideal gas laws than H_2.

22. (a) Identify the most important intermolecular force between the molecules of each of the following substances:

CO_2 H_2O HF I_2 ICl   He

(b) What is the strongest intermolecular attractive interaction or bond that must be broken when the following substances are melted?

Benzene, C_6H_6 Ethanol, C_2H_5OH C_2H_6
CaO Cl_2 HCl

23. For each of the following pairs, predict which substance would have the higher boiling point. In each case, explain your choice.

(a) Ca or Na
(b) SiH_4 or CH_4
(c) C_3H_8 or C_4H_{10}
(d) NH_3 or PH_3
(e) F_2 or Cl_2
(f) SO_2 or SiO_2

24. (a) In which of the following substances are there dipole-dipole interactions between the molecules?

CO H_2S CH_2Cl_2 CCl_4 CH_4 Cl_2

(b) In which of the following are there hydrogen-bonding interactions between the molecules?

CH_3OH CH_3-O-CH_3 CH_3F NH_3
CH_4 H_2N-NH_2 CH_3COOH H_2

25. Which substance in each of the following pairs is likely to be more soluble in water? Explain your choices. For the least soluble in each pair, suggest a better solvent.

(a) Hydrogen peroxide or benzene
(b) Ethane or ethanediol, $HOCH_2CH_2OH$
(c) Sugar or dodecane
(d) Chloroform (trichloromethane) or magnesium chloride
(e) Iodine or hydrogen iodide
(f) Lithium chloride or carbon tetrachloride (tetrachloromethane)
(g) Methanol, CH_3OH, or ethane

Reactions of Water

26. Write balanced equations for the reactions, if any, of each of the following substances with water. Classify each reaction as an acid-base reaction or an oxidation-reduction reaction, where appropriate. In each case, state whether the reaction proceeds at 25°C or only at high temperature.

(a) CaC_2
(b) SO_2
(c) SO_3
(d) Mg_3N_2
(e) CO
(f) $NaNH_2$
(g) P_4
(h) PCl_3
(i) PCl_5
(j) $COCl_2$
(k) Ca
(l) Na_2O
(m) F_2
(n) Br_2

27. Write balanced equations for two reactions in which water behaves as a Lewis base.

28. Write balanced equations for two reactions in which water behaves as a Brönsted-Lowry acid.

29. Write balanced equations for half-reactions showing water behaving as an oxidizing agent and as a reducing agent.

30. Write balanced equations for two reactions in which water behaves as a base.

31. Explain why water is described as amphoteric.

Colligative Properties

32. What mass of sugar, $C_{12}H_{22}O_{11}$, must be dissolved in one cup (250 mL) of water to raise the boiling point to 101.0°C at a pressure of 1.00 atm? 250 ml (1 atm)
$K_b = 0.52 \frac{°kg}{mol}$ = 250 ml (1 $\frac{3}{mol}$) = .25 kg

33. Either camphor, $C_{10}H_{16}O$, or naphthalene, $C_{10}H_8$, can be used to make mothballs. A 5.00-g sample of mothballs was dissolved in 100.0 g of ethanol. The solution had a boiling point of 78.89°C. The pure solvent has a boiling point of 78.41°C. Were the mothballs made of camphor or naphthalene?

34. What is the freezing point of each of the following?

(a) A 0.10-m solution of sugar in water
(b) A 0.10-m NaCl solution in water $K_f = 2(-1.86 \frac{°kg}{mol})$
(c) A 0.10-m $CaCl_2$ solution in water $K_f = 3(-1.86)$

35. A solution containing 1.00 g of aluminum bromide in 100 g of benzene has a freezing point 0.099°C below that of pure benzene. What is the molar mass of aluminum bromide in benzene? What is its molecular formula? Draw a Lewis structure for the molecule.

36. Down to what temperature is sodium chloride effective in melting ice on streets and sidewalks? The solubility of sodium chloride in water at 0°C is 5.8m.

37. A 1% by mass solution of gum arabic (empirical formula, $C_{12}H_{22}O_{11}$) was found to have an osmotic pressure of 7.2 mm Hg at 25°C. What is the average molar mass of gum arabic and the number of monomer units per polymer molecule?

38. Insulin is a protein that regulates carbohydrate metabolism by decreasing blood sugar levels. A deficiency of insulin leads to diabetes. An aqueous solution of insulin containing 5.00 g of insulin in 50 mL of solution was found to have an osmotic pressure of 0.427 atm at 25°C. What is the molecular mass of insulin?

$Ca C_2 + 2H_2O \rightarrow Ca(OH)_2 + C_2H_2$

$SO_2 + H_2O \rightarrow H_2SO_3$

$SO_3 + H_2O \rightarrow H_2SO_4$

$Mg_3N_2 + 6H_2O \rightarrow 3Mg(OH)_2 + 2NH_3$

(high Temp only) $CO + H_2O \rightarrow CO_2 + H_2$

$Na NH_2 + H_2O \rightarrow Na OH + NH_3$

$P_4 + H_2O \rightarrow$ no reaction

$PCl_3 + H_2O \rightarrow H_3PO_3 + 3HCl$

$PCl_5 + H_2O \rightarrow H_3PO_4 + 5HCl$

$COCl_2 + H_2O \rightarrow 2HCl + CO_2$

$Ca + 2H_2O \rightarrow Ca(OH)_2 + H_2$

$Na_2O + H_2O \rightarrow 2NaOH$

$2F_2 + 2H_2O \rightarrow 4HF + O_2$

$Br_2 + H_2O \rightarrow HBr + HOBr$

CHAPTER 14

CHEMICAL EQUILIBRIUM

Gas Phase Equilibria and Acid-Base Equilibria in Aqueous Solutions

A very important question concerning any reaction is the extent to which it proceeds under particular conditions. Does it proceed to completion, that is, until one of the reactants is completely used up? Does it proceed to only a limited extent, or perhaps not at all? In other words, do we obtain the maximum possible yield of the products, only a small yield, or perhaps no yield at all? For example, we have seen in Chapter 5 that if hydrogen and chlorine are exposed to a bright light, an explosive reaction occurs in which the hydrogen and chlorine combine completely; in other words, a quantitative yield of HCl is obtained. The reaction is said to proceed to completion. On the other hand, if hydrogen and iodine are heated together, the reaction is much less vigorous and does not go to completion; some hydrogen and iodine remain unreacted.

A strong acid such as HCl ionizes completely in water; that is, all the HCl molecules donate their protons to H_2O molecules to give H_3O^+ and Cl^-. In contrast, a weak acid such as HF does not ionize completely in water. The HF molecules, H_2O molecules, H_3O^+ ions, and F^- ions are all present in a state of equilibrium in which their concentrations do not change. But the fact that the concentrations do not change does not mean that the reaction of HF with H_2O stops. In fact, HF molecules continue to donate protons to H_2O molecules, but at the same time H_3O^+ ions donate protons to F^- ions. At equilibrium the rates of these two reactions are equal, so the concentrations of all of the species taking part in the reaction remain constant. We write the equation for such a reaction as follows:

$$HF(aq) + H_2O(l) \rightleftharpoons H_3O^+(aq) + F^-(aq)$$

where the double arrow indicates that the reaction is taking place in both directions simultaneously.

14.1 EQUILIBRIUM
CONSTANT

501

In this chapter we show how chemical equilibria can be treated in a quantitative manner. We will see that we can assign to a reaction a quantity called the *equilibrium constant* that enables us to predict how far a reaction will proceed toward completion before it comes to equilibrium. We will also see how the concentrations of the reactants and products that are present at equilibrium in a reaction depend on several factors, such as initial concentration of the reactants, pressure, and temperature. We will see how in an industrial process such as the synthesis of ammonia, it is possible to adjust the conditions to maximize the yield of ammonia.

Acid-base equilibria in aqueous solutions are particularly important in chemistry and biology, and we will consider them in some detail. We will discuss the extent of ionization of weak acids and bases, the changes in hydronium ion concentration in acid-base titrations, the use of indicators in titrations, and the use of equilibria to maintain a constant hydronium ion concentration in a solution.

14.1 EQUILIBRIUM CONSTANT

All chemical reactions carried out in a closed system eventually reach a state of **equilibrium** in which the concentrations of all the reactants and products do not change with time. By "in a closed system" we mean that the total mass of the reactants and products remains constant; in other words, as the reaction proceeds, none of the products are removed or allowed to escape from the reaction mixture, and no additions of any of the reactants are made. Some reactions proceed very rapidly; others are much slower. But in all cases a reacting system eventually reaches equilibrium. In some cases, such as the reaction of HF with H_2O, all the reactants and the products are present together in measurable concentrations when equilibrium is reached. In other reactions, such as that between H_2 and Cl_2 to give HCl, the concentrations of H_2 and Cl_2 that remain when equilibrium is reached are too small to be measured. In such a case we say that the reaction has gone to completion, and we usually write a single arrow between the reactants and the products in the equation for the reaction:

$$H_2(g) + Cl_2(g) \longrightarrow 2HCl(g)$$

In contrast, the reaction between H_2 and I_2 in the gas phase is another reaction in which there are measurable concentrations of all the reactants and products when equilibrium is reached. When I_2 vapor, which has a violet color, is mixed with colorless H_2, colorless hydrogen iodide, HI, is formed. As the reaction proceeds and I_2 is used up, the color of the I_2 vapor fades, eventually reaching a constant intensity that does not change further; the reaction has then reached equilibrium. The rate of the forward reaction between H_2 and I_2 is equal to the rate of the reverse reaction between HI molecules to give back H_2 and I_2, so the concentrations of all three species, H_2, I_2, and HI, remain constant:

$$H_2(g) + I_2(g) \rightleftharpoons 2HI(g)$$

If we start with equal concentrations of H_2 and I_2, then these concentrations decrease and the concentration of HI increases with time, until all three concentrations become constant when equilibrium is reached, as illustrated in Figure 14.1.

Equilibrium is reached whether we start with the reactants or the products. For example, if a colorless sample of HI is kept at 698 K, it gradually becomes violet as I_2 is formed. The concentration of HI decreases, and the concentrations of H_2 and I_2 increase. Eventually, the intensity of the violet color due to the I_2 becomes constant as the concentrations of H_2, I_2, and HI reach their equilibrium values. Thus whether we start with 1 mol of I_2 and 1 mol of H_2 or with 2 mol of HI in a vessel of the same volume, the same equilibrium mixture of H_2, I_2, and HI is obtained, as illustrated in Figure 14.1.

The reactants and the products do not always have the same equilibrium concentrations. In fact, equilibrium concentrations depend on the initial concentrations of the reactants and the products and on the temperature. But measurements have shown that there is a simple relationship between the concentrations of the reactants and the concentrations of the products when a reaction reaches equilibrium at a given temperature. Table 14.1 shows some different concentrations of H_2, I_2, and HI that are at equilibrium at 698 K. The fourth column gives values for the quantity

$$K = \frac{[\text{HI}]^2_{\text{eq}}}{[\text{H}_2]_{\text{eq}}[\text{I}_2]_{\text{eq}}}$$

calculated from these equilibrium concentrations. The $[\ ]_{\text{eq}}$ are the concentrations of the reactants and the products *at equilibrium*. We see that K is a constant that is independent of the actual concentrations of the reactants and the products at equilibrium. The constant K is called the **equilibrium constant** for the reaction. It has a value of 54.4 at 698 K for the H_2–I_2 reaction. It has a different value at another temperature, so the temperature must always be stated.

Experiments have shown that for any reaction at equilibrium a similar expression involving the concentrations of the products and the reactants at equilibrium has a constant value. We can write a general equation for a reaction as follows:

$$a\text{A} + b\text{B} + c\text{C} + \cdots \rightleftharpoons p\text{P} + q\text{Q} + r\text{R} + \cdots$$

In this equation A, B, C, and so on, are the reactants; and P, Q, R, and so on, are the products. The letters $a, b, c, \ldots, p, q, r, \ldots$ represent the *number of moles* of each substance involved in the balanced equation for the reaction. For this general reaction the equilibrium constant is

$$K = \left(\frac{[\text{P}]^p[\text{Q}]^q[\text{R}]^r \cdots}{[\text{A}]^a[\text{B}]^b[\text{C}]^c \cdots} \right)_{\text{eq}}$$

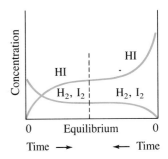

Figure 14.1 Concentration Changes During Approach to Equilibrium.

$$\text{H}_2(g) + \text{I}_2(g) \rightleftharpoons 2\text{HI}(g)$$

On the left, we start with equal initial concentrations of H_2 and I_2. As the reaction proceeds, the concentrations of H_2 and I_2 decrease and the concentration of HI increases, until each attains a constant value at equilibrium. *On the right*, we start with HI alone. As the reaction to give H_2 and I_2 proceeds, the concentration of HI decreases and the concentrations of H_2 and I_2 increase, until they all become constant when equilibrium is reached. In the figure the initial concentration of HI on the right is twice the initial concentration of H_2 and I_2 on the left. Under these circumstances the same equilibrium mixture is attained whether we start from the left or from the right.

Table 14.1 Equilibrium Concentrations in the H_2–I_2 System at 698 K

$[\text{I}_2]$ (mol L^{-1})	$[\text{H}_2]$ (mol L^{-1})	$[\text{HI}]$ (mol L^{-1})	$K = \left(\frac{[\text{HI}]^2}{[\text{H}_2][\text{I}_2]} \right)_{\text{eq}}$
0.47×10^{-3}	0.48×10^{-3}	3.5×10^{-3}	54.3
1.14×10^{-3}	1.14×10^{-3}	8.4×10^{-3}	54.3
1.71×10^{-3}	2.91×10^{-3}	16.5×10^{-3}	54.7
1.25×10^{-3}	3.56×10^{-3}	15.6×10^{-3}	54.7
0.74×10^{-3}	4.56×10^{-3}	13.5×10^{-3}	54.0
2.34×10^{-3}	2.25×10^{-3}	16.9×10^{-3}	54.2
3.13×10^{-3}	1.83×10^{-3}	17.7×10^{-3}	54.7
		Average value	54.4

where $[A]$, $[B]$, $[C]$, ..., $[P]$, $[Q]$, $[R]$, ... are the concentrations of the reactants and the products at equilibrium. To simplify the writing of the expression, we have dropped the subscript eq on each concentration term and have given the expression in brackets a subscript eq to remind us that the concentrations are equilibrium concentrations. In other words:

To obtain the expression for the equilibrium constant for any reaction, we raise the equilibrium concentration of each product to the power given by the number of moles of that product in the balanced equation for the reaction, and we multiply these. We then multiply the concentrations of each reactant, similarly raised to the power given by the number of moles of the reactant in the balanced equation. Finally, we divide the resulting expression for the products by that for the reactants.

For example, for the H_2–I_2 reaction $p = 2$, $a = 1$, and $b = 1$; so

$$K = \left(\frac{[HI]^2}{[H_2][I_2]} \right)_{eq}$$

For the reaction between N_2 and H_2 to give ammonia, NH_3,

$$N_2 + 3H_2 \rightleftharpoons 2NH_3$$

$p = 2$, $a = 1$, and $b = 3$; so

$$K = \left(\frac{[NH_3]^2}{[N_2][H_2]^3} \right)_{eq}$$

Some data for this reaction are given in Table 14.2. We see that K has a constant value independent of the concentrations of the reactants at equilibrium. No other expressions of this type involving different powers of the concentrations have a constant value unless these powers are all the same multiple of the powers in the expression above.

The expression for the equilibrium constant can be derived by the methods of thermodynamics and also by considering how the rates of the forward and reverse reactions depend on the concentrations of the reactants and products, as we will see in Chapter 18.

Example 14.1 What is the expression for the equilibrium constant of each of the following reactions?

$$2SO_2(g) + O_2(g) \rightleftharpoons 2SO_3(g)$$
$$2H_2(g) + O_2(g) \rightleftharpoons 2H_2O(g)$$
$$CH_4(g) + H_2O(g) \rightleftharpoons CO(g) + 3H_2(g)$$

Solution

$$K_1 = \left(\frac{[SO_3]^2}{[SO_2]^2[O_2]} \right)_{eq} \qquad K_2 = \left(\frac{[H_2O]^2}{[H_2]^2[O_2]} \right)_{eq} \qquad K_3 = \left(\frac{[CO][H_2]^3}{[CH_4][H_2O]} \right)_{eq}$$

Measurements of the equilibrium concentrations of the reactants and the products have given the following values for the equilibrium constants in Example 14.1:

$$K_1 = 287 \text{ mol}^{-1} \text{ L at } 1000°C$$
$$K_2 = 5.9 \times 10^{40} \text{ mol}^{-1} \text{ L at } 25°C$$
$$K_3 = 3.73 \times 10^{-3} \text{ mol}^2 \text{ L}^{-2} \text{ at } 1000 \text{ K}$$

Concentrations are normally expressed in mol L^{-1}, but the units of the equilibrium constant depend on the form of the equilibrium constant expression.

Table 14.2 Experimental Measurements of Equilibrium Concentrations in the $N_2 + 3H_2 \rightleftharpoons 2NH_3$ System at 500°C

$[H_2]$ (mol L^{-1})	$[N_2]$ (mol L^{-1})	$[NH_3]$ (mol L^{-1})	$K = \left(\dfrac{[NH_3]^2}{[N_2][H_2]^3}\right)_{eq}$ (mol^{-2} L^2)
1.15	0.75	0.261	5.98×10^{-2}
0.50	1.00	0.087	6.05×10^{-2}
1.35	1.15	0.412	6.00×10^{-2}
2.43	1.85	1.27	6.08×10^{-2}
1.47	0.750	0.376	5.93×10^{-2}
		Average value	6.0×10^{-2}

For reactions involving gases, we will often find it convenient to express the concentrations in terms of the partial pressures of the gases. Consider a mixture of n_A moles of A, n_B moles of B, ..., in a volume V. If the partial pressure of A is p_A, then from the ideal gas law we may write

$$p_A V = n_A RT$$

or

$$p_A = \frac{n_A}{V} RT$$

Now n_A/V is the molar concentration of A, that is, $[A]$. Hence $p_A = [A]RT$. Similarly, we may write $p_B = [B]RT$, and so on, for every gas in the mixture.

At a constant temperature RT is a constant, so $p_A \propto [A]$. Similarly, for every gas in the mixture the partial pressure is proportional to the concentration. Therefore the equilibrium constant expression written in terms of partial pressures is also a constant for the reaction at a given temperature. For example, for the synthesis of NH_3,

$$N_2(g) + 3H_2(g) \rightleftharpoons 2NH_3(g)$$

we have

$$K_p = \left\{\frac{(p_{NH_3})^2}{(p_{N_2})(p_{H_2})^3}\right\}_{eq}$$

where p_{NH_3} is the partial pressure of NH_3, and so on. The experimental value of K_p at 400 K is 55 atm^{-2}. Note that the units of K_p are pressure units. The value of K_p depends on the units used to express pressure, which are usually atmospheres. Note also that we write K_p with a subscript p to denote that the equilibrium constant has pressure units. Similarly, we may write K_c with a subscript c to denote that the equilibrium constant has concentration (mol L^{-1}) units.

An important use of the value of the equilibrium constant for a reaction is to estimate approximately where the position of equilibrium lies, that is, whether or not the reaction goes almost to completion. For example, for the reaction between H_2 and Cl_2 in the gas phase,

$$H_2(g) + Cl_2(g) \rightleftharpoons 2HCl(g) \qquad K_c = \left(\frac{[HCl]^2}{[H_2][Cl_2]}\right)_{eq}$$

the equilibrium constant K at 25°C has the very large value of 2.5×10^{33}. The large value of K indicates that at equilibrium the concentration of HCl is very large compared with the concentrations of the reactants H_2 and Cl_2. In other words, the reaction goes very nearly to completion. The equilibrium is said to lie "far to the right," and the equation for the reaction is normally written with a single arrow:

$$H_2(g) + Cl_2(g) \longrightarrow 2HCl(g)$$

In contrast, the equilibrium constant for the decomposition of water to H_2 and O_2,

$$2H_2O(g) \rightleftharpoons 2H_2(g) + O_2(g) \qquad K_c = \left(\frac{[H_2]^2[O_2]}{[H_2O]^2}\right)_{eq}$$

has the very small value of 1.7×10^{-41} mol L^{-1} at 25°C. This small value indicates that the equilibrium lies far to the left. At equilibrium there is a very large concentration of H_2O and only small concentrations of O_2 and H_2. Thus if we start with H_2 and O_2 instead of H_2O, then the reaction goes almost to completion. That is, at equilibrium there is a very large concentration of H_2O and only very small concentrations of H_2 and O_2. The *formation* of H_2O from H_2 and O_2,

$$2H_2(g) + O_2(g) \rightleftharpoons 2H_2O(g)$$

has the equilibrium constant

$$K_c = \left(\frac{[H_2O]^2}{[H_2]^2[O_2]}\right)_{eq}$$

If we call the equilibrium constant for the decomposition of water K_{decomp} and the equilibrium constant for the formation of water K_{form}, we see that

$$K_{form} = \left(\frac{[H_2O]^2}{[H_2]^2[O_2]}\right)_{eq} = \frac{1}{\left(\dfrac{[H_2]^2[O_2]}{[H_2O]^2}\right)_{eq}} = \frac{1}{K_{decomp}}$$

Therefore from the value for K_{decomp} we can find the value for K_{form}:

$$K_{form} = \frac{1}{K_{decomp}} = \frac{1}{1.7 \times 10^{-41} \text{ mol } L^{-1}} = 5.9 \times 10^{40} \text{ mol}^{-1} \text{ L}$$

As we expect, the equilibrium constant for the formation of water has a very large value; the reaction goes very nearly to completion.

In general, the equilibrium constant for a reaction as written from right to left is the reciprocal of the equilibrium constant for the reaction written from left to right:

$$K_{right\ to\ left} = \frac{1}{K_{left\ to\ right}}$$

Reaction Quotient

We can also use the equilibrium constant to predict in which direction a reaction will proceed for any given initial—that is, nonequilibrium—concentrations of reactants and products. Suppose we mix certain amounts of H_2, I_2, and HI. Will there be more or less HI when equilibrium is reached? For example, if 1.0×10^{-2} mol each of H_2 and I_2 and 2.0×10^{-3} mol of HI are placed in a 1-L flask at 698 K, will more HI or more H_2 and I_2 be formed?

To answer this question, we calculate the **reaction quotient**, Q. It has the same form as the expression for the equilibrium constant, but the concentrations of reactants and products are *not* the equilibrium concentrations. We first find the value of the reaction quotient,

$$Q = \frac{[HI]^2}{[H_2][I_2]}$$

for the *initial concentrations*. Substituting, we have

$$Q = \frac{(2.0 \times 10^{-3})^2 \, mol^2 \, L^{-2}}{(1.0 \times 10^{-2}) \, mol \, L^{-1} \times (1.0 \times 10^{-2}) \, mol \, L^{-1}} = 0.040$$

We then compare this value with the value of K_c:

$$K_c = \left(\frac{[HI]^2}{[H_2][I_2]} \right)_{eq} = 54.4$$

We see that Q is much smaller than K_c. If the value of

$$Q = \frac{[HI]^2}{[H_2][I_2]}$$

is to increase until it equals K_c, the concentration of HI must increase and the concentrations of H_2 and I_2 must decrease. Thus we conclude that if we start with these initial concentrations of reactants and products, more HI will be formed until equilibrium is reached.

If we increase the initial concentration of HI to $2.0 \times 10^{-1} \, mol \, L^{-1}$, then we have

$$Q = \frac{(2.0 \times 10^{-1})^2 \, mol^2 \, L^{-2}}{(1.0 \times 10^{-2}) \, mol \, L^{-1} \times (1.0 \times 10^{-2}) \, mol \, L^{-1}} = 400$$

In this case the reaction quotient is greater than K_c. So for equilibrium the concentration of HI must decrease and the concentrations of H_2 and I_2 must increase; more H_2 and I_2 will be formed until equilibrium is reached.

Finally, suppose the initial concentration of HI is $7.38 \times 10^{-2} \, mol \, L^{-1}$ and the H_2 and I_2 concentrations are each $1.00 \times 10^{-2} \, mol \, L^{-1}$. In this case the reaction quotient is

$$Q = \frac{(7.38 \times 10^{-2})^2}{(1.00 \times 10^{-2})^2} = 54.4$$

which is equal to K_c. Thus the system is at equilibrium, and there will be no change in the concentrations of H_2, I_2, or HI.

In short, if we know the initial concentrations of the reactants and the products, we can predict the direction in which a reaction will proceed, as follows:

- If $Q < K$, the concentrations of the products will increase and the concentrations of the reactants will decrease until equilibrium is reached.

- If $Q > K$, the concentrations of the reactants will increase and the concentrations of the products will decrease until equilibrium is reached.

- If $Q = K$, the reactants and the products are at equilibrium, and there will be no change in the concentrations of reactants or products.

Example 14.2 The equilibrium constant for the reaction

$$2SO_2(g) + O_2(g) \rightleftharpoons 2SO_3(g)$$

is $K_p = 0.14 \, atm^{-1}$ at 900 K. If a reaction vessel is filled with SO_3 at a partial pressure of 0.10 atm and with O_2 and SO_2 each at a partial pressure of 0.20 atm, is the reaction at equilibrium? If not, in what direction does it proceed?

Solution The value of the reaction quotient Q for the initial conditions is

$$Q = \frac{(p_{SO_3})^2}{(p_{SO_2})^2(p_{O_2})} = \frac{(0.10 \text{ atm})^2}{(0.20 \text{ atm})^2(0.20 \text{ atm})} = 1.25 \text{ atm}^{-1}$$

Since $Q > K_p$, the reaction is not at equilibrium. The reaction will proceed so as to decrease the value of Q. Thus more SO_2 and O_2 will be formed, and the amount of SO_3 will decrease. In other words, the reaction will proceed to the left.

Calculation of Equilibrium Concentrations

The most important use of the equilibrium constant is to calculate the concentrations of reactants and products that will be present at equilibrium starting with some initial concentrations. We illustrate this use of the equilibrium constant in Examples 14.3 and 14.4.

Example 14.3 Suppose we introduce 0.100 mol of H_2 and 0.100 mol of I_2 into a 10.0-L flask at 698 K. What are the concentrations of H_2, I_2, and HI at equilibrium?

Solution *The first step in solving any equilibrium problem* is to write the equation for the equilibrium reaction. In this case the reaction is

$$H_2 + I_2 \rightleftharpoons 2HI$$

The second step is to write the expression for the equilibrium constant and look up its value at the temperature of the reaction if the value is not given with the data for the problem. In this case the value $K_c = 54.4$ at 698 K has been given on page 503. Thus we can write

$$K_c = \left(\frac{[HI]^2}{[H_2][I_2]}\right)_{eq} = 54.4$$

The third step is to write expressions for the concentrations of each of the molecules present at equilibrium. We do not know these concentrations, but they are all related to each other through the balanced equation for the reaction. Suppose that when equilibrium is reached, x moles of H_2 have reacted. Then the equation shows that they must have combined with x moles of I_2 to form $2x$ moles of HI. We can conveniently write this information under the equation for the reaction:

	H_2	$+$ I_2	$\rightleftharpoons$ 2HI
Initial amounts	0.100 mol	0.100 mol	0 mol
Equilibrium amounts	$(0.100 - x)$ mol	$(0.100 - x)$ mol	$2x$ mol
Equilibrium concentrations	$\dfrac{(0.100 - x) \text{ mol}}{10.0 \text{ L}}$	$\dfrac{(0.100 - x) \text{ mol}}{10.0 \text{ L}}$	$\dfrac{2x \text{ mol}}{10.0 \text{ L}}$

The fourth step is to substitute these concentrations in the expression for K_c and to solve for x. We have

$$\frac{(2x/10)^2 \text{ mol}^2 \text{ L}^{-2}}{[(0.100 - x)/10 \text{ L}][(0.100 - x)/10 \text{ L}] \text{ mol}^2 \text{ L}^{-2}} = 54.4$$

Taking the square root of both sides of the equation and multiplying numerator and denominator by 10, we have

$$\frac{2x}{0.100 - x} = 7.38$$

Solving for x gives

$$x = 0.0787$$

Substituting this value of x in the equilibrium concentrations gives

$$[H_2] = \frac{(0.100 - 0.0787) \text{ mol}}{10.0 \text{ L}} = 0.002\,13 \text{ mol L}^{-1}$$

Similarly,

$$[I_2] = 0.002\,13 \text{ mol L}^{-1} \qquad [HI] = 0.0157 \text{ mol L}^{-1}$$

We can check these results by calculating K_c:

$$K_c = \frac{[HI]^2}{[H_2][I_2]} = \frac{(0.0157)^2}{(0.002\,13)^2} = 54.3$$

Example 14.4 A 10.0 L flask is filled with 0.20 mol of HI at 500°C. What are the concentrations of H_2, I_2, and HI at equilibrium?

Solution Following the same steps as in Example 14.3, we have:

First step

$$2HI \rightleftharpoons H_2 + I_2$$

Second step

$$K_c = \left(\frac{[H_2][I_2]}{[HI]^2} \right)_{eq}$$

What is the value of the equilibrium constant for this reaction? In Example 14.3 we used the value for the equilibrium constant K_{form} for the formation of HI,

$$K_{form} = \left(\frac{[HI]^2}{[H_2][I_2]} \right)_{eq} = 54.4$$

If we write K_{decomp} for the equilibrium constant for the decomposition of HI, we have

$$K_{decomp} = \frac{1}{K_{form}} = \frac{1}{54.4} = 0.0184 = K_c$$

For the *third step*, we assume that at equilibrium x moles of H_2 are formed. The equation for the reaction shows that x moles of I_2 are also formed and that $2x$ moles of HI have reacted, so the amount present at equilibrium is $0.20 - 2x$.

	2HI	$\rightleftharpoons$	H_2	+	I_2
Initial amounts	0.200 mol		0 mol		0 mol
Equilibrium amounts	$(0.200 - 2x)$ mol		x mol		x mol
Equilibrium concentrations	$\dfrac{(0.200 - 2x) \text{ mol}}{10.0 \text{ L}}$		$\dfrac{x \text{ mol}}{10.0 \text{ L}}$		$\dfrac{x \text{ mol}}{10.0 \text{ L}}$

The fourth step is to substitute these values in the expression for the equilibrium constant:

$$K_c = \frac{(x/10.0)^2 \text{ mol}^2 \text{ L}^{-2}}{[(0.200 - 2x)/10.0]^2 \text{ mol}^2 \text{ L}^{-2}} = 0.0184$$

Taking the square root of both sides and multiplying numerator and denominator by 10.0, we have

$$\frac{x}{0.200 - 2x} = 0.136$$

Solving for x gives

$$x = 0.0213$$

Since the volume is 10.0 L, the concentrations at equilibrium are

$$[H_2] = 0.002\ 13\ mol\ L^{-1} \qquad [I_2] = 0.002\ 13\ mol\ L^{-1}$$

$$[HI] = \frac{0.20 - 2(0.0213)}{10.0} = 0.0157\ mol\ L^{-1}$$

Note that the equilibrium concentrations are exactly the same as the concentrations we obtained in Example 14.3, which demonstrates again that the same equilibrium position can be approached from both sides. In other words, the equilibrium mixture is the same whether we start with 0.01 mol L^{-1} of H_2 and 0.01 mol of I_2 or with 0.02 mol L^{-1} of HI in 10.0 L.

14.2 EFFECT OF CONDITIONS: LE CHATELIER'S PRINCIPLE

The final equilibrium concentrations in any reaction depend on (1) the initial concentrations or partial pressures of the reactants and products and (2) the temperature, because a change in temperature changes the value of the equilibrium constant. We have already seen how, at a given temperature, we can calculate the concentrations of the reactants and the products at equilibrium for any initial concentrations if we know the value of the equilibrium constant for a given temperature. Even if we do not know how the equilibrium constant changes with temperature, we can make qualitative predictions about the effect of a change in conditions on an equilibrium by using Le Châtelier's principle.

Recall from Chapter 7 that **Le Châtelier's principle** states that:

If any one of the conditions affecting an equilibrium is changed, the equilibrium shifts so as to minimize the change.

Let us see how this principle enables us to make qualitative predictions about the effects of concentration, pressure, and temperature changes on an equilibrium.

Concentration Changes

We can restate Le Châtelier's principle for the special case of concentration changes:

When the concentrations of any of the reactants or products in a reaction at equilibrium are changed, the composition of the equilibrium mixture changes so as to minimize the change in concentration that was made.

For example, if to the reaction

$$H_2(g) + I_2(g) \rightleftharpoons 2HI(g)$$

at equilibrium we add H_2, the concentration of H_2 is increased so that it is no longer the equilibrium concentration. According to Le Châtelier's principle, a new equilibrium will be set up in which the concentration of H_2 is less than it was after the addition of H_2 but more than in the original mixture (see Figure 14.2). The adjustment of the equilibrium never fully compensates for the change in concentration that was made. We can summarize the changes that occur in the system by stating that the addition of H_2 causes the position of the equilibrium to shift to the right.

We could arrive at the same conclusion by considering the reaction quotient,

$$Q = \frac{[HI]^2}{[H_2][I_2]}$$

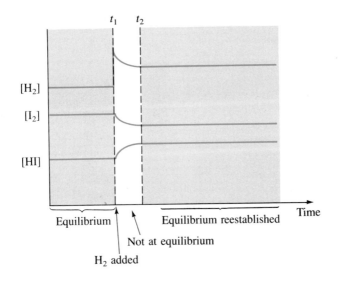

Figure 14.2 Le Châtelier's
**Principle: Effect of Concentration
Changes.** When H_2 is added to the
system $H_2 + I_2 \rightleftharpoons 2HI$ at equilibrium, at time t_1 the equilibrium
shifts to the right. So the concentrations of H_2 and I_2 both decrease
and the concentration of HI increases, until a new position of
equilibrium is reestablished at time
t_2. The final equilibrium concentration of H_2 is less than it
was after the addition of H_2 but
greater than it was in the original
equilibrium mixture.

After the addition of H_2 the concentration of H_2 is greater than the equilibrium
value, so Q is less than K. The concentration of HI will therefore increase and
the concentrations of H_2 and I_2 will decrease until $Q = K$ and equilibrium is
reestablished. Again, we conclude that the addition of H_2 causes the position
of equilibrium to shift to the right.

Similarly, we can predict that if we were to remove H_2 from the reaction
mixture at equilibrium, more H_2 would be formed to replace the H_2 removed.
A new equilibrium would be set up with a decreased concentration of HI and
an increased concentration of I_2.

Although removing H_2 from this particular reaction mixture would not be
very practical, the shift in an equilibrium caused by the removal of a product
often has great practical importance. When a product is removed, a reaction
that would otherwise come to equilibrium without going to completion can be
driven to completion. This circumstance is favored, for example, when one of the
products of a reaction is a gas or a substance that can easily be vaporized. We
saw in Chapter 7 that when sodium chloride reacts with concentrated sulfuric
acid,

$$NaCl(s) + H_2SO_4(\text{conc. aq}) \rightleftharpoons HCl(g) + NaHSO_4(aq)$$

the equilibrium is shifted to the right, because HCl is a gas that escapes from the
reaction mixture. Therefore the equilibrium concentration of HCl in the reaction mixture is never reached, and the reaction goes very nearly to completion.
The effects of concentration changes on an equilibrium are demonstrated in
Experiment 14.1.

Pressure Changes

The position of equilibrium of a reaction involving gases is affected by a pressure
change because the partial pressures, and therefore the concentrations, of the
reacting gases change correspondingly. Let us see why, in the Haber process for
the synthesis of ammonia (Chapter 3), a high pressure is used.

Suppose that the equilibrium

$$N_2(g) + 3H_2(g) \rightleftharpoons 2NH_3(g)$$

has been established in a vessel with a variable volume. What is the effect of

14.2 EFFECT OF
CONDITIONS: LE
CHÂTELIER'S PRINCIPLE

511

Le Châtelier's Principle

Left: The beaker on the right is an aqueous solution containing the Fe^{3+} and SCN^- (thiocyanate) ions. These two ions are in equilibrium with the ion $FeSCN^{2+}$, which has a red-brown color.

$$Fe^{3+}(aq) + SCN^-(aq) \rightleftharpoons FeSCN^{2+}(aq)$$

pale yellow   colorless   red-brown

Excess SCN^- has been added to the next beaker to the left. The intensity of the color increases because more $FeSCN^{2+}$ is formed as the equilibrium is shifted to the right. Excess Fe^{3+} has been added to the next beaker to the left, again shifting the equilibrium to the right. This increases the concentration of $FeSCN^{2+}$ and hence the intensity of the color. Silver nitrate solution has been added to the beaker at the far left. This causes a precipitate of insoluble $AgSCN$ to form, thus decreasing the concentration of SCN^- in solution. Hence the $FeSCN^{2+}$ ion dissociates and its concentration becomes too low to cause an observable color.

Right: The formation of $FeSCN^{2+}$ is an exothermic reaction. Thus when the solution is cooled, the equilibrium shifts to the right and the intensity of the color increases.

decreasing the volume of the vessel, thereby increasing the concentrations, or partial pressures, of all the gases and, therefore, the total pressure? (See Figure 14.3.) Le Châtelier's principle predicts that the position of equilibrium will shift so as to minimize the pressure increase. Thus more NH_3 will be formed, because the formation of two molecules of product ($2NH_3$) causes the disappearance of four molecules of reactants ($N_2 + 3H_2$) and therefore a decrease in the total number of molecules. Because the total pressure is proportional to the total number of molecules in a given volume, there will be a corresponding pressure decrease. The amount of NH_3 at equilibrium is therefore increased by increasing the pressure. Hence, the Haber process for the synthesis of ammonia employs a pressure in the range of 100–300 atm.

Consider now the decomposition (cracking) of ethane to ethene and hydrogen:

$$C_2H_6(g) \rightleftharpoons C_2H_4(g) + H_2(g)$$

There are more molecules on the right than on the left. Therefore in this case a pressure increase will shift the equilibrium to the left. In other words, the amounts of C_2H_4 and H_2 at equilibrium will be reduced by increasing the pressure. The reaction is therefore carried out at atmospheric pressure.

We can summarize the effect of a pressure increase on any gas phase equilibrium as follows:

$$\text{Reactants} \underset{\text{Fewer molecules on left}}{\overset{\text{Fewer molecules on right}}{\rightleftharpoons}} \text{Products}$$

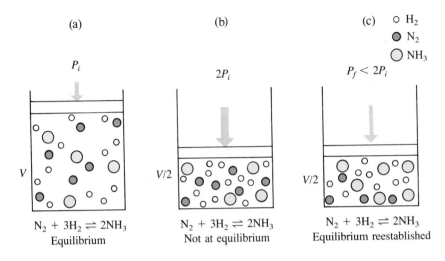

(a)

P_i

V

$N_2 + 3H_2 \rightleftharpoons 2NH_3$
Equilibrium

(b)

$2P_i$

$V/2$

$N_2 + 3H_2 \rightleftharpoons 2NH_3$
Not at equilibrium

(c) ○ H_2
● N_2
○ NH_3

$P_f < 2P_i$

$V/2$

$N_2 + 3H_2 \rightleftharpoons 2NH_3$
Equilibrium reestablished

Figure 14.3 Le Châtelier's Principle: Effect of Pressure. (a) In the reaction

$$N_2 + 3H_2 \rightleftharpoons 2NH_3$$

if the volume of the reaction vessel is decreased to half its original value, (b), the concentrations (partial pressures) of N_2, H_2, and NH_3 are all doubled, and the total pressure is therefore doubled. The equilibrium thus shifts to the right, (c), which decreases the total number of molecules and thereby decreases the total pressure. When equilibrium is reestablished, the final pressure P_f is less than $2P_i$ but greater than P_i.

If, as in the case of the equilibrium

$$H_2(g) + I_2(g) \rightleftharpoons 2HI(g)$$

the reaction is not accompanied by any change in the total number of molecules, a change of pressure has no effect on the position of the equilibrium.

Temperature Changes

A change in temperature affects the position of an equilibrium because it changes the value of the equilibrium constant. However, even if we do not know the values of the equilibrium constant at different temperatures, Le Châtelier's principle enables us to predict the direction in which an equilibrium will be shifted by a temperature change. To make this prediction we need only know whether the reaction is exothermic (ΔH has a negative value) or endothermic (ΔH has a positive value).

The formation of hydrogen iodide from hydrogen and iodine is an exothermic reaction. For the purpose of applying Le Châtelier's principle, we can conveniently write this equilibrium as

$$H_2(g) + I_2(g) \rightleftharpoons 2HI(g) + \text{Heat}$$

When this reaction proceeds to the right, energy is liberated in the form of heat, which raises the temperature of the reaction mixture. According to Le Châtelier's principle, changing the temperature causes the position of the equilibrium to shift in the direction that minimizes the temperature change. Thus if the temperature is decreased, the hydrogen-iodine equilibrium will shift to the right, increasing the concentration of hydrogen iodide, because the heat thus produced will tend to raise the temperature, thus counteracting the initial temperature decrease. Conversely, if the temperature is increased, the equilibrium will shift to the left, because in this direction the reaction is endothermic, and the heat absorbed will tend to lower the temperature of the system, thus counteracting the initial temperature increase.

We can come to the same conclusion if we know the value of the equilibrium constant at two temperatures. In this case the equilibrium constant,

$$K_c = \left(\frac{[HI]^2}{[H_2][I_2]}\right)_{eq}$$

decreases from 67.5 at 325°C to 50.0 at 450°C. Thus the yield of HI decreases with increasing temperature, as we predicted from Le Châtelier's principle.

The synthesis of ammonia is an exothermic reaction:

$$N_2(g) + 3H_2(g) \rightleftharpoons 2NH_3(g) \qquad \Delta H° = -92.38 \text{ kJ}$$

We may therefore write

$$N_2(g) + 3H_2(g) \rightleftharpoons 2NH_3(g) + \text{Heat}$$

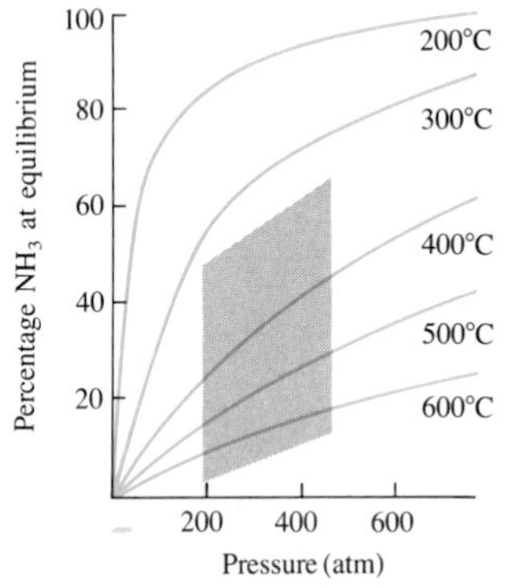

Figure 14.4 Effect of Temperature and Pressure on the Equilibrium $N_2 + 3H_2 \rightleftharpoons 2NH_3$. Increasing pressure increases the percentage of NH_3 at equilibrium; increasing temperature decreases the percentage of NH_3 at equilibrium. The range of conditions normally used in the Haber process is shown by the shaded area.

According to Le Châtelier's principle, raising the temperature shifts the equilibrium to the left and decreases the equilibrium concentration of ammonia. Thus for a maximum yield of ammonia the reaction should be carried out at as low a temperature as possible. A very low temperature cannot be used, however, because the reaction is then too slow. In practice, a catalyst is used to speed up the reaction, and even then a temperature in the range 400–500°C is necessary in order to obtain a reasonable reaction rate (Figure 14.4). *Catalysts do not change the position of an equilibrium, but they do increase the rate at which equilibrium is attained.* We will discuss catalysts in more detail in Chapter 18. The effects of a temperature change on the position of an equilibrium are demonstrated in Experiment 14.1

We can summarize the effect of *a temperature increase* on an equilibrium as follows:

Exothermic reactions	*Endothermic reactions*
Equilibrium concentration of products decreases	Equilibrium concentration of products increases
Position of equilibrium shifts to left	Position of equilibrium shifts to right
K decreases	K increases

The effects of a temperature decrease are just the reverse of the effects of a temperature increase.

Example 14.5 What conditions of pressure and temperature give a high equilibrium yield of products in the following reactions?

(a) $2SO_2(g) + O_2(g) \rightleftharpoons 2SO_3(g)$; $\Delta H° = -198$ kJ.

(b) $H_2(g) + CO_2(g) \rightleftharpoons H_2O(g) + CO(g)$; $\Delta H° = +41$ kJ.

Solution

(a) This reaction is accompanied by a decrease in the number of gaseous molecules; hence a high equilibrium concentration of the product, SO_3, will be favored by a high pressure. The reaction is exothermic, so the equilibrium yield of SO_3 will be increased by carrying out the reaction at a low temperature.

(b) This reaction is not accompanied by any changes in the number of molecules, so it is not affected by a change in pressure. The reaction is endothermic, so the yield of H_2O and CO will be increased at a high temperature.

14.3 HETEROGENEOUS EQUILIBRIA

In the equilibria that we have discussed so far in this chapter, all the reactants and products have been gases. Equilibria in which all the substances involved are in the same phase are called **homogeneous equilibria**. However, many equilibria involve solids and gases, solids and liquids, or liquids and gases. These equilibria are called **heterogeneous equilibria**.

An example is the decomposition of calcium carbonate, which when heated, gives calcium oxide and carbon dioxide.

$$CaCO_3(s) \rightleftharpoons CaO(s) + CO_2(g)$$

The equilibrium constant for this reaction is

$$K'_c = \left(\frac{[CaO][CO_2]}{[CaCO_3]} \right)_{eq}$$

But $CaCO_3$ and CaO are pure solids. What do we mean by the concentration of a pure solid? Concentration has the units moles per liter. If we multiply these units by the molar mass ($g\,mol^{-1}$), we obtain grams per liter, which is density. The density of a pure solid is constant, independent of the amount of the solid that is present. Therefore the "concentration" of a pure solid is also constant. So both $[CaO]$ and $[CaCO_3]$ are constant. We can rearrange the expression for K'_c to give

$$\frac{K'_c[CaCO_3]}{[CaO]} = [CO_2]$$

and since $[CaCO_3]$ and $[CaO]$ are constant, $K'_c[CaCO_3]/[CaO]$ is a constant, which we will write as K_c. Thus

$$K_c = [CO_2]_{eq}$$

or in terms of partial pressures,

$$K_p = (p_{CO_2})_{eq}$$

At 900°C, $K_p = 1.04$ atm; thus if $CaCO_3$ is heated in a closed vessel, it will decompose to CaO and CO_2 until the pressure of CO_2 reaches 1.04 atm. The system will then have reached equilibrium, and the amounts of $CaCO_3$, CaO, and CO_2 will remain constant. However, if $CaCO_3$ is heated in a vessel open to the atmosphere, the CO_2 diffuses away, and it never attains a partial pressure of 1.04 atm. So the reaction is driven to completion, and all the $CaCO_3$ decomposes to CaO (see Figure 14.5).

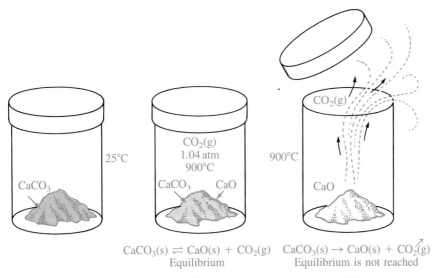

$$CaCO_3(s) \rightleftharpoons CaO(s) + CO_2(g) \qquad CaCO_3(s) \rightarrow CaO(s) + CO_2(g)$$
Equilibrium   Equilibrium is not reached

(a)   (b)   (c)

Figure 14.5 Heterogeneous Equilibrium. (a) At 25°C there is no observable tendency for solid $CaCO_3$ to decompose. (b) When solid $CaCO_3$ is heated in a closed vessel at 900°C, it decomposes to give solid CaO and gaseous CO_2. When the pressure of CO_2 reaches 1.04 atm, equilibrium is established. (c) When $CaCO_3$ is heated in an open vessel, the CO_2 diffuses away. The equilibrium pressure of CO_2 is never reached, and $CaCO_3$ is completely converted to CaO.

In general, for heterogeneous equilibria the concentrations of pure liquids and solids are not included in the equilibrium constant expression. They are constants and are incorporated in the equilibrium constant K.

Example 14.6 Write the expressions for the equilibrium constants for the following reactions in terms of concentrations and in terms of partial pressures:

(a) $C(s) + CO_2(g) \rightleftharpoons 2CO(g)$

(b) $FeO(s) + CO(g) \rightleftharpoons Fe(s) + CO_2(g)$

Solution

(a) $K_c = \left(\dfrac{[CO]^2}{[CO_2]}\right)_{eq} \qquad K_p = \left(\dfrac{(p_{CO})^2}{p_{CO_2}}\right)_{eq}$

(b) $K_c = \left(\dfrac{[CO_2]}{[CO]}\right)_{eq} \qquad K_p = \left(\dfrac{p_{CO_2}}{p_{CO}}\right)_{eq}$

14.4 EQUILIBRIA IN AQUEOUS SOLUTIONS OF ACIDS AND BASES

Some of the most important applications of equilibria are concerned with reactions in aqueous solution. In this section we will discuss the quantitative treatment of acid-base reactions. We consider precipitation reactions in Chapter 15 and oxidation-reduction reactions in Chapter 16.

Strong Acids and Bases

A strong acid such as hydrochloric acid, HCl, or nitric acid, HNO_3, reacts very rapidly and very nearly completely with water:

$$HCl + H_2O \longrightarrow H_3O^+ + Cl^-$$

The equilibrium lies far to the right; we say that the acid is fully ionized or that it is fully dissociated into hydronium ions and the conjugate base of the acid. The equilibrium constant,

$$K = \left(\frac{[H_3O^+][Cl^-]}{[H_2O][HCl]}\right)_{eq}$$

has a very large value, that is, $K \gg 1$.

Example 14.7 What are the concentrations of H_3O^+ and Cl^- in 250 mL of an aqueous HCl solution containing 0.01 mol of HCl?

Solution The equation shows that 0.01 mol of HCl is converted to 0.01 mol of H_3O^+ and 0.01 mol of Cl^-. We then convert these amounts to concentrations by using the volume of the solution. Note that there must be some HCl in equilibrium with H_3O^+ and Cl^-, but the amount is so small that it is effectively zero.

	HCl	+	H₂O	⟶	H₃O⁺	+	Cl⁻

$$HCl + H_2O \longrightarrow H_3O^+ + Cl^-$$

	HCl	H₃O⁺	Cl⁻
Initial amounts	0.01 mol	0 mol	0 mol
Equilibrium amounts	0 mol	0.01 mol	0.01 mol
Equilibrium concentrations	0 mol L⁻¹	$\left(\dfrac{0.01 \text{ mol}}{250 \text{ mL}}\right)\left(\dfrac{10^3 \text{ mL}}{1 \text{ L}}\right)$	$\left(\dfrac{0.01 \text{ mol}}{250 \text{ mL}}\right)\left(\dfrac{10^3 \text{ mL}}{1 \text{ L}}\right)$
	0 mol L⁻¹	0.04 mol L⁻¹	0.04 ml L⁻¹

Thus

$$[H_3O^+] = [Cl^-] = 0.04 \text{ mol L}^{-1}$$

Example 14.7 is simple but it illustrates two important points. First, a strong acid is quantitatively ionized. Second, $[H_3O^+] = [Cl^-]$. The second conclusion comes from the balanced equation, and it exemplifies an important principle: *All solutions are electrically neutral.* The sum of the charges on all the positive ions must always be equal to the sum of the charges on all the negative ions in a solution.

As we saw in Chapter 5, a strong base is either an ionic substance containing OH^- or an ion or molecule that is completely protonated in water, thus producing an equal amount of OH^-. The most common strong bases are soluble hydroxides such as NaOH and KOH, which are ionic compounds consisting of metal ions and hydroxide ions. They therefore give a quantitative amount of OH^- when dissolved in water. Two examples of species that are completely protonated in water are the oxide ion, O^{2-}, and the hydride ion, H^-:

$$O^{2-} + H_2O \longrightarrow OH^- + OH^-$$
$$H^- + H_2O \longrightarrow H_2 + OH^-$$

In both cases the position of equilibrium lies far to the right.

Weak Acids

Weak acids do not react completely with water. For example, in a solution of acetic acid in water the following equilibrium is established:

$$CH_3CO_2H + H_2O \rightleftharpoons CH_3CO_2^- + H_3O^+$$

The equilibrium constant expression for this reaction is

$$K = \left(\frac{[CH_3CO_2^-][H_3O^+]}{[CH_3CO_2H][H_2O]}\right)_{eq}$$

In solution all concentrations are measured in moles per liter.

We will be concerned only with dilute solutions of acids and bases in which there is an enormous excess of water compared with the solute species. The concentration of water is very large and is hardly changed by changing the amounts of the solute species. Thus in the above expression for the equilibrium constant we can take $[H_2O]$ to be constant. We can therefore modify this expression by multiplying both sides by $[H_2O]$:

$$K_c[H_2O] = \left(\frac{[H_3O^+][CH_3CO_2^-]}{[CH_3CO_2H]}\right)_{eq}$$

14.4 EQUILIBRIA IN AQUEOUS SOLUTIONS OF ACIDS AND BASES

Since $[H_2O]$ is constant in dilute solutions, the product $K_c[H_2O]$ is also constant. It is called the **acid dissociation constant** and is given the symbol K_a:

$$K_a(CH_3CO_2H) = K_c[H_2O] = \left(\frac{[H_3O^+][CH_3CO_2^-]}{[CH_3CO_2H]}\right)_{eq}$$

The expressions for the acid dissociation constants of other weak acids are written in the same way. For example, HF is a weak acid in water:

$$HF + H_2O \rightleftharpoons H_3O^+ + F^-$$

Thus the equilibrium constant $K_a(HF)$ is given by the expression

$$K_a(HF) = \left(\frac{[H_3O^+][F^-]}{[HF]}\right)_{eq}$$

In general, for any weak acid, HA, in water we may write

$$K_a = \left(\frac{[H_3O^+][A^-]}{[HA]}\right)_{eq}$$

Values of the acid dissociation constants of some weak acids are given in Table 14.3. Like other equilibrium constants, K_a values depend on the temperature. They are normally quoted at 25°C. These values show the extent to

Table 14.3 Dissociation Constants for Some Acids in Water at 25°C

	PROTON TRANSFER REACTION		K_a (mol L^{-1})	pK_a ($= -\log_{10} K_a$)
	ACID	CONJUGATE BASE		
Perchloric acid	$HClO_4$ + $H_2O \longrightarrow H_3O^+ + ClO_4^-$		Very large	—
Hydrochloric acid	HCl + $H_2O \longrightarrow H_3O^+ + Cl^-$		Very large	—
Sulfuric acid	H_2SO_4 + $H_2O \longrightarrow H_3O^+ + HSO_4^-$		Very large	—
Nitric acid	HNO_3 + $H_2O \longrightarrow H_3O^+ + NO_3^-$		Very large	—
Hydronium ion	H_3O^+ + $H_2O \rightleftharpoons H_3O^+ + H_2O$		1.0	0.0
Sulfurous acid[a]	SO_2 + $2H_2O \rightleftharpoons H_3O^+ + HSO_3^-$		1.5×10^{-2}	1.8
Hydrogen sulfate ion	HSO_4^- + $H_2O \rightleftharpoons H_3O^+ + SO_4^{2-}$		1.2×10^{-2}	1.9
Phosphoric acid	H_3PO_4 + $H_2O \rightleftharpoons H_3O^+ + H_2PO_4^-$		7.5×10^{-3}	2.1
Hydrofluoric acid	HF + $H_2O \rightleftharpoons H_3O^+ + F^-$		3.5×10^{-4}	3.5
Acetic acid	CH_3CO_2H + $H_2O \rightleftharpoons H_3O^+ + CH_3CO_2^-$		1.8×10^{-5}	4.7
Hydrated aluminum ion	$Al(H_2O)_6^{3+}$ + $H_2O \rightleftharpoons H_3O^+ + Al(H_2O)_5OH^{2+}$		7.2×10^{-6}	5.1
Carbonic acid[b]	CO_2 + $2H_2O \rightleftharpoons H_3O^+ + HCO_3^-$		4.3×10^{-7}	6.4
Hydrogen sulfide	H_2S + $H_2O \rightleftharpoons H_3O^+ + HS^-$		9.1×10^{-8}	7.0
Dihydrogen phosphate ion	$H_2PO_4^-$ + $H_2O \rightleftharpoons H_3O^+ + HPO_4^{2-}$		6.2×10^{-8}	7.2
Hypochlorous acid	$HOCl$ + $H_2O \rightleftharpoons H_3O^+ + OCl^-$		3.1×10^{-8}	7.5
Ammonium ion	NH_4^+ + $H_2O \rightleftharpoons H_3O^+ + NH_3$		5.6×10^{-10}	9.3
Hydrocyanic acid	HCN + $H_2O \rightleftharpoons H_3O^+ + CN^-$		4.9×10^{-10}	9.3
Hydrogen phosphate ion	HPO_4^{2-} + $H_2O \rightleftharpoons H_3O^+ + PO_4^{3-}$		2.1×10^{-13}	12.4
Water[c]	H_2O + $H_2O \rightleftharpoons H_3O^+ + OH^-$		1.8×10^{-16}	15.8
	CONJUGATE ACID	BASE		

[a] There is no evidence for the formation of the undissociated acid H_2SO_3.
[b] The equilibrium given here is the sum of the two equilibria $H_2O + CO_2 \rightleftharpoons H_2CO_3$ and $H_2CO_3 + H_2O \rightleftharpoons H_3O^+ + HCO_3^-$ because the amount of H_2CO_3 formed is small and not accurately known.
[c] $K_a(H_2O) = K_w/[H_2O]$

which the acids are dissociated in water. The strong acids have very large dissociation constants that cannot be measured. The weak acids are listed in order of decreasing strength. Thus phosphoric acid, $K_a = 7.5 \times 10^{-3}\ mol\ L^{-1}$, is much more extensively ionized than hydrocyanic acid, $K_a = 4.9 \times 10^{-10}\ mol\ L^{-1}$. Phosphoric acid is a stronger acid than hydrocyanic acid, although both are weak; neither is completely ionized. The position of the equilibrium lies to the left in both cases, but it is much further to the left for hydrocyanic acid than for phosphoric acid. From the value of the dissociation constant K_a of a weak acid, we can calculate the extent of its dissociation in water.

Calculating the Extent of Dissociation of a Weak Acid

As an example, we will calculate the concentration of H_3O^+ in a $0.20M$ solution of acetic acid.

The *first step* in solving any equilibrium problem is to write the equation for the equilibrium reaction. In this case the reaction is

$$CH_3CO_2H(aq) + H_2O(l) \rightleftharpoons H_3O^+(aq) + CH_3CO_2^-(aq)$$

The *second step* is to write the expression for the equilibrium constant and look up its value. From Table 14.3, K_a for acetic acid is $1.8 \times 10^{-5}\ mol\ L^{-1}$. Thus we can write

$$K_a = \left(\frac{[H_3O^+][CH_3CO_2^-]}{[CH_3CO_2H]}\right)_{eq} = 1.8 \times 10^{-5}\ mol\ L^{-1}$$

The *third step* is to write expressions for the concentrations of each of the species present at equilibrium. We first imagine that when acetic acid is added to water, no proton transfer occurs, so the *initial* concentration of acetic acid is $0.200M$. Then we imagine that the proton transfer occurs to give a certain concentration of H_3O^+, which we wish to find. Let us call the concentration of H_3O^+ $x\ mol\ L^{-1}$. The concentration of acetate ion, $CH_3CO_2^-$, is then also $x\ mol\ L^{-1}$, since 1 mol of $CH_3CO_2^-$ is formed for each mole of H_3O^+ formed. The concentration of undissociated acetic acid, CH_3CO_2H, that remains at equilibrium is therefore $(0.20 - x)\ mol\ L^{-1}$. Thus

	$CH_3CO_2H + H_2O \rightleftharpoons$	H_3O^+ +	$CH_3CO_2^-$	
Initial concentrations	0.20	0	0	$mol\ L^{-1}$
Equilibrium concentrations	$0.20 - x$	x	x	$mol\ L^{-1}$

The *fourth step* is to substitute these concentrations in the expression for K_a and to solve for x, the H_3O^+ concentration:

$$K_a = \left(\frac{[H_3O^+][CH_3CO_2^-]}{[CH_3CO_2H]}\right)_{eq} = \frac{(x\ mol\ L^{-1})(x\ mol\ L^{-1})}{(0.20 - x)\ mol\ L^{-1}}$$

$$= \frac{x^2}{0.20 - x}\ mol\ L^{-1} = 1.8 \times 10^{-5}\ mol\ L^{-1}$$

Since this expression can be rearranged to give a quadratic equation, it can be solved by the formula given in Appendix A. However, a simpler and much quicker approximate method can be used. Because the value of K_a is very small, we know that the position of the equilibrium is to the left; only a very small amount of acetic acid is dissociated. Therefore x, the H_3O^+ concentration, will be very small compared with the concentration of CH_3CO_2H. We will assume then that x is much smaller than 0.20 and can be neglected with respect to 0.20.

So we can write

$$0.20 - x \simeq 0.20$$

Using this approximation, we have

$$\frac{x^2}{0.20} \text{ mol L}^{-1} = 1.8 \times 10^{-5} \text{ mol L}^{-1}$$

$$x^2 = 0.20 \times 1.8 \times 10^{-5} = 0.36 \times 10^{-5} = 3.6 \times 10^{-6}$$

Taking the square root of both sides, we have $x = 1.9 \times 10^{-3}$, and

$$[H_3O^+] = 1.9 \times 10^{-3} \text{ mol L}^{-1}$$

We see that x is indeed much smaller than 0.20, so our approximation is justified. The obvious question is, How much smaller than the initial concentration of acid must $[H_3O^+] = x$ mol L^{-1} be for the approximation to be valid? The answer depends on the accuracy with which we wish to calculate the H_3O^+ concentration. For most purposes, a sufficiently accurate answer is obtained if $[H_3O^+]$ is 5% or less of the initial acid concentration. In this case $[H_3O^+]$ is

$$\frac{0.0019 \text{ mol L}^{-1}}{0.20 \text{ mol L}^{-1}} \times 100\% = 1\%$$

of the initial acid concentration. In all the examples given in this book, except for a few concerned with polyprotic acids, the approximation gives a sufficiently accurate result.

Example 14.8 What is the H_3O^+ concentration in a 0.20M solution of hydrocyanic acid, HCN? What is the percent dissociation of the acid?

Solution *First*, we write the equation for the equilibrium:

$$HCN(aq) + H_2O(l) \rightleftharpoons H_3O^+(aq) + CN^-(aq)$$

Second, we write the expression for the equilibrium constant and find the value of K_a from Table 14.3:

$$K_a = \left(\frac{[H_3O^+][CN^-]}{[HCN]} \right)_{eq} = 4.9 \times 10^{-10} \text{ mol L}^{-1}$$

Third, we let x mol L$^{-1} = [H_3O^+]$ at equilibrium. Then we have

$$HCN(aq) + H_2O \rightleftharpoons H_3O^+(aq) + CN^-(aq)$$

Initial concentrations	0.20	0	0 mol L^{-1}
Equilibrium concentrations	0.20 − x	x	x mol L^{-1}

Substituting into the expression for K_a, we have

$$K_a = \frac{x^2}{0.20 - x} \text{ mol L}^{-1} = 4.9 \times 10^{-10} \text{ mol L}^{-1}$$

Since K_a is very small, HCN is only very slightly dissociated and the concentration of H_3O^+ will be very small. Therefore we assume that x is much smaller than 0.20 and can be neglected with respect to 0.20. In other words, $0.20 - x \simeq 0.20$. We then have

$$\frac{x^2}{0.20} = 4.9 \times 10^{-10}$$

$$x^2 = 0.98 \times 10^{-10} = 98 \times 10^{-12}$$

Taking the square roots of both sides, we obtain

$$x = 9.9 \times 10^{-6}$$

Therefore $$[H_3O^+] = 9.9 \times 10^{-6} \text{ mol L}^{-1}$$

We see that our assumption that $x \ll 0.2$ is certainly justified.

Since the concentration of HCN that is dissociated is equal to the concentration of $[H_3O^+]$ that is formed, 9.9×10^{-6} mol L^{-1}, the percent dissociation of the acid is

$$\frac{9.9 \times 10^{-6} \text{ mol L}^{-1}}{0.20 \text{ mol L}^{-1}} \times 100\% = 0.005\%$$

At a concentration of $0.20M$, HCN is ionized to only a very small extent, namely, 0.005%. Thus it is an extremely weak acid. Only 5 molecules in 100,000 are ionized; the rest remain as unionized HCN molecules.

Weak Bases

A weak base is incompletely protonated by water:

$$B + H_2O \rightleftharpoons BH^+ + OH^-$$

The solution contains the base, B, the protonated base, BH^+, hydroxide ion, OH^-, and water in equilibrium. The equilibrium constant for this reaction is

$$K_c = \left(\frac{[BH^+][OH^-]}{[B][H_2O]} \right)_{eq}$$

As we did when considering weak acids, we assume that $[H_2O]$ is constant. Hence we may incorporate $[H_2O]$ into the equilibrium constant and write

$$K_c[H_2O] = K_b = \left(\frac{[BH^+][OH^-]}{[B]} \right)_{eq}$$

The constant K_b is called the **base dissociation constant**. A list of K_b values for some weak bases is given in Table 14.4.

Table 14.4 Dissociation Constants of Some Weak Bases at 25°C

	PROTON TRANSFER REACTION		K_b (mol L^{-1})	pK_b
	BASE	CONJUGATE ACID		
Hydride ion	H^-	$+ H_2O \rightleftharpoons OH^- + H_2$	Very large	
Amide ion	NH_2^-	$+ H_2O \rightleftharpoons OH^- + NH_3$	Very large	—
Oxide ion	O^{2-}	$+ H_2O \rightleftharpoons OH^- + OH^-$	Very large	—
Hydroxide ion	OH^-	$+ H_2O \rightleftharpoons OH^- + H_2O$	1.00	0.0
Phosphate ion	PO_4^{3-}	$+ H_2O \rightleftharpoons OH^- + HPO_4^{2-}$	4.8×10^{-2}	1.3
Carbonate ion	CO_3^{2-}	$+ H_2O \rightleftharpoons OH^- + HCO_3^-$	2.1×10^{-4}	3.7
Cyanide ion	CN^-	$+ H_2O \rightleftharpoons OH^- + HCN$	2.0×10^{-5}	4.7
Ammonia	NH_3	$+ H_2O \rightleftharpoons OH^- + NH_4^+$	1.8×10^{-5}	4.7
Hydrogen phosphate ion	HPO_4^{2-}	$+ H_2O \rightleftharpoons OH^- + H_2PO_4^-$	1.6×10^{-7}	6.8
Hydrogen carbonate ion	HCO_3^-	$+ H_2O \rightleftharpoons OH^- + H_2CO_3$	2.5×10^{-8}	7.6
Aniline	$C_6H_5NH_2$	$+ H_2O \rightleftharpoons OH^- + C_6H_5NH_3^+$	4.3×10^{-10}	9.4
Dihydrogen phosphate ion	$H_2PO_4^-$	$+ H_2O \rightleftharpoons OH^- + H_3PO_4$	1.3×10^{-12}	11.9
	CONJUGATE BASE	ACID		

We can find the OH^- concentration in an aqueous solution of a weak base by a method similar to the method we used to find the H_3O^+ concentration in a solution of a weak acid, as the following example shows.

Example 14.9 What is the concentration of OH^- in a $0.10M$ aqueous solution of ammonia?

Solution *The first step* is to write the equation for the equilibrium reaction:

$$NH_3(aq) + H_2O(l) \rightleftharpoons NH_4^+(aq) + OH^-(aq)$$

The second step is to write the expression for the equilibrium constant and look up its value. From Table 14.4 we see that K_b for NH_3 is 1.8×10^{-5} mol L^{-1}. Therefore

$$K_b = \left(\frac{[NH_4^+][OH^-]}{[NH_3]}\right)_{eq} = 1.8 \times 10^{-5} \text{ mol L}^{-1}$$

The third step is to write expressions for the concentration of each of the species in solution. We let the equilibrium concentration of OH^- be x mol L^{-1}. Thus

$$NH_3 + H_2O \rightleftharpoons NH_4^+ + OH^-$$

Initial concentrations	0.10	0	0	mol L^{-1}
Equilibrium concentrations	$0.10 - x$	x	x	mol L^{-1}

The fourth step is to substitute these values in the expression for the equilibrium constant and then solve for x. We have

$$K_b = \frac{[NH_4^+][OH^-]}{[NH_3]} = \frac{x^2}{0.10 - x} \text{ mol L}^{-1} = 1.8 \times 10^{-5} \text{ mol L}^{-1}$$

We could solve this equation for x by using the formula for a quadratic equation (Appendix A). However, because K_b is small, we know that the extent of ionization will be small and therefore that x is much less than 0.10. Hence we assume that $0.10 - x \simeq 0.10$, and the equation becomes

$$\frac{x^2}{0.10} = 1.8 \times 10^{-5}$$

$$x^2 = 1.8 \times 10^{-6}$$

$$x = 1.3 \times 10^{-3}$$

Hence

$$[OH^-] = 1.3 \times 10^{-3} \text{ mol L}^{-1}$$

Thus $[OH^-]$ is approximately 1% of the initial concentration of NH_3, 0.10 mol L^{-1}; hence the assumption that we made in solving the equation was justified.

14.5 SELF-IONIZATION OF WATER

We saw in Chapters 5 and 13 that water can act as a very weak acid and also as a very weak base. Very small concentrations of H_3O^+ and OH^- are therefore formed in water by the reaction

$$H_2O + H_2O \rightleftharpoons H_3O^+(aq) + OH^-(aq)$$

This reaction is called the **self-ionization, or autoprotolysis, of water**, and the equilibrium constant is

$$K_c = \left(\frac{[H_3O^+][OH^-]}{[H_2O]^2}\right)_{eq}$$

Because the concentration of water is very nearly constant in any dilute solution, we can multiply K_c by $[H_2O]^2$ to give a new constant, K_w, called the **ionic product constant of water**:

$$K_w = K_c[H_2O]^2 = ([H_3O^+][OH^-])_{eq}$$

Measurement of the electrical conductivity of carefully purified water has shown that at 25°C, $[H_3O^+] = [OH^-] = 1.00 \times 10^{-7}$ mol L^{-1}. Thus

$$K_w = (1.00 \times 10^{-7} \text{ mol L}^{-1})(1.00 \times 10^{-7} \text{ mol L}^{-1})$$
$$= 1.00 \times 10^{-14} \text{ mol}^2 \text{ L}^{-2} \text{ at } 25°C$$

This equilibrium constant applies not only to pure water but also to the self-ionization of water in any aqueous solution. Hydronium ions and hydroxide ions are present in any aqueous solution, and they are always in equilibrium with water molecules.

Let us calculate the concentration of OH^- in a 0.010M solution of HCl. The concentration of OH^- will be less than it is in pure water because the H_3O^+ from the ionization of the HCl will shift the position of the equilibrium of the self-ionization reaction to the left. If the concentration of OH^- is x mol L^{-1}, then the concentration of H_3O^+ *from the self-ionization of water* must also be x mol L^{-1}. The concentration of H_3O^+ from the ionization of HCl is 0.010 mol L^{-1}, so the total concentration of H_3O^+ is $(0.010 + x)$ mol L^{-1}. Thus

$$2H_2O \rightleftharpoons H_3O^+ \quad + \quad OH^-$$

| Equilibrium concentrations | $0.010 + x$ | x | mol L^{-1} |

Substituting into the equilibrium constant for the self-ionization of water, we have

$$K_w = ([H_3O^+][OH^-])_{eq} = (0.010 + x)(x) \text{ mol}^2 \text{ L}^{-2} = 1.0 \times 10^{-14} \text{ mol}^2 \text{ L}^{-2}$$

Since x must be very small, we can assume that $x \ll 0.010$, and therefore $0.010 + x \simeq 0.010$. So

$$0.010x = 1.0 \times 10^{-14}$$

Hence

$$x = 1.0 \times 10^{-12}$$
$$[OH^-] = 1.0 \times 10^{-12} \text{ mol L}^{-1}$$

We see that in a solution of an acid the concentration of H_3O^+ arising from the self-dissociation of water is very small, and the corresponding concentration of OH^- is also very small. Nevertheless, there *is* a very small concentration of OH^-, even in an acid solution. The approximation that we made, that $x \ll 0.010$, is clearly justified. In fact, *in an aqueous solution of an acid, the concentration of H_3O^+ from the self-dissociation of water is negligible compared with the concentration of H_3O^+ from the dissociation of the acid*, except when the concentration of H_3O^+ from the acid is extremely small, that is, less than 10^{-6} mol L^{-1}. The total H_3O^+ concentration can then be calculated only by a more detailed analysis that is beyond the scope of this book. The H_3O^+ from the ionization of an acid shifts the position of the self-ionization equilibrium to the left so that the H_3O^+ concentration from the self-ionization is much less than 10^{-7} mol L^{-1}. The acid is said to *repress* the self-ionization of water.

Similar considerations apply to solutions of bases in water. In an aqueous solution of a base, *the concentration of OH^- arising from the self-ionization of water is negligible compared with the concentration of OH^- due to the base*, except when the concentration due to the base, is extremely small.

14.5 SELF-IONIZATION
OF WATER

523

Example 14.10 What is the concentration of OH^- in a $0.000\,01\,M$ solution of nitric acid?

Solution Nitric acid is a strong acid and is therefore completely dissociated. The concentration of H_3O^+ from the acid is therefore $0.000\,01\,M = 10^{-5}\,M$. We can neglect the H_3O^+ from the self-ionization of water, so the total H_3O^+ concentration in the solution is also $10^{-5}\,\text{mol L}^{-1}$. Thus

$$K_w = ([H_3O^+][OH^-])_{eq} = 10^{-14}\,\text{mol}^2\,\text{L}^{-2}$$

Substituting $[H_3O^+] = 10^{-5}\,\text{mol L}^{-1}$ gives

$$(10^{-5}\,\text{mol L}^{-1})[OH^-] = 10^{-14}\,\text{mol}^2\,\text{L}^{-2}$$

Thus

$$[OH^-] = 10^{-9}\,\text{mol L}^{-1}$$

pH Scale

Solutions of acids and bases in water may have H_3O^+ and OH^- concentrations that vary over a very wide range, that is, from greater than $1\,\text{mol L}^{-1}$ down to $10^{-14}\,\text{mol L}^{-1}$ or lower. The relationship between the H_3O^+ and OH^- concentrations in aqueous solutions is shown in Table 14.5. *Acidic solutions* have $[H_3O^+]$ greater than $10^{-7}\,\text{mol L}^{-1}$ and $[OH^-]$ less than $10^{-7}\,\text{mol L}^{-1}$ at 25°C. *Basic (alkaline) solutions* have $[OH^-]$ greater than $10^{-7}\,\text{mol L}^{-1}$ and $[H_3O^+]$ less than $10^{-7}\,\text{mol L}^{-1}$ at 25°C. A *neutral solution*, such as pure water, has $[H_3O^+] = [OH^-] = 10^{-7}$

The concentration of H_3O^+ or OH^- in aqueous solutions has many important consequences. Plants tolerate only limited ranges of H_3O^+ concentration

Table 14.5 H_3O^+ and OH^- Concentrations of Common Substances and the pH Scale

pH	$[H_3O^+]$	$[OH^-]$	pH OF SOME COMMON SUBSTANCES
-1	10	10^{-15}	Concentrated HCl (37%)
0	1	10^{-14}	
1	10^{-1}	10^{-13}	
2	10^{-2}	10^{-12}	Stomach acid
			Lemon juice
3	10^{-3}	10^{-11}	Orange juice
			Wine
4	10^{-4}	10^{-10}	Soda water
5	10^{-5}	10^{-9}	Tomato juice
6	10^{-6}	10^{-8}	Rainwater
7	10^{-7}	10^{-7}	Milk
			Blood
8	10^{-8}	10^{-6}	Seawater
9	10^{-9}	10^{-5}	Baking soda solution
			Borax solution
10	10^{-10}	10^{-4}	Toilet soap
11	10^{-11}	10^{-3}	Limewater
12	10^{-12}	10^{-2}	Household ammonia
13	10^{-13}	10^{-1}	
14	10^{-14}	1	$1M$ NaOH solution
15	10^{-15}	10	Drain cleaner

in the soil. Thus an important factor determining the distribution of a plant is the soil acidity or basicity. Many reactions in living system occur only in very narrow ranges of H_3O^+ concentration. Blood has a constant H_3O^+ concentration very close to 2.5×10^{-7} mol L^{-1}. A solution with an H_3O^+ concentration of 10^{-3} mol L^{-1} has a sour or tart flavor, which is quite pleasant, especially if it is sweetened with sugar, as in many soft drinks. But a solution having an H_3O^+ concentration of 10^{-1} mol L^{-1} or higher is not only unpleasant to taste but also dangerous because it burns the skin (see Table 14.5).

In expressing the concentration of H_3O^+ in a solution, we can avoid the use of negative powers of 10 by using the pH scale. The pH of a solution is the *negative* logarithm to the base 10 of the hydrogen ion concentration, $[H_3O^+]$. Thus

$$pH = -\log_{10}[H_3O^+]$$

The units of $[H_3O^+]$ are moles per liter, but when the logarithm is taken, the units are dropped because it is impossible to take the logarithm of a quantity with units. Therefore pH has no units.

If, as in pure water, the H_3O^+ concentration is 10^{-7} mol L^{-1}, then

$$pH = -\log_{10}[H_3O^+] = -\log_{10}(10^{-7}) = -(-7.0) = +7.0$$

ACIDIC SOLUTIONS   In a $0.10M$ solution of a strong acid such as HCl, $[H_3O^+]$ is 1.0×10^{-1} mol L^{-1}. Therefore

$$pH = -\log_{10}(1.0 \times 10^{-1}) = -(-1.0) = +1.0$$

Example 14.11 What is the pH of each of the following aqueous solutions?

(a) $1 \times 10^{-3}M$ HCl   (b) $5 \times 10^{-3}M$ HCl   (c) $1M$ HCl

Solution Because HCl is a strong acid, we have the following concentrations:

(a) $[H_3O^+] = 1 \times 10^{-3}$ mol L^{-1}.

(b) $[H_3O^+] = 5 \times 10^{-3}$ mol L^{-1}.

(c) $[H_3O^+] = 1$ mol L^{-1}.

Dropping the units and taking the negative logarithm of $[H_3O^+]$, we have the following pH values:

(a) $pH = -\log[H_3O^+] = -\log(1 \times 10^{-3}) = 3.$

(b) $pH = -\log[H_3O^+] = -\log(5 \times 10^{-3}) = 2.3.$

(c) $pH = -\log[H_3O^+] = -\log(1) = 0.$

Example 14.12 What is the pH of a $0.20M$ solution of acetic acid?

Solution Acetic acid is a weak acid. We saw previously (page 520) that for a $0.20M$ solution of acetic acid $[H_3O^+]$ is 1.9×10^{-3} mol L^{-1}. So

$$pH = -\log_{10}(1.9 \times 10^{-3}) = -(-2.72) = 2.72$$

Example 14.13 What is the pH of a solution prepared by adding 25.0 mL of $0.20M$ NaOH to 40.0 mL of $0.15M$ HCl?

Solution An acid-base reaction occurs when we prepare this solution, so the first step is to find which reactant is limiting (see Example 2.17). Hence we calculate the number of moles of each reactant present initially:

$$25.0 \text{ mL NaOH solution} \left(\frac{0.20 \text{ mol NaOH}}{1 \text{ L NaOH solution}} \right) \left(\frac{1 \text{ L}}{1000 \text{ mL}} \right) = 5.0 \times 10^{-3} \text{ mol NaOH}$$

$$40.0 \text{ mL HCl solution} \left(\frac{0.15 \text{ mol HCl}}{1 \text{ L HCl solution}} \right) \left(\frac{1 \text{ L}}{1000 \text{ mL}} \right) = 6.0 \times 10^{-3} \text{ mol HCl}$$

Now HCl and NaOH react according to the following equation:

	HCl	+	NaOH	$\longrightarrow$	NaCl	+ H_2O	
Initial amounts	6.0×10^{-3}		5.0×10^{-3}		0		mol
Final amounts	1.0×10^{-3}		0		5.0×10^{-3}		mol

Thus NaOH is the limiting reactant, and 1×10^{-3} mol of HCl remains after the reaction.

Since HCl is a strong acid, it is fully ionized:

	HCl	+ H_2O $\longrightarrow$	H_3O^+	+	Cl^-	
Initial amounts	1.0×10^{-3}		0		0	mol
Equilibrium amounts	0		1.0×10^{-3}		1.0×10^{-3}	mol

In the solution there are therefore 1.0×10^{-3} mol of H_3O^+ in a total volume of $(25.0 + 40.0)$ mL $= 65.0$ mL of solution. The concentration of H_3O^+ in the solution is therefore

$$[H_3O^+] = \left(\frac{1.0 \times 10^{-3} \text{ mol } H_3O^+}{65.0 \text{ mL solution}} \right) \left(\frac{1000 \text{ mL}}{1 \text{ L}} \right) = 0.015 \text{ mol L}^{-1}$$

$$pH = -\log(0.015) = 1.82$$

BASIC SOLUTIONS   For any solution of a base we can calculate the OH^- concentration, as we saw in Example 14.9. The H_3O^+ concentration in basic solutions is very small, but it can be easily calculated from the OH^- concentration and the ionic product constant for water,

$$K_w = [H_3O^+][OH^-] = 1.00 \times 10^{-14} \text{ mol}^2 \text{ L}^{-2}$$

Thus we may also conveniently describe basic solutions in terms of their pH (see Table 14.5).

Example 14.14 Calculate the pH of the following aqueous solutions:

(a) $0.0100M$ NaOH   (b) $0.134M$ NaOH

Solution Since NaOH is a strong base, it is fully ionized to give Na^+ and OH^-.

(a) $$[OH^-] = 1.00 \times 10^{-2} \text{ mol L}^{-1}$$

and

$$[H_3O^+][OH^-] = 1.00 \times 10^{-14} \text{ mol}^2 \text{ L}^{-2}$$

So

$$[H_3O^+] = \frac{1.00 \times 10^{-14} \text{ mol}^2 \text{ L}^{-2}}{1.00 \times 10^{-2} \text{ mol L}^{-1}} = 1.00 \times 10^{-12} \text{ mol L}^{-1}$$

Hence
$$pH = 12.00$$

(b)
$$[OH^-] = 0.134 \text{ mol L}^{-1}$$

$$[H_3O^+] = \frac{1.00 \times 10^{-14} \text{ mol}^2 \text{ L}^{-2}}{0.134 \text{ mol L}^{-1}} = 7.46 \times 10^{-14} \text{ mol L}^{-1}$$

$$pH = -\log(7.46 \times 10^{-14}) = 13.13$$

In summary:

pH		$[H_3O^+]$	$[OH^-]$	
> 7.0	Basic	$< 10^{-7}$	$> 10^{-7}$	mol L^{-1}
$= 7.0$	Neutral	10^{-7}	10^{-7}	mol L^{-1}
< 7.0	Acidic	$> 10^{-7}$	$< 10^{-7}$	mol L^{-1}

There are no upper or lower limits to the pH scale, but for the vast majority of practical applications pH values are in the range 0 to 14.

Example 14.15 Aniline, $C_6H_5NH_2$, is a much weaker base than ammonia and has a dissociation constant of $K_b = 4.3 \times 10^{-10}$ mol L^{-1}. What is the pH of a $0.010M$ solution of aniline in water?

Solution We first write the equation for the equilibrium:

$$C_6H_5NH_2 + H_2O \rightleftharpoons C_6H_5NH_3^+ + OH^-$$

Next, we write the expression for the equilibrium constant:

$$K_b = \frac{[C_6H_5NH_3^+][OH^-]}{[C_6H_5NH_2]} = 4.3 \times 10^{-10} \text{ mol L}^{-1}$$

We then let the equilibrium concentration of OH$^-$ be x mol L^{-1} and write

	$C_6H_5NH_2 + H_2O \rightleftharpoons$	$C_6H_5NH_3^+ +$	OH^-	
Initial concentrations	0.010	0	0	mol L^{-1}
Equilibrium concentrations	$0.010 - x$	x	x	mol L^{-1}

We can now substitute these values in the equilibrium constant expression, so we have

$$\frac{x^2}{0.010 - x} \text{ mol L}^{-1} = 4.3 \times 10^{-10} \text{ mol L}^{-1}$$

Since the value of the equilibrium constant is very small, we can reasonably assume that x is very small compared with 0.010. Thus we have

$$\frac{x^2}{0.010} = 4.3 \times 10^{-10}$$

$$x^2 = 4.3 \times 10^{-12}$$

$$x = 2.1 \times 10^{-6}$$

$$[OH^-] = 2.1 \times 10^{-6} \text{ mol L}^{-1}$$

We see that x is indeed very small compared with 0.010, and therefore our assumption that $0.010 - x \simeq 0.010$ was justified.

Since $K_w = [H_3O^+][OH^-]$, we have

$$[H_3O^+] = \frac{K_w}{[OH^-]} = \frac{1.0 \times 10^{-14} \text{ mol}^2 \text{ L}^{-2}}{2.1 \times 10^{-6} \text{ mol L}^{-1}} = 4.8 \times 10^{-9} \text{ mol L}^{-1}$$

Hence

$$pH = -\log(4.8 \times 10^{-9}) = 8.32$$

14.5 SELF-IONIZATION
OF WATER

527

pK VALUES Because the dissociation constants of weak acids and bases are normally expressed in terms of negative powers of 10 (see Tables 14.3 and 14.4), it is convenient to define pK_a and pK_b in the same way that we defined pH:

$$pK_a = -\log K_a \quad \text{and} \quad pK_b = -\log K_b$$

Example 14.16 What is the pK_a value for hydrofluoric acid? From Table 14.3 we see that $K_a = 3.5 \times 10^{-4}$ mol L^{-1}.

Solution Since $K_a = 3.5 \times 10^{-4}$ mol L^{-1}, then

$$pK_a = -\log K_a = -\log(3.5 \times 10^{-4}) = 3.46$$

Relationship Between K_a and K_b for an Acid and Its Conjugate Base

We saw in Chapter 5 that the conjugate base of a strong acid in water has no basic properties in water, but the conjugate base of a weak acid is a weak base. For example, the position of the equilibrium

$$HCl + H_2O \rightleftharpoons H_3O^+ + Cl^-$$

lies far to the right. The reverse reaction does not occur to any significant extent; in other words, Cl^- is not protonated by H_3O^+ to any significant extent. Neither is it protonated by the weaker acid H_2O; in other words, it is not a base in water. The equilibrium

$$Cl^- + H_2O \rightleftharpoons HCl + OH^-$$

lies far to the left.

In contrast, the dissociation of a weak acid such as HF is not complete:

$$HF + H_2O \rightleftharpoons H_3O^+ + F^-$$

The reverse reaction does occur to a considerable extent; in other words, F^- is protonated to a considerable extent by H_3O^+. It is also protonated by H_2O to a significant extent:

$$F^- + H_2O \rightleftharpoons HF + OH^-$$

Thus HF is a weak acid, and its conjugate base, F^-, is a weak base.

The weaker an acid is, the greater is the strength of its conjugate base. Indeed, there is a quantitative relationship between the strength of an acid and the strength of its conjugate base. We can derive this relationship from the expressions for the acid and base dissociation constants, K_a and K_b. Consider the ionization of a weak acid, HA:

$$HA + H_2O \rightleftharpoons H_3O^+ + A^- \qquad K_a(HA) = \left(\frac{[H_3O^+][A^-]}{[HA]}\right)_{eq}$$

Its conjugate base, A^-, behaves as a weak base in water:

$$A^- + H_2O \rightleftharpoons HA + OH^- \qquad K_b(A^-) = \left(\frac{[HA][OH^-]}{[A^-]}\right)_{eq}$$

For convenience we omit the parentheses ( )$_{eq}$, but we must remember that these expressions for K_a and K_b are only valid when the [ ] refers to an *equilibrium* concentration.

If we now multiply $K_a(HA)$ and $K_b(A^-)$, we have

$$K_a(HA)K_b(A^-) = \frac{[H_3O^+][A^-]}{[HA]} \times \frac{[HA][OH^-]}{[A^-]} = [H_3O^+][OH^-]$$

Hence,

$$K_a(HA)K_b(A^-) = [H_3O^+][OH^-] = K_w = 10^{-14} \text{ mol}^2 \text{ L}^{-2} \text{ at } 25°C$$

In general, for any acid

$$K_a(\text{acid})K_b(\text{conjugate base}) = K_w = 10^{-14} \text{ mol}^2 \text{ L}^{-2}$$

Taking negative logarithms of both sides gives

$$pK_a(\text{acid}) + pK_b(\text{conjugate base}) = pK_w = 14$$

where we have written pK_w for $-\log K_w$. Thus the larger the K_a (that is, the stronger the acid), the smaller the K_b (that is, the weaker the conjugate base).

We do not need to list values of both K_a and K_b, because one can always be obtained from the other. But for convenience we have listed some common K_a and K_b values in Tables 14.3 and 14.4.

Example 14.17 What is the base dissociation constant for the fluoride ion, F^-? From Table 14.3 $K_a(HF) = 3.5 \times 10^{-4}$ mol L^{-1}.

Solution

$$K_a(HF)K_b(F^-) = K_w$$

$$K_b(F^-) = \frac{1.0 \times 10^{-14} \text{ mol}^2 \text{ L}^{-2}}{K_a(HF)}$$

$$= \frac{1.0 \times 10^{-14} \text{ mol}^2 \text{ L}^{-2}}{3.5 \times 10^{-4} \text{ mol L}^{-1}} = 2.9 \times 10^{-11} \text{ mol L}^{-1}$$

14.6 ACID-BASE PROPERTIES OF ANIONS, CATIONS, AND SALTS

When an acid reacts with a base to give a salt, the acid is often said to neutralize the base. Thus we might think that the solution of the salt that is formed is neutral, that is, that it has a pH of 7. Although many salts do give neutral solutions in water, a large number do not, because some cations and anions are acids or bases. The acid-base properties of some common anions and cations are summarized in Table 14.6.

Table 14.6 Acid-Base Properties of Some Common Ions

	CATIONS	ANIONS
Acidic	NH_4^+, H_3O^+, $Al(H_2O)_6^{3+}$, $Fe(H_2O)_6^{3+}$	HSO_4^-, $H_2PO_4^-$
Neutral	Mg^{2+}, Ca^{2+}, Sr^{2+}, Ba^{2+}, Li^+, Na^+, K^+, Rb^+, Cs^+, Ag^+	NO_3^-, ClO_4^-, Cl^-, Br^-, I^-
Basic	None	SO_4^{2-} (very weak, almost neutral), PO_4^{3-}, CO_3^{2-}, SO_3^{2-}, F^-, CN^-, OH^-, S^{2-}, $CH_3CO_2^-$, HCO_3^-

Anions

Anions (conjugate bases) of strong acids such as Cl^- have no basic properties in water. They give neutral solutions.

Anions (conjugate bases) of weak acids such as CN^- and CO_3^{2-} are weak bases. They give basic solutions in water. For example,

$$CN^- + H_2O \rightleftharpoons HCN + OH^-$$

Anions containing hydrogen that are derived from polyprotic acids may be either acids or bases. For example, HSO_4^- and $H_2PO_4^-$ are acids:

$$HSO_4^- + H_2O \rightleftharpoons SO_4^{2-} + H_3O^+$$

$$H_2PO_4^- + H_2O \rightleftharpoons HPO_4^{2-} + H_3O^+$$

But HPO_4^{2-} and HCO_3^- are bases:

$$HPO_4^{2-} + H_2O \rightleftharpoons H_2PO_4^- + OH^-$$

$$HCO_3^- + H_2O \rightleftharpoons H_2CO_3 + OH^-$$

If the acid dissociation constant of the anion is larger than the base dissociation constant (see Tables 14.3 and 14.4), then the anion is an acid. But if the base dissociation constant of the anion is larger than its acid dissociation constant, the anion is a base.

Cations

Metal ions are generally hydrated, as we have described in Chapters 9 and 13. Many hydrated metal ions behave as acids, particularly when the metal has a positive charge of 2 or greater. For example,

$$Al(H_2O)_6^{3+} + H_2O \rightleftharpoons H_3O^+ + Al(OH)(H_2O)_5^{2+}$$

The charge on the metal ion attracts electrons from the OH bonds, making the hydrogens considerably more acidic than they are in the free water molecule. The only common hydrated metal ions that do *not* behave as acids are Li^+, Na^+, K^+, Rb^+, Cs^+, Mg^{2+}, Ca^{2+}, Sr^{2+}, Ba^{2+}, and Ag^+. They give neutral solutions in water. All other hydrated metal ions give acidic solutions in water.

Cations (conjugate acids) of weak bases are weak acids. The most common example is the ammonium ion,

$$NH_4^+ + H_2O \rightleftharpoons H_3O^+ + NH_3$$

Most other acidic cations are derived from ammonia, for example, methylammonium, $CH_3NH_3^+$, anilinium, $C_6H_5NH_3^+$, and hydrazinium, $H_2NNH_3^+$.

Salts

We can now classify aqueous solutions of salts according to their acid-base properties (see Table 14.7).

Neutral salts contain a neutral cation and a neutral anion. They include salts of Li^+, Na^+, K^+, Rb^+, Cs^+, Mg^{2+}, Ca^{2+}, Sr^{2+}, Ba^{2+}, and Ag^+, with anions of strong acids, such as Cl^- and NO_3^-—for example, KCl, $BaCl_2$, and $AgNO_3$.

Acidic salts contain an acidic cation and a neutral anion or a neutral cation and an acidic anion. They include the following:

- Salts of metal cations, except Li^+, Na^+, K^+, Rb^+, Cs^+, Mg^{2+}, Sr^{2+}, Ba^{2+}, and Ag^+, with anions of strong acids, for example, $AlCl_3$ and $Fe_2(SO_4)_3$.

Table 14.7 Acid-Base Properties of Aqueous Solutions of Some Common Salts

BASIC SOLUTIONS, pH > 7	NEUTRAL SOLUTIONS, pH = 7	ACIDIC SOLUTIONS, pH < 7
Neutral cation Basic anion	Neutral cation Neutral anion	Acidic cation Neutral anion
NaCN	KCl	NH_4Cl
KF	$BaCl_2$	$Al(H_2O)_6Cl_3$
$Na(CH_3CO_2)$	$Ca(NO_3)_2$	$Fe(H_2O)_6(NO_3)_3$
Na_2CO_3	$Mg(ClO_4)_2$	$C_6H_5NH_3 \cdot Cl$
		Neutral cation Acidic anion
		$KHSO_4$

- Ammonium salts of strong acids, for example, NH_4Cl.
- Some salts of polyprotic acids, for example, $NaHSO_4$.

Basic salts contain a neutral cation and a basic anion. They include salts of Li^+, Na^+, K^+, Rb^+, Cs^+, Mg^{2+}, Ca^{2+}, Sr^{2+}, Ba^{2+}, and Ag^+, with anions of weak acids, such as CN^-, F^-, and CO_3^{2-}—for example, NaCN, KF, and Na_2CO_3.

For a salt of an acidic cation, such as NH_4^+, and a basic anion, such as CN^-, we cannot predict whether the solution will be acidic, basic, or neutral without knowing their acid and base dissociation constants.

Example 14.18 Predict whether the following salts give acidic, basic, or neutral solutions when dissolved in water: NaBr, K_2CO_3, $AlCl_3$, NH_4ClO_4, and $(NH_4)_2S$.

Solution

NaBr	Neutral cation, neutral anion	Therefore the solution is neutral.
K_2CO_3	Neutral cation, basic anion	Therefore the solution is basic.
$AlCl_3$	Acidic cation, neutral anion	Therefore the solution is acidic.
NH_4ClO_4	Acidic cation, neutral anion	Therefore the solution is acidic.
$(NH_4)_2S$	Acidic cation, basic anion	We cannot make a prediction without information on $K_a(NH_4^+)$ and $K_b(S^{2-})$.

Rather than just predict whether the solution of a salt is acidic or basic, we can calculate the pH of the solution, as the following examples show.

Example 14.19 What is the pH of a $0.10M$ solution of sodium cyanide?

Solution *The first step* is to write the equation for the equilibrium reaction. Because CN^- is the conjugate base of a weak acid, HCN, it is a weak base:

$$CN^- + H_2O \rightleftharpoons HCN + OH^-$$

The Na^+ ion is neither an acid nor a base.

The second step is to write the expression for the equilibrium constant and to look up the value of $K_b(CN^-)$ in Table 14.4 or to calculate it from $K_a(HCN)$:

$$K_b = \frac{[HCN][OH^-]}{[CN^-]} = 2.0 \times 10^{-5} \text{ mol L}^{-1}$$

The third step is to write an expression for the concentration of each of the species in solution. Because NaCN is a salt and is fully dissociated in aqueous solution, the initial concentrations of Na^+ and CN^- are both $0.10M$. If we let $[OH^-] = x$ mol L^{-1}, we have

$$CN^- + H_2O \rightleftharpoons HCN + OH^-$$

Initial concentrations	0.10	0	0	mol L^{-1}
Equilibrium concentrations	$0.10 - x$	x	x	mol L^{-1}

Hence

$$\frac{x^2}{0.10 - x} \text{ mol L}^{-1} = 2.0 \times 10^{-5} \text{ mol L}^{-1}$$

Since K_b is small, we assume that $x \ll 0.10$ and hence $0.10 - x \simeq 0.10$. Thus we have

$$\frac{x^2}{0.10} = 2.0 \times 10^{-5}$$

$$x^2 = 2.0 \times 10^{-6}$$

$$x = 1.4 \times 10^{-3}$$

$$[OH^-] = 1.4 \times 10^{-3} \text{ mol L}^{-1}$$

Since x is less than 5% of the initial concentration of CN^-, the assumption that $0.10 - x \simeq 0.10$ was justified. Thus

$$[H_3O^+] = \frac{1.0 \times 10^{-14} \text{ mol}^2 \text{ L}^{-2}}{[OH^-]} = \frac{1.0 \times 10^{-14} \text{ mol}^2 \text{ L}^{-2}}{1.4 \times 10^{-3} \text{ mol L}^{-1}}$$

$$= 7.1 \times 10^{-12} \text{ mol L}^{-1}$$

$$pH = 11.15$$

Example 14.20 What is the pH of a $0.20M$ solution of NH_4Cl?

Solution Ammonium chloride, NH_4Cl, is an ionic solid consisting of ammonium ions, NH_4^+, and chloride ions, Cl^-. The chloride ion is neither an acid nor a base, but the ammonium ion is a weak acid. We follow exactly the same steps as in Example 14.19. First, we write the equation for the reaction of ammonium ion with water:

$$NH_4^+ + H_2O \rightleftharpoons H_3O^+ + NH_3$$

The equilibrium constant is the acid dissociation constant, $K_a(NH_4^+)$. From Table 14.3 we see that $K_a(NH_4^+) = 5.6 \times 10^{-10}$ mol L^{-1}. Thus

$$K_a(NH_4^+) = \frac{[H_3O^+][NH_3]}{[NH_4^+]} = 5.6 \times 10^{-10} \text{ mol L}^{-1}$$

Now we obtain an expression for each of the concentrations by letting $[H_3O^+] = x$ mol L^{-1}:

$$NH_4^+ + H_2O \rightleftharpoons H_3O^+ + NH_3$$

Initial concentrations	0.20	0	0	mol L^{-1}
Equilibrium concentrations	$0.20 - x$	x	x	mol L^{-1}

Therefore

$$\frac{x^2}{0.20 - x} \text{ mol L}^{-1} = 5.6 \times 10^{-10} \text{ mol L}^{-1}.$$

Making the usual approximation that $x \ll 0.20$, we have

$$\frac{x^2}{0.20} = 5.6 \times 10^{-10}$$

Solving for x gives

$$x = 1.1 \times 10^{-5} \text{ and } [H_3O^+] = 1.1 \times 10^{-5} \text{ mol L}^{-1}$$

Therefore

$$pH = -\log(1.1 \times 10^{-5}) = 4.96$$

14.7 pH: APPLICATIONS AND MEASUREMENT

The measurement of the pH of aqueous solutions has many important applications. The rates of chemical reactions involved in biochemical processes are often very sensitive to the hydronium ion concentration of the medium. In fermentation, for example, control of the pH is very important. In fact, the concept of pH was invented in 1909 by the Danish chemist Søren Sørensen (1868–1939) while he was working at the Carlsberg Brewery in Copenhagen on problems connected with the brewing of beer. Many body fluids have well-defined pH's that must be maintained at these values if the body is to function in a normal way. For example, the fluid in the stomach has a pH of approximately 1.4. This rather high acidity is important for the proper digestion of food. But if the stomach fluid becomes much more acidic, we are soon made aware of it by the pain and discomfort. Blood has a constant pH of 7.4.

Pure water has a pH of 7.0. Ordinary rainwater and drinking water are normally very slightly acidic (pH < 7.0), mainly because they contain a small amount of dissolved CO_2. This dissolved CO_2, which reacts with water to some extent, produces H_3O^+ and hydrogen carbonate ion, HCO_3^-:

$$CO_2(aq) + 2H_2O(l) \rightleftharpoons H_3O^+(aq) + HCO_3^-(aq)$$

Rainwater also contains very small amounts of sulfuric and nitric acids. These acids are formed from SO_2 emitted by volcanoes and NO formed in lightning discharges. However, when we speak of *acid rain*, we mean rain that is much more acidic than it has normally been in the past or than it is in areas where the atmosphere is not polluted with SO_2 and other acid-producing substances resulting from industrial processes and automobile emissions (see Box 14.1).

Our sense of taste is remarkably sensitive to pH. Indeed, taste was one of the earliest ways in which acids and bases were distinguished. We are able to detect a sour, tart, or acidic taste in a solution with a pH between 4 and 5. Soda water, which is a solution of carbon dioxide in water, has a pH of about 4 and tastes quite definitely acidic. Most fruit juices and soft drinks have a pH in the range of 2–3. Their familiar acid taste arises from the weak acids that they contain. For example, lemon juice contains citric acid (see Chapter 19). A solution with a pH of 1 or less is not only unpleasant to taste but is dangerous because it burns the skin.

Basic solutions with a pH greater than 7 have a bitter taste and a characteristic slippery or soapy feel. They cause a characteristic wrinkling of the skin. A 2% solution of sodium hydrogen carbonate, $NaHCO_3$, is an effective mouthwash, but it has a very unpleasant taste. Very basic solutions of pH 12 or 13 are dangerous because they attack the skin quite rapidly. Some drain cleaners are a concentrated NaOH solution with added scent and coloring. They have a pH as high as 14 or 15 and should be handled with great care.

Indicators

A convenient way of determining the approximate pH of a solution is by the use of indicators. *An* **indicator** *is a weak acid that has a conjugate base with a different color from that of the acid.* Some naturally occurring colored substances

Rain is normally slightly acidic, with a pH of 5.6, because it contains dissolved carbon dioxide from the atmosphere. But the rain now falling in many parts of the industralized world, even in regions remote from industry, is often much more acidic. The average pH of rain in the northeastern United States has decreased over the past years and is now 4.3. Rain with a pH as low as 1.5—which is only three times less concentrated that the 0.1M acid often used for titrations in the laboratory—has been recorded at Wheeling, West Virginia. The Los Angeles basin routinely has fogs made up of suspended water droplets with a pH of 2.2 to 4.0.

Acid rain is caused by oxides of sulfur and nitrogen which are present in the atmosphere as a result of combustion processes. Coal and oil typically contain 1% to 3% sulfur and this is converted into SO_2 when the coal and oil are burned. Sulfur dioxide is also produced when sulfide ores are roasted to convert them to oxides that can be reduced to the metal (see Chapter 9). For example

$$Cu_2S + 2O_2 \longrightarrow 2CuO + SO_2$$

Sulfur dioxide is slowly oxidized to sulfur trioxide in the atmosphere and this dissolves in water droplets to give a dilute solution of sulfuric acid. Of course, some sulfur dioxide arises from natural sources. Scientists have estimated that the eruption of Mount St. Helens in the state of Washington in 1980 blew out approximately 400,000 tons of sulfur dioxide. But that is only about 1.5% of the estimated total sulfur dioxide from man-made sources in the United States for the same year. Nitrogen monoxide, NO, is also formed in combustion processes, particularly in internal combustion engines, as a result of the

combination of atmospheric nitrogen and oxygen. Nitrogen monoxide reacts with oxygen in the atmosphere to give nitrogen dioxide, NO_2, which in turn reacts with water in clouds and raindrops to give a solution of nitric acid.

$$3NO_2(g) + H_2O(l) \longrightarrow 2HNO_3(aq) + NO(g)$$

At least some of the harmful effects of acid rain have now been well-established. Because marble—a form of calcium carbonate—is soluble in acid (see Chapter 5), Greek and Italian monuments that have withstood centuries of natural weathering are now rapidly deteriorating. There are even more serious harmful ecological effects. In limestone areas acid rain is largely neutralized, but in other regions this does not happen and lakes and rivers have consequently become acidified. It is estimated that 20,000 lakes in Sweden have now become too acidic for fish and other life. The same effects are now being observed in lakes in the United States and Canada. Acid rain dissolves aluminum compounds from the soil and washes them into lakes where they poison the fish. Plant life is also affected because acid rain kills microorganisms in the soil that are responsible for nitrogen fixation, and it dissolves and washes away essential magnesium, calcium, and potassium compounds. Moreover, acid rain can dissolve the waxy coating that protects leaves from fungi and bacteria. The recent deterioration of some coniferious forests in Germany and Canada has been attributed to the effects of acid rain.

What can be done to combat the effects of acid rain? A local and limited solution that has been used in Sweden is to add lime, CaO, to lakes to neutralize the acid, but

have been used as indicators for a long time. Litmus, which is extracted from certain lichens, is an example. In basic solutions, in which it has a blue color, it is in its conjugate base form. In acid solutions, in which it has a red color, it is in its acid form. We saw in Experiment 5.10 that tea is an indicator, it has a brown color in basic solution and a yellow color in acid solution. Experiment 14.2 shows that colored substances present in flowers can also act as indicators. More commonly, in the laboratory we use various dyes such as phenolphthalein, which is colorless in acid solution and pink in basic solution, and methyl orange, which is red in acid solution and yellow orange in basic solution (see Figure 14.6).

The pH at which the color change occurs depends on the indicator. It is determined by the position of the equilibrium between the acid form of the indicator, denoted by HIn, and its conjugate base, denoted by In^-:

$$HIn + H_2O \rightleftharpoons In^- + H_3O^+$$

According to Le Châtelier's principle, the equilibrium is shifted to the left if the

Left: This photo taken in 1910 shows the effect of 400 years of weathering on a grotesque decorating Lincoln Cathedral in England. Right: In 1984, only 74 years later, acid rain and other atmospheric pollution have worn the figure to a barely recognizable remnant.

this is expensive and must be repeated annually. In the past, very high smoke stacks were built at smelting plants to disperse the sulfur dioxide as high in the atmosphere as possible. This certainly benefitted the vegetation around the smelting plants which in many cases had been completely killed for many miles around. However, sulfur dioxide is sometimes carried in air currents many hundreds of miles to be deposited, often in another country, as acid rain. The Swedes blame the British for their acid rain, and controversy rages between Canada and the United States over the acid rain believed by both countries to originate on the other side of the border. The obvious solution to the problem is to drastically reduce emissions of the oxides of sulfur and nitrogen. Sulfur dioxide could be removed from the gases produced by combustion and by smelting for example, by passing them over lime which reacts with sulfur dioxide to produce solid calcium sulfite.

Other solutions include removing sulfur from oil and coal or using only low sulfur content fuels. Or we could convert to alternative energy sources such as solar power or nuclear power. Most solutions are expensive or bring problems of their own—such as disposal of nuclear wastes. Because these solutions are expensive, and often unpopular with certain groups, such as coalminers who might lose their jobs, governments are reluctant to act. They use the fact that neither all the effects of acid rain, nor all the details of its formation are yet well understood, as an excuse to delay. Although scientists will continue to search for a better understanding of the phenomenon of acid rain and for better and less expensive solutions to the problems that it causes, we already know enough about its harmful effects to warrant drastic steps to reduce emissions of the oxides of sulfur and nitrogen.

$$CaO(g) + SO_2(g) \longrightarrow CaSO_3(s)$$

H_3O^+ concentration is increased, for example, by adding acid to the solution. If the indicator is phenolphthalein, it is then very largely in its colorless form, HIn. But if the solution is made basic, the H_3O^+ concentration is much reduced, the equilibrium shifts to the right, and therefore the indicator is converted almost entirely to the In^- form, which in the case of phenolphthalein is pink. Thus we can distinguish between an acidic solution and a basic solution by adding a small amount of phenolphthalein. The solution will be pink if it is basic but colorless if it is acidic.

The exact pH at which the color change of an indicator occurs can be found from the acid dissociation constant of the indicator:

$$K_a(HIn) = \frac{[In^-][H_3O^+]}{[HIn]}$$

Rearranging this equation, we have

$$[H_3O^+] = K_a \frac{[HIn]}{[In^-]}$$

A Natural Indicator

The compound that gives the color to a red rose can be dissolved in methanol to give a red solution, leaving the rose in the beaker on the left a pale pink color.

This solution can be used as an indicator. The small beakers from left to right contain solutions of pH 1, 3, 7, 9 and 11 to which a small amount of the red solution has been added,

Figure 14.6 The Indicators Phenolphthalein and Methyl Orange

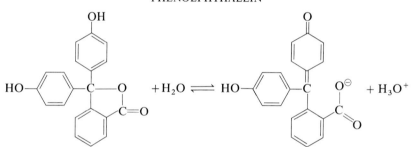

PHENOLPHTHALEIN

Acid form: Colorless   Conjugate base form: Pink

METHYL ORANGE

Acid form: Red   Conjugate base form: Yellow orange

Taking negative logarithms of both sides gives

$$-\log[H_3O^+] = -\log K_a - \log \frac{[HIn]}{[In^-]}$$

$$pH = pK_a - \log \frac{[HIn]}{[In^-]} = pK_a + \log \frac{[In^-]}{[HIn]}$$

The indicator will be half in its acid form and half in its base form—in other words, in the middle of its color change—when

$$[HIn] = [In^-] \qquad \text{or} \qquad \frac{[In^-]}{[HIn]} = 1$$

and the pH will then be

$$pH = pK_a + \log 1 = pK_a + 0 = pK_a$$

CHAPTER 14
CHEMICAL EQUILIBRIUM

536

Thus the pH at which an indicator changes color depends on its pK_a. For phenolphthalein $K_a = 3 \times 10^{-10}$, so the pH at which it is half in its base form and half in its acid form is

$$pH = pK_a = -\log(3 \times 10^{-10}) = 9.5$$

Thus we expect phenolphthalein to change color in the vicinity of pH 9.5. In fact, the color change occurs over a range of pH, and this range is partly determined by the sensitivity of the eye to the color change. In general, when $[In^-]/[HIn]$ or $[HIn]/[In^-]$ has a value of approximately 10, the eye cannot detect any further color change if the concentration ratio is further increased. Thus the visible color change occurs approximately over the range

$$pH = pK_a(HIn) \pm \log 10 \quad \text{or} \quad pH = pK_a(HIn) \pm 1$$

and the color change is half complete when $pH = pK_a(HIn)$. For example, methyl red has a pK_a of 5.2. It is clearly red in a solution of pH 4.2 and clearly yellow in a solution of pH 6.2. Thus as the acidity increases over this range, the indicator changes color from yellow through orange to red.

Table 14.8 lists several indicators together with their useful pH ranges. The colors of some of these indicators are shown in Figure 14.7. Notice that indicators do not in general change color in the immediate vicinity of pH 7. If we

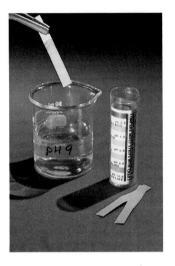

pH Paper

Table 14.8 Properties of Some Indicators

	pK_a	EFFECTIVE pH RANGE	COLOR	
			ACID FORM	BASE FORM
Methyl violet	1.6	0.0–3.0	Yellow	Violet
Methyl orange	4.2	2.1–4.4	Red	Yellow
Methyl red	5.0	4.2–6.2	Red	Yellow
Bromothymol blue	7.1	6.0–7.8	Yellow	Blue
Thymol blue	8.2	7.9–9.4	Yellow	Blue
Phenolphthalein	9.5	8.3–10.0	Colorless	Red
Alizarine yellow	11.0	10.1–12.1	Yellow	Red

Figure 14.7 Colors of Some Indicators. From left to right: Methyl red, phenophthalein, bromothymol blue, and universal.

14.7 pH: APPLICATIONS AND MEASUREMENT

Figure 14.8 pH Meter. A pH meter is an instrument that employs two electrodes to measure the potential difference between the solution to be tested and a standard solution of known pH. This potential difference is related to the pH of the solution being tested. The pH is read directly from the dial of the instrument.

compare the color of a solution of unknown pH containing a suitable indicator with the colors of that indicator in a number of solutions of known pH, we can make a fairly accurate determination of the pH. The smallest amount of indicator that will give a clearly observable color should be used. Adding too much indicator disturbs the acid-base equilibrium in the solution and changes the pH.

Universal indicator is a mixture of several indicators that changes color continuously over a wide pH range. With such a solution one can find the approximate pH of any solution within this range. So-called *pH paper* is impregnated with a universal indicator. When a strip of this paper is immersed in a solution, the pH can be judged from the resulting color.

For accurate pH measurements a pH meter such as the one shown in Figure 14.8 is used. The principles on which the pH meter operates are described in Chapter 16. Experiment 14.3 shows the pH of some common solutions.

14.8 BUFFER SOLUTIONS

In many chemical processes and in many reactions that occur in living systems, the pH must remain at a constant value. For example, the control of the pH is very important in the treatment of sewage, in electroplating, and in the manu-

EXPERIMENT 14.3

The pH of Some Common Solutions

The color of universal indicator shows that solutions of lime juice in water, soda water, and vinegar are all acidic.

The color of the universal indicator shows that Drāno, household ammonia, and Milk of Magnesia are all basic (alkaline).

pH can be measured more accurately with a pH meter. Vinegar has a pH of 2.3.

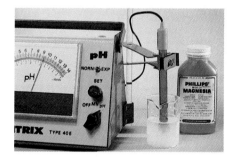

A suspension of Milk of Magnesia in water has a pH of 9.4.

facture of photographic materials. Maintaining a constant pH is important in many metabolic processes, because the function of proteins depends on their structures and their structures depend on the pH. Proteins are both acids and bases, and as their charge changes on gaining or losing protons, their shape changes. Their function depends critically on their shape, and therefore the pH must remain constant so that they maintain a particular shape. For example, the pH of the blood has to be close to 7.4; the pH of saliva is close to 6.8; and the enzymes in the stomach function only at a pH of approximately 1.5.

A solution that has the ability to maintain a very nearly constant pH even when moderate amounts of acid or base are added to it is called a **buffer solution**.

Buffer solutions contain either a weak acid and a salt of the weak acid, or a weak base and a salt of the weak base, in approximately equal proportions.

To understand how a solution acts as a buffer solution—that is, why its pH is insensitive to the addition of moderate amounts of acid or base—consider a solution containing both acetic acid, and its conjugate base, the acetate ion, in comparable amounts. Such a solution can be prepared by adding an acetate, such as sodium acetate, to an acetic acid solution. Because this solution contains both a weak acid, acetic acid, and a weak base, acetate ion, it can react with both added acids and bases. If H_3O^+ is added to the solution, it combines almost completely with $CH_3CO_2^-$ to form CH_3CO_2H:

$$CH_3CO_2^- + H_3O^+ \longrightarrow CH_3CO_2H + H_2O$$

Therefore there is no appreciable change in the H_3O^+ concentration. That this reaction does go almost to completion can be seen from the value of its equilibrium constant, which is the inverse of the acid dissociation constant of acetic acid:

$$K = \frac{[CH_3CO_2H]}{[CH_3CO_2^-][H_3O^+]} = \frac{1}{\dfrac{[H_3O^+][CH_3CO_2^-]}{[CH_3CO_2H]}} = \frac{1}{K_a(CH_3CO_2H)}$$

$$= \frac{1}{1.8 \times 10^{-5}} = 5.6 \times 10^4 \text{ mol}^{-1} \text{ L}$$

If OH^- is added to the acetic acid–acetate buffer solution, it combines almost completely with acetic acid, converting it to acetate ion:

$$CH_3CO_2H + OH^- \longrightarrow CH_3CO_2^- + H_2O$$

So again there is almost no change in the OH^- (or H_3O^+) concentration and therefore in the pH. That this reaction also goes very nearly to completion is shown by its very large equilibrium constant, which is the inverse of the base dissociation constant of $CH_3CO_2^-$:

$$K = \frac{[CH_3CO_2^-]}{[CH_3CO_2H][OH^-]} = \frac{1}{\dfrac{[CH_3CO_2H][OH^-]}{[CH_3CO_2^-]}} = \frac{1}{K_b(CH_3CO_2^-)}$$

$$= \frac{1}{5.6 \times 10^{-10} \text{ mol}^{-1} \text{ L}} = 1.8 \times 10^9 \text{ mol}^{-1} \text{ L}$$

Another way of looking at the consequences of adding acid or base to an acetic acid–acetate solution is to apply Le Châtelier's principle to the equilibrium:

$$CH_3CO_2H + H_2O \rightleftharpoons CH_3CO_2^- + H_3O^+$$

Addition of acid, that is, of H_3O^+, drives the equilibrium to the left. Since acetic acid is a weak acid, only a very small concentration of H_3O^+ can remain when the new equilibrium is reached. Therefore virtually all the added H_3O^+ is removed, and there is almost no decrease in the pH. If OH^- is added to the buffer solution, it reacts with H_3O^+ to form water. In accordance with Le Châtelier's principle more acetic acid must then dissociate to give more H_3O^+ to replace that which is removed. The overall result is the removal of the added OH^- and the simultaneous conversion of CH_3CO_2H to acetate ion, $CH_3CO_2^-$. Almost all the added OH^- is removed in this way, and there is almost no increase in the pH of the solution.

An acetic acid–acetate buffer solution will therefore maintain an almost constant pH provided that the amounts of added acid or base are not large enough to remove most of the $CH_3CO_2^-$ or CH_3CO_2H originally present in the buffer solution. As long as both $CH_3CO_2^-$ and CH_3CO_2H are present in appreciable amounts, a small and nearly constant equilibrium concentration of H_3O^+ is maintained, and there is little change in the pH.

A buffer solution composed of a weak base and a salt of the weak base behaves in a similar manner. For example, if we have a solution of ammonia and ammonium chloride, added acid reacts with the weak base, NH_3, and added base reacts with the ammonium ion, which is a weak acid:

$$NH_3 + H_3O^+ \longrightarrow NH_4^+ + H_2O$$
$$NH_4^+ + OH^- \longrightarrow NH_3 + H_2O$$

Again, there is no significant change in either the H_3O^+ or OH^- concentrations and thus no appreciable change in the pH of the solution.

Calculation of the pH of a Buffer Solution

To make up a buffer solution for some particular pH, we must know how to calculate the pH of any given buffer solution. As an example, we will calculate the pH of a solution of 0.050 mol of acetic acid and 0.050 mol of sodium acetate in 1 L.

The first step is to write the equation for the equilibrium in the solution. The solution contains both acetate ions and acetic acid molecules, and the equilibrium is

$$CH_3CO_2H + H_2O \rightleftharpoons CH_3CO_2^- + H_3O^+$$

This is the equilibrium for the ionization of acetic acid in water. However, the solution that we are considering differs from a solution of acetic acid in water in that it has a large concentration of acetate ion as well as acetic acid.

The second step is to write the expression for the equilibrium constant and look up its value. In this case we have

$$K_a = \frac{[H_3O^+][CH_3CO_2^-]}{[CH_3CO_2H]} = 1.8 \times 10^{-5} \ mol \ L^{-1}$$

The third step is to write an expression for the concentration of each of the species, remembering that 0.050 mol $CH_3CO_2^-$ comes from the sodium acetate:

	$CH_3CO_2H + H_2O \rightleftharpoons$	$CH_3CO_2^- +$	H_3O^+	
Initial concentrations	0.050	0.050	0	mol L^{-1}
Equilibrium concentrations	$0.050 - x$	$0.050 + x$	x	mol L^{-1}

The initial concentrations of the acetic acid and the acetate ion are both 0.050 mol L^{-1}, but a small amount of the acetic acid must ionize in order to establish the equilibrium with the acetate ion. We let the concentration of hydronium ion that is formed be x mol L^{-1}. Then at equilibrium the concentration of acetic acid will be $0.050 - x$ mol L^{-1}, and the concentration of acetate ion will be $0.050 + x$ mol L^{-1}. Hence

$$K_a = \frac{[H_3O^+][CH_3CO_2^-]}{[CH_3CO_2H]} = \frac{(0.050 + x)(x)}{0.050 - x} \text{ mol } L^{-1} = 1.8 \times 10^{-5} \text{ mol } L^{-1}$$

This is a quadratic equation and may be solved by using the formula in Appendix A. However, because acetic acid is a weak acid, we know that x will be small compared with 0.050, and we can see, by using Le Châtelier's principle, that the large concentration of acetate ion will further decrease the ionization of acetic acid; thus x must be very small indeed. Therefore we make the approximation that $0.050 - x \simeq 0.050 + x \simeq 0.050$. Thus

$$\frac{0.050x}{0.050} \text{ mol } L^{-1} = 1.8 \times 10^{-5} \text{ mol } L^{-1}$$

$$x = 1.8 \times 10^{-5}$$

$$[H_3O^+] = 1.8 \times 10^{-5} \text{ mol } L^{-1}$$

$$pH = -\log(1.8 \times 10^{-5}) = 4.74$$

We see that x is indeed very small compared with 0.050, and therefore our approximation was justified.

Now we want to calculate the effect of adding 0.001 mol of HCl, which is fully ionized to give H_3O^+ and Cl^-. The added H_3O^+ reacts almost completely with the acetate ion, so we may write

$$CH_3CO_2^- + H_3O^+ \longrightarrow CH_3CO_2H + H_2O$$

Before reaction with HCl	0.050	0.001	0.050	mol L^{-1}
After reaction with HCl	0.049	0	0.051	mol L^{-1}

Now we recalculate the pH, using the new concentrations:

$$CH_3CO_2H + H_2O \rightleftharpoons CH_3CO_2^- + H_3O^+$$

Initial concentrations	0.051	0.049	0	mol L^{-1}
Equilibrium concentrations	$0.051 - x$	$0.049 + x$	x	mol L^{-1}

Hence

$$K_a = 1.8 \times 10^{-5} \text{ mol } L^{-1} = \frac{[CH_3CO_2^-][H_3O^+]}{[CH_3CO_2H]} = \frac{(0.049 + x)(x)}{(0.051 - x)} \text{ mol } L^{-1}$$

To solve this quadratic equation, we assume $x \ll 0.049$. Thus $0.049 + x = 0.049$ and $0.051 - x = 0.051$. So

$$1.8 \times 10^{-5} = \left(\frac{0.049}{0.051}\right)x$$

$$x = \left(\frac{0.051}{0.049}\right)(1.8 \times 10^{-5}) = 1.9 \times 10^{-5}$$

$$[H_3O^+] = 1.9 \times 10^{-5} \text{ mol } L^{-1}$$

Therefore

$$pH = -\log(1.9 \times 10^{-5}) = 4.72$$

This pH differs by only 0.02 from the pH of the original buffer solution.

It may seem that we added only a small amount of HCl. But if the same amount of HCl had been added to a solution of pH 4.7 that was *not* a buffer solution, the pH would have changed by a much larger amount. For example, an HCl solution of concentration 0.000020 mol L^{-1} has a pH of 4.7. If we add 0.001 mol of HCl to 1 L of this solution, the total $[H_3O^+]$ is 0.001020, and the pH of the solution is 3.0. Thus in this case the pH changes from 4.7 to 3.0—a much greater change than we found for the buffer solution. Or if we add 0.001 mol of HCl to 1 L of pure water, the pH changes from 7 to 3.

Example 14.21 What is the pH change when 0.01 mol of HCl is added to a solution containing 0.05 mol of acetate ion and 0.05 mol of acetic acid in 1.0 L of solution?

Solution Following the previous argument, we find that the equilibrium concentrations after the addition of HCl are

$$[CH_3CO_2H] = 0.06 \text{ mol } L^{-1} \quad \text{and} \quad [CH_3CO_2^-] = 0.04 \text{ mol } L^{-1}$$

Hence

$$K_a = 1.8 \times 10^{-5} = \left(\frac{0.04}{0.06}\right)x$$

Therefore

$$x = 2.7 \times 10^{-5} \quad \text{and} \quad pH = 4.57$$

Thus the pH changes from 4.74 to 4.57.

We see that even on adding ten times as much HCl as in the preceding discussion, the change in pH is still very small. If this much HCl has been added to 1.0 L of solution of pH 4.7 that was *not* a buffer, the pH would have changed from 4.7 to 2.0.

Example 14.22 What is the pH of a buffer solution prepared from 1.00 mol of NH_3 and 0.40 mol of NH_4Cl in 1.0 L of solution?

Solution The equilibrium is between NH_3 and the NH_4^+:

$$NH_3 + H_2O \rightleftharpoons NH_4^+ + OH^-$$

The appropriate equilibrium constant is therefore the base dissociation constant of NH_3 (Table 14.4):

$$K_b = \frac{[NH_4^+][OH^-]}{[NH_3]} = 1.8 \times 10^{-5} \text{ mol } L^{-1}$$

We then obtain the concentrations of the various species in solution:

	NH_3	$+ H_2O \rightleftharpoons$	NH_4^+	$+ OH^-$	
Initial concentrations	1.0		0.4	0	mol L^{-1}
Equilibrium concentrations	$1.0 - x$		$0.4 + x$	x	mol L^{-1}

Substituting in the equilibrium expression gives

$$K_b = \frac{(0.4 + x)(x)}{(1.0 - x)} \text{ mol } L^{-1} = 1.8 \times 10^{-5} \text{ mol } L^{-1}$$

Making the usual assumption that x is very small compared with the initial concentrations, we have

$$\frac{0.4x}{1.0} = 1.8 \times 10^{-5}$$

$$x = 4.5 \times 10^{-5}$$

Thus

and since

$$[OH^-] = 4.5 \times 10^{-5} \text{ mol L}^{-1}$$

$$[H_3O^+][OH^-] = 1.0 \times 10^{-14} \text{ mol}^2 \text{ L}^{-2}$$

$$[H_3O^+] = 2.2 \times 10^{-10} \text{ mol L}^{-1}$$

$$pH = 9.66$$

We can derive a general equation for calculating the pH of any buffer solution that is a mixture of a weak acid and its conjugate base. We have

$$HA + H_2O \rightleftharpoons A^- + H_3O^+$$

$$K_a = \left(\frac{[H_3O^+][A^-]}{[HA]} \right)_{eq}$$

This equation can be rearranged to give

$$[H_3O^+] = \frac{K_a[HA]}{[A^-]}$$

Taking negative logarithms of both sides, we obtain

$$-\log[H_3O^+] = -\log K_a - \log \frac{[HA]}{[A^-]}$$

$$pH = pK_a + \log \left(\frac{[A^-]}{[HA]} \right)_{eq}$$

We have seen that the equilibrium concentrations of HA and A^- do not differ significantly from the concentrations of the weak acid and its conjugate base used to make up the buffer solution. Hence to a very good approximation we may write

$$pH = pK_a + \log \left(\frac{[A^-]}{[HA]} \right)_{initial}$$

or

$$pH = pK_a + \log \frac{[base]}{[acid]}$$

where [acid] and [base] are the concentrations of acid and conjugate base used to make up the buffer solution. This equation is called the **Henderson-Hasselbalch equation**.

A buffer is most efficient in resisting changes in pH when both the acid and its conjugate base are present in approximately equal amounts—in other words, when $[A^-] = [HA]$. In this case

$$pH = pK_a + \log \frac{[A^-]}{[HA]} = pK_a + \log 1 = pK_a$$

Thus we can choose a buffer for a given pH range on the basis of the pK_a value of the acid. For a buffer solution to operate around pH 9, we might choose an $NH_3 - NH_4^+$ buffer, since $pK_a(NH_4^+) = 9.3$. For a buffer to operate around a pH of 5, we might choose a $CH_3CO_2H–CH_3CO_2^-$ buffer, since $pK_a(CH_3CO_2H) = 4.7$.

The constant pH of blood is maintained by several conjugate acid-base pairs including $H_2PO_4^-$, HPO_4^{2-} and H_2CO_3, HCO_3^-. Several proteins also contribute to the buffering action (see Experiment 14.4).

The Buffer Action of Blood Plasma

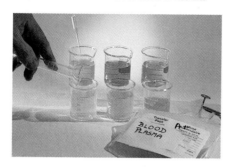

The top row of beakers contain water to which a few drops of bromothymol blue indicator has been added. The bottom row contain water to which blood plasma and bromothymol blue have been added. Right: No acid or base has been added. Top middle: A few drops of a 0.1 M NaOH solution have been added and the solution has become alkaline as shown by the blue color of the indicator. Bottom middle: Ten mL of 0.1 M NaOH solution have been added to the plasma solution. The indicator color does not change, showing that there is little change in the pH. Top left: A few drops of 0.1 M HCl solution have been added to the water, it becomes acidic as shown by the change in the color of the indicator. Bottom left: When 10 mL of 0.1 M HCl solution is added to the blood plasma in water, the color of the indicator does not change, showing that there is little change in the pH.

Example 14.23 Blood has a pH of 7.40. Use the Henderson-Hasselbalch equation to calculate the ratio of hydrogen carbonate ion, HCO_3^-, to carbonic acid, H_2CO_3, in blood; $pK_a(H_2CO_3) = 6.37$.

Solution The Henderson-Hasselbalch equation is

$$pH = pK_a + \log \frac{[\text{base}]}{[\text{acid}]}$$

Rearranging to solve for the ratio of concentrations, we obtain

$$\log \frac{[\text{base}]}{[\text{acid}]} = pH - pK_a$$

$$= 7.40 - 6.37 = 1.03$$

Therefore

$$\frac{[\text{base}]}{[\text{acid}]} = 10.7$$

Thus the ratio of HCO_3^- to H_2CO_3 must be 10.7 to 1.

Example 14.24 What is the pH of a buffer solution obtained by adding 25.00 mL of a $0.10M$ sodium hydroxide solution to 75.00 mL of a $0.10M$ solution of acetic acid? Use the Henderson-Hasselbalch equation.

Solution The NaOH and CH_3CO_2H react to form sodium acetate, CH_3CO_2Na. We need first to calculate the amounts of CH_3CO_2H and CH_3CO_2Na in the buffer solution. Initially, we have

$$\text{moles of NaOH} = (25 \text{ mL})(0.10 \text{ mol L}^{-1})\left(\frac{1 \text{ L}}{10^3 \text{ mL}}\right) = 2.5 \times 10^{-3} \text{ mol}$$

$$\text{moles of } CH_3CO_2H = (75 \text{ mL})(0.10 \text{ mol L}^{-1})\left(\frac{1 \text{ L}}{10^3 \text{ mL}}\right) = 7.5 \times 10^{-3} \text{ mol}$$

$$CH_3CO_2H + OH^- \longrightarrow CH_3CO_2^- + H_2O$$

Initial amounts	7.5×10^{-3}	2.5×10^{-3}	0	mol
Equilibrium amounts	5.0×10^{-3}	0	2.5×10^{-3}	mol

Note that to apply the Henderson-Hasselbalch equation, we do not need to change these amounts into concentrations, because the ratio of concentrations, [base]/[acid], is equal to the ratio of the number of moles. That is,

$$pH = pK_a + \log \frac{[\text{base}]}{[\text{acid}]} = 4.74 + \log \frac{2.5 \times 10^{-3} \text{ mol}}{5.0 \times 10^{-3} \text{ mol}}$$

$$= 4.74 + \log 0.5 = 4.44$$

14.9 ACID-BASE TITRATIONS

Acids and bases are such common and important substances that it is frequently necessary to determine the concentration of a solution of an acid or a base. Hydrochloric acid can be made by passing hydrogen chloride into water, but how do we find the concentration of the acid produced? In the industrial preparation of sulfuric acid, SO_3 and water are added continuously to 98% sulfuric acid at such a rate as to produce more 98% acid. How can the concentration of the acid being produced be checked?

Answers to questions such as these can be obtained by *volumetric analysis*, in which the volumes of solutions are measured. In a procedure called a **titration**, a solution of known concentration is used to determine the unknown concentration of another solution. A titration can be based on any reaction that is rapid and goes to completion, for example, the reaction between an acid and a base. A solution of known concentration of a base is added to a known volume of a solution of an acid of unknown concentration, and the volume of the solution of the base needed to just react completely with the acid is determined.

The calculation of the concentration of the acid solution from the measured volume of the base solution was explained in Chapter 5. We will be concerned here with how we find out when just enough base has been added to completely react with all the acid. The **equivalence point** of the titration will then have been reached. To determine the equivalence point in an acid-base reaction, we may use an *indicator*. To choose an indicator that changes color at the equivalence point, we need to know how the pH changes during an acid-base titration and, in particular, what the pH is at the equivalence point.

Titration of a Strong Acid with a Strong Base

We consider first the change in the pH during the titration of a strong acid with a strong base. As an example, we will calculate the change in pH during the titration of 25 mL of $0.100M$ HCl solution with $0.100M$ NaOH solution. The reaction is

$$HCl + NaOH \longrightarrow NaCl + H_2O$$

At the equivalence point the solution contains only NaCl; therefore the pH is 7. At any other point in the titration n_b moles of base have been added to n_a moles of acid. Since the acid and base are strong, $[H_3O^+] = n_a$ and $[OH^-] = n_b$. And since 1 mol of H_3O^+ reacts with 1 mol of OH^-, we have the following amounts of H_3O^+ and OH^- before the equivalence point is reached:

	H_3O^+	OH^-	
Before reaction	n_a	n_b	moles
Before equivalence point	$n_a - n_b$	0	moles
After equivalence point	0	$n_b - n_a$	moles

From the number of moles of H_3O^+ or OH^- and the total volume of the solution, we can calculate $[H_3O^+]$ and $[OH^-]$ and hence the pH, as shown in Table 14.9. The change in pH during the titration is shown graphically in Figure 14.9. There is a rapid change in the pH of the solution very close to the equivalence point, and any indicator with a pK_a between 4 and 10 would be suitable for detecting the equivalence point in this titration. Phenolphthalein is often used because of its definite color change from colorless to pink. However, many other indicators such as methyl red and bromothymol blue are equally suitable.

Titration of a Weak Acid with a Strong Base

The choice of indicator for the titration of a weak acid such as acetic acid with a strong base such as sodium hydroxide is more limited. The variation in the

Table 14.9 Change in pH During Titration of 25.0 mL of $0.100M$ HCl with $0.100M$ NaOH at 25°C

HCl (mL), V_a	NaOH (mL), V_b	TOTAL VOLUME (mL), $V_a + V_b$	n_a* (mmol)	n_b* (mmol)	$n_{H_3O^+}$ (mmol), $n_a - n_b$	$[H_3O^+]$ (mmol L^{-1})	pH
25.0	0.0	25.0	2.50	0.00	2.50	100.0	1.00
	10.0	35.0		1.00	1.50	42.9	1.37
	20.0	45.0		2.00	0.50	11.1	1.95
	24.5	49.5		2.45	0.05	1.0	3.00
	24.9	49.9		2.49	0.01	0.2	3.70
	25.0	50.0		2.50	—	10^{-7}	7.00[a]

					n_{OH^-} (mmol), $n_b - n_a$	$[OH^-]$ (mmol L^{-1})	
	25.1	50.1		2.51	0.01	0.2	10.30
	25.5	50.5		2.55	0.05	1.0	11.00
	30.0	55.0		3.00	0.50	9.1	11.96
	40.0	65.0		4.00	1.50	23.1	12.36
	50.0	75.0		5.00	2.50	33.3	12.52

Sample calculation

When 20.0 mL of $0.100M$ NaOH has been added, $n_a = 2.50 \times 10^{-3}$ mol and $n_b = 2.00 \times 10^{-3}$ mol. The moles of unreacted acid are given by $n_a - n_b = 0.50 \times 10^{-3}$ mol L^{-1}, and the total volume of the solution is $V_a + V_b = (25.0 + 20.0)$ mL $= 45.0$ mL. So

$$[H_3O^+] = \left(\frac{0.50 \times 10^{-3} \text{ mol}}{45.0 \text{ mL}} \right) \left(\frac{10^3 \text{ mL}}{1 \text{ L}} \right) = 0.0111 \text{ mol L}^{-1}$$

$$pH = 1.95$$

Note: n_a = initial millimoles of acid; n_b = millimoles of added base.

[a] Equivalence point: solution contains only NaCl.

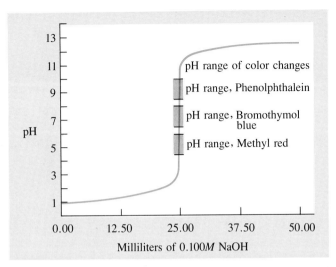

Figure 14.9 Titration of 25.00 mL of 0.100M HCl with 0.100M NaOH. Any one of these three indicators could be used to determine the equivalence point of this titration.

pH is not the same as in a strong-acid–strong-base titration, because at the equivalence point of a titration of acetic acid with sodium hydroxide, the solution contains sodium acetate. This solution is basic and has pH > 7.0.

We consider now the change in pH during the titration of 25 mL of 0.100M acetic acid solution with a 0.100M sodium hydroxide solution. We can conveniently consider four stages in the titration.

1. *Initially*, we have a 0.100M solution of CH_3CO_2H. We can calculate the pH of this solution as described in Example 14.12, and we find pH = 2.9.

2. *Between the initial acid solution and the equivalence point*, insufficient NaOH has been added to neutralize all the CH_3CO_2H, so the solution contains both CH_3CO_2H and CH_3CO_2Na. These solutions are therefore *buffer solutions*, and their pH may be calculated as shown in Examples 14.23 and 14.24.

3. *At the equivalence point* the addition of 25.00 mL of 0.100M NaOH solution to 25.00 mL of 0.100M CH_3CO_2H solution gives a 0.050M solution of CH_3CO_2Na. The pH of this solution is calculated to be 8.87 using the method given in Example 14.19.

4. *After the equivalence point is reached*, we are adding strong base, NaOH, to a solution of CH_3CO_2Na. The excess OH^- represses the ionization of $CH_3CO_2^-$ as a weak base:

$$CH_3CO_2^- + H_2O \rightleftharpoons CH_3CO_2H + OH^-$$

So the additional OH^- from this equilibrium is negligible compared with the added OH^-. Therefore the pH can be calculated simply from the concentration of OH^- added after the equivalence point.

The calculations of the pH for selected points in the titration of 25 mL of 0.100M CH_3CO_2H with 0.100M NaOH are summarized in Table 14.10. The complete curve is shown in Figure 14.10.

From the figure we see that methyl red changes color long before the equivalence point, so methyl red is of no use for detecting the equivalence point. In contrast, phenolphthalein changes color at the equivalence point. Furthermore, at this point the pH changes very rapidly with the volume of added NaOH. Therefore the volume of NaOH solution needed to reach the equivalence point can be determined accurately, that is, to within one drop of the added solution. Thus phenolphthalein, but not methyl red, is a suitable indicator for this titration. *The only suitable indicators are those that have a color change over a pH range that falls within the steep part of the titration curve close to the equivalence point.*

Table 14.10 Change in pH During Titration of 25.0 mL of 0.100M CH$_3$CO$_2$H with 0.100M NaOH at 25°C

CH$_3$CO$_2$H (mL), V_a	NaOH (mL), V_b	n_b (mmol OF SALT, CH$_3$CO$_2$Na)	n_a (mmol OF UNREACTED ACID)	$\dfrac{n_b}{n_a}$	[H$_3$O$^+$] (μmol L^{-1})*	pH
25.0	0.00	0.00	2.50	—	134.0	2.87[a]
	5.00	0.50	2.00	0.25	72.0	4.14[b]
	10.0	1.00	1.50	0.67	26.9	4.57[b]
	12.5	1.25	1.25	1.00	18.0	4.74[b]
	15.0	1.50	1.00	1.50	12.0	4.92[b]
	20.0	2.00	0.50	4.00	4.5	5.35[b]
	22.0	2.20	0.30	7.40	2.4	5.61[b]
	23.0	2.30	0.20	11.50	1.6	5.81[b]
	24.0	2.40	0.10	34.00	0.8	6.12[b]
	25.0	25.0	0.00	—	1.9×10^{-6}	8.72[c]

			m_a (mmol OF BASE)		[OH$^-$] (mmol L^{-1})	
	26.0	2.50	0.10		1.96	11.29[d]
	27.0	2.50	0.20		3.85	11.59[d]
	30.0	2.50	0.50		9.09	11.96[d]
	35.0	2.50	1.00		16.67	12.22[d]
	40.0	2.50	1.50		23.08	12.36[d]
	50.0	2.50	2.50		33.33	12.52[d]

*Note: 1 μmol = 10^{-6} mol.

[a] 0.100 M solutions of CH$_3$CO$_2$H.
[b] Buffer solutions.
[c] 0.050M solution of CH$_3$CO$_2$Na.
[d] Solutions containing excess base and CH$_3$CO$_2$Na.

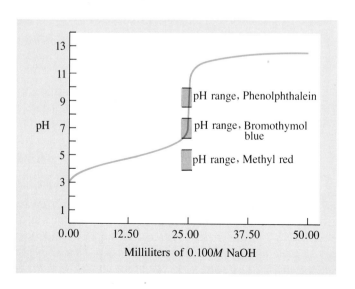

Figure 14.10 Titration of 25.00 mL of 0.100M CH$_3$CO$_2$H with 0.100M NaOH. Of the three indicators, only phenolphthalein could be used to accurately determine the equivalence point of this titration.

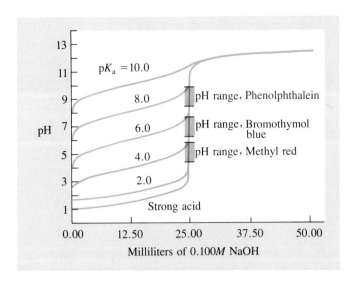

Figure 14.11 Titration of 25.00 mL of 0.100M Solutions of Weak Acids with 0.100M NaOH. In the titration of very weak acids, $K_a \leq 10^{-8}$, the pH changes only slowly in the vicinity of the equivalence point, so no indicator will give a sharp color change.

Figure 14.11 shows how the shape of the titration curve changes with the dissociation constant of the acid being titrated. When the acid is very weak, $K_a < 10^{-8}$, the slope of the curve in the vicinity of the equivalence point is not steep enough to cause any indicator to change color rapidly. So titration using an indicator is not a good method for determining the concentration of a solution of a very weak acid. But for strong acids and for many common weak acids and bases, titrations are a widely used method for determining the concentrations of solutions of acids and bases.

Titration of a Weak Base with a Strong Acid

The change in the pH during a titration of a weak base, such as NH_3, with a strong acid, such as HCl, can be calculated in just the same way as illustrated for the titration of a weak acid with a strong base. At the equivalence point we have a solution of NH_4Cl, and the pH is less than 7. A typical titration curve is shown in Figure 14.12.

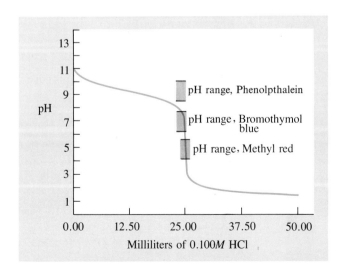

Figure 14.12 Titration of 25.00 mL of 0.100M NH_3 with 0.100M HCl.

14.10 POLYPROTIC ACIDS

An acid that can donate more than one proton to a base is called a **polyprotic acid**. In Chapter 7 we saw that sulfuric acid, H_2SO_4, is a *diprotic acid* and that phosphoric acid, H_3PO_4, is a *triprotic acid*. In Chapter 11 we saw that carbonic acid, H_2CO_3, is a diprotic acid.

Phosphoric acid has three distinct acid dissociation constants.

1. Dissociation of phosphoric acid:

$$H_3PO_4(aq) + H_2O(l) \rightleftharpoons H_3O^+(aq) + H_2PO_4^-(aq)$$

$$K_{a_1}(H_3PO_4) = \frac{[H_3O^+][H_2PO_4^-]}{[H_3PO_4]} = 7.1 \times 10^{-3} \text{ mol L}^{-1}$$

2. Dissociation of the dihydrogen phosphate ion:

$$H_2PO_4^-(aq) + H_2O(l) \rightleftharpoons H_3O^+(aq) + HPO_4^{2-}(aq)$$

$$K_{a_2}(H_3PO_4) = K_a(H_2PO_4^-) = \frac{[H_3O^+][HPO_4^{2-}]}{[H_2PO_4^-]} = 6.2 \times 10^{-8} \text{ mol L}^{-1}$$

3. Dissociation of the monohydrogen phosphate ion:

$$HPO_4^{2-}(aq) + H_2O(l) \rightleftharpoons H_3O^+(aq) + PO_4^{3-}(aq)$$

$$K_{a_3}(H_3PO_4) = K_a(HPO_4^{2-}) = \frac{[H_3O^+][PO_4^{3-}]}{[HPO_4^{2-}]} = 4.4 \times 10^{-13} \text{ mol L}^{-1}$$

Each successive acid dissociation constant of a polyprotic acid is approximately 10^{-5} times the value of the preceding one. Each successive proton is more difficult to remove because of the increased charge of the anion from which it is being removed.

Let us see now how we calculate the concentrations of all the species in a $0.100M$ aqueous solution of H_3PO_4. We consider first the first dissociation:

$$H_3PO_4(aq) + H_2O(l) \rightleftharpoons H_3O^+(aq) + H_2PO_4^-(aq)$$

Initial concentrations	0.100	0	0	mol L^{-1}
Equilibrium concentrations	$0.100 - x$	x	x	mol L^{-1}

$$K_{a_1}(H_3PO_4) = \frac{[H_3O^+][H_2PO_4^-]}{[H_3PO_4]} = \frac{x^2}{0.100 - x} \text{ mol L}^{-1} = 7.1 \times 10^{-3} \text{ mol L}^{-1}$$

Solving the quadratic equation for x (see Appendix A) gives $x = 0.023$, and therefore $[H_3O^+] = 0.023$ mol L^{-1}. Hence

$$[H_2PO_4^-] = 0.023 \text{ mol L}^{-1} \quad \text{and} \quad [H_3PO_4] = 0.077 \text{ mol L}^{-1}$$

Note that the approximation that $x \ll 0.1$ is not valid in this case because $x > (5\%)[H_3PO_4]$.

You might object that these concentrations cannot be correct because we have not taken into account that some of the $H_2PO_4^-$ is dissociated to HPO_4^{2-}, and in principle you would be justified. However, the differences in the dissociation constants K_{a_1}, K_{a_2}, and K_{a_3} are so great that the concentrations of HPO_4^{2-} and PO_4^{3-} formed are negligibly small, as we will now show.

We have

$$K_{a_2}(H_3PO_4) = K_a(H_2PO_4^{2-}) = \frac{[H_3O^+][HPO_4^{2-}]}{[H_2PO_4^-]} = 6.2 \times 10^{-8} \text{ mol L}^{-1}$$

$$H_2PO_4^- + H_2O \rightleftharpoons H_3O^+ + HPO_4^2$$

Initial concentrations	0.023	0.023	0	mol L^{-1}
Equilibrium concentrations	$0.023 - x$	$0.023 + x$	x	mol L^{-1}

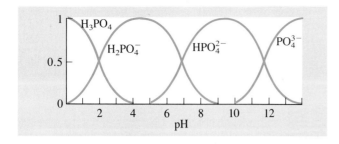

Figure 14.13 Distribution Diagram for H_3PO_4. This diagram shows how the amount of each species expressed as a fraction, α, of the total varies with the pH of the aqueous solution, for a total concentration of $0.10M$.

Since x is very small, we can write

$$0.023 - x \simeq 0.023 + x \simeq 0.023$$

Hence

$$[HPO_4^{2-}] = K_{a_2} \frac{[H_2PO_4^-]}{[H_3O^+]} = \frac{(6.2 \times 10^{-8})(0.023)}{0.023} \text{ mol L}^{-1}$$

$$= 6.2 \times 10^{-8} \text{ mol L}^{-1}$$

Thus we see that $[HPO_4^{2-}]$ is negligible compared with $[H_2PO_4^-] = 0.023$ mol L^{-1}.

Similarly, we have

$$[PO_4^{3-}] = \frac{K_{a_3}[HPO_4^-]}{[H_3O^+]} = \frac{(4.4 \times 10^{-13})(6.2 \times 10^{-8})}{0.023} \text{ mol L}^{-1}$$

$$= 1.2 \times 10^{-18} \text{ mol L}^{-1}$$

which is again negligible compared with $[H_2PO_4^-] = 0.023$.

Thus the concentrations of the various species in a $0.100M$ H_3PO_4 solution are

$$[H_3PO_4] = 0.077M \qquad [H_3O^+] = 0.023M \qquad [H_2PO_4^-] = 0.023M$$
$$[HPO_4^{2-}] = 6.2 \times 10^{-8}M \qquad [PO_4^{3-}] = 1.2 \times 10^{-18}M$$

The titration curve of phosphoric acid has three equivalence points corresponding to the three dissociations (see Figure 14.14). A simple way to show how the concentrations of the species in H_3PO_4 solution depend on the pH is by means of a *distribution diagram*, such as the one shown in Figure 14.13, in which the concentrations of the species are plotted against the pH. This diagram shows that although three separate dissociation equilibria are involved, only one of the dissociation equilibria is important at any given pH value, because the three acid dissociation constants have very different values.

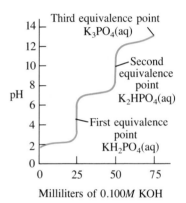

Figure 14.14 The titration of 25.00 mL of phosphoric acid, H_3PO_4, with 0.100M KOH solution. Phosphoric acid is a triprotic acid, so there are three equivalence points. The first equivalence point can be detected using methyl red as an indicator, and the second could be detected using phenolphthalein. The pH changes very slowly in the vicinity of the third equivalence point because the third dissociation of phosphoric acid is very weak. So this equivalence point cannot be accurately determined with an indicator.

IMPORTANT TERMS

The **acid dissociation constant** for a weak acid in water is given by the expression

$$K_a = \left(\frac{[H_3O^+][A^-]}{[HA]} \right)_{eq}$$

The **base dissociation constant** for a weak base in water is given by the expression

$$K_b = \left(\frac{[BH^+][OH^-]}{[B]} \right)_{eq}$$

A **buffer solution** has the ability to resist changes in pH. A buffer solution can be prepared from a weak acid and a salt of the weak acid or from a weak base and a salt of the weak base.

Equilibrium is said to have been reached in a chemical reaction when the rate of the forward reaction is equal to the rate of the back reaction and no further changes in the concentrations of the reactants or the products take place.

The **equilibrium constant** is an expression involving the concentrations of the reactants and the concentrations of the products; it has a constant value for a system at equilibrium. For the reaction

$$aA + bB + \cdots \rightleftharpoons pP + qQ + \cdots$$

the equilibrium constant expression is

$$K = \left(\frac{[P]^p[Q]^q \cdots}{[A]^a[B]^b \cdots}\right)_{eq}$$

An **indicator** is a weak acid or base whose color depends on the pH of the solution, because its acid and base forms have different colors.

The **ionic product constant of water**, K_w, is the product of the equilibrium concentrations of H_3O^+ and OH^-. $K_w = ([H_3O^+][OH^-])_{eq}$. This is also called the **autoprotolysis constant** of water.

Le Châtelier's principle states that if any of the conditions that affect a system at equilibrium are changed, the equilibrium shifts so as to minimize the change.

pH is a convenient method for expressing the concentration of H_3O^+ in a solution. It is defined as

$$pH = -\log_{10}[H_3O^+]$$

A **polyprotic acid** has more than one ionizable hydrogen.

The **reaction quotient** is a ratio of concentration terms that has the same form as the equilibrium constant expression, but the concentrations are *not* the equilibrium concentrations. Its value indicates the direction in which a reaction will proceed in order to establish equilibrium.

A **titration** is a procedure in which the volume of a solution needed to react completely with a known volume of another solution may be determined.

PROBLEMS

Gas Phase Equilibria

1. What are the equilibrium constant expressions K_c and K_p for each of the following reactions?

(a) $2N_2O_5(g) \rightleftharpoons 4NO_2(g) + O_2(g)$

(b) $2SO_2(g) + O_2(g) \rightleftharpoons 2SO_3(g)$

(c) $SO_2(g) + \frac{1}{2}O_2(g) \rightleftharpoons SO_3(g)$

(d) $P_4(g) + 5O_2(g) \rightleftharpoons P_4O_{10}(g)$

2. The reaction of nitrogen with oxygen to give nitrogen monoxide,

$$N_2 + O_2 \rightleftharpoons 2NO$$

has an equilibrium constant of 2.5×10^{-3} at 2100°C.

(a) What is the expression for the equilibrium constant?

(b) What are the units of the equilibrium constant?

3. The equilibrium constant for the equilibrium

$$N_2O_4(g) \rightleftharpoons 2NO_2(g)$$

at 100°C is 0.212 mol L^{-1}. What is the value of the equilibrium constant for the same reaction written as follows?

(a) $2NO_2(g) \rightleftharpoons N_2O_4(g)$

(b) $NO_2(g) \rightleftharpoons \frac{1}{2}N_2O_4(g)$

4. At 425°C, $K_c = 300 \text{ mol}^{-2} L^2$ for the reaction in which methanol, CH_3OH, is synthesized from hydrogen and carbon monoxide:

$$2H_2(g) + CO(g) \rightleftharpoons CH_3OH(g)$$

If the initial concentrations of H_2, CO, and CH_3OH are 0.10 mol L^{-1}, is the system at equilibrium?

5. When the reaction

$$2SO_2(g) + O_2(g) \rightleftharpoons 2SO_3(g)$$

had reached equilibrium, the concentrations of the reactants and the products were found to be $[SO_2] = 0.010M$, $[SO_3] = 0.100M$, and $[O_2] = 0.20M$. What is the value of K_c?

6. Dinitrogen tetraoxide dissociates to nitrogen dioxide according to the equation

$$N_2O_4(g) \rightleftharpoons 2NO_2(g)$$

In a mixture of the two gases at 100°C the concentrations were found to be $[N_2O_4] = 0.10M$ and $[NO_2] = 0.12M$.

(a) What is the value for the reaction quotient of the mixture?

(b) Given that $K_c = 0.212 \text{ mol } L^{-1}$ at 100°C, is the system at equilibrium?

(c) If not, will $[NO_2]$ increase or decrease as equilibrium is achieved?

(d) What are the final equilibrium concentrations of NO_2 and N_2O_4? What is the value of K_p at 100°C for this reaction?

7. What are the equilibrium expressions, K_c and K_p, for each of the following heterogeneous reactions?

(a) $C(s) + O_2(g) \rightleftharpoons CO_2(g)$

(b) $MgCO_3(s) \rightleftharpoons MgO(s) + CO_2(g)$

(c) $2NaHCO_3(s) \rightleftharpoons$
$$Na_2CO_3(s) + CO_2(g) + H_2O(g)$$

(d) $ZnO(s) + CO(g) \rightleftharpoons Zn(s) + CO_2(g)$

(e) $3Fe(s) + 4H_2O(g) \rightleftharpoons Fe_3O_4(s) + 4H_2(g)$

Le Châtelier's Principle

8. The value of the equilibrium constant for a reaction increases as the temperature is increased. Is the forward reaction exothermic or endothermic?

9. The dissociation of $PCl_5(g)$ to $PCl_3(g)$ and $Cl_2(g)$ is an endothermic reaction. What is the effect on the position of equilibrium of each of the following changes?

(a) Compressing the gaseous mixture.

(b) Increasing the volume of the gaseous mixture.

(c) Decreasing the temperature.

(d) Adding $Cl_2(g)$ to the equilibrium mixture at constant volume.

10. For each of the following reactions, calculate the standard enthalpy change for the forward reaction. What is the effect on the position of equilibrium of changing temperature and pressure? What combination of temperature and pressure will maximize the yield of products? (Calculate the enthalpy changes from the standard enthalpies of formation given in Appendix B.)

(a) $2NO(g) + Cl_2(g) \rightleftharpoons 2NOCl(g)$

(b) $2SO_2(g) + O_2(g) \rightleftharpoons 2SO_3(g)$

(c) $N_2(g) + 3H_2(g) \rightleftharpoons 2NH_3(g)$

(d) $CO(g) + H_2O(g) \rightleftharpoons CO_2(g) + H_2(g)$

11. What will be the effect on the equilibrium concentration of SO_3 in the equilibrium

$$2SO_2(g) + O_2(g) \rightleftharpoons 2SO_3(g) \qquad \Delta H° = -198 \text{ kJ}$$

of the following changes?

(a) Doubling the volume of the reaction vessel.

(b) Increasing the temperature without changing the volume of the reaction vessel.

(c) Adding more O_2 to the reaction vessel.

(d) Adding He to the reaction vessel.

Acid-Base Equilibria

12. What are the molar concentrations of the ions in each of the following solutions?

(a) $10^{-5}M \text{ HNO}_3$ (b) $0.0023M$ HCl

(c) $0.113M \text{ HClO}_4$ (d) $0.034M$ HBr

(e) $10^{-3}M$ NaOH   (f) $0.145M$ Ba(OH)$_2$

13. What are the H_3O^+ concentration and the percentage dissociation of a $0.010M$ solution of hydrocyanic acid $(K_a = 4.9 \times 10^{-10} \text{ mol L}^{-1})$?

14. Muscles may ache after strenuous exercise because lactic acid $(K_a = 8.4 \times 10^{-4} \text{ mol L}^{-1})$ is formed faster than it is metabolized to CO_2 and H_2O. What is the pH of the fluid in a muscle when the lactic acid concentration reaches $1.0 \times 10^{-3} \text{ mol L}^{-1}$?

15. What is the OH^- concentration and what is the percentage dissociation of a $0.010M$ solution of ammonia in water?

16. What is the pH value of each of the acidic solutions in Problem 12?

17. What is the percentage dissociation and the OH^- concentration of a $0.10M$ solution of aniline $(pK_b = 9.4)$?

18. What is the pH value of each of the basic solutions in Problem 12?

19. What is the pH of each of the following solutions?

(a) A solution prepared by adding 25.0 mL of $0.10M$ KOH to 50.0 mL of $0.080M$ HNO_3.

(b) A solution prepared by adding 25.0 mL of $0.13M$ HCl to 35.0 mL of $0.12M$ NaOH.

(c) A solution prepared by adding 35.0 mL of $0.050M$ HNO_3 to 70.0 mL of $0.025M$ NaOH.

*20. The pK_a values of acetic acid, monochloracetic acid, $ClCH_2CO_2H$, dichloroacetic acid, Cl_2CHCO_2H, and trichloroacetic acid, Cl_3CCO_2H, are 4.8, 2.9, 1.3, and 0.7, respectively. Arrange these acids in order of acid strength. What is the pH of a $0.10M$ solution of each acid?

21. The pH of $0.10M$ hypochlorous acid, HClO, is 4.2. What is the value of K_a for hypochlorous acid?

22. What is the pH of a $0.0050M$ aqueous solution of KF?

23. What are the equilibrium concentrations of NH_4^+, NH_3, H_3O^+, and OH^- in a $0.020M$ solution of NH_4Cl?

24. What is the pH of each of the following solutions?

(a) $0.010M$ acetic acid

(b) $0.10M$ hydrogen fluoride

(c) $0.0030M$ ammonia

(d) $0.10M$ sodium acetate

(e) $0.20M$ ammonium chloride

25. Suggest a suitable acid-base reaction for preparing each of the following salts. Would you expect $0.1M$ solutions of these salts to be acidic, basic, or neutral? Explain your answer in each case.

(a) Ammonium nitrate   (b) Ammonium chloride

(c) Calcium sulfate   (d) Potassium acetate

(e) Aluminum chloride   (f) Lithium iodide

26. Write balanced equations for the reactions that occur when the following solids are dissolved in water:

(a) Na_2O (b) K_2S (c) $NaNH_2$

Indicators

27. At what pH value is the ratio of the basic form, In^-, to the acidic form, HIn, of the indicator methyl orange $(pK_a = 4.2)$ (a) 5:1, (b) 1:1, (c) 1:5?

28. Estimate the pH of a colorless aqueous solution that turns yellow when methyl orange is added and yellow when bromothymol blue is added.

29. Estimate the pH of a colorless aqueous solution that remains colorless when phenolphthalein is added but turns blue when bromothymol blue is added.

30. The indicator thymol blue turns green over the pH range 7.9 to 9.4. Estimate the pK_a value for thymol blue.

Buffer Solutions

31. A buffer solution contains $0.30M$ ammonia and $0.25M$ ammonium chloride. What is its pH?

32. A buffer solution is made from equal volumes of $0.10M$ hydrogen cyanide and $0.10M$ sodium cyanide. What is its pH?

33. What ratio of masses of ammonia and ammonium chloride must be used to prepare a buffer solution with a pH of 9.0?

34. What is the pH of a solution prepared by adding 25 mL of $0.10M$ NaOH to 50 mL of $0.10M$ acetic acid?

35. What is the pH of a solution prepared by adding 15 mL of $0.010M$ ammonia to 25 mL of $0.010M$ ammonium chloride?

36. What is the change in pH when 1.00 mL of $1M$ NaOH is added to 0.10 L of a solution containing $0.18M$ NH_3 and $0.10M$ NH_4Cl?

37. Use the Henderson-Hasselbalch equation to find the ratio of acetic acid to acetate ion that must be used to prepare a buffer solution of pH 4.50.

38. A phosphate buffer is frequently used in biological experiments. What is the pH of the phosphate buffer solution prepared by dissolving 3.40 g of KH_2PO_4 and 3.55 g of NaH_2PO_4 in sufficient water to give 1.00 L of solution?

Titrations

39. Consider the titration of 10.0 mL of $0.010M$ hypochlorous acid, HOCl, with $0.010M$ potassium hydroxide. Calculate the pH of the solution after the following volumes of base have been added:

(a) 0.0 mL   (b) 1.0 mL   (c) 5.0 mL

(d) 9.9 mL   (e) 10.0 mL   (f) 10.1 mL

(g) 11.0 mL

40. Repeat Problem 39 for the titration of 10 mL of $0.010M$ ammonia with $0.010M$ hydrochloric acid.

41. From Table 14.8, select indicators suitable for detecting the endpoints of the following titrations:

(a) $0.1M$ NaOH with $0.1M$ HCl

(b) $0.1M$ acetic acid with $0.1M$ NaOH

(c) $0.1M$ ammonia with $0.1M$ HCl

Miscellaneous

42. Calculate the equilibrium concentrations of the ions CO_3^{2-}, HCO_3^-, OH^-, and H_3O^+ for a $0.20M$ solution of Na_2CO_3 in water (see Table 14.4 for data).

43. What is the pH of the following solutions?

(a) A solution prepared by adding 25.00 mL of $0.102M$ NaCl to 25.00 mL of $0.112M$ HCl.

(b) A solution prepared by adding 25.00 mL of $0.0120M$ NaOH to 10.00 mL of $0.0150M$ H_2SO_4.

(c) A solution prepared by adding 50.00 mL of $0.124M$ HCl to 25.00 mL of $0.084M$ NH_3.

44. What is the pH of $0.010M$ aqueous solutions of each of the following?

(a) KOH   (b) KCl   (c) NH_3

(d) HF   (e) KF   (f) HNO_3

45. What is the pH of a buffer solution prepared by mixing 25.0 mL of a $0.020M$ solution of aniline, $C_6H_5NH_2$, with 10.0 mL of $0.030M$ solution of the salt $C_6H_5NH_3^+Cl^-$? How would the pH of the solution change if you added 1.0 mL of $0.040M$ HNO_3 to the solution? How would the pH change if instead you added 2.0 mL of $0.030M$ KOH to the solution?

46. Which of the following solutions act as buffers?

(a) 25 mL of $0.10M$ HNO_3 and 25 mL of $0.10M$ $NaNO_3$.

(b) 25 mL of $0.10M$ acetic acid and 25 mL of $0.05M$ KOH.

(c) 25 mL of $0.10M$ acetic acid and 25 mL of $0.15M$ KOH.

47. What is the pH of the solution in the titration of 25.0 mL of $0.020M$ acetic acid after the following amounts of $0.050M$ KOH solution have been added?

(a) 0.0 mL   (b) 5.0 mL   (c) 9.9 mL

(d) 10.0 mL   (e) 12.0 mL

48. What is the solubility of phenol, C_6H_5OH, in water at 25°C if the pH of a saturated solution of phenol is 4.90? K_a(phenol) $= 1.6 \times 10^{-10}$ mol L^{-1}.

$$C_6H_5OH + H_2O \rightleftharpoons H_3O^+ + C_6H_5O^-$$

49. One liter of dinitrogen tetraoxide, N_2O_4, has a mass of 2.50 g at 60°C and 1.00 atm pressure.

(a) What is the percentage dissociation into NO_2?

(b) What is the value of K_p for the reaction

$$N_2O_4(g) \rightleftharpoons 2NO_2(g)?$$

50. Sulfur dioxide and oxygen, in the ratio 2 mol:1 mol, are allowed to reach equilibrium in the presence of a catalyst at a pressure of 5.00 atm. At equilibrium 33% of the SO_2 is converted to SO_3. What is the equilibrium constant, K_p, for the reaction?

$$2SO_2(g) + O_2(g) \rightleftharpoons 2SO_3(g)$$

CHAPTER 15
THE ALKALI AND ALKALINE EARTH METALS

Solubility and Precipitation Reactions

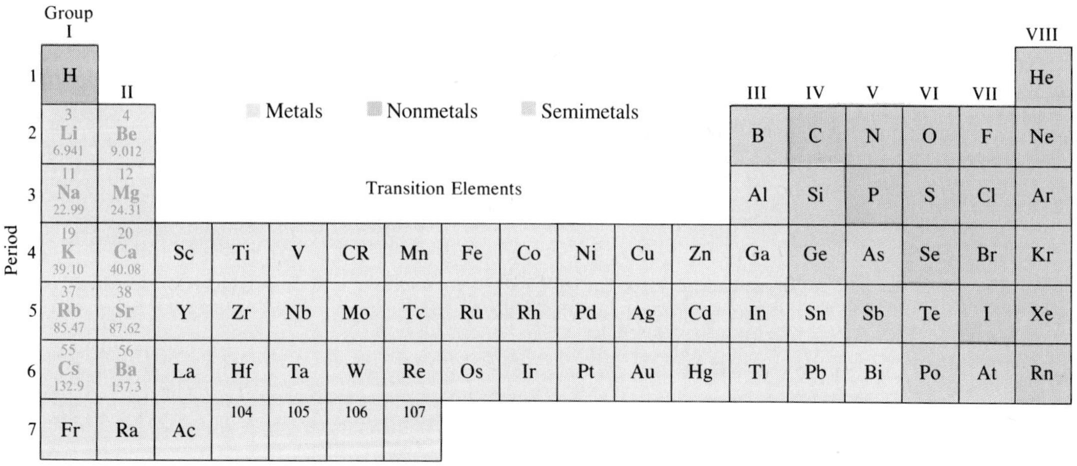

The elements of group I, the *alkali metals*, are lithium, sodium, potassium, rubidium, cesium, and francium. The elements of group II, the *alkaline earth metals*, are beryllium, magnesium, calcium, strontium, barium, and radium. Four of these elements, namely calcium, sodium, potassium, and magnesium, are among the ten most abundant in the earth's crust (Table 3.2). In this chapter we will concentrate mainly on these four elements.

We have previously mentioned that sodium chloride and limestone, $CaCO_3$, are two of the most important raw materials on which a large part of the chemical industry is based. They are used, for example, for the production of calcium oxide (lime), sodium hydroxide, and sodium carbonate, which are among the top twenty chemicals produced by the chemical industry (see Table 15.1) and which have a multitude of uses. Because they are soluble in water and are

15.1 THE ELEMENTS

555

Table 15.1 Top 20 Industrial Chemicals Produced in the United States (1983)

1. Sulfuric acid
2. Nitrogen
3. Calcium oxide (lime)
4. Oxygen
5. Ethene
6. Ammonia
7. Sodium hydroxide
8. Chlorine
9. Phosphoric acid
10. Sodium carbonate
11. Nitric acid
12. Propene
13. Ammonium nitrate
14. Urea
15. Dichloroethane
16. Benzene
17. Ethylbenzene
18. Carbon dioxide
19. Toluene
20. Styrene

readily available, the sodium and potassium salts of the common acids are widely used compounds. Potassium ion, K^+, is essential to nearly all forms of life, both plant and animal. Sodium ion, Na^+, also is essential to higher animals, for whom it is the main cation in the fluids outside the cells; K^+ is the main cation inside the cells.

In this chapter we consider the properties and reactions of the alkali and alkaline earth metals and their compounds. Almost all the salts of the alkali metals are soluble in water, but several common salts of the alkaline earth metals have rather low solubilities. To discuss these solubilities in a quantitative way, we use the ideas of chemical equilibrium from Chapter 14 to develop the concept of the solubility product.

15.1 THE ELEMENTS

Physical Properties

The alkali and alkaline earth metals have typical metallic properties. They are good conductors of heat, have high electrical conductivities, which increase with decreasing temperature, are mostly malleable and ductile, and have a shiny metallic luster when freshly cut (see Experiment 15.1). All the elements in these

EXPERIMENT 15.1

Physical Properties of the Alkali Metals.

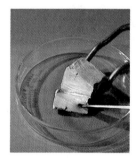

Sodium is soft enough to be cut with a knife. The bright metallic luster soon dulls as the surface is oxidized.

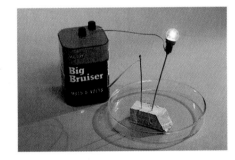

Sodium is good conductor of electricity.

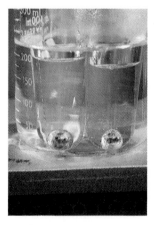

Sodium melts at 97.5°C and potassium melts at 62.3°C. The tubes contain sodium (left) and potassium (right) covered with paraffin to protect them from oxidation by the oxygen in the atmosphere. When the tubes are heated in boiling water, both sodium and potassium melt to form a shiny metallic ball that looks like mercury.

two groups, with the exception of magnesium and beryllium, are rather soft and mechanically weak, and they react with the atmosphere; consequently, they are of no importance as structural materials. However, magnesium is an important constituent of many lightweight structural alloys.

The alkali metals have several unusual properties. Unlike the alkaline earth metals and most other typical metals, the alkali metals have low densities and low melting points (see Tables 15.2 and 15.3). In fact, lithium is the lightest of all metals. Lithium, sodium, and potassium are all less dense than water, which is very unusual for a metal. The melting points of all the alkali metals are considerably lower than those of almost all other metals. The melting point of cesium is so low (29°C) that it becomes a liquid on a hot day. The only metal that has a lower melting point is mercury (-30°C). Alloys of two or more alkali metals have melting points that are even lower than those of the pure metals. Sodium-potassium alloys containing 40%–90% potassium are liquid at room temperature. They are used in certain types of nuclear reactors as coolants. Liquid metals are particularly useful as coolants because they have a high heat conductivity, a high heat capacity, and very high boiling points.

The alkali metals and barium have the same structure as iron, namely, the monatomic body-centered cubic structure (Chapters 9 and 10). In this structure each atom has only eight close neighbors, as compared with twelve in the cubic and hexagonal close-packed structures that are the other common structures for metals. The metallic bonding in the alkali metals is relatively weak, because they have only one valence electron per atom available for metallic bonding. In contrast, the alkaline earth metals have two valence electrons, and other metals have two or more valence electrons available for metallic bonding. In each period the atomic radii decrease from left to right with increasing core charge. Thus in a given period the alkali metal has the largest atom and therefore a low density.

When a metal is melted, the metallic bonds holding the atoms together are

Table 15.2 Properties of the Alkali Metals

ELEMENT	DENSITY (g cm^{-3})	MELTING POINT (°C)	BOILING POINT (°C)	ΔH_{fus} (kJ mol^{-1})	ΔH_{vap} (kJ mol^{-1})
Lithium	0.53	186	1336	2.9	148
Sodium	0.97	97.5	880	2.6	99
Potassium	0.86	62.3	760	2.4	79
Rubidium	1.53	38.5	700	2.2	76
Cesium	1.87	28.5	670	2.1	66

Table 15.3 Properties of the Alkaline Earth Metals

ELEMENT	DENSITY (g cm^{-3})	MELTING POINT (°C)	BOILING POINT (°C)	ΔH_{fus} (kJ mol^{-1})	ΔH_{vap} (kJ mol^{-1})
Beryllium	1.86	1280	2970	12	309
Magnesium	1.74	650	1100	9	128
Calcium	1.55	850	1490	8	151
Strontium	2.6	770	1380	9	139
Barium	3.6	710	1140	7.7	151

not broken, because the mobile electron cloud that holds the atomic cores together can move with the moving atoms. The melting of a metal therefore requires only a slight weakening of the bonds as the atoms move a little further apart than they were in the solid. For the alkali metals, which have relatively weak metallic bonding, melting requires only a small amount of energy, and their melting points and enthalpies of fusion are quite low (see Table 15.2). However, when a liquid metal is vaporized, individual atoms or small molecules are formed, and the majority of the metallic bonds are broken. Thus a rather large amount of energy is needed to vaporize a metal, and the boiling points and enthalpies of vaporization are always high (see Table 15.2), although lower for the alkali metals, because of their relatively weak bonding, than for other metals.

The alkali metals have only one electron in their valence shells, whereas the alkaline earth metals have two (Tables 15.4 and 15.5). The first column in Table 15.4 gives the energy needed to remove the most easily removed electron to give an M^+ ion. This energy is called the *first ionization energy*. For the alkali metals, the most easily removed electron is an s electron. The second column gives the energy needed to remove the next most easily removed electron to give an M^{2+} ion. This energy is called the *second ionization energy*. For an alkali metal atom this second electron must come from a completed inner shell, and therefore it has a much higher ionization energy. Thus one, but not two, electrons may be rather easily removed from an alkali metal atom. For the alkaline earth metals the two s electrons can be rather easily removed—the first and second ionization energies are relatively low. But a third electron can only be removed with much greater difficulty, because it must come from a completed inner shell. Thus the third ionization energy is much larger than the first and second (Table 15.5). Hence the alkali and alkaline earth metals are readily oxidized to M^+ and M^{2+}, respectively. The ionization energies decrease from the top to the bottom of each group as the size of the atoms increases and the valence electrons are therefore at an increasingly greater distance from the core. The increasing size of the atoms and the decreasing ionization energy with increasing atomic number means that the electrons in the mobile charge cloud are at a greater distance from the nuclei and are therefore less strongly attracted. Hence the strength of the metallic bonding and the melting and boiling points decrease from the top to the bottom of both groups.

Recall that when we previously discussed ionization energies in Chapter 6, we considered the removal of *one* electron only but not necessarily the most

Table 15.4 Properties of the Alkali Metal Atoms

| | IONIZATION ENERGIES (MJ mol^{-1}) | | | | |
| | 1ST | 2ND | ATOMIC RADIUS (pm) | IONIC RADIUS (M^+) (pm) | ELECTRON CONFIGURATION |
ELEMENT	$M \rightarrow M^+$	$M^+ \rightarrow M^{2+}$			
Lithium	0.54 (2s)	7.28 (1s)	155	90	[He] 2s^1
Sodium	0.51 (3s)	4.56 (2p)	190	116	[Ne] 3s^1
Potassium	0.42 (4s)	3.06 (3p)	235	152	[Ar] 4s^1
Rubidium	0.40 (5s)	2.65 (4p)	248	166	[Kr] 5s^1
Cesium	0.38 (6s)	2.42 (5p)	267	188	[Xe] 6s^1

Table 15.5 Properties of the Alkaline Earth Metal Atoms

ELEMENT	IONIZATION ENERGIES (MJ mol^{-1})			ATOMIC RADIUS (pm)	IONIC RADIUS (M^{2+}) (pm)	ELECTRON CONFIGURATION
	1ST $M \rightarrow M^+$	2ND $M^+ \rightarrow M^{2+}$	3RD $M^{2+} \rightarrow M^{3+}$			
Beryllium	0.90 (2s)	1.75 (2s)	14.81 (1s)	112	41	[He] 2s^2
Magnesium	0.74 (3s)	1.45 (3s)	7.72 (2p)	160	86	[Ne] 3s^2
Calcium	0.59 (4s)	1.15 (4s)	4.93 (3p)	197	114	[Ar] 4s^2
Strontium	0.55 (5s)	1.06 (5s)	4.14 (4p)	215	132	[Kr] 5s^2
Barium	0.50 (6s)	0.96 (6s)	3.47 (5p)	222	149	[Xe] 6s^2

easily removed electron. The ionization energies in Table 6.1 are for the removal of one electron from the various orbitals of a neutral atom. These neutral-atom ionization energies also show that removing the 3s electron of sodium (0.50 MJ mol^{-1}) is much easier than removing a 2p electron (3.67 MJ mol^{-1}). In Tables 15.4 and 15.5 the ionization energies are for the successive removal of electrons from the same atom. The most weakly held electron is removed first, then the next most weakly held electron, and so on. Thus the second electron is removed from an M^+ ion, the third from an M^{2+} ion, and so on.

Flame Tests

When the alkali and alkaline earth metals and their compounds are strongly heated in a flame, they impart characteristic bright colors to the flame (see Table 15.6). Only beryllium and magnesium do not give colored flames. As we saw in Chapter 6, the colors arise from atoms that have been raised to excited states at the high temperature of the flame; atoms in these excited states may lose their excess energy by emitting light of characteristic wavelengths. Compounds of these elements contain the metals as positive ions in the solid state; but when they are heated to a high temperature in a flame, they dissociate into gaseous atoms, not ions. For example, at high temperatures sodium chloride dissociates into Na and Cl atoms, not into Na$^+$ and Cl$^-$ ions. Hence the compounds exhibit the same characteristic flame colors as the elements. These colored flames provide a very convenient qualitative test for these elements in mixtures and compounds (see Experiments 6.1 and 15.2).

Occurrence

The alkali metals commonly occur as chlorides in enormous salt deposits that were formed by the evaporation of ancient seas. They are also present in vast

Table 15.6 Flame Colors Produced by the Alkali and Alkaline Earth Metals

METAL	FLAME COLOR	METAL	FLAME COLOR
Li	Red	Cs	Blue
Na	Yellow	Ca	Orange red
K	Lilac	Sr	Deep red
Rb	Purple	Ba	Pale green

Flame Tests for the Alkaline Earth Metals

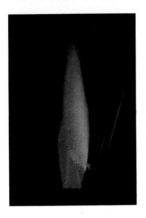

Calcium compounds give an orange red color.

Strontium compounds give a deep red color.

Barium compounds give a pale green color.

Table 15.7 Concentrations of Ions in Seawater

ELEMENT	CONCENTRATION (mol L^{-1})
Na^+	0.48
Mg^{2+}	0.05
Ca^{2+}	0.01
K^+	0.01
Cl^-	0.51
SO_4^{2-}	0.07
HCO_3^-	0.002

quantities, but in less accessible form, in the world's oceans (see Table 15.7). Some inland seas from which there is no outlet contain much higher concentrations of sodium chloride. For example, the Dead Sea between Israel and Jordan contains about 20% by mass of sodium chloride plus large amounts of other salts.

When an inland sea evaporates, the solid that is deposited is not a uniform mixture. Instead, because the least soluble salts are deposited first, followed by the more soluble ones, a typical salt deposit consists of several layers. Each of these layers consists of a relatively pure salt such as *rock salt*, NaCl, *sylvite*, KCl, and *carnallite*, $KMgCl_3 \cdot 6H_2O$. These salt deposits are mined either by conventional methods or by forcing water into them to form a concentrated salt solution, *brine*, which is pumped to the surface and evaporated.

Magnesium occurs not only as the chloride in seawater and in salt deposits but also as the carbonate in the minerals *magnesite*, $MgCO_3$, and *dolomite*, $MgCa(CO_3)_2$, and as the sulfate *epsomite*, $MgSO_4 \cdot 7H_2O$.

Calcium occurs widely and abundantly as the carbonate, phosphate, sulfate, and fluoride. Naturally occurring calcium carbonate, $CaCO_3$, exists in a wide variety of forms including *chalk, limestone,* and *marble.* Chalk has been subjected to the least pressure during its formation and is the softest $CaCO_3$ rock. Limestone is a more highly compressed form, while marble, which is the hardest, has been subjected to such high temperatures and pressures that it has melted and subsequently recrystallized. Marble is an impure form of the crystalline form of calcium carbonate, called *calcite.* Marble may contain many other minerals, and it consequently varies greatly in color and texture. Pure colorless crystals of calcite also occur naturally.

The enormous deposits of these forms of calcium carbonate that we find on the earth all had their origin in the sea. The weathering of calcium silicate rocks over millions of years by water and wind gradually changed insoluble calcium silicate into soluble calcium salts, which were carried by rivers into the sea. The calcium dissolved in the sea was used by small sea creatures to form their shells, which are chiefly calcium carbonate. As these animals died, deposits of calcium carbonate were laid down on the ocean floor. Eventually, the deposits were compressed by layers of other deposits above them to form sedimentary rock. Calcium carbonate deposits continue to be laid down on the ocean floor today. Phosphate rock, a mixture of *fluorapatite*, $Ca_5(PO_4)_3F$, *chlorapatite*, $Ca_5(PO_4)_3Cl$, and *hydroxyapatite*, $Ca_5(PO_4)_3(OH)$, occurs widely; as we saw in Chapter 7, it is the most important source of phosphorus. Hydroxyapatite is the principal constituent of bones in humans and animals.

The alkali and alkaline earth metals also occur widely as complex silicates and aluminosilicates. But because it is difficult to obtain the elements or other useful compounds of the alkali or alkaline earth metals from these minerals, they are not important as sources of the elements.

Preparation

Since the elements occur exclusively in the form of the M^+ and M^{2+} ions, these ions must be reduced in order to prepare the elements. The very low ionization energies of the alkali and alkaline earth metals indicate that they are very easily oxidized to their positive ions; thus the ions are difficult to reduce to the elemental form.

For many years sodium was obtained by reducing sodium carbonate with carbon at a high temperature:

$$Na_2CO_3(s) + 2C(s) \rightleftharpoons 2Na(l) + 3CO(g)$$

But today sodium is made entirely by electrolysis of molten sodium chloride, as described in Chapter 16. In electrolysis sodium ions are reduced by free electrons, which are the strongest possible reducing agents:

$$Na^+ + e^- \longrightarrow Na$$

Potassium can be made by electrolysis of molten potassium chloride, but it is nearly all made by using sodium as the reducing agent:

$$KCl(l) + Na(l) \longrightarrow NaCl(l) + K(g)$$

Molten sodium is introduced at the bottom of a column of molten potassium chloride, and potassium vapor escapes at the top. The reaction is driven to the right because potassium is more volatile than sodium (see Table 15.2). The alkaline earth metals are usually prepared by electrolysis (see Chapter 16).

Reactions

The chemistry of the alkali and alkaline earth metals is relatively simple. Because one or two electrons are easily removed from their atoms to form the M^+ and M^{2+} ions, respectively, there is only one stable oxidation state for each group, $+1$ for the alkali metals and $+2$ for the alkaline earth metals. With the exception of a very few compounds, primarily those of lithium and beryllium, all their compounds are ionic.

As we might expect from their low ionization energies, all these elements are very reactive. They react directly with many other elements and compounds. In Chapter 5 we saw that they react with the halogens to give ionic halides, such as NaCl and $CaBr_2$. Except for beryllium, they react with hydrogen on heating to give ionic hydrides, such as NaH and CaH_2. Lithium, magnesium, calcium, strontium, and barium even react with nitrogen to give ionic nitrides containing the N^{3-} ion, such as Li_3N and Ca_3N_2. The reactions with hydrogen and nitrogen, particularly, illustrate the great reactivity and strong reducing properties of these metals. Molecular nitrogen, which has a strong triple bond (see Table 12.2), is a rather unreactive substance that does not react directly with many elements or compounds.

The alkali and alkaline earth metals all react readily with oxygen, and except for beryllium and magnesium, they must be stored out of contact with the atmosphere, usually under kerosene or some other unreactive hydrocarbon. The reactions with oxygen are more complicated than we might have expected. Only lithium, beryllium, magnesium, and calcium give the expected oxides, Li_2O, BeO, MgO, and CaO, in a pure state when they react with excess oxygen under ordinary conditions. The other metals give peroxides such as Na_2O_2, which contain the O_2^{2-} ion and superoxides such as KO_2, which contain the O_2^- ion. The peroxide and superoxide ions are discussed in Chapter 17.

Some of the most important reactions of the alkali and alkaline earth metals are summarized in Tables 15.8 and 15.9. In each case the element is converted to an ionic compound containing an M^+ or M^{2+}, so that these are all oxidation-reduction reactions.

In the next two sections we discuss some of the more important compounds of the alkali and alkaline earth metals.

Table 15.8 Some Reactions of the Alkali Metals

With oxygen

$4Li(s) + O_2(g) \longrightarrow 2Li_2O(s)$

$2Na(s) + O_2(g) \longrightarrow Na_2O_2(s)$

$K(s) + O_2(g) \longrightarrow KO_2(s)$

$Rb(s) + O_2(g) \longrightarrow RbO_2(s)$

$Cs(s) + O_2(g) \longrightarrow CsO_2(s)$

With halogens[a]

$2M(s) + X_2(g) \longrightarrow 2MX(s)$

With water

$2M(s) + 2H_2O(l) \longrightarrow 2MOH(aq) + H_2(g)$

With hydrogen

$2M(s) + H_2(g) \longrightarrow 2MH(s)$

Note: Except for the reactions with water, most of these reactions only proceed at a reasonable rate at temperatures above room temperature.

[a] M denotes an alkali metal and X a halogen.

Table 15.9 Some Reactions of the Alkaline Earth Metals

With oxygen

$2M(s) + O_2(g) \longrightarrow 2MO(s)$

$Ba(s) + O_2(g) \longrightarrow BaO_2(s)$

With halogens

$M(s) + X_2 \longrightarrow MX_2(s)$

With water

$Mg(s) + H_2O(g) \longrightarrow MgO(s) + H_2(g)$

$M(s) + 2H_2O(l) \longrightarrow M(OH)_2(s) + H_2(g)$

With hydrogen

$M(s) + H_2(g) \longrightarrow MH_2(s)$

With nitrogen[a]

$3M(s) + N_2(g) \longrightarrow M_3N_2(s)$

Note: Except for the reactions of Ca, Sr, and Ba with water, most of these reactions only proceed at a reasonable rate at temperatures above room temperature. M denotes Mg, Ca, Sr, or Ba unless otherwise specified.

[a] Except Be.

Hydroxides

The alkali metals react with water to give hydrogen and solutions of the metal hydroxide. For example,

$$2Na(s) + 2H_2O(l) \longrightarrow 2Na^+(aq) + 2OH^-(aq) + H_2(g)$$

The reactions of the alkali metals with water are quite vigorous and become increasingly so from lithium to cesium (see Experiments 4.1 and 15.3). A small piece of sodium floats when dropped onto water and moves around on the surface in a spectacular way, propelled by the hydrogen evolved and the currents set up in the water by the large amount of heat produced. With potassium, which also floats on the surface, the reaction is even more spectacular, since the heat evolved is sufficient to ignite the hydrogen, which burns with a flame colored violet by the potassium. The reactions of cesium and rubidium are still more violent and are dangerous to demonstrate.

The alkali metal hydroxides can also be produced by reaction of the metal oxides with water; for example,

$$Li_2O(s) + H_2O(l) \longrightarrow 2Li^+(aq) + 2OH^-(aq)$$

This is an acid-base reaction, not an oxidation-reduction reaction, since there is no change in the oxidation number of any of the elements involved. If we write the equation in terms of ions, we see that the reaction is a simple transfer of a proton from a water molecule to an oxide ion:

$$O^{2-} + H_2O \longrightarrow 2OH^-$$

The oxide ion behaves as a strong base in water and is quantitatively converted to the hydroxide ion (Chapters 5 and 14).

However, because preparing the alkali metals and their oxides is expensive and difficult, the hydroxides are made commercially by the electrolysis of an

EXPERIMENT 15.3

Reactions of Some Alkali and Alkaline Earth Metals with Water

Lithium—the lightest of all metals—floats on water, reacting less vigorously than sodium and producing hydrogen.

Calcium is heavier than water and sinks to the bottom of the beaker. A rapid stream of hydrogen bubbles rises to the surface.

Calcium hydroxide, $Ca(OH)_2$, is the other product of the reaction. Because it is not very soluble, a white precipitate is soon formed.

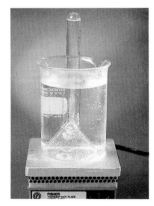

Magnesium does not react with cold water but it does react with hot water. The hydrogen produced collects at the top of the tube.

aqueous solution of the alkali metal chloride. This process is described in Chapter 16.

The alkali metal hydroxides are all white crystalline ionic solids containing the metal ion, M^+, and the hydroxide ion, OH^-. They are all *deliquescent*; that is, they absorb water very readily from the atmosphere to form a concentrated aqueous solution of the hydroxide. They behave as strong bases in water, dissolving to give a solution of M^+ and OH^- ions.

Sodium hydroxide is an important strong base. It is used extensively in the production of many chemicals, textiles, cleaners, soap, and paper and pulp and in petroleum refining. In industry and commerce sodium and potassium hydroxides are frequently called *caustic soda* and *caustic potash*.

Many salts of the alkali metals are easily and conveniently prepared by the reaction of an alkali metal hydroxide with the appropriate acid. For example, lithium nitrate may be prepared by the reaction between lithium hydroxide, $LiOH$, and nitric acid, HNO_3:

$$HNO_3(aq) + LiOH(aq) \longrightarrow LiNO_3(aq) + H_2O(l)$$
$$\text{Acid} \qquad\quad \text{Base} \qquad\qquad \text{Salt}$$

An important property of the alkali metal salts—in particular, the halides, sulfates, carbonates, nitrates, and the phosphates—is that they are all soluble in water. Since sodium and potassium salts are relatively inexpensive, they are the salts normally used to carry out reactions involving these anions in aqueous solution.

We now consider some of the more important alkali metal salts.

Halides

The alkali metal halides are all colorless, crystalline ionic compounds that are rather soluble in water. Some of the properties of the fluorides and chlorides were given in Table 5.4. All the alkali metal halides, except $CsCl$, $CsBr$, and CsI, have the sodium chloride structure. Cesium chloride, bromide, and iodide all have the cesium chloride structure. These structures were discussed in Chapter 10 (Figures 10.22 and 10.23).

The halides of the alkali metals are readily available compounds, most of which are found in natural salt deposits. They can be obtained by the direct reaction between the alkali metal and a halogen (see Experiment 5.4); for example,

$$2K(s) + Br_2(l) \longrightarrow 2KBr(s)$$

But they would not normally be made in this way, because the starting materials are expensive and the reactions are inconveniently vigorous. They may be more easily prepared by the reaction between the hydroxide and the appropriate acid; for example,

$$CsOH(aq) + HI(aq) \longrightarrow CsI(aq) + H_2O$$

Sodium chloride is the usual source of almost all other compounds containing sodium and chloride. Its most important uses are for the manufacture of hydrochloric acid (see Chapter 7) and of sodium hydroxide and chlorine (see Chapter 16). Sodium chloride is an essential component of the diet of humans and animals.

Hydrides

The hydrides of the alkali metals are white crystalline ionic compounds containing M^+ and H^- ions; they have the sodium chloride structure. They can be made by heating the element in hydrogen.

These hydrides are all powerful reducing agents. They react vigorously with water, liberating H_2 and forming OH^-:

$$H^- + H_2O \longrightarrow OH^- + H_2$$

The hydride ion, H^-, accepts a proton from water to become H_2. It is therefore a strong base in water. But this reaction is also an oxidation-reduction reaction. The hydride ion (oxidation number -1) is oxidized by the hydrogen in water (oxidation number $+1$) to give H_2 (oxidation number 0).

Nitrides

The nitrides of the alkali metals are white crystalline solids containing M^+ and N^{3-} ions. They can be made by heating the element in nitrogen. They react with water, forming ammonia:

$$Li_3N(s) + 3H_2O(l) \longrightarrow NH_3(aq) + 3LiOH(aq)$$

Nitride ion is a strong base in water and is completely protonated.

Carbonates

Potassium and sodium carbonates have been known since very early times. Potassium carbonate was obtained from the ashes left after the burning of wood and other land plants—hence the common name *potash*. Plants take up potassium from the soil, and in agriculture this must be replaced by adding potassium-containing fertilizers. Land plants make little use of sodium, but it is present in considerable quantities in marine plants. Thus the ashes from the burning of seaweed (kelp) were an early source of sodium carbonate (soda ash). They were also a source of iodine (see Chapter 5).

Sodium carbonate, Na_2CO_3, is a very important industrial chemical, which in 1983 ranked tenth among the industrial chemicals produced in the United States (see Table 15.1). It is used in the manufacture of glass, paper, detergents, and soap. In the home it is used for washing, for cleaning, and for softening "hard" water.

Although sodium carbonate occurs naturally, until recently the known deposits were insufficient to meet the enormous demand, and most sodium carbonate was made from limestone and sodium chloride by the *Solvay process*. There are several stages in the process, but the overall reaction is

$$CaCO_3(s) + 2NaCl(aq) \longrightarrow Na_2CO_3(aq) + CaCl_2(aq)$$

Recently, however, very large deposits of the mineral *trona*, $Na_5(CO_3)_2 \cdot (HCO_3) \cdot 2H_2O$, were discovered in Wyoming, and most of the sodium carbonate used in North America comes from this source. The trona is simply crushed and heated. Heating causes the reaction

$$2HCO_3^- \longrightarrow CO_3^{2-} + H_2O + CO_2$$

to proceed to the right as gaseous H_2O and CO_2 are evolved. Thus the equation for the overall reaction is

$$2Na_5(CO_3)_2(HCO_3) \cdot 2H_2O \longrightarrow 5Na_2CO_3 + CO_2 + 5H_2O$$

The crude sodium carbonate obtained in this way is dissolved in water, the insoluble impurities are filtered off, and the sodium carbonate is crystallized, filtered, and heated to give pure anhydrous sodium carbonate, Na_2CO_3.

At temperatures below 35.2°C sodium carbonate crystallizes from water as the hydrated form, $Na_2CO_3 \cdot 10H_2O$, and is familiar in this form under the name *soda crystals* or *washing soda*. Above this temperature the monohydrate $Na_2CO_3 \cdot H_2O$ crystallizes. On standing in the air, $Na_2CO_3 \cdot 10H_2O$ slowly loses water and crumbles to a white powder of $Na_2CO_3 \cdot H_2O$.

Because the carbonate ion is the anion of a weak acid, HCO_3^-, it is a weak base (Chapters 5 and 14):

$$CO_3^{2-} + H_2O \rightleftharpoons HCO_3^- + OH^-$$

Solutions of sodium carbonate are therefore basic (see Example 15.1).

Example 15.1 Suppose 100 g of washing soda is added to sufficient water to give 1.00 L of solution. What is its pH?

Solution Washing soda has the formula $Na_2CO_3 \cdot 10H_2O$ and a molar mass of 286.2 g. The number of moles of sodium carbonate in 1.00 L of solution is therefore

$$\frac{100 \text{ g}}{286.2 \text{ g mol}^{-1}} = 0.349 \text{ mol}$$

The concentration of carbonate ion in the solution is therefore

$$0.349 \text{ mol L}^{-1}$$

To find the pH of this solution, we follow the steps outlined in Chapter 14. We first write the equation for the equilibrium, which, since the carbonate ion is a weak base, is

$$CO_3^{2-} + H_2O \rightleftharpoons HCO_3^- + OH^-$$

We let the concentration of OH^- be x mol L^{-1}. Then we have

	$CO_3^{2-} + H_2O \rightleftharpoons$	HCO_3^- +	OH^-	
Initial concentrations	0.349	0	0	mol L^{-1}
Equilibrium concentrations	$0.349 - x$	x	x	mol L^{-1}

We find the value for the equilibrium constant, $K_b(CO_3^{2-}) = 2.1 \times 10^{-4}$ mol L^{-1}, in Table 14.4, and we substitute the values for the equilibrium concentrations. Thus

$$K_b = \frac{[OH^-][HCO_3^{2-}]}{[CO_3^{2-}]} = 2.1 \times 10^{-4} \text{ mol L}^{-1} = \frac{x^2}{0.349 - x} \text{ mol L}^{-1}$$

Assuming that $x \ll 0.349$, we have

$$x^2 = (0.349)(2.1 \times 10^{-4})$$
$$x = 7.3 \times 10^{-3}$$

and therefore

$$[OH^-] = 7.3 \times 10^{-3} \text{ mol L}^{-1}$$

So

$$[H_3O^+] = \frac{1.0 \times 10^{-14} \text{ mol}^2 \text{ L}^{-2}}{[OH^-]} = \frac{1.0 \times 10^{-14} \text{ mol}^2 \text{ L}^{-2}}{7.3 \times 10^{-3} \text{ mol L}^{-1}} = 1.4 \times 10^{-12} \text{ mol L}^{-1}$$

$$pH = -\log[H_3O^+] = 11.9$$

We see that this solution of washing soda is quite basic. In this calculation, we have neglected the fact that HCO_3^- is also a weak base and will form more OH^-. But since the concentration of HCO_3^- is quite small (9.2×10^{-3} mol L^{-1}), and since HCO_3^- is a considerably weaker base than CO_3^{2-}, the additional OH^- formed by the reaction of the HCO_3^- with water is negligible.

Sodium hydrogen carbonate, $NaHCO_3$ (sodium bicarbonate, bicarbonate of soda, or baking soda), can be made by passing CO_2 into a solution of NaOH:

$$CO_2(g) + NaOH(aq) \longrightarrow NaHCO_3(aq)$$

When $NaHCO_3(s)$ is heated, H_2O and CO_2 are driven off, and $Na_2CO_3(s)$ is formed:

$$2NaHCO_3(s) \longrightarrow Na_2CO_3(s) + H_2O(g) + CO_2(g)$$

Sodium hydrogen carbonate is used in cooking and in medicine and for the manufacture of baking powder, which is a mixture of sodium hydrogen carbonate and a weak acid, for example, sodium dihydrogen phosphate. Baking powder is a leavening agent used in the making of biscuits, cakes, and other baked goods. When it is mixed with water and warmed during cooking, carbon dioxide is formed:

$$H_3O^+ + HCO_3^- \rightleftharpoons CO_2 + 2H_2O$$

The equilibrium is driven to the right by the evolution of CO_2. The carbon dioxide causes the baked goods to "rise," giving them a light texture because of the many "holes" formed by the escaping carbon dioxide.

Solutions of $NaHCO_3$ in water are basic, but they have a lower pH than solutions of Na_2CO_3 because HCO_3^- is a weaker base than CO_3^{2-}.

Sulfates

Sodium sulfate, Na_2SO_4, and *sodium hydrogen sulfate*, $NaHSO_4$, are by-products of the manufacture of hydrogen chloride by the reaction of sulfuric acid with sodium chloride (Chapter 7):

$$NaCl + H_2SO_4 \longrightarrow NaHSO_4 + HCl$$
$$2NaCl + H_2SO_4 \longrightarrow Na_2SO_4 + 2HCl$$

Large amounts of sodium sulfate are used in the paper industry. Sodium hydrogen sulfate gives acidic solutions in water. It is used in the home as a toilet-bowl cleaner, for example.

15.3 COMPOUNDS OF THE ALKALINE EARTH METALS

The alkaline earth metals received this name because their oxides are very stable to heat and insoluble in water but react with water to give alkaline (basic) solutions. The alchemists referred to any nonmetallic substance that was insoluble in water and unaffected by strong heat as an *earth*. The oxides were therefore called *alkaline earths*. The name *alkaline earth metals* for the family of elements was then derived from the name for the oxides, which because of the difficulty of obtaining the metals, were known long before the elements themselves.

We will not consider the compounds of beryllium, because it is a rather rare element whose compounds are not of great practical importance. Moreover, like the first element in other groups, its properties and behavior are not always typical of the group as a whole.

Oxides and Hydroxides

The oxides of the alkaline earth metals are formed when the elements burn in air (Experiment 15.4), although for magnesium this reaction is accompanied by the formation of the nitride, Mg_3N_2:

$$2Mg(s) + O_2(g) \longrightarrow 2MgO(s)$$
$$3Mg(s) + N_2(g) \longrightarrow Mg_3N_2(s)$$

Both these reactions are highly exothermic, and energy is released as both heat and light (see Experiment 1.3). When barium is heated in air or oxygen at $500°-600°C$, the peroxide, BaO_2, is formed (see Chapter 17).

The oxides are normally made by decomposing the carbonates by heat; for example,

$$MgCO_3(s) \longrightarrow MgO(s) + CO_2(g)$$
$$CaCO_3(s) \longrightarrow CaO(s) + CO_2(g)$$

Calcium oxide is usually made in this reaction by heating limestone in a *limekiln* (Figure 15.1); the product, CaO, is known as *lime* (or *quicklime*).

Figure 15.1 Limekiln.

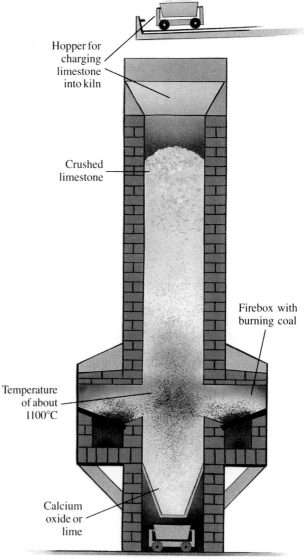

Hopper for charging limestone into kiln

Crushed limestone

Firebox with burning coal

Temperature of about 1100°C

Calcium oxide or lime

The oxides of magnesium, calcium, strontium, and barium all have the sodium chloride crystal structure. Magnesium oxide is insoluble in water, and its melting point is very high because of the strength of the interionic forces. The electrostatic force of attraction between an M^{2+} ion and an X^{2-} ion is four times the force between an M^+ ion and an X^- ion at the same distance apart.

The alkaline earth metal oxides are basic. They react with acids to give salts; for example

$$MgO + 2HCl \longrightarrow MgCl_2 + H_2O$$

Magnesium oxide reacts very slowly with water to give insoluble magnesium hydroxide:

$$MgO + H_2O \longrightarrow Mg(OH)_2$$

Calcium oxide reacts exothermically with water to give *calcium hydroxide*, $Ca(OH)_2$, or *slaked lime*, which has a low solubility in water.

$$CaO(s) + H_2O(l) \longrightarrow Ca(OH)_2(s) \qquad \Delta H^\circ = -65.2 \text{ kJ}$$

Calcium hydroxide can also be made by precipitation from a solution of a soluble calcium salt:

$$CaCl_2(aq) + 2NaOH(aq) \longrightarrow Ca(OH)_2(s) + 2NaCl(aq)$$

A third method of preparation is by the direct reaction of the element with water (see Experiment 15.3):

$$Ca(s) + 2H_2O(l) \longrightarrow Ca(OH)_2(s) + H_2(g)$$

A saturated solution of $Ca(OH)_2$ is called *limewater* and is commonly used as a test for CO_2. When CO_2 is passed through limewater, the solution turns cloudy, because insoluble $CaCO_3$ precipitates:

$$Ca(OH)_2(aq) + CO_2(aq) \longrightarrow CaCO_3(s) + H_2O(l)$$

EXPERIMENT 15.4

The Formation of Calcium Oxide

When calcium is strongly heated in the air it burns with a red flame.

The product is white solid calcium oxide.

But continued passage of CO_2 causes the precipitate to dissolve because of the formation of soluble $Ca(HCO_3)_2$:

$$CaCO_3(s) + CO_2(g) + H_2O(l) \longrightarrow Ca(HCO_3)_2(aq)$$

Calcium hydroxide is a strong base. It is completely dissociated into Ca^{2+} and OH^- in solution:

$$Ca(OH)_2(s) \rightleftharpoons Ca^{2+}(aq) + 2OH^-(aq)$$

Lime and slaked lime are very important commercial bases because of their ready availability and cheapness. Lime is used in making glass and cement. It is used to remove SiO_2 and other acidic oxides in metallurgical processes, such as the smelting of iron (Chapter 9), by combining with them to form a slag, such as calcium silicate, $CaSiO_3$.

Magnesium hydroxide, $Mg(OH)_2$, is only slightly soluble in water. It is conveniently made by adding a soluble hydroxide, such as NaOH, to a solution of a soluble magnesium salt, such as $MgCl_2$ or $MgSO_4$, to give a white precipitate of $Mg(OH)_2$:

$$MgSO_4(aq) + 2NaOH(aq) \longrightarrow Mg(OH)_2(s) + Na_2SO_4(aq)$$

Although magnesium reacts only very slowly with water, it reacts rapidly with steam to give the hydroxide:

$$Mg(s) + 2H_2O(g) \longrightarrow Mg(OH)_2(s) + H_2(g)$$

A burning piece of magnesium will continue to burn brightly in an atmosphere of steam (Experiment 1.3).

A suspension of magnesium hydroxide in water is called *milk of magnesia*; it is used as a medicine to correct excess acidity in the stomach (Box 15.1).

Carbonates

Calcium and magnesium carbonates are minerals and are the most important sources of these elements. Unlike the alkali metal carbonates (other than Li_2CO_3), they decompose when heated, giving the oxides and CO_2:

$$CaCO_3 \xrightarrow{\text{Heat}} CaO + CO_2$$

As we saw in Chapter 14, if $CaCO_3$ is heated in a closed system, it does not completely decompose but comes into equilibrium with CaO and CO_2. When it is heated in the atmosphere, however, the CO_2 escapes and never reaches the equilibrium pressure, so the reaction is driven to completion.

LIMESTONE CAVES, STALACTITES, AND STALAGMITES Large and beautiful caves are a feature of limestone country in many parts of the world. These caves have been formed by water running through the limestone in small streams and occasionally large underground rivers. Although $CaCO_3$ is insoluble in water, it is soluble in acids,

$$CaCO_3(s) + H_3O^+(aq) \longrightarrow Ca^{2+}(aq) + HCO_3^-(aq) + H_2O$$

which convert the insoluble $CaCO_3$ to soluble $Ca(HCO_3)_2$.

Natural water is slightly acidic because it contains dissolved carbon dioxide:

$$2H_2O(l) + CO_2(g) \rightleftharpoons H_3O^+(aq) + HCO_3^-(aq)$$

This acidity is sufficient to convert a small amount of calcium carbonate to soluble calcium hydrogen carbonate:

$$CaCO_3(s) + CO_2(aq) + H_2O(l) \rightleftharpoons Ca^{2+}(aq) + 2HCO_3^-(aq)$$

In an underground stream the dissolved calcium hydrogen carbonate is continually washed away. Thus equilibrium is never reached, so the calcium carbonate continues to dissolve, and over a very long time underground caves may be formed.

A common feature of limestone caves are the stalactites and stalagmites, which are often found in impressively large and beautiful shapes (Figure 15.2). They are formed by the reversal of the reaction that creates the caves. Water that has passed through limestone rock becomes saturated with $Ca(HCO_3)_2$, seeps through to the roof of a cave, and gathers as a drop hanging from the roof. The water slowly evaporates, and some of the CO_2 is lost from the solution. In accordance with Le Châtelier's principle, the equilibrium shifts to the left to produce more CO_2 and H_2O, and some of the Ca^{2+} in solution in redeposited as $CaCO_3$. The evaporation of many drops from the same place over many thousands of years gradually leads to the formation of a conical-shaped stalactite. Moreover, a drop hanging from the roof may fall to the floor, where it evaporates slowly, again depositing a very small amount of $CaCO_3$. If drops continue to hit the same spot, a stalagmite is built up. It may grow sufficiently to join the stalactite, forming a column.

HARDNESS OF WATER Water containing appreciable amounts of Ca^{2+} and Mg^{2+} is called *hard water*. The anions that are usually present with these cations are Cl^-, SO_4^{2-}, and HCO_3^-. Water containing only very small concentrations of Ca^{2+} and Mg^{2+} is called *soft water*.

Most natural water, particularly in limestone regions, is hard. The use of hard water for domestic purposes and in industry presents several problems. For example, soap consists of the sodium salts of long-chain carboxylic acids such as sodium stearate, $C_{17}H_{35}CO_2^-Na^+$. In hard water soap forms an insoluble scum of the insoluble calcium salts of these acids, which has no cleansing power.

Figure 15.2 Stalactites and Stalagmites in Limestone Cave.

15.3 COMPOUNDS OF THE ALKALINE EARTH METALS

When hard water is heated, the reaction by which $CaCO_3$ is dissolved in cave formation is reversed as CO_2 is driven off and $CaCO_3$ is precipitated:

$$Ca^{2+}(aq) + 2HCO_3^-(aq) \longrightarrow CaCO_3(s) + CO_2(aq) + H_2O(l)$$

Calcium sulfate, which is less soluble in hot water than in cold, is also precipitated. A mixture of $CaCO_3$ and $CaSO_4$ is the familiar "scale" that forms inside teakettles and in pipes in hot-water boilers and heating systems. After a sufficiently long time enough deposit may form on the inside of a pipe to block it completely.

Calcium and magnesium can be removed from water—in other words, the water can be "softened"—by adding Na_2CO_3 to precipitate the insoluble carbonates (Experiment 15.5):

$$Ca^{2+} + CO_3^{2-} \longrightarrow CaCO_3(s)$$
$$Mg^{2+} + CO_3^{2-} \longrightarrow MgCO_3(s)$$

If the hardness is due only to $Ca(HCO_3)_2$, it can be removed by simply boiling water to precipitate $CaCO_3$ or by adding enough lime to convert all the $Ca(HCO_3)_2$ to $CaCO_3$:

$$Ca^{2+} + 2HCO_3^- + Ca(OH)_2 \longrightarrow 2CaCO_3(s) + 2H_2O$$

Halides

The halides of the alkaline earth metals are all rather soluble in water, with the exception of CaF_2, SrF_2, and BaF_2, which have very low solubilities. Calcium fluoride occurs widely as the mineral *fluorite*, which is an important source of fluorine. The solubility of CaF_2 is only 0.000 22 mol L^{-1}, but when it is added to drinking water, it gives the low concentration of F^- that helps prevent tooth decay (Chapter 5).

Magnesium and calcium chlorides crystallize from water as $MgCl_2 \cdot 6H_2O$ and $CaCl_2 \cdot 6H_2O$. Anhydrous calcium chloride can be made by heating the hydrate. It combines strongly with water to re-form the hydrate, and it is often used for drying gases. Since calcium chloride is available in large quantities and at a low price as a by-product of the Solvay process, it is widely used to melt ice on the roads in winter in cold climates. However, when $MgCl_2 \cdot 6H_2O$ is heated, a hydrolysis reaction occurs, and the products are MgO and HCl:

EXPERIMENT 15.5

Hardness of Water.

Left: Two drops of soap solution form a thick lather in distilled water. Center: Two drops of soap solution in hard water give no lather, but the solution becomes cloudy because insoluble calcium salts of the soap are precipitated. Right: If sodium carbonate from the dish is added to the water, calcium ion is precipitated as calcium carbonate. Two drops of soap solution then give a small amount of lather.

$$MgCl_2 \cdot 6H_2O(s) \longrightarrow MgO(s) + 2HCl(g) + 5H_2O(g)$$

Because of the small size of Mg^{2+}, the $Mg(H_2O)_6^{2+}$ ion is relatively acidic and donates a proton to the Cl^- ion to give HCl, which is evolved as a gas, driving the reaction to the right. Hydrated aluminium chloride, $AlCl_3 \cdot 6H_2O$, and many other hydrated metal halides behave in the same way on heating.

Calcium, strontium, and barium fluoride have a relatively simple crystal structure known as the *calcium fluoride*, or *fluorite*, *structure* (Figures 10.26 and 15.3a). Like the sodium chloride structure, it is based on a face-centered cubic lattice. Magnesium chloride has a layer structure (Figures 10.21 and 15.3b).

Sulfates

Magnesium sulfate is soluble in water and crystallizes as the hydrate $MgSO_4 \cdot 7H_2O$. It occurs naturally in this form as *epsomite* and is commonly called *Epsom salts*. In medicine it is useful as a purgative.

Calcium sulfate occurs in nature as the mineral *gypsum*, $CaSO_4 \cdot 2H_2O$. When gypsum is heated to a little above 100°C, it loses three-quarters of its water of crystallization, giving a white powder of $CaSO_4 \cdot \frac{1}{2}H_2O$, which is called *plaster of paris*:

$$CaSO_4 \cdot 2H_2O \xrightarrow{\text{Heat}} CaSO_4 \cdot \tfrac{1}{2}H_2O + \tfrac{3}{2}H_2O$$

A paste of plaster of paris and a small amount of water rapidly sets to a solid, which consists of a mass of interlocking crystals of the dihydrate. The formation of the dihydrate is accompanied by an increase in volume, which is very valuable in making plaster casts, since the increase in volume forces the plaster into all the crevices of any mold into which it is poured; thus all the details are faithfully reproduced. Plaster of paris is used in the manufacture of wallboard, as a component of the plaster used for the surface of interior walls and ceilings, and for making blackboard crayons (erroneously called "chalk"). It was formerly used for making plaster casts to immobilize broken bones.

The solubilities of the alkaline earth metal sulfates decrease steadily from $BeSO_4$ and $MgSO_4$, which are very soluble in water, to $CaSO_4$, which is only slightly soluble, to $SrSO_4$ and $BaSO_4$, which are insoluble. Two important uses of $BaSO_4$ depend on its very low solubility. The formation of a precipitate of $BaSO_4$, when a solution of a soluble barium salt, such as $BaCl_2$, is added to a solution to be tested, indicates the presence of SO_4^{2-} in the solution (Experiment 15.6):

$$Ba^{2+}(aq) + SO_4^{2-}(aq) \longrightarrow BaSO_4(s)$$

Magnesium chloride

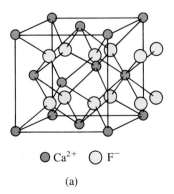

Ca^{2+} F$^-$

(a)

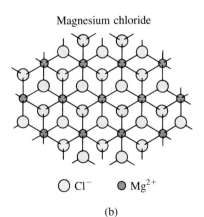

Cl$^-$ Mg^{2+}

(b)

Figure 15.3 Structures of Calcium Fluoride and Magnesium Chloride. (a) The calcium fluoride structure is based on a face-centered cubic lattice. The calcium ions are at the lattice points and have a cubic close-packed arrangement. The fluoride ions occupy all the tetrahedral holes. (b) Magnesium chloride has a layer structure. Each magnesium ion is surrounded by an octahedral arrangement of six chloride ions. The chloride ions form layers, with the magnesium ions sandwiched between them.

Some Insoluble Alkaline Earth Metal Compounds

A solution of ammonium carbonate added to a solution of calcium chloride produces a white precipitate of calcium carbonate.

A solution of sodium sulfate added to a solution of barium chloride produces a heavy white precipitate of barium sulfate.

A solution of sodium hydroxide added to a solution of magnesium chloride produces a gelatinous precipitate of magnesium hydroxide.

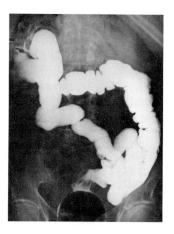

Figure 15.4 X-Ray Photograph of Large Intestine. The intestine has been coated with barium sulfate to render it opaque to X rays.

Like all elements of high atomic number, barium absorbs X rays strongly. If a suspension of barium sulfate is swallowed, it coats the alimentary canal. Then X ray photographs of the alimentary canal, which is normally transparent to X rays, can be obtained (Figure 15.4). Unlike calcium and magnesium, barium is poisonous, but the solubility of barium sulfate is so small that it passes through the body without being absorbed or having any harmful effects.

15.4 SOLUBILITY AND PRECIPITATION REACTIONS

Solubility Rules

We have seen that the hydroxides and all the salts of the alkali metals are soluble in water, whereas many of the hydroxides and salts of the alkaline earth metals have very small solubilities; these insoluble compounds include $Mg(OH)_2$, $CaCO_3$, CaF_2, $CaSO_4$, $BaCO_3$, and $BaSO_4$ (see Experiment 15.6). In previous chapters we encountered other insoluble salts such as $AgCl$, $PbSO_4$, and FeS and insoluble hydroxides such as $Cu(OH)_2$ and $Pb(OH)_2$. We can now give some rules for the solubilities of ionic compounds (Table 15.10). These rules apply to all the common cations that we have considered so far; namely, NH_4^+, Na^+, K^+, Mg^{2+}, Ca^{2+}, Sr^{2+}, Ba^{2+}, Al^{3+}, Fe^{2+}, Fe^{3+}, Cu^{2+}, and Pb^{2+}, as well as the cations Cr^{3+}, Mn^{2+}, Co^{2+}, Ni^{2+}, Ag^+, Hg^{2+}, and Hg_2^{2+}, which we will consider in Chapter 21, and to the common anions Cl^-, Br^-, I^-, NO_3^-, SO_4^{2-}, CO_3^{2-}, PO_4^{3-}, ClO_4^-, OH^-, and S^{2-}. Unfortunately, no simple and general explanations of these solubility differences exist. However, some observations may be helpful.

In general, as a consequence of Coulomb's law, ions having multiple charges are more strongly held together than ions having single charges, and smaller ions are more strongly held together than larger ions. Thus ionic substances with small, highly charged ions tend to be less soluble than ionic substances

Table 15.10 Solubilities of Some Common Salts and Hydroxides in Water at 25°C

| SOLUBLE[a] | EXCEPTIONS | |
	INSOLUBLE	SPARINGLY SOLUBLE
Na^+, K^+, and NH_4^+ salts		
Nitrates		
Perchlorates		
Chlorides, bromides, and iodides	Ag^+, Pb^{2+}, Hg_2^{2+}	$PbCl_2$, $PbBr_2$
Sulfates	Sr^{2+}, Ba^{2+}, Pb^{2+}	Ca^{2+}, Ag^+, Hg_2^{2+}
Acetates	Ag^+, Hg_2^{2+}	

| INSOLUBLE[a] | EXCEPTIONS | |
	SOLUBLE	SPARINGLY SOLUBLE
Carbonates[b]	Na^+, K^+, NH_4^+	
Phosphates[b]	Na^+, K^+, NH_4^+	
Sulfides	Na^+, K^+, NH_4^+, Mg^{2+}, Ca^{2+}, Sr^{2+}, Ba^{2+}	
Hydroxides	Na^+, K^+, NH_4^+, Ba^{2+}	Ca^{2+}, Sr^{2+}

[a] The classifications *soluble*, *sparingly soluble*, and *insoluble* are somewhat arbitrary. Their approximate significance is as follows: soluble ≥ 0.1 mol L^{-1}; sparingly soluble ≤ 0.1 mol L^{-1} and ≥ 0.01 mol L^{-1}; insoluble ≤ 0.01 mol L^{-1}.

[b] Many HCO_3^- and $H_2PO_4^-$ salts are soluble.

composed of large, singly charged ions. For example, the compounds of singly charged cations—the alkali metals and NH_4^+—are all soluble. And the salts of Cl^-, Br^-, I^-, NO_3^-, and ClO_4^- are all soluble, except for the salts of some M^{2+} ions and Ag^+. However, most hydroxides are insoluble. Also, we note that with few exceptions salts of the doubly charged ions CO_3^{2-}, S^{2-}, and PO_4^{3-} are insoluble. However, most sulfates are soluble. Note that although the salts of the multiply charged CO_3^{2-} and PO_4^{3-} ions are insoluble, many salts of the singly charged HCO_3^- and $H_2PO_4^-$ ions are soluble.

In addition to acid-base and oxidation-reduction reactions, a third important type of reaction that we have met many times in earlier chapters is a *precipitation reaction*. We can now use the solubility rules to make some predictions about precipitation reactions, as the following example shows.

Example 15.2 Use the solubility rules to predict whether a precipitate will form when the following pairs of aqueous solutions are mixed. Explain and write a net ionic equation for those cases in which a precipitate forms.

(a) $MgCl_2$ and Na_3PO_4

(b) $Al_2(SO_4)_3$ and $Ba(NO_3)_2$

(c) $Ba(NO_3)_2$ and $MgCl_2$

Solution

(a) Possible products are $Mg_3(PO_4)_2$ and NaCl. Since $Mg_3(PO_4)_2$ is insoluble, a precipitation reaction will occur:

$$3Mg^{2+} + 2PO_4^{3-} \longrightarrow Mg_3(PO_4)_2(s)$$

(b) Possible products are $Al(NO_3)_3$ and $BaSO_4$. Since $BaSO_4$ is insoluble, a precipitation reaction will occur:

$$Ba^{2+} + SO_4^{2-} \longrightarrow BaSO_4(s)$$

(c) Possible products are $BaCl_2$ and $Mg(NO_3)_2$. Both of these compounds are soluble, so there will be no reaction. The solution will consist of a mixture of Ba^{2+}, Cl^-, Mg^{2+}, and NO_3^-.

Solubility Product Constant

Strictly speaking, we should have answered parts (a) and (b) of Example 15.2 by stating that a precipitate *may* be formed, because if the solutions that were mixed were very dilute, a precipitate might not have been formed. What concentrations of ions will, in fact, give a precipitate? We can answer this question by considering the equilibrium between a solid salt and its ions in a saturated solution of the salt in a more quantitative manner.

Even if a salt is described as insoluble, there must be small, perhaps very small, concentrations of ions in equilibrium with the undissolved salt. For example, if we have a saturated solution of $CaSO_4$ in equilibrium with some undissolved $CaSO_4$, the Ca^{2+} and SO_4^{2-} ions are continually leaving the crystals and passing into solution, while ions from the solution are continually attaching themselves to the crystals. At equilibrium the rates of the two opposing processes are equal:

$$CaSO_4(s) \rightleftharpoons Ca^{2+}(aq) + SO_4^{2-}(aq)$$

The expression for the equilibrium constant is

$$K_c = \left(\frac{[Ca^{2+}][SO_4^{2-}]}{[CaSO_4(s)]} \right)_{eq}$$

The equilibrium between $CaSO_4(s)$ and $Ca^{2+}(aq)$ and $SO_4^{2-}(aq)$ is an example of a heterogeneous equilibrium because it involves both a solid and a solution. As we saw in Chapter 14, the concentration of a pure solid is a constant, which does not vary, and so it can be incorporated in the equilibrium constant to give another constant, K_{sp}, which is called the **solubility product constant**:

$$K_c[CaSO_4(s)] = K_{sp} = ([Ca^{2+}][SO_4^{2-}])_{eq}$$

In general, for an ionic compound with the formula A_xB_y the equilibrium in a saturated solution can be written as

$$A_xB_y(s) \rightleftharpoons xA^{m+}(aq) + yB^{n-}(aq)$$

The solubility product is then

$$K_{sp} = ([A^{m+}]^x[B^{n-}]^y)_{eq}$$

In other words, the solubility product constant is equal to the product of the concentrations of the ions involved in the equilibrium, each raised to the power of its coefficient in the equation for the equilibrium.

CALCULATING THE SOLUBILITY PRODUCT CONSTANT OF AN IONIC COMPOUND FROM ITS SOLUBILITY We can obtain the value of the solubility product constant for an ionic compound from the measured value of its solubility. For example, the solubility of $CaSO_4$ in water at 25°C is 4.9×10^{-3} mol L^{-1}. This is the concentration of the dissolved $CaSO_4$. But since the $CaSO_4$ is fully ionized to give equal amounts of Ca^{2+} and SO_4^{2-}, the concentration of each of these ions is also 4.9×10^{-3} mol L^{-1}. In other words,

$$[Ca^{2+}] = [SO_4^{2-}] = 4.9 \times 10^{-3} \text{ mol } L^{-1}$$

Therefore

$$K_{sp} = ([Ca^{2+}][SO_4^{2-}])_{eq} = (4.9 \times 10^{-3} \text{ mol } L^{-1})^2$$
$$= 2.4 \times 10^{-5} \text{ mol } L^{-2}$$

Solubility product constants for some common salts and hydroxides are given in Table 15.11. Very insoluble salts have very small values; more soluble salts have larger values. For example, the solubility product constant of the divalent metal hydroxides ranges from 5.0×10^{-3} for the relatively soluble $Ba(OH)_2$ to 1.6×10^{-19} for the much less soluble $Cu(OH)_2$. Many metal sulfides are very insoluble; for example, HgS has the solubility product constant 1.6×10^{-54}.

Table 15.11 Solubility Product Constants, K_{sp} of Some Slightly Soluble Salts at 25°C

Bromides			
$PbBr_2$	4.6×10^{-5}	$Fe(OH)_2$	1.8×10^{-15}
Hg_2Br_2	1.3×10^{-22}	$Fe(OH)_3$	1.0×10^{-38}
$AgBr$	5.0×10^{-13}	$Pb(OH)_2$	4.2×10^{-15}
Carbonates		$Mg(OH)_2$	8.9×10^{-12}
$BaCO_3$	1.6×10^{-9}	$Sr(OH)_2$	3.2×10^{-4}
$CaCO_3$	4.7×10^{-9}	$Zn(OH)_2$	4.5×10^{-17}
$CuCO_3$	2.5×10^{-10}	*Iodides*	
$FeCO_3$	2.1×10^{-11}	PbI_2	8.3×10^{-9}
$PbCO_3$	1.5×10^{-15}	Hg_2I_2	4.5×10^{-29}
$MgCO_3$	1×10^{-15}	AgI	8.5×10^{-17}
Ag_2CO_3	8.2×10^{-12}	*Phosphates*	
$SrCO_3$	7×10^{-10}	$Ba_3(PO_4)_2$	6×10^{-39}
$ZnCO_3$	2×10^{-10}	$Ca_3(PO_4)_2$	1.3×10^{-32}
Chlorides		Ag_3PO_4	1.8×10^{-18}
$PbCl_2$	1.7×10^{-5}	$Sr_3(PO_4)_2$	1×10^{-31}
Hg_2Cl_2	1.1×10^{-18}	*Sulfates*	
$AgCl$	1.7×10^{-10}	$BaSO_4$	1.5×10^{-9}
Fluorides		$CaSO_4$	2.4×10^{-5}
BaF_2	2.4×10^{-5}	$PbSO_4$	1.3×10^{-8}
CaF_2	3.4×10^{-11}	Ag_2SO_4	1.2×10^{-5}
PbF_2	4×10^{-8}	$SrSO_4$	7.6×10^{-7}
MgF_2	8×10^{-8}	*Sulfides*	
SrF_2	7.9×10^{-10}	CuS	8×10^{-37}
Hydroxides		FeS	6.3×10^{-18}
$Al(OH)_3$	5×10^{-33}	PbS	7×10^{-29}
$Ba(OH)_2$	5.0×10^{-3}	HgS	1.6×10^{-54}
$Ca(OH)_2$	1.4×10^{-6}	Ag_2S	5.5×10^{-51}
$Cu(OH)_2$	1.6×10^{-19}	ZnS	1.6×10^{-24}

15.4 SOLUBILITY AND PRECIPITATION REACTIONS

Example 15.3 The solubility of CaF_2 in water at 25°C is found to be 2.05×10^{-4} mol L^{-1}. What is the value of K_{sp} at this temperature?

Solution The equation for the equilibrium is

$$CaF_2(s) \rightleftharpoons Ca^{2+}(aq) + 2F^-(aq)$$

Hence each mole of CaF_2 that dissolves produces 1 mol of Ca^{2+} and 2 mol of F^-. Thus

$$[Ca^{2+}] = 2.05 \times 10^{-4} \text{ mol } L^{-1}$$
$$[F^-] = 4.10 \times 10^{-4} \text{ mol } L^{-1}$$

From the equation for the equilibrium we have

$$K_{sp}(CaF_2) = ([Ca^{2+}][F^-]^2)_{eq}$$

Hence

$$K_{sp} = (2.05 \times 10^{-4})(4.10 \times 10^{-4})^2 \text{ mol}^3 \text{ L}^{-3}$$
$$= 3.45 \times 10^{-11} \text{ mol}^3 \text{ L}^{-3}$$

CALCULATING SOLUBILITY FROM THE SOLUBILITY PRODUCT CONSTANT
We can calculate the solubility of an ionic compound if we know its solubility product constant, as the next example shows.

Example 15.4 The solubility product constant of lead(II) chloride, $PbCl_2$, is 1.7×10^{-5} mol^3 L^{-3} at 25°C. What is the solubility of $PbCl_2$ in water at this temperature?

Solution The equilibrium is

$$PbCl_2(s) \rightleftharpoons Pb^{2+}(aq) + 2Cl^-(aq)$$

We have to find the solubility of $PbCl_2$, so we will let this be x mol L^{-1}. The equation shows that every mole of $PbCl_2$ that dissolves gives 1 mol of Pb^{2+} and 2 mol of Cl^-. Therefore

$$[Pb^{2+}] = x \text{ mol } L^{-1} \quad \text{and} \quad [Cl^-] = 2x \text{ mol } L^{-1}$$

From the equation we can write

$$K_{sp} = ([Pb^{2+}][Cl^-]^2)_{eq}$$

And since we know that $K_{sp} = 1.7 \times 10^{-5}$ mol^3 L^{-3}, we have

$$[Pb^{2+}][Cl^-]^2 = 1.7 \times 10^{-5} \text{ mol}^3 \text{ L}^{-3}$$

Substituting gives

$$x(2x)^2 = 1.7 \times 10^{-5}$$
$$x^3 = 4.3 \times 10^{-6}$$

Taking the cube root of each side yields

$$x = 1.6 \times 10^{-2}$$
$$[Pb^{2+}] = 1.6 \times 10^{-2} \text{ mol } L^{-1}$$

Thus the solubility of $PbCl_2$ is 1.6×10^{-2} mol L^{-1}.

Example 15.4 is somewhat artificial because the solubility product constant is normally obtained from the measured solubility, as we saw in Example 15.3.

More important uses of the solubility product constant include predicting whether a precipitate will form when two solutions are mixed; how much of a given ion remains in solution after a precipitate has formed; and the solubility of a substance in the presence of another having a common ion. We take up these applications of the solubility product constant in the following sections.

Precipitation Reactions

We can use the solubility product to predict whether a precipitate will form when two solutions are mixed. To do this, we calculate the reaction quotient, Q, from the ion concentrations. Recall that Q is calculated from an expression that has the form of the equilibrium constant but in which the concentrations are the initial concentrations, *not* the equilibrium concentrations. In the present case Q has the form of the solubility product and is called the **ion product**. It is the product of the initial concentrations of the ions, each raised to the same power as in the K_{sp} expression. We then compare the value of Q with the value of the solubility product, K_{sp}. There are three possibilities:

- If $Q < K_{sp}$, the solution is not saturated, and no precipitate forms.
- If $Q = K_{sp}$, the solution is just saturated.
- If $Q > K_{sp}$, the solution is supersaturated, and a precipitate will form until the ion product becomes equal to K_{sp}.

For example, if we mix a $2 \times 10^{-4} M$ solution of $AgNO_3$ with an equal volume of a $2 \times 10^{-4} M$ solution of $NaCl$, will a precipitate form? Because the volume of the solution is doubled on mixing equal volumes, the concentrations of each of the ions will be halved. Therefore

$$[Ag^+] = 1 \times 10^{-4} \text{ mol L}^{-1} \quad \text{and} \quad [Cl^-] = 1 \times 10^{-4} \text{ mol L}^{-1}$$

Hence

$$Q = [Ag^+][Cl^-] = 1 \times 10^{-8} \text{ mol}^2 \text{ L}^{-2}$$

From Table 15.11 we find

$$K_{sp}(AgCl) = 1.7 \times 10^{-10} \text{ mol}^2 \text{ L}^{-2}$$

Thus $Q > K_{sp}$, so we conclude that a precipitate will form.

Example 15.5 Equal volumes of a $2 \times 10^{-3} M$ $Pb(NO_3)_2$ solution and a $2 \times 10^{-3} M$ NaI solution are mixed. Will a precipitate of PbI_2 form?

Solution Since equal volumes of the two solutions are mixed, the resulting concentration of Pb^{2+} will be $1 \times 10^{-3} M$ and the resulting concentration of I^- will be $1 \times 10^{-3} M$. Hence

$$Q = [Pb^{2+}][I^-]^2 = (1 \times 10^{-3})(1 \times 10^{-3})^2 = 1 \times 10^{-9} \text{ mol}^3 \text{ L}^{-3}$$

From Table 15.11

$$K_{sp} = 8.3 \times 10^{-9} \text{ mol}^3 \text{ L}^{-3}$$

Therefore

$$Q < K_{sp}$$

and we conclude that no precipitate will form.

Common-Ion Effect

Suppose we have a saturated solution of lead chloride in equilibrium with the solid salt:

$$PbCl_2(s) \rightleftharpoons Pb^{2+}(aq) + 2Cl^-(aq)$$

If we now add sodium chloride or some other soluble chloride to the solution and thus increase the concentration of chloride ion, then according to Le Châtelier's principle, the equilibrium will shift to the left. In other words, more lead chloride will precipitate, and the concentration of lead ions in solution will decrease. The solubility of lead chloride is less in a solution containing chloride ion than it is in pure water. The solubility of any salt is decreased in the presence of another salt that has a common ion. This decrease in solubility is an example of the **common-ion effect**. Any equilibrium in which ions are formed is shifted to the left in a solution that contains an ion common with one of those formed in the equilibrium.

Example 15.6 What is the solubility of $PbCl_2$ in a $1.00M$ HCl solution?

Solution Let the solubility of $PbCl_2$ in $1.00M$ HCl be x mol L^{-1}. The concentration of Pb^{2+} in the solution is then x mol L^{-1}, and the concentration of Cl^- is 1.00 mol L^{-1} (from the HCl) plus $2x$ (from the $PbCl_2$). Thus we have

$$PbCl_2(s) \rightleftharpoons Pb^{2+}(aq) + 2Cl^-(aq)$$

| Equilibrium concentrations | x | $1.00 + 2x$ | mol L^{-1} |

The solubility product expression is

$$K_{sp} = ([Pb^{2+}][Cl^-]^2)_{eq}$$

and the value for the solubility product, from Table 15.11, is 1.70×10^{-5} mol^3 L^{-3}. Substituting this value and the equilibrium concentrations, we have

$$x(1.00 + 2x)^2 \text{ mol}^3 \text{ L}^{-3} = 1.70 \times 10^{-5} \text{ mol}^3 \text{ L}^{-3}$$

Since x will be very small, we can assume that $2x \ll 1.00$ and therefore, to a good approximation, $1.00 + 2x \simeq 1.00$. Hence

$$x(1.00)^2 = 1.70 \times 10^{-5}$$
$$x = 1.70 \times 10^{-5}$$
$$[Pb^{2+}] = 1.70 \times 10^{-5} \text{ mol L}^{-1}$$

The Pb^{2+} concentration is decreased from $1.6 \times 10^{-2}M$ in water (see Example 15.4) to $1.70 \times 10^{-5}M$ in $1M$ HCl. Because of the common ion effect, the solubility of $PbCl_2$ in $1.00M$ HCl is reduced to about a thousandth of its solubility in pure water. Note that our assumption that $2x (= 3.4 \times 10^{-5}) \ll 1.0$ is justified.

The common-ion effect can be useful if we wish to ensure that we have almost completely removed an ion from solution. Radium, the heaviest of the alkaline earth metals, is very rare and highly radioactive. It could therefore be important to remove as much Ra^{2+} as possible from a solution. Since $RaSO_4$ has a very low solubility (6×10^{-6} mol L^{-1}) the concentration of Ra^{2+} could be decreased to 6×10^{-6} mol L^{-1} by precipitating $RaSO_4$. Although this solubility is small, it is not negligible, particularly in view of the fact that radium is highly radioactive. But the concentration of Ra^{2+} in solution is easily reduced to a much lower value by using an excess of SO_4^{2-} to precipitate $RaSO_4$.

Let us assume that we use sulfuric acid to precipitate $RaSO_4$ and that we add sufficient sulfuric acid so that the final concentration of SO_4^{2-} in solution is 1 mol L^{-1}. We can calculate the concentration of Ra^{2+} in solution if we know the solubility product constant for $RaSO_4$. We can find it from the solubility of $RaSO_4$ in water, which is 6.0×10^{-6} mol L^{-1}. Therefore

$$[Ra^{2+}] = [SO_4^{2-}] = 6.0 \times 10^{-6} \text{ mol } L^{-1}$$

Hence

$$K_{sp} = [Ra^{2+}][SO_4^{2-}] = (6.0 \times 10^{-6})^2 \text{ mol}^2 \text{ } L^{-2} = 3.6 \times 10^{-11} \text{ mol}^2 \text{ } L^{-2}$$

We have supposed that the final concentration of SO_4^{2-} is 1.00 mol L^{-1} after the $RaSO_4$ is precipitated by adding excess H_2SO_4. Therefore

$$[Ra^{+2}] = \frac{K_{sp}}{[SO_4^{2-}]} = \frac{3.6 \times 10^{-11} \text{ mol}^2 \text{ } L^{-2}}{1.00 \text{ mol } L^{-1}} = 3.6 \times 10^{-11} \text{ mol } L^{-1}$$

Because of the common-ion effect of the excess SO_4^{2-}, the solubility of $RaSO_4$ is reduced to 4×10^{-11} mol L^{-1}, which is approximately a hundred thousandth of its solubility in pure water.

Dependence of Solubility on pH

An insoluble precipitate may be dissolved by decreasing the concentration of one of the ions in equilibrium so that the ion product becomes less than the solubility product constant. One way of removing one of the ions is by the formation of a weak electrolyte. For example, insoluble carbonates, such as $CaCO_3$, are soluble in acids because CO_3^{2-} combines with the added H_3O^+ to form HCO_3^- and H_2CO_3:

$$CO_3^{2-} + H_3O^+ \rightleftharpoons HCO_3^- + H_2O$$
$$HCO_3^- + H_3O^+ \rightleftharpoons H_2CO_3 + H_2O$$

Then H_2CO_3 decomposes to give H_2O and CO_2:

$$H_2CO_3 \rightleftharpoons H_2O + CO_2$$

The combined effect of these reactions is to decrease the concentration of CO_3^{2-} so that the ion product, $Q = [Ca^{2+}][CO_3^{2-}]$, becomes smaller than the solubility product constant, $K_{sp}(CaCO_3)$. Hence the equilibrium

$$CaCO_3(s) \rightleftharpoons Ca^{2+}(aq) + CO_3^{2-}(aq)$$

shifts to the right. In other words, $CaCO_3$ dissolves until equilibrium is re-established with a higher concentration of Ca^{2+}. We can see from the distribution diagram for carbonic acid (Figure 15.5) that CO_3^{2-} is almost completely converted to HCO_3^- at pH = 8.0. Thus $CaCO_3$ is completely dissolved as

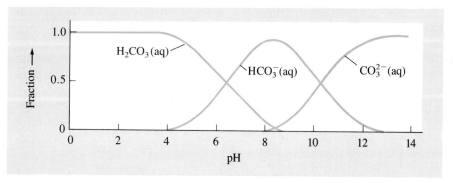

Figure 15.5 Distribution Diagram for Carbonic Acid. Carbonic acid, H_2CO_3, dissociates in water to produce the HCO_3^- and CO_3^{2-} ions. The distribution diagram shows how the concentration of each of these species, expressed as a fraction of the total concentration of all three species, varies with the pH of the solution.

$Ca(HCO_3)_2$ in a solution of pH ≤ 8.0:

$$CaCO_3(s) + 2H_3O^+(aq) \longrightarrow Ca^{2+}(aq) + 2HCO_3^-(aq) + 2H_2O$$

The solubility of a salt of any weak acid such as fluoride or a sulfide depends on the pH of the solution in a similar way. The solubility of a salt of a strong acid such as a chloride, a bromide, or a sulfate does not depend on the pH.

Example 15.7 Calcium fluoride, CaF_2, is insoluble in water but dissolves in dilute hydrochloric acid. Explain why, and write a net ionic equation for the reaction.

Solution In order for CaF_2 to dissolve, either the F^- or Ca^{2+} must be removed by reaction with H_3O^+ or Cl^-, thus shifting the equilibrium,

$$CaF_2(s) \rightleftharpoons Ca^{2+}(aq) + 2F^-(aq)$$

to the right. Because Ca^{2+} is not a base—it has no unshared electron pairs to accept a proton—it does not react with H_3O^+. But because HF is a weak acid, F^- is a weak base; therefore it reacts with H_3O^+ to form HF:

$$F^- + H_3O^+ \longrightarrow HF + H_2O$$

Removal of F^- from the solution causes the equilibrium

$$CaF_2(s) \rightleftharpoons Ca^{2+}(aq) + 2F^-(aq)$$

to shift to the right, and CaF_2 dissolves. Adding these two equations (after multiplying the first by 2) gives an overall equation for the reaction:

$$CaF_2(s) + 2H_3O^+(aq) \longrightarrow 2HF(aq) + 2H_2O(l) + Ca^{2+}(aq)$$

Selective Precipitation

Insoluble metal sulfides can similarly be dissolved in acid:

$$MS(s) + 2H_3O^+(aq) \longrightarrow M^{2+}(aq) + H_2S(g) + 2H_2O(l)$$

The more insoluble the sulfide—in other words, the smaller its solubility product constant—the higher the hydronium ion concentration that will be needed to dissolve it. Figure 15.6 shows the dependence of the solubilities of FeS ($K_{sp} = 6.3 \times 10^{-18}$) and ZnS ($K_{sp} = 1.6 \times 10^{-24}$) on the pH of the solution. If acid is added to a mixture of ZnS(s) and FeS(s), the FeS will dissolve first at a pH between 3 and 4. Or if we were to pass H_2S into a solution of Fe^{2+} and Zn^{2+} at a pH of 2.5, for example, only ZnS would be precipitated. If we had a solution of Fe^{2+} and Zn^{2+}, we could separate them by passing H_2S into the solution at pH = 2.5 to obtain a precipitate of ZnS, which could be filtered off, leaving Fe^{2+} in solution.

The solubility of metal hydroxides depends on the pH. The equilibrium between a solid hydroxide such as $Cu(OH)_2$ and its ions,

$$Cu(OH)_2(s) \rightleftharpoons Cu^{2+}(aq) + 2OH^-(aq)$$

is shifted to the right by the removal of OH^- on the addition of an acid. We can find how the solubility of $Cu(OH)_2$ depends on the pH of the solution as follows: The solubility equilibrium for $Cu(OH)_2$ is

$$Cu(OH)_2(s) \rightleftharpoons Cu^{2+}(aq) + 2OH^-(aq)$$

The solubility product constant is

$$K_{sp} = ([Cu^{2+}][OH^-]^2)_{eq} = 1.6 \times 10^{-19} \text{ mol}^3 \text{ L}^{-3}$$

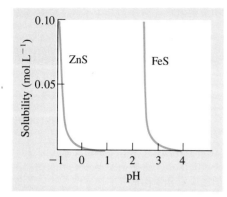

Figure 15.6 Solubilities of ZnS and FeS as a Function of pH. Iron(II) sulfate dissolves at a higher pH than that which is needed for ZnS to dissolve.

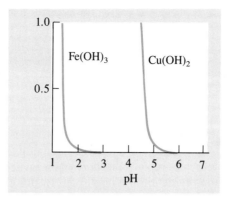

Figure 15.7 Solubilities of Cu(OH)$_2$ and Fe(OH)$_3$ as a Function of pH. Copper(II) hydroxide dissolves at a higher pH than that which is needed for Fe(OH)$_3$.

which we can rearrange as

$$[Cu^{2+}] = \frac{1.6 \times 10^{-19} \text{ mol}^3 \text{ L}^{-3}}{[OH^-]^2}$$

Now since $K_w = [H_3O^+][OH^-] = 1.0 \times 10^{-14} \text{ mol}^2 \text{ L}^{-2}$, then

$$[OH^-] = \frac{1.0 \times 10^{-14}}{[H_3O^+]}$$

Therefore

$$[Cu^{2+}] = \frac{(1.6 \times 10^{-19} \text{ mol}^3 \text{ L}^{-3})[H_3O^+]^2}{(1.0 \times 10^{-14})^2 \text{ mol}^4 \text{ L}^{-4}} = (1.6 \times 10^9 \text{ mol}^{-1} \text{ L})[H_3O^+]^2$$

At pH = 6, $[H_3O^+] = 10^{-6}$ mol L^{-1}, and therefore $[Cu^{2+}] = 1.6 \times 10^{-3}$ mol L^{-1}. In other words, the solubility of Cu(OH)$_2$ is only 1.6×10^{-3} mol L^{-1}. But at pH = 5.0, $[H_3O^+] = 10^{-5}$ mol L^{-1}, and therefore $[Cu^{2+}] = 1.6 \times 10^{-1}$ mol L^{-1}. The solubility of Cu(OH)$_2$ at pH 5 is therefore 0.16 mol L^{-1}, which is much greater than the solubility at pH = 6.0. Figure 15.7 shows a plot of the solubility of Cu(OH)$_2$ and of Fe(OH)$_3$ against the pH. We can see that if we were to adjust the pH to 4.0, Cu(OH)$_2$ would dissolve, but Fe(OH)$_3$, which has a smaller solubility product, would remain undissolved. In this way we could separate a mixture of Cu^{2+} and Fe^{3+}.

Example 15.8 What is the solubility of Fe(OH)$_3$ in an aqueous solution at pH = 3.0 and at pH = 1.5?

Solution For Fe(OH)$_3$ the solubility equilibrium is

$$Fe(OH)_3(s) \rightleftharpoons Fe^{3+}(aq) + 3OH^-(aq)$$

From Table 15.11 the solubility product constant is

$$K_{sp} = ([Fe^{3+}][OH^-]^3)_{eq} = 1.0 \times 10^{-38} \text{ mol}^3 \text{ L}^{-3}$$

Following the same argument we used for copper hydroxide, we deduce the equation

$$[Fe^{3+}] = (1.0 \times 10^4 \text{ mol}^{-2} \text{ L}^2)[H_3O^+]^3$$

At pH 3.0, $[H_3O^+] = 1.0 \times 10^{-3}$ mol L^{-1}, and therefore

$$[Fe^{3+}] = 1.0 \times 10^{-5} \text{ mol L}^{-1}$$

At pH 1.5, $[H_3O^+] = 3.2 \times 10^{-2}$, and therefore

$$[Fe^{3+}] = 0.32 \text{ mol L}^{-1}$$

Thus at pH 3.0 the solubility of $Fe(OH)_3$ is only 1.0×10^{-5} mol L^{-1}, but at pH 1.5 it is 0.32 mol L^{-1}.

We saw in Chapter 9 that a precipitate of $Cu(OH)_2$ will dissolve in concentrated aqueous ammonia to give a deep blue solution that contains the $Cu(NH_3)_4^{2+}$ ion. This ion is analogous to a hydrated ion, except that NH_3 molecules replace the water molecules. It is called a *complex ion* and is discussed in more detail in Chapter 21. In this case the equilibrium

$$Cu(OH)_2(s) \rightleftharpoons Cu^{2+}(aq) + 2OH^-(aq)$$

is shifted to the right. The $Cu(OH)_2$ dissolves because of the decrease in the concentration of $Cu^{2+}(aq)$ caused by the formation of $Cu(NH_3)_4^{2+}$ on the addition of NH_3 to the solution.

IMPORTANT TERMS

The **common-ion effect** is the effect on a system at equilibrium caused by the addition of a substance with an ion common with one involved in the equilibrium.

The **solubility product constant**, K_{sp}, for a slightly soluble ionic substance is the equilibrium constant for the equilib-

rium between the solid and the dissolved ions in a saturated solution. It is the product of the concentrations of the ions in the saturated solution raised to the powers indicated by the coefficients in the balanced equation for the equilibrium.

PROBLEMS

Alkali and Alkaline Earth Metals

1. Why does the reactivity of the alkali metals increase with increasing atomic number?

2. Write balanced equations to show the reactions of lithium and calcium with (a) bromine, (b) sulfur, (c) nitrogen.

3. Complete and balance the following equations:
  (a) $K(s) + Br_2(l) \longrightarrow$
  (b) $Li(s) + O_2(g) \longrightarrow$
  (c) $Na(s) + H_2(g) \longrightarrow$
  (d) $Li(s) + N_2(g) \longrightarrow$
  (e) $LiH(s) + H_2O(l) \longrightarrow$
  (f) $Li_3N(s) + H_2O(l) \longrightarrow$
  (g) $K(s) + H_2O(l) \longrightarrow$

4. Explain how a solution of sodium hydrogen carbonate can act as a buffer solution. What is a common name for sodium hydrogen carbonate?

5. What is lime? What happens when water is added to lime? Why is lime such an important industrial chemical?

6. Write balanced equations for the reaction of magnesium with (a) oxygen, (b) sulfur, (c) nitrogen.

* The asterisk denotes the more difficult problems.

7. Explain why the first ionization energy of sodium (0.51 MJ mol^{-1}) is lower than the first ionization energy of magnesium (0.74 MJ mol^{-1}), but the second ionization energy of sodium (4.56 MJ mol^{-1}) is much higher than the second ionization energy of magnesium (1.45 MJ mol^{-1}).

8. Which of the following ionic compounds of sodium react with water to give products that differ from the ions in the solid ionic compound? Write a balanced equation for each reaction.
  (a) NaCl (b) NaH (c) NaOH
  (d) Na_2O (e) Na_2SO_4 (f) Na_2CO_3

9. Complete and balance the following equations:
  (a) $Mg(s) + Cl_2(g) \longrightarrow$
  (b) $Ca(s) + O_2(g) \longrightarrow$
  (c) $Sr(s) + H_2(g) \longrightarrow$
  (d) $Mg(s) + N_2(g) \longrightarrow$
  (e) $Ca(s) + H_2O(l) \longrightarrow$

10. How do the solubilities of the hydroxides of the alkaline earth metals vary from top to bottom of the group? How do the solubilities of the sulfates vary?

11. Explain, with appropriate balanced equations, the formation of limestone caverns and the stalactites and stalagmites found in these caverns.

12. Write equations for the reactions that occur when the following compounds are heated:

    (a) $CaCO_3$ (b) $Ca(OH)_2$

    (c) $NaHCO_3$ (d) $MgCl_2 \cdot 6H_2O$

13. A 10.00-g sample of a mixture of $CaCO_3$, $Ca(HCO_3)_2$, and CaO was heated, and 0.200 g of H_2O and 1.500 g of CO_2 were obtained. What was the composition of the original mixture?

14. What volume of hydrogen gas at STP would be generated by the reaction of 2.00 g of calcium hydride with excess water?

15. Seawater contains 0.05 mol L^{-1} of Mg^{2+}. How many liters of water would have to be processed to yield 1 metric ton (10^3 kg) of magnesium if the extraction process is 70% efficient?

16. How much washing soda can be made from 1 metric ton (10^3 kg) of trona ore?

17. Write equations for the acid-base reactions that occur when soluble compounds of the following ions are added to water:

    (a) O^{2-} (b) H^- (c) N^{3-} (d) CO_3^{2-}

Solubility and Solubility Product Constant

18. Are the following compounds soluble or insoluble in water? Use the solubility rules to make your predictions.

    (a) $PbSO_4$ (b) AgI (c) Na_2CO_3

    (d) FeS (e) $AgNO_3$ (f) $Cu(OH)_2$

19. Are the following compounds soluble or insoluble in water? Use the solubility rules to make your predictions.

    (a) $AlCl_3$ (b) $CaSO_4$ (c) $CuSO_4$

    (d) $LiOH$ (e) $BaCO_3$ (f) Na_2S

20. Predict whether or not a precipitate will form when aqueous solutions of the following pairs of substances are mixed. If a reaction occurs, write the balanced equation for the reaction and also the corresponding net ionic equation.

    (a) $Na_2CO_3(aq) + CaCl_2(aq) \longrightarrow$

    (b) $NaNO_3(aq) + CaBr_2(aq) \longrightarrow$

    (c) $AgNO_3(aq) + NaI(aq) \longrightarrow$

    (d) $BaCl_2(aq) + MgSO_4(aq) \longrightarrow$

    (e) $HCl(aq) + Pb(NO_3)_2(aq) \longrightarrow$

21. Predict whether or not a precipitate will form when aqueous solutions of the following pairs of substances are mixed. If a reaction occurs, write the balanced equation for the reaction and also the corresponding net ionic equation.

    (a) $FeCl_3(aq) + NaOH(aq) \longrightarrow$

    (b) $NaClO_4(aq) + AgNO_3(aq) \longrightarrow$

    (c) $BaCl_2(aq) + KOH(aq) \longrightarrow$

    (d) $Pb(NO_3)_2(aq) + H_2SO_4(aq) \longrightarrow$

    (e) $AgNO_3(aq) + Na_2S(aq) \longrightarrow$

22. What are the solubility product constant expressions for $Fe(OH)_3$ and for $Ca_3(PO_4)_2$?

23. What are the solubility product constant expressions for each of the following?

    (a) $AgCl$ (b) BaF_2 (c) $Cr(OH)_3$ (d) Bi_2S_3

24. Write the solubility product constant expressions for the fluorides of silver, calcium, and lead, the hydroxides of silver, magnesium, and aluminum, and the sulfates of silver and strontium.

25. It is found that 0.060 g of $PbSO_4$ will dissolve in 2.0 L of water. What is the solubility product constant for $PbSO_4$?

26. The solubility of MgF_2 in water is 0.0012 mol L^{-1}. What is the solubility product constant for MgF_2?

27. The solubilities of PbS, CaF_2, and $Cr(OH)_3$ in water are 4.41 pg, 16.8 mg, and 5.62 μg per liter, respectively. What are the values of their solubility product constants?

28. Use the data in Table 15.11 to calculate the solubility in water at 25°C of each of the following:

    (a) $MgCO_3$ (b) $AgCl$ (c) $Al(OH)_3$

    (d) PbI_2 (e) Ag_3PO_4

29. The solubility product constant for AgCl is 1.7×10^{-10} mol^2 L^{-2}. Calculate the mass of AgCl that will dissolve in 250 mL of water.

30. The solubility product constant of $Pb(OH)_2$ is 4.2×10^{-15} mol^3 L^{-3}. What is the molar concentration of Pb^{2+} in a saturated solution of $Pb(OH)_2$?

31. Should a precipitate of $PbCl_2$ form when 50 mL of 0.10M $Pb(NO_3)_2$ solution is added to 100 mL of 0.05M NaCl solution?

32. The solubility products of $Fe(OH)_2$ and $Fe(OH)_3$ are 1.8×10^{-15} mol^3 L^{-3} and 6×10^{-38} mol^4 L^{-4}, respectively. Which of these two hydroxides is the least soluble?

33. Will a precipitate form if you mix 50 mL of 1.00M KOH(aq) with each of the following?

    (a) 50 mL of 0.0010M $BaCl_2(aq)$

    (b) 50 mL of 1.00M $BaCl_2(aq)$

*34. Suppose we mix 50 mL of 2.00M NaCl(aq) with 50 mL of 0.020M $AgNO_3(aq)$.

    (a) How many grams of AgCl will precipitate from the solution?

    (b) What concentration of $Ag^+(aq)$ remains in solution after the precipitate has formed?

*35. Is a precipitate formed when you mix 50 mL of 2.00M $Pb(NO_3)_2(aq)$ with 50.0 mL of $4.00 \times 10^{-3}M$ NaI(aq)? If a precipitate does form, what is it, and how many moles are obtained? What are the concentrations of Pb^{2+}, I^-, NO_3^-, and Na^+ that remain in the solution?

Common-Ion Effect

36. What is the solubility, at 25°C, of calcium fluoride in each of the following?

    (a) Pure water   (b) $0.01M$ $CaCl_2(aq)$

    (c) $0.10M$ $NaF(aq)$

37. What mass of BaF_2 will dissolve in 500 mL of a solution that is already $0.20M$ in Ba^{2+}?

38. An aqueous solution is prepared by dissolving 1×10^{-8} mol of $MgCO_3$ in 2 L of water. How many moles of the soluble salt $MgCl_2$ can be added to the solution before $MgCO_3$ begins to precipitate?

39. Predict whether the amount of $Fe(OH)_3$ in equilibrium with its saturated solution will increase or decrease under the following conditions:

    (a) An aqueous KOH solution is added.

    (b) An aqueous HCl solution is added.

40. What is the solubility of $Ca_3(PO_4)_2$ in each of the following?

    (a) An aqueous $0.10M$ $CaCl_2$ solution.

    (b) An aqueous $0.10M$ Na_3PO_4 solution.

    (c) Pure water.

41. An aqueous solution is $0.0010M$ in F^- and $0.010M$ in CO_3^{2-}. A concentrated aqueous solution of $MgCl_2$ is added. Which precipitates first, MgF_2 or $MgCO_3$?

42. A solution initially is $0.10M$ in both F^- and SO_4^{2-}. If Pb^{2+} is added (in the form of a soluble salt), which precipitates first, $PbSO_4$ or PbF_2? What percentage of one ion remains when the other begins to precipitate as the lead salt?

Solubility and pH

43. What is the minimum pH at which a precipitate of $Fe(OH)_2$ will form in a $0.005M$ $FeCl_2$ solution?

44. Milk of magnesia is an aqueous suspension of magnesium hydroxide, $Mg(OH)_2$. What is the pH of milk of magnesia?

45. For which of the following compounds does the solubility increase when the pH of the solution is decreased?

    (a) $CuS(s)$ (b) $AgI(s)$ (c) $MgCO_3(s)$

    (d) $Cu(OH)_2(s)$ (e) $CaF_2(s)$ (f) $PbSO_4(s)$

46. What is the solubility of $Pb(OH)_2$ in an aqueous solution buffered at pH = 8.0?

47. At pH = 8.0 the solubility of $Ni(OH)_2$ is $0.0020M$. What is K_{sp} for this hydroxide? What is the solubility of the hydroxide at pH = 7.0?

48. What are the solubilities of $Cu(OH)_2$ and $Pb(OH)_2$ in an aqueous solution buffered at pH = 5.5? Can $Cu(OH)_2$ be separated from $Pb(OH)_2$ at this pH?

***49.** A saturated solution of H_2S in water contains 0.10 mol L^{-1} of H_2S at 25°C. The acid dissociation constant for the equilibrium

$$H_2S + 2H_2O \rightleftharpoons 2H_3O^+ + S^{2-}$$

has a value of 1.0×10^{-19} mol^2 L^{-2} at 25°C. What are the maximum concentrations of Cu^{2+}, Fe^{2+}, Pb^{2+}, and Ag^+ ions that can exist in solutions saturated with H_2S and maintained at a pH of 4.00?

CHAPTER 16
ELECTROCHEMISTRY

Oxidation-reduction reactions involve the transfer of electrons. Since a flow of electrons constitutes an electric current, we can use chemical reactions to produce electrical energy and electrical energy to produce chemical reactions.

The batteries for flashlights, radios, pocket calculators, watches, and automobiles, to name just a few of their many applications, use chemical reactions to produce an electric current. The industrial production of many metals, such as sodium, aluminum, and copper, depends on the use of electrical energy to reduce the positive metal ions to the metal.

In this chapter we consider first some examples of the use of electrical energy to carry out chemical reactions. This process is called *electrolysis*. Then we discuss the use of chemical reactions for producing electricity in *electrochemical cells*. We will see that an understanding of electrochemical cells will improve our understanding of the chemistry of metals and allow us to answer a question such as, Why do iron and magnesium dissolve in dilute hydrochloric acid, whereas copper and silver do not, although they will dissolve in concentrated nitric acid? In fact, we will be able to put oxidation-reduction reactions on a quantitative basis, just as we did for acid-base reactions in Chapter 14. Corrosion of metals occurs as a result of oxidation-reduction reactions, and we will see that the chemistry of corrosion is closely related to the chemistry of electrochemical cells.

16.1 ELECTROLYSIS

We mentioned in Chapter 15 that metal ions, such as Na^+, K^+, and Ca^{2+}, are difficult to reduce to the metal, and that in many cases the most practical method is to use electrons directly, that is, to use an electric current. **Electrolysis** may be defined as a *process in which electrical energy is used to produce chemical change*. We have seen previously that molten ionic compounds and solutions of electrolytes in water conduct an electric current because the ions can move under the influence of an electric potential. But we have not discussed

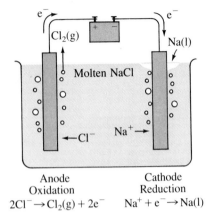

Anode
Oxidation
$2Cl^- \rightarrow Cl_2(g) + 2e^-$

Cathode
Reduction
$Na^+ + e^- \rightarrow Na(l)$

Figure 16.1 Electrolysis of Molten Sodium Chloride. The electric current outside the electrolytic cell is carried by the electrons, which are pushed around the circuit by the battery. Inside the molten electrolyte the current is carried by the movement of the positive and negative ions toward the electrodes. At the anode Cl^- ions give up electrons to the electrode, to give Cl atoms, which combine to give Cl_2 molecules. At the cathode Na^+ ions accept electrons from the electrode, to give Na atoms, which form liquid sodium metal.

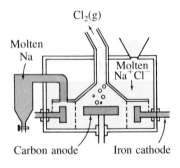

Figure 16.2 Cell for Commercial Production of Sodium. A section through the cell is shown. The iron cathode in the form of a ring surrounds the carbon anode. The cell is designed to prevent the sodium produced at the cathode from coming into contact with the chlorine formed at the anode to re-form sodium chloride.

how the electric current passes from the power supply and external circuit to the molten salt or electrolyte solution. To this end, we consider the preparation of sodium and chlorine by the electrolysis of molten sodium chloride.

Electrolysis of Molten Sodium Chloride

The electrolysis of molten sodium chloride is carried out in an **electrolytic cell**, which consists of a container to hold the molten sodium chloride and two *electrodes* that are made of a solid conducting material, usually a metal or graphite. When a battery or a direct current (DC) source is connected to the two electrodes by conducting wires, it pushes electrons into one electrode, which becomes negatively charged, and it withdraws electrons from the other electrode, which becomes positively charged (see Figure 16.1). The negatively charged electrode attracts positive sodium ions to its surface. Here each sodium ion acquires an electron from the electrode and is reduced to a sodium atom. These sodium atoms combine to form sodium metal, which rises to the surface of the molten sodium chloride. Negative chloride ions are attracted to the positive electrode, and each chloride ion gives up an electron to the electrode and is oxidized to a chlorine atom. These chlorine atoms combine to form chlorine molecules, which form bubbles of gas that rise to the surface.

Thus the reactions taking place at the electrode surfaces are as follows:

Positive electrode: anode $2Cl^- \longrightarrow Cl_2 + 2e^-$ Oxidation

Negative electrode: cathode $Na^+ + e^- \longrightarrow Na$ Reduction

The electrode at which oxidation occurs is called the **anode**. The electrode at which reduction occurs is called the **cathode**. To remember this, note that both *a*node and *o*xidation begin with a vowel, while both *c*athode and *r*eduction begin with a consonant.

The overall reaction that takes place in the electrolytic cell is called the *cell reaction*. The equation for the complete cell reaction is obtained by combining the anode half-reaction with the cathode half-reaction. It may be necessary to multiply either or both of the equations by appropriate coefficients so that the same number of electrons as are produced in the anode reaction are used up in the cathode reaction. Recall that we used this process before for balancing oxidation-reduction equations. In this case we must multiply the equation for the reduction half-reaction by 2 to give

$$
\begin{aligned}
2Na^+(l) + 2e^- &\longrightarrow 2Na(l) \\
2Cl^-(l) &\longrightarrow Cl_2(g) + 2e^- \\
\hline
2Na^+(l) + 2Cl^-(l) &\longrightarrow Cl_2(g) + 2Na(l)
\end{aligned}
$$

We saw in Chapter 5 that sodium and chlorine combine spontaneously in a highly exothermic reaction to give sodium chloride. The equilibrium

$$2Na(s) + Cl_2(g) \longrightarrow 2NaCl(s) \qquad \Delta H^\circ = -822 \text{ kJ}$$

lies far to the right. The reverse reaction, the decomposition of sodium chloride, is highly endothermic, and a considerable amount of energy is needed to carry it out. In the electrolysis of sodium chloride this energy is supplied by the external source of electric power.

The industrial preparation of sodium is carried out in a cell such as the one shown in Figure 16.2. The cell is designed to prevent the sodium and chlorine produced from coming into contact and re-forming sodium chloride. This process is the most important method for making sodium and is also a source of chlorine.

In the course of some of the earliest experiments on the effects of electric currents on substances, the elements sodium and potassium were first prepared by Humphrey Davy (Box. 16.1) in 1806 by the electrolysis of their molten carbonates.

In any electrolysis electric current is supplied by an external source. The electrons enter and leave the material being electrolyzed—the *electrolyte*—by means of suitable conducting electrodes. At the surface of each electrode, electrons are transferred from the electrode to the electrolyte, or from the electrolyte to the electrode, by means of chemical reactions. *Electrons are transferred to the electrolyte from the cathode in a reduction reaction*, for example, the reduction of Na^+ to Na. *Electrons are transferred from the electrolyte to the anode in an oxidation reaction*, for example, the oxidation of Cl^- to Cl_2. An electric current flows through the electrolyte. In order words, charge is carried across the

---

Box 16.1

Davy was brought up on a farm in Cornwall, England. His father had lost much of his money in a mining speculation and died when Davy was only sixteen. Forced to choose a profession that would enable him to help support the family, Davy was apprenticed to a surgeon and apothecary. During his apprenticeship he read widely and studied hard. He set himself a course of study that included theology, geography, medicine, logic, languages (English, French, Latin, Greek, Italian, Spanish, and Hebrew), physics, mechanics, rhetoric and oratory, history, and mathematics.

Just before his nineteenth birthday he began his study of chemistry by reading Lavoisier's famous *Traité Elementaire de Chimie* (Box 3.1). Almost immediately, without having had any formal training in chemistry, he began to test Lavoisier's theories with home-made apparatus. Among his many experiments during this period was one showing that plants absorb carbon dioxide in the presence of light and produce oxygen. Despite his youth he had the confidence to challenge established ideas, even Lavoisier's.

In 1797 Davy became an assistant at an institute for the study of the curative powers of the various gases that were being discovered at that time. Among these gases were oxygen, hydrogen, carbon dioxide, and carbon monoxide, not all of which had beneficial effects on the patients at the institute. Davy breathed all these gases and appears to have nearly killed himself with carbon monoxide. In early experiments he noticed that dinitrogen oxide, N_2O, had an intoxicating effect. It was commonly called "laughing gas," and laughing gas parties became fashionable among the rich. Davy suggested that it be used during operations, which at the time were carried out without anesthetics. Indeed, dinitrogen oxide later became widely used as the first chemical anesthetic.

In 1801, at the age of twenty-three, Davy was made director of the recently formed Royal Institution in London. His public lectures on chemistry were phenomenally successful, and Davy became a popular celebrity; he has been called the most handsome of all the great scientists.

In one of his greatest experiments, Davy prepared sodium and potassium by the electrolysis of their molten carbonates, using an enormous battery consisting of 250 metal plates.

In 1812, suffering from overwork and almost certainly from poisoning caused by tasting and smelling all the substances with which he worked and even breathing large quantities of gases, Davy needed a vacation. He resigned from the Royal Institution, married a rich Scottish widow, and set out on a grand tour of Europe, with Michael Faraday as his assistant (Box 16.3). As he traveled, he had discussions with many famous scientists, made geological and biological observations, and continued to do many experiments in chemistry and physics. He applied his ability to many practical problems. One of his most important inventions was the safety lamp for miners. It has been said, perhaps unjustly, that despite his many achievements his greatest discovery was Michael Faraday.

electrolyte not by the movement of free electrons but by the movement of positive ions toward the negative electrode, or cathode, and negative ions toward the positive electrode, or anode.

This type of conduction in which the charge carriers are ions is called **ionic conduction**. In the electrodes and in the external circuit (the wires connecting the electrodes to the source of power), the charge carriers are free electrons. This is *electronic, or metallic, conduction* (Chapter 9). The charge carriers change from electrons to ions, or vice versa, at the surfaces between the electrodes and the electrolyte. It is this change that produces the chemical reactions at the electrodes—oxidation at the anode and reduction at the cathode.

Preparation of Metals by Electrolysis

Magnesium, calcium, strontium, and barium are prepared by electrolysis of their molten chlorides. Magnesium is the most important of these metals and is produced in the largest quantities. Magnesium chloride is obtained from seawater as described in Chapter 13. It is then melted and electrolyzed to give magnesium and chlorine. The overall reaction is

$$MgCl_2(l) \xrightarrow{\text{Electrolysis}} Mg(l) + Cl_2(g)$$

Probably the most important electrolytic process for the preparation of a metal is that for aluminum (Chapter 9). This process is called the *Hall process* in North America after its inventor Charles M. Hall (Box 16.2). Aluminum

---

Box 16.2

CHARLES MARTIN HALL (1863–1914)

The process we use today for manufacturing aluminum was invented by a young American while he was still an undergraduate at Oberlin College. Inspired by a professor's remark that anyone who could invent a cheap process for producing aluminum would make a fortune, Charles Hall set out to try in 1885. At the time, aluminum cost $90 a pound and was more expensive than either silver or gold. It is said that the very rich flaunted their wealth by dining with aluminum knives and forks.

Hall worked in a woodshed using homemade and borrowed equipment. After about a year he found that Al_2O_3 dissolves in molten cryolite to give a conducting solution, from which aluminum could be deposited by passing an electric current. He used an iron frying pan as a container for the molten cryolite-alumina mixture, which he melted over a blacksmith's forge. The electric current came from electrochemical cells that he made from jars that his mother used to can fruit.

By an odd coincidence Paul Héroult, who was the same age as Hall, made the same discovery independently in France about the same time. As a result of the discovery of Hall and Héroult, the large-scale production of aluminum became economically feasible for the first time, and it became a common and familiar metal.

is produced by the electrochemical reduction of a molten mixture of Al_2O_3 and *cryolite*, Na_3AlF_6, an ionic compound composed of Na^+ and octahedral AlF_6^{3-} ions. Cryolite is a rare mineral found in appreciable quantities only in Greenland, but the large quantities required for the Hall process are now made synthetically. The aluminum oxide is obtained from the mineral bauxite, $Al_2O_3 \cdot xH_2O$, which generally contains only 35%–60% Al_2O_3 together with oxides of iron, silicon, and titanium as well as small amounts of clay and other silicates. The melting point of Al_2O_3 (2050°) is much too high to allow it to be conveniently used for electrolysis. But it dissolves in molten cryolite, Na_3AlF_6, at about 1000°C. A typical cell is shown in Figure 16.3. The anodes are graphite rods dipping into the molten electrolyte. The cathode is a steel vessel, lined with graphite, that contains the molten electrolyte. Neither the nature of the species present in solution nor the electrode reactions are completely understood. For simplicity we shall assume that the Al_2O_3 is ionized to Al^{3+} and O^{2-}. We can represent the reaction at the cathode by the equation

$$\text{Cathode} \qquad Al^{3+} + 3e^- \longrightarrow Al(l)$$

At the anode several reactions occur in which carbon dioxide, oxygen, and fluorine are formed. The principal reaction at this electrode may be approximately represented by the equation

$$\text{Anode} \qquad C(s) + 2O^{2-} \longrightarrow CO_2(g) + 4e^-$$

The approximate overall reaction is therefore

$$4Al^{3+}(l) + 6O^{2-}(l) + 3C(s) \longrightarrow 4Al(l) + 3CO_2(g)$$

The graphite electrodes are gradually converted to carbon dioxide and must be replaced from time to time. The energy consumption is high, and the process is economically feasible only when carried out near a cheap source of electric power, for example, at a site where hydroelectric power is produced.

Quantitative Aspects of Electrolysis

The principal cost in the electrolytic preparation of aluminum and other metals is the cost of electricity. Therefore it is important to know how much electricity is needed to produce a certain amount of metal. The amount of electricity needed can be calculated from the equation for the appropriate electrode reaction. In the case of the electrolysis of molten sodium chloride, the reactions are

$$\text{Cathode} \qquad Na^+(l) + e^- \longrightarrow Na(l)$$
$$\text{Anode} \qquad 2Cl^-(l) \longrightarrow Cl_2(g) + 2e^-$$

The passage of one electron produces one sodium atom; the passage of 1 mol of electrons produces 1 mol of sodium. The passage of two electrons produces one molecule of Cl_2; the passage of 2 mol of electrons produces 1 mol of Cl_2. In summary:

Electrode reaction	Number of moles of electrons	Product
$Na + e^- \longrightarrow Na$	1 mol electrons	1 mol Na = 23.0 g Na
$2Cl^- \longrightarrow Cl_2 + 2e^-$	2 mol electrons	1 mol Cl_2 = 70.9 g Cl_2

The charge on one electron is $1.602\,19 \times 10^{-19}$ coulomb (C). Therefore the charge on 1 mol, $(6.022\,05 \times 10^{23})$ electrons, is $(6.022\,05 \times 10^{23}$ electrons $\text{mol}^{-1})(1.602\,19 \times 10^{-19}$ C electron$^{-1}) = 96\,485$ C mol^{-1}. The charge of 1 mol

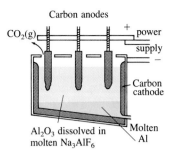

Figure 16.3 Electrolytic Cell for the Production of Aluminum by the Hall Process. A section through the cell is shown. The carbon anodes are oxidized to $CO_2(g)$ and as they are consumed they are lowered further into the molten mixture of Al_2O_3 and Na_3AlF_6. Aluminium ions are reduced to aluminum at the carbon cathode that lines the tank. Because molten aluminum is more dense than the molten Al_2O_3—Na_3AlF_6 mixture, aluminum collects at the bottom of the cell and may be run off.

of electrons ($96\,485\ \mathrm{C\ mol^{-1}}$) is called the Faraday constant, F, in honor of Michael Faraday, who discovered the quantitative laws governing electrolysis (Box 16.3):

$$F = 96\,485\ \mathrm{C\ mol^{-1}}$$

or $96\,500\ \mathrm{C\ mol^{-1}}$ when rounded off to three significant figures, which is accurate enough for most purposes. Thus we can find the charge Q of n moles of electrons by using the expression

$$Q = nF$$

Box 16.3

Michael Faraday was one of ten children of a blacksmith in London. He had only a basic elementary education before being apprenticed to a bookbinder at the age of fourteen. His employer, who was uncharacteristically lenient for the times, allowed Faraday to read the books in his shop, and Faraday educated himself. In 1812 he was given tickets to attend lectures given by Humphrey Davy (Box 16.1) at the Royal Institution. He wrote up careful, complete notes of the lectures and bound them in a book. Friends persuaded Faraday to send the notes to Davy in support of his application for a position as an assistant to Davy, who was director of the Royal Institution. Faraday obtained the position, beginning not only his prolific scientific career but also the very fruitful collaboration with Davy. Soon after he was appointed, Faraday left with Davy on a grand tour of Europe, which did much to broaden Faraday's scientific education and gave him the opportunity to meet many famous scientists.

When they were in Florence, Davy and Faraday were able to use a very large lens belonging to the Duke of Tuscany to prove conclusively that diamond consists only of carbon—an idea that was difficult for many scientists at that time to accept. Davy and Faraday used the lens to focus the suns rays on a diamond enclosed in a bulb of pure oxygen. After about an hour the diamond began to burn. "The diamond," Faraday wrote in his journal, "glowed brilliantly with a scarlet light and when placed in the dark continued to burn for about four minutes." They burned the diamond completely and showed that the bulb then contained nothing but carbon dioxide and excess oxygen.

In 1825 Faraday replaced Davy as director of the Royal Institution, and his reputation soon began to rival that of Davy. Faraday was the first to liquefy several gases, including carbon dioxide, hydrogen sulfide, hydrogen bromide, and chlorine; he discovered benzene and determined its composition; and he discovered the quantitative laws of electrolysis. The Faraday constant, F, was named in his honor. Faraday made even greater contributions to physics. He found that an electric current could be induced in a wire by a moving magnet. He provided both the experimental basis and basic ideas for the theory of electromagnetism, which was later developed by Maxwell. Albert Einstein ranked Faraday with Newton, Galileo, and Maxwell, as the greatest physicists of all time.

Faraday was a member of a religious sect so strict that for a time he was denied membership of the church because he had accepted an invitation to have lunch with Queen Victoria on a Sunday! In accordance with his religious beliefs, he tried to live a simple life, accepting rather reluctantly the many honors that came to him. His beliefs enabled him to solve without any uncertainty a moral problem that still faces scientists. During the Crimean War, in the 1850s, the British government asked him to head an investigation of the possibility of preparing large quantities of poison gas for use on the battlefield. Faraday refused to have anything to do with the project, and nothing came of the idea at that time.

Since the production of 1 mol, or 23.0 g, of sodium by reduction of sodium ions requires 1 mol of electrons, it requires an amount of charge equal to

$$1 \text{ mol electrons} \times F = (1 \text{ mol electrons})(96\ 500 \text{ C mol}^{-1})$$
$$= 96\ 500 \text{ C}$$

The production of 1 mol Cl_2 requires 2 mol of electrons and therefore an amount of charge equal to $2 \times 96\ 500$ C. In summary:

Electrode reaction	Charge	Product
$Na^+ + e^- \longrightarrow Na$	96 500 C	1 mol Na = 23.0 g Na
$2Cl^- \longrightarrow Cl_2 + 2e^-$	$2 \times 96\ 500$ C	1 mol Cl_2 = 70.9 g Cl_2

In practice, charge is usually determined by measuring a current flow for a given time. A charge of one coulomb passes a given point when a current of one ampere (A) flows for one second:

$$1 \text{ coulomb} = 1 \text{ ampere} \times 1 \text{ second}$$
$$1 \text{ C} = 1 \text{ A s}$$

Thus we can calculate, for example, that if molten NaCl is electrolyzed for 1.00 hour with a current of 50.0 A, the number of coulombs passed through it is

$$(50.0 \text{ A})\left(\frac{1 \text{ C}}{1 \text{ A s}}\right) \times 1 \text{ h}\left(\frac{60 \text{ min}}{1 \text{ h}}\right)\left(\frac{60 \text{ s}}{1 \text{ min}}\right) = 180\ 000 \text{ C}$$

Then we have

$$n = \frac{Q}{F} = \frac{180\ 000 \text{ C}}{96\ 500 \text{ C mol}^{-1}} = 1.87 \text{ mol electrons}$$

and 1.87 mol electrons produces 1.87 mol of Na (43.0 g) and $1.87/2 = 0.935$ mol of Cl_2 (66.3 g).

Example 16.1 What mass of aluminum will be produced in 1.00 h by the electrolysis of molten $AlCl_3$, using a current of 10.0 A?

Solution First, from the current and the time, we must calculate the number of coulombs:

$$10.0 \text{ A}\left(\frac{1 \text{ C}}{1 \text{ A s}}\right) \times 1.00 \text{ h}\left(\frac{3600 \text{ s}}{1 \text{ h}}\right) = 3.60 \times 10^4 \text{ C}$$

We then find the number of moles of electrons:

$$n = \frac{Q}{F} = \frac{3.60 \times 10^4 \text{ C}}{96\ 500 \text{ C mol}^{-1}} = 0.373 \text{ mol of electrons}$$

The half-reaction for the reduction of aluminum is

$$Al^{3+} + 3e^- \longrightarrow Al$$

Thus 3 mol of electrons is needed to produce 1 mol (27.0 g) of Al. Hence 0.373 mol of electrons produces

$$0.373 \text{ mol of electrons}\left(\frac{1 \text{ mol Al}}{3 \text{ mol electrons}}\right)\left(\frac{27.0 \text{ g Al}}{1 \text{ mol Al}}\right) = 3.36 \text{ g Al}$$

Example 16.2 What volume of chlorine at STP is produced when a current of 20.0 A is passed through molten sodium chloride for 2.00 h?

Solution

$$\text{Number of coulombs} = 20.0 \text{ A} \left(\frac{1 \text{ C}}{1 \text{ A s}}\right) \times 2.00 \text{ h} \times \frac{3600 \text{ s}}{1 \text{ h}}$$

$$= 1.44 \times 10^5 \text{ C}$$

$$\text{Number of moles of electrons} = \frac{1.44 \times 10^5 \text{ C}}{96\,500 \text{ C mol}^{-1}} = 1.49 \text{ mol electrons}$$

The anode reaction is

$$2Cl^- \longrightarrow Cl_2 + 2e^-$$

Therefore 2 mol of electrons produces 1 mol of Cl_2. Hence 1.49 mol of electrons produces

$$\frac{1.49 \text{ mol } Cl_2}{2} = 0.745 \text{ mol } Cl_2$$

From the ideal gas law, we know that 1 mol of an ideal gas occupies 22.4 L at STP. Therefore

$$\text{Volume of } Cl_2 \text{ at STP} = 0.745 \text{ mol } Cl_2 \left(\frac{22.4 \text{ L}}{1 \text{ mol } Cl_2}\right)$$

$$= 16.7 \text{ L}$$

Electrolysis of Aqueous Solutions

In any electrolysis there must be an oxidation reaction at the anode and a reduction reaction at the cathode. But these reactions are not necessarily the oxidation of the negative ions and the reduction of the positive ions in the solution. In aqueous solution, water may be oxidized or reduced or the electrodes may be oxidized or reduced. For the moment we will consider only inert electrodes. We first discuss the electrolysis of an aqueous sodium chloride solution, which is an important industrial process for the manufacture of sodium hydroxide and chlorine.

ELECTROLYSIS OF AQUEOUS SODIUM CHLORIDE If we electrolyze a concentrated aqueous sodium chloride solution, we find that chlorine is produced at the anode, but at the cathode we obtain hydrogen and not sodium, as in the electrolysis of molten sodium chloride (see Figure 16.4). Thus the anode reaction is

$$\text{Anode} \quad 2Cl^- \longrightarrow Cl_2(g) + 2e^- \quad \text{(Oxidation)}$$

Figure 16.4 Electrolysis of Aqueous Sodium Chloride Solution. The Cl^- ions are attracted to the anode, to which they give up electrons to form Cl atoms, which combine to form $Cl_2(g)$. At the cathode electrons are transferred to H_2O molecules; water is reduced to OH^- ions and H atoms, and the latter combine to give $H_2(g)$. The Na^+ ions attracted to the cathode are not reduced but keep the solution neutral; in other words, a solution of NaOH is formed at the cathode.

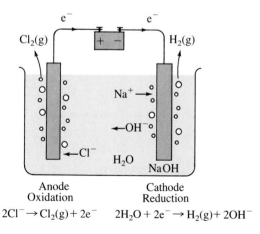

as in the electrolysis of molten sodium chloride. The alternative possibility is the oxidation of water:

$$2H_2O \longrightarrow O_2 + 4H^+ + 4e^- \quad \text{(Oxidation)}$$

But the production of chlorine rather than oxygen at the anode shows that Cl^- is more easily oxidized than H_2O.

At the cathode hydrogen and hydroxide ions are produced. Thus the cathode reaction is the reduction of water,

$$2H_2O + 2e^- \longrightarrow H_2 + 2OH^-$$

rather than the reduction of sodium ion to sodium metal:

$$Na^+ + e^- \longrightarrow Na$$

We conclude that water is more easily reduced than the sodium ion.

The overall equation for the cell reaction is

$$
\begin{array}{l}
2Cl^-(aq) \longrightarrow Cl_2(g) + 2e^- \\
2H_2O(l) + 2e^- \longrightarrow H_2(g) + 2OH^-(aq) \\
\hline
2H_2O(l) + 2Cl^-(aq) \longrightarrow Cl_2(g) + H_2(g) + 2OH^-(aq)
\end{array}
$$

Since the sodium ion is not reduced at the cathode, it takes no part in the reaction—it is a spectator ion. By adding sodium ion to both sides we can rewrite this equation in the form

$$2H_2O(l) + 2NaCl(aq) \longrightarrow Cl_2(g) + H_2(g) + 2NaOH(aq)$$

The products of the electrolysis of an aqueous solution of sodium chloride are chlorine at the anode and sodium hydroxide and hydrogen at the cathode.

ELECTROLYSIS OF AQUEOUS SODIUM SULFATE If an aqueous solution of sodium sulfate is electrolyzed, the products are oxygen at the anode and hydrogen at the cathode (see Figure 16.5). At the anode the possibilities are either that sulfate ion is oxidized or that water is oxidized. Since sulfur in the sulfate ion is in its highest oxidation state, that is, $+6$, it is not surprising that it is the water that is oxidized. The anode reaction is

$$\text{Anode} \quad 2H_2O(l) \longrightarrow O_2(g) + 4H^+(aq) + 4e^- \quad \text{(Oxidation)}$$

We have seen in discussing the electrolysis of aqueous sodium chloride that water, not sodium ion, is reduced at the cathode:

$$\text{Cathode} \quad 2H_2O(l) + 2e^- \longrightarrow H_2(g) + 2OH^-(aq) \quad \text{(Reduction)}$$

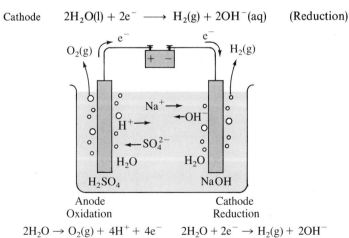

Anode
Oxidation
$$2H_2O \rightarrow O_2(g) + 4H^+ + 4e^-$$

Cathode
Reduction
$$2H_2O + 2e^- \rightarrow H_2(g) + 2OH^-$$

Figure 16.5 Electrolysis of Aqueous Sodium Sulfate Solution. Water molecules give up two electrons at the anode to give O atoms and H^+ ions; the former combine to give O_2 molecules. Water molecules are oxidized. Water molecules accept electrons at the cathode to give H atoms and OH^- ions; the former combine to give H_2 molecules. Water molecules are reduced. The Na^+ and SO_4^{2-} ions transport the current through the cell, and by their motion in opposite directions they keep the solution neutral at every point. Thus a solution of H_2SO_4 ($H_3O^+HSO_4^-$) is formed at the anode, and a solution of NaOH (Na^+OH^-) is formed at the cathode.

If we multiply this equation by 2 and add it to the equation for the anode reaction, we obtain the equation for the overall cell reaction:

$$6H_2O(l) \longrightarrow O_2(g) + 2H_2(g) + 4OH^-(aq) + 4H^+(aq)$$

This equation can be simplified, because H^+ and OH^- combine to give water. We can then subtract $4H_2O$ from each side to give

$$2H_2O(l) \longrightarrow O_2(g) + 2H_2(g)$$

This is the overall equation for the electrolysis of water.

Neither the sodium ion nor the sulfate ion participate in the electrode reactions. What then is their function? They serve to conduct the current between the electrodes and to maintain the electrical neutrality of every part of the solution. Hydronium (hydrogen) ions are produced at the anode, and the sulfate ions arriving at the anode neutralize the charge of the hydronium ions, so the solution remains neutral at all times. Hydroxide ions are produced at the cathode. Sodium ions arriving at the anode neutralize the charge of the hydroxide ions. The more concentrated the sodium sulfate solution, the more sulfate ions arrive at the anode in a given time, so the more quickly hydrogen ions can be produced. Electrolysis of a very dilute solution of sodium sulfate is very slow, because the electrolysis can only occur as fast as ions arrive at the electrodes. In principle, we could electrolyze pure water, but the concentration of ions from the self-dissociation of water is extremely small. So only a very small current would flow, and the electrolysis would be extremely slow.

ELECTROLYSIS OF AQUEOUS COPPER(II) SULFATE In the electrolysis of a copper sulfate solution, oxygen is evolved at the anode and copper is deposited on the cathode. Thus as we saw for sodium sulfate, sulfate ion is not oxidized at the anode; rather, water is oxidized to oxygen. The anode reaction is

$$\text{Anode} \qquad 2H_2O(l) \longrightarrow O_2(g) + 4H^+(aq) + 4e^- \text{(Oxidation)}$$

In solutions of sodium chloride and sodium sulfate, water is reduced at the cathode, rather than Na^+, but in the present case Cu^{2+} is reduced to copper. We conclude that Cu^{2+} is more easily reduced than water, although Na^+ is less easily reduced than water. The cathode reaction is

$$\text{Cathode} \qquad Cu^{2+}(aq) + 2e^- \longrightarrow Cu(s) \qquad \text{(Reduction)}$$

The equation for the overall cell reaction can be obtained by multiplying the equation for the cathode reaction by 2 and adding it to the equation for the anode reaction, and we get

$$2H_2O(l) + 2Cu^{2+}(aq) \longrightarrow O_2(g) + 4H^+(aq) + 2Cu(s)$$

The electrolysis of some aqueous solutions is demonstrated in Experiment 16.1.

Electrolytic Preparation of Chlorine and Sodium Hydroxide

The electrolysis of aqueous sodium chloride is an extremely important reaction because it is the principal method by which chlorine and sodium hydroxide are made industrially. Indeed, more than 90% of the world's chlorine is made by this process. Sodium hydroxide and chlorine are two of the basic chemicals produced by the chemical industry: In 1983 they ranked seventh and eighth,

Electrolysis of Some Aqueous Solutions

Electrolysis of a colorless aqueous solution of potassium iodide with inert (copper) electrodes produces brown I_3^- at the anode and H_2 and OH^- at the cathode. The bubbles of hydrogen can be seen rising from the cathode, and the OH^- ion causes phenophthalein in the solution to turn pink.

The beaker contains a solution of copper sulfate and a graphite electrode which is connected to the positive terminal of a battery. The other electrode is a strip of silver which is connected to the negative terminal of the battery.

When the silver electrode is dipped into the solution, electrolysis commences and bubbles of oxygen can be seen rising from the graphite electrode.

After a few minutes the silver electrode is removed from the solution. It can be seen to be coated with a layer of red-brown copper.

respectively, among all the chemicals produced in the United States. From the equation for the overall cell reaction,

$$2NaCl(aq) + 2H_2O(l) \longrightarrow Cl_2(g) + H_2(g) + 2NaOH(aq)$$

we see that hydrogen is another important product of the process, although as we saw in Chapter 3, there are other important methods for making hydrogen.

The chlorine gas must be kept separated from the hydrogen and from the sodium hydroxide so that it cannot react with them. The cell shown in Figure 16.6 accomplishes separation by using a porous diaphragm, usually made of asbestos, through which the solution can be made to flow but which prevents the passage of the gases produced at the electrodes. This diaphragm also prevents OH^- produced at the cathode from diffusing to the anode compartment,

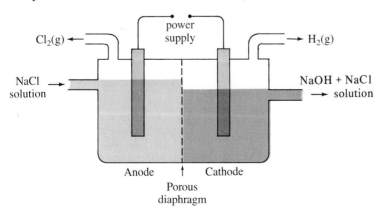

Figure 16.6 Diaphragm Cell for Production of Cl_2 by Electrolysis of Aqueous Sodium Chloride Solution. The anode and cathode compartments are separated by a porous asbestos diaphragm. A difference in level between the solutions in the anode and cathode compartments keeps the solution flowing from the anode to the cathode compartment, which prevents movement of OH^- ions formed at the cathode into the anode compartment, where they would react with the chlorine.

where it would react with the chlorine. The solution that flows out of the cathode compartment contains 11% NaOH and 16% NaCl. This solution is concentrated by evaporation. The less soluble NaCl crystallizes and can be filtered off. When the concentration of NaOH reaches 50% by mass, only about 1% NaCl remains in solution. For many applications this small impurity of NaCl is not important.

In the type of cell shown in Figure 16.7, a stream of mercury is used as the negative electrode. Hydrogen is not formed at this electrode, but sodium ions are reduced to sodium, which dissolves in the mercury. The mercury electrode is not inert; it takes part in the electrode reaction by dissolving the sodium that is formed by reduction of Na^+.

$$Na^+ + Hg + e^- \longrightarrow Na(Hg)$$

Sodium amalgam

The solution of sodium in mercury, which is called *sodium amalgam*, flows out of the cell and is subsequently decomposed by being mixed with water:

$$2Na\ (amalgam) + 2H_2O \longrightarrow 2Na^+ + 2OH^- + H_2 + (mercury)$$

Thus the hydrogen and the sodium hydroxide are produced separately from the chlorine, and very pure sodium hydroxide is obtained. The mercury is returned to the main cell and used again, but some mercury is always lost. A very small amount of mercury tends to find its way into the waste water from the plant, creating an environmental hazard (see Box 21.3).

Electrolytic Refining of Copper

In most of the examples of electrolysis that we have considered so far, the electrodes were inert; in other words, they did not react under the conditions of the electrolysis. However, not all electrodes are inert. We have already seen that in the manufacture of aluminum by the electrolysis of a solution of Al_2O_3 in molten Na_3AlF_6, the carbon anodes are oxidized to CO_2. Oxidation of the anode also occurs when copper is used as an anode in the electrolysis of an aqueous solution. Copper dissolves to give Cu^{2+}, rather than water being oxidized. The anode reaction is

$$\text{Anode} \qquad Cu(s) \longrightarrow Cu^{2+}(aq) + 2e^-$$

Figure 16.7 Mercury Cell for Production of Chlorine by Electrolysis of Aqueous Sodium Chloride Solution. At the anode Cl^- ions are oxidized to $Cl_2(g)$. At the mercury cathode Na^+ ions are reduced to sodium, which dissolves in the mercury. This Na/Hg amalgam is pumped into the tank at the top of the figure, where it comes into contact with water and reacts to form sodium hydroxide and hydrogen. The mercury is then pumped back into the electrolytic cell.

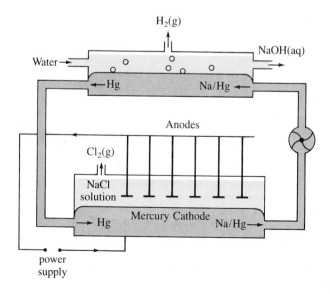

In a cell in which we have a copper anode in a solution of copper(II) sulfate, copper goes into solution at the anode, and at the cathode copper is deposited. The cathode reaction is

Cathode $Cu^{2+}(aq) + 2e^- \longrightarrow Cu(s)$

This type of cell is the basis for an important method for the purification (refining) of crude copper (see Figure 16.8). Impure copper is the anode of the cell. Unreactive metals that are not as easily oxidized as copper, such as silver and gold, fall to the bottom of the cell as a sludge. Other metals that are easily oxidized, such as zinc and iron, go into solution as their positive ions, together with the copper. But if the potential at which the cell operates is suitably adjusted, ions such as Fe^{2+} and Zn^{2+}, which are less readily reduced than Cu^{2+}, are not reduced to the metal at the cathode. Thus only copper is deposited at the cathode, which may be a thin sheet of pure copper or another metal on which a layer of pure copper is built up. The sludge is treated to isolate the less easily oxidized metals such as Ag, Au, and Pt which are valuable by-products of the process.

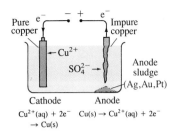

Figure 16.8 Electrolytic Refining of Copper. The impure copper dissolves at the anode. Pure copper is deposited at the cathode.

Electroplating

We have just seen that copper can be "plated out" on the cathode during electrolysis. Other metals can be deposited on a cathode in the same way. This process is called *electroplating*. The object to be plated is the cathode and the plating metal is the anode of an electrolytic cell. Many automobile parts such as bumpers and door handles are plated with chromium to improve their appearance and to protect them against corrosion (Experiment 16.2). Knives, forks, spoons, trays, jugs, and many other objects in the home are often silver-plated.

Example 16.3 How much copper would be deposited during the electrolysis of an aqueous copper sulfate solution if a current of 1.30 A was passed for 1.50 h?

Solution

$$\text{Number of coulombs} = (1.30\ A)\left(\frac{1\ C}{1\ A\ s}\right)(1.50\ h)\left(\frac{3600\ s}{1\ h}\right)$$

$$= 7.02 \times 10^3\ C$$

$$n = \frac{Q}{F} = \frac{7.02 \times 10^3\ C}{96\ 500\ C\ mol^{-1}}$$

$$= 0.0727\ \text{mol electrons}$$

Copper is produced at the cathode according to the equation

$$Cu^{2+} + 2e^- \longrightarrow Cu$$

$$\text{Moles of Cu} = 0.0727\ \text{mol of electrons}\left(\frac{1\ \text{mol Cu}}{2\ \text{mol of electrons}}\right)$$

$$= 0.0364\ \text{mol} = 2.31\ \text{g Cu}$$

Example 16.4 By measuring the amount of a metal deposited at the cathode by weighing the cathode before and after an electrolysis, we can find the amount of charge passed through the cell. An apparatus for measuring charge in this way is called a coulometer. If 0.5230 g of silver was deposited from a silver nitrate solution in a coulometer, how many coulombs of charge were passed through the coulometer?

Electroplating

A current is passed between strips of copper in an aqueous solution of dichromic acid $H_2Cr_2O_7$, obtained by adding chromium trioxide, CrO_3, to water.

After a few minutes the copper strip has become plated with a shiny layer of chromium.

Solution The equation for the cathode reaction is

$$Ag^+ + e^- \longrightarrow Ag$$

The 0.5230 g of Ag is

$$0.5230 \text{ g Ag} \left(\frac{1 \text{ mol Ag}}{107.87 \text{ g Ag}} \right) = 0.004\,848 \text{ mol Ag}$$

The 0.004 848 mol of Ag requires

$$0.004\,848 \text{ mol of electrons} = 0.004\,848 \text{ mol} \times 96\,485 \text{ C mol}^{-1}$$
$$= 467.8 \text{ C}$$

Summary of Electrode Reactions

There are three different types of reactions that may occur at the electrodes in electrolysis:

Reaction involves	Anode	Cathode
1. Electrolyte	Anions oxidized	Cations reduced
2. Solvent	Solvent oxidized	Solvent reduced
3. Electrode	Anode oxidized	Cathode reduced

1. The simplest situation is the electrolysis of a *molten* electrolyte with inert electrodes. In such cases the cation is always reduced and the anion is always oxidized. For example:

Electrolyte	Anode reaction	Cathode reaction
Molten NaCl	$2Cl^- \longrightarrow Cl_2 + 2e^-$	$Na^+ + e^- \longrightarrow Na$
Molten $MgCl_2$	$2Cl^- \longrightarrow Cl_2 + 2e^-$	$Mg^{2+} + 2e^- \longrightarrow Mg$

2. In the electrolysis of *aqueous* electrolyte solutions water is oxidized to oxygen at the anode if the anion is less easily oxidized than water; water is reduced to hydrogen at the cathode if the cation is less easily reduced than water. From the results of experiments we can classify the common anions and cations according to whether they are more or less easily oxidized than water:

Cations	Anions
Less easily reduced than water:	Less easily oxidized than water
$Na^+, K^+, Mg^{2+}, Ca^{2+}, Al^{3+}, Fe^{2+}, Pb^{2+}$	$F^-, SO_4^{2-}, NO_3^-, ClO_4^-, CO_3^{2-}, PO_4^{3-}$
More easily reduced than water:	More easily oxidized than water:
$Cu^{2+}, Ag^+, H_3O^+(H^+)$	Cl^-, Br^-, I^-

Some examples are as follows:

Electrolyte	Anode reaction	Cathode reaction
Aqueous Na_2SO_4	$2H_2O \longrightarrow O_2 + 4H^+ + 4e^-$	$2H_2O + 2e^- \longrightarrow H_2 + 2OH^-$
Aqueous KI	$2I^- \longrightarrow I_2 + 2e^-$	$2H_2O + 2e^- \longrightarrow H_2 + 2OH^-$
Aqueous $CuSO_4$	$2H_2O \longrightarrow O_2 + 4H^+ + 4e^-$	$Cu^{2+} + 2e^- \longrightarrow Cu$

Chloride ion is a borderline case in that it is not much more easily oxidized than water. In a concentrated solution, chloride ion is oxidized to chlorine; but in a dilute solution of Cl^- there is such a large excess of water that water is preferentially oxidized to oxygen.

3. Mostly we are concerned with electrolysis using inert electrodes, but there are a few important cases where the electrode undergoes reaction. For example, copper is more easily oxidized than water. So if a copper anode is used in the electrolysis of a solution having an anion that is not easily oxidized, the copper anode goes into solution as Cu^{2+}. In the electrolysis of molten Al_2O_3, oxide ions react with the graphite anode to give carbon dioxide. In the electrolysis of aqueous sodium chloride with a mercury cathode, sodium ions are reduced to sodium, which dissolves in the mercury electrode. Some examples are as follows:

Electrolyte	Anode material	Anode reaction
Aqueous Na_2SO_4	Cu	$Cu \longrightarrow Cu^{2+} + 2e^-$
Aqueous $CuSO_4$	Cu	$Cu \longrightarrow Cu^{2+} + 2e^-$
Molten Al_2O_3	C	$C + 2O^{2-} \longrightarrow CO_2 + 4e^-$
Aqueous NaCl	Inert	$2Cl^- \longrightarrow Cl_2 + 2e^-$

Electrolyte	Cathode material	Cathode reaction
Aqueous Na_2SO_4	Inert	$2H_2O + 2e^- \longrightarrow H_2 + 2OH^-$
Aqueous $CuSO_4$	Inert	$Cu^{2+} + 2e^- \longrightarrow Cu$
Molten Al_2O_3	Inert	$Al^{3+} + 3e^- \longrightarrow Al$
Aqueous NaCl	Hg	$Na^+ + Hg + e^- \longrightarrow Na/Hg(amalgam)$

Example 16.5 Predict the products of electrolysis of each of the following:

(a) A dilute aqueous solution of sulfuric acid with inert electrodes.

(b) Molten lead bromide with inert electrodes.

(c) An aqueous solution of silver nitrate with inert electrodes.

(d) An aqueous solution of silver nitrate with silver electrodes.

Solution

(a) At the cathode H^+ (aq) is more easily reduced than H_2O. Thus H^+(aq) is reduced to hydrogen:

$$2H^+(aq) + 2e^- \longrightarrow H_2$$

At the anode SO_4^{2-} is less easily oxidized than H_2O. Thus water is oxidized to oxygen:

$$2H_2O \longrightarrow O_2 + 4H^+ + 4e^-$$

(b) At the cathode Pb^{2+} is reduced to lead:

$$Pb^{2+} + 2e^- \longrightarrow Pb$$

At the anode Br^- is oxidized to bromine:

$$2Br^- \longrightarrow Br_2 + 2e^-$$

(c) At the cathode Ag^+ is more easily reduced than water, so silver is deposited:

$$Ag^+ \longrightarrow Ag(s) + e^-$$

At the anode water is more easily oxidized than NO_3^-. Thus water is oxidized to oxygen.

(d) At the cathode Ag^+ is more easily reduced than water, so silver is deposited. At the anode Ag is more easily oxidized than water, so the silver electrode dissolves to give $Ag^+(aq)$:

$$Ag(s) \longrightarrow Ag^+(aq) + e^-$$

16.2 ELECTROCHEMICAL CELLS

We have just seen how we can use an electric current to cause oxidation and reduction reactions to occur at electrodes in molten ionic compounds or in solutions of an electrolyte in water. Conversely, we can use oxidation-reduction reactions to produce an electric current if we set up an appropriate cell. To understand this, we consider the reaction that occurs when we place a piece of zinc in a solution of copper(II) sulfate. The zinc soon becomes coated with a reddish deposit of metallic copper, and the blue color of the solution due to the hydrated copper ion begins to fade. If enough zinc is present and enough time allowed, the solution eventually becomes colorless. These changes result from a reaction in which electrons are transferred from zinc to copper; each Cu^{2+} ion deposited as metallic copper takes two electrons from a zinc atom, which dissolves as Zn^{2+} ion (Experiment 16.3).

$$Zn(s) + Cu^{2+}(aq) \longrightarrow Zn^{2+}(aq) + Cu(s)$$
$$\text{Silvery} \qquad \text{Blue} \qquad\qquad \text{Colorless} \quad \text{Red}$$

In other words, zinc is oxidized to $Zn^{2+}(aq)$, while $Cu^{2+}(aq)$ is reduced to copper metal. This reaction can be used as the basis of an **electrochemical cell** to produce an electric current.

An electrochemical cell, which is also called a *voltaic cell*, or a *galvanic cell*, employs a chemical reaction to produce an electric current. In contrast, in an *electrolytic cell* an electric current is used to produce a chemical reaction. But when the reaction between zinc and copper ions occurs directly on the surface of the zinc, the flow of electrons cannot be used. The oxidation and reduction processes must be separated to get an electric current. This is done by setting up an electrochemical cell such as the one shown in Figure 16.9.

A piece of zinc, called the zinc electrode, is immersed in a solution of zinc sulfate, and a piece of copper, the copper electrode, is immersed in a solution of copper(II) sulfate. The two solutions are kept apart by a porous barrier that permits ions to move from one solution to the other but prevents the mixing of the two solutions by diffusion. A conducting metal wire connects the two electrodes. At the surface of the zinc electrode, Zn^{2+} ions are formed and pass into the solution, leaving excess electrons in the electrode. These electrons flow through the wire to the copper electrode, where they combine with Cu^{2+} ions arriving at the electrode surface. These Cu^{2+} ions are thereby converted to copper atoms, which are deposited on the electrode surface. (Any metal could be used for this electrode and it would become coated with a layer of copper.) The electrode at which oxidation occurs, in this case the zinc electrode, is called the *anode*. The electrode at which reduction occurs, in this case the copper

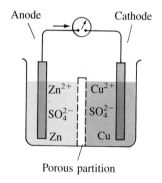

Figure 16.9 An Electrochemical Cell. This cell is based on the overall reaction

$$Zn(s) + Cu^{2+}(aq) \longrightarrow$$
$$Zn^{2+}(aq) + Cu(s).$$

The Reaction of Zinc with Aqueous Solution of Copper Sulfate

A strip of zinc and an aqueous solution of copper sulfate.

When the zinc strip is placed in the copper sulfate solution it rapidly becomes coated with copper.

When the zinc strip is removed from the solution, the characteristic red-brown color of the copper deposit is clearly seen.

electrode, is the *cathode*. These definitions are the same as those we gave previously for an electrolytic cell.

In the copper-zinc cell the electrode reactions are as follows:

| Anode | $Zn(s) \longrightarrow Zn^{2+}(aq) + 2e^-$ | (Oxidation) |
| Cathode | $Cu^{2+}(aq) + 2e^- \longrightarrow Cu(s)$ | (Reduction) |

The two halves of the cell are called half cells. The zinc electrode dipping into a zinc sulfate solution is one of the half cells, and the copper electrode dipping into a copper sulfate solution is the other half cell.

Neither half-cell reaction can take place by itself. Each half-cell reaction must be accompanied by another half-cell reaction that can use up or supply the necessary electrons. These electrons constitute the electric current that flows around the external circuit. Not only must the circuit be complete outside the cell, but it must be complete inside as well; that is, ions must be able to move from one half cell to the other. In the cell in Figure 16.9, the ions move through the porous partition.

Another method is to connect the two half cells by a *salt bridge*, as shown in Figure 16.10. If the two solutions are not in contact in this way, the electrode reactions would quickly cease. Zinc ions going into solution at the zinc electrode would give the solution in the zinc half cell a positive charge, and the attraction of this charge for the electrons in the zinc electrode would prevent them from leaving. Similarly, at the copper electrode the removal of copper ions from the solution would leave the solution with an overall negative charge. This negative charge would prevent more electrons from entering the copper electrode, so the electrode reaction would stop. When the porous partition is in place, sulfate ions can move from the copper half cell to the zinc half cell. The loss of the sulfate ions from the copper half cell keeps this solution neutral, and the sulfate ions that pass through the barrier neutralize the charge of the zinc ions that are formed in the zinc half cell, so this solution also remains neutral. A salt bridge serves the same purpose (see Figure 16.10). The movement of sulfate ions in one direction and of zinc and copper ions in the opposite direction constitutes the current that flows through the cell.

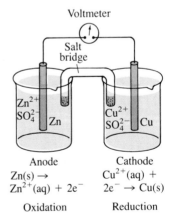

Anode $\qquad$ Cathode

$Zn(s) \rightarrow$ $\qquad$ $Cu^{2+}(aq) +$
$Zn^{2+}(aq) + 2e^-$ $\qquad$ $2e^- \rightarrow Cu(s)$

Oxidation $\qquad$ Reduction

Figure 16.10 Electrochemical Cell with Salt Bridge. The two half cells are connected by a salt bridge, which consists of an inverted U tube containing a conducting solution, such as $NaNO_3(aq)$ or $Na_2SO_4(aq)$. The salt bridge allows movement of ions between the two half cells but prevents the solution in the anode and cathode compartments from mixing.

A Simple Electrochemical Cell

Strips of copper and zinc inserted into a lemon form an electrochemical cell. The measured voltage is 0.9 V.

Experiment 16.4 shows that an electrochemical cell is formed whenever two different metals are inserted into a conducting medium.

Cell Potentials

The electric current produced by an electrochemical cell is the result of electrons being pushed out of the anode and around the external circuit to the cathode, where they are used up. When a current flows between two points, we say that there is a *potential difference* between the two points: The current flows from the higher potential to the lower potential. This potential difference is measured in volts and is often called the *voltage* of the cell. When we speak of a 6-volt (V) battery, we mean a battery that has a potential difference of 6 V between its terminals. The potential difference is a measure of the work that the battery can do by pushing charge around an external circuit. If 1 J of work is done by moving a charge of 1 C through a potential difference, the potential difference is 1 V:

$$1\ V = 1\ J\ C^{-1}$$

The potential between the electrodes of an electrochemical cell is called the *cell potential*, E_{cell}. The cell potential depends on the concentrations of the ions in the cell, the temperature, and the partial pressures of any gases that might be involved in the cell reactions. When all the concentrations are 1 mol L^{-1}, all partial pressures of gases are 1 atm, and the temperature is 25°C, the cell potential is called the **standard cell potential**, E_{cell}°. The cell potential can be measured by connecting a high-resistance voltmeter across the cell.

In principle, any oxidation—reduction reaction can be used as the basis of an electrochemical cell. For example, zinc dissolves in dilute aqueous acids:

$$Zn(s) + 2H^{+}(aq) \longrightarrow Zn^{2+}(aq) + H_2(g)$$

How do we set up an electrode involving hydrogen? We use a piece of platinum foil coated with a black layer of finely divided platinum that serves as a catalyst for the reaction

$$2H^{+}(aq) + 2e^{-} \longrightarrow H_2(g)$$

This electrode is immersed in a solution of H^{+}(aq) ions, for example, a sulfuric acid solution. Hydrogen ions combine with electrons from the metal to give hydrogen gas, which forms bubbles on the surface of the platinum. Thus a hydrogen electrode consists of bubbles of hydrogen gas on the surface of platinum (see Figure 16.11).

A cell based on the reaction between zinc and dilute acid is shown in Figure 16.12. It is composed of a half cell consisting of a hydrogen electrode in a solution of sulfuric acid and a half cell consisting of a zinc electrode in a solution of zinc sulfate. Zinc dissolves to give Zn^{2+} ions, and the electrons released flow around the wire of the external circuit and enter the platinum electrode, where they combine with hydrogen ions at its surface to give H_2. The standard cell potential is 0.76 V.

Rather than represent a cell by a detailed drawing, as in Figures 16.9, 16.10, and 16.12, we can use a **cell diagram**. For example, the cell in Figure 16.9 would be represented by

$$Zn(s)|ZnSO_4(aq)||CuSO_4(aq)|Cu(s)$$

and the cell shown in Figure 16.12 would be represented as

$$Zn(s)|ZnSO_4(aq)||H_2SO_4(aq)|H_2(g)\ Pt(s)$$

In these diagrams the single vertical line separates the electrode from the solution with which it is in contact, and the double vertical lines indicate a salt bridge or porous barrier between the two solutions. In such a cell diagram the anode—the electrode at which oxidation occurs—is on the left, and the cathode—the electrode at which reduction occurs—is on the right.

Just as we can think of the overall cell reaction as the sum of two half-reactions, we can think of the cell potential as the sum of the two half cell potentials: E_{ox} due to the oxidation half-reaction and E_{red} due to the reduction half-reaction:

$$E_{cell} = E_{ox} + E_{red}$$

Standard Reduction Potentials

It would be useful to have values for half-cell potentials, but no potential can be measured unless two half cells are connected to give a cell in which an overall reaction can occur. Therefore we cannot measure experimentally the potential associated with any individual half-reaction; we can measure only the *sum* of two half-cell potentials.

However, we can obtain relative values of standard half-cell potentials by making an arbitrary assumption about the potential of one particular half cell. The half cell chosen as the standard with which all other half-cell potentials are compared is the hydrogen half cell. This half cell is arbitrarily assigned a potential of 0 V.

$$E^\circ_{2H^+ + 2e^- \to H_2} = 0$$

The standard cell potential for the zinc-hydrogen cell is 0.76 V. Now since

$$E^\circ_{cell} = E^\circ_{ox} + E^\circ_{red}$$
$$0.76\ V = E^\circ_{Zn \to Zn^{2+} + 2e^-} + E^\circ_{2H^+ + 2e^- \to H_2}$$

And since we have assumed that

$$E^\circ_{2H^+ + 2e^- \to H_2} = 0$$
$$0.76\ V = E^\circ_{Zn \to Zn^{2+} + 2e^-} + 0$$

or

$$E^\circ_{Zn \to Zn^{2+} + 2e^-} = 0.76\ V$$

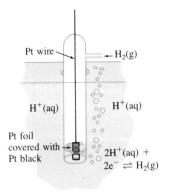

Pt wire — H₂(g)

H⁺(aq)   H⁺(aq)

Pt foil covered with Pt black $2H^+(aq) + 2e^- \rightleftharpoons H_2(g)$

Figure 16.11 A Hydrogen Electrode. This consists of a piece of platinum foil covered with a layer of finely divided platinum (platinum black), attached to a Pt wire, and surrounded by an atmosphere of hydrogen. The whole is immersed in an aqueous acidic solution.

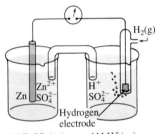

$H_2(g)$

Zn^{2+} H^+
$Zn\ SO_4^{2-}$ SO_4^{2-}

Hydrogen electrode

1M ZnSO₄(aq)   1M H⁺(aq)

Anode   Cathode

$Zn(s) \rightarrow$ $2H^+ + 2e^- \rightarrow$
$Zn^{2+} + 2e^-$ $H_2(g)$

Figure 16.12 Standard Electrochemical Cell Using Hydrogen Electrode. This cell is based on the reaction

$$Zn(s) + 2H^+(aq) \longrightarrow Zn^{2+}(aq) + H_2(g)$$

16.2 ELECTROCHEMICAL CELLS

605

Thus we see that the standard potential for the half-cell reaction

$$Zn(s) \longrightarrow Zn^{2+} + 2e^-$$

is $E^\circ_{ox} = 0.76$ V.

Now we can use this value to find the half-cell potential for the copper electrode by combining this electrode with the zinc electrode. We have already seen that the standard potential of the zinc-copper cell is 1.10 V. Therefore

$$1.10 \text{ V} = E^\circ_{ox} + E^\circ_{red}$$
$$= E^\circ_{Zn(s)\rightarrow Zn^{2+}+2e^-} + E^\circ_{Cu^{2+}+2e^-\rightarrow Cu(s)}$$
$$= 0.76 \text{ V} + E^\circ_{Cu^{2+}+2e^-\rightarrow Cu(s)}$$
$$E^\circ_{Cu^{2+}+2e^-\rightarrow Cu(s)} = 0.34 \text{ V}$$

Thus for the reaction

$$Cu^{2+} + 2e^- \longrightarrow Cu(s)$$

we have

$$E^\circ_{red} = 0.34 \text{ V}$$

We could alternatively have obtained the standard potential for the copper half cell by combining it with a hydrogen electrode. However, we know that copper does not dissolve in dilute aqueous acids. The equilibrium

$$Cu(s) + 2H^+ \rightleftharpoons Cu^{2+} + H_2(g)$$

lies far to the left. In other words, it is the reverse reaction that occurs when we combine a hydrogen half cell with a copper half cell; that is,

$$Cu^{2+} + H_2(g) \longrightarrow Cu(s) + 2H^+$$

The experimentally measured voltage of this cell is 0.34 V. Thus

$$0.34 \text{ V} = E^\circ_{ox} + E^\circ_{red}$$
$$= E^\circ_{H_2(g)\rightarrow 2H^++2e^-} + E^\circ_{Cu^{2+}+2e^-\rightarrow Cu(s)}$$
$$= 0 + E^\circ_{Cu^{2+}+2e^-\rightarrow Cu(s)}$$

So for the reaction

$$Cu^{2+} + 2e^- \longrightarrow Cu(s)$$

we have

$$E^\circ_{red} = 0.34 \text{ V}$$

This is the same value we obtained previously.

Example 16.6 Iron dissolves in dilute acid to give Fe^{2+}:

$$Fe(s) + 2H^+(aq) \longrightarrow Fe^{2+} + H_2(g)$$

The potential of a cell constructed from a standard hydrogen electrode and a half cell consisting of an iron electrode in a $1M$ solution of $FeSO_4$ is 0.44 V. What is the standard potential of the Fe half cell?

Solution

$$E^\circ_{cell} = E^\circ_{ox} + E^\circ_{red}$$
$$0.44 \text{ V} = E^\circ_{Fe\rightarrow Fe^{2+}+2e^-} + 0$$

Therefore

$$Fe \longrightarrow Fe^{2+} + 2e^- \qquad E^\circ_{ox} = 0.44 \text{ V}$$

By combining any half cell with a hydrogen electrode, we can determine the standard half-cell potential. Some of these potentials are for oxidation reactions, and others are for reduction reactions. It is convenient, in tabulating the values, to write them all the same way. By convention, *all standard half-cell potentials are listed as reductions* and are called **standard reduction potentials**. The signs for the potentials of oxidation reactions are reversed when they are written as reduction reactions. By convention, the most negative reduction potentials are listed at the top of the table, the most positive at the bottom. Table 16.1 gives the value of the standard reduction potentials for a number of half-reactions. Recall that these are the half-cell potentials when the partial pressures of any gases are 1 atm and the concentrations of ions are $1M$. The significance of a negative sign is that relative to H_2, the substances on the right tend to give up electrons and the half-reaction tends to proceed from right to left. The larger the negative value, the greater is the tendency of the reaction to proceed from right to left. The positive sign indicates that, relative to H^+, the substances on the left tend to acquire electrons, and the half-reaction tends to proceed from left to right. The larger the positive value, the greater is the tendency of the reaction to proceed to the right.

The substances in the table are listed in order of decreasing strength as reducing agents. At the top of the table the substances on the right, such as Li, K, and Mg, and H^-, are strong reducing agents. At the bottom of the table the substances on the right are very poor reducing agents, but those on the left are strong oxidizing agents, for example, NO_3^-, O_2, MnO_4^-, and Cl_2. The substances on the left are listed in order of increasing strength as oxidizing agents. Lithium is the strongest reducing agent listed in the table, and F_2 is the strongest oxidizing agent.

Notice that the table includes a number of half-reactions that do not involve the reduction of a metal ion to the corresponding metal, for example, the reduction of MnO_4^- to Mn^{2+}, the reduction of the halogens to the corresponding halide ions, and the reduction of Fe^{3+} to Fe^{2+}. In principle, at least, the electrode potentials corresponding to these reactions can be determined by setting up a half cell consisting of a solution of the ions (and gases) involved in the equilibrium and a platinum electrode to supply or remove the electrons. This half cell can then be combined with a hydrogen electrode or any other electrode of known potential.

Uses of Standard Reduction Potentials

The table of standard reduction potentials has a number of important uses.

RELATIVE STRENGTHS OF OXIDIZING AND REDUCING AGENTS We can easily predict the relative oxidizing or reducing strengths of any of the substances listed in the table. For example, since zinc comes above iron and has a more negative reduction potential, it is the stronger reducing agent. Since Br_2 comes below Fe^{3+} in the table and has a larger positive potential, it is the stronger oxidizing agent.

Example 16.7 Place the following in order of their strengths as oxidizing agents:

Cu^{2+}, MnO_4^-, Br_2, and Zn^{2+}

Table 16.1 Standard Reduction Potentials

OXIDIZING AGENTS	REACTION[a]	REDUCING AGENTS	$E°_{red}$ (V)
Acidic Solution			
Very weak	$Li^+ + e^- \longrightarrow Li(s)$	Very strong	-3.05
	$K^+ + e^- \longrightarrow K(s)$		-2.93
	$Ca^{2+} + 2e^- \longrightarrow Ca(s)$		-2.87
	$Na^+ + e^- \longrightarrow Na(s)$		-2.71
	$Mg^{2+} + 2e^- \longrightarrow Mg(s)$		-2.36
	$H_2(g) + 2e^- \longrightarrow 2H^-$		-2.25
	$Al^{3+} + 3e^- \longrightarrow Al(s)$		-1.66
	$2H_2O + 2e^- \longrightarrow H_2(g) + 2OH^-$		-0.83
	$Zn^{2+} + 2e^- \longrightarrow Zn(s)$		-0.76
	$Cr^{3+} + 3e^- \longrightarrow Cr(s)$		-0.74
	$Fe^{2+} + 2e^- \longrightarrow Fe(s)$		-0.44
	$Cr^{3+} + e^- \longrightarrow Cr^{2+}$		-0.41
	$V^{3+} + e^- \longrightarrow V^{2+}$		-0.26
	$Ni^{2+} + 2e^- \longrightarrow Ni(s)$		-0.25
	$Sn^{2+} + 2e^- \longrightarrow Sn$		-0.16
	$Pb^{2+} + 2e^- \longrightarrow Pb(s)$		-0.13
	$2H^+ + 2e^- \longrightarrow H_2(g)$		0
	$AgBr(s) + e^- \longrightarrow Ag(s) + Br^-$		$+0.10$
	$S(s) + 2H^+ + 2e^- \longrightarrow H_2S(aq)$		$+0.14$
	$Cu^{2+} + e^- \longrightarrow Cu^+$		$+0.15$
	$AgCl(s) + e^- \longrightarrow Ag(s) + Cl^-$		$+0.22$
	$Cu^{2+} + 2e^- \longrightarrow Cu(s)$		$+0.34$
	$Cu^+ + e^- \longrightarrow Cu(s)$		$+0.52$
	$I_2(s) + 2e^- \longrightarrow 2I^-$		$+0.54$
	$O_2(g) + 2H^+ + 2e^- \longrightarrow H_2O_2(aq)$		$+0.68$
	$Fe^{3+} + e^- \longrightarrow Fe^{2+}$		$+0.77$
	$Ag^+ + e^- \longrightarrow Ag(s)$		$+0.80$
	$2NO_3^- + 4H^+ + 2e^- \longrightarrow N_2O_4(g) + 2H_2O$		$+0.80$
	$2Hg^{2+} + 2e^- \longrightarrow Hg_2^{2+}$		$+0.92$
	$NO_3^- + 4H^+ + 3e^- \longrightarrow NO(g) + 2H_2O$		$+0.97$
	$Br_2 + 2e^- \longrightarrow 2Br^-$		$+1.09$
	$O_2(g) + 4H^+ + 4e^- \longrightarrow 2H_2O$		$+1.23$
	$Cr_2O_7^{2-} + 14H^+ + 6e^- \longrightarrow 2Cr^{3+} + 7H_2O$		$+1.33$
	$Cl_2(g) + 2e^- \longrightarrow 2Cl^-$		$+1.36$
	$MnO_4^- + 8H^+ + 5e^- \longrightarrow Mn^{2+} + 4H_2O$		$+1.49$
	$Au^{3+} + 3e^- \longrightarrow Au(s)$		$+1.50$
	$MnO_2 + 4H^+ + 2e^- \longrightarrow Mn^{2+} + 4H_2O$		$+1.61$
	$H_2O_2(aq) + 2H^+ + 2e^- \longrightarrow 2H_2O$		$+1.78$
	$Co^{3+} + e^- \longrightarrow Co^{2+}$		$+1.81$
Very strong	$F_2 + 2e^- \longrightarrow 2F^-$	Very weak	$+2.87$
Basic Solution			
	$2H_2O + 2e^- \longrightarrow H_2(g) + 2OH^-$		-0.83
	$Fe(OH)_3(s) + e^- \longrightarrow Fe(OH)_2(s) + OH^-$		-0.56
	$O_2(g) + e^- \longrightarrow O_2^-(aq)$		-0.56
	$O_2(g) + 2H_2O + 4e^- \longrightarrow 4OH^-$		$+0.40$

[a] All ions are in aqueous solution, and H_2O is in the liquid state.

Solution If we list the reduction potentials, we have

$$Zn^{2+} + 2e^- \longrightarrow Zn \qquad\qquad -0.76 \text{ V}$$
$$Cu^{2+} + 2e^- \longrightarrow Cu \qquad\qquad +0.34 \text{ V}$$
$$Br_2 + 2e^- \longrightarrow 2Br^- \qquad\qquad +1.09 \text{ V}$$
$$MnO_4^- + 8H^+ + 5e^- \longrightarrow Mn^{2+} + 4H_2O \qquad +1.49 \text{ V}$$

The substance with the highest positive potential is the strongest oxidizing agent. Therefore the order of oxidizing strength is

$$MnO_4^- > Br_2 > Cu^{2+} > Zn^{2+}$$

DIRECTION OF OXIDATION-REDUCTION REACTIONS   We can predict which way the overall reaction will proceed when we combine any two half-reactions. The reactions at the top of the table have a greater tendency to proceed to the left than those at the bottom. Therefore, when we combine any two half-reactions, the one that is higher in the table will go to the left, driving the other reaction to the right. Consider the combination of two half-reactions:

$$Zn^{2+} + 2e^- \longrightarrow Zn(s) \quad\text{and}\quad Fe^{2+} + 2e^- \longrightarrow Fe(s)$$

The zinc reaction is the higher in the table, so it will go to the left, forcing the other reaction to the right. Thus we must reverse the first reaction,

$$Zn(s) \longrightarrow Zn^{2+} + 2e^-$$

and then add it to the second,

$$Fe^{2+} + 2e^- \longrightarrow Fe(s)$$

to get the overall reaction:

$$Zn(s) + Fe^{2+} \longrightarrow Fe(s) + Zn^{2+}$$

Zinc is the stronger reducing agent and it reduces Fe^{2+} to Fe.

A simple way to remember how to combine two half-reactions is to write them down in the order in which they are listed in the table:

$$Zn^{2+} + 2e^- \longrightarrow Zn(s)$$
$$Fe^{2+} + 2e^- \longrightarrow Fe(s)$$

Then connect the two reactions by a counterclockwise arrow—it has the form of a letter *C* (combine counterclockwise)—that shows the direction in which the reactions proceed when they are combined. Thus if we wish to combine the two reactions

$$Cu^{2+} + 2e^- \longrightarrow Cu(s) \quad\text{and}\quad Ag^+ + e^- \longrightarrow Ag(s)$$

we write them in the order in which they are found in the table and we connect them with a ↻ to show us which way the reactions will proceed when combined. We then multiply the second reaction by 2 and add the two reactions to give

$$Cu(s) + 2Ag^+ \longrightarrow Cu^{2+} + Ag(s)$$

Copper reduces silver ion to silver.

We can similarly predict that iron will reduce Cu^{2+} to Cu, that zinc will reduce Pb^{2+} to Pb and will reduce Sn^{2+} to Sn (see Experiment 16.5).

CALCULATION OF CELL POTENTIALS   We can calculate the voltage of a cell made up of any two half cells by adding the half-cell voltages. Consider

16.2 ELECTROCHEMICAL CELLS

609

Metal Displacement Reactions

An iron nail dipped in copper sulfate solution becomes coated with a reddish brown layer of copper.

A coil of copper wire suspended in silver nitrate solution becomes covered with "whiskers" of silver, and the solution slowly turns blue as copper ions are formed.

A zinc rod in lead nitrate solution becomes coated with a spongy layer of lead.

Pieces of granulated zinc in an acidic solution of tin(II) chloride, $SnCl_2$ become coated with fine crystals of tin.

a cell made by combining the two half cells

$$Zn^{2+} + 2e^- \longrightarrow Zn(s) \quad \text{and} \quad Fe^{2+} + 2e^- \longrightarrow Fe(s)$$

We have seen that the first reaction will proceed in the reverse direction and that the overall reaction in the cell will be

$$Zn(s) + Fe^{2+} \longrightarrow Zn^{2+} + Fe(s)$$

The cell diagram is

$$Zn(s)|Zn^{2+}(aq)||Fe^{2+}(aq)|Fe(s)$$

The half-reactions and the corresponding potentials are

| Anode | $Zn(s) \longrightarrow Zn^{2+} + 2e^-$ | $E^\circ_{ox} = +0.76$ V |
| Cathode | $Fe^{2+} + 2e^- \longrightarrow Fe(s)$ | $E^\circ_{red} = -0.44$ V |

The potential for the oxidation of zinc is found by changing the sign of the potential for the reduction reaction given in Table 16.1. Adding the half-cell reactions and potentials gives, for the complete cell,

$$Zn(s) + Fe^{2+} \longrightarrow Zn^{2+} + Fe(s) \quad E^\circ_{cell} = +0.32 \text{ V}$$

Now consider the cell

$$Cu(s)|Cu^{2+}(aq)||Ag^+(aq)|Ag(s)$$

obtained by combining the two half cells

$$Cu^{2+} + 2e^- \longrightarrow Cu(s) \quad \text{and} \quad Ag^+ + e^- \longrightarrow Ag(s)$$

The copper half cell is higher in the table, and therefore the half-reaction must be reversed and written as an oxidation before combining it with the silver half-cell reaction:

Anode	$Cu(s) \longrightarrow Cu^{2+} + 2e^-$	$E_{ox}^\circ = -E_{red}^\circ = -0.34$ V
Cathode	$2[Ag^+ + e^- \longrightarrow Ag(s)]$	$E_{red}^\circ = +0.80$ V
Overall	$Cu(s) + 2Ag^+ \longrightarrow Cu^{2+} + 2Ag(s)$	$E_{cell}^\circ = +0.46$ V

Notice that when we multiply the equation for a half-reaction by an appropriate coefficient to obtain an equation for the overall reaction, we do *not* multiply the half-cell potential by this coefficient. The half-cell potential is always the value given in the table and is independent of any coefficient by which it may be necessary to multiply the corresponding equation in order to obtain a balanced equation for the overall reaction. The standard potential does not depend on the amounts of the reactants and the products but only on their concentrations, which are 1 mol L^{-1}. It therefore does not depend on how many electrons are transferred. A potential difference may be regarded as a difference in level between which the electrons flow; it does not depend on how many electrons flow between the levels.

Let us calculate the cell voltage for the cell

$$Ag(s)|Ag^+(aq)\| Cu^{2+}(aq)|Cu(s)$$

that is, for the reaction

$$2Ag(s) + Cu^{2+} \longrightarrow 2Ag^+ + Cu(s)$$

which is the reverse of the reaction that we have just considered. The half-cell reactions and their appropriate potentials are as follows:

Anode	$2[Ag(s) \longrightarrow Ag^+ + e^-]$	$E_{ox}^\circ = -E_{red}^\circ = -0.80$ V
Cathode	$Cu^{2+} + 2e^- \longrightarrow Cu(s)$	$E_{red}^\circ = +0.34$ V
Overall	$2Ag(s) + Cu^{2+} \longrightarrow 2Ag^+ + Cu(s)$	$E_{cell}^\circ = -0.46$ V

The overall cell voltage has a negative value, whereas in the previous case it had a positive value. This negative value indicates that the reaction does *not* proceed in the direction written but in the reverse direction. The reaction actually proceeds in the direction

$$2Ag^+ + Cu(s) \longrightarrow Cu^{2+} + 2Ag(s)$$

as we have seen previously. The *calculated* cell voltage for a reaction that proceeds spontaneously as written from left to right is always positive. A *calculated* cell voltage that is negative indicates that the reaction in fact proceeds in the reverse direction.

Example 16.8 What will be the spontaneous reaction when the following half-reactions are combined? What is the value of E_{cell}°?

(a) $Fe^{3+} + e^- \longrightarrow Fe^{2+}$

(b) $MnO_4^- + 8H^+ + 5e^- \longrightarrow Mn^{2+} + 4H_2O$

Solution From Table 16.1 we see that reaction (a) is higher in the table, so it proceeds to the left, driving reaction (b) to the right. We therefore reverse reaction (a) and change the sign of its potential. After multiplying by the appropriate coefficient, we add it to reaction (b). We then add the half-cell potentials to obtain E_{cell}°:

$5[Fe^{2+} \longrightarrow Fe^{3+} + e^-]$	$E_{ox}^\circ = -0.77$ V
$MnO_4^- + 8H^+ + 5e^- \longrightarrow Mn^{2+} + 4H_2O$	$E_{red}^\circ = +1.49$ V
$5Fe^{2+} + MnO_4^- + 8H^+ \longrightarrow 5Fe^{3+} + Mn^{2+} + 4H_2O$	$E_{cell}^\circ = +0.72$ V

The overall reaction is the reduction of MnO_4^- to Mn^{2+} and the oxidation of Fe^{2+} to Fe^{3+}. The cell voltage has a positive value, confirming that we combined the two half-cell reactions correctly.

PRODUCTS OF ELECTROLYSIS We saw earlier that in the electrolysis of aqueous solutions the reactions at the electrodes might be oxidation or reduction of ions, of water, or of the electrode. The table of standard reduction potentials enables us to predict which reactions will occur at the electrode, although, as we will see, there are some limitations to these predictions.

The substance that is reduced at the cathode will be the one that is most easily reduced. All those metals ions that are above the reaction for the reduction of water in the table are more difficult to reduce than water, so it is water that is preferentially reduced. Thus Na^+, Ca^{2+}, and Al^{3+}, for example, cannot be reduced in aqueous solution because water is preferentially reduced. But ions such as Cr^{3+}, Cu^{2+}, and Ag^+, which come below water in the table, are more easily reduced; they are therefore reduced to the metal in the electrolysis of aqueous solutions of their cations. Similarly, any reaction that comes above the reaction for the oxidation of water,

$$2H_2O \longrightarrow 4e^- + 4H^+ + O_2(g)$$

will occur more easily than the oxidation of water. Thus we see that Br^- will be preferentially oxidized to Br_2, and I^- will be oxidized to I_2. But F^- will not be oxidized to F_2, and Mn^{2+} will not be oxidized to MnO_4^-; instead, water will be oxidized.

However, there is a complication. We have already seen that chloride ion is oxidized to chlorine when a sodium chloride solution is electrolyzed, although the table predicts that water should be oxidized preferentially. This unexpected result is due to the fact that a higher voltage is sometimes needed for electrolysis than is indicated by the reduction potential. This additional voltage required to cause electrolysis is called *overvoltage*. Its origin is connected with the rates of the electrode reactions. If these reactions are slow, as is often the case when a gas is involved, then an additional voltage is required for the electrode reaction to take place. The overvoltage is considerably higher for the oxidation of water than for chloride ion, so chloride ion is more easily oxidized. As we will see, there is also a concentration effect. In very dilute solutions of chloride ion, water is more easily oxidized, and oxygen rather than chlorine is obtained.

Unfortunately, we cannot therefore always make reliable predictions about the products of electrolysis from standard reduction potentials. Nevertheless, we understand the general principles that govern which electrode reactions will occur and therefore why different electrode reactions are observed. Overvoltages are of considerable practical importance, but further discussion of them is beyond the scope of this book.

Nonstandard Conditions: Nernst Equation

The values of the standard reduction potentials listed in Table 16.1 are for solutions in which the dissolved species all have concentrations of $1M$ and the partial pressures of any gases involved are 1 atm. The greater the departure of the actual concentrations from these values, the greater is the deviation of the actual potential from the value listed. We can deduce the direction in which the potential changes with changing concentrations and pressures by applying Le Châtelier's principle.

Consider the equilibrium

$$Fe^{2+}(aq) + 2e^- \rightleftharpoons Fe(s) \qquad E^\circ_{red} = -0.44 \text{ V}$$

If the concentration of Fe^{2+} is decreased below $1M$, the equilibrium shifts to the left. The reduction has less tendency to proceed, and so E_{red} becomes more negative than -0.44 V. Similarly, if the concentration of Zn^{2+} is greater than $1M$, then the equilibrium

$$Zn(s) \rightleftharpoons Zn^{2+} + 2e^- \qquad E^\circ_{ox} = +0.76 \text{ V}$$

shifts to the left, and the potential is less than $+0.76$ V. The zinc-iron cell has a potential of $+0.32$ V if the concentrations of Fe^{2+} and Zn^{2+} are initially $1M$, but as the reaction

$$Zn(s) + Fe^{2+} \longrightarrow Zn^{2+} + Fe(s)$$

proceeds, the concentration of Fe^{2+} decreases and the concentration of Zn^{2+} increases. Thus both E_{red} and E_{ox} decrease, and therefore the potential of the cell ($E_{cell} = E_{red} + E_{ox}$) decreases. Eventually, the cell potential becomes zero—when the cell reaction has come to equilibrium and the reaction no longer has any tendency to proceed. Since all chemical reactions eventually reach equilibrium, the voltage of any cell decreases with time as equilibrium is approached and becomes zero when equilibrium is reached. We then say that a battery is "dead" or that a battery has "run down"; the cell reaction has proceeded to equilibrium or so close to equilibrium that the corresponding cell potential is very nearly zero.

The quantitative relationship between the cell potential and the concentrations of the reactants is called the **Nernst equation**. It was first derived by German chemist Walter Nernst (1864–1941) in 1899. The potential E of a cell is given by the expression

$$E = E^\circ - \frac{RT}{nF} \ln Q$$

In this equation, E° is the standard potential; Q is the reaction quotient; that is, an expression that has the same form as the expression for the equilibrium constant but the concentrations are not the equilibrium concentrations (see Chapter 14). In this equation, F is the Faraday constant and n is the number of moles of electrons transferred in the reaction. Note that ln means logarithm to the base e, that is, the natural logarithm, not logarithm to the base 10, which we write as log. The relationship between $\log x$ and $\ln x$ is $\ln x = 2.303 \log x$. At 25°C

$$\frac{RT}{F} = \frac{(8.314 \text{ J K}^{-1})(298.2 \text{ K})}{96\,500 \text{ C mol}^{-1}} = 0.0257 \text{ J C}^{-1} = 0.0257 \text{ V}$$

Thus at 25°C the Nernst equation becomes

$$E = E^\circ - \frac{0.0257}{n} \ln Q \qquad \text{or} \qquad E = E^\circ - \frac{0.0592}{n} \log Q$$

Using this equation, we can calculate the potential of a cell if we know the concentrations of the reactants and the standard cell potential, E°. Consider, as an example, a cell based on the reaction

$$Zn(s) + Cu^{2+}(aq) \longrightarrow Zn^{2+}(aq) + Cu(s)$$

for which $E^\circ = 1.10$ V. In this reaction 2 mol of electrons are transferred for each mole of Cu^{2+} that is reduced, so $n = 2$. Hence the Nernst equation becomes

$$E = 1.10 - \frac{0.0257}{2} \ln \frac{[Zn^{2+}]}{[Cu^{2+}]}$$

Recall from Chapter 14 that Q includes the concentrations of species in solution but not the solids taking part in a reaction.

Consider a case in which the cell started to operate with $[Zn^{2+}] = [Cu^{2+}] = 1.00M$, and after a certain time the concentrations had changed so that $[Cu^{2+}] = 0.10M$ and $[Zn^{2+}] = 1.90M$. Then

$$E = 1.10 - \frac{0.0257}{2} \ln \frac{1.9}{0.1} = 1.10 - \frac{0.0257}{2} \ln 19$$

$$= 1.10 - 0.038 = 1.06 \text{ V}$$

We see that the cell voltage has decreased but only by a rather small amount.

Let us calculate the voltage after a further period of time when the concentrations have reached the values $[Cu^{2+}] = 0.001$ and $[Zn^{2+}] = 1.999$:

$$E = 1.10 - \frac{0.0257}{2} \ln \frac{1.999}{0.001}$$

$$= 1.10 - 0.10 = 1.00 \text{ V}$$

Thus as we deduced quantitatively by using Le Châtelier's principle, the cell voltage continually decreases as the cell reaction proceeds.

EQUILIBRIUM CONSTANTS FROM THE NERNST EQUATION The potential of a cell becomes zero when the cell reaction has attained equilibrium. The reaction quotient, Q, is then equal to the equilibrium constant, K, and the Nernst equation may be written in the form

$$E = E° - \frac{RT}{nF} \ln K = 0$$

This can be rearranged to

$$\ln K = \frac{nFE°}{RT} = \frac{nE°}{0.0257} \qquad \text{or} \qquad \log K = \frac{nE°}{0.0592} \qquad \text{at } 25°C$$

This equation enables us to calculate the equilibrium constant for any oxidation-reduction reaction in aqueous solution from the corresponding standard cell potential, which in turn can often be obtained from the tabulated standard reduction potentials of the two half-cell reactions (Table 16.1).

Example 16.9 What is the value of the equilibrium constant for the oxidation of Fe^{2+} to Fe^{3+} by MnO_4^- in aqueous acid solution?

Solution The two half-cell reactions and the corresponding standard reduction potentials, from Table 16.1, are as follows:

$$5[Fe^{2+}(aq) \longrightarrow Fe^{3+}(aq) + e^-] \qquad E°_{ox} = -0.77 \text{ V}$$

$$MnO_4^- + 8H^+(aq) + 5e^- \longrightarrow Mn^{2+}(aq) + 4H_2O \qquad E°_{red} = +1.49 \text{ V}$$

Adding the two half-cell reactions and the corresponding potentials gives the overall cell reaction and the corresponding cell voltage:

$$5Fe^{2+}(aq) + MnO_4^-(aq) + 8H^+(aq) \longrightarrow 5Fe^{3+}(aq) + Mn^{2+}(aq) + 4H_2O$$

$$E°_{cell} = +0.72 \text{ V}$$

The positive cell voltage shows that the reaction proceeds as written.

Next, we can calculate the equilibrium constant from the expression

$$\ln K = \frac{nE^\circ}{0.0257}$$

In this reaction 5 mol of electrons are transferred, so $n = 5$. Thus

$$\ln K = \frac{5(0.72)}{0.0257} = 1.4 \times 10^2$$

$$K = 6.3 \times 10^{60}$$

This value is extremely large, and we therefore know that this reaction goes essentially to completion.

This example illustrates an important use of the Nernst equation, namely, to obtain values of equilibrium constants for reactions that go essentially to completion—that is, for reactions that have very large (or very small) equilibrium constants—and that are difficult to determine in any other way. Solubility product constants, which often have very small values, can be similarly obtained by using the Nernst equation.

Example 16.10 Calculate a value for the solubility product constant of AgCl at 25°C from standard reduction potentials.

Solution The overall reaction for which we require the cell potential is

$$AgCl(s) \longrightarrow Ag^+ + Cl^-$$

We can obtain this reaction by adding the half-reactions:

$$Ag(s) \longrightarrow Ag^+ + e^- \qquad E^\circ_{ox} = -0.80 \text{ V}$$
$$AgCl(s) + e^- \longrightarrow Ag(s) + Cl^- \qquad E^\circ_{red} = +0.22 \text{ V}$$

Adding the half-cell potentials gives the standard cell potential:

$$AgCl(s) \longrightarrow Ag^+ + Cl^- \qquad E^\circ = -0.58 \text{ V}$$

The negative potential shows that the reaction is spontaneous in the reverse direction from that written. We know that addition of chloride ion to a solution of silver ion gives a precipitate of silver chloride. The equilibrium position of the reaction lies far to the left. We can calculate the equilibrium constant from the expression

$$\ln K = \frac{nE^\circ}{0.0257} = \frac{1(-0.58)}{0.0257} = -22.6$$

Hence

$$K = [Ag^+][Cl^-] = 1.5 \times 10^{-10} = K_{sp}(AgCl)$$

pH METER Another important application of the Nernst equation is for determining the concentration of an ion in solution by measuring the potential of a suitably designed cell. Consider a cell composed of a copper electrode and a hydrogen electrode:

$$Pt, H_2(g) | H^+(aq) || Cu^{2+}(aq) | Cu(s)$$

The cell reaction is

$$Cu^{2+}(aq) + H_2(g) \longrightarrow Cu(s) + 2H^+(aq)$$

and the Nernst equation for this reaction is

$$E = E° - \frac{0.0257}{2} \ln \frac{[H^+]^2}{[Cu^{2+}]p_{H_2}}$$

If the concentration of Cu^{2+} is 1 mol L^{-1} and the pressure of H_2 is 1 atm, then

$$E = E° - \frac{0.0257}{2} \ln [H^+]^2$$

or

$$E = E° - 0.0257 \ln [H^+]$$

Hence

$$E = E° - 0.0592 \log [H^+] = E° + 0.0592(pH)$$

Thus the emf of the cell is proportional to the pH. By measuring the emf of the cell, we can obtain the pH of the solution in which the hydrogen electrode is immersed. A pH meter is an instrument based on such a cell, although it does not use a hydrogen electrode but a more practical electrode called a *glass electrode*. This electrode makes use of the fact that a potential difference is developed across a thin glass membrane that depends on the concentrations of hydrogen ions on the two sides of the membrane. If the hydrogen ion concentration inside a thin-walled glass bulb is kept constant, the hydrogen ion concentration of the solution into which the electrode is dipped can be measured from the potential developed across the thin glass wall (Figure 16.13).

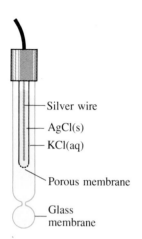

Silver wire
AgCl(s)
KCl(aq)
Porous membrane
Glass membrane

Figure 16.13 Glass Electrode. The potential of this electrode depends on the concentration of hydrogen ion in any solution into which it is inserted.

Batteries

A **battery** is an electrochemical cell that is used as an energy source. It stores energy in the form of oxidizing and reducing agents and then releases the energy as electricity when it is needed. In principle, any oxidation-reduction reaction can be used as the basis of an electrochemical cell, but there are many limitations to the use of most reactions as the basis of a practical battery. A useful battery should be reasonably light and compact and easily transported; it should have a reasonably long life, both when it is being used and when it is not. Also, it is essential for many applications that the voltage of the cell stay constant during use.

There are two main types of batteries: *primary cells*, in which the reaction occurs only once and the battery is then dead and cannot be used again, and *secondary cells*, which can be recharged by passing a current through them so that they can be used again and again.

PRIMARY CELLS The most familiar battery is the type that is widely used, for example, in portable radios and flashlights (Figure 16.14). It is called a *Leclanché* cell, after its inventor, or a *dry cell*. The anode consists of a zinc can; the cathode is a graphite rod surrounded by powdered MnO_2, which is a conductor. The space between the electrodes is filled with a moist paste of NH_4Cl and $ZnCl_2$. Strictly speaking, the cell is not dry; it contains a moist paste in place of an aqueous solution. The electrode reactions are complex but they may be written approximately as

Anode $\qquad\qquad Zn(s) \longrightarrow Zn^{2+} + 2e^-$

Cathode $\qquad MnO_2 + NH_4^+ + e^- \longrightarrow MnO(OH) + NH_3$

In the cathode reaction manganese is reduced from the +4 oxidation state to the +3 oxidation state. Ammonia is not liberated as a gas, but it combines with Zn^{2+} to form the ion, $Zn(NH_3)_4^{2+}$. The dry cell does not have an indefinite life, even when not used, because the acidic NH_4Cl corrodes the zinc can.

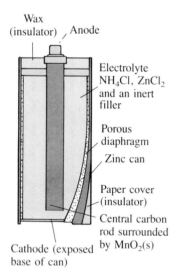

Wax (insulator)
Anode
Electrolyte NH_4Cl, $ZnCl_2$ and an inert filler
Porous diaphragm
Zinc can
Paper cover (insulator)
Central carbon rod surrounded by $MnO_2(s)$
Cathode (exposed base of can)

Figure 16.14 Dry Cell.

An improved version of this battery, known as an *alkaline battery*, has a zinc rod anode and a cathode of MnO_2. The electrolyte is KOH in the form of a thick gel. The battery is enclosed in a steel container. The alkaline battery has a longer life than the dry cell but is more expensive. Both cells have a potential of 1.5 V.

Another important, although more expensive, battery is the *mercury cell*. It can be made in very small sizes and has many uses, for example, in hearing aids, watches, and cameras. A zinc-mercury amalgam is the anode; a paste of HgO and carbon is the cathode. The electrolyte is a paste of KOH and ZnO (Figure 16.15). The electrode reactions are as follows:

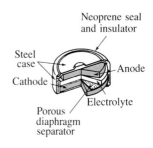

Anode $Zn(amalgam) + 2OH^- \longrightarrow ZnO(s) + H_2O + 2e^-$

Cathode $HgO(s) + H_2O + 2e^- \longrightarrow Hg(l) + 2OH^-$

Overall $Zn(amalgam) + HgO(s) \longrightarrow ZnO(s) + Hg(l)$

Figure 16.15 Mercury Cell.

The overall cell reaction does not involve any ions in solution whose concentrations can change. As a result, this battery has the advantage of maintaining a constant potential (1.34 V) throughout its life.

SECONDARY CELLS The best known and most important secondary cell is the *lead storage battery* (Figure 16.16). The anode is a grid of lead alloy packed with finely divided spongy lead. The cathode is a grid of lead alloy packed with lead(IV) oxide, PbO_2. The electrolyte is a solution of sulfuric acid (38% by mass). At the anode lead is oxidized to Pb^{2+} and insoluble $PbSO_4$ is formed. At the cathode PbO_2 is reduced to Pb^{2+} and $PbSO_4$ is formed. The reactions are as follows:

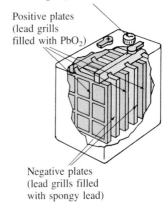

Capped hole for testing and replenishing electrolyte of H_2SO_4 and distilled water.

Positive plates (lead grills filled with PbO_2)

Negative plates (lead grills filled with spongy lead)

Anode $Pb(s) + SO_4^{2-} \longrightarrow PbSO_4(s) + 2e^-$

Cathode $PbO_2(s) + SO_4^{2-} + 4H^+ + 2e^- \longrightarrow PbSO_4(s) + 2H_2O$

Overall $Pb(s) + PbO_2(s) + 2H_2SO_4(aq) \rightleftharpoons 2PbSO_4(s) + 2H_2O(l)$

Figure 16.16 Lead Storage Battery.

Thus lead sulfate is formed at each electrode, and sulfuric acid is used up. The sulfuric acid solution thereby becomes more dilute, and the voltage falls. The extent to which the battery has been discharged can be checked by measuring the density of the acid, which decreases with decreasing concentration (see Figure 16.17).

An important advantage of a secondary cell is that it can be recharged. A potential that is larger than the cell potential is applied to the cell so as to force a current through the cell and reverse the electrode reactions. Therefore during recharging the $PbSO_4$ is reduced to Pb at one electrode, and at the other electrode Pb is oxidized to PbO_2. The overall reaction during charge is

$$2PbSO_4(s) + 2H_2O(l) \longrightarrow Pb(s) + PbO_2(s) + 2H_2SO_4(aq)$$

A single lead cell has a potential of 2.0 V. Normally, three or six such cells are connected in series to give a 6-V or a 12-V battery. The lead storage battery has proved to be a very practical source of electric power, particularly for applications in which it is used many times but can be easily recharged between. It is used, for example, to start the engine of an automobile and then is recharged by the generator as soon as the engine is running.

If a lead storage battery is recharged too rapidly, however, hydrogen and oxygen may be formed at the electrodes:

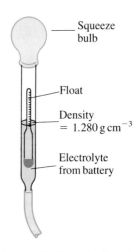

Figure 16.17 Measuring Density of Sulfuric Acid Electrolyte in Lead Storage Battery by Using Hydrometer. The depth to which the float is submerged depends on the density of the liquid.

Anode $2H_2O(l) \longrightarrow O_2(g) + 4H^+(aq) + 4e^-$

Cathode $2H^+(aq) + 2e^- \longrightarrow H_2(g)$

16.2 ELECTROCHEMICAL CELLS

617

There is then some danger of an explosion because of the possible reaction

$$2H_2(g) + O_2(g) \longrightarrow 2H_2O(g)$$

The oxygen for this reaction may be the oxygen produced in the cell or may come from the air. Thus sparks and flames should not be brought near a lead storage battery when it is being charged. The anode reaction that produces oxygen uses up water, which may also be lost by evaporation. Thus water must be added to the battery from time to time. Maintenance-free batteries use lead-calcium alloy electrodes at which the evolution of hydrogen and oxygen is very slow. Because very little water is electrolyzed, there is no need to add water to the cell.

The *nickel-cadmium* (Nicad) *battery* is more expensive than the lead battery, but it is smaller and lighter and it can be made as a sealed unit. It is used in portable, cordless appliances such as electric razors and calculators. It consists of a cadmium anode and a cathode that is a metal grid containing $NiO(OH)$. The electrolyte is KOH. The reactions during discharge and charge are

$$\text{Anode} \qquad Cd(s) + 2OH^- \underset{\text{Charge}}{\overset{\text{Discharge}}{\rightleftharpoons}} Cd(OH)_2(s) + 2e^-$$

$$\text{Cathode} \qquad NiO(OH)(s) + H_2O + e^- \underset{\text{Charge}}{\overset{\text{Discharge}}{\rightleftharpoons}} Ni(OH)_2 + OH^-$$

The overall reaction is

$$Cd(s) + 2NiO(OH)(s) + 2H_2O \underset{\text{Charge}}{\overset{\text{Discharge}}{\rightleftharpoons}} Cd(OH)_2(s) + 2Ni(OH)_2(s)$$

Since all the reactants and products in the overall reaction are in the solid state, there is no change in the concentration of any ions in solution during the discharge reaction. Therefore the voltage of the cell (1.35 V) remains very constant.

In recent years much research has been done to develop secondary cells that will deliver large currents for a lengthy period. Such a cell would be particularly useful for an electrically powered automobile. However, a suitable cell for this purpose has not yet been developed.

FUEL CELLS The oxidizing and reducing agents in a secondary cell can be regenerated by recharging, but batteries can also be made in which the reactants are fed continuously to the electrodes. Such batteries are called **fuel cells**. One of the most successful fuel cells uses the reaction of hydrogen with oxygen to form water. The cell is illustrated in Figure 16.18. The electrodes are hollow tubes made of porous compressed carbon impregnated with a catalyst. The electrolyte is KOH. At the electrodes hydrogen is oxidized to water, and oxygen is reduced to the -2 oxidation state in water:

$$\text{Anode} \qquad 2[H_2(g) + 2OH^-(aq) \longrightarrow 2H_2O(l) + 2e^-]$$

$$\text{Cathode} \qquad O_2(g) + 2H_2O(l) + 4e^- \longrightarrow 4OH^-(aq)$$

$$\text{Overall} \qquad 2H_2(g) + O_2(g) \longrightarrow 2H_2O(l)$$

Such a cell runs continuously as long as the reactants are supplied. Because fuel cells convert the energy of a fuel directly to electricity, they are potentially more efficient than conventional methods of generating electricity on a large scale by burning hydrocarbon fuels or by using nuclear reactors. In both these cases heat is produced to turn water into steam to drive turbines, which in turn drive electric generators. These indirect methods of generating electricity are not very efficient. Only about 40% of the energy produced by burning fossil fuels is converted to electricity. At present fuel cells are too expensive for large-scale use, but their importance is growing. They have been used very successfully in some special applications, for example, in spacecraft.

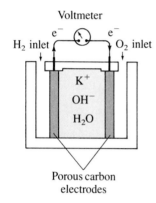

Figure 16.18 Fuel Cell. This cell is based on the reaction $2H_2 + O_2 \rightarrow 2H_2O$. The hydrogen and oxygen gases are fed continuously to the cell.

Corrosion is the deterioration of metals as a result of their reactions with the environment. The rusting of iron, the tarnishing of silver, the development of a green coating on copper, brass, and bronze are all familiar examples of corrosion. It causes enormous damage to bridges, ships, and cars, for example. The damage and the efforts taken to prevent it cost billions of dollars a year.

Corrosion occurs because most metals are relatively easily oxidized by the loss of electrons to oxygen in the air to form metal oxides. Oxygen is adsorbed on the metal surface, where it reacts to form an oxide layer. The formation of oxide is not always harmful, however. Aluminum forms a tough, dense transparent layer of Al_2O_3, which protects the metal underneath from further oxidation. Iron, on the other hand, forms an oxide coating that is relatively porous and easily cracks as it thickens. As a result, oxygen and moisture can continue to reach the metal, and it continues to oxidize until it is completely destroyed.

Rust is a hydrated oxide of iron(III), $Fe_2O_3 \cdot xH_2O$. It forms only in the presence of oxygen and water. Rusting is an electrochemical process. The reactions involved are complex and not completely understood, but the main steps are thought to be as follows: The iron in one part of an iron object behaves as an anode, and the iron is oxidized to Fe^{2+}:

$$Fe(s) \longrightarrow Fe^{2+} + 2e^- \qquad E^\circ_{ox} = 0.44 \text{ V}$$

The electrons produced in this half-reaction flow through the metal to another part of the object, which acts as a cathode. Here atmospheric oxen is reduced in the presence of $H^+(aq)$ supplied by H_2CO_3 formed from dissolved CO_2:

$$O_2 + 4H^+ + 4e^- \longrightarrow 2H_2O \qquad E^\circ_{red} = 1.23 \text{ V}$$

Figure 16.19 shows the formation of rust in the vicinity of a drop of water on an iron surface.

The circuit is completed by the movement of ions through water on the surface of the iron. This explains why rusting is particularly rapid in salt water, which has a high concentration of ions. Salt spread on roads in the winter similarly speeds up the rusting of cars. The overall reaction is the sum of the cathode and anode reactions:

$$2Fe(s) + O_2(g) + 4H^+ \longrightarrow 2Fe^{2+} + 2H_2O \qquad E^\circ_{cell} = 1.67 \text{ V}$$

The Fe^{2+} is further oxidized by atmospheric oxygen to rust:

$$4Fe^{2+} + O_2 + 4H_2O \longrightarrow 2Fe_2O_3(s) + 8H^+$$

Note that this reaction also provides hydrogen ions that are needed in the reduction of oxygen.

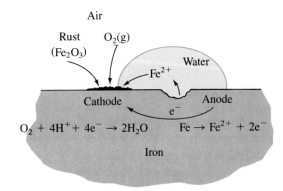

Figure 16.19 The Rusting of Iron. Iron in contact with water forms the anode, where iron is oxidized to Fe^{2+}. Iron in contact with air forms the cathode, where oxygen is reduced to water. Thus a cell is set up and iron continues to go into solution as Fe^{2+}. Where the Fe^{2+} solution is in contact with the air, the Fe^{2+} is oxidized to insoluble Fe_2O_3 (rust).

619

Inhibiting Corrosion

The most obvious way of minimizing corrosion is to cover the surface of the metal with a coating that keeps out air and moisture. This coating can be provided in several ways. Steel used in bridges and buildings is often painted with red lead paint, which contains Pb_3O_4. The Pb_3O_4 apparently oxidizes the surface of the iron to form a tough continuous layer of the oxide, which resists further oxidation.

Another method of protecting iron is to coat it with a layer of another metal. Zinc, tin, and chromium are often used for this purpose. The use of a zinc coating to protect iron is called *galvanizing*. Since zinc is above iron in the table of reduction potentials, it is more easily oxidized. It would not appear therefore to be very useful for protecting iron. However, zinc is oxidized to $Zn(OH)_2$, which reacts with the CO_2 in the atmosphere to form a tough layer of $Zn(OH)_2 \cdot xZnCO_3$, which strongly adheres to the surface, protecting the metal beneath. Even if the zinc layer is cracked or broken, it continues to protect the iron, as it becomes the anode in a cell in which the iron is the cathode. Hence zinc is oxidized rather than iron (see Figure 16.20).

A layer of tin, which becomes coated with a layer of oxide, can also be used to protect iron, as, for example, in tin cans. Tin comes below iron in the table of standard reduction potentials and is therefore less easily oxidized. But if the tin coating is cracked or scratched, the tin becomes the cathode in a cell and the iron becomes the anode. The oxidation of Fe to Fe^{3+}—rusting—will then be speeded up.

CATHODIC PROTECTION We have seen that iron corrodes by becoming the anode of an electrochemical cell. However, if we connect the iron to a more easily oxidized metal such as zinc or magnesium, then this metal becomes the anode of a cell and *it* corrodes instead of the iron. The iron becomes the cathode at which oxygen is reduced to water. This method of preventing the corrosion of iron is called **cathodic protection** (Experiment 16.6). For example, if a magnesium or zinc rod is connected to an underground tank or steel pipe, then the magnesium or zinc is oxidized instead of the pipe (see Figure 16.21). The more easily oxidized metal is called a *sacrificial anode*. Although it is gradually dissolved, it is easier to replace than the pipe or the tank. This method also provides excellent protection for bridge foundations and steel ships.

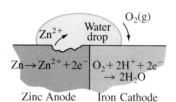

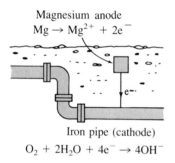

Figure 16.20 Galvanized Iron. Cathodic protection of iron in contact with zinc is provided by galvanizing.

Magnesium anode
$$Mg \rightarrow Mg^{2+} + 2e^-$$

Iron pipe (cathode)
$$O_2 + 2H_2O + 4e^- \rightarrow 4OH^-$$

Figure 16.21 Cathodic Protection. When a block of magnesium or zinc is attached to an iron pipe or tank buried underground, the magnesium forms the anode in an electrolytic cell and the iron becomes the cathode. Thus the magnesium or zinc is slowly oxidized and goes into solution as Mg^{2+} or Zn^{2+}. Oxygen is reduced at the iron cathode leaving the iron unattacked.

Cathodic Protection

Left: An iron nail placed in a beaker of warm water containing phenophthalein soon becomes covered in rust, and the indicator turns pink due to the hydroxide ions that are formed. Right: If the nail is wrapped in magnesium ribbon, the magnesium, which is more easily oxidized, becomes the anode in an electrochemical cell and the iron becomes the cathode. Thus magnesium, rather than iron, is oxidized and no rust forms.

IMPORTANT TERMS

An **anode** is the electrode at which oxidation occurs.

A **battery** is an electrochemical cell that is used as an energy source.

A **cathode** is the electrode at which reduction occurs.

Cathodic protection is a method of protecting a metal from corrosion by making it the cathode of a cell in which some other, more easily oxidized metal acts as the anode and is oxidized preferentially.

The **coulomb** (C) is the SI unit of electric charge. It is the quantity of electric charge carried by 1 ampere (A) in 1 second (s): $1 \text{ C} = 1 \text{ A s}$.

An **electrochemical cell** is a device for producing an electric current. It consists of two half cells, in one of which a substance is oxidized and in the other a substance is reduced.

An **electrode** is a strip of metal, or other conducting material, by means of which electrons are transferred to or from a solution in an electrochemical or electrolytic cell.

Electrolysis is a process in which electrical energy is used to produce chemical change.

An **electrolytic cell** is an apparatus in which electrolysis occurs, that is, in which a chemical reaction is carried out by using an external source of electrical energy.

The **Faraday constant** is the charge associated with 1 mol of electrons, $96\,485 \text{ C mol}^{-1}$.

A **fuel cell** is an electrochemical cell in which the substances that are oxidized and reduced are fed continuously to the anode and the cathode, respectively.

In **ionic conduction**, an electric current is carried by the movement of ions.

The **Nernst equation** is an expression that gives the relation between the voltage of a cell and the concentrations of the reactants.

The **standard cell potential**, $E°_{\text{cell}}$, is the potential of a cell when all the concentrations of solutes are 1 mol L$^-$, all the partial pressures of gases are 1 atm, and the temperature is 25°C.

The **standard reduction potential**, $E°$, is the reduction potential of a half-reaction with all the ions at $1M$ concentration and all the gases at 1 atm, measured relative to the hydrogen electrode, for which $E°$ is exactly 0 V at 25°C.

PROBLEMS

Electrolysis

1. What is meant by electrolysis and by an electrolyte? How does ionic conduction differ from electronic conduction?

2. How many coulombs are required for the following reductions?

    (a) 1 mol of Cu^{2+} to Cu

    (b) 1 mol of Fe^{3+} to Fe^{2+}

    (c) 1 mol of MnO_4^- to Mn^{2+}

    (d) 1 mol of ClO_3^- to Cl^-

3. How many coulombs are required for the following oxidations?

    (a) 1 mol H_2O to O_2 (b) 1 mol Cl_2 to ClO_3^-

    (c) 1 mol Pb to PbO_2 (d) 1 mol FeO to Fe_2O_3

4. How many coulombs are required to produce the following?

    (a) 50.0 mL of O_2 (g) (at STP) from Na_2SO_4 (aq).

    (b) 50.0 kg of aluminum from molten Al_2O_3.

    (c) 20.0 g of calcium from molten $CaCl_2$.

    (d) 5.00 g of silver from an aqueous $AgNO_3$ solution.

5. How long will it take to produce 1.00 kg of nickel by electrolysis of an aqueous solution containing Ni^{2+} if a current of 2.00 A is used?

6. What mass of NaOH and what mass of Cl_2 will be produced by electrolysis of an aqueous sodium chloride solution for 3.00 h, using a current of 0.200 A?

7. What volumes of H_2 and O_2 gases are produced by the electrolysis of an aqueous sodium sulfate solution for 30 min, using a current of 4.00 A, if the atmospheric pressure and temperature are 740 torr and 27°C?

8. How long would it take to deposit 16.0 g of silver from a solution of $AgNO_3$, using a current of 6.00 A?

9. How long would it take to plate a metal tray with a coating of silver of thickness 0.020 mm, using a current of 7.65 A, if the tray is of dimensions 24 cm × 12 cm? Neglect the amount of silver needed to coat the edges of the tray (density of Ag $= 10.54 \text{ g cm}^{-3}$).

10. For how long would an electric current of 1.50 A have to be passed through a solution containing Cr^{3+} ions in order for a steel object of surface area 0.10 m^2 to be coated with a layer of chromium 0.1 mm thick (density of Cr $= 7.1 \text{ g cm}^{-3}$).

11. The same quantity of electricity that causes 10.0 g of copper to be plated out from a solution of $CuSO_4$ is passed through a solution of $NiCl_2$. What mass of nickel will be plated out?

12. Write equations for the electrode reactions in the electrolysis of (a) molten $AlCl_3$ and (b) a dilute aqueous solution of $AlCl_3$. Why are the products of the electrolysis of (b) different from those of (a)?

13. What are the products of the electrolysis of the following with inert electrodes?

    (a) Molten $MgBr_2$ (b) Aqueous $Cu(NO_3)_2$

    (c) Aqueous HI   (d) Molten $FeCl_3$

14. Sketch a cell for the electrolysis of an aqueous HBr solution using platinum electrodes. Label the anode and the cathode. Write equations for the electrode reactions. Show the directions in which the ions are moving.

15. Sketch a cell for the electrolysis of a $CuSO_4$ solution using copper electrodes. Label the anode and the cathode. Show the directions in which the ions are moving. Write equations for the electrode reactions.

16. When a current of 1.487 A was passed through an aqueous HI solution for 1 h 0 min 1.5 s, 7.2428 g of I_2 was formed at the anode. Determine a value for the Faraday constant from these data. Compare this value with the accepted value.

17. A Hall cell for the production of aluminum from alumina, Al_2O_3, operates at a current of 130 000 A. What mass of aluminum will be produced in 1 min? If, in practice, 100 such cells are connected in series, how much aluminum can be produced per day? What mass of carbon will be consumed at the anodes per day?

18. When a current of 0.500 A was passed through a solution of an M^{2+} ion for exactly 2 h, 1.98 g of the metal M was deposited. What is the atomic mass of the metal M? What is the metal?

Standard Reduction Potentials

19. Which of the following will be oxidized by an acid solution of the dichromate ion, $Cr_2O_7^{2-}$?

F^- Cl^- Br^- I^-

Hg_2^{2+} Mn^{2+} Fe^{2+} H_2S (to S)

20. Use standard half-cell reduction potentials to predict which of the following reactions will occur in acid solution with all soluble substances at a concentration of $1M$. Complete and balance the equation for each reaction that is predicted to occur.

    (a) $H_2O_2 + Cu^{2+} \longrightarrow Cu + O_2$

    (b) $Ag^+ + Fe^{2+} \longrightarrow Ag + Fe^{3+}$

    (c) $I^- + NO_3^- \longrightarrow I_2 + NO$

    (d) $H_2SO_3 + H_2S \longrightarrow S$

21. From a consideration of the appropriate standard reduction potentials, predict whether or not dichromate ion in acid solution will oxidize water to oxygen, O_2. Why does an acidic solution of dichromate ion in water remain unchanged for a long period?

22. Which of the following reactions will occur in the forward direction under standard conditions? Balance the equations.

    (a) $Mg + Cr^{3+} \longrightarrow Mg^{2+} + Cr$

    (b) $I_2(aq) + H^+ \longrightarrow H_2 + I^-$

    (c) $Cu + NO_3^- \longrightarrow Cu^{2+} + NO_2 + H_2O$

in acid solution

23. Arrange the following metals in the order in which they displace each other from solution:

Aluminum   Copper   Iron   Magnesium   Zinc

24. Which of the following species are reduced by Fe^{2+}?

Ag^+ Cr^{3+} Zn^{2+} I_2 Br_2 MnO_4^-

25. Which of the following species are oxidized by manganese dioxide?

Br^- Ag^+ I^- Cl^-

26. Use standard reduction potentials to predict the reaction, if any, that occurs between the following:

    (a) $Fe^{3+}(aq)$ and $I^-(aq)$

    (b) $Ag^+(aq)$ and $Cu(s)$

    (c) $Fe^{3+}(aq)$ and $Br^-(aq)$

    (d) $Ag(s)$ and $Fe^{3+}(aq)$

    (e) $Br_2(aq)$ and $Fe^{2+}(aq)$

27. Use standard reduction potentials to predict whether each of the following reactions will occur in acid solution with all soluble substances at a concentration of $1M$. Complete and balance the equation for each reaction that occurs.

    (a) $Mn^{2+} + Cr_2O_7^{2-} \longrightarrow MnO_4^- + Cr^{3+}$

    (b) $O_2 + Br^- \longrightarrow Br_2$

    (c) $Au + Cl_2 \longrightarrow Au^{3+} + Cl^-$

Electrochemical Cells

28. Sketch an electrochemical cell in which the reaction is

$$Zn(s) + 2Ag^+(aq) \longrightarrow Zn^{2+}(aq) + 2Ag(s)$$

    (a) Show the cathode and the anode, the directions in which the ions move, the direction in which the electrons move in the external circuit and the electrode reactions.

    (b) Calculate the standard cell voltage.

29. A cell based on the reaction

$$5Fe^{2+}(aq) + MnO_4^-(aq) + 8H^+(aq) \longrightarrow$$
$$5Fe^{3+}(aq) + Mn^{2+}(aq) + 4H_2O(l)$$

can be set up as follows: A platinum electrode is inserted into a beaker containing a solution of $KMnO_4$ and H_2SO_4. Another platinum electrode is inserted into a solution of $FeSO_4$ in another beaker. A salt bridge is used to join the beakers. The two electrodes are connected to a voltmeter.

    (a) What are the reactions occurring at the anode and the cathode?

(b) In what direction do the electrons move through the external circuit?

(c) In which direction do the ions move through the solution?

(d) What is the emf of the cell under standard conditions?

30. Write equations for the reactions occurring at the anode and the cathode of each of the following cells. What is the standard potential for each cell? Draw a cell diagram for each cell, label the anode and the cathode, and show the direction in which electrons flow in the external circuit.

(a) A lead wire dipping into a $1M$ $PbCl_2$ solution and a copper wire dipping into a $1M$ $CuSO_4$ solution.

(b) Chlorine gas bubbling over a platinum wire in a $1M$ $NaCl$ solution and a silver wire coated with $AgCl$ dipping into the same solution.

(c) Two platinum wires dipping into a $1M$ solution of HI. One electrode has $H_2(g)$ bubbling over it at a pressure of 1 atm.

Nernst Equation

31. Predict the effect of increasing the concentration of iodide ion on the emf of an electrochemical cell utilizing the following reaction:

$$Cl_2(aq) + 2I^-(aq) \longrightarrow 2Cl^-(aq) + I_2(s)$$

32. Calculate the equilibrium constant for the reaction

$$Cl_2(g) + 2H_2O \longrightarrow H_3O^+(aq) + Cl^-(aq) + HOCl(aq)$$

from the appropriate standard reduction potentials.

$$(2HOCl + 2H^+ + 2e^- \longrightarrow Cl_2 + 2H_2O$$
$$E^\circ = 1.63 \text{ V})$$

33. An electrochemical cell is constructed on the basis of the following reaction:

$$Sn^{2+}(aq) + Pb(s) \longrightarrow Pb^{2+}(aq) + Sn(s)$$

The concentration of Sn^{2+} in the cathode compartment is $1.00M$, and the emf of the cell is 0.22 V at 25°C. What is the concentration of Pb^{2+} in the anode compartment? If $[SO_4^{2-}] = 1.00M$ in the anode compartment, what is the K_{sp} of $PbSO_4$?

34. A piece of iron is placed in an aqueous solution in $1M$ Fe^{2+}, and a piece of copper is placed in a separate container in which there is a $1M$ Cu^{2+} solution. If the solutions are connected by a tube containing an aqueous salt solution and the metals are connected by a conducting wire, predict which metal will dissolve and which will increase in mass. What potential (voltage) will develop between the electrodes? Which of the two solutions will become more concentrated? Will this potential increase or decrease as the reaction proceeds?

Fuel Cells

35. Write the two half-reactions for the oxidation of $C_2H_6(g)$ by $O_2(g)$ in a fuel cell to give $CO_2(g)$ and H_2O in acidic solution. Calculate how many liters of $C_2H_6(g)$ at STP would be needed at the anode to generate a current of 0.50 A for 6.0 h. How many liters of $O_2(g)$ at STP would be needed at the cathode?

Miscellaneous

36. A possible method for producing power for a heart pacemaker is to implant a zinc and a platinum electrode into the body tissues. These electrodes in the oxygen-containing body fluid will form a "biogalvanic" cell in which zinc metal is oxidized and oxygen is reduced.

(a) Write equations for the anode and the cathode reaction.

(b) Make an estimate of the voltage that such a cell would generate.

(c) If a current of 40 μA is drawn from the cell, how often will a 5.0-g zinc electrode need replacing?

CHAPTER 17

FURTHER CHEMISTRY OF NITROGEN AND OXYGEN

In earlier chapters we described some of the simpler compounds of the very common and very important elements nitrogen and oxygen. But not all aspects of their chemistry could be discussed adequately in terms of the ideas and theories described in those chapters. Nitrogen monoxide and nitrogen dioxide are the most common and most important of the oxides of nitrogen. Nitrogen monoxide, NO, is formed from N_2 and O_2 by lightning discharges in the atmosphere and then reacts with oxygen to form NO_2. The same reaction occurs in an internal combustion engine, and NO and NO_2 have become well known in recent years as serious air pollutants. Yet we cannot write a conventional Lewis structure for either of these oxides in which all the atoms obey the octet rule. Unlike the vast majority of molecules, both NO and NO_2 have an odd number of electrons.

In this chapter we first describe the oxides of nitrogen and discuss their structures. Then we discuss the oxoacids, nitrous acid, HNO_2, and nitric acid, HNO_3. Nitric acid is a very important industrial chemical that is used in the manufacture of explosives, fertilizers, dyes, and plastics.

Other compounds of nitrogen that we discuss in this chapter include hydrazine, N_2H_4, which is an important rocket fuel; hydroxylamine, NH_2OH, and hydrazoic acid, HN_3.

Oxygen, too, has some unusual and unexpected compounds. Some of the alkali and alkaline earth metals react with oxygen to form peroxides, such as $(Na^+)_2O_2^{2-}$, and superoxides, such as $K^+O_2^-$. The superoxide ion, O_2^-, is similar to NO in that it has an odd number of electrons and we cannot write a normal Lewis structure for it. When treated with acids, peroxides give hydrogen peroxide, H_2O_2. This is a strong and reactive oxidizing agent that in dilute solution in water is used as an antiseptic and as a bleach. The pure substance is important as a rocket propellant. We also discuss ozone, O_3, an allotrope of oxygen, which is a more reactive oxidizing agent than oxygen and is an important constituent of the upper atmosphere.

From its position in group V of the periodic table, and from its electron configuration $1s^2 2s^2 2p^3$ we except nitrogen to have a valence of 3 and therefore to form the oxide N_2O_3. This oxide can be made at low temperatures, but it readily decomposes to NO and NO_2, which are much more stable. The oxides of nitrogen and some of their physical properties are listed in Table 17.1. Nitrogen exhibits positive oxidation states only in its compounds with oxygen and fluorine, because only these elements are more electronegative than nitrogen. In its oxides nitrogen has oxidation numbers from $+1$ to $+5$ (Table 17.2).

Nitrogen Monoxide, NO

This can also be called *nitrogen(II) oxide*, and it is often called by its older name of *nitric oxide*. It is a colorless, reactive gas that is only slightly soluble in water.

PREPARATION   Nitrogen combines directly with oxygen slowly and at very high temperatures to give a small equilibrium yield of NO:

$$N_2(g) + O_2(g) \rightleftharpoons 2NO(g) \qquad \Delta H^\circ = 180 \text{ kJ}$$

The reaction is endothermic, and the equilibrium constant is only 10^{-30} at 25°C. In a ___ with Le Châtelier's principle, the equilibrium shifts to the right w___ ___t even at 2400 K the equilibrium constant is only 2.___ ___concentration of NO is small.

___h oxygen to give NO in lightning discharges
ir ___ the discharge, it is rapidly cooled to ordinary
t___ ___decomposition of NO are then very slow, so
t___ ___k to the left to any appreciable extent. In other
___ ___is "frozen" at the high-temperature value.
___t of NO is formed by the reaction between the
___ air at the high temperature (≈ 2300°C) of the
___l combustion engines. Automobiles are therefore
___ution of the atmosphere caused by NO and NO_2

Table 17.1 ___des of Nitrogen

		NAME	PHYSICAL STATE AT 25°C	mp (°C)	bp (°C)
		Dinitrogen pentaoxide	White solid	30	47
		Dinitrogen tetraoxide	White solid	−11.2	21.2
		Nitrogen dioxide	Brown gas		
+3	N_2O_3	Dinitrogen trioxide	Deep blue liquid (−80°C)	−102	—
+2	NO	Nitrogen monoxide (nitric oxide)	Colorless gas	−90.8	−88.5
+1	N_2O	Dinitrogen monoxide (nitrous oxide)	Colorless gas	−163.7	−151.8

Table 17.2 Oxidation States of Nitrogen

OXIDATION STATE	EXAMPLE
+5	HNO_3, N_2O_5
+4	NO_2, N_2O_4
+3	HNO_2, N_2O_3
+2	NO
+1	N_2O
0	N_2
−1	NH_2OH
−2	N_2H_4
−3	NH_3

Box 17.1
SMOG

An unpleasant and dangerous combination of smoke and fog that we now call smog was a feature of many industrial cities in Europe for hundreds of years. It was a product of a damp climate and smoke and sulfur dioxide produced by the burning of coal for heat and power. This type of smog has become much less common in recent years, since the widespread use of coal has decreased, but it has been replaced by a different type of smog called photochemical smog, which is primarily the product of a combination of automobile exhaust and a sunny climate. The process of photochemical smog formation is complex and all its details are not yet well understood. It begins with the combination of atmospheric nitrogen and oxygen to form NO at the high temperature of an automobile engine or the furnaces of coal- or oil-burning electric power plants. When NO is emitted into the atmosphere, it is slowly oxidized by oxygen to NO_2. The concentration of NO_2 is sometimes high enough for it to be clearly visible as a brown layer in the atmosphere, particularly from an airplane.

Although NO_2 is a health hazard, primarily because it attacks the lungs, its concentration usually does not reach a dangerous level. However, as we have explained in Box 14.1, it leads to the formation of acid rain. Moreover, it initiates a complex series of reactions that lead to other, more dangerous atmospheric pollutants. The NO_2 is dissociated by

ultraviolet radiation ($\lambda \leq 392$ nm) from the sun to give NO molecules and O atoms:

$$NO_2 \longrightarrow NO + O$$

Free oxygen atoms are extremely reactive and combine with molecular oxygen to give ozone:

$$O + O_2 \longrightarrow O_3$$

Because ozone is a powerful oxidizing agent, it reacts destructively with many materials, including rubber, paint, and vegetation. Moreover, it is very irritating to the eyes and it may damage the lungs. It also reacts with hydrocarbons that are released into the atmosphere from automobile exhausts as a result of the incomplete combustion of gasoline. These reactions produce a variety of organic molecules such as organic peroxides and peroxoacetylnitrates (abbreviated PAN), which

are strong eye irritants and damaging to plant life.

Los Angeles was the first city to achieve notoriety because of its photochemical smog, which results from its huge number of automobiles, its sunny climate, and its geographic situation. Normally, the temperature of the air decreases with increasing altitude (Chapter 3), and as dirty, hot air rises from the surface, cleaner, cool air descends to take its place. In certain locations such as Los Angeles a condition known as a temperature inversion can arise. Cool air flows down from the nearby mountains to form a pool on the surface underneath a mass of hot air. Here it is trapped, and atmospheric pollutants accumulate in this pool of cool air rather than being removed by rising hot air currents.

In recent years serious efforts have been made to reduce the smog problem in many cities. For example, standards have been set for automobile exhaust emissions. Automobile manufacturers have attempted to reduce the concentrations of undesirable components of automobile exhaust in two ways: first, by altering the conditions of combustion to reduce NO formation and the concentration of unburnt fuel and, second, by adding catalytic converters to the exhaust system to remove unburnt hydrocarbons and to decompose NO to N_2 and O_2. But a really efficient catalyst for this decomposition has still to be found.

Nitrogen monoxide is prepared on a large scale in the first step of the *Ostwald process* for the manufacture of nitric acid. Ammonia and oxygen are passed over a platinum-rhodium wire gauze that has been heated to approximately 900°C by an electric current. The hot wire gauze catalyzes the oxidation of ammonia to nitrogen monoxide:

$$4NH_3(g) + 5O_2(g) \longrightarrow 4NO(g) + 6H_2O(g) \qquad \Delta H° = -906 \text{ kJ}$$

Once the reaction has started, it supplies the heat needed to keep the catalyst at the operating temperature (see Experiment 17.1).

Nitrogen monoxide can be conveniently prepared in the laboratory by reducing *dilute* nitric acid with copper (see Experiment 17.2):

$$3Cu(s) + 8HNO_3(aq) \longrightarrow 3Cu(NO_3)_2(aq) + 2NO(g) + 4H_2O(l)$$

PROPERTIES   Because the nitrogen in NO is in a low ($+2$) oxidation state, NO behaves as a good reducing agent. With oxygen it forms nitrogen dioxide, NO_2:

$$2NO(g) + O_2(g) \longrightarrow 2NO_2(g)$$

Nitrogen monoxide reduces ozone to oxygen and is oxidized to NO_2,

$$NO(g) + O_3(g) \longrightarrow O_2(g) + NO_2(g)$$

and it reduces CO_2 to CO:

$$NO(g) + CO_2(g) \xrightarrow{\text{Heat}} CO(g) + NO_2(g)$$

With chlorine it forms nitrosyl chloride, NOCl,

$$2NO(g) + Cl_2(g) \longrightarrow 2NOCl(g)$$

an orange-yellow gas that contains nitrogen in the $+3$ state. When bubbled into aqueous solutions of oxidizing agents, NO is usually oxidized to nitrate ion. For example, it reduces iodine to iodide ion,

$$2NO(g) + 3I_2(s) + 4H_2O \longrightarrow 2NO_3^-(aq) + 8H^+(aq) + 6I^-(aq)$$

and MnO_4^- to Mn^{2+}.

Because there are stable lower oxidation states of nitrogen, NO will also react as an oxidizing agent. For example, it is reduced to nitrogen by hydrogen, which is oxidized to water,

$$2NO(g) + 2H_2(g) \longrightarrow N_2(g) + 2H_2O(l)$$

and by phosphorus, which is oxidized to P_4O_6,

$$P_4(s) + 6NO(g) \longrightarrow P_4O_6(s) + 3N_2(g)$$

EXPERIMENT 17.1

Preparation of NO by Catalytic Oxidation of NH_3

When a coil of copper wire is heated in a flame and then suspended over a warm, concentrated, aqueous solution of ammonia, the copper coil continues to glow brightly for a long time as the ammonia is oxidized on the metal surface in a strongly exothermic reaction. The heat produced in the reaction is usually sufficient to melt the copper wire so that molten copper drops into the ammonia solution producing a blue color.

Nitrogen Monoxide, NO

Nitrogen monoxide may be prepared by reacting copper with 6 M nitric acid. In this experiment a jar containing 6 M nitric acid has been inverted over some pieces of copper in a trough of water. Colorless NO is evolved and collects in the jar as it is only slightly soluble in water. The Cu^{2+} ion formed causes the nitric acid to turn blue.

A jar of colorless NO.

When the stopper is removed, the NO reacts with oxygen of the atmosphere to form brown NO_2.

If the jar of NO_2 is then inverted over water containing a little methyl red indicator, the water rises rapidly in the jar as the NO_2 dissolves to form a solution of HNO_2 and HNO_3, which changes the color of the indicator from yellow to red.

Nitrogen Dioxide, NO_2

The product of the reaction between NO and oxygen is nitrogen dioxide, NO_2. It is a red-brown gas. It may also be called nitrogen (IV) oxide. Two molecules of NO_2 combine with each other to form colorless *dinitrogen tetraoxide*, N_2O_4. At ordinary temperatures the gas is an equilibrium mixture of NO_2 and N_2O_4:

$$2NO_2 \rightleftharpoons N_2O_4 \qquad \Delta H° = -57 \text{ kJ}$$

Because the formation of N_2O_4 is an exothermic reaction, the equilibrium shifts to the right with decreasing temperature, and more N_2O_4 is formed. At 21°C the gas condenses to a deep red-brown liquid, which is a mixture of NO_2 and N_2O_4 in equilibrium. As the temperature is decreased, the color of the liquid mixture becomes less intense. Finally, at $-11°C$ it freezes to a white solid that consists entirely of N_2O_4 molecules (see Experiment 17.3). The equilibrium between NO_2 and N_2O_4 in the gas phase can be rather easily studied by measuring the intensity of the color of the mixture.

Example 17.1 Calculate the equilibrium constant at 25°C for the equilibrium $N_2O_4 \rightleftharpoons 2NO_2$ from the following experimental data:

Equilibrium concentrations at 25°C (mol L^{-1})

NO_2	N_2O_4
0.0508	0.555
0.0457	0.452
0.0480	0.491
0.0524	0.592

Nitrogen Dioxide and Dinitrogen Tetraoxide

The equilibrium $N_2O_4 \leftrightarrows 2NO_2$ shifts to the left with decreasing temperature. A large amount of dark brown NO_2 vapor can be seen in the tube on the left which is in hot water. The tube in the middle, which is in ice, is much paler because the equilibrium has been shifted to the left and there is therefore less NO_2 vapor in the tube. The yellow-brown liquid at the bottom of the tube consists of N_2O_4 containing a little NO_2.
The tube on the right has been kept in a freezing mixture at a temperature of approximately $-15°C$. It contains white solid N_2O_4.

Solution We write the expression for the equilibrium constant,

$$K_c = \left(\frac{[NO_2]^2}{[N_2O_4]}\right)_{eq}$$

Substituting the first set of data gives

$$K_c = \frac{(0.0508)^2 \text{ mol}^2 \text{ L}^{-2}}{0.555 \text{ mol L}^{-1}} = 4.65 \times 10^{-3} \text{ mol L}^{-1}$$

For the other sets of data we get $K_c = 4.62 \times 10^{-3}$ mol L^{-1}, 4.69×10^{-3} mol L^{-1}, and 4.64×10^{-3} mol L^{-1}. We see that the equilibrium expression is indeed constant within experimental error. We may take the average value of 4.65×10^{-3} mol L^{-1} as the equilibrium constant for the reaction at 25°C.

Example 17.2 Will the equilibrium constant for the equilibrium

$$N_2O_4 \rightleftharpoons 2NO_2$$

increase or decrease with increasing temperature?

Solution The reaction

$$N_2O_4 \rightleftharpoons 2NO_2 \qquad \Delta H° = +57 \text{ kJ mol}^{-1}$$

is endothermic, and therefore the equilibrium shifts to the right as the temperature is raised. The equilibrium amount of NO_2 increases and the equilibrium amount of N_2O_4 decreases with increasing temperature. Therefore the equilibrium constant

$$K_c = \left(\frac{[NO_2]^2}{[N_2O_4]}\right)_{eq}$$

increases with increasing temperature.

Nitrogen dioxide can be conveniently made in the laboratory by reducing *concentrated* nitric acid with copper:

$$Cu(s) + 4HNO_3(aq) \longrightarrow Cu(NO_3)_2(aq) + 2NO_2(g) + 2H_2O(l)$$

It can also be produced by heating certain metal nitrates (see Experiment 17.4). For example,

$$2Pb(NO_3)_2(s) \longrightarrow 2PbO(s) + 4NO_2(g) + O_2(g)$$

17.1 OXIDES OF NITROGEN

Preparation of Nitrogen Dioxide

Nitrogen dioxide may be prepared by heating lead nitrate, $Pb(NO_3)_2$. Here the NO_2 may be seen condensing as a pale brown liquid in the U-tube cooled in ice. Oxygen, which is also produced in the reaction, ignites a glowing splint held at the exit tube.

Since nitrogen is in a high ($+4$) oxidation state, NO_2 is a strong oxidizing agent. For example, it oxidizes SO_2 to SO_3:

$$SO_2(g) + NO_2(g) \longrightarrow SO_3(g) + NO(g)$$

It oxidizes CO to CO_2,

$$CO(g) + NO_2(g) \longrightarrow CO_2(g) + NO(g)$$

and I^- to I_2, but it reduces MnO_4^- to Mn^{2+} (see Experiment 17.5).

Structures of NO and NO_2: Free Radicals

Nitrogen monoxide and nitrogen dioxide are unusual molecules in that they have an odd number of valence electrons: 11 for NO and 17 for NO_2. The vast majority of stable molecules have an even number of electrons, which may be described either as shared (bonding) pairs or as unshared (nonbonding) pairs.

NITROGEN MONOXIDE Because NO and NO_2 have an odd number of valence electrons, we cannot write normal Lewis structures for them in which

Properties of Nitrogen Dioxide

A colorless aqueous solution of KI turns brown when NO_2 gas is passed into it, The NO_2 oxidizes colorless I^- to I_2, forming brown I_3^-.

A pink aqueous solution of $KMnO_4$ becomes colorless as NO_2 gas is passed into it. The NO_2 reduces MnO_4^- to Mn^{2+} and is itself oxidized to NO_3^-.

all the electrons are paired and in which all the atoms have an octet of electrons in their valence shells. For example, we can write the following structure for NO:

$$:\overset{\cdot}{N}=\overset{\cdot\cdot}{O}:$$

Here oxygen has eight electrons in its valence shell, but nitrogen has only seven, and the odd or unpaired electron is located on nitrogen. We can write another possible Lewis structure for NO by placing the odd electron on oxygen rather than on nitrogen, so oxygen has only seven electrons in its valence shell and nitrogen has eight:

$$:\overset{\ominus\cdot\cdot}{N}=\overset{\cdot\,\oplus}{O}:$$

This second structure has formal charges; it is therefore somewhat less important than the previous structure. The two structures may be regarded as resonance structures:

$$:\overset{\cdot}{N}=\overset{\cdot\cdot}{O}: \longleftrightarrow :\overset{\ominus\cdot\cdot}{N}=\overset{\cdot\,\oplus}{O}:$$

Thus the odd electron is found on both the nitrogen atom and the oxygen atom. Hence we can alternatively describe NO by assuming that the odd electron is shared between the nitrogen atom and the oxygen atom:

$$:\overset{\cdot}{N}\!\doteq\!\overset{\cdot}{O}:$$

In this structure both the nitrogen atom and the oxygen atom have a valence shell of eight electrons. The bond may be described as consisting of two shared pairs and a single shared electron. It has an order of 2.5. The NO bond length is consistent with this description, since it is intermediate between that of the triple bond in N_2 and that of the double bond in O_2:

	$:N\equiv N:$	$:\overset{\cdot}{N}\!\doteq\!\overset{\cdot}{O}:$	$:\overset{\cdot\cdot}{O}=\overset{\cdot\cdot}{O}:$
Bond length	109 pm	115 pm	121 pm
Bond order	3	2.5	2

Example 17.3 What are the formal charges on the nitrogen and oxygen atoms in the following structure?

$$:\overset{\cdot}{N}\!\doteq\!\overset{\cdot}{O}:$$

Solution

$$:\overset{\ominus\frac{1}{2}\quad\oplus\frac{1}{2}}{N}\!\doteq\!\overset{}{O}:$$

Nitrogen has three electrons of its own and a half share in the five bonding electrons. Hence it has a total of $5\frac{1}{2}$ electrons and has a formal charge of $-\frac{1}{2}$. Oxygen similarly has a total of $5\frac{1}{2}$ electrons; since its core charge is $+6$, it has a formal charge of $+\frac{1}{2}$.

FREE RADICALS Molecules containing an odd number of electrons are quite rare. They are usually exceedingly reactive and have only very short lifetimes because they react with almost any other substance with which they come in contact. Odd-electron molecules are called **free radicals**. Examples include the methyl radical, and the hydroxyl radical, $H\!-\!\overset{\cdot\cdot}{\underset{\cdot\cdot}{O}}\!\cdot$. Although free radicals are often formed in small amounts during the course of a reaction, they get used up again—they are said to be intermediates—and they cannot generally be isolated in appreciable amounts. The methyl radical, $CH_3\cdot$, for example, reacts rapidly with many other molecules. In the absence of other molecules it combines with another methyl radical to give ethane, C_2H_6.

$$\overset{\textstyle H}{\underset{\textstyle H}{H\!-\!C\!\cdot}}$$

Although NO reacts with oxygen and with some other substances, it is relatively unreactive for an odd-electron molecule or free radical. At ordinary temperatures two NO molecules do not combine with each other. That the odd electron in NO is shared between the nitrogen and oxygen atoms rather than localized on one atom, as in the methyl radical, may be the reason that NO is considerably less reactive than most other free radicals.

NITROGEN DIOXIDE   The NO_2 molecule is also an odd-electron molecule or free radical; it has 17 valence electrons. We can write several Lewis structures for NO_2. There are two structures with the odd electron on nitrogen and 7 in the valence shell of nitrogen:

$$:O{=}\overset{\cdot}{\overset{\oplus}{N}}{-}\overset{\ominus}{O}: \quad\longleftrightarrow\quad :\overset{\ominus}{O}{-}\overset{\cdot}{\overset{\oplus}{N}}{=}O:$$

There are also two structures with the odd electron on oxygen and only 7 electrons in the valence shell of one of the oxygen atoms:

$$:\overset{\cdot}{O}{-}N{=}O: \quad\longleftrightarrow\quad :O{=}N{-}\overset{\cdot}{O}:$$

Again, the odd electron can be considered to be shared between oxygen and nitrogen, and we can then write the following two structures in which all the atoms have octets:

$$:\overset{\cdot}{O}{-}\overset{\cdot}{N}{=}O: \quad\longleftrightarrow\quad :O{=}\overset{\cdot}{N}{-}\overset{\cdot}{O}:$$

The molecule is angular with a bond length of 119 pm and a bond angle of 134°. Because a single unshared electron occupies a smaller volume than an unshared pair of electrons, the bond angle is considerably larger than the ideal angle of 120° for an AX_2E molecule.

MOLECULAR ORBITALS   The difficulty of formulating satisfactory Lewis structures for NO and NO_2 again illustrates the limitations of Lewis structures. They are very convenient for the large number of molecules in which all the electrons can be described with reasonable accuracy as localized pairs, either bonding or nonbonding. However, for molecules in which some of the electrons are delocalized or for molecules with odd numbers of electrons, such as NO and NO_2, Lewis structures are less satisfactory. In such cases molecular orbital theory provides an alternative to the use of the concept of resonance to improve the description of molecules in terms of Lewis structures. In **molecular orbital theory** electrons are not described as localized bonding and nonbonding pairs; instead, they are allocated to orbitals, called **molecular orbitals**, that are associated with the entire molecule. (See Chapter 25.)

Dinitrogen Tetraoxide, N_2O_4, and Dinitrogen Trioxide, N_2O_3

Two NO_2 molecules combine to give dinitrogen tetraoxide, N_2O_4, a white solid that melts at $-11°C$ to give an equilibrium mixture of NO_2 and N_2O_4. The strong oxidizing power of N_2O_4 has made it useful as a rocket fuel. A Lewis structure for N_2O_4 may be written by combining two NO_2 structures in which the odd electron is on nitrogen. A total of four resonance structures can be obtained in this way (see Figure 17.1).

$$\text{O}_2\text{N} \xrightarrow[134°]{178\ \text{pm}} \text{N} \xleftarrow{118\ \text{pm}} \text{O}_2$$

(a)

(b)

Figure 17.1 Structure of Dinitrogen Tetraoxide. Dinitrogen tetra-oxide is a planar molecule with the dimensions shown in (a). It can be represented by the four resonance structures in (b).

Nitrogen monoxide and nitrogen dioxide also combine at low temperature by pairing their odd electrons to give *dinitrogen trioxide*, N_2O_3 (see Figure 17.2). This deep blue liquid decomposes completely below room temperature to NO and NO_2 (see Experiment 17.6).

$$114\ \text{pm}\ \text{O} \!-\! \text{N} \xrightarrow{186\ \text{pm}} \text{N} \begin{smallmatrix} \text{O}\ 122\ \text{pm} \\ \text{O}\ 120\ \text{pm} \end{smallmatrix}$$

(a)

(b)

Figure 17.2 Structure of Dinitrogen Trioxide. (a) Dimensions; (b) resonance structures.

EXPERIMENT 17.6

Dinitrogen Trioxide, N_2O_3

The tube contains pale brown liquid N_2O_4 at 0°C.

When colorless NO is bubbled into the N_2O_4 it is converted to deep blue liquid N_2O_3.

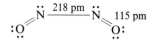

Figure 17.3 Structure of N_2O_2 in the Solid State.

At very low temperatures two NO molecules also combine in the same way to give the N_2O_2 molecule in the solid state (see Figure 17.3). This molecule is very unstable and is not known in the liquid or gaseous states. In other words, at room temperature the equilibrium lies far to the right:

$$N_2O_2(g) \rightleftharpoons 2NO(g)$$

In each of the molecules N_2O_4, N_2O_3, and N_2O_2 the N—N bond is much longer than the value of 140 pm predicted from the covalent radius of nitrogen (Figure 4.7), and the bond length increases from N_2O_4 to N_2O_3 to N_2O_2. With increasing length the N—N bond becomes weaker and the molecules therefore dissociate with increasing ease. No simple theory can describe satisfactorily the bonding in these molecules.

Dinitrogen Monoxide, N_2O

When solid ammonium nitrate is gently heated, it melts and decomposes to give N_2O, dinitrogen monoxide,

$$\overset{-3}{N}H_4\overset{+5}{N}O_3(l) \longrightarrow \overset{+1}{N_2}O(g) + 2H_2O(g) \qquad \Delta H° = -37 \text{ kJ}$$

which is often called by its older name, *nitrous oxide*. In this oxidation-reduction reaction, nitrogen in the -3 oxidation state in NH_4^+ is oxidized by nitrogen in the $+5$ state in the nitrate ion, to give N_2O in which nitrogen is in the $+1$ oxidation state. Since the reaction forms 3 mol of gaseous products and is exothermic, it can occur with explosive violence if the solid is heated too strongly (see Box 12.1).

Dinitrogen monoxide may be described by two resonance structures:

$$:\overset{..}{O}=\overset{\oplus}{N}=\overset{..}{N}:^{\ominus} \longleftrightarrow {}^{\ominus}:\overset{..}{O}-\overset{\oplus}{N}\equiv N:$$

Dinitrogen monoxide is isoelectronic with carbon dioxide,

$$:\overset{..}{O}=C=\overset{..}{O}:$$

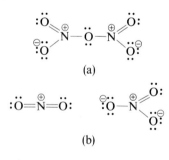

(a)

(b)

Figure 17.4 Structure of Dinitrogen Pentaoxide. (a) In the gas state, dinitrogen pentaoxide consists of N_2O_5 molecules. One of four equivalent resonance structures is shown here. (b) In the solid state, dinitrogen pentaoxide is an ionic compound composed of the ions NO_2^+ and NO_3^-.

and has the same linear AX_2 structure.

In contrast to NO and NO_2, dinitrogen monoxide has an even number of electrons and is a rather unreactive substance. It is a colorless gas with a pleasant smell and a sweet taste. When breathed in small amounts, it acts as a mild intoxicant. Because of this property, it has been called laughing gas. In large amounts N_2O acts as a general anesthetic and is sometimes used for this purpose in dental and other minor surgery. Dinitrogen monoxide also has the useful property of being rather soluble in fats; a property exploited in making self-whipping cream. Cream is packaged with N_2O under pressure to increase its solubility. When the pressure is released, some of the N_2O escapes to form tiny bubbles, which produce whipped cream.

Dinitrogen Pentaoxide, N_2O_5

The oxide N_2O_5 is made by dehydrating 100% nitric acid, HNO_3, with phosphorus (V) oxide, P_4O_{10}:

$$4HNO_3(l) + P_4O_{10}(s) \longrightarrow 2N_2O_5(g) + 4HPO_3(s)$$

It has the structure shown in Figure 17.4. Because it is formed by the removal of water from nitric acid, dinitrogen pentaoxide may be described as the *anhydride of nitric acid*. In general, an *acid anhydride* is the oxide obtained by removing water from an oxoacid.

$$2HNO_3 \longrightarrow N_2O_5 + \qquad H_2O$$
$$\text{Combines with} P_4O_{10}$$

Dinitrogen pentaoxide is a white volatile solid that decomposes slowly at room temperature to NO_2 and O_2:

$$2N_2O_5(s) \longrightarrow 4NO_2(l) + O_2(g)$$

It reacts readily with water to give nitric acid.

Dinitrogen pentaoxide is an interesting example of a substance that has a different structure in the solid and gaseous states. In the gas phase dinitrogen pentaoxide consists of covalent N_2O_5 molecules. In contrast, the white solid is an ionic compound composed of nitronium ions, NO_2^+, and nitrate ions, NO_3^-. It might therefore be called nitronium nitrate. The nitronium ion is isoelectronic with CO_2 and N_2O and has a linear AX_2 structure (see Figure 17.4). We have previously seen in Chapter 9 that aluminum trichloride is ionic in the solid state but consists of covalent Al_2Cl_6 molecules in the gaseous state.

17.2 OXOACIDS OF NITROGEN

There are two important oxoacids of nitrogen: nitrous acid, HNO_2, and nitric acid, HNO_3. They contain nitrogen in the $+3$ and $+5$ oxidation states, respectively. Nitric acid is a very important industrial chemical that ranks eleventh in annual U.S. production (see Table 15.1). It is used in the manufacture of explosives, dyes, and fertilizers. It is both a strong acid and a strong oxidizing agent.

Nitric Acid, HNO_3

Nitric acid is prepared for industrial use from ammonia by the **Ostwald process**, which was devised by German chemist Wilhelm Ostwald (1853–1932). The first step in this process is the catalyzed oxidation of NH_3 to NO (see page 626). The NO then reacts with excess oxygen to form NO_2,

$$2NO(g) + O_2(g) \longrightarrow 2NO_2(g)$$

which is passed into water, with which it reacts to give nitric acid and NO, which is recycled:

$$3NO_2(g) + H_2O(l) \longrightarrow 2HNO_3(aq) + NO(g)$$

In this reaction nitrogen in the $+4$ oxidation state is converted to nitrogen in the $+5$ state in nitric acid and nitrogen in the $+2$ state in NO. In other words, nitrogen is both oxidized and reduced. This reaction is another example of self-oxidation-reduction, or *disproportionation* (see Chapter 9).

The aqueous solution of nitric acid prepared in this way contains dissolved NO and NO_2, which are blown out by a stream of air and recycled. The acid is normally sold as "concentrated nitric acid," which has a concentration of 16M, or 68% HNO_3 by mass. Concentrated nitric acid is colorless, but with time it usually develops a slight yellow color, because it decomposes when exposed to light, forming yellow-brown NO_2:

$$4HNO_3(aq) \xrightarrow{hv} 4NO_2(g) + O_2(g) + 2H_2O(l)$$

Anhydrous (100%) HNO_3 is obtained by mixing aqueous nitric acid with concentrated sulfuric acid and distilling off the volatile HNO_3. It can also be made by heating a metal nitrate with concentrated sulfuric acid (see Experiment 17.7):

Nitric Acid

Nitric acid may be prepared by heating sodium nitrate with concentrated sulfuric acid. Nitric acid is volatile and distills over to the tube on the right, where it can be seen condensing as a pale yellow liquid. Pure liquid nitric acid is colorless. Here it is colored yellow by NO_2 formed by some decomposition of the hot HNO_3 vapor.

$$NaNO_3(s) + H_2SO_4(l) \rightleftharpoons NaHSO_4(s) + HNO_3(g)$$

The equilibrium is continually shifted to the right by the removal of the volatile HNO_3 from the reaction mixture.

PROPERTIES   Pure nitric acid is a colorless liquid that boils at 84.1°C and freezes at −41.6°C. The nitric acid molecule can be represented by two resonance structures:

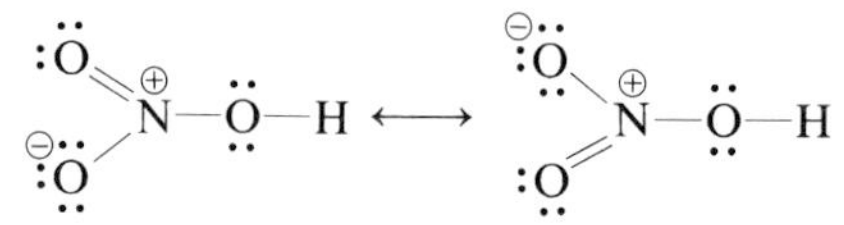

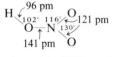

Dimensions of the HNO_3 molecule

Nitric acid is a strong acid that is 100% dissociated in dilute aqueous solution (see Chapter 7):

$$HNO_3(l) + H_2O(l) \longrightarrow H_3O^+(aq) + NO_3^-(aq)$$

NITRATES   There are many salts of nitric acid and almost all of them are soluble in water. The nitrate ion, NO_3^-, is isoelectronic with the carbonate ion and has the same planar triangular shape. The bonding in the nitrate ion can be described by three resonance structures:

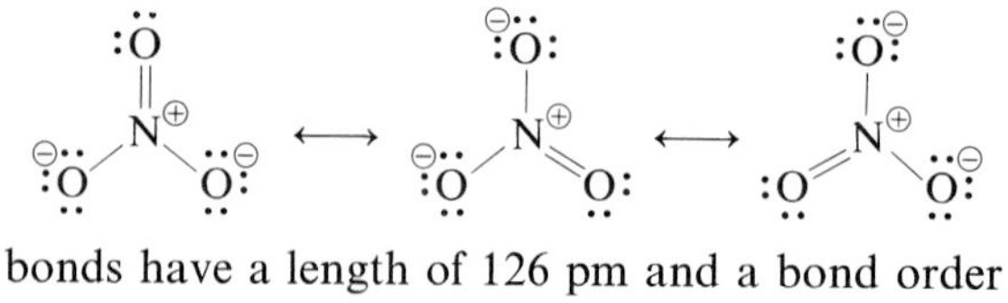

All three NO bonds have a length of 126 pm and a bond order of $1\frac{1}{3}$.

OXIDIZING PROPERTIES   Nitric acid and the nitrate ion both contain nitrogen in the +5 oxidation state and are good oxidizing agents. Depending on the concentration of nitric acid and the nature of the reducing agent, they may be reduced to any of the other oxidation states of nitrogen. The corresponding half-reactions are given in Table 17.3.

The two most common of these reactions and their corresponding standard

CHAPTER 17
FURTHER CHEMISTRY OF
NITROGEN AND OXYGEN

636

Table 17.3 Half-Reactions for Reduction of Nitrate Ion

REACTION	OXIDATION STATE OF NITROGEN IN REDUCTION PRODUCT
$NO_3^-(aq) + 2H^+(aq) + e^- \longrightarrow NO_2(g) + H_2O(l)$	+4
$NO_3^-(aq) + 3H^+(aq) + 2e^- \longrightarrow HNO_2(aq) + H_2O(l)$	+3
$NO_3^-(aq) + 4H^+(aq) + 3e^- \longrightarrow NO(g) + 2H_2O(l)$	+2
$2NO_3^-(aq) + 10H^+(aq) + 8e^- \longrightarrow N_2O(g) + 5H_2O(l)$	+1
$2NO_3^-(aq) + 12H^+(aq) + 10e^- \longrightarrow N_2(g) + 6H_2O(l)$	0
$NO_3^-(aq) + 10H^+(aq) + 8e^- \longrightarrow NH_4^+(aq) + 3H_2O(l)$	−3

reduction potentials are as follows:

$$NO_3^-(aq) + 2H^+(aq) + e^- \longrightarrow NO_2(g) + H_2O(l) \qquad E^\circ = 0.78 \text{ V}$$

$$NO_3^-(aq) + 4H^+(aq) + 3e^- \longrightarrow NO(g) + 2H_2O(l) \qquad E^\circ = 0.96 \text{ V}$$

Concentrated aqueous nitric acid oxidizes sulfur to sulfuric acid and carbon to carbon dioxide and is reduced to nitrogen dioxide:

$$S(s) + 6HNO_3 \text{ (conc. aq)} \longrightarrow H_2SO_4(aq) + 6NO_2(g) + 2H_2O(l)$$

$$C(s) + 4HNO_3 \text{ (conc. aq)} \longrightarrow CO_2(g) + 4NO_2(g) + 2H_2O(l)$$

Aqueous nitric acid oxidizes H_2S to sulfur and SO_2 to sulfuric acid. In these reactions, nitric acid is reduced to NO:

$$3H_2S(aq) + 2HNO_3(aq) \longrightarrow 3S(s) + 2NO(g) + 4H_2O(l)$$

$$3SO_2(aq) + 2HNO_3(aq) + 2H_2O(l) \longrightarrow 3H_2SO_4(aq) + 2NO(g)$$

In the table of standard reduction potentials (Table 16.1) the half-reactions for most of the metals are above the half-reactions for the reduction of nitric acid to NO_2 and NO. Thus we expect nitric acid to dissolve all these metals, including metals such as copper and silver which are not oxidized by the H_3O^+ ion. Only a few metals such as gold and platinum, which have more positive reduction potentials than NO_3^- has, are not dissolved by nitric acid.

We can see from these half-reactions that the oxidizing strength of NO_3^- depends on the hydrogen ion concentration. According to Le Châtelier's principle, increasing the hydrogen ion concentration should cause the reactions to proceed more to the right. In other words, the oxidizing strength of NO_3^- should increase with increasing hydrogen ion concentration. Thus nitric acid is a much stronger oxidizing agent than a neutral solution of a metal nitrate. We can also arrive at this conclusion by using the Nernst equation (see Example 17.5).

Because of the variety of reduction products of nitric acid, its reactions with reducing agents are often quite complicated and give several reduction products. The relative amounts of the possible products depend on the temperature, the concentration of the nitric acid, and the nature of the reducing agent. As we saw in Chapter 9, when copper reacts with concentrated nitric acid, the main reduction product is NO_2, in which nitrogen is in the +4 oxidation state; with dilute nitric acid the main reduction product is NO, in which nitrogen is in the +2 oxidation state:

$$Cu(s) + 4HNO_3(\text{conc. aq}) \longrightarrow Cu(NO_3)_2(aq) + 2NO_2(g) + 2H_2O$$

$$3Cu(s) + 8HNO_3 \text{ (dil. aq)} \longrightarrow 3Cu(NO_3)_2(aq) + 2NO(g) + 4H_2O$$

17.2 OXOACIDS OF NITROGEN

637

The standard reduction potentials show that the reduction of NO_3^- with copper should not stop at NO_2 but should continue to NO. However, the reaction of concentrated nitric acid with copper is very rapid, and NO_2 escapes from the solution before it can be reduced to NO.

Example 17.4 Rather pure NO can be prepared by reducing dilute nitric acid with a solution of iron(II) sulfate. Write the balanced equation for this reaction. What is the standard potential for a cell based on this reaction?

Solution The unbalanced equation is

$$NO_3^- + Fe^{2+} \longrightarrow NO + Fe^{3+}$$

Since nitric acid is fully ionized, we have written NO_3^- rather than HNO_3. We can assume that Fe^{2+} is oxidized to Fe^{3+} (Chapter 9). The half equation for the reduction of NO_3^- to NO can be obtained as described in Example 9.1. It is the third equation in Table 17.3:

$$NO_3^- + 4H^+ + 3e^- \longrightarrow NO + 2H_2O$$

The half-reaction for the oxidation of Fe^{2+} to Fe^{3+} is

$$Fe^{2+} \longrightarrow Fe^{3+} + e^-$$

Multiplying this equation by 3 and adding the result to the previous equation gives the overall equation

$$3Fe^{2+} + NO_3^- + 4H^+ \longrightarrow 3Fe^{3+} + NO + 2H_2O$$

To find the cell potential, we add the half-cell potentials from Table 16.1:

$$
\begin{aligned}
NO_3^- + 4H^+ + 3e^- &\longrightarrow NO + 2H_2O & E_{ox}^\circ &= 0.96 \text{ V} \\
Fe^{2+} &\longrightarrow Fe^{3+} + e^- & E_{red}^\circ &= -0.77 \text{ V} \\
& & E_{cell}^\circ &= 0.19 \text{ V}
\end{aligned}
$$

Example 17.5 Use the Nernst equation to calculate the half-cell potential for

$$NO_3^- + 4H^+ + 3e^- \longrightarrow NO + 2H_2O$$

for (a) $[H^+] = 10$ mol L^{-1} and (b) a neutral solution with $[H^+] = 10^{-7}$ mol L^{-1}, assuming $[NO_3^-] = 1$ mol L^{-1} and $p_{NO} = 1$ atm.

Solution

(a) $\quad E = E^\circ - \dfrac{0.0257}{n} \ln \dfrac{p_{NO}}{[NO_3^-][H^+]^4} = 0.96 - \dfrac{0.0257}{3} \ln \dfrac{1}{(1)(10^4)}$

$\quad\quad = 0.96 + 0.08 = 1.04$ V

(b) $\quad E = E^\circ - \dfrac{0.0257}{3} \ln \dfrac{1}{(1)(10^{-7})^4} = 0.96 - 0.55 \text{ V} = 0.41$ V

We see that the standard half-cell potential (0.96 V when $[H^+] = 1M$) is increased by increasing the hydrogen ion concentration to 10 mol L^{-1} and is very much reduced in a neutral solution. Thus nitrate ion in neutral solution is a much weaker oxidizing agent than aqueous nitric acid.

USES OF NITRIC ACID   Much of the nitric acid produced is combined with ammonia to give ammonium nitrate:

$$NH_3(g) + HNO_3(l) \longrightarrow NH_4NO_3(s)$$

Ammonium nitrate is widely used as a fertilizer. Sodium nitrate and potassium nitrate are also important fertilizers. Extensive deposits of sodium nitrate are found in the desert regions of Chile.

Nitric acid is also used in the manufacture of explosives. Two important explosives are trinitrotoluene (TNT), and nitroglycerin. Nitroglycerin, the explosive component of dynamite (Box 12.1), is made by the reaction of glycerol (1,2,3-propanetriol) with nitric acid:

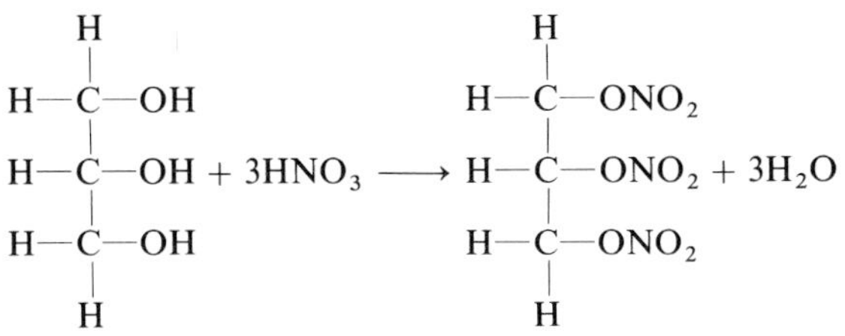

Trinitrotoluene is made by heating toluene with a mixture of concentrated nitric and sulfuric acids. Nitric acid is protonated by concentrated sulfuric acid:

$$HNO_3(l) + H_2SO_4(l) \longrightarrow H_2NO_3^+(soln) + HSO_4^-(soln)$$

Sulfuric acid is a stronger acid than nitric acid, and in the absence of water, nitric acid acts as a base toward sulfuric acid. The ion $H_2NO_3^+$ is then dehydrated by sulfuric acid, producing the nitronium ion, NO_2^+:

$$H_2NO_3^+(l) + H_2SO_4(conc) \longrightarrow NO_2^+(soln) + H_3O^+(soln) + HSO_4^-(soln)$$

Thus the overall reaction of nitric acid with concentrated sulfuric acid is

$$HNO_3 + 2H_2SO_4 \longrightarrow NO_2^+ + H_3O^+ + 2HSO_4^-$$

The nitronium ion reacts with toluene to give trinitrotoluene:

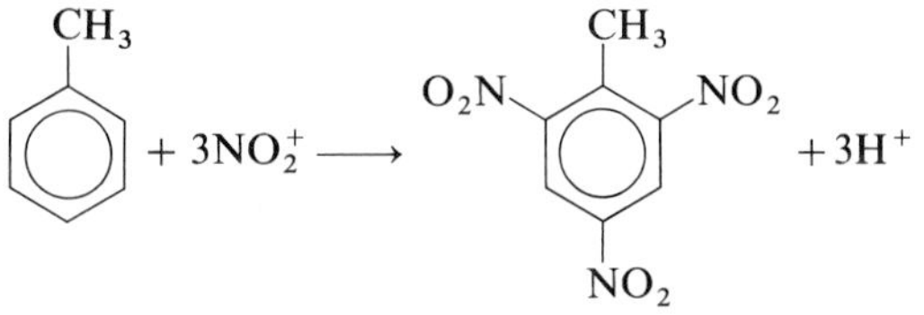

When TNT and nitroglycerin are heated or subjected to shock, they decompose rapidly, giving a large quantity of gaseous products in exothermic reactions. For example, nitroglycerin decomposes according to the following equation:

$$4C_3H_5N_3O_9(l) \longrightarrow 6N_2(g) + 12CO_2(g) + 10H_2O(g) + O_2(g)$$

Nitrous Acid, HNO_2

Unlike nitric acid, nitrous acid is a weak acid in water. It can be obtained by dissolving either N_2O_3 or a mixture of NO and NO_2 in water:

$$N_2O_3 + H_2O \longrightarrow 2HNO_2$$

Dinitrogen trioxide, N_2O_3, is the anhydride of nitrous acid. Only dilute solutions of nitrous acid can be obtained. If we attempt to concentrate a solution by heating, the nitrous acid decomposes to give NO and NO_2:

$$2HNO_2(aq) \longrightarrow NO(g) + NO_2(g) + H_2O(l)$$

The Lewis structure of nitrous acid is

$$:\ddot{O}=\ddot{N}-\ddot{O}-H$$

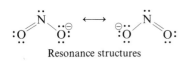

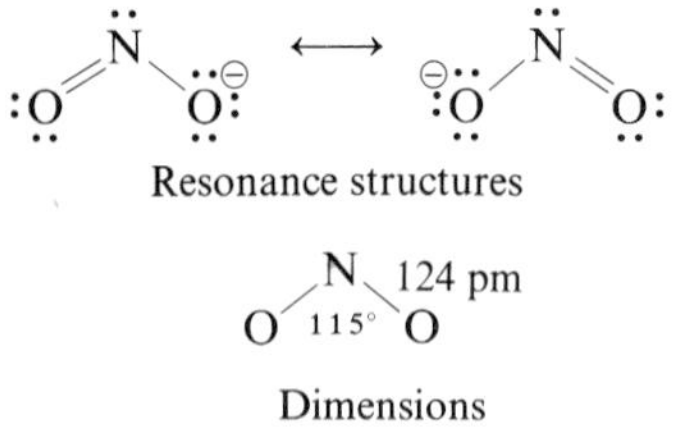

Resonance structures

N 124 pm
O 115° O

Dimensions

Figure 17.5 Structure of the Nitrite Ion.

The salts of nitrous acid are the *nitrites* (Figure 17.5). Sodium nitrite can be made by passing a mixture of NO and NO_2 into aqueous sodium hydroxide solution:

$$NO(aq) + NO_2(g) + 2NaOH(aq) \longrightarrow 2NaNO_2(aq) + H_2O(l)$$

Nitrites can also be made by heating alkali metal nitrates above their melting point:

$$2NaNO_3(l) \longrightarrow 2NaNO_2(l) + O_2(g)$$

Because they are readily oxidized to nitric acid and nitrate ion, nitrous acid and the nitrites are good reducing agents:

$$NO_2^- + H_2O \longrightarrow NO_3^- + 2e^- + 2H^+$$

Sodium nitrite and sodium nitrate have long been used in the preservation of meat. Meat darkens when it is stored because the blood is oxidized. Sodium nitrate and nitrite are reduced to NO by compounds in the meat. The NO combines with the hemoglobin and retards its oxidation, thus maintaining a fresh red appearance to the meat. Nitrites and nitrates also prevent the growth of a bacterium that causes botulism, a dangerous and sometimes fatal form of food poisoning. In the past few years, however, concern has risen about the possibility of a reaction between nitrite ion and certain organic compounds in the meat or in the digestive system to form nitrosamines, which are highly carcinogenic. But we should remember that some form of food preservation is necessary. Fish has been preserved by smoking in many parts of the world for thousands of years. But recently nitrosamines have been found in small amounts in smoked fish.

17.3 HYDRAZINE, HYDROXYLAMINE, AND HYDRAZOIC ACID

Hydrazine, N_2H_4

It is useful to think of hydrazine as being derived from ammonia by replacement of a hydrogen atom by an NH_2 group:

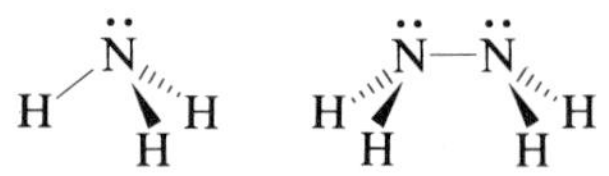

Ammonia   Hydrazine

Both nitrogen atoms in N_2H_4 have the expected pyramidal AX_3E geometry.

An aqueous solution of hydrazine can be prepared by oxidizing ammonia with a solution of sodium hypochlorite:

$$2NH_3(aq) + OCl^-(aq) \longrightarrow N_2H_4(aq) + H_2O(l) + Cl^-(aq)$$

The chloramines H_2NCl and $HNCl_2$ are also produced in this reaction. They are both toxic and explosive. Thus it is dangerous to mix household bleach and an ammonia cleaning solution.

Hydrazine is a colorless liquid that melts at 2°C and boils at 114°C. It is a weak base, which is protonated to give the $N_2H_5^+$ and $N_2H_6^{2+}$ ions. Hydrazine has an endothermic enthalpy of formation ($\Delta H_f^\circ = 51$ kJ mol^{-1}); it is therefore a somewhat unstable compound. It burns in air with a considerable evolution of heat. It reacts vigorously with strong oxidizing agents such as nitric acid, dinitrogen tetraoxide, and hydrogen peroxide with the evolution of a large amount of heat and the generation of a large volume of gaseous products:

$$2N_2H_4(l) + N_2O_4(l) \longrightarrow 3N_2(g) + 4H_2O(g)$$
$$N_2H_4(l) + 2H_2O_2(l) \longrightarrow N_2(g) + 4H_2O(g)$$

The oxidation of N_2H_4 or of dimethylhydrazine, $N_2H_2(CH_3)_2$, with one of these strong oxidizing agents has been used to power rockets. On the Apollo missions to the moon, the rocket engines of the command module and the lunar landing vehicles used N_2O_4 as an oxidizer for the fuel, which was a mixture of hydrazine and dimethylhydrazine. Hydrazine also reduces Fe^{3+} to Fe^{2+}, MnO_4^- to Mn^{2+}, and I_2 to I^-.

Example 17.6 Write a balanced equation for the reaction between hydrazine and Fe^{3+} in which hydrazine is oxidized to N_2.

Solution The oxidation number of each nitrogen in hydrazine is -2. Therefore the oxidation half-reaction involves 4 electrons and is

$$N_2H_4(aq) \longrightarrow N_2(g) + 4H^+(aq) + 4e^-$$

The reduction half-reaction is

$$Fe^{3+}(aq) + e^- \longrightarrow Fe^{2+}(aq)$$

Combining the equations for the two half-reactions gives the equation for the overall reaction.

$$4Fe^{3+}(aq) + N_2H_4(aq) \longrightarrow 4Fe^{2+}(aq) + N_2(g) + 4H^+(aq)$$

Example 17.7 What are the enthalpy changes for the following reactions? Use the enthalpies of formation given in Table 12.1 and $\Delta H_f^\circ(H_2O_2) = -188$ kJ, and $\Delta H_f^\circ(N_2H_4) = 51$ kJ.

(a) $N_2H_4(l) + 2H_2O_2(l) \longrightarrow N_2(g) + 4H_2O(g)$

(b) $2N_2H_4(l) + N_2O_4(l) \longrightarrow 3N_2(g) + 4H_2O(g)$

Solution

(a) $\Delta H^\circ = \Delta H_f^\circ(N_2) + 4\,\Delta H_f^\circ(H_2O) - \Delta H_f^\circ(N_2H_4) - 2\,\Delta H_f^\circ(H_2O_2)$

$= 0 + 4(-242) - 51 - 2(-188)\ kJ = -643\ kJ$

(b) $\Delta H^\circ = 3\,\Delta H_f^\circ(N_2) + 4\,\Delta H_f^\circ(H_2O) - 2\,\Delta H_f^\circ(N_2H_4) - \Delta H_f^\circ(N_2O_4)$

$= 0 + 4(-242) - 2(51) - (-20)\ kJ = -1050\ kJ$

We see that both these reactions are strongly exothermic.

Hydroxylamine, NH_2OH

We can think of hydroxylamine as resulting from the replacement of one H of an NH_3 molecule by an OH group or one hydrogen of a water molecule by an NH_2 group (see Figure 17.6). Hydroxylamine is a white solid that melts at 33°C. It slowly decomposes above 0°C. Therefore it is normally used as a solution in water, in which it is relatively stable. An aqueous solution of hydroxylamine can be prepared by reducing nitric acid with several oxidizing agents,

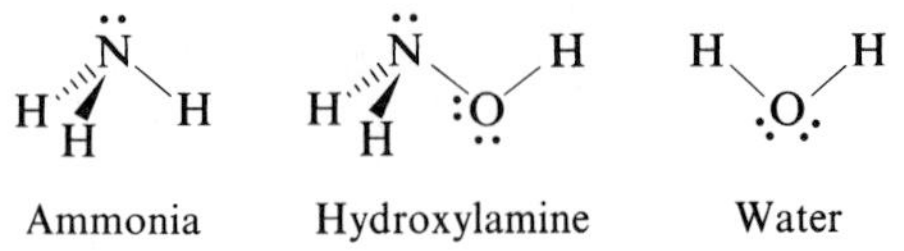

Figure 17.6 Structure of Hydroxylamine. The structure of hydroxylamine can be derived from the
structure of ammonia by replacing
an H by OH, or from the structure
of water by replacing an H by NH_2.
The geometry at the nitrogen atom
is AX_3E.

Ammonia   Hydroxylamine   Water

including tin and SO_2, under carefully controlled conditions:

$$HNO_3(aq) + 6e^- + 6H^+(aq) \longrightarrow NH_2OH(aq) + 2H_2O(l)$$

Hydroxylamine is a strong reducing agent, and like ammonia, it is a weak base:

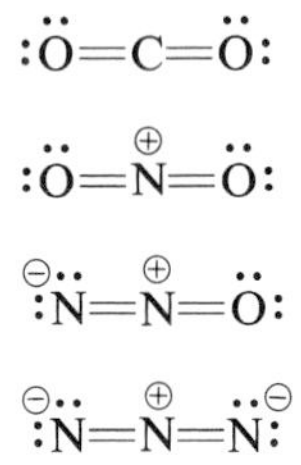

$$NH_2OH(aq) + H_2O(l) \rightleftharpoons NH_3OH^+(aq) + OH^-(aq)$$

Hydrazoic Acid, HN_3

Hydrazoic acid is a liquid that boils at 37°C. It is a weak acid in water. It is dangerously explosive. Salts of hydrazoic acid are called *azides*. The azides of heavy metals such as lead, mercury, and barium explode on being struck sharply and are used as detonators.

The azide ion, N_3^-, is isoelectronic with CO_2, N_2O, and NO_2^+, and it has the expected linear AX_2 structure with a bond length of 116 pm. Hydrazoic acid has the structure shown in Figure 17.7. It can be represented by the following two resonance structures:

124 pm

N—N—N
113 pm   110°   H

116 pm 116 pm
[N—N—N]⁻

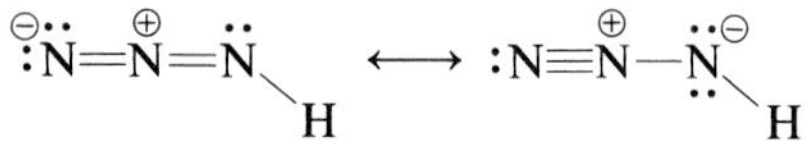

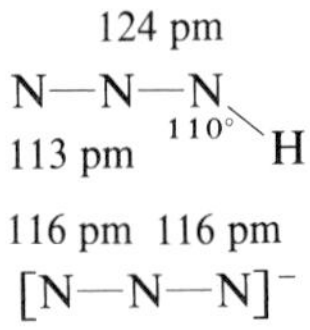

Figure 17.7 Structures of Hydrazoic Acid and the Azide Ion.

17.4 OZONE

Ozone is an allotrope of oxygen. It consists of O_3 molecules. It is a pale blue gas formed by passing an electric discharge through gaseous oxygen. Its characteristic odor may be noticed during electrical storms and in the vicinity of electrical equipment such as electric motors. It is an important constituent of the stratosphere (Box 17.2). It condenses to a deep blue liquid at −112°C. The liquid and even the gas, if it is not at low pressure or diluted with an inert gas such as nitrogen, may decompose explosively in a strongly exothermic reaction:

$$2O_3 \longrightarrow 3O_2 \qquad \Delta H° = -284 \text{ kJ}$$

Ozone is isoelectronic with the nitrite ion, NO_2^-, and is an angular AX_2E molecule (see Figure 17.8).

We might have expected ozone to be described by the Lewis structure

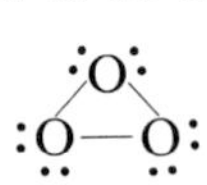

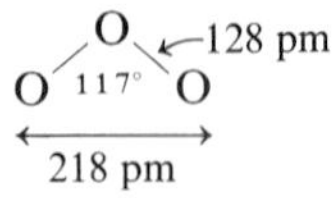

However, this structure implies bond angles of 60° and equal distances between all the oxygen atoms, which is not in agreement with the experimentally determined structure (Figure 17.8). Since the covalent radius of oxygen is 66 pm (Figure 4.7), the length of an oxygen-oxygen single bond is predicted to be 132 pm. The distance of 218 pm between two of the oxygen atoms in ozone is so much larger than this that there can be no bond between these two oxygen atoms. Ozone is therefore represented by the following two resonance structures:

$$:\ddot{O}=\overset{\oplus}{\underset{..}{O}}-\underset{..}{\overset{..}{O}}:^{\ominus} \longleftrightarrow {}^{\ominus}:\underset{..}{\overset{..}{O}}-\overset{\oplus}{\underset{..}{O}}=\ddot{O}:$$

Ozone is an extremely powerful oxidizing agent. The only common oxidizing agent that is stronger is fluorine. Ozone can be used for destroying bacteria in water by oxidation. Unlike chlorine, it leaves no taste in the water.

17.5 HYDROGEN PEROXIDE, PEROXIDES, AND SUPEROXIDES

Hydrogen peroxide, H_2O_2, can be thought of as being derived from water by replacement of a hydrogen atom by an OH group (see Experiment 17.8). Its structure is shown in Figure 17.9.

Although the angular geometry at each oxygen atom can be predicted by VSEPR theory, there is no simple theory that enables one to predict the overall shape of the molecule—that is, whether it will be planar or nonplanar. In the solid state it is nonplanar (see Figure 17.9). The O—O bond (147 pm) is somewhat longer than the single-bond distance of 132 pm predicted from the covalent radius of 66 pm (Table 4.8), which suggests, that the O—O bond is rather weak and therefore rather easily broken to give two $\cdot\ddot{O}H$ radicals. Thus H_2O_2 is a reactive substance.

In the laboratory hydrogen peroxide can be prepared by stirring solid barium peroxide, BaO_2 with a cold aqueous solution of sulfuric acid. Insoluble barium

$$H-\ddot{O}-\ddot{O}-H$$
(a)

(b)

Figure 17.9 Structure of Hydrogen Peroxide. (a) Lewis structure. (b) Dimensions. The two OH bonds in the H_2O_2 molecule are not in the same plane. In the solid state, there is an angle of 112° between them.

EXPERIMENT 17.8

Hydrogen Peroxide

An interesting although not very practical method of preparing a dilute aqueous solution of hydrogen peroxide is to allow a hydrogen flame to play on an ice cube. As the flame is rapidly cooled by the ice, the hydrogen reacts with oxygen to produce H_2O_2 as well as H_2O. The H_2O_2 dissolves in the water formed by the melting ice so that a dilute solution of H_2O_2 collects in the dish.

That the solution thus formed contains hydrogen peroxide can be demonstrated by adding it to a pink aqueous solution of $KMnO_4$. The $KMnO_4$ is decolorized as the hydrogen peroxide reduces the MnO_4^- ion to Mn^{2+} and is itself oxidized to oxygen.

When an aqueous solution of hydrogen peroxide is added to solid black lead sulfide, PbS, the PbS is oxidized to white $PbSO_4$ and the H_2O_2 is reduced to H_2O.

Ozone is an important constituent of the stratosphere, principally at altitudes between 15 and 25 km, where it may reach a concentration of 10 ppm, compared with 0.04 ppm in the troposphere (Chapter 3). Ozone is formed by the decomposition of oxygen by ultraviolet radiation from the sun having wavelengths shorter than 240 nm:

$$O_2 \xrightarrow{h\nu} 2O$$

At a height of 200 km or more above the earth's surface, the pressure is extremely low, and the concentration of oxygen atoms is so low that they rarely recombine to form O_2 molecules but remain as oxygen atoms, making up the atomic oxygen layer of the homosphere (Chapter 3). The presence of atomic oxygen in the upper atmosphere was clearly demonstrated in 1956 in an experiment in which 8 kg of NO was released from a rocket. As a result of the reaction between NO and O to produce NO_2, an orange-red glow appeared in the sky. More recently, the concentrations of the various atoms and molecules present in the atmosphere at high altitudes have been measured by mass spectrometers carried by satellites.

When oxygen atoms are produced by the dissociation of oxygen molecules at lower altitudes, they immediately combine with oxygen molecules to form ozone:

$$O + O_2 \longrightarrow O_3$$

But this ozone absorbs ultraviolet light, with wavelengths between 240 and 310 nm, and is decomposed back to O and O_2:

$$O_3 \xrightarrow{\text{UV light } (\lambda = 240-310 \text{ nm})} O + O_2$$

Thus an equilibrium is set up. The concentration of ozone remains constant at its equilibrium value, but ultraviolet light is continuously absorbed and is converted to kinetic energy of O atoms and O_2 molecules, that is, into heat. As a consequence, most of the ultraviolet radiation from the sun is absorbed before it reaches the earth's surface. Since living cells can be destroyed by ultraviolet radiation, the ozone layer protects us from its damaging effects. If there were no ozone in the stratosphere, the intensity of the ultraviolet radiation reaching the surface would be such that life as we know it would not be possible. Even a small decrease in the concentration of ozone could lead to an increased incidence of skin cancer.

Although too much ozone in the troposphere, as in photochemical smog, is harmful to us, too little in the stratosphere could be even more serious. There has been concern in recent years that certain atmospheric pollutants may be decreasing the concentration of ozone in the ozone layer. Supersonic aircraft (SSTs) fly at altitudes up to 18 km, that is, at the height of the ozone

---

sulfate is precipitated and can be filtered off. If stoichiometric amounts of BaO_2 and sulfuric acid are used, a pure aqueous solution of hydrogen peroxide can be obtained:

$$BaO_2(s) + H_2SO_4(aq) \longrightarrow BaSO_4(s) + H_2O_2(aq)$$

Hydrogen peroxide is a colorless, viscous liquid that boils at 150°C. Like water, it is strongly associated by hydrogen bonding (Chapter 13). When the pure liquid is heated, it decomposes rapidly and even explosively in a disproportionation reaction:

$$2H_2O_2(l) \longrightarrow 2H_2O(l) + O_2(g) \qquad \Delta H° = -196 \text{ kJ}$$

Normally, it is used as an aqueous solution (for example, 30% by mass for use in the laboratory and 3% by mass for pharmaceutical use). Such aqueous solutions decompose very slowly at room temperature, but the decomposition is catalyzed by many different substances, including Fe(II) salts, manganese dioxide, powdered platinum, and blood (Experiment 1.5). Hydrogen peroxide is a very weak acid in aqueous solution.

Hydrogen peroxide is an important industrial chemical that has many applications; for example, it is used as a bleaching agent for textiles and for wood pulp and waste paper in paper making. Its use for bleaching hair is well known. Its bleaching action is due to its strong oxidizing properties. The oxidation number of oxygen in hydrogen peroxide is -1. It is reduced to H_2O, in which

layer. The NO they emit in their exhaust reduces ozone to oxygen:

$$NO + O_3 \longrightarrow O_2 + NO_2$$

The NO_2 combines with oxygen atoms to give NO and O_2:

$$NO_2 + O \longrightarrow NO + O_2$$

The net reaction is

$$O_3 + O \longrightarrow 2O_2$$

Thus ozone is decomposed and NO is regenerated. Here NO is behaving as a catalyst: It increases the rate of decomposition of O_3 but is regenerated in the reaction and undergoes no overall change. Thus the release of NO by supersonic planes could apparently reduce the ozone concentration. However, this conclusion is by no means certain, because the chemistry of the upper atmosphere is very complex and not well understood.

Another cause of concern has been the release of chlorofluoromethanes, principally CF_2Cl_2 and $CFCl_3$, into the atmosphere. These substances have been widely used as propellants in spray cans and as refrigerant gases. They are unreactive compounds and do not appear to undergo any reactions in the lower atmosphere. They presumably diffuse into the stratosphere, where they are subject to ultraviolet radiation, which breaks a C—Cl bond, giving a free Cl atom:

$$CF_2Cl_2 \xrightarrow{h\nu} CF_2Cl + Cl$$

These Cl atoms can catalyze the decomposition of ozone in a manner analogous to NO:

$$O_3 + Cl \longrightarrow O_2 + ClO$$

$$ClO + O \longrightarrow O_2 + Cl$$

Chlorine atoms are regenerated, and the overall reaction is

$$O + O_3 \longrightarrow 2O_2$$

The ozone layer may also be perturbed by the release of N_2O into the atmosphere from the bacterial decomposition of nitrate fertilizers. They are being used in increasing amounts, and so increasing amounts of N_2O are being formed. Dinitrogen monoxide is unreactive, but in the stratosphere NO is produced by the reaction

$$N_2O + O \longrightarrow 2NO$$

This NO could then catalyze the decomposition of ozone. However, our understanding of all these reactions is not yet sufficient to enable us to make any certain predictions about their effects on the ozone layer. Many more observations and laboratory studies are needed.

oxygen has an oxidation number of -2. The corresponding half-reaction is

$$H_2O_2(aq) + 2H^+(aq) + 2e^- \longrightarrow 2H_2O(l) \qquad E° = +1.77 \text{ V}$$

The large positive value of the standard reduction potential shows that it is a stronger oxidizing agent than NO_3^- or MnO_4^-. Hydrogen peroxide oxidizes Fe^{2+} to Fe^{3+}, I^- to I_2, SO_2 (or SO_3^{2-}) to SO_4^{2-}, and PbS to $PbSO_4$ (see Experiment 17.8). This last reaction is used to restore the original white color to old paintings in which white lead pigment, $Pb_3(OH)_2(CO_3)_2$, has become converted to dark brown PbS in an urban atmosphere containing H_2S.

Because it can be oxidized to oxygen, in which the oxidation number of oxygen is zero, H_2O_2 can also behave as a reducing agent:

$$H_2O_2 \longrightarrow O_2 + 2H^+ + 2e^-$$

For example, H_2O_2 will reduce MnO_4^- to Mn^{2+}.

Example 17.8 Write balanced equations for (a) the oxidation of PbS to $PbSO_4$ with H_2O_2 and (b) the reduction of MnO_4^- to Mn^{2+} with H_2O_2.

Solution

(a) When S^{2-} is oxidized to SO_4^{2-}, the oxidation number of sulfur is increased from -2 to $+6$. The equation for the corresponding half-reaction is therefore

$$S^{2-} + 4H_2O \longrightarrow SO_4^{2-} + 8H^+ + 8e^-$$

17.5 HYDROGEN PEROXIDE, PEROXIDES, AND SUPEROXIDES

We combine this equation with the equation for the reduction of H_2O_2 to H_2O,

$$H_2O_2 + 2H^+ + 2e^- \longrightarrow 2H_2O$$

to give the overall equation

$$S^{2-} + 4H_2O_2 \longrightarrow SO_4^{2-} + 4H_2O$$

Adding Pb^{2+} to both sides gives

$$PbS + 4H_2O_2 \longrightarrow PbSO_4 + 4H_2O$$

(b) The equation for the reduction of MnO_4^- to Mn^{2+} is

$$MnO_4^- + 8H^+ + 5e^- \longrightarrow Mn^{2+} + 4H_2O$$

The equation for the oxidation of H_2O_2 to O_2 is

$$H_2O_2 \longrightarrow O_2 + 2H^+ + 2e^-$$

Multiplying the first equation by 2 and the second by 5 and combining them gives

$$2MnO_4^- + 5H_2O_2 + 6H^+ \longrightarrow 2Mn^{2+} + 5O_2 + 8H_2O$$

When sodium is heated in a limited supply of air, it forms the expected oxide Na_2O. But in an excess of air it forms the pale yellow peroxide Na_2O_2, which contains the peroxide ion, O_2^{2-}:

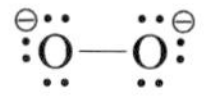

Barium behaves in the same way to give barium peroxide, BaO_2.

Potassium, rubidium, and cesium give the expected oxides K_2O, Rb_2O, and Cs_2O in a limited amount of air or oxygen. But when they are heated in excess oxygen, they form the oxides KO_2, RbO_2, and CsO_2. These orange-red compounds contain the superoxide ion, O_2^-, which has 13 electrons and is a stable, free-radical ion. It can be represented by two resonance structures,

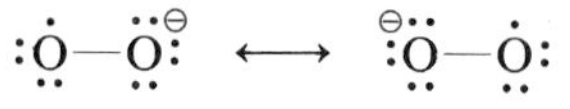

or by one structure in which the odd electron is shared between the two oxygen atoms, as in NO:

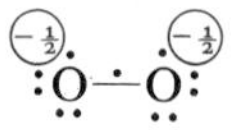

As the bond order increases from 1 in O_2^{2-} to 1.5 in O_2^- and to 2 in the oxygen molecule, O_2, the bond length decreases correspondingly:

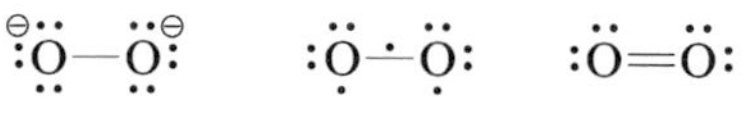

149 pm	133 pm	121 pm

IMPORTANT TERMS

An **acid anhydride** is the oxide obtained from an oxoacid by removing water.

A **free radical** is a molecule or ion containing an odd number of electrons.

PROBLEMS

Names and Formulas

1. What is the formula of each of the following compounds?

(a) Nitric acid   (e) Dinitrogen pentaoxide

(b) Nitrous acid   (f) Hydrazine

(c) Potassium nitrite   (g) Sodium azide

(d) Nitrogen monoxide   (h) Hydroxylamine

2. What is the formula of each of the following substances?

(a) Ozone   (b) Sodium peroxide

(c) Barium peroxide   (d) Potassium superoxide

3. What are the names of the following compounds?

(a) $NaNO_3$ (b) KNO_2 (c) N_2O_4

(d) HN_3 (e) NH_2OH (f) BaO_2

(g) KO_2 (h) N_2H_4 (i) LiN_3

(j) Li_3N

4. Write the formula for the anhydride of each of the following acids:

(a) HNO_3 (b) HNO_2 (c) H_3PO_3

(d) $HClO_4$ (e) H_3BO_3 (f) H_2SO_4

(g) H_2CO_3

Reactions

5. Write balanced equations for the following reactions by which nitrogen may be prepared in the laboratory. Indicate clearly in each case which substances have been oxidized and which reduced.

(a) The reaction of ammonia with copper(II) oxide to give nitrogen, water, and copper.

(b) The reaction of a solution of sodium nitrite with a solution of ammonium chloride, on gentle heating, to give nitrogen and sodium chloride.

(c) The reaction of nitric oxide with ammonia, when passed over a red-hot copper catalyst, to give nitrogen and water.

(d) The decomposition of solid ammonium dichromate, $(NH_4)_2Cr_2O_7$, on heating, to give Cr_2O_3, nitrogen, and water.

6. Write balanced equations and identify the oxidation number of nitrogen in the products formed when the following nitrates are heated.

(a) Potassium nitrate   (b) Ammonium nitrate

(c) Lead(II) nitrate

7. Write balanced equations for the following reactions:

(a) The decomposition of nitrous acid to nitric acid and nitrogen monoxide in aqueous solution.

(b) The decomposition of nitric acid to nitrogen dioxide and oxygen.

(c) The reaction of benzene with a mixture of sulfuric acid and nitric acid to give nitrobenzene.

8. Write balanced equations for the reactions of nitric acid (conc., dil., 100%) with the following:

(a) Cu   (b) P_4O_{10} (c) H_2SO_4

(d) NH_3 (e) $Mg(OH)_2$

9. Write a balanced equation for each of the following reactions. State whether each reaction is an acid-base reaction or an oxidation-reduction reaction. For the acid-base reactions, indicate which species is the acid and which is the base. For the oxidation-reduction reactions, state which species is oxidized and which is reduced.

(a) The reaction of sodium nitrate with concentrated sulfuric acid.

(b) The photochemical decomposition of nitric acid.

(c) The decomposition of lead(II) nitrate on heating.

(d) The reaction of nitric acid with concentrated sulfuric acid.

(e) The reaction of nitrous acid with water.

(f) The reaction of ammonia with hypochlorite ion in aqueous solution.

(g) The catalyzed decomposition of hydrogen peroxide.

(h) The reaction of barium peroxide with sulfuric acid.

10. Describe, using suitable equations, how nitric acid is prepared from ammonia.

11. When sulfur dioxide is bubbled through a strongly acidic solution of barium nitrate, a white precipitate is obtained. What is this white precipitate? Explain, by means of suitable equations, how it is formed.

Lewis Structures and Molecular Geometry

12. Draw Lewis structures for each of the following:

(a) NH_4^+ (b) $N_2H_5^+$ (c) HNO_2

(d) H_2O_2 (e) O_2^{2-} (f) $NOCl$

13. Draw Lewis structures for the following molecules and ions, and indicate in each case whether a single Lewis structure is sufficient for a good description of the bonding or if several resonance structures are necessary to adequately describe the bonding.

(a) Nitric acid   (b) Nitrate ion

(c) Nitrous acid   (d) Nitrite ion

(e) Nitrogen monoxide   (f) Dinitrogen tetraoxide

(g) Gaseous dinitrogen pentaoxide

14. Using VSEPR theory, categorize each of the following in terms of the AX_nE_m nomenclature, and predict the approximate molecular geometry in each case. In each case the central atom is nitrogen.

(a) NO_2 (b) NO_2^+ (c) NO_2^- (d) HNO_3

(e) N_2O (f) $NOCl$ (g) NF_3 (h) HNO_2

(i) CH_3NO_2

15. Draw the Lewis structure of the oxide that is the anhydride of each of the following acids:

(a) HNO_3 (b) HNO_2

(c) H_3PO_4 (d) H_2SO_4

16. Which of the following molecules has a dipole moment?

(a) NO (b) NO_2 (c) N_2

(d) O_3 (e) N_2O (f) N_2O_4

(g) H_2O_2 (h) N_2H_4 (i) HNO_3

17. Which three species mentioned in this chapter are isoelectronic with CO_2?

18. Explain why ozone is not given the Lewis structure

$$:\!O\!:$$
$$:\!O\!-\!O\!:$$

Stoichometry

19. Calculate the molar concentration of nitric acid in each of the following solutions.

(a) A concentrated solution of nitric acid containing 65.3% HNO_3 by mass and having a density of 1.40 g mL^{-1}.

(b) A solution of nitric acid prepared by starting with 100 g of a solution containing 24.84% HNO_3 by mass and diluting it with water to give a total volume of 500 mL.

20. What volume of nitrogen measured at $60°C$ and 750 torr is formed by the thermal decomposition of 10.0 g of ammonium nitrite?

21. What mass of ammonia would be required to neutralize 1.00 metric ton of nitric acid to give nitrogen and water?

22. An impure sample of $NaNO_3$ having a mass of 1.354 g was strongly heated, and the evolved oxygen was collected over water at $25°C$ and a pressure of 760 mm Hg. Calculate the percentage purity of the sample if 131.8 mL of oxygen was collected (vapor pressure of H_2O at $25°C$ is 20.8 mm Hg). Assume that all the oxygen comes from the $NaNO_3$.

Equilibrium

23. In the gas phase nitrogen dioxide is in equilibrium with dinitrogen tetraoxide. What will be the effect on the concentration of N_2O_4 of the following changes?

(a) Increasing the total pressure of the system by introducing an inert gas.

(b) Increasing the volume of the system at constant temperature.

(c) Decreasing the temperature.

24. Calculate the pH of the following:

(a) A 0.02 molar solution of nitrous acid

(b) A 0.02 molar solution of sodium nitrite.

(c) A solution that is 0.01 molar in sodium nitrite and 0.01 molar in nitrous acid. ($pK_a = 3.14$).

Thermochemistry

25. Calculate $\Delta H°$ for the reaction

$$4NH_3(g) + 5O_2(g) \longrightarrow 4NO(g) + 6H_2O(g)$$

from the standard enthalpies of formation in Table 12.1.

26. The standard enthalpy of formation of ozone, O_3, is $+142 \text{ kJ mol}^{-1}$. The dissociation energy of the O_2 molecule is 498 kJ mol^{-1}. What is the average bond energy of the two bonds in O_3?

27. Calculate the enthalpy change for the reaction

$$CO(g) + NO_2(g) \longrightarrow CO_2(g) + NO(g)$$

using the standard enthalpies of formation in Table 12.1.

28. Use the standard enthalpies of formation given in Table 12.1 and the dissociation energy of the oxygen molecule (498 kJ mol^{-1}) to find the enthalpy change for both steps of the conversion of O_3 to O_2 catalyzed by NO:

$$NO(g) + O_3(g) \longrightarrow NO_2(g) + O_2(g)$$
$$NO_2(g) + O(g) \longrightarrow O_2(g) + NO(g)$$

Oxidation-Reduction

29. What is the oxidation state of nitrogen in each of the following?

(a) NH_4^+ (b) N_2H_4 (c) KNO_2

(d) NO_2 (e) NH_2OH (f) N_2O

(g) NH_4NO_3 (h) LiN_3 (i) Li_3N

(j) NF_3

30. Balance the following oxidation-reduction equations:

(a) $H_2S + NO_3^- \longrightarrow S + NO$; acidic solution.

(b) $Zn + NO_3^- \longrightarrow$
$$Zn(OH)_2 + NH_3\text{ ; basic solution.}$$

(c) $P_4 + NO_3^- \longrightarrow H_3PO_4 + NO$; acidic solution.

31. Explain why hydrogen peroxide can act as both an oxidizing agent and a reducing agent. Write balanced equations for the following reactions:

(a) The oxidation of SO_2 to SO_4^{2-} by H_2O_2 in acidic solution.

(b) The reduction of O_3 to H_2O by H_2O_2 in acidic solution.

(c) The reduction of I_2 to I^- by H_2O_2 in acidic solution.

(d) The oxidation of $Cr(OH)_3$ to CrO_4^{2-} by H_2O_2 in basic solution.

(e) The oxidation of NO_2^- to NO_3^- by H_2O_2 in acidic solution.

CHAPTER 18
RATES OF CHEMICAL REACTIONS

There is evidence all around us that some chemical reactions are very fast while others are extremely slow. The explosive reaction between oxygen and a hydrocarbon in the cylinder of an automobile, the explosive decomposition of TNT, and the reaction between a strong acid and a strong base, which occurs as fast as acid is added to base, are all examples of very fast reactions. Although slower than explosions, the reactions that lead to the contraction of muscles and that transmit nervous impulses are still very rapid. In contrast, the rusting of iron in air is rather slow, although usually it is not as slow as we would like. The weathering of rocks is still slower and may take millions of years to complete. Although we would like to be able to further slow up reactions such as the rusting of iron and the deterioration of foods, we are more often concerned with how we could speed up slow reactions so that we can obtain a useful amount of a desired product in a reasonable time. The rates of two familiar everyday reactions are demonstrated in Experiment 18.1.

In order to be able to control the rates of reactions, we need to understand the factors that influence the rate. In this chapter we consider the effect of the concentrations of the reactants, the temperature, and the presence of catalysts on the rate of a reaction. We will see that studying the various factors that influence the rate of a reaction can provide us with information on how a reaction occurs. In order for a reaction to take place between two molecules, the molecules must collide. The probability of such a collision increases with increasing concentrations of the reactants. Thus we expect reaction rates to depend on the concentrations of the reactants. However, we will see that if every collision led to a reaction, then almost all reactions would be extremely fast. In fact, in most cases only a very small fraction of the collisions actually lead to a reaction, and we will discuss why this is so. We will see that increasing the temperature increases the rate of almost all reactions because it increases the number of effective collisions.

Rates of Oxidation

A freshly cut apple is exposed to the atmosphere.

In a few minutes the surface becomes brown as it is rapidly oxidized by the oxygen of the atmosphere.

When steel wool is placed in water it is slowly oxidized and within one to two hours becomes coated with a layer of brown rust.

Many reactions of industrial importance employ catalysts, and most reactions in living organisms are catalyzed. We will discuss how a catalyst can increase the rate of a reaction without being used up.

Many reactions, including many catalyzed reactions, take place at the surface between two different phases, for example, a solid and a gas or a solid and a liquid. These are called hetereogeneous reactions, and their rates depend on the area of the surface between the two phases. Increasing the surface area increases the rate.

The effect of temperature on the rate of a reaction is illustrated in Experiment 18.2.

Before discussing in more detail how the concentrations of the reactants, the temperature, and the catalysts affect the rates of reactions, we need to consider just how we express and measure the rate of a reaction.

The Effect of Temperature on Reaction Rate

Marble chips ($CaCO_3$) dissolve in dilute hydrochloric acid producing bubbles of carbon dioxide. When the reaction is carried out at 0°C as on the left, the reaction is slow. Only a few bubbles of CO_2 can be seen and only a small amount of CO_2 has collected in the jar since the reaction was started. On the right the reaction is being carried out at a temperature of approximately 60°C. There is a very rapid evolution of CO_2 and a large amount of gas has been collected in the same amount of time.

CHAPTER 18
RATES OF CHEMICAL
REACTIONS

The rate (or speed) of a chemical reaction is defined in the same way that we define the speed of an automobile or any other object. The speed of an automobile is equal to the change in its position, or the distance traveled, divided by the time taken to travel the distance between its initial and final positions:

$$\text{Speed of automobile} = \frac{\text{Change in position}}{\text{Time for the change}} = \frac{\text{Distance traveled}}{\text{Time needed}}$$

The **rate of a chemical reaction** is defined as the change in the concentration of a reactant (or product) in a certain time interval. Consider the reaction between nitrogen monoxide and ozone that may be responsible for decreasing the concentration of ozone in the ozone layer of the upper atmosphere (Box 17.2):

$$NO(g) + O_3(g) \longrightarrow NO_2(g) + O_2(g)$$

If the concentration of NO_2 at time t_1 is $[NO_2]_1$, and if at time t_2 it is $[NO_2]_2$, then the concentration of NO_2 has changed by

$$\Delta[NO_2] = [NO_2]_2 - [NO_2]_1$$

in the time interval $\Delta t = t_2 - t_1$. The rate of the reaction is then

$$\frac{\text{Change in concentration of } NO_2}{\text{Time needed for change}} = \frac{\Delta[NO_2]}{\Delta t}$$

The rate of the reaction is given by the change in the concentration of any of the reactants or products in a certain time interval. Thus we have

$$\text{Rate of reaction} = \frac{\Delta[NO_2]}{\Delta t} = \frac{\Delta[O_2]}{\Delta t} = -\frac{\Delta[NO]}{\Delta t} = -\frac{\Delta[O_3]}{\Delta t}$$

The rate is given by the change in the concentration divided by the time needed for the change. Since the concentrations of the reactants decrease, $\Delta[\text{reactant}]$ is a negative quantity. Therefore if we measure the rate in terms of the change in concentration of one of the reactants, we define the rate as $-\Delta[\text{reactant}]/\Delta t$ so that the rate is always a positive quantity.

For a reaction involving different numbers of moles of reactants and products, we must be careful to specify which species is being used to measure the rate of the reaction. For example, in the reaction

$$N_2(g) + 3H_2(g) \longrightarrow 2NH_3(g)$$

two ammonia molecules are formed for every nitrogen molecule that is used up. In other words, the rate at which ammonia is formed is twice the rate at which N_2 disappears, and hydrogen disappears three times as fast. Therefore if we express the rate of reaction in terms of the rate at which nitrogen disappears, we have

$$\text{Rate of reaction} = -\frac{\Delta[N_2]}{\Delta t} = \frac{1}{2}\frac{\Delta[NH_3]}{\Delta t} = -\frac{1}{3}\frac{\Delta[H_2]}{\Delta t}$$

Speed has the units of distance divided by time and may be expressed, for example, as miles per hour (miles h^{-1}), kilometers per hour (km h^{-1}), or meters per second (m s^{-1}). Reaction rate has the units of concentration divided by time. We express concentrations in moles per liter (mol L^{-1}), but time may be given in any convenient unit—seconds, minutes, hours, days, or possibly years. Therefore the units of reaction rate may be mol $L^{-1} s^{-1}$, mol $L^{-1} min^{-1}$, mol $L^{-1} h^{-1}$, and so on.

Example 18.1 The rate of the reaction

$$2N_2O_5(g) \longrightarrow 4NO_2(g) + O_2(g)$$

can be expressed as $\Delta[O_2]/\Delta t$. Write an expression for the rate in terms of each of the other molecules involved in the reaction.

Solution The equation shows that the N_2O_5 is used up twice as fast as O_2 is formed. Therefore

$$\text{Rate} = \frac{\Delta[O_2]}{\Delta t} = -\frac{1}{2}\frac{\Delta[N_2O_5]}{\Delta t}$$

The equation also shows that NO_2 is formed four times as fast as oxygen is formed. Therefore

$$\text{Rate} = \frac{\Delta[O_2]}{\Delta t} = -\frac{1}{2}\frac{\Delta[N_2O_5]}{\Delta t} = \frac{1}{4}\frac{\Delta[NO_2]}{\Delta t}$$

To measure the rate of a reaction experimentally, we can select any one of the reactants or products and measure the change in its concentration with time. Any convenient method of determining concentration can be used. A small sample could be withdrawn from the reaction mixture and the reaction in the sample stopped, for example, by lowering the temperature or by diluting it in a large amount of solvent. The sample could then be analyzed for one or more of the products or reactants.

A measurement that can be made directly on the reaction mixture is often more convenient. For example, in the reaction between NO and O_3, the change in the concentration of NO_2 can be followed by a measurement of the increase in the intensity of the red-brown color of NO_2. If the hydrogen ion concentration changes during a reaction, this can be followed by measuring the pH, for example, with a pH meter. Reactions involving gases are often accompanied by a change in volume or pressure. For example, in the reaction of N_2 with H_2 to give NH_3, the total number of molecules decreases with time, so the pressure decreases if the reaction is carried out in a vessel of constant volume. Thus the rate of the reaction can be measured by the rate of change of the pressure.

Table 18.1 shows the concentrations of the reactants at 10-s intervals for the

Table 18.1 Rate Data for the Reaction $CO(g) + NO_2(g) \rightarrow CO_2(g) + NO(g)$

[CO] (mol L^{-1})	[NO$_2$] (mol L^{-1})	t (s)	RATE $= -\dfrac{\Delta[CO]}{\Delta t} = -\dfrac{\Delta[NO_2]}{\Delta t}$ (mol L^{-1} s^{-1})	SLOPE = RATE AT TIME t (mol L^{-1} s^{-1})
0.100	0.100	0		4.9×10^{-3}
			3.3×10^{-3}	
0.067	0.067	10		2.2×10^{-3}
			1.7×10^{-3}	
0.050	0.050	20		1.2×10^{-3}
			1.0×10^{-3}	
0.040	0.040	30		0.8×10^{-3}
			0.7×10^{-3}	
0.033	0.033	40		0.5×10^{-3}
			0.3×10^{-3}	
0.017	0.017	100		0.1×10^{-3}
0.002	0.002	1000		

reaction

$$CO + NO_2 \longrightarrow CO_2 + NO$$

We can find the reaction rate for each 10-s interval by dividing the change in concentration over the 10-s interval by 10 s. For example, for the interval $t = 0$ to $t = 10$ s the rate is given by

$$-\frac{\Delta[CO]}{\Delta t} = -\frac{0.067 - 0.100}{10 \text{ s}} \text{ mol L}^{-1} = 3.3 \times 10^{-3} \text{ mol L}^{-1} \text{ s}^{-1}$$

Values of the rate obtained in this way are given in column 4 of Table 18.1. Because the reaction rate is changing with time, these values are only average rates over each 10-s interval.

We can obtain a more exact value for the rate by reducing the size of the time interval. In the limit when t is very small, $-\Delta[CO]/\Delta t$ is the slope of the tangent to the curve at the time t as shown in Figure 18.1. Thus

$$\text{Rate} = \text{Limiting value}\left(-\frac{\Delta[CO]}{\Delta t}\right) = -\text{Slope of curve}$$

This is the *instantaneous rate* at a particular time t. Values of the rate obtained from the slope of the curve at a given time are given in column 5 of Table 18.1. Notice from columns 4 and 5 that the rate of the reaction continually decreases as the reaction proceeds because the concentrations of the reactants are continually decreasing.

18.2 RATE LAWS AND REACTION ORDER

We have mentioned that we expect the rate of a reaction to depend on the concentrations of the reactants. For the reaction

$$NO(g) + O_3(g) \longrightarrow NO_2(g) + O_2(g)$$

experiment shows that

Rate of reaction is proportional to $[NO][O_3]$

or

Rate of reaction $= k[NO][O_3]$

where k is a constant called the **rate constant**. At a given temperature k is a constant characteristic of the reaction. Its value is independent of the concentrations of the reactants, although it does depend on the temperature. For the reaction between NO and O_3, $k = 1.6 \times 10^7 \text{ mol}^{-1} \text{ L s}^{-1}$ at $25°$. The rate constant k is a measure of the intrinsic rate of the reaction: It is the rate when the concentrations of all the reactants are 1 mol L^{-1}. Fast reactions have large k values, while slow reactions have small k values.

An expression of this type, which relates the rate of a reaction to the concentrations of the reactants, is called a **rate law**. In general, for a reaction

$$aA + bB + cC + \cdots \longrightarrow \text{Products}$$

the rate law often has the form

$$\text{Rate} = k[A]^x[B]^y[C]^z \cdots$$

where x is called the **order** of the reaction with respect to A, y is the order with respect to B, z is the order with respect to C, and the sum, $x + y + z + \cdots$, is called the **overall order**.

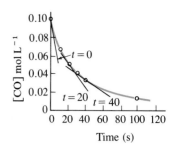

Figure 18.1 Concentration of CO as Function of Time for the Reaction

$$CO(g) + NO_2(g) \longrightarrow CO_2(g) + NO(g)$$

For the reaction between NO and O_3 the rate law is

$$\text{Rate} = k[\text{NO}][\text{O}_3]$$

We say that the reaction is first-order with respect to NO, first-order with respect to O_3, and second-order overall. This means that the rate of the reaction is directly proportional to the concentration of NO and also directly proportional to the concentration of O_3. If we double the concentration of either NO or O_3, the rate will double. If we double the concentrations of both NO and O_3, the rate increases by four times.

The rate law for any reaction must be determined experimentally, because, as we will see, *there is no necessary connection between the order with respect to any given reactant and the coefficient for the reactant in the balanced equation for the reaction.* In other words, x, y, z, ... are not necessarily the same as a, b, c, ...

How do we find experimentally how the rate depends on the concentration of each of the reactants? For a reaction in which there is only one reactant, the problem is fairly simple. Consider, for example, the decomposition of dinitrogen pentaoxide:

$$2\text{N}_2\text{O}_5(g) \longrightarrow 4\text{NO}_2(g) + \text{O}_2(g)$$

Table 18.2 and Figure 18.2 show some experimental data for this reaction. The reaction was followed by measuring the increase in pressure accompanying the reaction, from which the partial pressure of N_2O_5 (column 2) and hence the concentration of N_2O_5 (column 3) were calculated. The values of the rate given in column 4 were obtained by drawing tangents to the curve in Figure 18.2. By plotting the rate against the concentration of N_2O_5 (Figure 18.3), we obtain a straight line, showing that the rate is proportional to the concentration of N_2O_5. Therefore the rate law is

$$\text{Rate} = k[\text{N}_2\text{O}_5]$$

We can obtain the value of the rate constant from the slope of the straight line:

$$k = \frac{\text{Rate}}{[\text{N}_2\text{O}_5]} = \text{Slope} = 3.0 \times 10^{-4}\ \text{min}^{-1}$$

In contrast, a plot of the rate against $[\text{N}_2\text{O}_5]^2$ does *not* give a straight line (Figure 18.4), showing that the rate law is *not*

$$\text{Rate} = k[\text{N}_2\text{O}_5]^2$$

Table 18.2 Rate Data for Decomposition of $\text{N}_2\text{O}_5(g)$

TIME (min)	$P_{\text{N}_2\text{O}_5}$ (mm Hg)	$[\text{N}_2\text{O}_5]$ (mol L^{-1})	RATE (mol L^{-1} min^{-1})
0	340.2	0.0172	
10	224.8	0.0113	3.4×10^{-4}
20	166.7	0.0084	2.5×10^{-4}
30	123.2	0.0062	1.8×10^{-4}
40	92.2	0.0046	1.3×10^{-4}
50	69.1	0.0035	1.0×10^{-4}
60	51.1	0.0026	0.8×10^{-4}
70	37.5	0.0019	0.6×10^{-4}
80	27.4	0.0014	0.4×10^{-4}

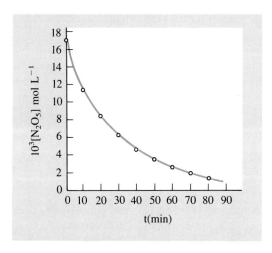

Figure 18.2 Concentration of N_2O_5 as Function of Time for Decomposition of N_2O_5.

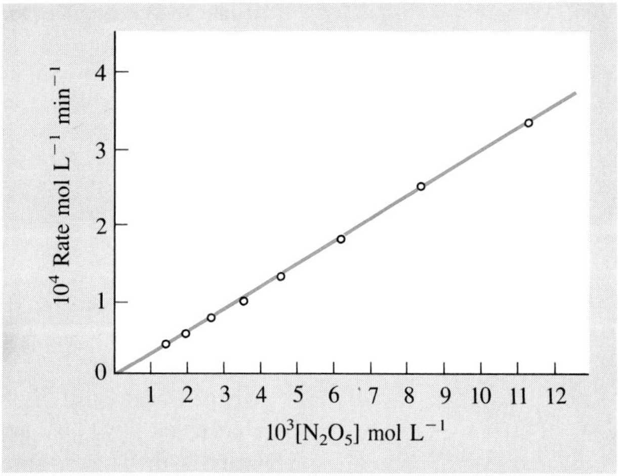

Figure 18.3 Rate of Decomposition of N_2O_5 as Function of N_2O_5 Concentration.

In other words, the reaction is *not* second-order. Notice that the form of the rate law does not follow from the balanced equation for the reaction.

Initial-Rate Method

For most reactions there is more than one reactant, all of whose concentrations are changing, so the previous method cannot be used. In such cases a simple method is the *initial-rate method*. In this method the rate of the reaction is measured at the beginning of the reaction, that is, before the concentrations have had time to change significantly. The initial concentration of only one reactant is then changed, while all the other initial concentrations are kept constant, and the rate is measured again. This procedure is repeated for each reactant, and the initial rate is measured each time.

For example, the data in Table 18.3 were obtained for the reaction between NO and Cl_2:

$$2NO(g) + Cl_2(g) \longrightarrow 2NOCl(g)$$

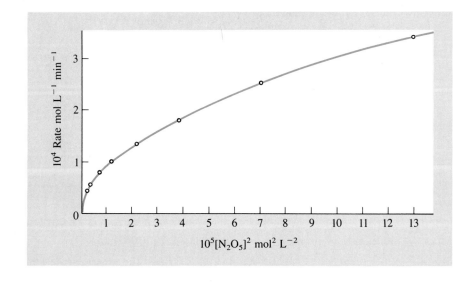

Figure 18.4 Rate of Decomposition of N_2O_5 as Function of $[N_2O_5]^2$.

Table 18.3 Rate Data for the Reaction $2NO(g) + Cl_2(g) \rightarrow 2NOCl(g)$ at 300 K

	INITIAL CONCENTRATIONS (mol L^{-1})		
EXPERIMENT	[NO]	[Cl$_2$]	INITIAL RATE (mol L^{-1} s^{-1})
1	0.010	0.010	1.2×10^{-4}
2	0.010	0.020	2.3×10^{-4}
3	0.020	0.020	9.6×10^{-4}

The rate law can be written in the general form

$$\text{Rate} = k[NO]^x[Cl_2]^y$$

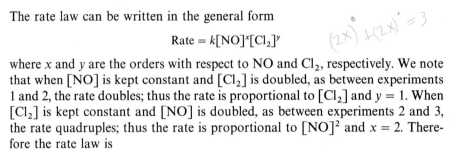

where x and y are the orders with respect to NO and Cl_2, respectively. We note that when [NO] is kept constant and [Cl$_2$] is doubled, as between experiments 1 and 2, the rate doubles; thus the rate is proportional to [Cl$_2$] and $y = 1$. When [Cl$_2$] is kept constant and [NO] is doubled, as between experiments 2 and 3, the rate quadruples; thus the rate is proportional to [NO]^2 and $x = 2$. Therefore the rate law is

$$\text{Rate} = k[NO]^2[Cl_2]$$

The reaction is second-order in NO, first-order in Cl$_2$, and third-order overall.

Example 18.2 What is the value of the rate constant at 300 K for the following reaction?

$$2NO(g) + Cl_2(g) \longrightarrow 2NOCl(g)$$

Solution We have seen that the rate law for this reaction is

$$\text{Rate} = k[NO]^2[Cl_2]$$

Therefore

$$k = \frac{\text{Rate}}{[NO]^2[Cl_2]}$$

Substituting the concentrations from experiment 1 (Table 18.3), we have

$$\frac{1.2 \times 10^{-4} \text{ mol L}^{-1}\text{s}^{-1}}{(0.010 \text{ mol L}^{-1})^2(0.010 \text{ mol L}^{-1})} = 1.2 \times 10^2 \text{ mol}^{-2}\text{L}^2\text{s}^{-1}$$

Substituting the concentrations and rates for experiments 2 and 3 also gives $k = 1.2 \times 10^2 \text{ mol}^{-2}\text{L}^2\text{s}^{-1}$ in both cases.

Example 18.3 What is the rate of the reaction in Example 18.2 when [NO] = 0.030 mol L^{-1} and [Cl$_2$] = 0.040 mol L^{-1}?

Solution We have seen that the rate law is rate = $k[NO]^2[Cl_2]$ and that k is $1.2 \times 10^2 \text{ mol}^{-2}\text{L}^2\text{s}^{-1}$. Therefore, if

$$[NO] = 0.030 \text{ mol L}^{-1} \quad \text{and} \quad [Cl_2] = 0.040 \text{ mol L}^{-1}$$

then

$$\text{Rate} = (1.2 \times 10^2 \text{ mol}^{-2}\text{L}^2\text{s}^{-1})(0.030 \text{ mol L}^{-1})^2(0.040 \text{ mol L}^{-1})$$

$$= 4.3 \times 10^{-3} \text{ mol L}^{-1}\text{s}^{-1}$$

Example 18.4 For the reaction between nitrogen monoxide and hydrogen,

$$2NO(g) + 2H_2(g) \longrightarrow N_2(g) + 2H_2O(g)$$

the following initial-rate data were obtained:

Experiment	Initial pressures (mm Hg)		Initial rate (mm Hg s^{-1})
	NO	H$_2$	
1	359	300	1.50
2	300	300	1.03
3	152	300	0.25
4	300	289	1.00
5	300	205	0.71
6	300	147	0.51

What is the order of the reaction with respect to NO and with respect to H$_2$?

Solution In determining the order we are concerned only with *ratios* of concentrations; so we can use the pressure data directly, without converting them to concentrations, because at constant V and T pressure is directly proportional to concentration. If we consider experiments 2 and 3, we see that when the pressure of NO is reduced by a factor of approximately two, the rate decreases by a factor of approximately four. We conclude that the reaction is second-order with respect to NO. By considering experiments 4 and 6, we see that reducing the pressure of H$_2$ by a factor of 2 reduces the rate from 1.00 to 0.51 mm Hg s^{-1}, that is, by a factor of approximately two. Thus the rate is directly proportional to the H$_2$ concentration; in other words, the reaction is first-order with respect to H$_2$. Therefore the overall rate law is

$$\text{Rate} = k[NO]^2[H_2]$$

Integrated Rate Law Method

Another generally useful method is the integrated rate law method. We have seen that we can determine the rate at different times from the slope of a plot of the concentration of a reactant or a product against time; and then by plotting the rate against concentration, we can find the order. It is generally more convenient, however, to determine the order directly from a suitable plot of concentration against time. This can be done using an integrated rate law. A rate law describes how the rate varies with the concentrations of the reactants. By the methods of the calculus, the rate law can be transformed to an *integrated rate law*, which *shows how the concentrations of the reactants vary with time.*

For a *first-order reaction*, if $[A]$ is the concentration of a reactant A and k_1 is the rate constant, then

$$\text{Rate} = -\frac{\Delta[A]}{\Delta t} = k_1[A]$$

From the methods of the calculus, it can be shown that

$$\ln[A]_t = -k_1 t + \ln[A]_0$$

where A$_t$ is the concentration of A at time t and A$_0$ is the initial concentration of A (ln is logarithm to the base e). The equation has the form of an equation for a straight line,

$$y = ax + b$$

where a is the slope of the line. Thus, for a first-order reaction, a plot of ln $[A]$ against t is a straight line with a slope of $-k_1$.

Table 18.4 gives data for the decomposition of N_2O_5 at 67°C, and a plot of $\ln [N_2O_5]$ against t is given in Figure 18.5. Since this plot is a straight line, we conclude that the reaction is first-order. From the slope of the line, we get the rate constant $k_1 = 0.345$ min^{-1}.

For a *second-order reaction* that follows the rate law,

$$\text{Rate} = k_2[A]^2$$

it can be shown from the calculus that the concentration of A varies with time according to the equation

$$\frac{1}{[A]_t} = k_2 t + \frac{1}{[A]_0}$$

This again has the form of an equation for a straight line. In this case a plot of $1/[A]$ against t gives a straight line. The slope of this line is k_2, where k_2 is the rate constant.

Half-life

A simple method for determining the order of a reaction is to measure the *half-life*. The half-life of a reaction, $t_{1/2}$, is the time required for the concentration of a reactant to decrease to half its initial value. If the initial concentration is $[A_0]$, then $[A] = \frac{1}{2}[A]_0$ at $t = t_{1/2}$.

The integrated rate law for a first-order reaction,

$$\ln [A]_t = -k_1 t + \ln [A]_0$$

can be written in the form

$$\ln \left(\frac{[A]_0}{[A]_t} \right) = k_1 t$$

If we substitute $[A] = \frac{1}{2}[A]_0$, we get

$$\ln \left(\frac{[A]_0}{\frac{1}{2}[A]_0} \right) = k_1 t_{1/2}$$

$$\ln 2 = k_1 t_{1/2}$$

Table 18.4 Decomposition of N_2O_5

t (min)	$[N_2O_5]$ (mol L^{-1})	$\ln [N_2O_5]$
0	1.00	0.000
1	0.705	−0.350
2	0.497	−0.699
3	0.349	−1.053
4	0.246	−1.402
5	0.173	−1.755

Figure 18.5 A First-Order Reaction. Plot of $\ln [N_2O_5]$ versus time for the decomposition of N_2O_5.

Slope $= -k_1 = -0.345$ min^{-1}
$k_1 = 0.345$ min^{-1}

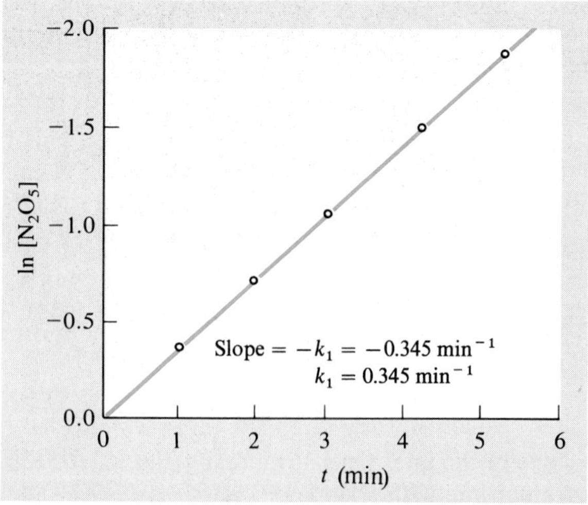

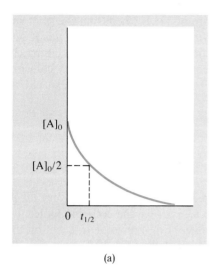

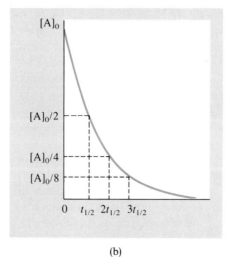

Figure 18.6 Half-Life of a First-Order Reaction. **The half-life** is independent of the initial concentration. In (a) the initial concentration of the reactant A_0 is less than it is in (b), but the time $t_{1/2}$ needed for the initial concentration to decrease to half its initial value is the same. The time needed for the initial concentration to decrease to a quarter of its initial value is $2t_{1/2}$ and to an eighth of its initial value is $3t_{1/2}$.

(a)   (b)

or

$$t_{1/2} = \frac{\ln 2}{k_1} = \frac{0.693}{k_1}$$

For a first-order reaction, $t_{1/2}$ is independent of the concentration of A. Thus if the concentration of the reactant decreases to half its initial value in 10 min, then it will decrease again by a factor of 2 in the next 10 min, so after 20 min it will be a fourth of its initial value, and so on (see Figure 18.6). In fact, at any time t during a first-order reaction, the time needed for the concentration at that time to drop to half its value is equal to the half-life, $t_{1/2}$ (see Figure 18.6).

For a second-order reaction, by substituting $[A] = \frac{1}{2}[A]_0$ at $t = t_{1/2}$, we find that

$$t_{1/2} = \frac{1}{k_2[A]_0}$$

so $t_{1/2}$ is *not* independent of the initial concentration. If we double the initial concentration, the half-life is halved.

Example 18.5 The hydrolysis of sucrose to glucose and fructose is catalyzed by the enzyme sucrase. It is first-order in the concentration of sucrose. If the half-life is 80 min at 20°C, what proportion of the initial sucrose will remain after 160 and 320 min?

Solution Since the reaction is first-order, the concentration is halved each 80 min. And since 160 min is two half-life periods, the initial concentration is reduced to $\frac{1}{2} \times \frac{1}{2} = \frac{1}{4}$ the original value. After 320 min, or four half-life periods, the concentration is reduced to $(\frac{1}{2})^4 = \frac{1}{16}$ its original value.

18.3 FACTORS INFLUENCING REACTION RATES

As we have stated earlier, a reaction can only occur when the reacting molecules collide. Thus the rate of a reaction depends on the rate at which collisions occur and on what fraction of these collisions are effective in leading to reaction.

Collision Rates

The rate at which molecules of two substances, A and B, collide in the gas phase can be calculated from a quantitative treatment of the molecular kinetic theory of gases (Chapter 3). If we have two gases, A and B, each at a pressure of 0.5 atm, in a total volume of 1 L and at a temperature of 298 K, then from the ideal gas equation we can calculate that there is 0.02 mol each of gas A and of gas B. Kinetic theory shows that under these conditions there will be approximately 10^7 mol of collisions per second—a truly enormous number. If every collision of an A molecule with a B molecule led to reaction, the reaction would be complete in approximately

$$\frac{0.02 \text{ mol}}{10^7 \text{ mol s}^{-1}} \approx 10^{-9} \text{ s}$$

In other words, the reaction would be over almost instantaneously if products were formed on every collision.

In fact, very few reactions occur this rapidly. Most reactions are very much slower because *most collisions do not result in reaction*; the colliding molecules frequently just bounce off each other unchanged. For a reaction to occur on the collision of two molecules, the molecules must collide (1) with a certain orientation and (2) with sufficient energy.

Orientation, or Steric, Effect

The necessity for two molecules to collide with the correct relative orientation in order that the reaction can occur is described as a **steric**, or **orientation, effect.** We can see the importance of the correct orientation of the reacting molecules by considering again the reaction of nitrogen monoxide with ozone:

$$NO(g) + O_3(g) \longrightarrow NO_2(g) + O_2(g)$$

In this reaction an oxygen atom must be transferred from an ozone molecule to a nitrogen monoxide molecule. Therefore it is reasonable to assume that for reaction to occur during a collision, one of the end oxygen atoms of an O_3 molecule must collide with the nitrogen of the NO molecule, as shown in Figure 18.7. We would not expect collisions between the central oxygen atom of the

Figure 18.7 The Orientation Effect. (a) A collision between an NO and an O_3 molecule in this orientation leads to reaction. (b) A collision between an NO and an O_3 molecule in this orientation does *not* lead to reaction.

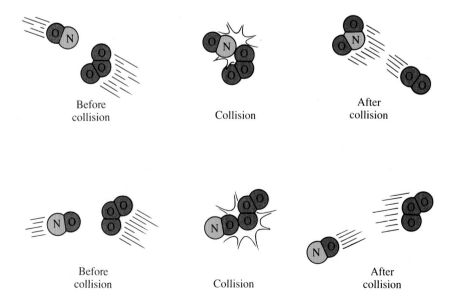

Before collision   Collision   After collision

Before collision   Collision   After collision

O_3 molecule and the nitrogen atom of the NO molecule, or between the O_3 molecule and the oxygen atom of the NO molecule, to lead readily to the transfer of an oxygen atom to the nitrogen atom. In other words, in such collisions the reaction would be unlikely to occur.

In many reactions, the required orientation of the molecules occurs only in a very small fraction of the collisions, particularly if the molecules are large and complicated.

Energy of Activation

The second reason why not every collision leads to a reaction is that a collision must occur with a certain minimum energy before a reaction can take place. When the atoms of the reactant molecules are rearranged to form the new product molecules, bonds must be broken and new bonds formed. For example, in the reaction of NO with O_3, one of the OO bonds in the O_3 molecule is broken, and a new NO bond is formed to give the NO_2 molecule (see Figure 18.7).

The OO bond must first be stretched—that is, the OO bond must be partially broken—before the oxygen atom can begin to form a bond with the nitrogen atom of NO. Energy must be supplied to accomplish this stretching. Partial formation of the new NO bond liberates some energy, and this partially compensates for the energy needed to break the OO bond. A point is reached eventually at which the energy liberated by the formation of the new NO bond is greater than the energy needed to further stretch the OO bond, and there is an overall liberation of energy.

Figure 18.8 shows how the total potential energy of the system changes as

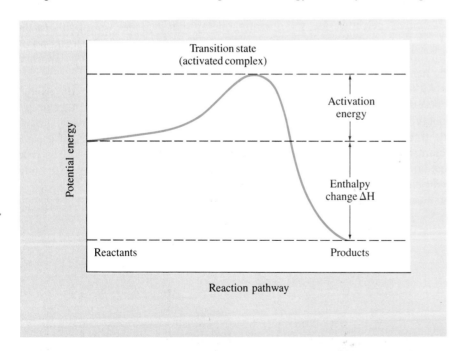

Figure 18.8 Change of Potential Energy in a Collision of O_3 with NO to Give O_2 and NO_2.

an oxygen atom moves from the O_3 molecule to the NO molecule. At first, the energy of the system increases as the OO bond stretches. The energy eventually reaches a maximum and finally decreases again as the new NO bond is formed.

The intermediate state of the reacting system at which its energy is a maximum is called the **transition state**, or **activated complex**, of the reaction. The energy needed to pass from the reactants to the transition state is the **activation energy**, E_a. It is the difference in energy between the reactants and the transition state.

The activation energy is provided by the kinetic energy of the reacting molecules. Unless the kinetic energy of their relative motion is at least equal to E_a, no reaction will occur, and the molecules simply bounce apart unchanged. The activation energy represents a barrier that must be overcome in order for the reaction to occur.

As an analogy, imagine the movement of a marble on the surface shown in Figure 18.9. The movement of the marble from left to right represents the progress of the reaction, and the barrier corresponds to the activation energy. As the marble climbs up the barrier, its potential energy increases. If it is rolled at a low velocity and therefore has low kinetic energy, it will climb only partway up the barrier and then roll back again (Figure 18.9a). Its kinetic energy is too low, and its energy is all converted to potential energy before it reaches the top of the barrier. This situation corresponds to a collision in which the kinetic energy of the reacting molecules is too small for reaction to occur; the molecules simply bounce apart unchanged. The marble must have sufficient initial kinetic energy to enable it to roll up the side of the barrier and reach the top, which in this analogy corresponds to the formation of the transition state. It may then roll down the other side and finish on the right-hand side, which corresponds to the completion of the reaction (Figure 18.9b). As the marble rolls down the other side, the potential energy that it gained on moving to the top of the energy barrier is reconverted to kinetic energy. The corresponding situation in a chemical reaction is when the product molecules move apart as the reaction is completed.

Although energy is always needed to reach the transition state, the overall reaction may be exothermic or endothermic, depending on whether more or less energy is needed to reach the transition state than is obtained in passing from the transition state to the products (see Figure 18.10). The energy of the reactant molecules depends on the temperature, and as we will see in the next section, raising the temperature increases the number of molecules with sufficient energy to react—and therefore increases the reaction rate.

Figure 18.9 Motion of Marble on Potential Energy Surface. (a) Here the marble has insufficient energy to climb the potential barrier. (b) Here the marble has sufficient energy to climb the potential barrier.

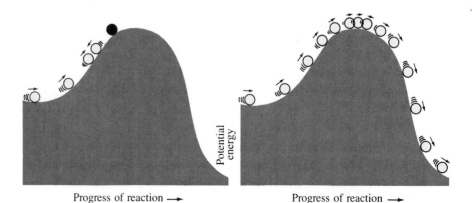

Progress of reaction →
(a)

Potential energy

Progress of reaction →
(b)

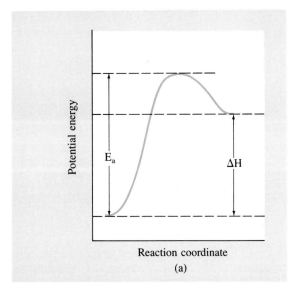

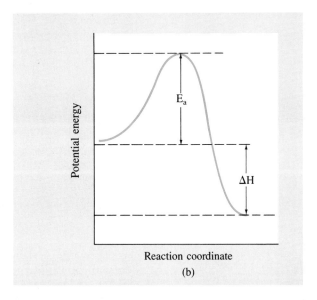

Figure 18.10 Change in Potential Energy During a Reaction.
(a) An endothermic reaction; (b) an exothermic reaction.

Temperature and Reaction Rate: The Arrhenius Equation

The rates of almost all chemical reactions increase with increasing temperature. The effect is quite marked, and the rate often increases by a factor of between 2 and 4 for every 10 K rise in temperature. Many industrial processes, such as the synthesis of ammonia, are carried out at a rather high temperature in order to speed up the reaction.

The average kinetic energy of the molecules in a gas increases with increasing temperature. Not all the molecules are moving at the same speed, but at a given temperature they have a certain average speed and a corresponding average kinetic energy. As we saw in Chapter 3, the distribution of speeds and kinetic energies of the molecules has the form shown in Figure 18.11(a). Only a very few molecules have very low energies or very high energies; the majority have energies close to the average. At a higher temperature the distribution of energies becomes wider, the number of molecules having rather high energies increases, and the average energy also increases.

The energies of the collisions between the molecules of a gas have a distribution similar to the kinetic energies of the molecules of a gas (see Figure 18.11b). In some cases the total energy of two colliding molecules will be small, and in a few cases it will be very high. In a large number of collisions the molecules will have an energy close to the average value. If the minimum energy for a reaction to occur—that is, the activation energy, E_a—is much higher than the average energy of the colliding molecules, only a very small fraction of the collisions can lead to reaction, and the reaction will be very slow. If the activation energy is small, however, a large fraction of the collisions that have a suitable orientation will lead to reaction, and the reaction will be fast. Thus the rate constant of a reaction depends on the activation energy for the reaction.

We see from Figure 18.11(b) that the fraction of the total collisions that can lead to reaction increases rather rapidly with increasing temperature. Thus the rate of most reactions increases greatly as the temperature is raised. Swedish

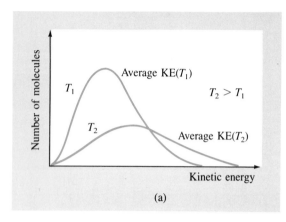

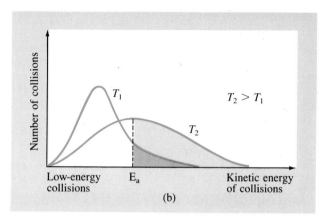

(a)   (b)

Figure 18.11 Kinetic Energy and Collision Energy Distributions.
(a) Distribution of kinetic energies of molecules at temperatures T_1 and T_2; (b) distribution of collision energies at temperatures T_1 and T_2. The shaded areas show that the fraction of molecules with kinetic energies of collision greater than the activation energy, E_a, increases with an increase in temperature.

chemist Svante Arrhenius (1859–1927) (Figure 13.23) first proposed the equation that shows how the rate constant of a reaction depends on the temperature. The **Arrhenius equation** is

$$\log\left(\frac{k_2}{k_1}\right) = \left(\frac{E_a}{2.30R}\right)\left(\frac{1}{T_1} - \frac{1}{T_2}\right)$$

where k_1 and k_2 are the rate constants at the absolute temperatures T_1 and T_2, E_a is the activation energy, and R is the gas constant ($8.31\ \mathrm{J\ K^{-1}\ mol^{-1}}$). We can use this equation to find the activation energy for a reaction if we know the rate constant at two different temperatures, as the next example illustrates.

Example 18.6 When cyclopropane is heated, it is converted to propene:

$$\mathrm{CH_2}\overset{\displaystyle CH_2}{\underset{\displaystyle CH_2}{\diagup\ \diagdown}}\quad \longrightarrow\quad CH_2{=}CH{-}CH_3$$

The rate constant k is $2.41 \times 10^{-10}\ \mathrm{s^{-1}}$ at 300°C. At 400°C, $k = 1.16 \times 10^{-6}\ \mathrm{s^{-1}}$. What is the activation energy for this reaction?

Solution First, we convert the temperatures to the Kelvin scale:

$$T_1 = 273 + 300 = 573\ \mathrm{K} \qquad T_2 = 273 + 400 = 673\ \mathrm{K}$$

Then we substitute the values for k and T in the Arrhenius equation:

$$\log\left(\frac{k_2}{k_1}\right) = \left(\frac{E_a}{2.30R}\right)\left(\frac{1}{T_1} - \frac{1}{T_2}\right)$$

$$\log\left(\frac{1.16 \times 10^{-6}\ \mathrm{s^{-1}}}{2.41 \times 10^{-10}\ \mathrm{s^{-1}}}\right) = \left(\frac{E_a}{2.30 \times 8.31\ \mathrm{J\ K^{-1}\ mol^{-1}}}\right)\left(\frac{1}{573\ \mathrm{K}} - \frac{1}{673\ \mathrm{K}}\right)$$

$$E_a = 2.71 \times 10^5\ \mathrm{J\ mol^{-1}}$$

$$= 271\ \mathrm{kJ\ mol^{-1}}$$

Example 18.7 What is the rate constant at 450°C for the reaction in Example 18.6?

Solution We again use the equation

$$\log\left(\frac{k_2}{k_1}\right) = \left(\frac{E_a}{2.30R}\right)\left(\frac{1}{T_1} - \frac{1}{T_2}\right)$$

This time we substitute $E_a = 271$ kJ mol^{-1} and $k_1 = 1.16 \times 10^{-6}$ s^{-1} at 400°C:

$$\log\left(\frac{k_2}{1.16 \times 10^{-6}\,\text{s}^{-1}}\right) = \left(\frac{2.71 \times 10^5\,\text{J mol}^{-1}}{2.30 \times 8.31\,\text{J K}^{-1}\,\text{mol}^{-1}}\right)\left(\frac{1}{723\,\text{K}} - \frac{1}{673\,\text{K}}\right)$$

$$= 1.46$$

$$\frac{k_2}{1.16 \times 10^{-6}} = 28.8$$

$$k_2 = 3.34 \times 10^{-5}\,\text{s}^{-1}$$

18.4 REACTION MECHANISMS

The study of the factors that influence the rate of a reaction, such as temperature and concentration, is important if we wish to control the rates of reactions—to speed them up or slow them down as necessary. But the study of reaction rates is also important in helping us understand in detail how reactions occur. The balanced equation does not tell us *how* a reaction occurs; it merely summarizes the numbers of moles of reactants and products. But how exactly are the reactants transformed to the products? It is often difficult to answer this question. But we can get much useful information, at least for some reactions, by studying their rates and how the rates are influenced by concentration and temperature changes.

Single-Step Reactions

The simplest reactions are those that occur in a single step. They may be classified according to the number of molecules that are involved in the single step.

BIMOLECULAR REACTIONS The most common type of single-step reactions are those that occur on the collision of two reactant molecules. These are called **bimolecular reactions**. Examples include

$$NO(g) + O_3(g) \longrightarrow NO_2(g) + O_2(g)$$

and

$$NOCl(g) + NOCl(g) \longrightarrow 2NO(g) + Cl_2(g)$$

As we have seen, we expect the rate of collision between two different molecules to be proportional to the concentrations of each of the molecules. That is,

$$\text{Rate} = k[NO][O_3] \quad \text{and} \quad \text{Rate} = k[NOCl][NOCl] = k[NOCl]^2$$

In other words, we expect these bimolecular reactions to have second-order rate laws. Second-order rate laws are, in fact, observed for these reactions.

TERMOLECULAR REACTIONS If a reaction occurs by the simultaneous collision of three molecules, we call it a **termolecular reaction**. The probability of a simultaneous collision between three molecules is much smaller than the probability of a simultaneous collision between two molecules; therefore such reactions are uncommon. The simultaneous collision of four molecules would be so rare that it is safe to say that no reaction actually occurs in this way. If the equation for a reaction shows four molecules of reactants, as, for example, in

$$2NO(g) + 2H_2(g) \longrightarrow N_2(g) + 2H_2O(g)$$

we presume that the reaction does not occur by the simultaneous collision of four molecules. Instead, we assume that it occurs in two or more simpler steps. However, before considering multistep reactions, we must discuss one further important type of single-step reaction—unimolecular reactions.

UNIMOLECULAR REACTIONS   Not all reactions are the direct result of collisions between molecules. A reaction may simply be the consequence of the decomposition or internal rearrangement of a single molecule. Such a reaction is called **unimolecular**.

We mentioned one example of a unimolecular reaction in Example 18.6: the conversion of cyclopropane to propene. Another example of a unimolecular reaction is the conversion of *cis*-2-butene to *trans*-2-butene:

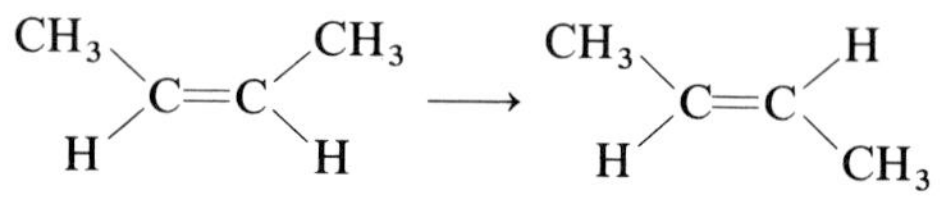

$$\begin{array}{c} CH_3 \\ \diagdown \\ H \diagup \end{array} C{=}C \begin{array}{c} CH_3 \\ \diagup \\ \diagdown H \end{array} \longrightarrow \begin{array}{c} CH_3 \\ \diagdown \\ H \diagup \end{array} C{=}C \begin{array}{c} H \\ \diagup \\ \diagdown CH_3 \end{array}$$

This reaction requires only a twisting or rotation around the double bond so that a CH_3 group exchanges places with the hydrogen atom on the same carbon atom. This is a unimolecular reaction because only one molecule is involved.

Considerable energy is needed to bring about this rotation. For one half of the molecule to rotate with respect to the other half, one component of the double bond must be broken, leaving only a single bond, around which rotation can easily occur. Thus the reaction has a considerable activation energy of 262 kJ mol^{-1}.

Energy is distributed among molecules as a result of their random collisions. So, as we have seen, a very small fraction of the molecules have a much higher energy than the average. However, for a unimolecular reaction to occur, it is not sufficient that the molecule have a high total energy. The total energy of a molecule is made up of its kinetic energy of motion in space (translational and rotational motion) and the kinetic and potential energy associated with its vibrations (see Figure 18.12). There are random fluctuations of energy between the various vibrations. Only when sufficient energy is concentrated in the vibration that causes one end of the molecule to twist with respect to the other end will one of the bonds of the double bond of a *cis*-2-butene molecule be broken so that it can be converted to a *trans*-2-butene molecule. Even when a *cis*-2-butene molecule has a large total vibrational energy, only very occasionally will a large enough fraction of its energy be concentrated in the twisting motion by which the reaction occurs. Therefore in a given period only a very small fraction of the *cis* molecules will be converted to *trans* molecules.

Now suppose that we double the concentration of *cis*-2-butene. There will then be twice as many molecules in each liter. The *fraction* of the total number of molecules that have sufficient energy in the twisting vibration to react during a given time interval is always the same, but with twice as many molecules, twice

Figure 18.12 Some Vibrations of the *cis*-2-Butene Molecule.
The arrows show the directions in which the atoms move during the vibration: (a) a stretching motion, (b) a bending motion, (c) a twisting motion.

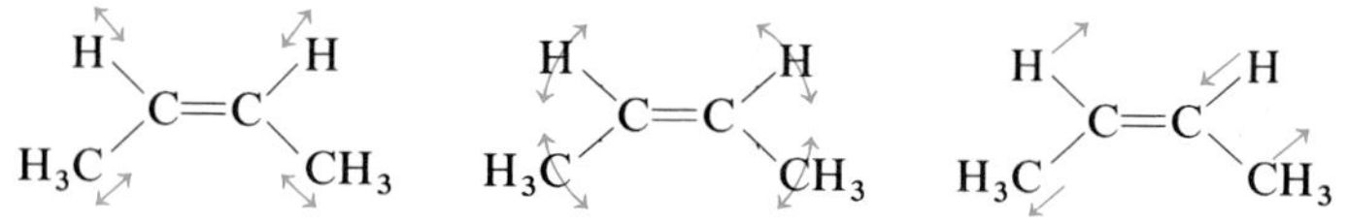

(a)   (b)   (c)

as many will be converted to the *trans* isomer in a given time. Twice as much *cis*-2-butene will be converted to *trans*-2-butene. In other words, the reaction rate is proportional to the concentration of *cis*-2-butene:

$$\text{Rate} = k[\textit{cis}\text{-2-butene}]$$

These arguments apply for any unimolecular reaction,

$$A \longrightarrow \text{Products}$$

so that the rate law is

$$\text{Rate} = k[A]$$

The rate is proportional simply to the concentration of one species; this is a *first-order rate law*.

Thus for single-step reactions:

- A unimolecular reaction has a first-order rate law.
- A bimolecular reaction has a second-order rate law.
- A termolecular reaction has a third-order rate law.

Multistep Reactions

If a reaction occurs in a single step, the rate law corresponds to the balanced equation for the reaction. However, the rate law for many reactions does not correspond to the balanced equation for the reaction.

For example, the rate law for the reaction

$$2N_2O_5(g) \longrightarrow 4NO_2(g) + O_2(g)$$

is found by experiment to be

$$\text{Rate} = k[N_2O_5]$$

Thus although the reaction could, in principle, take place by collisions between two N_2O_5 molecules, we can conclude that it does not, because such a single-step bimolecular reaction would lead to a second-order rate law:

$$\text{Rate} = k[N_2O_5]^2$$

It must therefore proceed by a more complicated process in which there are two or more successive steps.

The decomposition of N_2O_5 is, in fact, believed to proceed as follows:

$$2[N_2O_5 \longrightarrow NO_2 + NO_3] \qquad \text{(Slow)}$$
$$NO_2 + NO_3 \longrightarrow NO + NO_2 + O_2 \qquad \text{(Fast)}$$
$$NO + NO_3 \longrightarrow 2NO_2 \qquad \text{(Fast)}$$

Each of the steps is called an *elementary process*. Elementary processes are almost always either unimolecular or bimolecular. Occasionally, they may be termolecular. All the steps taken together constitute the **mechanism of the reaction**. The sum of the equations for the separate consecutive steps must give the overall equation for the reaction.

The different steps of a multistep reaction do not all have the same rate. They may have very different rates; in the simple cases that we will consider, one of the steps is usually much slower than any of the others. The overall rate of a reaction cannot be greater than the rate of the slowest step. Therefore if one of the steps in a reaction mechanism is much slower than the others, it is called the **rate-determining step**, or the **rate-limiting step**. In such cases the rate law is determined by the rate law for the slowest step of the reaction. For example, in the decomposition of N_2O_5 the rate law for the slow unimolecular step is

$$\text{Rate} = k[N_2O_5]$$

which is the rate law for the overall reaction.

Even when the exponents of the rate law match the coefficient of the balanced equation, we cannot be certain that the reaction occurs in one step. In fact, if the balanced equation involves more than two molecules of reactants, it probably occurs in several steps. An example is the reaction between NO and Cl_2,

$$2NO(g) + Cl_2(g) \longrightarrow 2NOCl(g)$$

for which the rate law is

$$\text{Rate} = k[NO]^2[Cl_2]$$

Thus, the reaction could be termolecular. But the reaction might occur by the following two steps: first, a rapid equilibrium with an equilibrium constant K_1,

$$2NO \overset{K_1}{\rightleftharpoons} N_2O_2 \qquad \text{(Fast equilibrium)}$$

and second, a slow bimolecular reaction with a rate constant k_2,

$$N_2O_2 + Cl_2 \overset{k_2}{\longrightarrow} 2NOCl \qquad \text{(Slow)}$$

The rate law for the slow second step is

$$\text{Rate} = k_2[N_2O_2][Cl_2] \tag{1}$$

In this form the rate law is not very useful because it involves N_2O_2 molecules, which are present in only a very small concentration during the reaction and is not one of the reactants nor one of the final products. We require the rate law in a form that can be measured experimentally, in other words, in a form that involves only the concentrations of the reactants—in this case NO and Cl_2.

We can eliminate the concentration of N_2O_2 from the rate law by using the fact that the first step of the mechanism is an equilibrium for which

$$K_1 = \left(\frac{[N_2O_2]}{[NO]^2}\right)_{eq}$$

where K_1 is the equilibrium constant. Hence

$$[N_2O_2] = K_1[NO]^2 \tag{2}$$

Substituting (2) in (1) gives

$$\text{Rate} = k_2K_1[NO]^2[Cl_2]$$

If we write $k_2K_1 = k$, then

$$\text{Rate} = k[NO]^2[Cl_2]$$

This is the observed rate law. Therefore this two-step mechanism is consistent with the rate law. But we cannot conclude with complete certainty that this *is* the mechanism of the reaction. We need to test other plausible mechanisms to make sure that they are not also consistent with the observed rate law.

Another possible mechanism is

$$2NO \xrightarrow{k_1} N_2O_2 \qquad \text{(Slow)}$$

$$N_2O_2 + Cl_2 \xrightarrow{k_2} 2NOCl \qquad \text{(Fast)}$$

The rate law for the first step is

$$Rate = k_1[NO]^2$$

Since this step is the slow, rate-limiting step, this rate law is also the rate law for the overall reaction:

$$Rate = k[NO]^2$$

But this is not the observed rate law, so we conclude that this mechanism is not the actual mechanism of the reaction.

Example 18.8 What is the rate law for the following mechanism?

$$NO + Cl_2 \xrightarrow{k_1} NOCl + Cl \qquad \text{(Slow)}$$

$$NO + Cl \xrightarrow{k_2} NOCl \qquad \text{(Fast)}$$

Solution For the first step, the rate law is

$$Rate = k_1[NO][Cl_2]$$

Since this step is the slow, rate-limiting step, this rate law is also the rate law for the overall reaction:

$$Rate = k[NO][Cl_2]$$

Again, this differs from the experimentally observed rate law, so we can conclude that this also is not the mechanism of the reaction between NO and Cl_2.

Another reaction that is more complicated than we might expect from the rate law is the gas phase reaction between hydrogen and iodine to give hydrogen iodide:

$$H_2(g) + I_2(g) \longrightarrow 2HI(g)$$

The second-order rate law,

$$Rate = k[H_2][I_2]$$

was first established experimentally in 1894. For over seventy years chemists thought that the reaction occurred by the simple bimolecular collision of an H_2 molecule with an I_2 molecule. But in 1967 it was shown that the reaction is considerably speeded up when exposed to an intense visible light.

Because the iodine molecule has a relatively weak bond (bond energy = 149 kJ mol^{-1}), visible light is capable of dissociating iodine molecules to iodine atoms:

$$I_2 \xrightarrow{h\nu} 2I$$

This suggests that iodine atoms are involved in the reaction, which is now believed to occur by the following sequence of steps:

$I_2 \rightleftharpoons 2I$	(Fast equilibrium)	(1)
$I + H_2 \rightleftharpoons H_2I$	(Fast equilibrium)	(2)
$H_2I + I \longrightarrow 2HI$	(Slow)	(3)

In the first reaction a small equilibrium concentration of iodine atoms is established by the dissociation of iodine molecules. Some of these iodine atoms react rapidly with hydrogen molecules in the second step to form a small equilibrium concentration of H_2I molecules. Finally, iodine atoms and H_2I molecules react relatively slowly to give HI. As the H_2I and I are used up in reaction (3) they are replenished by the shifting of the equilibria (1) and (2).

Step (3) is the slowest, so it is the rate-limiting reaction. It determines the rate of formation of HI and therefore the rate of the overall reaction. Since the third reaction is a bimolecular reaction, its rate law is

$$\text{Rate} = k_3[H_2I][I] \tag{4}$$

and because reaction (3) is the rate-determining step, this rate law is also the rate law for the overall reaction. Although it is not expressed in terms of the reactants H_2 and I_2, we may relate the concentrations of both H_2I and I to the concentrations of H_2 and I_2 by means of the equilibrium constants for reactions (1) and (2). These are

$$K_1 = \left(\frac{[I]^2}{[I_2]}\right)_{eq} \qquad K_2 = \left(\frac{[H_2I]}{[I][H_2]}\right)_{eq}$$

Hence

$$[I]^2 = K_1[I_2] \quad \text{and} \quad [H_2I] = K_2[I][H_2]$$

Substituting for $[H_2I]$ in the rate law (4)

$$\text{Rate} = k_3 K_2[I][H_2][I]$$

Then substituting for $[I]^2$ gives

$$\text{Rate} = k_3 K_2 K_1[H_2][I_2]$$

If we then define a new constant $k = k_3 K_2 K_1$, we may write

$$\text{Rate} = k[H_2][I_2]$$

This is the observed rate law. It shows that the proposed mechanism is a possible mechanism for the reaction. Since it also accounts for the effect of visible light on the reaction, which increases the concentration of iodine atoms and therefore speeds up the reaction, it is now the accepted mechanism.

The determination of the rate law for a reaction is a very important part of the study of any reaction. Although several mechanisms may be consistent with the rate law, there will be many that are not consistent and can therefore be eliminated. Further experiments can then often be made to obtain the additional information necessary to decide between those mechanisms that are consistent with the rate law. However, we can only say that a given mechanism is *consistent* with the rate law and with all the other information about the reaction and is therefore an acceptable mechanism. It is conceivable that another mechanism will be discovered that also is consistent with the rate law and all other available experimental data. New experimental data will then be needed to decide between these two possible mechanisms.

Reaction Intermediates

In most multistep reactions, species formed in one or more of the steps are used up in the following steps and do not appear in the final products. Such species are called **reaction intermediates**. They are often unfamiliar molecules such as N_2O_2, NO_3, and H_2I or free atoms such as I. Frequently, as in the cases of NO_3,

H_2I, and I, they are odd-electron species, that is, free radicals. They are usually very reactive and play an important role in reactions, but they are normally present only in very small concentrations during the reaction.

Equilibrium Constant and Reaction Mechanism

When a reacting system reaches equilibrium, the rate of the forward reaction equals the rate of the reverse reaction. For the single-step bimolecular reaction

$$NO(g) + O_3(g) \rightleftharpoons NO_2(g) + O_2(g)$$

the rate laws for the forward and back reactions are

$$\text{Rate forward reaction} = k_f[NO][O_3]$$
$$\text{Rate reverse reaction} = k_r[NO_2][O_2]$$

At equilibrium the rate of the forward reaction equals the rate of the reverse reaction. Thus

$$k_f([NO][O_3])_{eq} = k_r([NO_2][O_2])_{eq}$$

so that

$$\left(\frac{[NO_2][O_2]}{[NO][O_3]}\right)_{eq} = \frac{k_f}{k_r}$$

Since k_f/k_r is a constant, which we may write as K, we have

$$\frac{k_f}{k_r} = K = \left(\frac{[NO_2][O_2]}{[NO][O_3]}\right)_{eq}$$

We recognize this expression as the equilibrium constant expression for the reaction. Thus the equilibrium constant expression for a single-step reaction is a consequence of the rate laws for the forward and back reactions and the fact that at equilibrium the rates of the forward and the back reactions are equal.

We have seen in Chapter 14 that we can always write the correct expression for the equilibrium constant from the balanced equation for a reaction, although we cannot always deduce the rate law from the balanced equation. For example, from the balanced equation for the reaction

$$H_2 + I_2 \rightleftharpoons 2HI$$

we may write

$$K = \frac{[HI]^2}{[H_2][I_2]}$$

If we were to assume, incorrectly, that the reaction is a single-step bimolecular reaction, then we could write

$$\text{Rate forward reaction} = k_f[H_2][I_2]$$
$$\text{Rate reverse reaction} = k_r[HI]^2$$

At equilibrium

$$k_f([H_2][I_2])_{eq} = k_r([HI]^2)_{eq}$$
$$\frac{k_f}{k_r} = \left(\frac{[HI]^2}{[H_2][I_2]}\right)_{eq} = K$$

This is the correct expression for the equilibrium constant, even though we did not use the correct mechanism. Let us now derive the equilibrium constant expression from the accepted mechanism.

When equilibrium is reached between the reactants and the final products of a reaction, each step in the reaction mechanism must also have reached equilibrium. In the present case three equilibria will be established:

$$I_2 \rightleftharpoons 2I \qquad\qquad (1)$$

$$I + H_2 \rightleftharpoons H_2I \qquad\qquad (2)$$

$$H_2I + I \rightleftharpoons 2HI \qquad\qquad (3)*$$

The three equilibrium constants for the three successive steps are

$$K_1 = \left(\frac{[I]^2}{[I_2]}\right)_{eq} \qquad K_2 = \left(\frac{[H_2I]}{[I][H_2]}\right)_{eq} \qquad K_3 = \left(\frac{[HI]^2}{[H_2I][I]}\right)_{eq}$$

Since the sum of equations (1), (2), and (3) is the overall equation

$$H_2 + I_2 \rightleftharpoons 2HI$$

the product of the equilibrium constants $K_1K_2K_3$ is the equilibrium constant for the overall reaction (see Chapter 14):

$$K_1K_2K_3 = \frac{[I]^2[H_2I][HI]^2}{[I_2][I][H_2][H_2I][I]} = \frac{[HI]^2}{[I_2][H_2]} = K$$

Thus we see that taking the individual steps of the reaction into account gives the same expression for the equilibrium constant as is obtained from the balanced equation for the equilibrium between the reactants and the final products of the reaction. In fact, for any reaction *the equilibrium constant expression is independent of the actual mechanism by which the reaction occurs*. Thus we can always write the correct expression for the equilibrium constant for a reaction simply from the balanced equation for the reaction, whether or not we know the detailed mechanism by which the reaction occurs.

18.5 CATALYSIS

In previous chapters we have met many examples of the use of catalysts to increase the rate of a reaction. We are now in a position to look at how a catalyst works. A **catalyst** is a substance that increases the rate of a chemical reaction although it is not used up in the reaction. In other words, its final concentration is equal to its initial concentration. It does not affect the equilibrium position of a reaction, but it does increase the rate at which equilibrium is reached. A catalyst does not appear in the overall equation for the reaction because it is regenerated during the reaction. It may be regarded as both a reactant and a product, and it therefore cancels out of the balanced equation. Experiment 18.3 demonstrates a reaction that is catalyzed by water.

As an example of how a catalyst operates, we will consider the conversion of *cis*-2-butene to *trans*-2-butene. In the absence of a catalyst this reaction is very slow and has an activation energy of 262 kJ mol^{-1}. Iodine is a catalyst for this reaction. It greatly speeds up the reaction by lowering the activation energy to 115 kJ mol^{-1}. Let us see how it can lower the activation energy.

In the presence of iodine the reaction is believed to proceed by the following

* Step (3) is written as an equilibrium here, whereas in deriving the rate law only the forward reaction was considered. The rate law thus applies only to the early stages of the reaction, when it is far from equilibrium. Derivation of the rate law for the system as it approaches equilibrium is complex and beyond the scope of this book.

Catalysis

The upper jar contains SO_2 gas. The lower jar contains H_2S gas. The two gases are separated by a glass plate.

The plate is removed but no reaction is observed. A few drops of water are then added and the pair of jars is inverted a few times to mix the two gases quickly.

Water catalyzes the reaction between H_2S and SO_2 so that the sides of the jars are rapidly coated with yellow sulfur.

steps: First, there is a dissociation of an iodine molecule to two iodine atoms:

$$I_2 \rightleftharpoons 2I$$

Then an iodine atom attacks the double bond to give the free radical C_4H_8I, in the slow, rate-limiting step of the reaction:

$$I + cis\text{–}C_4H_8 \longrightarrow C_4H_8I \quad \text{(Slow)}$$

Because there is now a single bond between the two carbon atoms, rotation around this bond readily gives another conformation, as shown in Figure 18.13. Loss of an iodine atom then gives the trans isomer in a fast step:

$$C_4H_8I \longrightarrow trans\text{-}C_4H_8 + I \quad \text{(Fast)}$$

Because the iodine atom is regenerated, a given iodine atom can successively

Figure 18.13 Catalysis of the Reaction *cis*-2-Butene → *trans*-2-Butene by Iodine. The double bond is broken by the addition of an iodine atom to the *cis* isomer in (a) to give the radical intermediate in (b). Rotation around the single C—C bond can then occur easily to give (c). This is followed by the loss of an iodine atom to give the *trans* isomer in (d)

lead to the conversion of many cis-C_4H_8 molecules to the trans isomer. Eventually, the iodine atoms recombine,

$$2I \rightleftharpoons I_2$$

and the catalyst I_2 is regenerated.

The function of the iodine is to form a reaction intermediate, C_4H_8I, in which rotation around the CC bond is much easier than it is in C_4H_8. The much lower activation energy for the catalyzed reaction is made up approximately as follows: The bond dissociation energy of the iodine molecule is 149 kJ mol^{-1}, so 74.5 kJ mol^{-1} is required for the formation of 1 mol of iodine atoms. The formation of C_4H_8I requires 28 kJ mol^{-1}, and 12 kJ mol^{-1} is needed for the twisting around the C—C single bond in C_4H_8I. Thus a total of 115 kJ mol^{-1} is needed to form the activated complex, which is very much less than the 262 kJ required for the uncatalyzed reaction; hence the catalyzed reaction is much faster than the uncatalyzed reaction. The catalyst provides an alternative pathway that has a lower activation energy (Figure 18.14).

Example 18.9 The rate law for the catalyzed conversion of cis-2-butene to $trans$-2-butene is

$$\text{Rate} = k[cis\text{-}C_4H_8][I_2]^{1/2}$$

Show that the proposed mechanism is consistent with this rate law.

Solution The rate-limiting reaction is

$$I + cis\text{-}C_4H_8 \longrightarrow C_4H_8I$$

from which we can write

$$\text{Rate} = k'[I][cis\text{-}C_4H_8]$$

For the equilibrium $I_2 \rightleftharpoons 2I$ we have

$$K = \left(\frac{[I]^2}{[I_2]}\right)_{eq}$$

Hence

$$[I] = K^{1/2}[I_2]^{1/2}$$

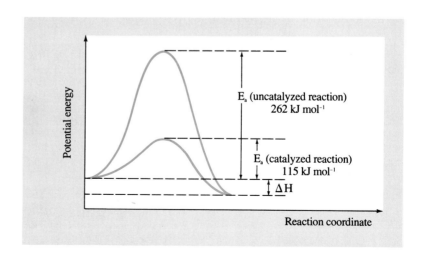

Figure 18.14 Activation Energy of Iodine-Catalyzed Isomerization of cis-2-Butene to $trans$-2-Butene.

Potential energy

E_a (uncatalyzed reaction)
262 kJ mol^{-1}

E_a (catalyzed reaction)
115 kJ mol^{-1}

ΔH

Reaction coordinate

Therefore by substituting for $[I]$ in the rate equation, we have

$$\text{Rate} = k'K^{1/2}[cis\text{-}C_4H_8][I_2]^{1/2}$$
$$= k[cis\text{-}C_4H_8][I_2]^{1/2}$$

Notice that the catalyst appears in the rate law although it does not appear in the balanced equation for the overall reaction. Notice also that a rate law may sometimes involve a fractional power of a concentration. In this case, $[I_2]^{1/2}$ appears in the rate law because iodine atoms rather than iodine molecules are involved in the reaction.

Homogeneous Catalysis

In the conversion of *cis*-butene to *trans*-butene the reactants and the catalyst are all in the same phase. They are either both gases or both in solution. This kind of catalysis in which the catalyst is in the same phase as the reactants is called **homogeneous catalysis**. In previous chapters we have met several examples of this kind of catalysis. In Chapter 17 we discussed the catalysis of the decomposition of ozone by nitrogen monoxide. We saw that the catalyzed reaction has the following mechanism:

$$NO + O_3 \longrightarrow NO_2 + O_2$$
$$NO_2 + O \longrightarrow NO + O_2$$

The overall reaction is

$$O_3 + O \longrightarrow 2O_2$$

The rate of this reaction is increased by NO, which is used up in the first step and regenerated in the second step.

Nitrogen monoxide also catalyzes the reaction between SO_2 and O_2 to give SO_3 in the gas phase. The mechanism of the catalyzed reaction is

$$O_2 + 2NO \longrightarrow 2NO_2$$
$$2[NO_2 + SO_2 \longrightarrow NO + SO_3]$$

and the sum of these two reactions is

$$O_2 + 2SO_2 \longrightarrow 2SO_3$$

The NO that is used up in the first step is regenerated in the second step.

Heterogeneous Catalysis

Many important industrial processes depend on a different kind of catalysis called **heterogeneous catalysis**. In this case the catalyst is in a different phase than the reactants. Usually, the catalyst is a solid, the reactants are gases, and the rate-limiting step occurs on the surface of the solid catalyst. Hence heterogeneous catalysis is often also called **surface catalysis**.

Although heterogeneous catalysis is of enormous importance, the detailed mechanism by which many heterogeneous catalysts work is not well understood. But we do know that reactant molecules become attached to the surface of the catalyst; this weakens some of the bonds in the reactant molecules and enables them to take part much more easily in a reaction with another molecule.

An important example is the use of the surface of a metal, such as platinum or nickel, as a catalyst for the addition of hydrogen to a carbon-carbon double bond, as in the conversion of ethene to ethane. In the absence of a catalyst this

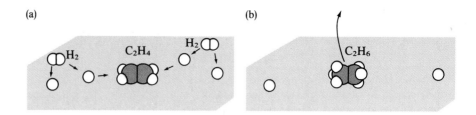

reaction does not occur readily. But in the presence of finely divided nickel or platinum, it occurs rapidly at room temperature and at a high pressure of several hundred atmospheres:

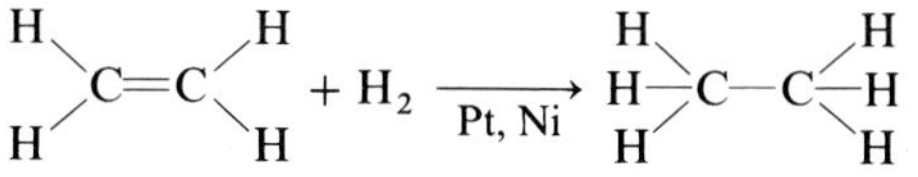

$$\underset{H}{\overset{H}{>}}C=C\underset{H}{\overset{H}{<}} + H_2 \xrightarrow[\text{Pt, Ni}]{} H\underset{H}{\overset{H}{-}}C-C\underset{H}{\overset{H}{-}}H$$

Both the C_2H_4 and the H_2 molecules become attached to the metal surface; they are said to be *adsorbed* (see Figure 18.15). The H_2 molecules then dissociate into H atoms, which remain attached to platinum atoms on the surface. But these hydrogen atoms can move rather easily across the metal surface from one platinum atom to another. When they encounter an ethene molecule, they combine with it readily. The successive addition of two hydrogen atoms gives the final product, ethane, which is not very strongly adsorbed and escapes from the surface.

Important industrial processes that use heterogeneous catalysts include the Ostwald process for the synthesis of nitric acid (Chapter 17); the Haber process for the synthesis of ammonia (Chapter 14); the cracking of hydrocarbons (Chapter 12); and the oxidation of SO_2 to SO_3 in the manufacture of sulfuric acid (Chapter 7). An important application of heterogeneous catalysts is in catalytic converters in the exhaust systems of automobiles. They are used to increase the rate of oxidation of CO and unburnt hydrocarbons to CO_2 and to increase the rate of reduction of NO to N_2.

Enzymes

Enzymes, which catalyze reactions in living organisms, are another very important class of catalysts. They are complex proteins that are very specific catalysts for biochemical reactions. For example, a solution of sugar at 37°C does not oxidize at a significant rate. Yet in the body at this temperature sugar is rapidly oxidized to CO_2 and H_2O:

$$C_{12}H_{22}O_{11}(aq) + 12O_2(aq) \longrightarrow 12CO_2(aq) + 11H_2O(l)$$

This reaction is accomplished by means of an enzyme catalyst. Enzymes are amazingly efficient catalysts that increase the rates of reactions by factors as great as 10^{20} so that they occur rapidly at body temperature. They are discussed in more detail in Chapter 23.

18.6 CHAIN REACTIONS

We have seen in Chapter 5 that the reaction between hydrogen and chlorine at room temperature is extremely slow; no measurable amount of hydrogen chloride is formed, even over a long period. However, if the mixture of gases is exposed to a bright light, a very rapid reaction occurs, which we observe as an explosion. This reaction is an example of an important class of reactions called **chain reactions**. A chain reaction is a multistep reaction in which a free radical is formed and the free radical initiates a series of two or more reactions in which the products are formed and the free radical is regenerated. Because one free radical may thus cause the formation of a large number of product molecules, chain reactions may be very fast. As an example, we will consider the mechanism of the reaction between hydrogen and chlorine.

Absorption of light of a certain minimum frequency (see Example 6.4) causes chlorine molecules to dissociate to atoms:

$$Cl_2 \xrightarrow{hv} 2Cl \qquad\qquad (1)$$

This first step is called *chain initiation*. Chlorine atoms are very reactive, and a Cl atom reacts rapidly with an H_2 molecule, producing an HCl molecule and an H atom:

$$Cl + H_2 \longrightarrow HCl + H \qquad\qquad (2)$$

The very reactive H atom then attacks a Cl_2 molecule, forming an HCl molecule and regenerating a Cl atom:

$$H + Cl_2 \longrightarrow HCl + Cl \qquad\qquad (3)$$

The Cl atom can attack another H_2 molecule, as in step 2, producing another H atom, which can react with a Cl_2 molecule, as in step 3. These two reactions can repeat themselves many times, giving a chain of successive reactions: They are called *chain-propagating steps*. The overall reaction

$$H_2 + Cl_2 \longrightarrow 2HCl$$

is the sum of reactions (2) and (3). These reactions continue indefinitely until the Cl atoms are removed in some way, for example, by combining with each other:

$$Cl + Cl \longrightarrow Cl_2$$

This reaction is called a *chain termination reaction*. Because the concentration of Cl atoms is very low, this reaction is relatively slow. Thus a Cl atom may cause the rapid formation of many thousands of HCl molecules before it is removed from the reacting system (see Figure 18.16).

The mechanism of the H_2–Cl_2 reaction can therefore be summarized by the following equations:

$$Cl_2 \longrightarrow 2Cl \qquad \text{(Chain initiation)}$$

$$\left.\begin{array}{l} Cl + H_2 \longrightarrow HCl + H \\ H + Cl_2 \longrightarrow HCl + Cl \end{array}\right\} \quad \text{(Chain propagation)}$$

$$2Cl \longrightarrow Cl_2 \qquad \text{(Chain termination)}$$

Such a sequence of reactions is typical of a chain reaction. A chain reaction is initiated by a relatively slow step that produces a highly reactive intermediate. In subsequent steps this reactive intermediate attacks a reactant molecule to form a product molecule and either regenerates itself or produces another reactive intermediate, which can attack another reactant molecule to form the product, and so on. Eventually, the reactive intermediate is removed in some chain termination step.

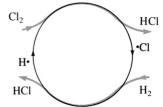

Figure 18.16 Hydrogen-Chlorine Chain Reaction. A Cl atom reacts with an H_2 molecule to form an HCl molecule and an H atom. The H atom reacts with a Cl_2 molecule to form another HCl molecule and another Cl atom, which reacts with another H_2 molecule, and so on. One Cl atom can cause the formation of a very large number of HCl molecules.

Chain reactions can lead to *explosions*, as in the hydrogen-chlorine reaction, if the products of the reaction are gases and if the chain-propagating steps produce heat faster than it can be conducted away. In such a case the increasing temperature continually increases the rate of the reaction, which produces heat still faster, and so on. The extremely rapid increase in the volume of the gaseous products then produces an explosion (Box 12.1).

A very important application of chain reactions is in *polymerization*, the type of reaction used in the manufacture of plastics, such as polyethylene (Chapter 23).

IMPORTANT TERMS

The **activated complex** is the particular arrangement of atoms that has the highest energy during the rearrangement of the atoms of the reactants to those of the products. It is also called the *transition state*.

The **activation energy** is the minimum energy that the reactant molecules must possess in order to be able to react. It is the difference in energy between the reactants and the transition state.

The **Arrhenius equation** is

$$\log\left(\frac{k_2}{k_1}\right) = \left(\frac{E_a}{2.30\,R}\right)\left(\frac{1}{T_1} - \frac{1}{T_2}\right)$$

where k_1 is the rate constant at temperature T_1, k_2 is the rate constant at temperature T_2, and E_a is the activation energy.

A **bimolecular reaction** is a single-step reaction between two molecules.

A **catalyst** is a substance that increases the rate of a reaction without being used up in the reaction. A catalyst changes the mechanism of a reaction and decreases the activation energy, E_a.

A **chain reaction** is a multistep reaction that begins with an initiation step, which produces a reactive intermediate, continues with chain-propagating steps, which form the products and regenerate the reactive intermediate, and is completed by one or more chain termination steps, in which the reactive intermediate is removed.

The **half-life** of a reaction, $t_{1/2}$, is the time required for the concentration of a reactant to decrease to half its initial value.

A **heterogeneous catalyst** is in a different phase from the reactants.

A **homogeneous catalyst** is in the same phase as the reactants.

The **order of a reaction** with respect to a given reactant is the exponent of the concentration of that reactant in the rate law.

The **orientation effect** describes the fact that colliding molecules must have a suitable relative orientation in order to react.

The **overall order of a reaction** is the sum of the exponents of the concentration terms in the rate law for the reaction.

The **rate constant** is the proportionality constant, k, that appears in a rate law: rate $= k[A]^x[B]^y[C]^z \ldots$.

The **rate law** is the equation that relates the rate of a reaction to the concentrations of the reactants; it has the form rate $= k[A]^x[B]^y[C]^z \ldots$.

A **rate-limiting reaction** (or *rate-determining reaction*) is the slowest step in the series of reaction steps that constitute the mechanism of a reaction. It is the step that determines the rate of the overall reaction.

Reaction intermediates are species formed in one step of a multistep reaction and used up in a following step, so they do not appear in the final products.

A **reaction mechanism** is the series of successive reaction steps (elementary processes) by which reactants are converted to products in a reaction.

The **reaction rate** is measured by the change in the concentration of one of the reactants, or one of the products, of a reaction in a given time interval.

A **surface catalyst** provides a reactive surface on which the rate-limiting step of a reaction can take place. It is a heterogeneous catalyst.

A **termolecular reaction** is a reaction that occurs by the simultaneous collision of three molecules.

In a **unimolecular reaction** a single molecule undergoes decomposition or rearrangement.

Reaction Rates

1. The reaction

$$CH_3OH(aq) + HCl(aq) \longrightarrow CH_3Cl(aq) + H_2O(l)$$

was followed by measuring the hydrogen ion concentration as it changed with time. The results in the accompanying table were obtained. Find the average rate of the reaction for each time interval.

TIME (min)	$[H^+]$ (mol L^{-1})
0	1.85
80	1.66
159	1.53
314	1.31
628	1.02

2. When gaseous SO_2Cl_2 is heated, it decomposes to SO_2 and Cl_2. The accompanying table gives data for an experiment at 320°C. The concentration of the SO_2Cl_2 remaining was determined at certain time intervals, and the results are given in the accompanying table. Find the average rate for each time interval.

t (min)	$[SO_2Cl_2]$ (mol L^{-1})
0	0.010 0
20	0.009 70
50	0.009 28
100	0.008 61
200	0.007 41
400	0.005 49
700	0.003 50
1000	0.002 23

3. For each of the following reactions, express the rate in terms of the change in concentration of the reactant and relate this to the rate of formation of each product:

(a) $2HI(g) \longrightarrow H_2(g) + I_2(g)$

(b) $2NOCl(g) \longrightarrow 2NO(g) + Cl_2(g)$

(c) $2N_2O_5(g) \longrightarrow 4NO_2(g) + O_2(g)$

Rate Laws

4. The rate constant for a first-order reaction is $3.7 \times 10^{-1} s^{-1}$. What is the initial rate in (a) mol $L^{-1} s^{-1}$, (b) mol $L^{-1} h^{-1}$, for an initial concentration of reactant of 0.04 mol L^{-1}?

* The asterisk denotes the more difficult problems.

5. The accompanying data were obtained for the decomposition of nitrosyl bromide,

$$2NOBr(g) \longrightarrow 2NO(g) + Br_2(g)$$

$[NOBr]$ (mol L^{-1})	RATE (mol $L^{-1} s^{-1}$)
0.14	0.50
0.31	1.90
0.45	4.00
0.54	5.70
0.71	9.80

What is the order of the reaction with respect to NOBr?

6. Plot the rates that you found in Problem 2 against the concentration of SO_2Cl_2 to find the order of the reaction.

7. For the reaction

$$4NH_3(g) + 3O_2(g) \longrightarrow 2N_2(g) + 6H_2O(g)$$

it was found that at a certain time N_2 was being formed at a rate of 0.27 mol $L^{-1} s^{-1}$.

(a) At what rate was water being formed?

(b) At what rate was NH_3 being used up?

(c) At what rate was O_2 being used up?

8. On what factors does the rate of a reaction depend?

9. Distinguish clearly between the rate of a reaction, the rate law for the reaction, and the rate constant of the reaction.

10. The initial rate of a second-order reaction in which the initial concentrations of two reacting substances were each 0.2 mol L^{-1} was found to be 8.1×10^{-3} mol L^{-1} min^{-1}. What is the value of the rate constant in (a) mol$^{-1} L s^{-1}$ (b) molecules^{-1} mL min^{-1}?

11. In the gas phase reaction between chlorine and nitrogen monoxide, $2NO + Cl_2 \rightarrow 2NOCl$, doubling the concentrations of both reactants increases the reaction rate by a factor of 8, but doubling the concentration of chlorine alone only doubles the rate. What is the order of the reaction with respect to (a) nitrogen monoxide, (b) chlorine, and (c) overall?

12. When ethanal (acetaldehyde), CH_3CHO, is heated, it decomposes to methane and carbon monoxide. If the initial concentration of CH_3CHO is doubled (without changing the temperature), the reaction rate increases to 2.83 times its initial value. What is the order of the reaction?

13. For the reaction

$$2NO(g) + H_2(g) \longrightarrow N_2O(g) + H_2O(g)$$

the data in the accompanying table were obtained. What is the rate law for the reaction? What is the value of the rate constant?

[NO] (mol L^{-1})	[H$_2$] (mol L^{-1})	INITIAL RATE (mol L^{-1} min^{-1})
0.150	0.80	0.500
0.075	0.80	0.125
0.150	0.40	0.250

14. In the reaction of H_2 with Br_2, doubling the concentration of H_2 increases the rate by a factor of 2, but increasing the concentration of Br_2 three times only increases the rate by a factor of 1.73. What are the orders of the reaction with respect to H_2 and Br_2? Is the reaction a one-step bimolecular reaction?

15. The decomposition of an aqueous solution of hydrogen peroxide to oxygen and water is a first-order reaction. In a certain experiment the half-life of hydrogen peroxide was found to be 17.0 min. What fraction of the initial hydrogen peroxide would have been left after (a) 51.0 min and (b) ten half-life periods?

16. In another experiment at a higher temperature than in Problem 15, a quarter of the initial H_2O_2 was found to remain after 8.0 min. What is the half-life of H_2O_2 under these conditions? How much longer will it take for the H_2O_2 concentration to decrease to $\frac{1}{32}$ of its initial value?

17. The rate law for a reaction is found to be

$$\text{Rate} = k[B]^2[C]$$

where $k = 4.0 \times 10^{-3}$ mol^{-2} L^2 s^{-1}. What is the reaction rate when $[B] = [C] = 0.01M$?

18. The following initial rates were found for the reaction

$$A + B \longrightarrow \text{Products}$$

EXPERIMENT	[A]	[B]	RATE
1	0.20M	0.10M	0.00340M s^{-1}
2	0.20M	0.30M	0.01020M s^{-1}
3	0.40M	0.30M	0.04080M s^{-1}

Calculate the rate law for the reaction and the rate constant k. What will be the rate of the reaction when $[A] = [B] = 0.50M$?

19. In problem 2, data was given for the variation of $[SO_2Cl_2]$ with time during an experiment in which SO_2Cl_2 decomposed to SO_2 and Cl_2. Determine whether this reaction is first order or second order in SO_2Cl_2 by plotting

$\ln[SO_2Cl_2]$ against t and by plotting $1/[SO_2Cl_2]$ against t. When you have decided which order applies, use the appropriate plot to determine the value of the rate constant.

20. The rate of decomposition of hydrogen peroxide in aqueous solution was measured by titrating samples of the solution with potassium permanganate solution at certain time intervals. The following results were obtained.

t(min)	0	10	20
volume KMnO$_4$ solution (mL)	22.8	13.8	8.3

Confirm that the reaction is first order in hydrogen peroxide and find its half-life and rate constant.

Activation Energy

21. The rate constants for the reaction $2HI \rightarrow H_2 + I_2$ are given in the accompanying table at several temperatures. Use the Arrhenius equation to obtain the activation energy for this reaction. Calculate the rate constant at 400°C.

TEMPERATURE (°C)	k (mol^{-1} L s^{-1})
302	1.18×10^{-6}
356	3.33×10^{-5}
374	8.96×10^{-5}
410	5.53×10^{-4}
427	1.21×10^{-3}

22. The activation energy for a reaction is found to be 80 kJ mol^{-1}. Calculate the temperature at which the rate will be ten times that at 0°C.

23. To what temperature must a reaction mixture be raised if the rate constant for the reaction is to be exactly twice the value at 27°C, given that the activation energy for the reaction is 100 kJ mol^{-1}?

24. If the rate of a reaction at 50°C is three times that at 25°C, what is the energy of activation?

25. Nitrogen dioxide decomposes to nitrogen monoxide and oxygen at high temperature in the gas phase. The values in the accompanying table were obtained for the rate constant of this reaction. Find the activation energy for this reaction.

k (L mol^{-1} s^{-1})	T (K)
3.16	650
28.2	730
158	800
1120	900
5010	1000

26. The rate constant for a reaction is 0.040 at 22°C and 0.050 at 30°C. Calculate the activation energy for the reaction and the value of the rate constant at 40°C.

27. The rate of a certain reaction at 50°C is found to be six times that at -10°C. Calculate the activation energy for this reaction and the temperature at which its rate will be twice that at -10°C.

Mechanisms

28. The reaction of hydrogen with bromine has the following mechanism:

$$Br_2 \rightleftharpoons 2Br \qquad \text{(Equilibrium)}$$
$$Br + H_2 \longrightarrow HBr + H \qquad \text{(Slow)}$$
$$H + Br_2 \longrightarrow HBr + Br \qquad \text{(Fast)}$$

Deduce the rate law expected for this mechanism in terms of the concentrations of the reactants H_2 and Br_2, assuming that the second step is rate limiting.

29. Carbon monoxide and chlorine react to form phosgene:

$$Cl_2 + CO \longrightarrow COCl_2$$

Deduce the rate law if the reaction proceeds by the following mechanism:

$$Cl_2 \rightleftharpoons 2Cl \qquad \text{(Equilibrium)}$$
$$Cl + CO \rightleftharpoons COCl \qquad \text{(Equilibrium)}$$
$$COCl + Cl_2 \longrightarrow COCl_2 + Cl \qquad \text{(Rate-limiting step)}$$

30. The reaction

$$NO_2(g) + CO(g) \longrightarrow NO(g) + CO_2(g)$$

is believed to occur by the two following bimolecular reactions:

$$NO_2(g) + NO_2(g) \longrightarrow NO_3(g) + NO(g) \qquad \text{(Slow)}$$
$$NO_3(g) + CO(g) \longrightarrow NO_2(g) + CO_2(g) \qquad \text{(Fast)}$$

What is the rate law corresponding to this mechanism? What would the rate law be if the reaction occurred directly in a single step?

31. The conversion of ozone to molecular oxygen,

$$2O_3(g) \longrightarrow 3O_2(g)$$

occurs via the following mechanism:

$$O_3 \rightleftharpoons O_2 + O \qquad \text{(Equilibrium)}$$
$$O + O_3 \longrightarrow 2O_2 \qquad \text{(Slow)}$$

Deduce the rate law for the reaction. Explain in words why the reaction rate decreases as the concentration of O_2 increases; in other words, why O_2 appears in the rate law with a negative order.

32. The rate law for the reaction $2NO(g) + O_2(g) \rightarrow 2NO_2(g)$ is

$$\text{Rate} = k[NO]^2[O_2]$$

Devise two mechanisms for the reaction that are consistent with this rate law but that do *not* involve the simultaneous collision of three molecules.

33. Explain why the fact that visible light increases the rate of the $H_2 + I_2 \rightarrow 2HI$ reaction cannot be explained by the formation of H atoms.

*34. Bromide ions are oxidized to bromine by hydrogen peroxide in acidic aqueous solution.

    (a) Write the balanced equation for the reaction.

    (b) In separate experiments where the initial concentrations of H_2O_2 and Br^- were in turn doubled, keeping the other concentrations constant, the rate was found to also double. When the pH of the solution was decreased by 0.60, the rate of the reaction increased fourfold. Deduce the rate law for the reaction.

    (c) If under certain conditions the rate of disappearance of Br^- is 7.2×10^{-3} mol L^{-1} s^{-1}, what is the rate of disappearance of H_2O_2? What is the rate of appearance of Br_2?

    (d) What is the effect on the rate constant, k, of increasing the pH?

    (e) If an initial solution was diluted with water so that its volume doubled, what would be the effect on the value of the rate constant? What would be the effect on the initial rate of the reaction?

    (f) Suggest a possible mechanism for the reaction.

Catalysis

35. The decomposition of nitramide, NH_2NO_2, in aqueous solution is catalyzed by hydroxide ion. Show that the following mechanism,

$$NH_2NO_2(aq) \rightleftharpoons NHNO_2^-(aq) + H^+(aq)$$
$$\text{(Fast equilibrium)}$$
$$NHNO_2^-(aq) \longrightarrow OH^-(aq) + N_2O(g) \qquad \text{(Slow)}$$
$$OH^-(aq) + H^+(aq) \rightleftharpoons H_2O(1) \qquad \text{(Fast equilibrium)}$$

leads to the rate law

$$\text{Rate} = \frac{k[NH_2NO_2]}{[H^+]}$$

Explain why decreasing the concentration of H^+ increases the rate.

Chain Reactions

36. Describe each step in the following mechanism for the bromination of methane as chain initiation, chain propagation, or chain termination:

$$Br_2 \longrightarrow 2Br$$
$$Br + CH_4 \longrightarrow CH_3 + HBr$$
$$CH_3 + Br_2 \longrightarrow CH_3Br + Br$$
$$2Br \longrightarrow Br_2$$

CHAPTER 19
ORGANIC CHEMISTRY

In Chapter 11 we described several types of hydrocarbons; Figure 19.1 gives some examples. Hydrocarbons are the starting materials for the manufacture of an enormous number of *organic compounds* which are all compounds of carbon. A very large number of organic compounds are found in nature, and they are essential constituents of living matter. Organic compounds were at one time obtained almost entirely from plants and animals, but petroleum and natural gas are now the main sources of organic compounds. Although we tend to think of hydrocarbons primarily as fuels, they are no less important as the source of the enormous variety of organic compounds used in the manufacture of many materials essential to modern life, such as synthetic fabrics, rubber, plastics, medicines, drugs, and solvents. Substances derived from petroleum are frequently referred to as *petrochemicals*, and the industry that produces them is called the *petrochemical industry*.

Figure 19.1 Hydrocarbons.

Aliphatic hydrocarbons

Methane
Alkane

Ethene
Alkene

Ethyne
Alkyne

Aromatic hydrocarbon

Benzene

19.1 PETROCHEMICAL INDUSTRY

683

Organic compounds either are hydrocarbons or can be conveniently regarded as derived from hydrocarbons by replacing one or more of the hydrogen atoms with other atoms or groups of atoms called **functional groups**. For example, replacing a hydrogen atom in ethane with an OH group gives ethanol:

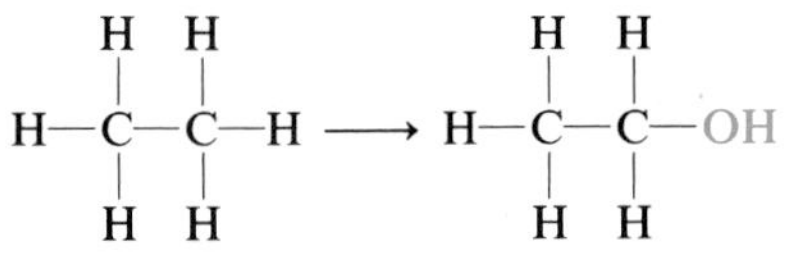

$$
H-\underset{\underset{H}{|}}{\overset{\overset{H}{|}}{C}}-\underset{\underset{H}{|}}{\overset{\overset{H}{|}}{C}}-H \longrightarrow H-\underset{\underset{H}{|}}{\overset{\overset{H}{|}}{C}}-\underset{\underset{H}{|}}{\overset{\overset{H}{|}}{C}}-OH
$$

Organic compounds containing an OH group are called **alcohols** (or phenols if they are derived from an arene). Alcohols have many common properties associated with the OH group.

The vast majority of organic compounds contain oxygen, nitrogen, or both, in addition to carbon and hydrogen; the most important functional groups are those containing oxygen or nitrogen or both (see Table 19.1). We may conveniently consider organic compounds as composed of a hydrocarbon fragment, denoted as R, and one or more functional groups, such as those in Table 19.1. Thus the general formula for an alcohol is written as ROH. The alkane part of an organic molecule is relatively unreactive, and most of the reactions of organic molecules involve the formation or removal of functional groups and the conversion of one functional group into another.

In the following sections we consider the characteristic properties and reactions of several important functional groups. First, however, we take a brief look at two basic processes of the petrochemical industry.

Table 19.1 Functional Groups

FUNCTIONAL GROUP	GENERAL FORMULA FOR COMPOUND	NAME	SUFFIX USED IN SYSTEMATIC NAME
—OH	R—OH[a]	Alcohol (phenol)	-ol
—OR	R—O—R	Ether	alkoxy (prefix)
—CHO	R—CHO	Aldehyde	-al
—COR	R—COR	Ketone	-one
—COOH	R—COOH	Carboxylic acid	-oic acid
—COOR	R—COOR	Ester	-oate
—X[b]	R—X	Haloalkane	halo (prefix)
—NH$_2$	R—NH$_2$	Amine	amino (prefix)
—CONH$_2$	R—CONH$_2$	Amide	-amide

[a] R is an alkyl or an aryl group. ROH is an alcohol when R is an alkyl group and a phenol when R is an aryl group.
[b] X is F, Cl, Br, or I.

19.1 PETROCHEMICAL INDUSTRY

The raw materials of the petrochemical industry are natural gas and petroleum. Natural gas consists primarily of methane and small amounts of ethane, propane, and the butanes, the exact amounts of which vary with the source of the natural gas. Petroleum consists of a complex mixture of alkanes containing five or more carbon atoms (Chapter 11). In the petrochemical industry these hydrocarbons are converted into a variety of important organic compounds. Two primary processes of the petrochemical industry are thermal cracking and the manufacture of synthesis gas.

Thermal Cracking

Thermal cracking is a process in which a mixture of alkanes from natural gas or petroleum is heated to approximately 1400°C for a short time. At this temperature longer-chain alkanes are broken down to shorter-chain alkanes, alkenes, and hydrogen. The reactions that occur are complex. When ethane is a major component of the hydrocarbon mixture, an important reaction is the decomposition of ethane to ethene and hydrogen.

$$CH_3{-}CH_3 \longrightarrow CH_2{=}CH_2 + H_2$$

This reaction is believed to take place by a free-radical mechanism (see Figure 19.2).

The important products of the thermal cracking of alkane mixtures are ethene, C_2H_4, and propene, $CH_3{-}CH{=}CH_2$, together with hydrogen, methane, ethane, propane, and other hydrocarbons, depending on the composition of the original mixture. The products are separated by low-temperature fractional distillation. The higher–boiling point fractions are further treated to make gasoline, and the ethane and methane are recycled. Ethene and propene are important primary materials of the petrochemical industry that are used for the preparation of many other important compounds.

Synthesis Gas

Methane and other alkanes are converted to a mixture of hydrogen and carbon monoxide by heating them with steam at 700°–800°C in the presence of a nickel catalyst; for example,

$$CH_4(g) + H_2O(g) \longrightarrow CO(g) + 3H_2(g)$$
$$C_9H_{20}(g) + 9H_2O(g) \longrightarrow 9CO(g) + 19H_2(g)$$

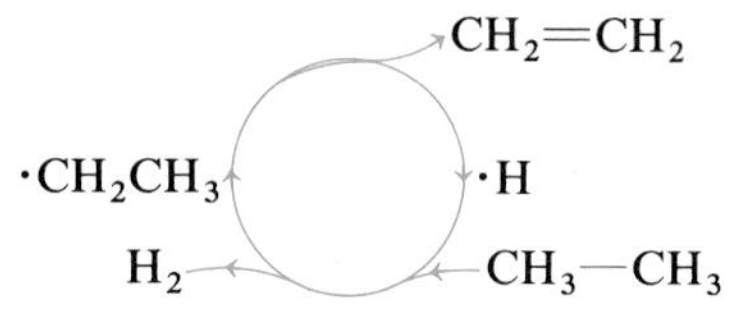

$$\left.\begin{array}{rl} CH_3{-}CH_3 & \longrightarrow \cdot CH_3 + \cdot CH_3 \\ \cdot CH_3 + H{-}CH_2{-}CH_3 & \longrightarrow CH_4 + \cdot CH_2CH_3 \end{array}\right\} \text{Chain initiation}$$

$$\left.\begin{array}{rl} \cdot CH_2CH_3 & \longrightarrow \cdot H + H_2C{=}CH_2 \\ \cdot H + CH_3{-}CH_3 & \longrightarrow H_2 + \cdot CH_2{-}CH_3 \end{array}\right\} \text{Chain propagation}$$

$$\left.\begin{array}{rl} \cdot H + \cdot H & \longrightarrow H_2 \\ \cdot CH_3 + \cdot CH_3 & \longrightarrow C_2H_6 \end{array}\right\} \text{Chain termination}$$

Figure 19.2 Thermal Cracking of Ethane. The cracking of ethane is a chain reaction involving free-radical intermediates.

The mixture of carbon monoxide and hydrogen obtained by this process is called **synthesis gas**. It is used as a fuel for domestic and industrial heating and as a starting material for the manufacture of many organic compounds. The major organic product made from synthesis gas is methanol.

19.2 ALCOHOLS

Industrial Preparation

When synthesis gas having the composition $2H_2:1CO$ is heated at $240°-260°C$ under a pressure of $50-100$ atm in the presence of a $ZnO/CuO/Al_2O_3$ catalyst, a good yield of methanol is obtained:

$$CO(g) + 2H_2(g) \rightleftharpoons CH_3OH(g) \qquad \Delta H° = -91 \text{ kJ}$$

The position of the equilibrium is very favorable at room temperature, but because the reaction is highly exothermic, the equilibrium constant decreases with increasing temperature (Table 19.2). However, even in the presence of a catalyst, the reaction is very slow at ordinary temperatures. The operating temperature of approximately $250°C$ is as low as can be used to obtain a reasonable reaction rate, even with a catalyst. High pressure increases the yield because the reaction is accompanied by a decrease in the number of molecules.

Methanol is a primary product of the petrochemical industry and is used in the manufacture of many other substances including plastics, fertilizers, and pharmaceuticals. A minor but familiar use is as a component of windshield washer liquid, because it is miscible with water and has a low freezing point of $-98°C$. Methanol is also called *wood alcohol*, because it can be obtained by heating wood in the absence of air in a process called *destructive distillation*.

Methanol can be regarded as being derived from methane by replacement of a hydrogen atom by a hydroxyl group, OH:

Methane   Methanol

It is a member of the class of compounds called alcohols, which contain the OH functional group and have the general formula ROH.

As we mentioned in Chapter 11, alkenes readily undergo *addition reactions*. An important reaction of this type is the addition of water to a double bond. Addition of water to ethene gives ethanol, according to the reaction

Addition of water to propene gives 2-propanol according to the reaction

$$CH_3-CH=CH_2 + H_2O \longrightarrow CH_3-CH-CH_3$$
$$\qquad\qquad\qquad\qquad\qquad\qquad | $$
$$\qquad\qquad\qquad\qquad\qquad\quad OH$$

although some 1-propanol, $CH_3-CH_2-CH_2-OH$, is produced as well. These reactions are extremely slow at room temperature, so a temperature of approxi-

Table 19.2 Temperature Variation of the Equilibrium Constant for the Reaction $CO(g) + 2H_2(g) \rightleftharpoons CH_3OH(g)$

TEMPERATURE (°C)	K (mol^{-2} L^2)
0	5.27×10^5
100	1.08×10
200	1.70×10^{-2}
300	2.31×10^{-4}
400	1.09×10^{-5}

mately 300°C and a catalyst, such as phosphoric acid, are used to obtain a sufficiently rapid reaction. In both reactions there is a decrease in the number of molecules, so a high pressure of 70 atm is used to increase the equilibrium yield.

Ethanol is commonly called *ethyl alcohol*, or simply alcohol. It can be imagined as being derived from ethane by replacing a H atom by a hydroxyl, —OH, group. Only one alcohol is obtained from ethane because replacement of any of the H atoms gives the same molecule, C_2H_5OH.

However, two *different* propanols can be derived from propane; in other words, there are two isomers of propanol:

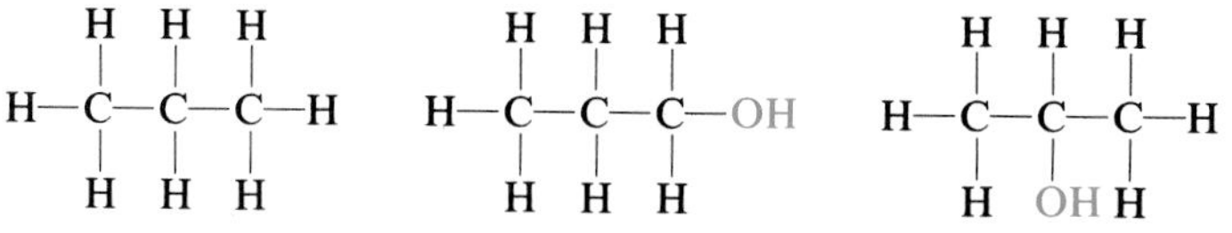

Propane   1-Propanol   2-Propanol

Nomenclature

Alcohols are named by replacing the -e in the name of the alkane from which they are derived by the ending -ol. For example,

$$\text{Methane} \longrightarrow \text{Methanol}$$

$$\text{Ethane} \longrightarrow \text{Ethanol}$$

More complicated structures are dealt with by the same rules that we described for the hydrocarbons (Chapter 11). The longest carbon chain in a molecule is numbered in the direction that gives the substituent groups the smallest possible numbers. Thus $CH_3CH_2CH_2OH$ is called 1-propanol, not 3-propanol.

There are four alcohols, C_4H_9OH. They can be derived from the two isomeric alkanes, C_4H_{10},

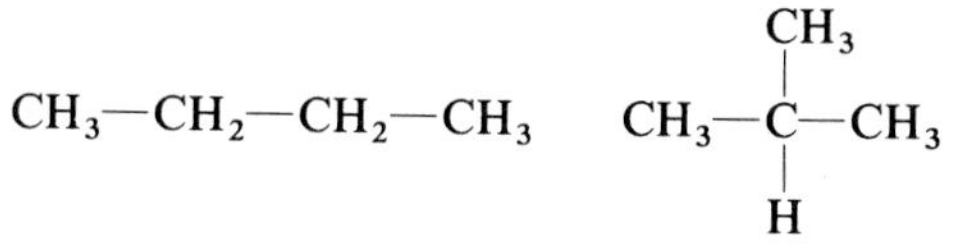

$$CH_3{-}CH_2{-}CH_2{-}CH_3$$

1-Butane

$$CH_3{-}\overset{\overset{\displaystyle CH_3}{|}}{\underset{\underset{\displaystyle H}{|}}{C}}{-}CH_3$$

2-Methylpropane

They are

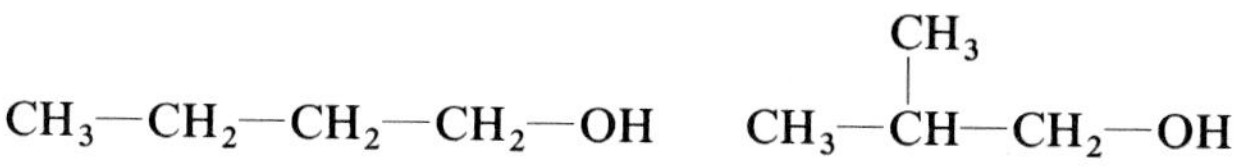

$$CH_3{-}CH_2{-}CH_2{-}CH_2{-}OH$$

1-Butanol

$$CH_3{-}\overset{\overset{\displaystyle CH_3}{|}}{CH}{-}CH_2{-}OH$$

2-Methyl-1-propanol

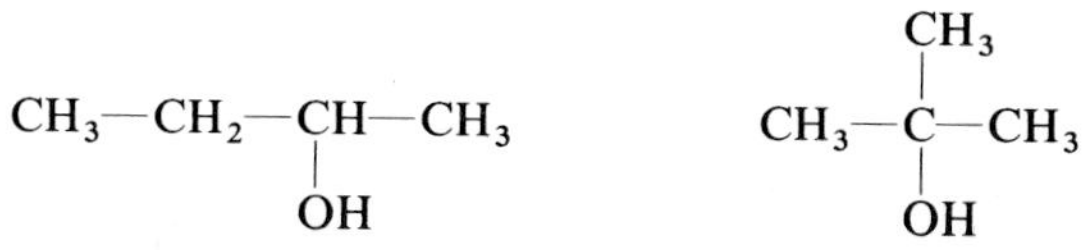

$$CH_3{-}CH_2{-}\underset{\underset{\displaystyle OH}{|}}{CH}{-}CH_3$$

2-Butanol

$$CH_3{-}\overset{\overset{\displaystyle CH_3}{|}}{\underset{\underset{\displaystyle OH}{|}}{C}}{-}CH_3$$

2-Methyl-2-propanol

An alcohol containing a —CH_2OH group is known as a *primary alcohol*. An alcohol containing a $>$CHOH group is a *secondary alcohol*, and an alcohol containing a $\geqslant$COH group is a *tertiary alcohol*.

Example 19.1 Classify the four alcohols, C_4H_9OH, as primary, secondary, and tertiary.

Solution

Primary $CH_3CH_2CH_2CH_2—OH$ $$CH_3—\overset{\displaystyle CH_3}{\underset{\displaystyle H}{C}}—CH_2OH$$

                          1-Butanol   2-Methyl-1-propanol

Secondary $$CH_3CH_2\underset{\displaystyle OH}{C}HCH_3$$

                          2-Butanol

Tertiary $$CH_3—\overset{\displaystyle CH_3}{\underset{\displaystyle OH}{C}}—CH_3$$

                          2-Methyl-2-propanol

Fermentation

In the past ethanol was always made by fermentation of carbohydrates and sugars. Beer and wine are still made in this way. **Fermentation** is a reaction, catalyzed by certain enzymes found in yeast, in which the complex molecules of sugars and carbohydrates are broken down into ethanol and carbon dioxide (Experiment 19.1). For example,

$$C_6H_{12}O_6 \xrightarrow{\text{Yeast}} 2C_2H_5OH + 2CO_2$$
$$\text{Glucose} \qquad\qquad\quad \text{Ethanol}$$

If we begin with an aqueous solution of sugar, fermentation proceeds until the ethanol concentration reaches 13%, at which point the yeast no longer functions and fermentation ceases. Wine contains up to about 13% ethanol and beer usually around 4%–6%. If wine is left exposed to the air, bacteria enter and catalyze the oxidation of ethanol to acetic acid (vinegar), CH_3CO_2H:

$$C_2H_5OH + O_2 \longrightarrow CH_3CO_2H + H_2O$$

This deterioration of wine is prevented if the ethanol content is increased to approximately 20%. The ethanol content of fortified wines such as port, sherry, and vermouth is increased to this level by adding more alcohol after fermentation is complete. Beverages such as whisky that contain more alcohol than fortified

Ancient Egyptians also made wine by fermentation.

Preparation of Ethanol by Fermentation

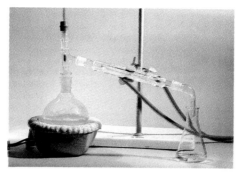

The flask contains a aqueous solution of sugar to which a little yeast has been added. The enzymes in the yeast catalyze the decomposition of sugar to ethanol and carbon dioxide. The CO_2 can be observed bubbling through lime water, a solution of $Ca(OH)_2$, producing a precipitate of insoluble calcium carbonate. After several days the evolution of CO_2 ceases and the reaction is complete.

When the aqueous solution produced in this way is distilled, a more concentrated ethanol solution is obtained by collecting the fraction boiling at approximately 78°C.

When this concentrated ethanol solution is poured into a dish it burns when ignited.

wines are made by distillation, which raises the ethanol content to about 50% (Experiment 19.1). The concentration of ethanol in alcoholic beverages is often expressed as *proof*, which is simply twice the percentage of ethanol by volume. Thus 100-proof whisky contains 50% alcohol.

The intoxicating effect of ethanol is well known. Prolonged and excessive consumption of ethanol can lead to permanent liver damage. Methanol is considerably more toxic; small amounts can lead to blindness and even death.

Diols and Triols

Alcohols containing two or more OH groups are called *diols*, *triols*, and so on. Two important compounds of this type are 1,2-ethanediol, also known as *ethylene glycol*, or simply *glycol*,

$$\underset{\underset{\text{OH}}{|}}{\text{CH}_2}\!-\!\underset{\underset{\text{OH}}{|}}{\text{CH}_2}$$

and 1,2,3-propanetriol, which is also known as *glycerol*, or *glycerine*,

$$\underset{\underset{\text{OH}}{|}}{\text{CH}_2}\!-\!\underset{\underset{\text{OH}}{|}}{\text{CH}}\!-\!\underset{\underset{\text{OH}}{|}}{\text{CH}_2}$$

1,2-Ethanediol is the main component of permanent antifreeze used in automobiles. Another important use is for the manufacture of polyester fibers. 1,2,3-Propanetriol is used in the manufacture of the explosives nitroglycerine and dynamite (Chapter 12), in cosmetics, as a sweetening agent, and in the manufacture of plastics and synthetic fibers.

19.2 ALCOHOLS

689

Properties

Under normal conditions methanol, ethanol, and other alcohols containing up to about twelve carbon atoms are liquids. The boiling point increases regularly as the length of the carbon chain increases; the strength of the intermolecular forces increases as the number of carbon atoms increases (see Table 19.3). The boiling points are all much higher than the boiling points for the corresponding alkanes. For example, methane boils at $-164°C$, whereas methanol boils at $65°C$; ethane boils at $-89°C$, but ethanol boils at $78°C$. The higher boiling points of the alcohols result from hydrogen bonding. The hydrogen atom bonded to the highly electronegative oxygen atom of one molecule is hydrogen-bonded to the oxygen of a neighboring molecule:

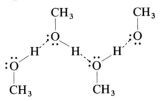

Methanol, ethanol, and the propanols are soluble in water at room temperature in all proportions. In contrast, the higher alcohols have only a limited solubility. Although the —OH group forms a hydrogen bond with water, as the hydrocarbon chains become longer, they progressively decrease the solubility.

Reactions

Like water, alcohols behave as very weak acids and bases. They are too weakly acidic to donate a proton to water, but they ionize in concentrated aqueous hydroxide solutions and in the presence of other strong bases, such as NH_2^-, to give the ethoxide ion:

$$C_2H_5OH + OH^- \rightleftharpoons C_2H_5O^- + H_2O$$
$$C_2H_5OH + NH_2^- \longrightarrow C_2H_5O^- + NH_3$$

Ethoxide ion

The reaction of sodium with ethanol is similar to its reaction with water (Experiment 19.2). The products are hydrogen and sodium ethoxide, $C_2H_5O^-Na^+$:

$$2C_2H_5OH + 2Na \longrightarrow 2C_2H_5O^-Na^+ + H_2$$

As bases, alcohols are similar in strength to water and are protonated by strong acids. For example,

$$CH_3-\overset{..}{\underset{..}{O}}-H + H-Br \longrightarrow CH_3-\overset{H}{\underset{\oplus}{O}}-H + Br^-$$

Table 19.3 Properties of Some Alcohols

ALCOHOL	NAME	bp (°C)	SOLUBILITY (g/100 g H$_2$O), 25°C
CH_3OH	Methanol	65 ⎫	
CH_3CH_2OH	Ethanol	78 ⎬ Soluble in all proportions, miscible	
$CH_3CH_2CH_2OH$	1-Propanol	97 ⎭	
$CH_3CH_2CH_2CH_2OH$	1-Butanol	117	9.0
$CH_3CH_2CH_2CH_2CH_2OH$	1-Pentanol	138	2.7
$CH_3CH_2CH_2CH_2CH_2CH_2OH$	1-Hexanol	158	0.6

Reaction of Sodium with Ethanol

Sodium reacts with ethanol producing sodium ethoxide and hydrogen. When sodium is added to ethanol it sinks to the bottom and reacts less vigorously than it reacts with water. The hydrogen produced may be easily collected as shown. Because ethoxide ion is basic, a few drops of phenolphthalein added to the ethanol color it pink.

Most alcohols can be dehydrated by heating them with concentrated H_2SO_4 or H_3PO_4 to give alkenes:

$$C_2H_5OH(l) \xrightarrow[\text{Heat}]{H_2SO_4(\text{conc})} CH_2{=}CH_2 + H_2O \; (+ H_2SO_4 \rightarrow H_3O^+ HSO_4^-)$$

This reaction is the reverse of the formation of an alcohol from an alkene, that is, the addition of H_2O to a double bond. The reaction is reversible, and the direction in which it proceeds is determined by the conditions. Concentrated sulfuric acid removes water and drives the reaction to the right.

Another important reaction of alcohols is oxidation to give aldehydes and ketones. This reaction is discussed later in this chapter.

Alcohols as Fuels

Complete oxidation of a hydrocarbon gives carbon dioxide and water. The alcohols represent an intermediate or incomplete stage of oxidation of a hydrocarbon, so their enthalpies of combustion, although high, are not as high as those of the corresponding alkanes:

$$CH_4(g) + 2O_2(g) \longrightarrow CO_2(g) + 2H_2O(l) \qquad \Delta H^\circ = -891 \text{ kJ mol}^{-1}$$
$$CH_3OH(l) + 1\tfrac{1}{2}O_2(g) \longrightarrow CO_2(g) + 2H_2O(l) \qquad \Delta H^\circ = -726 \text{ kJ mol}^{-1}$$
$$C_2H_6(g) + 3\tfrac{1}{2}O_2(g) \longrightarrow 2CO_2(g) + 3H_2O(l) \qquad \Delta H^\circ = -1560 \text{ kJ mol}^{-1}$$
$$C_2H_5OH(l) + 3O_2(g) \longrightarrow 2CO_2(g) + 3H_2O(l) \qquad \Delta H^\circ = -1367 \text{ kJ mol}^{-1}$$

Because of an increase in the cost of petroleum products in recent years, there has been interest in replacing gasoline with alcohol produced by fermentation of agricultural waste. Mixtures of ethanol and gasoline—gasohol—are now sold as automobile fuel in some parts of the United States and in other countries, and methanol, which has a high octane rating of 110, is used in racing cars. Ethanol is used as a fuel in some camp stoves and in the home for heating and cooking food at the table. Unlike hydrocarbons, it normally does not form explosive mixtures with air, and any fire caused by burning alcohol can be easily extinguished with water.

Phenols

When a hydrogen atom in an arene (aromatic hydrocarbon) is replaced by an OH group, the resulting compound is called a **phenol**, not an alcohol. Replacing one of the hydrogen atoms of benzene with an -OH group gives the compound C_6H_5OH, which is called *benzenol* but is commonly known as *phenol*. It has the structure

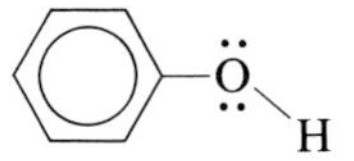

Benzenol
(phenol)

Phenol is a stronger acid than the alcohols, although it is still quite weak ($K_a = 1.3 \times 10^{-10}$ mol L^{-1} at 25°C). It is a much weaker base than the alcohols.

Originally called *carbolic acid*, phenol is used as an antiseptic, a disinfectant, and a starting material in the manufacture of dyes and plastics. It is very effective in killing bacteria and was first used in medicine by Sir Joseph Lister in 1867.

19.3 ETHERS

Ethers have the general formula

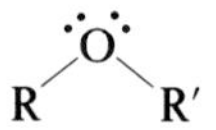

Where R and R′ are alkyl groups. They may be prepared by dehydrating alcohols; for example,

$$C_2H_5OH + HOC_2H_5 \longrightarrow C_2H_5\text{—}O\text{—}C_2H_5 + H_2O$$

Diethyl ether

This reaction can be carried out by using a dehydrating agent such as concentrated sulfuric acid. We have just seen that with excess sulfuric acid ethene is obtained, but with excess ethanol the main product is diethyl ether.

Another method of preparing ethers is by the reaction of a sodium salt of an alcohol with an alkyl halide. For example,

$$C_2H_5\text{—}O^-Na^+ + CH_3\text{—}Br \longrightarrow C_2H_5OCH_3 + Na^+Br^-$$

The common names for ethers are based on the two alkyl groups attached to the oxygen atom. For example, $CH_3OC_2H_5$ is methyl ethyl ether. The systematic name is based on the alkyl group that has the longest carbon chain. The alkyl group with the shorter carbon chain and the oxygen atom are then called an *alkoxy group*; for example, $CH_3OC_2H_5$ is methoxy ethane. These systematic names are not widely used.

Both alcohols and ethers are related to water and have the same angular structure at the oxygen atom. They are all examples of angular AX_2E_2 molecules. For example,

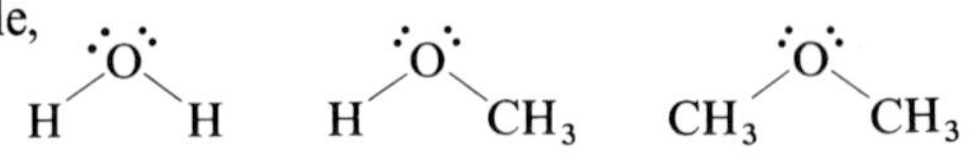

Water	Methanol	Dimethyl ether
(bp = 100°C)	(bp = 65°C)	(bp = −25°C)

Both methanol and water are liquids, but dimethyl ether is a gas, because unlike water and methanol, dimethyl ether cannot form hydrogen bonds. In fact, the boiling point of dimethyl ether is not very different from that of a simple alkane with a similar molecular mass (CH_3—O—CH_3, bp = −25°C; CH_3—CH_2—CH_3, bp = −42°C).

Diethyl ether (ethoxy ethane), $C_2H_5OC_2H_5$ (bp = 35°C), has only a limited solubility in water and is less dense than water. It is very widely used as a solvent for organic compounds and for extracting organic compounds from water and other solvents. Many organic compounds are considerably more soluble in ether than in water. Thus when an aqueous solution of an organic compound is shaken with ether, most of the organic compound is transferred from the water to the ether. It is said to be *extracted*. The ether rapidly separates and forms a layer on top of the water. The ether layer containing the dissolved organic compound is then easily separated from the water layer. Because of its low boiling point, diethyl ether is readily distilled from the dissolved organic compound at a temperature so low that even relatively unstable substances are not decomposed. But diethyl ether is very flammable and must be used with caution. At one time it was widely used as an anesthetic. However, since mixtures with air are very explosive, it has now been replaced as an anesthetic by compounds that are safer to handle.

19.4 ALDEHYDES AND KETONES

Primary and secondary alcohols decompose at 550°–600°C in the presence of a suitable catalyst (such as copper or silver) to form hydrogen and aldehyde or ketone. This is called a **dehydrogenation reaction** and is an industrial method of preparation for aldehydes and ketones. Some examples are

$$\begin{array}{c} H \\ | \\ H-C-OH \\ | \\ H \end{array} \longrightarrow \begin{array}{c} H \\ \diagdown \\ \diagup \ C=O \\ H \end{array} + H_2$$

Methan*ol*   Methan*al*
(formaldehyde)
(bp = −21°C)

$$\begin{array}{cc} H & H \\ | & | \\ H-C-C-OH \\ | & | \\ H & H \end{array} \longrightarrow \begin{array}{cc} H & H \\ | & | \\ H-C-C=O \\ | \\ H \end{array} + H_2$$

Ethan*ol*   Ethan*al*
(acetaldehyde)
(bp = 21°C)

EXPERIMENT 19.3

Oxidation of Ethanol to Ethanal and the Identification of Ethanal.

When ethanol is heated with an acidified aqueous solution of $K_2Cr_2O_7$ it is oxidized to ethanal and the orange dichromate ion is reduced to dark green Cr^{3+}. If the ethanal that is produced is bubbled into a cold solution of 2,4-dinitrophenylhydrazine, a yellow precipitate of the hydrazone of ethanal soon begins to crystallize from the solution. If this is filtered off and dried, its melting point can be determined. A melting point of 168°C confirms that the aldehyde formed was ethanal.

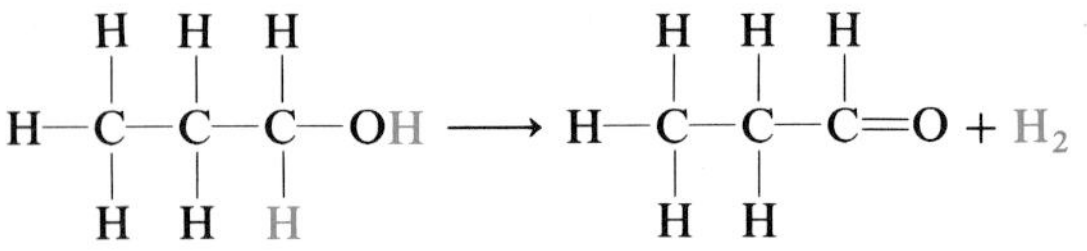

$$\underset{\text{1-Propanol}}{\overset{\overset{\displaystyle H \quad H \quad H}{|\quad\ |\quad\ |}}{H-C-C-C-OH}} \longrightarrow \underset{\text{Propanal}}{\overset{\overset{\displaystyle H \quad H \quad H}{|\quad\ |}}{H-C-C-C=O}} + H_2$$

1-Propanol   Propanal
(bp = 50°C)

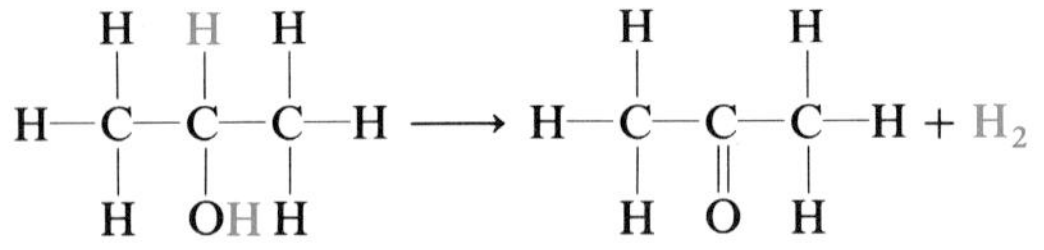

$$\underset{\text{2-Propanol}}{\overset{\overset{\displaystyle H \quad H \quad H}{|\quad\ |\quad\ |}}{H-C-C-C-H}} \longrightarrow \underset{\text{2-Propanone}}{H-C-C-C-H} + H_2$$

2-Propanol   2-Propanone
(acetone)
(bp = 56°C)

A primary alcohol, $-CH_2OH$, gives an aldehyde, $-C\overset{\displaystyle H}{\underset{\displaystyle O}{}}$

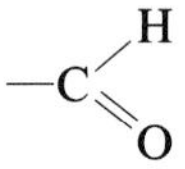

while a secondary alcohol, $>C\overset{\displaystyle H}{\underset{\displaystyle OH}{}}$

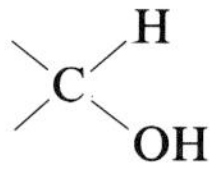

gives a ketone, $>C=O$

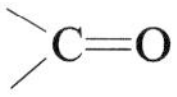

A tertiary alcohol cannot be dehydrogenated in this way, because it has no H atom on the C atom to which the OH group is attached.

Methanal, ethanal, and propanal all contain the

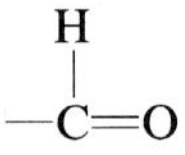

$$-\overset{\displaystyle H}{\underset{}{C}}=O$$

group and are the simplest members of the series of compounds known as **aldehydes**, which have the general formula

$$R-\overset{\displaystyle H}{\underset{}{C}}=O$$

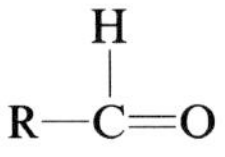

2-Propanone (acetone) is the simplest of a series of compounds known as **ketones**, all of which contain the

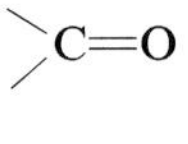

$$>C=O$$

group and have the general formula

$$\overset{\displaystyle R}{\underset{\displaystyle R'}{>}}C=O$$

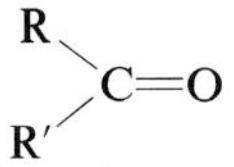

Some examples are given in Figure 19.3.

Preparation

As we have seen, aldehydes and ketones can be prepared by the dehydrogenation of alcohols. They can also be prepared by oxidation of alcohols with oxygen, which removes the hydrogen as water. For example,

$$2CH_3OH + O_2 \longrightarrow 2CH_2O + 2H_2O$$

Figure 19.3 Aldehydes and Ketones.

$$\underset{H}{\overset{H}{\diagdown}}C=O$$

Methanal
(formaldehyde)

$$\underset{CH_3}{\overset{H}{\diagdown}}C=O$$

Ethanal
(acetaldehyde)

$$\underset{CH_3CH_2}{\overset{H}{\diagdown}}C=O$$

Propanal

$$\underset{R}{\overset{H}{\diagdown}}C=O$$

General formula for an aldehyde, where R is an alkyl group

$$\underset{CH_3}{\overset{CH_3}{\diagdown}}C=O$$

Propanone
(acetone)

$$\underset{CH_3CH_2}{\overset{CH_3}{\diagdown}}C=O$$

2-Butanone

$$\underset{R'}{\overset{R}{\diagdown}}C=O$$

General formula for a ketone, where R and R′ represent alkyl groups

In the laboratory the oxidation of alcohols is usually carried out by using an acidic solution of potassium permanganate, $KMnO_4$, or potassium dichromate, $K_2Cr_2O_7$, as oxidizing agents, which are reduced to Mn^{2+} and Cr^{3+}, respectively (see Experiment 19.3 on page 693):

$$CH_3CH_2OH \xrightarrow{K_2Cr_2O_7, \, H_3O^+} CH_3CHO$$

Experiment 19.4 demonstrates the oxidation of ethanol to ethanal using copper oxide, CuO, as the oxidizing agent.

Nomenclature

Aldehydes are named by replacing the -*e* of the corresponding alkane by -*al*. Since aldehydes are derived from primary alcohols, the

$$\underset{}{\overset{H}{\underset{|}{-C}}}=O$$

group must always be at the end of the chain, so its C atom must always be carbon atom 1. Therefore we do not need to indicate its position in the name.

Ketones are named by replacing the -*e* in the name of the corresponding

EXPERIMENT 19.4

Oxidation of Ethanol to Ethanal with Copper Oxide

A piece of copper sheet is heated in a bunsen flame.

It rapidly becomes coated with a black layer of copper(II) oxide, CuO.

When the hot, oxide-coated copper is dipped into ethanol, it rapidly oxidizes the ethanol to ethanal and is itself reduced back to shiny metallic copper.

alkane by *-one*. The longest chain of C atoms is numbered from the end that gives the C atom of the C=O group the lowest possible number. A $>$C=O group is called a *carbonyl group*.

The common names *formaldehyde*, *acetaldehyde*, and *acetone* are very commonly used and should be remembered.

Structures

Ketones and aldehydes are examples of AX_3 molecules and therefore have a planar geometry around the carbon atom of the C=O group, with approximately 120° bond angles (Figure 19.4). Because oxygen is more electronegative than carbon, the oxygen atom of the carbonyl group carries a small negative charge, and the carbon of the C=O group carries a small positive charge. In other words, the carbonyl group is polar:

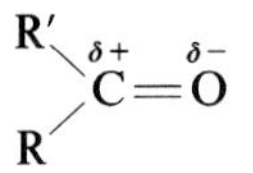

$$\begin{array}{c} R' \\ \diagdown \overset{\delta+}{C} = \overset{\delta-}{O} \\ \diagup \\ R \end{array}$$

The polarity of the carbonyl group can also be represented by the two resonance structures

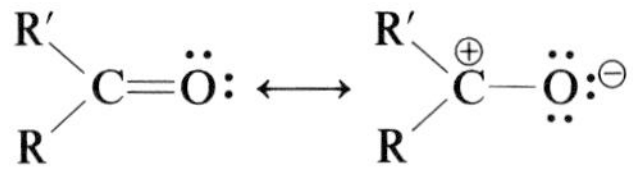

$$\begin{array}{c} R' \\ \diagdown \\ \diagup C = \ddot{O}: \\ R \end{array} \longleftrightarrow \begin{array}{c} R' \\ \diagdown \overset{\oplus}{C} - \ddot{O}: \overset{\ominus}{} \\ \diagup \\ R \end{array}$$

Properties

Methanal (formaldehyde) is a gas with a penetrating odor and is very soluble in water. It is normally stored and sold as a 40% aqueous solution known as *formalin*. This is used as a disinfectant and for preserving biological specimens. A major use of formaldehyde is for the manufacture of resins, plastics, and adhesives.

Ethanal (acetaldehyde), CH_3CHO, is a liquid with a low boiling point of 21°C. It is used largely for the synthesis of other organic compounds.

Propanone (acetone), CH_3COCH_3, is a volatile liquid boiling at 56°C. It is a good solvent for many compounds, and it is completely miscible with water.

Many aromatic aldehydes have pleasant odors (Figure 19.5).

Figure 19.4 Structures of Some Aldehydes and Ketones. These are all AX_3 molecules with a planar triangular structure and approximately 120° bond angles.

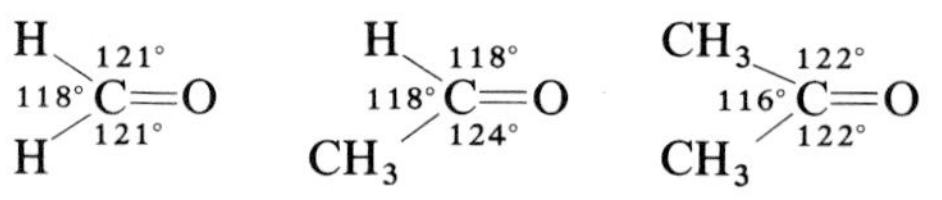

Methanal   Ethanal   Propanone

Figure 19.5 Some Aromatic Aldehydes.

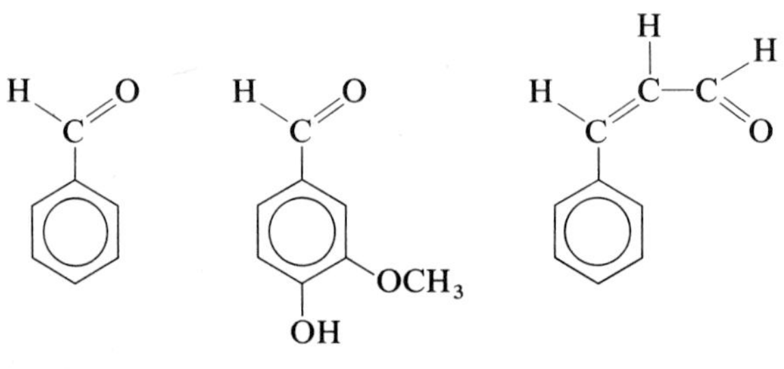

Benzaldehyde   Vanillin   Cinnamaldehyde
(bitter almonds)   (vanilla)   (cinnamon)

Reactions

Many of the reactions of aldehydes and ketones involve addition to the C=O double bond. Two important examples are the addition of H_2 and of hydrazine.

ADDITION OF H_2 (HYDROGENATION) This can be carried out by using a metal catalyst (Ni or Pt) and hydrogen under pressure:

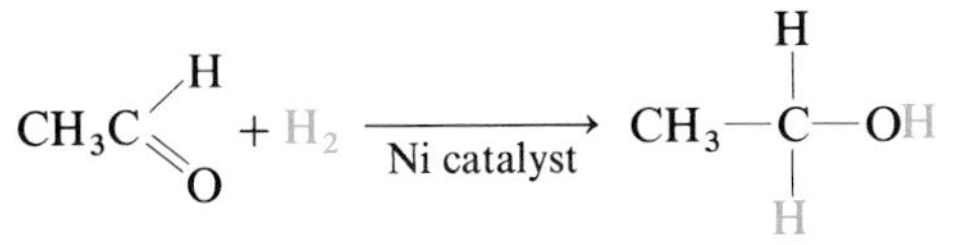

$$CH_3C\overset{H}{\underset{O}{\diagdown}} + H_2 \xrightarrow[\text{Ni catalyst}]{} CH_3 - \underset{\underset{H}{|}}{\overset{\overset{H}{|}}{C}} - OH$$

Ethanal (aldehyde) Ethanol (primary alcohol)

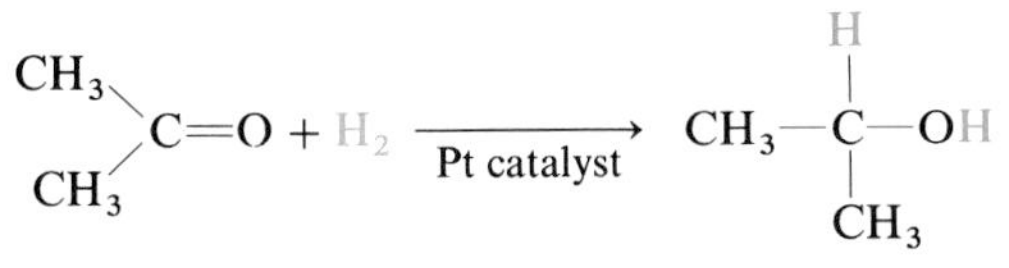

$$\underset{CH_3}{\overset{CH_3}{\diagdown}}C=O + H_2 \xrightarrow[\text{Pt catalyst}]{} CH_3 - \underset{\underset{CH_3}{|}}{\overset{\overset{H}{|}}{C}} - OH$$

Propanone (ketone) 2-Propanol (secondary alcohol)

These reactions are the reverse of the dehydrogenation reactions that are used to prepare aldehydes and ketones from alcohols.

ADDITION OF HYDRAZINE AND ITS DERIVATIVES Because the determination of the melting point is a simple and convenient method for identifying a substance, a liquid is often converted to a solid crystalline derivative to identify it. Many common aldehydes and ketones are liquids with low boiling points, and the reaction with hydrazine or a derivative of hydrazine is used to convert them into crystalline solids for the purposes of identification.

Hydrazine adds to the C=O bond of a ketone or an aldehyde to form an intermediate that loses water to give a compound containing a C=N double bond, called a *hydrazone*:

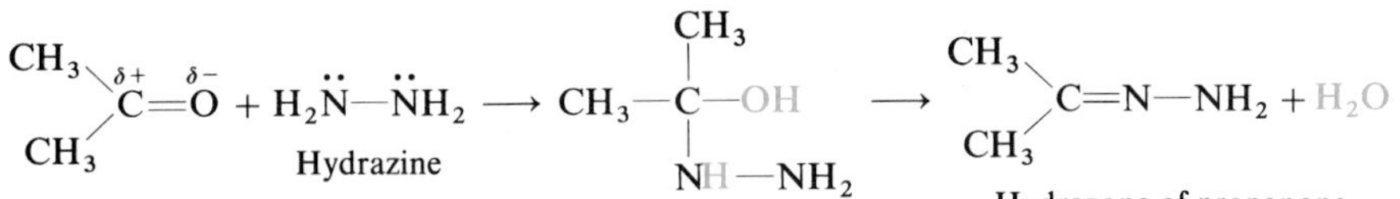

$$\underset{CH_3}{\overset{CH_3}{\diagdown}}\overset{\delta+}{\underset{\delta-}{C=O}} + H_2\overset{..}{N}-\overset{..}{N}H_2 \longrightarrow CH_3 - \underset{\underset{NH-NH_2}{|}}{\overset{\overset{CH_3}{|}}{C}} - OH \longrightarrow \underset{CH_3}{\overset{CH_3}{\diagdown}}C=N-NH_2 + H_2O$$

Hydrazine

Hydrazone of propanone

A common reagent for preparing derivatives of ketones and aldehydes is 2,4-dinitrophenylhydrazine, which gives crystalline compounds with convenient characteristic melting points. For example,

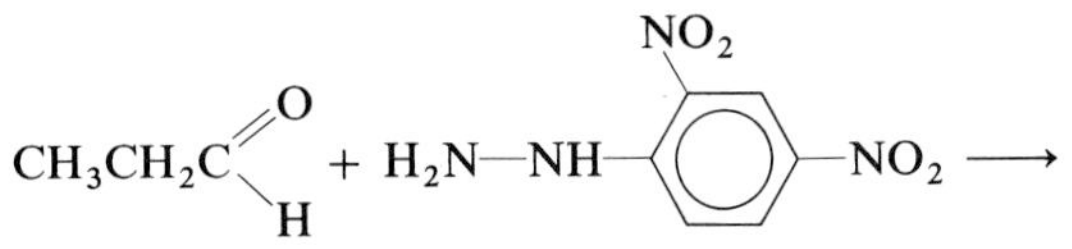

$$CH_3CH_2C\overset{O}{\underset{H}{\diagdown}} + H_2N-NH-\underset{NO_2}{\underset{|}{\bigcirc}}-NO_2 \longrightarrow$$

Propanal 2,4-Dinitrophenylhydrazine

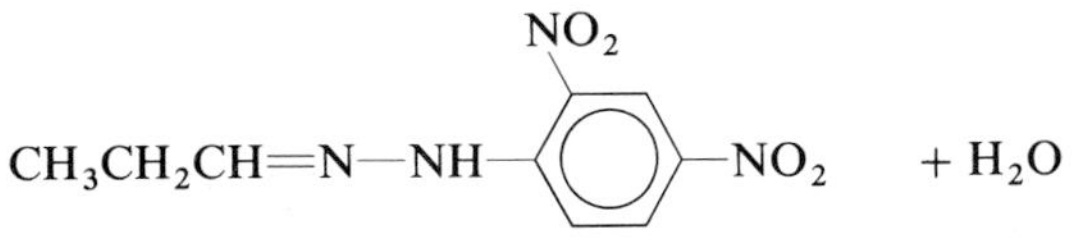

$$CH_3CH_2CH=N-NH-\underset{NO_2}{\underset{|}{\bigcirc}}-NO_2 + H_2O$$

2,4-Dinitrophenylhydrazone of propanal (mp = 155°C)

(See Experiment 19.3.)

OXIDATION OF ALDEHYDES   Aldehydes are readily oxidized to give carboxylic acids; these reactions are discussed in the next section (see Experiment 19.5). Ketones are not easily oxidized.

19.5 CARBOXYLIC ACIDS

In Chapters 5 and 14 we discussed acetic acid as an example of a weak acid in aqueous solution. It is a member of the class of compounds known as **carboxylic acids**, all of which have the general formula

$$\begin{array}{c} O \\ \parallel \\ R-C-OH \end{array}$$

For example, acetic acid is

$$\begin{array}{c} O \\ \parallel \\ CH_3-C-OH \end{array}$$

The $\begin{array}{c} O \\ \parallel \\ -C-OH \end{array}$ group is called the *carboxyl group*.

Carboxylic acids are named systematically by replacing the *-e* in the name of the parent alkane by *-oic acid*. Thus methane becomes methanoic acid, ethane becomes ethanoic acid, propane becomes propanoic acid, and so on. Examples are given in Table 19.4. These systematic names, however, have not been widely accepted, and the common names continue to be used extensively.

Table 19.4 Carboxylic Acids

HYDROCARBON	CARBOXYLIC ACID	COMMON NAME OF CARBOXYLIC ACID
$$\begin{array}{c} H \\ \mid \\ H-C-H \\ \mid \\ H \end{array}$$ Methane	$$\begin{array}{c} O \\ \parallel \\ H-C-OH \end{array}$$ Methanoic acid	Formic acid
CH_3-CH_3 Ethane	$$\begin{array}{c} O \\ \parallel \\ CH_3-C-OH \end{array}$$ Ethanoic acid	Acetic acid
$CH_3-CH_2-CH_3$ Propane	$$\begin{array}{c} O \\ \parallel \\ CH_3-CH_2-C-OH \end{array}$$ Propanoic acid	Propionic acid
$CH_3-CH_2-CH_2-CH_3$ Butane	$$\begin{array}{c} O \\ \parallel \\ CH_3-CH_2-CH_2-C-OH \end{array}$$ 1-Butanoic acid	Butyric acid
$$\begin{array}{c} CH_3 \\ \mid \\ CH_3-C-CH_3 \\ \mid \\ H \end{array}$$ 2-Methylpropane	$$\begin{array}{c} CH_3\;\;O \\ \mid\;\;\;\;\parallel \\ CH_3-C-C-OH \\ \mid \\ H \end{array}$$ 2-Methylpropanoic acid	Isobutyric acid

Preparation

A general method for preparing carboxylic acids in the laboratory is the oxidation of the appropriate aldehyde with an oxidizing agent such as potassium dichromate, $K_2Cr_2O_7$, in acid solution (see Experiment 19.5):

$$CH_3CH_2C\overset{O}{\underset{H}{\diagdown}} \xrightarrow{K_2Cr_2O_7,\ H_3O^+} CH_3CH_2C\overset{O}{\underset{OH}{\diagdown}}$$

Propanal   Propanoic acid

In this reaction Cr in the $+6$ oxidation state ($K_2Cr_2O_7$) is reduced to the $+3$ oxidation state (Cr^{3+}). Experiment 19.5 shows the oxidation of ethanal to ethanoic acid with hot copper(II) oxide.

Since primary alcohols may be oxidized to aldehydes, which in turn are readily oxidized to carboxylic acids, primary alcohols may be oxidized directly to carboxylic acids. Aldehydes can therefore only be obtained by the oxidation of alcohols if a limited amount of oxidizing agent is used. Use of excess oxidizing agent gives the corresponding carboxylic acid:

$$CH_3CH_2OH \nearrow \xrightarrow[\text{Limited amount}]{K_2Cr_2O_7,\ H_3O^+,} CH_3\overset{H}{\underset{}{C}}=O$$

Ethanol   Ethanal

$$\searrow \xrightarrow[\text{Excess}]{K_2Cr_2O_7,\ H_3O^+,} CH_3\overset{OH}{\underset{}{C}}=O$$

Ethanoic acid

Even a rather weak oxidizing agent, such as Ag^+, can oxidize an aldehyde to a carboxylic acid. Silver ion is reduced to metallic silver, which under the right conditions forms a shiny mirror on the sides of the reaction tube. This reaction is a useful test for aldehydes and serves to distinguish them from ketones and other related compounds (Experiment 19.6).

EXPERIMENT 19.5

Oxidation of Ethanal to Ethanoic Acid

The colorless liquid at the bottom of the tube is ethanal (acetaldehyde), which has a boiling point of 21°C. At the top of the tube black copper(II) oxide, CuO, is held in place by a plug of glasswool. When the copper oxide is heated, ethanal vapor is oxidized to ethanoic (acetic) acid and the copper oxide is reduced to copper. Blue litmus paper held at the mouth of the tube is turned red by the acid vapor, and the copper oxide can be seen to have been reduced to red-brown copper.

Silver Mirror Test for Aldehydes

A clean beaker (or test tube) contains a solution of silver nitrate in aqueous ammonia. An aqueous solution of ethanal is added from a dropper and the solution is stirred.

The solution rapidly darkens in color as ethanal is oxidized to ethanoic acid and Ag^+ is reduced to silver.

The inside of the beaker finally becomes coated with a shiny metallic silver mirror. Other aldehydes will give the same result.

ACETIC ACID   One industrial method for preparing acetic acid is the oxidation of ethanal (acetaldehyde) with air at 60°–80°C:

$$2CH_3CHO + O_2 \longrightarrow 2CH_3CO_2H$$

Acetic acid can also be made by the enzyme-catalyzed oxidation of ethanol by air. This process has been used, since ancient times, to produce vinegar from wine and cider, but today much of the vinegar sold is made by diluting manufactured acetic acid. Vinegar is an approximately 5% solution of acetic acid in water. The oxidation of ethanol occurs in two stages; the second stage is more rapid than the first:

$$CH_3-CH_2-OH \xrightarrow{O_2} CH_3-\overset{H}{\underset{}{C}}=O \xrightarrow{O_2} CH_3-\overset{OH}{\underset{}{C}}=O$$

Ethanol   Ethanal   Ethanoic acid
                (acetaldehyde)   (acetic acid)

FORMIC ACID   The first member of the carboxylic acid series is *formic acid* (methanoic acid), HCO_2H:

$$H-C\overset{\ddot{O}:}{\underset{\ddot{O}-H}{\diagdown}}$$

It was first prepared by the distillation of ants (Latin, *formica*). The swelling and irritation associated with ant bites and bee stings is partly due to formic acid.

One method for the industrial preparation of formic acid is the reaction of carbon monoxide with sodium hydroxide at 200°C to give sodium formate. This salt is then acidified with HCl, and the formic acid is distilled off:

$$NaOH + CO \xrightarrow{200°C} H-\overset{\overset{\displaystyle O}{\|}}{C}-O^-Na^+$$

$$H-\overset{\overset{\displaystyle O}{\|}}{C}-O^-Na^+ + HCl \longrightarrow H-\overset{\overset{\displaystyle O}{\|}}{C}-OH + Na^+Cl^-$$

$$(bp = 100.5°C)$$

OTHER CARBOXYLIC ACIDS   In addition to formic acid, many other carboxylic acids can be isolated from natural sources. *Butyric acid (1-butanoic acid)*, $CH_3CH_2CH_2CO_2H$, is obtained from butterfat and is responsible for the odor of rancid butter. The long-chain acids *palmitic acid*, $CH_3(CH_2)_{14}CO_2H$, and *stearic acid*, $CH_3(CH_2)_{16}CO_2H$, are obtained from many animal and vegetable fats. Their sodium salts are important ingredients of soap.

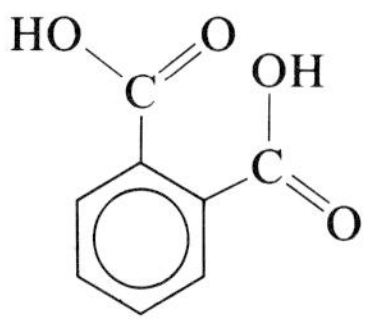

There are also important carboxylic acids that contain more than one carboxyl group. The simplest is *oxalic acid (ethanedioic acid)*, $(CO_2H)_2$, which consists of two carboxyl groups joined together (see Experiment 19.7). Oxalic acid occurs in the leaves of rhubarb and related plants and is poisonous. It removes calcium from the body as insoluble calcium oxalate, $Ca^{2+}(C_2O_4)^{2-}$.

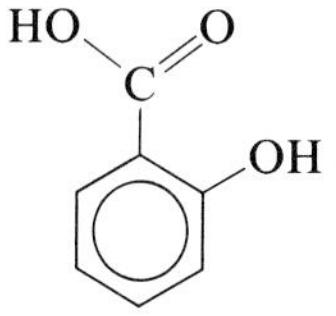

Other important carboxylic acids contain hydroxyl groups as well as the carboxylic acid group. *Lactic acid* (2-hydroxypropanoic acid), $CH_3CH(OH)CO_2H$, is found in sour milk. It accumulates in the muscles of the body after strenuous exercise and is responsible for the soreness in the muscles. *Citric acid* is found in the juice of citrus fruits. The simplest aromatic carboxylic acid is *benzoic acid*, $C_6H_5CO_2H$. It is made by the oxidation of methylbenzene (toluene) with oxygen in the presence of a catalyst:

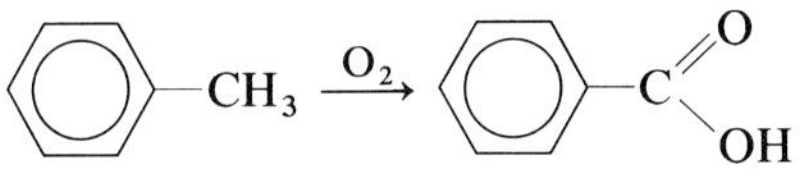

Phthalic acid

Salicylic acid

Benzoic acid

The dicarboxylic acid $C_6H_4(CO_2H)_2$ is called *phthalic acid*. It is a relatively strong acid ($pK_a = 3.0$). Phthalic acid is used in the synthesis of dyes and in the manufacture of paints and varnishes. The aromatic hydroxycarboxylic acid $C_6H_4(OH)(CO_2H)$ is known as *salicylic acid*. Aspirin is acetylsalicylic acid (see Box 19.1).

EXPERIMENT 19.7

Properties of Oxalic Acid

Oxalic acid is a white solid that is soluble in water. Bromothymol blue indicator has a yellow color in an aqueous solution of oxalic acid (left). The aqueous solution reacts with calcium carbonate (center) to produce carbon dioxide, and with magnesium (right) to produce hydrogen.

From early times people have sought substances that would relieve pain. Ethanol, opium, cocaine, and marijuana were all used by early societies for this purpose. Such pain-relieving substances are generally known as *analgesics*. In the eighteenth century it was found that an extract of willow bark was a good analgesic. In 1860 salicylic acid was isolated from this willow bark extract and found to be an efficient analgesic and an anti-inflammatory drug. About the same time chemists discovered that salicylic acid could be synthesized conveniently and cheaply by heating the sodium salt of phenol with carbon dioxide under pressure:

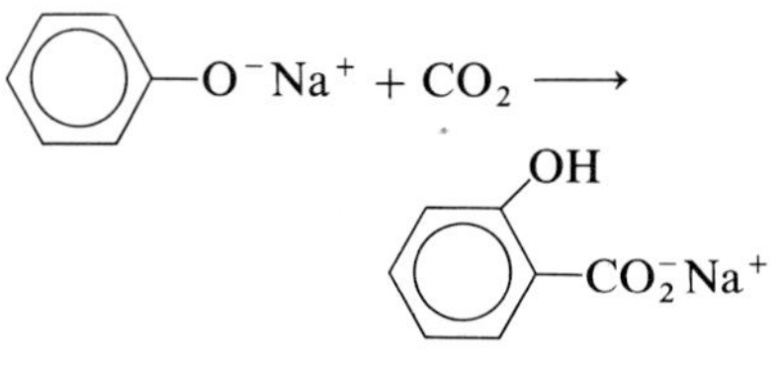

$$\text{\includegraphics{benzene}}-\text{O}^-\text{Na}^+ + \text{CO}_2 \longrightarrow$$

$$\text{\includegraphics{salicylate}}-\text{CO}_2^-\text{Na}^+$$

Salicylic acid then began to be widely used. But salicylic acid has a very sour taste and damages the tissue of the mouth and throat. Acetylsalicylic acid—aspirin—was found to be equally effective and not to have the disadvantages of salicylic acid. It was put into large-scale production in 1899 and soon became the largest-selling drug in the world, a position it retains today.

Aspirin not only relieves pain but also reduces fever and local inflammation. The only undesirable side effect is gastrointestinal bleeding. While the bleeding is usually slight and unimportant, in some persons it may be serious. Other persons have an allergic hypersensitivity to aspirin. Aspirin also inhibits the clotting of blood and should not be used by persons facing surgery or women awaiting childbirth.

More than $100 million is spent every year in the United States on thousands of tons of aspirin; three times as much is spent on products that are combinations of other drugs with aspirin. These products generally contain small amounts of the stimulant caffeine and other pain relievers that are probably no more effective than aspirin but are considerably more expensive. Despite its long use and intensive investigation, only in the past few years have we begun to understand how aspirin works.

Structure

The carboxylic acid group has a planar AX_3 geometry around the carbon atom (see Figure 19.6).

Properties

Carboxylic acids are weak acids in water; for example,

$$\underset{\text{Acetic acid}}{CH_3-\overset{\displaystyle O}{\overset{\|}{C}}-OH} + H_2O \rightleftharpoons H_3O^+ + \underset{\text{Acetate ion}}{CH_3-\overset{\displaystyle O}{\overset{\|}{C}}-O^-}$$

Table 19.5 gives pK_a values for some carboxylic acids. They are much stronger acids than the corresponding alcohols.

The acid strength of the carboxylic acids can be attributed to the presence of the polar carbonyl group, $C{=}O$. The electronegative oxygen atom attracts

Figure 19.6 Structure of Methanoic (Formic) Acid. The CO_2H group has a planar AX_3 geometry with bond angles that are close to 120°.

CHAPTER 19
ORGANIC CHEMISTRY

Table 19.5 Melting Points, Boiling Points, and pK_a Values for Some Carboxylic Acids

ACID	FORMULA	mp (°C)	bp (°C)	pK_a
Methanoic (formic)	HCO_2H	8.4	110.5	3.77
Ethanoic (acetic)	CH_3CO_2H	16.6	118	4.76
Propanoic (propionic)	$CH_3CH_2CO_2H$	−22	141	4.88
1-Butanoic (butyric)	$CH_3CH_2CH_2CO_2H$	−5	163	4.82
Ethanedioic (oxalic)	$(CO_2H)_2$	187	—	1.46
2-Hydroxypropanoic (lactic)	$CH_3CH(OH)CO_2H$	18	—	3.87
Benzoic	$C_6H_5CO_2H$	122	249	4.17
Phthalic	$C_6H_4(CO_2H)_2$	200[a]	—	3.00
Salicylic	$C_6H_4(OH)(CO_2H)$	159	—	3.00

[a] Decomposes.

electrons away from the carbon atom to which it is attached, making this carbon atom more electronegative. The carbon atom in turn attracts electrons from the OH group, making the O—H bond more polar than it is in an alcohol and facilitating the donation of the proton to a water molecule (see Figure 19.7).

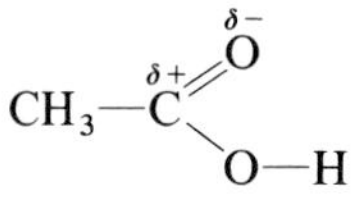

In the anion, $CH_3CO_2^-$, the negative charge is delocalized over both oxygen atoms. This delocalization of charge contributes to the stability of the anion and, therefore, to the tendency of the acid to ionize:

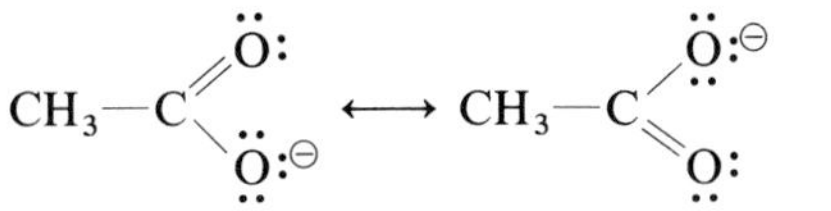

A similar delocalization of negative charge is not possible in the anion of an alcohol, such as CH_3O^-, and therefore alcohols are weaker acids than carboxylic acid. Both CO bonds in the acetate ion are expected to have a bond order of $1\frac{1}{2}$. Their length is consistent with this bond order, as can be seen from the data in Table 19.6.

Important salts of carboxylic acids are sodium benzoate, which is used as a food preservative, and monosodium glutamate (MSG), which is used as a flavor enhancer (Figure 19.8).

Figure 19.7 Acidity of Carboxylic Acids. The acidity of the OH group is enhanced by the presence of the polar $C{=}O$ group.

Table 19.6 Bond Lengths and Bond Orders in Some Molecules Containing CO Bonds

MOLECULE	BOND ORDER	BOND LENGTH (pm)
$CH_3{-}OH$	1.00	143
CO_3^{2-}	1.33	129
$CH_3CO_2^-$	1.50	127
H_2CO	2.00	122
CO	3.00	113

Figure 19.8 Monosodium Glutamate. This compound is used as a meat tenderizer and flavor enhancer.

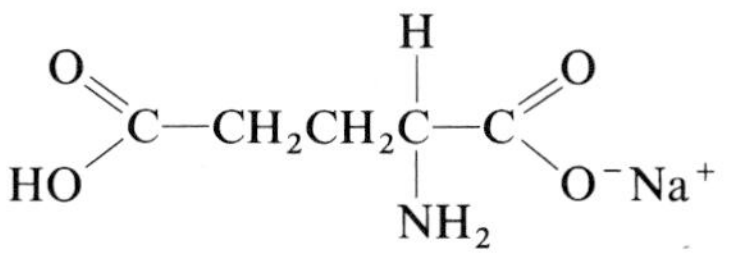

19.6 ESTERS

Carboxylic acids react with alcohols in the presence of an acid as a catalyst to form compounds called **esters**. For example, acetic acid reacts with ethanol to give ethyl acetate (ethyl ethanoate) and water:

$$CH_3\overset{\overset{\displaystyle O}{\|}}{C}-OH + H-OC_2H_5 \longrightarrow CH_3\overset{\overset{\displaystyle O}{\|}}{C}-O-C_2H_5 + H_2O$$

Butanoic acid reacts with ethanol to give *ethylbutanoate*:

$$CH_3-CH_2-CH_2-\overset{\overset{\displaystyle O}{\|}}{C}-OH + H-O-C_2H_5 \longrightarrow$$

$$CH_3-CH_2-CH_2-\overset{\overset{\displaystyle O}{\|}}{C}-O-C_2H_5 + H_2O$$

In general, a carboxylic acid RCO_2H reacts with an alcohol $R'OH$ to give the ester

$$R-\overset{\overset{\displaystyle O}{\|}}{C}-OR'$$

The systematic names for esters are obtained from the names of the acid and the alcohol from which they may be prepared. The ending *-oic* of the acid is replaced by *-oate* and the name is preceded by that of the alkyl group in the alcohol. For example, *methylpropanoate* is prepared from methanol and propanoic acid:

$$CH_3-CH_2-\overset{\overset{\displaystyle O}{\|}}{C}-OH + HO-CH_3 \longrightarrow CH_3-CH_2-\overset{\overset{\displaystyle O}{\|}}{C}-OCH_3 + H_2O$$

Example 19.2 Write an equation for the preparation of pentylbutanoate from the appropriate alcohol and carboxylic acid.

Solution

$$CH_3CH_2CH_2CO_2H + HOCH_2CH_2CH_2CH_2CH_3 \longrightarrow$$

Butanoic acid   1-Pentanol

$$CH_3CH_2CH_2\overset{\overset{\displaystyle O}{\|}}{C}OCH_2CH_2CH_2CH_2CH_3 + H_2O$$

Pentylbutanoate

Many esters have pleasant fruity odors. Indeed, they play an important role in determining the taste and fragrance of many fruits and flowers (see Table 19.7). In general, the flavor and odor of a fruit is due to a combination of many substances, not one compound. For example, it is estimated that more than 250 substances determine the flavor of strawberries. These include alcohols, carboxylic acids, esters, aldehydes, and ketones. Synthetic esters are widely used in perfumes and for artificial flavoring in foods and candies.

Since salicylic acid has both a carboxylic acid group and an —OH group, it can form esters in two ways. It can react with another carboxylic acid such

Table 19.7 Odors of Some Esters

FORMULA	NAME	ODOR
$CH_3\overset{\displaystyle O}{\overset{\|}{C}}—OCH_2CH_2CH_2CH_2CH_3$	Pentylethanoate (amyl acetate)	Banana
$CH_3CH_2CH_2\overset{\displaystyle O}{\overset{\|}{C}}OCH_2CH_3$	Ethylbutanoate (ethyl butyrate)	Pineapple
$CH_3\overset{\displaystyle O}{\overset{\|}{C}}OCH_2(CH_2)_6CH_3$	Octylethanoate	Orange
$CH_3CH_2CH_2\overset{\displaystyle O}{\overset{\|}{C}}OCH_2CH_2CH_2CH_2CH_3$	Pentylbutanoate (amyl butyrate)	Apricot
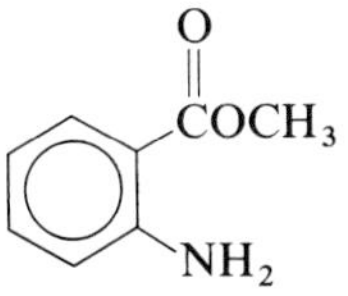	Methyl anthranilate	Grape
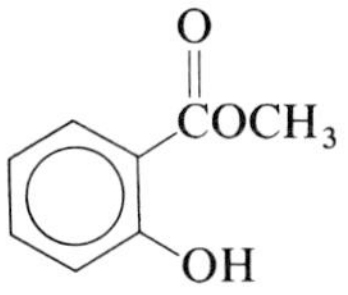	Methyl salicylate	Oil of wintergreen

as acetic acid to give the ester acetylsalicyclic acid, which is aspirin (see Box 19.1):

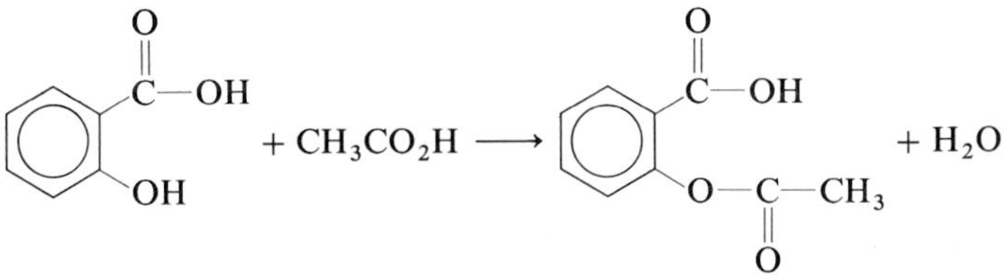

Salicylic acid   Aspirin

Salicyclic acid can also react with alcohols. With methanol the methyl ester, methyl salicylate, is formed. This is also known as *oil of wintergreen* and is used in the manufacture of perfumes and artificial flavors:

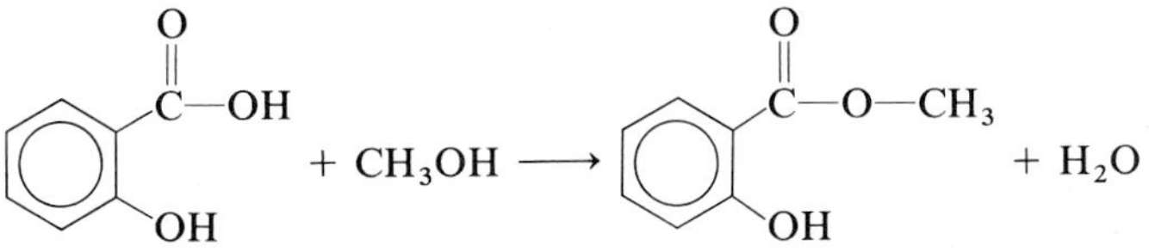

Methyl salicylate (oil of wintergreen)

The synthesis of aspirin and oil of wintergreen starting with the hydrocarbon benzene is a simple illustration of the important procedure called *organic synthesis* in which functional groups are added to a hydrocarbon and modified in a step-by-step procedure until the final desired product is obtained (Box 19.2).

Box 19.2

The production of organic compounds from readily available materials usually requires a number of steps. Each of these steps normally involves either the insertion of a functional group into a hydrocarbon or the conversion of one functional group to another. Each step produces an intermediate product that is used in the next step. The final step produces the desired product.

A simple example of a multistep organic synthesis is provided by the synthesis of aspirin and oil of wintergreen starting from methylbenzene (toluene). In this scheme we have not written a balanced equation for each reaction but we have merely shown each organic substance and the other reactant and conditions needed to bring about the reaction.

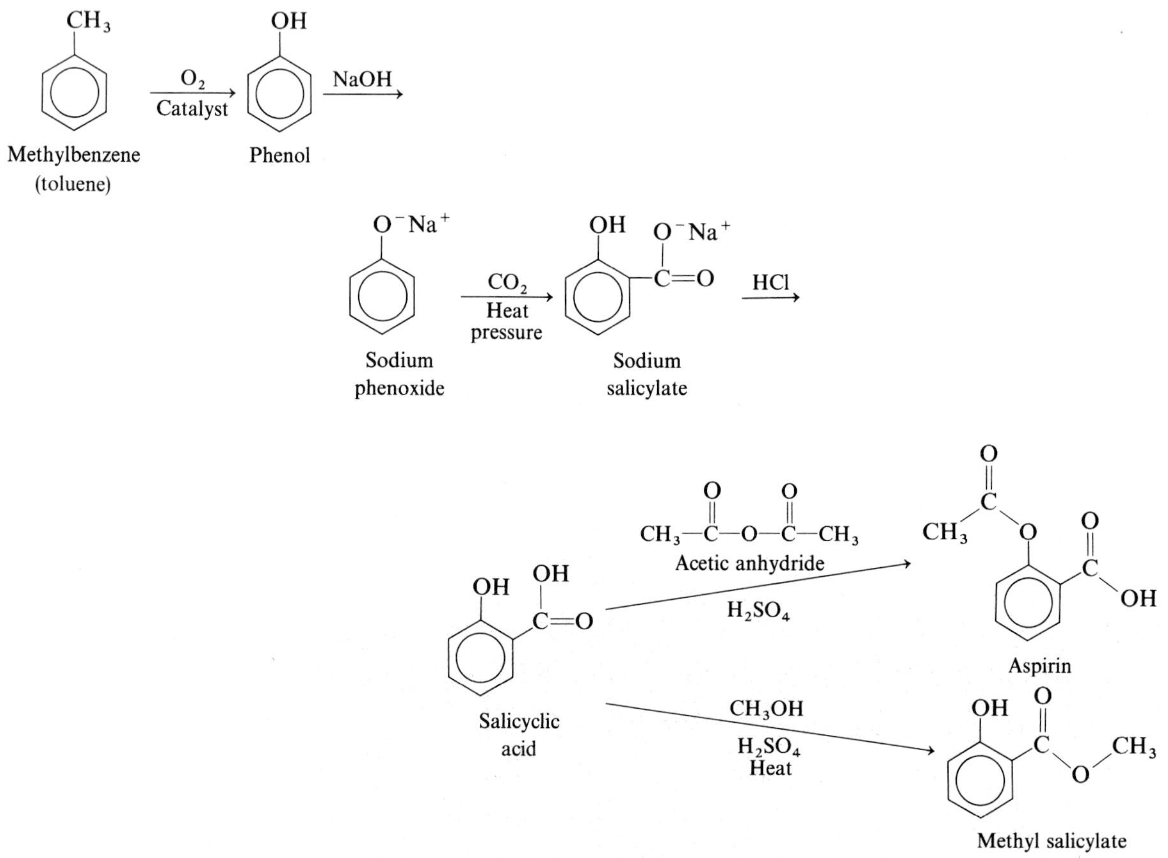

The synthesis of large, complex organic molecules occurring in plants and animals may require a very large number of steps.

There may be many different possible routes to a given compound, and the best choice has to be made considering the following factors:

1. The cost and availability of the starting materials.
2. The number of separate steps needed.
3. The yield and the purity of the product obtained in each step.
4. The ease of separating and purifying the product of each step.

It has sometimes been said that organic synthesis is more of an art than a science. Designing a good organic synthesis, like all creative activity, requires imagination, extensive background knowledge, and experience. It is a challenging field for the chemist and one that has contributed enormously to our material well-being. Chemists have synthesized a truly enormous number of organic compounds—well over 2 million are known today and new ones are being made every day. These include plastics, synthetic fibers, paints, dyes, pesticides, medicines, drugs, perfumes, flavors, preservatives, lubricants, and solvents, all of which are prepared by synthesis from coal, petroleum, and natural gas.

The esters of diols and triols with carboxylic acids are an important group of natural compounds. Animal fats (such as lard and butter) and vegetable oils (such as corn oil, soybean oil, linseed oil, olive oil, and peanut oil) are all esters of 1,2,3-propanetriol (glycerol) and are called *glycerides*. They have the general formula

$$RCO_2CH_2$$
$$R'CO_2CH$$
$$R''CO_2CH_2$$

The solid glycerol esters are called *fats*, and the liquid glycerol esters are called *oils*. The R, R', and R'' groups are hydrocarbon chains with from 3 to 21 carbon atoms. These R groups may be saturated alkyl groups or they may contain one or more C=C double bonds. In general, animal fats contain saturated hydrocarbon chains, whereas vegetable oils have unsaturated hydrocarbon chains with one or more double bonds. In the manufacture of margarine from vegetable oils, some of the double bonds are converted to single bonds by adding hydrogen to them. This raises the melting point and converts an oil to a fat.

Inorganic acids also form esters with alcohols. For example, glycerol reacts with nitric acid to form the nitrate ester glyceroltrinitrate, commonly called nitroglycerine (Box 12.1). The esters of adenosine with di- and triphosphoric acid, ADP and ATP, are important energy storage compounds in biochemical processes (see Figure 19.9).

Figure 19.9 Adenosine Triphosphate. ATP is an ester of triphosphoric acid and adenosine (an alcohol).

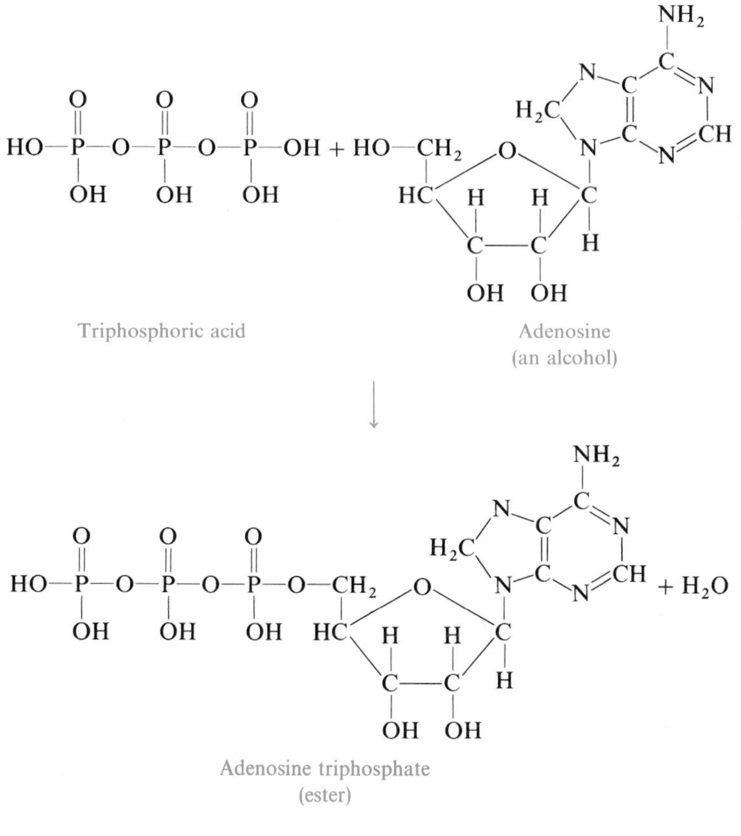

Triphosphoric acid   Adenosine
                                  (an alcohol)

Adenosine triphosphate
(ester)

19.7 AMINES

In the same way that alcohols and ethers may be regarded as derivatives of water, there are analogous compounds derived from ammonia. They are called **amines**. We can distinguish primary, secondary, and tertiary amines, depending on the number of alkyl groups attached to the nitrogen atom; for example,

| Ammonia | Primary amine | Secondary amine | Tertiary amine |

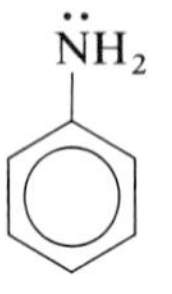

Examples are given in Figure 19.10. All the amines have the same pyramidal AX_3E geometry around the nitrogen as the ammonia molecule has.

The simplest aromatic amine is *aminobenzene*, $C_6H_5NH_2$, commonly known as *aniline*. It is the starting material for the preparation of a number of important dyes called azo dyes.

The methylamines are obtained industrially by the reaction of methanol with ammonia at 400°C in the presence of aluminum oxide as a catalyst:

$$CH_3OH + NH_3 \longrightarrow CH_3NH_2 + H_2O$$
$$CH_3OH + CH_3NH_2 \longrightarrow (CH_3)_2NH + H_2O$$
$$CH_3OH + (CH_3)_2NH \longrightarrow (CH_3)_3N + H_2O$$

Most amines have unpleasant odors. The stench of decaying flesh is due to amines such as putrescine produced by the decomposition of proteins. Some examples of amines are given in Figure 19.11.

The simple aliphatic amines are generally soluble in water, but aniline is insoluble. Like ammonia, they are weak bases (see Table 19.8):

$$CH_3{-}NH_2 + H_2O \rightleftharpoons CH_3{-}NH_3^+ + OH^-$$
$$\text{Methylammonium ion}$$

Aminobenzene
(aniline)

Figure 19.10 Primary, Secondary, and Tertiary Amines.

Methylamine
(primary amine)

Dimethylamine
(secondary amine)

Trimethylamine
(tertiary amine)

Figure 19.11 Some Amines.

$NH_2CH_2CH_2CH_2CH_2NH_2$ — Putrescine

$NH_2CH_2CH_2CH_2CH_2CH_2NH_2$ — Cadaverine

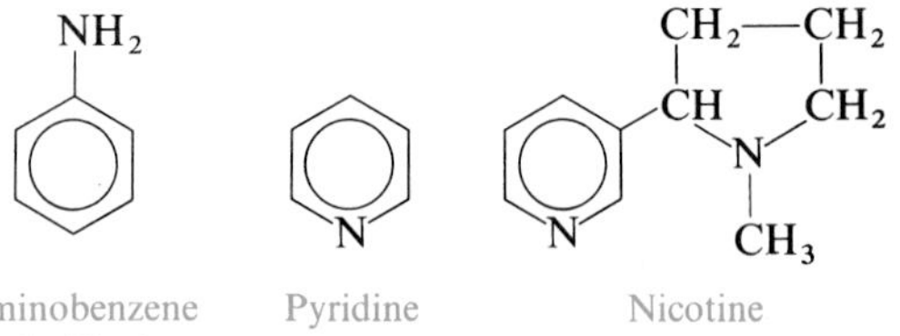

Aminobenzene
(Aniline)   Pyridine   Nicotine

Table 19.8 Properties of Ammonia and Some Amines

COMPOUND	FORMULA	mp (°C)	bp (°C)	pK_b
Ammonia	NH_3	−77.7	−33.4	4.75
Methylamine	CH_3NH_2	−92.5	−6.5	3.43
Dimethylamine	$(CH_3)_2NH$	−96	7.4	3.27
Trimethylamine	$(CH_3)_3N$	−124	3.5	4.19
Aniline	$C_6H_5NH_2$	−6	184	9.37

Basic Properties of Aniline

Aniline is a colorless liquid, but the slightly impure commercial product has a pale yellow-brown color. When it is added to water in the tube on the left, it forms a layer at the bottom of the tube because it is immiscible with water and it has a higher density. Aniline is a weak base. If a small amount of concentrated hydrochloric acid is added, the aniline begins to dissolve (middle tube). When sufficient acid has been added (right tube), a colorless solution of anilinium chloride, $C_6H_5NH_3^+Cl^-$, is obtained.

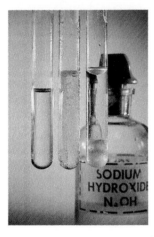

Anilinium ion is a weak acid. When sodium hydroxide solution is added to the colorless solution of anilinium chloride (left tube), the anilinium ion loses its proton and aniline separates out again (middle and right tubes).

They react with acids, such as HCl, to give salts (see Experiment 19.8):

$$C_6H_5NH_2 + HCl \longrightarrow C_6H_5NH_3^+Cl^-$$

Anilinium chloride

19.8 AMIDES

Replacement of the OH group of a carboxylic acid by an NH_2 group gives a compound called an **amide**:

Carboxylic acid   Amide

The simplest members of this family of compounds are

Methanamide   and   Ethanamide
(formamide)   (acetamide)

Amides can be prepared by the reaction of a carboxylic acid with ammonia or an amine to form a salt, which, when heated eliminates water to give the amide. For example,

$$CH_3COOH + NH_3 \longrightarrow CH_3CO_2^-NH_4^+ \longrightarrow CH_3CONH_2 + H_2O$$

Acetic acid   Ammonia   Ammonium acetate   Acetamide

$$CH_3COOH + CH_3NH_2 \longrightarrow CH_3CO_2^-CH_3NH_3^+ \longrightarrow CH_3CONHCH_3 + H_2O$$

Methylamine   Methylammonium acetate   N-Methylacetamide

The amide group

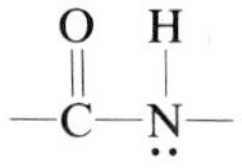

is the key functional group in the structure of proteins and some important polymers. They are discussed in Chapter 23.

An important feature of the amide group is that it has a planar geometry at both the carbon and the nitrogen atoms (see Figure 19.12). We expect a planar AX_3 geometry around the carbon atom, but we might have expected a pyramidal AX_3E geometry at the nitrogen atom. The reason that the geometry is planar is that the unshared pair of electrons on nitrogen is partially used to form a double bond to the adjacent carbon atom:

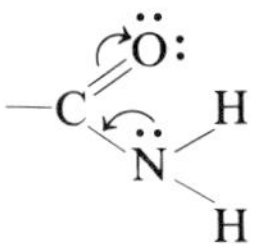

In other words, the amide group may be best represented by two resonance structures:

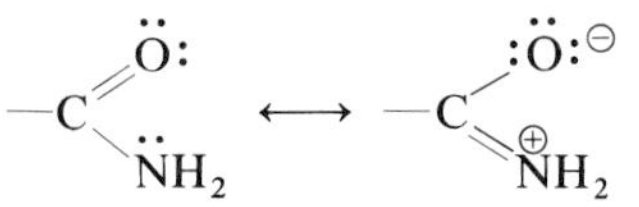

Hence both the carbon and the nitrogen have a planar AX_3 geometry. The C—N bond length of 130 pm in methanamide is intermediate between the C—N single bond length of 147 pm and the C=N double bond length of 124 pm. This short CN bond length is further evidence that the amide group is best described by the two resonance structures above.

Urea,

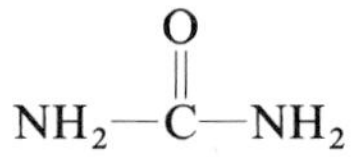

is a major animal waste product that is found in urine. In 1828 German chemist Friedrich Wöhler (1800–1882) made urea by heating ammonium cyanate:

$$NH_4^+OCN^- \xrightarrow{\text{Heat}} NH_2CONH_2$$

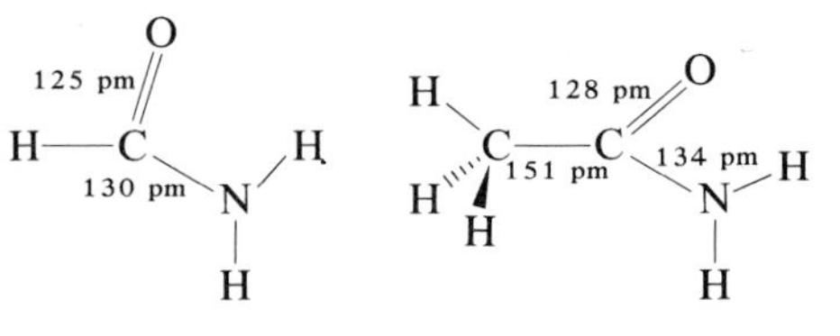

Figure 19.12 Structures of Methanamide and Ethanamide. All the atoms of the amide group CONH$_2$ are in the same plane. Both the C and N atoms of the amide group have a planar AX_3 geometry.

This experiment led to the overthrow of the idea, widely accepted at the time, that organic compounds could only be obtained from living matter.

Urea ranks fifteenth among the industrial chemicals produced in the United States. It is made by heating ammonia with carbon dioxide at 175°–200°C and a pressure of 200 to 400 atm:

$$2NH_3 + CO_2 \longrightarrow H_2N\overset{\overset{O}{\|}}{C}NH_2 + H_2O$$

Because of its high nitrogen content (46% by mass), urea is an important fertilizer. In the soil it decomposes slowly to ammonia. It is also used in manufacturing plastics, foams, and adhesives.

19.9 HALOGEN DERIVATIVES OF ALKANES AND ALKENES

In Chapter 11 we saw that alkanes are relatively unreactive compounds. Their carbon atoms have complete valence shells and cannot accept additional electron pairs. They also have no unshared electron pairs to donate to other atoms. For an alkane to undergo reaction, C—H bonds must be broken; but because a C—H bond is strong (Chapter 12), a large amount of energy is needed to break it. However, at high temperatures the alkanes do undergo some important reactions. Combustion is one; it was discussed in Chapter 11. Another very important reaction is with the halogens. Particularly important is the reaction of methane with chlorine.

Chloromethanes

In the industrial process the reaction of methane and chlorine is carried out at a temperature between 350° and 750°C:

$$CH_4 + Cl_2 \longrightarrow CH_3Cl + HCl$$

The product is *chloromethane* (methyl chloride), CH_3Cl. However, chloromethane is more reactive than methane, so the reaction continues to give *dichloromethane* (methylene dichloride), CH_2Cl_2, *trichloromethane* (chloroform), $CHCl_3$, and *tetrachloromethane* (carbon tetrachloride), CCl_4:

$$CH_3Cl + Cl_2 \longrightarrow CH_2Cl_2 + HCl$$
$$CH_2Cl_2 + Cl_2 \longrightarrow CHCl_3 + HCl$$
$$CHCl_3 + Cl_2 \longrightarrow CCl_4 + HCl$$

These reactions are chain reactions (see Figure 19.13). The relative amounts of the various chloromethanes in the product can be adjusted by changing the proportions of the reactants. The HCl is removed by dissolving it in water, and the chloromethanes are then condensed and separated by fractional distillation. The boiling points of the chloromethanes increase from 24°C for CH_3Cl to 79° for CCl_4, with increasing molecular size and the consequent increase in the strength of the intermolecular (London) forces.

The main use of chloromethane is in the preparation of chloromethyl silanes, such as $(CH_3)_2SiCl_2$, which are important intermediates in the manufacture of silicones (see Chapter 22).

Dichloromethane is a very good solvent, and nearly all its commercial uses depend on this property. For example, it is an important component of paint strippers. In the past *trichloromethane* (*chloroform*) was widely used as an

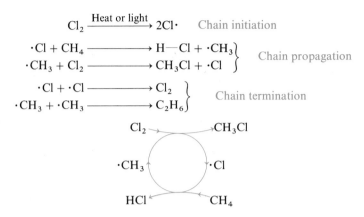

Figure 19.13 Reaction Between Methane and Chlorine. This is a chain reaction involving free-radical intermediates. It can be initiated by heat or light.

anesthetic, but it is no longer an important anesthetic because it is believed to be carcinogenic. Its main use is as an intermediate in the manufacture of chlorofluoromethanes.

Tetrachloromethane (*carbon tetrachloride*) was formerly used as a dry-cleaning agent and in fire extinguishers. It has been replaced by other substances for these purposes because it is toxic and at high temperatures it reacts with the oxygen of the air to form the highly toxic gas phosgene, $Cl_2C{=}O$ (Chapter 11).

Other Chloroalkanes and Chloroalkenes

Chloroethane, C_2H_5Cl, can be made by the chlorination of ethane, but it is more commonly made by adding HCl to ethene:

$$H_2C{=}CH_2 + HCl \longrightarrow C_2H_5Cl$$

A major use of chloroethane is for the manufacture of tetraethyl lead, $Pb(C_2H_5)_4$, which is added to gasoline as an antiknock agent, although this use is now declining.

Addition of chlorine to ethene gives *1,2-dichloroethane*:

$$H_2C{=}CH_2 + Cl_2 \longrightarrow ClCH_2{-}CH_2Cl$$

1,2-Dichloroethane ranks sixteenth among the industrial chemicals produced in the United States. Almost all of it is used to produce *chloroethene* (*vinyl chloride*), another very important industrial chemical. When dichloroethane is heated, HCl is eliminated.

$$ClCH_2{-}CH_2Cl \xrightarrow[\text{Heat}]{} H_2C{=}CHCl + HCl$$
$$\text{Chloroethene}$$
$$\text{(vinyl chloride)}$$

Vinyl chloride, which ranks nineteenth among industrial chemicals, is used mainly for the manufacture of polyvinyl chloride (PVC) (see Chapter 23).

Several *chlorofluoromethanes* and *chlorofluoroethanes* have important uses as refrigerator fluids and as aerosol propellants in cans of spray polish, shaving cream, and so on. These compounds are commonly called Freons and include CF_2Cl_2 (bp $= -30°C$) and $CClF_2CClF_2$ (bp $= 24°C$). As we discussed in Chapter 17, there has been considerable controversy over whether these substances are decreasing the concentration of ozone in the upper atmosphere.

Tetrachloroethene, $Cl_2C{=}CCl_2$, has replaced CCl_4 in "dry" cleaning because of its lower volatility and lower toxicity. *Tetrafluoroethene*, C_2F_4, on polymerization yields the plastic Teflon, which is widely used because of its inertness and resistance to heat. Teflon-lined nonstick pans are familiar in the kitchen.

CHAPTER 19
ORGANIC CHEMISTRY

712

In this final section we review the functional groups and some of the reactions by which they are inserted into alkanes and by which they may be interconverted.

The simple functional groups OH and OR may be thought of as being derived from water and NH_2 as being derived from ammonia. Table 19.9 shows that five more important functional groups contain the carbonyl group, $C=O$. The carbonyl group may be combined with H or R to give aldehydes and ketones and with the simple functional groups OH, OR, and NH_2 to give carboxylic acids, esters, and amides, respectively.

The various reactions by which simple one, two, and three-carbon compounds containing these functional groups may be obtained from methane, ethene, and propene are summarized in Figures 19.14 and 19.15.

Table 19.9 Relationships Between Functional Groups

—H		$\overset{\overset{\textstyle O}{\|\|}}{-C-H}$	Aldehyde
—R	Alkyl	$\overset{\overset{\textstyle O}{\|\|}}{-C-R}$	Ketone
—OH	Alcohol	$\overset{\overset{\textstyle O}{\|\|}}{-C-OH}$	Carboxylic acid
—OR	Ether	$\overset{\overset{\textstyle O}{\|\|}}{-C-OR}$	Ester
—NH_2	Amine	$\overset{\overset{\textstyle O}{\|\|}}{-C-NH_2}$	Amide

Figure 19.14 Some Products Made from Methane and Methanol.

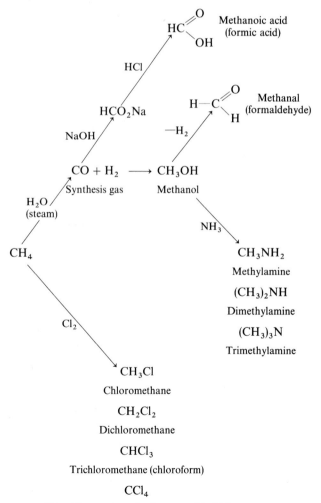

Figure 19.15 Some Products Made from Ethene and Propene.

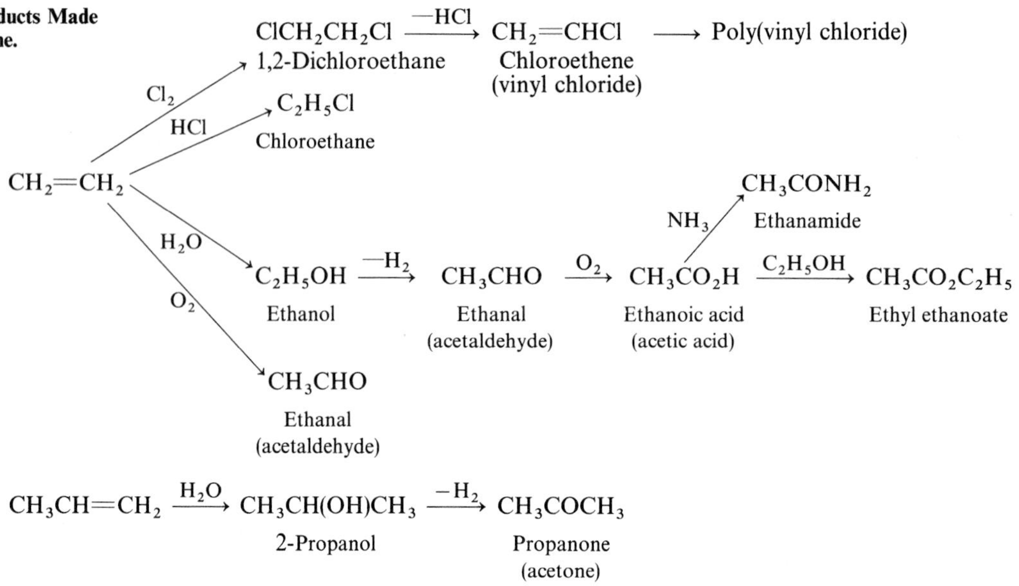

$$ClCH_2CH_2Cl \xrightarrow{-HCl} CH_2{=}CHCl \longrightarrow Poly(vinyl\ chloride)$$

1,2-Dichloroethane → Chloroethene (vinyl chloride)

$$\xrightarrow{Cl_2}$$

$$\xrightarrow{HCl} C_2H_5Cl$$
Chloroethane

$$CH_2{=}CH_2$$

$$\xrightarrow{H_2O}$$

$$C_2H_5OH \xrightarrow{-H_2} CH_3CHO \xrightarrow{O_2} CH_3CO_2H \xrightarrow{C_2H_5OH} CH_3CO_2C_2H_5$$
Ethanol → Ethanal (acetaldehyde) → Ethanoic acid (acetic acid) → Ethyl ethanoate

$$\xrightarrow{NH_3} CH_3CONH_2$$
Ethanamide

$$\xrightarrow{O_2} CH_3CHO$$
Ethanal (acetaldehyde)

$$CH_3CH{=}CH_2 \xrightarrow{H_2O} CH_3CH(OH)CH_3 \xrightarrow{-H_2} CH_3COCH_3$$
2-Propanol → Propanone (acetone)

IMPORTANT TERMS

A **dehydrogenation reaction** is one in which a hydrogen molecule is eliminated from a larger molecule.

Fermentation is a reaction catalyzed by certain enzymes in which the complex molecules of sugars and carbohydrates are broken down into ethanol and carbon dioxide.

A **functional group** is an atom or group of atoms that replaces a hydrogen atom of a hydrocarbon; for example, the hydroxyl group, OH, and the carboxyl group, CO_2H, are functional groups.

Synthesis gas is a mixture of hydrogen and carbon monoxide that is obtained by heating methane with steam at 700–800°C in the presence of a nickel catalyst.

Thermal cracking is a process in which alkanes are broken down on heating to give shorter-chain alkanes, alkenes, and hydrogen.

PROBLEMS

Names and Formulas

1. Give the IUPAC names for each of the following alcohols:

(a) CH_3CHCH_2OH
     |
    CH_3

(b) CH_3CHCH_2Cl
     |
    OH

(c) $CH_3CH_2CHC(CH_3)_3$
       |
      OH

(d) $CH_3CHCH_2CHCH_3$
     | |
    OH OH

2. Write structural formulas for each of the following compounds:

(a) 2-Methyl-2-butanol
(b) 5-Methyl-2-hexanol
(c) 2-Chloro-1-propanal
(d) 3-Chlorocyclohexanol

3. Classify the following alcohols as primary, secondary, or tertiary:

(a) $CH_3CH_2CH_2OH$

(b) $CH_3CH_2C{-}CH_2CH_3$
        CH_3 (top) |
        OH

(c) $CH_3{-}CHCH_2CH_3$
        |
      OH

(d) $CH_3CH_2C{-}H$
       CH_3 (top) |
       OH

4. Give the IUPAC names for each of the following aldehydes and ketones:

(a) CH_3CH_2CHO

(b) CH_3CH_2CHCHO
      CH_3 |

(c) $CH_3CH_2\overset{O}{\overset{\|}{C}}CH_3$

(d) $CH_3CH{-}\overset{O}{\overset{\|}{C}}{-}CH_2CH_3$
    CH_3 |

5. Write structural formulas for each of the following aldehydes and ketones:

(a) Butanal
(b) 2-Pentanone
(c) 3-Methyl-2-butanone
(d) 3,3-Dimethylhexanal

6. Write structural formulas for each of the following:

(a) 1,2,2-Trichlorobutane

(b) 1,3,5-Tribromobenzene

(c) 1,2-Dichlorocyclohexane

7. Write structural formulas and give the systematic names of the isomers of $C_3H_6Cl_2$ and C_4H_9Br.

8. Identify and name the functional groups in the following molecules:

(a) H_3C-CHO

(b) $ClCH_2CH_2\underset{\underset{\displaystyle OH}{|}}{\overset{\overset{\displaystyle H}{|}}{C}}-CH_2-CH_2-\overset{\overset{\displaystyle O}{\|}}{C}-OH$

(c)
$$\underset{}{\overset{OH}{\bigcirc}}\overset{\overset{H}{|}}{\underset{}{C}}=O$$

(d) $CH_3\overset{\overset{\displaystyle O}{\|}}{C}-CH_2-CH_2-CO_2H$

(e) $CH_3O-\overset{\overset{\displaystyle O}{\|}}{C}-CH_2-CH_2-CH_3$

(f) $CH_3CH_2\underset{\underset{\displaystyle NH_2}{|}}{C}=O$

9. Name the following compounds:

(a) CH_3OH

(b) CH_3CH_2OH

(c) HCO_2H

(d) $CH_3CH_2\underset{\underset{\displaystyle OH}{|}}{C}HCH_3$

(e) CH_3CH_2CHO

(f) $\underset{\underset{\displaystyle OH}{|}}{C}H_2-\underset{\underset{\displaystyle OH}{|}}{C}H_2$

(g) CH_3COCH_3

(h) $CH_3-\underset{\underset{\displaystyle CH_3}{|}}{C}H-CHO$

(i) $\underset{\underset{\displaystyle Cl}{|}}{C}H_2-\underset{\underset{\displaystyle Br}{|}}{C}H-CH_3$

Molecular Geometry

10. In terms of VSEPR theory, classify the following molecules and ions in terms of the geometry around the starred atoms as AX_2, AX_3, AX_4, AX_3E, or AX_2E_2, and draw a diagram to show the geometry in each case.

(a) $\overset{*}{C}H_3OH$

(b) $CH_3\overset{*}{O}CH_3$

(c) $H\overset{*}{C}O_2H$

(d) $H\overset{*}{C}O_2$

(e) $CH_3\overset{*}{C}ONH_2$

(f) $(CH_3)_2\overset{*}{C}O$

(g) $CH_3\overset{*}{N}H_2$

(h) $[(CH_3)_4\overset{*}{N}]^+$

(i) $(CH_3)_3\overset{*}{N}$

11. Explain why all five atoms in the amide group $-CONH_2$ are in the same plane.

Reactions

12. The following aldehydes and ketones may be prepared by oxidation of a suitable alcohol. In each case, name the alcohol, and draw its structural formula.

(a) Ethanal   (b) Propanone

(c) 2-Methylpropanal   (d) 2-Pentanone

13. Draw the structures of the esters formed from the following:

(a) Ethanoic acid (acetic acid) and 2-propanol.

(b) Ethanoic acid (acetic acid) and 1-butanol.

(c) Benzoic acid and methanol.

(d) Methanoic acid (formic acid) and ethanol.

14. (a) Describe, giving a suitable balanced equation, how synthesis gas is made.

(b) Describe, giving a suitable equation, how methanol is made from synthesis gas.

15. Complete and balance the following equations:

(a) $C_2H_5OH(l) + O_2(g) \xrightarrow{\text{Catalyst}}$

(b) $CH_3CH_2CH_2OH(l) + NaOH(s) \longrightarrow$

(c) $CH_3OH(l) + Na(s) \longrightarrow$

(d) $C_2H_5OH(l) + HBr(g) \longrightarrow$

(e) $C_2H_5OH(l) + H_2SO_4(conc) \longrightarrow$

(f) $C_2H_5OH(g) + O_2(g) \xrightarrow{\text{Heat}}$

(g) $CH_3CH=CH_2(g) + H_2O(g) \xrightarrow[\text{Catalyst}]{\text{Heat, pressure}}$

16. Complete and balance the following equations:

(a) $C_2H_5ONa + CH_3Br \longrightarrow$

(b) $C_2H_5OH \xrightarrow{\text{Heat, catalyst (Cu or Ag)}}$

(c) $CH_3\underset{\underset{\displaystyle OH}{|}}{C}HCH_3 \xrightarrow{K_2Cr_2O_7(H_3O^+),\ heat}$

(d) $CH_3CH_2COCH_3 + H_2 \xrightarrow{\text{Pt catalyst, heat}}$

(e) $CH_3CH_2COCH_3 + H_2NNH_2 \longrightarrow$

17. Complete and balance the following equations:

(a) $CH_3CH_2CHO \xrightarrow{K_2Cr_2O_7(H_3O^+),\ heat}$

(b) $CO + NaOH \xrightarrow{200°C}$

(c) $CH_3CH_2CO_2H + (CH_3)_2CHOH \xrightarrow{\text{Heat}}$

(d) $CH_3OH + NH_3 \xrightarrow{\text{Heat}}$

(e) $CH_3NH_2 + H_2O(l) \longrightarrow$

(f) $CH_3NH_2 + HCl(aq) \longrightarrow$

(g) $CH_3CO_2H + C_2H_5NH_2 \xrightarrow{\text{Heat}}$

18. (a) What is the overall equation for the production of methanol from methane and water via synthesis gas?

(b) The excess H_2, which appears as a product in the equation in part (a), can be used to convert CO_2 to CO via the reaction

$$H_2 + CO_2 \longrightarrow CO + H_2O$$

This CO then reacts with the remaining H_2 to produce more methanol. Deduce what fraction of the H_2 should be caused to react with CO_2 to use all the H_2 and CO in synthesis gas to form methanol.

19. Write an equation for the reaction by which the ester methyl methanoate (methyl formate) can be produced from the appropriate acid and alcohol.

20. Give the structural formula and the name of the product of the following reactions:

(a) $CO(g) + 2H_2(g) \xrightarrow{\text{ZnO catalyst}}$

(b) $\begin{matrix} CH_3 \\ \diagdown \\ CH_3 \diagup \end{matrix} C=C \begin{matrix} CH_3 \\ \diagup \\ \diagdown H \end{matrix} + Br_2 \longrightarrow$

(c) $\begin{matrix} CH_3 \\ \diagdown \\ CH_3 \diagup \end{matrix} C=C \begin{matrix} CH_3 \\ \diagup \\ \diagdown H \end{matrix} + H_2 \longrightarrow$

(d) $CH_3OH + C_2H_5OH \xrightarrow{H_2SO_4}$

(e) $CH_3CH_2CH_2OH + Na \longrightarrow$

21. How would you prepare vinyl chloride (chloroethene) starting from ethene and chlorine?

22. Describe, with suitable equations, how methane can be converted to methanoic acid (formic acid).

23. Describe, with suitable equations, how ethene may be converted to ethyl ethanoate.

24. Outline the steps by which aspirin may be prepared from toluene (methylbenzene).

25. Each of the following compounds can be prepared from an alkene or an alkyne and another reactant. In each case, give the name and structure of the alkene or alkyne and the other reactant.

(a) CH_3CHCH_3 (b) $CH_3CBr_2CBr_2CH_3$
     |
     OH

(c) $CH_3CH{=}CHCH_3$

(d) $CH_3C{=}CH_2$
     |
     Br

26. How would you prepare the following?

(a) 1,2-Dichloroethane from ethene

(b) Propanoic acid from 1-propanol

(c) Ethyl ethanoate from ethanal

27. Propane reacts with bromine when irradiated with light of a suitable wavelength. Write a series of equations to describe the mechanism of this reaction.

Miscellaneous

28. A compound is found, on analysis, to have the molecular formula $C_5H_{12}O$. On oxidation it is converted to a compound with the molecular formula $C_5H_{10}O$, which has the reactions of a ketone. Give two or more reasonable structures for the compound $C_5H_{12}O$.

29. (a) What is the pH of a 0.10M solution of methylamine ($pK_b = 3.34$)? What is the pK_a of the $CH_3NH_3^+$ ion?

(b) What is the pH of a solution obtained by adding 24.5 mL of 0.090M methylamine to 10.00 mL of a solution containing 0.10M hydrogen chloride?

30. Calculate the enthalpies of combustion of methanol and ethanol from the standard enthalpies of formation in Table 12.1 and ΔH_f° $CH_3OH = -239$ kJ mol^{-1}, and ΔH_f° $C_2H_5OH = -277$ kJ mol^{-1}.

31. An organic compound was found to contain carbon, hydrogen, and oxygen and analysis gave 52.1% C and 13.1% H. A mass of 0.230 g of the compound occupied a volume of 153 mL at 100°C and a pressure of 1.00 atm. Calculate the empirical formula and molecular formula of the compound, and draw possible structures. The compound was found to react with sodium to give hydrogen and a salt. Write a balanced equation for this reaction, and name the compound. What volume of hydrogen would result from the reaction of 0.250 g of the compound with excess sodium at 25°C and a pressure of 750 torr?

32. Explain the differences in the following boiling points: water, 100°C; methanol, 65°C; ethanol, 78°C; methane, -162°C; ethane, -89°C; dimethyl ether, -25°C.

33. The representative alkane of gasoline is nonane, C_9H_{20}, which has a density of 0.72 g cm^{-3} and a ΔH_f° of -275 kJ mol^{-1}. Ethanol has a density of 0.79 g cm^{-3} and a ΔH_f° of -277 kJ mol^{-1}. Compare the heat energy obtained by burning gasoline with that obtained by burning ethanol (a) on the basis of equal volumes and (b) on the basis of equal masses.

34. Would you expect the structural isomers dimethyl ether, $CH_3{-}O{-}CH_3$, and ethanol, CH_3CH_2OH, to differ significantly in their boiling points? Why? What simple chemical test could be used to distinguish between them?

35. A 156-mg sample of an organic compound was burned in excess oxygen to give 512 mg of CO_2 and 209 mg of H_2O. At 100°C, 156 mg of this substance occupied a volume of 93.3 mL at a pressure of 882 torr. Suggest a suitable structure for the compound, and identify the functional group.

CHAPTER 20

THE NOBLE GASES: MORE CHEMISTRY OF THE HALOGENS

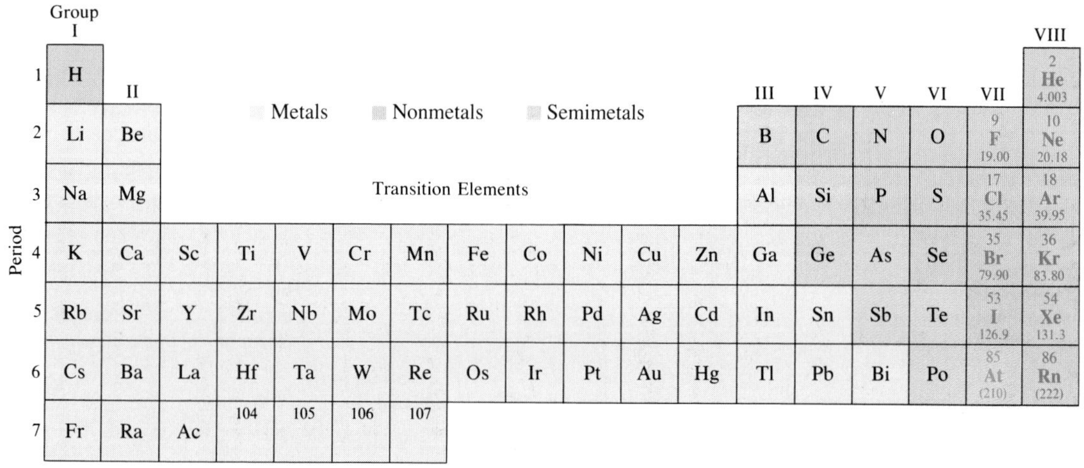

One of the most startling discoveries in chemistry in recent times occurred in 1962, when Neil Bartlett demonstrated that xenon is capable of forming compounds. This discovery greatly surprised most chemists, since it had become generally accepted that the noble gases did not form compounds. In fact, the lack of compound formation by the noble gases, all of which, except helium, have eight electrons in their valence shells, had led to the idea that a valence shell of eight electrons is a particularly stable arrangement. This idea was the basis of Lewis's octet rule. Following the general acceptance of Lewis's ideas, very few attempts were made to prepare compounds of the noble gases, although a few chemists predicted that compounds of at least the heavier noble gases should exist.

The chemistry of the noble gases is closely related to the chemistry of the halogens in their higher oxidation states. In Chapter 5 we restricted our attention almost exclusively to the -1 oxidation state of the halogens, in particular, the halide ions and the hydrogen halides. We now take up some further aspects of the chemistry of the halogens, including important compounds such as the chlorates, perchlorates, and iodates. In fact, by extrapolating from the known compounds of the halogens and the elements of groups V and VI in their higher oxidation states, some chemists had predicted the probable existence of compounds of the noble gases.

20.1 HALIDES OF NONMETALS IN THEIR HIGHER OXIDATION STATES

In Chapter 7 we discussed the higher oxidation states of phosphorus and sulfur, namely, the $+5$ state of phosphorus and the $+4$ and $+6$ states of sulfur. These higher oxidation states are possible because these elements can have more than eight electrons in their valence shells. Recall that compounds containing the element in these higher oxidation states can be regarded as being formed from a valence state in which one or more electrons are promoted to the 3d orbitals (see Figure 20.1). Only the most electronegative ligands such as oxygen, fluorine, and chlorine can form these higher–oxidation state compounds because only these ligands are sufficiently electronegative to pull one or more electrons away from s and p orbitals into 3d orbitals. In addition to the -1 and $+1$ oxidation states that arise from the ground state electron configuration, chlorine, bromine, and iodine have $+3$, $+5$, and $+7$ oxidation states, which correspond to valence state electron configurations in which one, two, or three electrons have been promoted to d orbitals (see Figure 20.1 and Table 20.1).

Fluorine is such a strong oxidizing agent that it can oxidize most nonmetals to their highest oxidation states. Thus fluorine reacts with phosphorus to give PF_3 and PF_5 and with sulfur to give SF_4 and SF_6. Fluorine oxidizes even the other halogens, chlorine, bromine, and iodine, to higher oxidation states. It

Figure 20.1 Ground State and Valence State Electron Configurations for Phosphorus, Sulfur, and Chlorine.

				Number of unpaired electrons	Example compounds
P	Ground state			3	PCl_3, PH_3
	Valence state			5	PCl_5, H_3PO_4
S	Ground state			2	SCl_2, H_2S
	Valence state			4	SF_4, SO_2
	Valence state			6	SF_6, SO_3, H_2SO_4
Cl	Ground state			1	ClF, $HOCl$
	Valence state			3	ClF_3, $HClO_2$
	Valence state			5	ClF_5, $HClO_3$
	Valence state			7	$HClO_4$

Table 20.1 Classification of Compounds of the Halogens by Oxidation State

OXIDATION STATE	TYPICAL COMPOUNDS
-1	NaCl, HCl, HI
0	F_2, Cl_2, Br_2, I_2
$+1$	HClO, NaClO, ClF
$+3$	$HClO_2, NaClO_2, ClF_3, BrF_3$
$+5$	$HClO_3, NaClO_3, BrF_5, IF_5$
$+7$	$HClO_4, NaClO_4, H_5IO_6, IF_7$

oxidizes chlorine and bromine to the $+1$, $+3$, and $+5$ oxidation states in ClF, ClF_3, and ClF_5, and BrF, BrF_3, and BrF_5; and iodine to IF_3 and IF_5, and to the $+7$ state in IF_7. Chlorine reacts with phosphorus to give PCl_3 and PCl_5 (see Experiment 20.1), but since it is a weaker oxidizing agent than fluorine, it gives chlorides of sulfur only in lower oxidation states, namely, S_2Cl_2 and SCl_2 (Chapter 7). It oxidizes iodine to ICl and ICl_3 but not to the $+5$ and $+7$ states (see Experiment 20.2). These compounds of fluorine and chlorine with the other halogens are known as **interhalogen compounds**.

The nonmetal halides are typical covalent substances, consisting of covalent molecules and having relatively low melting points and boiling points. Some of their physical properties are summarized in Table 20.2. Phosphorus pentachloride resembles $AlCl_3$ in that it is ionic in the solid state but covalent in the gas phase (Chapter 9). It consists of covalent PCl_5 molecules in the gas phase but is an ionic crystal composed of PCl_4^+ and PCl_6^- ions in the solid state (Figure 20.2).

Properties

Most nonmetal halides are rather reactive compounds. They almost all react very readily with water to give a hydrogen halide and either an oxoacid or an

---

EXPERIMENT 20.1

Preparation of Phosphorus Pentachloride, PCl₅

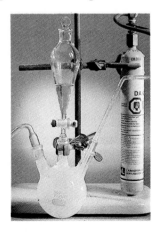

Phosphorus pentachloride may be prepared from the reaction of phosphorus trichloride with chlorine. Here, colorless liquid PCl_3 is being slowly dripped into a flask into which gaseous chlorine is being passed. White (very pale yellow) PCl_5 is formed immediately and may be seen coating the sides of the flask.

20.1 HALIDES OF
NONMETALS IN THEIR
HIGHER OXIDATION STATES

Preparation of Iodine Monochloride and Iodine Trichloride

Chlorine is passed into a flask containing solid iodine.

Red liquid ICl and a small amount of orange solid ICl_3 are rapidly formed.

Table 20.2 Halides of Phosphorus, Sulfur, and the Halogens

Phosphorus		
PF_3	Colorless gas	bp = $-95°C$
PF_5	Colorless gas	bp = $-84°C$
PCl_3	Colorless liquid	bp = $76°C$
PCl_5	White solid	Sublimes at $163°C$
Sulfur		
SF_4	Colorless gas	bp = $-38°C$
SF_6	Colorless gas	bp = $-64°C$
S_2Cl_2	Yellow liquid	bp = $138°C$
SCl_2	Red liquid	bp = $59°C$
Chlorine		
ClF	Colorless gas	bp = $-100°C$
ClF_3	Colorless gas	bp = $12°C$
ClF_5	Colorless gas	bp = $-14°C$
Bromine		
BrF	Pale brown gas	bp = $20°C$
BrF_3	Yellow liquid	bp = $126°C$
BrF_5	Colorless liquid	bp = $41°C$
Iodine		
IF_3	Yellow solid	bp = $28°C$ (decomposes)
IF_5	Colorless liquid	bp = $100°C$
IF_7	Colorless gas	Solid sublimes at $5°C$
ICl	Red solid	mp = $27°C$
ICl_3	Orange solid	Sublimes at $64°C$

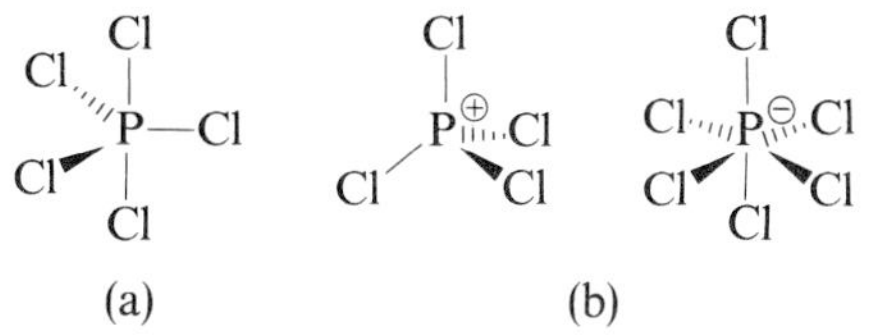

Figure 20.2 Structure of Phosphorus Pentachloride. (a) In the gas phase PCl_5 consists of PCl_5 molecules. (b) In the solid state it is an ionic crystal composed of tetrahedral PCl_4^+ ions and octahedral PCl_6^- ions.

oxide; for example,

$$PCl_5(s) + 4H_2O(l) \longrightarrow H_3PO_4(aq) + 5HCl(aq)$$

$$SF_4(g) + 2H_2O(l) \longrightarrow SO_2(aq) + 4HF(aq)$$

$$ClF_5(g) + 3H_2O(l) \longrightarrow HClO_3(aq) + 5HF(aq)$$

One exception is SF_6, which is extremely inert and is not attacked by water even at 500°C. It appears that six fluorine atoms completely fill all the space around the sulfur atom. There is therefore no space for a water molecule to approach close enough to the sulfur atom to donate an electron pair into its valence shell. In its inertness to water SF_6 resembles CCl_4 and CF_4, two other nonmetal halides that do not react with water. Four chlorine or fluorine atoms completely surround a carbon atom, leaving no room for attack by a water molecule.

Another property of many nonmetal halides is their ability either to lose or to gain halide ions to form cations and anions; for example,

$$PCl_5 \longrightarrow PCl_4^+ + Cl^-$$

$$PCl_5 + Cl^- \longrightarrow PCl_6^-$$

The formation of the ionic solid $PCl_4^+PCl_6^-$ from gaseous PCl_5 molecules can be regarded as a transfer of a chloride ion from one PCl_5 molecule to another. Other examples include

$$BrF_3(l) + KF(s) \longrightarrow K^+BrF_4^-(s)$$

$$ICl_3(s) + AlCl_3(s) \longrightarrow ICl_2^+AlCl_4^-(s)$$

$$ICl_3(s) + KCl(s) \longrightarrow K^+ICl_4^-(s)$$

$$ICl(s) + KCl(s) \longrightarrow K^+ICl_2^-(s)$$

Anions such as BrF_4^-, ICl_4^-, and ICl_2^- are called **polyhalide ions**. Other anions of this type are I_3^-, which is formed by the addition of I^- to I_2, and Br_3^-, which is formed by the similar reaction of Br^- with Br_2. The formation of I_3^- was demonstrated in Experiment 5.6.

Molecular Shapes

In Chapter 8 we discussed the VSEPR theory, which shows how the shapes of molecules depend on the number of electron pairs in the valence shell of the central atom. In that chapter we concentrated on valence shells containing two, three, and four pairs of electrons. The Lewis structures of some nonmetal halides are given in Figure 20.3.

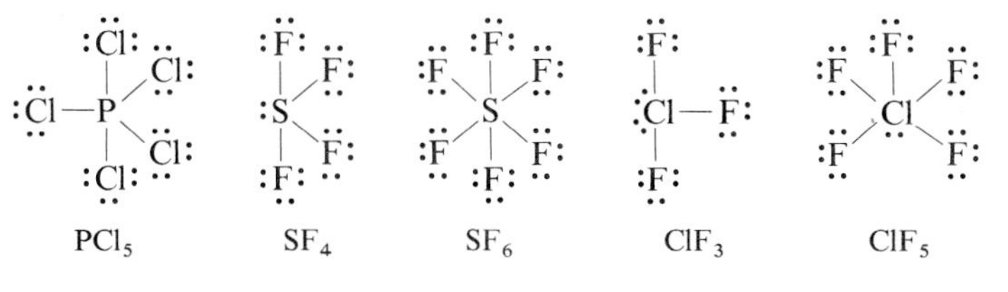

Figure 20.3 Lewis Structures of Some Nonmetal Halides.

Table 20.3 Shapes of Molecules Based on Valence Shells Containing Five and Six Electron Pairs

NUMBER OF ELECTRON PAIRS	NUMBER OF UNSHARED PAIRS	TYPE OF MOLECULE	MOLECULAR SHAPE	EXAMPLE
5	0	AX_5	Trigonal bipyramid	PF_5, PCl_5
	1	AX_4E	Disphenoid	SF_4
	2	AX_3E_2	T shape	ClF_3
	3	AX_2E_3	Linear	ICl_2^-, XeF_2
6	0	AX_6	Octahedron	SF_6
	1	AX_5E	Square pyramid	BrF_5, IF_5
	2	AX_4E_2	Square	ICl_4^-, XeF_4

To understand the shapes of the molecules of the halides of the nonmetals in their higher oxidation states, we need to extend our previous discussion of VSEPR theory (Chapter 8) and to consider the shapes of molecules that are based on a central atom having five or six electron pairs in its valence shell. These shapes are summarized in Table 20.3.

VALENCE SHELLS CONTAINING SIX ELECTRON PAIRS: AX_6, AX_5E, AND AX_4E_2 MOLECULES The most probable arrangement of six electron pairs is at the corners of an octahedron. An AX_6 molecule therefore has an octahedral shape in which all six bonds are equivalent and make angles of 90° with their nearest neighbors. Sulfur hexafluoride, SF_6, and the anion PCl_6^- are examples of octahedral AX_6 species (Figure 20.4). Recall that many hydrated ions such as $Al(H_2O)_6^{3+}$ and $Mg(H_2O)_6^{2+}$ also have this shape.

If one of the electron pairs of an octahedral arrangement is a nonbonding pair E, the molecule is of the AX_5E type and has a *square pyramidal shape*, as Figure 20.5 illustrates. The halogen pentafluorides, ClF_5, BrF_5, and IF_5, are examples of this type of molecule. Because the nonbonding electron pair is

Figure 20.4 Two Octahedral AX_6 Species, SF_6 and PCl_6^-.

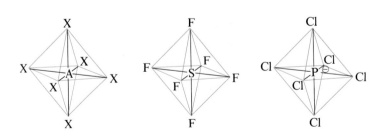

Figure 20.5 Some Square Pyramidal AX_5E Molecules, ClF_5, BrF_5, and IF_5.

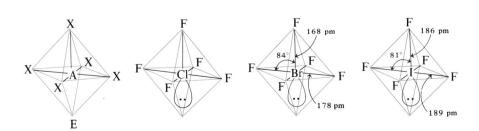

722

larger than the bonding pairs (Chapter 8), the molecules are slightly distorted from the ideal square pyramidal shape. All the bond angles are slightly smaller than the ideal angle of 90°, and the bonds in the square base are slightly longer than the axial bond. The large nonbonding pair pushes adjacent bonding pairs away from the core, thus increasing these bond lengths, but the nonbonding pair is too far away from the axial bond pair to affect it appreciably.

In an AX_4E_2 molecule the two nonbonding pairs are at opposite corners of the octahedron so that the molecule has a *square planar shape*. Because of their large size, the unshared pairs tend to keep as far apart as possible. They therefore occupy opposite (trans) positions of the octahedron rather than adjacent (cis) positions. An example is provided by ICl_4^- (see Figure 20.6.)

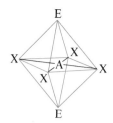

VALENCE SHELLS CONTAINING FIVE ELECTRON PAIRS: AX_5, AX_4E, AX_3E_2, AND AX_2E_3 MOLECULES

The most probable arrangement of five electron pairs is at the corners of a trigonal bipyramid. An AX_5 molecule therefore has a trigonal bipyramidal shape. Phosphorus pentachloride and phosphorus pentafluoride are examples of trigonal bipyramidal AX_5 molecules (see Figure 20.7).

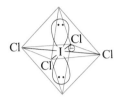

Figure 20.6 Square Planar AX_4E_2 Molecule: ICl_4^-.

Of the five bonds, three are equivalent *equatorial* bonds in the same plane, and the other two are equivalent *axial* bonds. The equatorial bonds make angles of 120° with each other, and the axial bonds are colinear and make angles of 90° with the equatorial bonds. The axial and equatorial bonds are not geometrically equivalent. The two axial bonds are longer than the three equatorial bonds, because the axial positions, which have three nearest neighbors at 90°, are more crowded than the equatorial positions, which have only two nearest neighbors at 90°. The axial pairs therefore tend to be forced further away from the central core than the equatorial pairs (Figure 20.7).

Because the equatorial positions are less crowded than the axial positions, there is a strong preference for larger nonbonding pairs of electrons to occupy the equatorial rather than the axial positions of a trigonal bipyramidal arrangement. Thus in AX_4E, AX_3E_2, and AX_2E_3 molecules the nonbonding pairs of electrons always occupy the equatorial positions. As a result, as Figure 20.8 shows, AX_4E molecules have the shape of a disphenoid, which may be approximately described as a squashed tetrahedron; AX_3E_2 molecules may be described as T-shaped; and AX_2E_3 molecules are linear.

Because nonbonding pairs take up more space than bonding pairs, the observed bond angles in AX_4E and AX_3E_2 molecules are smaller than the 90° and 120° angles in the ideal shapes. The difference in the lengths of the axial and equatorial bonds that is observed whenever there are five electron pairs in the valence shell of the central atom is accentuated in the AX_3E_2 and AX_4E molecules because the nonbonding pair is closer to the axial bonding pairs than to the equatorial pairs.

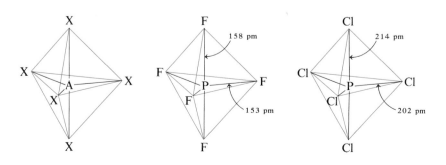

Figure 20.7 Trigonal Bipyramidal AX_5 Molecules, PF_5 and PCl_5.

20.1 HALIDES OF NONMETALS IN THEIR HIGHER OXIDATION STATES

723

Figure 20.8 AX$_4$E, AX$_3$E$_2$, and AX$_2$E$_3$ Molecules.

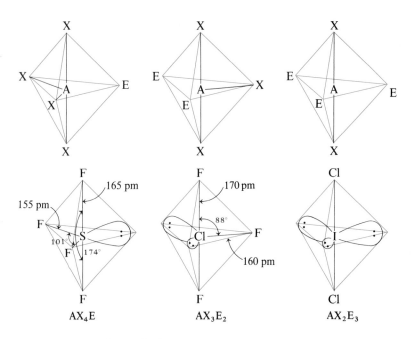

The three nonbonding pairs in AX$_2$E$_3$ molecules have a symmetrical arrangement, so they do not distort the predicted linear geometry, although they cause the A—X bonds to be longer than they would be in the absence of the unshared pairs.

20.2 OXIDES AND OXOACIDS OF THE HALOGENS

The most important compounds containing the halogens in positive oxidation states are the oxoacids. These are summarized in Table 20.4

Recall that when there are only two oxoacids of a given element, the *-ous* ending is used to indicate the lower–oxidation-state acid and *-ic* the higher–oxidation-state acid. For elements such as the halogens that form oxoacids in

Table 20.4 Oxoacids and Oxoanions of the Halogens

OXIDATION	FORMULA	NAME	FORMULA	NAME	FORMULA	NAME
Oxoacids						
+1	HClO	Hypochlorous	HBrO	Hypobromous	HIO	Hypoiodous
+3	HClO$_2$	Chlorous				
+5	HClO$_3$	Chloric	HBrO$_3$	Bromic	HIO$_3$	Iodic
+7	HClO$_4$	Perchloric	HBrO$_4$	Perbromic	HIO$_4$	Periodic
					H$_5$IO$_6$	Paraperiodic
Anions						
+1	ClO$^-$	Hypochlorite	BrO$^-$	Hypobromite	IO$^-$	Hypoiodite
+3	ClO$_2^-$	Chlorite				
+5	ClO$_3^-$	Chlorate	BrO$_3^-$	Bromate	IO$_3^-$	Iodate
+7	ClO$_4^-$	Perchlorate	BrO$_4^-$	Perbromate	IO$_4^-$	Periodate
					IO$_6^{5-}$	Paraperiodate

as many as four oxidation states, this simple terminology must be modified by the addition of prefixes to the names. The prefix *hypo-* signifies a lower oxidation state and *per-* a higher oxidation state, as Table 20.5 shows. In the IUPAC system the oxidation state is designated by a Roman numeral, and the names of all oxoacids end in *ic*, as shown in Table 20.5. It would seem logical to use the systematic IUPAC nomenclature for the halogen oxoacids, but the traditional names continue to be widely used, and it will be a long time before the change to the new names is complete. In this book we use the traditional names because they are used much more frequently at present than the IUPAC names.

Oxoacids of Chlorine

The oxoacids of chlorine and their salts are particularly important and have several useful applications. There are oxoacids corresponding to each of the positive oxidation states of chlorine (Table 20.4).

HYPOCHLOROUS ACID AND HYPOCHLORITES Hypochlorous acid is a weak acid in water ($K_a = 3.1 \times 10^{-8}$ mol L^{-1}). It exists only in dilute aqueous solution. The pure acid, or even concentrated solutions, cannot be made because the acid decomposes when dilute solutions are concentrated:

$$2HOCl(aq) + 2H_2O(aq) \longrightarrow 2H_3O^+(aq) + 2Cl^-(aq) + O_2(g)$$

In this reaction hypochlorous acid, in which chlorine is in the $+1$ oxidation state, oxidizes water to oxygen and is itself reduced to Cl$^-$ (oxidation number $= -1$). Even fairly dilute solutions of hypochlorous acid are not completely stable; they decompose slowly—more rapidly in sunlight.

Hypochlorous acid is formed in equilibrium with molecular chlorine in solutions of chlorine in water:

$$Cl_2 + 2H_2O \rightleftharpoons HOCl + H_3O^+ + Cl^-$$

About a third of the chlorine in a $0.1M$ solution is converted to HOCl. We can increase the concentration of HOCl in the solution by removing the Cl$^-$ and thus shifting the equilibrium to the right. Chloride ion can be removed by adding silver oxide, $(Ag^+)_2O^{2-}$. The Ag$^+$ ion reacts with Cl$^-$ to form insoluble AgCl, which can be removed by filtration, and the O^{2-} neutralizes the H$_3$O$^+$. The overall equation for this reaction is

$$2Cl_2(g) + Ag_2O(s) + H_2O \longrightarrow 2AgCl(s) + 2HOCl(aq)$$

Hypochlorites can be made by passing chlorine into a cold solution of a soluble hydroxide:

$$Cl_2 + 2OH^- \longrightarrow Cl^- + ClO^- + H_2O$$

Table 20.5 Names of Chlorine Oxoacids

COMMON NAME	FORMULA	SYSTEMATIC NAME
Hypochlorous	HClO	Chloric(I) acid
Chlor*ous*	HClO$_2$	Chloric(III) acid
Chlor*ic*	HClO$_3$	Chloric(V) acid
*Per*chloric	HClO$_4$	Chloric(VII) acid

This is essentially the same as the reaction of Cl_2 with water. The equilibrium

$$Cl_2 + 2H_2O \rightleftharpoons HOCl + H_3O^+ + Cl^-$$

is shifted to the right by removal of HOCl and H_3O^+ by reaction with the hydroxide ion, which converts them to OCl^- and H_2O, respectively. The reactions of chlorine with water and with OH^- are disproportionation reactions in which chlorine in the zero oxidation state is converted to chlorine in the $+1$ and -1 oxidation states.

Hypochlorite solutions are prepared on a large scale by electrolyzing a *cold* aqueous solution of sodium chloride. The solution is vigorously stirred to mix the chlorine produced at the anode,

$$2Cl^- \longrightarrow Cl_2 + 2e^-$$

with the hydroxide ion produced at the cathode,

$$2H_2O + 2e^- \longrightarrow H_2 + 2OH^-$$

so that the reaction

$$Cl_2 + 2OH^- \longrightarrow Cl^- + ClO^- + H_2O$$

takes place.

Hypochlorous acid and hypochlorites are strong oxidizing agents:

$$OCl^-(aq) + 2H^+(aq) + 2e^- \longrightarrow Cl^-(aq) + H_2O(l)$$

Sodium hypochlorite, NaOCl, and calcium hypochlorite, $Ca(OCl)_2$, have many important uses. Calcium hypochlorite is known as bleaching powder, and solutions of sodium hypochlorite are familiar as household bleach (Experiment 20.3). Hypochlorites are used in industry to bleach cotton and linen materials and paper pulp. They kill microorganisms and are therefore used as disinfectants and to purify water.

CHLORIC ACID AND CHLORATES In aqueous solution hypochlorite ion decomposes slowly in two ways. As we have stated, it decomposes to H_3O^+, Cl^-, and O_2, and it also disproportionates to chlorate ion, ClO_3^-, and Cl^-:

$$3ClO^- \longrightarrow 2Cl^- + ClO_3^-$$

The rate of this reaction increases if the solution is warmed. Chlorates can therefore be prepared by passing chlorine into a *hot* hydroxide solution. The overall equation for the reaction is

$$3Cl_2 + 6OH^- \longrightarrow 5Cl^- + ClO_3^- + 3H_2O$$

Changing the temperature of the reaction between chlorine and an aqueous solution of sodium hydroxide thus alters the products of the reaction. In the commercial preparation a hot solution of potassium chloride is electrolyzed with vigorous stirring so that the chlorine produced at the anode mixes with the hydroxide produced at the cathode. Potassium chlorate has a solubility of only 3 g in 100 g of water at 0°C, but potassium chloride has a solubility of 28 g in 100 g. The solubility of $KClO_3$ is further decreased by the common-ion effect (Chapter 15) of the potassium chloride produced in the reaction. Hence when the solution resulting from the electrolysis is cooled, potassium chlorate crystallizes.

Chloric acid, $HClO_3$, is a strong acid in water, but neither the pure substance nor concentrated aqueous solutions can be obtained. If solutions of $HClO_3$ are evaporated, the acid decomposes, sometimes explosively.

Properties of Sodium Hypochlorite

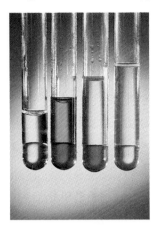

Left to right: (a) Tetrachloroethene, C_2Cl_4 is added to a colorless aqueous solution of potassium iodide. The C_2Cl_4 is immiscible with water and sinks to the bottom of the tube, forming a separate layer. (b) A few drops of a dilute aqueous solution of sodium hypochlorite, NaOCl, are added. This oxidizes iodide ion to iodine, which forms a purple solution in the C_2Cl_4 and a brown solution of I_3^- in the aqueous layer. (c) When the tube is shaken, all the iodine is extracted into the C_2Cl_4. (d) When an excess of sodium hypochlorite solution is added, iodine is further oxidized to iodate ion, IO_3^-, and when the tube is shaken, both layers again become colorless.

Household bleach is a solution of sodium hypochlorite. Here its ability to bleach different samples of colored paper is shown. The bleaching action results from the oxidation of the colored dye to a colorless compound.

The chlorate ion may be described by three equivalent resonance structures. It is an AX_3E molecule and has the expected triangular pyramidal structure (see Figure 20.9).

Chloric acid and chlorates are powerful oxidizing agents. Chlorates form sensitive explosive mixtures with reducing agents such as carbon, sulfur, and many organic compounds, which are converted in exothermic reactions to gaseous products such as CO_2 and SO_2. Potassium chlorate finds extensive use in matches, fireworks, and various types of explosives because of its strength as an oxidizing agent. Sodium chlorate destroys microorganisms and bacteria in the soil by oxidation. It is therefore used to kill weeds and other vegetation. However, this use is potentially hazardous, because if wood or clothing are saturated with a solution of sodium chlorate, they may ignite spontaneously.

When potassium chlorate is heated it decomposes in two ways:

(1) The reaction

$$2KClO_3(s) \longrightarrow 2KCl(s) + 3O_2(g)$$

:O̤=C̤l—O̤:⊖   :O̤=C̤l=O̤:   ⊖:O̤—C̤l=O̤:
　　‖　　　　　　　│　　　　　　　　‖
　　:O̤　　　　　　:O̤:⊖　　　　　　:O̤

Resonance structures

　　　Cl
O⟋　║ˎˎˎ148 pm
　　　│　　O
　　　O
　　　106°

Dimensions

Figure 20.9 Structure of the Chlorate Ion: An AX_3E Molecule.

20.2 OXIDES AND OXOACIDS OF THE HALOGENS

is catalyzed by various substances; manganese dioxide is frequently used (see Experiment 20.4). In the past this reaction has been widely used for the laboratory preparation of oxygen. However, since impurities in the catalyst, such as carbon, can sometimes lead to explosions, suitable precautions must be taken.

(2) In the absence of a catalyst, gently heating potassium chlorate gives potassium perchlorate as the main product:

$$4KClO_3(s) \longrightarrow 3KClO_4(s) + KCl(s)$$

This reaction is an example of a disproportionation, in which potassium chlorate may be said to oxidize itself. Although almost all perchlorates are rather soluble in water, the solubility of potassium perchlorate is only 0.75 g in 100 g of water at 0°C; therefore it is easily separated from the KCl.

PERCHLORIC ACID AND PERCHLORATES   Perchlorates are strong oxidizing agents but are somewhat less sensitive and less dangerous than chlorates. Potassium perchlorate and ammonium perchlorate mixed with carbon or various organic compounds are used as explosives and as rocket fuels. The solid-booster rockets of the space shuttle use a solid propellant that consists of 70% ammonium perchlorate, 16% aluminum powder, and a polymer known as PBAN (a polybutadiene–acrylic acid–acrylonitrile copolymer), which binds the mixture into a fairly hard material that looks and feels like a hard rubber eraser. This propellant is quite safe to prepare and handle. Perchlorates are also used extensively in fireworks and flares.

The perchlorate ion, ClO_4^-, may be represented by four equivalent resonance structures. It is an AX_4 species and therefore has a tetrahedral structure (see Figure 20.10). It is isoelectronic with the tetrahedral PO_4^{3-} and SO_4^{2-} ions. The observed bond length (144 pm) is much shorter than the sum of the single-bond, covalent radii of chlorine and oxygen, which is 165 pm. This bond length is consistent with the bond order of 1.75 implied by the resonance structures.

Anhydrous perchloric acid, $HClO_4$, can be prepared by distilling a mixture of potassium perchlorate and concentrated sulfuric acid under reduced pressure. Perchloric acid is a strong acid in water. It is a very powerful oxidizing agent that explodes violently on contact with organic matter such as wood and cloth. The preparation of anhydrous perchloric acid is therefore hazardous and must always be carried out in scrupulously clean apparatus with precautions against explosions. However, dilute aqueous solutions of perchloric acid are

One of 4 equivalent
resonance structures

Dimensions

Figure 20.10 Structure of the Perchlorate Ion: An AX₄ Molecule.

EXPERIMENT 20.4

Thermal Decomposition of Potassium Chlorate

When white potassium chlorate, $KClO_3$, is heated in the presence of black manganese dioxide, MnO_2, which acts as a catalyst, it decomposes to potassium chloride and oxygen. Here we see the oxygen being collected over water.

quite safe. Although aqueous solutions of the perchlorate ion are potentially good oxidizing agents, the reactions are usually quite slow.

All the halogen oxoacids and their anions are good oxidizing agents. Their oxidizing ability and, particularly, the rates at which they react depend very much on the conditions. Their oxidizing strengths do not simply increase with increasing number of oxygen atoms.

CHLOROUS ACID AND CHLORITES There is a fourth oxoacid of chlorine, chlorous acid, $HClO_2$, which is known only in aqueous solution, in which it behaves as a weak acid in water ($K_a = 1.1 \times 10^{-2}$ mol L^{-1}). Its salts are the chlorites. They are good oxidizing agents but are not very important. The chlorite ion is an angular AX_2E_2 molecule. It can be represented by two resonance structures (see Figure 20.11). In the series of anions ClO_2^-, ClO_3^-, ClO_4^-, the bond orders obtained from the resonance structures are 1.50, 1.66, and 1.75, and the bond lengths decrease correspondingly from 157 to 148 to 144 pm.

Resonance structures

Dimensions

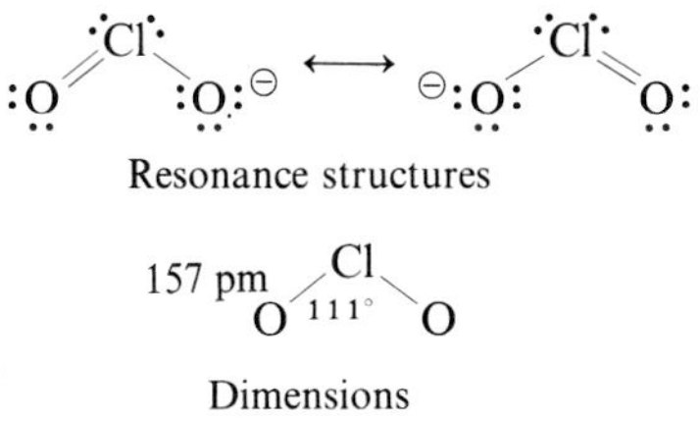

Figure 20.11 Structure of the Chlorite Ion: An AX_2E_2 Molecule.

STRENGTHS OF CHLORINE OXOACIDS The oxoacids of chlorine form an interesting series,

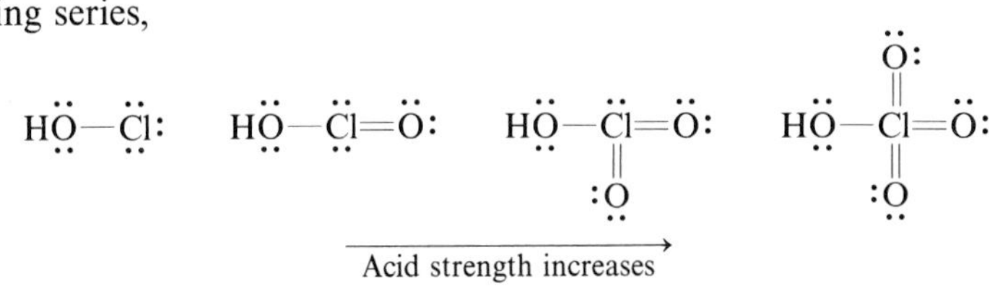

Acid strength increases

in which the nonbonding pairs are successively replaced by oxygen atoms. This causes a steady increase in the acid strength from hypochlorous acid, which is the weakest, to chlorous acid, which is still a weak acid, and to chloric acid and perchloric acid, which are both strong acids. We have previously seen that sulfuric acid, H_2SO_4, is stronger than sulfurous acid, H_2SO_3, and that nitric acid, HNO_3, is stronger than nitrous acid, HNO_2. The increasing number of oxygen atoms in the series HOCl, HOClO, HOClO_2, and HOClO_3 increases the effective electronegativity of the chlorine atom by withdrawing electrons from the chlorine, thereby increasing the ability of chlorine to withdraw electrons from the OH group (Chapter 7). In addition, the increasing number of oxygen atoms allows more delocalization of the negative charge of the anion, thereby stabilizing the anion and decreasing its base strength (Chapter 8).

We have seen previously (Chapter 7) that if the formulas of oxoacids are written in the form $XO_m(OH)_n$ the strength of the acid increases with the value of m (see Table 20.6). According to this classification, perchloric acid is a stronger acid than sulfuric acid, nitric acid, or chloric acid. However, all four acids are completely ionized in aqueous solution. They are all strong acids, and we cannot distinguish between their strengths by studying their solutions in water. Measurements on the ionization of these acids in solvents that are less basic than water have shown that perchloric acid is indeed a stronger acid than nitric acid or sulfuric acid.

Chlorine Oxides

The oxides of chlorine, unlike those of phosphorus and sulfur, are all somewhat unstable and explosive. They include Cl_2O, Cl_2O_3, ClO_2, Cl_2O_6, and Cl_2O_7. The structures of some of these oxides are shown in Figure 20.12. Of

Table 20.6 Strengths of Some Oxoacids $XO_m(OH)_n$

	0 $X(OH)_n$	1 $XO(OH)_n$	2 $XO_2(OH)_n$	3 $XO_3(OH)_n$
	ClOH	ClO(OH)	ClO$_2$(OH)	ClO$_3$(OH)
	BrOH	PO(OH)$_3$	SO$_2$(OH)$_2$	
	IOH	IO(OH)$_5$	NO$_2$(OH)	
	Si(OH)$_4$	NO(OH)		
	B(OH)$_3$	HPO(OH)$_2$		
		CO(OH)$_2$		
Strength in water	Very weak	Weak	Strong	Very strong
K_a (mol L^{-1})	$< 10^{-7}$	10^{-2}–10^{-4}	> 1	$\gg 1$

these only chlorine dioxide, ClO_2, is of any importance. It is prepared by re-ducing $NaClO_3$ with SO_2 in acid solution:

$$2NaClO_3(aq) + SO_2(g) + H_2SO_4(aq) \longrightarrow 2ClO_2(g) + 2NaHSO_4(aq)$$

Chlorine dioxide is a yellow-brown gas that condenses to a red liquid at $-59°C$. The ClO_2 molecule, which has 33 electrons, is an odd-electron mole-cule. Like NO_2, it is a stable free radical, but it shows much less tendency to dimerize than NO_2. It can be represented by the Lewis structures in Figure 20.12. They indicate that the odd electron is delocalized over the chlorine and oxygen atoms. It is an angular AX_2E_2 molecule.

Despite its instability and the fact that it can form explosive mixtures with air, ClO_2 is produced in large quantities. It is used as a bleaching agent for cellulose in the pulp and paper industry and in the manufacture of white flour. It is also used for water purification.

The chlorine oxides are unusual compounds that are still not well under-stood. No one has yet explained satisfactorily the stability of either ClO_2 or of NO_2.

Oxoacids of Bromine and Iodine

The oxoacids of bromine are HOBr, $HBrO_3$, and $HBrO_4$. They resemble the corresponding chlorine oxoacids. Bromous acid, $HBrO_2$, is not known.

The oxoacids of iodine are hypoiodous acid, HOI, iodic acid, HIO_3, and the periodic acids, HIO_4 and H_5IO_6. Sodium iodate occurs naturally in small amounts in sodium nitrate, deposits of which are found in certain desert regions of Chile. Iodine is obtained from sodium iodate by reduction with sulfur dioxide

Figure 20.12 Structures of Some Oxides of Chlorine.

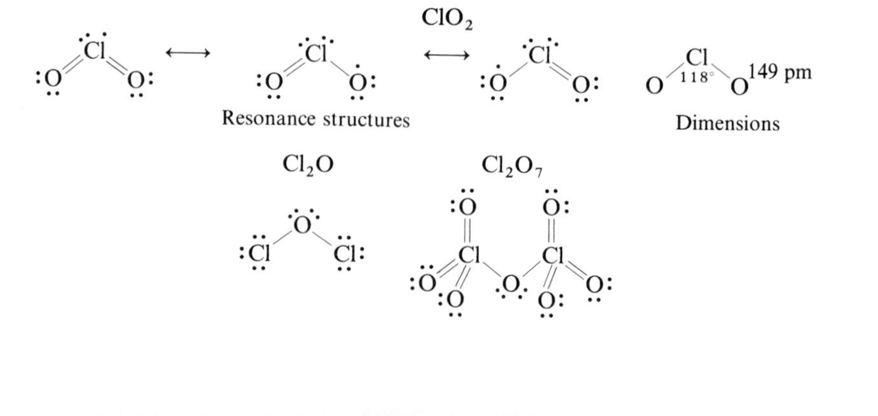

or with sodium hydrogen sulfite, $NaHSO_3$:

$$2IO_3^-(aq) + 5SO_2(g) + 4H_2O(l) \longrightarrow I_2(s) + 8H^+(aq) + 5SO_4^{2-}(aq)$$

Experiment 20.5 shows how potassium iodate is reduced to iodine by sulfur dioxide.

Iodine is insoluble in water but dissolves in a solution of hydroxide ion with the formation of hypoiodite ion, IO^-. This reaction is quite similar to the reaction of Cl_2 with OH^-:

$$I_2(s) + 2OH^-(aq) \longrightarrow IO^-(aq) + I^-(aq) + H_2O(l)$$

However, hypoiodite ion is considerably less stable than hypochlorite ion, and it disproportionates rather rapidly:

$$3IO^-(aq) \longrightarrow IO_3^-(aq) + 2I^-(aq)$$

This reaction is further accelerated by warming the solution. Thus when iodine is dissolved in *warm* hydroxide solution, iodate and iodide are formed:

$$3I_2(s) + 6OH^-(aq) \longrightarrow 5I^-(aq) + IO_3^-(aq) + 3H_2O(l)$$

We have seen that chloric acid reacts similarly to give chlorate and chloride in hot NaOH solution.

Whereas chloric acid is unstable and cannot be isolated in a pure form, iodic acid is a stable, white crystalline solid. It can be made by oxidizing iodine with nitric acid (see Experiment 20.6):

$$I_2(s) + 10HNO_3(aq) \longrightarrow 2HIO_3(aq) + 10NO_2(g) + 4H_2O(l)$$

Like the chlorate and bromate ions, iodic acid and the iodate ion have a triangular pyramidal AX_3E geometry (see Figure 20.13).

Periodic acid and the periodates differ from perchloric acid and perbromic acid in that they exist in two forms (see Figure 20.14). Periodic acid is a solid that, when crystallized from water, has the formula H_5IO_6. This form is known as *paraperiodic acid*. It is a weak acid (see Table 20.6). If it is heated to 80°C

Reduction of Potassium Iodate

Sulfur dioxide is passed into a colorless solution of potassium iodate

The potassium iodate is rapidly reduced to black solid iodine.

20.2 OXIDES AND OXOACIDS OF THE HALOGENS

Preparation of Iodic Acid

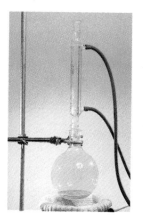

Iodine can be oxidized to iodic acid, HIO_3, by concentrated nitric acid. Here, pale yellow concentrated nitric acid has been added to solid iodine in the bottom of the flask.

On warming, the flask is filled with brown NO_2 vapor resulting from the reduction of nitric acid.

After some time, the iodic acid formed can be seen as a white crystalline solid in the bottom of the flask and coating the sides.

under vacuum, it loses water to give HIO_4, which is known as *metaperiodic acid*:

$$H_5IO_6 \longrightarrow HIO_4 + 2H_2O$$

Salts of both forms of the acid are known, such as $NaIO_4$, $Na_2H_3IO_6$, and Ag_5IO_6.

Figure 20.13 Structures of Iodic Acid and the Iodate Ion.

Iodate ion, IO_3^-

Iodic acid, HIO_3

Figure 20.14 Structures of Periodic Acid and the Periodate Ion.

Paraperiodic acid

(six equivalent resonance structures)

Paraperiodate

(four equivalent resonance structures)

Metaperiodate

Occurrence and Uses

Helium was the first noble gas to be discovered. It is formed in the sun by the fusion of hydrogen nuclei (protons); this reaction provides the sun with its energy (Chapter 24). Indeed, helium was identified on the sun before it was discovered on the earth. Certain lines in the spectrum of the sun were attributed to an element not known on earth; hence it was named helium after the Greek word *helios*, for "sun." The discovery of the other noble gases is described in Box 20.1.

The noble gases occur in the atmosphere as monatomic gases. Argon constitutes approximately 1% of the atmosphere, neon only about 0.002%, and the other noble gases are present in still smaller amounts. Helium, formed by the disintegration of uranium and other radioactive elements, is found trapped in uranium minerals, from which it can be liberated by heating. Some of the helium formed in this way has gradually diffused into pockets of natural gas, which often contains as much as 1% helium. Although helium is rather rare on the earth, it is the second most abundant element in the sun and in the solar system as a whole. Like hydrogen, it is too light to have remained in appreciable amounts in the earth's atmosphere (Chapter 3).

The fact that the noble gases occur in the uncombined monatomic state in the atmosphere is a clear indication of their lack of reactivity. All their important uses depend on this lack of reactivity.

Helium is the second lightest of all the elements and is therefore used for filling balloons and airships in place of the much more dangerous hydrogen (Box 3.4). It is also used, mixed with oxygen, to provide a gas for divers to breathe. If they breathe ordinary air, nitrogen dissolves in the blood under the high pressures at which divers work. When a diver comes to the surface, the dissolved nitrogen is released into the blood in the form of bubbles, a painful and dangerous condition. A helium-oxygen mixture can be breathed safely because helium is much less soluble in the blood than is nitrogen.

The most important use for neon is in neon signs. When an electric current is passed through neon at low pressure, excited neon atoms emit the characteristic atomic spectrum of neon, which has strong lines in the red part of the spectrum (Chapter 6).

Argon, by far the most abundant of the noble gases, has some important applications. It is used to increase the useful life of incandescent light bulbs and to permit the filament to be heated to a higher temperature and thus produce a whiter light. Argon is extensively used in industry to provide a chemically inert atmosphere for welding and in the manufacture of metals and alloys.

No significant use has been found for krypton and xenon. Somewhat surprisingly, xenon is a good anesthetic, but it is too expensive for general use. The radioactive gas radon is used in the treatment of cancer. Radon is sealed into a small tube and placed close to the cancerous tissues to be destroyed.

Compounds of the Noble Gases

After the discovery of the noble gases, many attempts were made to prepare compounds of these elements. But none were successful, and the elements became known as the *inert gases*. In fact, no compound of any of these elements was made until 1962, when Neil Bartlett prepared the first compound of xenon

Box 20.1
DISCOVERY OF THE NOBLE GASES

The story of the discovery of the noble gases illustrates the importance of paying attention to minor discrepancies in experimental results. For over a hundred years it had been thought that air consisted solely of nitrogen and oxygen, together with small variable amounts of water vapor and carbon dioxide. Then in 1785 English scientist Henry Cavendish (1731–1810) made some new investigations of the atmosphere. He added oxygen to the air and passed an electric spark through the mixture. This caused the formation of NO_2:

$$N_2 + O_2 \longrightarrow 2NO \quad \text{and} \quad 2NO + O_2 \longrightarrow 2NO_2$$

Cavendish continued the sparking until no further change in the volume occurred, that is, until all the nitrogen had been removed. He then removed the NO_2 by reacting it with a solution of KOH, and he removed the excess of oxygen by allowing it to react with a solution of potassium pentasulfide:

$$2K_2S_5 + 3O_2 \longrightarrow 2K_2S_2O_3 + 6S$$

After this treatment a small bubble of gas remained; it had a volume of not more than $\frac{1}{120}$ the original volume of the air. Cavendish concluded that if there were another component of the air, other than nitrogen and oxygen,

it constituted less than 1% by volume. But most chemists apparently assumed that the remaining bubble was simply nitrogen that had not been removed, perhaps because the sparking had not been carried on for a sufficient time, and Cavendish's experiment was forgotten.

More than a hundred years later Lord Rayleigh (1842–1919), an English physicist, was making careful measurements of the densities of several gases, including nitrogen. He prepared nitrogen from ammonia by passing a mixture of oxygen with excess ammonia over red hot copper; the copper catalyzes the oxidation of ammonia to nitrogen:

$$4NH_3 + 3O_2 \longrightarrow 6H_2O + 2N_2$$

After it was dried, the nitrogen was found to have a density of 1.2505 g L^{-1} at 0°C and 1 atm. Rayleigh also made nitrogen by passing dry air over red-hot copper, which removed the oxygen by combining with it to form copper oxide. The remaining gas, after it was dried, was presumed to be nitrogen. But it was found to have a density of 1.2672 g L^{-1} at 0°C and 1 atm. In other words, nitrogen prepared from air had a density 0.5% greater than that of nitrogen prepared from ammonia. Rayleigh was convinced that this difference was greater than the experi-

(see Box. 20.2). Since that time more compounds of xenon have been made, as well as a few compounds of krypton and radon, and we can no longer regard these elements as truly inert. They are now called *noble gases*, or, because of their low abundance in the atmosphere, the *rare gases*.

Although before 1962 it had become firmly established in most chemists' minds that the inert gases were incapable of forming compounds, a few chemists continued to believe that synthesizing compounds of the noble gases should be possible. In 1933 Linus Pauling (Box 20.3) went so far as to predict the formulas of several compounds of the noble gases that he felt should exist.

A consideration of the periodic table shows us that Pauling's predictions were quite logical. We saw in Chapter 4 that if we consider the normal valences of the nonmetals, we conclude that the noble gases would have a valence of zero (see Table 20.7). A valence of zero is consistent with their inertness. But the nonmetals in the third and subsequent periods also have higher valences, because they can expand their valence shells beyond 8 electrons; there are d

Table 20.7 Predicted Valences of Nonmetals Based on a Valence Shell of Eight Electrons

Group	IV	V	VI	VII	VIII
Valence	4	3	2	1	0
	CF_4	NF_3	OF_2	FF	Ne
	SiF_4	PF_3	SF_2	ClF	Ar

mental error in his measurements, and he suggested several tentative explanations. For example, he proposed that atmospheric nitrogen might contain N_3 molecules analogous to O_3 molecules.

William Ramsay (1852–1916), professor of chemistry at University College, London, then became interested in the problem. He believed that nitrogen prepared from the atmosphere might contain a small amount of another, unidentified gas, as suggested by Cavendish's experiment. He therefore repeated Cavendish's experiment, using several different methods to remove both the oxygen and the nitrogen from the air. Like Cavendish, he found that approximately 1% of the air remained after both the oxygen and the nitrogen were removed. By examining the atomic spectrum of the residual gas, Ramsay then showed that it was not nitrogen but a new element and that it had a considerably higher density than nitrogen. Ramsay tried to cause this new element to react with many other substances, but he had no success. He therefore called the element *argon*, after the Greek word for "lazy."

Ramsay also identified the gas given off by some radioactive minerals as helium, whose atomic spectrum had previously been observed in the light from the sun.

Because helium, like argon, proved to be inert, Ramsay concluded that there must be a whole family of inert, monatomic gases like argon and helium. He considered it likely that these gases were also present in the atmosphere, so he set out with his assistant, William Travers, to find them. They prepared large amounts of liquid air and then fractionally distilled it. In this way they separated argon and the new element krypton in May 1898, followed by neon in June, and by xenon in July of the same year.

Both Cavendish's and Rayleigh's experiments illustrate the importance of knowing the accuracy of measurements, so that discrepancies greater than the possible experimental error can be recognized. Rayleigh's attention to a small discrepancy that was greater than his experimental error led to the discovery of the whole family of noble gases.

Although Cavendish had isolated a new component of the atmosphere, he had no means of identifying the very small amount of inert gas that he obtained. Ramsay was able to identify the new element, argon, because he could observe its atomic spectrum; but Cavendish could not do this because the technique was not known in his time.

orbitals in these shells that can be used in compound formation. For example, in some fluorides of the nonmetals the central atom has a valence shell of 12 electrons and in others a valence shell of 10 electrons (see Tables 20.8 and 20.9). Comparison of the formulas of these compounds leads to the prediction that

Table 20.8 Nonmetal Fluorides with a Valence Shell of Twelve Electrons

KNOWN COMPOUNDS		PREDICTED BY EXTRAPOLATION
GROUP IV	GROUP VII	GROUP VIII
SF_6	ClF_5	(ArF_4), still unknown
SeF_6	BrF_5	(KrF_4), still unknown
TeF_6	IF_5	XeF_4, now known

Table 20.9 Nonmetal Fluorides with a Valence Shell of Ten Electrons

KNOWN COMPOUNDS			PREDICTED BY EXTRAPOLATION
GROUP V	GROUP VI	GROUP VII	GROUP VIII
PF_5	SF_4	ClF_3	(ArF_2), still unknown
AsF_5	SeF_4	BrF_3	KrF_2, now known
SbF_5	TeF_4	IF_3	XeF_2, now known

Box 20.2
DISCOVERY OF NOBLE GAS COMPOUNDS

The first compound of the noble gases was prepared in 1962 at the University of British Columbia by Neil Bartlett, who was born in England in 1932. At that time Bartlett was studying the reactions of platinum hexafluoride, PtF_6, which is a red gas. He found that it reacts rapidly with oxygen at room temperature to form the yellow ionic solid $O_2^+ PtF_6^-$. This surprising reaction implies that PtF_6 is a very strong oxidizing agent that can remove an electron from an oxygen molecule to give the ion O_2^+. Bartlett argued that if PtF_6 could remove an electron from O_2, it should be able to remove an electron from Xe, since the ionization energy of xenon (1.167 MJ mol^{-1}) is slightly lower than the ionization energy of O_2 (1.176 MJ mol^{-1}). He therefore predicted that xenon and PtF_6 would react to form $Xe^+PtF_6^-$:

$$Xe + PtF_6 \longrightarrow Xe^+PtF_6^-$$

Bartlett carried out the reaction and found that xenon does indeed react with the red gas PtF_6 to give a yellow solid, which Bartlett believed to be $Xe^+PtF_6^-$.

Bartlett's discovery caused great interest and excitement. His reaction was immediately repeated by others, and other reactions of xenon were tried. In particular, xenon was found to react directly with fluorine under pressure to give XeF_2, XeF_4, and XeF_6.

It is interesting to note that William Ramsay (Box 20.1) sent a sample of argon to the French chemist Henri Moissan (1852–1907), who had recently discovered fluorine, to see whether it would react with fluorine. Moissan did not succeed. Unfortunately, he was unable to try the reaction between fluorine and xenon because Ramsay had prepared such a small amount of xenon that he was unable to send a sample to Moissan. Had Moissan been able to attempt this reaction, the compounds of xenon might have been discovered 60 years earlier. Such a discovery might have greatly influenced the development of

the theory of chemical bonding; the octet rule might not have then attained such a prominent place in the development of ideas on valence and the chemical bond. The fact that the noble gases were found by Ramsay and others to be completely inert led Lewis to formulate his octet rule. The success of this rule reinforced the idea that the noble gases were unreactive and, indeed, gave some apparent theoretical foundation to the concept of the inertness of the noble gases. Students learned that the noble gases did not form compounds because they had a stable octet of electrons, and only a very few chemists, such as Linus Pauling, continued to think about the possible existence of noble gas compounds.

Bartlett not only made a logical and apparently simple, but nevertheless brilliant, deduction from his experiment with oxygen and PtF_6, but he also had sufficient confidence in the correctness of his argument to question the orthodox view that the noble gases did not form compounds.

As a postscript to this story, it is interesting to point out that the true composition of the xenon compound prepared by Bartlett has never been established with certainty. Probably it was not the simple compound $Xe^+PtF_6^-$ but rather a complex mixture containing the compounds $XeF^+PtF_6^-$ and $XeF^+Pt_2F_{11}^-$. Despite the apparent logic of Bartlett's argument that PtF_6 should be able to oxidize Xe to Xe^+, there is no firm evidence that Xe^+ exists as a stable species. Because Xe^+ is a free radical, we expect it to be very reactive, and if it is formed in the reaction with PtF_6, it probably combines immediately with a fluorine atom to give XeF^+. Nevertheless, Bartlett's brilliantly conceived experiment opened a new field of chemistry—the chemistry of the noble gases. Had it not been for Bartlett's experiment, the chemistry of the noble gases might still be a closed book to us.

---

argon, krypton, and xenon should form tetrafluorides and difluorides. Of these possible compounds only XeF_4, XeF_2, and KrF_2 have so far been prepared.

Similarly, we might predict the formation of the trioxides and tetraoxides of the noble gases by extrapolation from the known formulas of the isoelectronic anions of the preceding nonmetals (see Table 20.10). Of these possible compounds only XeO_3 and XeO_4 have so far been prepared.

Since helium has a valence shell filled with two electrons and neon one with eight electrons, we do not expect to find any compounds of these two elements. However, argon, krypton, xenon, and radon can all have more than eight electrons in their valence shells, and thus they could, in principle, form compounds. No compounds of argon have so far been made, and although there appears to be at least one compound of radon, the chemistry of this element

Table 20.10 Isoelectronic Oxoanions and Oxides

KNOWN ANIONS			PREDICTED BY EXTRAPOLATION
GROUP V	GROUP VI	GROUP VII	GROUP VIII
—	SO_3^{2-}	ClO_3^-	(ArO_3), still unknown
AsO_3^{3-}	SeO_3^{2-}	BrO_3^-	(KrO_3), still unknown
SbO_3^{3-}	TeO_3^{2-}	IO_3^-	XeO_3, now known
PO_4^{3-}	SO_4^{2-}	ClO_4^-	(ArO_4), still unknown
AsO_4^{3-}	SeO_4^{2-}	BrO_4^-	(KrO_4), still unknown
—	—	IO_4^-	XeO_4, now known

is very difficult to study because of its radioactivity. Our knowledge of the chemistry of the noble gases is therefore confined to krypton and xenon.

The noble gases in their ground states do not have any unpaired electrons, that is, electrons in singly occupied orbitals. Consequently, they can form compounds only via a valence state in which one or more s or p electrons is promoted to an unoccupied d orbital, as illustrated in Figure 20.15.

Only the most electronegative elements—in particular, fluorine and oxygen—exert sufficient attraction on the valence shell electrons to move some of them into the d orbitals. Except for one compound that contains Xe—N bonds, all the known compounds of krypton and xenon have Kr—F, Xe—F, and Xe—O bonds. In fact, fluorine is the only element that will react directly with xenon and krypton.

FLUORIDES   When xenon and fluorine are heated together, XeF_2 and XeF_4 are formed; the relative amounts of these compounds depend on whether xenon or fluorine is in excess. If xenon is heated with an excess of fluorine under pressure, XeF_6 is obtained:

$$Xe(g) + F_2(g) \longrightarrow XeF_2(s)$$
$$Xe(g) + 2F_2(g) \longrightarrow XeF_4(s)$$
$$Xe(g) + 3F_2(g) \longrightarrow XeF_6(s)$$

Xenon difluoride can also be prepared more simply by exposing a glass bulb containing equal amounts of fluorine and xenon gases to strong sunlight or

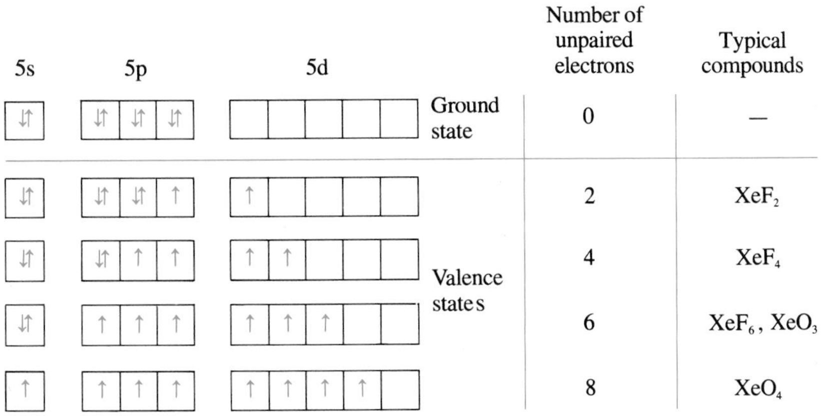

Figure 20.15 Ground State and Valence State Electron Configurations for Xenon.

Linus Pauling was born in Portland, Oregon, in 1901. He graduated from Oregon State College in 1922 and obtained his Ph.D. in chemistry in 1925 from the California Institute of Technology. After a period of study in Europe, he became a professor at the California Institute of Technology in 1927, and he remained there for the rest of his academic career.

Pauling did much to develop our understanding of the chemical bond. Among the important concepts that he introduced were electronegativity and resonance structures. He was one of a very small number of chemists who seriously considered the possibility that the noble gases might form compounds. In 1939 he published his ideas on chemical bonding in a book entitled *The Nature of the Chemical Bond*. This book proved to be one of the most influential chemistry books of the present century. In 1954 Pauling was awarded the Nobel Prize in chemistry for his work on molecular structure.

In the 1950s, Pauling turned to studying the structures of biopolymers (Chapter 23). He was one of the first to suggest that protein molecules have a helical shape. Pauling's claim that large doses of Vitamin C are effective in preventing the common cold has attracted attention since 1970.

After World War II, Pauling became a passionate supporter of nuclear disarmament. In 1962 he was awarded the Nobel Peace Prize, thus becoming the second person in history to win two Nobel Prizes.

ultraviolet light. Over a few hours, colorless crystals of xenon difluoride form on the sides of the bulb (Figure 20.16).

The fluorides XeF_2, XeF_4, and XeF_6 are all stable, colorless, crystalline, but very reactive, solids. They are all strong oxidizing agents. For example, a solution of XeF_2 in water rapidly oxidizes HCl to Cl_2 and OH^- to O_2:

$$2HCl(aq) + XeF_2(aq) \longrightarrow Xe(g) + Cl_2(g) + 2HF(aq)$$
$$4OH^-(aq) + 2XeF_2(aq) \longrightarrow 2Xe(s) + O_2(g) + 4F^-(aq) + 2H_2O(l)$$

In each case the XeF_2 is reduced to xenon. Only one fluoride of krypton, KrF_2, is known.

OXIDES   The oxides of xenon, XeO_3 and XeO_4, cannot be prepared by the direct reaction of xenon with oxygen. The trioxide, which is a white explosive solid, is obtained by the reaction of either XeF_4 or XeF_6 with water. Xenon hexafluoride reacts very rapidly with water to form xenon trioxide:

$$XeF_6(s) + 3H_2O(l) \longrightarrow XeO_3(s) + 6HF(aq)$$

This is a typical reaction of a nonmetal halide with water to give a hydrogen halide and a nonmetal oxide or oxoacid. For example, it is analogous to the

reaction of PCl$_5$ with water:

$$PCl_5(s) + 4H_2O(l) \longrightarrow H_3PO_4(aq) + 5HCl(aq)$$

Hydrolysis of XeF$_4$ does not yield XeO$_2$ but rather XeO$_3$ and Xe by a disproportionation reaction:

$$6XeF_4(s) + 12H_2O(l) \longrightarrow 2XeO_3(s) + 4Xe(g) + 3O_2(g) + 24HF(aq)$$

Xenon tetraoxide is obtained in a complex disproportionation of an alkaline aqueous solution of XeO$_3$. It is a highly unstable and explosive gas.

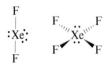

Figure 20.17 Structures of XeF$_2$, an AX$_2$E$_3$ Molecule, and XeF$_4$, an AX$_4$E$_2$ Molecule.

STRUCTURES OF NOBLE GAS COMPOUNDS   The structures of the noble gas compounds provide some interesting examples of the application of the VSEPR theory. Xenon difluoride is a linear AX$_2$E$_3$ molecule, while XeF$_4$ is a square planar AX$_4$E$_2$ molecule with the two nonbonding electron pairs in trans positions (see Figure 20.17). Xenon hexafluoride, XeF$_6$, has seven electron pairs in the valence shell of xenon and is an AX$_6$E molecule, a type we have not yet discussed. Because of the presence of the nonbonding pair, the arrangement of the six bonding pairs would not be expected to be octahedral. It may be described as a distorted octahedron. The presence of the unshared electron pair forces the fluorine atoms surrounding the unshared pair further apart than in a regular octahedron (see Figure 20.18).

The two oxides XeO$_3$ and XeO$_4$ are isoelectronic with iodate and metaperiodate and have the same pyramidal AX$_3$E and tetrahedral AX$_4$ structures, as Figure 20.19 shows. Xenon tetraoxide, which is a nonpolar molecule, is a gas, while XeO$_3$, which is polar, is a solid.

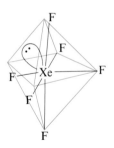

Figure 20.18 Structure of XeF$_6$: An AX$_6$E Distorted Octahedral Molecule.

Figure 20.19 The structures of XeO$_3$ and XeO$_4$.

IMPORTANT TERM

An **interhalogen compound** is a compound formed between two different halogens. Examples are ClF and BrF$_5$.

PROBLEMS

Molecular Geometry

1. Draw orbital box diagrams for the following:

(a) The $+1$, $+3$, $+5$, and $+7$ oxidation states of iodine.

(b) The $+2$, $+4$, and $+6$ oxidation states of Xe.

2. On the basis of VSEPR theory, what are the expected shapes of AX_5 and AX_6 molecules? Give two examples of molecules having these shapes. What are the expected shapes of SeF_4, BrF_3, XeF_4, and ICl_2^-?

3. Compare the structures of $PCl_5(g)$ and $PCl_5(s)$.

4. Why is it possible for the elements of the third and subsequent periods to expand their valence shells in reactions with fluorine? Describe the shapes of the molecules formed between phosphorus and fluorine and between sulfur and fluorine.

5. The molecular formula of the trichloride of iodine is I_2Cl_6 rather than ICl_3. By analogy with the structures of Al_2Cl_6 and $AlCl_3$, predict the structure of I_2Cl_6.

6. Chlorine, bromine, and iodine form covalent oxofluorides of general formula AO_2F and AO_3F. Predict the shapes of these molecules.

7. The fluorides XeF_4 and XeF_6 react with SbF_5 to give the ionic compounds $XeF_3^+SbF_6^-$ and $XeF_5^+SbF_6^-$. Draw a Lewis structure for each of the ions, and use the VSEPR theory to predict their shapes. Comment on the expected bond angles.

8. Which of the following molecules will have a dipole moment?

(a) PF_5 (b) SF_4 (c) ClF_3

(d) BrF_5 (e) XeF_2 (f) XeF_4

9. Use VSEPR theory to predict the shapes of the following ions:

(a) SiF_6^{2-} (b) SiF_5^- (c) PF_6^-

(d) SeF_5^- (e) BrF_4^- (f) IF_4^+

10. Use VSEPR theory to predict the shapes of the following oxofluorides of xenon:

(a) $XeOF_4$ (b) XeO_2F_2 (c) XeO_3F_2

Halogens

11. Arrange the following oxoacids in order of increasing acid strength, and explain your choice of order: $HClO$, H_3PO_3, $HClO_4$, H_2SO_4.

12. Write balanced equations for the following:

(a) The oxidation of a solution of chromium(III) sulfate, $Cr_2(SO_4)_3$, with sodium hypochlorite to give sodium chromate, Na_2CrO_4, and sodium chloride in basic aqueous solution.

(b) The oxidation of potassium iodide by sodium chlorate to give iodine and sodium chloride in basic aqueous solution.

(c) The oxidation of iodide by iodate ion in acidic aqueous solution.

13. Write equations for the following reactions.

(a) The formation of perbromate ion and fluoride ion by the reaction of bromate ion and fluorine gas in a basic aqueous solution.

(b) The oxidation of bromate ion to perbromate ion, xenon, and fluoride ion by XeF_2 in a basic aqueous solution.

What is the Lewis structure and the expected geometry of the perbromate ion?

14. Write equations for the reactions that occur when chlorine is bubbled into (a) a cold aqueous solution of $NaOH$ and (b) a hot aqueous solution of $NaOH$.

15. What reaction occurs if $KClO_3$ is gently heated? What reaction occurs if MnO_2 is present when $KClO_3$ is heated? Write balanced equations for both reactions.

16. Write balanced equations for the reactions of each of the following with water:

(a) PCl_5 (b) IF_5 (c) SF_4 (d) Br_2

17. Describe, by means of suitable equations, the preparation of a dilute aqueous solution of hypochlorous acid. What happens if an attempt is made to concentrate the solution by evaporation?

18. Write balanced equations for each of the following reactions:

(a) Bromine is added to a hot aqueous solution of sodium hydroxide.

(b) Bromine is added to an aqueous hydrogen peroxide solution and oxygen is liberated.

(c) Potassium bromide is heated with concentrated phosphoric acid.

(d) Sulfur dioxide is bubbled through an aqueous solution of sodium iodate.

19. Name each of the following compounds:

(a) $HOBr$ (b) $Ca(OCl)_2$ (c) $KBrO_3$

(d) $KClO_2$ (e) $Mg(ClO_4)_2$ (f) HIO_3

(g) $HBrO_4$ (h) H_5IO_6

When it is heated, iodic acid is at first dehydrated to give the oxide I_2O_5, and then it further decomposes to oxygen and iodine. Write balanced equations for these two reactions. Draw what you consider would be the most probable Lewis structure for I_2O_5.

21. Write equations for the half-reactions describing the behavior of ClO_3^-, ClO_2^-, and ClO^- as oxidizing agents in which they are reduced to Cl^- in acid solution.

22. Explain why sulfur forms the fluoride SF_4 but oxygen does not form the corresponding fluoride OF_4.

23. Iodine reacts with liquid chlorine at $-40°C$ to give an orange compound containing 54.5% iodine and 45.5% chlorine. What is the empirical formula of the orange compound?

24. What is the oxidation number of the halogen in each of the following molecules and ions?

    (a) ClO_3^- (b) BrF_3 (c) $HClO_4$

    (d) ClO_2^- (e) ClO_2 (f) H_5IO_6

25. Write the equation for the half-reaction for the reduction of ClO_3^- to Cl_2. This reduction can be carried out with $Fe^{2+}(aq)$. Write the equation for the complete oxidation-reduction reaction.

26. What is a disproportionation reaction? Give two examples from the chemistry of the halogens and another from the chemistry of the noble gases.

27. Iodine is prepared from sodium iodate by adding a controlled amount of sodium hydrogen sulfite, $NaHSO_3$, which is oxidized to sulfate. Why is it important not to add an excess of $NaHSO_3$? How much sodium hydrogen sulfite should be used to prepare iodine from 50.0 kg of $NaIO_3$?

Noble Gases

28. What is the oxidation number of xenon in each of the following molecules and ions?

    (a) XeF_2 (b) XeF_4 (c) XeO_2F_2

    (d) XeF_3^+ (e) XeO_3 (f) XeO_4

29. Why do we not expect to find compounds of neon analogous to those formed by krypton and xenon?

30. Colorless crystals of a compound A containing only Xe and F reacted with a limited amount of water to give compound B, a colorless liquid containing Xe, F, and O. Compound B reacted with excess water to give a colorless solution from which compound C, a colorless crystalline solid containing only Xe and O, was isolated when the solution was evaporated to dryness. Analysis and molar mass determinations gave the results in the accompanying table. Determine the molecular formulas of compounds A, B, and C. Explain the reactions, and predict the shapes of A, B, and C on the basis of VSEPR theory.

COMPOUND	PERCENT F	PERCENT O	MOLAR MASS $(g\ mol^{-1})$
A	46.5	—	245
B	34.0	7.2	223
C	—	26.8	179

31. The $\Delta H°$ for the following reaction is $-402\ kJ\ mol^{-1}$:

$$XeO_3(s) \longrightarrow Xe(g) + \tfrac{3}{2}O_2(g)$$

The enthalpy of sublimation of XeO_3 is $80\ kJ\ mol^{-1}$. Calculate the $Xe=O$ bond energy. Compare this value with some of the bond energies given in Table 12.2, and comment on the instability of XeO_3. $\Delta H_f° (O, g) = 249.1\ kJ\ mol^{-1}$.

32. Xenon trioxide added to an aqueous solution of Mn^{2+} oxidizes it immediately to MnO_4^- and is itself reduced to xenon. Write a balanced equation for this reaction.

CHAPTER 21

THE TRANSITION METALS

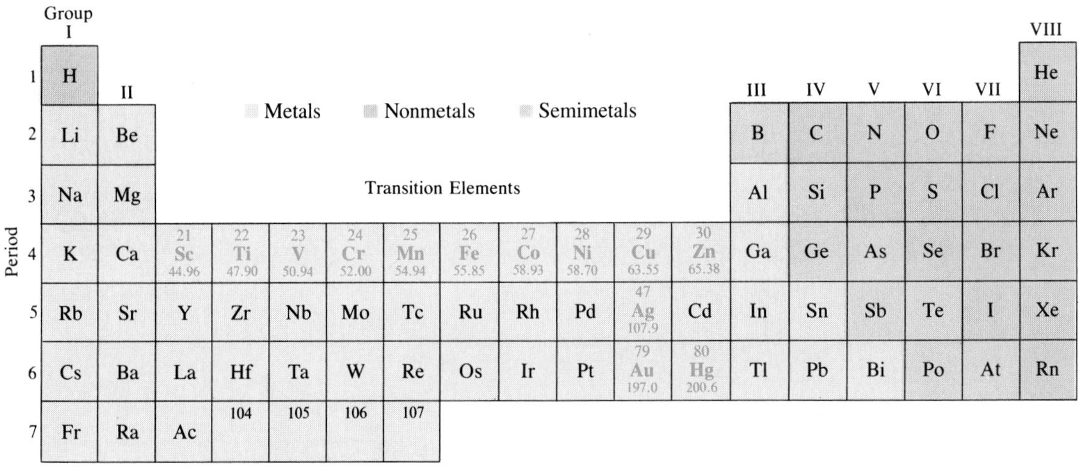

More than three-quarters of the elements are metals. Among these are 30 elements known as the **transition metals**. They occupy the middle portion of the periodic table between the main groups II and III in periods 4, 5, and 6. They include some very important metals, such as iron and copper, that we discussed in Chapter 9, as well as vanadium, chromium, cobalt, nickel, silver, gold, and mercury.

They exhibit a wide variety of properties, but they have several common characteristics. These include:

1. *Multiple oxidation states*: Unlike the metals in groups I and II and aluminum in group III, the majority of the transition metals form compounds in several different oxidation states.

2. *Complex ions*: The transition metals form a large number of complex ions. In this respect they again differ from the metals of groups I and II, which form relatively few complex ions.

3. *Color*: The majority of the compounds of the transition metals are colored. We have already met purple $KMnO_4$, yellow-brown $FeCl_3$, and blue $CuSO_4 \cdot 5H_2O$. Again, they differ from the main group metals, the majority of which form colorless compounds.

Many of the transition metals have important practical applications. Some of them are very hard and strong and are used in structural metals. The prime example is iron and the many different steels, which are alloys of iron with other transition metals such as vanadium, chromium, manganese, cobalt, and nickel, as well as other elements such as carbon. Other transition metals such as silver, gold, and platinum are important for their lack of reactivity and resistance to corrosion.

Several of the transition metals are important as trace elements in living systems. In addition to H, C, N, O, P, S, Na, K, and Ca, which are required by the human body in relatively large amounts, we also require smaller amounts of other elements including several of the transition metals. Iron is an essential part of hemoglobin, the oxygen carrier in the blood. Iron, copper, and zinc and some other transition metals are essential constituents of some enzymes.

The three rows of transition elements are often called the first, second, and third transition series. We will restrict our discussion mainly to the first series because this includes most of the familiar and more important of the transition metals, although we will briefly discuss silver, gold, and mercury.

Some transition metal compounds. Clockwise from left: Chromium (3 samples), cobalt (2 samples), copper (2 samples), nickel (1 sample), and iron (1 sample).

21.1 PROPERTIES OF THE TRANSITION METALS

We begin by reviewing the electron configurations of the transition metals because they determine their characteristic properties. We will see that they have an important feature not found in the electron configurations of the main group elements.

Electron Configurations and Ionization Energies

The electron configurations of the first 38 elements are given in Table 21.1. Recall from Chapter 6 that for some elements the sublevels of the $n = 3$ shell and the $n = 4$ shell overlap (see Figure 21.1). In particular, this is the case for potassium, calcium, scandium, and the succeeding elements. Thus for potassium and calcium the 4s level is lower in energy than the 3d level and is therefore filled before the 3d level. Then at scandium the 3d level begins to fill. Scandium is the first element in the first transition series, and zinc, which has both 3d and 4s levels filled, is the last element. In the second series the 4d level is filled, and in the third series the 5d level is filled.

As we saw in Chapter 6, we can obtain considerable information on electron configurations from ionization energies. Table 21.2 gives the first ionization energy, that is, the ionization energy of the most easily removed electron, for the first 36 elements. Let us compare the ionization energies for the third- and fourth-period elements.

In the third period the ionization energy increases from Na ($3s^1$) to Mg ($3s^2$) and then decreases for Al ($3s^2 3p^1$), despite the increased core charge. The reason for the decrease is that in Al there is an electron in a 3p level that has a higher energy than the electron in the 3s level, and this electron therefore has a lower ionization energy. In the fourth period there is a similar increase in ionization

Table 21.1 Electron Configurations of the First 38 Elements

		n = 1	2		3			4				5
PERIOD		s	s	p	s	p	d	s	p	d	f	s
1	H	$1s^1$										
	He	$1s^2$										
2	Li	$1s^2$	$2s^1$									
	Be	$1s^2$	$2s^2$									
	B	$1s^2$	$2s^2$	$2p^1$								
	C	$1s^2$	$2s^2$	$2p^2$								
	N	$1s^2$	$2s^2$	$2p^3$								
	O	$1s^2$	$2s^2$	$2p^4$								
	F	$1s^2$	$2s^2$	$2p^5$								
	Ne	$1s^2$	$2s^2$	$2p^6$								
3	Na	$1s^2$	$2s^2$	$2p^6$	$3s^1$							
	Mg	$1s^2$	$2s^2$	$2p^6$	$3s^2$							
	Al	$1s^2$	$2s^2$	$2p^6$	$3s^2$	$3p^1$						
	Si	$1s^2$	$2s^2$	$2p^6$	$3s^2$	$3p^2$						
	P	$1s^2$	$2s^2$	$2p^6$	$3s^2$	$3p^3$						
	S	$1s^2$	$2s^2$	$2p^6$	$3s^2$	$3p^4$						
	Cl	$1s^2$	$2s^2$	$2p^6$	$3s^2$	$3p^5$						
	Ar	$1s^2$	$2s^2$	$2p^6$	$3s^2$	$3p^6$						
4	K	$1s^2$	$2s^2$	$2p^6$	$3s^2$	$3p^6$		$4s^1$				
	Ca	$1s^2$	$2s^2$	$2p^6$	$3s^2$	$3p^6$		$4s^2$				
	Sc	$1s^2$	$2s^2$	$2p^6$	$3s^2$	$3p^6$	$3d^1$	$4s^2$				
	Ti	$1s^2$	$2s^2$	$2p^6$	$3s^2$	$3p^6$	$3d^2$	$4s^2$				
	V	$1s^2$	$2s^2$	$2p^6$	$3s^2$	$3p^6$	$3d^3$	$4s^2$				
	Cr	$1s^2$	$2s^2$	$2p^6$	$3s^2$	$3p^6$	$3d^5$	$4s^1$				
	Mn	$1s^2$	$2s^2$	$2p^6$	$3s^2$	$3p^6$	$3d^5$	$4s^2$				
	Fe	$1s^2$	$2s^2$	$2p^6$	$3s^2$	$3p^6$	$3d^6$	$4s^2$				
	Co	$1s^2$	$2s^2$	$2p^6$	$3s^2$	$3p^6$	$3d^7$	$4s^2$				
	Ni	$1s^2$	$2s^2$	$2p^6$	$3s^2$	$3p^6$	$3d^8$	$4s^2$				
	Cu	$1s^2$	$2s^2$	$2p^6$	$3s^2$	$3p^6$	$3d^{10}$	$4s^1$				
	Zn	$1s^2$	$2s^2$	$2p^6$	$3s^2$	$3p^6$	$3d^{10}$	$4s^2$				
	Ga	$1s^2$	$2s^2$	$2p^6$	$3s^2$	$3p^6$	$3d^{10}$	$4s^2$	$4p^1$			
	Ge	$1s^2$	$2s^2$	$2p^6$	$3s^2$	$3p^6$	$3d^{10}$	$4s^2$	$4p^2$			
	As	$1s^2$	$2s^2$	$2p^6$	$3s^2$	$3p^6$	$3d^{10}$	$4s^2$	$4p^3$			
	Se	$1s^2$	$2s^2$	$2p^6$	$3s^2$	$3p^6$	$3d^{10}$	$4s^2$	$4p^4$			
	Br	$1s^2$	$2s^2$	$2p^6$	$3s^2$	$3p^6$	$3d^{10}$	$4s^2$	$4p^5$			
	Kr	$1s^2$	$2s^2$	$2p^6$	$3s^2$	$3p^6$	$3d^{10}$	$4s^2$	$4p^6$			
5	Rb	$1s^2$	$2s^2$	$2p^6$	$3s^2$	$3p^6$	$3d^{10}$	$4s^2$	$4p^6$			$5s^1$
	Sr	$1s^2$	$2s^2$	$2p^6$	$3s^2$	$3p^6$	$3d^{10}$	$4s^2$	$4p^6$			$5s^2$

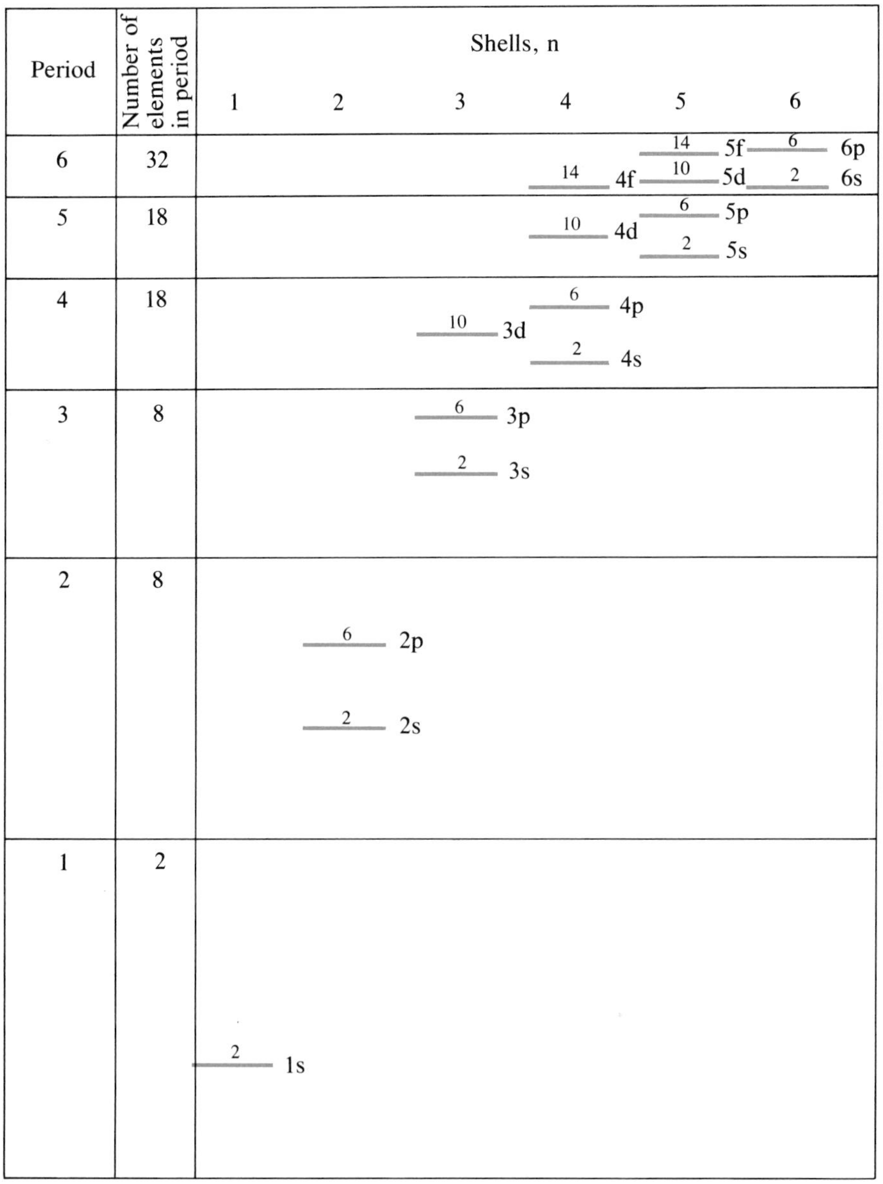

Period	Number of elements in period	Shells, n					
		1	2	3	4	5	6
6	32				4f 14	5f 14 · 5d 10	6p 6 · 6s 2
5	18				4d 10	5p 6 · 5s 2	
4	18			3d 10	4p 6 · 4s 2		
3	8			3p 6 · 3s 2			
2	8		2p 6 · 2s 2				
1	2	1s 2					

Figure 21.1 Energy Level Diagrams. This diagram is not to scale and it is not meant to represent the energy levels for any particular element. By filling each of the levels in order of increasing energy, we can, however, correctly predict the electron configurations of most of the elements. In particular, it shows that for the neutral atoms of transition elements, the 4s level is below the 3d level and is therefore filled first. However, the 4s level is not below the 3d level for all the elements or, in fact, for the ions of the transition metals.

energy from K ($4s^1$) to Ca ($4s^2$), but at Sc there is a small increase in the ionization energy rather than the decrease observed for Al in period 3. The increase strongly indicates that Sc is not analogous to aluminum in electron configuration. As we have seen, the 3d level begins to be occupied at this point, and it has an energy very close to that of the 4s level. Scandium has the electron configuration $3d^1 4s^2$. The ionization energy then increases slowly with increasing nuclear charge up to Zn. Only then do we observe a marked decrease in the ionization energy at Ga ($3d^{10} 4s^2 4p^1$), when the 4p level starts to fill.

Densities, Melting Points, Radii, and Metallic Bonding

Some properties of potassium, calcium, and the first series of transition metals are listed in Table 21.3. Since the atoms in a metallic crystal are packed closely together and are touching each other, the distance between any two adjacent

Table 21.2 First Ionization Energies for the First 36 Elements

Z	ELEMENT	IONIZATION ENERGY (MJ mol^{-1})	Z	ELEMENT	IONIZATION ENERGY (MJ mol^{-1})
1	H	1.312	19	K	0.419
2	He	2.373	20	Ca	0.590
3	Li	0.520	21	Sc	0.631
4	Be	0.899	22	Ti	0.658
5	B	0.809	23	V	0.650
6	C	1.086	24	Cr	0.653
7	N	1.400	25	Mn	0.717
8	O	1.314	26	Fe	0.759
9	F	1.680	27	Co	0.758
10	Ne	2.080	28	Ni	0.737
11	Na	0.496	29	Cu	0.745
12	Mg	0.732	30	Zn	0.906
13	Al	0.578	31	Ga	0.579
14	Si	0.786	32	Ge	0.762
15	P	1.012	33	As	0.946
16	S	1.000	34	Se	0.941
17	Cl	1.251	35	Br	1.139
18	Ar	1.521	36	Kr	1.303

atoms is twice the radius of the atom. These radii, which are called **metallic radii**, are listed in Table 21.3 and are also shown in Figure 21.2.

We recall from Chapter 9 that we expect the melting point of a metal to increase with increasing strength of the metallic bond. The strength of the bonding in a metal is determined primarily by the number of electrons that contribute to the metallic bonding. Potassium ($4s^1$) has only one electron available for metallic bonding, while calcium ($4s^2$) has two electrons available for metallic bonding. Thus calcium has a higher melting point than potassium (Table 21.3 and Figure 21.3). Since scandium has a still higher melting point, and the melting point continues to increase up to vanadium, we conclude that the strength of

Table 21.3 Properties of Potassium, Calcium, and the First Series of Transition Metals

METAL	ATOMIC NUMBER	ATOMIC MASS	MELTING POINT (K)	DENSITY (g cm^{-3})	$r_{metallic}$ (pm)
K	19	39.10	337	0.86	235
Ca	20	40.08	1110	1.55	197
Sc	21	44.96	1812	3.0	187
Ti	22	47.90	1941	4.51	147
V	23	50.94	2173	6.1	134
Cr	24	52.00	2148	7.19	128
Mn	25	54.94	1518	7.43	127
Fe	26	55.85	1809	7.86	126
Co	27	58.93	1768	8.9	125
Ni	28	58.70	1726	8.9	125
Cu	29	63.55	1356	8.96	128
Zn	30	65.38	693	7.14	134

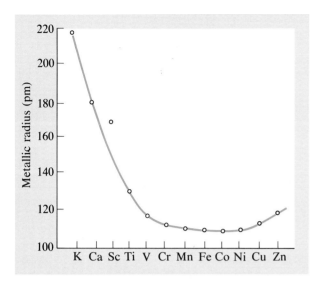

Figure 21.2 Metallic Radii of the Elements K to Zn.
These radii are obtained by taking one-half the distance
between adjacent atoms in the crystal of the metal.

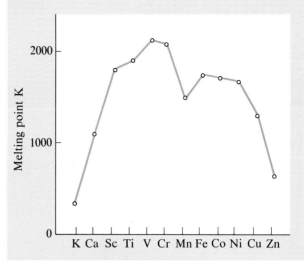

Figure 21.3 Melting Points of the Elements from K to Zn.

the metallic bonding also increases up to vanadium. An increasing number of electrons, therefore, must contribute to the metallic bonding; these must be the 4s and 3d electrons. Presumably, scandium ($3d^1 4s^2$) has three metallic bonding electrons, titanium ($3d^2 4s^2$) four, and vanadium ($3d^3 4s^2$) five. However, after vanadium the melting point decreases slightly to nickel and then more markedly to copper and zinc. Manganese has an unexpectedly low melting point, which reflects the fact that it has an unusual structure that is not close-packed.

There is also a marked increase in the density from 0.86 g cm^{-3} for potassium to 7.19 g cm^{-3} for chromium. The density depends on the mass of the atoms and the volume that they occupy. The atomic mass increases by only 20% from potassium to chromium but the density increases by over 800%, so the volume that the atoms occupy must be decreasing. The increasingly strong metallic bonds pull the atoms more closely together, so their radii decrease from potassium to chromium (see Table 21.3 and Figure 21.2). The metallic radius is then almost constant from chromium to nickel and finally increases slightly to copper and then zinc.

We conclude from the melting points and the densities (or metallic radii) that after vanadium, which can have a maximum of five metallic bonding electrons, or chromium, which can have a maximum of six, the number of metallic bonding electrons remains constant at five or six up to nickel, despite the fact that the total number of 3d electrons is increasing. After nickel the number of metallic bonding electrons decreases. We expect all the electrons to be more tightly held with increasing atomic number and core charge. Apparently the 3d electrons become so tightly held after vanadium or chromium that not all of them are used to form metallic bonds. We will see that we reach a similar conclusion about the availability of these electrons for bonding from considering the oxidation states of these elements.

Oxidation States

We saw in the previous section that the 3d electrons contribute to the metallic bonding in the metals of the first transition series. Not surprisingly, therefore,

Table 21.4 Common Oxidation States of the First Series of Transition Metals

	Sc	Ti	V	Cr	Mn	Fe	Co	Ni	Cu	Zn	
Total number of 3d and 4s electrons	3	4	5	6	7	8	9	10	11	12	
	—	—	—	—	—	—	—	—	+1	—	⎫
	—	+2	+2	+2	+2	+2	+2	+2	+2	+2	⎬ Ionic compounds
	+3	+3	+3	+3	—	+3	+3				⎭
	—	+4	+4	—	+4	—					⎫
			+5	—	—	—					⎬ Covalent compounds
				+6	+6	(+6)					
					+7						⎭

the 3d electrons are also used in compound formation by these elements, the 4s electrons are always used, and one or more of the 3d electrons may be used. This possibility of using different numbers of electrons in compound formation is responsible for the variety of oxidation states exhibited by these elements. The common oxidation states of the transition metals are summarized in Table 21.4.

The maximum observed oxidation state increases from $+3$ for scandium to $+7$ for manganese, but thereafter it decreases. Although the $+6$ oxidation state is known for iron, its most common oxidation states are $+2$ and $+3$. For nickel, copper, and zinc, the highest common oxidation state is only $+2$. The variation in the maximum oxidation state shows that in compound formation all the 3d electrons are available for bonding up to manganese, but for the following elements, although the total number of 3d electrons increases, the number of 3d electrons available for bonding decreases from 4 in iron, in the $+6$ oxidation state, to only one in the $+3$ oxidation state of cobalt, and to none in the $+2$ oxidation state of zinc. We see again that as the core charge increases, the number of 3d electrons that are available for bonding decreases. In fact, for zinc the 3d electrons are not used at all; they may be considered to have become part of the core. The $+2$ oxidation state is found for all the elements except scandium. It corresponds to the loss of the two 4s electrons to form dipositive ions such as Mn^{2+}, Fe^{2+}, and Ni^{2+}.

21.2 FIRST SERIES OF TRANSITION ELEMENTS

Many transition metals, including most of the elements of the first series, are very important in industry and are produced in very large amounts, either as the pure metals or as alloys. The most important alloys are the many varieties of steel, the principal constituent of which is iron, together with one or more other metals, often those of the first transition series, and small amounts of some nonmetals such as carbon, silicon, and phosphorus.

In this section we consider some of the more important compounds of the first series of transition elements. We will discuss briefly their preparation, properties, and important uses, and we will relate their properties to their electron configurations. In their highest oxidation states these elements have high electronegativities and behave more like nonmetals than metals. Thus the chromate ion, CrO_4^{2-}, is in some ways like the sulfate ion, SO_4^{2-}, and the permanganate ion, MnO_4^-, resembles the perchlorate ion, ClO_4^-.

Scandium ($3d^1 4s^2$)

Scandium is widely distributed in nature, principally in the form of Sc_2O_3. The three valence electrons are easily removed to give Sc^{3+}. All the compounds of scandium contain this ion; they resemble the corresponding Al^{3+} compounds. They have no important uses.

Titanium ($3d^2 4s^2$)

Titanium is the tenth most abundant element in the earth's crust. After iron it is the second most abundant transition metal. The principal ores are *rutile*, TiO_2, and *ilmenite*, $FeTiO_3$.

PREPARATION AND USES The element is difficult to prepare in the pure state because at high temperatures it reacts readily with carbon, hydrogen, nitrogen, and oxygen. The most important method for preparing the metal involves heating rutile or ilmenite with coke and chlorine. The volatile covalent compound $TiCl_4$ distills off, leaving behind involatile, ionic $FeCl_2$:

$$FeTiO_3(s) + 3C(s) + 3Cl_2(g) \longrightarrow FeCl_2(s) + 3CO(g) + TiCl_4(g)$$

The $TiCl_4$ is condensed, purified, and then vaporized into a vessel filled with red-hot magnesium shavings in an atmosphere of argon in which it is reduced to titanium metal. Most of the liquid $MgCl_2$ is drained off, and the remainder is removed by washing it out with water or by vaporizing it in a vacuum at about 900°C:

$$TiCl_4(g) + 2Mg(s) \longrightarrow 2MgCl_2(l) + Ti(s)$$

Titanium is becoming an increasingly important metal because it is very strong, has a low density, is highly resistant to corrosion, and has a high melting point. It has many important uses in rockets, jet engines, and aircraft, where its very high melting point makes it much more useful than aluminum. It is also widely used in chemical plants because of its great resistance to corrosion; for example, it is not attacked by any of the common acids or by chlorine. Steel containing titanium is particularly hard and very resistant to abrasion.

OXIDATION STATES Titanium can use two, three, or four valence electrons to give compounds containing titanium in the $+2$, $+3$, and $+4$ oxidation states (see Table 21.5). The $+4$ state is the most stable oxidation state. Compounds of titanium in the $+3$ and $+2$ states are very strong reducing agents that are readily oxidized to the $+4$ state.

COMPOUNDS The most important Ti(IV) compound is the dioxide TiO_2, which occurs naturally as the mineral rutile. Titanium dioxide is very important as a white pigment that is widely used in paints, paper, and many other products. Titanium dioxide is an acidic oxide that reacts with molten metal hydroxides to give titanates such as Na_2TiO_3.

Table 21.5 Oxidation States of Titanium

OXIDATION STATE	UNUSED VALENCE SHELL ELECTRONS	TYPICAL EXAMPLES
$+4$	d^0	TiO_2, $TiCl_4$
$+3$	d^1	Ti^{3+}
$+2$	d^2	$TiCl_2$

21.2 FIRST SERIES OF TRANSITION ELEMENTS

Titanium(IV) chloride consists of covalent $TiCl_4$ molecules. It is a colorless liquid that boils at 136°C and has a pungent odor. It fumes strongly in moist air, forming a dense white cloud of titanium dioxide (Experiment 21.1):

$$TiCl_4 + 2H_2O \longrightarrow TiO_2 + 4HCl$$

It is used in producing smoke screens and in skywriting. In its volatility and reaction with water, it resembles a nonmetal halide such as $SiCl_4$ or PCl_3. Titanium tetrachloride, $TiCl_4$, can be prepared by heating the element in chlorine or by heating the more readily available dioxide with carbon and chlorine:

$$TiO_2(s) + 2C(s) + 2Cl_2(g) \longrightarrow TiCl_4(g) + 2CO(g)$$

Titanium(III) chloride, $TiCl_3$, can be made by heating $TiCl_4$ with hydrogen:

$$2TiCl_4 + H_2 \longrightarrow 2TiCl_3 + 2HCl$$

It is an ionic compound that is soluble in water to give violet solutions containing the hydrated titanium(III) ion, $Ti(H_2O)_6^{3+}$.

In the $+3$ oxidation state titanium has one unused valence electron; that is, it has the valence shell electron arrangement $3d^1$, which is usually written simply as d^1. We emphasize that titanium uses all its valence electrons in the $+4$ state by describing it as a d^0 state (see Table 21.5).

When $TiCl_4$ is heated with titanium, titanium(II) chloride, $TiCl_2$, can be prepared:

$$TiCl_4 + Ti \longrightarrow 2TiCl_2$$

All Ti(II) compounds are very strong reducing agents. For example, they reduce water to hydrogen:

$$TiCl_2 + 2H_2O \longrightarrow TiO_2 + 2HCl + H_2$$

There is therefore no aqueous solution chemistry of Ti^{2+}.

Vanadium ($3d^3 4s^2$)

OCCURRENCE AND USES Although vanadium is very widely distributed, it constitutes only 0.02% of the earth's crust. It occurs mainly in the form of the vanadate ion, VO_4^{3-}. Like titanium, pure vanadium is very difficult to prepare

EXPERIMENT 21.1

Hydrolysis of Titanium Tetrachloride

When dry air is bubbled through colorless liquid titanium tetrachloride, $TiCl_4$, a dense white smoke of TiO_2 is formed as soon as the $TiCl_4$ vapor comes into contact with atmospheric moisture.

because it combines readily at high temperature with oxygen, nitrogen, and carbon. Because it is used mainly as a component of alloy steels, it is usually made in the form of ferrovanadium, an iron-vanadium alloy, rather than as the pure metal.

COMPOUNDS   The oxidation states of vanadium are summarized in Table 21.6. In its highest ($+5$) oxidation state vanadium behaves more like a nonmetal than a metal.

The most important compound of vanadium is vanadium(V) oxide, V_2O_5, an orange-red solid that can be prepared by heating vanadium metal in oxygen. It is important as a catalyst in the contact process for the manufacture of sulfuric acid (Chapter 7). Vanadium(V) oxide is amphoteric. It dissolves in a strongly basic solution to give the vanadate ion, VO_4^{3-}:

$$V_2O_5(s) + 6OH^-(aq) \longrightarrow 2VO_4^{3-}(aq) + 3H_2O$$

It dissolves in acids to give $VO_2^+(aq)$:

$$V_2O_5(s) + 2H_3O^+(aq) \longrightarrow 2VO_2^+(aq) + 3H_2O$$

Vanadate, VO_4^{3-}, is a colorless ion with a tetrahedral structure like the phosphate ion. Many polymeric forms of the vanadate ion are known; for example, $(VO_3^-)_n$ is like the metaphosphate ion $(PO_3^-)_n$ and consists of an infinite chain of VO_4 tetrahedra that share corners:

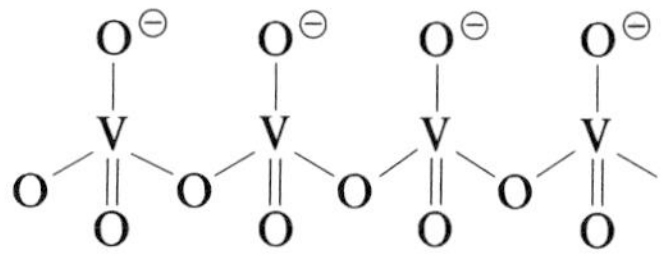

The VO_2^+ ion has a linear structure like the NO_2^+ ion (Chapter 17).

If an acidic solution of $VO_2^+(aq)$ is treated with a reducing agent such as zinc or Fe^{2+}, a blue solution of oxovanadium(IV) ion, $VO(H_2O)_5^{2+}$, is obtained.

If vanadium is heated in chlorine, another vanadium(IV) compound, vanadium tetrachloride, VCl_4, is obtained. This is a red-brown liquid with a boiling point of 154°C, and like $TiCl_4$, it reacts vigorously with water:

$$VCl_4(l) + 6H_2O(l) \longrightarrow VO(H_2O)_5^{2+}(aq) + 2H^+(aq) + 4Cl^-(aq)$$

If a $VO(H_2O)_5^{2+}$ solution is further reduced with zinc or Zn-Hg amalgam, the color of the solution changes from bright blue to the green of $V(H_2O)_6^{3+}$. With an excess of zinc, V^{3+} can be further reduced to a violet solution of $V(H_2O)_6^{2+}$ (see Experiment 21.2). Both V^{2+} and V^{3+} are strong reducing agents, and their aqueous solutions are oxidized by the air to vanadium(IV).

Chromium ($3d^5 4s^1$)

OCCURRENCE AND USES   Chromium is familiar as the protective coating applied to automobile parts, such as fenders, and to many other steel objects to

Table 21.6 Oxidation States of Vanadium

OXIDATION STATE	UNUSED VALENCE SHELL ELECTRONS	TYPICAL EXAMPLES
$+5$	d^0	V_2O_5, VO_4^{3-}, VO_2^+
$+4$	d^1	VO^{2+}
$+3$	d^2	V^{3+}
$+2$	d^3	V^{2+}

Oxidation States of Vanadium

A yellow acidified solution of ammonium vanadate, $(NH_4)_3VO_4$, in which vanadium is in the $+5$ oxidation state, can be reduced with zinc amalgam first to a blue solution of VO^{2+}, in which vanadium is in the $+4$ oxidation state, then to a green solution of V^{3+}, and finally to a violet solution of V^{2+}.

If a layer of purple solution of potassium permanganate is gently added on top of a violet solution of V^{2+}, it slowly oxidizes the V^{2+} over a period of a few hours, forming successive layers of the different oxidation states of vanadium. From bottom to top: violet V^{2+}, green V^{3+}, blue VO^{2+}, yellow VO_4^{3-}, brown MnO_2 (produced by reduction of MnO_4^-), and purple $KMnO_4$.

improve their appearance and increase their resistance to corrosion. The commonest ore of chromium is *chromite*, $FeCr_2O_4$. This ore is reduced by carbon in an electric furnace to give ferrochrome, an iron-chromium alloy, that is used to make a variety of stainless steels.

OXIDATION STATES The principal oxidation states of chromium are summarized in Table 21.7.

The unexpected electron configuration $3d^54s^1$, rather than $3d^44s^2$, again emphasizes that the 3d and 4s levels are very close in energy and that alternative electron arrangements may differ only very slightly in energy. When the 3d and 4s levels are sufficiently close, the repulsion between the two electrons in the 4s orbital can cause one of them to move into a 3d orbital.

The $+6$ oxidation state is the maximum possible oxidation state for chromium, because in this state it uses all six valence electrons in the formation of bonds. Chromium(VI) compounds are generally strong oxidizing agents and are readily reduced to the $+3$ state, which is the most stable oxidation state of chromium. In the $+2$ state chromium is a reducing agent and is readily oxidized to the $+3$ state.

THE $+6$ OXIDATION STATE In its highest oxidation state chromium has a high electronegativity and behaves like a nonmetal. Chromium(VI) compounds are predominately covalent. The yellow chromate ion, CrO_4^{2-}, resembles the

Table 21.7 Oxidation States of Chromium

OXIDATION STATE	UNUSED VALENCE SHELL ELECTRONS	TYPICAL EXAMPLES
+6	d^0	CrO_3
+3	d^3	Cr_2O_3, Cr^{3+}
+2	d^4	Cr^{2+}

sulfate ion and has the same tetrahedral structure. Metal chromates often resemble the corresponding sulfates. For example, like $BaSO_4$ and $PbSO_4$, $BaCrO_4$ and $PbCrO_4$ are insoluble in water.

The corresponding acid, chromic acid, H_2CrO_4, cannot be made, because when H_3O^+ is added to a yellow chromate solution, the solution becomes orange-red owing to the formation of the dichromate ion, $Cr_2O_7^{2-}$ (see Experiment 21.3):

$$2CrO_4^{2-}(aq) + 2H_3O^+(aq) \longrightarrow Cr_2O_7^{2-}(aq) + 3H_2O(l)$$

We can think of this reaction as occurring by the protonation of the chromate ion to give the hydrogen chromate ion, $HCrO_4^-$, followed by elimination of water from two $HCrO_4^-$ ions:

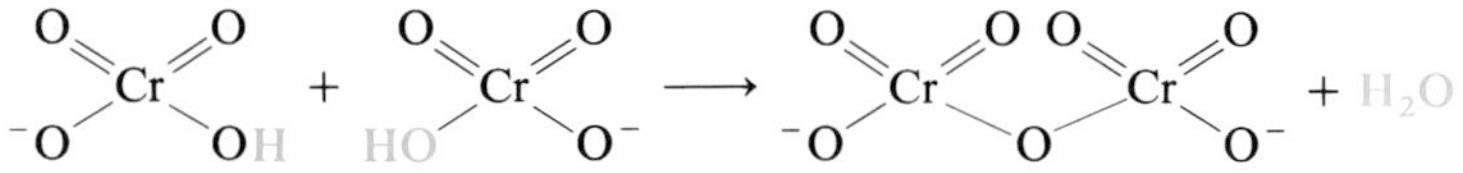

This is an example of a condensation reaction (Chapter 7). The dichromate ion has a structure similar to that of the disulfate ion (Chapter 7). Potassium dichromate, $K_2Cr_2O_7$, crystallizes from solution as bright orange-red crystals. The formation of the dichromate ion is reversible, and if base is added to a dichromate solution, the color changes back to the yellow color of the chromate ion:

$$Cr_2O_7^{2-}(aq) + 2OH^-(aq) \longrightarrow 2CrO_4^{2-}(aq) + H_2O$$

Orange-red   Yellow

If potassium dichromate is dissolved in concentrated sulfuric acid, it forms dichromic acid, which is dehydrated to give red solid chromium(VI) oxide (chromium trioxide), CrO_3:

$$Cr_2O_7^{2-}(aq) + 2H^+(aq) \rightleftharpoons H_2Cr_2O_7(aq) \longrightarrow 2CrO_3(s) + H_2O$$

EXPERIMENT 21.3

Chromium (VI) Compounds

Left to right: If water is added to dark red solid CrO_3, an orange solution of dichromic acid, $H_2Cr_2O_7$, is formed. When sodium hydroxide is added to this solution, the dichromate ion, $Cr_2O_7^{2-}$, is converted to the chromate ion, CrO_4^{2-}, and the color changes to yellow. When aqueous hydrochloric acid is added, the orange dichromate ion, $Cr_2O_7^{2-}$, is reformed.

Figure 21.4 Structure of
Chromium Trioxide.

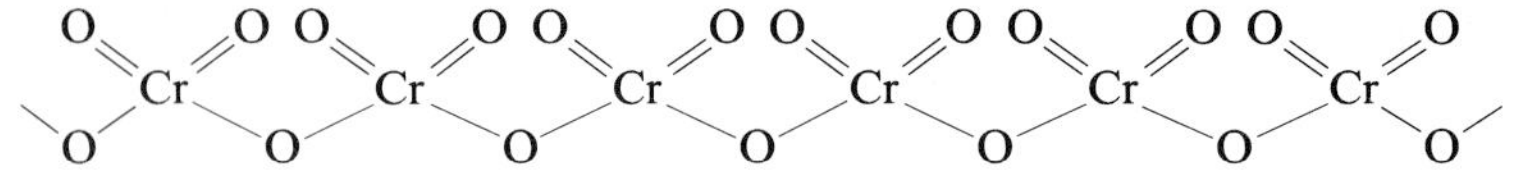

Figure 21.4 shows that chromium trioxide has a chain structure like that of polymeric sulfur trioxide (Chapter 7).

Chromium in the $+6$ oxidation state is a strong oxidizing agent, and potassium dichromate and chromium trioxide are widely used as oxidizing agents. In particular, chromium trioxide and the dichromate ion are often used to oxidize alcohols to ketones and aldehydes (Chapter 19). For example, 2-propanol is oxidized by CrO_3 or $Cr_2O_7^{2-}$ to propanone (acetone):

$$3CH_3\!-\!\underset{\underset{\displaystyle OH}{|}}{CH}\!-\!CH_3 + 2CrO_3 + 6H^+ \longrightarrow 3CH_3\!-\!\underset{\underset{\displaystyle O}{\|}}{C}\!-\!CH_3 + 2Cr^{3+} + 6H_2O$$

And 1-propanol is oxidized to propanal:

$$CH_3CH_2CH_2OH \xrightarrow{\;Cr_2O_7^{2-},\,H^+\;} CH_3CH_2CHO$$

$$\text{1-Propanal} \qquad\qquad\qquad \text{Propanal}$$

Although $Cr_2O_7^{2-}$ and CrO_3 are important oxidizing agents in the laboratory, they are too expensive to be used on an industrial scale. For large-scale preparations oxygen is the cheapest and most readily available oxidizing agent, although high temperatures and a catalyst are usually needed to obtain convenient and economical reaction rates.

Example 21.1

(a) Write a balanced equation for the half-reaction in which $Cr_2O_7^{2-}$ is reduced to Cr^{3+} in acid solution.

(b) Then write a balanced equation for the oxidation of Fe^{2+} to Fe^{3+} by $Cr_2O_7^{2-}$ in acid solution.

Solution

(a) First, we write the known reactants and products and assign oxidation numbers:

$$\overset{+6}{Cr_2}O_7^{2-} \longrightarrow 2\overset{+3}{Cr^{3+}}$$

Two chromium atoms are each reduced from the $+6$ state to the $+3$ state, so the total change in oxidation number is -6. This requires six electrons, so we can write

$$Cr_2O_7^{2-} + 6e^- \longrightarrow 2Cr^{3+}$$

Next, we balance for charges by adding H^+ or OH^- ions as appropriate. Since the reaction is in acid solution and we have 8 negative charges on the left and 6 positive charges on the right, we can balance charges by adding $14H^+$ to the left side:

$$Cr_2O_7^{2-} + 14H^+ + 6e^- \longrightarrow 2Cr^{3+}$$

Now we have 7 O atoms and 14 H atoms on the left side that do not appear on the right. If we add 7 water molecules to the right side, the equation is balanced:

$$Cr_2O_7^{2-} + 14H^+ + 6e^- \longrightarrow 2Cr^{3+} + 7H_2O$$

(b) The equation for the oxidation of Fe^{2+} to Fe^{3+} is

$$Fe^{2+} \longrightarrow Fe^{3+} + e^-$$

To combine this equation with the equation for the reduction of dichromate, we must first multiply it by 6 so that when the two half-reactions are added, the electrons cancel:

$$6Fe^{2+} \longrightarrow 6Fe^{3+} + 6e^-$$

Adding this equation to that for the reduction of $Cr_2O_7^{2-}$ gives the overall equation:

$$Cr_2O_7^{2-} + 14H^+ + 6Fe^{2+} \longrightarrow 2Cr^{3+} + 6Fe^{3+} + 7H_2O$$

THE +3 OXIDATION STATE The Cr^{3+} ion, which has a violet color in aqueous solution, forms many salts. These include $CrCl_3 \cdot 6H_2O$, $Cr_2(SO_4)_3 \cdot 18H_2O$, and *chrome alum*, $KCr(SO_4)_2 \cdot 12H_2O$, which forms large, violet octahedral crystals. In these salts Cr^{3+} is hydrated by six water molecules to give the octahedral ion $Cr(H_2O)_6^{3+}$.

When an aqueous solution of ammonia or sodium hydroxide is added to a solution of a chromium(III) salt, a grey-green precipitate of chromium(III) hydroxide, $Cr(OH)_3$, is obtained (Experiment 21.4):

$$Cr^{3+}(aq) + 3OH^-(aq) \longrightarrow Cr(OH)_3(s)$$

Chromium(III) hydroxide is amphoteric and it is dissolved by an excess of hydroxide ion or by dilute acids:

$$Cr(OH)_3(s) + OH^-(aq) \longrightarrow Cr(OH)_4^-(aq)$$
$$Cr(OH)_3(s) + 3HCl(aq) \longrightarrow CrCl_3(aq) + 3H_2O$$

EXPERIMENT 21.4

Reactions of Chromium and its Compounds

Chromium dissolves in dilute hydrochloric acid to give hydrogen and a blue solution of $CrCl_2$, which is rapidly oxidized by oxygen in the air to a green solution of $CrCl_3$.

When sodium peroxide, Na_2O_2, solution is added to a green solution of $CrCl_3$, it is oxidized to a yellow solution of CrO_4^{2-}.

When aqueous sodium hydroxide is added to a green solution of $CrCl_3$, a grey-green precipitate of $Cr(OH)_3$ is formed.

If the hydroxide is strongly heated, it is dehydrated to give chromium(III) oxide, Cr_2O_3:

$$2Cr(OH)_3 \longrightarrow Cr_2O_3 + 3H_2O$$

Chromium(III) oxide can also be made conveniently in a rather spectacular reaction by heating ammonium dichromate:

$$(NH_4)_2Cr_2O_7(s) \longrightarrow N_2(g) + 4H_2O(g) + Cr_2O_3(s)$$

In this reaction the dichromate ion oxidizes the ammonium ion to nitrogen and is itself reduced to Cr_2O_3 (see Experiment 21.5).

Chromium(III) oxide is a rather inert substance of high melting point. When exposed to the air, chromium becomes coated with a very thin, but very hard, unreactive layer of Cr_2O_3, which prevents any further attack on the metal; thus chromium is used as a protective and decorative coating for other metals. The oxide is used as a green pigment in paint.

Acidified Cr^{3+} solutions can be reduced with zinc or other reducing agents to Cr^{2+}. In aqueous solution this ion is hydrated, $Cr(H_2O)_6^{2+}$, and is blue in color (see Experiment 21.4). The chromium(II) ion is a very strong reducing agent, and its solutions must be protected from oxygen, which oxidizes it rapidly to Cr^{3+}. Oxygen can be conveniently removed from a mixture of gases by bubbling it through a chromium(II) solution.

Manganese ($3d^5 4s^2$)

OCCURRENCE AND USES   The most important ore of manganese is *pyrolusite*, MnO_2, from which the metal is obtained by reduction with carbon or carbon monoxide. Since the pure metal is not widely used, it is mostly prepared

EXPERIMENT 21.5

Thermal Decomposition of Ammonium Dichromate: The Volcano Experiment

A pile of orange ammonium dichromate can be ignited with a flame.

It then burns spontaneously, producing a shower of sparks, in a very exothermic reaction.

The pile becomes red hot, giving the appearance of a volcano in eruption. The product is dark green Cr_2O_3, which is a light, fluffy powder with a considerably greater volume than the original ammonium dichromate.

in the form of ferromanganese, an iron-manganese alloy, by the reduction of a mixture of MnO_2 and iron oxides. Manganese is used to produce steel, containing 12% Mn and 1% C, that is very hard and tough and resists abrasion extremely well. It is used, for example, for naval armor plate, bulldozer blades, and dredger buckets.

The principal oxidation states of manganese are summarized in Table 21.8.

COMPOUNDS The formula of manganese(IV) oxide is usually written as MnO_2, but careful analysis shows that it is usually closer to $MnO_{1.85}$. Although it has considerable covalent character, this oxide may be approximately described as consisting of Mn^{4+} ions and oxide ions. The unusual variable composition arises because some of the oxide ions are missing from the crystal structure, leaving a number of holes. A corresponding number of Mn^{4+} ions are replaced by Mn^{3+} ions in order to compensate for the deficiency of negative charge caused by the missing oxide ions. In this structure an oxide ion can move from its position in the lattice to an adjacent hole, which thus creates a new hole into which another oxide ion can move, and so on. Thus oxide ions can move through solid MnO_2, and it is therefore an electrical conductor. The fact that MnO_2 is an electrical conductor as well as an oxidizing agent is the basis for its use in the dry cell (Chapter 16).

A compound that has a formula such as $MnO_{1.85}$, in which the subscripts are not whole numbers, is called a **nonstoichiometric compound**. Such compounds are only possible in the solid state and only when a metal has at least two stable oxidation states. Nonstoichiometric compounds are common among the oxides and other compounds of the transition metals in their lower oxidation states.

Because manganese in MnO_2 is in an intermediate oxidation state, it behaves as both an oxidizing agent and a reducing agent. It oxidizes hydrochloric acid to chlorine and is reduced to manganese(II) ion, Mn^{2+}:

$$MnO_2(s) + 4HCl(aq) \longrightarrow Cl_2(g) + Mn^{2+}(aq) + 2Cl^-(aq) + 2H_2O(l)$$

This reaction provides a convenient method for preparing chlorine in the laboratory. Hydrated $MnCl_2 \cdot 6H_2O$ can be crystallized from this solution, which contains the pale pink hydrated Mn^{2+} ion, $Mn(H_2O)_6^{2+}$. There are many other salts of Mn^{2+} such as $MnSO_4 \cdot 7H_2O$ and $Mn(NO_3)_2 \cdot 6H_2O$, which are all pale pink and also contain the hydrated ion $Mn(H_2O)_6^{2+}$.

When hydroxide ion is added to an aqueous solution containing Mn^{2+}, a white precipitate of $Mn(OH)_2$ is obtained. In the presence of air this precipitate rapidly turns brown as it is oxidized to $MnO(OH)$, one of the few compounds that contain manganese in the +3 oxidation state (see Experiment 21.7).

When H_2S is passed into a basic solution of an Mn^{2+} salt, a pale pink precipitate of MnS, manganese(II) sulfide, is obtained:

$$Mn^{2+}(aq) + S^{2-}(aq) \longrightarrow MnS(s)$$

Table 21.8 Oxidation States of Manganese

OXIDATION STATE	UNUSED VALENCE SHELL ELECTRONS	TYPICAL EXAMPLES
+7	d^0	MnO_4^-
+6	d^1	MnO_4^{2-}
+4	d^3	MnO_2
+3	d^4	$MnO(OH)$
+2	d^5	Mn^{2+}

21.2 FIRST SERIES OF
TRANSITION ELEMENTS

EXPERIMENT 21.6

Preparation of Potassium Manganate, K_2MnO_4, and Potassium Permanganate, $KMnO_4$

The dish on the left contains a mixture of white pellets of KOH and black solid MnO_2. When this mixture is heated, a very dark green molten mass of K_2MnO_4 is obtained (right).

A solution of dark green K_2MnO_4 is obtained when water is added to the K_2MnO_4 in the dish. This dark green solution is shown in the left test tube. When dilute HCl is added to the dark green K_2MnO_4, the K_2MnO_4 disproportionates to pink $KMnO_4$ and black MnO_2. In the middle tube the reaction is incomplete and some dark green K_2MnO_4 solution remains at the bottom of the tube. In the right tube the reaction has gone to completion, and the MnO_2 can be seen at the bottom of the tube below the pink solution of $KMnO_4$.

If MnO_2 is heated with solid KOH in the presence of air, it reduces oxygen from the 0 to the -2 oxidation state and is itself oxidized to the manganate ion, MnO_4^{2-}, which contains manganese in the $+6$ oxidation state (see Experiment 21.6):

$$2MnO_2(s) + 4KOH(s) + O_2(g) \longrightarrow 2K_2MnO_4(s) + 2H_2O(l)$$

Potassium manganate is an ionic compound containing the green manganate ion, MnO_4^{2-}. Salts of the manganate ion are the only known compounds containing manganese in the $+6$ oxidation state.

The manganate ion, MnO_4^{2-}, is stable only in basic solution. If a solution of manganate, MnO_4^{2-}, is acidified, it gives permanganate, MnO_4^-, and manganese dioxide, MnO_2. In this reaction the $+6$ oxidation state disproportionates to the $+7$ and $+4$ states:

$$3MnO_4^{2-} + 4H^+ \rightleftharpoons 2MnO_4^- + MnO_2 + 2H_2O$$

$$\text{Green} \qquad\qquad \text{Purple} \quad \text{Black}$$

The permanganate ion, MnO_4^-, has a deep purple color. This reaction is reversible. If OH^- is added to the purple solution of MnO_4^- containing black insoluble MnO_2, a clear green solution of MnO_4^{2-} is obtained. Manganate ion can also be oxidized to permanganate ion electrolytically or with chlorine:

$$2MnO_4^{2-} + Cl_2 \longrightarrow 2MnO_4^- + 2Cl^-$$

A solution of potassium permanganate is a powerful oxidizing agent that is used as a disinfectant and bleaching agent and for purifying water. Potassium

Reactions of Manganese(II)

Left to right: (a) An aqueous solution of the manganese(II) cation, Mn^{2+}, is very pale pink. (b) When aqueous sodium hydroxide solution is added, a gelatinous precipitate of $Mn(OH)_2$ is obtained. (c) On standing, this slowly becomes brown as it is oxidized to the manganese(III) compound, $MnO(OH)$. (d) If hydrogen peroxide is added to the $Mn(OH)_2$ precipitate it is rapidly oxidized to dark brown manganese(IV) oxide, MnO_2.

permanganate is used extensively as an oxidizing agent in analytical chemistry. Under acidic conditions MnO_4^- is reduced to Mn^{2+}:

$$MnO_4^- + 5e^- + 8H^+ \longrightarrow Mn^{2+} + 4H_2O$$

Example 21.2 Write balanced equations for the following oxidations by MnO_4^- in acid solution:

(a) Oxidation of I^- to I_2.

(b) Oxidation of Fe^{2+} to Fe^{3+}.

(c) Oxidation of oxalic acid, $(COOH)_2$, to CO_2.

Solution In each case we write the balanced oxidation half-reaction and then add it to the permanganate reduction half-reaction, after multiplying both half-reactions by the appropriate coefficient so as to eliminate the electrons on each side of the balanced equation.

(a)
$$5[2I^- \longrightarrow I_2 + 2e^-] \quad \text{(Oxidation)}$$
$$\underline{2[MnO_4^- + 5e^- + 8H^+ \longrightarrow Mn^{2+} + 4H_2O]} \quad \text{(Reduction)}$$
$$10I^- + 2MnO_4^- + 16H^+ \longrightarrow 5I_2 + 2Mn^{2+} + 8H_2O$$

(b)
$$5[Fe^{2+} \longrightarrow Fe^{3+} + e^-] \quad \text{(Oxidation)}$$
$$\underline{MnO_4^- + 5e^- + 8H^+ \longrightarrow Mn^{2+} + 4H_2O} \quad \text{(Reduction)}$$
$$5Fe^{2+} + MnO_4^- + 8H^+ \longrightarrow 5Fe^{3+} + Mn^{2+} + 4H_2O$$

(c)
$$5[(COOH)_2 \longrightarrow 2CO_2 + 2e^- + 2H^+] \quad \text{(Oxidation)}$$
$$\underline{2[MnO_4^- + 5e^- + 8H^+ \longrightarrow Mn^{2+} + 4H_2O]} \quad \text{(Reduction)}$$
$$5(COOH)_2 + 2MnO_4^- + 6H^+ \longrightarrow 10CO_2 + 2Mn^{2+} + 8H_2O$$

Example 21.3 What is the concentration of an acidic potassium permanganate solution if exactly 25.0 mL of it is completely reduced to Mn^{2+} by 73.2 mL of a $0.106M$ solution of $FeSO_4$?

21.2 FIRST SERIES OF TRANSITION ELEMENTS

Solution The number of moles of $FeSO_4$ in 73.2 mL of solution is

$$73.2 \text{ mL} \left(\frac{0.106 \text{ mol } FeSO_4}{1 \text{ L}} \right) \left(\frac{1 \text{ L}}{1000 \text{ mL}} \right) = 7.76 \times 10^{-3} \text{ mol } FeSO_4$$

From the balanced equation in Example 21.2(b) we have

$$5 \text{ mol } FeSO_4 \text{ reacts with 1 mol } KMnO_4$$

Therefore

$$7.76 \times 10^{-3} \text{ mol } FeSO_4 \left(\frac{1 \text{ mol } KMnO_4}{5 \text{ mol } FeSO_4} \right) = 1.55 \times 10^{-3} \text{ mol } KMnO_4$$

Thus the concentration of the $KMnO_4$ solution is

$$\left(\frac{1.55 \times 10^{-3} \text{ mol } KMnO_4}{25.0 \text{ mL}} \right) \left(\frac{1000 \text{ mL}}{1 \text{ L}} \right) = 0.0620 M$$

Iron, Cobalt, and Nickel

OCCURRENCE AND USES Iron was considered in Chapter 9. Cobalt is a rather scarce element that occurs principally as the sulfide CoS. It is usually found together with nickel sulfide, NiS, and copper sulfide, CuS. In the extraction of the metal, the ore is roasted in air to give the impure oxide Co_2O_3, which is then dissolved in sulfuric acid, and the hydroxide, $Co(OH)_3$, is precipitated by adding $Ca(OH)_2$ or Na_2CO_3. The hydroxide is heated to give the oxide, which is then reduced with hydrogen or carbon. Cobalt is used in the manufacture of a number of special alloys, for example, Alnico, an alloy of Co, Ni, Al, and Cu, which can be very strongly magnetized and is therefore used for making magnets.

Nickel occurs as NiS together with CuS and FeS. It occurs in sufficient concentrations to be worth mining in only a few places. The most important nickel deposits are in Sudbury, Ontario, and central Africa. The metal is prepared by roasting the ore to give the oxide and then reducing the oxide with carbon to give impure nickel containing copper and iron. Pure nickel is obtained by passing CO(g) over the impure nickel at 80°C. The nickel reacts to give nickel tetracarbonyl, $Ni(CO)_4$, which is a volatile covalent compound that boils at 43°C and is purified by distillation. It is then decomposed to pure nickel and CO(g) by heating at 200°C:

$$Ni(s) + 4CO(g) \rightleftharpoons Ni(CO)_4(g)$$

The properties of nickel are much like those of cobalt; it is almost entirely used in the manufacture of stainless steel and special alloys. An important alloy is Monel (72% Ni, 25% Cu, and 3% Fe), which is resistant to oxidation and corrosion and therefore has important applications in chemical plants. Monel is also the sheath metal on electric heating elements in kitchen ranges and ovens. Another important alloy is coinage metal, which is an alloy of 70% Cu and 30% Ni. Nichrome (60% Ni, 25% Fe, and 15% Cr) has a high melting point and a relatively high resistance and is widely used for electric resistance heaters.

COMPOUNDS OF IRON The common oxidation states of iron are $+2$ and $+3$. In Chapter 9 we described several compounds of iron in these oxidation states. There are a few compounds of iron in the $+4$ and $+6$ oxidation states. For example, if Fe_2O_3 is treated with chlorine and a strong base, a solution of the ferrate ion, FeO_4^{2-}, can be obtained. As we would expect, FeO_4^{2-} is a powerful oxidizing agent.

As we described in Chapter 9, iron forms three oxides, FeO, Fe_2O_3, and Fe_3O_4. Like MnO_2, these oxides have a variable composition and are non-stoichiometric. The oxide FeO has the sodium chloride structure and may be regarded as a close-packed arrangement of oxide ions with Fe^{2+} ions occupying the octahedral holes (Chapter 10). If a small number of Fe^{2+} ions were replaced by two-thirds as many Fe^{3+} ions, we would have an iron-deficient but electrically neutral crystal. The actual composition of iron(II) oxide is usually $Fe_{0.95}O$. Conversion of three-quarters of the Fe^{2+} to Fe^{3+} would give the composition $FeO \cdot Fe_2O_3$, or Fe_3O_4. Finally, replacement of all the Fe^{2+} by two-thirds as many Fe^{3+} gives the composition $Fe_{0.67}O$, that is, Fe_2O_3.

COMPOUNDS OF COBALT   The common oxidation states of cobalt are $+2$ and $+3$. Most of the simple compounds of cobalt are cobalt(II) compounds. But many complex ions of cobalt exist in which cobalt is in the $+3$ oxidation state; they are described later in this chapter.

The cobalt(II) ion is hydrated in aqueous solution and has the formula $Co(H_2O)_6^{2+}$. It has a pale pink color. Most salts crystallize from solution in a hydrated form and are pink or red. Examples are $CoCl_2 \cdot 6H_2O$, $Co(NO_3)_2 \cdot 6H_2O$, and $CoSO_4 \cdot 6H_2O$. They all contain $Co(H_2O)_6^{2+}$.

Addition of sodium hydroxide solution to a solution of a cobalt(II) compound gives a pink precipitate of cobalt(II) hydroxide, $Co(OH)_2$ (see Experiment 21.8). If $Co(OH)_2$ is heated, it loses water to give the olive green oxide CoO, an ionic compound that has the sodium chloride structure. It is a basic oxide that dissolves in acids to give cobalt(II) salts.

If a solution of ammonium sulfide is added to a solution of a cobalt(II) salt, or if H_2S is passed into a basic solution of a cobalt(II) salt, a black precipitate of cobalt(II) sulfide, CoS, is obtained:

$$Co^{2+}(aq) + S^{2-}(aq) \longrightarrow CoS(s)$$

COMPOUNDS OF NICKEL   All the common compounds of nickel contain the element in the $+2$ oxidation state. Nickel(II) salts such as $NiCl_2 \cdot 6H_2O$ and $NiSO_4 \cdot 7H_2O$ contain the green hydrated nickel ion, $Ni(H_2O)_6^{2+}$. Addition

EXPERIMENT 21.8

Reactions of Cobalt(II)

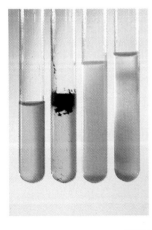

Left to right: (a) A pink solution of $CoCl_2$. (b) When a solution of ammonium sulfide is added, a black precipitate of cobalt(II) sulfide, CoS, is obtained. (c) When excess aqueous NaOH solution is added, a pink precipitate of $Co(OH)_2$ is obtained. (d) When a solution of hydrogen peroxide is added on top of the suspended $Co(OH)_2$ precipitate, the $Co(OH)_2$ is oxidized to $Co(OH)_3$, which forms a brown ring on the interface between the two solutions.

of NaOH or KOH to a solution of an Ni^{2+} salt gives a green precipitate of $Ni(OH)_2$ (see Experiment 21.9). Aqueous ammonia also gives the same precipitate, but with excess ammonia this precipitate dissolves to give a deep blue solution containing the complex ion $Ni(NH_3)_6^{2+}$.

Copper

The element following nickel in the first transition metal series is copper, which we discussed in Chapter 9. Its electron configuration is $3d^{10}4s^1$. As with chromium, this differs from the electron configuration of most of the other transition metals of the first series in that it only has one 4s electron. Copper may lose the single 4s electron to give Cu(I) compounds in which it has a $3d^{10}$ electron arrangement and therefore has a completed $n = 3$ shell. In addition, it may also lose one of the 3d electrons to give the Cu(II) oxidation state, which has a d^9 electron arrangement.

Zinc

The next element, and the last in the first transition metal series, is zinc. The principal ore of zinc is zinc sulfide, ZnS, which is known as *sphalerite*. The structure of ZnS was described in Chapter 10. When ZnS is roasted in a furnace, it is converted to the oxide ZnO, which is then reduced by strongly heating it with carbon (coke).

Zinc has only one oxidation state, the $+2$ state. Its valence shell electron configuration is $3d^{10}4s^2$. It can lose the two 4s electrons to give the Zn^{2+} ion, which has a d^{10} arrangement of unused valence electrons. Typical salts of Zn^{2+} are $ZnCl_2 \cdot H_2O$, $ZnSO_4 \cdot 7H_2O$, and $Zn(NO_3)_2 \cdot 6H_2O$. In aqueous solution the zinc(II) ion is in the hydrated form, $Zn(H_2O)_6^{2+}$.

Addition of an alkali metal hydroxide or aqueous ammonia to a solution of a zinc(II) salt gives a white gelatinous precipitate of zinc(II) hydroxide, $Zn(OH)_2$:

$$Zn^{2+}(aq) + 2OH^-(aq) \longrightarrow Zn(OH)_2(s)$$

EXPERIMENT 21.9

Reactions of Nickel(II)

Left to right: (a) A solution of nickel(II) chloride, $NiCl_2$, is green. (b) When a solution of ammonium sulfide is added, a black precipitate of nickel(II) sulfide, NiS, is obtained. (c) When NaOH(aq) is added, a pale green precipitate of $Ni(OH)_2$ is obtained.

The precipitate dissolves in excess hydroxide because of the formation of the $Zn(OH)_4^{2-}$ ion,

$$Zn(OH)_2(s) + 2OH^-(aq) \longrightarrow Zn(OH)_4^{2-}(aq)$$

and in excess ammonia because of the formation of the $Zn(NH_3)_4^{2+}$ ion.

If hydrogen sulfide is passed into a basic solution of a zinc(II) salt, a white precipitate of zinc sulfide, ZnS, is obtained.

Trends in Properties

Now that we have considered the chemistry of the elements from scandium to zinc, we can review the trends in the properties of these elements. The highest oxidation state increases from +3 for scandium to +7 for manganese. The compounds of these elements in oxidation states of +4 or higher are predominantly covalent, and in these oxidation states the elements behave more like nonmetals than metals. This is because they have high core charges in these high oxidation states and therefore have high electronegativities. Their oxides are acidic, and they form oxoacids and oxoanions as the elements in the main groups IV to VII, do—for example, titanates, TiO_3^{2-} (like silicates, SiO_3^{2-}), vanadates, VO_4^{3-} (like phosphates, PO_4^{3-}), and chromates, CrO_4^{2-} (like sulfates, SO_4^{2-}). However, after manganese the theoretically possible maximum oxidation states, which would be +8 for iron, +9 for cobalt, and so on, are not known. Because the core charge increases from +4 for titanium, to +7 for manganese, and to +9 for cobalt, the 3d electrons are increasingly strongly held, so that after manganese not all of them are available for compound formation.

Thus from titanium to manganese, the highest oxidation state becomes increasingly less stable, and the oxoanions become increasingly strong oxidizing agents. As the higher oxidation states become less stable, the lower oxidation states become more stable. Thus whereas Ti^{2+} is a strong reducing agent that is easily oxidized to Ti(IV) compounds, Ni^{2+}, Cu^{2+}, and Zn^{2+} are the stable oxidation states of these elements, and they cannot easily be oxidized to higher oxidation states. In these lower oxidation states these elements show typical metal behavior; they form basic oxides and hydroxides, such as FeO and NiO and $Fe(OH)_2$, and $Ni(OH)_2$, that dissolve in acids to form ionic salts.

For the elements beyond manganese, not all the 3d electrons can be regarded as belonging to the valence shell. At zinc the 3d shell is filled with ten electrons, and these electrons have no tendency to take part in compound formation. They cannot, therefore, be considered as part of the valence shell; rather, they may be thought of as core electrons. For the elements between manganese and zinc— that is, iron, cobalt, nickel, and copper—the 3d electrons cannot be clearly identified as either valence shell electrons or as core electrons. This special nature of the 3d electrons is responsible for most of the unusual properties of the transition metals.

An obvious question that comes to mind is, why are the 4s electrons always used before the 3d electrons in compound formation, although we stated earlier in discussing electron configurations that the 4s level has a lower energy than the 3d level? The reason is that although the 3d level is above the 4s level for the neutral atoms of the transition metals, the 3d level is below the 4s level for their ions. The greater resultant charge acting on the electrons in the ions than in the neutral atom decreases the energy of the 3d level more than that of the 4s level; as a result, in the ions the 3d level has a slightly lower energy than

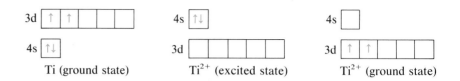

Figure 21.5 Orbital Energies for Ti and Ti^{2+}. Although the 3d orbitals have a lower energy than the 4s orbital in the neutral Ti atom, they have a higher energy in the Ti^{2+} ion. Thus the ground state configuration of Ti^{2+} is $3d^24s^0$, not $4s^23d^0$. If we imagine that two electrons are removed from the 3d orbital of Ti($3d^24s^2$), we obtain an excited state of Ti^{2+}, $3d^04s^2$, which reverts to the ground state, $3d^24s^0$.

Ti (ground state)   Ti^{2+} (excited state)   Ti^{2+} (ground state)

the 4s level. Thus when electrons are removed from a neutral transition metal atom, the first two always *appear to have come* from the 4s level (see Figure 21.5). Thus the electron configurations of the first transition series metal ions can be obtained from the electron configurations of the elements by first removing the 4s electrons and then as many d electrons as necessary.

21.3 INORGANIC COMPLEXES

A characteristic property of transition metals is their ability to form a large number of stable complexes. An inorganic molecule or ion that contains several atoms, including one or more metal atoms, is called a **complex**; it most commonly consists of a single metal atom surrounded by several atoms or groups of atoms. When the complex carries a charge, it is called a **complex ion**. Hydrated ions such as $Al(H_2O)_6^{3+}$ and $Fe(H_2O)_4^{2+}$ and the ions $AlCl_4^-$, $Al(OH)_4^-$, and $FeCl_4^-$, which we discussed in Chapter 9, are all examples of complex ions. In this chapter we have encountered a variety of other complex ions such as $Cr(H_2O)_6^{3+}$, $Ni(H_2O)_6^{2+}$, $Ni(NH_3)_6^{2+}$, $Zn(NH_3)_4^{2+}$, and $Zn(OH)_4^{2-}$.

We have seen in Chapter 9 that a metal ion, particularly if it is small and multiply charged, strongly attracts an unshared electron pair in a molecule such as H_2O or NH_3 or an anion such as Cl^- or OH^-. A molecule or ion that becomes bonded to a metal atom in this way is called a **ligand**. The transition metals form complexes with a wide variety of molecules and anions. The ligands that are most commonly found in transition metal complexes are listed in Table 21.9.

As we discussed in Chapter 9, the bond between a metal atom and a ligand is intermediate in character between an ionic bond and a covalent bond. In some complexes it is largely ionic, while in others it is largely covalent. At the ionic extreme the bond can be considered to be due to the electrostatic attraction between the metal ion and the negative charge of the anion or the negative end of a polar molecule. At the covalent extreme the originally unshared pair

Table 21.9 Some Common Ligands

	LIGAND	NAME IN COMPLEX
Water	H_2O	Aqua
Ammonia	NH_3	Ammine[a]
Bromide	Br^-	Bromo
Chloride	Cl^-	Chloro
Cyanide	CN^-	Cyano
Hydroxide	OH^-	Hydroxo
Oxalate	$C_2O_4^{2-}$	Oxalato
Ethylenediamine	$H_2NCH_2CH_2NH_2$	Ethylenediamine (en)

[a] Note the double *mm*, rather than a single *m*, as in the amines RNH_2 (Chapter 19).

of the ligand is shared with the metal atom. If a considerable amount of charge is donated from the ligand to the metal atom, then the bond is largely covalent. But if only a small amount of charge is donated from the ligand to the metal atom, then the bond is largely ionic (see Figure 21.6).

The ligands are said to be *coordinated* to the metal atom. The number of bonds formed to ligands is called the **coordination number** of the metal atom. Complex compounds are often called **coordination compounds**. The bond between a ligand and a metal is sometimes called a coordinate bond. But a coordinate bond consists of a pair of shared electrons, like any other single bond, and like other single bonds, it may be predominantly covalent or predominantly ionic.

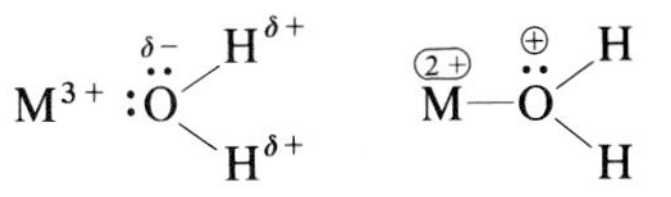

The coordination number of transition metal atoms is very frequently 6; less commonly, it is 4. Other coordination numbers such as 2, 3, 5, 8, and 9 are sometimes found. If the ligand is a negative ion, then the charge on the complex ion differs from the charge on the uncoordinated metal ion; for example,

$$Fe^{3+} + 4Cl^- \longrightarrow FeCl_4^-$$

In a case like this, each chloride ion shares an electron pair with the Fe atom and therefore becomes covalently bound to the Fe atom, so that it can no longer be regarded as a Cl^- ion. A complex ion may have both neutral molecules and negative ions as ligands. For example, Cr^{3+} forms the ion $Cr(H_2O)_4Cl_2^+$.

The tendency of the transition metals to form complex ions is so strong that in aqueous solution they are always either hydrated or in the form of some other complex ion. Because of the large number and variety of these complex ions, the transition metals often have a more complicated and more interesting aqueous solution chemistry than that of the main group metals. For example, when ammonium chloride and aqueous ammonia are added to a solution of cobalt(II), and the solution is treated with an oxidizing agent such as air, iodine, or hydrogen peroxide and then acidified with hydrochloric acid, four compounds can be obtained: (1) a yellow compound with the composition $CoCl_3 \cdot 6NH_3$, (2) a purplish red compound $CoCl_3 \cdot 5NH_3$, (3) a green compound $CoCl_3 \cdot 4NH_3$, and (4) a violet compound with the same composition, $CoCl_3 \cdot 4NH_3$.

For a long time the bewildering variety of similar, closely related compounds of cobalt and many other transition metals posed a difficult problem. The structures of these compounds were not understood until Alfred Werner, a 26-year-old Swiss chemist (1866–1919), suggested in 1893 that ammonia molecules, chloride ions, and other molecules and ions could be strongly attached to a metal ion to give complex ions. Werner proposed that all these compounds of cobalt chloride and ammonia should be formulated as ionic compounds containing complex ions of cobalt.

Werner based his theory on a great variety of experimental evidence. One example of the evidence that he used to support his theory is provided by the reaction of these cobalt compounds with excess silver nitrate. He found the following results:

- One mole of $CoCl_3 \cdot 6NH_3$ reacts immediately with 3 mol of $AgNO_3$ to give a precipitate of 3 mol of AgCl.
- One mole of $CoCl_3 \cdot 5NH_3$ reacts immediately with only 2 mol of $AgNO_3$ to give a precipitate of 2 mol of AgCl.
- One mole of $CoCl_3 \cdot 4NH_3$ reacts immediately with only 1 mol of $AgNO_3$ to give a precipitate of 1 mol of AgCl.

Figure 21.6 Metal-Ligand (Coordinate) Bond.

He concluded that $CoCl_3 \cdot 6NH_3$ contains three Cl^- ions, whereas $CoCl_3 \cdot 5NH_3$ contains only two Cl^- ions and $CoCl_3 \cdot 4NH_3$ contains only one Cl^- ion. In order to account for these results, he proposed that all these compounds contain complex ions. The first compound contains the complex ion $[Co(NH_3)_6]^{3+}$ and should therefore be formulated as $[Co(NH_3)_6]^{3+}(Cl^-)_3$. The second contains the complex ion $[Co(NH_3)_5Cl]^{2+}$ and should therefore be formulated as $[Co(NH_3)_5Cl]^{2+}(Cl^-)_2$, in which there are only two chloride ions and one chlorine that is covalently bound to the metal atom and does not, therefore, give a precipitate with silver chloride. Finally, $CoCl_3 \cdot 4NH_3$ contains the complex ion $[Co(NH_3)_4Cl_2]^+$ and should be formulated as $[Co(NH_3)_4Cl_2]^+Cl^-$, in which there is only one chloride ion and two chlorine atoms covalently bound to the cobalt atom.

Werner's brilliant proposal was made a long time before it was possible to directly determine the structures of complex ions by X ray crystallography, and he was awarded the Nobel Prize in chemistry in 1913 for this major contribution to inorganic chemistry.

The formation of $Co(NH_3)_6^{3+}$ and the similar complex ammines $Ni(NH_3)_6^{2+}$ and $Cu(NH_3)_4^{2+}$ is demonstrated in Experiment 21.10.

Notice that in order to indicate clearly which ions or molecules are attached to the metal atom, we enclose the formula of the complex ion in square brackets. The charge on the complex ion is written outside the square brackets. These brackets should not be confused with the brackets that we used in previous chapters to indicate the concentration of a species.

We still have to discuss why there are two compounds with the formula $[Co(NH_3)_4Cl_2]Cl$. This we do in the next section.

Geometry of Complex Ions

The most common coordination numbers observed in complex ions are 2, 4, and 6. According to the VSEPR theory (Chapters 8 and 20), AX_2, AX_4, and AX_6 molecules, having respectively two, four, and six ligands and no unshared pairs of electrons, are linear, tetrahedral, and octahedral, respectively. Most transition metal complex ions have these structures, and the octahedral geom-

EXPERIMENT 21.10

Copper(II), Cobalt(III), and Nickel(II) Ammines

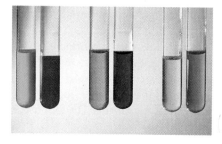

Left to right: (a) When an excess of aqueous ammonia is added to a blue solution of $Cu^{2+}(aq)$, the very deep blue $Cu(NH_3)_4^{2+}$ complex ion is formed. (b) When an excess of ammonia is added to a pink solution of $Co^{2+}(aq)$, it is oxidized to Co^{3+} by the oxygen in the air and becomes deep red in color as the $Co(NH_3)_6^{3+}$ complex ion is formed. (c) A green solution of $Ni^{2+}(aq)$ becomes violet in color when an excess of aqueous ammonia is added because the $Ni(NH_3)_6^{2+}$ complex ion is formed.

etry is particularly common (see Figure 21.7). However, we must remember that in most cases there are nonbonding d electrons in the valence shell of the metal ion. Unlike the unshared pairs of the main group elements, these nonbonding electrons often have only a negligible effect on the geometry of the complex ions. This is not surprising, since in many cases these electrons are associated more with the core than with the valence shell. In some cases, however, the nonbonding d electrons do distort the basic shapes. For example, a few AX_4 complexes, such as $Ni(CN)_4^{2-}$ and $Cu(NH_3)_4^{2+}$, have a square planar rather than a tetrahedral shape.

We saw in Chapter 11 that isomers in which all the atoms are connected in the same way and that differ only with respect to the positions of identical groups in space are called *geometric isomers*. Geometric isomers are possible for octahedral complexes containing two or more different ligands. The simplest case is a complex of the type MX_4Y_2 with four ligands X and two ligands Y, such as the complex ion $[Co(NH_3)_4Cl_2]^+$, which we have seen exists in green and violet forms. These two forms have the structures

cis (violet)   *trans* (green)

In the cis isomer the two chloro ligands are in positions adjacent to each other ($Cl\widehat{Co}Cl = 90°$); in the trans isomer they are opposite to each other ($Cl\widehat{Co}Cl = 180°$). The explanation of the existence of two isomers of the compound $CoCl_3 \cdot 4NH_3$ was one of the important achievements of Werner's theory.

Chelates

The atom in a ligand that is bound directly to the metal is called the *donor atom*. For example, oxygen is the donor atom in $Cr(H_2O)_6^{3+}$. Most of the ligands in Table 21.9 have only one donor atom. They are called **monodentate ligands**. But two ligands in the table have two donor atoms. Ethylenediamine

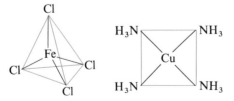

H_3N—Ag—NH_3

(a) Linear $[Ag(NH_3)_2]^+$ (b) Tetrahedral $[FeCl_4]^-$; Square planar $[Cu(NH_3)_4]^{2+}$

(c) Octahedral $[Co(NH_3)_6]^{3+}$

Figure 21.7 Some Common Shapes for Inorganic Complexes.
(a) AX_2 complexes are linear.
(b) AX_4 complexes are tetrahedral but occasionally square planar.
(c) AX_6 complexes are octahedral.

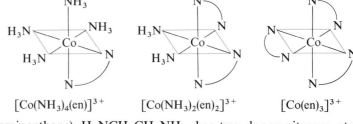

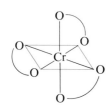

Figure 21.8 Some Chelate Complexes of Cobalt Formed by the Bidentate Ligand H₂NCH₂CH₂NH₂. The ligand $H_2NCH_2CH_2NH_2$(en) is shown as N—N.

$$[Co(NH_3)_4(en)]^{3+} \qquad [Co(NH_3)_2(en)_2]^{3+} \qquad [Co(en)_3]^{3+}$$

(1,2-diaminoethane), $H_2NCH_2CH_2NH_2$, has two donor nitrogen atoms, and in the oxalate ion, $C_2O_4^{2-}$, two of the oxygen atoms behave as donor atoms. These ligands are called **bidentate ligands**.

In an octahedral complex ethylenediamine (en) occupies two adjacent octahedral positions. It forms complexes such as those shown in Figure 21.8. The formation of some ethylenediamine complexes of Ni(II) is demonstrated in Experiment 21.11. The oxalate ion, $C_2O_4^{2-}$, forms similar complexes (Figure 21.9).

Other ligands can occupy three or more coordination sites. They are called **polydentate ligands**. An interesting example is the ethylenediaminetetraacetate ion, $EDTA^{4-}$ (Figure 21.10), which has six donor atoms and can therefore occupy all six sites in an octahedrally coordinated ion, as in the complex ion $Co(EDTA)^-$. Complexes containing bidentate or polydentate ligands are often known as **chelates**, a name derived from the Greek word for "claw," because the polydentate ligands appear to grasp the metal atom between two or more donor atoms. Such ligands are called *chelating agents*.

In general, polydentate ligands form more stable complexes than do related monodentate ligands. This property makes chelating agents very useful for removing metal ions from solution. For example, Ca^{2+} and Mg^{2+} ions, which cause hardness in water (Chapter 15), can be removed by forming complex ions with the triphosphate anion:

$$^{\ominus}O-\underset{\underset{O_{\ominus}}{\overset{\overset{O}{\parallel}}{|}}}{P}-O-\underset{\underset{O_{\ominus}}{\overset{\overset{O}{\parallel}}{|}}}{P}-O-\underset{\underset{O_{\ominus}}{\overset{\overset{O}{\parallel}}{|}}}{P}-O^{\ominus}$$

Certain metal ions such as Hg^{2+} and Pb^{2+}, which tend to accumulate in the body and act as poisons, can be removed by forming a chelate with EDTA, thus converting them into harmless forms that are eliminated from the body. Various complexes of transition metals with chelating agents play vital roles in reactions in living systems. Box 21.1 discusses one example.

Figure 21.9 Chelate Complex $Cr(C_2O_4)^{3-}$. The oxalate ion,

is abbreviated as O—O.

EXPERIMENT 21.11

Ethylenediamine Complexes of Ni(II)

When diaminoethane (ethylene diamine), $H_2NCH_2CH_2NH_2$, is added to a green aqueous solution of Ni^{2+}, the solution changes first to green-blue, then to purple-blue, and finally to violet as the chelate complexes $Ni(H_2O)_4(en)^{2+}$, $Ni(H_2O)_2(en)_2^{2+}$, and $Ni(en)_3^{2+}$ are successively formed.

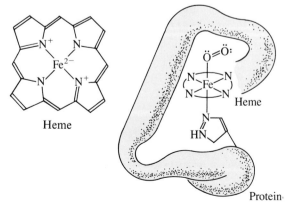

Figure 21.10 Chelate Complex of Cobalt Formed by the Hexadentate Ligand Ethylenediaminetetraacetate.

Formation Constants

The stabilities of complex ions in aqueous solution are measured by determining the equilibrium constant for the reaction of the hydrated ion with the ligand. For example, for the reaction

$$Co(H_2O)_6^{2+}(aq) + 6NH_3(aq) \rightleftharpoons Co(NH_3)_6^{2+}(aq) + 6H_2O$$

the expression for the equilibrium constant is

$$K_c = \frac{[Co(NH_3)_6^{2+}][H_2O]^6}{[Co(H_2O)_6^{2+}][NH_3]^6}$$

Box 21.1

HEMOGLOBIN

Among the transition metals that are essential to life, iron is extremely important as an essential component of the oxygen carrier *hemoglobin*. If our diet is deficient in iron, we may suffer from anemia and feel tired and weak. Hemoglobin contains four complex protein molecules, each of which has a heme molecule as part of its structure. Each heme molecule contains an iron atom coordinated by four nitrogen atoms in a square. Octahedral coordination around the iron is completed by a nitrogen atom from another part of the protein molecule and an oxygen molecule or a water molecule:

Hemoglobin picks up oxygen and forms *oxyhemoglobin* in the lungs. Oxygen is released in the tissues, where it is needed for cell metabolism, and replaced by a water molecule to form *deoxyhemoglobin*.

Carbon monoxide also behaves as a ligand. It forms a complex with hemoglobin that is about two hundred times as strong as the complex with oxygen and makes the hemoglobin useless as an oxygen carrier. If we breathe carbon monoxide, some of the oxyhemoglobin in our blood is converted to carboxyhemoglobin; breathing air containing only 0.1% CO converts about 60% of the hemoglobin to carboxyhemoglobin in a few hours. If a substantial part of the hemoglobin in the blood is converted to carboxyhemoglobin, the body suffers from acute oxygen deficiency and death soon results.

Long exposure to even small concentrations of carbon monoxide can have serious consequences. For example, air inhaled through a lighted cigarette contains about 400 ppm of carbon monoxide. Consequently, heavy smokers have as much as 6% of the hemoglobin in their blood constantly converted to carboxyhemoglobin. As a result, their blood is not as efficient at carrying oxygen as the blood of nonsmokers, and their hearts must work harder. This is probably a contributing factor in heart disease and heart attacks.

Since for dilute solutions $[H_2O]$ is very nearly constant, we can rearrange this equation as

$$K_f = \frac{K_c}{[H_2O]^6} = \frac{[Co(NH_3)_6^{2+}]}{[Co(H_2O)_6^{2+}][NH_3]^6}$$

where K_f is the **formation constant** of the complex ion.

For $Co(NH_3)_6^{2+}$, $K_f = 1 \times 10^5$ mol^{-6} L^6. Table 21.10 gives values of formation constants for some other complex ions. A large value for K_f indicates a very stable complex. For example, the formation constant of $Ni(NH_3)_6^{2+}$ is 6×10^8 mol^{-6} L^6, whereas that for chelate $Ni(en)_3^{2+}$ is 4×10^{18} mol^{-3} L^3, indicating that the bidentate ligand $NH_2CH_2CH_2NH_2$ forms a considerably more stable complex than NH_3.

Nomenclature

The rules for the systematic naming of complex compounds are as follows:

1. The common ligands have the names given in Table 21.9.

2. The number of any particular ligand is specified by di = 2, tri = 3, tetra = 4, penta = 5, hexa = 6, and so on. When confusion might result from the use of these prefixes, the alternative prefixes bis = 2, tris = 3, tetrakis = 4, and so on, are used.

3. The name of a negative (anionic) complex always ends in the suffix -*ate*, which is appended to the name of the metal or the stem of the name of the metal. For some metals the Latin stem is used. For example, in a negative complex iron is named ferrate and lead is named plumbate.

4. The oxidation state of the metal is indicated by a Roman numeral in parenthesis following the name of the metal.

Some examples are given in Table 21.11.

Table 21.10 Formation Constants for Some Complex Ions at 25°C

ION	K_f	ION	K_f
$Cu(NH_3)_4^{2+}$	1×10^{12}	$Cu(en)_2^{2+}$	1.6×10^{20}
$Ni(NH_3)_6^{2+}$	6×10^8	$Ni(en)_3^{2+}$	4×10^{18}
$Zn(NH_3)_4^{2+}$	5×10^8	$Co(en)_3^{3+}$	8×10^{13}
$Ag(NH_3)_2^{+}$	1×10^8	$Zn(en)_3^{2+}$	1.2×10^{13}
$Co(NH_3)_6^{2+}$	1×10^5	$Fe(en)_3^{3+}$	4×10^9
$Fe(CN)_6^{3-}$	1×10^{31}	$Mn(en)_3^{2+}$	5×10^5
$Fe(CN)_6^{4-}$	1×10^{24}	$Al(EDTA)^-$	1.4×10^{16}
$Ni(CN)_4^{2-}$	1×10^{30}	$Co(EDTA)^-$	2×10^{16}
$Zn(CN)_4^{2-}$	5×10^{16}	$Cu(EDTA)^{2-}$	6.3×10^{18}
$Hg(CN)_4^{2-}$	4×10^{41}	$Fe(EDTA)^{2-}$	2.1×10^{14}
$Ag(CN)_2^-$	1×10^{21}	$Fe(EDTA)^-$	1.3×10^{25}
$Al(C_2O_4)_2^-$	1.0×10^{13}	$Mn(EDTA)^{2-}$	1.1×10^{14}
$Co(C_2O_4)_2^{2-}$	1.3×10^7	$Hg(EDTA)^{2-}$	6.3×10^{21}
$Cu(C_2O_4)^{2-}$	2.0×10^{10}	$Ni(EDTA)^{2-}$	4.2×10^{18}
$Fe(C_2O_4)_2^{2-}$	4.0×10^9	$Ag(EDTA)^{3-}$	2.1×10^7
$Mn(C_2O_4)_2^{2-}$	6.3×10^5	$Zn(EDTA)^{2-}$	3.2×10^{16}
$Ni(C_2O_4)_2^{2-}$	3.3×10^6		
$Zn(C_2O_4)_2^{2-}$	2.3×10^7		

Table 21.11 Names of Some Complexes

COMPLEX	NAME
$[Co(H_2O)_6]^{3+}$	Hexaaquocobalt(III) ion
$[CoCl_6]^{3-}$	Hexachlorocobaltate(III) ion
$[Co(NH_3)_4Cl_2]^+$	Dichlorotetraamminecobalt(III) ion
$[Ag(NH_3)_2]^+$	Diamminesilver(I) ion
$[Ag(CN)_2]^-$	Dicyanoargentate(I) ion
$[Cr(NH_3)_3Cl_3]$	Trichlorotriamminechromium(III)

WHY ARE THERE SO MANY COMPLEXES OF THE TRANSITION METALS? The transition metals form many more complexes than the main group metals, and those that they form are more stable. Why? Most main group metals have completed inner shells of electrons beneath the valence shell. Therefore the core charge is equal to the charge on the ion, for example, $+1$ for the alkali metals $+2$ for the alkaline earth metals, and $+3$ for aluminum. In many cases, however the transition metals have higher core charges than the charge on the ions that they form. Consider, for example, the Cr^{3+} ion. It has the electron configuration d^3. If we count these electrons as part of the valence shell, then the core charge of chromium is $+6$. If we think of the d electrons as belonging partly to the core and partly to the valence shell, we conclude that the core charge is between $+3$ and $+6$. The Fe^{2+} ion has the valence shell configuration d^6. Its core charge is, therefore, somewhere between $+2$ and $+8$.

The core charge increases with increasing atomic number, but this increase is counterbalanced by the increasing tendency of the d electrons to form part of the core. Thus from scandium to zinc the core charge at first increases with increasing atomic number, but then decreases again as more d electrons are drawn into the core, until at zinc the core charge is only $+2$. For most of the transition metals, however, the core charge is greater than $+3$. This large core charge exerts a strong attraction on an unshared electron pair of a ligand. Thus the transition metal ions form much stronger complexes than do the main group metal ions.

Colors of Transition Metal Complexes

We have seen that most transition metal compounds are colored. For example, hydrated Cu^{2+} salts are blue, hydrated Ni^{2+} salts are green, MnO_4^- is purple, and Fe_2O_3 is a deep red brown. In contrast, most of the compounds of the main group metals are colorless. Substances are colored because they absorb certain wavelengths of visible light while transmitting (or reflecting) others. Leaves of plants are green because they contain the molecule chlorophyll, which absorbs red and blue light but transmits (or reflects) green and yellow light. Figure 21.11 shows the absorption spectrum of chlorophyll.

A substance absorbs light when one or more of its electrons are excited from their ground state energy levels to higher energy levels. If the difference in energy of the levels between which the electron is excited is between 170 and 290 kJ mol^{-1}, then as we saw in Chapter 6, the substance absorbs visible light, that is, light of wavelengths between 700 and 400 nm. For many substances this energy difference is greater than 290 kJ mol^{-1}, so they absorb ultraviolet light, rather than visible light, and are therefore colorless.

In contrast to the spectra of free atoms, which, as we saw in Chapter 6, consist of sharp lines, the spectra of molecules are broad bands. Molecules absorb

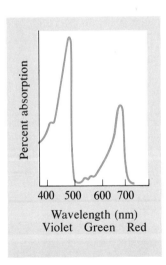

Figure 21.11 Absorption Spectrum of Chlorophyll. This shows the percentage of the incident light that is absorbed at each wavelength. Chlorophyll absorbs red and blue light; therefore the transmitted (or reflected) light is green.

not one wavelength but a range of wavelengths (see Figure 21.11) because the energy levels of an atom in a molecule are affected by the surrounding atoms. And because atoms in molecules are in constant motion, vibrating around their equilibrium position, their effect on the energy levels of an atom to which they are bonded is constantly varying. Thus the transition of an electron when it absorbs light occurs not between energy levels of precisely fixed energy but between energy levels that have a range of energies.

In transition metal compounds, the transitions of the d electrons to higher energy levels are usually responsible for the absorption of visible light. In a free atom all five d orbitals have the same energy. But in a compound the energies of the orbitals of an atom are affected by the atoms to which it is bonded. As a result, not all the d orbitals in a transition metal complex have the same energy.

The transitions of an electron between d orbitals of different energies are responsible for the colors of most transition metal compounds. Thus Sc^{3+} (d^0), which has no d electrons, and Zn^{2+} (d^{10}), which has all its d orbitals completely filled, are colorless because no transitions of electrons between d levels are possible. That the ligands have a profound effect on the energy levels of the d electrons in a metal atom is shown clearly by the great differences in color caused by changing the ligands on a particular metal atom. Thus $[Cu(H_2O)_4]^{2+}$ is pale blue, but $[Cu(NH_3)_4^{2+}]$ is an intense deep blue; $[Fe(H_2O)_6]^{3+}$ is a pale violet, but $[Fe(H_2O)_5SCN]^{2+}$ is deep red; and $[Co(NH_3)_6]^{3+}$ is yellow, but $[Co(NH_3)_5Cl]^{2+}$ is purple. The greater the covalent character of the bond between the ligand and the metal ion, the greater the perturbation of the energy of the d orbitals and the greater the energy differences between the d orbitals. The colors of some cobalt complexes are given in Table 21.12. A ligand such as F^- forms a very ionic bond with Co^{3+}. Therefore the difference in energy between the d orbitals is only small, and so $[CoF_6]^{3-}$ absorbs at long wavelengths. In contrast, a ligand such as CN^-, which forms strong covalent bonds with the Co atom, causes the energy difference between the d orbitals to be greater, and so $[Co(CN)_5Br]^{3-}$ absorbs short-wavelength light.

But the colors of compounds are not due only to the transitions of d electrons. Recall from Chapter 12 that the polarizability of a molecule is a measure of the ease with which its electrons can be displaced and therefore of the ease with which they can be excited to higher energy levels. Easily polarizable molecules are more likely to have transitions in the visible region than less easily polarizable molecules. Thus whereas fluorine is a very pale yellow, chlorine is a deeper yellow green, bromine is a deep red brown, and iodine is a very dark, almost black, violet. Similarly, AgCl is white, AgBr a pale yellow, and AgI a deeper yellow.

Table 21.12 Colors of Some Cobalt Complexes

Co^{3+} COMPLEX	WAVELENGTH ABSORBED (nm)	COLOR ABSORBED	COLOR OF COMPLEX
CoF_6^{3-}	700	Red	Green
$[Co(CO_3)_3]^{3-}$	640	Red orange	Green blue
$[Co(H_2O)_6]^{3+}$	600	Orange	Blue
$[Co(NH_3)_5Cl]^{2+}$	535	Yellow	Purple
$[Co(NH_3)_5OH]^{2+}$	500	Blue green	Red
$[Co(NH_3)_6]^{3+}$	475	Blue	Yellow orange
$[Co(CN)_5Br]^{3-}$	415	Violet	Yellow

We discuss the second and third series of transition elements in periods 5 and 6 only very briefly, considering in detail only silver, gold, and mercury, which are metals of significant historical and practical importance.

In the fifth period, after two electrons enter the 5s subshell to give the alkali metal rubidium, Rb ($5s^1$), and the alkaline earth metal strontium, Sr ($5s^2$), the elements from yttrium, Y, to cadmium, Cd, constitute a *second series of ten transition metals* in which the 4d subshell is filled with electrons (Table 21.13).

The second series of ten transition elements is followed by six main group elements, In, Sn, Sb, Te, I, and Xe, completing the eighteen elements in the fifth period. Xenon, the last element in the period, has the electron arrangement [Kr] $4d^{10}5s^25p^6$. Neither the $n = 4$ shell nor the $n = 5$ shell is yet complete, since the 4f subshell and the 5d, 5f, and 5g subshells are still empty. In the two elements Cs and Ba, the 6s shell is completed. The 4f shell is then filled, which gives rise to a series of 14 elements beginning with lanthanum, La, that have no counterparts in the previous periods of the periodic table. They are called **lanthanides**. The filling of the 4f shell completes the $n = 4$ shell, which then has a maximum of 32 electrons. The lanthanides are followed by the third series of transition metals in which the 5d subshell is filled. The most familiar of these metals are tungsten, platinum, gold, and mercury.

Silver and Gold

Copper, silver, and gold are in the same group of transition metals and are known as the *coinage metals*. They all have a $d^{10}s^1$ outer-electron configuration. Because of the high core charges of these elements, we expect only a few of the d electrons to be available for compound formation. In fact, all three elements form compounds in the +1 oxidation state, in which only the single s electron has been used, leaving a complete d^{10} subshell. However, these elements also form compounds in higher oxidation states. Indeed, the most common oxidation state of copper is Cu(II). Silver forms both Ag(II) and Ag(III) compounds, but these compounds are rather uncommon. Gold has a common Au(III) oxidation state in addition to Au(I).

Both silver and gold are rare elements, but they have long been known in their elemental forms. Some of their properties are summarized in Table 21.14. Like copper, they are relatively soft metals that are very malleable and ductile, and they have relatively low melting points compared with many other metals. Silver, like most metals, has a silvery white color, but the red color of copper and the yellow of gold are unusual. Because they occur in the free state in nature, because they are easily worked into different shapes, and because they are rather inert, they have had many uses since ancient times (Box 9.1). Gold, because it is completely resistant to the atmosphere and because it has a fine yellow color and luster and is rare, has always been valued for jewelry and other ornamental purposes. It is the most malleable and most ductile of all metals. It can, for example, be hammered into sheets only 10 nm (10^{-6} cm) thick. Very thin sheets of gold—gold leaf—have been used for many purposes, from decorating books to decorating large buildings.

The three metals, copper, silver, and gold, are known as the coinage metals because of their traditional use for coins. However, as gold and silver have become more valuable, their use for this purpose has greatly decreased. Except for special commemorative coins, gold and silver coins are no longer in current use. Up to the mid 1960s U.S. silver coins contained 90% silver and 10% copper.

Roman coins were made from a natural mixture of gold and silver.

Table 21.13 Electron Configurations for Elements from Rb to Rn

PERIOD		$n=$ 4 s	p	d	f	5 s	p	d	f	6 s	p
5	Rb	$4s^2$	$4p^6$			$5s^1$					
	Sr	$4s^2$	$4p^6$			$5s^2$					
	Y	$4s^2$	$4p^6$	$4d^1$		$5s^2$					
	Zr	$4s^2$	$4p^6$	$4d^2$		$5s^2$					
	Nb	$4s^2$	$4p^6$	$4d^4$		$5s^1$					
	Mo	$4s^2$	$4p^6$	$4d^5$		$5s^1$					
	Tc	$4s^2$	$4p^6$	$4d^5$		$5s^2$					
	Ru	$4s^2$	$4p^6$	$4d^7$		$5s^1$					
	Rh	$4s^2$	$4p^6$	$4d^8$		$5s^1$					
	Pd	$4s^2$	$4p^6$	$4d^{10}$							
	Ag	$4s^2$	$4p^6$	$4d^{10}$		$5s^1$					
	Cd	$4s^2$	$4p^6$	$4d^{10}$		$5s^2$					
	In	$4s^2$	$4p^6$	$4d^{10}$		$5s^2$	$5p^1$				
	Sn	$4s^2$	$4p^6$	$4d^{10}$		$5s^2$	$5p^2$				
	Sb	$4s^2$	$4p^6$	$4d^{10}$		$5s^2$	$5p^3$				
	Te	$4s^2$	$4p^6$	$4d^{10}$		$5s^2$	$5p^4$				
	I	$4s^2$	$4p^6$	$4d^{10}$		$5s^2$	$5p^5$				
	Xe	$4s^2$	$4p^6$	$4d^{10}$		$5s^2$	$5p^6$				
6	Cs	$4s^2$	$4p^6$	$4d^{10}$		$5s^2$	$5p^6$			$6s^1$	
	Ba	$4s^2$	$4p^6$	$4d^{10}$		$5s^2$	$5p^6$			$6s^2$	
	La	$4s^2$	$4p^6$	$4d^{10}$		$5s^2$	$5p^6$	$5d^1$		$6s^2$	
	Ce	$4s^2$	$4p^6$	$4d^{10}$	$4f^2$	$5s^2$	$5p^6$			$6s^2$	
	Pr	$4s^2$	$4p^6$	$4d^{10}$	$4f^3$	$5s^2$	$5p^6$			$6s^2$	
	Nd	$4s^2$	$4p^6$	$4d^{10}$	$4f^4$	$5s^2$	$5p^6$			$6s^2$	
	Pm	$4s^2$	$4p^6$	$4d^{10}$	$4f^5$	$5s^2$	$5p^6$			$6s^2$	
	Sm	$4s^2$	$4p^6$	$4d^{10}$	$4f^6$	$5s^2$	$5p^6$			$6s^2$	
	Eu	$4s^2$	$4p^6$	$4d^{10}$	$4f^7$	$5s^2$	$5p^6$			$6s^2$	
	Gd	$4s^2$	$4p^6$	$4d^{10}$	$4f^7$	$5s^2$	$5p^6$	$5d^1$		$6s^2$	
	Tb	$4s^2$	$4p^6$	$4d^{10}$	$4f^9$	$5s^2$	$5p^6$			$6s^2$	
	Dy	$4s^2$	$4p^6$	$4d^{10}$	$4f^{10}$	$5s^2$	$5p^6$			$6s^2$	
	Ho	$4s^2$	$4p^6$	$4d^{10}$	$4f^{11}$	$5s^2$	$5p^6$			$6s^2$	
	Er	$4s^2$	$4p^6$	$4d^{10}$	$4f^{12}$	$5s^2$	$5p^6$			$6s^2$	
	Tm	$4s^2$	$4p^6$	$4d^{10}$	$4f^{13}$	$5s^2$	$5p^6$			$6s^2$	
	Yb	$4s^2$	$4p^6$	$4d^{10}$	$4f^{14}$	$5s^2$	$5p^6$			$6s^2$	
	Lu	$4s^2$	$4p^6$	$4d^{10}$	$4f^{14}$	$5s^2$	$5p^6$	$5d^1$		$6s^2$	
	Hf	$4s^2$	$4p^6$	$4d^{10}$	$4f^{14}$	$5s^2$	$5p^6$	$5d^2$		$6s^2$	
	Ta	$4s^2$	$4p^6$	$4d^{10}$	$4f^{14}$	$5s^2$	$5p^6$	$5d^3$		$6s^2$	
	W	$4s^2$	$4p^6$	$4d^{10}$	$4f^{14}$	$5s^2$	$5p^6$	$5d^4$		$6s^2$	
	Re	$4s^2$	$4p^6$	$4d^{10}$	$4f^{14}$	$5s^2$	$5p^6$	$5d^5$		$6s^2$	
	Os	$4s^2$	$4p^6$	$4d^{10}$	$4f^{14}$	$5s^2$	$5p^6$	$5d^6$		$6s^2$	
	Ir	$4s^2$	$4p^6$	$4d^{10}$	$4f^{14}$	$5s^2$	$5p^6$	$5d^7$		$6s^2$	
	Pt	$4s^2$	$4p^6$	$4d^{10}$	$4f^{14}$	$5s^2$	$5p^6$	$5d^9$		$6s^1$	
	Au	$4s^2$	$4p^6$	$4d^{10}$	$4f^{14}$	$5s^2$	$5p^6$	$5d^{10}$		$6s^1$	
	Hg	$4s^2$	$4p^6$	$4d^{10}$	$4f^{14}$	$5s^2$	$5p^6$	$5d^{10}$		$6s^2$	
	Tl	$4s^2$	$4p^6$	$4d^{10}$	$4f^{14}$	$5s^2$	$5p^6$	$5d^{10}$		$6s^2$	$6p^1$
	Pb	$4s^2$	$4p^6$	$4d^{10}$	$4f^{14}$	$5s^2$	$5p^6$	$5d^{10}$		$6s^2$	$6p^2$
	Bi	$4s^2$	$4p^6$	$4d^{10}$	$4f^{14}$	$5s^2$	$5p^6$	$5d^{10}$		$6s^2$	$6p^3$
	Po	$4s^2$	$4p^6$	$4d^{10}$	$4f^{14}$	$5s^2$	$5p^6$	$5d^{10}$		$6s^2$	$6p^4$
	At	$4s^2$	$4p^6$	$4d^{10}$	$4f^{14}$	$5s^2$	$5p^6$	$5d^{10}$		$6s^2$	$6p^5$
	Rn	$4s^2$	$4p^6$	$4d^{10}$	$4f^{14}$	$5s^2$	$5p^6$	$5d^{10}$		$6s^2$	$6p^6$

Table 21.14 Some Properties of Copper, Silver, and Gold

	ATOMIC NUMBER	ATOMIC MASS	DENSITY (g cm^{-3})	MELTING POINT (°C)	$r_{metallic}$ (pm)
Copper	29	63.54	8.97	1083	128
Silver	47	107.87	10.54	960	144
Gold	79	196.97	19.42	1063	144

This composition is known as coin silver. Sterling silver is an alloy of 92% silver and 8% copper used in tableware. "Silver" coins are now made from copper-nickel alloy. Pure gold is too soft for normal use in jewelry, so it is alloyed with copper, silver, and other metals.

COMPOUNDS OF SILVER   Silver occurs not only as the free metal but also as *argentite*, Ag_2S, and *chlorargyrite*, $AgCl$. Extracting silver from the sulfide calls for treating the ore with a solution of sodium cyanide, which dissolves the silver as the complex ion $Ag(CN)_2^-$. Addition of zinc to this solution then precipitates metallic silver:

$$Ag_2S(s) + 4CN^-(aq) \longrightarrow 2Ag(CN)_2^-(aq) + S^{2-}(aq)$$

$$2Ag(CN)_2^-(aq) + Zn(s) \longrightarrow 2Ag(s) + Zn(CN)_4^{2-}(aq)$$

Silver tarnishes in the atmosphere because of the presence of traces of hydrogen sulfide, which reacts with silver in the presence of oxygen to form a thin film of black silver sulfide:

$$4Ag(s) + 2H_2S(g) + O_2(g) \longrightarrow 2Ag_2S(s) + 2H_2O(l)$$

Sulfur-containing proteins in eggs and other foods stain silver spoons by a similar reaction.

Silver, like copper, is insoluble in dilute hydrochloric and sulfuric acids but dissolves in dilute nitric acid to give silver nitrate, $AgNO_3$. This is a colorless crystalline salt and is the most common compound of silver. Silver nitrate is easily reduced to metallic silver by organic matter. If a drop of silver nitrate solution is spilled on the skin, it leaves a brown-black stain of finely divided metallic silver. Silver ion can be used to oxidize aldehydes to carboxylic acids (Chapter 19).

Silver oxide, Ag_2O, is obtained as a dark brown precipitate when sodium hydroxide is added to a solution of silver nitrate:

$$2Ag^+(aq) + 2OH^-(aq) \rightleftharpoons Ag_2O(s) + H_2O(l)$$

It is slightly soluble in water, producing a weakly basic solution because it is in equilibrium with Ag^+ and OH^-.

When Cl^-, Br^-, or I^- is added to an aqueous solution of silver ion, a precipitate of the insoluble silver halide is obtained. Silver chloride is white, $AgBr$ is pale yellow, and AgI is bright yellow. In contrast, silver fluoride, AgF, is very soluble in water. The formation of these precipitates is often used as a test for Cl^-, Br^-, and I^- and also for Ag^+ (see Chapters 7 and 15). The test can be confirmed by noting the solubility of the precipitates in aqueous ammonia solution. Silver chloride and AgBr both dissolve because of the formation of the $Ag(NH_3)_2^+$ complex ion, but AgI remains insoluble.

Silver chloride, bromide, and iodide slowly turn black when exposed to ordinary daylight, because they undergo a photochemical decomposition to give

The Photographic Process

A key is placed on a filter paper covered with powdered white silver chloride.

The silver chloride is then exposed to a strong light. The silver chloride slowly darkens as grey-black silver is formed.

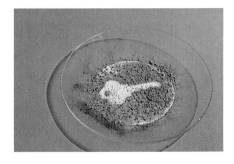

When the key is removed, a white image of the key is left because the silver chloride underneath the key was protected from the light.

silver and the corresponding halogen:

$$2AgBr \xrightarrow{h\nu} 2Ag + Br_2$$

This reaction is the basis of the photographic process (Experiment 21.12 and Box 21.2). Thirty percent of the silver used by industry in the United States goes into the manufacture of photographic film.

COMPOUNDS OF GOLD Gold is a very unreactive metal, and most of its compounds can be reduced or decomposed back to gold rather easily. It does not react with oxygen under any conditions or with any of the common acids. Gold does dissolve in a mixture of concentrated nitric and hydrochloric acids, which has long been known as *aqua regia*, forming a yellow solution of the strong acid $HAuCl_4$, which is fully ionized to the H_3O^+ ion and the complex $AuCl_4^-$ ion. Yellow crystals of the hydrate $(H_3O^+)(AuCl_4^-) \cdot 3H_2O$ can be obtained from this solution. When heated, this hydrate loses water and HCl to give yellow $Au(III)Cl_3$, which then loses Cl_2 to give yellow $Au(I)Cl$ and finally elemental gold.

When $OH^-(aq)$ is added to a solution of $HAuCl_4$, $Au(OH)_3$ is precipitated. When $Au(OH)_3$ is heated, it gives brown Au_2O_3, which decomposes at 150°C to gold and oxygen.

Effect of Complex-Ion Formation on Solubility

Silver bromide and silver chloride are soluble in an aqueous ammonia solution, whereas silver iodide is insoluble because of the much lower solubility in water of silver iodide than silver chloride and bromide. Let us look at this difference in solubility in a quantitative way.

The equilibria that we are concerned with are

$$AgCl(s) \rightleftharpoons Ag^+(aq) + Cl^-(aq) \qquad K_{sp} = 1.7 \times 10^{-10} \text{ mol}^2 \text{ L}^{-2} \text{ (see Table 15.11)}$$

Black and white photographic film is composed of very small crystals of a silver halide (usually silver bromide) suspended in gelatin—called photographic emulsion—coated on a cellulose acetate plastic. When the film is briefly exposed to light, a very small fraction of the silver ions in some of the silver halide crystals undergo photochemical decomposition, producing silver atoms. This process is called *photosensitization*. The crystals that are sensitized in this way are much more easily reduced completely to silver than the crystals that have not been sensitized. The reason sensitized crystals are much more easily reduced is not well understood, but it is a very important feature of the photographic process.

The process of reducing the sensitized crystals is known as *developing*, and the reducing agent is the *developer*. Organic reducing agents are used that penetrate the gelatin and rapidly reduce the sensitized crystals but only reduce the nonsensitized crystals very slowly. A common developer is 1,4-dihydroxybenzene (hydroquinone), which is oxidized to quinone, a ketone containing two carbonyl groups:

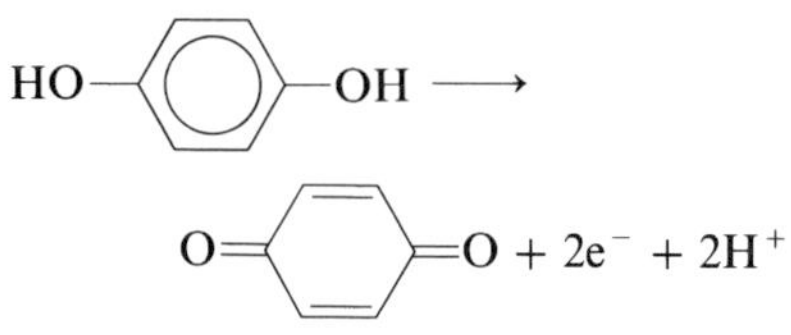

When the sensitized film is developed, the sensitized crystals are reduced to silver; the amount of elemental silver is thereby increased by a factor of about 10^{10}. The image on the film is therefore intensified by the same factor. In the next step, called *fixing*, undeveloped crystals of silver halide are dissolved by a solution of sodium thiosulfate, commonly called *hypo*. Without this step nonsensitized crystals would be slowly reduced by light, and the whole film would eventually turn black. The sodium thiosulfate solution dissolves the nonsensitized silver halide crystals by forming the complex ion $Ag(S_2O_3)_2^{3-}$:

$$AgBr(s) + 2S_2O_3^{2-} \longrightarrow Ag(S_2O_3)_2^{3-} + Br^-$$

After the film has been thoroughly washed, all that remains on the film is metallic silver. It is denser and therefore darker in the parts of the film that were exposed to the most light; it is less dense, and therefore not so dark, in those parts of the film that were exposed to less light. The film is then called a *negative*. When light is shone through the negative onto photographic paper—that is, paper covered with photographic emulsion—a positive print is obtained by essentially the same process.

and

$$Ag^+(aq) + 2NH_3(aq) \rightleftharpoons Ag(NH_3)_2^+(aq) \qquad K_f = 1 \times 10^8 \text{ mol}^{-2} \text{ L}^2 \text{ (see Table 21.10)}$$

These equations may be added to give the equation for the overall equilibrium:

$$AgCl(s) + 2NH_3(aq) \rightleftharpoons Ag(NH_3)_2^+(aq) + Cl^-(aq)$$

The corresponding equilibrium constant is obtained by multiplying the equilibrium constants K_{sp} and K_f. Therefore we have

$$K = K_{sp} \cdot K_f = \left(\frac{[Ag^+][Cl^-][Ag(NH_3)_2^+]}{[Ag^+][NH_3]^2}\right)_{eq} = \left(\frac{[Ag(NH_3)_2^+][Cl^-]}{[NH_3]^2}\right)_{eq}$$

$$= (1.7 \times 10^{-10} \text{ mol}^2 \text{ L}^{-2})(1 \times 10^8 \text{ mol}^{-2} \text{ L}^2) = 1.7 \times 10^{-2}$$

If silver chloride is added to a $1M$ NH_3 solution, and if we let x mol L^{-1}

21.4 SECOND AND THIRD SERIES OF TRANSITION METALS

be the concentration of $Ag(NH_3)_2^+$ that is formed, we may write

$$AgCl(s) + 2NH_3(aq) \rightleftharpoons Ag(NH_3)_2^+(aq) + Cl^-(aq)$$

Initial concentrations	1	0	0	mol L^{-1}
Equilibrium concentrations	$1 - 2x$	x	x	mol L^{-1}

$$K = \left(\frac{[Ag(NH_3)_2^+][Cl^-]}{[NH_3]^2}\right)_{eq} = \frac{x^2}{(1 - 2x)^2} = 1.7 \times 10^{-2}$$

Taking the square root of both sides, we have

$$\frac{x}{1 - 2x} = 0.13$$

Solving for x gives

$$x = 0.10 \quad \text{or} \quad [Ag(NH_3)_2^+] = 0.10 \text{ mol } L^{-1}$$

The concentration of the complex ion $Ag(NH_3)_2^+$ in equilibrium with solid silver chloride is quite high, showing that silver chloride has a considerable solubility in ammonia. We can find the concentration of Ag^+ in solution before the addition of ammonia as follows:

$$[Ag^+][Cl^-] = K_{sp} = 1.7 \times 10^{-10} \text{ mol}^2 L^{-2}$$

But $[Ag^+] = [Cl^-]$. Therefore,

$$[Ag^+]^2 = 1.7 \times 10^{-10} \text{ mol}^2 L^{-2}$$
$$[Ag^+] = 1.3 \times 10^{-5} \text{ mol } L^{-1}$$

The concentration of silver in solution as Ag^+ is only 1.3×10^{-5} mol L^{-1}, but after ammonia is added, the concentration of Ag in solution as $Ag(NH_3)_2^+$ is 0.10 mol L^{-1}.

We can make a similar calculation for silver iodide. We have

$$AgI(s) \rightleftharpoons Ag^+(aq) + I^-(aq) \qquad K_{sp} = 8 \times 10^{-17} \text{ mol}^2 L^{-2} \text{ (see Table 15.11)}$$

$$Ag^+(aq) + 2NH_3(aq) \rightleftharpoons Ag(NH_3)_2^+(aq) \qquad K_f = 1 \times 10^8 \text{ mol}^{-2} L^2 \text{ (see Table 21.10)}$$

Hence

$$AgI(s) + 2NH_3(aq) \rightleftharpoons Ag(NH_3)_2^+(aq) + I^-(aq)$$

and

$$K = K_{sp} \cdot K_f = (8 \times 10^{-17} \text{ mol}^2 L^{-2})(1 \times 10^8 \text{ mol}^{-2} L^2) = 8 \times 10^{-9}$$

This small equilibrium constant shows that the equilibrium lies well to the left; that is, silver iodide does not dissolve in ammonia.

If we again let the concentration of $Ag(NH_3)_2^+$ in equilibrium with solid AgI be x and the concentration of NH_3 be $1M$, we have

$$\left(\frac{[Ag(NH_3)_2^+][I^-]}{[NH_3]^2}\right)_{eq} = \frac{x^2}{(1 - 2x)^2} = 8 \times 10^{-9}$$

Hence,

$$\frac{x}{1 - 2x} = 9 \times 10^{-5}$$

Therefore

$$x = 9 \times 10^{-5} \quad \text{or} \quad [Ag(NH_3)_2^+] = 9 \times 10^{-5} \text{ mol } L^{-1}$$

The concentration of silver in solution in the form of the complex ion $Ag(NH_3)_2^+$ is very small. In other words, no significant amount of silver iodide dissolves in $1M$ ammonia solution.

In general, an insoluble precipitate of a metal salt will dissolve in a solution of a complexing agent if the formation constant of the complex is large enough, that is, if the overall equilibrium between the insoluble salt and the complexing agent to give the complex ion lies toward the right. The equilibrium constant for this reaction depends on the relative magnitudes of the solubility product constant of the insoluble salt and the formation constant of the complex. A very insoluble salt will be dissolved only by a complexing agent that forms a very strong complex. Thus ammonia forms a strong enough complex with the silver ion to dissolve silver chloride and bromide, but it will not dissolve the much less soluble silver iodide, which has a much smaller solubility product. Similarly, cyanide ion is used in extracting silver from the ore argentite, Ag_2S. Although Ag_2S is very insoluble in water, it can be dissolved in a solution of CN^- because of the formation of the complex ion $Ag(CN)_2^-$.

Mercury

Mercury is in the same group of transition metals as zinc and cadmium. Some of the properties of zinc, cadmium, and mercury are summarized in Table 21.15. We discussed zinc earlier in this chapter. Cadmium is sometimes used as a protective coating for iron and steel. It is also used for the control rods in nuclear reactors (Chapter 24) and as a component of the nickel-cadmium cell (Chapter 16). The chemistry of cadmium resembles that of zinc rather closely. The three elements Zn, Cd, and Hg have the same $d^{10}s^2$ outer-electron configuration. Only the s electrons are used in compound formation. Thus zinc and cadmium form the Zn^{2+} and Cd^{2+} ions and many complexes of these ions. Mercury also forms a similar Hg^{2+} ion as well as the unusual $[Hg\!-\!Hg]^{2+}$ ion in which two mercury atoms are joined by a covalent bond.

The symbol for mercury comes from the Latin *hydrargyrum*, meaning "liquid silver." Mercury is the only metal that is liquid at room temperature (mp $= -38.9°C$), although the melting points of two other metals are only just above room temperature (Cs, mp $= 28.5°C$; Ga, mp $= 29.8°C$). Some of mercury's many uses depend on this property. It is used, for example, in electric switches, thermometers, and barometers. It is also used in mercury vapor lamps and in fluorescent lighting tubes (Chapter 6). Mercury vapor and soluble mercury compounds are toxic (see Box 21.3).

Mercury forms an alloy with almost every metal except iron. These mercury alloys are known as **amalgams**. The metal in these amalgams is often less reactive than it is in the free state. For example, sodium amalgam reacts only slowly with water (Experiment 21.13). A mercury amalgam called dental amalgam is

Table 21.15 Some Properties of Zinc, Cadmium, and Mercury

	ATOMIC NUMBER	ATOMIC MASS	DENSITY (g cm^{-3})	MELTING POINT (°C)	$r_{metallic}$ (pm)
Zinc	30	65.37	7.14	419	138
Cadmium	48	112.41	8.64	321	154
Mercury	80	200.59	13.55	-39	157

21.4 SECOND AND THIRD SERIES OF TRANSITION METALS

Box 21.3

MERCURY: POISONING AND POLLUTION

Mercury, lead, and arsenic are commonly known to be toxic, but there are many popular misconceptions about the extent and dangers of environmental pollution by mercury and other toxic elements. It must be emphasized that the toxicity of the element and of various compounds of the element may be very different. For mercury to exert its toxic effect, it must enter the blood, either in the digestive system or in the lungs. Mercury amalgams, because of their insolubility, have long been used in dentistry without any harmful effects. Mercury has only a very small vapor pressure, but if mercury vapor is breathed over a long time, small amounts enter the body through the lungs, and a toxic concentration may be slowly built up. More volatile mercury compounds such as dimethyl mercury are much more dangerous. Soluble mercury compounds are equally dangerous because they can enter the body through the digestive system.

Mercury(II) nitrate, which is soluble in water, was at one time used to treat the fur used to make felt hats. Persons working in this trade who had long exposure to the mercury(II) nitrate often displayed symptoms of mercury poisoning, which include the loss of teeth and hair, and nervous disorders such as tremors, loss of memory, paralysis, and insanity. Hence the origin of the phrase "mad as a hatter," familiar from the mad hatter in *Alice in Wonderland*. But many mercury compounds are rather insoluble and are therefore less dangerous. For example, mercury(I) chloride, Hg_2Cl_2, is insoluble in water and has in the past actually been used as a cathartic (purgative).

In recent years the pollution of the environment by mercury has aroused concern. Mercury is present in the earth's crust as the ore cinnabar, which is very insoluble and therefore presents no direct danger. It is also believed that volcanoes emit considerable quantities of mercury compounds. Until the materials contained in these emissions become widely dispersed, they may be somewhat hazardous. In addition, humans have liberated considerable amounts of mercury into the environment. The burning of coal, which contains about 1 ppm Hg, releases about 5000 tons of mercury into the atmosphere every year. In the past paper mills have used a mercury-containing fungicide to prevent the growth of fungi during pulping. Mercury compounds therefore occurred in the waste products of the process and even in small amounts in the paper. Large amounts of mercury are used in the production of sodium hydroxide and chlorine by elec-

used to fill teeth. It consists of about 50% mercury and 50% "dental alloy," which is mainly silver and tin.

The most important ore of mercury is the sulfide *cinnabar*, HgS. Mercury is easily obtained from the ore by heating it in the air:

$$HgS(s) + O_2(g) \longrightarrow Hg(l) + SO_2(g)$$

Although mercury is a very rare element, it has been known since ancient times, because it is so easily obtained from its ores.

Mercury does not react with dilute acids, but it does react with concentrated nitric acid and, if heated, with concentrated sulfuric acid to give mercury(II) nitrate and mercury(II) sulfate, respectively:

$$Hg(l) + 4HNO_3(concd) \longrightarrow Hg(NO_3)_2(aq) + 2NO_2(g) + 2H_2O(l)$$
$$Hg(l) + 2H_2SO_4(concd) \longrightarrow HgSO_4(aq) + SO_2(g) + H_2O(l)$$

Mercury(II) chloride, $HgCl_2$, can be made by heating mercury(II) sulfate with sodium chloride:

$$HgSO_4(s) + 2NaCl(s) \longrightarrow Na_2SO_4(s) + HgCl_2(g)$$

The volatile $HgCl_2$ distills off. Mercury(II) chloride is a covalent compound that even in aqueous solution is present mainly as un-ionized covalent molecules. Since there are only two bonding electron pairs in the valence shell of the mercury atom, $HgCl_2$ is a linear AX_2 molecule.

Mercury combines with oxygen when heated in the air to give mercury(II)

trolysis of aqueous sodium chloride (see Chapter 16). Although in principle this mercury is recovered, small amounts find their way into the waste products of the process, which are frequently discharged into rivers.

It was once believed that the discharge of mercury compounds into rivers and lakes was safe because they are mostly insoluble and were thought to be converted slowly to very insoluble mercury(II) sulfide. However, it is now clear that this practice constitutes a very considerable hazard. Elemental mercury and mercury(I) compounds become slowly converted to mercury(II) compounds. Bacteria in the water then convert the Hg(II) compounds to CH_3Hg^+ and $(CH_3)_2Hg$. These compounds gradually accumulate in the plants and small organisms growing in the water. Fish feed on these plants and organisms and in turn tend to accumulate the mercury. This mercury then passes into larger fish, which eat the small fish, and in fish such as sharks and swordfish mercury can reach dangerously high levels. Before this danger was realized, there were several cases of mass mercury poisoning. For example, at Minimata, Japan, industrial mercury wastes accumulated in the sediment at the bottom of an ocean bay over many years. Over a period of ten years more than fifty people in Minimata died of mercury poisoning as a result of eating fish caught in the bay. Many others suffered from the mercury poisoning, and many children were born with serious deformities and brain damage.

We should note, however, that high levels of mercury in fish are not necessarily always to be attributed to human carelessness. Natural sources account for the mercury found in fish in some isolated mountain lakes and in arctic regions.

Because mercury is eliminated from the body, we can tolerate small amounts. The half-life of inorganic mercury compounds in the body is about six days; that is, half of the mercury that is ingested is eliminated in about six days. If only small amounts enter the body at infrequent intervals, a dangerous concentration of mercury is not built up. In contrast, the half-life of organomercury compounds, such as $(CH_3)_2Hg$, in the body averages about 70 days and is even longer in some organs such as the brain. Thus even if a person regularly ingests even very small amounts of organomercury compounds, the concentration of mercury in the body continues to rise for some months until it reaches a constant level, which may be a toxic level.

oxide:

$$2Hg(l) + O_2(g) \longrightarrow 2HgO(s)$$

However, when HgO is heated more strongly, it decomposes to mercury and oxygen (Experiment 21.14). Mercury(II) oxide can also be obtained by adding hydroxide ion to a solution of an Hg^{2+} salt:

$$Hg^{2+}(aq) + 2OH^-(aq) \longrightarrow HgO(s) + H_2O(l)$$

When H_2S is passed into a solution of a mercury(II) salt, insoluble black HgS is precipitated.

A very unusual property of mercury is that it forms the Hg_2^{2+} ion in which mercury atoms are joined by a covalent bond. Cadmium is the only other metal known to form an analogous ion. This ion can be obtained, for example by the reaction between mercury and a solution of mercury(II) nitrate:

$$Hg(l) + Hg(NO_3)_2(aq) \longrightarrow Hg_2(NO_3)_2(aq)$$

The Hg_2^{2+} ion contains mercury in the $+1$ oxidation state and is called the mercury(I) ion.

If chloride ion is added to a solution of mercury(I) nitrate, a white insoluble precipitate of mercury(I) chloride, Hg_2Cl_2, is obtained. Like mercury(II) chloride, Hg_2Cl_2 is a covalent molecule with a linear structure in which both mercury atoms have the expected linear AX_2 geometry.

Dimethyl mercury, CH_3—Hg—CH_3, is a volatile covalent compound. It is an example of an **organometallic compound**, a compound in which alkyl groups,

Cl—Hg—Hg—Cl

Mercury(I) chloride

21.4 SECOND AND THIRD SERIES OF TRANSITION METALS

Sodium Amalgam

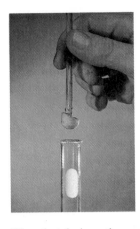

A piece of sodium floats on the surface of mercury, which has a much higher density.

When the tube is gently warmed, a solid compound of sodium and mercury (sodium amalgam) is formed. Here this solid amalgam is being lifted out of the tube.

Sodium amalgam reacts considerably more slowly with water than does sodium itself to form hydrogen and an NaOH solution.

or substituted alkyl groups, are attached to a metal atom. Dimethyl mercury can be prepared by the reaction of mercury-sodium amalgam with chloromethane:

$$Hg(l) + 2Na(l) + 2CH_3Cl(g) \longrightarrow Hg(CH_3)_2(l) + 2NaCl(s)$$

The Thermal Decomposition of Mercury(II) Oxide

When red mercury(II) oxide is heated it darkens and decomposes to give oxygen and mercury, which condenses as silver droplets on the cooler part of the tube.

When the tube is allowed to cool, the mercury oxide returns to its original red color.

CHAPTER 21
THE TRANSITION METALS

It is a strong-smelling, very toxic, volatile substance (bp = 96°C). It has a linear covalent structure. Reaction of $(CH_3)_2Hg$ with $HgCl_2$ gives methylmercury chloride, CH_3HgCl, another linear covalent molecule:

$$(CH_3)_2Hg + HgCl_2 \longrightarrow 2CH_3HgCl$$

If the Cl is replaced by sulfate or nitrate, an ionic salt that contains the covalent methylmercury cation, $CH_3\!-\!Hg^+$, is obtained, for example, $CH_3Hg^+NO_3^-$ (see Box 21.3).

We have previously mentioned another organometallic compound, namely, tetraethyl lead, $Pb(C_2H_5)_4$, (Box 11.1). Many other metals form organometallic compounds, but discussion of these interesting compounds is beyond the scope of this book.

IMPORTANT TERMS

An **amalgam** is a liquid or solid alloy of mercury.

A **bidentate ligand** is a molecule or ion containing two atoms that can each donate an electron pair to a metal atom, such as ethylenediamine and the oxalate ion.

A **chelate** is a ligand with two or more donor atoms.

A **complex (coordination compound)** is a molecule containing one or more ligands bonded to a metal atom.

A **complex ion** is a complex that has an overall charge.

A **coordinate bond** is the bond between a ligand and a metal atom in a complex. It does not differ from an ordinary single bond.

The **coordination number** is the number of donor atoms bonded to a metal atom.

The **formation constant** is the equilibrium constant for the equilibrium between a hydrated ion and a complex ion in aqueous solution.

The **lanthanides** are the 14 metals in period 5 beginning with lanthanum.

A **ligand** is a molecule or an ion that is attached to a central metal atom in a complex.

A **monodentate** ligand is a ligand with only one donor atom.

A **nonstoichiometric compound** is a solid with a formula in which one of the subscripts is nonintegral, for example, $MnO_{1.85}$.

An **organometallic compound** is a compound containing metal-carbon bonds.

A **polydentate ligand** is a ligand in which the number of donor atoms is greater than two, for example, $(EDTA)^{4-}$, which is hexadentate.

The **transition metals** are a group of thirty elements that occupy the middle portion of the periodic table between Groups II and III in periods 4, 5, and 6.

PROBLEMS

Electron Configurations

1. What are the ground state electron configurations of each of the following ions?

Cr^{3+} Ni^{2+} Cu^+ Co^{2+} Au^{3+} Fe^{3+}

2. What are the ground state electron configurations of the following?

Ti   Cr   Fe   Ni   Zn
Au^{3+} Hg^{2+} Cr^{3+} Co^{2+}

3. How many 3d electrons are there in each of the following ions?

Cu^{2+} Fe^{3+} Mn^{2+} Ag^+ V^{3+} Ti^{2+}

Oxidation States

4. What are the oxidation numbers of the transition metals in the following compounds and ions?

$KMnO_4$ $K_2Cr_2O_7$ $Ag(CN)_2^-$ $[Co(NH_3)_4Cl_2]^+$
$K_3[Cr(CN)_6]$ Na_2CoCl_4 K_2MnO_4 $MnO(OH)$
VO_2Cl $TiO\cdot SO_4$

5. What is the oxidation state of the transition metal in each of the following complex ions?

$FeCl_4^-$ $Co(H_2O)_6^{3+}$ $Al(OH)_4^-$
$Ag(NH_3)_2^+$ $Fe(CN)_6^{4-}$ $Fe(CN)_6^{3-}$

6. By writing out their electronic configurations, decide what oxidation states are, in principle, possible for the elements of atomic numbers 22, 25, and 27.

7. What is the oxidation state of the transition metal in each of the following ions?

CrO_4^{2-} $Cr_2O_7^{2-}$ MnO_4^{2-}
VO_4^{3-} VO^{2+} FeO_4^{2-}

Complexes

8. Consider the complex ion $[Co(NH_3)_3(H_2O)_2Cl]^+$.

(a) Identify the ligands and their charges.

(b) What is the oxidation state of cobalt?

(c) What would be the charge on the complex if chloride ions replaced the water molecules?

9. What is the coordination number of the central metal atom in each of the following compounds?

$[Zn(NH_3)_4]Cl_2$ $[Co(NH_3)_3Cl_3]$ $[Co(NH_3)_5Cl]Cl_2$
$[Cr(en)_2Cl_2]^+$ $K_2[FeCl_4]$

10. Draw diagrams of the structures of each of the following complex ions:

$trans[Cr(NH_3)_4Cl_2]^+$ $[Co(C_2O_4)_3]^{3-}$
$[Cr(C_2O_4)Br_4]^{3-}$ $cis[Pt(en)_2(CN)_2]^{2-}$

11. Name each of the complex ions in Problem 10.

12. Addition of an aqueous silver nitrate solution to a solution of each of the compounds of platinum listed in the accompanying table was found by Werner to give the amounts of silver chloride given in the table. How did Werner account for these observations?

EMPIRICAL FORMULA	MOLES OF AgCl OBTAINED FOR EACH MOLE OF PLATINUM COMPOUND
$PtCl_4 \cdot 6NH_3$	4
$PtCl_4 \cdot 5NH_3$	3
$PtCl_4 \cdot 4NH_3$	2
$PtCl_4 \cdot 3NH_3$	1
$PtCl_4 \cdot 2NH_3$	0

13. The compound $Co(NH_3)_5SO_4Br$ exists in a red form and a violet form. Solutions of the red compound give a precipitate of AgBr on addition of an $AgNO_3$ solution but no precipitate on addition of a $BaCl_2$ solution. Solutions of the violet compound give a white precipitate of $BaSO_4$ on addition of a $BaCl_2$ solution but no precipitate on addition of an $AgNO_3$ solution. From this evidence, draw structures for the two compounds.

14. Write balanced chemical equations to represent the following observations:

(a) Solid silver chloride dissolves in a solution of sodium thiosulfate.

(b) The green compound $[Cr(en)_2Cl_2]Cl$ reacts slowly with water to give an orange-brown compound. When a solution of this compound is treated with an aqueous silver nitrate solution, 3 mol of AgCl is obtained per mole of the orange-brown compound.

(c) Insoluble $Ni(OH)_2$ dissolves in an excess of aqueous ammonia.

(d) A pink solution of $CoSO_4$ turns deep blue on the addition of concentrated hydrochloric acid.

15. Explain why calcium oxalate, which is insoluble in water, will dissolve in a solution of EDTA.

16. Draw a Lewis structure for the tetrahydroxozincate-(II) ion, $Zn(OH)_4^{2-}$. What is the formal charge on the zinc? What is the shape of this ion?

17. Draw a Lewis structure for the molecule CrO_2Cl_2, and predict its geometry.

Properties of Transition Metals

18. How does the acidic character of an oxide of a transition metal vary with its oxidation state? Illustrate your answer by means of the oxides of a particular transition metal.

19. How does the ionic-covalent character of the compounds of a transition metal vary with its oxidation state? Illustrate your answer by describing some of the compounds of one particular transition metal.

20. Explain why the melting points of the first series of transition metals rise to a maximum at Cr and then decrease again.

21. Name the oxide of chromium that is most likely to be:

(a) Basic and ionic (b) Acidic and covalent

(c) An oxidizing agent (d) A reducing agent

Reactions

22. Why is the solution made by dissolving the pale violet solid $Fe(NO_3)_3 \cdot 6H_2O$ in water yellow? When an aqueous nitric acid solution is added to the yellow solution, it becomes pale violet again. Explain why.

23. In what form does copper exist in an aqueous copper(II) sulfate solution? After the addition of concentrated hydrochloric acid to this solution? After the addition of concentrated aqueous ammonia to this solution? After the addition of aqueous potassium cyanide to this solution?

24. Dilute sulfuric acid is added to samples of nickel and cobalt in separate beakers. Describe the reactions that occur and the color of the resulting solutions. Describe what occurs if each solution is made basic by the addition of aqueous NaOH.

25. Consider the elements Cr, Mn, Fe, Co, Ni, Cu, and Zn.

(a) Which of these elements form ions that give a colorless solution in water?

(b) Which oxidation state is common to all these elements?

(c) Identify the following hydroxides of general formula $M(OH)_2$: (i) blue, (ii) white, (iii) green.

(d) Identify the ion that gives a pale green solution and forms a red-brown precipitate on the addition of OH^- and H_2O_2.

(e) Identify the ion that gives a pale pink solution that becomes blue on the addition of excess hydrochloric acid.

(f) Which of the elements form compounds in the $+6$ oxidation state? Give examples of compounds of this oxidation state for two different elements.

(g) A solution containing manganese has a pale pink color. It might be a solution of Mn^{2+} or a very dilute solution of MnO_4^-. How would you decide between these two possibilities?

26. Write equations for the reduction of dichromate ion in acidic solution by the following:

(a) Sulfur dioxide.

(b) Ethanol, which is oxidized to ethanal.

(c) Iodide ion, which is oxidized to triiodide ion, I_3^-.

Miscellaneous

27. A 0.540-g sample of pure iron powder was heated with powdered sulfur. A reaction occurred to give a grey-black compound, and after excess sulfur was burned off, the mass of the resulting grey-black compound was found to be 0.905 g. Calculate the formula of the sulfide of iron that is formed. Do you think this result shows a large experimental error? Or could the experimental result be accurate? Explain why you consider the result either accurate or inaccurate.

28. What volume of $Ni(CO)_4$, measured at 80°C and 2 atm, would result from the complete reaction of 2.5 g of nickel with carbon monoxide? What volume of carbon monoxide, measured at 25°C and 1 atm, would result from the decomposition of this amount of $Ni(CO_4)$ at 200°C?

29. A 0.50-g sample of brass, a zinc-copper alloy, was treated with dilute sulfuric acid. After the reaction was complete, the volume of hydrogen collected over water at 25°C and a pressure of 756 mm Hg was found to be 102.8 mL. What is the percentage composition (by mass) of brass? (The vapor pressure of water at 25°C is 23.8 mm Hg.)

30. An important type of stainless steel contains 18% by mass of nickel. How much nickel sulfide ore, NiS, is required to produce 1 tonne of stainless steel?

31. Vitamin B_{12} is a coordination compound of cobalt with the molecular formula $C_{63}H_{90}O_{14}N_{14}PCo$. What is the percentage of cobalt in vitamin B_{12}? If the minimum daily requirement of vitamin B_{12} is 1 μg, how much cobalt does the human body require each day to stay healthy?

32. The solubility of $Cr(OH)_3$ in water at 25°C is 5.6 $\mu g\ L^{-1}$. What is the value of the solubility product constant?

33. For how long would a current of 1.5 A have to be passed through a solution containing $Cr^{3+}(aq)$ ions in order for a metal object of surface area 1 m^2 to be coated with a layer of Cr metal 0.1 mm thick? Why do you suppose that the thickness of the coating on chromium-plated steel is generally very thin?

34. The K_{sp} for Ag_3PO_4 is 1.8×10^{-18}. What is solubility of Ag_3PO_4 in water and in an 0.0010 M solution of $AgNO_3$?

CHAPTER 22

BORON AND SILICON: TWO SEMIMETALS

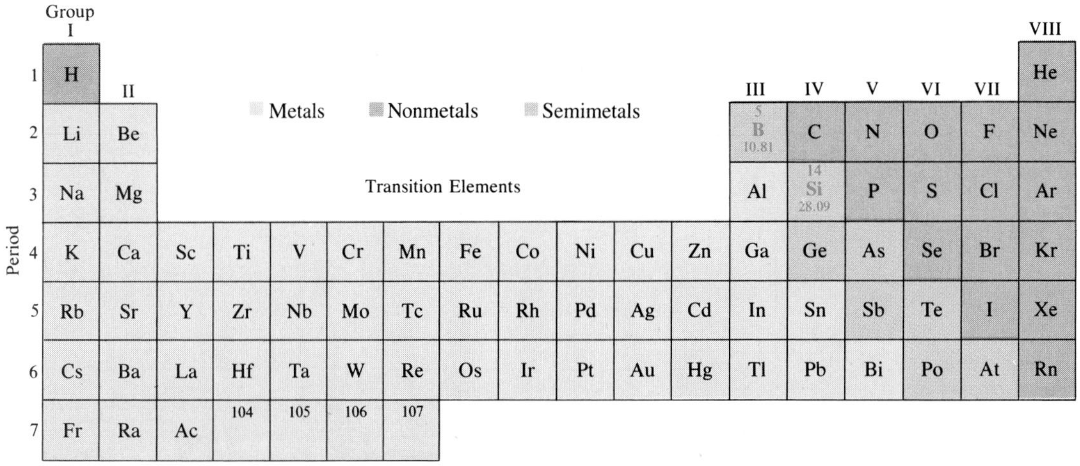

	Group I																	VIII
1	H	II											III	IV	V	VI	VII	He
2	Li	Be		Metals	Nonmetals	Semimetals							5 B 10.81	C	N	O	F	Ne
3	Na	Mg			Transition Elements								Al	14 Si 28.09	P	S	Cl	Ar
4	K	Ca	Sc	Ti	V	Cr	Mn	Fe	Co	Ni	Cu	Zn	Ga	Ge	As	Se	Br	Kr
5	Rb	Sr	Y	Zr	Nb	Mo	Tc	Ru	Rh	Pd	Ag	Cd	In	Sn	Sb	Te	I	Xe
6	Cs	Ba	La	Hf	Ta	W	Re	Os	Ir	Pt	Au	Hg	Tl	Pb	Bi	Po	At	Rn
7	Fr	Ra	Ac	104	105	106	107											

Period

Silicon is the second most abundant element in the earth's crust. *Silicates*, which are compounds of silicon, oxygen, and many different metals, comprise 95% of the rocks that make up the earth's crust. It is not surprising, therefore, that silicates have played a very significant role in history. One feature that distinguishes *Homo sapiens* from all other mammals is a highly developed ability to use tools. The first tools were pebbles of flint, a form of silicon dioxide, some of which were fashioned into sharp cutting edges for axes and hunting weapons. The Latin word for flint is *silex*, and from it comes our words *silicon*, *silica*, *silicate*, *silicone*, and so on. Another important step in human development occurred when pottery was first made by heating clay, which is a mixture of certain types of silicates. Pottery making ranks with brewing as one of the oldest chemical technologies developed.

As civilization developed, silicate glazes and enamels for decorating pottery and metal objects, and glass for making bottles and ornamental objects were developed. These are all made from molten silicates, which do not crystallize readily on cooling but, instead, form a hard, transparent, amorphous solid that we call a glass. The chemistry of silicates is, however, only one aspect of the chemistry of silicon. Today the element itself is assuming great importance because the silicon chip is the basis of the microcomputer revolution.

In contrast to silicon, boron is rare; it constitutes only 0.0003% of the earth's crust. However, because it occurs in rather concentrated deposits in a few areas, it is a relatively well-known element with some important uses. It also has some rather unusual chemistry, which will remind us that although we have discussed a rather considerable amount of chemistry in this book, we have only scratched the surface of the subject.

Boron and silicon are neighboring elements in the periodic table, and although boron is in Group III and silicon is in Group IV, they have many similar properties. Both are **semimetals**, or *metalloids*, that is, elements with some of the properties of a metal and some of a nonmetal. The semimetals are found in a band running diagonally across the periodic table from top left to bottom right.

22.1 SILICON

Silicon follows carbon in Group IV. Whereas carbon is a typical nonmetal, silicon and the next element, germanium, are semimetals, and they are followed by tin and lead, which are metals. Unlike the elements in the groups on the left and right sides of the periodic table, those in the main groups in the middle of the table show a considerable variation in properties on descending the group. Silicon is a shiny, silvery solid that looks like a metal but has only a low electrical conductivity; moreover, its conductivity increases with increasing temperature, whereas the conductivity of a metal decreases with increasing temperature (Chapter 9). Substances that have an electrical conductivity in the solid state that increases appreciably with increasing temperature are known as **semiconductors**. We will discuss them and their important role in modern technology at the end of this chapter.

Silicon occurs not only as silicates but also as *silicon dioxide*, SiO_2, which has been known for centuries as *silica*. The silicates and silica are found as rocks, clay, sand, and so on.

The element can be prepared from silica by heating it with coke to a temperature of about 3000°C in an electric arc furnace:

$$SiO_2(s) + 2C(s) \longrightarrow Si(l) + 2CO(g)$$

The reactants are added continuously at the top of the furnace. Carbon monoxide escapes from the furnace and burns to give carbon dioxide, while the molten silicon (mp = 1414°C) runs out from the bottom of the furnace and solidifies. This silicon is pure enough for many purposes, such as the manufacture of alloys with metals, but ultrapure silicon is needed in solid-state electronic devices.

Ultrapure silicon is obtained by first converting impure silicon to silicon tetrachloride by heating it in chlorine:

$$Si(s) + 2Cl_2(g) \longrightarrow SiCl_4(l) \qquad (bp = 57.6°C)$$

The resulting $SiCl_4$ is then purified by distillation and reduced to silicon by

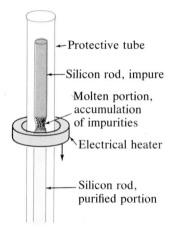

← Protective tube

← Silicon rod, impure

Molten portion,
accumulation
of impurities

Electrical heater

Silicon rod,
purified portion

Figure 22.1 Purification of Silicon by Zone Melting. A rod of impure silicon is drawn very slowly through the electric heater. The molten zone in which the impurities dissolve is thus moved to one end of the rod, where, after solidifying, it can be cut off.

heating with hydrogen or magnesium:

$$SiCl_4 + 2H_2 \longrightarrow Si + 4HCl$$
$$SiCl_4 + 2Mg \longrightarrow Si + 2MgCl_2$$

The magnesium chloride is removed from the silicon by washing it out with hot water. The silicon can be further purified by **zone refining** (Figure 22.1). In this process, a short segment of a rod of silicon is heated until it melts. The impurities are more soluble in the molten silicon than in the solid and therefore concentrate in the liquid. The rod is slowly moved through the heater so that the molten zone traverses the length of the rod, removing impurities as it moves. When the impure molten zone reaches the end of the rod, it is allowed to solidify and is cut off. This process is repeated as often as necessary. In the laboratory silica can be reduced to silicon by heating it with magnesium (Experiment 22.1).

Silicon has the diamond structure (Chapter 10), but the Si—Si bonds are longer and weaker than the C—C bonds, and so the melting point and boiling point of silicon are lower than those of carbon (see Table 22.1). Silicon has no allotropic form analogous to graphite. This is consistent with the fact that a third-period element such as silicon has a much smaller tendency to form multiple bonds than a second-row element such as carbon; in graphite each carbon atom forms two single and one double bond.

Silicon is a rather unreactive element and is not attacked by acids. It reacts with hot, concentrated aqueous hydroxide solutions and with molten hydroxides to give silicates; it reacts with oxygen at high temperatures to give silicon dioxide.

EXPERIMENT 22.1

Reduction of Silica to Silicon

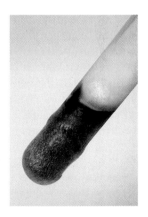

When silica, SiO_2 (sand), is heated with magnesium powder, a strongly exothermic reaction occurs and the mixture begins to glow brightly.

It continues to glow very brightly even when the tube is removed from the bunsen burner.

When the tube is allowed to cool, it is seen to have become distorted by the heat of the reaction and the inside is coated with a shiny metallic-looking deposit of silicon.
The other product of the reaction, magnesium oxide, can be seen as a white powder higher up in the tube.

Table 22.1 Properties of Boron, Carbon, and Silicon

	ATOMIC NUMBER	ATOMIC MASS	DENSITY (g cm⁻³)	MELTING POINT (°C)	BOILING POINT (°C)	COVALENT RADIUS (pm)
B	5	10.81	2.54	2300	2550	90
C (diamond)	6	12.01	3.52	3550	4827	77
Si	14	28.08	2.36	1414	2355	117

22.2 SILICON DIOXIDE, SiO₂

Silicon dioxide, or *silica*, occurs in three different crystalline forms, as the minerals *quartz*, *cristobalite*, and *tridymite*, and in several amorphous forms, such as agate and flint. Quartz is the best known of the silica minerals. It is one of the commonest minerals in the earth's crust and is often found in the form of colorless, transparent crystals, which are sometimes beautifully formed and of enormous size; crystals having a mass of 100 kg or more have been found. Quartz is a constituent of many rocks, which are often complex mixtures of different minerals. Granite, for example, is a mixture of the three silicon minerals, quartz, mica, and feldspar. Some forms of quartz that contain traces of impurities are beautifully colored and are often used as semiprecious gemstones. Amethyst, for example, is quartz that is colored violet by traces of Fe(III). Onyx, jasper, carnelian, and flint are colored forms of noncrystalline silicon dioxide.

Silicon dioxide also forms the cell walls of tiny one-cell plants called *diatoms*, which are the most abundant components of marine plankton. When the diatom dies, the silica cell walls sink to the bottom of the ocean to form deposits of a material called *diatomaceous earth*. This is used in many commercial preparations, including polishes and paint removers, and also for decolorizing oils.

The structures of all three crystalline forms of silicon dioxide are based on the SiO₄ tetrahedron. Each silicon atom forms four Si—O bonds, which have a tetrahedral AX₄ arrangement around the silicon. Each oxygen is bonded to another silicon atom to give an infinite three-dimensional structure. The structures of the three different forms differ in the ways that the tetrahedra are arranged. Cristobalite has the simplest structure, closely related to that of diamond (see Figure 22.2). Each silicon atom occupies the position of a carbon atom in the diamond structure. But the silicon atoms are not connected directly to each other as in elemental silicon; rather, they are connected via an oxygen atom, which is situated between each pair of silicon atoms (see Figure 22.2). In quartz the SiO₄ tetrahedra are joined in a more complicated manner.

The contrast between silicon dioxide and carbon dioxide is striking. Carbon dioxide is a molecular substance consisting of simple triatomic CO_2 molecules and is a gas at room temperature. Silicon dioxide has an infinite three-dimensional structure and is a hard crystalline solid. As previously noted, second-period elements, such as carbon, readily form multiple bonds, as in the CO_2 molecule, whereas silicon, a third-period element, has less tendency to form multiple bonds, and thus SiO_2 has an infinite three-dimensional network structure that involves only single bonds. Because of the considerable difference in electronegativity between silicon and oxygen, the SiO bonds are polar, and the structure may be regarded as being intermediate between a purely covalent three-dimensional network and an ionic crystal consisting of Si^{4+} and O^{2-} ions.

If any form of silica is melted (mp ≈ 1600°C) and the liquid is then cooled,

Group of quartz crystals

SiO₄ tetrahedron

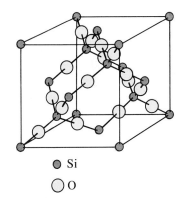

● Si
○ O

Figure 22.2 Structure of Silicon Dioxide. The structures of the three forms of silicon dioxide, quartz, cristobalite, and tridymite, all have a tetrahedral arrangement of four Si—O bonds around each silicon atom. The structure of cristobalite is based on a face-centered cubic lattice. Each silicon atom occupies the position of a carbon atom in the structure of diamond. An oxygen atom is located between each pair of silicon atoms.

it normally does not crystallize at the original melting point; instead, it becomes more and more viscous as the temperature is lowered. At about 1500°C it is so viscous that it no longer flows, having many of the characteristics of a solid. It is an amorphous rather than a crystalline solid, and it is called *silica glass*. It has a structure related to that of quartz and the other crystalline forms of silicon dioxide in that each silicon is surrounded by a tetrahedral arrangement of four oxygen atoms and each tetrahedron is joined to four others by the oxygen atoms at their corners. However, the tetrahedra have a random arrangement rather than one of the regular arrangements of quartz, cristobalite, or tridymite (see Figure 22.3).

Silica glass is transparent and has many of the properties of ordinary glass, but it is more difficult to work; it becomes soft at a much higher temperature than ordinary glass and has a smaller range of temperature over which it remains soft and able to be shaped. Silica glass expands very little when it is heated, and it can therefore be heated and cooled very rapidly without breaking. For this reason it is useful for certain laboratory apparatus. In addition, it is transparent to ultraviolet light and is therefore used for making mercury vapor ultraviolet lamps and for windows and lenses in instruments that employ ultraviolet light.

Silicon dioxide is an acidic oxide that reacts with NaOH or Na_2CO_3 on heating to give silicates

$$SiO_2(s) + 4NaOH(s) \longrightarrow Na_4SiO_4(s) + 2H_2O(g)$$

$$SiO_2(s) + Na_2CO_3(s) \longrightarrow Na_2SiO_3(s) + CO_2(g)$$

These are only simplified representations of the reactions; there are many silicates with complex structures, as we will see in the next section.

A concentrated solution of sodium silicate in water is called *water glass*. It is used for fireproofing wood and cloth, for adhesives, and for preserving eggs. It can be used to prepare other metal silicates, as demonstrated in Experiment 22.2.

EXPERIMENT 22.2

Formation of Metal Silicates: A Chemical Garden

The bottle contains an aqueous solution of sodium silicate to which crystals of various colored transition metal salts have been added. The metal ions react with the sodium silicate to form a thin membrane of insoluble metal silicate. Water penetrates the membrane by osmosis, causing it to expand and then burst. A new membrane then forms and the process repeats, causing a column of colored metal silicates to form. Here the crystals are resting on a layer of sand, and the columns growing up from the crystals form what is sometimes called a "chemical garden."

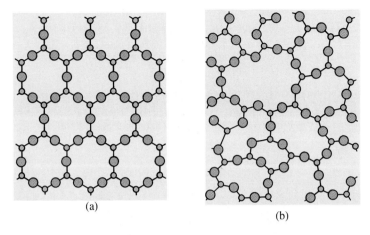

(a)

(b)

Figure 22.3 Two-Dimensional Representation of Structures of Crystalline SiO$_2$ and Amorphous SiO$_2$.
(a) In crystalline SiO$_2$, the SiO$_4$ tetrahedra have a regular arrangement.
(b) In amorphous SiO$_2$ (silica glass), the SiO$_4$ tetrahedra have an irregular, random arrangement.

22.3 SILICIC ACID AND SILICATES

Silicon dioxide is the anhydride of silicic acid. When finely powdered silicon dioxide is shaken with water for a long time, a very slightly acidic solution is obtained, which is due to the formation of a very small amount of silicic acid, Si(OH)$_4$:

$$SiO_2(s) + 2H_2O(l) \rightleftharpoons Si(OH)_4(aq)$$

The very small solubility of silicon dioxide in water is increased at high pressure and temperature, which accounts for the deposits of silica found around many hot geysers. The silicon dioxide that is dissolved in water at high pressure below the surface comes out of solution when the water rises to the surface, where the pressure and temperature are much lower.

More concentrated solutions of silicic acid can be obtained by the reaction of silicon tetrachloride with water (see Experiment 5.3),

$$SiCl_4(l) + 4H_2O(l) \longrightarrow Si(OH)_4(aq) + 4HCl(aq)$$

or by the reaction of an aqueous solution of sodium silicate with hydrochloric acid,

$$Na_2SiO_3(aq) + H_2O(l) + 2HCl(aq) \longrightarrow Si(OH)_4(aq) + 2NaCl(aq)$$

However, the product of these reactions is not simply a solution of Si(OH)$_4$. A gelatinous solid is formed that consists of polymeric silicic acids formed by condensation reactions such as

$$\underset{\underset{OH}{|}}{\overset{\overset{OH}{|}}{HO-Si-OH}} + \underset{\underset{OH}{|}}{\overset{\overset{OH}{|}}{HO-Si-OH}} \longrightarrow \underset{\underset{OH}{|}}{\overset{\overset{OH}{|}}{HO-Si-O}}-\underset{\underset{OH}{|}}{\overset{\overset{OH}{|}}{Si-OH}} + H_2O$$

Continuation of such reactions leads to a complex mixture of many polymeric acids. In Chapter 7 we noted a similar tendency of phosphoric acid to form polymeric acids by condensation, although the tendency is less marked for phosphorus than for silicon. Phosphoric acid gives stable solutions in water that contain only H$_3$PO$_4$ and its ions, whereas Si(OH)$_4$, although probably present

$XO_m(OH)_n$	$X(OH)_4$	$XO(OH)_3$	$XO_2(OH)_2$	$XO_3(OH)$
$m =$	0	1	2	3

$$HO-\underset{\underset{OH}{\vert}}{\overset{\overset{OH}{\vert}}{Si}}-OH \qquad HO-\underset{\underset{OH}{\vert}}{\overset{\overset{O}{\vert\vert}}{P}}-OH \qquad O=\underset{\underset{OH}{\vert}}{\overset{\overset{O}{\vert\vert}}{S}}-OH \qquad O=\underset{\underset{O}{\vert\vert}}{\overset{\overset{O}{\vert\vert}}{Cl}}-OH$$

Very weak   Very strong

Acid strength increases ⟶

Tendency to polymerize decreases

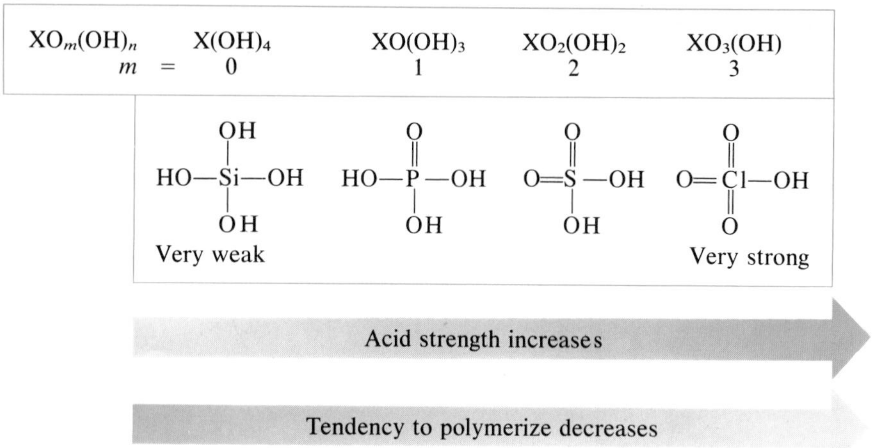

in aqueous solutions, cannot be separated from the complicated mixture of the condensed forms of the acid that is always formed.

Silicic acid is very weak ($K_a = 1 \times 10^{-10}$ mol L^{-1}). It is a member of the class of very weak acids such as $HOCl$ and $B(OH)_3$, which have the general formula $X(OH)_n$ (Chapter 20). It is the first member of the series of third-period oxoacids that increase in strength from the very weak silicic acid to the very strong perchloric acid (see Figure 22.4).

If the mixture of polymeric silicic acids is heated, causing further condensation reactions to occur, a hard, granular, translucent substance called *silica gel* is obtained (see Experiment 22.3). It has a large surface area and readily adsorbs water and other substances. It is widely used as a drying agent and as a catalyst. Small bags of silica gel are frequently packed with delicate scientific apparatus, cameras, and electronic equipment to protect them from damage by moisture during transport.

Silicates

Over one thousand silicates occur naturally. Their number and complexity result from the many different ways in which SiO_4 tetrahedra can be linked

EXPERIMENT 22.3

A Dessicator Containing Silica Gel

A dessicator is often used in the laboratory for storing substances that are sensitive to moisture. The dessicator contains a substance such as silica gel, which strongly absorbs water. The silica gel used here is colored with cobalt(II) chloride which is blue when it is anhydrous but becomes pink when it is hydrated, showing that the silica gel has absorbed so much water that it is no longer active.

together. The simplest silicates contain the anion SiO_4^{4-}, derived from the acid $Si(OH)_4$. *Olivine* is an important mineral of this type; it is the principal component of the earth's mantle, which lies between the core and the crust (see Box 22.1). Olivine is an iron magnesium silicate that is often represented by the formula $FeMgSiO_4$, although its composition may vary from Mg_2SiO_4 to Fe_2SiO_4. The SiO_4^{4-} tetrahedra are packed together in a compact way in olivine, giving it a greater density than most other silicate minerals.

The silicates are the salts of the many polymeric forms of silicic acid. The condensation of two silicic acid molecules gives the acid $H_6Si_2O_7$:

$$HO-\underset{\underset{OH}{|}}{\overset{\overset{OH}{|}}{Si}}-OH + HO-\underset{\underset{OH}{|}}{\overset{\overset{OH}{|}}{Si}}-OH \longrightarrow HO-\underset{\underset{OH}{|}}{\overset{\overset{OH}{|}}{Si}}-O-\underset{\underset{OH}{|}}{\overset{\overset{OH}{|}}{Si}}-OH + H_2O$$

The anion derived from this acid is $Si_2O_7^{6-}$. It consists of two SiO_4 tetrahedra

---

Box 22.1

MINERALS AND THE COMPOSITION AND STRUCTURE OF EARTH

The earth consists principally of three concentric layers. There is a central sphere consisting mainly of iron and nickel, which has a radius of approximately 3500 km. Surrounding the core is a mantle of silicate rock, mostly olivine, $FeMgSiO_4$, approximately 3000 km thick. On top of the mantle is a very thin crust ranging from about 5 km in thickness under the oceans to as much as 100 km under the continents. The temperature of the mantle is such that the rocks are soft and somewhat plastic and flow slowly. As a result, pieces of the crust, called *plates*, can be regarded as "floating" on the mantle, like icebergs in the oceans. Collisions between these plates produce mountain ranges, volcanic activity, and earthquakes.

Soon after the formation of our planet, most of the interior was molten; thus the heaviest substances—in particular, iron and nickel—sank to the center, forming the core. Around this core various metal silicates stratified according to their density. One of the most dense of the silicates, olivine, now constitutes the mantle. The less dense silicates floated to the top and built the crust.

The substances that constitute the earth's crust are called *minerals*. Minerals may be classified into three major groups; native elements, silicate minerals, and nonsilicate minerals. The most important of the native elements are copper, silver, gold, and sulfur. Some of the more important nonsilicate minerals that have been mentioned in earlier chapters are the following:

- The oxides Al_2O_3 (bauxite), Cu_2O (cuprite), and Fe_2O_3 (hematite).
- The carbonates $CaCO_3$ (limestone, calcite) and $MgCO_3$ (magnesite).
- The sulfates $BaSO_4$ (barite) and $CaSO_4 \cdot 2H_2O$ (gypsum).
- The sulfides Cu_2S (chalcocite) and FeS_2 (pyrite).
- The halides $NaCl$ (halite, rock salt) and KCl (sylvite).
- The phosphate $Ca_5(PO_4)_3F$ (fluorapatite).

Despite their abundance, the silicate minerals are not useful as sources of metals, because it is more difficult to obtain a metal from a silicate than from other compounds such as oxides and sulfides. Thus we rely on these much rarer minerals as sources of metals. The silicates are used only for making pottery, glass, and cement.

sharing a corner:

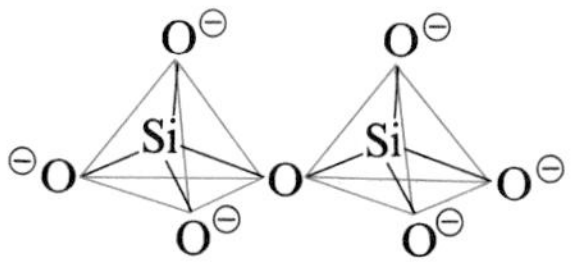

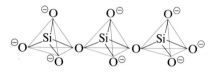

The condensation of three silicic acid molecules gives the acid $H_8Si_3O_{10}$:

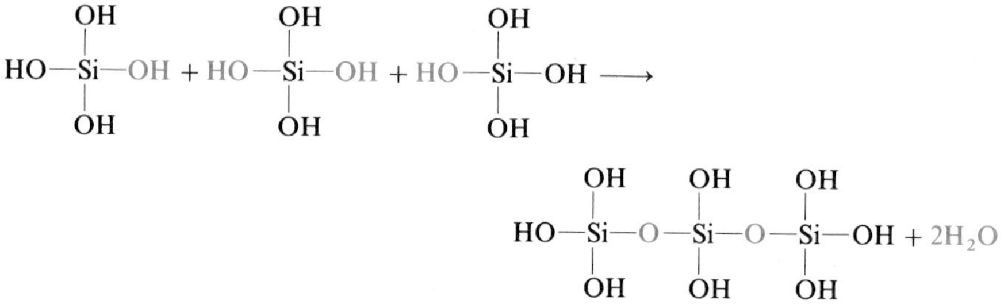

$$HO-\underset{\underset{OH}{|}}{\overset{\overset{OH}{|}}{Si}}-OH + HO-\underset{\underset{OH}{|}}{\overset{\overset{OH}{|}}{Si}}-OH + HO-\underset{\underset{OH}{|}}{\overset{\overset{OH}{|}}{Si}}-OH \longrightarrow$$

$$HO-\underset{\underset{OH}{|}}{\overset{\overset{OH}{|}}{Si}}-O-\underset{\underset{OH}{|}}{\overset{\overset{OH}{|}}{Si}}-O-\underset{\underset{OH}{|}}{\overset{\overset{OH}{|}}{Si}}-OH + 2H_2O$$

The anion derived from this acid is $Si_3O_{10}^{8-}$, and it consists of three SiO_4 tetrahedra sharing corners:

Continued condensation reactions of this type can give still longer chains, which may also join head to tail to give rings. A typical example is the cyclic anion $Si_6O_{18}^{12-}$ found in the mineral *beryl*, $Be_3Al_2Si_6O_{18}$ or $(Be^{2+})_3(Al^{3+})_2(Si_6O_{18}^{12-})$. In this anion there are six SiO_4 tetrahedra, each of which shares two corners with neighboring SiO_4 tetrahedra.

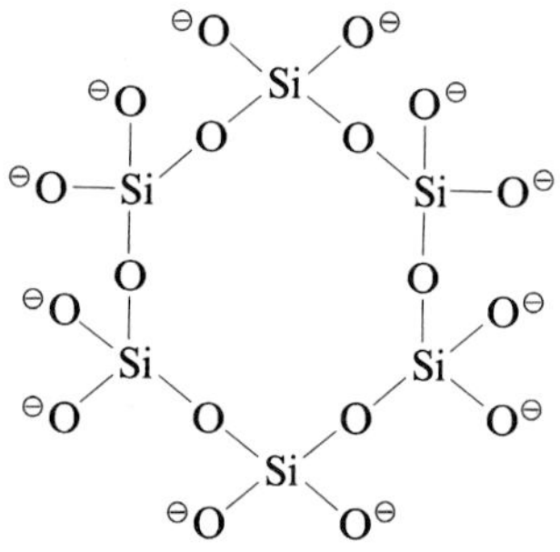

Each of the oxygen atoms that is not shared with another silicon carries a negative charge, giving a total charge of 12− to the ion. Rings of other sizes are also known, as well as the infinitely long chain,

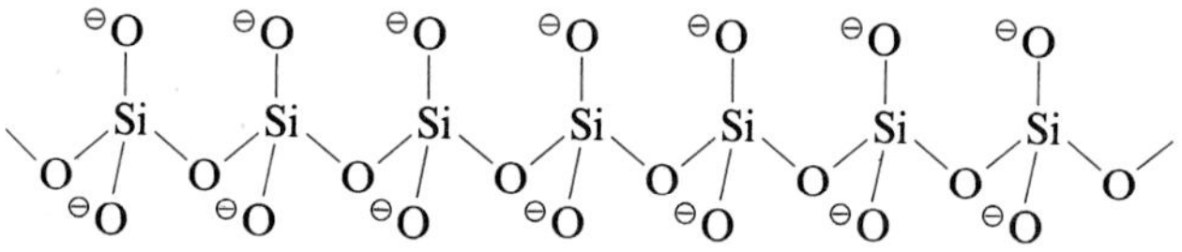

In this chain, each silicon is bonded to four oxygens, two of which are not shared and two of which are shared with other silicon atoms. Thus there is a total of 3 oxygen atoms $[(2 \times 1) + (2 \times \frac{1}{2})]$ for each silicon atom. The empirical formula for the chain is SiO_3^{2-} since each of the unshared oxygen atoms carries a negative charge. Silicates that contain this infinite-chain anion are called *pyroxenes*.

An example is *diopside*, $CaMg(SiO_3)_2$, which is composed of Ca^{2+} and Mg^{2+} ions and the infinite-chain anion $(SiO_3^{2-})_n$ (see Figure 22.5).

Two $(SiO_3^{2-})_n$ chains may be joined together by sharing oxygen atoms on alternate tetrahedra, giving a double chain with the empirical formula $Si_4O_{11}^{6-}$ (see Figure 22.5). Silicates that contain this double-chain anion are called *amphiboles*.

The mineral *tremolite*, $Ca_2Mg_5(Si_4O_{11})_2(OH)_2$ is an example of an amphibole. Tremolite is composed of Ca^{2+} ions, Mg^{2+} ions, OH^- ions, and the infinite, double-chain anion $(Si_4O_{11}^{6-})_n$. It is one of a group of minerals called *asbestos*. Asbestos minerals have a fibrous structure that reflects the infinite-chain structure of the anion. Fibers of asbestos are very strong and have many uses. Asbestos is a very effective insulating and fire-resistant material. The fibers can be woven into blankets that are used for insulation and extinguishing fires and into cloth for making protective clothing. Unfortunately, small particles of asbestos suspended in the air and in water supplies are a health hazard (see Box 22.2).

If a large number of chains are joined side by side, a layer of SiO_4 tetrahedra results (see Figure 22.5). Each tetrahedron then shares three corners with neighboring tetrahedra. Each silicon atom is bonded to three oxygen atoms that it shares with other silicon atoms and to one oxygen atom that carries a negative

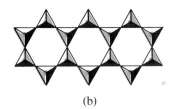
(a)

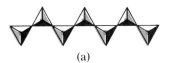
(b)

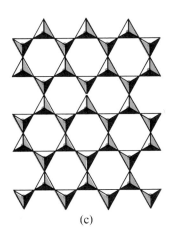
(c)

Figure 22.5 Polymeric Silicate Anions. (a) Pyroxenes have single chains of linked SiO_4 tetrahedra. (b) Amphiboles have double chains of linked SiO_4 tetrahedra. (c) Clays and mica have sheets of linked SiO_4 tetrahedra. Each tetrahedron represents an SiO_4 group.

Box 22.2

ASBESTOS

The ready availability of the fibrous silicate minerals known as asbestos and their nonflammability, flexibility, mechanical strength, and inertness to chemical attack have led to their widespread use. Asbestos has many applications in the building industry as a fire retardant and as a heat and sound insulating material. It is an important component of solid materials used for pipes, brake pads, and many other applications.

Asbestos fibers magnified 100 times

Crude asbestos fibers

When airborne asbestos particles reach the lungs, they do not dissolve and are not expelled: They tend to remain, irritating and scarring the lungs. Extensive exposure to airborne asbestos in appreciable concentrations can lead to the disease asbestosis and to a greatly increased risk of cancer, especially for those who smoke. Many workers who have been exposed to asbestos in its mining and processing have contracted asbestosis, and there is concern that the general public is being exposed to levels of asbestos that after 20 to 30 years may give some of them the same health problems experienced by asbestos workers.

charge. Thus each silicon atom is associated with $2\frac{1}{2}$ oxygen atoms $[(1 \times 1) + (3 \times \frac{1}{2})]$, one of which carries a negative charge. Therefore the empirical formula of this sheet anion is $Si_2O_5^{2-}$. Several minerals contain this sheet anion, for example, *talc*, $Mg_3(Si_2O_5)_2(OH)_2$, which is used for making talcum powder, and *kaolinite*, $Al_2Si_2O_5(OH)_4$, which is a type of clay that is particularly valuable for pottery making. Talc and kaolinite consist of layers, each of which is composed of two $(Si_2O_5^{2-})_n$ sheet anions together with hydroxide ions and sufficient cations to neutralize the total charge of the $(Si_2O_5^{2-})_n$ anions and the hydroxide ions (Figure 22.6). These neutral layers slide easily over each other, giving these minerals a characteristic soft, soapy feel. Talc is very soft and has a hardness of only 1 on the Mohs scale (Box 10.1). *Soapstone* is talc in compact form, an easily carved material. Soapstone objects made by the Inuit inhabitants of Canada and Alaska have become well known.

The layer structure of a clay such as kaolinite can be penetrated by water molecules, which provide a lubricant between the layers, enabling them to slip easily over each other. Thus wet clay is pliable and slippery, but when it is baked or fired in a kiln to 1100°C or more, the water molecules are driven out, and the layers lock into a rigid structure.

Because molecules can be inserted between the layers of a clay mineral, these substances have enormous effective surface areas. They can absorb large amounts of water and many organic substances. A dried clay called Fuller's earth is often used for removing stains and grease from fabrics, for removing some of the natural oil from wool in the textile industry, and for clarifying and deodorizing oils.

We have seen then that when SiO_4 tetrahedra share two corners, we get chain and ring silicates; and when they share three corners, we get sheet silicates. When SiO_4 tetrahedra share all four corners, the various three-dimensional structures of silicon dioxide result.

Aluminosilicates

In many silicate minerals, some of the silicon atoms in the SiO_4 tetrahedra are replaced by aluminum atoms to give *aluminosilicates*. But to form four bonds like the silicon atom it replaces, the aluminum atom must have four electrons; in other words, Si is replaced by an Al^- ion. Replacement of some of the Si atoms in neutral silicon dioxide by Al^- ions gives an infinite three-dimensional aluminosilicate anion. There are numerous aluminosilicate anions. We consider just a few examples.

In the mineral *albite*, $NaAlSi_3O_8$, one-quarter of the silicon atoms in silicon dioxide are replaced by Al^-, giving the three-dimensional framework anion $(AlSi_3O_8^-)_n$. Albite consists of this anion and sodium ions, Na^+. Minerals in which up to one-half of the silicon atoms in silicon dioxide are replaced by aluminum are *feldspars*. Feldspar minerals are one of the essential constituents of granite.

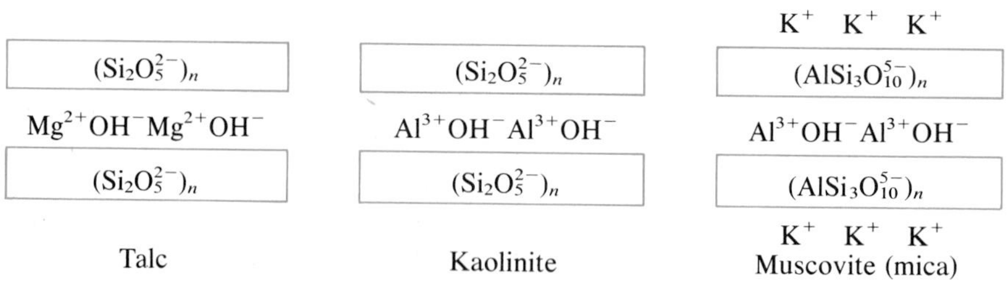

Figure 22.6 Structures of Some Layer Silicates and Aluminosilicates. Edge view of layer stacking in talc, $Mg_3(Si_2O_5)_2(OH)_2$, kaolinite, $Al_2(Si_2O_5)(OH)_4$, and muscovite (mica), $KAl_2(AlSi_3O_{10})$.

		K^+ K^+ K^+
$(Si_2O_5^{2-})_n$	$(Si_2O_5^{2-})_n$	$(AlSi_3O_{10}^{5-})_n$
$Mg^{2+}OH^-Mg^{2+}OH^-$	$Al^{3+}OH^-Al^{3+}OH^-$	$Al^{3+}OH^-Al^{3+}OH^-$
$(Si_2O_5^{2-})_n$	$(Si_2O_5^{2-})_n$	$(AlSi_3O_{10}^{5-})_n$
		K^+ K^+ K^+
Talc	Kaolinite	Muscovite (mica)

Example 22.1 The mineral *anorthite* is a feldspar mineral containing Ca^{2+} ions and an anion formed by replacing half of the silicon atoms in SiO_2 by aluminum. What is the empirical formula of this mineral?

Solution If we double the empirical formula SiO_2 and then replace one of the Si atoms by Al^-, we obtain the empirical formula $AlSiO_4^-$ for the infinite, three-dimensional aluminosilicate anion. For electrical neutrality, two of these empirical formula units are required by each Ca^{2+}, so the empirical formula of the mineral anorthite is $Ca(AlSiO_4)_2$.

Aluminosilicates can also be derived from silicate anions by replacing some of the Si atoms by Al^- ions. Muscovite, $KAl_2(AlSi_3O_{10})(OH)_2$, is a *mica* mineral based on the amphibole layer anion with one-quarter of the silicon atoms replaced by aluminum. If we double the empirical formula $Si_2O_5^{2-}$ of the amphibole layer anion, we obtain $Si_4O_{10}^{4-}$. Then replacing one of the silicon atoms by Al^- gives the empirical formula $AlSi_3O_{10}^{5-}$. In addition, there are two hydroxide ions, giving a total of seven negative charges, which are neutralized by one K^+ ion and two Al^{3+} ions. Thus the mica mineral muscovite consists of infinite layer $(AlSi_3O_{10})^{5-}$ anions together with OH^- ions, K^+ ions, and Al^{3+} ions. Note that in this mineral aluminum occurs both as Al^{3+} ions and as the central atom in AlO_4 tetrahedra in the $(AlSi_3O_{10}^{5-})_n$ anion.

In muscovite and other mica minerals, layer anions are held strongly together by cations (see Figure 22.6). These layers do not slide over each other as readily as the neutral layers in talc, and so mica is not as soft as talc. Mica cleaves readily into thin transparent sheets that are used for windows in stoves and furnaces and for insulation in electrical equipment.

Glass

The manufacture and use of **glass** is another chemical technology that was developed early in history. We have already mentioned that when molten silicon dioxide is cooled, it does not crystallize but gradually becomes more viscous and eventually gives a glass or amorphous solid, in which the SiO_4 tetrahedra are arranged in a random manner. Many silicates and aluminosilicates have the same property. Since they usually melt at lower temperatures and stay soft over a wider temperature range, they are easier to work with than silicon dioxide. For example, ordinary window and bottle glass is a mixture of sodium and calcium silicates.

The composition of glass is normally expressed in terms of the oxides of the elements that it contains, even though these elements are actually present in the form of complex sodium and calcium silicates. Ordinary glass has the approximate composition by mass of 12% Na_2O, 12% CaO, and 76% SiO_2. It is made by heating together sodium carbonate, calcium oxide (lime) or calcium carbonate (limestone), and silicon dioxide in the form of white sand. Sodium carbonate and calcium carbonate decompose on being heated to give Na_2O and CaO. These oxides combine with SiO_2 in reactions such as

$$CaO + SiO_2 \longrightarrow CaSiO_3$$
$$Na_2O + SiO_2 \longrightarrow Na_2SiO_3$$

which exemplify the reaction of an acidic oxide, SiO_2, with a basic oxide, CaO or Na_2O, to give a salt, $CaSiO_3$ or Na_2SiO_3 (Chapter 9). Numerous other oxides can be added to glass to give it special properties. For example, adding boron oxide, B_2O_3, gives borosilicate glass, which expands very little on being heated.

This type of glass is used in cooking utensils and laboratory ware and is sold under trade names such as Pyrex. Addition of PbO gives a glass with a high refractive index and a high density; it is used for the manufacture of cut glass and for lenses.

Glass may also be colored by the addition of other oxides. Indeed, ordinary glass often contains a small amount of Fe(II), from an impurity in the sand, which gives it a green color. This color can be removed by the addition of MnO_2. Addition of CoO gives a blue glass; addition of Cr_2O_3 gives a deep green glass; and addition of SnO_2 gives a white opaque glass.

A recent development has been that of "photochromic" glass for use in sunglasses. This glass darkens on exposure to bright sunlight but becomes clear again in a subdued indoor light. The glass contains AgCl (or AgBr), which is decomposed by light (Chapter 21) to give finely divided black silver:

$$AgCl \xrightarrow{h\nu} Ag + Cl$$

Since the Ag and Cl atoms are trapped in adjacent positions in the solid glass, they recombine in the absence of light to re-form AgCl.

Glass provides the glaze on pottery that makes it impervious to water. A glaze is simply a coating of glass that is formed by covering the pottery with a thin paste of the appropriate oxides and heating to a high temperature. A similar coating of glass, called *vitreous enamel*, can also be applied to metal surfaces in much the same way. Ornamental objects made of enameled metal were found in the tomb of Tutankahamen, who died in 1350 B.C.

Cement

Cement is an aluminosilicate made by heating a powdered mixture of limestone ($CaCO_3$), sand (SiO_2), and clay to about 1500°C. The resulting solid mass is powdered and mixed with a little gypsum, $CaSO_4 \cdot 2H_2O$. Cement has the approximate composition 60% CaO, 20% SiO_2, 10% Al_2O_3, and 10% other oxides. When cement is mixed with water, a plastic mass is obtained, which slowly hardens as interlocking crystals of hydrated aluminosilicates are formed. Cement is usually mixed with sand, gravel, or crushed rock to give a hard, strong material known as *concrete*.

22.4 SILICONES

The wide range of applications of the mineral silicates depends on their great thermal stability and inertness toward other substances. These properties are related to the great strength of the silicon-oxygen bond, which has an average bond energy (430 kJ mol^{-1}) that is much greater than the silicon-silicon bond energy (190 kJ mol^{-1}) or even the carbon-carbon, single-bond energy (340 kJ mol^{-1}). Chemists have been able to combine the strength and inertness of the silicon-oxygen bond with some of the useful properties of organic polymers (Chapter 23) in the synthetic polymers known as *silicones*.

The simplest silicones are chain polymers that have the general formula $(R_2SiO)_n$, where R is an alkyl or aryl group such as CH_3, C_2H_5, or C_6H_5 (see Figure 22.7). There are also cyclic polymers (see Figure 22.7) and cross-linked polymers in which chains are held together by sharing oxygen atoms.

Silicones are prepared by heating silicon with chloroalkanes such as chloromethane, CH_3Cl. The main product of the reaction with CH_3Cl is dimethyl

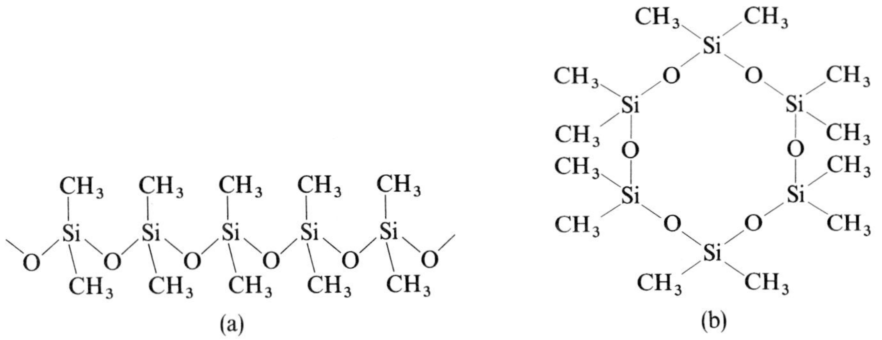

Figure 22.7 Structures of Some Silicone Polymers. (a) A long-chain polymer, or (b) A cyclic polymer.

silicon dichloride, $(CH_3)_2SiCl_2$:

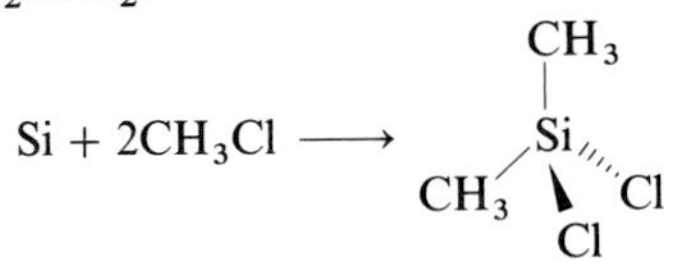

But small amounts of $(CH_3)_3SiCl$ and CH_3SiCl_3 are also formed. The reaction of dimethyl silicon dichloride with water gives dimethyl silicon dihydroxide, $(CH_3)_2Si(OH)_2$:

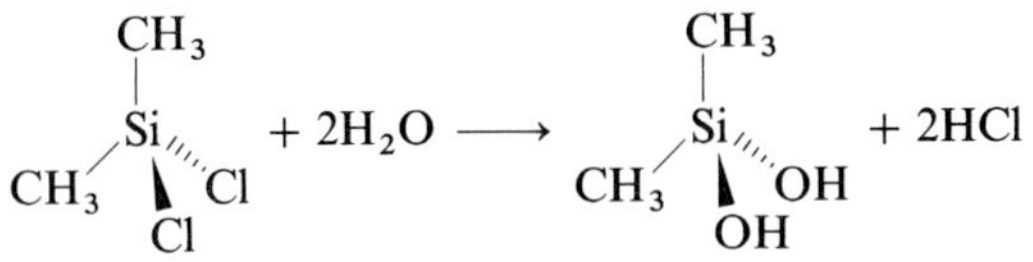

This polymerizes in a condensation reaction to give a chain polymer:

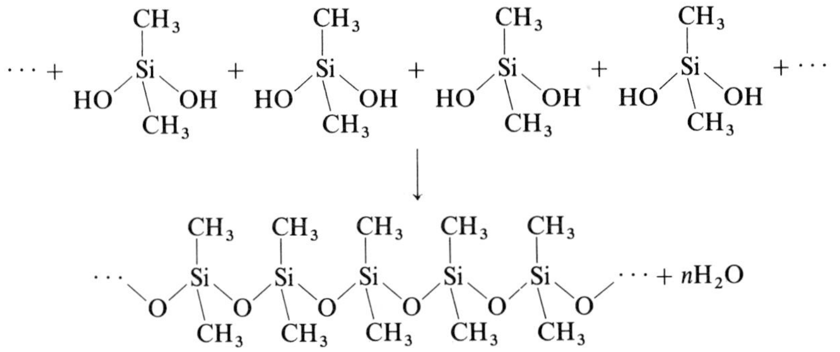

The length of the polymer chain can be controlled by introducing a certain amount of $(CH_3)_3SiCl$ into the reaction mixture to produce $(CH_3)_3SiOH$. When $(CH_3)_3SiOH$ condenses with another Si—OH group, it terminates the chain:

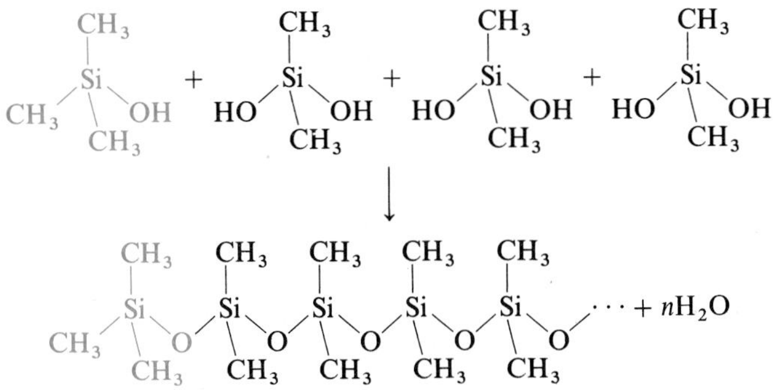

Silicones consisting of chains with up to about ten silicon atoms are liquids, while those with longer chains are greases and waxes.

22.4 SILICONES

799

If a certain amount of CH_3SiCl_3 is introduced into the reaction mixture, cross-links between the chains are produced. The reaction of CH_3SiCl_3 with water gives $CH_3Si(OH)_3$. When this product condenses with other Si—OH groups, it can form three Si—O—Si links and it can therefore join the SiOSiOSi chains together (see Figure 22.8). In this way a variety of solids can be produced, ranging from hard materials to soft, rubbery materials.

Silicones have several useful properties. They are excellent electrical insulators, good lubricants, water-repellent, and nontoxic. Because they are very resistant to heat, they can be used as replacements for rubber and lubricants in many high-temperature applications. The viscosity of the liquid silicones changes very little with temperature, so they retain their fluidity down to very low temperatures. They are often used in place of hydrocarbon oils for low-temperature applications. The solids also retain their rubbery properties at low temperatures and do not become hard and brittle, like many rubbers based on organic molecules.

Silicones with the formula $(R_2SiO)_n$ are the silicon analogues of ketones. However, whereas ketones are simple molecules with C=O double bonds, the corresponding silicones are all polymers in which there are only single bonds. As we have seen before, carbon, a second-period element, has a much stronger tendency to form multiple bonds than does silicon, a third-period element.

22.5 OTHER SILICON COMPOUNDS

Silanes

The great difference between the chemistry of silicon and the chemistry of carbon is also illustrated by the hydrides of silicon, which are called *silanes*. In their general formula Si_nH_{2n+2}, the silanes are analogous to the alkanes, C_nH_{2n+2}, but only a few silanes are known. In contrast to the alkanes, they are extremely reactive. They are spontaneously flammable in the air (see Experiment 22.4):

$$SiH_4(g) + 2O_2(g) \longrightarrow SiO_2(s) + 2H_2O(l)$$

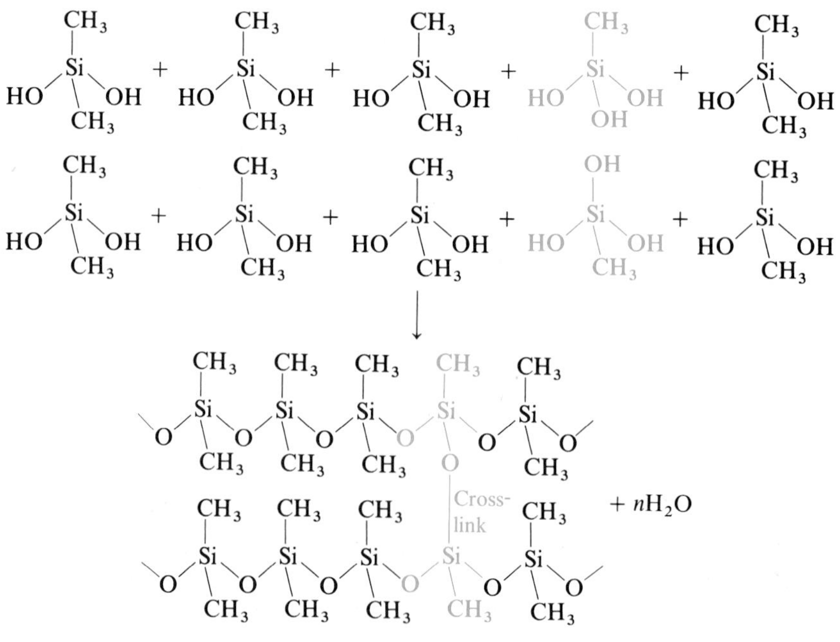

Figure 22.8 Formation of Cross-Linked Silicone Polymer. Introduction of a small amount of $CH_3Si(OH)_3$ into $(CH_3)_2Si(OH)_2$ causes the formation of cross-links between chains when it is polymerized.

Preparation of Silane, SiH_4

A mixture of silica, SiO_2, is heated with magnesium in a 1:2 mole ratio.

The reaction is highly exothermic and the tube glows brightly even when it is removed from the bunsen burner. The product contains magnesium silicide, Mg_2Si.

When the magnesium silicide is tipped into dilute hydrochloric acid it reacts to form silane, SiH_4. When the bubbles of silane reach the surface of the solution, they ignite spontaneously with an explosive pop.

In contrast, the alkanes burn only when ignited. The silanes react with basic aqueous solutions to give hydrogen:

$$SiH_4(g) + OH^-(aq) + 3H_2O(l) \longrightarrow SiO(OH)_3^-(aq) + 4H_2(g)$$

Silicon Halides

Silicon tetrachloride, which is a colorless liquid (bp = 57°C), can be made by passing chlorine through a red-hot mixture of sand and coke:

$$SiO_2(s) + 2C(s) + 2Cl_2(g) \longrightarrow SiCl_4(l) + 2CO(g)$$

Silicon tetrafluoride, which is a colorless gas, can be made by the reaction of silicon with fluorine,

$$Si(s) + 2F_2(g) \longrightarrow SiF_4(g)$$

or by heating calcium fluoride with silicon dioxide (sand) and concentrated sulfuric acid:

$$2CaF_2(s) + 2H_2SO_4(conc) + SiO_2(s) \longrightarrow SiF_4(g) + 2CaSO_4(s) + 2H_2O(l)$$

In this reaction, hydrogen fluoride is produced by the reaction of calcium fluoride with concentrated sulfuric acid, and then HF reacts with silicon dioxide:

$$SiO_2 + 4HF \rightleftharpoons SiF_4 + 2H_2O$$

This reaction is an equilibrium and is driven to the right by the removal of water by sulfuric acid.

Both $SiCl_4$ and SiF_4 are covalent tetrahedral AX_4 molecules. In contrast to the carbon compounds CF_4 and CCl_4, which are stable in the presence of water, both $SiCl_4$ and SiF_4 react rapidly with water to give silicic acid, $Si(OH)_4$, and

22.5 OTHER SILICON COMPOUNDS

801

Figure 22.9 Formation of SiF_5^- and SiF_6^{2-} Ions. Because it can accept more electrons into its valence shell, the silicon atom in SiF_4 can combine with one or two F^- ions to form the SiF_5^- and SiF_6^{2-} ions. In these reactions SiF_4 is behaving as a Lewis acid.

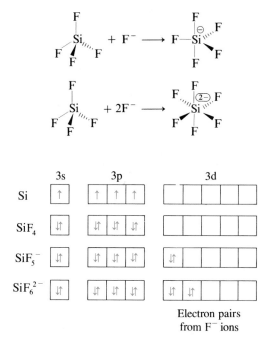

a hydrogen halide:

$$SiCl_4 + 4H_2O \longrightarrow Si(OH)_4 + 4HCl$$
$$SiF_4 + 4H_2O \longrightarrow Si(OH)_4 + 4HF$$

Silicon tetrafluoride reacts with fluoride ion to form the ions SiF_5^- and SiF_6^{2-}. These ions are isoelectronic with PF_5 and SF_6 and have the same trigonal bipyramidal AX_5 and octahedral AX_6 structures (see Figure 22.9).

The corresponding acid H_2SiF_6 is not known in the pure state, but in water it is a strong acid existing in the fully ionized form $(H_3O^+)_2SiF_6^{2-}$. This acid is also a product of the reaction of SiF_4 with water, the HF that is formed reacting with more SiF_4 and water:

$$SiF_4(g) + 2HF(aq) + 2H_2O(l) \longrightarrow 2H_3O^+(aq) + SiF_6^{2-}(aq)$$

22.6 COMPARISON OF CARBON AND SILICON

We have mentioned that there are considerable differences in the chemistry of the elements in the second period of the periodic table and the elements of the same groups in the third period. These differences are particularly evident for carbon and silicon. In this section we review some of these differences and give some explanations for them.

The alkanes and other hydrocarbons form the basic framework for hundreds of thousands of organic compounds, whereas there are very few silanes and they are of very limited usefulness. Chains of carbon atoms are rather unreactive at ordinary temperatures, whereas chains of silicon atoms are very reactive. We may attribute these reactivity differences to the difference in the C—C and Si—Si bond energies and to the fact that the valence shell of silicon can accommodate more than four electron pairs. In other words, silicon can use 3d orbitals in bond formation.

Table 22.2 Bond Energy Values (kJ mol^{-1})

B—B	C—C	N—N	O—O	F—F
301	347	159	143	159
	Si—Si	P—P	S—S	Cl—Cl
	196	197	266	242
	Ge—Ge	As—As	Se—Se	Br—Br
	163	177	193	193
	Sn—Sn	Sb—Sb	Te—Te	I—I
	152	142	126	151
B—O	C—O	N—O	O—O	F—O
523	360	113	143	213
	Si—O	P—O		Cl—O
	464	368		207

As we saw in Chapter 12, one useful measure of bond strength is the bond energy. Table 22.2 lists bond energies for boron and most of the elements in Groups IV to VII. The C—C bond is indeed the strongest of all the single covalent bonds between like atoms, an important factor in the lack of reactivity of compounds containing chains and rings of carbon atoms.

Another important difference between silicon and carbon is that the valence shell of carbon is completely filled in all its compounds, whereas silicon has the possibility of adding extra electrons to its valence shell by utilizing its 3d orbitals. Carbon tetrafluoride does not combine with fluoride ion, but SiF_4 forms SiF_5^- and SiF_6^{2-}, because silicon can exceed the octet by making use of its 3d orbitals to accept F$^-$ ions (see Figure 22.9). Similarly, SiH_4, $SiCl_4$, and SiF_4 react rapidly with water, whereas the alkanes and CCl_4 and CF_4 do not. The empty 3d orbitals on the silicon atom provide an easy route for the water molecule to attack a silicon atom by donating one of its unshared electron pairs to a vacant 3d orbital to form the activated complex (transition state).

Although the Si—Si bond is not so strong as the C—C bond (Table 22.2), the Si—O bond is stronger than the C—O bond and indeed second only to the B—O bond in strength among the energies of bonds formed by oxygen (Table 22.2). The strength of the Si—O bond is certainly consistent with the great stability of silicon dioxide and the mineral silicates, but it does not completely explain why the silicates lack reactivity, since the bond energy of the Si—F bond in SiF_4 is even greater (598 kJ mol^{-1}) and yet SiF_4 reacts rapidly with water. The reasons for reactivity or lack of reactivity of compounds are many and complex, and they cannot be fully discussed in terms of a few simple concepts such as bond energies. Further discussion of these interesting questions would, however, take us well beyond the scope of this book.

Finally, it is interesting that a complete lack of success in making compounds containing Si=Si double bonds had led chemists to conclude that such a bond does not form. However, Robert West found in 1981 that compounds containing Si=Si bonds can be made if the groups attached to the silicon atoms are large enough that they prevent other molecules from approaching the very reactive silicon atoms (see Figure 22.10). Thus we see that the ability of silicon to accept extra electron pairs into its valence shell is responsible for the lack of compounds containing Si=Si bonds. The Si=Si bond can be formed, but molecules with an Si=Si bond are too reactive to be isolated unless the molecule has a structure that prevents other reactants, such as water molecules, from coming close to the silicon atoms.

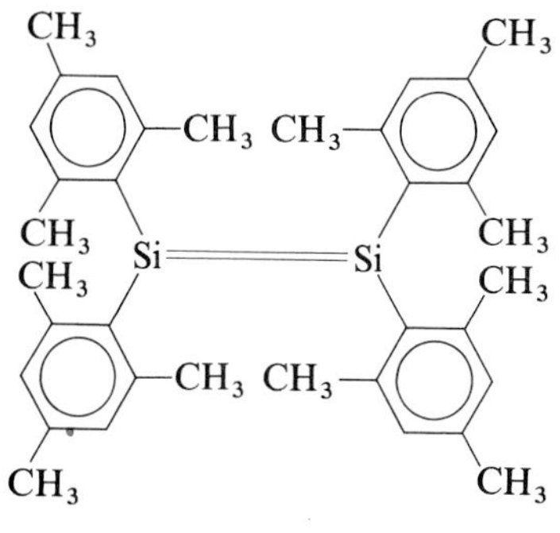

Figure 22.10 The First Compound Containing an Si=Si Double Bond. It was prepared by Robert West at the University of Michigan in 1981. The bulky organic groups prevent other molecules from reaching and reacting with the Si=Si double bond.

Boron is the first element in Group III. It differs in many of its properties from the next element in the group, aluminum, which is a metal. Boron is a very hard, black, shiny solid. Although it is somewhat metallic in appearance, it is a very poor conductor of electricity. It is best regarded as a semimetal, like silicon.

Boric Acid and Borates

Boric acid and the borates are among the simplest and most important of the compounds of boron. Boric acid, $B(OH)_3$, is a stable, colorless crystalline compound that forms thin, platelike crystals. It consists of planar molecules with an equilateral triangular AX_3 geometry around boron. The molecules are held together in flat sheets by hydrogen bonds, as shown in Figure 22.11.

Boric acid is a weak monoprotic acid ($K_a = 6.0 \times 10^{-10}$ mol L^{-1}). It ionizes in water in an unusual way. Instead of donating one of its hydrogen atoms to a water molecule, it removes an OH^- from a water molecule, leaving an H^+ ion, which combines with another water molecule to give an H_3O^+ ion:

$$B(OH)_3(aq) + 2H_2O(l) \rightleftharpoons B(OH)_4^-(aq) + H_3O^+(aq)$$

Thus $B(OH)_3$ behaves as a Lewis acid (Chapter 9). The boron atom has a vacant 2p orbital, which can accept an electron pair from a water molecule. This water molecule, simultaneously or in a subsequent step, loses a proton to another water molecule:

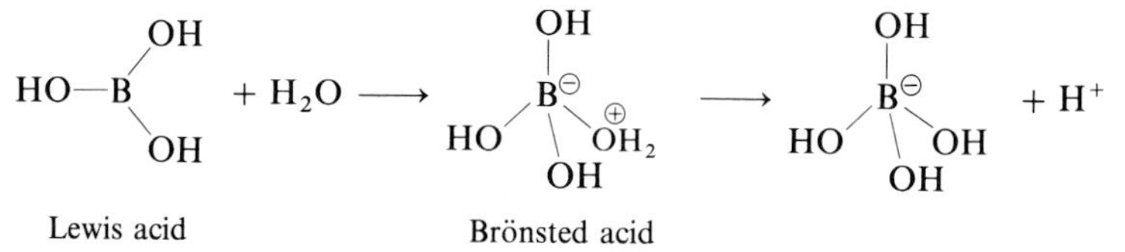

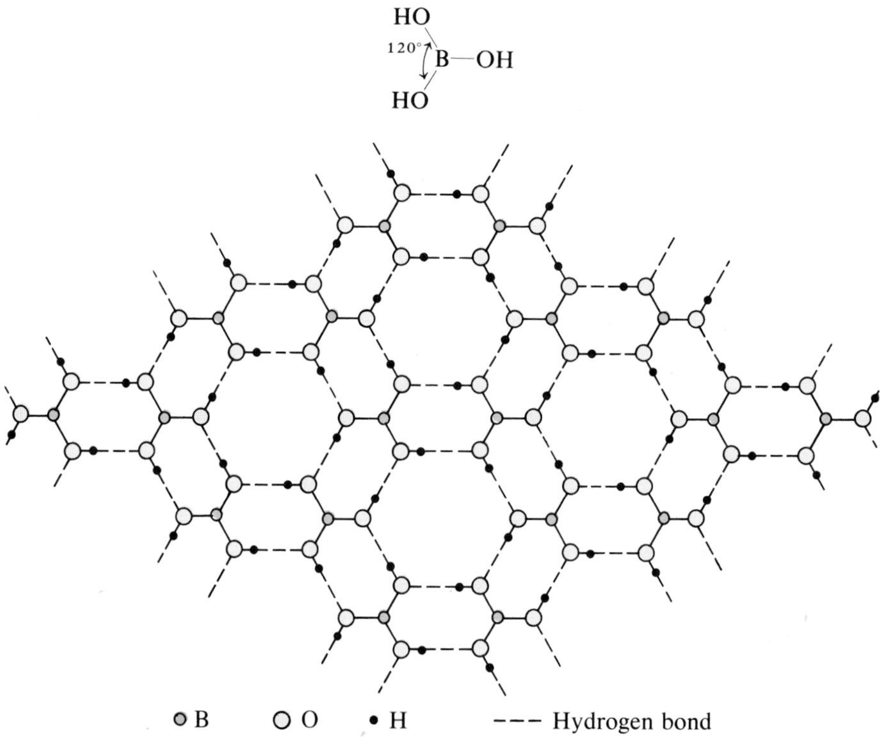

Figure 22.11 Structure of Boric Acid. The planar triangular AX_3 molecules are held together in flat sheets by hydrogen bonds.

○ B   ○ O   • H   --- Hydrogen bond

In this way, boron completes its octet by forming the tetrahedral borate ion $B(OH)_4^-$.

When it is heated, boric acid loses water to form various condensed boric acids, such as the cyclic metaboric acid shown in Figure 22.12. Further heating gives boric oxide, B_2O_3, which, like silicon dioxide, is often formed in an amorphous form, or glass. The crystalline form has a complex framework structure based on planar BO_3 groups.

A few of the salts of boric acid, the borates, contain the simple anion BO_3^{3-}, but most are derived from condensed forms of the acid. The most familiar and most important is sodium tetraborate, $Na_2[B_4O_5(OH)_4]\cdot 8H_2O$, which is commonly called *borax* and for which the formula is often written as $Na_2B_4O_7\cdot 10H_2O$. The structure of the tetraborate anion, $B_4O_5(OH)_4^{2-}$, is shown in Figure 22.13. Most borax is obtained directly from dry lakes such as Searles Lake in California. Aqueous solutions of borax are basic, because tetraborate is the anion of a weak acid and is therefore a weak base (Experiment 22.5):

$$B_4O_5(OH)_4^{2-} + 5H_2O \rightleftharpoons 4H_3BO_3 + 2OH^-$$

Because borax is a weak base, and because the borates of calcium and magnesium are insoluble, it can remove Ca^{2+} and Mg^{2+} from water. Thus borax is used as a water softener and as a component of washing powders.

Boron Halides

The boron halides are typical covalent nonmetal halides. Boron trifluoride, BF_3, and boron trichloride, BCl_3, are gases at room temperature; BBr_3 is a liquid, and BI_3 is a solid. They all consist of molecules with the expected AX_3 planar triangular structure. Although the electronegativity difference between boron and fluorine is 2.1, boron trifluoride is a covalent molecular compound with polar B—F bonds rather than an ionic crystal containing B^{3+} and F^- ions. We see again that the electronegativity difference between two elements is not always a reliable guide to whether they will form an ionic crystal or a molecular compound (Chapter 5).

Because of the presence of the vacant 2p orbital on the boron atom in the boron halides, they are rather reactive compounds, unlike the carbon tetrahalides. For example, BF_3 reacts with an F^- ion to form BF_4^- in which the valence shell of boron is completed. Boron trifluoride reacts in a similar way

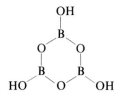

Figure 22.12 Structure of Metaboric Acid, HBO_2.

Figure 22.13 Structure of Tetraborate Anion, $B_4O_5(OH)_4^{2-}$, in Sodium Tetraborate (Borax). Two of the boron atoms have a tetrahedral AX_4 geometry, and the other two have a planar AX_3 geometry. The two tetrahedrally coordinated boron atoms each have a formal negative charge.

EXPERIMENT 22.5

Boric Acid and Borax (Sodium Tetraborate)

Boric acid, H_3BO_3, is a white solid which behaves as a weak acid in water. It therefore causes bromothymol blue indicator to turn yellow. Sodium tetraborate is a salt of a condensed form of boric acid. The tetraborate ion is therefore a weak base and gives a basic solution, which causes bromothymol blue indicator to turn blue.

22.7 BORON AND ITS COMPOUNDS

with many other molecules and ions that have unshared electron pairs that can be donated to its vacant 2p orbital. Thus it forms compounds with ammonia and other amines and with ethers and alcohols. In all these cases BF_3 is behaving as a Lewis acid, that is, as an electron pair acceptor (Chapter 9):

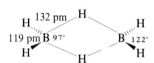

The boron halides react with water to give boric acid and the hydrogen halides. For example,

$$BCl_3 + 3H_2O \longrightarrow B(OH)_3 + 3HCl$$

This reaction is similar to the reaction of $SiCl_4$ with water. It occurs readily because the vacant 2p orbital on the boron can accept an electron pair from a water molecule. In contrast, the carbon tetrahalides have no vacant orbital, and they do not react with water.

Boranes

At least seventeen different compounds of boron and hydrogen have been prepared. They are known as the *boranes*. They all have unexpected formulas such as B_2H_6, B_4H_{10}, B_5H_9, and $B_{10}H_{14}$. We would expect the formula of the simplest borane to be BH_3, but this molecule is not known as a stable species. The simplest borane that can be isolated is diborane, B_2H_6; its structure is shown in Figure 22.14.

The unusual and unexpected feature of this structure is that there are two hydrogen atoms, called *bridging hydrogens*, shared between the two borons. However, there are not enough electrons for each of the lines shown in the structure to represent an electron pair. Each boron contributes 3 electrons and each hydrogen 1 electron, making a total of 12 electrons, or six pairs, for the molecule. Thus there can be a maximum of only six ordinary covalent bonds, whereas the structure appears to have eight bonds. Because B_2H_6 has too few electrons for all the atoms to be held together by normal two-center bonds, it is often described as an **electron deficient molecule**. The bonding in diborane is best described as involving two **three-center bonds**, in which one electron pair holds together three rather than two nuclei. The arrangement of the electron pairs is shown in Figure 22.15. Each boron atom is surrounded by four electron pairs which have the expected tetrahedral arrangement. But two of these electron pairs form three center bonds in which one electron pair holds together two boron nuclei and a hydrogen nucleus. In the following diagram of the structure the two electron pairs forming the three center bonds are denoted by ⊙:

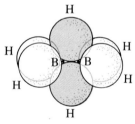

Figure 22.14 Structure of Diborane, B_2H_6.

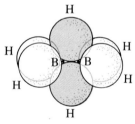

○ Two-center bonds

● Three-center bonds

Figure 22.15 Electron Pair Sphere Model of Diborane. There is a tetrahedral arrangement of electron pairs around each of the boron atoms. Each three-center bond is formed by one electron pair shared between two boron atoms and a hydrogen atom.

The following alternative but entirely equivalent representation of the bonding in diborane emphasizes its close relationship to the bent bond model of ethene:

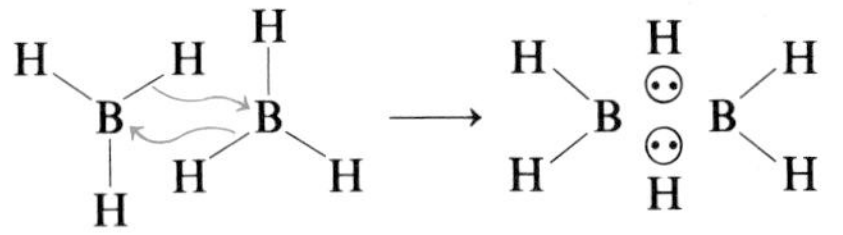

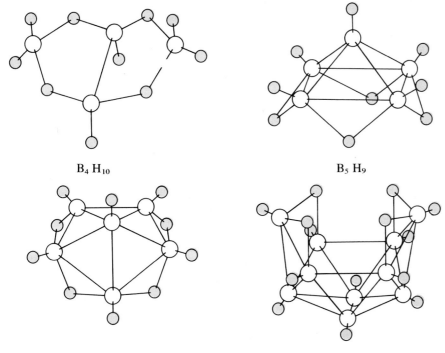

Figure 22.16 Formation of B₂H₆ from Two BH₃ Molecules. The molecule BH₃ is not known under ordinary conditions. Two BH₃ molecules combine to give a B₂H₆ molecule.

The structure of B_2H_6 can be thought of as resulting from the fact that in BH_3 boron has a vacant 2p orbital. Two BH_3 molecules can then combine as each boron attracts the electrons of a B—H bond in the other BH_3 molecule, converting a B—H bonding electron pair into a three-center bonding pair (see Figure 22.16). Another way to represent the structure of this molecule is by means of the two resonance structures

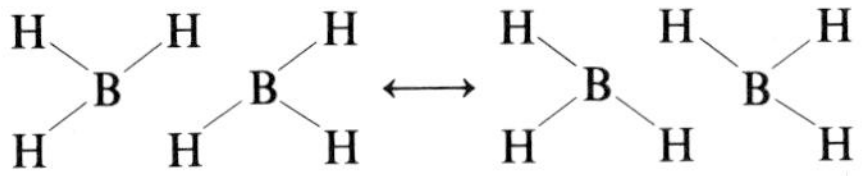

A molecule of diborane readily adds two hydride ions to give two borohydride ions, BH_4^-. This ion has the expected tetrahedral AX_4 structure and ordinary two-center, electron pair bonds:

$$B_2H_6 + 2H^- \longrightarrow 2BH_4^-$$

Salts such as lithium borohydride, $LiBH_4$, and sodium borohydride, $NaBH_4$, are good reducing agents with many applications in organic chemistry.

The higher boranes have unusual and fascinating structures (see Figure 22.17) that cannot be explained in terms of the simple bonding theories that we have described. It is necessary to make use of three-center and four-center bonds or to use molecular orbital theory.

There are also many anions derived from the boranes that are more complicated than BH_4^-. A particularly fascinating example is $B_{12}H_{12}^{2-}$ (Figure 22.18).

$B_4 H_{10}$

$B_5 H_9$

$B_6 H_{10}$

$B_{10} H_{14}$

Figure 22.17 Structures of Some Higher Boranes. They are all electron-deficient molecules in which there are not enough electrons to form ordinary single bonds between adjacent atoms. The bonding in these molecules can only be described in terms of three-center and other multicenter bonds.

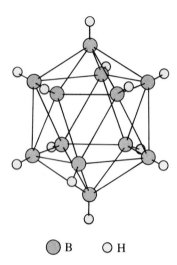

⬤ B   ◯ H

Figure 22.18 Structure of $B_{12}H_{12}^{2-}$ Anion. Each boron atom is at one of the corners of an icosahedron. A hydrogen atom is attached to each boron atom. The icosahedron is a regular solid with 12 equivalent vertices, 20 equilateral triangular faces, and 30 equivalent edges.

It has the shape of a regular icosahedron, which has 12 equivalent vertices, 20 equilateral triangular faces, and 30 equivalent edges. The bonding in the electron-deficient $B_{12}H_{12}^{2-}$ can be satisfactorily described only by using molecular orbital theory. The same icosahedron of 12 boron atoms is found in the different allotropes of elemental boron. This arrangement of boron atoms is very stable and unreactive. Crystalline boron resists oxidation to a much higher temperature than diamond.

The structures of these remarkable compounds serve to illustrate the fact that although the simple ideas of chemical bonding presented in this book enable us to understand the structures of many compounds, they are inadequate for many others. A more detailed, quantum mechanical treatment of the chemical bond is needed to be able to understand such compounds. Although our understanding of chemical bonding has progressed considerably since Lewis first proposed the idea of the shared electron pair, many aspects of this subject are still not fully understood. Chemists continue to prepare new compounds, the structures of which present a challenge for even the most sophisticated theories.

22.8 SEMICONDUCTORS

We are in the midst of a technological revolution, the computer revolution, which is no less significant than the Industrial Revolution of the nineteenth century. The computer revolution is based on the semiconductor properties of silicon and other semimetals. A *semiconductor* is not as conducting as a metal, but it does have a significant electrical conductivity, which unlike the conductivity of a metal, increases with increasing temperature.

Semiconductors have electrical properties that are intermediate between those of nonconducting covalent solids and highly conducting metals. In order to understand semiconductors, let us first review our picture of the bonding in metals. We have previously discussed the properties of metals in terms of the charge cloud or electron gas model. In this model we imagine that each metal atom provides one or more of its valence electrons to a mobile charge cloud that can move freely between the positively charged metal atom cores. These mobile electrons hold the positively charged cores together and constitute the metallic bond. In a covalent solid all the valence electrons are held tightly by the atoms in localized bonds or as unshared pairs. They are not free to move through the crystal, and a covalent crystal is a nonconductor. We can think of a semiconductor as a covalent substance in which the valence electrons are not held very strongly; as a consequence, a few of them have enough energy to break away from the atom with which they were associated and move freely through the crystal. With increasing temperature more electrons acquire sufficient energy to break away from their atoms, and so the conductivity increases with increasing temperature.

This is a very oversimplified picture of a semiconductor; a full understanding of semiconductors requires a knowledge of quantum mechanics and a treatment that is beyond the scope of this book. Nevertheless, we can go a little more deeply into the properties of many semiconductors by starting with an alternative view of the bonding in metals.

Energy Levels in Solids

When atoms are brought as close together as they are in a crystal, they interact with each other so that the energies of their outer orbitals are slightly altered.

Thus in a crystal of sodium, for example, the 3s orbital of an isolated sodium atom is replaced by a whole set of closely spaced energy levels equal in number to the number of atoms in the crystal. The set of energy levels is called a *band*, in this case a 3s band (see Figure 22.19). Since each orbital, in accordance with the Pauli exclusion principle, may contain two electrons, and since each sodium atom contributes only one electron, only half the levels are occupied. Thus there are many empty levels into which an electron may be excited, and the small amount of energy provided by the application of an electric potential to the metal can move some electrons into these empty orbitals. The movement of electrons from one orbital to another constitutes an electric current; thus sodium is an electric conductor.

If we apply the same arguments to magnesium, which has two electrons in a 3s orbital, we conclude that the 3s band will be completely filled. There are no empty orbitals in the 3s band into which electrons can be excited by the application of an electric field. In other words, an electric field cannot cause any movement of the electrons, and therefore magnesium might be expected to be a nonconductor. However, the 3p orbitals also form a band of closely spaced energy levels, and this 3p band overlaps the 3s band (see Figure 22.19). Thus the application of an electric field can move electrons into this empty band, which is called a *conduction band*. The filled 3s band is called the *valence band*.

In an insulator such as diamond the valence band is completely filled, and the conduction band is at a much higher energy; so the electrons from the valence band cannot move into the conduction band, and diamond is a non-

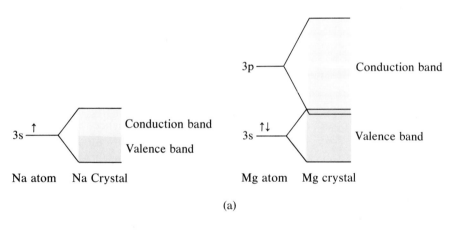

(a)

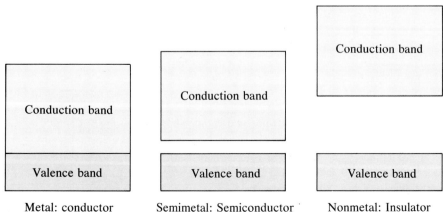

Metal: conductor   Semimetal: Semiconductor   Nonmetal: Insulator

(b)

Figure 22.19 Valence and Conduction Bands in Metals, Semiconductors, and Insulators.

(a) In a sodium or a magnesium crystal, the single 3s orbital of an isolated metal atom is replaced by numerous closely spaced energy levels, which are said to form a band. In sodium, this band is half-filled. In magnesium, this band is full but overlaps the empty 3p band. (b) The separation of the valence band and the conduction band increases from a metal, to a semiconductor, to an insulator.

22.8 SEMICONDUCTORS

conductor. In a semiconductor such as silicon the conduction band does not overlap the valence band as in a metal but is sufficiently close to the valence band that a few electrons have sufficient thermal energy to jump the gap and occupy the conduction band. Thus silicon is a conductor, but because there are only very few electrons in the conduction band, it is a much poorer conductor than a metal.

The conductivity of a semiconductor increases with increasing temperature because an increasing number of electrons have sufficient thermal energy to occupy the conduction band. Thus the difference between a metal, a semiconductor, and a nonconductor depends on the energy gap between the valence band and the conduction band. Figure 22.20 shows the energy gaps for carbon, silicon, germanium, and tin. Carbon is an insulator; silicon and germanium are semiconductors; and tin is a metal. In a metal the conduction band overlaps the valence band or is very close to the valence band.

Applications of Semiconductors

The important applications of semiconductors depend on the very important discovery that the conductivity of semiconductors can be changed in a dramatic but controllable way by the addition of very small amounts of certain impurities. If a very small amount of phosphorus or arsenic is added to crystalline silicon, the phosphorus or arsenic atoms are incorporated into the silicon structure. However, only four electrons are needed to form bonds to the surrounding four silicon atoms, so one electron is left over. This electron occupies an energy level in the conduction band, and under the influence of an applied electric potential it can move through the crystal. The number of electrons that move into the conduction band and, therefore, the conductivity are proportional to the number of phosphorus or arsenic atoms that are added.

If a boron atom is added to silicon, it contributes only three electrons to the structure. Because it needs four electrons to form four bonds, the boron steals an electron from an adjacent silicon atom, leaving this atom short of one electron and therefore positively charged. This vacancy in the valence shell of silicon is called a *positive hole*. An electron from an adjacent silicon atom can move into this hole, thus transferring the positive hole to this adjacent atom, and so on. Under the influence of an applied potential the positive hole moves through

Figure 22.20 Valence and Conduction Bands for Carbon, Silicon, Germanium, and Tin.

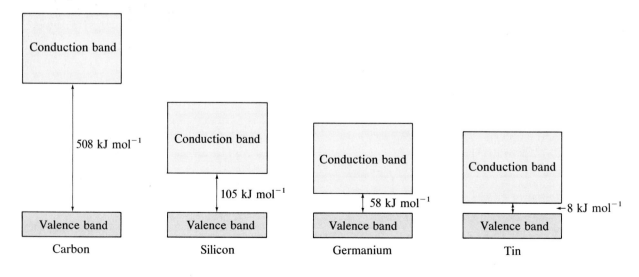

Carbon   Silicon   Germanium   Tin

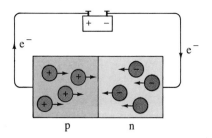 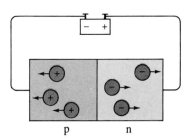

Figure 22.21 An *n–p* **Semiconductor Junction.** Current is carried in a *p*-type semiconductor by the motion of holes and in an *n*-type semiconductor by the motion of electrons. Left: With the potential in this direction, a large current can flow. Right: With the potential applied in this direction, little current flows.

the crystal. Because the charge is carried by the positive holes, this type of conductor is called a *p*-type conductor. In contrast, if arsenic or phosphorus is added to the silicon, the conductor is called an *n*-type conductor, because conduction is due to negative electrons moving in the conduction band.

If a *p*-type semiconductor is placed in contact with an *n*-type semiconductor, a very useful device called a *p–n* junction is obtained. Figure 22.21 shows what happens if a potential is applied across a *p–n* junction. If the *n*-type conductor is made negative, electrons flow into the conductor and continue toward the junction. Electrons are simultaneously removed from the *p*-type conductor, producing more holes, which move toward the junction. At the junction the electrons fall into the holes—the holes and electrons may be said to neutralize each other—and the current continues to flow.

If the potential is applied in the opposite direction, electrons flow out of the *n*-type conductor and electrons enter the *p*-type conductor, thus removing the holes, which flow away from the junction. Therefore in the region near the junction there are no longer holes in the *p* conductor nor any free electrons in the *n* conductor. Since there is no mechanism for producing more holes or electrons, the current ceases to flow.

Thus a *p–n* junction conducts current in only one direction. It can therefore convert alternating current to direct current. Such a device is called a *rectifier*. The *p–n* junction has replaced the vacuum tube diode in electronics for this purpose and is often called a diode.

In addition to rectifying current, *n* and *p* junctions can be combined into devices, called *transistors*, that amplify currents and voltages. Since their invention in 1948, transistors and other semiconductor devices have completely revolutionized the electronics industry.

The most important development in recent years has been the production of *integrated circuits* (Figure 22.22). A small thin wafer of *n*- or *p*-type silicon, with dimensions of only a few millimeters, called a *silicon chip*, is baked in oxygen to cover its surface with a layer of silicon dioxide. The silicon dioxide is coated with a photosensitive emulsion, and a tiny mask scaled down from a circuit diagram is photographed on the chip. It is then exposed to ultraviolet light, which hardens the silicon dioxide that is exposed. The softer silicon dioxide that was not exposed to the ultraviolet light is then etched away with acid. Semiconductive material of opposite type to that of the chip is then diffused into the channels. The chip can then again be coated with silicon dioxide, a further mask applied, another ultraviolet treatment and an acid treatment carried out, and so on. Finally, a conducting layer of aluminum is applied, which is again etched to give a complete circuit diagram. In this way a complete circuit consisting of thousands of resistors, transistors, rectifiers, and capacitors is built up, all on a single piece of silicon having dimensions of no more than a few millimeters. Integrated circuit chips are at the heart of digital wristwatches, handheld calculators, and personal computers.

Figure 22.22 Integrated Circuit on Silicon Chip. A chip this size can contain hundreds of transistors.

IMPORTANT TERMS

An **electron-deficient molecule** is a molecule in which there are insufficient electrons for all the atoms to be held together by two-center, electron pair bonds.

A **glass** is a solid that has an amorphous rather than a crystalline structure. One of its characteristic properties is that it softens over a temperature range rather than melting sharply at a well-defined temperature.

A **semiconductor** is a solid that has an appreciable electrical conductivity, which increases with increasing tempera-

ture rather than decreasing, as in the case of a metal.

A **semimetal** (or *metalloid*) is an element that has properties intermediate between those of a metal and a nonmetal.

In a **three-center bond** one pair of electrons holds three nuclei together.

Zone refining is a technique for purifying solids in which a molten zone is moved along a bar of a solid material. Impurities concentrate in the molten zone and are thus removed to one end of the bar, which can then be cut off.

PROBLEMS

Silicon

1. What is the fundamental structural unit of the silicates? How is this unit modified in the aluminosilicates?

2. Give an example of a silicate mineral with a layer structure. Explain the empirical formula of the silicate anion in this structure.

3. What is an amphibole mineral? What is the empirical formula of the silicate anion in the amphiboles?

4. What are the three principal components of glass?

5. Which properties of glass are similar to those of a crystalline solid? Which properties are similar to those of a liquid?

6. What is meant by photochromic glass? Explain how it works.

7. The solid-state structure of silicon is the same as that of diamond. Draw a sketch of the unit cell. How many atoms are in the unit cell? How many nearest neighbors does each silicon atom have? The edge of the unit cell has a length of 545 pm. Calculate the density of silicon, the length of the Si—Si bond, and the radius of the silicon atom.

8. Draw the Lewis structure for the cyclic $Si_4O_{12}^{8-}$ anion.

9. Write a reasonable ionic formulation for the following minerals, and describe the structure of the silicate or aluminosilicate anion.

    (a) Diopside, $CaMgSi_2O_6$

    (b) Orthoclase, $KAlSi_3O_8$

    (c) Hardystonite, $Ca_2ZnSi_2O_7$

    (d) Denitoite, $BaTiSi_3O_9$

    (e) Pyrophyllite, $AlSi_2O_5(OH)$

    (f) Anorthite, $CaAl_2Si_2O_8$

10. Draw Lewis structures for the molecules Si_2Cl_6 and Si_2Cl_6O. Which of these molecules would be expected to have a dipole moment?

11. Draw the structure of the anion $Si_6O_{18}^{12-}$. Which silicate mineral contains this anion?

12. Explain what is meant by *zone refining*.

13. Why does silicon not have an allotrope like graphite?

14. What is a silicone? Draw a diagram of the structure of a typical silicone. What are some of the important properties of silicones?

15. What carbon compounds are the analogues of the silicones? How do their structures differ from those of the silicones? Explain this difference in structure.

16. Why are silanes much more reactive than alkanes?

17. Why are so few compounds known containing Si=Si double bonds?

Boron

18. A sample of a gaseous borane has a density of 1.23 $g\,L^{-1}$ at STP. What is the molar mass of the compound? What is a plausible molecular formula?

19. Explain the differences in the melting points and boiling points of the fluorides and chlorides of boron and aluminum.

	mp (°C)	bp (°C)
BF_3	−128	−102
AlF_3	—	1257 (sublimes)
BCl_3	−107	12
$AlCl_3$	—	180 (sublimes)

20. Explain why $B(OH)_3$ is a weak acid, whereas $Al(OH)_3$ is amphoteric in aqueous solution.

21. Write three balanced equations showing compounds of boron behaving as Lewis acids.

22. What is borax? Draw a diagram of the structure of its anion. Why can borax be used as a water softener?

Miscellaneous

23. What is the chemical composition of each of the following minerals: limestone, gypsum, silica, bauxite, pyrite, beryl, and talc?

24. Explain the increase in acid strength in the following series: $Si(OH)_4 < OP(OH)_3 < O_2S(OH)_2 < O_3ClOH$

25. Explain, with the aid of an appropriate example, what is meant by the term *three-center bond*.

CHAPTER 23

POLYMERS: SYNTHETIC AND NATURAL

We cannot go through a single day without using a dozen or more materials based on *synthetic polymers*. The materials that we commonly call **plastics**, which are used for cups and dishes, combs, telephones, pens, bags, pipes, paints, synthetic fibers, kitchen counter tops, and so on, are all composed of synthetic polymers. The names of many of these materials are well known to most of us: polyethylene, polystyrene, polyurethane, Teflon, Formica, Saran, and so on. **Polymers** are very large molecules that are formed by the combination of a very large number of relatively small molecules called **monomers**. In synthetic polymers there is often only one type of monomer unit—or at the most a small number of different monomer units. In Chapter 11 we mentioned polyethylene, which is a polymer of composition $(CH_2)_n$ formed from ethene. Nylon is a polymer formed from two different monomers, hexanedioic acid and diaminohexane. There are also very many natural polymers, often called *biopolymers*, such as carbohydrates and proteins, that often contain many different monomer units.

Polymers may be made from both inorganic and organic molecules. We have previously mentioned a number of inorganic polymers such as $(SO_3)_n$ and metaphosphoric acid, $(HPO_3)_n$ (Chapter 7), and the silicates and silicones (Chapter 22). Polymer molecules may have many different forms; in particular, they may be chains, like polyethylene, or sheets, like talc and mica, or three-dimensional giant molecules, like quartz. But most synthetic polymers are long-chain organic molecules that typically contain thousands of monomer units. Such molecules have very high masses, of the order of 100 000 u or more. They are also often called **macromolecules**.

Although today polymers are familiar to almost everybody, only in the past 40 years have chemists learned how to synthesize them. Their enormous importance at the present time can be judged from the fact that half of the professional organic chemists employed by industry in the United States are engaged in research and development related to polymers. But nature has been using polymers, often far more complex than those synthesized by chemists, since the beginning of life. In this chapter we consider first a few examples of synthetic polymers and then some natural polymers, or biopolymers.

Polyethylene and Other Addition Polymers

Polyethylene is formed by joining a large number of ethene molecules to form a long chain:

$$\cdots CH_2{=}CH_2 + CH_2{=}CH_2 + CH_2{=}CH_2 \cdots \longrightarrow$$
$$-CH_2-CH_2-CH_2-CH_2-CH_2-CH_2-$$

This repeated addition of small molecules to each other to form a polymer is called *polymerization*.

Ethene molecules are not easily polymerized, even at high temperature and pressure; the activation energy for the polymerization is high. The reaction must be initiated by introducing a reactive free radical (Chapters 17 and 18), usually provided by thermal decomposition of an organic peroxide, $R{-}O{-}O{-}R$, to give two free radicals, $R{-}O\cdot$. The first step in the polymerization is then

$$R{-}O\cdot + CH_2{=}CH_2 \longrightarrow R{-}O{-}CH_2{-}CH_2\cdot$$

The product is a free radical that can react with another ethene molecule to give another free radical,

$$R{-}O{-}CH_2{-}CH_2\cdot + CH_2{=}CH_2 \longrightarrow R{-}O{-}CH_2{-}CH_2{-}CH_2{-}CH_2\cdot$$

and so on. The chain can continue to grow in this way until a termination reaction occurs, in which, for example, two chains add to each other or another free radical adds to the chain. The length of the chains and the degree to which the chains are branched can be controlled by varying the conditions of the polymerization and using different catalysts. Thus polyethylenes with a variety of different properties can be obtained. They may be viscous liquids or solids of varying degrees of hardness. The solid polymers are generally amorphous, but they may be at least partly crystalline (see Figure 23.1).

The polymerization of ethene can be written as

$$nCH_2{=}CH_2 \longrightarrow {+}CH_2CH_2{+}_n$$

The formula ${+}CH_2CH_2{+}_n$ means that the group in brackets is repeated n times; the value of n is usually several thousand or larger and varies from one chain to another. If we measure the chain length or the molecular mass of a polymer, we obtain an average value. The chains must be terminated by some other group or groups, but since there is only one such group for several

Figure 23.1 Structure of Polyethylene.

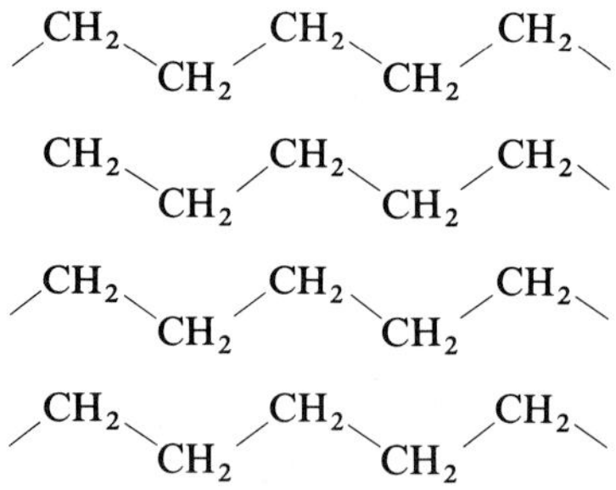

Crystalline polyethylene

(a)

Crystalline and amorphous regions in polyethylene

(b)

thousand monomer units, these end groups have no appreciable effect on the composition or the properties.

A large number of different polymers can be prepared by polymerizing various substituted ethenes. Some examples are given in Table 23.1. Polymers of this type are prepared by the successive addition of monomer units and are called **addition polymers**.

Example 23.1 What is the molar mass of polystyrene if a single molecule contains 2000 monomer units? Styrene has the molecular formula $C_6H_5\text{—}CH\text{=}CH_2$.

Solution The molar mass of styrene is 104.2 g mol^{-1}. The polymer molecule has the molecular formula $(C_8H_8)_{2000}$. The molar mass is therefore

$$(2000)(104.2) \text{ g mol}^{-1} = 2.084 \times 10^5 \text{ g mol}^{-1}$$

Table 23.1 Some Addition Polymers Produced from Substituted Ethenes

MONOMER	POLYMER	TYPICAL USES
$CH_2\text{=}CH_2$ Ethene	$\text{—(}CH_2\text{—}CH_2\text{)}_n$ Polyethylene	Containers, pipes, bags, toys, wire insulation
$CH_2\text{=}CHCH_3$ Propene	$\text{—(}CH_2\text{—}CH\text{)}_n$ | CH_3 Polypropylene	Fibers for carpets, artificial turf, rope, fishing nets
$CH_2\text{=}CHCl$ Chloroethene (vinyl chloride)	$\text{—(}CH_2\text{—}CH\text{)}_n$ | Cl Polyvinyl chloride (PVC)	Garden hoses, floor tiles, plumbing, records, laboratory tubing
$CH_2\text{=}CHCN$ Propenenitrile (acrylonitrile)	$\text{—(}CH_2\text{—}CH\text{)}_n$ | CN Polyacrylonitrile (Orlon, Acrilan)	Fibers for cloth, carpets, upholstery
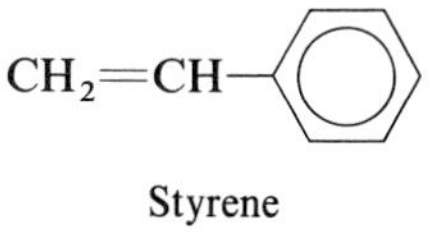 $CH_2\text{=}CH\text{—}\bigcirc$ Styrene	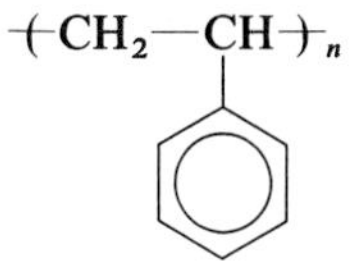 $\text{—(}CH_2\text{—}CH\text{)}_n$ Polystyrene	Styrofoam, hot-drink cups, insulation
$CF_2\text{=}CF_2$ Tetrafluoroethene	$\text{—(}CF_2\text{—}CF_2\text{)}_n$ Teflon	Non-stick coating for kitchen utensils
$CH_2\text{=}\overset{\displaystyle CH_3}{\underset{\displaystyle }{C}}\text{—}CO_2CH_3$ Methyl methacrylate	$\text{—(}CH_2\text{—}\overset{\displaystyle CH_3}{\underset{\displaystyle CO_2CH_3}{C}}\text{)}_n$ Polymethyl methacrylate	Plexiglass, Lucite, headlamp lenses, sunglasses, aircraft windows

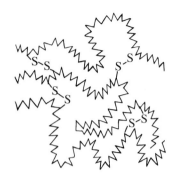

Natural and Synthetic Rubber

Natural rubber is an addition polymer of 2-methyl-1,3-butadiene (isoprene) with an average chain length of about 5000 monomer units:

$$n\ CH_2\text{=}CH\text{—}\underset{\underset{CH_3}{|}}{C}\text{=}CH_2 \longrightarrow \text{+}CH_2\text{—}CH\text{=}\underset{\underset{CH_3}{|}}{C}\text{—}CH_2\text{+}_n$$

Rubber was discovered by the native peoples of America. It was brought back to Europe by Columbus, and it was named rubber by Joseph Priestley (Chapter 3) because he found that it could be used to rub out pencil marks.

Natural rubber is an inconveniently sticky material. In 1846 American scientist Charles Goodyear found that this disadvantage could be overcome by heating rubber with sulfur in a process called *vulcanization*. The sulfur forms short chains of sulfur atoms that link the hydrocarbon chains together—the polymer is then said to be **cross-linked** (Figure 23.2). In a polymer the chains are normally tangled up with each other, rather like a bowl of spaghetti. When solid rubber is stretched, the chains straighten out to some extent. When the tension on the rubber is released, the chains tend to coil up again (see Experiment 23.1). The straighter, more ordered chains have a lower entropy than the coiled up, more disordered chains. When rubber is cross-linked by sulfur chains, the extent to which the chains can be straightened is limited, and they have a greater tendency to resume their original shape when the tension is released. Thus vulcanized rubber is stronger, harder, less sticky, and more "rubbery" than natural rubber.

Synthetic rubber is made from butadiene and other substituted butadienes:

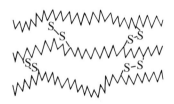

$$n(CH_2\text{=}CH\text{—}CH\text{=}CH_2) \longrightarrow \text{+}CH_2\text{—}CH\text{=}CH\text{—}CH_2\text{+}_n$$
Butadiene   Polybutadiene

A material like rubber that returns to its original shape after stretching is said to be elastic. Polymers that are elastic are called **elastomers**.

Figure 23.2 Vulcanized Rubber. The hydrocarbon chains are held together by cross-linking chains of sulfur atoms. When tension is applied, the chains can straighten out, but they cannot slip past each other because of the polysulfide bridges. Thus rubber can be stretched only to a limited extent. When the tension is removed, the chains tend to coil up again, and the rubber resumes its original shape.

Example 23.2 Draw the structure of the addition polymer (neoprene rubber) formed from 2-chloro-1,3-butadiene (chloroprene).

EXPERIMENT 23.1

Rubber

A sharp pointed needle can be pushed slowly through an inflated rubber balloon without causing it to burst because the long flexible rubber molecules move around the hole and seal it.

Solution The structure of 2-chloro-1,3-butadiene is

$$CH_2{=}\overset{\displaystyle Cl}{\underset{\displaystyle |}{C}}{-}CH{=}CH_2$$

so the polymer has the structure

$$\cdots{-}CH_2\overset{\displaystyle Cl}{\diagdown}{C}{=}{C}\overset{\displaystyle H}{\diagup}CH_2{-}CH_2\overset{\displaystyle Cl}{\diagdown}{C}{=}{C}\overset{\displaystyle H}{\diagup}CH_2{-}CH_2\overset{\displaystyle Cl}{\diagdown}{C}{=}{C}\overset{\displaystyle H}{\diagup}CH_2{-}\cdots$$

Neoprene rubber

Condensation Polymers

Addition polymers contain all the atoms of the original monomers. Another important class of polymers are the **condensation polymers**, so-called because they are formed through condensation reactions in which monomers are joined into polymer chains by the elimination of small molecules such as water.

In Chapter 19 we saw that when a carboxylic acid and an amine are heated together, a condensation reaction occurs to give an amide and water:

$$\underset{\text{Carboxylic acid}}{R{-}\overset{\displaystyle O}{\overset{\displaystyle ||}{C}}{-}OH} + \underset{\text{Amine}}{H{-}\overset{\displaystyle H}{\underset{\displaystyle |}{N}}{-}R'} \longrightarrow \underset{\text{Amide}}{R{-}\overset{\displaystyle O}{\overset{\displaystyle ||}{C}}{-}\overset{\displaystyle H}{\underset{\displaystyle |}{N}}{-}R'} + \underset{\text{Water}}{H_2O}$$

We also saw that an ester is formed from a carboxylic acid and an alcohol in a condensation reaction:

$$R{-}CO_2H + R'{-}OH \longrightarrow R{-}\overset{\displaystyle O}{\overset{\displaystyle ||}{C}}{-}OR' + H_2O$$

Polyamides and *polyesters* are among the most important condensation polymers.

NYLON   Nylon-66 is a **polyamide** that is prepared by heating 1,6-hexanedioic acid (adipic acid) and 1,6-hexanediamine at 270°C under pressure:

$$\cdot\;\underset{\text{Hexanedioic acid}}{HO{-}\overset{\displaystyle O}{\overset{\displaystyle ||}{C}}{-}(CH_2)_4{-}\overset{\displaystyle O}{\overset{\displaystyle ||}{C}}{-}OH} + \underset{\text{1,6-Hexanediamine}}{H_2N{-}(CH_2)_6{-}NH_2} + HO{-}\overset{\displaystyle O}{\overset{\displaystyle ||}{C}}{-}(CH_2)_4{-}\overset{\displaystyle O}{\overset{\displaystyle ||}{C}}{-}OH$$

$$\downarrow$$

$${-}\overset{\displaystyle H}{\underset{\displaystyle |}{N}}{-}\overset{\displaystyle O}{\overset{\displaystyle ||}{C}}{-}(CH_2)_4{-}\overset{\displaystyle O}{\overset{\displaystyle ||}{C}}{-}\overset{\displaystyle H}{\underset{\displaystyle |}{N}}{-}(CH_2)_6{-}\overset{\displaystyle H}{\underset{\displaystyle |}{N}}{-}\overset{\displaystyle O}{\overset{\displaystyle ||}{C}}{-}(CH_2)_4{-}\overset{\displaystyle O}{\overset{\displaystyle ||}{C}}{-}\overset{\displaystyle H}{\underset{\displaystyle |}{N}}{-} + nH_2O$$

The structure of Nylon-66 may be written more briefly as

$$\left(\overset{\displaystyle H}{\underset{\displaystyle |}{N}}{-}\overset{\displaystyle O}{\overset{\displaystyle ||}{C}}{-}(CH_2)_4{-}\overset{\displaystyle O}{\overset{\displaystyle ||}{C}}{-}\overset{\displaystyle H}{\underset{\displaystyle |}{N}}{-}(CH_2)_6\right)_n$$

Nylon produced in this way has a molecular mass of about 10 000 u and a melting point of 250°C. While molten, it can be extruded into fibers. Hydrogen bonds between the NH groups in one chain and the CO groups in an adjacent chain hold the molecules together strongly enough to give nylon considerable tensile strength but not so strongly that it cannot be pulled out into thin fibers.

Nylon is used in hosiery and other clothing. It resembles silk in its structure and properties, but being cheaper to produce, it has almost entirely replaced silk. The name nylon-66 refers to the fact that it is made from a six-carbon-atom carboxylic acid and a six-carbon-atom amine. Many other nylons have been made from other dicarboxylic acids and diamines (see Experiment 23.2).

POLYESTERS When a diol reacts with a dicarboxylic acid, a **polyester** is formed. For example, Dacron is a condensation polymer formed from 1,2-ethanediol and terephthalic acid (1,4-benzene dicarboxylic acid):

$$HOCH_2CH_2OH + HO\!-\!\overset{\overset{\displaystyle O}{\|}}{C}\!-\!\bigcirc\!-\!\overset{\overset{\displaystyle O}{\|}}{C}\!-\!OH + HOCH_2CH_2OH + \cdots$$

$$-CH_2CH_2\!-\!O\!-\!\overset{\overset{\displaystyle O}{\|}}{C}\!-\!\bigcirc\!-\!\overset{\overset{\displaystyle O}{\|}}{C}\!-\!O\!-\!CH_2CH_2\!-\!O\!-\!\overset{\overset{\displaystyle O}{\|}}{C}\!-\!\bigcirc\!-\!\overset{\overset{\displaystyle O}{\|}}{C}\!-\!O\!-$$

$$+\, n\,H_2O$$

The structure of Dacron may be written

$$\left(\!O\!-\!CH_2\!-\!CH_2\!-\!O\!-\!\overset{\overset{\displaystyle O}{\|}}{C}\!-\!\bigcirc\!-\!\overset{\overset{\displaystyle O}{\|}}{C}\!\right)_{\!n}$$

Dacron forms strong fibers. It is used as a blend with cotton in clothing, and it has specialized uses, such as for seat belts and sails. In the form of thin sheets, this polymer is called Mylar.

Example 23.3 Kodel is a polyester made from terephthalic acid (1,4-benzene dicarboxylic acid) and the diol

$$HO\!-\!CH_2\!-\!CH\!\!\overset{\displaystyle CH_2\!-\!CH_2}{\underset{\displaystyle CH_2\!-\!CH_2}{\diagup\!\!\diagdown}}\!\!CH\!-\!CH_2\!-\!OH$$

EXPERIMENT 23.2

Synthesis of Nylon-610

When a solution of 1,6-diaminohexane, $H_2N(CH_2)_6NH_2$, in aqueous sodium hydroxide is poured gently onto a solution of decanedioyl chloride, $COCl(CH_2)_8COCl$, a white film of nylon-610 forms between the two layers. The film can be grasped with tweezers and pulled up as a nylon string, which can be wound on a glass rod as shown.

What is the structure of Kodel?

Solution

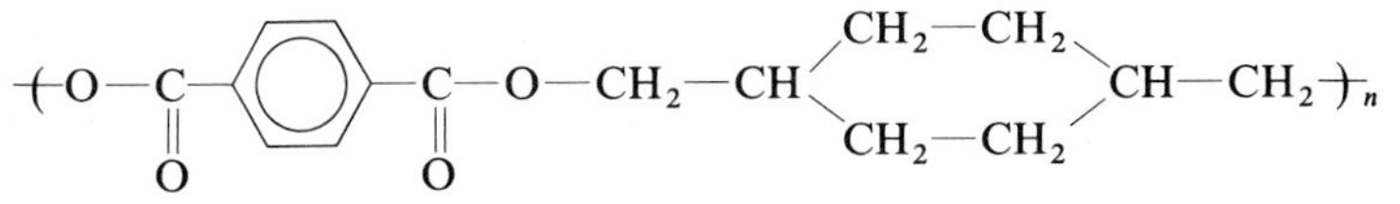

$$\left(\!\!\!O\!-\!\overset{\displaystyle C}{\underset{\displaystyle O}{\|}}\!-\!\!\!\bigcirc\!\!\!-\!\overset{\displaystyle C}{\underset{\displaystyle O}{\|}}\!-\!O\!-\!CH_2\!-\!CH\overset{\displaystyle CH_2-CH_2}{\underset{\displaystyle CH_2-CH_2}{\diagup\diagdown}}CH\!-\!CH_2\!\right)_n$$

Example 23.4 Poly-4-methyl-1-pentene is a solid transparent polymer used in the manufacture of laboratory ware such as flasks and beakers. Is this an addition or a condensation polymer? Draw the structure of the polymer.

Solution It is an addition polymer

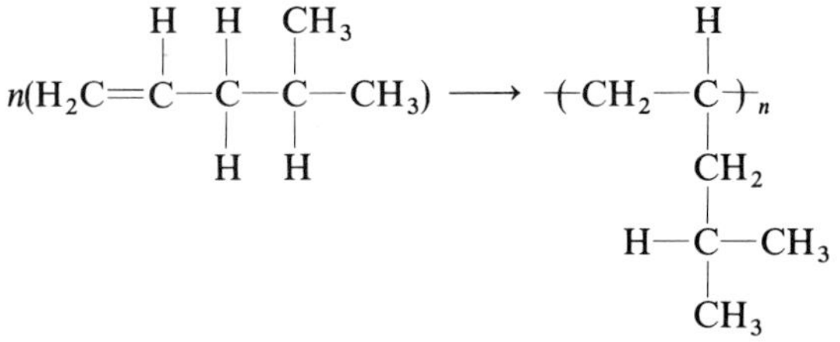

$$n(H_2C\!=\!\overset{\displaystyle H}{\underset{\displaystyle H}{C}}\!-\!\overset{\displaystyle H}{\underset{\displaystyle H}{C}}\!-\!\overset{\displaystyle CH_3}{C}\!-\!CH_3) \longrightarrow \left(\!CH_2\!-\!\overset{\displaystyle H}{\underset{\displaystyle \underset{\displaystyle \underset{\displaystyle CH_3}{H-C-CH_3}}{CH_2}}{C}}\!\right)_n$$

23.2 BIOPOLYMERS

Many biologically important substances are polymers. In this section we consider carbohydrates, proteins, and nucleic acids. Carbohydrates serve as energy sources and as the structural material of plants. Proteins are found in all parts of the body, and they have an enormous variety of functions. Some proteins are the structural components of skin, muscle, and hair; others control the transmission of nerve impulses; still others are enzymes that catalyze reactions. The nucleic acid, DNA, is the molecule in which an organism stores genetic information and through which it passes this information from generation to generation.

Carbohydrates

Carbohydrates are synthesized by green plants from CO_2 and H_2O in the presence of sunlight in a process called *photosynthesis*:

$$xCO_2 + yH_2O \longrightarrow C_xH_{2y}O_y + xO_2$$

Because their empirical formulas can be written as $C_x(H_2O)_y$, carbohydrates were originally thought to be hydrates of carbon (hence the name). Carbohydrates can be classified into three main groups: monosaccharides, disaccharides, and polysaccharides.

About twenty monosaccharides occur naturally. They are commonly known as *sugars*. Two important examples are *glucose* and *fructose*. They both have the molecular formula $C_6H_{12}O_6$, but they have different structures. Glucose has a six-membered ring of five carbon atoms and an oxygen atom and has five OH groups, which accounts for its solubility in water (see Figure 23.3). It exists in two forms called α-*glucose* and β-*glucose*, which differ only in the orientation of one of the OH groups with respect to the ring. Fructose has a five-membered ring of four carbon atoms and one oxygen atom (see Figure 23.4).

Figure 23.3 Structure of Glucose.
The α and β forms of glucose differ in the orientation of the OH group on C–1. In α-glucose this OH group is perpendicular to a plane through the ring. In β-glucose this OH group lies approximately in a plane through the ring.

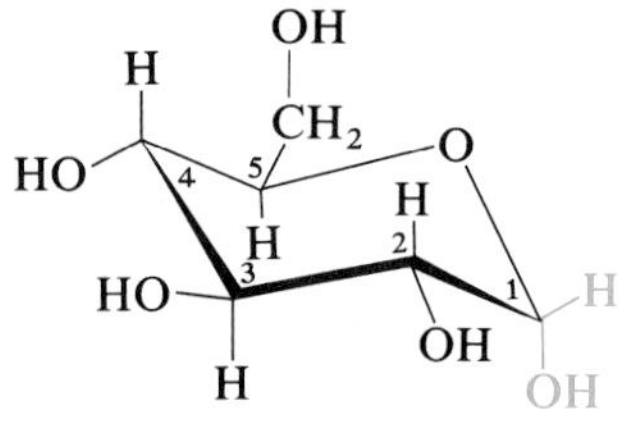

α-Glucose

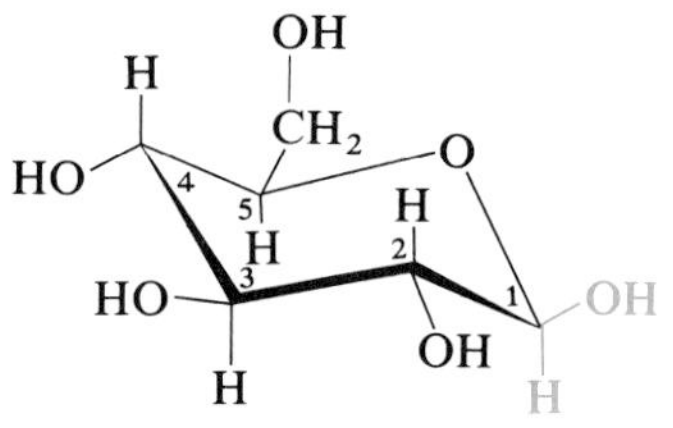

β-Glucose

A common disaccharide is sucrose (common table sugar). It is formed from a glucose molecule and a fructose molecule condensed together with the elimination of a water molecule (see Figure 23.5). The two monosaccharide units are joined by an ether linkage. The condensation reaction by which sucrose is formed from glucose and fructose is reversed in the stomach in a reaction that is catalyzed by the enzyme *sucrase*. Thus when we digest sucrose, glucose and fructose are formed, which are absorbed into the blood. The oxidation of glucose in living cells (aerobic metabolism) is an important source of energy for all animals. It occurs in many steps catalyzed by enzymes and ultimately results in the formation of CO_2 and water:

$$C_6H_{12}O_6 + 6O_2 \longrightarrow 6CO_2 + 6H_2O \qquad \Delta H° = -2880 \text{ kJ}$$

Two important polysaccharides are starch and cellulose. *Starch* is a mixture of polymers of α-glucose, and cellulose is a polymer of β-glucose. Starch consists mainly of *amylose*, which is a straight-chain (unbranched) polymer of α-glucose containing an average of 200 glucose units (Figure 23.6). Starch is broken down in the digestive tract in a series of steps that are catalyzed by enzymes.

Cellulose is a straight-chain polymer of β-glucose containing on average about 3000 glucose units (Figure 23.7). Cellulose is the major structural component of wood and other plants. It accounts for more than one-half of all living matter. Humans do not possess the enzymes necessary to break down cellulose into glucose. Thus we are unable to digest cellulose. Animals such as cows and deer have intestinal bacteria that have the necessary enzymes for breaking cellulose down to glucose. If chemists could find a simple way to break cellulose down to glucose, we would have another important source of food.

Example 23.5 Write equations for the hydrolysis of sucrose and starch.

Solution

$$C_{12}H_{22}O_{11} + H_2O \longrightarrow C_6H_{12}O_6 + C_6H_{12}O_6$$

Sucrose \qquad\qquad Glucose \quad Fructose

$$(C_6H_{10}O_5)_n + nH_2O \longrightarrow nC_6H_{12}O_6$$

Starch \qquad\qquad Glucose

Figure 23.4 Structure of Fructose.

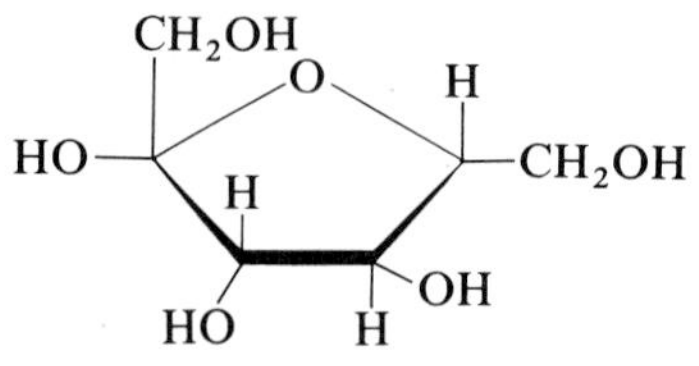

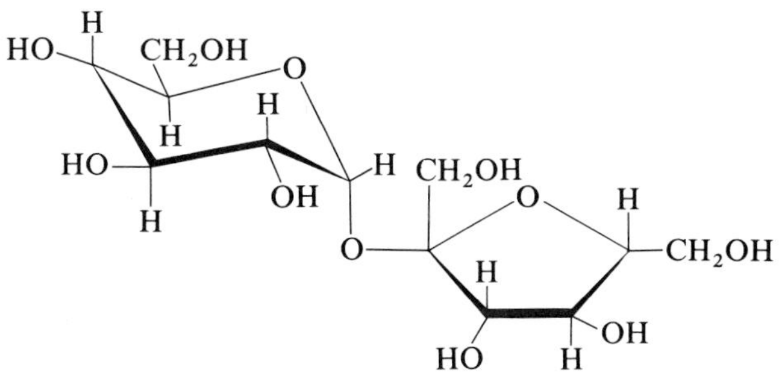

Figure 23.5 Structure of Sucrose, a Disaccharide.

Example 23.6 Amylose has a molar mass of about 3.0×10^5 g mol^{-1}. Approximately how many glucose units does an amylose molecule contain?

Solution Each glucose unit in amylose has the formula $C_6H_{12}O_6 - H_2O = C_6H_{10}O_5$ and therefore has a formula mass of 162 g mol^{-1}. Hence

$$\text{Number of glucose units} = \frac{3.00 \times 10^5 \text{ g mol}^{-1}}{162 \text{ g mol}^{-1}} = 1900$$

Semisynthetic Polymers from Cellulose

Cellulose contains a very large number of —OH groups that can react with carboxylic acids to give esters. Especially important are the *acetates*, such as cellulose triacetate, in which all three —OH groups of each glucose unit are replaced by acetate,

$$—O—\overset{\displaystyle O}{\overset{\|}{C}}—CH_3$$

groups. This class of polymers is referred to collectively as *acetate polymers.* An example is Arnel, which is a strong fiber used in clothing, fabrics, and electric insulators.

Cellulose in the form of cotton or wood pulp also reacts with nitric acid to give *nitrocellulose.* Depending on the conditions of the reaction, the —OH groups of each glucose unit may be partially or fully replaced by nitrate groups, $—ONO_2$. The partially nitrated form, with camphor added to make it softer and more malleable, is called *celluloid.* It was patented as early as 1869 and was the first synthetic plastic. It was used for many years to make articles as diverse as baby rattles, shirt collars, and photographic film, but its use has diminished in recent years due to its flammability. Fully nitrated cellulose is

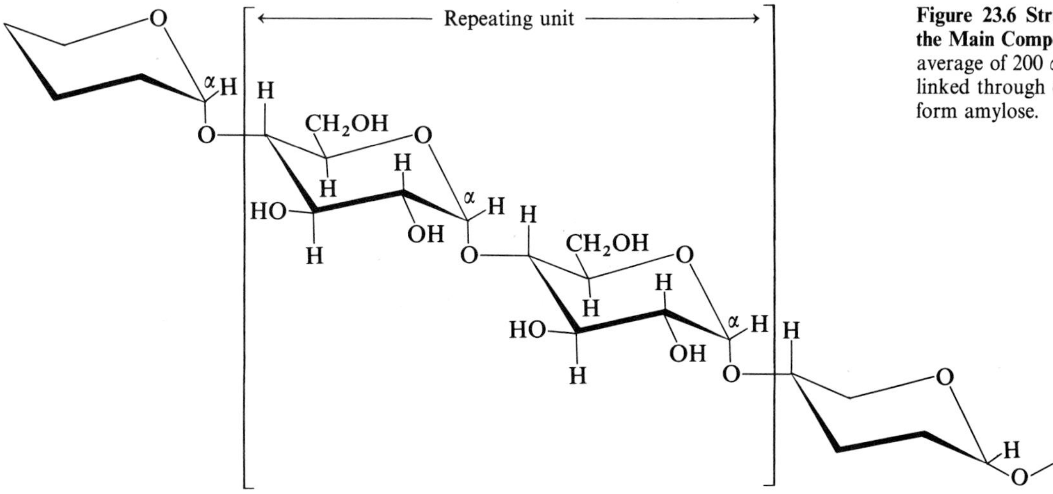

Figure 23.6 Structure of Amylose, the Main Component of Starch. An average of 200 α-glucose units are linked through oxygen atoms to form amylose.

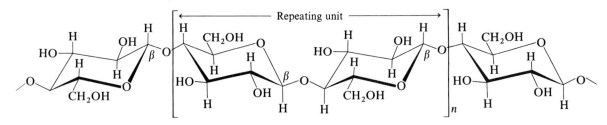

Figure 23.7 Structure of Cellulose. On average about 3000 β-glucose units are linked through oxygen atoms to form a cellulose molecule.

known as *guncotton* and is used as an explosive and a propellant. Nitrocellulose dissolved in solvents such as ether-alcohol mixtures is used as a base for quick-drying paints.

Proteins

Thousands of different proteins go into the makeup of a living cell. They take part in the thousands of reactions that take place in a living cell. **Proteins** are very complex, giant molecules, and one of the greatest achievements of modern science has been determining the structure of numerous proteins. But the exact details of all the complex processes in which they are involved will defy our understanding for some time to come. Here we can only give a glimpse of this fascinating and challenging area of chemistry.

Proteins are polyamides formed from amino acids. **Amino acids** have both an NH_2 group and a carboxyl group, CO_2H. The amino acids in proteins are called **α-amino acids** because they have the NH_2 group on the same carbon atom as the carboxylic acid group.

An amide is formed by condensation of an NH_2 group with a carboxylic acid group. Because an amino acid has both these functional groups, a polymer—a *polyamide*—can be formed by condensation:

$$H_2NCHC \overset{O}{\underset{R_1}{\diagup}} OH + H_2NCHC \overset{O}{\underset{R_2}{\diagup}} OH + H_2NCHC \overset{O}{\underset{R_3}{\diagup}} OH$$

$$\downarrow$$

$$-NH-CH-\overset{O}{\overset{\|}{C}}-NH-CH-\overset{O}{\overset{\|}{C}}-NH-CH-\overset{O}{\overset{\|}{C}}-NH- + n H_2O$$
$$\qquad R_1 \qquad\qquad R_2 \qquad\qquad R_3$$

The $-\overset{O}{\overset{\|}{C}}-NH-$ groups linking the R groups in the polymer are called peptide links, and a polyamide is also called a *polypeptide*. Proteins are naturally occurring polypeptides. The general formula of a protein as we have written it looks simple, but there are 20 amino acids that are commonly found in nature and a given protein may contain many or all of them. These different amino acids are listed in Table 23.2. They differ in the nature of the group R. They are often denoted by the three-letter abbreviations given in Table 23.2 such as *ala* for alanine and *gly* for glycine. Since a protein commonly contains as many as fifty or more monomer units, the number of possibilities for different proteins is truly enormous. Fortunately, in nature we find only a very small fraction of all these possibilities.

Table 23.2 Biologically Important α-Amino Acids

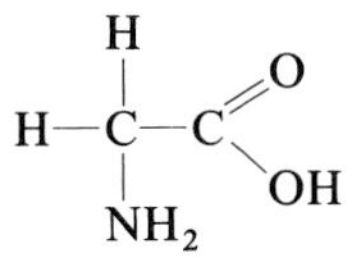

Glycine, gly

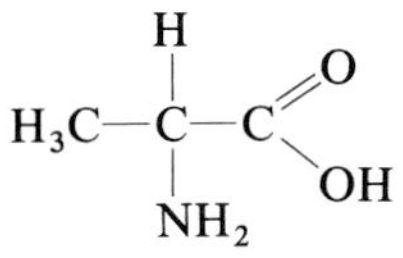

Alanine, ala

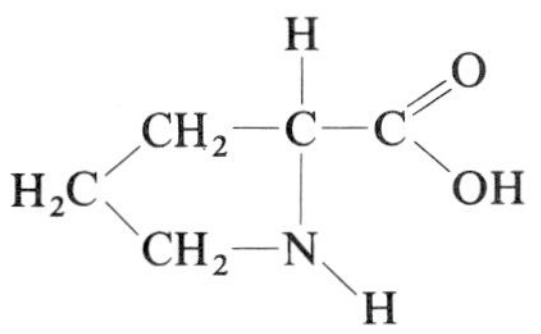

Proline, pro

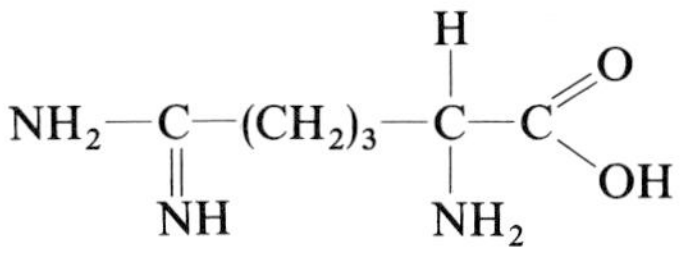

Arginine, arg

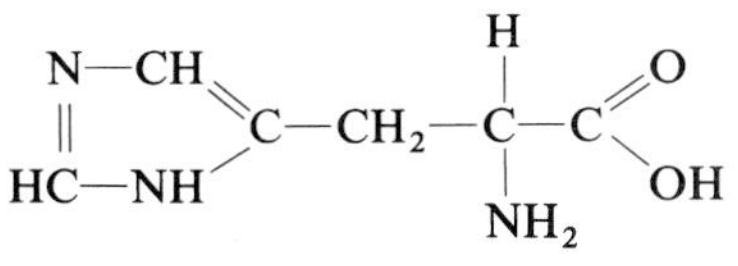

Histidine, his

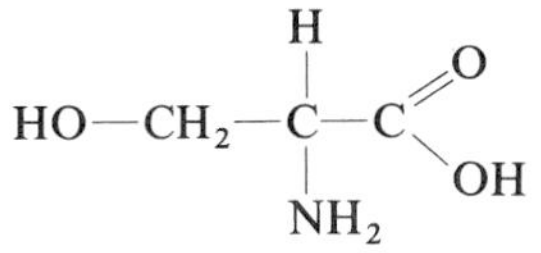

Serine, ser

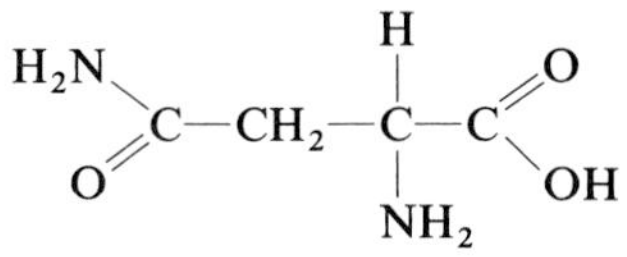

Aspargine asn

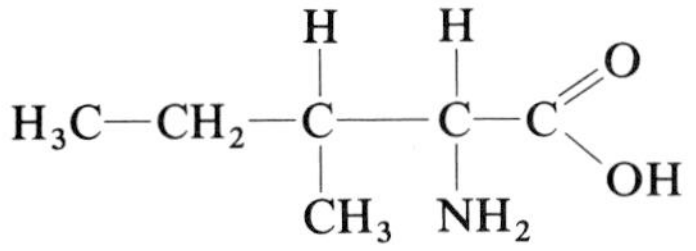

Isoleucine, ile

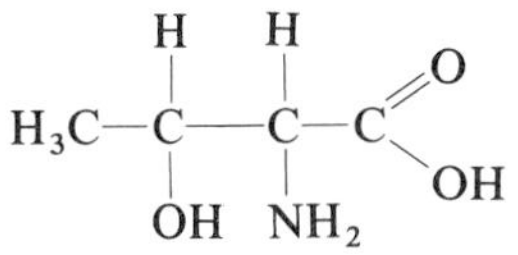

Threonine, thr

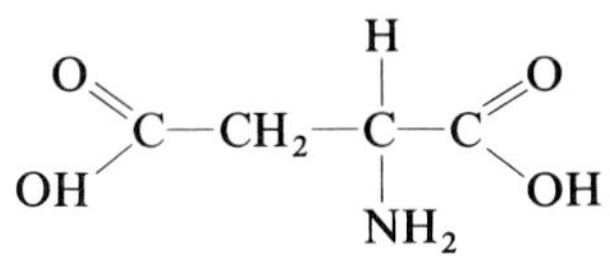

Aspartic acid, asp

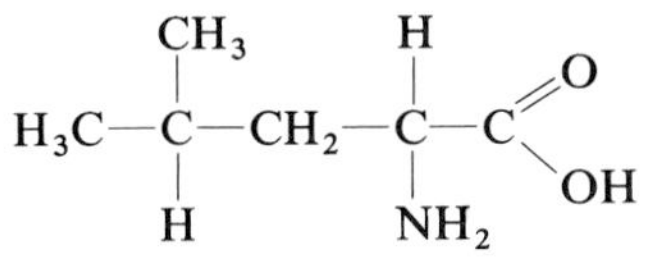

Leucine, leu

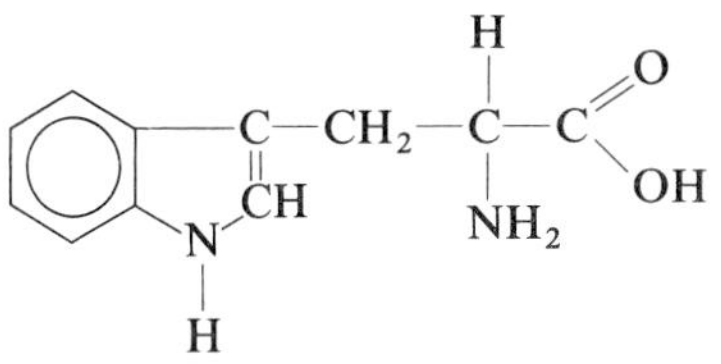

Tryptophan, trp

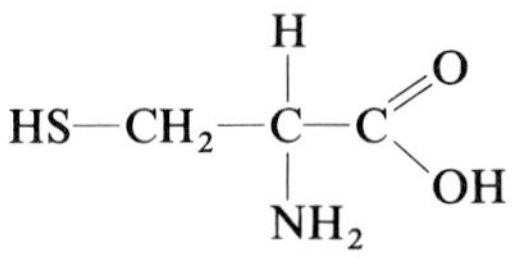

Cysteine, cys

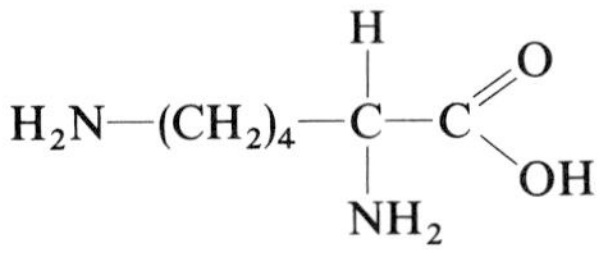

Lysine, lys

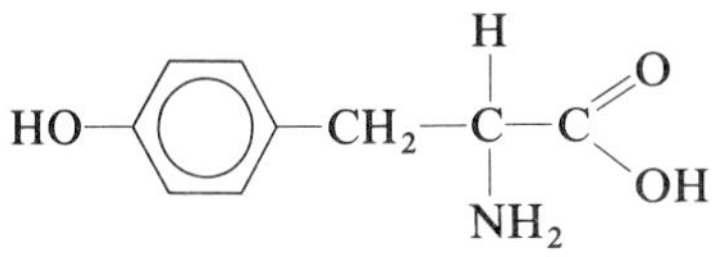

Tyrosine, tyr

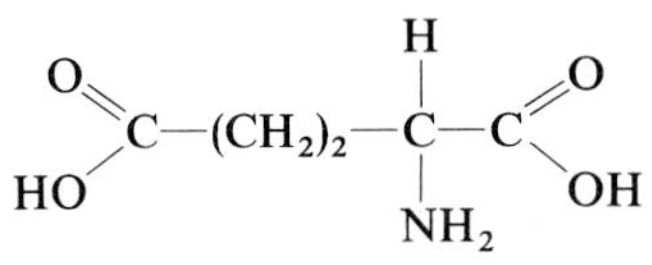

Glutamic acid, glu

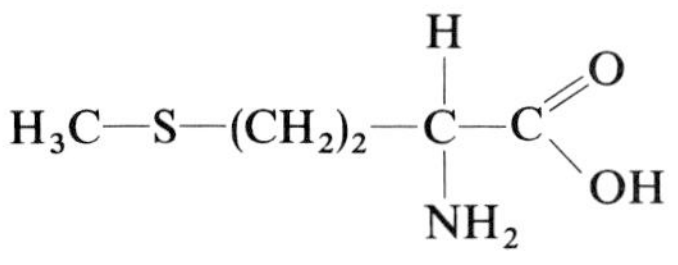

Methionine, met

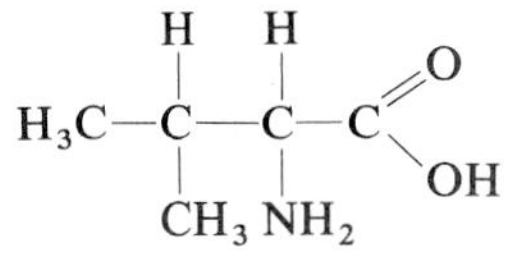

Valine, val

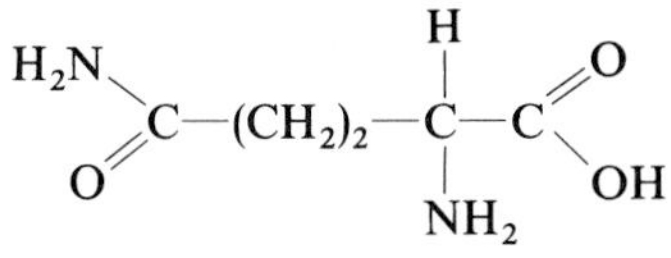

Glutamine, gln

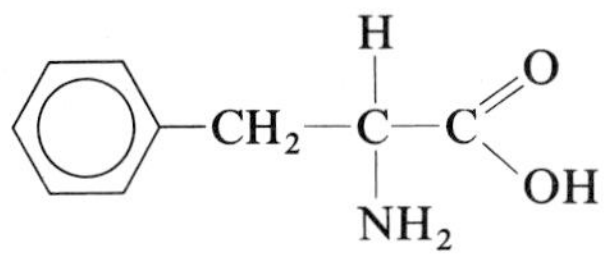

Phenylalanine, phe

Example 23.7 What amides can result from the condensation of a valine molecule with a cysteine molecule?

Solution

$$\underset{\text{Cysteine}}{\text{HS}-\overset{\overset{\displaystyle H}{|}}{\underset{\underset{\displaystyle H}{|}}{C}}-\overset{\overset{\displaystyle H}{|}}{\underset{\underset{\displaystyle NH_2}{|}}{C}}-\overset{\overset{\displaystyle O}{\|}}{C}-OH} + \underset{\text{Valine}}{H-\overset{\overset{\displaystyle H}{|}}{N}-\overset{\overset{\displaystyle H}{|}}{\underset{\underset{\displaystyle CO_2H}{|}}{C}}-\overset{\overset{\displaystyle CH_3}{|}}{\underset{\underset{\displaystyle H}{|}}{C}}-CH_3} \longrightarrow$$

$$\text{HS}-\overset{\overset{\displaystyle H}{|}}{\underset{\underset{\displaystyle H}{|}}{C}}-\overset{\overset{\displaystyle H}{|}}{\underset{\underset{\displaystyle NH_2}{|}}{C}}-\overset{\overset{\displaystyle O}{\|}}{C}-\overset{\overset{\displaystyle H}{|}}{N}-\overset{\overset{\displaystyle H}{|}}{\underset{\underset{\displaystyle CO_2H}{|}}{C}}-\overset{\overset{\displaystyle CH_3}{|}}{\underset{\underset{\displaystyle H}{|}}{C}}-CH_3 + H_2O$$

$$\underset{\text{Cysteine}}{\text{HS}-\overset{\overset{\displaystyle H}{|}}{\underset{\underset{\displaystyle H}{|}}{C}}-\overset{\overset{\displaystyle CO_2H}{|}}{\underset{\underset{\displaystyle H}{|}}{C}}-\overset{\displaystyle H}{N}-H} + \underset{\text{Valine}}{HO-\overset{\overset{\displaystyle O}{\|}}{C}-\overset{\overset{\displaystyle H}{|}}{\underset{\underset{\displaystyle NH_2}{|}}{C}}-\overset{\overset{\displaystyle CH_3}{|}}{\underset{\underset{\displaystyle H}{|}}{C}}-CH_3} \longrightarrow$$

$$\text{HS}-\overset{\overset{\displaystyle H}{|}}{\underset{\underset{\displaystyle H}{|}}{C}}-\overset{\overset{\displaystyle CO_2H}{|}}{\underset{\underset{\displaystyle H}{|}}{C}}-\overset{\displaystyle H}{N}-\overset{\overset{\displaystyle O}{\|}}{C}-\overset{\overset{\displaystyle H}{|}}{\underset{\underset{\displaystyle NH_2}{|}}{C}}-\overset{\overset{\displaystyle CH_3}{|}}{\underset{\underset{\displaystyle H}{|}}{C}}-CH_3 + H_2O$$

The order in which the amino acids occur in a protein is called the *primary structure*. The first primary structure of a protein was determined by British chemist Frederick Sanger in 1953. Since then the primary structures of several hundred proteins have been determined (Box 23.1). The primary structure of

---

Box 23.1

The complete sequence of amino acid units in several hundred protein molecules is now known. An example is *insulin*, the hormone produced in the pancreas that is essential to the metabolism of carbohydrates in the body and the lack of which leads to diabetes. Diabetes is a serious and widespread disease, and diabetics have to be injected with insulin daily to regulate their condition. Insulin contains 51 amino acids arranged in two chains and cross-linked in two places by the disulfide bond of cysteine. One chain contains 21 amino acid units and the other has 30 amino acid units (Figure 23.8). The polymer has a molecular mass of 5733. The amino acid sequence was determined by British biochemist Frederick Sanger, who received a Nobel Prize for the work in 1958, and insulin was syn-thesized in the laboratory for the first time in 1963.

The principal source of insulin for medical use since the 1920s, when it was discovered by Canadian scientist Frederick Banting (1891–1941) at the University of Toronto, had been the pancreases of cattle, from which it had to be extracted. Today, however, synthetic insulin is available. Very recently, it has become possible to make insulin by the methods of *genetic engineering*. In this technique, genes are made—either synthetically or by modifying existing genes—and then added to simple organisms, such as bacteria, that can use the information from the genetic code to "manufacture" proteins. In 1978 chemically synthesized genes were added to the bacteria *E. coli*, and human insulin resulted.

beef insulin is shown in Figure 23.8. It has 51 amino acids that are linked into two polypeptide chains held together by S—S links. Some proteins consist of only a single polypeptide chain.

A very long protein chain can have an enormous number of different conformations. But a protein does not have a floppy structure that is continually changing; rather, it adopts a very definite conformation, called the *secondary structure*. This structure is very largely a consequence of hydrogen bonding. If two protein chains are laid parallel to each other but running in opposite directions, a very large number of N—H- - -O=C hydrogen bonds can be formed between them. In fact, numerous parallel chains can be bonded together in this way to form a sheet (Figure 23.9). These sheets are then stacked one upon another to form a three-dimensional structure. Silk has a structure of this type, with the protein chains running in the direction of the silk fibers. This structure gives silk its characteristic mechanical properties. Since stretching a silk fiber means pulling against the covalent bonds in the protein chains, silk fibers are not very elastic. However, they bend easily, because when a fiber bends, the protein sheets slide over each other. This sliding occurs readily because the sheets are held together only by relatively weak intermolecular forces. A stack of sheets of paper cannot be stretched but can be rather easily bent because the sheets can slide over each other.

Wool and hair have a different type of structure, in which hydrogen bonds are formed between CO and NH groups in a single chain. These hydrogen bonds cause the chain to coil up into a spiral, called an α-helix, in which there are 3.6 amino acids for each turn of the helix (Figure 23.10). Three α-helices are then twisted together as in a rope to give a structure called a *protofibril*. The protofibrils are then packed together in parallel bundles in a wool or hair fiber. Wool fibers are stretchy because only the weak hydrogen bonds must be broken in order to allow the helix to increase in length, much as a spring stretches when it is pulled. Then when the tension is released, the hydrogen bonds re-form, pulling the fiber back into its original helical shape.

Other proteins, such as myoglobin and hemoglobin in the blood and those proteins that behave as enzymes, have a still more complex structure. They con-

NH₂	NH₂
Gly	Phe
He	Val
Val	Asn
Glu	Gln
Gln	His
Cys	Leu
Cys — S — S — Cys	
Ala	Gly
Ser	Ser
Val	His
Cys	Leu
Ser	Val
Leu	Glu
Tyr	Ala
Gln	Leu
Leu	Tyr
Glu	Leu
Asn	Val
Tyr	Cys
Cys — S — S	Gly
Asn	Glu
	Arg
	Gly
	Phe
	Phe
	Tyr
	Thr
	Pro
	Lys
	Ala

Figure 23.8 Primary Structure of Beef Insulin. There are two polypeptide chains held together by disulfide linkages between cysteine residues.

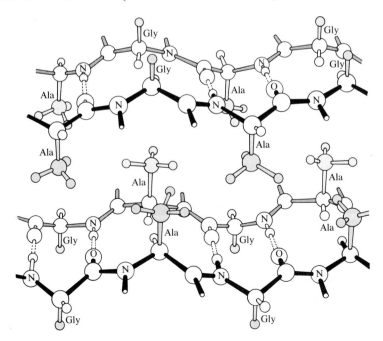

Figure 23.9 Sheets of Hydrogen-Bonded Protein Chains as Found in Silk.

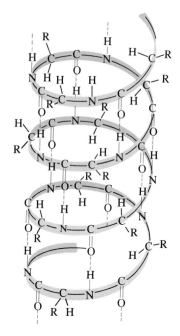

Figure 23.10 α-Helix Structure of Protein. The helix conformation of the molecule is called the secondary structure.

sist of helical chains, but the chain is folded up in a complex way to give a much more compact structure. These proteins are called *globular proteins*. Parts of the chain have a helical structure, but at the bends in a chain the regular helical structure is disrupted (Figure 23.11). The protein is held in this folded shape mainly by interactions between the side groups R, some of which are ionic or polar, and also by disulfide bridges —S—S— between cysteine residues. The form of the folded chain is called the *tertiary* structure of the protein. Different globular proteins have very different and very characteristic folded shapes.

Enzymes are globular proteins that catalyze chemical reactions in living systems. More than a thousand enzymes have been identified, and the amino acid sequence in over a hundred of them has been determined. Two remarkable properties of enzymes are their extraordinary specificity—each enzyme catalyzes only one reaction or one group of closely related reactions—and their amazing efficiency—they may speed up reactions by factors of up to 10^{20}. Enzymes provide a very effective method for the control of reactions in living systems. Biochemical reactions do not take place in the body at an appreciable rate in the absence of the appropriate enzyme catalyst. Thus the presence or absence of the enzyme at a particular site enables reactions to be switched on or off.

The mechanism by which an enzyme acts has been the subject of intense research. A simple and popular theory is the lock-and-key theory, according to which the reactant molecule or molecules, called the *substrate*, fit into a pocket or cavity in the complex folded structure of the enzyme (Figure 23.12). The pocket in any particular enzyme has a very specific shape that can only accommodate one particular molecule or group of similar molecules. When the reactant molecules are held in the correct orientation for the reaction to occur, the reaction is much more rapid than it would be if the correct orientation were achieved only in a small percentage of random collisions. In other words, the enzyme behaves as a catalyst.

The lack of even one of the many enzymes in the body can cause serious disease. For example, some mentally retarded children suffer from the disease known as PKU (phenylketonuria). They lack the enzyme that converts phenylalanine to tyrosine:

$$\text{Phenylalanine} \xrightarrow{\text{Enzyme}} \text{Tyrosine}$$

Phenylalanine   Tyrosine

Figure 23.11 Tertiary Structure of the Globular Protein Myoglobin.

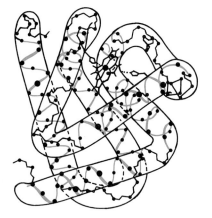

Instead, phenylalanine is converted to phenylpyruvic acid, and high levels of phenylpyruvic acid can lead to mental retardation in some way that is not well understood:

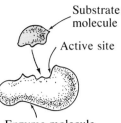

Figure 23.12 Lock-and-Key Theory of Enzyme Action. The substrate fits the active site of an enzyme as a key fits a lock. The bond-breaking and bond-making processes that transform a substrate (reactant) to products occur while the substrate is bound to the active site of the enzyme.

Phenylpyruvic acid is easily detected in the urine, and infants are now routinely tested for its presence. If it is found in large amounts, the child can be given a special diet low in phenylalanine, and so the mental retardation can be prevented.

Nucleic Acids

One of the most amazing aspects of life is the ability of living organisms to transmit their characteristics to their progeny. The observation that organisms reproduce their own species is widely known and self-evident. Yet the mystery of how this happens is one of the most challenging problems facing science today. Although we have a fair understanding of this process, our knowledge of many of its details is far from complete. Differences in species appear to result from differences in proteins. For example, the hemoglobin of cats differs slightly in amino acid sequence from the hemoglobin of mice and of humans.

How does an organism synthesize correctly its own characteristic proteins? We know that the information that is necessary to guide the correct synthesis of proteins is stored in the molecule **deoxyribonucleic acid**—usually abbreviated as **DNA**—which is found in the nuclei of all cells. Deoxyribonucleic acid is an example of a **nucleic acid**. Nucleic acids are polymers of nucleotides. A **nucleotide** is made by the condensation of a molecule of phosphoric acid, a molecule of a sugar—deoxyribose—and a molecule of a nitrogen compound called a *nitrogen base*. Four different nitrogen bases are found in DNA; they are adenine, guanine, cytosine, and thymine (see Figure 23.13). Thus there are four different nucleotides in DNA. These nucleotides are then condensed into a polynucleotide, which therefore consists of a sugar-phosphate backbone with a nitrogen base attached to each sugar (see Figure 23.14). Numerous different sequences of

Figure 23.13 The Four Nitrogen Bases Found in the Polynucleotide DNA.

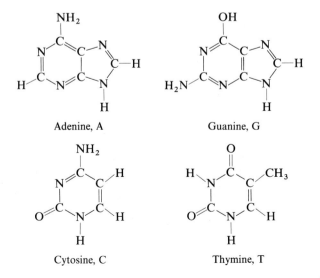

Adenine, A

Guanine, G

Cytosine, C

Thymine, T

Figure 23.14 Structure of a Nucleic Acid.

nitrogen bases are possible. In a typical small DNA polymer strand containing 1500 nucleotide units, there are 4^{1500}, or 10^{900}, different possible sequences.

The key to understanding how DNA works lies in its three-dimensional structure. In 1953 James Watson, an American biologist, and Francis Crick, an English biophysicist, working together in Cambridge, England, proposed that DNA consists of two polynucleotides in the form of a **double helix** (see Figure 23.15). They based this suggestion on two observations: (1) X ray diffrac-

Figure 23.15 Double-Helix Structure of DNA. (a) This model is a three-dimensional, space-filling model. (b) This is a simplified representation that shows the hydrogen bonds between the two helices.

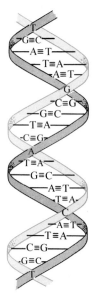

tion patterns of DNA indicated that it has a helical structure; (2) chemical analysis of DNA had shown that although the amounts of the different nitrogen bases in DNA vary from species to species, the amount of adenine is always equal to the amount of thymine and the amount of guanine is always equal to the amount of cytosine. Thus it seemed that these bases must somehow always be paired together.

Working with ball-and-stick molecular models (see Figure 23.16), they found that adenine, A, and thymine, T, were just the right size and shape to be linked together by two hydrogen bonds and that guanine, G, and cytosine, C, can join together at exactly the same distance (1.1 nm) by forming three hydrogen bonds (see Figure 23.17). Held together by hydrogen bonds in this way, the two strands have a constant separation of 1.1 nm. No other nitrogen base pairs have the right size and shape to form hydrogen bonds with the same separation between the chains. The two chains are said to be *complementary* to each other because the sequence of nitrogen bases in one chain completely determines the sequence in the other chain. Thus if the hydrogen bonds break, the helix can uncoil, and each of the separate chains can then act as a template for the formation of a new complementary chain. Thus we have an explanation of how DNA can replicate itself.

The sequence of bases in the DNA molecule is a code for the synthesis of all the proteins characteristic of a given organism. Each sequence of three bases along a DNA chain—for example, CGT—is the code for the synthesis of a particular amino acid—in this case alanine. Thus the particular sequence of bases in a segment of DNA corresponds to the sequence of amino acids in a particular protein. Each of the proteins needed by an organism is coded into the DNA in this way. The triplet AAA is the code for phenylalanine, so the sequence CGTAAA corresponds to an ala-phe segment of a polypeptide, and so on. Each segment of the DNA molecule that codes the synthesis of one particular protein is called a *gene*. The exact mechanism by which the nitrogen base sequence along a DNA strand is used to build up a protein is complicated but reasonably well understood. However, consideration of this mechanism would take us too deeply into molecular biology and must be left for other courses.

Mutations in DNA can cause errors in the biosynthesis of proteins with serious consequences for the organism. A well-known example is sickle-cell anemia. The red blood cells of persons afflicted with this disease have an unusual shape. This unusual shape results simply from the replacement of one glutamic

Figure 23.16 Ball-and-Stick Model of DNA Constructed by Watson and Crick.
Watson (*left*) and Crick are examining the model of DNA that they built at Cambridge in 1953. James Watson was born in 1928 in Chicago, and he graduated from the University of Chicago at the age of nineteen. He obtained his Ph.D. in zoology from the University of Indiana in 1950, when he was only twenty-two. In 1951 he went to the University of Cambridge, where he worked with Francis Crick on the structure of DNA. Francis Crick was born in Northampton, England, in 1916. He graduated in physics, later joining a group of physicists and other scientists at Cambridge who had turned their attention to solving the challenging problems of the new science of molecular biology. Watson and Crick's proposal of the double-helix structure of DNA is regarded as one of the most significant break-throughs in science in recent times. For their work they were awarded the Nobel Prize in medicine and physiology in 1962. Watson later wrote a popular and highly successful account of the work leading up to their discovery in his book *The Double Helix*.

Figure 23.17 The Hydrogen Bonds that Determine Shape of the DNA Double Helix.

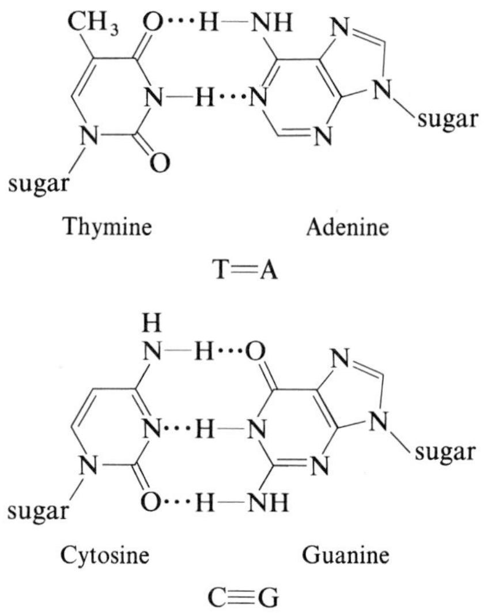

acid monomer in the hemoglobin protein by a valine monomer. The unusual
shape of these cells causes them to clump together and to block capillaries, thus
preventing oxygen-carrying cells from reaching the tissues. Sickle-cell anemia
is one example of many molecular diseases that are now being identified.

IMPORTANT TERMS

An **addition polymer** is a polymer that contains all of the atoms of the monomer units from which it is composed.

An **α-amino acid** is an amino acid in which the —CO_2H and —NH_2 groups are attached to the same carbon atom.

An **amino acid** is an organic molecule containing both a carboxy group and an amino group.

The **amino acid sequence** is the sequence of amino acids found in proteins (polypeptides).

A **carbohydrate** is a compound of carbon, hydrogen, and oxygen in which the ratio of H to O is the same as in water.

Cellulose is a polysaccharide polymer formed by the condensation of β-glucose units.

A **condensation polymer** is a polymer formed from monomers by a condensation reaction in which molecules of small molecular mass, such as H_2O, are eliminated.

A **cross-linked polymer** is a polymer in which polymer chains are joined by cross-links.

DNA, deoxyribonucleic acid, is a polynucleotide consisting of two polynucleotide strands forming a double helix.

The **double helix** is the structure of DNA in which two helical monomer strands (polynucleotides) are held together by hydrogen bonding.

An **elastomer** is a polymer with elastic properties.

A **macromolecule** is a polymer.

A **monomer** is the basic repeating unit of a polymer.

A **nucleic acid** is a polymer of nucleotides.

A **nucleotide** is the basic building block of nucleic acids. It is formed by the condensation of an organic nitrogen base, a sugar, and a phosphoric acid molecule.

Plastic is the common name for a synthetic polymer.

A **polyamide** is a condensation polymer formed from an amino acid or from a dicarboxylic acid and a diamine.

A **polyester** is a condensation polymer formed from a dicarboxylic acid and a diol.

A **polymer** is a molecule consisting of a number of repeating units called monomers.

A **protein** is naturally occurring polyamide (polypeptide).

PROBLEMS

Synthetic Polymers

1. Define each of the following terms and give one example of each:
   (a) An addition polymer
   (b) A condensation polymer
   (c) An α-amino acid
   (d) A sugar
   (e) A polypeptide

2. From what monomers are the following formed?
   (a) Teflon   (b) Saran   (c) PVC
   (d) Nylon   (e) Dacron

3. What are the structures of the polymers formed from the following monomers?
   (a) Propene
   (b) $CH_2=CH-CH=CH_2$
   (c) 1,6-hexane diamine and 1,4-hexanedioic acid
   (d) 1,2-Ethanediol and terephthalic acid

4. The trans isomer of poly-2-methyl-1,3-butadiene is a hard natural material known as gutta-percha. What is the structure of this polymer?

5. Orlon has the polymeric chain structure:

$$-CH_2-CH-CH_2-CH-CH_2-CH-$$
$$\quad\quad\quad | \quad\quad\quad\quad | \quad\quad\quad\quad |$$
$$\quad\quad\quad CN \quad\quad\quad CN \quad\quad\quad CN$$

What is the monomer from which this can be made?

6. Draw the structure of the repeating unit in a condensation polymer made by the elimination of methanol, CH_3OH, from

H   H
| |
H—C—C—H   and $CH_3O-\overset{O}{\overset{||}{C}}-CH_2-\overset{O}{\overset{||}{C}}-OCH_3$
| |
OH OH

1,2-Ethanediol   Dimethylpropanedioate
                                  (dimethylmalonate)

7. Nylon stockings dissolve rapidly in concentrated hydrochloric or sulfuric acids. Suggest a probable explanation.

8. What structural feature must be present in an organic molecule for it to undergo addition polymerization?

Biopolymers

9. What molecular units combine to give a nucleotide?

10. Write the structure of each of the following α-amino acids. Which are dicarboxylic acids?
    (a) Alanine   (b) Glycine
    (c) Aspargine   (d) Glutamic acid

11. What is the difference between α-glucose and β-glucose?

12. What special characteristics of adenine, thymine, guanine, and cytosine make them important as constituents of DNA?

13. In what way does starch differ in structure from cellulose?

14. How many glucose residues are there in a starch molecule of molecular mass 2.57×10^5 u?

15. If six different amino acids formed all the possible tripeptides, how many would there be?

16. What is meant by the primary, secondary, and tertiary structures of a protein?

17. Explain how an enzyme increases the rate of a reaction.

Madame Curie

CHAPTER 24

NUCLEAR AND RADIOCHEMISTRY

We discussed the structure of the nuclei of atoms very early in this book because the charge of the nucleus—the atomic number Z—determines the number of surrounding electrons and therefore the chemical properties of the atom. Subsequently, however, we have paid little attention to the nucleus because nuclei remain unchanged in chemical reactions. Any process in which a nucleus undergoes a change is called a **nuclear reaction**, and such changes are usually considered to form part of physics rather than chemistry. Nevertheless, nuclear reactions have some very important applications in chemistry as well as being sources of enormous amounts of energy. Life as we know it exists because of the nuclear reactions that provide the energy for the sun. In this chapter we discuss nuclear reactions and some of their important applications. Nuclear reactions can be used in destructive ways, but electric power generation, medical diagnosis, and a variety of industrial applications have enriched our lives.

Some nuclei are unstable and spontaneously change into other nuclei by emitting electrons, positrons, or other particles, such as helium nuclei (α particles). Such nuclei are said to be radioactive. Many stable nuclei can also be transformed into unstable radioactive nuclei by bombarding them with other particles such as α particles and neutrons.

24.1 RADIOACTIVITY

All elements with atomic numbers of 83 or less, with the exception of technetium ($Z = 43$) and promethium ($Z = 61$), have one or more stable isotopes, but the nuclei of all the isotopes of the elements with atomic numbers greater than 83 (bismuth) are unstable. About 260 stable isotopes are found in nature, but over 1100 unstable radioactive nuclei are known. Of these, about 65 are found in nature and the rest have been made by nuclear reactions. The spontaneous disintegration of a nucleus is called **radioactivity**, and an unstable nucleus that decomposes spontaneously is said to be **radioactive**. For example, uranium-238 nuclei emit helium nuclei, which for historical reasons are called α particles. An uranium nucleus is thus transformed into a thorium nucleus:

$$^{238}_{92}\text{U} \longrightarrow {}^{234}_{90}\text{Th} + {}^{4}_{2}\text{He}$$

In such nuclear reactions the total number of nucleons (protons and neutrons), and therefore their total charge, remains constant. The sum of the mass numbers and the sum of the atomic numbers of the products must equal the mass number and the atomic number of the disintegrating nucleus. Thus whenever a helium nucleus is emitted, the mass number of the disintegrating nucleus decreases by 4 and the charge (atomic number) decreases by 2.

Not all radioactive nuclei emit α particles; some emit electrons. In the early studies of radioactivity, before these particles were identified as electrons, they were called β particles and the emission of electrons is still often called β emission. Thus thorium-234, which is produced by the radioactive disintegration of uranium-238, emits electrons and is transformed into protactinium-234:

$$^{234}_{90}\text{Th} \longrightarrow {}^{234}_{91}\text{Pa} + {}^{0}_{-1}\text{e}$$

In equations for nuclear reactions the electron is written as ${}^{0}_{-1}\text{e}$. The superscript refers to the very small mass of the electron relative to that of a proton or a neutron, and the subscript refers to the charge on the electron. Hydrogen has a radioactive isotope called tritium, ${}^{3}_{1}\text{H}$, which emits an electron to give an isotope of helium:

$$^{3}_{1}\text{H} \longrightarrow {}^{3}_{2}\text{He} + {}^{0}_{-1}\text{e}$$

When an electron is emitted, the mass number does not change, but the atomic number increases by 1. There are no electrons in nuclei. The emission of an electron results from the transformation of a neutron into a proton, which can be represented by the equation

$$^{1}_{0}\text{n} \longrightarrow {}^{1}_{1}\text{H} + {}^{0}_{-1}\text{e}$$

Other nuclei emit positrons. A **positron** is a particle with the same mass as an electron but with a positive charge. The symbol for a positron in an equation for a nuclear reaction is ${}^{0}_{1}\text{e}$. Two examples of positron emission are

$$^{39}_{19}\text{K} \longrightarrow {}^{39}_{18}\text{Ar} + {}^{0}_{1}\text{e}$$
$$^{11}_{6}\text{C} \longrightarrow {}^{11}_{5}\text{B} + {}^{0}_{1}\text{e}$$

There are no positrons in nuclei. The emission of a positron can be considered to result from the conversion of a proton to a neutron:

$$^{1}_{1}\text{H} \longrightarrow {}^{1}_{0}\text{n} + {}^{0}_{1}\text{e}$$

Positrons exist only for a very short time. Within about 10^{-9} s they combine with an electron and are converted to high-energy radiation called γ-radiation that has a shorter wavelength than X rays.

Some nuclei undergo radioactive transformation without emitting any particles. Instead, one of the inner electrons of an atom, for example a 1s electron, enters the nucleus. This is called an **electron capture process**. Rubidium-81 is transformed in this way to krypton 81:

$$^{81}_{37}\text{Rb} + {}^{0}_{-1}\text{e} \longrightarrow {}^{81}_{36}\text{Kr}$$

Positron emission and electron capture lead to the same result; they both decrease the nuclear charge by 1 and leave the mass number unchanged.

The new nucleus formed in a radioactive-decay process may be in an excited state—in other words, its constituent neutrons and protons do not have their most stable arrangement—and it decays to the ground state by emitting high-energy γ radiation. For example, in the decay of ${}^{234}_{92}\text{U}$ to ${}^{230}_{90}\text{Th}$ by α- particle emission, 77% of the ${}^{230}_{90}\text{Th}$ nuclei are produced in the ground state by the emission of an α- particle with an energy of 6.69×10^{-13} J, but 23% are produced in an excited state by the emission of an α-particle with an energy of only 6.61×10^{-13} J. This excited state thorium atom, ^{230}Th, then emits γ radiation of energy 8×10^{-15} J and returns to the ground state (Figure 24.1).

The radioactive isotope cobalt-60 is used in the treatment of cancer. It emits an electron to give $^{60}_{28}$Ni in an excited state which decays to the ground state by two successive γ-ray emissions

$$^{60}_{27}\text{Co} \longrightarrow {}^{60}_{28}\text{Ni}\left(\begin{array}{c}\text{excited}\\\text{state 1}\end{array}\right) + {}_{-1}^{0}\text{e}$$

$$^{60}_{28}\text{Ni}\left(\begin{array}{c}\text{excited}\\\text{state 1}\end{array}\right) \longrightarrow {}^{60}_{28}\text{Ni}\left(\begin{array}{c}\text{excited}\\\text{state 2}\end{array}\right) + \gamma(8.7 \times 10^{-13}\text{ J})$$

$$^{60}_{28}\text{Ni}\left(\begin{array}{c}\text{excited}\\\text{state 2}\end{array}\right) \longrightarrow {}^{60}_{28}\text{Ni}\left(\begin{array}{c}\text{ground}\\\text{state}\end{array}\right) + \gamma(2.1 \times 10^{-13}\text{ J})$$

Table 24.1 summarizes the various types of radioactive decay.

Example 24.1 Complete the following nuclear equations:

(a) $^{32}_{15}\text{P} \longrightarrow ? + {}_{-1}^{0}\text{e}$

(b) $^{43}_{19}\text{K} \longrightarrow {}^{43}_{20}\text{Ca} + ?$

(c) $^{210}_{84}\text{Po} \longrightarrow {}^{206}_{82}\text{Pb} + ?$

(d) $^{17}_{9}\text{F} \longrightarrow ? + {}_{1}^{0}\text{e}$

Solution

(a) The emission of an electron does not change the mass number but it increases Z by 1, so the nucleus that is formed is $^{32}_{16}\text{S}$.

(b) The mass number remains unchanged but the atomic number increases by 1, so the particle emitted must be an electron, $_{-1}^{0}\text{e}$.

(c) The mass number decreases by 4 and the atomic number decreases by 2, so the particle emitted must be a helium nucleus, $^{4}_{2}\text{He}$ (α particle).

(d) The emission of a positron means that the mass number does not change but the charge decreases by 1. Therefore the nucleus that is formed is $^{17}_{8}\text{O}$.

A variety of methods can be used to detect the emissions from radioactive materials. Photographic film and plates are sensitive not only to light but also to the energetic α particles, electrons, and γ rays emitted by radioactive substances. The greater the exposure to radioactive emissions, the greater is the

Figure 24.1 Production of Gamma Rays. In the decay of $^{234}_{92}$U by α particle emission, α particles of two different energies are produced. Emission of α particles of energy 6.61×10^{-13} J leads to the formation of $^{230}_{90}$Th in an excited state. This excited state decays to the ground state by emission of a γ ray photon of energy 8×10^{-15} J.

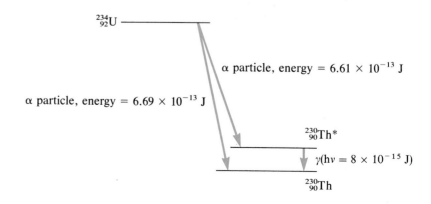

Table 24.1 Radioactive-Decay Processes

PARTICLE EMITTED		CHANGE IN		EXAMPLE
		MASS NUMBER	ATOMIC NUMBER	
Helium nucleus (α particle)	^4_2He	Decreases by 4	Decreases by 2	$^{238}_{92}\text{U} \longrightarrow {}^{234}_{90}\text{Th} + {}^4_2\text{He}$
Electron (β particle)	$^0_{-1}\text{e}$	No change	Increases by 1	$^{14}_{6}\text{C} \longrightarrow {}^{14}_{7}\text{N} + {}^0_{-1}\text{e}$
Positron	$^0_{+1}\text{e}$	No change	Decreases by 1	$^{64}_{29}\text{Cu} \longrightarrow {}^{64}_{28}\text{Ni} + {}^0_{+1}\text{e}$
Electron capture		No change	Decreases by 1	$^{195}_{79}\text{Au} + {}^0_{-1}\text{e} \longrightarrow {}^{195}_{78}\text{Pt}$
γ ray photon	γ	No change	No change	$^{87}_{38}\text{Sr*} \longrightarrow {}^{87}_{38}\text{Sr} + \gamma$

* The asterisk denotes a nucleus in an excited state.

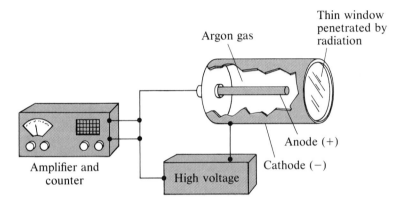

Figure 24.2 Schematic Representation of Geiger Counter.

blackening of the negative. Persons who work with radioactive substances carry a film badge to record the extent of their exposure to radiation.

An important instrument for detecting and measuring radioactivity is the **Geiger counter**. A Geiger counter (Figure 24.2) consists of a metal tube filled with a gas such as argon. One end of the tube has a thin window that allows fast-moving electrons, α particles and γ rays, to pass through. In the center of the tube is a wire electrode. A potential difference of about 1000 V is maintained between the metal tube and the central wire. If a high-energy electron, α particle, or γ ray photon enters the tube through the window, it knocks electrons out of the atoms in its path. The electrons and ions that are thus formed are accelerated to high speeds by the voltage between the central wire and the tube, and they in turn ionize other atoms, producing more electrons and ions, which in turn ionize more atoms, and so on. Thus a single, high-energy α particle or γ ray photon entering the tube produces an avalanche of ions and electrons. These give a brief pulse of electric current in the external circuit. This current pulse is amplified and recorded or is made audible as a click. The number of clicks in a given time is a direct measure of the number of particles entering the Geiger counter. Because a Geiger counter can record individual particles, it is an extremely sensitive device.

24.2 NUCLEAR STABILITY

What makes some nuclei stable and others unstable? If we plot the number of neutrons against the number of protons, that is, the atomic number, for all the known stable nuclei, we find that they all fall in a narrow band called the *band of stability* (see Figure 24.3). For stable nuclei with low atomic numbers the

Marie Curie (1867–1934) and her daughter, Irene (1897–1956). Marie Curie shared the 1903 Nobel Prize in physics with her husband, Pierre, and Antoine Becquerel for their research on radioactivity. In 1911 she won the Nobel Prize in chemistry for the discovery of the elements polonium and radium. Irene Curie shared the 1935 Nobel Prize in chemistry with her husband, Frederic Joliot.

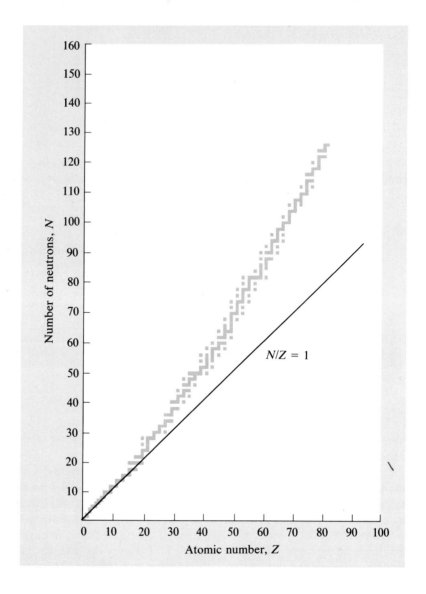

number of protons is equal to the number of neutrons, but as Z increases, the number of neutrons exceeds the number of protons and the ratio of the number of neutrons to the number of protons reaches a value of approximately 1.5 for the heaviest stable nuclei. The distances between the particles in nuclei are very small—less than 10^{-15} m. At these distances the electrostatic repulsion between the protons is extremely large. Nuclei are only stable because there are very strong attractive forces between the nucleons. We call these attractive forces **nuclear forces**. Their nature has been under investigation for many years. These forces are believed to be caused by the exchange of particles called π mesons. As the number of protons increases, the electrostatic repulsion between them increases, and a higher proportion of neutrons is needed to provide additional attractive forces to hold the nucleus together. But after bismuth, $Z = 83$, no increase in the number of neutrons is sufficient to hold the nucleus together, and all the nuclei with more protons than the bismuth nucleus has are unstable.

Nuclei that lie outside the band of stability are unstable and decompose to give a nucleus with a more stable neutron-to-proton ratio. For example, a nucleus that lies above the band of stability must either gain protons or lose neutrons to become more stable. Thus we can understand why a nucleus such as ^{14}C, which lies above the band of stability, decays by the emission of an electron, because this process converts a neutron into a proton:

$$\,^1_0n \longrightarrow \,^1_1H + \,^0_{-1}e$$

For ^{14}C we have

$$\,^{14}_6C \longrightarrow \,^{14}_7N + \,^0_{-1}e$$

Nuclei located below the band of stability must increase their neutron/proton ratio to achieve stability. They can do so either by positron emission or by electron capture. Positron emission converts a proton into a neutron. An example is the decay of ^{11}C:

$$\,^{11}_6C \longrightarrow \,^{11}_5B + \,^0_1e$$

Electron capture similarly converts a proton to a neutron. For example,

$$\,^7_4Be + \,^0_{-1}e \longrightarrow \,^7_3Li$$

The electron is captured from an inner shell of the atom, which leaves the resulting atom in an excited state; but an electron quickly drops from the valence shell to fill the vacant orbital, and a corresponding amount of energy in the form of X rays is emitted. An example is

$$\,^{40}_{19}K + \,^0_{-1}e \longrightarrow \,^{40}_{18}Ar$$

Nuclei with atomic numbers greater than 83 cannot achieve stability by electron or positron emission or electron capture. As a result, they often decay by emission of a helium nucleus (α particle), which removes two protons and two neutrons simultaneously. We have already mentioned this type of decay for uranium. Another possibility for heavy elements is fission, which we discuss later.

We have seen that ^{238}U decays to ^{234}Th, which is also radioactive and decays to ^{234}Pa. This isotope is unstable and decays to ^{234}U, which is also radioactive, and so on. Such a series of radioactive disintegrations continues until a stable (nonradioactive) isotope of an element is formed. Such a series of nuclear reactions is called a **radioactive-decay series**. For example, ^{238}U decays in a series of 14 nuclear reactions that eventually lead to the stable isotope ^{206}Pb.

24.3 RADIOACTIVE-DECAY RATES

Half-Life

We cannot predict when an individual radioactive nucleus will decay, but each nucleus of the same kind in a sample has the same probability of decaying in a certain interval of time as any other. As a result, the rate of decay of a sample of radioactive material, that is, the number of disintegrations per unit time, is directly proportional to the number of radioactive nuclei, N, that the sample contains. The rate of decay of a given sample of radioactive material is therefore not constant, but it decreases with time. A characteristic property of each radioactive isotope is the time needed for half a given sample to disintegrate. This is called the **half-life**, $t_{1/2}$, of the particular radioactive isotope (Table 24.2). Some radioactive isotopes have very long half-lives. Others have very short half-lives. For example, $t_{1/2}$ for uranium-238 is 4.5×10^9 years, that of radon-222 is 3.8

Table 24.2 Half-Lives of Some Radioisotopes

ISOTOPE	HALF-LIFE	MODE OF DECAY
$^{214}_{84}\text{Po}$	164 s	α
$^{25}_{11}\text{Na}$	1.0 min	β
$^{131}_{53}\text{I}$	8.0 days	β
$^{222}_{86}\text{Rn}$	3.8 days	α
$^{32}_{15}\text{P}$	14.3 days	β
$^{60}_{27}\text{Co}$	5.3 years	β
$^{90}_{38}\text{Sr}$	28.8 years	β
$^{14}_{6}\text{C}$	5730 years	β
$^{230}_{94}\text{Pu}$	2.4×10^4 years	α
$^{40}_{19}\text{K}$	1.3×10^9 years	α
$^{238}_{92}\text{U}$	4.5×10^9 years	α
$^{232}_{90}\text{Th}$	1.4×10^{10} years	α

days, while that of sodium-25 is only 1.0 min. Thus if we have a sample of 1 mg of radon-222—a noble gas—in a container, $\frac{1}{2}$ mg will remain after 3.8 days. After another 3.8 days one-half of the $\frac{1}{2}$ mg—or $\frac{1}{4}$ mg—will be left. Thus after 7.6 days only $\frac{1}{4}$ mg radon will remain, and after 3(3.8) days = 11.4 days, $\frac{1}{8}$ mg will remain, and so on. Figure 24.4 shows a plot of the decay of $^{222}_{86}\text{Rn}$. Since the number of disintegrations in a given time is proportional to the number of radioactive nuclei present, radioactive decay is an example of a first-order rate process (Chapter 18).

If there are N_0 nuclei at $t = 0$, and N at time t, then $\Delta N = N_0 - N$ nuclei have disintegrated in a time interval $\Delta t = t - t_0$. So

$$\text{Rate of decay} = \frac{\Delta N}{\Delta t} = kN$$

where k is a first-order rate constant. By comparison with the analogous equation for a first-order rate process that we discussed in Chapter 18, the number of nuclei N remaining after a time t is given by the equation

$$\ln \frac{N_0}{N} = kt$$

Figure 24.4 Plot of Amount of Radon 222 Against Time for Sample with Initial Mass of 1.0 mg. Radon 222 decays to polonium-218 by α particle emission with a half-life of 3.8 days.

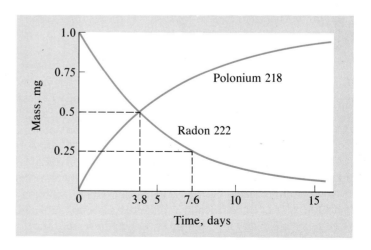

where N_0 is the number of nuclei at $t = 0$. The time $t_{1/2}$ needed for half of the radioactive nuclei in any sample to decay—that is, the half-life—can be found by writing $N = \frac{1}{2}N_0$ at $t = t_{1/2}$, which gives

$$\ln \frac{N_0}{\frac{1}{2}N_0} = kt_{1/2} \quad \text{or} \quad kt_{1/2} = \ln 2$$

Thus

$$t_{1/2} = \frac{0.693}{k}$$

Example 24.2 The half-life of $^{222}_{86}\text{Rn}$ is 3.8 days. How much $^{222}_{86}\text{Rn}$ will remain after 8.5 days in a sample initially containing 45 μg of $^{222}_{86}\text{Rn}$?

Solution We determine the rate constant k from the expression

$$k = \frac{0.693}{t_{1/2}} = \frac{0.693}{3.8} = 0.18 \text{ day}^{-1}$$

Since for any given radioactive nucleus the number of atoms is proportional to the mass, we have

$$\ln \frac{N_0}{N} = \ln \frac{\text{Initial mass Rn}}{\text{Mass Rn after 8 days}} = kt = 0.18 \text{ day}^{-1} \times 8.5 \text{ days} = 1.5$$

$$\ln \frac{45 \ \mu\text{g Rn}}{\text{Mass Rn after 8 days}} = 1.5, \quad \text{hence} \quad \frac{45 \ \mu\text{g Rn}}{\text{Mass Rn after 8 days}} = 4.5$$

Therefore, mass Rn after 8 days $= \dfrac{45 \ \mu\text{g Rn}}{4.5} = 10 \ \mu\text{g Rn}$

Example 24.3 The radioisotope $^{131}_{53}\text{I}$ is used in studies and tests on the thyroid gland. A sample that originally contained 1.00 mg of $^{131}_{53}\text{I}$ contained 0.32 mg of $^{131}_{53}\text{I}$ after 13.3 days. What is the half-life of $^{131}_{53}\text{I}$?

Solution

$$\ln \frac{N_0}{N} = kt = \frac{0.693t}{t_{1/2}}$$

$$\ln \frac{1.00 \text{ mg } ^{131}_{53}\text{I}}{0.32 \text{ mg } ^{131}_{53}\text{I}} = \frac{0.693 \times 13.3 \text{ days}}{t_{1/2}}$$

$$t_{1/2} = \frac{0.693 \times 13.3}{1.14} = 8.08 \text{ days}$$

You may wonder how we can possibly measure the half-life of a nucleus that disintegrates as slowly as uranium-238 ($t_{1/2} = 4.51 \times 10^9$ years) because the change in the mass of a sample of uranium over any measurable period of time, even as long as 10 years, is completely negligible. But because an instrument such as a Geiger counter is so sensitive that it can record individual particles, we can measure the half-life of uranium by measuring the rate at which it emits α particles. Let us assume that we have a 1.0-mg sample of uranium. This sample contains

$$\left(\frac{1.0 \times 10^{-3} \text{ g}}{238 \text{ g mol}^{-1}} \right) (6.022 \times 10^{23} \text{ nuclei mol}^{-1}) = 2.5 \times 10^{18} \text{ nuclei}$$

The first-order rate constant for the disintegration of uranium is

$$k = \frac{0.693}{t_{1/2}} = \frac{0.693}{4.51 \times 10^9 \text{ years}}$$

$$= \frac{0.693}{4.51 \times 10^9 \times 365 \times 24 \times 3600 \text{ s}}$$

$$= 4.9 \times 10^{-18} \text{ s}^{-1}$$

$$\text{Rate of disintegration} = \frac{\Delta N}{\Delta t} = kN$$

$$= 4.9 \times 10^{-18} \text{ s}^{-1} \times 2.5 \times 10^{18} \text{ nuclei}$$

$$= 12 \text{ nuclei s}^{-1}$$

Thus we see that the disintegration of uranium proceeds at a rate which can be relatively easily measured with a Geiger counter.

Radiochemical Dating

An important application of radioactive decay is the dating of rocks, fossils, and ancient objects. Naturally occurring radioactive uranium-238 decays in a series of steps, the first of which gives ^{234}Th by α particle emission. This process has a half-life of 4.51×10^9 years and is by far the slowest of the steps that finally lead to the stable isotope ^{206}Pb. If we measure the amount of ^{206}Pb, for example with a mass spectrometer, we can find how much uranium-238 was present initially and, therefore, knowing the rate constant, how long it has taken this much ^{206}Pb to form. If we assume that the lead begins to accumulate once the rock has formed and solidified, the time it has taken the lead to form is equal to the age of the solid rock.

For rocks that contain potassium the reaction

$$^{40}_{19}\text{K} \xrightarrow{\text{Electron capture}} {}^{40}_{18}\text{Ar} \qquad t_{1/2} = 1.3 \times 10^9 \text{ years}$$

can be used for dating.

Example 24.4 A sample of rock is found to contain 13.2 μg of uranium-238 and 3.42 μg of lead-206. If the half-life of $^{238}_{92}\text{U}$ is 4.51×10^9 years, what is the age of the rock?

Solution We first need to find how many grams of uranium-238 have decayed. The number of micrograms of $^{238}_{92}\text{U}$ transformed into $^{206}_{82}\text{Pb}$ is

$$(3.42 \ \mu\text{g} \ ^{206}\text{Pb})\left(\frac{238 \text{ g U mol}^{-1}}{206 \text{ g Pb mol}^{-1}}\right) = 3.95 \ \mu\text{g} \ ^{238}\text{U}$$

The initial amount of uranium-238 in the ore was 13.2 μg plus the 3.95 μg that has been transformed into lead, that is,

$$13.2 \ \mu\text{g} + 3.95 \ \mu\text{g} = 17.2 \ \mu\text{g}$$

Then we have
$$\ln \frac{N_0}{N} = \frac{0.693}{t_{1/2}} t$$

$$\ln \frac{17.2 \text{ g}}{13.2 \text{ g}} = \frac{0.693}{4.51 \times 10^9 \text{ years}} t$$

$$t = \frac{0.26 \times 4.51 \times 10^9 \text{ years}}{0.693} = 1.7 \times 10^9 \text{ years}$$

Thus we conclude that the uranium mineral crystallized from the molten magma 1.7 billion years ago. The oldest rocks on earth analyzed by this method solidified about 3.6 billion years ago. Clearly the earth is older than this, and other information indicates that the age of the earth is about 4.5 billion years.

Another important radiodating method is that based on carbon-14, which decays by the reaction

$$^{14}_{6}\text{C} \longrightarrow {}^{14}_{7}\text{N} + {}^{0}_{-1}\text{e}$$

The half-life of carbon-14 is 5730 years. Carbon-14 is produced in the atmosphere by bombardment of nitrogen with cosmic rays that are high-energy particles such as protons and neutrons that originate from the sun and other parts of the universe. Carbon-14 is formed by the reaction

$$^{14}_{7}\text{N} + {}^{1}_{0}\text{n} \longrightarrow {}^{14}_{6}\text{C} + {}^{1}_{1}\text{H}$$

Because ^{14}C is produced in the upper atmosphere at a constant rate and because it decays at a constant rate, there is a small but constant concentration of $^{14}\text{CO}_2$ in the atmosphere. Plants use atmospheric CO_2 to make carbohydrates in photosynthesis, so there is the same small concentration of carbon-14 in all living plants and animals. But when a plant or animal dies, it no longer incorporates carbon-14, so the amount of carbon-14 decreases. Thus by measuring the amount of carbon-14 left in any formerly living material, we can determine the time that has passed since it died.

Because the amount of carbon-14 in living matter is extremely small, it would be very difficult to measure the amount present with any accuracy. It is much simpler and much more accurate to measure the rate at which the carbon-14 disintegrates by counting the number of disintegrations per second per gram of material. In the atmosphere and in all living organisms, there are 15.3 disintegrations of carbon-14 per minute per gram of carbon. When the organism dies, this rate of disintegration decreases with a half-life of 5730 years. The rate of disintegration, R, at time t is proportional to N, the number of radioactive nuclei at time t. Thus we can transform the equation

$$t = \frac{t_{1/2}}{0.693} \ln \frac{N_0}{N}$$

to the form

$$t = \frac{t_{1/2}}{0.693} \ln \frac{R_0}{R}$$

For carbon, $R_0 = 15.3$ disintegrations per minute per gram and $t_{1/2} = 5730$ years. So we have

$$t = \frac{5730}{0.693} \ln \frac{15.3}{R} \text{ years} = 8.27 \times 10^3 \ln \frac{15.3}{R} \text{ years}$$

The time that has elapsed since any living material died can be determined by using this equation, as illustrated in the following example.

Example 24.5 A sample of charcoal from one of the earliest Polynesian settlements in Hawaii had a disintegration rate of 13.6 disintegrations per minute per gram. What is the age of the charcoal?

Solution The time since the tree providing the charcoal was cut can be obtained from the equation

$$t = 8.27 \times 10^3 \ln \frac{15.3}{R} \text{ years}$$

In this case $R = 13.6$ disintegrations per minute per gram, so that

$$t = 8.27 \times 10^3 \ln \frac{15.3}{13.6} \text{ years} = 974 \text{ years}$$

This result suggests that the Polynesians first arrived in Hawaii around the year 1010 A.D.

24.4 ARTIFICIAL RADIOISOTOPES

About one-half of the uranium-238 present when the earth's crust solidified still remains, the rest having been transformed into lead-206. But most radioactive nuclei decay much more rapidly, so even if they were present when the earth was formed, they would have completely disappeared many years ago and we would know nothing about them. However, a large number of radioactive nuclei have been made by nuclear reactions in recent times.

Synthesis of Radioisotopes

Rutherford was the first to carry out a nuclear reaction in the laboratory. He observed that when he bombarded nitrogen with a beam of α particles, he produced $^{17}_8O$ and protons by the reaction

$$^{14}_7N + {}^4_2He \longrightarrow {}^{17}_8O + {}^1_1H$$

This nuclear reaction gives the stable isotope $^{17}_8O$, but many radioactive nuclei can be produced by similar reactions. For example, $^{27}_{13}Al$ can be transformed to $^{30}_{15}P$ by bombardment with α particles:

$$^{27}_{13}Al + {}^4_2He \longrightarrow {}^{30}_{15}P + {}^1_0n$$

In these experiments α particles were obtained by the disintegration of radioactive elements such as uranium. But α particles obtained in this way do not have enough energy to react with many heavy nuclei. The high charge of a heavy nucleus repels a positively charged α particle so that it cannot enter the nucleus. If α particles are to react with heavy nuclei, their energy must be increased by accelerating them to a high velocity in a machine called a particle accelerator, such as a cyclotron. For example, uranium-238 is converted to plutonium-239 when it is bombarded with high-energy α particles:

$$^{238}_{92}U + {}^4_2He \longrightarrow {}^{239}_{94}Pu + 3{}^1_0n$$

An alternative method for producing new nuclei is to use neutrons as the bombarding particles. Because they are neutral, neutrons are not repelled by nuclei and they do not therefore need to be moving at a high speed to enter a nucleus. Neutrons can be obtained from neutron-producing reactions such as the conversion of $^{27}_{13}Al$ to $^{30}_{15}P$ or the conversion of $^{238}_{92}U$ to $^{239}_{94}Pu$. Nuclear reactors, in which many neutron–producing reactions occur, are often used for bombarding nuclei with neutrons to produce new nuclei. For example, cobalt-60, used in radiation therapy for cancer, is produced by the reaction:

$$^{59}_{27}Co + {}^1_0n \longrightarrow {}^{60}_{27}Co$$

In discussing the halogens in Chapters 5 and 20, we only very briefly mentioned the last element in the halogen family, astatine, because it has been much less studied than the other halogens. Astatine is radioactive and can only be produced in very small amounts. It was first made by the reaction

$$^{209}_{83}Bi + {}^4_2He \longrightarrow {}^{211}_{85}At + 2{}^1_0n$$

The isotope $^{211}_{85}At$ has a half-life of only 7.5 h, and even the most stable isotope, $^{210}_{85}At$, has a half-life of only 8.5 h, so large amounts cannot be accumulated.

Another radioactive element that is not found in nature is technetium, Tc, which is in the second series of transition metals below manganese. Technetium can be made by neutron bombardment of molybdenum:

$$^{98}_{42}Mo + ^{1}_{0}n \longrightarrow ^{99}_{42}Mo$$

$$^{99}_{42}Mo \longrightarrow ^{99}_{43}Tc + ^{0}_{-1}e$$

Uses of Radioisotopes

Many artificially produced radioisotopes have important applications in medicine, agriculture, oil exploration, and many other fields. Because the radiation emitted from a radioisotope is easily detected, the movement of an element in the body is easily followed. The radiation emitted by a radioisotope can give an image of any organ in which the radioisotope concentrates. Thus sodium-24 is used to follow blood circulation, technetium-99 is used for brain and liver scans, and iodine-123 is used for thyroid imaging.

In chemistry one of the important applications of radioisotopes is the study of reaction mechanisms. We saw in Chapter 18 that most reactions take place in a series of steps. Radioisotopes can often be used to help work out the exact sequence of steps in a complex reaction mechanism. If a very small percentage of the atoms of an element are exchanged for a radioactive isotope, the element is said to be *labeled*. Since a radioactive isotope has the same chemical properties as the stable isotopes of an element, the radioactivity of the labeled element can be used to detect the movement of the element through a complex series of reactions.

Melvin Calvin (born 1911, St. Paul, Minnesota) and his fellow researchers kept growing plants in an atmosphere of carbon dioxide labeled with radioactive carbon-14 for just a few seconds. They then extracted and separated as many of the compounds in the plant as possible. Those compounds that were found to contain carbon-14 could then be supposed to be involved in the early stages of photosynthesis. By much painstaking work from 1949 to 1957 they were able to work out a mechanism for the very complex reaction by which plants convert carbon dioxide and water to sugars and carbohydrates in the presence of light.

Until the development in the 1940s of accelerators for producing high-speed particles, the last element in the periodic table was element 92, uranium. Since that time the periodic table has been extended up to at least element 106. These transuranium elements have been produced by nuclear reactions. We have already mentioned the formation of plutonium by α particle bombardment of uranium. Some other examples are given in Table 24.3. Some of these transuranium elements can be produced in small but, nevertheless, commercially

Table 24.3 Synthesis of Some Transuranium Elements.

ATOMIC NUMBER	NAME	SYMBOL	REACTION
93	Neptunium	Np	$^{238}_{92}U + ^{1}_{0}n \longrightarrow ^{239}_{93}Np + ^{0}_{-1}e$
94	Plutonium	Pu	$^{238}_{92}U + ^{2}_{1}H \longrightarrow ^{238}_{93}Np + 2^{1}_{0}n$
			$^{238}_{93}Np \longrightarrow ^{238}_{94}Pu + ^{0}_{-1}e$
95	Americium	Am	$^{239}_{94}Pu + ^{1}_{0}n \longrightarrow ^{240}_{95}Am + ^{0}_{-1}e$
96	Curium	Cm	$^{239}_{94}Pu + ^{4}_{2}He \longrightarrow ^{242}_{96}Cm + ^{1}_{0}n$

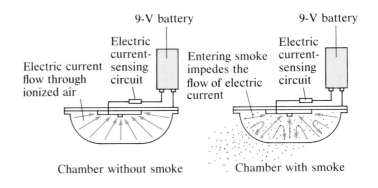

Figure 24.5 Home Smoke Detector. The ionization chamber contains a small quantity of americium-241, which decays by α particle emission with a half-life of 432 years. The α particles ionize the air in the ionization chamber. The ions are accelerated by a potential provided by a battery so that an electric current flows across the ionization chamber. This current can be detected in an external circuit. When smoke enters the chamber, the smoke particles impede the movement of ions, and the current is reduced. This decrease in current is detected electronically.

useful quantities. Americium-241 is used in one type of home smoke detector (Figure 24.5). The α particles emitted by americium-241, like all α particles produced by radioactive nuclei, have a very low penetrating power, so they do not escape from the detector and thus do not constitute a health hazard.

Effects of Radiation

The increasing use of radioisotopes in medicine and industry and the increasing number of nuclear reactors have led to increased concern over the biological effects of radiation. Electrons, α particles, and γ rays emitted by radioactive nuclei have energies far in excess of that needed to break chemical bonds. When these high-energy particles and gamma rays pass through matter, they break up molecules, forming free radicals and ions. For example, water may be split into hydrogen atoms and hydroxyl radicals:

$$H_2O \longrightarrow H\cdot + OH\cdot$$

Many of these free radicals and ions are very reactive. In a biological system free radicals may disrupt the normal operation of the cell and may even kill the cell. Indeed, γ rays are used routinely to destroy cancerous cells.

The damage caused by a radiation source outside the body depends on the penetrating ability of the radiation. Gamma rays are particularly dangerous because, like X rays, they penetrate human tissue very effectively. In contrast, α particles are stopped by the skin, and electrons do not penetrate far below the skin. However, if a radiation source enters the body, it can be particularly dangerous. For example, α emitters are generally the nuclei of heavy elements that tend to concentrate in the bones where they may cause considerable damage to the bone and surrounding tissues. However, we should be aware that we are subject to radiation at all times. Radioactive minerals, such as those of uranium, have been present on the earth since its formation. Cosmic radiation causes nuclear reactions in the atmosphere which produce radioactive nuclei such as those of carbon and potassium. Since these are essential elements in living organisms, humans and all other organisms are continually subjected to radiation from the disintegration of these nuclei. The normal low level of radiation to which we are all exposed is called background radiation. This background radiation causes mutations in living cells and thus has actually contributed to the mechanism by which the great variety of living organisms has been produced. It is only exposure to radiation that is many times in excess of normal background radiation that is dangerous.

We saw in Chapter 2 that the mass of a helium-4 atom is not equal to the mass of its constituents. The mass of two protons and two neutrons is 4.031 88 u, but the mass of one $^4_2$He nucleus is only 4.001 50 u. The difference of 0.030 38 u arises because a very large amount of energy is released when protons and neutrons combine to form a nucleus. The amount of energy is so large that it has a significant mass equivalent, given by the Einstein equation

$$E = mc^2$$

where E is the energy, m is the mass, and c is the velocity of light. For a change in mass, Δm, the energy change is $\Delta E = c^2 \, \Delta m$. Thus the energy change for the formation of a helium nucleus from two protons and two neutrons is

$$E = (0.030\ 38\ \text{u})(2.998 \times 10^8\ \text{m s}^{-1})^2 \left(\frac{1.000\ \text{g}}{6.022 \times 10^{23}\ \text{u}}\right)\left(\frac{1\ \text{kg}}{1000\ \text{g}}\right)$$

$$= 4.534 \times 10^{-12}\ \text{kg m}^2\,\text{s}^{-2} = 4.534 \times 10^{-12}\ \text{J}$$

The formation of 1 mol of helium-4 nuclei would produce an enormous amount of energy:

$$\left(\frac{6.022 \times 10^{23}\ \text{nuclei}}{1\ \text{mol nuclei}}\right)(4.534 \times 10^{-12}\ \text{J nuclei}^{-1}) = 2.730 \times 10^{12}\ \text{J mol}^{-1}$$

Conversely, this amount of energy would be needed to break 1 mol of helium nuclei into their constituent protons and neutrons. The energy required to decompose a nucleus into protons and neutrons is called the **binding energy** of the nucleus, it is the energy needed to overcome the very strong nuclear forces that hold the nucleus together.

For comparison of the binding energies of different nuclei, it is convenient to quote the value of the *binding energy per nucleon*:

$$\text{Binding energy per nucleon} = \frac{\text{Binding energy}}{\text{Number of nucleons}} = \frac{\text{Binding energy}}{\text{Mass number}}$$

Example 24.6 The mass of $^{56}_{26}$Fe is 55.920 66 u. What are the binding energy and the binding energy per nucleon?

Solution The masses of the proton and the neutron are given in Table 2.1. They are mass $^1_1$H = 1.007 28 and mass $^1_0$n = 1.008 66 u. The mass difference between $^{56}_{26}$Fe and its constituent particles (26 protons and 30 neutrons) is

$$\Delta m = (26 \times 1.007\ 28\ \text{u}) + (30 \times 1.008\ 66\ \text{u}) - 55.920\ 66\ \text{u}$$

$$= 56.4491\ \text{u} - 55.920\ 66\ \text{u} = 0.5284\ \text{u}$$

Hence

$$\Delta E = c^2\,\Delta m = (2.998 \times 10^8\ \text{m s}^{-1})^2(0.5284\ \text{u})\left(\frac{1.000\ \text{g}}{6.022 \times 10^{23}\ \text{u}}\right)\left(\frac{1\ \text{kg}}{1000\ \text{g}}\right)$$

$$= 7.884 \times 10^{-11}\ \text{kg m}^2\,\text{s}^{-2} = 7.887 \times 10^{-11}\ \text{J}$$

This is the binding energy of $^{56}_{26}$Fe. There are 56 nucleons (26 protons and 30 neutrons) in $^{56}_{26}$Fe, so the binding energy per nucleon is

$$\frac{7.884 \times 10^{-11}\ \text{J}}{56\ \text{nucleons}} = 1.408 \times 10^{-12}\ \text{J nucleon}^{-1}$$

Figure 24.6 Plot of Binding Energy per Nucleon Against Mass Number. The most stable nucleus is $^{56}_{29}$Fe. If lighter nuclei are combined in a fusion reaction, energy is released. If nuclei heavier than $^{56}_{29}$Fe are split in a fission reaction, energy is released.

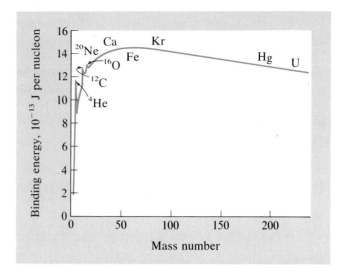

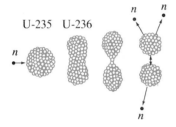

U-235   U-236

Figure 24.7 Neutron-Induced Fission of Uranium-235. Absorption of a neutron produces the $^{236}_{92}$U nucleus in an excited state. This deforms and splits in two in the same way as a vibrating drop of liquid might split in two. Simultaneously, several neutrons are emitted.

Figure 24.8 Branching Chain Reaction Produced When Uranium-235 Undergoes Fission. If each fission produces two neutrons, these neutrons cause the fission of two uranium-235 nuclei, each of which produces two neutrons. The four neutrons thus produced cause the fission of four more nuclei, thus producing eight neutrons, which can cause the fission of eight nuclei, and so on. The number of fission reactions increases very rapidly, resulting in an explosive release of energy.

Figure 24.6 shows a plot of the binding energy per nucleon against the mass number. The binding energy per nucleon increases up to ^{56}Fe and then decreases slowly. Thus ^{56}Fe is the most stable nucleus. This curve shows that energy is released when heavy nuclei are broken up into lighter nuclei. This process is called **fission**. A fission reaction was first discovered during the search for transuranium elements in the 1930s. If a uranium-235 nucleus absorbs a neutron, it breaks into two nuclei and at the same time several neutrons are ejected (Figure 24.7). Uranium-235 nuclei may split into several different pairs of nuclei. Two of these reactions are

$$^{235}_{92}U + {}^{1}_{0}n \quad \begin{array}{c} \nearrow \quad {}^{137}_{52}Te + {}^{97}_{40}Zr + 2{}^{1}_{0}n \\ \searrow \quad {}^{142}_{56}Ba + {}^{91}_{36}Kr + 3{}^{1}_{0}n \end{array}$$

The average number of neutrons produced in the fission of an uranium-235 nucleus is 2.4. Suppose that two of these neutrons each cause the fission of a uranium nucleus. These two fissions produce four neutrons, which can cause the fission of four more nuclei, which produces eight more neutrons, which can cause the fission of eight nuclei, and so on (Figure 24.8). The number of fissions

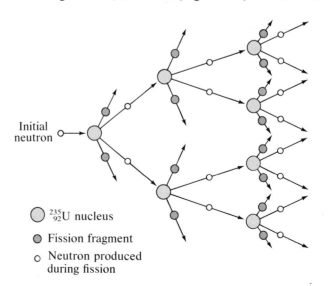

Initial neutron

◯ $^{235}_{92}$U nucleus

⬤ Fission fragment

○ Neutron produced during fission

rapidly escalates, and this process is called a **branching chain reaction**. The associated liberation of a large amount of energy causes an enormous explosion. But for a chain reaction to occur, the sample of uranium must have a minimum mass. Otherwise, neutrons escape from the surface before they have a chance to cause fission. The minimum mass of fissionable material needed to ensure that each fission process causes one further fission is called the **critical mass**. The critical mass depends on the shape and purity of a particular sample. If the critical mass is exceeded, a branching chain reaction occurs. One of the ways that a critical mass can be obtained is to very rapidly combine two pieces of fissionable material, each of which has less than the critical mass, for example, by firing one piece into the other by using a conventional explosive. This method is used in one type of nuclear (atom) bomb (Figure 24.9).

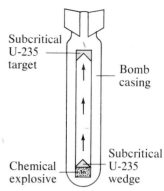

Figure 24.9 Diagram of Atomic Bomb. One piece of uranium-235 is fired into the other piece by means of a conventional explosive. When the two masses collide, the critical mass is exceeded, and a neutron chain reaction ensues with a corresponding release of energy as an explosion.

Nuclear Reactors

In a nuclear reactor the fission reaction is carried out in a controlled manner so that just one of the 2.4 neutrons emitted in a fission is captured by another fissionable nucleus. Achieving this precise control is one of the main problems of nuclear reactors. Some neutrons, of course, are lost through the reactor surface and by absorption into the material of which the reactor is constructed. But most important is the fact that natural uranium contains only 0.7% uranium-235. The much more abundant uranium-238 readily captures the fast neutrons liberated in fission but does not usually undergo fission itself. The neutrons absorbed by ^{238}U are therefore wasted. However, ^{238}U has little ability to capture *slow* neutrons, which readily cause fission in ^{235}U. Thus if the neutrons are slowed down they will not be absorbed by ^{238}U and therefore wasted, and more will be available to cause fission in ^{235}U.

A substance called a *moderator* is used to slow down the neutrons. When a fast neutron collides with a nucleus, some of its kinetic energy is imparted to the nucleus. When two particles collide, the greatest transfer of energy occurs if the particles have equal masses. Thus hydrogen nuclei would be the most efficient moderator for neutrons because they have very nearly the same mass. Indeed, ordinary ("light") water is used as a moderator in one type of nuclear reactor. The water is also the coolant in such reactors. However, protons also combine with neutrons to form deuterons

$$^1_1H + ^1_0n \longrightarrow ^2_1H$$

so that many neutrons are lost in this way. Thus in such a reactor it is necessary to use uranium enriched to about 3% in ^{235}U. Uranium is enriched by converting it to uranium hexafluoride, UF_6, and using gaseous diffusion to separate $^{235}UF_6$ from $^{238}UF_6$ (see Chapter 3). This process requires a large plant and much energy and is therefore expensive.

Another important type of nuclear reactor uses "heavy" water, D_2O, as a moderator because deuterium nuclei have a much smaller tendency to combine with neutrons than with protons. In this type of reactor natural uranium can be used as a fuel. The production of heavy water is an energy-consuming and expensive process, but once the reactor has been constructed only small amounts are needed to make up for small losses.

The fuel of a nuclear reactor consists of uranium oxide, UO_2, pellets packed into a zirconium alloy tube to form what is called a fuel rod. A number of these fuel rods are then packed together to form the reactor core. Control rods made of substances such as cadmium or boron, which strongly absorb neutrons, are inserted into spaces between the fuel rods. The control rods are moved in and out of the reactor core to adjust the rate of the fission reaction. The energy of the fission process appears as heat. The heat is transferred outside the reactor by means of a suitable fluid—often water. In some types of reactors it may be another liquid such as sodium, or a sodium-potassium alloy, or a gas such as carbon dioxide or helium. This heat is used to boil water to form steam which is used to drive a turbine as in a conventional power plant (Figure 24.10).

Figure 24.10 A Nuclear Reactor.
The heat produced in the core is transferred to water, which flows in a closed loop through a vessel containing water that is heated to its boiling point, forming steam. The steam is used to drive a turbine, which produces electricity. The steam from the turbine is condensed to water in a cooling tower, and the resulting water is pumped back into the steam generator.

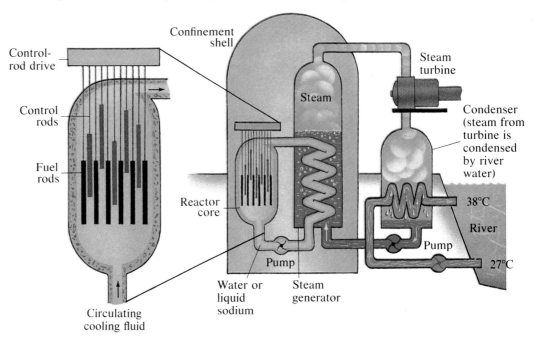

A nuclear power plant

Nuclear Fusion

We can see from Figure 24.6 that the conversion of very light nuclei to heavier nuclei also results in a release of large amounts of energy, even larger, in fact, than the energy obtained from fission reactions. Such reactions are called **fusion reactions** because small nuclei fuse to form a large nucleus.

The basic energy-producing process in stars—and hence the source of nearly all the energy in the universe—is the fusion of hydrogen nuclei into helium nuclei. In the sun the following reactions predominate:

$$\ce{^1_1H + ^1_1H \longrightarrow ^2_1H + ^0_1e}$$
$$\ce{^1_1H + ^2_1H \longrightarrow ^3_2He}$$
$$\ce{^3_2He + ^3_2He \longrightarrow ^4_2He + 2^1_1H}$$

A very high temperature of approximately 10^7 K is needed for fusion reactions to occur in such quantity that a substantial amount of energy is produced. The high temperature is needed so that the nuclei collide with sufficient kinetic energy to overcome their mutual electrostatic repulsion.

In order to produce a fusion reaction, we must somehow reproduce on earth the extremely high temperature of the sun. In the hydrogen or thermonuclear bomb this high temperature is produced by a fission reaction. In one such device a fission bomb is surrounded by lithium deuteride, $\text{Li}^2\text{H(LiD)}$. Neutrons from the fission reaction are captured by lithium nuclei:

$$\ce{^6_3Li + ^1_0n \longrightarrow ^4_2He + ^3_1H}$$

And at the very high temperature produced by the fission reaction, tritium, ^3_1H, undergoes fusion with deuterium:

$$\ce{^3_1H + ^2_1H \longrightarrow ^4_2He + ^1_0n}$$

But controlling nuclear fusion so that it can be used for the production of usable power is a problem that has challenged scientists and engineers for over twenty years. A fundamental difficulty is that the temperature is too high for the fusion reaction to be confined to any solid container. One approach involves using powerful laser beams to heat tiny pellets containing deuterium to a very high temperature to produce what are in effect miniature hydrogen bomb explosions. A succession of such explosions could furnish a steady supply of energy. Another method that has been extensively studied is to contain an extremely hot mass of gaseous atoms stripped of their electrons—called a plasma—by means of a very strong magnetic field. Intensive research continues, but practical fusion reactors are probably still far in the future.

IMPORTANT TERMS

The **binding energy** of a nucleus is the energy needed to overcome the very strong nuclear forces that hold the nucleus together.

A **branching chain reaction** is a fission reaction caused by a neutron, which produces more than one neutron, which can in turn cause the fission of more nuclei, and so on.

The **critical mass** is the minimum mass of fissionable material for which a branching chain reaction can occur.

Electron capture is a process in which a nucleus absorbs an inner electron and is thus transformed to a new nucleus.

A **fission reaction** is a nuclear reaction in which a heavy nucleus splits into two lighter nuclei with the production of a large amount of energy.

A **fusion reaction** is a nuclear reaction in which two light nuclei combine to form a heavier nucleus with the production of a large amount of energy.

A **Geiger counter** is an instrument for detecting the α particles and γ rays emitted by radioactive nuclei.

The **half-life** of a radioactive nucleus is the time in which one-half of the nuclei in a given sample have disintegrated.

Nuclear forces are the strong attractive forces that hold the protons and the neutrons together in a nucleus.

A nuclear reaction is a process in which nuclei are changed into other nuclei.

A positron is a particle with the same mass as an electron but with a positive charge.

A radioactive-decay series is the series of nuclear reactions by which an unstable radioactive nucleus decays to a stable nucleus.

Radioactivity is the spontaneous decomposition of unstable nuclei.

PROBLEMS

Composition of Nuclei

1. How many protons and how many neutrons are there in each of the following nuclei?

$$^6_3\text{Li} \qquad ^{13}_6\text{C} \qquad ^{94}_{40}\text{Zr} \qquad ^{137}_{56}\text{Ba}$$

2. How many protons and how many neutrons are there in each of the following nuclei?

$$^{22}\text{Ne} \qquad ^{88}\text{Sr} \qquad ^{92}\text{Sr} \qquad ^{180}\text{W} \qquad ^{242}\text{Cm}$$

Radioactivity

3. An $^{80}_{35}\text{Br}$ nucleus can decay by electron emission, by positron emission, or by electron capture. What nucleus is formed in each case?

4. The nucleus $^{233}_{90}\text{Th}$ undergoes two successive electron emissions to become an isotope of uranium. Which isotope of uranium is formed?

5. Complete the following equations for radioactive decay:

(a) $^{32}_{15}\text{P} \longrightarrow ? + ^{\;\;0}_{-1}\text{e}$

(b) $^{15}_{8}\text{O} \longrightarrow ^{15}_{7}\text{N} + ?$

(c) $^{52}_{26}\text{Fe} \longrightarrow ^{52}_{25}\text{Mn} + ?$

(d) $^{218}_{87}\text{Fr} \longrightarrow ? + ^4_2\text{He}$

(e) $^{50}_{26}\text{Fe} \longrightarrow ? + ^{\;\;0}_{-1}\text{e}$

(f) $^{122}_{53}\text{I} \longrightarrow ^{122}_{54}\text{Xe} + ?$

Nuclear Reactions

6. Fill in the missing symbols in the following equations for nuclear reactions:

(a) $^{35}_{17}\text{Cl} + ? \longrightarrow ^{32}_{16}\text{S} + ^4_2\text{He}$

(b) $^{15}_{7}\text{N} + ? \longrightarrow ^{12}_{6}\text{C} + ^5_2\text{He}$

(c) $^{12}_{6}\text{C} + ^{12}_{6}\text{C} \longrightarrow ? + ^1_1\text{H}$

(d) $? + ^4_2\text{He} \longrightarrow ^7_4\text{Be} + \gamma$

7. Complete and balance the following equations for nuclear reactions:

(a) $^{32}_{16}\text{S} + ^1_0\text{n} \longrightarrow ^1_1\text{H} + ?$

(b) $^7_4\text{Be} + ^{\;\;0}_{-1}\text{e} \text{ (1s electron)} \longrightarrow ?$

(c) $^{81}_{37}\text{Rb} \longrightarrow ? + ^{\;0}_{+1}\text{e}$

(d) $^1_1\text{H} + ^{11}_{5}\text{B} \longrightarrow 3?$

(e) $^{235}_{92}\text{U} + ^1_0\text{n} \longrightarrow ^{135}_{54}\text{Xe} + ? + 2^1_0\text{n}$

(f) $^{249}_{98}\text{Cf} + ^{18}_{8}\text{O} \longrightarrow ? + 4^1_0\text{n}$

Nuclear-Decay Rates and Half-Lives

8. Germanium-66 decays by positron emission with a half-life of 2.5 h. Write the equation for the nuclear reaction. How much ^{66}Ge remains from a 50.0-mg sample after 12.5 h?

9. A sample of wood containing 50.0 g of carbon from an ancient casket had a disintegration rate of 95 disintegrations per minute. Approximately when did the person buried in that casket die?

10. Cesium-137 is produced by fission reactions in nuclear reactors. It has a half-life of 30.2 years. How long will it take for its activity to decrease to 1.0% of its value after the spent fuel rods have been removed from the reactor?

11. A sample of charcoal from the Lascaux cave in France had a count rate of 2.4 disintegrations per minute per gram. Assuming that the fire that produced the charcoal was lit by the artists of the renowned cave paintings, when did these artists live?

Binding Energy

12. The mass of the $^{35}_{17}\text{Cl}$ nucleus is 34.9689 u. Calculate the binding energy and the binding energy per nucleon.

13. Calculate the total binding energy and the binding energy per nucleon for each of the following nuclei:

(a) $^{20}_{10}\text{Ne}$ (atomic mass 19.9924 u)

(b) $^{64}_{30}\text{Zn}$ (atomic mass 63.929 14)

(c) $^{61}_{28}\text{Ni}$ (atomic mass 60.930 06 u)

(d) $^{226}_{88}\text{Ra}$ (atomic mass 226.0254 u)

Nuclear Fission and Fusion

14. A possible nuclear reaction for a controlled fusion process is the conversion of deuterium ^2_1H to the helium isotope ^3_2He:

$$^2_1\text{H} + ^2_1\text{H} \longrightarrow ^3_2\text{He} + ^1_0\text{n}$$

Using the atomic masses $^2_1\text{H} = 2.014\,10$ u and $^3_2\text{He} = 3.026\,03$ u and the mass of the neutron given in Table 2.1, find the energy released per mole of deuterium. How much energy could be obtained by burning 1 mol of deuterium to form heavy water, $^2_1\text{H}_2\text{O}$?

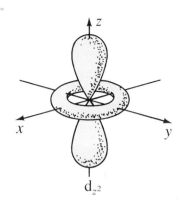

d$_{z^2}$

CHAPTER 25

QUANTUM MECHANICS AND
THE CHEMICAL BOND

In this chapter we take our discussion of the chemical bond a little further than we did in Chapters 6 and 8. We have seen that the behavior of atoms is determined by the electrons that surround the nucleus and, in particular, by the valence shell electrons. We have also seen that the behavior of electrons cannot be described by classical (Newtonian) mechanics. Complete understanding of the behavior of atoms and molecules therefore requires the detailed application of *quantum mechanics*, which is an abstract and rather mathematical subject that is mostly beyond the scope of this book. Fortunately, for an introduction to chemistry we need use only a few relatively simple ideas of quantum mechanics, and we can avoid almost all the associated mathematics. Although the material in this chapter is not essential for understanding the chemistry presented in this book, it will interest those who wish to delve a little more deeply into the fascinating subject of chemical bonding, and it will serve as an introduction to more advanced courses.

In 1925 Austrian physicist Erwin Schrödinger (1887–1961) (Figure 25.1) proposed an equation for describing the behavior of electrons. This equation is often called the *Schrödinger wave equation* because it has the same general form as the equations used to describe the behavior of waves, such as the waves on water and the waves formed by a violin string. The amplitude of the wave described by the Schrödinger equation is given the symbol ψ (psi). When applied to an atomic system, the solutions of the Schrödinger equation give the possible energy levels of the system, and the square of the amplitude, ψ^2, gives the *probability* of finding an electron at any given point in the atom. In other words, it gives the density of the electron charge cloud at a particular point. Recall that the intensity of a light wave is proportional to (amplitude)^2.

Before we consider the results obtained by applying the Schrödinger equation to atomic systems, we will find it helpful, as an analogy, to consider the vibrations of some familiar mechanical systems.

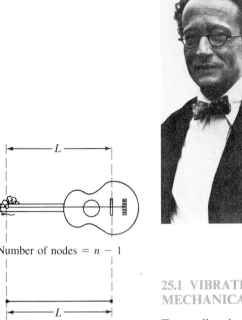

Figure 25.1 Erwin Schrödinger (1887–1961).
Schrödinger was born and educated in Austria, obtaining his Ph.D. at the University of Vienna in 1910. After serving in the Austrian army in World War I, he held positions at several universities. While at the University of Zurich, in 1925, he published his famous wave equation that made such an important contribution to the new field of quantum mechanics, which was also being developed at the same time by several other physicists, including Heisenberg and Pauli. In 1928 he succeeded Planck as professor of theoretical physics at the University of Berlin. Subsequently, his life reflected the troubled times in Europe. In 1933, disgusted with the advent of Nazism, he left Germany and went to Oxford. In 1936 he returned to Austria, but only two years later Hitler annexed Austria to Germany and he was forced to move again. He arrived in Rome with only the possessions that he could carry in a backpack and found asylum in the Vatican. From there he went to the Institute of Advanced Studies in Dublin, where he remained until 1955, when he retired to his native Austria.

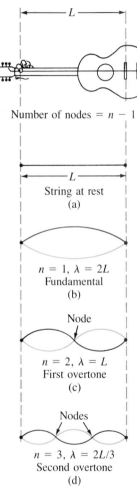

Number of nodes $= n - 1$

L

String at rest
(a)

$n = 1, \lambda = 2L$
Fundamental
(b)

Node

$n = 2, \lambda = L$
First overtone
(c)

Nodes

$n = 3, \lambda = 2L/3$
Second overtone
(d)

Figure 25.2 Vibrations of a String, such as a Guitar or Violin String, Fixed at Both Ends. (a) String of length L at rest. (b) The lowest energy vibration ($n = 1$) is called the fundamental. (c) The next-highest energy vibration ($n = 2$) is called the first overtone. (d) The next-highest energy vibration ($n = 3$) is called the second overtone. In general, there are $n - 1$ points of zero amplitude (nodes) for each value of n, and the wavelength λ is given by $2L/n$.

25.1 VIBRATIONS OF SOME MECHANICAL SYSTEMS

For a vibrating string fixed at both ends, such as a violin or guitar string, only certain modes of vibration are possible. A few of these vibrations are illustrated in Figure 25.2. Only integral or half-integral wavelengths can fit into the length of the string. The points at which the amplitude of the wave is zero are called **nodes**. Excluding the fixed ends of the string, the lowest energy vibration has zero nodes; the next lowest has one node; and so on. The possible vibrations are designated by a number n. For the lowest energy vibration (called the fundamental), $n = 1$, and there are zero nodes. For the vibration of next-highest energy (called the first overtone), $n = 2$, and there is one node. The second overtone, $n = 3$, has two nodes, and so on. In general, a vibration has $n - 1$ nodes.

The vibrations of a violin string are in one dimension. Now consider the vibrations of a circular drumhead, which are in two dimensions. They are illustrated in Figure 25.3. The nodes are not points but lines on the surface. They are of two kinds: straight lines and circles. The total number of nodes is again given by $n - 1$, while the number of straight-line nodes is given by a second number, ℓ, which cannot be greater than the total number of nodes, $n - 1$. Therefore the value of ℓ is limited to $0, 1, 2, \ldots, n - 1$. For example, for $n = 2$ the possible values of ℓ are 0 or 1; thus there are two different types of vibration possible, one having a circular node (for which $\ell = 0$) and the other a linear node ($\ell = 1$). The $\ell = 1$ vibration that has a linear node can occur in two mutually perpendicular directions. There are only two such vibrations because a vibration in any other direction can always be represented as a sum of the two independent vibrations.

Finally, consider a three-dimensional vibrating system such as a pulsating spherical dust cloud or a gaseous star, like the sun. In this case the circular and linear nodal lines of the two-dimensional vibrations are replaced by nodal surfaces, which may be spherical or planar. The lower-energy vibrations are shown in Figure 25.4. The total number of nodes for each vibration is again $n - 1$. The number of planar nodes is given by the number ℓ; as before, ℓ can have all the integral values from 0 to $n - 1$. In this three-dimensional case the $\ell = 1$ vibra-

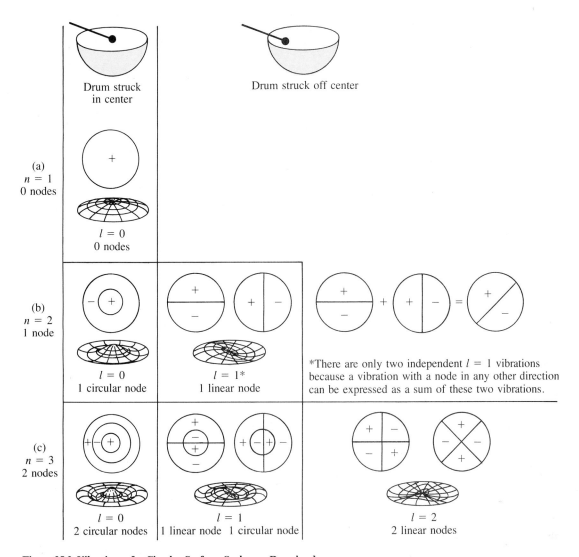

(a)
$n = 1$
0 nodes

Drum struck in center

$l = 0$
0 nodes

(b)
$n = 2$
1 node

$l = 0$
1 circular node

$l = 1*$
1 linear node

Drum struck off center

*There are only two independent $l = 1$ vibrations because a vibration with a node in any other direction can be expressed as a sum of these two vibrations.

(c)
$n = 3$
2 nodes

$l = 0$
2 circular nodes

$l = 1$
1 linear node   1 circular node

$l = 2$
2 linear nodes

Figure 25.3 Vibrations of a Circular Surface, Such as a Drumhead.
(a) There is only one vibration ($n = 1$) with no nodes; this is the lowest energy vibration.
(b) There are three vibrations with one node ($n = 2$); one has a circular node and the other two have linear nodes. There are only two vibrations with a linear node, because a vibration with a node in any other direction can be expressed by the appropriate sum of the two independent vibrations with nodes at right angles. (c) There are five vibrations with two nodes ($n = 3$); one has two circular nodes, two have one circular and one linear node, and two have two linear nodes.

tion, which has one planar node, can have three independent orientations in which the planar node is perpendicular to the x, y, and z axes respectively. Any other orientation that is not perpendicular to one of these axes can be obtained by taking an appropriate combination of these three. The number of possible different independent orientations depends on the form of the vibration. For any given vibration the number of independent orientations is given by the possible values of a third number m, which can have any integral value from $-\ell$ through 0 to $+\ell$. Hence for $\ell = 0$, m must equal 0; there is only one distinguishable orientation for a spherical $\ell = 0$ vibration. For $\ell = 1$ (one planar node) there are three orientations, as we have seen: m may equal -1, 0, or $+1$. For $\ell = 2$ there are five orientations: m may equal -2, -1, 0, $+1$, or $+2$.

25.1 VIBRATIONS OF SOME MECHANICAL SYSTEMS

853

Figure 25.4 Vibrations of a Sphere Such as a Gaseous Star. (a) There is only one vibration ($n = 1$) with no nodes. (b) There are four vibrations with one node—one has a spherical node; the other three have a planar node and differ only in that their nodes are in three mutually perpendicular planes. (c) There are nine vibrations with two nodes ($n = 3$)—one has two spherical nodes, three have one spherical and one planar node, and five have two planar nodes.

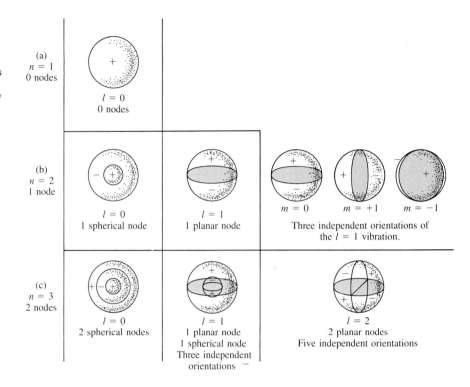

(a)
$n = 1$
0 nodes

$l = 0$
0 nodes

(b)
$n = 2$
1 node

$l = 0$
1 spherical node

$l = 1$
1 planar node

$m = 0$ $m = +1$ $m = -1$

Three independent orientations of the $l = 1$ vibration.

(c)
$n = 3$
2 nodes

$l = 0$
2 spherical nodes

$l = 1$
1 planar node
1 spherical node
Three independent orientations

$l = 2$
2 planar nodes
Five independent orientations

25.2 HYDROGEN ATOMIC ORBITALS

We have seen that the possible vibrational modes—that is, the possible energy states—for a vibrating system can be described by a set of numbers, each of which can take certain integral values. The possible energy states of the hydrogen atom and the corresponding forms of the wave function, ψ, that are obtained by solving the Schrödinger equation for the hydrogen atom are very similar to the energy states of the vibrating sphere. In fact, the set of integers used to describe the various possible states of the hydrogen atom are identical to the set of integers used to describe the vibrations of a pulsating sphere. In the case of the hydrogen atom the numbers n, ℓ, and m are called **quantum numbers**.

The possible values of the quantum numbers for the hydrogen atom are summarized in Table 25.1. The wave function ψ corresponding to any particular set of quantum numbers is called an **orbital**. The number and form of the nodes determine how the wave function varies in the space around the hydrogen nucleus; that is, they determine the shape of the orbital.

An orbital for which $\ell = 0$ is called an s orbital. Orbitals corresponding to $\ell = 1$ are p orbitals; orbitals corresponding to $\ell = 2$ are d orbitals; and orbitals corresponding to $\ell = 3$ are f orbitals. By prefixing the letter s, p, d, or f with the value of the quantum number n, we obtain a useful designation of the orbital. Thus the $n = 2$, $\ell = 1$ orbital is called a 2p orbital.

The forms of these orbitals are analogous to the vibrations of the pulsating sphere. All s orbitals are spherical. The 1s orbital has no nodes; the 2s orbital has one spherical node; and the 3s orbital has two spherical nodes. All p orbitals have one planar node. The diagrams of these orbitals in Figure 25.5 show only the regions where ψ, the amplitude of the wave function, has a positive value and where it has a negative value. We can refine this description by indicating

Table 25.1 Quantum Numbers for Hydrogen Atom

| QUANTUM NUMBERS | | | ORBITAL SYMBOL | NODES $(n-1)$ | | NUMBER OF ORBITALS n^2 |
n	ℓ	m		SPHERICAL $n-\ell-1$	PLANAR ℓ	
1	0	0	1s	0	0	1
2	0	0	2s	1	0	1 ⎱ 4
	1	0, ±1	2p	0	1	3 ⎰
3	0	0	3s	2	0	1 ⎱
	1	0, ±1	3p	1	1	3 ⎬ 9
	2	0, ±1, ±2	3d	0	2	5 ⎰
4	0	0	4s	3	0	1 ⎱
	1	0, ±1	4p	2	1	3 ⎟
	2	0, ±1, ±2	4d	1	2	5 ⎬ 16
	3	0, ±1, ±2, ±3	4f	0	3	7 ⎰

more exactly how ψ varies throughout space. Since s orbitals are spherical, the variation in ψ is exactly the same in any direction from the nucleus. In the 1s orbital ψ has its maximum value at the nucleus and decreases with increasing distance from the nucleus, eventually reaching an infinitesimally small value. In the 2s orbital ψ again has its maximum value at the nucleus; but since there is a spherical node, ψ passes through a value of zero, then decreases to a minimum value, and finally increases again to an infinitesimally small negative value.

The overall shape of the 2p orbital may be approximately described as a sphere sliced in half by the planar node through the nucleus. In one hemisphere ψ has a positive value, and in the other hemisphere it has a negative value;

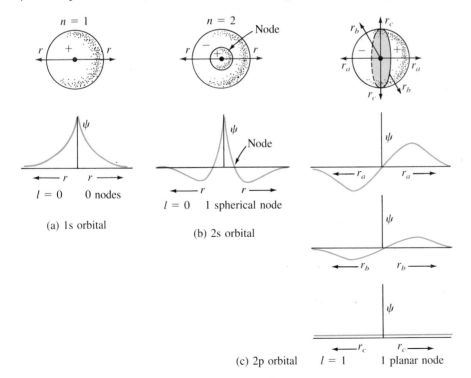

Figure 25.5 Hydrogen Atom Orbitals. (a) The spherical 1s orbital and the variation of ψ in any direction from the nucleus in a 1s orbital. (b) The spherical 2s orbital and the variation of ψ in any direction from the nucleus in a 2s orbital. (c) The 2p orbital and the variation of ψ in the directions r_a, r_b, and r_c in the 2p orbital.

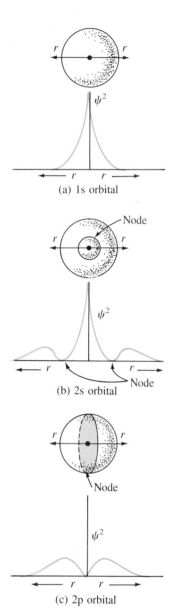

(a) 1s orbital

(b) 2s orbital

(c) 2p orbital

Figure 25.6 Variation of ψ^2 (Probability of Finding an Electron) in 1s, 2s, and 2p Orbitals of the Hydrogen Atom. (a) The 1s orbital; (b) the 2s orbital; (c) the 2p orbital

at the nucleus the value of ψ is zero. In the positive half of the orbital ψ increases to a maximum value and then decreases to an infinitesimally small value. In the negative half of the orbital ψ decreases to a minimum value and then increases to an infinitesimally small negative value. The orbital is not truly spherical, and ψ does not attain the same maximum value in every direction. The largest maximum and minimum values of ψ are along the directions (r_a) perpendicular to the nodal plane. In any other direction the maxima and minima in ψ are smaller, and in any direction along the nodal plane ψ is always zero. There are three independent 2p orbitals with their nodal planes perpendicular to the x, y, and z axes, respectively; they are designated $2p_x$, $2p_y$, and $2p_z$.

Since ψ has no physical significance, but ψ^2 is a measure of the probability of finding the electron at some point in the atom, we are usually more interested in the variation of ψ^2 in space than the variation of ψ. The variation of ψ^2 is quite similar to the variation of ψ except that ψ^2 is everywhere either positive or zero. We can use diagrams similar to those in Figure 25.5 to show the variation of ψ^2, but we can leave out the positive and negative signs because ψ^2 is everywhere positive except at the nodes, where it is zero, as shown in Figure 25.6.

We can conveniently think of the electron as spread out in the form of a negative charge cloud surrounding the nucleus. The density of the charge cloud is not uniform; the value of ψ^2 at any point represents the density of the charge cloud at that point. We can represent the charge cloud by a dot density diagram in which the density of dots in any particular small volume represents the density of the charge cloud, that is, the value of ψ^2 in the small volume (see Figure 25.7). In the 1s orbital the density of the charge cloud is greatest at the nucleus and decreases to an infinitesimally small value with increasing distance from the nucleus. In the 2s orbital the maximum density is again at the nucleus, but it decreases to zero at the node and then increases again and passes through a maximum before decreasing to an infinitesimally small value.

The 2p orbitals have a different shape, as shown in Figure 25.7. The charge density for a $2p_z$ orbital consists of two clouds that have the shape of flattened spheres, one above and one below the nodal plane. The density of these charge clouds is again not uniform, but they are more dense toward the center of each cloud, the point of maximum density being on the z axis above and below the nodal plane. The charge density distributions for the other 2p orbitals have the same shape; but in the $2p_y$ orbital the two maximum density regions are along the y axis, and in the $2p_x$ orbital they are along the x axis.

Although the electron density of an isolated atom in principle extends to infinity, it rapidly becomes insignificant at quite a small distance from the nucleus. It is therefore convenient to choose a surface of constant, but small, electron density to represent the size and shape of the orbital. This boundary surface is usually chosen so that it encloses 90% of the electron density. It is therefore a good representation of the overall size and shape of the electron density distribution. As Figure 25.8 shows, an s orbital has a spherical boundary surface and a p orbital has a boundary surface that can be approximately described as two flattened spheres, one on each side of the nodal plane. The representations of orbitals are often still simplified further by using the two-dimensional shape obtained by taking a planar slice through the orbital that passes through the nucleus. Thus an s orbital is often represented by a circle and a p orbital by two roughly ellipsoidal or squashed circle shapes, one on each side of the nodal plane (see Figure 25.9).

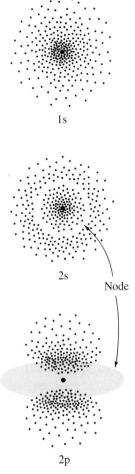

1s

2s

Node

2p

Figure 25.7 Dot Density Diagrams Showing Charge Clouds for 1s, 2s, and 2p Orbitals of the H Atom.

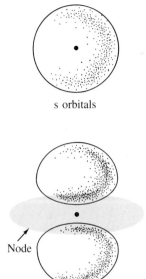

s orbitals

Node

p orbitals

Figure 25.8 Boundary Surfaces for s and p Orbitals. All s and p orbitals (1s, 2s, 3s, . . . , and 2p, 3p, . . .) are often represented by these simple boundary surfaces, and any nodes inside the surface are not shown.

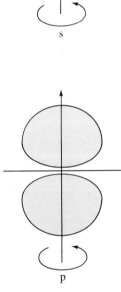

s

p

Figure 25.9 Two-Dimensional Representations of Boundary Surfaces of s and p Orbitals. The three-dimensional boundary surface can be generated from such a diagram by rotating it through 180° about an axis passing through the nucleus, which in the case of a p orbital is also perpendicular to the nodal plane.

The orbitals that we have described represent the probability of finding the electron at any point in the space around the nucleus, or the distribution of electron density around the nucleus, for the single electron of the hydrogen atom in the ground (1s) state and for the various excited states of the hydrogen atom. The electron density is closest to the nucleus in the ground (1s) state. In the excited states there is more electron density further away from the nucleus, and there are more nodes in the distribution. An important difference between s and p orbitals is that the s orbitals are spherical, whereas p orbitals have their electron density more concentrated along one direction in space.

25.3 POLYELECTRONIC ATOMS

The Schrödinger equation cannot be solved exactly for any system containing more than one electron, that is, for any atom other than hydrogen or for any molecule. In fact, the equations governing the behavior of *any* system involving more than two interacting bodies cannot be solved exactly, although with sufficient work very good approximations can be obtained. This applies not only

25.3 POLYELECTRONIC ATOM

857

to atomic systems but also, for example, to the problem of computing the simultaneous gravitational interaction of the earth, the moon, and an orbiting spacecraft. Thus the Schrödinger equation for the helium atom—in which account must be taken of the interaction of the nucleus with two electrons as well as the repulsion between the two electrons—cannot be solved exactly, although very good approximations have been obtained for this relatively simple system. With increasing numbers of electrons, however, the complexity of the problem rapidly escalates. For a copper atom with 29 electrons there are 29 attractions between the electrons and the nucleus and 406 repulsions between pairs of electrons to be considered. For many purposes, however, an adequate approximation can be obtained by assuming that the electrons occupy orbitals that have the same general form as the orbitals of the hydrogen atom. Such approximate orbitals in a polyelectronic atom are often called **hydrogenlike orbitals**.

How, then, are the electrons of a polyatomic atom distributed among these hydrogenlike orbitals? This distribution is determined by the number of electrons that can occupy a single orbital and the energy levels of the hydrogenlike orbitals.

Recall from Chapter 6 that according to the *Pauli exclusion principle*, two electrons of the same spin keep apart in space, whereas two electrons of opposite spin may occupy the same region of space. Within the limits of the approximation that polyelectronic atoms may be described in terms of hydrogen-like orbitals, this means that *two electrons of the opposite spin may occupy the same orbital, but two electrons of the same spin must occupy separate orbitals*.

Whereas the energy of an orbital in the hydrogen atom depends only on the value of the quantum number n, the energy of a hydrogenlike orbital in a polyelectronic atom depends on both n and ℓ. *For a given value of n, the greater the value of ℓ, the higher is the energy*. This reflects differences in the distribution of electron density in orbitals with different ℓ values.

For $n = 1$, ℓ can only have the value 0. Thus the single electron in the hydrogen atom in its ground state is in a 1s orbital. Helium has two electrons with opposite spins in a hydrogenlike 1s orbital. For lithium, with three electrons, two electrons can occupy a hydrogenlike 1s orbital, but the third electron must enter the $n = 2$ shell, which contains 2s, $2p_x$, $2p_y$ and $2p_z$ hydrogenlike orbitals. Of these, the 2s orbital has the lowest ℓ value ($\ell = 0$), so the third electron enters this orbital rather than one of the 2p orbitals ($\ell = 1$). Let us examine the situation more closely to see why this should be so.

To a first approximation, 2s and 2p electrons have their density located very largely outside the density of the 1s electrons, which are close to the nucleus of charge $+Z$. Thus they are attracted by a nuclear charge of $+Z$ but repelled by the -2 charge of the two electrons in the filled 1s orbital; in other words, to a first approximation, they experience the attraction of a *core charge* of $+Z - 2$.

However, this is not quite exact, because 2s and 2p electrons have some probability of being "inside" the 1s shell, so part of their density experiences the attraction of the full nuclear charge of $+Z$. As a result, the average charge acting on all the electron density is somewhat greater than the core charge of $+Z - 2$ and is different for 2s and 2p electrons because of their different electron density distributions. We call the actual charge that an electron in an orbital of a polyelectronic atom experiences the *effective nuclear charge*. For a 2s electron there is some probability of it being "inside" the 1s shell very close to, and indeed right at, the nucleus. In contrast, a 2p electron has a node at the nucleus and only a very small electron density inside the 1s shell. Hence the effective nuclear charge for a 2s electron is somewhat greater than $+Z-2$, whereas for a 2p electron it is only slightly greater than $+Z-2$ and smaller than that for a 2s electron. Hence the energy of a 2s electron in a polyelectronic atom is a little lower than that of a 2p electron (see Figure 25.10).

We have seen that the $2p_x$, $2p_y$, and $2p_z$ orbitals are identical, except for their relative orientations in space. Thus they have the same energy. The set of three 2p orbitals of the same energy is referred to as a *degenerate set*. A set of degenerate orbitals constitutes a *subshell*. The $n = 2$ shell consists of two sub-shells with somewhat different energies: the 2s subshell consisting of one orbital and the 2p subshell consisting of three orbitals. The total number of orbitals in any shell is n^2, while each subshell consists of $2\ell + 1$ orbitals of the same energy. Thus the $n = 3$ shell consists of a total of nine orbitals in three subshells, one 3s orbital, three 3p orbitals, and five 3d orbitals. The 3d orbitals have even less density close to the nucleus than the 3p orbitals; thus the effective nuclear charges for the subshells of the $n = 3$ shell increase in the order 3d < 3p < 3s, and the order of the energies of the orbitals is 3s < 3p < 3d.

In general, the energies of the subshells in any shell increase in the order s < p < d < $\cdots$. But as noted in Chapters 6 and 21, the energies of the sub-shells in the $n = 3$ shell overlap with those in the $n = 4$ shell for some elements. In such cases the 3d orbitals are at a slightly higher energy than the 4s orbital (Figure 21.1). There is a similar overlap of the energies of all the higher shells, for example, between the $n = 4$ shell and the $n = 5$ shell.

25.4 THE COVALENT BOND

We can use the shapes of s and p orbitals to give a somewhat more detailed account of the covalent bond than we gave in Chapter 6. We consider first the changes in electron distribution that occur in the formation of the hydrogen molecule. As two hydrogen atoms approach each other, the electron density of each H atom experiences an attractive force exerted by the nucleus of the other hydrogen atom. In other words, both electrons are simultaneously attracted by both nuclei. Since each hydrogen atom can accommodate two electrons in its 1s orbital, provided they have opposite spins, both electrons can occupy both orbitals simultaneously. Alternatively, we can say that the electron clouds of the two hydrogen atoms can overlap each other. Constructive interference can then occur, leading to a reinforcement of the electron density in the overlap region. We can think of the two electrons now as belonging to the molecule rather than one to each separate atom. The two electrons may be said to occupy a **molecular orbital** that surrounds both nuclei.

Although the Schrödinger equation cannot be solved exactly for any mole-cule, we can obtain a reasonable approximation to the ground state wave func-tion for the hydrogen molecule, ψ_{molecule}, by simply adding the two hydrogen 1s wave functions, which we will designate as ψ_a and ψ_b,

$$\psi_{\text{molecule}} = \psi_a + \psi_b$$

just as we can add the amplitudes of two waves to get the resultant amplitude.

Overlap of the atomic orbitals—that is, the constructive interference be-tween the wave functions—causes an increase in the electron density in the overlap region. In other words, electron density moves into this region (Figure 25.11). This increased electron density in the internuclear region provides the force that holds the two nuclei together.

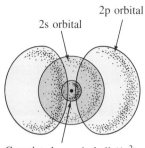

2s orbital   2p orbital

Completed $n = 1$ shell $(1s^2)$

Figure 25.10 Effective Nuclear Charge. A 2s electron penetrates inside the 1s shell and has some probability of being found right at the nucleus. This part of the electron density experiences a greater attractive force than the rest of the electron density, which is "outside" the 1s shell. Thus a 2s electron experiences the attraction of a greater charge than the core charge of $Z - 2$—this charge is called the effective nuclear charge. In contrast, the 2p electron has no density at the nucleus and only a very small amount of density "inside" the 1s shell. Thus it experiences the attraction of a charge only very slightly greater than $Z - 2$. A 2s electron experiences a greater effective nuclear charge than a 2p electron does, so an electron in a 2s orbital has a slightly lower energy than one in a 2p orbital.

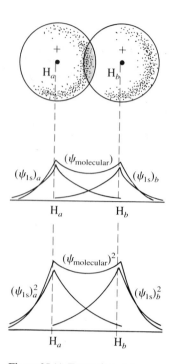

Figure 25.11 Formation of the Hydrogen Molecule. Because of constructive interference between the two wave functions where they overlap, there is a concentration of electron density in the region between the nuclei. This additional electron density provides the attractive force holding the nuclei together.

Any two atomic orbitals can be combined in the same way as we have combined two hydrogen 1s orbitals to give a molecular orbital. A molecular orbital that is restricted to two nuclei is often called a **localized molecular orbital**, or a **bonding orbital.** Like an atomic orbital, a molecular orbital may contain only two electrons of opposite spin in accordance with the Pauli exclusion principle. For example, the F atom has the electron configuration $1s^2 2s^2 2p_x^2 2p_y^2 2p_z^1$; it has a 2p orbital that contains only one electron. This 2p orbital can be combined with the 1s orbital of a hydrogen atom to give a localized molecular orbital that provides an approximate description of the electron distribution of the bonding electron pair in the HF molecule (see Figure 25.12). Any shared pair of electrons—that is, any single bond—can be described in a similar manner. For example, the 2p orbital of an F atom may be combined with the 2p orbital of another F atom to give an approximate description of the electron distribution of the bonding electron pair in the F_2 molecule, as shown in Figure 25.12.

25.5 ATOMIC ORBITAL APPROACH TO MOLECULAR GEOMETRY

Instead of using the VSEPR approach described in Chapter 8, we can discuss molecular geometry in terms of the directed character of p, d, ... hydrogenlike atomic orbitals. For example, consider the H_2O molecule. Each of the bonds may be approximately described by a localized molecular orbital formed from a singly occupied oxygen 2p orbital and a hydrogen 1s orbital. For a given O—H distance maximum overlap of the two orbitals, and therefore maximum increase of electron density in the bonding region, is obtained if the hydrogen 1s orbital is located on the axis of the oxygen 2p orbital, as in Figure 25.13. In other words, the strongest bond is obtained if it is formed in the direction of the axis of the 2p orbital.

In general, a bond is formed in the direction that gives maximum overlap of the two atomic orbitals; this statement is called the *criterion of maximum overlap.* Thus we predict that the two bonds formed by an oxygen atom will be at right angles to each other; that is, the water molecule is expected to be angular with a bond angle of 90° (see Figure 25.13). Similarly, the nitrogen atom has three singly occupied 2p orbitals. Each of these 2p orbitals can be combined with hydrogen 1s orbitals to give three localized molecular orbitals at right angles to each other. Thus the ammonia molecule is predicted to be pyramidal with 90° bond angles (see Figure 25.14). We recall that the VSEPR method

Figure 25.12 Formation of Localized Molecular Orbitals to Describe Bonds in HF and F_2 Molecules.

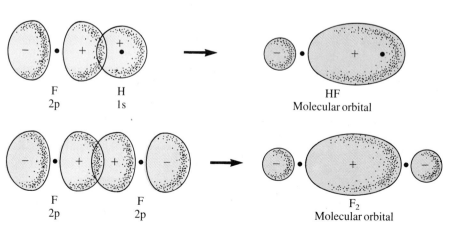

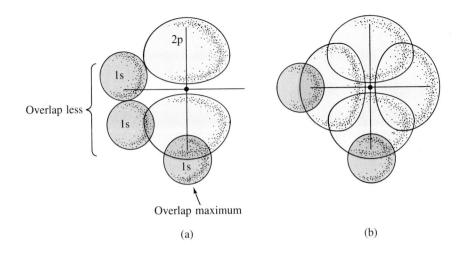

Figure 25.13 Formation of Localized Molecular Orbitals from Oxygen 2p Orbitals and Hydrogen 1s Orbitals to Describe the O—H Bonds in the Water Molecule.
(a) The overlap between the H (1s) and the O (2p) orbital is greatest when the 1s orbital is on the axis of the 2p orbital. (b) The two O—H bonds are predicted to be at 90°.

predicts bond angles of slightly less than 109.5° for these two molecules, which is in better agreement with the observed angles of 104.5° and 107.5° than with the angle of 90° predicted by the criterion of maximum overlap.

However, the structures of AX_n molecules in which the central atom uses an s orbital for bonding cannot easily be predicted on the basis of the criterion of maximum overlap, since the s orbital has no directional character. Thus for the $2s2p^3$ configuration of the carbon atom, a simple application of the criterion of maximum overlap predicts that the three p orbitals will form three bonds at right angles and that a fourth bond will be formed by the s orbital in a direction that cannot be predicted. We see that it does not predict the observed tetrahedral geometry (see Figure 25.15).

For the $2s2p^2$ configuration of the boron atom in a BX_3 molecule, the criterion of maximum overlap predicts two bonds at right angles and a third bond in some unspecified direction: It does not predict the observed equilateral triangular structure (Figure 25.16). Finally, in the case of the $2s2p$ configuration of a beryllium atom, we can make no prediction about the angle between the two bonds in a BeX_2 molecule, because the direction of the bond formed by the s orbital cannot be predicted (Figure 25.17).

These difficulties arise primarily because we are basing our description of a polyelectronic atom on hydrogenlike orbitals, which, strictly speaking, are accurate only for single-electron atoms. When there is more than one electron, they interact with each other, and the distribution of any one electron is not independent of the distribution of the other. If the distribution of an electron in a 2p orbital were independent of that in the 2s orbital, there would be a probability of finding both electrons at the same point wherever the two orbitals overlap. But according to the Pauli exclusion principle, there is a zero probability of finding two electrons with the same spin at the same point in space. Thus we see that the distribution of an electron in a 2s orbital cannot be independent of the distribution of an electron in a 2p orbital. To find the most probable arrangement of the four electrons in the $2s2p^3$ configuration of the carbon atom, we must consider them all together and take proper account of the Pauli exclusion principle. How this is done is beyond the scope of this book, but the result is that the most probable arrangement of four electrons is at the corners of a tetrahedron. This arrangement is just what we predicted in Chapter 8 simply on the basis of the Pauli exclusion principle, without allocating the electrons to any specific orbitals. The same considerations apply to the 2s2p

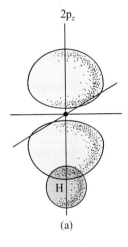

(a)

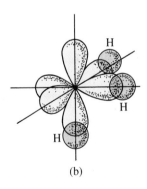

(b)

Figure 25.14 Formation of Localized Molecular Orbitals from N (2p) Orbitals and H (1s) Orbitals.
(a) Overlap of an H (1s) and an N (2p) orbital. (b) Here the 2p orbitals are drawn in a distorted shape so that all three can be distinguished. The three N—H bonds are predicted to be at 90°.

25.5 ATOMIC ORBITAL APPROACH TO MOLECULAR GEOMETRY

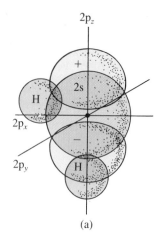

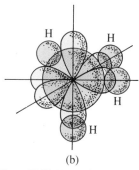

(a)

(b)

Figure 25.15 Formation of Localized Molecular Orbitals from C (2s) and C (2p) Orbitals.
(a) Overlap of the C (2p$_z$) and H (1s) orbitals and of the C (2s) and H (1s) orbitals. (b) The shape of the 2p orbitals has been distorted so that all three can be seen. This model predicts three C—H bonds at 90° and a fourth bond—formed by the 2s orbital—in an unspecified direction.

state of the beryllium atom, for which the most probable arrangement of the two electrons is with one on either side of the nucleus at an angle of 180° to each other. For the 2s2p^2 configuration of the boron atom, the most probable arrangement of the three electrons is at the corners of an equilateral triangle.

Hybrid Orbitals

We have seen that hydrogenlike orbitals are not very useful for describing the bonds in molecules such as BeCl$_2$, BF$_3$, and CH$_4$. But these hydrogenlike orbitals are not unique. We may obtain other orbitals, which are equivalent to a given set of 2s and 2p orbitals, by making combinations of the 2s and 2p orbitals. These combinations are called **hybrid orbitals**. In fact, there is an infinite number of ways that a set of 2s and 2p orbitals may be combined to give new orbitals. However, since we know that in molecules such as BeCl$_2$, BF$_3$, and CH$_4$ each of the bonds to the central atom has the same length and strength, we choose combinations of the 2s and 2p orbitals that give **equivalent orbitals**. For the orbitals to be equivalent, each orbital must be composed of equal amounts of the 2s and 2p orbitals. For one 2s and one 2p orbital there are only two such combinations:

$$(\psi_{sp})_1 = \frac{1}{\sqrt{2}}(\psi_{2s} + \psi_{2p}) \qquad (\psi_{sp})_2 = \frac{1}{\sqrt{2}}(\psi_{2s} - \psi_{2p})$$

They are analogous to the in-phase and out-of-phase combinations of two waves. These two orbitals are called sp *hybrid orbitals*. Their shapes may be obtained by simple algebraic addition of the 2s and 2p orbitals, as shown in Figure 25.18. These sp hybrid orbitals indicate the correct relative distribution

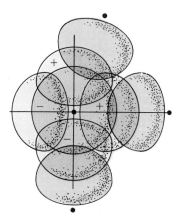

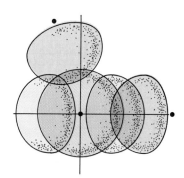

Figure 25.16 Formation of Localized Molecular Orbitals in BCl$_3$ from Boron Atomic Orbitals. The overlap of the Cl (3p) with the B (2p$_x$) and B (2p$_y$) orbitals is a maximum along their axes. The overlap of the Cl (3p) orbital with the B (2s) orbital is the same in all directions. Thus two B—Cl bonds are predicted to be at 90° to each other, and the direction of the third B—Cl bond cannot be predicted. For simplification, only half of the Cl (3p) orbital is shown.

Figure 25.17 Formation of Localized Molecular Orbitals in BeCl$_2$ from Beryllium Atomic Orbitals. The overlap of the Cl (3p) orbital with the Be (2p) orbital is a maximum along the axis of the Be (2p) orbital. The overlap of the Cl (3p) orbital with the Be (2s) orbital is the same in all directions. Therefore no prediction can be made about the shape of the BeCl$_2$ molecule. For simplification, only half of the Cl (3p) orbital is shown.

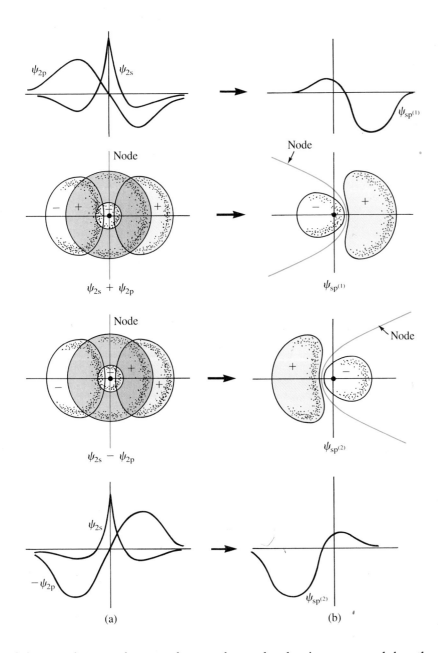

Figure 25.18 Formation of sp Hybrid Orbitals. By taking the sum and difference of a 2s and a 2p orbital, we can construct two equivalent hybrid (sp) orbitals:

(a) $(\psi_{sp})_1 = \dfrac{1}{\sqrt{2}}(\psi_{2s} + \psi_{2p})$;

(b) $(\psi_{sp})_2 = \dfrac{1}{\sqrt{2}}(\psi_{2s} - \psi_{2p})$;

of the two electrons, because they overlap each other in space much less than the 2s and 2p orbitals, and therefore they take better account of the Pauli exclusion principle. By combining each of these sp orbitals with the 3p orbital of a chlorine atom, we obtain two localized bonding orbitals that describe the bonds in $BeCl_2$.

Similarly, one 2s and two 2p orbitals may be combined to give three equivalent sp^2 hybrid orbitals, as shown in Figure 25.19. They are oriented at $120°$ to each other, and they may be used to form localized molecular orbitals to describe the bonds in a molecule such as BF_3.

A set of four equivalent hybrid orbitals can be constructed from one 2s and three 2p orbitals; they are called sp^3 hybrid orbitals. They have the same general shape as the sp and sp^2 hybrid orbitals, but are oriented in the tetrahedral directions, as Figure 25.20 shows. Again, their mutual overlap is considerably

25.5 ATOMIC ORBITAL APPROACH TO MOLECULAR GEOMETRY

**Figure 25.19 Formation of sp²
Hybrid Orbitals.** By taking
appropriate combinations of 2s,
$2p_x$, and $2p_y$ orbitals, we can obtain
a set of equivalent sp² orbitals.

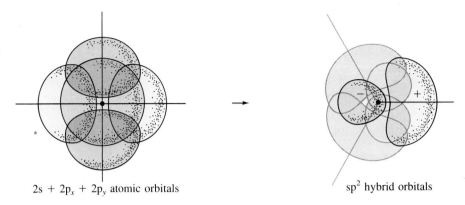

$2s + 2p_x + 2p_y$ atomic orbitals $\qquad\qquad$ sp² hybrid orbitals

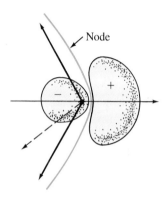

Node

**Figure 25.20 The sp³ Hybrid
Orbitals.** The shape of only one
orbital is shown; the directions of
the maxima of the other three are
shown by the arrows.

less than the overlap of the individual 2s and 2p orbitals. Therefore they make
reasonably adequate allowance for the Pauli exclusion principle, so they correctly indicate the most probable arrangement of the four electrons, which is
tetrahedral. Each of these orbitals can be used to form a localized molecular
orbital with, for example, the 1s orbital of a hydrogen atom to describe the
bonds in the methane molecule.

We cannot emphasize too strongly that the formation of hybrid orbitals—
hybridization—is *not* a phenomenon. It does not change the total electron density in an atom. For example, the total electron density on an atom having one
electron in a 2s orbital and another electron in a 2p orbital is

$$\Psi^2 = \psi_{2s}^2 + \psi_{2p}^2$$

The total electron density for an atom with an electron in each of two sp hybrid
orbitals is the same; that is,

$$
\begin{aligned}
\Psi^2 &= (\psi_{sp})_1^2 + (\psi_{sp})_2^2 \\
&= \tfrac{1}{2}(\psi_{2s} + \psi_{2p})^2 + \tfrac{1}{2}(\psi_{2s} - \psi_{2p})^2 \\
&= \tfrac{1}{2}(\psi_{2s}^2 + \psi_{2p}^2 + 2\psi_{2s}\psi_{2p}) + \tfrac{1}{2}(\psi_{2s}^2 + \psi_{2p}^2 - 2\psi_{2s}\psi_{2p}) \\
&= \psi_{2s}^2 + \psi_{2p}^2
\end{aligned}
$$

We can similarly show that the total electron density for an atom with three
electrons in a 2s and two 2p orbitals is exactly the same as the total density
for three electrons in three sp² hybrid orbitals (see Figure 25.21). The total electron density for four electrons in one 2s and three 2p orbitals is exactly the
same as the total electron density for an electron in each of the four sp³ hybrid
orbitals (see Figure 25.22): It is spherical in both cases. The two different descriptions of an atom—namely, the hydrogenlike orbital description and the hybrid
orbital description—simply represent two different, but equivalent, ways of
dividing up the total electron density into separate parts, one for each electron.

**Figure 25.21 Total Electron
Density for Three Electrons in a Set
of sp² Hybrid Orbitals.** The total
electron density has the shape of a
flattened sphere and is exactly the
same as the total electron density
for three electrons in the 2s, $2p_x$
and $2p_y$ orbitals.

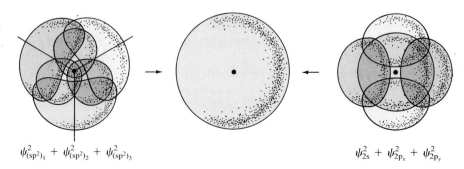

$\psi_{(sp^2)_1}^2 + \psi_{(sp^2)_2}^2 + \psi_{(sp^2)_3}^2$ $\qquad\qquad\qquad$ $\psi_{2s}^2 + \psi_{2p_x}^2 + \psi_{2p_y}^2$

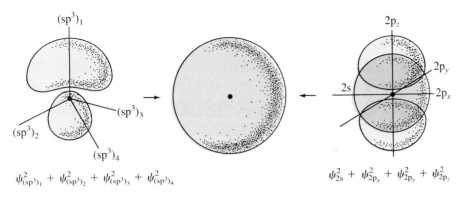

$$\psi^2_{(sp^3)_1} + \psi^2_{(sp^3)_2} + \psi^2_{(sp^3)_3} + \psi^2_{(sp^3)_4}$$

$$\psi^2_{2s} + \psi^2_{2p_x} + \psi^2_{2p_y} + \psi^2_{2p_z}$$

Figure 25.22 Total Electron Density for Four Electrons in a Set of sp³ Hybrid Orbitals. The total electron density is spherical and is exactly the same for four electrons in the 2s, $2p_x$, $2p_y$, and $2p_z$ orbitals.

In short, a set of hydrogenlike orbitals and a set of hybrid orbitals constructed from them are equally valid descriptions of the atom. We may choose whichever description is the most convenient for a particular purpose. If we wish to describe the bonds in the methane molecule in terms of localized bonding orbitals, then we must form these orbitals from the four equivalent sp³ orbitals. The basic reason for the tetrahedral geometry of the methane molecule is, however, the Pauli exclusion principle, which determines that the most probable arrangement of four pairs of electrons around a central core is the tetrahedral arrangement.

According to the very simple electron-pair-sphere model that we described earlier (Chapter 8), the four bonding pairs of electrons in the methane molecule are described as occupying four localized spherical regions of space, which, in order to get as close as possible to the positively charged core, naturally adopt a tetrahedral arrangement. The picture of the distribution of electrons given by this model corresponds very closely to that provided by the hybrid orbital model according to which the four pairs of electrons occupy four localized molecular orbitals constructed from the carbon sp³ hybrid orbitals and hydrogen 1s orbitals. The electron-pair-sphere model, however, has the advantage that is simpler to understand and to use.

The concept of hybrid orbitals is widely used in chemistry, but it is sometimes misused and misunderstood. Both hydrogenlike atomic orbitals and hybrid orbitals are often depicted in very approximate forms, which may give the incorrect impression that the overall electron distribution in a set of hybrid orbitals is different from the set of hydrogenlike orbitals from which the hybrid orbitals were constructed.

Figure 25.23(a) is a common approximate representation of a set of p orbitals, showing the correct relative orientations of the orbitals but giving only an approximate picture of the orbital shape. The shapes are often distorted in this way so as to reduce the overlap of the orbitals and thus facilitate the drawing of diagrams. This practice is convenient, but if these diagrams are taken literally, they tend to give a false impression. For example, the sum of the electron densities of three electrons in the three p orbitals is spherical, but this spherical shape is not obvious from the representation of the shapes of p orbitals in Figure 25.23(a).

Similarly, approximate representations of sp³ hybrid orbitals (Figure 25.23b) can give the incorrect impression that the sum of the electron densities of four electrons occupying sp³ orbitals is not spherical and that it differs from the sum of the electron densities of four electrons occupying the s and three p orbitals from which the sp³ hybrid orbitals were constructed. Of course, when a localized molecular orbital is constructed, the electron distribution changes due to the concentration of electron density in the bonding region. Thus in

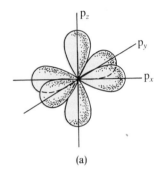

(a)

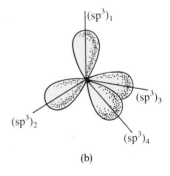

(b)

Figure 25.23 Approximate Representations of p Orbitals and sp³ Hybrid Orbitals. So that the orbitals can be more clearly represented in a diagram, their shapes are distorted by reducing the extent to which they spread out in space, thus emphasizing their directional character. It is not apparent from these diagrams that (a) a set of three p electrons (or three p electrons and an s electron) or (b) a set of four sp³ electrons have a total electron density that is spherical.

the methane molecule, the electron density distribution is no longer spherical, as it is in the carbon atom, but is more concentrated along the C—H bonds. The hybridization of s and p orbitals to give sp^3 hybrid orbitals is not, however, what causes the change in the electron distribution; rather, it is the formation of four C—H bonds that causes a concentration of electron density along the four tetrahedral directions.

It is sometimes implied, incorrectly, that hybrid orbitals provide an explanation for molecular geometry. In fact, molecular geometry is determined by the operation of the Pauli exclusion principle. *Hybrid orbitals are constructed so as to agree with the observed geometry. They do not themselves explain the geometry.* A suitable set of hybrid orbitals can be constructed to correspond to many different possible geometries. Chemists often loosely refer to a carbon atom in a molecule such as methane as sp^3-hybridized. This description is convenient but not strictly correct; it really means that it is convenient to describe the bonding in terms of localized molecular orbitals constructed from a set of equivalent sp^3 hybrid orbitals.

Hybrid Orbital Descriptions of Multiple Bonds

π **ORBITALS** An important use of hybrid orbitals is in describing multiple bonds, a method that is an alternative to that given in Chapter 8. Consider, for example, the ethene molecule, which has the following planar geometry:

$$\begin{array}{c} H \\ \end{array} \overset{122°}{\underset{116°}{}} C = C \begin{array}{c} H \\ \end{array}$$

The bonds may be approximately described by using a set of three equivalent sp^2 hybrid orbitals on each of the carbon atoms to form localized molecular orbitals with two hydrogen 1s orbitals and with an sp^2 orbital on the other carbon atom (see Figure 25.24). A single electron remains in the $2p_z$ orbital on each carbon atom. These two orbitals are combined in a "sideways" overlap. The result is a localized molecular orbital that has a nodal plane in the plane of the molecule. It is called a π **(pi) orbital**. Localized molecular orbitals that are formed by "end-on" overlap of atomic orbitals are symmetrical around the bond axis and are called σ **(sigma) orbitals**. This is only an approximate description of the bonding because it corresponds to bond angles of 120°.

DOUBLE BONDS The C=C double bond is thus approximately described as consisting of a σ orbital formed by "end-on" overlap of sp^2 orbitals and a π orbital formed by "sideways" overlap of two 2p orbitals. According to this description, the double bond consists of a σ bond and a π bond. In contrast, our previous description suggested that the double bond consists of two equivalent bent bonds. Neither description is exact and, as we will see, they are in fact very similar.

We may put our previous description of the double bond, which was based on a tetrahedral arrangement of four electron pairs around each carbon atom, into orbital language by using four equivalent tetrahedral sp^3 hybrid orbitals on each carbon atom. Two of these hybrid orbitals are combined with hydrogen 1s orbitals to describe the C—H bonds, and two are combined with similar orbitals on the other carbon atom to form the double bond orbitals (see Figure 25.25). In this description the double bond consists of two charge clouds, one on either side of the C—C axis. In contrast, in the alternative description the double bond consists of an axially symmetrical σ charge cloud and a π charge

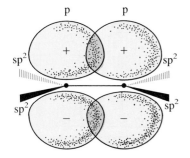

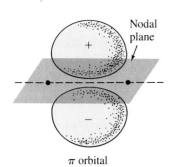

π orbital

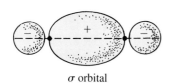

σ orbital

Figure 25.24 Description of the Ethene Molecule in Terms of σ and π Orbitals. The C—H and C—C σ bond orbitals formed by a set of sp^2 hybrid orbitals on each carbon atom are shown here only as lines. The "sideways" overlap of two p orbitals gives a π orbital. A σ orbital is symmetrical around the bond direction, but a π orbital has a planar node in the bond direction.

cloud that overlaps the σ charge cloud. Despite the apparent difference, these two descriptions are very similar. The similarity becomes apparent if we combine the σ and π orbitals as we previously combined s and p orbitals to get sp hybrids (see Figure 25.26). Combining them in this way gives the two localized "hybrid" orbitals, $\psi_1 = (1/\sqrt{2})(\sigma + \pi)$ and $\psi_2 = (1/\sqrt{2})(\sigma - \pi)$, which are largely concentrated one on each side of the C—C axis.

The total electron density in the double bond is obtained by taking the sum of both two-electron charge clouds. This sum gives a four-electron charge cloud that has a maximum along the C—C axis and in which separate regions, corresponding to separate electron pairs, cannot be distinguished (Figure 25.26). Any division of this charge cloud into separate orbitals, each representing an electron pair, is arbitrary and is made only for convenience.

TRIPLE BONDS A triple bond, such as that in ethyne, is frequently described by using a set of two sp hybrid orbitals and a $2p_y$ and a $2p_z$ orbital (see Figure 25.27). The sp hybrid orbitals are used to form the localized orbitals describing the C—H and C—C bonds, and the $2p_y$ and $2p_z$ orbitals are combined with similar orbitals on the other carbon atom to form two π-type orbitals. Again, however, this description is not unique. Using a set of sp^3 orbitals on each carbon atom is equally valid. One of these orbitals is used to form the C—H bonding orbital, and the other three are used to form three bent bonds between the carbon

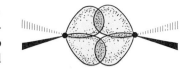

Figure 25.25 Bent-Bond Description of Ethene Molecule. Here the double bond is represented by two bent bonds, each of which is formed by the overlap of a pair of sp^3 orbitals.

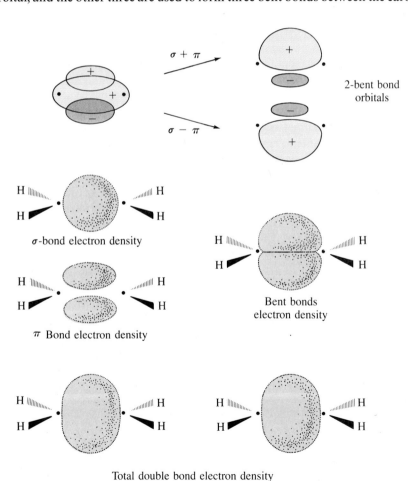

σ + π

2-bent bond orbitals

σ − π

σ-bond electron density

π Bond electron density

Bent bonds electron density

Total double bond electron density
(4-electron charge cloud)

Figure 25.26 Equivalence of $\sigma-\pi$ and Bent-Bond Models for Double Bonds. These two models correspond to the same total electron density for the double bond. They are simply two different ways of dividing the total electron density.

25.5 ATOMIC ORBITAL
APPROACH TO MOLECULAR
GEOMETRY

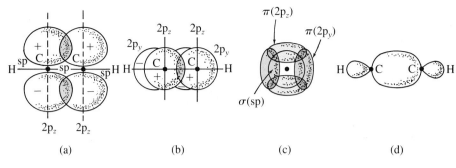

Figure 25.27 The σ–π Description of the Triple Bond in Ethyne.
(a) The H—C—C—H σ bonds are formed by overlap of carbon sp hybrid orbitals with the H 1s orbitals and with each other, leaving the $2p_z$ and $2p_y$ orbitals on each carbon atom for π orbital formation. The overlap of the $2p_z$ orbitals to form a π orbital is shown. (b) The $2p_y$ orbitals overlap to form a second π orbital. (c) A view of the σ and π orbitals down the C—C axis. (d) The cylindrical shape of the total electron density in the triple bond.

atoms (Figure 25.28). This description corresponds closely to the electron-pair-sphere model described in Chapter 8.

Molecular Orbitals

For planar molecules such as the carbonate ion, CO_3^{2-}, the σ–π model provides an alternative to the resonance structure description given in Chapter 8. If sp^2 hybrid orbitals are used on the carbon atom and each of the three oxygen atoms, three electron pairs can be placed in three C—O σ bonding orbitals and six more pairs can be placed in sp^2 nonbonding orbitals on each oxygen atom (Figure 25.29). Eighteen of the 24 electrons in the carbonate ion are thus accounted for. The remaining six can be accommodated in three *molecular orbitals* that extend over the whole molecule. These orbitals may be constructed from the $2p_z$ orbitals of the carbon and oxygen atoms. They all have a node in the plane of the molecule. For the orbital of lowest energy, there are no additional nodes. The two next-highest energy orbitals are equivalent and have an additional node perpendicular to the molecular plane. The six delocalized electrons occupy these three molecular orbitals, as also shown in Figure 25.29.

Another important example is provided by the benzene molecule. In this case three sp^2 hybrid orbitals on each carbon atom can be used to form one C—H bond and two C—C bonds, thus constructing a hexagonal framework of σ-bonds (see Figure 25.30). One electron is thus left in the $2p_z$ orbital on each carbon atom. These $2p_z$ orbitals can be used to construct a set of six π molecular orbitals that cover all six carbon atoms. The three orbitals of lowest energy are each occupied by two electrons. This description of the six delocalized electrons is an alternative to the use of resonance structures (Chapter 11). This molecular orbital method of describing benzene, which is also appropriate for many related hydrocarbons and their derivatives, is important and widely used, but further discussion is beyond the scope of this book.

Figure 25.28 Bent-Bond Model of the Triple Bond in Ethyne.
Here the triple bond is represented as three bent bonds. (a) The bent bonds are depicted by a simple stick model. (b) The orbitals corresponding to the three bent bonds are shown. (c) A view is shown of the same three orbitals looking down the C—C axis. (d) The total electron density due to the six electrons of the triple bond is illustrated.

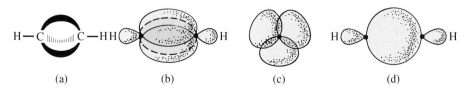

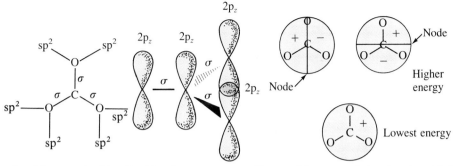

σ bond framework of localized bonds
and nonbonding pairs formed
from sp² hybrid orbitals

Shapes of three occupied molecular orbitals,
each of which contains two electrons

Figure 25.29 Partial Molecular Orbital Description of the Carbonate Ion. Eighteen of the electrons occupy localized σ bond orbitals or nonbonding orbitals formed from sp² hybrid orbitals on the carbon and oxygen atoms. The six delocalized electrons occupy three molecular orbitals formed by combining the $2p_z$ orbitals on each of the oxygen atoms and the carbon atom. In addition to the nodes shown, they each have a node in the plane of the four atoms.

Hybrid Orbitals in Water and Ammonia Molecules

The bond angles in H_2O and NH_3 are closer to the tetrahedral value predicted by the VSEPR rules than to the angle of 90° that is expected if localized bonding orbitals are constructed from 2p orbitals on the nitrogen and oxygen atoms. The bonds may be better described by using a set of four sp³ hybrid orbitals. Two of these orbitals are occupied by pairs of electrons and therefore describe the nonbonding electron pairs, and two are singly occupied and may be combined with hydrogen 1s orbitals to form localized O—H bonding orbitals (see Figure 25.31). Similarly, the ammonia molecule may be described by a set of sp³ hybrid orbitals; one is occupied by a pair of electrons, the nonbonding pair, and the others are singly occupied and may be used to construct the localized N—H bonding orbitals by combining them with the H 1s orbitals (Figure 25.31).

Since, in fact, the bond angles in both these molecules are slightly smaller than the tetrahedral angle of 109.5°, more accurate bonding orbitals would be based on hybrid orbitals that have a little more p character than the sp³ orbitals and therefore a slightly smaller interorbital angle. The nonbonding pair orbital(s) would therefore have slightly more s character than an sp³ orbital. Since the s orbital is spherically symmetrical, an orbital with more s character is more spread out—that is, is less localized—and takes up more space in the valence shell of an atom than an orbital with more p character, as discussed in Chapter 8.

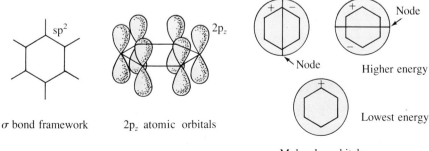

σ bond framework   2p_z atomic orbitals

Molecular orbitals

Figure 25.30 The σ Bond Framework and Three Lowest-Energy Molecular Orbitals for the Benzene Molecule. These three orbitals are occupied by the six delocalized electrons. In addition to the nodes shown, each orbital has a node in the plane of the ring.

25.5 ATOMIC ORBITAL
APPROACH TO MOLECULAR
GEOMETRY

869

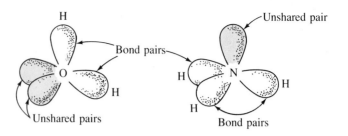

AX₅ and AX₆ Molecules: Hybrid Orbitals Involving d Orbitals

To describe the bonds in molecules with more than eight electrons in the valence shell of the central atom—for example, the AX_5 and AX_6 molecules that we discussed in Chapter 20—we must construct a set of five or six appropriate hybrid orbitals that make use of d orbitals on the central atom in addition to the s and p orbitals. The shapes of the five 3d orbitals are shown in Figure 25.32. The 3d orbitals for which $n = 3$ and $\ell = 2$ have two planar nodes, just as we have seen for the vibrations of a sphere (Figure 25.4). There are five possible independent orientations of the set of two planar nodes, which give rise to the orbital shapes in Figure 25.32.

A set of five hybrid orbitals for describing the bonding in a trigonal bipyramidal AX_5 molecule can be constructed from one s, three p, and the d_{z^2} orbitals. They are called sp^3d hybrids. A set of six hybrid orbitals for describing the bond-

Figure 25.32 Approximate Representations of the Five 3d Orbitals.
(a) The d_{xy} orbital lies in the xy plane; it has planar nodes in the xz and yz planes.
(b) The d_{yz} orbital lies in the yz plane; it has planar nodes in the xz and yz planes.
(c) The d_{xz} orbital lies in the xz plane; it has planar nodes in the xz and yz planes.
(d) The $d_{x^2-y^2}$ orbital lies in the xy plane; it has planar nodes that are perpendicular to the xy plane and bisect the angles between the x and y axes. (e) The d_{z^2} orbital, which is a combination of $d_{x^2-z^2}$ and $d_{y^2-z^2}$, has a more complicated shape: It has two conical nodes surrounding the z axis, each with its apex at the nucleus.

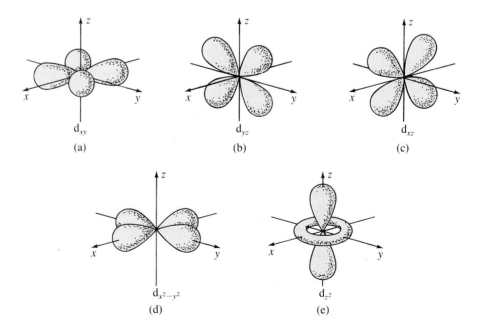

ing in an octahedral AX_6 molecule can be constructed from one s orbital, three p orbitals, and the $d_{x^2-y^2}$ and d_{z^2} orbitals. They are called sp^3d^2 hybrids (Figure 25.33).

Table 25.2 on page 872 summarizes the geometry of molecules with up to six electron pairs in the valence shell and the corresponding hybrid orbitals that may be used to describe the bonds in terms of localized molecular orbitals.

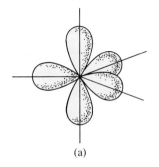

(a)

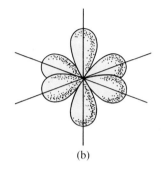

(b)

Figure 25.33 The sp^3d and sp^3d^2 Hybrid Orbitals. (a) The five trigonal bipyramidal $sp^3d_{z^2}$ hybrid orbitals. (b) The six octahedral $sp^3d_{x^2-y^2}d_{z^2}$ hybrid orbitals.

IMPORTANT TERMS

Equivalent orbitals are a set of hybrid orbitals that are chosen so that all the orbitals in the set are equivalent. An example is the set of sp^3 hybrid orbitals.

A hybrid orbital is a combination of two or more atomic orbitals. Hybrid orbitals are constructed so as to agree with the observed geometry of the bonds around an atom.

Hydrogenlike orbitals are orbitals having the same general form as the orbitals of a hydrogen atom. They are used to give an approximate description of a polyelectronic atom.

A localized molecular orbital is a molecular orbital that extends over only two atomic nuclei.

Molecular orbitals are orbitals that extend over two or more nuclei and often over all the nuclei in a molecule.

A node is a point, line, or surface at which the amplitude of a wave or the wave function, ψ, is zero.

An orbital is the wave function corresponding to a particular set of quantum numbers.

A pi (π) orbital has a nodal plane through the atoms of a planar molecule. Its approximate shape may be obtained by combining two or more atomic p orbitals in a "sideways" manner.

Quantum numbers are the integers used to describe the orbitals of the hydrogen atom.

A sigma (σ) orbital is a localized molecular (or bonding) orbital that is symmetrical around the straight line joining two atoms. It may be obtained by combining two atomic or hybrid orbitals in an "end-on" manner.

PROBLEMS

1. For the quantum number $n = 3$ of a hydrogen atom (or a pulsating, spherical dust cloud), how many spherical and how many planar nodal surfaces are present for $\ell = 0$? For $\ell = 2$? Is $\ell = 3$ possible here? How are the $n = 3$, $\ell = 0$ and $n = 3$, $\ell = 2$ orbitals usually designated?

2. What is the relationship between the number of planar nodes in an atomic orbital and the quantum number ℓ.

3. Explain the relationship between the quantum number ℓ and the shape of the orbitals for $n = 3$.

4. Using the atomic orbital approach, predict the shapes of PH_3 and H_2S.

* The asterisk denotes the more difficult problems.

5. Give a hybrid orbital description of the bonds in carbon tetrachloride, CCl_4.

6. Give a hybrid orbital description of the bonds in nitrogen trifluoride, NF_3.

7. Give a hybrid orbital description of the bonds in hydrogen peroxide, H_2O_2.

8. Give a hybrid orbital description of the bonds in propyne, $CH_3-C\equiv CH$.

9. Give a hybrid orbital description of the bonds in ethanal, CH_3CHO.

10. Give a hybrid orbital description of the bonds in propene, $CH_3CH=CH_2$.

Table 25.2 Molecular Shapes and Hybrid Orbitals

NUMBER OF ELECTRON PAIRS IN VALENCE SHELL OF CENTRAL ATOM	ARRANGEMENT OF ELECTRON PAIRS	TYPE OF MOLECULE IN TERMS OF VSEPR THEORY	SHAPE OF MOLECULE	HYBRID ORBITALS ON CENTRAL ATOM USED TO DESCRIBE BONDS AND NONBONDING PAIRS[a]	EXAMPLES		
					AX_n	AX_mY_n	AX_mE_n
2	Linear	AX_2	Linear	sp	$BeCl_2$, $HgCl_2$ $Ag(CN)_2^-$, CO_2	CH_3HgCl, C_2H_2	
3	Equilateral triangular	AX_3	Equilateral triangular	sp^2	BCl_3, BF_3	BF_2Cl, C_2H_4	
		AX_2E	Angular				$SnCl_2$
4	Tetrahedral	AX_4	Tetrahedral	sp^3	CH_4, $SiCl_4$ $Zn(OH)_4^{2-}$	$CHCl_3$	
		AX_3E	Triangular pyramidal				NH_3, H_3O^+
		AX_2E_2	Angular				H_2O, $(CH_3)_2O$
5	Trigonal bipyramidal	AX_5	Trigonal bipyramidal	sp^3d	PCl_5		
		AX_4E	Disphenoid				SF_4
		AX_3E_2	T-shaped				ClF_3
		AX_2E_3	Linear				XeF_2
6	Octahedral	AX_6	Octahedral	sp^3d^2	SF_6		
		AX_5E	Square pyramidal				IF_5
		AX_4E_2	Square planar				ICl_4^-

[a] For AX_mE_n and AX_mY_n molecules, these hybrid orbital descriptions are only approximate. Bonds to different ligands and to unshared pairs can be more accurately described by orbitals that are intermediate in character between sp, sp^2, sp^3, sp^3d, and sp^3d^2 orbitals.

11. Describe the bonding in the nitrate ion, NO_3^-, in terms of localized σ orbitals and π molecular orbitals.

12. Describe the bonds and unshared electron pairs in ClF_3 and SF_4 in terms of hybrid orbitals.

13. Explain, in terms of the $\sigma-\pi$ model, why there is no free rotation around a double bond.

14. Give hybrid orbital descriptions of the bonds in the molecules $AlCl_3$ and Al_2Cl_6.

15. Give hybrid orbital descriptions of CO, CN^-, and C_2^{2-}.

16. (a) Give an atomic orbital description of the bonding in $SnCl_2$.

(b) Give a hybrid orbital description of the bonding in $SnCl_2$.

(c) Predict the geometry of $SnCl_2$ by using the VS-EPR method. Which is in better agreement with the expected bond angle, the hybrid orbital description or the atomic orbital description of the bonding?

17. Describe a double bond and a triple bond in terms of σ and π orbitals.

18. Describe the methanoate (formate) ion, HCO_2^-, (a) in terms of two resonance structures and (b) in terms of a σ bond framework and two π molecular orbitals.

19. What is the qualitative relationship between the number of nodes in an atomic or molecular orbital and its energy? What is the relationship between the number of nodes of an atomic orbital and the quantum number n?

*20. The five independent 3d orbitals can be represented in several different ways. Sketch the $3d_{x^2-z^2}$ and $3d_{y^2-z^2}$ orbitals and show that $3d_{x^2-z^2} - 3d_{y^2-z^2} = 3d_{x^2-y^2}$. Thus only two of the $3d_{x^2-y^2}$, $3d_{x^2-z^2}$, and $3d_{y^2-z^2}$ are independent. The two orbitals that are normally chosen are $3d_{x^2-y^2}$ and $3d_{z^2}$. Draw a diagram to show that the sum of the $3d_{x^2-z^2}$ and $3d_{y^2-z^2}$ orbitals has the shape of the $3d_{z^2}$ orbital.

CHAPTER 26

THERMODYNAMICS AND THE DRIVING FORCE OF CHEMICAL REACTIONS

Entropy, Gibbs Free Energy, and the Second Law of Thermodynamics

One of the greatest achievements in history was the discovery of how sources of energy other than our own muscles or those of animals can be used. The development of this knowledge created modern civilization. Most of our energy comes from the burning of fossil fuels and from nuclear reactions; these processes supply energy as heat. But most energy is needed in the form of mechanical energy to do work in order to operate a machine or propel a vehicle. The transformation of heat to work is achieved by the use of devices called engines, which use the energy stored in a fuel to perform mechanical work to drive, for example, an electric generator, an automobile, or a ship.

The science of energy and its transformations is known as *thermodynamics.* It grew out of the Industrial Revolution and the need to understand how to convert as much as possible of the heat obtained by burning a fuel into mechanical energy—in other words, how to improve the efficiency of an engine. In engineering the main use of thermodynamics is therefore in connection with engines. In chemistry, however, thermodynamics has another very important application; it helps us to answer the question "Why do chemical reactions occur?" Why do some reactions proceed spontaneously to completion, others proceed only partially, and still others not at all? This question is clearly one of the most important in chemistry, but it is not one of the easiest to answer. Because some of the basic ideas of thermodynamics are rather abstract and difficult to understand fully at first, a full appreciation of thermodynamics comes only with using it over a period of time. All we can do here is to make a start.

We do not attempt in this chapter to give a complete introduction to thermodynamics, since to do so would require concepts and mathematics that are well beyond the scope of this book. We will not derive any equations; we will simply show how they can be used to answer the question of why certain reactions proceed spontaneously and others do not.

We will see that the spontaneity of a reaction is determined by both the change in energy and the change in the *entropy* during the reaction. We have previously briefly mentioned that entropy is a measure of the disorder in a system (Chapter 12). In this chapter we consider entropy in more detail, and then show how a quantity called the *Gibbs free energy*, which combines the energy change and the entropy change for a reaction, enables us to predict whether or not the reaction will be spontaneous under some given conditions.

26.1 SPONTANEOUS PROCESSES

Some changes in nature occur spontaneously, and some do not. A gas expands to fill the space available to it, but it does not spontaneously contract to a smaller volume. A hot object cools down to the temperature of the environment, but an object does not spontaneously get hotter than its environment. Many chemical reactions proceed spontaneously in one direction only. Hydrogen combines explosively with oxygen to give water when the reaction is initiated by a spark, but water does not decompose spontaneously to hydrogen and oxygen. Burning diamonds in oxygen gives carbon dioxide, but carbon dioxide, even when strongly heated, does not decompose to give diamonds and oxygen. Clearly, something causes some chemical reactions and certain physical processes to proceed in one direction only. Of course, these changes can be reversed: A gas *can* be compressed to a smaller volume; an object *can* be cooled to below the temperature of its environment; and water *can* be decomposed to hydrogen and oxygen by passing an electric current through it. However, in each case work must be done to accomplish these changes. They do not occur spontaneously; rather, it is the reverse process that occurs spontaneously. What is it that causes all these changes to occur spontaneously in one particular direction?

Common experience shows that spontaneous processes involving macroscopic objects proceed with a decrease in potential energy. A rock spontaneously rolls down a mountain if it is dislodged, but it never rolls up to the top again. A skier slides spontaneously down a ski run, but work must be done by the skier, or by the ski lift, if the skier is to reach the top of the run again.

Although we take it as almost self-evident that an object tends to achieve a position of minimum energy, we must remember that if something loses energy spontaneously, something must at the same time gain energy, because according to the first law of thermodynamics, the total energy of a system and its surroundings remains constant (Chapter 12). The potential energy of the skier descending the ski run is converted to kinetic energy as his speed increases and to heat because of friction between the skis and the snow. Finally, when he stops, voluntarily or not, all his kinetic energy is converted to heat. Although the skier has lost energy, the surroundings have gained energy, so the total energy has not decreased. Instead, there has been a transfer of energy from the system to the surroundings.

From what we have said, we might expect that a chemical reaction would proceed spontaneously if the reacting system decreases in energy by transferring heat to its surroundings. In other words, we might expect spontaneous reactions to be exothermic, and this is frequently the case. But there are many endothermic chemical reactions and other processes that proceed spontaneously. For example, ammonium chloride dissolves spontaneously in water in an endothermic change (see Experiment 13.3):

$$NH_4Cl(s) \xrightarrow{H_2O} NH_4^+(aq) + Cl^-(aq) \qquad \Delta H° = 14.0 \text{ kJ}$$

Dinitrogen pentoxide decomposes spontaneously at room temperature into NO_2 and O_2 in an endothermic reaction (Chapter 17):

$$2N_2O_5(s) \longrightarrow 4NO_2(s) + O_2(g) \qquad \Delta H° = 219.0 \text{ kJ}$$

Water left in an open container spontaneously evaporates, although the vaporization of water is an endothermic process:

$$H_2O(l) \longrightarrow H_2O(g) \qquad \Delta H° = 44.0 \text{ kJ}$$

In all these examples the enthalpy of the system increases rather than decreases. Recall from Chapter 12 that the change in internal energy of a system, ΔE, is not in general exactly equal to the change in the enthalpy, ΔH. But in almost all cases the difference between ΔE and ΔH is quite small, and it can be neglected for our present purposes. Clearly, the direction of spontaneous change is not always determined by the tendency of a system to achieve minimum energy. Thus the idea that a system spontaneously tends to decrease its energy is too simplistic; the real situation is more complicated.

It may help us to zero in on the real reason for spontaneous processes if we consider a process in which there is no energy change at all. An ideal gas expands spontaneously at constant temperature into any space made available to it. Since the energy of an ideal gas depends only on the temperature, its energy does not change during the process, and therefore no heat is evolved or absorbed. Moreover, if the gas expands into a vacuum, no work is done by, or on, the system (see Figure 26.1). That this spontaneous expansion of a gas should occur does not surprise us; indeed, we would be very surprised if it did not occur. But *why* does it occur? Because, according to the kinetic molecular theory the molecules of a gas are moving randomly and at high speeds. They move through the valve connecting the two flasks in Figure 26.1 and thus move into the empty flask. Some molecules eventually return to the original flask, but more move from left to right than from right to left, until there are equal numbers of molecules in both flasks and equilibrium is established. Thus the spontaneity of this process is associated with the molecular nature of matter and specifically with the random motions of large numbers of molecules. This random molecular motion causes the molecules of a gas to spread out in space as much as possible. Not only does a single gas diffuse into an empty space, but two gases diffuse into each other to form a uniform mixture. They become completely mixed up with each other as a result of their random thermal motions.

In general, because of the random motions of molecules, there is a tendency for all molecules to become more dispersed in space and more mixed up with each other. In other words, there is a tendency for them to become more disordered. If we start with two separate gases and allow them to mix, the final state of the system is more mixed-up—that is, is more disordered—than the original state. This is quite similar to the shuffling of a pack of cards that was initially ordered according to the four suits. The process of shuffling mixes them up—that is, it increases their disorder—and continued shuffling is very unlikely to return them to the original order. The mixing up of the cards is a

Stopcock closed

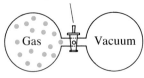

Stopcock opened

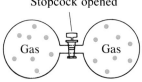

Figure 26.1 The Expansion of an Ideal Gas is a Spontaneous Process.

CHAPTER 26
THERMODYNAMICS AND
THE DRIVING FORCE OF
CHEMICAL REACTIONS

876

"spontaneous" process that occurs in only one direction; we never observe the "unmixing" of the cards. Thus a spontaneous process in an isolated system is one in which there is an increase in disorder in the system.

For a system that is not isolated from its surroundings, we must also take into account what happens in the surroundings. For example, in an exothermic process the total energy of the system and its surroundings is constant. But energy is transferred as heat from the system to the surroundings; that is, it is dispersed in the surroundings. For example, a hot block of metal spontaneously cools down as energy is transferred as heat from the metal to its surroundings. The random thermal motion of the molecules of the surroundings is increased by an amount corresponding to the heat transferred from the system to the surroundings. The energy that was originally concentrated in the metal block becomes much more dispersed because of the very large—essentially infinite—size of the surroundings. In this spontaneous process there is no change in the total energy of the system and its surroundings, but there is an increase in the disorder of the system and its surroundings.

The example of the book that is dropped on the table, which we discussed at the end of Chapter 12, also illustrates that a spontaneous process is accompanied by an increase in disorder. When the book hits the table, the ordered motion of its molecules, which are all moving toward the table as it falls, is replaced by the chaotic, random motion of the molecules of the book and the table, which both become a little warmer. It would be quite consistent with the first law of thermodynamics for the book and table to become a little cooler and for the book to jump back off the table! We know that this never happens. The change that occurs when the book drops to the table and its energy is converted to heat is a spontaneous process, whereas the reverse process is not spontaneous. There has been no change in the total energy of the book and the table in this process. What has changed is the way the energy is distributed. The energy that was originally concentrated in the book has become dispersed over the table and the book as random thermal motions. The reverse process—which would involve the spontaneous ordering of the random motions of the molecules to cause them all to move in the same direction so that the book jumps off the table—is so improbable that it is never observed, although it would not be contrary to the law of conservation of energy.

If we consider a system and its surroundings, then whether or not a change in the system is spontaneous does not depend on energy changes, because the total energy of the system and its surroundings is constant, but it depends on the change in the disorder of the system and its surroundings. *In any spontaneous process the total disorder in the system and its surroundings increases.*

26.2 ENTROPY

To put the above ideas on a quantitative basis, we need a quantity that is a measure of the disorder of the particles (atoms and molecules) that make up the system and the dispersal of energy associated with these particles. This quantity is called **entropy** and it is given the symbol S. The disorder in a system depends only on the conditions that determine the state of the system, such as composition, temperature, and pressure. The change in entropy therefore depends only on the initial and final states of the system. *Entropy*, like enthalpy, *is a state function.*

The Second Law of Thermodynamics

> **In any spontaneous process the total entropy of a system and its surroundings increases.**

This law of nature is called the **second law of thermodynamics**. It follows that *the entropy of the universe is increasing.* This is an alternative statement of the second law. An analogous statement of the *first law* is that *the energy of the universe is constant.* For any spontaneous process we may write

$$\Delta S_{universe} = \Delta S_{system} + \Delta S_{surroundings} > 0$$

In other words, the total entropy change must be positive.

Standard Molar Entropies

So far we have given only a qualitative description of the quantity we have called entropy. Although we cannot discuss here how quantitative values for the entropies of substances can be obtained, we can increase our understanding of the concept of entropy by considering the results of the measurement of the entropy of a substance as a function of temperature.

The entropy of oxygen as a function of temperature is shown in Figure 26.2. At 0 K the entropy of any pure crystalline substance is zero—the molecules have a perfectly regular arrangement and they have no thermal motion. As the temperature increases, the molecules begin to vibrate about their mean positions—their random thermal motion begins to increase—and therefore the entropy of oxygen increases. At the melting point the regular arrangement of the molecules in the solid changes to the more random arrangement of the molecules in the liquid, so there is an abrupt increase in the entropy. Then the entropy again steadily increases with increasing temperature until the liquid boils. There is then a large increase in the volume of the oxygen and therefore in the random distribution of the molecules in space. Therefore there is a correspondingly large, abrupt increase in the entropy. The entropy of gaseous oxygen then continues to increase slowly with increasing temperature.

Values of the standard molar entropies of some substances at 25°C are given in Table 26.1. (Additional values are given in Appendix B.) The **standard molar entropy**, $S°$, is the entropy of 1 mol of the substance in its standard state—

Figure 26.2 Entropy of Oxygen as a Function of Temperature.

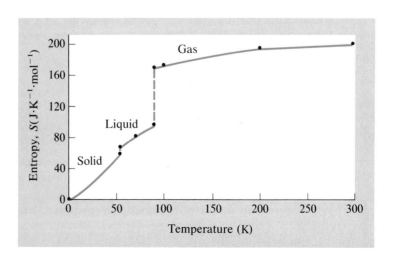

Table 26.1 Standard Molar Entropies, $S°$, at 25°C (J K^{-1} mol^{-1})*

SOLIDS		LIQUIDS		GASES	
C (diamond)	2.4	H$_2$O	70.0	H$_2$	130.6
C (graphite)	5.8	Hg	75.9	N$_2$	191.5
Fe	27.3	Br$_2$	152.2	F$_2$	202.7
S (rhombic)	32.0	CH$_3$OH	126.8	O$_2$	205.0
Cu	33.2	HNO$_3$	155.6	Cl$_2$	223.0
P (white)	41.1	CH$_3$CO$_2$H	159.8	I$_2$	260.6
Ag	42.6	C$_2$H$_5$OH	160.7	CH$_4$	86.1
I$_2$	116.1	CH$_2$Cl$_2$	177.8	HCl	186.8
MgO	27.0	CHCl$_3$	201.7	H$_2$O	188.7
CaO	38.1	CCl$_4$	216.4	NH$_3$	192.7
SiO$_2$ (quartz)	41.5	SiCl$_4$	239.7	H$_2$S	205.6
NaF	51.3	C$_5$H$_{12}$^{a}	263.3	HI	206.5
NaCl	72.5	C$_6$H$_6$^{b}	172.2	CO$_2$	213.7
NaBr	87.2	C$_6$H$_{14}$^{a}	295.9	C$_2$H$_6$	229.5
Fe$_2$O$_3$	87.4	C$_7$H$_{16}$^{a}	328.5	C$_3$H$_8$	269.9
CaCO$_3$	92.9	C$_8$H$_{18}$^{a}	361.2	C$_4$H$_{10}$^{a}	310.1
AgCl	96.2			C$_5$H$_{12}$^{a}	348.9
NaI	98.5			C$_6$H$_6$^{b}	269.2
Glucose (C$_6$H$_{12}$O$_6$)	182.4				
P$_4$O$_{10}$	231				
Sucrose (C$_{12}$H$_{22}$O$_{11}$)	360				

* Additional values are given in Tables B.2 and B.3.
^{a} Straight-chain alkanes. ^{b} Benzene.

that is, at 1 atm pressure—and at 25°C. The units of entropy are joules per kelvin per mol (J K^{-1} mol^{-1}). Note that entropy values are normally given in joules, not kilojoules.

The data in Table 26.1 shows that gases in general have larger entropies than solids. In a gas, the molecules have a more random motion than in a solid, and their energy is spread over a much larger volume; that is, the energy of the molecules of a gas is more dispersed than the energy associated with the motions of the molecules in a solid. The entropies of liquids, in general, fall between those of gases and solids. We note also that substances with large molecules usually have higher entropies than substances with smaller molecules, because energy is shared between more atoms and is therefore more dispersed.

Entropy Changes in Reactions

The **standard entropy change** for a reaction is easily calculated from the standard molar entropies, using the expression

$$\Delta S° = \sum S°(\text{products}) - \sum S°(\text{reactants})$$

As an example, let us calculate the standard entropy change for the rusting of iron:

$$4\text{Fe(s)} + 3\text{O}_2(\text{g}) \longrightarrow 2\text{Fe}_2\text{O}_3(\text{s})$$

For this reaction we may write

$$\Delta S^\circ = 2S^\circ(Fe_2O_3) - [4S^\circ(Fe) + 3S^\circ(O_2)]$$

Using the values given in Table 26.1, we obtain

$$\begin{aligned}
\Delta S^\circ &= (2 \text{ mol})(87.4 \text{ J K}^{-1} \text{ mol}^{-1}) - [(4 \text{ mol})(27.3 \text{ J K}^{-1} \text{ mol}^{-1}) \\
&\quad + (3 \text{ mol})(205.0 \text{ J K}^{-1} \text{ mol}^{-1})] \\
&= -549.4 \text{ J K}^{-1}
\end{aligned}$$

Example 26.1 Use the data in Table 26.1 to calculate ΔS° for the reaction

$$CaCO_3(s) \longrightarrow CaO(s) + CO_2(g)$$

Solution

$$\begin{aligned}
\Delta S^\circ &= S^\circ(CaO) + S^\circ(CO_2) - S^\circ(CaCO_3) \\
&= (1 \text{ mol})(38.1 \text{ J K}^{-1} \text{ mol}^{-1}) + (1 \text{ mol})(213.7 \text{ J K}^{-1} \text{ mol}^{-1}) \\
&\quad - (1 \text{ mol})(92.9 \text{ J K}^{-1} \text{ mol}^{-1}) \\
&= 158.9 \text{ J K}^{-1}
\end{aligned}$$

We see that there is a large decrease in entropy in the rusting of iron because the highly dispersed oxygen gas reacts to form a compact, ordered solid. Conversely, in the decomposition of calcium carbonate there is a large increase in entropy because a gas is produced from a solid. Whenever a reaction involves one or more gases, we can easily make a qualitative prediction of the entropy change for the reaction. If the number of moles of gas increases during a reaction, the entropy change for the reaction will be positive. But if the number of moles of gas decreases during a reaction, the entropy change will be negative.

Example 26.2 Predict whether the entropy increases or decreases in the following reaction:

$$N_2(g) + 3H_2(g) \longrightarrow 2NH_3(g)$$

Solution Two moles of gas are produced from 4 mol, so we expect a decrease in the entropy in this reaction; in other words, we predict that ΔS will be negative.

26.3 GIBBS FREE ENERGY

The rusting of iron is accompanied by a large decrease in the entropy of the system. Nevertheless, the rusting of iron is a spontaneous process. To understand why, we must take into account the entropy change in the surroundings, which results from the exchange of energy as heat between the system and its surroundings. An exothermic reaction transfers energy to its surroundings in the form of heat. This energy becomes dispersed in the surroundings; therefore the entropy of the surroundings is increased. Conversely, an endothermic reaction withdraws energy in the form of heat from its surroundings. Therefore the entropy of the surroundings decreases.

To predict whether or not a reaction will occur spontaneously, we must find the entropy change for the system *and* the surroundings. How do we find the entropy change for the surroundings? Thermodynamics provides a quantitative relationship between the energy transferred as heat and the entropy change of the surroundings at constant pressure. The energy transferred as heat from the system to the surroundings is $-\Delta H_{system}$, and the relationship is

$$\Delta S_{surroundings} = -\frac{\Delta H_{system}}{T}$$

where ΔH_{system} is the enthalpy change in the reacting system.

For the rusting of iron the enthalpy change for the reaction

$$4Fe(s) + 3O_2(g) \longrightarrow 2Fe_2O_3(s)$$

is simply twice the molar enthalpy of formation of $Fe_2O_3(s)$:

$$\Delta H° = 2 \Delta H_f°(Fe_2O_3) = (2 \text{ mol})(-824.2 \text{ kJ mol}^{-1}) = -1648.4 \text{ kJ}$$

The release of this energy to the surroundings increases the entropy of the surroundings by the amount $-\Delta H_{system}/T$:

$$\Delta S_{surroundings} = -\frac{\Delta H_{system}}{T} = -\frac{(-1648.4 \text{ kJ})}{298.15 \text{ K}}$$

$$= +5529 \text{ J K}^{-1}$$

This large increase in the entropy of the surroundings completely outweighs the entropy decrease in the reacting system, which we calculated in Section 26.2, so there is a large net increase in the total entropy:

$$\Delta S_{total}° = \Delta S_{system}° + \Delta S_{surroundings}°$$

$$= -549 \text{ J K}^{-1} + 5529 \text{ J K}^{-1} = +4980 \text{ J K}^{-1}$$

Thus although in the rusting of iron there is a decrease in the entropy of the reacting system, there is a larger increase in the entropy of the surroundings. The rusting of iron is therefore a spontaneous process. We can never hope to stop this process, only to slow it up as much as possible.

Instead of calculating ΔS_{system} and $\Delta S_{surroundings}$ in order to predict whether or not a reaction is spontaneous, it would be much more convenient if we could make the prediction on the basis of a property of the system only. If fact, we can do so by making use of the relationship $\Delta S_{surroundings} = -\Delta H_{system}/T$. For a spontaneous reaction

$$\Delta S_{total} = \Delta S_{system} + \Delta S_{surroundings} > 0$$

Substituting $\Delta S_{surroundings} = -\Delta H_{system}/T$ gives

$$\Delta S_{total} = \Delta S_{system} - \frac{\Delta H_{system}}{T}$$

Since all the quantities in this equation refer to the system, we can drop the qualification "system" and write

$$\Delta S_{total} = \Delta S - \frac{\Delta H}{T}$$

Multiplying through by $-T$, we have

$$-T \Delta S_{total} = \Delta H - T \Delta S$$

The quantity $-T\,\Delta S_{total}$ is given the symbol ΔG, $\Delta G = -T\,\Delta S_{total}$. Therefore

$$\Delta G = \Delta H - T\,\Delta S$$

Since H and S are state functions of the system, $G = H - TS$ is also a state function of the system. It is called the *Gibbs function*, the *Gibbs energy*, or, more usually, the **Gibbs free energy**, after J. Willard Gibbs (Figure 12.8), who was the first to recognize its importance. It is one of the most important equations in thermodynamics. It enables us to predict whether or not a given reaction will occur spontaneously. Since $\Delta G = -T\,\Delta S_{total}$, and since ΔS_{total} is greater than zero—that is, is positive—for a spontaneous reaction, ΔG *must be negative for a spontaneous reaction*. In other words, the Gibbs free energy must decrease during a spontaneous process.

For many reactions ΔH is much larger than $T\,\Delta S$, so as a rough approximation, $\Delta G \approx \Delta H$. Since ΔG is negative in a spontaneous process, we see why so many spontaneous reactions are exothermic (ΔH is negative). Nevertheless, there are some endothermic reactions that are spontaneous because they are accompanied by a large increase in the entropy of the system. Many thermal decompositions that form gases, such as the decomposition of calcium carbonate, are of this type.

We can recognize four possible combinations of ΔH and ΔS:

	ΔH	ΔS	ΔG	*Change*
(1)	$-$	$+$	$-$	Spontaneous
(2)	$-$	$-$	$-$ at low T	Spontaneous
			$+$ at high T	Not spontaneous
(3)	$+$	$-$	$+$	Not spontaneous
(4)	$+$	$+$	$+$ at low T	Not spontaneous
			$-$ at high T	Spontaneous

1. This first case is an exothermic reaction (ΔH is negative) in which there is an increase in entropy in the system (ΔS is positive). The heat emitted to the surroundings causes an increase in the entropy of the surroundings, and since the entropy of the system also increases, the total entropy increases and the process is spontaneous.

2. If the entropy of the system decreases (ΔS is negative), then at a sufficiently high temperature $T\,\Delta S$ will be large and negative and therefore $\Delta H - T\,\Delta S$ will be positive, and the process will not be spontaneous. At low temperature, however, $T\,\Delta S$ is small. In this case $\Delta G = \Delta H - T\,\Delta S$ is negative, so the process is spontaneous. At low temperature the surroundings are relatively ordered, so the transfer of a given amount of heat causes a relatively large increase in the disorder. But at high temperature the surroundings are more disordered, and the same amount of heat produces a much smaller increase in the entropy. Thus at low temperature a given amount of heat transferred to the surroundings produces a large entropy increase in the surroundings, $\Delta S_{surroundings} = -\Delta H/T$, which counterbalances the entropy decrease in the system, so the total entropy increase in the universe is positive. However, the transfer of the same amount of heat at high temperature causes only a small increase in the entropy of the surroundings, which is insufficient to overcome the negative entropy change in the system.

3. A reaction that is endothermic and has a negative entropy change can never be spontaneous because ΔG is always positive.

4. But an endothermic reaction may be spontaneous if it is accompanied by a large enough increase in the entropy of the system. Since the term $T\,\Delta S$ increases with increasing T, the reaction is more likely to be spontaneous at high temperature than at low temperature. In an endothermic reaction the surroundings lose heat,

so the entropy of the surroundings must decrease, and this entropy change for a given amount of heat is larger at lower temperatures than at higher temperatures, when the system is more highly disordered. Thus only at some sufficiently high temperature is the negative entropy change in the surroundings sufficiently small that it is outweighed by the positive entropy change in the system.

Notice that a reaction for which ΔG is positive and that is therefore not spontaneous in the forward direction—that is, from left to right—has a negative ΔG in the reverse direction—that is, from right to left. Thus a reaction that does not proceed spontaneously in the forward direction is spontaneous in the reverse direction. Finally, if $\Delta G = 0$ the reaction has no tendency to proceed in either direction; in other words, the system must be at equilibrium. In summary:

ΔG	Reaction
Negative	Spontaneous
Positive	Not spontaneous (spontaneous in reverse direction)
Zero	At equilibrium

Calculating Standard Free Energy Changes

The **standard free energy change**, $\Delta G°$, for any reaction may be found from the standard free energies of formation, $\Delta G_f°$, of the reactants and products in just the same way as a standard enthalpy change is calculated.

$$\Delta G° = \sum \Delta G_f°(\text{products}) - \sum \Delta G_f°(\text{reactants})$$

The **standard free energy of formation**, $\Delta G_f°$, is the free energy change for the formation of 1 mol of a compound from its elements in their standard states. The standard free energies of formation, $\Delta G_f°$, of the elements in their standard states are taken to be zero, as for $\Delta H_f°$ values. But notice that the standard entropies of the elements are *not* zero. As we have seen, absolute values of S can be obtained, but only relative, not absolute, values of G and H can be obtained. Values of $\Delta G_f°$ can be calculated from the corresponding $\Delta H_f°$ and $S°$ values. However, it is a great convenience not to have to do these calculations each time a $\Delta G°$ value is required but to simply refer to a table of $\Delta G_f°$ values, such as Table 26.2.

We will now calculate the entropy, enthalpy, and free energy changes associated with the following reactions:

1. $CaCO_3(s) \longrightarrow CaO(s) + CO_2(g)$
2. $N_2(g) + 3H_2(g) \longrightarrow 2NH_3(g)$
3. $H_2(g) + Cl_2(g) \longrightarrow 2HCl(g)$

First, we calculate the entropy changes for these reactions, using the data in Table 26.1.

1.
$$\Delta S° = S°(CaO) + S°(CO_2) - S°(CaCO_3)$$
$$= (1 \text{ mol})(38.1 \text{ J K}^{-1} \text{ mol}^{-1}) + (1 \text{ mol})(213.7 \text{ J K}^{-1} \text{ mol}^{-1})$$
$$\quad - (1 \text{ mol})(92.9 \text{ J K}^{-1} \text{ mol}^{-1})$$
$$= 158.9 \text{ J K}^{-1}$$

2.
$$\Delta S° = 2S°(NH_3) - [S°(N_2) + 3S°(H_2)]$$
$$= (2 \text{ mol})(192.7 \text{ J K}^{-1} \text{ mol}^{-1}) - [(1 \text{ mol})(191.5 \text{ J K}^{-1} \text{ mol}^{-1})$$
$$\quad + (3 \text{ mol})(130.6 \text{ J K}^{-1} \text{ mol}^{-1})]$$
$$= -197.9 \text{ J K}^{-1}$$

Table 26.2 Standard Free Energies of Formation, ΔG_f°, at 25°C

	ΔG_f° (kJ mol^{-1})		ΔG_f° (kJ mol^{-1})
AgCl(s)	−109.8	CH$_3$CO$_2$H(l)	−390
CO(g)	−137.2	C$_6$H$_6$(l)	124.7
CO$_2$(g)	−394.4	Fe$_2$O$_3$(s)	−742.2
CH$_4$(g)	−50.8	H$_2$S(g)	−33.4
C$_2$H$_6$(g)	−32.9	HCl(g)	−95.3
C$_3$H$_8$(g)	−23.4	HI(g)	1.6
C$_4$H$_{10}$(g)^a	−17.2	HNO$_3$(l)	−80.8
C$_5$H$_{12}$(l)^a	−9.6	H$_2$SO$_4$(l)	−690.1
C$_6$H$_{14}$(l)^a	−4.4	H$_2$O(l)	−237.2
C$_7$H$_{16}$(l)^a	1.0	H$_2$O(g)	−228.6
C$_8$H$_{18}$(l)^a	6.4	NH$_3$(g)	−16.4
CH$_2$O(g)	−113	NO(g)	86.6
CH$_3$OH(l)	−116.4	NO$_2$(g)	51.3
C$_2$H$_5$OH(l)	−174.9	N$_2$O$_4$(g)	97.8
CH$_2$Cl$_2$(l)	−67.3	NaCl(s)	−384.3
CH$_3$Cl(g)	−57.4	NaBr(s)	−349.1
CCl$_4$(l)	−65.3	NaI(s)	−282.4
C$_6$H$_{12}$O$_6$(s) (glucose)	−919.2	CaO(s)	−603.5
		CaCO$_3$(s)	−1128.8

^{a} Straight-chain alkanes.

3.
$$\Delta S^\circ = 2S^\circ(\text{HCl}) - [S^\circ(\text{H}_2) + S^\circ(\text{Cl}_2)]$$
$$= (2 \text{ mol})(186.8 \text{ J K}^{-1}\text{mol}^{-1}) - [(1 \text{ mol})(130.6 \text{ J K}^{-1}\text{mol}^{-1})$$
$$+ (1 \text{ mol})(223.0 \text{ J K}^{-1}\text{mol}^{-1})]$$
$$= 20.0 \text{ J K}^{-1}$$

Note that in the third reaction the entropy change is small because 2 mol of gaseous reactants form 2 mol of gaseous products.

Let us now find the enthalpy changes in these reactions so that we can finally calculate the free energy changes and thus predict whether these reactions will occur spontaneously. Using the data in Tables 12.1, B.2, and B.3, we have

1.
$$\Delta H^\circ = \Delta H_f^\circ(\text{CaO}) + \Delta H_f^\circ(\text{CO}_2) - \Delta H_f^\circ(\text{CaCO}_3)$$
$$= (1 \text{ mol})(-635.1 \text{ kJ mol}^{-1}) + (1 \text{ mol})(-393.5 \text{ kJ mol}^{-1})$$
$$- (1 \text{ mol})(-1206.9 \text{ kJ mol}^{-1})$$
$$= 178.3 \text{ kJ}$$

2.
$$\Delta H^\circ = 2 \Delta H_f^\circ(\text{NH}_3) - [\Delta H_f^\circ(\text{N}_2) - 3 \Delta H_f^\circ(\text{H}_2)]$$
$$= (2 \text{ mol})(-45.9 \text{ kJ mol}^{-1}) - (0 + 0)$$
$$= -91.8 \text{ kJ}$$

3.
$$\Delta H^\circ = 2 \Delta H_f^\circ(\text{HCl}) - [\Delta H_f^\circ(\text{H}_2) + \Delta H_f^\circ(\text{Cl}_2)]$$
$$= (2 \text{ mol})(-92.3 \text{ kJ mol}^{-1}) - (0 + 0)$$
$$= -184.6 \text{ kJ}$$

We see that the first reaction, the thermal decomposition of calcium carbonate, is endothermic; ΔH has a positive value. This reaction can only be spontaneous if it is accompanied by a sufficiently large entropy increase. As we have seen, the entropy change for the reaction is indeed positive. For the reaction to be spontaneous, the entropy increase in the system must be greater than the decrease in the entropy of the surroundings caused by the loss of heat to the system.

The second reaction, the synthesis of ammonia, is exothermic, but it is accompanied by a large negative entropy change. This reaction will proceed spontaneously only if the heat given off to the surroundings produces an increase in the entropy of the surroundings that is larger than the decrease in the entropy of the system.

The synthesis of hydrogen chloride from hydrogen and chlorine is also a strongly exothermic reaction, and since the entropy change is positive, although small, we expect that this reaction will proceed spontaneously.

Let us now calculate the free energy changes for these reactions so that we can examine our qualitative predictions more closely. We can calculate the free energy change either from the expression

$$\Delta G^\circ = \Delta H^\circ - T\,\Delta S^\circ$$

or directly from the values of the free energies of formation listed in Table 26.2.

For reaction (1), $\Delta S^\circ = 159 \text{ J K}^{-1} = 0.159 \text{ kJ K}^{-1}$. Therefore

$$\Delta G^\circ = 178.3 \text{ kJ} - (298 \text{ K})(0.159 \text{ kJ K}^{-1})$$
$$= 130.9 \text{ kJ}$$

Alternatively,

$$\Delta G^\circ = \Delta G_f^\circ(\text{CaO}) + \Delta G_f^\circ(\text{CO}_2) - \Delta G_f^\circ(\text{CaCO}_3)$$
$$= (1 \text{ mol})(-603.5 \text{ kJ mol}^{-1}) + (1 \text{ mol})(-394.4 \text{ kJ mol}^{-1})$$
$$- (1 \text{ mol})(-1128.8 \text{ kJ mol}^{-1})$$
$$= 130.9 \text{ kJ}$$

This large positive ΔG° value indicates that this reaction has no tendency to proceed at 25°C under standard conditions.

For reaction (2), $\Delta S^\circ = -198 \text{ J K}^{-1} = -0.198 \text{ kJ K}^{-1}$. Therefore

$$\Delta G^\circ = -91.8 \text{ kJ} - (298 \text{ K})(-0.198 \text{ kJ K}^{-1})$$
$$= -32.8 \text{ kJ}$$

or alternatively,

$$\Delta G^\circ = 2\,\Delta G_f^\circ(\text{NH}_3) - [\Delta G_f^\circ(\text{N}_2) + 3\,\Delta G_f^\circ(\text{H}_2)]$$
$$= -(2 \text{ mol})(-16.4 \text{ kJ mol}^{-1}) - 0 + 0$$
$$= -32.8 \text{ kJ}$$

In this case the negative ΔG° value indicates that the reaction will, in principle, proceed spontaneously at room temperature. However, we should be careful to note that the statement that a reaction proceeds spontaneously does not mean that it necessarily proceeds rapidly. In fact, a spontaneous reaction may be very slow. For example, the synthesis of ammonia is a very slow reaction at room temperature. In order to speed up the reaction, one must increase the temperature and employ a catalyst (see Chapters 3 and 18).

For reaction (3), $\Delta S° = 20 \text{ J K}^{-1} = 0.020 \text{ kJ K}^{-1}$. Therefore

$$\Delta G° = -184.6 \text{ kJ} - (298 \text{ K})(0.020 \text{ kJ K}^{-1})$$
$$= -190.6 \text{ kJ}$$

Or alternatively,

$$\Delta G° = 2 \Delta G_f°(\text{HCl}) - [\Delta G_f°(\text{H}_2) + \Delta G_f°(\text{Cl}_2)]$$
$$= (2 \text{ mol})(-95.3 \text{ kJ mol}^{-1}) - (0 + 0)$$
$$= -190.6 \text{ kJ}$$

This large negative $\Delta G°$ value indicates that this reaction has a strong tendency to proceed at room temperature. As we saw in Chapters 5 and 18, if this reaction is initiated by a spark or by light of suitable frequency, it proceeds explosively to give a nearly quantitative yield of the products.

Example 26.3 Using the data in Table 26.2, calculate $\Delta G°$ for the reaction

$$\text{CH}_3\text{OH(l)} + \tfrac{3}{2}\text{O}_2(\text{g}) \longrightarrow \text{CO}_2(\text{g}) + 2\text{H}_2\text{O(l)}$$

Solution We have

$$\Delta G° = \Delta G_f°(\text{CO}_2) + 2 \Delta G_f°(\text{H}_2\text{O}) - [\Delta G_f°(\text{CH}_3\text{OH}) + \tfrac{3}{2} \Delta G_f°(\text{O}_2)$$
$$= (1 \text{ mol})(-394.4 \text{ kJ mol}^{-1}) + (2 \text{ mol})(-237.2 \text{ kJ mol}^{-1})$$
$$- [(1 \text{ mol})(-116.4 \text{ kJ mol}^{-1}) + (\tfrac{3}{2} \text{ mol})(0 \text{ kJ mol}^{-1})]$$
$$= -752.4 \text{ kJ}$$

Temperature Dependence of $\Delta G°$

We have seen that the reaction

$$\text{CaCO}_3(\text{s}) \longrightarrow \text{CaO(s)} + \text{CO}_2(\text{g})$$

has a large positive $\Delta G°$ value at room temperature, indicating that it has no tendency to proceed under these conditions; indeed, it tends to proceed in the reverse direction. However, we can see from the equation

$$\Delta G° = \Delta H° - T \Delta S°$$

that because $\Delta S°$ is positive, $\Delta G°$ will have a negative value if the temperature is increased sufficiently. Let us calculate the value of $\Delta G°$ for $T = 1000 \text{ K}$, assuming that ΔS and ΔH do not change very much with temperature:

$$\Delta G° = 178.3 \text{ kJ} - (1000 \text{ K})(0.159 \text{ kJ mol}^{-1}) = 19 \text{ kJ}$$

The value is still positive. But at 1200 K

$$\Delta G° = 178.3 \text{ kJ} - (1200 \text{ K})(0.159 \text{ kJ K}^{-1}) = -13 \text{ kJ}$$

The value of $\Delta G°$ is negative, so we expect the reaction to proceed spontaneously at this temperature.

This prediction agrees with predictions made previously on the basis of Le Châtelier's principle. Since the decomposition of calcium carbonate is an endothermic reaction, Le Châtelier's principle predicts that the equilibrium is shifted to the right if the temperature is raised; that is, the products are favored. At higher temperatures the transfer of a given amount of energy as heat from the surroundings to the system produces a smaller decrease in the entropy of the surroundings than at lower temperatures. At a sufficiently high temperature the entropy decrease in the surroundings becomes smaller than the entropy increase in the reacting system, and the reaction becomes spontaneous.

Since $\Delta G°$ is negative at 1200 K but positive at 1000 K, we could guess that at a temperature of approximately 1100 K, $\Delta G°$ would be zero. Thus at this temperature the reaction would have no tendency to proceed in either direction; it would be at equilibrium. We can easily calculate this temperature from the equation

$$\Delta G° = \Delta H° - T \, \Delta S°$$

by setting $\Delta G° = 0$. In this case we have

$$0 = 179.3 - T(0.159 \text{ kJ K}^{-1})$$

$$T = 1128 \text{ K} = 855°C$$

Thus at 855°C and at standard conditions—that is, a pressure of 1 atm—the system is at equilibrium. At a higher temperature it proceeds to the right; at a lower temperature it proceeds to the left. However, such a calculation is only approximate, because it is based on the assumption that $\Delta H°$ and $\Delta S°$ do not vary with temperature, which is only a rough approximation.

Many thermal decompositions are of this type. Solids that are stable at room temperature often decompose in an endothermic reaction at high temperature, if the decomposition produces one or more gases and is therefore accompanied by a large increase in entropy of the system. For example, nitrates of many metals decompose on heating to give gaseous NO_2 and O_2 (Chapters 9 and 17):

$$2Pb(NO_3)_2(s) \longrightarrow 2PbO(s) + 4NO_2(g) + O_2(g)$$

Ammonium nitrate decomposes on heating to give N_2O and water (Chapter 17):

$$NH_4NO_3(s) \longrightarrow N_2O(g) + 2H_2O(g)$$

Gibbs Free Energy and Equilibrium Constant

In Chapter 14 we introduced the reaction quotient Q, which for the gas-phase reaction $aA + bB + \cdots \rightarrow pP + qQ + \cdots$ can be written in the form

$$Q = \frac{(p_P)^p(p_Q)^q \cdots}{(p_A)^a(p_B)^b \cdots}$$

where $p_A, p_B, \cdots$ are the partial pressures of A, B, $\cdots$ and so on, and in general are *not* the equilibrium partial pressures. We also saw in Chapter 14 that:

- If $Q < K_p$ or $Q/K_p < 1$, the reaction proceeds from left to right; in other words, the reaction is *spontaneous*.
- If $Q > K_p$ or $Q/K_p > 1$, the reaction proceeds from right to left; in other words, it is *not spontaneous* in the direction written (left to right), but it is spontaneous in the reverse direction.
- If $Q = K_p$ or $Q/K_p = 1$, the reaction is at *equilibrium*.

We expect therefore that there will be a relationship between Q/K_p and ΔG. Thermodynamics shows that this relationship is

$$\Delta G = RT \ln Q/K_p$$

or

$$\Delta G = RT \ln Q - RT \ln K_p$$

If all the reactants and products are in their standard states—that is, 1 atm pressure—$\Delta G = \Delta G°$ and $Q = 1$, or $\ln Q = 0$, so that

$$\Delta G° = -RT \ln K_p$$

This equation shows that the equilibrium constant is determined by the standard Gibbs free energy change for the reaction. If $\Delta G°$ is negative, $K_p > 1$ and the reaction is spontaneous under standard conditions. If $\Delta G°$ is positive, $K_p < 1$ and the reaction is not spontaneous under standard conditions. If $\Delta G° = 0$, $K_p = 1$ and the reaction is at equilibrium under standard conditions.

As an example, let us calculate the equilibrium constant for the formation of NO(g) from the elements at 25°C. The equation for the reaction is

$$N_2(g) + O_2(g) \longrightarrow 2NO(g)$$
$$\Delta G° = 2\,\Delta G_f°(NO, g) - [\Delta G_f°(N_2, g) + \Delta G_f°(O_2, g)]$$
$$= (2\text{ mol})(86.6\text{ kJ mol}^{-1}) - (0 + 0) = 173.2\text{ kJ mol}^{-1}$$

Then we can find K_p from the expression

$$\ln K_p = \frac{-\Delta G°}{RT} = -\frac{173.2\text{ kJ mol}^{-1}}{(0.00831\text{ kJ mol}^{-1}\text{ K}^{-1})(298\text{ K})} = -69.9$$
$$K_p = 4.25 \times 10^{-31}$$

This very small equilibrium constant shows that the reaction has very little tendency to proceed at room temperature. At standard conditions, because all the gas pressures are 1 atm, $Q = 1$, so $Q \gg K$, or $Q/K \gg 1$. Thus the reaction is not spontaneous in the direction written but proceeds spontaneously in the reverse direction. But at 25°C this reverse reaction is exceedingly slow.

Example 26.4 From standard free energies of formation, calculate the equilibrium constant, K_p, at 25°C for the reaction

$$N_2(g) + 3H_2(g) \rightleftharpoons 2NH_3(g)$$

Solution We have

$$\Delta G° = (2\text{ mol})[\Delta G_f°(NH_3, g)] - \{(1\text{ mol})[\Delta G_f°(N_2, g)] + (3\text{ mol})[\Delta G_f°(H_2, g)]\}$$

Substituting values from Table 26.2 gives

$$\Delta G° = (2\text{ mol})(-16.4\text{ kJ mol}^{-1}) - (0 + 0) = -32.8\text{ kJ}$$

Also,

$$\Delta G° = -RT \ln K_p$$

Therefore,

$$\ln K_p = \frac{-\Delta G°}{RT} = \frac{32.8\text{ kJ mol}^{-1}}{(0.00831\text{ kJ mol}^{-1}\text{ K}^{-1})(298\text{ K})} = 13.25$$
$$K_p = 5.68 \times 10^5\text{ atm}^{-2}$$

This large value of the equilibrium constant tells us that the position of the equilibrium is far to the right; that is, at equilibrium there will be a high pressure (or concentration) of NH_3 and low pressures (or concentrations) of N_2 and H_2. We can reach essentially the same conclusion from the negative $\Delta G°$ value, which tells us that at standard conditions—all the gases at 1 atm pressure—the reaction is spontaneous to the right.

Example 26.5 Calculate ΔG for the reaction, at 298 K,

$$N_2(g) + 3H_2(g) \rightleftharpoons 2NH_3(g)$$

if the reaction mixture consists of 10 atm N_2, 10 atm H_2, and 1 atm NH_3.

Solution

$$Q = \frac{(p_{NH_3})^2}{(p_{N_2})(p_{H_2})^3} = \frac{1^2}{10 \times 10^3} = \frac{1}{10^4} \quad \text{and} \quad K_p = 5.7 \times 10^5 \text{ atm}^{-2} \text{ (see Example 26.4)}$$

$$\Delta G = -RT \ln \frac{Q}{K_p} = (8.31 \text{ J K}^{-1})(298 \text{ K})\left(\ln \frac{1}{5.7 \times 10^9}\right)$$

$$= (8.31 \text{ J K}^{-1})(298 \text{ K})(-22.5) = -55.7 \text{ kJ}$$

Increasing the pressure of both N_2 and H_2 from 1 to 10 atm changes ΔG from -32.8 to -55.7 kJ. Thus the reaction has a still greater tendency to proceed to the right, as we could have predicted using Le Châtelier's principle.

Gibbs Free Energy and Work

Another very useful property of free energy is that the free energy change for a process is equal to the maximum possible work that can be derived from the process. Any process that occurs spontaneously can in principle be utilized for the performance of work. The greater the free energy change, the greater is the tendency for the reaction to occur spontaneously and the greater is the amount of work that can be obtained from the process.

If we have some gas compressed in a cylinder fitted with a piston, then the spontaneous expansion of the gas will drive the piston, which in turn could be used to turn a wheel, for example. In an automobile engine the spontaneous combustion of gasoline produces a large amount of gas, which drives the pistons of the engine. The amount of work that can be derived from a particular process depends on how it is carried out. If gasoline is burned in an open container, some of the energy of the reaction is used to push away the atmosphere, and the rest is converted to heat so that no work can be obtained. But if gasoline is burned in an automobile engine, some work can be obtained. In practice, the theoretically possible maximum amount of work can never be obtained—all engines are somewhat inefficient. In an automobile engine only about 20% of the maximum possible work is obtained. Nevertheless, it is very useful to know the maximum amount of work that can theoretically be obtained from any chemical reaction or other process.

In our bodies we oxidize foodstuffs to obtain energy to do work. For the oxidation of 1 mol of glucose,

$$\Delta H° = -2808 \text{ kJ} \quad \text{and} \quad \Delta G° = -2870 \text{ kJ}$$

Hence 2870 kJ is the maximum amount of work that can be done by a person as a result of metabolizing 1 mol (180.2 g) of glucose. The work that a person must do in climbing a height h is given by $w = mgh$, where m is the mass of the person and g is the acceleration due to gravity. Thus in order to climb a height of 100 m, a 60-kg woman would need to do $(60 \text{ kg})(9.8 \text{ m s}^{-2})(100 \text{ m}) = 60\,000 \text{ J} = 60 \text{ kJ}$ of work. In order to do this much work, she would need to metabolize a minimum of $(60 \text{ kJ}/2870 \text{ kJ})(180.2 \text{ g}) = 3.8 \text{ g}$ of glucose. In practice, because the conversion of energy to work in the body is not 100% efficient, she would need more than this amount of glucose.

The energy of the hydrogen-oxygen reaction can be converted to work by generating electricity in the hydrogen-oxygen fuel cell (Chapter 16). The reaction is

$$2H_2(g) + O_2(g) \longrightarrow 2H_2O(l)$$

for which

$$\Delta H° = -572 \text{ kJ} \qquad \Delta G° = -474 \text{ kJ} \qquad \Delta S° = -326 \text{ J K}^{-1}$$

The value of $\Delta H° = -572$ kJ means that 572 kJ of heat is evolved during the reaction if no work is done. The value of $\Delta G° = -474 \text{ kJ mol}^{-1}$ means that only 474 kJ of this energy can be converted to work; that is, 474 kJ is the maximum work that can be obtained from the reaction. The difference between 572 and 474—that is, 98 kJ mol^{-1}—must be released to the surroundings to increase the entropy of the surroundings to compensate for the decrease in entropy of the system during the reaction in which 3 mol of gas is converted to 2 mol of liquid. In principle, therefore, $(474/572)(100) = 83\%$ of the energy of this reaction could be converted to electrical energy in a fuel cell. In practice, somewhat less than this amount of work would be obtained.

The reaction between hydrogen and oxygen to produce water is a spontaneous process that can be utilized to do work. The reverse reaction, the decomposition of water to hydrogen and oxygen, is a nonspontaneous reaction with a positive $\Delta G°$. Work must be done to decompose water to hydrogen and oxygen. For the decomposition of 1 mol of water at 25°C and 1 atm pressure, 474 kJ of work must be done, by passing an electric current through the water, for example. Another way in which work can be done on a reacting system in order to drive a reaction in the nonspontaneous direction is to couple the reaction with another reaction that can do work, that is, one that has a negative ΔG.

Coupled Reactions

Consider, for example, the extraction of copper from the ore Cu_2S. For the decomposition

$$Cu_2S(s) \longrightarrow 2Cu(s) + S(s) \tag{1}$$

$\Delta H° = +79.5$ kJ and $\Delta S° = -22.4 \text{ J K}^{-1}$. The reaction is strongly endothermic and has a negative entropy change. Therefore $\Delta G°$ is positive, and the reaction is nonspontaneous at all temperatures. At 25°C we calculate the value of $\Delta G° = \Delta H° - T \Delta S°$ to be $+86.2 \text{ kJ mol}^{-1}$. For this reaction to proceed, work must be done on the system in some way. We can couple it with another reaction so that the overall process is spontaneous. Consider, for example, the reaction

$$S(s) + O_2(g) \longrightarrow SO_2(g) \tag{2}$$

for which

$$\Delta H° = -296.8 \text{ kJ} \qquad \text{and} \qquad \Delta S° = 11.1 \text{ kJ K}^{-1}$$

This reaction is exothermic and is accompanied by an increase in entropy. It is therefore spontaneous under all conditions. At 25°C, $\Delta G° = -300.1$ kJ.

By adding the equations (1) and (2), we obtain the equation for the overall reaction. By adding the values of $\Delta H°$ and $\Delta G°$ for the two reactions, we obtain values for the overall reaction:

$$\begin{array}{llll} \text{Cu}_2\text{S(s)} \longrightarrow 2\text{Cu(s)} + \text{S(s)} & \Delta G^\circ = +86.2\,\text{kJ} & \Delta H^\circ = +79.5\,\text{kJ} \\ \underline{\text{S(s)} + \text{O}_2\text{(g)} \longrightarrow \text{SO}_2\text{(g)}} & \underline{\Delta G^\circ = -300.1\,\text{kJ}} & \underline{\Delta H^\circ = -296.8\,\text{kJ}} \\ \text{Cu}_2\text{S(s)} + \text{O}_2\text{(g)} \longrightarrow 2\text{Cu(s)} + \text{SO}_2\text{(s)} & \Delta G^\circ = -213.9\,\text{kJ} & \Delta H^\circ = -217.3\,\text{kJ} \end{array}$$

We see that the overall reaction is exothermic. More importantly, it has a negative ΔG value and it is spontaneous, because the negative free energy change for the second reaction is larger than the positive free energy change for the first reaction. The second reaction may be said to drive the first reaction.

The coupling of reactions to cause a nonspontaneous reaction to occur is very important in biochemical systems. Many of the reactions that are essential to life do not occur spontaneously in the human body. These reactions are made to occur by coupling them with reaction that are spontaneous. The energy for these nonspontaneous processes is obtained primarily by the metabolism of foods. For example, when glucose is oxidized in the body, a substantial amount of energy is released:

$$\text{C}_6\text{H}_{12}\text{O}_6\text{(s)} + 6\text{O}_2\text{(g)} \longrightarrow 6\text{CO}_2\text{(g)} + 6\text{H}_2\text{O(l)}$$
$$\Delta G^\circ = -2870\,\text{kJ} \qquad \Delta H^\circ = -2816\,\text{kJ}$$

This energy is employed to drive nonspontaneous reactions. It is not coupled to these reactions directly but by way of an intermediate reaction involving the molecules ADP and ATP:

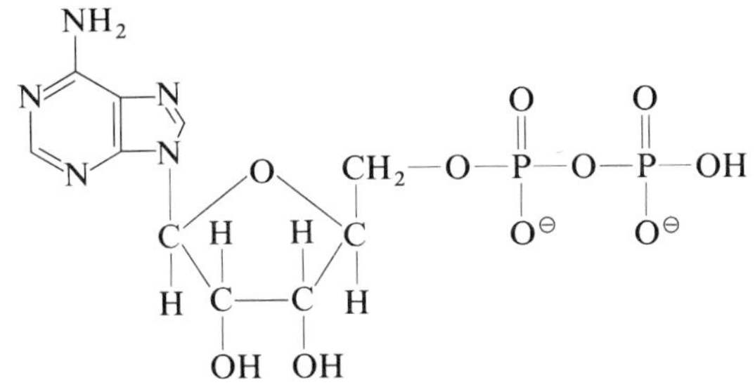

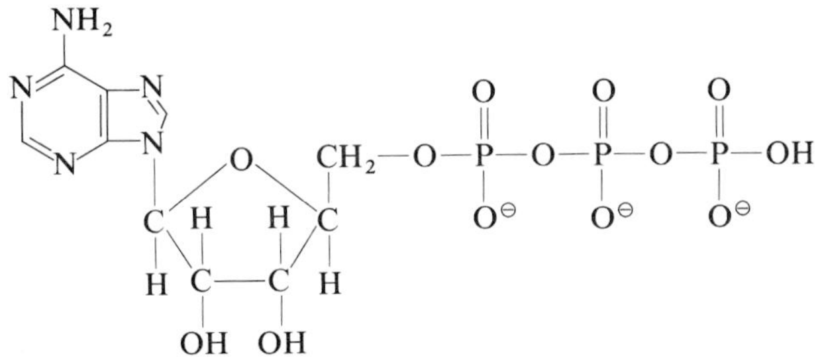

ATP hydrolyzes readily to ADP:

$$\text{ATP} + \text{H}_2\text{O} \longrightarrow \text{ADP} + \text{H}_2\text{PO}_4^- + 2\text{H}^+$$

Under the conditions in the body the entropy, enthalpy, and free energy changes for this reaction are

$$\Delta G^\circ = -30\,\text{kJ} \qquad \Delta H^\circ = -20\,\text{kJ} \qquad \Delta S^\circ = +34\,\text{J K}^{-1}$$

The hydrolysis of ATP can therefore be used to drive any reaction for which $\Delta G < +30\,\text{kJ}$. For example, the biosynthesis of sucrose from glucose and fructose has a $\Delta G°$ of $+23$ kJ. Hence this reaction can be driven by the hydrolysis of ATP:

$$\text{Glucose + Fructose + ATP} \longrightarrow \text{Sucrose + ADP} \qquad \Delta G° = -7\text{ kJ}$$

Gibbs Free Energy and Cell Voltage

We saw in Chapter 16 that the calculated voltage of a cell based on an oxidation-reduction reaction provides another criterion for the spontaneity of a reaction. If the cell voltage is calculated to be positive, then the reaction is spontaneous. But if it is calculated to be negative, the reaction is not spontaneous. If the cell voltage is zero, the reaction is at equilibrium. We therefore expect that there is a relationship between the cell voltage E and the free energy change for the reaction. This relationship is

$$\Delta G° = -nFE°$$

where F is the Faraday constant.

Example 26.6 What is the free energy change associated with the standard zinc-copper cell for which $E° = 1.10$ V.

Solution The reaction in the cell is

$$\text{Zn(s) + Cu}^{2+}\text{(aq)} \longrightarrow \text{Zn}^{2+}\text{(aq) + Cu(s)} \qquad n = 2$$

So

$$\Delta G° = -2(96\,500\text{ C})(1.10\text{ V}) = 212\,000\text{ C V}$$

Since

$$1\text{ V} = \frac{1\text{ J}}{1\text{ C}}$$

then

$$\Delta G° = -212\,000\text{ J} = -212\text{ kJ}$$

This very large negative ΔG value indicates that the reaction has a very strong tendency to proceed from left to right.

The free energy change for a cell reaction is equal to the maximum work that the cell can do in, for example, driving an electric motor. Thus the standard zinc-copper cell can do 212 kJ of work for every mole of zinc that reacts.

The various criteria that we can use to determine the spontaneity of a reaction are summarized in Table 26.3.

Table 26.3 Criteria for Spontaneity of Reactions

	REACTION QUOTIENT EQUILIBRIUM CONSTANT Q/K	GIBBS FREEE ENERGY CHANGE, ΔG	CELL VOLTAGE, E
Spontaneous	<1	$-$	$+$
Not spontaneous	>1	$+$	$-$
Equilibrium	$=1$	0	0

For standard conditions $Q = 1$, $\Delta G = \Delta G°$, $E = E°$

	EQUILIBRIUM CONSTANT, K	STANDARD GIBBS FREE ENERGY CHANGE, $\Delta G°$	STANDARD CELL VOLTAGE $E°$
Spontaneous	>1	$-$	$+$
Not spontaneous	<1	$+$	$-$
Equilibrium	$=1$	0	0

IMPORTANT TERMS

Entropy, S, is a state function that measures the extent of disorder or randomness in a system and the dispersal of energy in the system.

The **Gibbs free energy**, G, is a state function related to the enthalpy, H, the temperature, T, and the entropy, S, of a system: $G = H - TS$. It is a measure of the spontaneity of a reaction.

The **second law of thermodynamics** states that the entropy of the universe (a system plus surroundings) increases in any spontaneous process—or simply that the entropy of the universe is increasing.

The **standard entropy change**, $\Delta S°$, of a reaction is the change in entropy under standard conditions (1 atm and a specified temperature, usually 25°C); it is given by

$$\Delta S° = \sum S°(\text{products}) - \sum S°(\text{reactants})$$

The **standard free energy change**, $\Delta G°$, for a reaction is the change in free energy under standard conditions (1 atm and a specified temperature, usually 25°C); it is given by

$$\Delta G° = \sum \Delta G_f°(\text{products}) - \sum \Delta G_f°(\text{reactants})$$

The **standard free energy of formation**, $\Delta G_f°$, is the free energy change when 1 mol of a substance is formed from its elements in their standard states.

The **standard molar entropy**, $S°$, of a substance is the entropy of 1 mol of the substance in its standard state (1 atm) and at a specified temperature, usually 25°C.

PROBLEMS†

Entropy

1. In each case, use qualitative reasoning to decide which of the following will have the largest entropy.

(a) A mole of ice at 0°C, or a mole of water at the same temperature.

(b) Solid ammonium chloride, or a solution of NH_4Cl in water.

(c) A collection of jigsaw pieces, or the completed puzzle.

(d) One kilogram of raw rubber, or 1 kg of vulcanized rubber.

(e) A pack of cards arranged in suits, or a pack of cards randomly shuffled.

* The asterisk denotes the more difficult problems.

† For data not given in Tables 12.1, 26.1, and 26.2, refer to Tables B.2 and B.3.

2. Will the entropy change in each of the following processes be positive or negative? Does the degree of disorder in each process increase or decrease?

(a) The evaporation of 1 mol of ethanol.

(b) $2Mg(s) + O_2(g) \longrightarrow 2MgO(s)$

(c) $XeO_3(s) \longrightarrow Xe(g) + \frac{3}{2}O_2(g)$

(d) $N_2(g) + 3H_2(g) \longrightarrow 2NH_3(g)$

(e) $BaCl_2 \cdot H_2O(s) \longrightarrow BaCl_2(s) + H_2O(g)$

(f) The dilution of an aqueous solution of NaCl.

3. Predict the sign of the entropy change for each of the following reactions:

(a) $CaCO_3(s) \longrightarrow CaO(s) + CO_2(g)$

(b) $NH_3(g) + HCl(g) \longrightarrow NH_4Cl(s)$

(c) $BaO(s) + CO_2(g) \longrightarrow BaCO_3(s)$

(d) $NaCl(s) \longrightarrow Na^+(aq) + Cl^-(aq)$

4. Does an aqueous solution of Al^{3+} ions have a larger entropy before or after hydration of the ions with water molecules? Why, then, are the ions hydrated?

5. Under what conditions are the following statements true?

(a) "In a spontaneous process the system moves toward a state of lower energy."

(b) "In a spontaneous process the entropy of the system increases."

6. Predict the sign of the entropy change for each of the following reactions:

(a) $2CO(g) + O_2(g) \longrightarrow 2CO_2(g)$
(b) $Mg(s) + Cl_2(g) \longrightarrow MgCl_2(s)$
(c) $I_2(s) \longrightarrow I_2(g)$
(d) $2C_2H_6(g) + 7O_2(g) \longrightarrow 4CO_2(g) + 6H_2O(g)$
(e) $CH_4(g) + 2O_2(g) \longrightarrow CO_2(g) + 2H_2O(l)$
(f) $Al_2Cl_6(g) \longrightarrow 2AlCl_3(g)$

7. What is the entropy change associated with each of the following reactions at 298 K?

(a) $C(s, graphite) + O_2(g) \longrightarrow CO_2(g)$
(b) $C_2H_5OH(l) + 3O_2(g) \longrightarrow 2CO_2(g) + 3H_2O(l)$
(c) $C_6H_{12}O_6(s) + 6O_2(g) \longrightarrow 6CO_2(g) + 6H_2O(l)$
(d) $H_2(g) + I_2(s) \longrightarrow 2HI(g)$

8. What is the entropy change associated with each of the following reactions at 298 K? Explain the sign of each of the entropy changes by comparing qualitatively the molecular disorder in reactants and products.

(a) $CaCO_3(s) \longrightarrow CaO(s) + CO_2(g)$
(b) $Br_2(l) + 3F_2(g) \longrightarrow 2BrF_3(g)$
(c) $2CO(g) + O_2(g) \longrightarrow 2CO_2(g)$
(d) $C(s, graphite) + H_2O(l) \longrightarrow CO(g) + H_2(g)$
(e) $2Na(s) + Cl_2(g) \longrightarrow 2NaCl(s)$

9. What is the entropy change, $\Delta S°$, associated with each of the following?

(a) $S(s, rhombic) + O_2(g) \longrightarrow SO_2(g)$
(b) $N_2(g) + O_2(g) \longrightarrow 2NO(g)$
(c) $P_4(s) + 5O_2(g) \longrightarrow P_4O_{10}(s)$
(d) $4Fe(s) + 3O_2(g) \longrightarrow 2Fe_2O_3(s)$
(e) $2H_2(g) + O_2(g) \longrightarrow 2H_2O(l)$

Gibbs Free Energy

10. Calculate the standard enthalpy change, $\Delta H°$, the standard entropy change, $\Delta S°$, and the standard free energy change, $\Delta G°$, for the reaction

$$Fe_2O_3(s) + 3C(s, graphite) \longrightarrow 2Fe(s) + 3CO(g)$$

from standard data. Is this reaction spontaneous at 25°C? Verify that $\Delta G° = \Delta H° - T \Delta S°$, and discuss whether the

enthalpy change and the entropy change, respectively, work for or against the spontaneity of the reaction. Which factor dominates?

11. Calculate the standard free energy change at 25°C for the reaction

$$H_2(g) + Cl_2(g) \longrightarrow 2HCl(g)$$

Is the reaction as written spontaneous? What are the relative contributions of enthalpy and entropy to the spontaneity of the reaction? Which effect predominates?

12. What is the value of $\Delta G°$ for each of the following reactions? Which of these reactions is spontaneous under standard conditions?

(a) $C_3H_8(g) + 5O_2(g) \longrightarrow 3CO_2(g) + 4H_2O(g)$
(b) $N_2O_4(g) \longrightarrow 2NO_2(g)$
(c) $CH_4(g) + CCl_4(l) \longrightarrow 2CH_2Cl_2(l)$

13. The $\Delta G°$ values for $SO_2(g)$, $H_2S(g)$, and $NO_2(g)$ are, respectively, -300.1, -33.4, and $+51.3$ kJ mol^{-1}. Which of these gases shows the greatest tendency to decompose to its elements at 298 K?

14. Calculate the standard free energy change for each of the following reactions. Comment on the relative oxidizing powers of F_2, Cl_2, and Br_2 and on Zn and Fe as reducing agents.

(a) $2NaF(s) + Cl_2(g) \longrightarrow 2NaCl(s) + F_2(g)$
(b) $2NaBr(s) + Cl_2(g) \longrightarrow 2NaCl(s) + Br_2(l)$
(c) $PbO_2(s) + 2Zn(s) \longrightarrow Pb(s) + 2ZnO(s)$
(d) $Al_2O_3(s) + 2Fe(s) \longrightarrow 2Al(s) + Fe_2O_3(s)$

Gibbs Free Energy and Equilibrium Constant

15. What is the equilibrium constant, K_p, for the reaction

$$2SO_2(g) + O_2(g) \longrightarrow 2SO_3(g)$$

at 298 K? Is this reaction spontaneous at 298 K?

16. What is the equilibrium constant, K_p, for the reaction

$$C(s, graphite) + CO_2(g) \longrightarrow 2CO(g)$$

at 700 K? (Assume $\Delta H°$ and $\Delta S°$ are independent of temperature.)

17. What is the equilibrium constant, K_p, at 298 K for the following reaction?

$$PCl_5(g) \longrightarrow PCl_3(g) + Cl_2(g)$$

18. What is the equilibrium constant, K_p, at 298 K for the following reaction?

$$N_2O_4(g) \longrightarrow 2NO_2(g)$$

19. Consider the reaction

$$CH_4(g) + 2O_2(g) \longrightarrow CO_2(g) + 2H_2O(g)$$

(a) According to its calculated free energy change, is this reaction spontaneous?

(b) What is the value of the equilibrium constant at 25°C?

(c) How do you account for the fact that a mixture of methane and oxygen can remain mixed for a very long time without any detectable reaction?

20. Calculate the free energy change associated with the combustion of liquid methanol:

$$2CH_3OH(l) + 3O_2(g) \longrightarrow 2CO_2(g) + 4H_2O(l)$$

(a) Is the reaction spontaneous under standard conditions?

(b) What is the value of the equilibrium constant at 298 K?

(c) Does the equilibrium constant favor formation of reactants or products?

(d) What effect would increased pressure have on the spontaneity of the reaction?

(e) What effect would increased temperature have on the spontaneity of the reaction?

21. The standard free energy change for the reaction

$$2C(s, \text{graphite}) + H_2(g) \longrightarrow C_2H_2(g)$$

is 209 kJ.

(a) Is this reaction a practical route for the synthesis of ethyne, C_2H_2, at room temperature?

(b) Would the reaction proceed spontaneously at a high temperature?

(c) Calculate the equilibrium constant at 1200 K. (Assume that $\Delta H°$ and $\Delta S°$ are independent of temperature.)

22. From your answer to Problem 18, deduce whether at 25°C a mixture in which the partial pressure of $NO_2(g)$ is 0.020 atm and the partial pressure of $N_2O_4(g)$ is 0.040 atm is at equilibrium. If not, in which direction will the reaction proceed?

Miscellaneous

23. What is ΔG for the reaction

$$2CO(g) + O_2(g) \longrightarrow 2CO_2(g)$$

for an initial gas mixture in which the partial pressure of each gas is 0.020 atm? Is ΔG smaller or larger than $\Delta G°$? In what direction will the reaction proceed in this mixture?

24. What is $\Delta G°$ for each of the following reactions? If these reactions were coupled, what would be the overall reaction, and what would be its $\Delta G°$ value? Would the coupled reaction be spontaneous?

(a) $2CO_2(g) + 4H_2O(l) \longrightarrow 2CH_3OH(l) + 3O_2(g)$

(b) $2C(s) + 2O_2(g) \longrightarrow 2CO_2(g)$

25. Explain, in terms of the changes in the entropy of the system and of the surroundings, why endothermic reactions are favored by an increase of temperature.

26. Explain, in terms of the natural tendency for energy to disperse, why chemical reactions take place spontaneously in the direction corresponding to a decrease in the Gibbs free energy.

*27. The standard Gibbs free energies of formation of CO and CO_2 at 1500°C are

$$\Delta G_f°(CO) = -250 \text{ kJ mol}^{-1}$$

and

$$\Delta G_f°(CO_2) = -380 \text{ kJ mol}^{-1}$$

On the basis of the following information, discuss the feasibility of reducing Al_2O_3, FeO, PbO, and CuO to the metal with carbon at 1500°C.

$4Al + 3O_2 \longrightarrow 2Al_2O_3$	$\Delta G°(1500°C) = -2250 \text{ kJ mol}^{-1}$
$2Fe + O_2 \longrightarrow 2FeO$	$\Delta G°(1500°C) = -250 \text{ kJ mol}^{-1}$
$2Pb + O_2 \longrightarrow 2PbO$	$\Delta G°(1500°C) = -120 \text{ kJ mol}^{-1}$
$2Cu + O_2 \longrightarrow 2CuO$	$\Delta G°(1500°C) = 0 \text{ kJ mol}^{-1}$

APPENDIX A
MATHEMATICAL REVIEW

A.1 SCIENTIFIC (EXPONENTIAL) NOTATION

As we have seen in Chapter 1, pages 15 and 16, it is usual in chemical calculations to express all numbers in a standard form. In scientific notation, all numbers, however large or small, are expressed as a number between $1.000\ldots$ and $9.999\ldots$, multiplied or divided by 10 an appropriate number of times. For example

$$138.42 = 1.3842 \times 10 \times 10$$

is written in the form 1.3842×10^2, where 10^2 means 10×10. Here 2 is the power, or exponent, to which 10 is raised.

In general in scientific notation a number is expressed in the form

$$a.bcd \ldots \times 10^n$$

where $a.bcd\ldots$ is a number between $1.000\ldots$ and $9.999\ldots$ and n is a number (not necessarily a single digit) called an *exponent*.

To express a number smaller than $1.000\ldots$ in scientific notation, the appropriate number between $1.000\ldots$ and $9.999\ldots$ is divided by 10 an appropriate number of times. For example

$$0.000\,138\,42 = \frac{1.3842}{10 \times 10 \times 10 \times 10} = 1.3842 \times 10^{-4}$$

Here the exponent -4 means that the number 1.3842 has been divided by 10 four times or, in other words, multiplied by $\frac{1}{10}$ four times. In general

$$10^{-n} = \frac{1}{10^n}$$

To transform a number larger than $9.999\ldots$ to scientific notation, the decimal point is moved to the left until there is only one nonzero digit before the decimal point. If the decimal point is moved x places, the exponent $n = x$.

Thus in transforming 138.42 to scientific notation the decimal point is moved to the left two places

$$138.42$$

so the exponent $n = 2$ and we can write

$$138.42 = 1.3842 \times 10^2$$

To transform a number smaller than $1.000\ldots$ to scientific notation, the decimal point is moved to the right until there is one nonzero digit before the decimal point. If the decimal point is moved y places the exponent $n = -y$.

Thus in transforming 0.000 138 42 to scientific notation the decimal point is moved to the right four places

$$0.000\ 138\ 42$$

so that the exponent $n = -4$ and we can write

$$0.000\ 138\ 42 = 1.3842 \times 10^{-4}$$

To transform a number from scientific notation to the ordinary form, we must move the decimal point in the opposite direction. If n is positive, the number is larger than $9.9999\ldots$ and the decimal point is moved to the right. Thus to transform 4.21×10^4 we move the decimal point four places to the right

$$4.2100$$

to give 42 100. If n is negative, the number is smaller than $1.000\ldots$ so the decimal point is moved to the left. Thus to transform 6.2×10^{-3} we move the decimal point three places to the left

$$0\,006.2$$

to give 0.0062. Note that if the number we are dealing with is negative, the negative sign is retained when the number is transformed to or from scientific notation. Thus

$$-42100 = -4.21 \times 10^4$$
$$-0.00394 = -3.94 \times 10^{-3}$$

To add or subtract numbers in scientific notation, the power of 10—that is, the exponent, n—must be the same in both numbers. For example, to add 6.234×10^4 and 1.203×10^3 we must first express both numbers so that they have the same exponent, n. If we transform 1.203×10^3 to 0.1203×10^4 we can then add the two numbers

$$6.234 \times 10^4 + 0.1203 \times 10^4 = 6.354 \times 10^4.$$

To multiply two numbers in scientific notation we make use of the relation

$$(10)^x(10)^y = 10^{x+y}$$

In other words, we add the exponents. For example

$$(3.025 \times 10^3)(6.217 \times 10^{-6}) = 18.81 \times 10^{3-6}$$
$$= 18.81 \times 10^{-3}$$
$$= 1.881 \times 10^{-2}$$

To divide two numbers in scientific notation we make use of the relation

$$\frac{10^x}{10^y} = 10^{x-y}$$

In other words, we subtract the exponent of the number in the denominator from the exponent of the number in the numerator. For example

$$\frac{3.81 \times 10^{12}}{6.22 \times 10^{23}} = \frac{3.81}{6.22} \times 10^{12-23}$$

$$= 0.613 \times 10^{-11}$$

$$= 6.13 \times 10^{-12}$$

To raise a number to a power we make use of the relation

$$(10^x)^y = 10^{xy}$$

In other words, we multiply the exponents. For example

$$(3.142 \times 10^3)^4 = (3.142)^4 \times 10^{12}$$

$$= 97.46 \times 10^{12}$$

$$= 9.746 \times 10^{13}$$

To take a root of a number we make use of the relation

$$\sqrt[y]{10^x} = (10^x)^{1/y} = 10^{x/y}$$

In other words, we divide the exponents. For example

$$\sqrt[3]{6.22 \times 10^{23}} = (6.22 \times 10^{23})^{1/3}$$

$$= (0.622 \times 10^{24})^{1/3}$$

$$= (0.622)^{1/3} \times 10^8$$

$$= 0.854 \times 10^8$$

Note that the number must be written in such a form that the exponent is divisible by 3. In this case, therefore, we transformed 6.22×10^{23} to 0.622×10^{24}.

A.2 USING A CALCULATOR

It is essential to be able to enter numbers expressed in scientific (exponential) notation into your calculator. Most calculators have an EXP or EE button to assist with such operations. To enter a number such as 2.5×10^7, first make the three touches required for 2.5, then press the EXP key, and finally press the key for 7. In other words, the keys you press after you've touched EXP correspond to the exponent of 10.

Notice that the "EXP" or "10" does not appear in the readout on your calculator. The blank spaces left between the 2.5 and the 7 (or 07) are meant to imply that the latter number is an exponent of 10.

Practice by entering 6.20×10^9, and 3.83×10^6 into your calculator.

Once a number has been entered in this manner, it can be used as a complete unit in any arithmetic operation. For example, multiplying the numbers 6.2×10^9 and 3.826×10^6 together should yield 2.37×10^{16}; most calculators use the following sequence:

6.20
EXP (or EE)
9
×
3.83
EXP (or EE)
6
=

To enter a number that has a *negative* exponent of 10, use the $+/-$ key on your calculator. For example, to enter 9.7×10^{-11}, first press the keys for 9.7, then EXP (or EE), then the $+/-$ key which changes the sign of the exponent from $+$ to $-$, and finally press the keys for 11. (Some calculators allow you to press the $+/-$ key after the numerical part of the exponent has been punched in.) To divide 3.92×10^{-6} by 4.44×10^{-8} use the following sequence:

$$3.92$$
$$\text{EXP}$$
$$+/-$$
$$6$$
$$\div$$
$$4.44$$
$$\text{EXP}$$
$$+/-$$
$$8$$
$$=$$

The answer is 88.3.

A.3 SIGNIFICANT FIGURES

Introduction

Two types of numbers are used in chemistry

1. Exact numbers
2. Inexact numbers

Examples of exact numbers are numbers of things, such as 10 beakers or 12 pencils, and numbers whose values are precisely fixed by definition (60 minutes = 1 hour; the mass of exactly 12 atomic mass units for 1 atom of carbon-12).

Every measurement (other than that of counting) gives an inexact number because every such measurement is uncertain to some extent. The precision of a measurement is indicated by the number of figures used to record it. The digits in a properly recorded measurement are known as *significant figures*.

Use the following rules to determine the proper number of significant figures to be recorded for the result of a calculation.

Location of the Decimal Point

Zeros which are used to locate the decimal point, that is, zeros *before* the first nonzero digit, are *not* significant. For example, suppose that the distance between two points is measured as 3 cm. This measurement could also be expressed as 0.03 m.

$$3 \text{ cm} = 0.03 \text{ m.}$$

Both values contain only one significant figure. The zeros in the second value are not significant since they only serve to locate the decimal point.

Zeros that arise as part of a measurement as significant. For example,

1. The number 106.540 has 6 significant figures
2. The number 0.0005030 has 4 significant figures
3. The number 6.02×10^{23} has 3 significant figures

Numbers are often expressed in scientific (exponential) notation, using the appropriate number of significant figures, in order to avoid ambiguity. Thus, a number ending in zero such as 1580 is best expressed as 1.580×10^3 or 1.58×10^3 depending upon whether or not the measurement merits three or four significant figures.

Rounding Off

If the calculated answer to a problem contains more figures than are significant, it should be rounded off:

1. If the figures following the last number to be retained are 4999... or less, they are discarded and the last number is left unchanged. For example,

$$3.624 \text{ is } 3.62 \text{ to 3 significant figures.}$$

2. If the figures following the last number to be retained are 5000... or greater, they are discarded and the last number is increased by one. For example,

$$7.635 \text{ becomes } 7.64 \text{ if there are 3 significant figures.}$$
$$28.7257 \text{ becomes } 28.726 \text{ if there are 5 significant figures.}$$

Calculations

ADDITION AND SUBTRACTION The result of an addition or subtraction should be reported to the same number of decimal places as that of the term with the least number of decimal places. For example, the answer to the addition

$$
\begin{array}{r}
28.16 \\
5.423 \\
0.0004 \\
\hline
33.5834
\end{array}
$$

should be reported as 33.58 (four significant figures).

MULTIPLICATION AND DIVISION The result of a multiplication or division is rounded off to the number of significant figures in the least precise term used in the calculation. The result of the multiplication

$$52.064 \times 1.24 = 64.5594$$

would be reported as 64.6 since the least precise term (1.24) has three significant figures. The result of the division

$$\frac{0.24}{1.346} = 0.1783$$

would be reported as 0.18 since the least precise term (0.24) has only two significant figures.

The presence of exact numbers in a mathematical expression does not affect the number of significant figures in the answer. Thus the result of the following calculation, in which the exact number 2 is used,

$$\frac{2.58 \times 0.1056 \times 2}{0.0267} = 20.4081$$

would be expressed as 20.4, that is, to three significant figures in keeping with the data. In other words, an exact number is considered to have an infinite number of significant figures.

MULTISTEP CALCULATIONS When a calculation involves several steps, small errors often are introduced by rounding off at the intermediate stages. Sometimes these errors become important if several rounding off corrections all happen to be in the same direction. Such errors may be avoided by doing all the numerical calculations in a problem in one operation, or by carrying an extra digit on the intermediate figures in a calculation, and rounding off only the final answer.

A.4 LOGARITHMS, EXPONENTS, AND EXPONENTIALS

Certain relationships in the study of chemistry involve exponents, exponentials, and/or logarithms. Here we review some properties of these operations.

If a number y is expressed in the form a^b, then we say that the exponent b is the *logarithm of y to the base a.* Thus,

if $$y = a^b$$

then $$\log_a y = b$$

The most common base used in science is the natural number $e(= 2.718\ldots)$; logarithms to base e often are abbreviated ln. Thus,

if $$y = e^b$$

then $$\log_e y = \ln y = b$$

Values for the ln (natural logarithm) of numbers can be obtained from your calculator by first entering the number and then pressing the LN button.

> **Exercise** Obtain ln (38.43), and ln (8.4×10^{-2})
> **Answers** 3.649, -2.48

Given the value x for the ln of a number y, it is possible, using your calculator, to establish the value of the number y; $-y$ is called the antilogarithm of x. If your calculator has an e^x button, enter the value (x) of the logarithm and then press this button. For example, entering 3.649 and pressing e^x yields 38.436. Thus we have reversed the process used in the exercise above. If your calculator has no e^x button, the value can be obtained by entering x and pressing INV then LN; that is, by "inverting" the ln operation.

> **Exercise** Given that ln $y = -3.58$, find the value of y
> **Answer** 2.79×10^{-2}

In some applications it is required to raise some number y (which we'll call the base) to the exponent of ("power of") another number x, for example, $0.5^{3.2}$. If your calculator has a y^x button, first enter the base y (0.5), press the y^x button, then enter 3.2 and press =. For example, by this procedure $0.5^{3.2}$ is found to equal 0.109.

In some applications, the value b of the exponent of a term in an equation is unknown; it is then convenient to rewrite the equation in logarithmic form in order to solve it. This can be done readily by realizing that if

$$y = a^b$$

then it follows that

$$\ln y = \ln (a^b)$$

In other words, we can equate the logarithms of the two sides of any equation. Further, recall that

$$\ln (a^b) = b \ln a$$

Thus our equation relating $\ln y$ to $\ln a$ can be transformed to

$$\ln y = b \ln a$$

Solving for b, we obtain

$$b = \frac{\ln y}{\ln a}$$

The value of b is then obtained by dividing the numerical value for $\ln y$ by the value for $\ln a$ (but *not* the $\ln$ of y/a!).

For example, let us find the value of the exponent b for which the following equation is true.

$$1.342 = 1.800^b$$

First rewrite the equation in logarithmic form:

$$\ln 1.342 = b \ln 1.800$$

Therefore

$$b = \frac{\ln 1.342}{\ln 1.800} = \frac{0.2942}{0.5878} = 0.50$$

In other applications we may need to relate the logarithm of a product of two numbers to the logs of the individual numbers:

If

$$y = ab$$

then

$$\ln y = \ln a + \ln b$$

Similarly

if

$$y = ae^x$$

then

$$\ln y = \ln a + \ln e^x$$
$$= \ln a + x$$

Similarly

if

$$y = c/d$$

then

$$\ln y = \ln c - \ln d$$

Thus if we encounter equations which involve sums of differences of logarithmic terms, it often is convenient to re-express them as logarithms of products or quotients. We could rewrite

$$\ln a + \ln b \quad \text{as} \quad \ln (ab),$$

and

$$\ln c - \ln d \quad \text{as} \quad \ln (c/d).$$

In chemistry, some quantities are defined as the logarithms to base 10 of other quantities. For example, $pH = -\log_{10} [H^+]$. If a number, y, is expressed as a power of 10, that is, as

$$y = 10^b$$

then the logarithm to base 10 of y is b, thus

$$\log_{10} y = b$$

The value of the $\log_{10}$ of any number can be obtained by entering the number in your calculator and pressing the LOG button; on some calculators this is a second function and another button must first be pressed—see your instructions.

Exercise Evaluate $\log_{10}$ of 38.43 and of 8.4×10^{-2}
Answers 1.585, -1.076

In the expression

$$\log_{10} z = x$$

x is called the logarithm of z and z is called the antilogarithm of x. Thus for the expression

$$\log_{10} z = -1.076$$

which is equivalent to

$$z = 10^{-1.076}$$

if we wish to evaluate z, we say that we need the antilog of -1.076. To find this, we enter -1.076 in the calculator and press the 10^x or INV LOG button. This gives the result

$$8.4 \times 10^{-2}$$

Exercise Given that $\log_{10} z = -1.076$, evaluate z.
Answer 8.4×10^{-2}

In summary, it is useful to memorize these relationships:

If $y = e^x$, $\log_e y = \ln y = x$

If $y = 10^x$, $\log_{10} y = x$

If $y = a^b$, $\ln y = b \ln a$, $\log_{10} y = b \log_{10} a$

If $y = ab$, $\ln y = \ln a + \ln b$, $\log_{10} y = \log_{10} a + \log_{10} b$

If $y = c/d$, $\ln y = \ln c - \ln d$, $\log_{10} y = \log_{10} c - \log_{10} d$

A.5 SOLVING MATHEMATICAL EQUATIONS

Many quantitative problems in introductory chemistry involve solving an algebraic equation. Many students, particularly those who have not dealt with arithmetic and algebra for several years, initially have difficulties with equation solving. If you follow the guidelines below, however, you should gain proficiency in the required techniques.

An equation may be simple in appearance,

$$4y = 15 \qquad \textbf{(1)}$$

or rather complicated

$$ay^2 = be^{-cx} \qquad \textbf{(2)}$$

In most cases, however, solving an equation requires only two steps:

1. Rearrange the equation. On the left side, isolate the variable (y or a or b or ...) whose numerical value is *not* known and for which you are trying to determine a value. After this step, the equation will have the form

Left side		Right side
unknown variable	=	algebraic expression involving numbers and variables whose values *are* known to you

2. Substitute numerical values for variables in the right-hand side of the new equation. Using your calculator determine the value for the unknown variable.

For example consider equation (1) above. We can isolate the unknown, y, on the left side by dividing both sides of the equation by the constant on the left, that is, by 4

$$\frac{4y}{4} = \frac{15}{4}$$

thus,

$$y = 15/4$$

or

$$y = 3.75$$

Next consider equation (2). Let us assume first that the quantity whose value we wish to determine is b. To obtain it on the left side, switch sides of the equation:

$$be^{-cx} = ay^2$$

Then divide both sides by the terms *other* than b on the left, that is, divide by e^{-cx}:

$$b = \frac{ay^2}{e^{-cx}}$$

If we are supplied with numerical values for a, y, c, and x, the value of b can be computed.

As another example, suppose that the unknown variable in equation (2) is y rather than b. Dividing both sides of the original equation by a isolates y^2 on the left:

$$y^2 = \frac{be^{-cx}}{a}$$

To obtain y, take the square root of both sides:

$$y = \sqrt{\frac{be^{-cx}}{a}}$$

Finally, consider a case in which the unknown variable in equation (2) is x. We can isolate (on the left) the exponential term involving x by switching sides of the equation

$$be^{-cx} = ay^2$$

and then dividing both sides by b to obtain

$$e^{-cx} = ay^2/b$$

Since x occurs in an exponent, this equation must be manipulated further so that it takes the form

$$x = \ldots\ldots\ldots$$

To accomplish this, take the logarithm to base e, that is, ln, of both sides of the equation:

Since $\qquad\qquad\qquad\qquad \ln(e^{-cx}) = -cx$

Therefore $\qquad\qquad\qquad -cx = \ln(ay^2/b)$

Dividing by $-c$, we obtain our final result

$$x = \frac{-\ln(ay^2/b)}{c}$$

As an example of these procedures, consider the equation

$$0.5^x A = Q^2(1 - 1/T)$$

Suppose that we must solve for A, given that $X = 2$, $Q = 0.0048$, and $T = 4.8$. To obtain an equation for A, divide both sides by 0.5^x:

$$A = \frac{Q^2(1 - 1/T)}{0.5^x}$$

Now substitute the values for the variables on the right side:

$$A = \frac{(0.0048)^2(1 - 1/4.8)}{0.5^2}$$

Evaluating the two terms on top and the one on bottom gives

$$A = \frac{(0.0000230)(0.792)}{0.25}$$

which yields

$$A = 0.000073$$

Next let us suppose that we must solve for T, given the values $A = 4.20$, $Q = 3.90$, and $x = 3$. First switch sides of the equation to obtain T on the left:

$$Q^2(1 - 1/T) = 0.5^x A$$

The term involving T is isolated by dividing both sides by Q^2:

$$1 - 1/T = \frac{0.5^x A}{Q^2}$$

Subtracting 1 from both sides yields a term on the left which involves T only:

$$-1/T = \frac{0.5^x A}{Q^2} - 1$$

Rather than inverting the entire equation to obtain T, it is easier to evaluate the right side first, otherwise a complicated equation will result

$$-1/T = \frac{0.5^3 \times 4.20}{(3.90)^2} - 1$$

$$-1/T = -0.965$$

Inverting both sides gives

$$-T = -1/0.965$$
$$= -1.04$$

Multiplying both sides by -1 eliminates the negative signs:

$$T = +1.04$$

Finally, let us suppose that we must solve the equation for x, given that $A = 6.8$, $Q = 0.37$, and $T = 3.5$. Dividing through by A isolates the term involving x on the left:

$$0.5^x = \frac{Q^2}{A}(1 - 1/T)$$

To remove x from the exponent and have it appear on the line, take the logarithm of both sides:

$$\ln 0.5^x = \ln\left[\frac{Q^2(1 - 1/T)}{A}\right]$$

Now by the properties of logarithms

$$\ln 0.5^x = x \ln 0.5$$

we then obtain

$$x \ln 0.5 = \ln\left[\frac{Q^2(1 - 1/T)}{A}\right]$$

Dividing both sides by $\ln 0.5$, we obtain an equation for x

$$x = \frac{\ln\left[\dfrac{Q^2(1 - 1/T)}{A}\right]}{\ln 0.5}$$

Now we substitute values for Q, T, and A

$$x = \frac{\ln\left[\dfrac{0.37^2(1 - 1/3.5)}{6.8}\right]}{\ln 0.5}$$

$$= \frac{\ln 0.0144}{\ln 0.5}$$

$$= \frac{-4.24}{-0.693}$$

$$= +6.1$$

Solving Quadratic Equations

In some equilibrium calculations, you may be required to solve equations of the form

$$\frac{y^2}{M - y} = K \quad \text{and also} \quad \frac{y(N + y)}{M - y} = K$$

By multiplying both sides by $(M - y)$ and rearranging, either type can be recast into the standard form for a quadratic equation, that is,

$$ay^2 + by + c = 0$$

Recall that the general solutions to this equation are

$$y = \frac{-b \pm \sqrt{b^2 - 4ac}}{2a}$$

We'll only encounter situations where y and b are positive, and thus the only root of interest is the positive one, that is

$$y = \frac{\sqrt{b^2 - 4ac} - b}{2a}$$

To solve for y, given values of a, b, and c, follow the following steps:

1. evaluate $b^2 - 4ac$
2. obtain the square root of $b^2 - 4ac$ (press $\sqrt{}$ on your calculator)
3. from this result, subtract b
4. divide your answer by 2, then by a

For example, to solve

$$\frac{y^2}{10^{-2} - y} = 5 \times 10^{-4}$$

we multiply both sides by $10^{-2} - y$ and obtain

$$y^2 = 5 \times 10^{-4}(10^{-2} - y)$$

that is

$$y^2 = 5 \times 10^{-6} - 5 \times 10^{-4}y$$

All terms are now brought to the left side:

$$y^2 + 5 \times 10^{-4}y - 5 \times 10^{-6} = 0$$

Comparing to the standard form, we make the following identifications:

$$a = 1$$
$$b = 5 \times 10^{-4}$$
$$c = -5 \times 10^{-6}$$

Thus

$$b^2 - 4ac = 25 \times 10^{-8} + 20 \times 10^{-6} = 2.025 \times 10^{-5}$$
$$\sqrt{b^2 - 4ac} = 4.50 \times 10^{-3}$$
$$\sqrt{b^2 - 4ac} - b = 4.00 \times 10^{-3}$$

Therefore

$$y = (\sqrt{b^2 - 4ac} - b)/2a = 2.00 \times 10^{-3}$$

Similarly, solving

$$\frac{y^2}{2 \times 10^{-4} - y} = 2 \times 10^{-6}$$

yields

$$y = 1.90 \times 10^{-5}$$

APPENDIX B

TABLES

Table B.1 Atomic and Molar Masses of the Elements[a]

ELEMENT	SYMBOL	ATOMIC NUMBER	ATOMIC MASS (u) MOLAR MASS (g mol^{-1})	ELEMENT	SYMBOL	ATOMIC NUMBER	ATOMIC MASS (u) MOLAR MASS (g mol^{-1})
Actinium	Ac	89	227.0278[b]	Iodine	I	53	126.9045
Aluminum	Al	13	26.981 54	Iridium	Ir	77	192.22
Americium	Am	95	(243)[c]	Iron	Fe	26	55.847
Antimony	Sb	51	121.75	Krypton	Kr	36	83.80
Argon	Ar	18	39.948	Lanthanum	La	57	138.9055
Arsenic	As	33	74.9216	Lawrencium	Lr	103	(260)[c]
Astatine	At	85	(210)[c]	Lead	Pb	82	207.2
Barium	Ba	56	137.33	Lithium	Li	3	6.941
Berkelium	Bk	97	(247)[c]	Lutetium	Lu	71	174.967
Beryllium	Be	4	9.012 18	Magnesium	Mg	12	24.305
Bismuth	Bi	83	208.9804	Manganese	Mn	25	54.9380
Boron	B	5	10.81	Mendelevium	Md	101	(258)[c]
Bromine	Br	35	79.904	Mercury	Hg	80	200.59
Cadmium	Cd	48	112.41	Molybdenum	Mo	42	95.94
Calcium	Ca	20	40.08	Neodymium	Nd	60	144.24
Californium	Cf	98	(249)[c]	Neon	Ne	10	20.179
Carbon	C	6	12.011	Neptunium	Np	93	237.0482[b]
Cerium	Ce	58	140.12	Nickel	Ni	28	58.69
Cesium	Cs	55	132.9054	Niobium	Nb	41	92.9064
Chlorine	Cl	17	35.453	Nitrogen	N	7	14.0067
Chromium	Cr	24	51.996	Nobelium	No	102	(259)[c]
Cobalt	Co	27	58.9332	Osmium	Os	76	190.2
Copper	Cu	29	63.546	Oxygen	O	8	15.9994
Curium	Cm	96	(247)[c]	Palladium	Pd	46	106.42
Dysprosium	Dy	66	162.50	Phosphorus	P	15	30.973 76
Einsteinium	Es	99	(252)[c]	Platinum	Pt	78	195.08
Erbium	Er	68	167.26	Plutonium	Pu	94	(244)[c]
Europium	Eu	63	151.96	Polonium	Po	84	(209)[c]
Fermium	Fm	100	(257)[c]	Potassium	K	19	39.0983
Fluorine	F	9	18.998 403	Praseodymium	Pr	59	140.9077
Francium	Fr	87	(223)[c]	Promethium	Pm	61	(145)[c]
Gadolinium	Gd	64	157.25	Protactinium	Pa	91	231.0359[b]
Gallium	Ga	31	69.72	Radium	Ra	88	226.0254[b]
Germanium	Ge	32	72.59	Radon	Rn	86	(222)[c]
Gold	Au	79	196.9665	Rhenium	Re	75	186.207
Hafnium	Hf	72	178.49	Rhodium	Rh	45	102.9055
Helium	He	2	4.002 60	Rubidium	Rb	37	85.4678
Holmium	Ho	67	164.9304	Ruthenium	Ru	44	101.07
Hydrogen	H	1	1.007 94	Samarium	Sm	62	150.36
Indium	In	49	114.82	Scandium	Sc	21	44.9559

[a] The atomic masses of many elements are not invariant but depend on the origin and treatment of the material; the values given here apply to elements as they exist naturally on the earth and to certain artificial elements.

[b] For these radioactive elements the mass given is that for the longest-lived isotope.

[c] Atomic masses for these radioactive elements cannot be quoted precisely without knowledge of the origin of the elements; the value given is the atomic mass number of the isotope of that element of longest known half-life.

Table B.1 (continued)

ELEMENT	SYMBOL	ATOMIC NUMBER	ATOMIC MASS (u) MOLAR MASS (g mol^{-1})	ELEMENT	SYMBOL	ATOMIC NUMBER	ATOMIC MASS (u) MOLAR MASS (g mol^{-1})
Selenium	Se	34	78.96	Thulium	Tm	69	168.9342
Silicon	Si	14	28.0855	Tin	Sn	50	118.69
Silver	Ag	47	107.8682	Titanium	Ti	22	47.88
Sodium	Na	11	22.989 77	Tungsten	W	74	183.85
Strontium	Sr	38	87.62	Uranium	U	92	238.0289
Sulfur	S	16	32.06	Vanadium	V	23	50.9415
Tantalum	Ta	73	180.9479	Xenon	Xe	54	131.29
Technetium	Tc	43	(98)	Ytterbium	Yb	70	173.04
Tellurium	Te	52	127.60	Yttrium	Y	39	88.9059
Terbium	Tb	65	158.9254	Zinc	Zn	30	65.38
Thallium	Tl	81	204.37	Zirconium	Zr	40	91.22
Thorium	Th	90	232.0381[a]				

Table B.2 Thermodynamic Data: Elements and Inorganic Compounds

COMPOUND	ΔH_f° kJ mol^{-1}	S° J K^{-1} mol^{-1}	ΔG_f° kJ mol^{-1}	COMPOUND	ΔH_f° kJ mol^{-1}	S° J K^{-1} mol^{-1}	ΔG_f° kJ mol^{-1}
Ag(s)	0.0	42.6	0.0	Na(s)	0.0	51.3	0.0
AgCl(s)	−127.1	96.2	−109.8	NaF(s)	−573.7	51.3	−546.3
Al$_2$O$_3$(s)	−1676	50.9	−1582	NaCl(s)	−411.2	72.5	−384.3
B$_5$H$_9$(s)	73.2	276	175	NaBr(s)	−361.1	87.2	−349.1
B$_2$O$_3$(s)	−1273.5	54.0	−1194.4	NaI(s)	−287.8	98.5	−282.4
Br$_2$(l)	0.0	152.2	0.0	NaOH(s)	−425.6	64.5	−379.7
BrF$_3$(g)	−255.6	292.4	−229.5	Na$_2$O$_2$(s)	−511.7	104	−451.0
CaO(s)	−635.1	38.1	−603.5	NH$_3$(g)	−46.2	192.7	−16.4
CaCO$_3$(s) (calcite)	−1206.9	92.9	−1128.8	NO(g)	90.3	210.6	86.6
Cl$_2$(g)	0.0	223.0	0.0	NO$_2$(g)	33.2	240.0	51.3
Cu(s)	0.0	33.2	0.0	HNO$_3$(l)	−174.1	155.6	−80.8
F$_2$(g)	0.0	202.7	0.0	NOCl(g)	51.7	261.6	66.1
Fe(s)	0.0	27.3	0.0	O$_2$(g)	0.0	205.0	0.0
Fe$_2$O$_3$(s) (hematite)	−824	87.4	−742.2	O$_3$(g)	142.7	238.8	163.2
H(g)	218.0	114.6	203.3	P(s) (white)	0.0	41.1	0.0
H$_2$(g)	0.0	130.6	0.0	P$_4$O$_{10}$(s)	−3010	231	−2724
HCl(g)	−92.3	186.8	−95.3	PCl$_3$(g)	−287.0	311.7	−267.8
HI(g)	26.4	206.5	1.6	PCl$_5$(g)	−374.9	364.5	−305.0
H$_2$O(g)	−241.8	188.7	−228.6	PbO$_2$(s)	−277.4	68.6	−217.4
H$_2$O(l)	−258.8	70.0	−237.2	S(s) (rhombic)	0.0	32.0	0.0
Hg(l)	0.0	75.9	0.0	H$_2$S(g)	−20.6	205.6	−33.4
I$_2$(s)	0.0	116.1	0.0	SiO$_2$ (quartz)	−910.7	41.5	−856.3
I$_2$(g)	62.4	260.6	19.4	SiCl$_4$(l)	−687.0	239.7	−619.9
MgO(s)	−601.5	27.0	−569.2	SO$_2$(g)	−296.8	248.1	−300.1
MnO$_2$(s)	−520.0	53.1	−465.2	SO$_3$(g)	−395.7	256.6	−371.1
N$_2$(g)	0.0	191.5	0.0	H$_2$SO$_4$(l)	−814.0	145.9	−690.1
N$_2$O$_4$(g)	9.3	304.2	97.8	ZnO(s)	−350.5	43.6	−320.5

Table B.3 Thermodynamic Data: Carbon and Carbon Compounds

COMPOUND	ΔH_f° kJ mol^{-1}	S° J K^{-1} mol^{-1}	ΔG_f° kJ mol^{-1}
C(g)	716.7	158.0	671.3
C (graphite)	0.0	5.8	0.0
C (diamond)	1.9	2.4	2.9
CO(g)	−110.5	197.6	−137.2
CO$_2$(g)	−393.5	213.7	−394.4
CH$_4$(g)	−74.5	186.1	−50.8
C$_2$H$_2$(g)	228.0	200.8	209.2
C$_2$H$_4$(g)	52.3	219.4	68.1
C$_2$H$_6$(g)	−84.7	229.5	−32.9
C$_3$H$_6$(g) (cyclopropane)	53.3	237	104
C$_3$H$_8$(g)	−103.8	269.9	−23.4
C$_4$H$_8$(g) (cyclobutane)	28.4	265	100
C$_4$H$_{10}$(g) (n-butane)	−126.1	310.1	−17.2
C$_5$H$_{10}$(g) (cyclopentane)	−78.4	293	39
C$_5$H$_{12}$(g) (n-pentane)	−146.4	348.9	−8.4
C$_5$H$_{12}$(l) (n-pentane)	−173.2	263.3	−9.6
C$_6$H$_6$(l) (benzene)	49.0	172.2	124.7
C$_6$H$_{12}$(g) (cyclohexane)	−123.3	298	32
n-C$_6$H$_{14}$(l)	−198.6	295.9	−4.4
n-C$_7$H$_{16}$(l)	−224.0	328.5	1.0
n-C$_8$H$_{18}$(l)	−250.0	361.2	6.4
CH$_2$O(g)	−108.7	218.7	−113
CH$_3$OH(l)	−239.1	126.8	−166.4
C$_2$H$_5$OH(l)	−277.1	160.7	−174.9
CH$_3$CO$_2$H(l)	−484.3	159.8	−390
C$_6$H$_{12}$O$_6$(s) (glucose)	−1273.3	182.4	−919.2
C$_{12}$H$_{12}$O$_{11}$(s) (sucrose)	−2226.1	360	
CH$_2$Cl$_2$(l)	−124.1	177.8	−67.3
CHCl$_3$(l)	−135.1	201.7	−73.7
CCl$_4$(l)	−129.6	216.4	−65.3

The International System of Units (SI)

Table B.4 SI Base Units

PHYSICAL QUANTITY	NAME	SYMBOL
Length	meter	m
Mass	kilogram	kg
Time	second	s
Electric current	ampere	A
Thermodynamic temperature	kelvin	K
Amount of substance	mole	mol
Luminous intensity	candela	cd

The above base units are defined as follows.

1. The **meter** is the length equal to 1 650 763.73 wavelength in vacuum of the radiation corresponding to the transition between the levels 2p$_{10}$ and 5d, of the krypton-86 atom.

2. The **kilogram** is the unit of mass; it is equal to the mass of the international prototype of the kilogram.

3. The **second** is the duration of 9 192 631 770 periods of the radiation corresponding to the transition between the two hyperfine levels of the ground state of the cesium-133 atom.

4. The **ampere** is that constant current which, if maintained in two straight parallel conductors of infinite length, of negligible curcular cross section, and placed 1 meter apart in a vacuum, would produce between these conductors a force equal to 2×10^{-7} newton per meter of length.

5. The **kelvin** unit of thermodynamic temperature is the fraction 1/273.16 of the thermodynamic temperature of the triple point of water.

6. The **mole** is the amount of substance of a system which contains as many elementary entities as there are atoms in 0.012 kilogram of carbon-12. When the mole is used, the elementary entities must be specified and may be atoms, molecules, ions, electrons, other particles, or specified groups of such particles.

7. The **candela** is the luminous intensity, in the perpendicular direction, of a surface of 1/600 000 square meter of black body at the temperature of freezing platinum under a pressure of 101 325 newtons per square meter.

Table B.5 SI Derived Units

PHYSICAL QUANTITY	NAME	SYMBOL	DEFINITION
Frequency	hertz	Hz	s^{-1}
Energy	joule	J	$kg\, m^2\, s^{-2}$
Force	newton	N	$kg\, m\, s^{-2} = J\, m^{-1}$
Power	watt	W	$kg\, m^2\, s^{-3} = J\, s^{-1}$
Pressure	pascal	Pa	$kg\, m^{-1}\, s^{-2} = N\, m^{-2} = J\, m^{-3}$
Electric charge	coulomb	C	$A\, s$
Electric potential difference	volt	V	$kg\, m^2\, s^{-3}\, A^{-1} = J\, A^{-1}\, s^{-1}$
Electric resistance	ohm	Ω	$kg\, m^2\, s^{-3}\, A^{-2} = V\, A^{-1}$

CHAPTER 1

1 (a) hydrogen H   carbon C.   oxygen O   nitrogen N
chlorine Cl   sodium Na   potassium K   silicon Si   iron Fe
nickel Ni   chromium Cr   krypton Kr   barium Ba   lead Pb
uranium U   (b) He Helium   Ne Neon   F Fluorine
Mg Magnesium   Al Aluminum   P Phosphorus   S Sulfur
Ca Calcium   Fe Iron   Mn Manganese   Co Cobalt
Cu Copper   Zn Zinc   As Arsenic   K Potassium
Br Bromine   Ag Silver   Pt Platinum   Au Gold

3.

MATERIAL	COLOR	STATE	ELECTRICAL CONDUCTIVITY
Water	colorless	liquid	very little (when pure)
Sugar	white	solid	none
Mercury	silver	liquid	high
Copper	red-brown	solid	high
Maple syrup	light brown	liquid	very little
Oxygen	colorless	gas	none
Glass	colorless	solid	none
Bromine	orange-brown	liquid	none

5. (a) The pure substances are nitrogen, iron, carbon, sodium chloride, nylon, carbon dioxide, oxygen, diamond, and distilled water; the rest are mixtures.   (b) The pure substances nitrogen, iron, carbon and oxygen are elements; the rest are compounds. The heterogeneous mixtures are milk, cottage cheese, vegetable soup, wet sand, salad dressing, and smog. The homogeneous mixtures are amalgam, black coffee, iodized table salt, filtered sea water, gasoline, a dime, and 14 carat gold.   **9.** (a) Molecular formula is As_4; empirical formula is As.   (b) Molecular formula is C_3H_6; empirical formula is CH_2.   (c) Molecular formula is P_4O_{10}; empirical formula is P_2O_5.   (d) Molecular formula is XeF_4; empirical formula is XeF_4.
11. $d_{HH} = 190$ pm.   **13.** H_2O is *angular*. XeF_2 is *linear*. PH_3 is *pyramidal*.   **15.** $d_{OH} = 96.9$ pm.   **17.** (a) 300   (b) 116200
(c) 0.0048   (d) -0.0644.   **19.** (a) 9×10^{-1} (b) 6.099×10^{23}
(c) 1.0×10^6 (d) 2.0×10^{10} (e) 3.233×10^2
21. (a) 5.84×10^{-9} kg   (b) 5.434×10^4 kg   (c) 3.54×10^{-4} kg
(d) 1.673×10^{-24} g   **23.** (a) 3.9×10^7 m   (b) 68 kg
(c) 9.8×10^3 kg m^{-2} (d) 2.7×10^{-2} km sec^{-1}
25. 9.2 L per 100 km   **29.** 3.8×10^{13} km 2.4×10^{13} miles.
31. (a) 3   (b) -3 (c) -2 (d) -12 (e) -6
35. (a) 1.84 kg   (b) 0.196 kg   (c) 0.943 kg   (d) 0.0354 kg
37. 50.3 cm^3 **39.** 0.789 g cm^{-3} **41.** 5.61 g cm^{-3}
43. 0.05 kg acetic acid   **45.** 26 cm^3

CHAPTER 2

1. ATOM

ATOM	ELECTRONS AND PROTONS	NUMBER OF NEUTRONS
^2_1H	1	1
$^{19}_9F$	9	10
$^{40}_{20}Ca$	20	20
$^{112}_{48}Cd$	48	64
$^{117}_{50}Sn$	50	67
$^{131}_{54}Xe$	54	77

3.

ATOM SYMBOL	MASS NUMBER	ATOMIC NUMBER	NUMBER OF PROTONS	NUMBER OF ELECTRONS	NUMBER OF NEUTRONS
$^{24}_{12}Mg$	24	12	12	12	12
$^{106}_{47}Ag$	106	47	47	47	59
$^{137}_{56}Ba$	137	56	56	56	81

5. (a) $^{40}_{19}K$ (b) $^{30}_{14}Si$ (c) $^{40}_{18}Ar$ (d) $^{15}_7N$ **7.** $^1H^1H^{16}O$;
$^1H^1H^{17}O$; $^1H^1H^{18}O$; $^2H^1H^{16}O$; $^2H^1H^{17}O$; $^2H^1H^{18}O$; $^2H^2H^{16}O$;
$^2H^2H^{17}O$; $^2H^2H^{18}O$ **9.** 10.81 u   **11.** 69.5% and 30.5%,
respectively.   **13.** 0.7%   **15.** 3.00×10^{21} atoms Hg.
13. 0.7%   **15.** 3.00×10^{23} atoms Hg.
17. (a) 0.4 moles cells.   (b) 1.7×10^{27} molecules H_2O.
19. Salt   **21.** (a) 4.37×10^{-4} SO_2 (b) 192.2 g SO_2.
23. 78.12 u; 6.78×10^{21} molecules C_6H_6; 8.14×10^{22} atoms.
25. NH_3: N = 82.2%; H = 17.8% Cl_2: Cl = 100%
NaOH: Na = 57.48%; O = 40.00%; H = 2.52%
C_2H_6O: C = 52.13%; H = 13.1%; O = 34.72%
$C_6H_5NO_2$: C = 58.53%; H = 4.10%; N = 11.38%; O = 25.99%
27. 21.20% nitrogen by mass. 623 g $(NH_4)_2SO_4$.
29. CBr_2; empirical formula mass of 171.8 u.
31. C_3OF_6; empirical formula mass 166.0 u.   **33.** $C_4H_5N_2O$
35. $C_3H_8O_3$ **37.** Empirical formula: C_5H_7N;
molecular formula: $(C_5H_7N)_2$, i.e. $C_{10}H_{14}N_2$.
39. Red copper oxide: Cu_2O; Black copper oxide: CuO.
41. 45.9%   **43.** $x = 1$ **45.** $C_{12}H_4Cl_6$
47. (a) $2S + 3O_2 \longrightarrow 2SO_3$ (b) $2C_2H_2 + 3O_2 \longrightarrow$
$4CO + 2H_2O$ (c) $Na_2CO_3 + Ca(OH)_2 \longrightarrow$
$2NaOH + CaCO_3$ (d) $Na_2SO_4 + 4H_2 \longrightarrow Na_2S + 4H_2O$
(e) $Cu_2S + 2O_2 \longrightarrow 2CuO + SO_2$ (f) $2Cu_2O + Cu_2S \longrightarrow$
$6Cu + SO_2$ (g) $Cu + 3H_2SO_4 \longrightarrow CuSO_4 + 2H_2O + SO_2$
(h) $4B + 3SiO_2 \longrightarrow 3Si + 2B_2O_3$
49. (a) $Na_2SO_4(s) + 2C(s) \longrightarrow Na_2S(s) + 2CO_2(g)$
(b) $Cl_2(aq) + H_2O(l) \longrightarrow HCl(aq) + HOCl(aq)$

(c) $PCl_3(l) + 3H_2O(l) \longrightarrow H_3PO_3(aq) + 3HCl(aq)$
(d) $3NO_2(g) + H_2O(l) \longrightarrow 2HNO_3(aq) + NO(g)$
(e) $Mg_3N_2(s) = 6H_2O(l) \longrightarrow 3Mg(OH)_2(s) + 2NH_3(g)$
51. 3.29 g NaOH; 4.81 g NaCl   **53.** 22.55 g P   **55.** BaO_2
57. $B_2O_3 + 6Mg \longrightarrow 3MgO + Mg_3B_2$; 3.831 g B_4H_{10}
59. $2AgNO_3(aq) + CaCl_2(aq) \longrightarrow 2AgCl(s) + Ca(NO_3)_2(aq)$;
1.4 g AgCl   **61.** 3.53 g Sb_2S_3 **63.** $Ba(NO_3)_2$ is limiting, and
7.14 g of BaSO is produced
65. $Mg(OH)_2$ is limiting, and 10.5 g of $Mg_3(PO_4)_2$ is produced
67. 0.03797 M **69.** 4.8 mL solution   **71.** (a) 45 g KOH
(b) 12 g (of the 50% solution)   (c) 4.3×10^2 g H_3PO_4
73. (a) 8.7×10^{-4} g   (b) 7.8×10^{10} J   (c) 5.95×10^4 g H_2O

CHAPTER 3

3. (a) Oxygen and silicon   (b) Aluminum   (c) Hydrogen
5. (a) $Fe_2O_3 + 3CO \longrightarrow 2Fe + 3CO_2$ (b) $Fe_2O_3 + 3H_2 \longrightarrow$
$2Fe + 3H_2O$ (c) $CuO + CO \longrightarrow Cu + CO_2$
(d) $Mg + H_2O \longrightarrow MgO + H_2$ **7.** (a) $N_2 + 3H_2 \longrightarrow$
$2NH_3$ (b) $N_2 + O_2 \longrightarrow 2NO$ (c) $3Mg + N_2 \longrightarrow Mg_3N_2$
9. Carbon monoxide, CO, has a very low solubility in water while
carbon dioxide, CO_2, is readily soluble. Hydrogen also has a very
low solubility in water, so the CO_2 formed may be readily removed
from the mixture by "scrubbing" (dissolving it in water).
11. (a) $2N_2 + 5O_2 + 2H_2O \longrightarrow 4HNO_3$ (b) 0.38 g HNO_3
13 2.5   atm   **15.** 0.28   atm   **17.** 186.5K   or $-86.5°C$
19 10.2   L   **21.** 2.11   atm   **23.** 0.615   g   **25.** 1.2   L
27. 0.0860   g   **29.** 0.731   g L^{-1} **31.** 4.90   L   **33.** O_3
35. 31.7 g mol^{-1} **37.** 34.6 g mol^{-1} **39.** (a) 116 g mol^{-1}
(b) $C_6H_{12}O_2$ **41.** (a) $P_{He} = 5.5$ atm; $P_{H_2} = 10.8$ atm
(b) 16.3 atm   **43.** The effusion ratio = 1.07:1. Therefore, N_2
effuses faster than O_2.   **45.** 1.004:1   **47.** 0.686 m from the
end at which the ammonia is entering   **49.** (a) 2.00 L
(b) 9.50 L   **51.** 2.59 L   **53.** 18.2 g   **55.** C_3H_6
57. 1630

CHAPTER 4

1. (a) Helium, nonmetal   (b) Phosphorus, nonmetal   (c) Potassium, metal   (d) Calcium, metal   (e) Tellurium, semimetal
(f) Bromine, nonmetal   (g) Aluminum, metal   (h) Tin, metal
3. Metal; titanium.   **5.** (a) $Mg(s) + H_2O(g) \longrightarrow$
$MgO(s) + H_2(g)$ (b) $H_2(g) + S(s) \longrightarrow H_2S(g)$ (c) $2Na(s) +$
$I_2(s) \longrightarrow 2NaI(s)$ (d) $2K(s) + 2H_2O(l) \longrightarrow 2KOH(aq) + H_2(g)$
(e) $H_2(g) + Cl_2(g) \longrightarrow 2HCl(g)$ (f) $Ne(g) + H_2O(l) \longrightarrow$ no
reaction   **7.** FrH; FrCl; SnH_4; $SnCl_4$; AtH; AtCl; Radon will
not form a hydride or chloride.   **9.** (a) Li_2S; BeS; B_2S_3; CS_2;
N_2S_3; SO; F_2S; (Ne)   (b) Li_3N; Be_3N_2 BN; C_3N_4; N_2; S_3N_2; NF_3;
(Ne)   **13.** Element A is calcium; m = 40.0.   **15.** The valence
shell is the outermost shell of a neutral atom that contains electrons. 3 for B, 7 for a halogen, 2 for He, 8 for Ne and 2 for Mg.
17. 4 for C; 2 for Mg and Mg^{2+}; 4 for Si; 6 for O, O^{2-}, and S^{2-}.
19. Ne > F > Na.   **21.** 0.022 MJ.   **25.** C 77;   Cl 99;
Br 117;   I 138; Br_2 234;   BrCl 216; I_2 276 pm.

27. K̈   Ċȧ   ·Ḃ·   ·S̈ṅ·   ·S̈ḃ·   ·T̈e: ·B̈r: :Ẍe: ·Ȧs·   ·Ḡe·

29. (a) Li^+ $:\ddot{\underset{..}{Cl}}:^-$

(b) $(Na^+)_2$ $:\ddot{\underset{..}{O}}:^{2-}$

(c) Al^{3+} $(:\ddot{\underset{..}{F}}:^-)_3$

(d) $Ca^{2+}:\ddot{\underset{..}{S}}:^{2-}$

(e) $Mg^{2+}(:\ddot{\underset{..}{Br}}:^-)_2$

31. (a) Li_2S (b) CaO   (c) $MgBr_2$ (d) NaH   (e) AlI_3
33. (a) $(NH_4)_3PO_4$ (b) Fe_2O_3 (c) Cu_2O (d) $Al_2(SO_4)_3$
39. (a) $\underset{H}{\overset{H}{\diagdown}}C{=}\ddot{\underset{..}{O}}:$ (b) $:P{\equiv}P:$ (c) $H{-}\ddot{N}{=}\ddot{N}{-}H$

(d) $H{-}\ddot{N}{-}\ddot{N}{-}H$
$\qquad\quad | \quad\ \ |$
$\qquad\ \ \ H\ \ \ H$

CHAPTER 5

1. (a) $2P + 3Cl_2 \longrightarrow 2PCl_3$ (b) (i) $S + Cl_2 \longrightarrow SCl_2$
(b) (ii) $2S + Cl_2 \longrightarrow S_2Cl_2$ (c) $C + 2F_2 \longrightarrow CF_4$
(d) $2As + 3Br_2 \longrightarrow 2AsBr_3$
5. Electronegativity increases from left to right along each period,
since the core charge increases in that direction. Electronegativity
decreases from top to bottom in a group, since although the core
charge is constant, the distance of the valence shell from the nucleus
increases.
7. Cl_2: covalent; PCl_3, ClF: polar covalent;   LiCl, $MgCl_2$: ionic.
(The structure for S_2Cl_2 is Cl—S—S—Cl. Its S—S bond is covalent,
but each S—Cl bond is polar covalent.)
9. I_2, H_2 are nonpolar; HBr, ClF are polar.   **13.** (a) K^+
(b) S^{2-} (c) Cl^- (d) Na^+ (e) I^- **15.** (a) CaI_2 (b) BeO
(c) Al_2S_3 (d) $MgBr_2$ (e) Rb_2Se (f) BaO

19. $Cl{-}\overset{\displaystyle |}{\underset{\displaystyle |}{\overset{..}{\underset{..}{P}}}}{-}Cl$ $:\ddot{\underset{..}{Cl}}{-}\ddot{\underset{..}{F}}:$ $[Li^+]:\ddot{\underset{..}{F}}:^-$ $:\ddot{\underset{..}{F}}{-}\overset{\displaystyle :\ddot{F}:}{\underset{\displaystyle :\ddot{F}:}{Si}}{-}\ddot{\underset{..}{F}}:$
$\qquad\ \, \overset{\displaystyle |}{Cl}$

21. (a) $Cl_2 + 2KI(aq) \longrightarrow I_2 + 2KCl(aq)$ (b) NR
(c) $Br_2 + 2NaI(aq) \longrightarrow I_2 + 2NaBr(aq)$
(d) $2F_2 + 2H_2O \longrightarrow 4HF + O_2$
23. (a) Rb is oxidized and therefore acts as the reducing agent. I_2
gains electrons, is reduced, and therefore acts as the oxidizing agent.
(b) Al is the reducing agent and is oxidized; O_2 is the oxidizing
agent and is reduced.   (c) Cu^{2+} is reduced, and acts as the oxidizing
agent. Two I^- ions lose electrons, are therefore oxidized, and act
as the reducing agents.   (d) Zn is the reducing agent and is oxidized;
S is the oxidizing agent and is reduced.   (e) Mg is the reducing
agent and is oxidized. H_3O^+ is reduced and acts as the oxidizing
agent.   **27.** $AgNO_3$ dissolves, but precipitates AgCl.
29. Action on an indicator; action on a carbonate; action on a
metal. One would not use taste.
31. $Na_2O + H_2O \longrightarrow 2Na^+ + 2OH^-$;
$KOH \xrightarrow{\text{water}} K^+ + OH^-$; $NH_3 + H_2O \rightleftharpoons NH_4^+ + OH^-$;
$KNH_2 \longrightarrow K^+ + NH_2^-$
$\qquad\qquad\qquad\quad \downarrow {\scriptstyle H_2O}$
$\qquad\quad NH_3 + OH^-$;   Of these bases, only NH_3 is

weak. **33.** $HZ + H_2O \longrightarrow Z^- + H_3O^+$. The conjugate base Z^- has negligible base strength in water. A solution of NaZ will contain only Na^+ and Z^- ions. The conjugate base Y^- is weak in water: $HY + H_2O \rightleftharpoons Y^- + H_3O^+$. An aqueous solution of NaY will contain not only Na^+ and Y^-, but also HY and OH^- due to the reaction of Y^- with H_2O. **35.** 8.6 g
37. (a) NH_4^+ (b) $CH_3NH_3^+$ (c) H_2O (e) H_3O^+
41. Acid: H_3O^+; base: OH^- $LiH + H_2O \longrightarrow LiOH + H_2$
$CaH_2 + 2H_2O \longrightarrow Ca(OH)_2 + 2H_2$
$Li_2O + H_2O \longrightarrow 2LiOH$ $CaO + H_2O \longrightarrow Ca(OH)_2$
43. 21.6 mL   **45.** (a) Solid (b) Ionization energy smaller, size larger (c) AtBr (d) Strong (e) Pyramidal (f) Ionic
(g) $2Na + At_2 \longrightarrow 2NaAt$ $Ca + At_2 \longrightarrow CaAt_2$
$2P + 3At_2 \longrightarrow 2PAt_3$; with excess At_2, $2P + 5At_2 \longrightarrow 2PAt_5$
$H_2 + At_2 \longrightarrow 2HAt$ **47.** $Cl_2 + 2NaOH \longrightarrow$
$NaOCl + NaCl + H_2O$; 3.72 g
49. (a) Oxidation-reduction. I^- is the reducing agent, Cl_2 is the oxidizing agent. (b) Acid-base. HCl is the acid, H_2O acts here as the base. (c) Oxidation-reduction. Zn is the reducing agent, H_3O^+ is the oxidizing agent. (d) Acid-base. H_3O^+ is the acid, HCO_3^- is the base.

CHAPTER 6

1. 3.19 m; 405 m   **3.** 11.0 m   **5.** 4.74×10^{14} Hz
7. 2.97×10^{-19} J; 179 kJ   **11.** 102 kJ mol^{-1} **13.** (a) 19×10^{-21} J (b) 4.23×10^{-19} J (c) 4.21×10^{-19} J   **15.** 4.09×10^{-19} J; 487 nm   **17.** (a) 2.18×10^{-18} J atom^{-1}; 91.2 nm
(b) 5.45×10^{-19} J atom^{-1}; 365 nm   **19.** 1.32×10^3 m s^{-1}
21. 1.1×10^{-34} m   **23.** 1.03×10^{-18} J   **25.** 1p, 2d
27. (b), (d)   **29.** (a) $_{19}K$: $1s^2 2s^2 2p^6 3s^2 3p^6 4s^1$
(b) $_{13}Al$: $1s^2 2s^2 2p^6 3s^2 3p^1$ (c) $_{17}Cl$: $1s^2 2s^2 2p^6 3s^2 3p^5$
(d) $_{22}Ti$: $1s^2 2s^2 2p^6 3s^2 3p^6 4s^2 3d^2$
(e) $_{30}Zn$: $1s^2 2s^2 2p^6 3s^2 3p^6 3d^{10} 4s^2$
(f) $_{33}As$: $1s^2 2s^2 2p^6 3s^2 3p^6 3d^{10} 4s^2 4p^3$

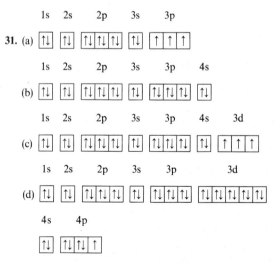

33. (b); (e)   **39.** (a) Li (c) O (e) Na (g) P

CHAPTER 7

1. Orthorhombic: brilliant yellow crystals, contain S_8 molecules. The most stable form of S.   Monoclinic: long needle crystals; slowly reverts to orthorhombic form. Made by cooling molten sulfur. Plastic: brown, rubbery solid; slowly reverts to orthorhombic form. Made by pouring molten S into cold water.   **3.** (a) With concentrated sulfuric acid, even without heating $CaSO_4 \cdot 2H_2O \longrightarrow CaSO_4 + 2H_2O$ (absorbed by H_2SO_4). In addition, we expect $CaSO_4 + H_2SO_4 \longrightarrow Ca(HSO_4)_2$. (b) $H_2SO_4 + NaOH \longrightarrow NaHSO_4 + H_2O$ and $NaHSO_4 + NaOH \longrightarrow Na_2SO_4 + H_2O$. These acid-base reactions occur whether the acid is dilute or concentrated, hot or cold. (c) $BaCl_2(aq) + H_2SO_4 \longrightarrow BaSO_4(s) + 2HCl$. This reaction should proceed regardless of the concentration and temperature of the acid. (d) $H_2SO_4 + C_2H_5OH \longrightarrow C_2H_4 + H_3O^+ + HSO_4^-$. This reaction requires high temperature and concentrated acid. (e) $NaF + H_2SO_4 \longrightarrow HF(g) + NaHSO_4$. This reaction should proceed at high temperature with concentrated acid. (f) $NaNO_3 + H_2SO_4 \longrightarrow HNO_3 + NaHSO_4$. This reaction requires high temperature and concentrated acid. (g) $C(s) + 2H_2SO_4 \longrightarrow CO_2(g) + 2H_2O + SO_2(g)$ This reaction requires high temperatures and concentrated acid.

5. valence = 2

valence = 4

valence = 6

Sulfur cannot achieve a valence of 8 since there are no further valence-shell lone pairs from which one electron can be promoted to the 3d level. Oxygen cannot achieve valences of 4 and 6 since it does not have low-lying d orbitals into which 2s or 2p electrons could be promoted.   **13.** $4PH_3 + 8O_2 \longrightarrow P_4O_{10} + 6H_2O$
17. $P_4S_3 + 8O_2 \longrightarrow P_4O_{10} + 3SO_2$; 0.0505 L SO_2 results
19. PH_3; $PH_3 + HI \longrightarrow PH_4I$; $2AlP + 3H_2SO_4 \longrightarrow Al_2(SO_4)_3 + 2PH_3$ **21.** $+5$: H_3PO_4, P_4O_{10}, PO_4^{3-}; $+3$: H_3PO_3, P_4O_6, HPO_3^{2-}; 0: elemental (all forms) -3: PH_3, Ca_3P_2, AlP, PH_4^+ **23.** (a) $+2$ (b) $+4$ (c) $+5$
(d) $+6$ (e) -1 (f) $+3$ **25.** $2Br^- + H_2SO_4 + 2H^+ \longrightarrow Br_2 + SO_2 + 2H_2O$ $Br_2 + SO_2 + 2H_2O \longrightarrow 2Br^- + H_2SO_4 + 2H^+$

27. (a), 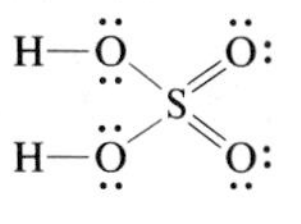 (b), 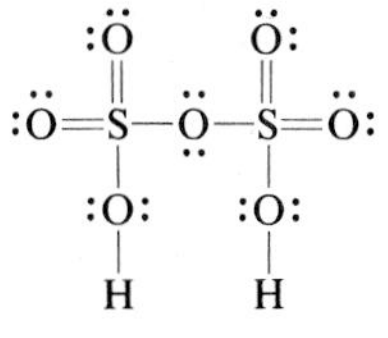 (c), 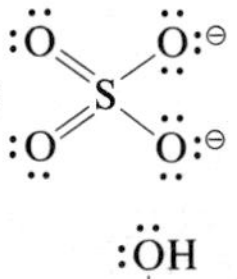 (d), 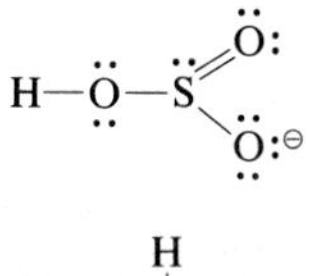 **29.** (a), 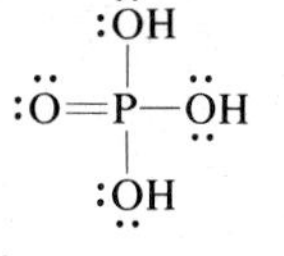 (b), 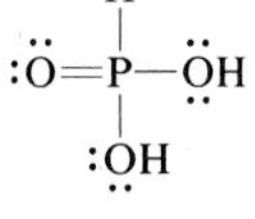 (c) 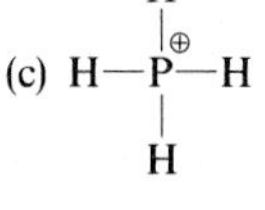 structures shown

A-19

(d)
$$O = \overset{\overset{\displaystyle :\ddot{O}H}{|}}{\underset{\underset{\displaystyle :\ddot{O}:^{\ominus}}{|}}{P}} - \ddot{O}:^{\ominus}$$
(e)
$$Cl - \overset{\overset{\displaystyle :\ddot{O}:^{\ominus}}{|}}{\underset{\underset{\displaystyle Cl}{|}}{P}} - Cl$$

39. (a) $CaSO_4$ (b) PBr_5 (c) PI_3 (d) $(NH_4)_2HPO_4$ (e) Ca_3N_2
(f) SF_4 (g) $CrCl_3$ **41.** (a) Potassium selenate (b) Hydrogen telluride (c) Disodium tetrasulfide (d) Iron(II) disulfide (e) Rubidium hydrogen sulfate (f) Tetraphosphorus hexaoxide
43. (a) $HNO_3 + NaOH \longrightarrow H_2O + NaNO_3$ sodium nitrate
(b) $H_2SO_4 + NaOH \longrightarrow H_2O + NaHSO_4$ sodium hydrogen-sulfate
$NaHSO_4 + NaOH \longrightarrow H_2O + Na_2SO_4$ sodium sulfate
(c) $H_3PO_4 + NaOH \longrightarrow H_2O + NaH_2PO_4$ sodium dihydrogenphosphate
$NaH_2PO_4 + NaOH \longrightarrow H_2O + Na_2HPO_4$ sodium hydrogenphosphate
$Na_2HPO_4 + NaOH \longrightarrow H_2O + Na_3PO_4$ sodium phosphate
(d) $H_3PO_3 + NaOH \longrightarrow H_2O + NaH_2PO_3$ sodium dihydrogenphosphite
$NaH_2PO_3 + NaOH \longrightarrow H_2O + Na_2HPO_3$ sodium hydrogenphosphite
(e) $H_4P_2O_7 + NaOH \longrightarrow H_2O + NaH_3P_2O_7$ sodium trihydrogendiphosphate
$NaH_3P_2O_7 + NaOH \longrightarrow H_2O + Na_2H_2P_2O_7$ (di)sodium dihydrogendiphosphate
$Na_2H_2P_2O_7 + NaOH \longrightarrow H_2O + Na_3HP_2O_7$ (tri)sodium hydrogendiphosphate
$Na_3HP_2O_7 + NaOH \longrightarrow H_2O + Na_4P_2O_7$ (tetra)sodium diphosphate
45. (a) Acid-base; H_2SO_4 (acid), HSO_4^- (conj. base); CO_3^{2-} (base), HCO_3^- (conj. acid). (b) Acid-base; HSO_4^- (acid), SO_4^{2-} (conj. base); H_2O (base), H_3O^+ (conj. acid). (c) Oxidation-reduction; Cu is oxidized, H_2SO_4 is reduced. (d) Oxidation-reduction; Br^- is oxidized, H_2SO_4 is reduced. (e) Acid-base; CO_3^{2-} (base), HCO_3^- (conj. acid); H_3O^+ (acid), H_2O (conj. base). (f) Oxidation-reduction; Mg is oxidized; H_3O^+ is reduced. (g) Acid-base; H_2O (acid), OH^- (conj. base); P^{3-} (base), PH_3 (conj. acid).

CHAPTER 8

1. (a) $^{\ominus}:\ddot{S} - \ddot{S}:^{\ominus}$ (b) $\ddot{O} = S - \ddot{O}:^{\ominus}$ with $:\ddot{O}:^{\ominus}$ below

(c) $:\ddot{Cl} - \ddot{O}:$ (d) $\ddot{O} = Cl - \ddot{O}:^{\ominus}$ with $:\overset{:O:}{O}:$ above and $:\ddot{O}:$ below

3. (a) $:\ddot{O} = \overset{\overset{\displaystyle :\ddot{O}:^{\ominus}}{}}{S} = \ddot{O}:$ with $:\ddot{S}:_{\ominus}$ below (b) $:\ddot{O} = S = \ddot{O}:$ with $:\ddot{O}:^{\ominus}$ below (c) $:\ddot{F} - \overset{\overset{\displaystyle :\ddot{F}:}{}}{S} - \ddot{F}:$ with $:\ddot{O}:$ below

(d) $\ddot{O} = \overset{\overset{\displaystyle :\ddot{F}:}{}}{S} = \ddot{O}$ with $:\ddot{F}:$ below (e) $:\ddot{F} - \overset{\overset{\displaystyle :\ddot{F}:}{}}{S} - \ddot{F}:$ with $\overset{\|}{N}$ below

A-20

---

7. (a) $^{\ominus}:\ddot{N} = \overset{\overset{\textcircled{2+}}{}}{\ddot{O}} = \ddot{N}:^{\ominus}$ (b) $^{\ominus}:\ddot{N} = \overset{\oplus}{N} = \ddot{O}:$
Structure (b) is preferred since it has fewer formal charges.

9. (a) $X - A - X$ Bond angles = $180°$

(b) $X - A \overset{X}{\underset{X}{\diagdown}}$ (c) $X \cdots A \cdots X$ with X top and X bottom

Bond angles = $120°$ Bond angles = $109.5°$

11. H_2O: AX_2E_2, angular; H_3O^+: AX_3E, triangular pyramid; PCl_3: AX_3E, triangular pyramid; BCl_3: AX_3, triangular planar; SiH_4: AX_4, tetrahedron; BH_4^-: AX_4, tetrahedron; H_2S: AX_2E_2, angular; SCl_2: AX_2E_2, angular; NH_4^+: AX_4, tetrahedron; BeH_2: AX_2, linear; BeH_4^{2-}: AX_4, Tetrahedron **13.** $109.5°$.
15. (a) Bond order 1.33 (3 equivalent structures). (b) Bond order 1.15 (2 equivalent structures). (c) Bond order 1.25 (4 equivalent structures). (d) Bond order 1.5 (6 equivalent structures).
(e) Bond order 1.75 (4 equivalent structures). **19.** Dipole moment: NF_3, SCl_2, ICl, $CHCl_3$, CH_2Cl_2 Zero dipole moment: BF_3, $BeCl_2$, I_2, CCl_4.
21.

	SO_2	CO_2	H_2O	NH_3	SO_3	BeH_2
	$\overset{\delta+\delta-}{S-O}$	$\overset{\delta+\delta-}{C-O}$	$\overset{\delta-\delta+}{O-H}$	$\overset{\delta-\delta+}{N-H}$	$\overset{\delta+\delta-}{S-O}$	$\overset{\delta+\delta-}{Be-H}$
	$\mu \neq 0$	0	$\neq 0$	$\neq 0$	0	0

1. carbon, carbon monoxide, aluminum, free electrons (in electrolysis). Copper: $Cu_2O + CO \longrightarrow 2Cu + CO_2$ Iron: $Fe_2O_3 + 2Al \longrightarrow 2Fe + Al_2O_3$ Aluminum: $Al_2O_3 \longrightarrow 2Al + 3/2\ O_2$ Lead: $PbO + C \longrightarrow Pb + CO$ **3.** 4.23×10^5 g.
5. (a) 4.8×10^6 metric tons of ore. (b) 210 kg iron/ton ore. (c) 68 kg coke. **9.** The bonding in the Na_2 molecule is covalent. In sodium metal, the bonding is metallic. The essential difference between the Na_2 molecule and the metal is that in the former the electrons are localized between specific atoms, whereas in the metal they are not. **15.** Both iron, Fe, and zinc, Zn, react with dilute hydrochloric acid solution. The equations are $Fe(s) + 2HCl(aq) \longrightarrow FeCl_2(aq) + H2(g)$ (iron(II) chloride and hydrogen gas) and $Zn(s) + 2HCl(aq) \longrightarrow ZnCl_2(aq) + H_2(g)$ (zinc(II) chloride and hydrogen gas). Neither copper, Cu, nor lead, Pb, reacts with dilute hydrochloric acid.
17. (a) $5SO_3^{2-}(aq) + 2MnO_4^-(aq) + 6H^+ \longrightarrow$
$5SO_4^{2-}(aq) + 2Mn^{2+}(aq) + 3H_2O$
(b) $5H_2O_2(aq) + 2MnO_4^-(aq) + 6H^+ \longrightarrow$
$2Mn^{2+} + 5O_2(g) + 8H_2O$
(c) $4Zn(s) + 2NO_3^-(aq) + 10H^+ \longrightarrow$
$4Zn^{2+}(aq) + N_2O(g) + 5H_2O$
(d) $3P(s) + 5NO_3^-(aq) + 2H^+ + 2H_2O \longrightarrow$
$3H_2PO_4^-(aq) + 5NO(g)$
(e) $10NO_3^-(aq) + I_2(s) + 8H^+ \longrightarrow$
$2IO_3^-(aq) + 10NO_2(g) + 4H_2O$

19. $\overset{+3\ -2}{Al_2O_3} \cdot xH_2O$ $\overset{+1\ -2}{AlCl_4^-}$ $\overset{+3\ -1}{SiO_2}$ $\overset{+4\ -2}{SiO_3^{2-}}$ $\overset{+2\ -2+1}{Pb(OH)_4^{2-}}$
$\overset{+4\ -1}{PbCl_4}$ $\overset{+3\ -2}{Fe_2O_3}$ $\overset{+2\ -3+1}{Cu(NH_3)_4^{2+}}$

21. $K_2O + H_2O \longrightarrow 2KOH$ (basic);
$SrO + H_2O \longrightarrow Sr(OH)_2$ (basic);
$SO_2 + 2H_2O \rightleftharpoons H_3O^+ + HSO_3^-$ (acidic);
$SO_3 + H_2O \longrightarrow H_2SO_4$ (acidic);
$CO_2 + 2H_2O \rightleftharpoons H_3O^+ + HCO_3^-$ (acidic);
$P_4O_6 + 6H_2O \longrightarrow 4H_3PO_3$ (acidic);
$Cl_2O_7 + H_2O \longrightarrow 2HClO_4$ (acidic);
23. (a) $Li_2O + SiO_2 \longrightarrow Li_2SiO_3$
(b) $Na_2O + N_2O_5 \longrightarrow 2NaNO_3$
(c) $6CaO + P_4O_{10} \longrightarrow 2Ca_3(PO_4)_2$
27. Add HCl to a portion of each solution. Only the $Pb(NO_3)_2$ solution forms a precipitate (white $PbCl_2$). Add NaOH to portions of the other solutions. The $FeSO_4$ solution gives a pale green precipitate of $Fe(OH)_2$ that turns brown on standing because it is air oxidized to $Fe(OH)_3$. The $CuuSO_4$ solution gives a pale blue precipitate of $Cu(OH)_2$ that dissolves on addition of NH_4OH to give a deep blue solution ($Cu(NH_3)_4^{2+}$). The $Al_2(SO_4)_3$ solution gives a white precipitate of $Al(OH)_3$ that dissolves in excess NaOH to give a colorless solution of $Al(OH)_4^-$. Additional tests: H_2S gives a black precipitate of metal sulfides with Pb(II), Cu(II), and Fe(II) but not with Al(III). I^- gives a yellow precipitate of PbI_2 with Pb(II) and a white precipitate of CuI and soluble brown I_3^- with Cu(II). **29.** HNO_3: $Cu(NO_3)_2$, Copper(II) nitrate; $Fe(NO_3)_2$, Iron(II) nitrate; $Fe(NO_3)_3$, Iron(III) nitrate H_2SO_4: Cu_2SO_4, Copper(I) sulfate; $CuSO_4$, Copper(II) sulfate; $FeSO_4$, Iron(II) sulfate; $Fe_2(SO_4)_3$, Iron(III) sulfate H_3PO_4: Cu_3PO_4, Copper(I) phosphate; $Cu_3(PO_4)_2$, Copper(II) phosphate; $Fe_3(PO_4)_2$, Iron(II) phosphate; $FePO_4$, Iron(III) phosphate. **33.** In the blast furnace process to produce iron, the slag is chiefly calcium silicate, $CaSiO_3$, formed by the reaction of calcium oxide, CaO, with the silica impurities in the iron ore. It serves a useful purpose by forming a protective layer on top of the molten iron, preventing reoxidation of the iron. **35.** The empirical formula is $AlCl_3$. Molecular formula is Al_2Cl_6.

CHAPTER 10

1. Each carbon in diamond forms covalent single bonds to four neighboring carbons. The geometry about each carbon is exactly tetrahedral. A giant 3-D structure of carbon atoms is thereby produced. **3.** The double bond electrons in graphite are delocalized, giving a type of metallic bonding in each sheet. Thus each sheet of carbon atoms in graphite conducts electricity. Since the sheets are only weakly bonded to each other, they slide easily over each other, thus making graphite a good lubricant. In diamond all the atoms are held together by strong covalent bonds in a 3-D arrangement and there are no delocalized electrons. Thus diamond is a nonconductor. It is hard, not soft like graphite, and it is not a lubricant. **5.** To melt a molecular solid, it is necessary only to break the weak intermolecular forces between the molecules. Thus they melt at a much lower temperature than do network solids, for which (strong) covalent or ionic bonds between atoms must be broken before the atoms are sufficient mobile to slide by each other. **9.** (a) Silica, SiO_2, contains Si—O single bonds in a three-dimensional network solid. In contrast, carbon dioxide consists of discrete O=C=O molecules which are only weakly attracted to each other. Thus SiO_2 is a high-melting solid, whereas CO_2 is a gas. (b) Oxygen consists of O_2 molecules, and since they attract each other weakly, the element is a gas. Sulfur is a solid consisting of S_8 rings, which because they contain 8 relatively heavy atoms attract

each other fairly strongly. **11.** A space lattice is a regular, repeating arrangement of points in space. **13.** The primitive cubic unit cell contains eight points, one at each corner. The body-centered cube is identical, except for the addition of one point in the center of the cube. The face-centered cubic has six points more than the primitive, one in the center of each of the six faces. (See Fig. 10.14b of the text.) **15.** The three common structures of metals are the (monatomic) body-centered cubic, the (monatomic) face-centered cubic (better known as cubic close-packed), and hexagonal close-packed. **17.** KCl and BaO have the sodium chloride structure—they are based upon a face-centered cubic lattice, with the anions halfway along each cell edge. CuCl has the sphalerite (ZnS) structure, which is also based upon the face-centered cubic lattice but with anions at the lattice points and cations one-quarter the distance along each body diagonal. **21.** 195 a
23. (a) 578 pm   (b) 250 pm   (c) 70 pm
27. 6.07×10^{23} pairs/mole   **29.** 4

CHAPTER 11

1. Conditions: T = high temperature, C = catalyst.
(a) $CO + 2H_2 \longrightarrow CH_3OH$ (T, C) (b) $2CO + O_2 \longrightarrow 2CO_2$ (T) (c) $CO + Cl_2 \longrightarrow COCl_2$ (T, C)
(d) $CO + H_2O \longrightarrow CO_2 + H_2$ (T) (e) $3CO + Fe_2O_3 \longrightarrow 2Fe + 3CO_2$ (T).
3. (a) Ö=C=Ö (b) :C≡O: (c) :C≡N: (d) :C≡C:
(e) H—C≡N: (f) Ö=C⟨Cl: / Cl:
5. S̈=C=S̈ (preferred structure); :S̈—C≡S̈: ; C=S̈
7. $COBr_2$; :O=C⟨Br: / Br   Planar triangular
11. (a) CN, $(CN)_2$ (b) NCCN (c) :N≡C—C≡N: Linear about each carbon atom (d) No dipole moment **13.** A cracking reaction occurs when the alkanes are heated to a sufficiently high temperature by themselves, and they decompose to give a mixture of products including smaller alkanes, other hydrocarbons containing unsaturated linkages, and hydrogen. **15.** An addition reaction is a reaction in which two molecules combine to give a third molecule. An elimination reaction involves the decomposition of a molecule into two molecules, one of which, normally the smaller of the two, is said to be eliminated. An elimination reaction is the opposite of an addition reaction. **17.** The ring bonds in cyclopropane and the double bonds in alkenes can be thought of as strained (bent); the CC bonds in alkanes can not.
19. (a) $CH_3—CH_2—CH_3 + 5O_2 \longrightarrow 3CO_2 + 4H_2O$
(b) $CH_2=CH_2 + Cl_2 \longrightarrow CH_2Cl—CH_2Cl$
(c) $CH_3—CH_3 + Cl_2 \longrightarrow CH_3—CH_2Cl + HCl$
(d) $CaC_2 + 2H_2O \longrightarrow CH≡CH + Ca(OH)_2$
(e) $n(CH_2=CH_2) \xrightarrow{catalyst} (—CH_2—CH_2—)n$
(f) $CH_3—CH_3 \xrightarrow{heat} CH_2=CH_2 + H_2$
21. (a) *Cracking* is the process whereby alkanes, when heated to a sufficiently high temperature, decompose to form a mixture of molecules including smaller alkanes, unsaturated hydrocarbons, and hydrogen. (b) *Polymerization* A polymerization reaction is one in which a large number of small molecules (monomers) combine to form a very large molecule (polymer). (c) *Addition Reaction* A reaction in which two molecules combine to give a third

molecule. In organic chemistry they are normally reactions of the double or triple bond. (d) *Cyclic Hydrocarbon* A hydrocarbon containing a closed chain or ring of carbon atoms. (e) *Alkyl Group* A substituent group formed from an alkane by the removal of one hydrogen atom. In the name, the *-ane* ending is replaced by *-yl*. (f) *Aromatic Hydrocarbon* The name given to members of the arene series of hydrocarbons—hydrocarbons containing the benzene ring or fused benzene ring systems. The name originally referred to the pleasant odors associated with many of these compounds. (g) *Conformation* Conformations are the different arrangements of the atoms and groups that are possible as a result of rotations about single bonds. **23.** When two tetrahedra share an edge (double bond) the remaining four corners (F atoms) are in the same plane.

25. A 1,3-butadiene

B 2-butyne $CH_3—C\equiv C—CH_3$

C cyclobutene

D 1-butyne $CH\equiv C—CH_2—CH_3$

29. A saturated hydrocarbon contains only single covalent bonds. An unsaturated hydrocarbon contains double and/or triple bonds. **33.** No free rotation is possible about the double bond in 2-butene. Hence, *cis* and *trans* isomers are possible. But 2-butyne is linear and no different spatial arrangements of groups or atoms are possible.

35. (a)

(b)

(c)

(d)

(e) $CH_2=CH—CH=CH_2$ (f)

(g)

39. (a) 3-Methylpentane (b) 2,2-Dimethylbutane (c) Pentane (d) Correct **41.** Ten **43.** (a) Addition reaction, and also an oxidation-reduction (b) Oxidation-reduction (c) Addition reaction, and also an oxidation-reduction (d) Oxidation-reduction (e) Acid-base (f) Elimination **45.** (a) A mixture of gaseous

hydrocarbons, mainly methane (b) Liquid mixture of hydrocarbons (c) Coal tar is a liquid distillate formed when coal is heated in the absence of air. (d) Synthesis gas is a mixture of carbon monoxide and hydrogen formed by the action of steam on methane in the presence of a catalyst.

CHAPTER 12

1. $+170$ kJ **3.** -676.5 kJ **5.** -49 kJ
7. $+2803$ kJ mol^{-1}; 15.56 kJ g^{-1} glucose **9.** -227 kJ
11. -924.7 kJ **13.** -57.1 kJ for NO $\longrightarrow$ NO$_2$;
-98.9 kJ for SO$_2$ $\longrightarrow$ SO$_3$; $+142.7$ kJ for O$_2$ $\longrightarrow$ O$_3$
15. -128.0 kJ

17.

ALCOHOL	ΔH PER MOLE
CH_3OH	-711
C_2H_5OH	-1342
C_3H_7OH	-2012
1-C_4H_9OH	-2635
2-C_4H_9OH	-2604

19. -104 kJ mol^{-1} **21.** -85 kJ mol^{-1} **23.** BE(OH) = -463.5 kJ mol^{-1}, 143.7 kJ mol^{-1}. The O—O bond strength of 143.7 is much less than half the O=O value. **25.** E(C=O) = 692 kJ Since the average C=O value in CO$_2$ is 804.2 kJ we conclude that the CO$_2$ value is greater. **27.** (a) -78 kJ (b) -218 kJ (c) -27 kJ **29.** ΔH_f° (C$_4$H$_8$, g) = -85 kJ; ΔH_f° (C$_6$H$_{12}$, g) = -128 kJ The discrepancy for cyclobutane arises because it is strained (has bent bonds). **31.** 119 kJ **33.** -5450 kJ mol^{-1} **35.** -66.5 kJ/mol **37.** 657.5 kJ

CHAPTER 13

5. (a) 726 torr. (b) 324 mL. **7.** 1.73 L; 1.35 M; 1.52 L.
11. (a) BrF (b) BrCl (c) KBr (d) K **13.** (a) London forces (b) Ionic (c) London forces (d) Dipole-dipole (e) Hydrogen bonds **17.** Boiling point, density, solubility, and viscosity. **19.** (a) Induced dipole-induced dipole (London) (b) Induced dipole-induced dipole (London) (c) Ionic (d) Ionic (e) Hydrogen bonds (f) Hydrogen bonds (g) Dipole-dipole (h) Induced dipole-induced dipole (London) **21.** (a) Ice is less dense water; thus any solid water is found at the top and floats on the more dense liquid layer. (b) *n*-Octane molecules are long and thin. They can approach each other more closely and consequently experience greater London forces of attraction than can the spherical, branched octane molecules. (c) NH$_3$ molecules attract each other by dipole-dipole forces, whereas H$_2$ molecules are attracted only by much weaker London forces. Thus NH$_3$ should show a greater deviation from ideal gas behavior than H$_2$, with respect both to deviations caused by intermolecular attraction, and also to size, since NH$_3$ is larger than H$_2$. **23.** (a) Ca (b) SiH$_4$ (c) C$_4$H$_{10}$ (d) NH$_3$ (e) Cl$_2$ (f) SiO$_2$ **25.** (a) H$_2$O$_2$ (b) Ethanediol (c) Sugar (d) Magnesium chloride (e) HI (f) LiCl (g) CH$_3$OH **27.** H$_2$O + SO$_3$ $\longrightarrow$ H$_2$SO$_4$; H$_2$O + CO$_2$ $\rightleftharpoons$ H$_2$CO$_3$ **29.** 2e$^-$ + 2H$_2$O $\longrightarrow$ 2OH$^-$ + H$_2$; 2H$_2$O $\longrightarrow$ 4H$^+$ + O$_2$ + 4e$^-$ **31.** Water is described as amphoteric because it can behave as an acid or as a base depending upon what is dissolved in it. **33.** Naphthalene **35.** 515 g mol^{-1}; Al$_2$Br$_6$ **37.** 2.6×10^4 g mol^{-1}; 76 monomer units in each polymer molecule

CHAPTER 14

1. (a) $K_c = [NO_2]^4[O_2]/[N_2O_5]^2$; $K_p = p_{NO_2}^4/p_{O_2}p_{N_2O_5}^2$
(b) $K_c = [SO_3]^2/[SO_2]^2[O_2]$; $K_p = p_{SO_3}^2/p_{SO_2}p_{O_2}^2$
(c) $K_c = [SO_3]/[SO_2][O_2]^{1/2}$; $K_p = p_{SO_3}/p_{SO_2}p_{O_2}^{1/2}$
(d) $K_c = [P_4O_{10}]/[P_4][O_2]^5$; $K_p = p_{P_4O_{10}}/p_{P_4}p_{O_2}^5$
3. (a) $4.72 \ mol^{-1} \ L$ (b) $2.17 \ mol^{-\frac{1}{2}} \ L^{-\frac{1}{2}}$ **5.** $500 \ M^{-1}$
9. (a) shifts to the left; (b) shifts to the right; (c) shifts to the left;
(d) shifts to the left; **11.** (a) decreases; (b) decreases;
(c) increases; (d) no effect; **13.** 2.2×10^{-6} M; 0.022%
15. 4.2×10^{-4} M; 4.2% **17.** 6.3×10^{-6} M; 6.3×10^{-3}%
19. (a) 1.70 (b) 12.20 (c) 7.00 **21.** 4.0×10^{-8}
23. $[H_3O^+] = [NH_3] = 3.3 \times 10^{-6}$ M; $[NH_4^+] = 0.020$ M;
$[OH^-] = 3.0 \times 10^{-9}$ M
25. (a) $NH_3 + HNO_3 \longrightarrow NH_4^+ + NO_3^-$ (acidic)
(b) $NH_3 + HCl \longrightarrow NH_4^+ + Cl^-$ (acidic)
(c) $Ca(OH)_2 + H_2SO_4 \longrightarrow 2H_2O + Ca^{2+} + SO_4^{2-}$ (basic)
(d) $CH_3CO_2H + KOH \longrightarrow CH_3CO_2K + H_2O$ (basic)
(e) $Al(OH)_3 + 3HCl \longrightarrow AlCl_3 + 3H_2O$ (acidic)
(f) $LiOH + HI \longrightarrow LiI + H_2O$ (neutral)
27. (a) 4.9 (b) 4.2 (c) 3.5 **29.** 8.3 ± 1.2 **31.** 9.4
33. 1/5.6 **35.** 9.1 **37.** 1.7 **39.** (a) 4.75 (b) 6.55
(c) 7.51 (d) 9.50 (e) 9.60 (f) 9.70 (g) 10.68
41. (a) bromothymol blue (b) thymol blue or phenolphthalein
(c) methyl red **43.** (a) 1.25 (b) 7.0 (c) 1.26
45. 4.86; 4.76; 5.00 **47.** (a) 3.22 (b) 4.74 (c) 6.72 (d) 8.45
(e) 11.43 **49.** (a) 35.5% (b) 0.579 atm

CHAPTER 15

1. The reactivity increases with increasing atomic number because the ionization energies (for $M \longrightarrow M^+ + e^-$) decrease.
3. (a) $2K(s) + Br_2(l) \longrightarrow 2KBr(s)$
(b) $4Li(s) + O_2(g) \longrightarrow 2Li_2O(s)$
(c) $2Na(s) + H_2(g) \longrightarrow 2NaH(s)$
(d) $6Li(s) + N_2(g) \longrightarrow 2Li_3N(s)$
(e) $LiH(s) + H_2O(l) \longrightarrow H_2(g) + Li^+(aq) + OH^-(aq)$
(f) $Li_3N(s) + 3H_2O(l) \longrightarrow NH_3(g) + 3Li^+(aq) + 3OH^-(aq)$
(g) $2K(s) + 2H_2O(l) \longrightarrow H_2(g) + 2K^+(aq) + 2OH^-(aq)$
9. (a) $Mg(s) + Cl_2(g) \longrightarrow MgCl_2(s)$
(b) $2Ca(s) + O_2(g) \longrightarrow 2CaO(s)$
(c) $Sr(s) + H_2(g) \longrightarrow SrH_2(s)$
(d) $3Mg(s) + N_2(g) \longrightarrow Mg_3N_2(s)$
(e) $Ca(s) + 2H_2O(l) \longrightarrow H_2(g) + Ca^{2+}(aq) + 2OH^-(aq)$
13. $Ca(HCO_3)_2$ 18.0%; $CaCO_3$ 11.9%; CaO 70.1%
15. 1.2×10^6 L **17.** (a) $O^{2-} + H_2O \longrightarrow 2OH^-$
(b) $H^- + H_2O \longrightarrow H_2 + OH^-$
(c) $N^{3-} + 3H_2O \longrightarrow NH_3 + 3OH^-$.
(d) $CO_3^{2-} + H_2O \longrightarrow HCO_3^- + OH^-$. **19.** (a) soluble
(b) sparingly soluble (c) soluble (d) soluble (e) insoluble
(f) soluble **21.** (a) $FeCl_3(aq) + 3NaOH(aq) \longrightarrow 3NaCl(aq) + Fe(OH)_3(s)$ (b) no precipitate (c) no precipitate
(d) $Pb(NO_3)_2(aq) + H_2SO_4(aq) \longrightarrow 2HNO_3(aq) + PbSO_4(s)$
(e) $2AgNO_3(aq) + Na_2S(aq) \longrightarrow 2NaNO_3(aq) + Ag_2S(s)$
23. (a) $[Ag^+][Cl^-]$ (b) $[Ba^{2+}][F^-]^2$ (c) $[Cr^{3+}][OH^-]^3$
(d) $[Bi^{3+}]^2[S^{2-}]^3$ **25.** 1.0×10^{-8}
27. $3.4 \times 10^{-28} \ mol^2 \ L^{-2}$; $4.0 \times 10^{-11} \ mol^3 \ L^{-3}$;
$2.4 \times 10^{-28} \ mol^4 \ L^{-4}$ **29.** 4.7×10^{-4} g **31.** Yes
33. (a) No (b) Yes **35.** Yes. 9.5×10^{-4} mol PbI_2; $[Pb^{2+}] = 1.00$ M; $[I^-] = 9 \times 10^{-6}$ M; $[Na^+] = 2 \times 10^{-3}$ M; $[NO_3^-] =$

2.00 M **37.** 0.48 g **39.** (a) increase (b) decrease
41. $MgCO_3$ **43.** 7.78 **45.** (a); (c); (d); (e) **47.** 2.0×10^{-15}
$mol^2 \ L^{-2}$; 0.20 M **49.** 8.0×10^{-25} M; 6.3×10^{-7} M;
7.0×10^{-17} M; 7.4×10^{-20} M

CHAPTER 16

1. Electrolysis is a process in which electrical energy is used to produce a chemical change. An electrolyte is a substance which, when melted or dissolved in a solvent, will carry an electrical current. In ionic conduction, the electrical current is carried by ions whereas in electronic conduction it is carried by electrons. **3.** (a) 9.6×10^4 C (b) 9.65×10^5 C (c) 3.86×10^5 C **5.** 1.64×10^6 s
7. O_2: 0.472 L; H_2: 0.943 L **9.** 1420 s **11.** 9.24 g
13. (a) Mg and Br_2 (b) Cu and O_2 (c) H_2 and I_2 (d) Fe and
Cl_2 **15.** $Cu^{2+}(aq) + 2e^- \longrightarrow Cu(s)$; $2H_2O(l) \longrightarrow O_2(g) + 4H^+(aq) + 4e^-$ **17.** 1.05×10^8 g Al; 3.50×10^7 g C
19. Br^-, I^-, Hg_2^{2+}, Fe^{2+}, and H_2S **23.** Mg, Al, Zn, Fe, Cu
25. Cl^-, Br^-, I^- **27.** (a) The balanced reaction will *not* occur.
$6Mn^{2+} + 5Cr_2O_7^{2-} + 22H^+ \longrightarrow 6MnO_4^- + 10Cr^{3+} + 11H_2O$.
(b) The balanced overall reaction, which will proceed, is
$O_2 + 4Br^- + 4H^+ \longrightarrow 2H_2O + Br_2$. (c) The overall reaction
will *not* proceed: $2Au + 3Cl_2 \longrightarrow 2Au^{3+} + 6Cl^-$.
29. (a) $MnO_4^- + 8H^+ + 5e^- \longrightarrow Mn^{2+} + 4H_2O$ (cathode)
$Fe^{2+} \longrightarrow Fe^{3+} + e^-$ (anode) (b) from the Pt electrode in
$FeSO_4$ to that dipping into MnO_4^-. (c) The Fe^{2+} ions move
toward the anode and the MnO_4^- to the cathode. (d) $+0.72$ V
31. The emf will increase. **33.** 1.7×10^{-8} M;
$1.7 \times 10^{-8} \ mol^2 \ L^{-2}$
35. $C_2H_6 + 4H_2O \longrightarrow 2CO_2 + 14H^+ + 14e^-$;
$O_2 + 4H^+ + 4e^- \longrightarrow 2H_2O$; 0.18 L C_2H_6; 0.63 L O_2

CHAPTER 17

1. (a) HNO_3 (b) HNO_2 (c) KNO_2 (d) NO (e) N_2O_5
(f) N_2H_4 (g) NaN_3 (h) NH_2OH **3.** (a) Sodium nitrate
(b) Potassium nitrite (c) Dinitrogen tetraoxide
(d) Hydrazoic acid (e) Hydroxylamine (f) Barium peroxide
(g) Potassium superoxide (h) Hydrazine (i) Lithium azide
(j) Lithium nitride
7. (a) $3HNO_2 \longrightarrow HNO_3 + 2NO + H_2O$
(b) $4HNO_3 \longrightarrow 4NO_2 + 2H_2O + O_2$
(c) $C_6H_6 + HNO_3 + H_2SO_4 \longrightarrow C_6H_5NO_2 + H_3O^+$
9. (a) $NaNO_3 + H_2SO_4 \longrightarrow HNO_3 + NaHSO_4$; acid-base;
H_2SO_4-acid, NO_3^--base (b) $4HNO_3 \xrightarrow{h\nu} 4NO_2 + O_2 + 2H_2O$; oxidation-reduction; N is reduced, O is oxidized
(c) $2Pb(NO_3)_2 \longrightarrow 2PbO + 4NO_2 + O_2$; oxidation-reduction;
N is reduced, O is oxidized. (d) $HNO_3 + 2H_2SO_4 \longrightarrow NO_2^+ + H_3O^+ + 2HSO_4^-$; acid-base; H_2SO_4-acid, HNO_3-base.
(e) $HNO_2 + H_2O \longrightarrow NO_2^- + H_3O^+$; acid-base; HNO_2-acid,
H_2O-base. (f) $2NH_3 + OCl^- \longrightarrow N_2H_4 + H_2O + Cl^-$;
oxidation-reduction; N is oxidized, Cl is reduced.
(g) $2H_2O_2 \longrightarrow 2H_2O + O_2$; oxidation-reduction; O is oxidized
and reduced. (h) $BaO_2 + H_2SO_4 \longrightarrow BaSO_4 + H_2O_2$;
acid-base; H_2SO_4-acid, O_2^{2-}-base.

13. (a)

(b) $\Theta:\ddot{O}-\overset{\oplus}{N}\overset{:\ddot{O}:}{\underset{\ddot{O}:^\Theta}{}} \longleftrightarrow \Theta:\ddot{O}-\overset{\oplus}{N}\overset{:\ddot{O}:^\Theta}{\underset{\ddot{O}:}{}} \longleftrightarrow :\ddot{O}=\overset{\oplus}{N}\overset{:\ddot{O}:^\Theta}{\underset{\ddot{O}:^\Theta}{}}$

(c) $H-\ddot{O}-\ddot{N}=\ddot{O}:$

(d) $\Theta:\ddot{O}-\ddot{N}=\ddot{O}: \longleftrightarrow :\ddot{O}=\ddot{N}-\ddot{O}:^\Theta$

(e) $:\dot{N}=\ddot{O}: \longleftrightarrow \Theta:\ddot{N}=\ddot{O}:^\oplus$

17. NO_2^+; N_2O; N_3^- **19.** (a) 14.5 M   (b) 0.788 M
21. 0.27 metric ton   **23.** (a) Increasing the total pressure by adding an inert gas has no effect.   (b) Decreases $[N_2O_4]$
(c) Increases $[N_2O_4]$.   **25.** -9048 kJ mol^{-1} **27.** -225.9 kJ
29. (a) -3 (b) -2 (c) $+3$ (d) $+4$ (e) -1 (f) $+1$
(g) $-3, +5$ (h) $-1/3$ (i) -3 (j) $+3$ **31.** O in H_2O_2 is in the -1 oxidation state. It can be oxidized to $O_2(0)$ or reduced to $H_2O(-2)$.   (a) $H_2O_2 + SO_2 \longrightarrow SO_4^{2-} + 2H^+$
(b) $3H_2O_2 + O_3 \longrightarrow 3O_2 + 3H_2O$
(c) $I_2 + H_2O_2 \longrightarrow O_2 + 2H^+ + 2I^-$
(d) $2Cr(OH)_3 + 3H_2O_2 + 4OH^- \longrightarrow 2CrO_4^{2-} + 8H_2O$
(e) $H_2O_2 + NO_2^- \longrightarrow NO_3^- + H_2O$

CHAPTER 18

1. Average rate $= -\Delta[H^+]/\Delta t$: 0.0024; 0.0016; 0.0014; 0.0009

3. (a) Rate $= \dfrac{\Delta[I_2]}{\Delta t} = \dfrac{\Delta[H_2]}{\Delta t} = -\dfrac{1}{2}\dfrac{\Delta[HI]}{\Delta t}$

(b) Rate $\dfrac{\Delta[Cl_2]}{\Delta t} = \dfrac{1}{2}\dfrac{\Delta[NO]}{\Delta t} = -\dfrac{1}{2}\dfrac{\Delta[NOCl]}{\Delta t}$

(c) Rate $= \dfrac{\Delta[O_2]}{\Delta t} = \dfrac{1}{4}\dfrac{\Delta[NO_2]}{\Delta t} = -\dfrac{1}{2}\dfrac{\Delta[N_2O_5]}{\Delta t}$

5. Second order   **7.** (a) 0.81 mol L^{-1} s^{-1}
(b) 0.54 mol L^{-1} s^{-1} (c) 0.41 mol L^{-1} s^{-1} **11.** First order in $[Cl_2]$; second order in $[NO]$; the overall reaction is third order.
13. Rate $= k[NO]^2[H_2]$; $k = 28$ mol^{-2} L^2 min^{-1} **15.** (a) $\frac{1}{8}$
(b) 1/1024   **17.** 4.0×10^{-9} mol L^{-1} s^{-1} **19.** first order;
$k = 0.0015$ min^{-1} **21.** $E_a = 186{,}000$ J mol^{-1}; 3.4×10^{-4}
mol^{-1} L s^{-1} **23.** 305 K   **25.** 108 kJ mol^{-1}
27. 21.1 kJ mol^{-1}; $+10°$C   **29.** Rate $= k[CO][Cl_2]^{3/2}$
31. Rate $= k[O_3]^2[O_2]$ **33.** Visible light is unable to dissociate H_2 since the energy of one photon of visible light is less than the H—H bond energy.

CHAPTER 19

1. (a) 2-methyl-1-propanol   (b) 1-chloro-2-propanol
(c) 2,2-dimethyl-3-pentanol   (d) 2,4-pentanediol
3. (a) primary   (b) tertiary   (c) secondary   (d) secondary

5. (a) $\underset{H}{\overset{O}{\underset{\|}{}}}C-CH_2-CH_2-CH_3$ (b) $CH_3\overset{O}{\underset{\|}{C}}CH_2CH_2CH_3$

(c) $CH_3-\underset{\underset{CH_3}{|}}{CH}-\overset{O}{\underset{\|}{C}}CH_3$

(d) $\underset{H}{\overset{O}{\underset{\|}{}}}C-CH_2-\underset{\underset{CH_3}{|}}{\overset{\overset{CH_3}{|}}{C}}-CH_2-CH_2-CH_3$

7. $CHCl_2-CH_2-CH_3$ (1,1-dichloropropane);
$CH_3-CCl_2-CH_3$ (2,2-dichloropropane);
$CH_2Cl-CHCl-CH_3$ (1,2-dichloropropane);
$CH_2Cl-CH_2-CH_2Cl$ (1,3-dichloropropane);

$CH_2Br-CH_2-CH_2-CH_3$ (1-bromobutane);

$CH_3-CHBr-CH_2-CH_3$ (2-bromobutane);

$CH_2Br-\underset{\underset{CH_3}{|}}{CH}-CH_3$ (1-bromo-2-methylpropane);

$CH_3-\underset{\underset{CH_3}{|}}{\overset{\overset{CH_3}{|}}{C}}Br-CH_3$ (2-bromo-2-methylpropane)
9. (a) Mehanol   (b) Ethanol   (c) Formic acid   (d) 2-Butanol
(e) Propanal   (f) 1,2-ethanediol   (g) Acetone
(h) 2-methylpropanal   (i) 1-chloro-2-bromopropane
15. (a) $CH_3CH_2OH \longrightarrow CH_3CHO + H_2$
(b) $CH_3CH_2CH_2OH + NaOH \rightleftharpoons CH_3CH_2CH_2O^-Na^+ +$
H_2O (c) $2CH_3OH + 2Na \longrightarrow 2CH_3O^-Na^+ + H_2$
(d) $C_2H_5OH + HBr \longrightarrow C_2H_5OH_2^+ + Br^-$
(e) $CH_3CH_2OH + H_2SO_4 \longrightarrow CH_2{=}CH_2 + H_3O^+ + HSO_4^-$
(g) $CH_3CH{=}CH_2 + H_2O \longrightarrow CH_3CH(OH)CH_3$

19. $H-\overset{O}{\underset{\|}{C}}-OH + HO-CH_3 \longrightarrow H-\overset{O}{\underset{\|}{C}}-O-CH_3 + H_2O$

21. $CH_2{=}CH_2 + Cl_2 \longrightarrow CH_2Cl-CH_2Cl$;
$CH_2Cl-CH_2Cl \xrightarrow{\text{heat}} CH_2{=}CHCl + HCl$
27. $Br_2 \xrightarrow{h\nu} 2Br$; $Br + C_3H_8 \longrightarrow C_3H_7 + HBr$;
$C_3H_7 + Br_2 \longrightarrow C_3H_7Br + Br$ **29.** (a) pH $= 11.77$;
$pK_a = 10.57$ (b) pH $= 10.65$
31. $2CH_3CH_2OH + 2Na \longrightarrow 2CH_3CH_2ONa + H_2$; 67.2 mL
35. The molecular formula is C_2H_4O; the most likely structure is CH_3CHO, an aldehyde.

CHAPTER 20

1. (a) $+1$ [Kr]4d^{10} $\boxed{\uparrow\downarrow}$ $\boxed{\uparrow\downarrow|\uparrow\downarrow|\uparrow}$ $\boxed{\quad|\quad|\quad}$

$+3$ [Kr]4d^{10} $\boxed{\uparrow\downarrow}$ $\boxed{\uparrow\downarrow|\uparrow|\uparrow}$ $\boxed{\uparrow|\quad|\quad}$

$+5$ [Kr]4d^{10} $\boxed{\uparrow\downarrow}$ $\boxed{\uparrow|\uparrow|\uparrow}$ $\boxed{\uparrow|\uparrow|\quad}$

$+7$ [Kr]4d^{10} $\boxed{\uparrow}$ $\boxed{\uparrow|\uparrow|\uparrow}$ $\boxed{\uparrow|\uparrow|\uparrow}$

(b) 2 [Kr]4d^{10} $\boxed{\uparrow\downarrow}$ $\boxed{\uparrow\downarrow|\uparrow\downarrow|\uparrow}$ $\boxed{\uparrow|\quad|\quad}$

4 [Kr]4d^{10} $\boxed{\uparrow\downarrow}$ $\boxed{\uparrow\downarrow|\uparrow|\uparrow}$ $\boxed{\uparrow|\uparrow|\quad}$

6 [Kr]4d^{10} $\boxed{\uparrow\downarrow}$ $\boxed{\uparrow|\uparrow|\uparrow}$ $\boxed{\uparrow|\uparrow|\quad}$

3. In the gas-phase, PCl_5 consists of individual PCl_5 molecules, each of which has the trigonal bipyramid geometry. In the solid state, however, it consists of tetrahedral PCl_4^+ and octahedral PCl_6^-

ions. **5.** We expect all the atoms in I_2Cl_6 to lie in the same plane, with angles at I and the bridging Cl's of about $90°$

9.

ION	ELECTRON PAIRS ON A	TYPE	GEOMETRY
SiF_6^{2-}	6	AX_6	Octahedral
SiF_5^-	5	AX_5	Trigonal bipyramid
PF_6^-	6	AX_6	Octahedral
SeF_5^-	6	AX_5E	Square pyramid
BrF_4^-	6	AX_4E_2	Square planar
IF_4^+	5	AX_4E	Disphenoid

11. $HClO$, H_3PO_3, H_2SO_4, $HClO_4$
13. (a) $F_2 + BrO_3^- + 2OH^- \longrightarrow 2F^- + BrO_4^- + H_2O$
(b) $XeF_2 + BrO_3^- + 2OH^- \longrightarrow Xe + 2F^- + H_2O + BrO_4^-$
15. With gentle heating, $4KClO_3 \longrightarrow 3KClO_4 + KCl$; in the presence of a catalyst $2KClO_3 \longrightarrow 2KCl + 3O_2$
17. $2Cl_2 + Ag_2O + H_2O \longrightarrow 2HOCl + 2AgCl$; If an attempt is made to further concentrate the solution, the acid decomposes.
19. (a) hypobromous acid (b) calcium hypochlorite
(c) potassium bromate (d) potassium chlorite
(e) magnesium perchlorate (f) iodic acid (g) perbromic acid
(h) paraperiodic acid
21. $ClO_3^- + 6H^+ + 6e^- \longrightarrow Cl^- + 3H_2O$
$ClO_2^- + 4H^+ + 4e^- \longrightarrow Cl^- + 2H_2O$
$ClO^- + 2H^+ + 2e^- \longrightarrow Cl^- + H_2O$ **23.** ICl_3
25. $2ClO_3^- + 10Fe^{2+} + 12H^+ \longrightarrow Cl_2 + 10Fe^{3+} + 6H_2O$
27. Excess $NaHSO_3$ reduces I_2 to I^-. 6.57 kg $NaHSO_3$.
31. $BE(Xe{=}O) = 88$ kJ

1. $[Ar]3d^3$ $[Ar]3d^8$ $[Ar]3d^{10}$ $[Ar]3d^7$ $[Xe]4f^{14}5d^8$
$[Ar]3d^5$ **3.** 9; 5; 5; 10; 2; 2 **5.** $+3$ for Fe; $+3$ for Co; $+3$ for Al; $+1$ for Ag; $+2$ for Fe; $+3$ for Fe **7.** Cr is $+6$ in CrO_4^{2-}; Cr is $+6$ in $Cr_2O_7^{2-}$; Mn is $+6$ in MnO_4^{2-}; V is $+5$ in VO_3^{3-}; V is $+4$ in VO^{2+}; Fe is $+6$ in FeO_4^{2-} **9.** 4 in $[Zn(NH_3)_4]Cl_2$; 6 in $[Co(NH_3)_3Cl_3]$; 6 in $[Co(NH_3)_5Cl]Cl_2$; 6 in $[Cr(en)_2Cl_2]$; 4 in $K_2[FeCl_4]$ **11.** *trans*-Dichlorotetramminechromium(III) ion; Trioxalatocobaltate(III) ion; Tetrabromooxalatochromate(V) ion; *cis*-Dicyanodiethylenediamineplatinate(O)

17.

tetrahedral

19. With increasing oxidation state, more valence shell electrons participate in the bonding, the atomic core becomes less shielded, and the electronegativity of the atom increases. Thus, the ionic character of the bonds, in the sense $\overset{\delta+}{M}{-}\overset{\delta-}{X}$ decreases with increasing oxidation state. **23.** $Cu(H_2O)_6^{2+}$; When concentrated HCl is added, Cl^- ions can replace one or more of the H_2O ligands; $Cu(NH_3)_4^{2+}$; $Cu(CN)_4^{2-}$ **27.** The closest formula is $Fe_{11}S_{13}$. Presumably the compound is a nonstoichiometric version of FeS, with some of the Fe sites occupied by holes, since this type of compound is known to be formed by iron. Thus it is unlikely that the experiment itself is inaccurate, though this can't be ruled out completely. **29.** $Zn = 53\%$ and $Cu = 47\%$

1. SiO_4^{4-}; In aluminosilicates, some of the Si atoms are replaced by Al^-. **3.** An amphibole is a silicate which contains two $(SiO_3^{2-})_n$ chains joined by sharing oxygen atoms on alternate tetrahedra, giving a double chain in which the silicate anion's empirical formula is $Si_4O_{11}^{6-}$. **5.** Like a solid, glass does not flow to any significant extent. In it, each silicon atom is surrounded by a tetrahedron of oxygen atoms, and each tetrahedron is joined to four others via the oxygen atoms. However, as in a liquid the tetrahedra have a random arrangement. **9.** (a) $Ca^{2+}Mg^{2+}(SiO_3^{2-})_2$ silicate chain anion (b) $K^+(AlSi_3O_8)^-$ aluminosilicate 3D anion (c) $(Ca^{2+})_2Zn^{2+}(Si_2O_7^{6-})$ disilicate anion (d) $Ba^{2+}Ti^4(Si_3O_9^{6-})$ cyclic silicate anion (e) $Al^{3+}(Si_2O_5^-)OH^-$ silicate sheet anion (f) $Ca^{2+}(Al_2Si_2O_8^{2-})$ aluminosilicate 3D anion **13.** Silican has little tendency to form double bonds. If they are formed they are very reactive. **15.** The carbon analogues of silicones are ketones. Ketones are small molecules that contain $C{=}O$ bonds, whereas silicones are polymers that contain $Si{-}O$ bonds, since carbon has a much greater tendency to form double bonds than does silicon.
19. BF_3 and BCl_3 are covalent molecules between which there are only London forces. They therefore have low mp's and bp's, the differences being due largely to the greater polarizability of Cl than F. $AlCl_3$ and AlF_3 are ionic solids with relatively high sublimation temperatures, although the formation of $Al_2Cl_6(g)$ suggests that $AlCl_3(s)$ is less ionic than $AlF_3(s)$.

21. $F_3B + :NH_3 \longrightarrow F_3\overset{\ominus}{B}{-}\overset{\oplus}{N}H_3$;

$F_3B + :OR_2 \longrightarrow F_3\overset{\ominus}{B}{-}\overset{\oplus}{O}R_2$; $F_3B + F^- \longrightarrow BF_4^-$

23. $CaCO_3$; $CaSO_4 \cdot 2H_2O$; SiO_2; Al_2O_3; FeS_2; $Be_3Al_2Si_6O_{18}$; $Mg_3(Si_2O_5)_2(OH)_2$ **25.** In a three-center bond, one electron pair holds together three (rather than the usual two) nuclei. An example is the B_2H_6 molecule, in which two hydrogen atoms bridge the BH_2 groups; each BH_{bridge} B interaction is a three-center bond.

CHAPTER 23

1. (a) An addition polymer results from the successive addition of monomer units. Examples include polyethylene and polystyrene. (b) Condensation polymers are formed by condensation reactions in which monomers are joined into polymer chains by the elimination of small molecules such as water. Examples include the polyamides and polyesters—nylon, Dacron (Mylar), and Kodel. (c) An α-amino acid is a molecule with a CO_2H group and an NH_2 group on the same carbon atom. Examples include all those in Table 23.3. (d) A sugar is a monosaccharide or a disaccharide. Examples include glucose, fructose, and sucrose. (e) A polypeptide is a polyamide, which has the repeated unit $-NH-CHR-C(=O)-$. Proteins are examples. **5.** $CH_2{=}CH(CN)$ **9.** A nucleotide is prepared by the condensation of phosphoric acid, a sugar, and a nitrogen base. **11.** α- and β-glucose differ only in the orientation of one OH group with respect to the six-membered ring; see Figure 23.3. **13.** Starch is a mixture of polymers of α-glucose, whereas cellulose is a straight chain polymer of β-glucose. **17.** The reactant molecules fit into a cavity in the structure of the enzyme protein; since the reactant molecules are held in the correct orientation for reaction, the reaction occurs rapidly. The enzyme behaves as a catalyst.

CHAPTER 24

1. Li has 3 protons and 3 neutrons; C has 6 and 7 respectively; Zr has 40 and 54; and Ba has 56 and 81. **5.** $^{32}_{16}S$ (b) $^{0}_{1}e$ (c) $^{0}_{1}e$ (d) $^{214}_{85}At$ (e) $^{50}_{27}Co$ (f) $_{-1}^{0}e$ **7.** (a) $^{32}_{15}P$ (b) $^{7}_{3}Li$ (c) $^{81}_{36}Kr$ (d) $^{4}_{2}He$ (e) $^{99}_{38}Sr$ (f) Isotope mass 263, Z = 106 **9.** 17 000 years ago **11.** 15 000 years ago **13.** (a) 2.49×10^{-11} J; 1.25×10^{-12} J nucleon^{-1} (b) 8.7×10^{-11} J; 1.36×10^{-12} J nucleon^{-1} (c) 8.35×10^{-11} J; 1.37×10^{-12} J nucleon^{-1} (d) 2.70×10^{-10} J; 1.20×10^{-12} J nucleon.

CHAPTER 25

5. In CCl_4 there are four equivalent C—Cl bonds in a tetrahedral orientation so the carbon atom uses sp^3 hybrid orbitals. **7.** At each oxygen there are four electron pairs—two bonding and two nonbonding. Each oxygen can be considered to be sp^3 hybridized, with two hybrid orbitals used for bonding (one to H, one to the other oxygen) and two used for lone pairs. **9.** The hybridization at the methyl group is sp^3. Since the central carbon forms sigma bonds to methyl, H, and O it is sp^2 hybridized. The remaining 2p orbital on carbon and the parallel 2p on oxygen overlap to form a pi bond; hence the CO bond is a double bond. **13.** The $2p_z$ orbitals overlap sideways to form a pi bond. If the CH_2 group is rotated 90°, the 2p orbitals of the pi bond no longer overlap and the pi bond is destroyed. Thus there is no free rotation about a double bond because it would completely destroy the pi bond if it occurred. **17.** A double bond consists of one sigma and one pi bond. A triple bond consists of one sigma and two pi bonds. Each pi bond is the combination of sideways-overlapping p orbitals, one on each atom. **19.** In general, the more nodes an orbital has, the less stable it is. For an atomic orbital, there are n − 1 nodes.

CHAPTER 26

1. (a) Water (b) A solution (c) The collection (d) Raw rubber (e) Shuffled cards **3.** (a) $\Delta S > 0$ (b) $\Delta S < 0$ (c) $\Delta S < 0$ (d) $\Delta S > 0$ **7.** (a) 2.9 J K^{-1} mol^{-1} (b) 138.3 J K^{-1} mol^{-1} (c) 289.8 J K^{-1} mol^{-1} (d) 166.3 J K^{-1} mol^{-1} **11.** $\Delta G° = -190.6$ kJ mol^{-1}; the reaction is spontaneous. The enthalpy $\Delta H°$ makes a much larger contribution to $\Delta G°$ than does the entropy term—$T \Delta S°$. **13.** NO_2 **15.** 7.9×10^{24} atm^{-1} **17.** 3.1×10^{-7} atm **19.** (a) -800.8 kJ (b) 10^{140} (c) The activation energy must be very high. **21.** (a) No (b) Possibly (c) $K_p = 1.3 \times 10^{-7}$ **23.** $\Delta G° = -514.4$ kJ; $\Delta G =$. -504.7 kJ. Reaction will proceed in forward direction. **27.** The FeO, PbO, and CuO reductions are all feasible but the Al_2O_3 reduction is not.

INDEX

Boldface page numbers indicate end-of-chapter definitions; *t* indicates a table; *E* indicates an experiment.